11–19 PROGRESSION

Edexcel GCSE (9-1)

Mathematics

Higher

Student Book

Confidence • Fluency • Problem-solving • Reasoning

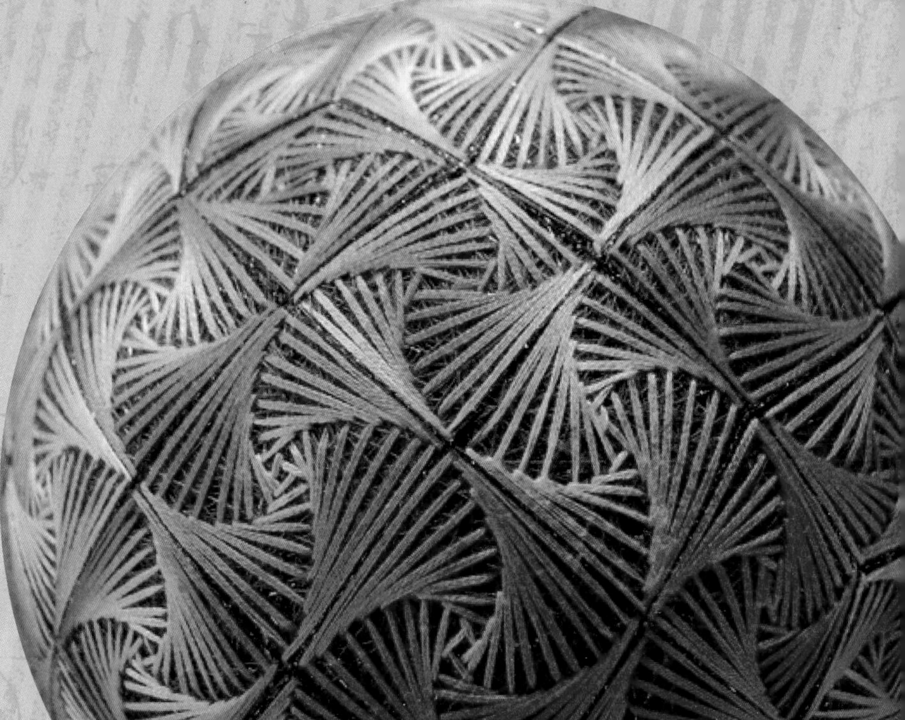

Series Editors:
Dr Naomi Norman • Katherine Pate

ALWAYS LEARNING

PEARSON

Published by Pearson Education Limited, 80 Strand, London, WC2R 0RL.

Registered in England 872828

www.pearsonschoolsandfecolleges.co.uk

Copies of official specifications for all Edexcel qualifications may be found on the website: www.edexcel.com

The rights of Jack Barraclough, Chris Baston, Ian Bettison, Sharon Bolger, Ian Boote, Judith Chadwick, Ian Jacques, Catherine Murphy, Su Nicholson, Naomi Norman, Diane Oliver, Katherine Pate, Jenny Roach, Peter Sherran and Robert Ward-Penny to be identified as authors of this work have been asserted by them in accordance with the Copyright, Designs and Patents Act 1988.

First published 2015

18 17 16 15
10 9 8 7 6 5

British Library Cataloguing in Publication Data
A catalogue record for this book is available from the British Library

ISBN 978 1 447 98020 9

Printed in Slovakia by Neografia

Acknowledgements
We would like to thank Glyn Payne for his work on this book and Amanda Hill, Kath Hipkiss, Mel Muldowney and Pietro Tozzi for their feedback on the ordering of this book.

The publisher would like to thank the following for their kind permission to reproduce their photographs:

(Key: b-bottom; c-centre; l-left; r-right; t-top)

123RF.com: fikmik 239, kurhan 30; **Alamy Images**: Graham Dunn 121, South West Images Scotland 296; **Getty Images**: Getty Images for Aegon 203, Rich Legg 531, Richard I'Anson 399, Rob Monk / Procycling Magazine 159; **Masterfile UK Ltd**: 280, photolibrary.com 186l; **Pearson Education Ltd**: Ben Nicholson 210; **Robert Harding World Imagery**: Last Refuge 501; **Science Photo Library Ltd**: Mischa Keijser / CULTURA 557; **Shutterstock.com**: Croisy 339, crolique 305, Garsya 265t, Happy Together 97, Hot Photo Pie 81, Hunor Focze 471, isak55 587, Joe Gough 265b, Kkulikov 186r, Kuttelvaserova Stuchelova 426, leungchopan 439, Lightspring 610, Lisa S. 325, MARGRIT HIRSCH 363, Monkey Business Images 61, Orla 548, pixeldreams.eu 456; **Sozaijiten**: 1

Cover images: Front: Created by Fusako, Photography by NanaAkua

All other images © Pearson Education

Every effort has been made to trace the copyright holders and we apologise in advance for any unintentional omissions. We would be pleased to insert the appropriate acknowledgement in any subsequent edition of this publication.

A note from the publisher
In order to ensure that this resource offers high-quality support for the associated Edexcel qualification, it has been through a review process by the awarding organisation to confirm that it fully covers the teaching and learning content of the specification or part of a specification at which it is aimed, and demonstrates an appropriate balance between the development of subject skills, knowledge and understanding, in addition to preparation for assessment.

While the publishers have made every attempt to ensure that advice on the qualification and its assessment is accurate, the official specification and associated assessment guidance materials are the only authoritative source of information and should always be referred to for definitive guidance.

Edexcel examiners have not contributed to any sections in this resource relevant to examination papers for which they have responsibility.

No material from an endorsed resource will be used verbatim in any assessment set by Edexcel.

Endorsement of a resource does not mean that the resource is required to achieve this qualification, nor does it mean that it is the only suitable material available to support the qualification, and any resource lists produced by the awarding organisation shall include this and other appropriate resources.

Contents

1	**Number**	**1**
	Prior knowledge check	1
1.1	Number problems and reasoning	2
1.2	Place value and estimating	4
1.3	HCF and LCM	6
1.4	Calculating with powers (indices)	8
1.5	Zero, negative and fractional indices	11
1.6	Powers of 10 and standard form	14
1.7	Surds	16
	Problem-solving	19
	Check up	20
	Strengthen	22
	Extend	25
	Knowledge check	27
	Unit test	28
2	**Algebra**	**30**
	Prior knowledge check	30
2.1	Algebraic indices	31
2.2	Expanding and factorising	33
2.3	Equations	35
2.4	Formulae	37
2.5	Linear sequences	40
2.6	Non-linear sequences	42
2.7	More expanding and factorising	46
	Problem-solving	48
	Check up	50
	Strengthen	51
	Extend	55
	Knowledge check	58
	Unit test	59
3	**Interpreting and representing data**	**61**
	Prior knowledge check	61
3.1	Statistical diagrams 1	63
3.2	Time series	67
3.3	Scatter graphs	70
3.4	Line of best fit	72
3.5	Averages and range	75
3.6	Statistical diagrams 2	78
	Problem-solving: Pollution particulates	81
	Check up	82
	Strengthen	84
	Extend	89
	Knowledge check	94
	Unit test	94
4	**Fractions, ratio and percentages**	**97**
	Prior knowledge check	97
4.1	Fractions	98
4.2	Ratios	101
4.3	Ratio and proportion	103
4.4	Percentages	105
4.5	Fractions, decimals and percentages	108
	Problem-solving	110
	Check up	111
	Strengthen	112
	Extend	115
	Knowledge check	118
	Unit test	119
5	**Angles and trigonometry**	**121**
	Prior knowledge check	121
5.1	Angle properties of triangles and quadrilaterals	122
5.2	Interior angles of a polygon	126
5.3	Exterior angles of a polygon	128
5.4	Pythagoras' theorem 1	131
5.5	Pythagoras' theorem 2	134
5.6	Trigonometry 1	136
5.7	Trigonometry 2	139
	Problem-solving	143
	Check up	144
	Strengthen	147
	Extend	153
	Knowledge check	155
	Unit test	157
6	**Graphs**	**159**
	Prior knowledge check	159
6.1	Linear graphs	161
6.2	More linear graphs	164
6.3	Graphing rates of change	166
6.4	Real-life graphs	170
6.5	Line segments	174
6.6	Quadratic graphs	176
6.7	Cubic and reciprocal graphs	180
6.8	More graphs	182
	Problem-solving: Profit parabolas	186
	Check up	187
	Strengthen	190
	Extend	195
	Knowledge check	198
	Unit test	200
7	**Area and volume**	**203**
	Prior knowledge check	203
7.1	Perimeter and area	204
7.2	Units and accuracy	207
7.3	Prisms	210

7.4	Circles	213
7.5	Sectors of circles	216
7.6	Cylinders and spheres	220
7.7	Pyramids and cones	222
	Problem-solving	225
	Check up	227
	Strengthen	229
	Extend	232
	Knowledge check	235
	Unit test	237
8	**Transformations and constructions**	**239**
	Prior knowledge check	239
8.1	3D solids	240
8.2	Reflection and rotation	242
8.3	Enlargement	245
8.4	Transformations and combinations of transformations	249
8.5	Bearings and scale drawings	253
8.6	Constructions 1	256
8.7	Constructions 2	259
8.8	Loci	262
	Problem-solving: Under construction	265
	Check up	266
	Strengthen	268
	Extend	273
	Knowledge check	275
	Unit test	277
9	**Equations and inequalities**	**280**
	Prior knowledge check	280
9.1	Solving quadratic equations 1	281
9.2	Solving quadratic equations 2	282
9.3	Completing the square	284
9.4	Solving simple simultaneous equations	287
9.5	More simultaneous equations	289
9.6	Solving linear and quadratic simultaneous equations	291
9.7	Solving linear inequalities	293
	Problem-solving: Overtaking	296
	Check up	297
	Strengthen	298
	Extend	300
	Knowledge check	302
	Unit test	303
10	**Probability**	**305**
	Prior knowledge check	305
10.1	Combined events	307
10.2	Mutually exclusive events	310
10.3	Experimental probability	312
10.4	Independent events and tree diagrams	314
10.5	Conditional probability	318
10.6	Venn diagrams and set notation	321
	Problem-solving: Drug testing	325
	Check up	326
	Strengthen	328
	Extend	333
	Knowledge check	335
	Unit test	336
11	**Multiplicative reasoning**	**339**
	Prior knowledge check	339
11.1	Growth and decay	340
11.2	Compound measures	343
11.3	More compound measures	346
11.4	Ratio and proportion	348
	Problem-solving	351
	Check up	352
	Strengthen	354
	Extend	357
	Knowledge check	359
	Unit test	360
12	**Similarity and congruence**	**363**
	Prior knowledge check	363
12.1	Congruence	364
12.2	Geometric proof and congruence	367
12.3	Similarity	370
12.4	More similarity	374
12.5	Similarity in 3D solids	378
	Problem-solving	382
	Check up	383
	Strengthen	385
	Extend	390
	Knowledge check	394
	Unit test	395
13	**More trigonometry**	**399**
	Prior knowledge check	399
13.1	Accuracy	400
13.2	Graph of the sine function	402
13.3	Graph of the cosine function	405
13.4	The tangent function	408
13.5	Calculating areas and the sine rule	411
13.6	The cosine rule and 2D trigonometric problems	415
13.7	Solving problems in 3D	418
13.8	Transforming trigonometric graphs 1	421
13.9	Transforming trigonometric graphs 2	423
	Problem-solving: Muddy tracks	426
	Check up	427
	Strengthen	429
	Extend	433
	Knowledge check	435
	Unit test	437

14	**Further statistics**	**439**
	Prior knowledge check	439
14.1	Sampling	440
14.2	Cumulative frequency	443
14.3	Box plots	446
14.4	Drawing histograms	449
14.5	Interpreting histograms	450
14.6	Comparing and describing populations	453
	Problem-solving: Brain training	456
	Check up	457
	Strengthen	458
	Extend	463
	Knowledge check	466
	Unit test	468
15	**Equations and graphs**	**471**
	Prior knowledge check	471
15.1	Solving simultaneous equations graphically	472
15.2	Representing inequalities graphically	475
15.3	Graphs of quadratic functions	478
15.4	Solving quadratic equations graphically	482
15.5	Graphs of cubic functions	485
	Problem-solving	490
	Check up	491
	Strengthen	492
	Extend	496
	Knowledge check	498
	Unit test	499
16	**Circle theorems**	**501**
	Prior knowledge check	501
16.1	Radii and chords	502
16.2	Tangents	504
16.3	Angles in circles 1	506
16.4	Angles in circles 2	509
16.5	Applying circle theorems	512
	Problem-solving	515
	Check up	517
	Strengthen	519
	Extend	523
	Knowledge check	526
	Unit test	527
17	**More algebra**	**531**
	Prior knowledge check	531
17.1	Rearranging formulae	532
17.2	Algebraic fractions	533
17.3	Simplifying algebraic fractions	535
17.4	More algebraic fractions	537
17.5	Surds	539
17.6	Solving algebraic fraction equations	541
17.7	Functions	543
17.8	Proof	545
	Problem-solving: Surface gravity	548
	Check up	549
	Strengthen	550
	Extend	553
	Knowledge check	555
	Unit test	556
18	**Vectors and geometric proof**	**557**
	Prior knowledge check	557
18.1	Vectors and vector notation	558
18.2	Vector arithmetic	560
18.3	More vector arithmetic	563
18.4	Parallel vectors and collinear points	566
18.5	Solving geometric problems	568
	Problem-solving	572
	Check up	574
	Strengthen	576
	Extend	579
	Knowledge check	582
	Unit test	584
19	**Proportion and graphs**	**587**
	Prior knowledge check	587
19.1	Direct proportion	588
19.2	More direct proportion	590
19.3	Inverse proportion	592
19.4	Exponential functions	595
19.5	Non-linear graphs	598
19.6	Translating graphs of functions	602
19.7	Reflecting and stretching graphs of functions	605
	Problem-solving: Modelling outbreaks	610
	Check up	611
	Strengthen	613
	Extend	615
	Knowledge check	619
	Unit test	621
	Answers	624
	Index	714

Welcome to Edexcel GCSE (9-1) Mathematics Higher Student Book

This Student Book is packed full of features to help you enjoy and feel confident in maths as well as preparing you for your GCSE.

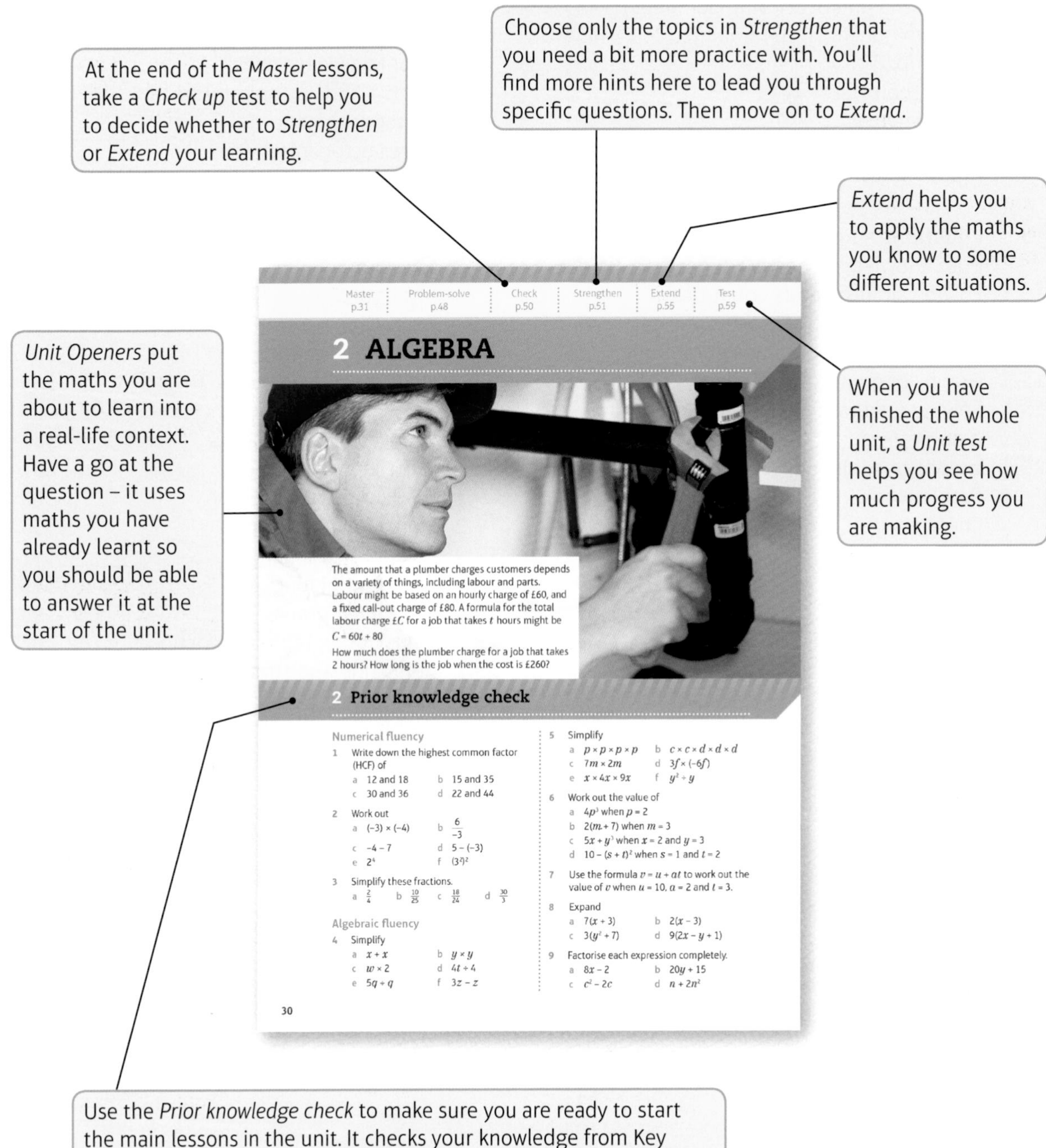

At the end of the *Master* lessons, take a *Check up* test to help you to decide whether to *Strengthen* or *Extend* your learning.
Choose only the topics in *Strengthen* that you need a bit more practice with. You'll find more hints here to lead you through specific questions. Then move on to *Extend*.
Extend helps you to apply the maths you know to some different situations.
Unit Openers put the maths you are about to learn into a real-life context. Have a go at the question – it uses maths you have already learnt so you should be able to answer it at the start of the unit.
When you have finished the whole unit, a *Unit test* helps you see how much progress you are making.
Use the *Prior knowledge check* to make sure you are ready to start the main lessons in the unit. It checks your knowledge from Key Stage 3 and from earlier in the GCSE course. Your teacher has access to worksheets if you need to recap anything.

Master p.31 | Problem-solve p.48 | Check p.50 | Strengthen p.51 | Extend p.55 | Test p.59

2 ALGEBRA

The amount that a plumber charges customers depends on a variety of things, including labour and parts. Labour might be based on an hourly charge of £60, and a fixed call-out charge of £80. A formula for the total labour charge £C for a job that takes t hours might be

$C = 60t + 80$

How much does the plumber charge for a job that takes 2 hours? How long is the job when the cost is £260?

2 Prior knowledge check

Numerical fluency

1 Write down the highest common factor (HCF) of
a 12 and 18 b 15 and 35
c 30 and 36 d 22 and 44

2 Work out
a $(-3) \times (-4)$ b $\frac{6}{-3}$
c $-4 - 7$ d $5 - (-3)$
e 2^4 f $(3^2)^2$

3 Simplify these fractions.
a $\frac{2}{4}$ b $\frac{10}{25}$ c $\frac{18}{24}$ d $\frac{30}{3}$

Algebraic fluency

4 Simplify
a $x + x$ b $y \times y$
c $w \times 2$ d $4t \div 4$
e $5q + q$ f $3z - z$

5 Simplify
a $p \times p \times p \times p$ b $c \times c \times d \times d \times d$
c $7m \times 2m$ d $3f \times (-6f)$
e $x \times 4x \times 9x$ f $y^2 \div y$

6 Work out the value of
a $4p^3$ when $p = 2$
b $2(m + 7)$ when $m = 3$
c $5x + y^3$ when $x = 2$ and $y = 3$
d $10 - (s + t)^2$ when $s = 1$ and $t = 2$

7 Use the formula $v = u + at$ to work out the value of v when $u = 10$, $a = 2$ and $t = 3$.

8 Expand
a $7(x + 3)$ b $2(x - 3)$
c $3(y^2 + 7)$ d $9(2x - y + 1)$

9 Factorise each expression completely.
a $8x - 2$ b $20y + 15$
c $c^2 - 2c$ d $n + 2n^2$

30

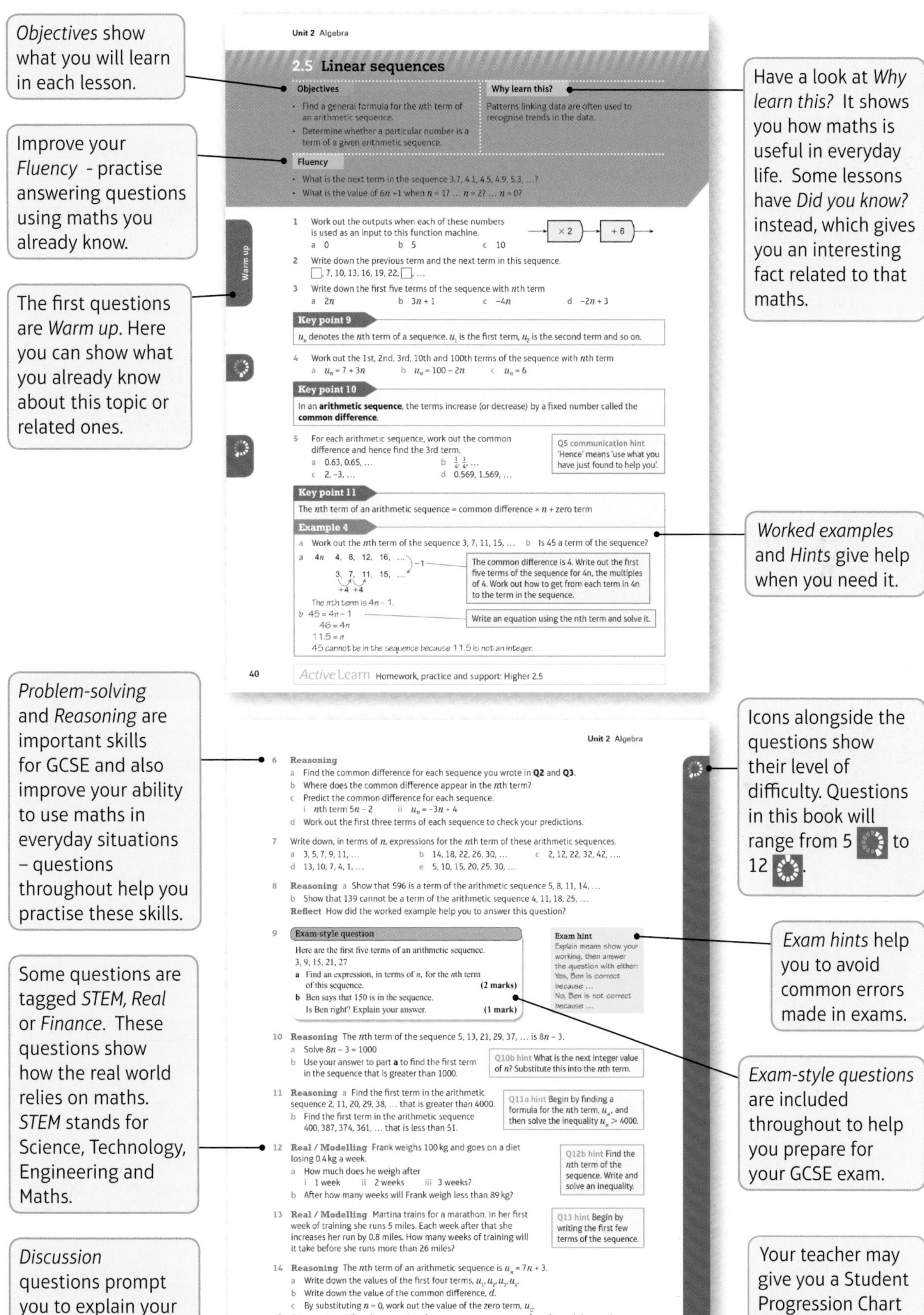

Objectives show what you will learn in each lesson.
Improve your *Fluency* - practise answering questions using maths you already know.
The first questions are *Warm up*. Here you can show what you already know about this topic or related ones.
Problem-solving and *Reasoning* are important skills for GCSE and also improve your ability to use maths in everyday situations – questions throughout help you practise these skills.
Some questions are tagged *STEM, Real* or *Finance*. These questions show how the real world relies on maths. *STEM* stands for Science, Technology, Engineering and Maths.
Discussion questions prompt you to explain your reasoning or explore new ideas with a partner.
Have a look at *Why learn this?* It shows you how maths is useful in everyday life. Some lessons have *Did you know?* instead, which gives you an interesting fact related to that maths.
Worked examples and *Hints* give help when you need it.
Icons alongside the questions show their level of difficulty. Questions in this book will range from 5 to 12.
Exam hints help you to avoid common errors made in exams.
Exam-style questions are included throughout to help you prepare for your GCSE exam.
Your teacher may give you a Student Progression Chart to help you see your progression through the units.

Unit 2 Algebra

2.5 Linear sequences

Objectives

- Find a general formula for the nth term of an arithmetic sequence.
- Determine whether a particular number is a term of a given arithmetic sequence.

Why learn this?

Patterns linking data are often used to recognise trends in the data.

Fluency

- What is the next term in the sequence 3.7, 4.1, 4.5, 4.9, 5.3, …?
- What is the value of $6n + 1$ when $n = 1$? … $n = 2$? … $n = 0$?

Warm up

1 Work out the outputs when each of these numbers is used as an input to this function machine.
a 0 b 5 c 10
→ × 2 → + 6 →

2 Write down the previous term and the next term in this sequence.
☐, 7, 10, 13, 16, 19, 22, ☐, …

3 Write down the first five terms of the sequence with nth term
a $2n$ b $3n + 1$ c $-4n$ d $-2n + 3$

Key point 9

u_n denotes the nth term of a sequence. u_1 is the first term, u_2 is the second term and so on.

4 Work out the 1st, 2nd, 3rd, 10th and 100th terms of the sequence with nth term
a $u_n = 7 + 3n$ b $u_n = 100 - 2n$ c $u_n = 6$

Key point 10

In an **arithmetic sequence**, the terms increase (or decrease) by a fixed number called the **common difference**.

5 For each arithmetic sequence, work out the common difference and hence find the 3rd term.
a 0.63, 0.65, … b $\frac{1}{4}, \frac{3}{4}, …$
c 2, −3, … d 0.569, 1.569, …

Q5 communication hint 'Hence' means 'use what you have just found to help you'.

Key point 11

The nth term of an arithmetic sequence = common difference × n + zero term

Example 4

a Work out the nth term of the sequence 3, 7, 11, 15, … b Is 45 a term of the sequence?

a $4n$ 4, 8, 12, 16, … −1
3, 7, 11, 15, …
+4 +4
The nth term is $4n - 1$.

The common difference is 4. Write out the first five terms of the sequence for $4n$, the multiples of 4. Work out how to get from each term in $4n$ to the term in the sequence.

b $45 = 4n - 1$
$46 = 4n$
$11.5 = n$
45 cannot be in the sequence because 11.5 is not an integer.

Write an equation using the nth term and solve it.

40 ActiveLearn Homework, practice and support: Higher 2.5

Unit 2 Algebra

6 **Reasoning**
a Find the common difference for each sequence you wrote in **Q2** and **Q3**.
b Where does the common difference appear in the nth term?
c Predict the common difference for each sequence.
i nth term $5n - 2$ ii $u_n = -3n + 4$
d Work out the first three terms of each sequence to check your predictions.

7 Write down, in terms of n, expressions for the nth term of these arithmetic sequences.
a 3, 5, 7, 9, 11, … b 14, 18, 22, 26, 30, … c 2, 12, 22, 32, 42, ….
d 13, 10, 7, 4, 1, … e 5, 10, 15, 20, 25, 30, …

8 **Reasoning** a Show that 596 is a term of the arithmetic sequence 5, 8, 11, 14, …
b Show that 139 cannot be a term of the arithmetic sequence 4, 11, 18, 25, …
Reflect How did the worked example help you to answer this question?

9 **Exam-style question**
Here are the first five terms of an arithmetic sequence.
3, 9, 15, 21, 27
a Find an expression, in terms of n, for the nth term of this sequence. **(2 marks)**
b Ben says that 150 is in the sequence.
Is Ben right? Explain your answer. **(1 mark)**

Exam hint Explain means show your working, then answer the question with either: Yes, Ben is correct because … No, Ben is not correct because …

10 **Reasoning** The nth term of the sequence 5, 13, 21, 29, 37, … is $8n - 3$.
a Solve $8n - 3 = 1000$
b Use your answer to part **a** to find the first term in the sequence that is greater than 1000.

Q10b hint What is the next integer value of n? Substitute this into the nth term.

11 **Reasoning** a Find the first term in the arithmetic sequence 2, 11, 20, 29, 38, … that is greater than 4000.
b Find the first term in the arithmetic sequence 400, 387, 374, 361, … that is less than 51.

Q11a hint Begin by finding a formula for the nth term, u_n, and then solve the inequality $u_n > 4000$.

12 **Real / Modelling** Frank weighs 100 kg and goes on a diet losing 0.4 kg a week.
a How much does he weigh after
i 1 week ii 2 weeks iii 3 weeks?
b After how many weeks will Frank weigh less than 89 kg?

Q12b hint Find the nth term of the sequence. Write and solve an inequality.

13 **Real / Modelling** Martina trains for a marathon. In her first week of training she runs 5 miles. Each week after that she increases her run by 0.8 miles. How many weeks of training will it take before she runs more than 26 miles?

Q13 hint Begin by writing the first few terms of the sequence.

14 **Reasoning** The nth term of an arithmetic sequence is $u_n = 7n + 3$.
a Write down the values of the first four terms, u_1, u_2, u_3, u_4.
b Write down the value of the common difference, d.
c By substituting $n = 0$, work out the value of the zero term, u_0.
Discussion What do you notice about your answers to parts **b** and **c**, and the numbers that appear in the formula, $u_n = 7n + 3$? What can you say about the zero terms of the sequences in **Q4**?

41

Problem-solving lessons

As well as problem-solving and reasoning throughout, this book includes a problem-solving lesson in every unit. There are two types:

- Some, such as the Unit 4 one below, give you strategies to approach problem-solving questions, for example using bar models. You are given a worked example which talks you through answering a question using the strategy and then a number of questions to practise on.
- Others, such as the Unit 3 one below, give you problem-solving questions in a real-life context to help you see how mathematical problem-solving is a part of many real-life activities.

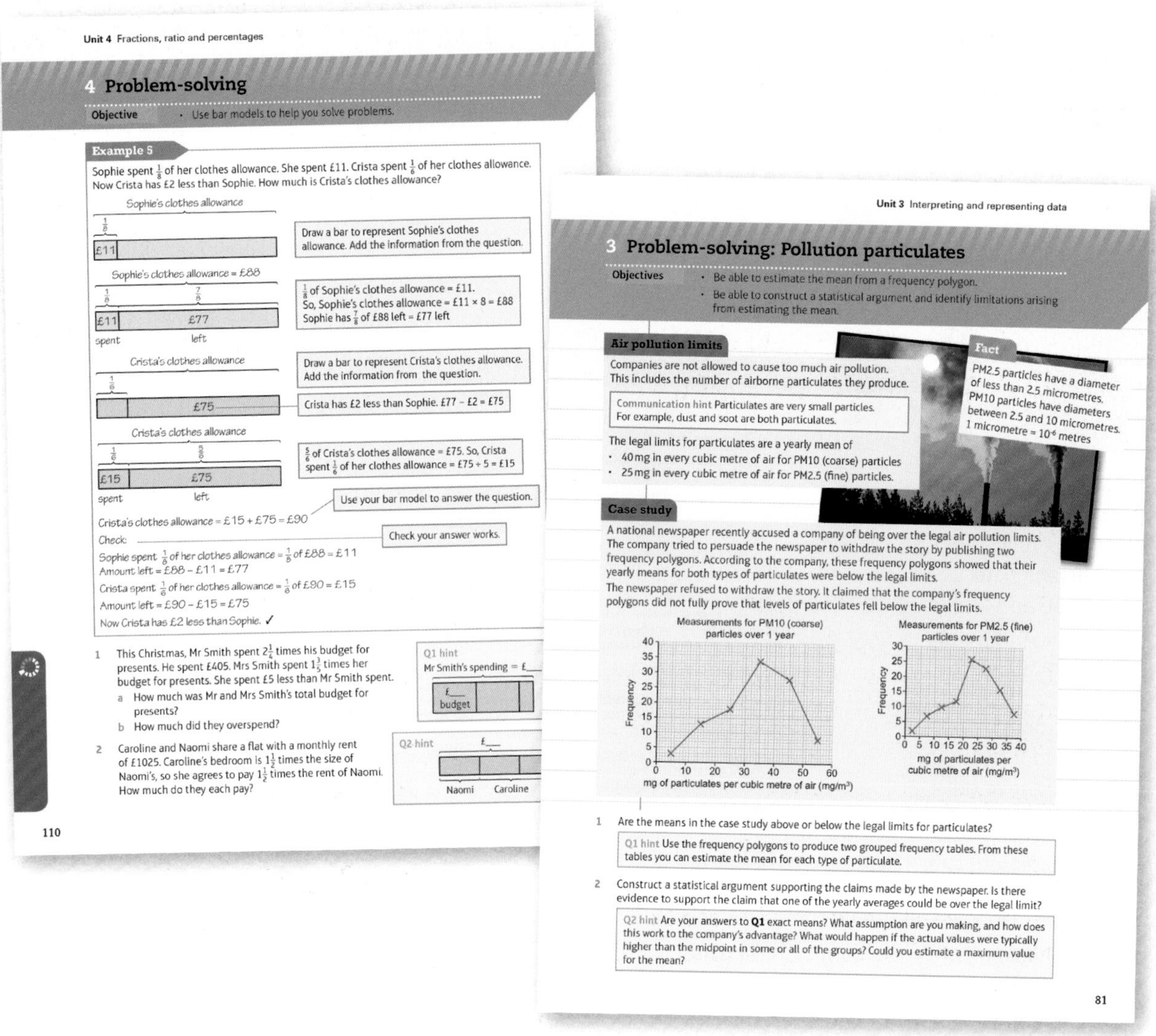

Unit 4 Fractions, ratio and percentages

4 Problem-solving

Objective · Use bar models to help you solve problems.

Example 5

Sophie spent $\frac{1}{8}$ of her clothes allowance. She spent £11. Crista spent $\frac{1}{6}$ of her clothes allowance. Now Crista has £2 less than Sophie. How much is Crista's clothes allowance?

Draw a bar to represent Sophie's clothes allowance. Add the information from the question.

$\frac{1}{8}$ of Sophie's clothes allowance = £11. So, Sophie's clothes allowance = £11 × 8 = £88. Sophie has $\frac{7}{8}$ of £88 left = £77 left

Draw a bar to represent Crista's clothes allowance. Add the information from the question.

Crista has £2 less than Sophie. £77 − £2 = £75

$\frac{5}{6}$ of Crista's clothes allowance = £75. So, Crista spent $\frac{1}{6}$ of her clothes allowance = £75 ÷ 5 = £15

Use your bar model to answer the question.

Crista's clothes allowance = £15 + £75 = £90

Check: Check your answer works.

Sophie spent $\frac{1}{8}$ of her clothes allowance = $\frac{1}{8}$ of £88 = £11

Amount left = £88 − £11 = £77

Crista spent $\frac{1}{6}$ of her clothes allowance = $\frac{1}{6}$ of £90 = £15

Amount left = £90 − £15 = £75

Now Crista has £2 less than Sophie. ✓

1 This Christmas, Mr Smith spent $2\frac{1}{4}$ times his budget for presents. He spent £405. Mrs Smith spent $1\frac{3}{5}$ times her budget for presents. She spent £5 less than Mr Smith spent.
 a How much was Mr and Mrs Smith's total budget for presents?
 b How much did they overspend?

Q1 hint Mr Smith's spending = £

2 Caroline and Naomi share a flat with a monthly rent of £1025. Caroline's bedroom is $1\frac{1}{2}$ times the size of Naomi's, so she agrees to pay $1\frac{1}{2}$ times the rent of Naomi. How much do they each pay?

Q2 hint

110

Unit 3 Interpreting and representing data

3 Problem-solving: Pollution particulates

Objectives
- Be able to estimate the mean from a frequency polygon.
- Be able to construct a statistical argument and identify limitations arising from estimating the mean.

Air pollution limits

Companies are not allowed to cause too much air pollution. This includes the number of airborne particulates they produce.

Communication hint Particulates are very small particles. For example, dust and soot are both particulates.

The legal limits for particulates are a yearly mean of
- 40 mg in every cubic metre of air for PM10 (coarse) particles
- 25 mg in every cubic metre of air for PM2.5 (fine) particles.

Fact

PM2.5 particles have a diameter of less than 2.5 micrometres. PM10 particles have diameters between 2.5 and 10 micrometres. 1 micrometre = 10^{-6} metres

Case study

A national newspaper recently accused a company of being over the legal air pollution limits. The company tried to persuade the newspaper to withdraw the story by publishing two frequency polygons. According to the company, these frequency polygons showed that their yearly means for both types of particulates were below the legal limits.

The newspaper refused to withdraw the story. It claimed that the company's frequency polygons did not fully prove that levels of particulates fell below the legal limits.

Measurements for PM10 (coarse) particles over 1 year

Measurements for PM2.5 (fine) particles over 1 year

1 Are the means in the case study above or below the legal limits for particulates?

Q1 hint Use the frequency polygons to produce two grouped frequency tables. From these tables you can estimate the mean for each type of particulate.

2 Construct a statistical argument supporting the claims made by the newspaper. Is there evidence to support the claim that one of the yearly averages could be over the legal limit?

Q2 hint Are your answers to **Q1** exact means? What assumption are you making, and how does this work to the company's advantage? What would happen if the actual values were typically higher than the midpoint in some or all of the groups? Could you estimate a maximum value for the mean?

81

Further support

You can easily access extra resources that tie in to each lesson – look for the *ActiveLearn Homework, practice and support* references on the first page of each lesson. This is online practice that is clearly mapped to the lessons and provides interactive exercises with lots of extra support for when you are working independently.

The Practice, Problem-solving and Reasoning Books are full of extra practice for key questions and will help you reinforce your learning and track your own progress.

1 NUMBER

Estimate the amount of money taken by a football club on match day.

1 Prior knowledge check

Numerical fluency

1 Work out

a 5×0.3 b 97×0.02
c $6 \div 0.2$ d $27 \div 0.09$
e $4.2 \div 0.1$ f 0.4×0.6
g 0.9×0.02 h 0.09×0.09
i $0.4 \div 0.2$ j $0.9 \div 0.03$
k $0.45 \div 0.3$ l $0.88 \div 0.04$

2 Choose the correct sign, < or >.

a 2.7 □ 2.5 b 3.04 □ 3.3
c −2.9 □ −2.8 d −5.16 □ −5.5

3 a Write down all the factors of 12 and 18.
b Make a list of the common factors.
c Write down the highest common factor.

4 a Copy and complete the Venn diagram to show the factors of 16 and 20.

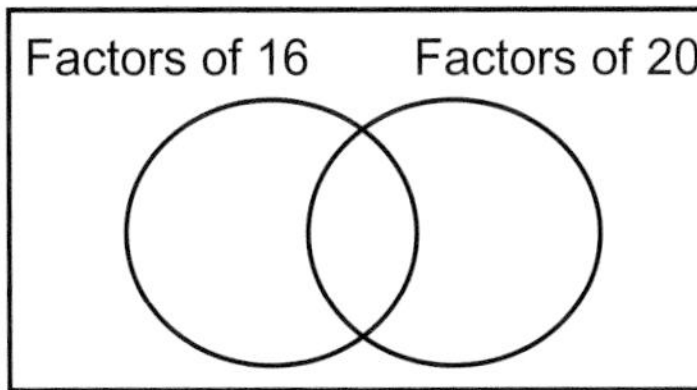

b Write down the highest common factor.

5 a Write down the first 10 multiples of 6 and 9.
b Make a list of the common multiples.
c Write down the lowest common multiple.

6 a Copy and complete the Venn diagram to show the first 10 multiples of 4 and 10.

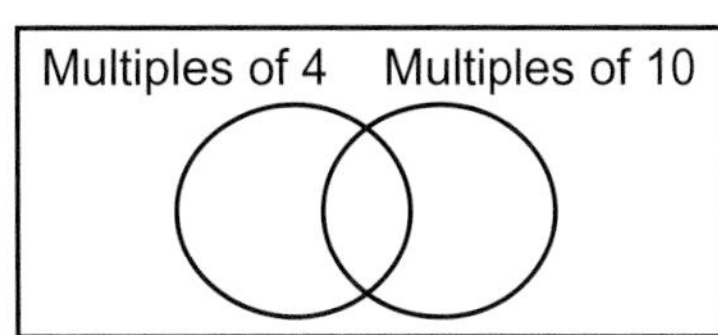

b Write down the lowest common multiple of 4 and 10.

7 Work out

a $8 - 2 \times 3$ b $(8 - 2) \times 3$
c $7 - (4 - 1) \times 6$ d $24 \div (8 - 2)$
e $4^2 + 1$ f $(-6)^2$

8 Insert brackets to make this calculation correct.
$9 + 18 \div 3 = 9$

9 Estimate

a 7.3×8.94 b $47 \div 2.1$
c $5.2 + 4.9$ d $7.9 - 2.4$

10 Write the positive and negative square roots of these numbers.

a 36 b 1 c 64

11 Work out

a $4 \times 9 \times 25$ b 102×48

c 182×99 d $27 \times 6 + 27 \times 4$

12 Copy and complete.

a $6 \times 6 = 6^{\square}$ b $3 \times 3 \times 3 \times 3 = 3^{\square}$

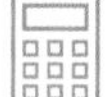

13 Work out

a 4^2 b 1^4

c 11^3 d 2^7

* Challenge

14 How many different ways can these cards be arranged?

What about now?

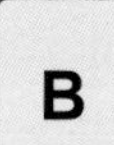

1.1 Number problems and reasoning

Objectives

- Work out the total number of ways of performing a series of tasks.

Why learn this?

5! in maths means 'five factorial' and is equal to $5 \times 4 \times 3 \times 2 \times 1$.

Fluency

Work out

- $4 \times 4 \times 4$
- $5 \times 4 \times 3$
- $10 \times 9 \times 8$
- $4 \times 3 \times 2 \times 1$

Warm up

1 a Copy and complete this list of all possible outcomes for rolling a dice and flipping a coin.

H, 1 H, 2 …

T, 1 … …

b How many outcomes are there altogether?

2 a Copy and complete this list of all possible outcomes for spinner A and spinner B.

2, 1 4, 1 6, 1

2, 3 4, 3 …

2, 5 … …

b How many outcomes are there altogether?

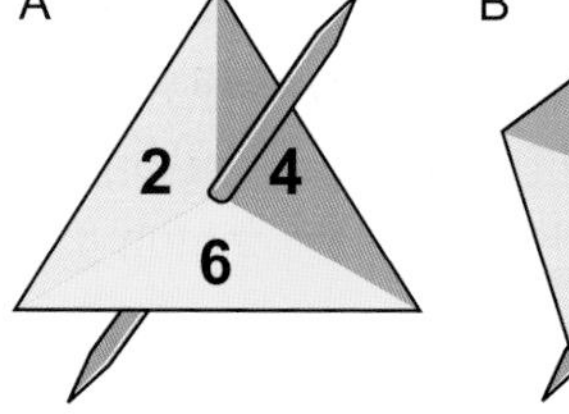

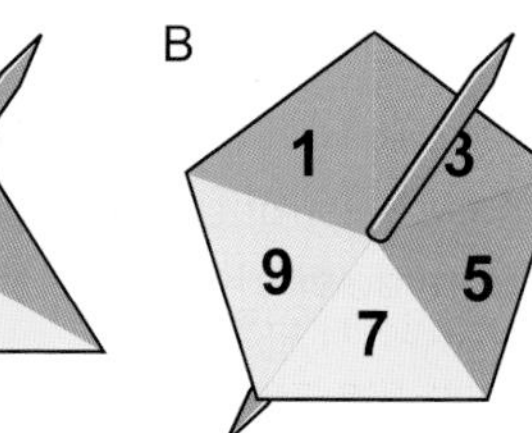

3 How many possible outcomes are there when

a rolling a dice

b flipping a coin

c spinning A in **Q2**

d spinning B in **Q2**?

Discussion What do you notice about your answers to

i **Q1b** and **Q3a** and **b** above

ii **Q2b** and **Q3c** and **d** above?

Questions in this unit are targeted at the steps indicated.

4 A restaurant offers a set menu for birthday parties.

a Write down all possible combinations of starters and main courses.

b **Reflect** How did you order your list to make sure you didn't miss any starters or mains?

The restaurant decides to offer fish (F) as a main course.

c How many possible combinations are there now?

d Copy and complete.

3 starters and 4 mains: ☐ combinations

3 starters and 5 mains: ☐ combinations

n starters and m mains: ☐ combinations

A different restaurant offers 2 starters, 4 mains and 3 desserts.

e How many possible combinations are there now?

Q4a hint Use letters for combinations, for example VP for vegetable soup, pizza.

Starters
Vegetable Soup (V)
Salad (S)
Melon (M)

Mains
Pizza (P)
Spaghetti Bolognaise (B)
Curry (C)
Lasagne (L)

Key point 1

When there are m ways of doing one task and n ways of doing a second task, the total number of ways of doing the first task then the second task is $m \times n$.

5 **Exam-style question**

Jess has a 4-digit password for her mobile phone.
Each digit can be between 0 and 9 **inclusive**.

a How many choices are possible for each digit of the code?

b What is the total number of 4-digit passwords that Jess can create?

Jess would like to choose an even number.
The code can start with a zero.

c How many different ways are possible now? **(5 marks)**

Q5 communication hint **Inclusive** means that the end numbers are also included.

6 Three people, A, B and C, enter a race.

a Write down the different orders in which they can finish first, second and third.

Harry says that there are 3 possible winners, but then only 2 possibilities for second place and only one person left for third place.

Discussion Is Harry correct? Explain your answer.

b How many different ways can people finish in

i a 4-person race ii a 6-person race iii a 10-person race?

Q6 communication hint
A factorial is the result of multiplying a sequence of descending integers.
For example '4 factorial' = $4! = 4 \times 3 \times 2 \times 1$.
Make sure you know how to use the factorial button on your calculator.

7 **Problem-solving** Eddie needs to choose a 6-digit code for his computer password.

a How many codes can Eddie create using

i 6 numbers

ii 4 numbers followed by 2 letters

iii 1 number followed by 5 letters?

Eddie decides that he does not want to repeat a digit or a letter.

b How many ways are possible in parts **i** to **iii** now?

1.2 Place value and estimating

Objectives

- Estimate an answer.
- Use place value to answer questions.

Why learn this?

Builders use estimates to give their clients an idea of how much the work will cost.

Fluency

Which two whole numbers does each square root lie between?

$\sqrt{3}$ $\sqrt{17}$

Warm up

1 Write each number to

  i 1 significant figure

  ii 2 significant figures.

  a 873 209 **b** 2019 **c** 0.007 059

2 Work out

  a $9 \times (4 + 7)$ **b** $5 + 3 \times 8$ **c** $7 \times 5 - 4 \times 2$

  d $30 - 5 \times 8$ **e** $72 - 9$ **f** $\sqrt{29 - 4}$

3 Work out the mean of 3, 6, 7, 9, 15 and 20.

4 Work out

  a 32×6 **b** 16×12 **c** 8×24 **d** 4×48

  Discussion What do you notice about your answers? Why has this happened?

5 $3.7 \times 9.86 = 36.482$

  Use this fact to work out the calculations below.

  Check your answers using an approximate calculation.

  a 37×9.86

  b 3.7×0.0986

  c 0.0037×98.6

  d $36.482 \div 9.86$

  e $3648.2 \div 98.6$

  f $364.82 \div 370$

Q5a hint Compare with the given calculation.

$\times\square$ $3.7 \times 9.86 = 36.482$ $\times\square$

$37 \times 9.86 = \square$

Q5d hint Rewrite the given calculation as a division.

6 **Reasoning** $54.8 \times 7.29 = 399.492$

  a Write down three more calculations that have the same answer.

  b Write down a division that has an answer of 54.8.

  c Write down a division that has an answer of 0.729.

  d Charlie says that $54.8 \times 72.9 = 3989.44$. Explain why Charlie must be wrong.

7 **a** Write down the value of $\sqrt{4}$ and $\sqrt{9}$.

  b Estimate the value of $\sqrt{5}$, $\sqrt{6}$, $\sqrt{7}$ and $\sqrt{8}$.
  Round each estimate to 1 decimal place.

  c Use a calculator to check your answer to part **b**.

Q7b hint Use a number line to help.

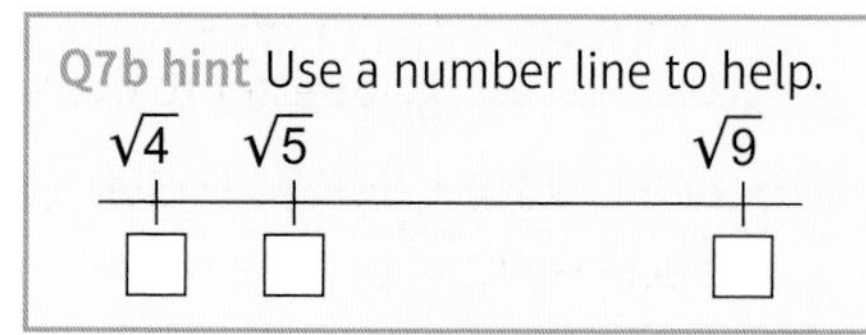

8 Estimate the value to the nearest tenth.

  a $\sqrt{47}$ **b** $\sqrt{22}$

  c $\sqrt{84}$ **d** $\sqrt{127}$

  e $\sqrt{10}$ **f** $\sqrt{40}$

Q8a hint You can write your answer

$\sqrt{47} \approx \square$

$\approx$ means 'is approximately equal to'

ActiveLearn Homework, practice and support: Higher 1.2

9 **Problem-solving** A mosaic uses 150 square tiles.
The total area is 3000 cm^2.

a Estimate the side length of a tile.

b Use a calculator to check your answer.

10 a Write down the value of 8^2 and 9^2.

b Estimate the value of 8.3^2 and 8.8^2. Round each estimate to the nearest whole number.

c Use a calculator to check your answer to part **b**.

11 Estimate to the nearest whole number.

a 3.2^2 b 4.7^2 c 1.7^2

d 7.1^2 e 6.3^2 f 9.8^2

Q11a hint Use a number line to help.

3^2 3.2^2 4^2

□ □ □

12 a Estimate answers to these.

i $(11.2 - \sqrt{50.3}) \times 4.08$ ii $(1.98 \times 3.14)^2 \div 8.85$

iii $\dfrac{88.72 - 21.9}{\sqrt{35.5}}$ iv $\sqrt{\dfrac{27.3 - 1.85}{3.93 \times 5.42}}$

Q12a iv hint The whole of the expression is being square rooted. So estimate the numerator and denominator before square rooting.

b Use your calculator to work out each answer.
Give your answers correct to one decimal place.

c **Reflect** How did you decide what to round each number to?
For **iii** and **iv** does it matter if you round the numerator or the denominator first?

13 The sum of the values on these cards is 12.

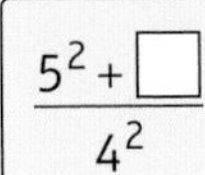

$\dfrac{5^2 + \square}{4^2}$ $\dfrac{80 - \sqrt{64}}{2 \times 4}$

Q13 hint Work out the value of the card on the right first.

Work out the missing number.

14 **Problem-solving** A large dice has a side length of 9.2 cm.
Estimate the surface area of the cube.

15 **Problem-solving** The area of a square is 80 cm^2.
Estimate the perimeter of the square.

16 **Problem-solving** Pieces of turf are 1 m long by 0.5 m wide. Each piece costs £3.79.

a Estimate the cost of turf required to cover these spaces.

i 9.6 m by 2.4 m

ii 6.2 m by 1.9 m

iii 4.4 m by 2.1 m

b Use a calculator to work out each answer.
How good were your estimates?

Discussion Is it better to overestimate or underestimate a cost?

17 **STEM** Robert uses a spreadsheet to record his runs for 10 innings.
His scores are in cells A1 to J1.
His mean score is in cell K1.

	A	B	C	D	E	F	G	H	I	J	K
1	78	12	4	15	0	35	0	7	12	21	11.8

a Use estimates to show that Robert's mean is wrong.

b Work out Robert's correct mean to the nearest tenth.

1.3 HCF and LCM

Objectives

- Write a number as the product of its prime factors.
- Find the HCF and LCM of two numbers.

Why learn this?

Astronomers use the lowest common multiple of patterns in the orbits of the Sun and the Moon to predict solar eclipses.

Fluency

Work out

- $2 \times 3 \times 5^2$
- $2^3 \times 3^2$
- $5 \times 5 \times 5 = 5^{\square}$
- $7 \times 7 \times 7 \times 7 \times 7 = 7^{\square}$

Warm up

1 a Write down all the factors of 20.
b Which of these factors are prime numbers?

2 a Write down all the prime numbers between 1 and 20.
b Write down all the factors of 24.
c Copy and complete this Venn diagram.

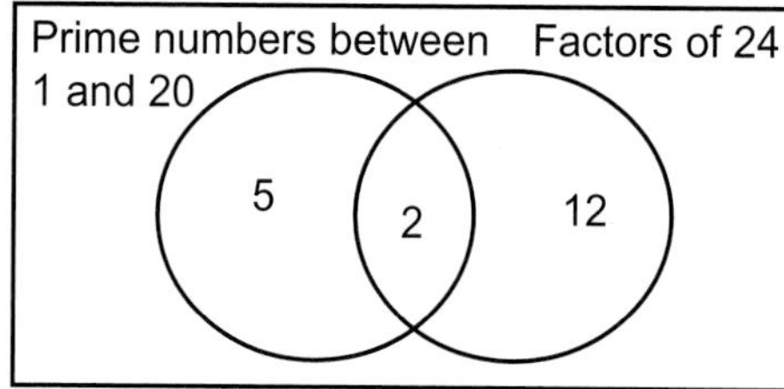

3 a Copy and complete this factor tree for 40.

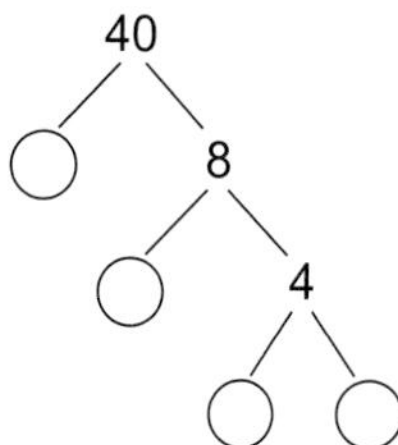

b Write 40 as a product of its prime factors.
$40 = \square \times \square \times \square \times \square = \square^{\square} \times \square$

Q3b hint Circle the prime factors in your factor tree.

4 Write 75 as a product of its prime factors.

Q4 hint Use the method from **Q3**.

5 Steve and Ian are asked to find 60 as a product of its prime factors.
Steve begins by writing $60 = 5 \times 12$
Ian begins by writing $60 = 6 \times 10$
a Work out a final answer for Steve. b Work out a final answer for Ian.
Discussion What do you notice about the two answers?
c Start the **prime factor decomposition** of 48 in two different ways: 6×8 and 12×4.
Discussion Does your first step in a prime factor decomposition affect your final answer?

6 **Reflect** a Write down your own short mathematical definition of these words.
i prime ii factor iii decomposition
b Use your definition to write down (in your own words) the meaning of prime factor decomposition.

ActiveLearn Homework, practice and support: Higher 1.3

7 Write each number as a product of its prime factors in **index form**.

a 18 b 42
c 25 d 36
e 24 f 80

> **Q7 communication hint** In **index form** means to write a number to a power or an index.
> 2^3 is written in index form. 3 is the power or index.

8 120 can be written as a product of its prime factors in the form $2^m \times n \times p$.
Work out m, n and p.

Example 1

Find the highest common factor and lowest common multiple of 24 and 60.

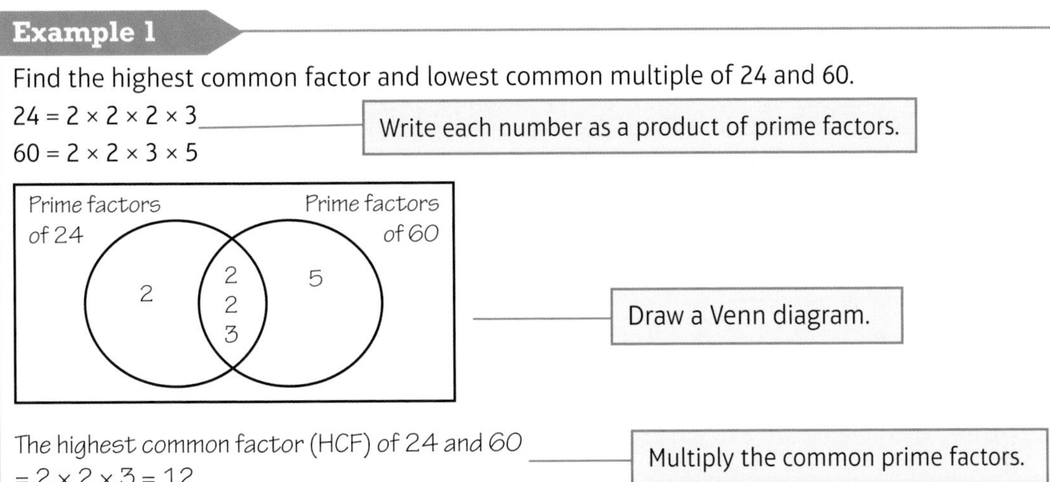

The highest common factor (HCF) of 24 and 60
$= 2 \times 2 \times 3 = 12$ — Multiply the common prime factors.

The lowest common multiple (LCM) of 24 and 60
$= 2 \times 2 \times 2 \times 3 \times 5 = 120$ — Multiply all the prime factors.

9 Find the HCF and LCM of

a 24 and 30 b 20 and 42
c 8 and 18 d 15 and 45
e 27 and 36 f 33 and 66

> **Q9 hint** Draw a Venn diagram for each question to help you.

10 **Real / Problem-solving** One bus leaves the bus station every 15 minutes.
Another bus leaves every 12 minutes.
At 2:30 pm both buses leave the bus station.
At what time will this next happen?

11 **Real / Problem-solving** Amber wants to tile her bathroom. It measures 1.2 m by 2.16 m.
She finds square tiles with a side length of 10 cm, 12 cm or 18 cm.
Which of these tiles will fit the wall exactly?

Discussion How do you know whether to find the HCF or LCM for **Q10** and **Q11**?

12 **Problem-solving** The HCF of two numbers is 2.
Write down three possible pairs of numbers.

> **Q12 hint** First choose two numbers where 2 is a factor. Is 2 the highest common factor of these numbers?

13 **Problem-solving** The LCM of two numbers is 18.
One of the numbers is 18.

a Write down all the possibilities for the other number.
b Describe the set of numbers you have created.

14 $48 = 2^4 \times 3$ and $36 = 2^2 \times 3^2$
Write down, as a product of its prime factors,

a the HCF of 48 and 36
b the LCM of 48 and 36.

> **Q14 hint** You could draw a Venn diagram.

15 **Exam-style question**

Given that $A = 2^3 \times 3^4 \times 5^2$ and $B = 2^2 \times 3^6 \times 5$

Write down, as a product of its prime factors,

a the HCF of A and B

b the LCM of A and B. **(2 marks)**

Exam hint
'Write as a product of its prime factors' means you don't have to calculate the number.

16 Write 80 as a product of its prime factors.

Discussion How can you use the prime factor decomposition of 80 to quickly work out the prime factor decomposition of 160? What about 40?

17 **Problem-solving** The prime factor decomposition of 2100 is $2^2 \times 3 \times 5^2 \times 7$.

Write down the prime factor decompositions of

a 75 b 24 c 12 d 30

18 a Harry says the prime factors of 75 appear in the prime factor decomposition of 2100, so 2100 is divisible by 75.
Is 2100 divisible by 24, 12 or 30?

b Use prime factors to show that 792 is divisible by 12.

c Is 792 divisible by 132? Explain your answer.

d Is 792 divisible by 27? Explain your answer.

19 In prime factor form, $700 = 2^2 \times 5^2 \times 7$ and $1960 = 2^3 \times 5 \times 7^2$

a What is the HCF of 700 and 1960?
Give your answer in prime factor form.

b What is the LCM of 700 and 1960?
Give your answer in prime factor form.

c Which of these are factors of 350 and 1960?

i $2 \times 5 \times 7$

ii 49

iii 20

iv $2^2 \times 5 \times 7^2$

Q19c hint What factors do 700 and 1960 have in common? Any factor of this number will be a factor of 350 and 1960.

d Which of these are multiples of 350 and 1960?

i $2^3 \times 5 \times 7^3$

ii $2^6 \times 5^2 \times 7^2$

iii $2^2 \times 5 \times 7$

Q19d hint What multiples do 700 and 1960 have in common? Any multiple of this number will be a multiple of 350 and 1960.

1.4 Calculating with powers (indices)

Objectives

- Use powers and roots in calculations.
- Multiply and divide using index laws.
- Work out a power raised to a power.

Why learn this?

A googol is a 1 followed by 100 zeros. It can be written as 10^{100}.

Fluency

Work out

- 6^2
- $(-4)^2$
- 2^4
- 1^5

ActiveLearn Homework, practice and support: Higher 1.4

Warm up

1 Work out

a 3^3 b $(-1)^3$ c 4×4^2 d $3^2 \times 5$

e $2^3 \times 10^2$ f 0.2^3 g $3 \times \sqrt{16}$ h $\sqrt{81} \times \sqrt{64}$

2 Work out

a $\dfrac{4 \times 4 \times 4 \times 4}{4 \times 4}$ b $\dfrac{3 \times 3 \times 3 \times 3 \times 3 \times 3}{3 \times 3 \times 3}$

3 Copy and complete.

a $2^{\square} = 16$ b $\square^3 = 64$

c $5^{\square} = 25$ d $\square^3 = 27$

Q3a hint $2 \times 2 \times \ldots = 16$

Q3b hint $\square \times \square \times \square = 64$

Key point 2

The inverse of a cube is the cube root.
$2^3 = 8$, so the cube root of 8 is $\sqrt[3]{8} = 2$

4 Work out

a $\sqrt[3]{27}$ b $\sqrt[3]{-1}$ c $\sqrt[3]{1000}$ d $\sqrt[3]{-125}$

Discussion Why it is possible to find the cube root of a negative number, but not the square root?

5 Work out these. Use a calculator to check your answers.

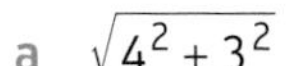

a $\sqrt{4^2 + 3^2}$ b $\sqrt[3]{10^2 + 5^2}$

c $43 - \sqrt[3]{-27}$ d $33 - \sqrt[3]{-8} - (-4)^2$

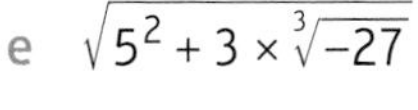

e $\sqrt{5^2 + 3 \times \sqrt[3]{-27}}$ f $\dfrac{5^2 \times \sqrt[3]{-27}}{\sqrt[3]{-8} - \sqrt{9}}$

g $\dfrac{-3^3}{\sqrt{9}} \times \dfrac{-\sqrt{64}}{\sqrt[3]{-1}}$ h $\dfrac{0.2^2 \times \sqrt[3]{-125}}{\sqrt[3]{8}}$

Q5a hint Use the priority of operations.

Q5e hint The square root applies to the whole calculation. Work out the cube root first.

6 Work out

a $[(3^3 - 5^2) \times 2)]^3$

b $20 - [3 \times 4^2 - (2^2 \times 3^2)]$

c $[72 \div (7 - 5)^3 - 3] \div \sqrt{9}$

Q6 communication hint Square brackets [ ] make the inner and outer brackets easier to see.

7 Work out

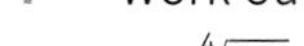

a $\sqrt[4]{16}$ b $\sqrt[4]{81}$ c $\sqrt[5]{100\ 000}$

Q7a hint

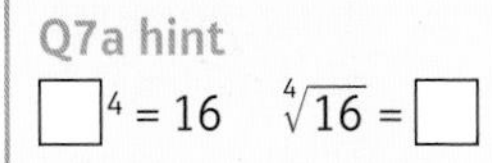

$\square^4 = 16$ $\sqrt[4]{16} = \square$

8 a Work out

i $10^3 \times 10^2$ ii 10^5 iii $10^6 \times 10^2$ iv 10^8

b How can you work out the answers to part **a** by using the indices of the powers you are multiplying?

c Check your rule works for

i $10^3 \times 10^4$ ii $10^5 \times 10$ iii $10^{-2} \times 10^{-3}$

Q8c ii hint $10 = 10^1$

9 Write each product as a single power.

a $3^2 \times 3^4$ b $4^2 \times 4^8$ c $9^3 \times 9^4$

Key point 3

To multiply, add the indices.
$x^m \times x^n = x^{m+n}$

10 Find the value of a.

a $8^4 \times 8^a = 8^7$ b $6^5 \times 6^a = 6^7$ c $2^3 \times 2^a = 2^{10}$

11 Write these calculations as a single power. Give your answers in index form.

a $27 \times 3^5 = 3^{\square} \times 3^5 = 3^{\square}$ b $4^3 \times 64$ c 5×125

d 32×4 e $8 \times 8 \times 8$ f $9 \times 27 \times 3$

Key point 4

You can only add the indices when multiplying powers of the same number.

12 **Reasoning**

a i Work out $\frac{5 \times 5 \times 5 \times 5 \times 5}{5 \times 5}$ by cancelling. Write your answer as a power of 5.

ii Copy and complete. $5^5 \div 5^2 = 5^{\square}$

b Copy and complete. $4^6 \div 4^2 = \frac{4 \times 4 \times 4 \times 4 \times 4 \times 4}{4 \times 4} = 4^{\square}$

c Work out $6^5 \div 6^4$

Discussion How can you quickly find $7^9 \div 7^3$ without writing all the 7s?

13 Work out

a $7^6 \div 7^2$ b $4^5 \div 4^3$ c $3^6 \div 3^5$

Key point 5

To divide powers, subtract the indices.
$x^m \div x^n = x^{m-n}$

14 Find the value of a.

a $9^6 \div 9^a = 9^4$ b $4^5 \div 4^a = 4$ c $7^6 \div 7^a = 7^9$

15 **Problem-solving**

a Yu multiplies three powers of 9 together.
$9^{\square} \times 9^{\square} \times 9^{\square} = 9^{12}$
What could the three powers be when

i all three powers are different

ii all three powers are the same?

b Harvey divides two powers of 5.
$5^{\square} \div 5^{\square} = 5^6$
What could the two powers be when

i both numbers are greater than 5^{20}

ii the power of one number is double the power of the other number?

16 Work out these. Write each answer as a single power.

a $5^3 \times 5^7 \div 5^4$ b $6^3 \div 6^2 \times 6^7$ c $\frac{5^4 \times 5^3}{5^5}$ d $\frac{8^2 \times 8^6}{8^7}$

17 **Real / STEM** The hard drive of Tom's computer holds 2^{38} bytes of data.
He buys a USB memory stick that holds 2^{34} bytes of data.

a How many memory sticks does he need to back up his computer?

He buys an external hard drive that holds 2^{39} bytes of data.

b What fraction of the external hard drive does he use when backing up his computer?

18 Copy and complete.

a $(2^3)^5 = 2^3 \times \square^{\square} \times \square^{\square} \times \square^{\square} \times \square^{\square} = 2^{\square}$

b $(6^4)^3 = \square^{\square} \times \square^{\square} \times \square^{\square} = 6^{\square}$

c $(8^7)^2 = \square^{\square} \times \square^{\square} = 8^{\square}$

Discussion What do you notice about the powers in the question and the powers in the final answer?

Key point 6

To work out a power to another power, multiply the powers together.
$(x^m)^n = x^{mn}$

19 Write as a single power.

a $(2^3)^4$ b $(6^2)^5$ c $(4^2)^{-3}$ d $(5^{-2})^{-6}$

20 **Problem-solving** Write each calculation as a single power.

a $8 \times 32 \times 8$ b $\frac{5^8}{125}$ c $\frac{16 \times 64 \times 16}{4^4}$

1.5 Zero, negative and fractional indices

Objectives

- Use negative indices.
- Use fractional indices.

Why learn this?

The smallest known time measurement is approximately 10^{-43} seconds. Scientists call this unit one Planck time, after Max Planck.

Fluency

- Work out $\sqrt{25}$ $\sqrt[3]{27}$ $\sqrt[4]{16}$
- Convert 0.3 to a fraction.

1 Work out

a 6^2 b 2^3 c 3^4 d 5^3

2 Write each calculation as a single power.

a $3^4 \times 3^6$ b $2^5 \div 2^3$ c 16×8 d $7^3 \times 7^5$

3 Work out

a $\frac{1}{\sqrt{25}}$ b $\sqrt[3]{\frac{-8}{27}}$ c $\frac{-3}{\sqrt[3]{27}}$ d $\sqrt[4]{81}$

4 Work out the value of n.

a $40 = 5 \times 2^n$ b $3^n \times 3^n = 3^8$

c $5^{2n} \div 5^n = 5^6$ d $\frac{1}{2} \times 4^n = 32$

Q4a hint

$40 = 5 \times \square$

How do you write this number as $2^{\square}$?

5 a Use a calculator to work out

i 2^{-1} ii 4^{-1} iii 5^{-1} iv 10^{-1}

b Write your answers to part **a** as fractions.

c Use a calculator to work out

i 2^{-2} ii 4^{-2} iii 5^{-2} iv 10^{-2}

d Write your answers to part **c** as fractions.

e Work out

i $\left(\frac{1}{2}\right)^{-1}$ ii $\left(\frac{3}{4}\right)^{-2}$

Discussion What is the rule for writing negative indices as fractions?

Warm up

6 a Match the equivalent cards.

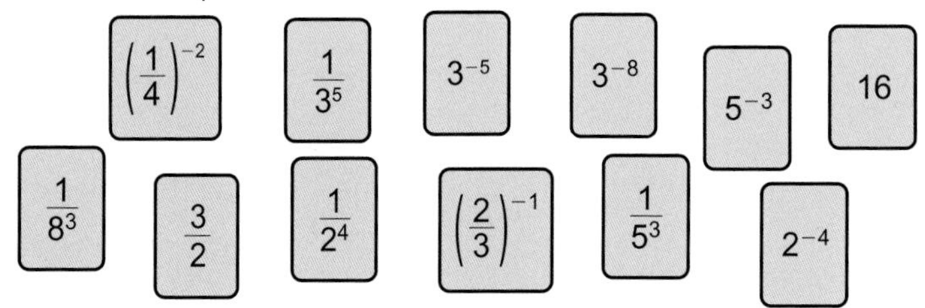

b Write a matching tile for the two tiles that are left over.

c Copy and complete. $\left(\frac{2}{3}\right)^{-1} = \square$, so $\left(\frac{a}{b}\right)^{-1} = \left(\frac{\square}{\square}\right)$

Key point 7

$x^{-n} = \frac{1}{x^n}$ for any number n, $x \neq 0$

7 Work out these. Write each answer as a single power.

a $6^2 \div 6^{-3} \times 6^7$ b $\frac{5^4 \times 5^{-3}}{5^5}$ c $\frac{8^{-2} \times 8^{-6}}{8^{-7}}$

8 **Problem-solving**

a Copy and complete. $2^3 \div 2^3 = 2^{\square}$

b Write down 2^3 as a whole number.

c $2^3 \div 2^3 = 8 \div \square = \square$

d Copy and complete using parts **a** and **c**. $2^3 \div 2^3 = 2^{\square} = \square$

e Repeat parts **a** and **b** for $7^5 \div 7^5$.

f Write down a rule for a^0, where a is any number.

Key point 8

$x^0 = 1$, where x is any non-zero number.

9 Work out

a 3^{-1} b 2^{-4} c 10^{-5}

d $\left(\frac{3}{4}\right)^{-1}$ e $\left(\frac{4}{5}\right)^{-3}$ f $\left(1\frac{1}{4}\right)^{-1}$

g $\left(2\frac{3}{4}\right)^{-2}$ h $(0.7)^{-1}$ i $(0.1)^{-5}$

j $(0.4)^{-3}$ k $(5^{-1})^0$ l $(7^{-1})^{-1}$

Q9f strategy hint Convert mixed numbers to improper fractions.

Q9h strategy hint Convert decimals to fractions.

10 a Use a calculator to work out

i $49^{\frac{1}{2}}$ ii $16^{\frac{1}{2}}$ iii $121^{\frac{1}{2}}$ iv $\left(\frac{4}{25}\right)^{\frac{1}{2}}$

b Copy and complete. $a^{\frac{1}{2}}$ is the same as the ________ ________ of a.

c Work out

i $27^{\frac{1}{3}}$ ii $1000^{\frac{1}{3}}$ iii $-1^{\frac{1}{3}}$ iv $\left(\frac{1}{1000}\right)^{\frac{1}{3}}$

d Copy and complete. $a^{\frac{1}{3}}$ is the same as the ________ ________ of a.

e Copy and complete.

i $625 = 5^{\square}$ so $625^{\frac{1}{4}} = \square$ ii $32 = \square^5$ so $32^{\frac{1}{5}} = \square$

11 Evaluate

a $36^{\frac{1}{2}}$ b $81^{\frac{1}{2}}$ c $\left(\frac{1}{9}\right)^{\frac{1}{2}}$

d $\left(\frac{16}{25}\right)^{\frac{1}{2}}$ e $\left(\frac{64}{49}\right)^{\frac{1}{2}}$ f $-8^{\frac{1}{3}}$

g $\left(\frac{1}{27}\right)^{\frac{1}{3}}$ h $\left(\frac{-64}{125}\right)^{\frac{1}{3}}$

Q11 communication hint Evaluate means 'work out the value of'.

Key point 9

$x^{\frac{1}{n}} = \sqrt[n]{x}$

12 Work out

a $25^{-\frac{1}{2}}$ b $64^{-\frac{1}{3}}$ c $\left(\frac{9}{25}\right)^{-\frac{1}{2}}$

Q12 hint $x^{-n} = \frac{1}{x^n}$

so $25^{-\frac{1}{2}} = \frac{1}{25^{\square}} = \frac{1}{\square}$

Example 2

Work out the value of **a** $27^{\frac{2}{3}}$ **b** $16^{-\frac{3}{4}}$

a $27^{\frac{2}{3}} = (27^{\frac{1}{3}})^2 = 3^2 = 9$

Use the rule $(x^m)^n = x^{mn}$. Work out the cube root of 27 first. Then square your answer.

b $16^{-\frac{3}{4}} = \frac{1}{16^{\frac{3}{4}}} = \frac{1}{(16^{\frac{1}{4}})^3} = \frac{1}{2^3} = \frac{1}{8}$

Use $x^{-n} = \frac{1}{x^n}$

13 Work out

a $64^{\frac{2}{3}}$ b $10\,000^{\frac{3}{4}}$ c $16^{\frac{3}{2}}$

d $\left(\frac{4}{9}\right)^{\frac{3}{2}}$ e $27^{-\frac{2}{3}}$ f $-81^{-\frac{3}{4}}$

Q13a hint
$64^{\frac{2}{3}} = \left(64^{\frac{1}{3}}\right)^2 = \left(\sqrt[3]{64}\right)^2 = \square^2 = \square$

Key point 10

$x^{\frac{n}{m}} = (\sqrt[m]{x})^n$

14 Work out

a $27^{-\frac{1}{3}} \times 9^{\frac{3}{2}}$ b $\left(\frac{4}{25}\right)^{-\frac{3}{2}} \times \left(\frac{8}{27}\right)^{\frac{1}{3}}$ c $\left(\frac{81}{16}\right)^{\frac{3}{4}} \times \left(\frac{9}{25}\right)^{-\frac{3}{2}}$

Q14a hint First work out $27^{-\frac{1}{3}}$.
Then work out $9^{\frac{3}{2}}$.
Then multiply these numbers together.

15 Find the value of n.

a $16 = 2^n$ b $\sqrt[3]{27} = 27^n$ c $\frac{1}{100} = 10^n$

d $\sqrt{\frac{4}{9}} = \left(\frac{9}{4}\right)^n$ e $(\sqrt{3})^7 = 3^n$ f $(\sqrt[4]{8})^7 = 8^n$

16 **Problem-solving / Reasoning** Will says that $25^{-\frac{1}{2}} \times 64^{\frac{2}{3}} = 80$

a Show that Will is wrong.

b What mistake did he make?

17 **Problem-solving / Reasoning**

Match the expressions with indices to their values.

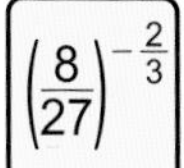

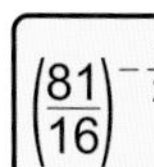

$\frac{1}{4}$

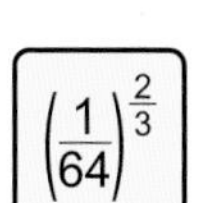

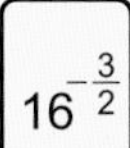

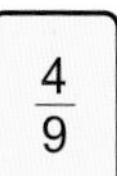

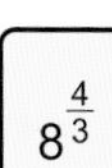

$\frac{9}{4}$ $\frac{1}{16}$ 16 $32^{-\frac{2}{5}}$ $\frac{1}{64}$ 8 $16^{\frac{3}{4}}$

1.6 Powers of 10 and standard form

Objectives

- Write a number in standard form.
- Calculate with numbers in standard form.

Why learn this?

Scientists use standard form to write very small or very large numbers.

Fluency

- Work out
  4.5×1000 0.0063×100 69.4×0.1 845.3×0.001
- Which of these are the same as $\div 10$?
  $\times \frac{1}{10}$ $\times 0.01$ $\times 10^{-1}$

Warm up

1 Copy and complete. If your answer is a fraction, write it as a decimal too.

a $10^0 =$ b $10^{-1} =$ c $10^{-2} =$
d $10^{-3} =$ e $10^{-4} =$ f $10^{-5} =$

2 Write down the value of x.

a $10^x = 1000$ b $10^5 = x$ c $10^x = 100\,000\,000$
d $10^{-1} = x$ e $10^x = 0.0001$ f $10^{-6} = x$

3 Copy and complete.

a 5 670 000 = □ million b 15 800 000 = □ million c 4 908 340 000 = □ billion

4 Copy and complete the table of **prefixes**.

Prefix	Letter	Power	Number
tera	T	10^{12}	1 000 000 000 000
giga	G	10^9	
mega	M		1 000 000
kilo	k	10^3	
deci	d		0.1
centi	c	10^{-2}	
milli	m		0.001
micro	μ	10^{-6}	
nano	n		0.000 000 001
pico	p	10^{-12}	

Q4 communication hint **Prefix** is the beginning part of a word.

Q4 communication hint μ, the letter for the prefix micro, is the Greek letter mu.

Key point 11

Some powers of 10 have a name called a prefix. Each prefix is represented by a letter. For example, kilo means 10^3 and is represented by the letter k, as in kg for kilogram.

5 Convert

a 15 mg into grams b 7 nm into metres
c 1.7 g into kg d 7.3 ps into seconds.

Q5a hint Use a number line.

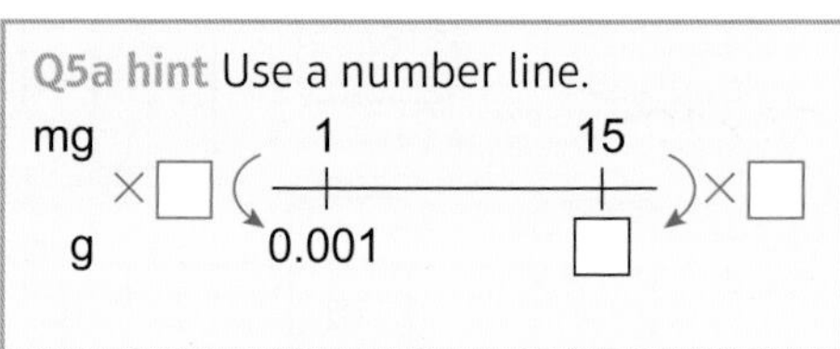

6 **STEM** Write these measurements in metres.

a The size of the influenza virus is about 1.2 μm.
b The radius of a hydrogen atom is 25 pm.
c A fingernail grows about 0.9 nm every second.

ActiveLearn Homework, practice and support: Higher 1.6

7 Copy and complete.

a $45\,000 = 4.5 \times \square$

b $10\,000 = 10^{\square}$

c $45\,000 = 4.5 \times 10^{\square}$

Key point 12

A number is in **standard form** when it is in the form $A \times 10^n$, where $1 \leqslant A < 10$ and n is an integer.
For example, 6.3×10^4 is written is standard form because 6.3 is between 1 and 10.
63×10^4 is *not* in standard form because 63 does not lie between 1 and 10.
Standard form is sometimes also called **scientific notation**.

8 Which of these numbers are in standard form?

A 4.5×10^7

B 13×10^4

C 0.9×10^{-2}

D 9.99×10^{-3}

E 4.5 billion

F 2.5×10

9 Write these numbers in standard form.

a 87 000

b 1 042 000

c 1 394 000 000

d 0.007

e 0.000 002 84

f 0.000 100 3

Q9 hint Write the number between 1 and 10 first.
Then multiply by a power of 10.

10 Write these as **ordinary numbers**.

a 4×10^5

b 3.5×10^2

c 6.78×10^3

d 6.2×10^{-2}

e 8.93×10^{-5}

f 4.04×10^{-3}

11 **STEM / Reasoning**

a The distance from the Sun to Neptune is 4 500 000 000 000 m.

i Write this number in standard form.

ii Enter the ordinary number in your calculator and press the = key.
Compare your calculator number with the standard form number.
Explain how your calculator displays a number in standard form.

b The thickness of a sheet of a paper is 0.000 07 m.

i Write this number in standard form.

ii Enter the ordinary number in your calculator and press the = key.
Compare your calculator number with the standard form number.

c **Reflect** Why do you think that scientists use standard form for very large and very small numbers?

Example 3

Work out $(5 \times 10^3) \times (7 \times 10^6)$

$5 \times 7 \times 10^3 \times 10^6$ — Rewrite the multiplication grouping the numbers and the powers.

35×10^9 — Simplify using multiplication and the index law $x^m \times x^n = x^{m+n}$.
This is not in standard form because 35 is not between 1 and 10.

$35 = 3.5 \times 10^1$ — Write 35 in standard form.

$35 \times 10^9 = 3.5 \times 10^1 \times 10^9 = 3.5 \times 10^{10}$ — Work out the final answer.

12 Work out these. Use a calculator to check your answers.

a $(3 \times 10^2) \times (2 \times 10^5)$
b $(5 \times 10^3) \times (4 \times 10^7)$
c $(8 \times 10^{-2}) \times (6 \times 10^7)$
d $(8 \times 10^6) \div (4 \times 10^3)$
e $(9 \times 10^{-2}) \div (3 \times 10^6)$
f $(2 \times 10^3) \div (8 \times 10^7)$
g $(5 \times 10^3)^2$
h $(4 \times 10^{-2})^3$

Q12g hint $(5 \times 10^3)^2 = (5 \times 10^3) \times (5 \times 10^3)$

13 **STEM / Problem-solving** The Sun is a distance of 1.5×10^8 km from the Earth.
Light travels at a speed of 3×10^5 km per second.
How many seconds will it take for light from the Sun to reach the Earth?

14 **STEM / Problem-solving** A water molecule has a mass of 3×10^{-29} kg.
A bottle contains 1.7×10^{28} molecules of water.
Calculate the mass of water in the bottle.

15 a Write these numbers as ordinary numbers.
i 8×10^4
ii 3×10^2
b Work out $(8 \times 10^4) + (3 \times 10^2)$, giving your answer in standard form.

Q15b hint Use your answers from part **a** to write the answer as an ordinary number. Then convert this to standard form.

16 Work out these. Give your answers in standard form.

a $3.4 \times 10^5 + 6.7 \times 10^4$
b $9.8 \times 10^4 - 2.2 \times 10^2$
c $7.2 \times 10^2 + 6.2 \times 10^{-1}$
d $8.3 \times 10^5 - 7 \times 10^{-1}$

17 **Exam-style question**

$(7 \times 10^x) + (7 \times 10^y) + (7 \times 10^z) = 700\,070.07$

Write down a possible set of values for x, y and z. **(3 marks)**

Exam hint
Don't just write down the possible values – give your working to show how you worked out the values.

1.7 Surds

Objectives

- Understand the difference between rational and irrational numbers.
- Simplify a surd.
- Rationalise a denominator.

Why learn this?

Surds are used to express irrational numbers in exact form.

Fluency

- What does the dot above the 1 mean in $0.\dot{1}$?
- What are the missing numbers?
  $147 = 3 \times \square$ $125 = \square \times 25$ $180 = 5 \times \square$ $96 = \square \times 16$

Warm up

1 Work out

a $\frac{3}{5} \times \frac{2}{4}$
b $\frac{7}{8} \times \frac{2}{3}$

2 Write each number as a fraction in its simplest form.

a 0.6
b 0.85
c 1.625
d 4.25
e $0.\dot{3}$
f $1.\dot{5}$

3 Write to 2 decimal places

a $\sqrt{5}$ b $\sqrt{7}$ c $\sqrt{19}$ d $\sqrt{53}$

Discussion Which is more exact, the square root or the decimal?

Key point 13

A **surd** is a number written exactly using square or cube roots.
For example $\sqrt{3}$ and $\sqrt[3]{5}$ are surds. $\sqrt{4}$ and $\sqrt[3]{27}$ are not surds, because $\sqrt{4} = 2$ and $\sqrt[3]{27} = 3$

4 a Work out

i $\sqrt{2} \times \sqrt{3}$ ii $\sqrt{6}$

b Work out

i $\sqrt{3}\sqrt{5}$ ii $\sqrt{15}$

c What do you notice about your answers to parts **a** and **b**?

d Find the missing numbers.

i $\sqrt{2} \times \sqrt{6} = \sqrt{\square}$ ii $\sqrt{2} \times \sqrt{\square} = \sqrt{10}$ iii $\sqrt{\square}\sqrt{7} = \sqrt{35}$

Key point 14

$\sqrt{mn} = \sqrt{m}\sqrt{n}$

5 Find the value of the integer k to simplify these surds.

a $\sqrt{150} = \sqrt{\square}\sqrt{6} = k\sqrt{6}$

b $\sqrt{40} = \sqrt{\square}\sqrt{10} = k\sqrt{10}$

c $\sqrt{128} = k\sqrt{2}$

d $\sqrt{108} = k\sqrt{3}$

Q5 communication hint An integer is a positive or negative whole number or zero.

6 Simplify these surds.

a $\sqrt{20}$ b $\sqrt{300}$ c $\sqrt{44}$

d $\sqrt{250}$ e $4\sqrt{50}$ f $6\sqrt{56}$

Q6 hint Find a factor that is also a square number.

7 Use a calculator to work out $\sqrt{75}$

a as a simplified surd

b as a decimal.

Q7 hint Make sure you know how to switch between surd form and decimals on your calculator.

8 a A surd simplifies to $4\sqrt{5}$. What could the original surd be?

b **Reflect** How did you find the surd?

Key point 15

$\sqrt{\frac{m}{n}} = \frac{\sqrt{m}}{\sqrt{n}}$

9 Simplify

a $\sqrt{\frac{7}{4}} = \frac{\sqrt{7}}{\sqrt{4}} =$ b $\sqrt{\frac{5}{9}}$ c $\sqrt{\frac{12}{49}} = \frac{\sqrt{12}}{\sqrt{49}} =$ d $\sqrt{\frac{18}{25}}$

Key point 16

Rational numbers can be written as a fraction in the form $\frac{a}{b}$, where a and b are integers and $b \neq 0$.

2 is rational as it can be written as $\frac{2}{1}$.

$0.\dot{2}$ is rational as it can be written as $\frac{2}{9}$.

$\sqrt{2}$ is irrational.

10 Copy and complete the table using the numbers below.

Rational	Irrational

$\sqrt[3]{6}$ $\quad \frac{3}{8}$ $\quad \sqrt{6.25}$ $\quad -4$ $\quad -\sqrt{8}$ $\quad \sqrt{17}$ $\quad 1.\dot{4}$ $\quad \sqrt{\frac{4}{49}}$ $\quad 0.3$

11 Solve the equation $x^2 - 90 = 0$, giving your answer as a surd in its simplest form.

Discussion Can you solve the equation $x^2 + 90 = 0$ in the same way? Explain your answer.

Q11 hint $x^2 = \square$
$x = \pm\sqrt{\square}$
$x = \pm\square\sqrt{\square}$

12 Solve these equations, giving your answer as a surd in its simplest form.

a $4x^2 = 200$ b $\frac{1}{2}x^2 = 80$ c $3x^2 = 36$ d $2x^2 - 14 = 42$

13 The area of a square is 60 cm^2.
Find the length of one side of the square.
Give your answer as a surd in its simplest form.

14 a Work out

i $5\sqrt{2} \times 4\sqrt{27}$ ii $4\sqrt{5} \times 6\sqrt{12}$
iii $9\sqrt{10} \times 4\sqrt{5}$ iv $8\sqrt{12} \times 3\sqrt{3}$

b Use a calculator to check your answers to parts **i** to **iv**.

Q14a hint Multiply the integers together and the surds together. Simplify.

Key point 17

To **rationalise the denominator** of $\frac{a}{\sqrt{b}}$, multiply by $\frac{\sqrt{b}}{\sqrt{b}}$. Then the fraction will have an integer as the denominator.

Example 4

Rationalise the denominator.

a $\frac{1}{\sqrt{2}}$

b $\frac{5}{\sqrt{75}}$

a $\frac{1}{\sqrt{2}} = \frac{1}{\sqrt{2}} \times \frac{\sqrt{2}}{\sqrt{2}}$
$= \frac{\sqrt{2}}{\sqrt{4}} = \frac{\sqrt{2}}{2}$

Multiplying by $\frac{\sqrt{2}}{\sqrt{2}}$ is the same as multiplying by 1, so this does not change the value.

b $\sqrt{75} = \sqrt{25}\sqrt{3} = 5\sqrt{3}$ — First simplify $\sqrt{75}$

$\frac{5}{\sqrt{75}} = \frac{5}{5\sqrt{3}} = \frac{1}{\sqrt{3}} \times \frac{\sqrt{3}}{\sqrt{3}} = \frac{\sqrt{3}}{\sqrt{9}} = \frac{\sqrt{3}}{3}$ — Simplify the fraction before rationalising.

15 Rationalise the denominators. Simplify your answers if possible.

a $\frac{1}{\sqrt{7}}$ b $\frac{1}{\sqrt{3}}$ c $\frac{1}{\sqrt{5}}$ d $\frac{1}{\sqrt{20}}$

e $\frac{2}{\sqrt{8}}$ f $\frac{3}{\sqrt{15}}$ g $\frac{32}{\sqrt{40}}$ h $\frac{11}{\sqrt{11}}$

16 Reasoning / Problem-solving

Ben types $\frac{1}{\sqrt{7}}$ into his calculator. His display shows $\frac{\sqrt{7}}{7}$.

a Show that $\frac{1}{\sqrt{7}} = \frac{\sqrt{7}}{7}$.

b Use your calculator to check your answers from **Q14**.

17 The area of a rectangle is $20\,\text{cm}^2$. The length of one side is $\sqrt{5}$ cm.
Work out the length of the other side. Give your answer as a surd in its simplest form.

18 Work out the area of these shapes.

a
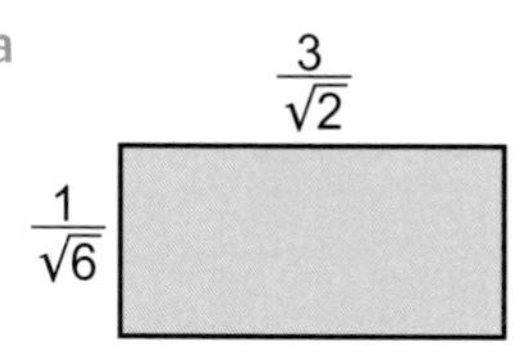

b
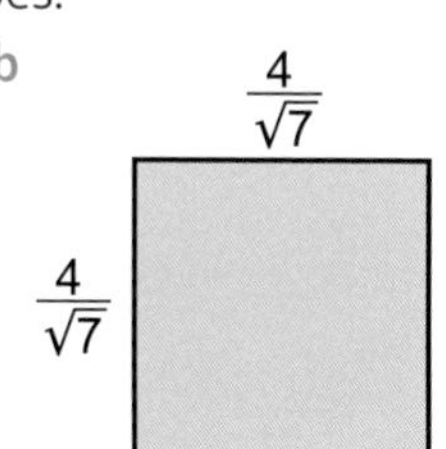

c
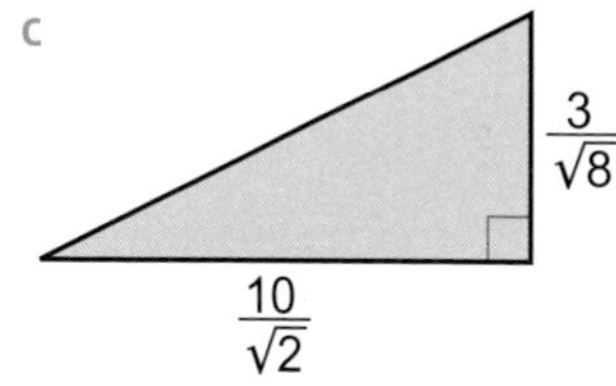

Give your answer as a surd in its simplest form.

Q18 hint Multiply the numerators and denominators separately. Then rationalise the denominator.

1 Problem-solving

Objective

- Use pictures or lists to help you solve problems.

Example 5

A furniture maker orders 22 metal legs. He uses the legs to make three-legged stools and four-legged chairs. Describe the different ways he can use all the legs.

| | | | | | | | | | | | | | | | | | | | | | | — Draw a picture to represent 22 metal legs.

1 stool

Circle 3 legs for one stool. Can you use the remaining 19 legs to make complete chairs? No.

2 stools 4 chairs

Circle another 3 legs for two stools. Can you use the remaining 16 legs to make complete chairs? Yes.

6 stools 1 chair

Draw the diagram again. Are there any other ways you can circle the legs to make complete chairs and stools?

He can use 22 legs to make 2 stools and 4 chairs or 6 stools and 1 chair. — Write your answer.

An alternative method is to use a list.

Number of stools	1	2	3	4	5	6
Number of chairs	4	4	3	2	1	1
Legs left over	3		1	2	3	

1 There are 20 chairs in a conference room. The conference organiser can put 4, 5 or 6 chairs at a table.
   a Describe the different ways the room can be arranged so that all the chairs are used.
   b What is the maximum number of tables required in the room?

Q1 hint Draw a picture or use a list.

2 A play park is 18 m wide and 31.5 m long. The council plans to enclose it with a fence, using a supporting post every 2.25 m. How many posts does the council need?

Q2 hint Draw a picture to help you see what you do to solve the problem.

3 When two plant stakes are placed end to end, their total length is 1.45 m.
When the two stakes are placed side by side, one is 0.15 m longer than the other.
What lengths are the stakes? Give your answer in cm.

Q3 hint Draw one picture to represent the first sentence. Draw another picture to represent the second sentence.

4 **Finance** In a canteen, a starter costs £0.80, a main costs £2.40 and a dessert costs £1.20.
Three friends bought lunch and paid £10 in total. They each had at least two courses.
How many starters, mains and desserts did they buy?

Q4 hint Find numbers that add to £10.

5 **Finance** A bicycle shop hires road bikes for £25 per day and tandems for £40 per day.
One day a family pays £155.
   a Which type of bicycles did they hire?
   b How many people are in the family?

6 A tour company offers three different walking tours.
The landmark tour leaves every 15 minutes.
The parks tour leaves every 20 minutes.
The museum tour leaves every 45 minutes.
All walking tours start at 9 am. When do the landmark, parks and museum tours next leave at the same time?

Q6 hint List all the different times each tour leaves.

7 **Reflect** How can you solve **Q6** without making a list?
Discussion Does it matter how you solve a maths problem?

1 Check up

Log how you did on your Student Progression Chart.

Calculations, factors and multiples

1 16.7 × 9.2 = 153.64
Use this fact to work out the calculations below.
Check your answers using an approximate calculation.
   a 167 × 9.2
   b 1.5364 ÷ 1.67

2 Estimate the value of $\sqrt{54}$ to the nearest tenth.

3 a Estimate i $(\sqrt{65.1} - 6.17) \times 1.98$ ii $\dfrac{\sqrt{8.19 \times 6.43}}{6.84 \times \sqrt{3.97}}$

   b Use your calculator to work out each answer.
   Give your answers correct to 1 decimal place.

ActiveLearn Homework, practice and support: Higher 1 Check up

4 Write 90 as a product of its prime factors in index form.

5 Find the highest common factor (HCF) and lowest common multiple (LCM) of 14 and 18.

6 In prime factor form, $2450 = 2 \times 5^2 \times 7^2$ and $68\,600 = 2^3 \times 5^2 \times 7^3$.
 a What is the HCF of 2450 and 68 600? Give your answer in prime factor form.
 b What is the LCM of 2450 and 68 600? Give your answer in prime factor form.

Indices and surds

7 Copy and complete.
 a $10^{\square} = 1000$ b $4^3 = \square$ c $2^{\square} = 16$ d $5^{\square} = 1$

8 Work out
 a $\sqrt[3]{\frac{9^2 - 1^2}{10}}$ b $[(6^2 - 2^5) \times 3]^2$ c $\sqrt[3]{\sqrt{81} + (-2)^3}$

9 Write each product as a single power.
 a $9^{-3} \times 9^7$ b $27 \times 9 \times 27$ c $5^7 \div 5^2$
 d $\frac{2^8 \times 16}{2^5}$ e $(2^4)^3$ f $(4^2)^{-1}$

10 Work out
 a 2^{-4} b $25^{\frac{3}{2}}$ c $\left(\frac{16}{81}\right)^{\frac{3}{4}}$ d $16^{-\frac{1}{2}}$

11 Simplify
 a $\sqrt{54}$ b $5\sqrt{1000}$

12 Rationalise the denominators. Simplify your answers if possible.
 a $\frac{1}{\sqrt{10}}$ b $\frac{4}{\sqrt{8}}$

Standard form

13 Write these numbers in standard form.
 a 32 040 000 b 0.0007

14 Write these as ordinary numbers.
 a 5.6×10^4 b 1.09×10^{-3}

15 Work out these. Give your answers in standard form.
 a $(5 \times 10^4) \times (9 \times 10^7)$ b $(3 \times 10^8) \div (6 \times 10^5)$ c $(8 \times 10^3) + (6 \times 10^2)$

16 How sure are you of your answers? Were you mostly
 Just guessing Feeling doubtful  Confident 
 What next? Use your results to decide whether to strengthen or extend your learning.

★ Challenge

17 The diagram shows a warehouse (W) and five destinations (A, B, C, D, E), and the times it takes to drive between each of them.
 A delivery driver has to deliver packages to A, B, C, D and E. He starts and ends at W.
 a Which is the quickest route?
 b Write your own delivery driver question.

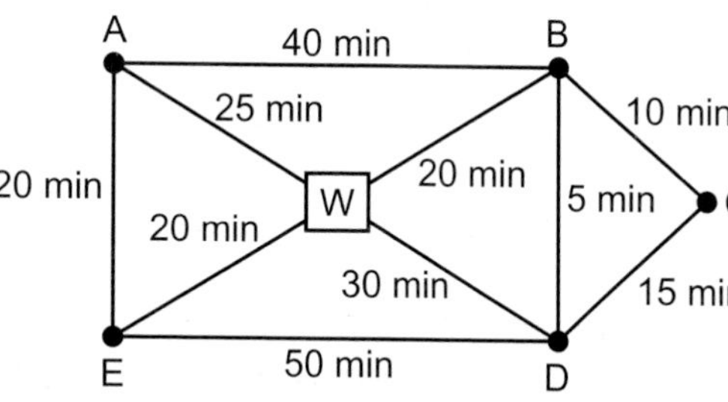

1 Strengthen

Calculations, factors and multiples

1 Copy and complete these number patterns.

a $0.38 \times 29.4 = 11.172$
$3.8 \times 29.4 = \square$
$38 \times 29.4 = \square$
$380 \times 29.4 = \square$
$3800 \times 29.4 = \square$

b $6011.545 \div 94.67 = 63.5$
$60\,115.45 \div 94.67 = \square$
$601\,154.5 \div 94.67 = \square$
$6\,011\,545 \div 94.67 = \square$
$60\,115\,450 \div 94.67 = \square$

Q1a hint $3.8 = 0.38 \times 10$
$3.8 \times 29.4 = 11.172 \times 10$

2 $8.9 \times 7.21 = 64.169$
Use this fact to work out the calculations below.
Check your answers using an approximate calculation.

a 8.9×72.1 b 8.9×7210 c 0.89×0.721
d 0.089×0.721 e $64.169 \div 72.1$ f $6416.9 \div 7.21$

Q2 hint Write out a number pattern to help you.

3 Copy and complete this square root number line.

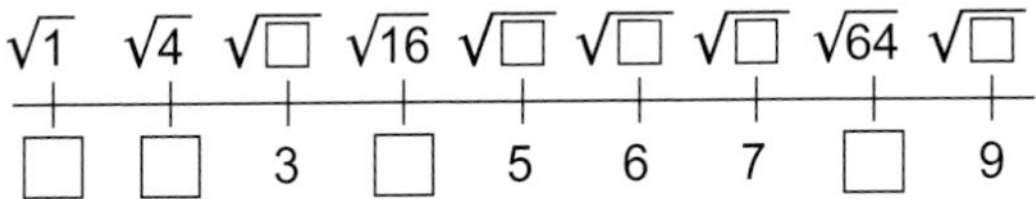

Q4a hint Use your number line from **Q3**. Which two square roots does $\sqrt{52}$ lie between? Which is it closer to?

4 Estimate the value to the nearest tenth.

a $\sqrt{52}$ b $\sqrt{60}$ c $\sqrt{75}$

5 a Estimate

i $\sqrt{4.09 \times 8.96}$
ii $25.76 - \sqrt{4.09 \times 8.96}$
iii $\sqrt[3]{26.64} + \sqrt{80.7}$
iv $\dfrac{\sqrt{6.91 \times 9.23}}{3.95^2 \div 2.03^3}$

b Use a calculator to work out each answer.
Give your answer correct to 1 decimal place.

Q5a i hint Round each number to the nearest whole number. Which square root is it closest to on your square root number line?

6 Copy and complete these calculations in index form

a $2 \times 2 \times 2 \times 3 \times 3 = 2^3 \times 3^{\square}$
b $2 \times 2 \times 3 \times 5 = 2^{\square} \times \square \times \square$
c $3 \times 3 \times 3 \times 3 \times 7 \times 7 =$

7 a Copy and complete this factor tree for 60 until you end up with just prime factors.

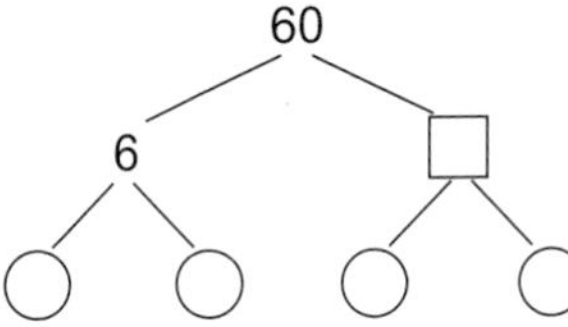

b Write 60 as a product of its prime factors.
c Write your answer to part **b** in index form.

Q7b hint Write all the prime factors from your tree multiplied together.

8 Write each number as a product of its prime factors in index form.

a 24 b 80 c 45
d 30 e 16 f 72

9 a Write 18 as a product of its prime factors.

b Write 45 as a product of its prime factors.

45 = □ × □ × □

c Copy and complete this diagram.

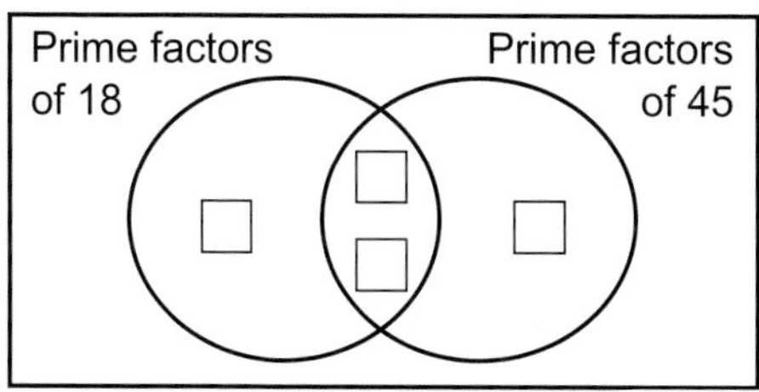

i Put any prime factors of both numbers where circles overlap.

ii Put the remaining prime factors of 18 in the left-hand part of the left circle.

iii Put the remaining prime factors of 45 in the right-hand part of the right circle.

d Work out the HCF.

e Work out the LCM.

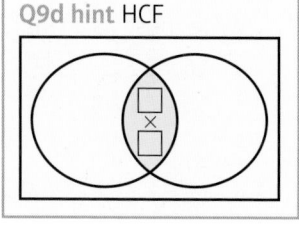

Q9d hint HCF

Q9e hint LCM

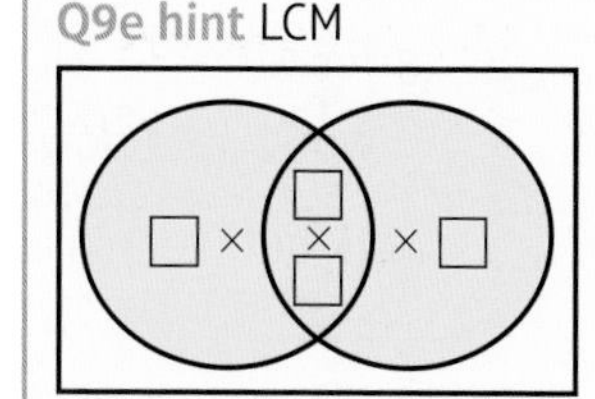

10 Find the highest common factor (HCF) and lowest common multiple (LCM) of

a 20 and 30 b 21 and 28

c 15 and 25 d 44 and 36

Q10 hint Use the method in **Q9**.

Indices and surds

1 Copy and complete.

a $2 \times 2 \times 2 = 2^{\square} = \square$ b $5 \times 5 = 5^{\square} = \square$ c $-3 \times -3 \times -3 = (-3)^{\square} = \square$

2 Work out

a $2^5 = \square$ b $5^3 = \square$

c $4^{\square} = 64$ d $10^{\square} = 1000$

Q2c hint
$4^1 = 4$
$4^2 = 4 \times 4 = \square$
$4^3 = 4 \times 4 \times 4 = \square$

3 Work out

a $3 \times \sqrt{81}$ b $2^3 \times \sqrt{16}$

c $\sqrt[3]{-27} \times \sqrt{64}$ d $10^2 \div -\sqrt{25}$

e $(-3)^2 \times \sqrt{49}$ f $\sqrt{9} \times \sqrt[3]{-125} \times \sqrt[3]{-1000}$

Q3 hint First work out any powers or roots. Then multiply or divide.

4 Work out

a $5 \times (4^2 - 3^2) - 2^3$

b $[(9^2 \div 3^2) + 2^2]^2$

c $\sqrt{2\sqrt{49} + \sqrt[3]{8}}$

Q4b hint First work out the round brackets. Then work out the square brackets.

5 Copy and complete.

a $(3 \times 3 \times 3 \times 3 \times 3) \times (3 \times 3 \times 3) = 3^{\square} \times 3^{\square} = 3^{\square}$

b $(4 \times 4 \times 4 \times 4) \times (4 \times 4 \times 4 \times 4 \times 4 \times 4) = 4^{\square} \times 4^{\square} = 4^{\square}$

c $\dfrac{⑥ \times ⑥ \times 6 \times 6}{⑥ \times ⑥} = \dfrac{6^4}{6^2} = 6^{\square}$

d $\dfrac{7 \times 7 \times 7 \times 7 \times 7 \times 7}{7 \times 7 \times 7} = \dfrac{7^{\square}}{7^{\square}} = 7^{\square}$

e To multiply powers, _____ the indices. To divide powers, _____ the indices.

6 Work out

a $5^6 \times 5^3 =$ b $7^2 \times 7^9 =$

c $5^8 \div 5^3 =$ d $9^8 \div 9^2 =$

e $8^4 \times 8^{-6} =$ f $7^3 \div 7^{-4} =$

Q6 hint Use the rules from **Q5**.

7 Write these as a single power of a prime number.

a $16 \times 8 = 2^{\square} \times 2^{\square} =$ b $25 \times 125 \times 25$

c $16 \times 64 \times 8$ d $27 \times 27 \times 9$

8 Copy and complete.

a $(4^2)^3 = 4^2 \times 4^2 \times 4^2 = 4^{\square}$

b $(6^3)^4 = 6^3 \times 6^3 \times 6^3 \times 6^3 = 6^{\square}$

c $(7^5)^2 =$

d $(8^3)^7 =$

e To work out a power raised to a power, __________ the indices.

9 a Work out using a calculator.

i 5^0 ii 7^0 iii 192^0 iv $(-3)^0$

b Use your answer to part **a** to work out these without a calculator.

i 12^0 ii $(-6)^0$ iii 2456^0 iv 10^0

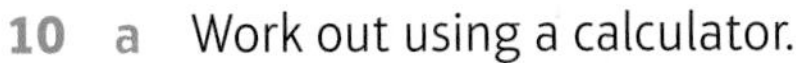

10 a Work out using a calculator.

i $\sqrt{169}$ ii $169^{\frac{1}{2}}$

b Use your answer to part **a** to work these out without a calculator.

i $64^{\frac{1}{2}}$ ii $25^{\frac{1}{2}}$ iii $81^{\frac{1}{2}}$ iv $144^{\frac{1}{2}}$

c Work out using a calculator.

i $\sqrt[3]{512}$ ii $512^{\frac{1}{3}}$

d Use your answer to part **c** to work these out without a calculator.

i $125^{\frac{1}{3}}$ ii $27^{\frac{1}{3}}$ iii $1000^{\frac{1}{3}}$ iv $8^{\frac{1}{3}}$

e Copy and complete.

$16^{\frac{1}{4}} = \sqrt[\square]{16} = \square$

Q10e hint Use what you have learned in parts **a** to **d**.

11 a Work out

i $64^{\frac{1}{3}}$ ii $64^{\frac{2}{3}}$

b Use your answer to **Q10d** to help you work out

i $125^{\frac{2}{3}}$ ii $27^{\frac{2}{3}}$ iii $1000^{\frac{2}{3}}$ iv $8^{\frac{2}{3}}$

c Use your answer to **Q10e** to help you work out $16^{\frac{3}{4}}$.

Q11a i hint $64^{\frac{2}{3}} = \left(64^{\frac{1}{3}}\right)^2$ Work out $64^{\frac{1}{3}}$ and square your answer.

Q11c hint $16^{\frac{3}{4}} = \left(16^{\frac{1}{4}}\right)^{\square}$ Use the same strategy as in part **a**.

12 a Copy and complete.

i $4^{-3} = \dfrac{1}{4^{\square}}$ ii $\dfrac{1}{10^5} = 10^{\square}$ iii $\dfrac{1}{2} = 2^{\square}$

iv $3^{-\frac{1}{3}} = \dfrac{1}{3^{\square}}$ v $\left(\dfrac{7}{6}\right)^{\square} = \dfrac{36}{49}$

b Work out

i 4^{-1} ii 10^{-2} iii $36^{-\frac{1}{2}}$ iv $125^{-\frac{1}{3}}$

13 Copy and complete.

a $50 = \square \times 2$, so $\sqrt{50} = \sqrt{\square} \times \sqrt{2} = \square\sqrt{2}$

b $84 = \square \times 21$, so $\sqrt{84} = \sqrt{\square} \times \sqrt{\square} = \square\sqrt{\square}$

Simplify

c $\sqrt{96}$ d $\sqrt{175}$ e $\sqrt{128}$

Q13c hint Write the square numbers up to 100. Find a square number that is a factor of 96.

14 Work out

a $\sqrt{4} \times \sqrt{4}$ b $\sqrt{25} \times \sqrt{25}$ c $\sqrt{17} \times \sqrt{17}$ d $\sqrt{21} \times \sqrt{21}$

15 Rationalise the denominator.
Simplify your answer if possible.

a $\frac{1}{\sqrt{17}}$ b $\frac{3}{\sqrt{21}}$

c $\frac{1}{\sqrt{8}}$ d $\frac{6}{\sqrt{20}}$

Q15a hint Multiply both the numerator and denominator by $\sqrt{17}$.

Q15c hint First rewrite the denominator.
$\sqrt{8} = \square\sqrt{2}$

Standard form

1 Are these numbers in standard form?
If not, give reasons why.

a 9.004×10^{-3}

b 32×10^{5}

c 7.3 million

d 0.8×10^{7}

Q1 hint A number written in standard form looks like this.

$A \times 10^{n}$

number between 1 and 10 — times sign — power of 10

2 Write each number using standard form.

a $68\,000 = 6.8 \times 10^{\square}$

b 94 000 000

c 801 000

d 0.000 004

e 0.0039

f 0.000 000 053

Q2a hint 6.8 lies between 1 and 10.
Multiply by 10 how many times to get 68 000?

Q2d hint 4 lies between 1 and 10.
Divide by 10 how many times to get 0.000 004?

3 Work out

a $(4 \times 10^{2}) \times (2 \times 10^{7}) = \square \times 10^{\square}$

b $(3 \times 10^{9}) \times (2 \times 10^{5})$

c $(6 \times 10^{4}) \times (1 \times 10^{-2})$

d $(6 \times 10^{8}) \times (8 \times 10^{4})$

e $(7 \times 10^{3}) \times (8 \times 10^{6})$

f $(8 \times 10^{-4}) \times (6 \times 10^{-2})$

Q3a hint

$\underbrace{4 \times 2}_{\square} \times \underbrace{10^{2} \times 10^{7}}_{10^{\square}}$

Q3d hint $48 = 4.8 \times 10^{\square}$

4 a Write 2.5×10^{4} and 1.3×10^{-2} as ordinary numbers.

b Use your answers to part **a** to help you work out
$(2.5 \times 10^{4}) + (1.3 \times 10^{-2})$

Q4a hint $10^{4} = 10\,000$
$10^{-2} = \frac{1}{100}$

1 Extend

1 **Problem-solving** Square A has a side length of 9.2 cm.
Square B has a perimeter of 34.4 cm.
Square C has an area of 80 cm^{2}.

a Which square has the greatest perimeter?

b Which square has the smallest area?

2 Show that $27^2 = 9^3 = 3^6$.

Q2 communication hint 'Show that' means show your working.

3 **Exam-style question**

Here are some properties of a number.

- It is a common factor of 216 and 540.
- It is a common multiple of 9 and 12.

Write two numbers with these properties. **(6 marks)**

Exam hint
There are 6 marks so most of them are for showing your working.

4 a Write 48, 90 and 150 as products of their prime factors.

b Use a Venn diagram to work out the HCF and LCM of 48, 90 and 150.

Discussion Explain how the diagram can be used to find the HCF and LCM of any two of the numbers.

Q4b hint

48 90 150

Put prime factors of all three in the very centre first.

5 **Real** A new school is deciding whether their lessons should be 30, 50 or 60 minutes.

Each length of lesson fits exactly into the total teaching time of the school day.

How long is the teaching time of the school day?

Discussion Ryan says there is more than one answer to this question. Is Ryan correct? Explain your answer.

6 **Reasoning**

a Use prime factors to explain why numbers ending in a zero must be divisible by 2 and 5.

b How many zeros are there at the end of $2^4 \times 3^7 \times 5^6 \times 7^2$?

c Use prime factors to work out $32 \times 9 \times 3125$. Write your answer as an ordinary number and in standard form.

7 Write each of these as a simplified product of powers.

a $10^5 \times 2^3 \times 5^4 = (2 \times 5)^5 \times 2^3 \times 5^4 = 2^{\square} \times 5^{\square} \times 2^3 \times 5^4 = 2^{\square} \times 5^{\square}$

b $6^3 \times 2^4 \times 3^3$

c $15^3 \times 10^4 \times 6^2$

d $30^4 \times 24^2 \times 15^3$

8 Estimate the value of 5.1^4

Q8 hint $5.1^4 = 5.1^2 \times 5.1^2$

9 **STEM** Write each answer

i as an ordinary number ii in standard form.

a Saturn has a diameter of 120 536 000 m.
Convert this to kilometres.

b The distance from the Sun to Mars is 227 900 000 km.
Convert this distance to metres.

c The diameter of a grain of sand is 4 μm.
Convert this to metres.

d The wavelength of an X-ray is 0.1 nm.
Convert this to metres.

10 Every six months, new licence plates are issued in the UK.

A licence plate consists of two letters, then two numbers, then three letters.

The numbers are fixed, but the letters vary.

a If all letters can be used, how many possible combinations are there?

b If only 21 letters can be used, how may possible combinations are there?

11 **STEM / Problem-solving** A container ship carries 1.8×10^8 kg.
An aeroplane can carry 3.8×10^5 kg.
What is the difference in their mass? Write your answer in standard form.

12 **Exam-style question**

Work out

a $\frac{5}{\sqrt{2}} + \frac{8}{\sqrt{32}}$ **b** $\frac{7}{\sqrt{72}} - \frac{3}{\sqrt{8}}$

Write each answer in the form $a\sqrt{2}$. **(3 marks)**

> **Q12 strategy hint** To add and subtract fractions you need to wite them with a common denominator.

13 Write $3^{-\frac{1}{2}}$ as a surd and rationalise the denominator.

14 **Exam-style question**

A restaurant offers 5 starters, 7 mains and 3 desserts.
A customer can choose

- just one course
- any combination of two courses
- all three courses.

Show that a customer has 191 options altogether. **(3 marks)**

> **Exam hint** Show your working clearly.

15 Estimate the value of

a $(3.1 \times 10^3)^2$ b $\sqrt{62 \times 10^4}$
c $(1.9 \times 10^{-2})^3$

> **Q15a hint** $(3.1 \times 10^3)^2 = 3.1 \times 10^3 \times 3.1 \times 10^3$

16 Estimate to the nearest whole number

a 2.7^3 b 1.4^3 c 2.1^4
d $\sqrt[3]{40}$ e $\sqrt[3]{12}$ f $\sqrt[5]{30}$

> **Q16d hint** Use a number line to help.

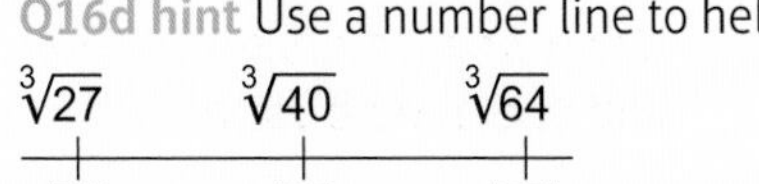

1 Knowledge check

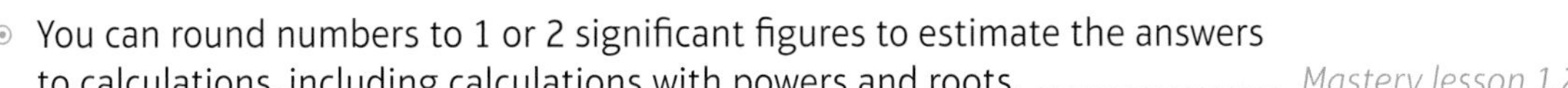

- When there are m ways of doing one task and n ways of doing a different task, the total number of ways the two tasks can be done is $m \times n$. .... *Mastery lesson 1.1*
- You can round numbers to 1 or 2 significant figures to estimate the answers to calculations, including calculations with powers and roots. .... *Mastery lesson 1.2*
- You can use a **prime factor tree** to write a number as the product of its **prime factors**. .... *Mastery lesson 1.3*
- You can use a **Venn diagram** of prime factors to work out the **highest common factor** and **lowest common multiple** of two numbers. .... *Mastery lesson 1.3*
- The **prime factor decomposition** of a number is the number written as the product of its prime factors. It is usually written in index form. .... *Mastery lesson 1.3*
- When multiplying powers, add the indices: $x^m \times x^n = x^{m+n}$
  When dividing powers, subtract the indices: $x^m \div x^n = x^{m-n}$
  To raise a power to another power, multiply the indices.
  $x^{-n} = \frac{1}{x^n}$ $x^{\frac{1}{n}} = \sqrt[n]{x}$ $x^{\frac{m}{n}} = (\sqrt[n]{x})^m$ .... *Mastery lesson 1.4, 1.5*
- A number in **standard form** is written in the format $A \times 10^n$, where A is a number between 1 and 10 and n is an integer. .... *Mastery lesson 1.6*

- To write a number in standard form:
  - work out the value of A
  - work out how many times A must be multiplied or divided by 10. This is the value of n. .... *Mastery lesson 1.6*
- To simplify a **surd**, identify any factors that are square numbers. .... *Mastery lesson 1.7*
- To **rationalise a denominator**, multiply the numerator and the denominator by the surd in the denominator and simplify. .... *Mastery lesson 1.7*

Reflect

For each statement A, B and C, choose a score:
1 – strongly disagree; 2 – disagree; 3 – agree; 4 – strongly agree
A I always try hard in mathematics
B Doing mathematics never makes me worried
C I am good at mathematics
For any statement you scored less than 3, write down two things you could do so that you agree more strongly in the future.

1 Unit test

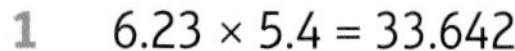

Log how you did on your Student Progression Chart.

1 $6.23 \times 5.4 = 33.642$
 a Write down two more multiplications with an answer of 33.642.
 b Write down a division with an answer of 0.623. *(3 marks)*

2 **Exam-style question**

List these numbers in order, starting with the smallest.
Show your working.

3.2^2 $\sqrt[3]{27}$ $\sqrt{69}$ 13.74 **(3 marks)**

3 a Estimate $(17.9 - \sqrt{36.13}) \times 3.89$
 b Use a calculator to work out the answer. Give your answer correct to 1 decimal place. *(2 marks)*

4 a Write 42 as a product of its prime factors.
 b Use your answer to write 84^3 as a product of its prime factors in index form. *(4 marks)*

5 Work out the HCF and LCM of 75 and 30. *(3 marks)*

6 **Real** Ben and Sadie are doing a sponsored walk around a circuit.
Ben takes 25 minutes to do one circuit and Sadie takes 45 minutes.
They start together at 9:30 am.
When will they next cross the start line together? *(2 marks)*

7 Find the value of a.
 a $5^3 \times 5^a = 5^9$ b $6^a \div 6^{-5} = 6^8$ c $8^a \times 8^a = 8^4$ *(3 marks)*

8 Write $(3^8)^4$ as a single power. *(1 mark)*

9 Use prime factors to determine whether 2520 is divisible by 18. *(2 marks)*

10 Write each number in standard form.
 a 0.000 000 65 b 0.9 million c 320×10^7 *(3 marks)*

11 Write $\left(\frac{4}{3}\right)^{-2}$ as a fraction in its simplest form. *(2 marks)*

ActiveLearn Homework, practice and support: Higher 1 Unit test

12 Let $x = 6 \times 10^5$ and $y = 8 \times 10^4$. Work out

a $x + y$ b $x - y$ c xy d $\frac{y}{x}$

Write your answers in standard form. *(4 marks)*

13 $9^{18} = 27^x$

Work out the value of x. *(2 marks)*

14 Work out the area of this shape.

Write your answer as a simplified surd.

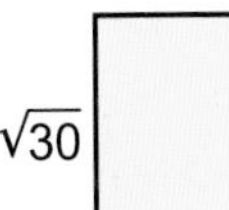

(3 marks)

15 **Exam-style question**

How many different 4-digit odd numbers are there, where the first digit is not zero? **(3 marks)**

16 Rationalise the denominator. $\frac{8}{\sqrt{6}}$ *(2 marks)*

Sample student answer

The maths is correct, but the student will only get 2 marks. Why?

Exam-style question

One sheet of paper is 9×10^{-3} cm thick.

Mark wants to put 500 sheets of paper into the paper tray of his printer.

The paper tray is 4 cm deep.

Is the paper tray deep enough for 500 sheets of paper?

You must explain your answer. **(3 marks)**

June 2013, Q15, 1MA0/1H

Student answer

$9 \times 10^{-3} \times 500$

0.009×500

$9 \times 500 = 4500 \div 1000 = 4.5$

2 ALGEBRA

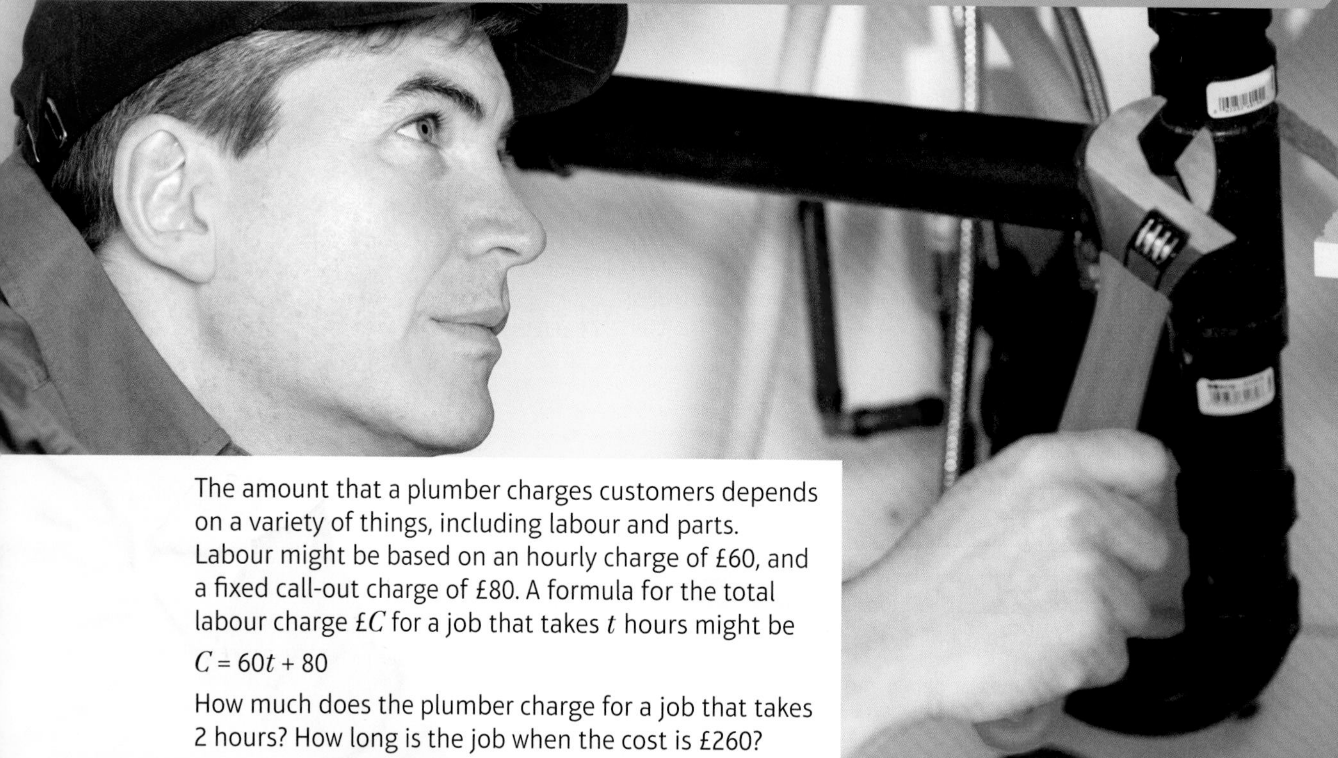

The amount that a plumber charges customers depends on a variety of things, including labour and parts. Labour might be based on an hourly charge of £60, and a fixed call-out charge of £80. A formula for the total labour charge £C for a job that takes t hours might be

$C = 60t + 80$

How much does the plumber charge for a job that takes 2 hours? How long is the job when the cost is £260?

2 Prior knowledge check

Numerical fluency

1 Write down the highest common factor (HCF) of
- a 12 and 18
- b 15 and 35
- c 30 and 36
- d 22 and 44

2 Work out
- a $(-3) \times (-4)$
- b $\frac{6}{-3}$
- c $-4 - 7$
- d $5 - (-3)$
- e 2^4
- f $(3^2)^2$

3 Simplify these fractions.
- a $\frac{2}{4}$
- b $\frac{10}{25}$
- c $\frac{18}{24}$
- d $\frac{30}{3}$

Algebraic fluency

4 Simplify
- a $x + x$
- b $y \times y$
- c $w \times 2$
- d $4t \div 4$
- e $5q \div q$
- f $3z - z$

5 Simplify
- a $p \times p \times p \times p$
- b $c \times c \times d \times d \times d$
- c $7m \times 2m$
- d $3f \times (-6f)$
- e $x \times 4x \times 9x$
- f $y^2 \div y$

6 Work out the value of
- a $4p^3$ when $p = 2$
- b $2(m + 7)$ when $m = 3$
- c $5x + y^3$ when $x = 2$ and $y = 3$
- d $10 - (s + t)^2$ when $s = 1$ and $t = 2$

7 Use the formula $v = u + at$ to work out the value of v when $u = 10$, $a = 2$ and $t = 3$.

8 Expand
- a $7(x + 3)$
- b $2(x - 3)$
- c $3(y^2 + 7)$
- d $9(2x - y + 1)$

9 Factorise each expression completely.
- a $8x - 2$
- b $20y + 15$
- c $c^2 - 2c$
- d $n + 2n^2$

10 Solve these equations.
Show your working.

a $x + 7 = 5$ b $5x - 1 = 19$

c $5(x - 3) = 10$ d $4(x + 1) = 36$

11 Find the value of x in the formula $R = 2ax - b$ when $R = 23$, $a = 3$ and $b = 7$.

12 Write an equation and use it to find the value of x in the diagram.

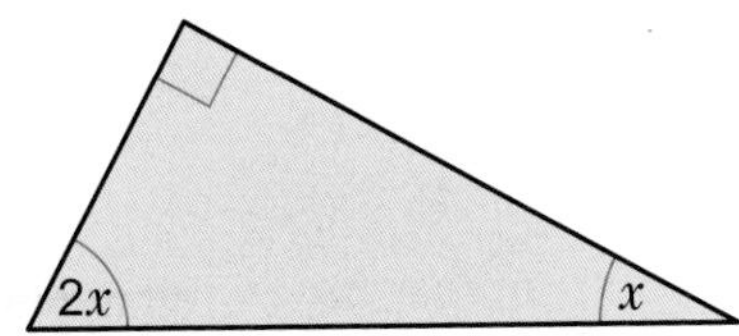

13 Make x the subject of each.

a $x - 5 = y$ b $4x = y$

14 a Work out the output of this function machine when the input is 4.

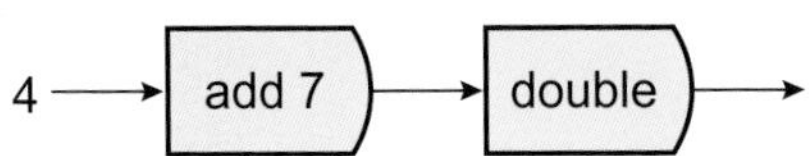

b By using an inverse function machine, or otherwise, find the input when the output is 48.

15 Use these position-to-term rules to work out the first four terms of each sequence.

a $7n + 2$ b $20 - 6n$

16 Write down the term-to-term rule and the next two terms of each sequence.

a 2, 11, 20, …. b −1, −3, −9, …

c 6, 2, −2, … d 2, 0.2, 0.02, …

17 Which of the sequences in **Q16** are

a arithmetic

b geometric

c ascending

d descending?

18 Write down the first four terms of the sequence defined by

a first term = 4
term-to-term rule is 'add 7'

b first term = 3
term-to-rule is 'multiply by 2'

* Challenge

19 **a** Work out

i $1 + 2$

ii $1 + 2 + 3$

iii $1 + 2 + 3 + 4$

iv $1 + 2 + 3 + 4 + 5$

b By substituting $n = 2, 3, 4$ and 5 into the formula $\frac{1}{2}n(n + 1)$, verify that this formula produces the sum of the first n positive whole numbers.

c Use the formula in part **b** to work out the sum of the first 100 whole numbers.

d Work out

i $1^3 + 2^3$

ii $1^3 + 2^3 + 3^3$

iii $1^3 + 2^3 + 3^3 + 4^3$

iv $1^3 + 2^3 + 3^3 + 4^3 + 5^3$

e By comparing your answers with those for part **a**, write down a formula for the sum of the first n cube numbers.

2.1 Algebraic indices

Objective

- Use the rules of indices to simplify algebraic expressions.

Did you know?

Algebraic functions have the same rules that you used with numbers in Unit 1.

Fluency

Evaluate $\sqrt{4} \times \sqrt{4}$ $\sqrt{25} \times \sqrt{25}$ $\sqrt[3]{8} \times \sqrt[3]{8} \times \sqrt[3]{8}$

Warm up

1 Write as a power of 2.

a $2^3 \times 2^4$ b $2^5 \div 2^2$ c $(2^3)^4$ d $\frac{1}{2}$

2 Write as a power of a single number.

a $\dfrac{10^4 \times 10^3}{10^2}$ b $(5^{-2})^3 \times 5^9$ c $\dfrac{(3^{10})^{\frac{1}{2}}}{3^2}$

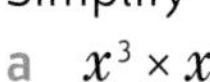

Questions in this unit are targeted at the steps indicated.

3 Simplify

a $x^3 \times x^4$ b $x^2 \times x^5$

c $a^7 \times a^4$ d $y^2 \times y^3 \times y^4$ e $m^{\frac{1}{2}} \times m^{\frac{3}{2}}$

Q3a hint $x^3 \times x^4 = \overbrace{x \times x \times x}^{3} \times \overbrace{x \times x \times x \times x}^{4} = x^{\square}$

Q3b hint $x^m \times x^n = x^{m+n}$

4 Simplify

a $2a^3 \times 3a^5$ b $4c \times 2c^5$

c $4n^2 \times 10n^5$ d $v^3 \times 7v^2$

e $5s^2t \times 3s^3t^5$ f $2pq^2 \times 5p^2q^3 \times 3p^3q$

Q4a hint $2a^3 \times 3a^5 = 2 \times 3 \times a^3 \times a^5 = \square a^{\square}$

Q4e hint $5s^2t \times 3s^3t^5 = 5 \times 3 \times s^2 \times s^3 \times t \times t^5$

5 Simplify

a $x^5 \div x^3$ b $x^7 \div x^4$

c $\dfrac{p^8}{p^5}$ d $y^7 \div y$

e $\dfrac{r^{10}}{r^9}$ f $\dfrac{t^3 \times t^5}{t^6}$

Q5a hint $x^5 \div x^3 = \dfrac{x^5}{x^3} = \dfrac{\cancel{x} \times \cancel{x} \times \cancel{x} \times x \times x}{\cancel{x} \times \cancel{x} \times \cancel{x}} = x^{\square}$

Q5b hint $x^m \div x^n = x^{m-n}$

6 Simplify

a $\dfrac{14g^{10}}{7g^8}$ b $\dfrac{6f^5}{2f}$ c $6x^4 \div 2x^2$ d $12w^7 \div 4w^5$

7 Simplify

a $(x^3)^2$ b $(x^6)^3$ c $(t^3)^3$ d $(j^2)^9$

Discussion Which of these expressions are equivalent?

$9x^2 \times x^3$ $(3x^2)^3$ $(3x^3)^2$ $27x^6$ $(-3x^3)^2$ $3x^3 \times 9x^2$

Q7a hint $(x^3)^2 = x^3 \times x^3 = x^{\square}$

8 Simplify

a $(2r^2)^3$ b $(3f^4)^2$ c $\left(\dfrac{b^2}{2}\right)^3$

Q8a hint $(2r^2)^3 = 2^3 \times (r^2)^3$
$= 2^3 \times r^{2 \times 3}$

9 Multiply or divide each pair of expressions connected by a line in this diagram. Divide in the direction of the arrow.

Reflect Which pair of expressions was easiest to multiply/divide? Why? Which pair was hardest? Why?

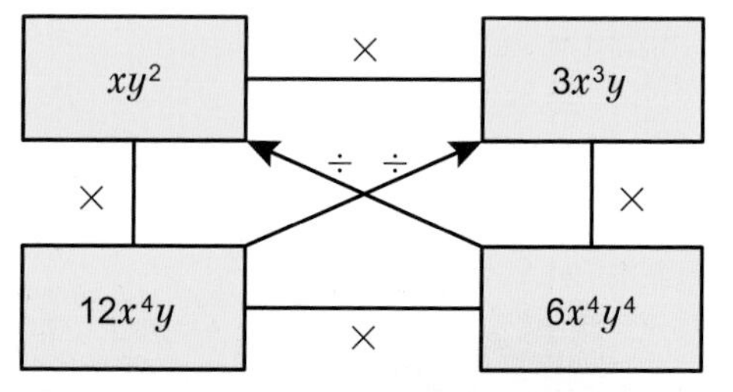

10 Simplify

a $(2x^2y^3)^3$ b $(6x^5y^2)^2$ c $(3x^2y)^4$ d 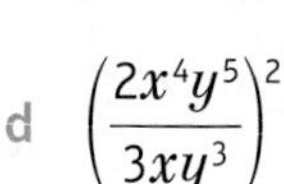 $\left(\dfrac{2x^4y^5}{3xy^3}\right)^2$

11 **Reasoning** Copy and complete

a $x^3 \div x^3 = x^{\square - \square} = x^{\square}$

$x^3 \div x^3 = \dfrac{x^3}{x^3} = \square$

Therefore $x^{\square} = \square$

b $x^3 \div x^4 = x^{\square - \square} = x^{\square}$

$x^3 \div x^4 = \dfrac{x \times x \times x}{x \times x \times x \times x} = \dfrac{\square}{\square}$

Therefore $x^{\square} = \dfrac{\square}{\square}$

c $x^3 \div x^5 = x^{\square - \square} = x^{\square}$

$x^3 \div x^5 = \dfrac{x \times x \times x}{x \times x \times x \times x \times x} = \dfrac{\square}{\square}$

Therefore $x^{\square} = \dfrac{\square}{\square}$

Q11b hint x has a negative power.

Key point 1

$x^0 = 1$ and $x^{-m} = \frac{1}{x^m}$

12 Simplify

a b^{-1} b h^{-3}

c p^0 d r^{-6}

Q12b hint $h^{-3} = (h^3)^{-1} = \frac{1}{\square^{\square}}$

13 **Exam-style question**

a Simplify $3c^2d^3 \times 4cd^{-2}$ **(3 marks)**

b $x^4 \times x^n = x^7$

Work out n. **(1 mark)**

Exam hint
In part **a** first multiply the numbers then simplify $c^2 \times c$ and $d^3 \times d^{-2}$.

14 Simplify

a $(t^2)^{-3}$ b $(x^{-1})^{-2}$ c $(q^2)^0$ d $(w^{-1})^{-1}$

Q14b hint
negative × negative = positive

15 Simplify

a $(x^7y^2)^0$ b $(e^2f^3)^{-1}$ c $(2p^5q)^{-2}$ d $\left(\frac{2u^4}{5v^3}\right)^{-1}$

16 Simplify

a $\sqrt{x^4}$ b $\sqrt{9x^2}$ c $\sqrt{4x^8}$ d $\sqrt{16x^4y^6}$

Q16a hint $x^{\square} \times x^{\square} = x^4$

17 **Reasoning** Copy and complete

a $x^{\frac{1}{2}} \times x^{\frac{1}{2}} = x^{\square+\square} = x^{\square} = \square$

$\sqrt{x} \times \sqrt{x} = \square$

Therefore $x^{\frac{1}{2}} = \square$

b $x^{\frac{1}{3}} \times x^{\frac{1}{3}} \times x^{\frac{1}{3}} = x^{\square+\square+\square} = x^{\square} = \square$

$\sqrt[3]{x} \times \sqrt[3]{x} \times \sqrt[3]{x} = \square$

Therefore $x^{\frac{1}{3}} = \square$

Key point 2

$x^{\frac{1}{n}} = \sqrt[n]{x}$

18 Simplify

a $(3pq^{-4})^{-2}$ b $(16c^6)^{\frac{1}{2}}$ c $(4x^{-2}y^8)^{-\frac{1}{2}}$ d $(32x^{10}y^{-5})^{-\frac{1}{5}}$

Q18 hint Begin by writing $(3pq^{-4})^{-2} = 3^{-2}p^{-2}(q^{-4})^{-2}$

2.2 Expanding and factorising

Objectives

- Expand brackets.
- Factorise algebraic expressions.

Why learn this?

Writing expressions in more than one way helps you to work with them in the easiest way.

Fluency

Simplify $y \times y$ $4x \times x$ $-2p \times p$ $x \times y$ $s \times r$

Warm up

1 Expand

a $4(x + 2)$ b $3(q - 5)$ c $7(2m + 1)$ d $-2(y + 6)$

2 Simplify by collecting like terms.

a $4a + 2 + 5a + 3$ b $3x + 5 - 2x - 1$ c $4r + 2s - 4r + 3s$

3 Find the highest common factor (HCF) of

a 12 and 10   b 18 and 27   c $9x$ and 15

4 **Reasoning** a Write down an expression containing brackets for the area of the rectangle.

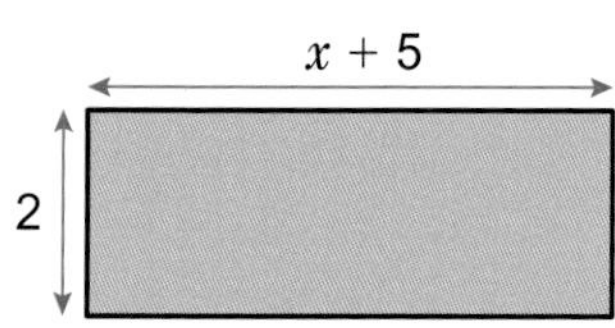

b Copy and complete this diagram to show the areas of the two small rectangles.

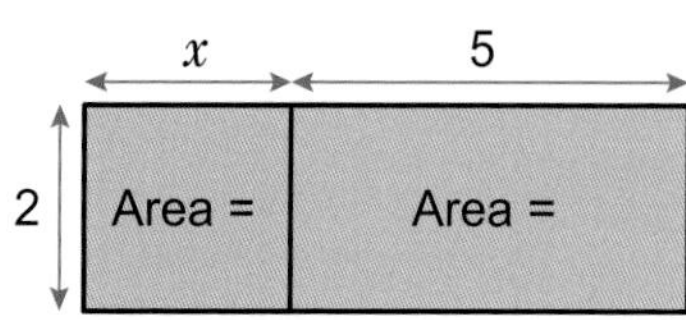

c What do you notice about your answers to parts **a** and **b**?

Key point 3

When the two sides of a relation such as $2(x + 5) = 2x + 10$ are equal for all values of x it is called an **identity** and we write $2(x + 5) \equiv 2x + 10$ using the '$\equiv$' symbol.
An **equation**, such as $2x = 6$, is only true for certain values of x (in this case $x = 3$).

5 **Reasoning** State whether each relation is an equation or an identity. Rewrite the identities using $\equiv$.

a $x \times x = x^2$ b $3x + 4x - x = 6x$ c $3x - 1 = 2x + 1$ d $\frac{6x}{3} = 2x$

6 **Reasoning** By drawing rectangles show that

a $3x(x + 4) \equiv 3x^2 + 12x$ b $x(2y + z) \equiv 2xy + xz$ c $x(y + z) \equiv xy + xz$

Key point 4

To **expand** a bracket, multiply each term inside the brackets by the term outside the brackets.

7 a Expand i $5x(y + 4)$ ii $3y(x + 2)$

b Use your answers to part **a** to expand and simplify $5x(y + 4) + 3y(x + 2)$

Q7b hint Add the two expansions in part **a** and collect like terms.

8 Expand and simplify

a $6(e + 3) + 2e$ b $6y + 2(y + 7)$ c $3(x + 9) + x$
d $6(m + 2) + 3(m + 5)$ e $2a + 5b + 3(a + b)$ f $2(5x + y) + 3(x + 2y)$

Q8 hint Expand all brackets first. Then collect like terms.

9 Expand and simplify

a $x(x - 2)$ b $4(y - 3) + 7y$ c $7t + 3(t - 2)$
d $2p(p + q) - q(p - q)$ e $2w - w(1 - 3w)$ f $5e(e + f) - 2f(e - f)$

10 Find the HCF of

a $4x$ and $6xy$ b $3xy$ and $5x$
c $8xy$ and $12y$ d $5x^2y$ and $10xy^2$

Q10 hint What is the HCF of the numbers? Which letters are common factors?

11 Factorise completely

a $2x + 12$ b $4x + 6xy = \square(\square + \square)$
c $3ab - 5b$ d $7xy + 7xz$
e $ab - abc$ f $t^3 + 2t^2$
g $6p^2q - 9pq$ h $3x^2z + 12xz$
i $20jk^2 - 15j^2k$ j $12pqr - 10pqs$

Q11b hint Write the HCF outside the brackets. Use your answer to **Q10a**.

Q11 Strategy hint
Expand brackets to check.

12 a What is the HCF of $4(s + 2t)^2$ and $8(s + 2t)$?

b Copy and complete

$4(s + 2t)^2 - 8(s + 2t)$

$= \square(s + 2t)[(s + 2t) - \square]$

$= \square(s + 2t)(s + 2t - \square)$

Q12b hint Use your answer to part **a**.

13 Factorise completely

a $14(p + 1)^2 + 21(p + 1)$

b $5(c + 1)^2 - 10(c + 1)$

c $12(y + 4)^2 - 8(y + 4)$

d $(a + 3b)^2 - 2(a + 3b)$

e $5(f + 5) + 10f(f + 5)$

f $5(a + b)^2 - 10(a + b)$

Communication hint **Consecutive integers** are one after the other.

Example 1

Show algebraically that the product of any two consecutive integers is divisible by 2.

One of these two numbers must be even, so it can be written as $2m$ for some whole number, m.

Numbers 1, 2, 3, 4, 5, … are odd, even, odd, even, odd, etc. so a pair of consecutive numbers must contain one odd and one even. If a number is even it is in the 2 times table.

If the other number is n then their product is $2m \times n = 2mn$. $2mn$ has a factor of 2 so it is divisible by 2.

14 **Communication / Reasoning**
Show algebraically that the product of three consecutive integers is divisible by 6.
Discussion What happens for four consecutive integers? Can you use algebra to show it?

Q14 hint When the numbers are consecutive at least one of them is even and one of them is a multiple of 3.

15 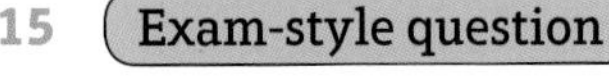
Exam-style question

a Expand $4x(2x - 5y)$ **(1 mark)**

b Factorise completely $4cp - 6cp^2$ **(1 mark)**

c Simplify $\sqrt{(9m^4n^6)}$ **(2 marks)**

Exam hint
Make sure that your final answer cannot be factorised further.

16 **Reflect** In this lesson you have learned about expanding, simplifying and factorising.
Why do you think these methods have these names?

2.3 Equations

Objectives

- Solve equations involving brackets and numerical fractions.
- Use equations to solve problems.

Why learn this?

You can use an equation to work out the distances travelled of a car journey.

Fluency

I think of a number, double it and add 1. The answer is 9. What number did I think of?

Warm up

1 Solve these equations.

a $4x - 5 = 23$ b $3(7x + 4) = 33$ c $9 = 3(7 - 2x)$

2 Write down the lowest common multiple (LCM) of

a 2 and 3 b 6 and 8 c 2, 3 and 12

3 Show that $x = 3$ is a solution of the equation $x^3 - 2x = 21$

4 Expand and simplify

a $2(4x + 3)$ b $2(3x + 1) + 3(5x - 2)$ c $2(2x + 1) - 3(4x - 5)$

5 a Copy and complete to begin to solve the equation.

$$3x + 1 = 5x - 9$$
$$3x + 1 - \square = 5x - 9 - \square$$
$$\square = \square x - 9$$

b Solve the equation.

Q5a hint Subtract $3x$ from both sides of the equation. Then simplify the expression on both sides of the equation.

6 Solve

a $2x + 4 = x + 9$ b $5x + 3 = 7x - 5$ c $x - 5 = 3x - 25$ d $11x - 7 = 9x - 11$

7 a Expand

i $4(3x - 4)$ ii $7(x - 3)$

b Use your answers to part a to solve $4(3x - 4) = 7(x - 3)$

8 a Expand and simplify $2(3x + 5) - 3(x - 2)$

b Use your answer to part **a** to solve $2(3x + 5) - 3(x - 2) = 25$

9 Solve these equations.

a $2(3x - 1) + 5(x + 3) = 24$ b $2(x - 1) - (3x - 4) = 3$

Key point 5

Unless a question asks for a decimal answer, give non-integer solutions to an equation as exact fractions.

10 Solve

a $2(4x - 1) + 3(x + 2) = 1$ b $4(2x + 3) = 5(3x - 2)$ c $3(2x + 9) = 2(4x - 1)$

d $9x - 2(3x - 5) = 6$ e $5(4x - 3) - (6 - 5x) = 0$ f $7(3 - 5x) = 2(x - 6)$

Discussion In part **a**, why is the fraction solution more accurate than the decimal?

11 Simplify these expressions by cancelling.

a $\frac{14x}{7}$ b $\frac{4y}{8}$ c $\frac{27z}{3}$ d $\frac{24w}{4}$

Q11a hint $\frac{14}{7} \times x = \square x$

12 a Copy and complete to begin to solve the equation.

$$\frac{7x - 1}{4} = 5$$
$$\frac{7x - 1}{4} \times \square = 5 \times \square$$
$$7x - 1 = \square$$

b Solve the equation.

Q12a hint Multiply both sides of the equation by 4. Then cancel.

13 a Copy to begin to solve the equation:

$$\frac{10}{x - 4} = 3$$
$$\frac{10}{x - 4} \times (\square) = 3 \times (\square)$$
$$10 = \square x - \square$$

b Solve the equation.

Q13a hint Multiply both sides by $x - 4$. Then cancel the left-hand side and expand the right-hand side

14 a By multiplying both sides of the equation $\frac{2x+1}{3} = \frac{x-5}{9}$ by 9, and cancelling, show that $3(2x + 1) = x - 5$. Then solve the equation.

b By multiplying both sides of the equation $\frac{x}{2} - \frac{x}{3} = \frac{7}{12}$ by 12, and cancelling, show that $6x - 4x = 7$. Then solve the equation.

Discussion How can you choose the number to multiply by?

> **Q14b hint** $\frac{x}{2} \times 12 = \frac{12x}{2} = \square x$
>
> $-\frac{x}{3} \times 12 = \frac{-12x}{3} = \square x$

Key point 6

To solve an equation involving fractions, multiply each term on both sides by the LCM of the denominators.

15 Solve these equations.

a $\frac{b-4}{2} = \frac{b+1}{4}$ b $\frac{n}{2} - \frac{n}{5} = \frac{3}{10}$ c $\frac{c-1}{4} + \frac{c+1}{8} = \frac{3}{2}$

d $\frac{2}{3x+1} = 5$ e $\frac{x-1}{3} + \frac{x+1}{2} + \frac{x}{6} = 7$

> **Q15a hint** Begin by multiplying both sides by the LCM of 2 and 4.

16 **Problem-solving** Find the size of the smallest angle in the triangle.

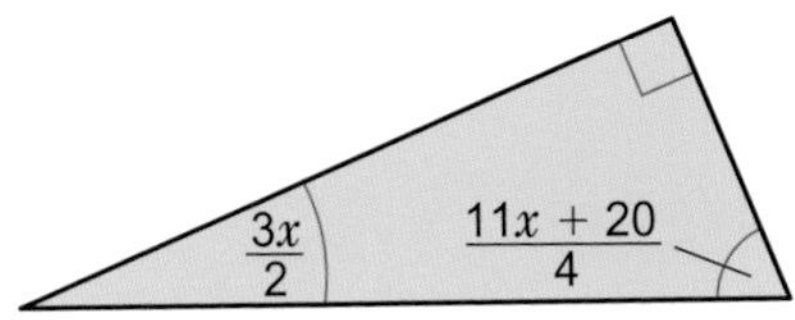

> **Q16 strategy hint** What fact do you know about angles in a triangle?

17 **Real / Reasoning** Bert drove from his house to Bolton at an average speed of 60 mph. He drove back at an average speed of 45 mph. His total driving time was 7 hours.

a Bert lives x miles from Bolton. Write down an expression for the time of his outward journey.

b Write down an expression for the time of his return journey.

c Write down an equation for both parts of the journey in terms of x.

d Solve the equation to work out how far Bert lives from Bolton.

> **Q17a hint**
>
> 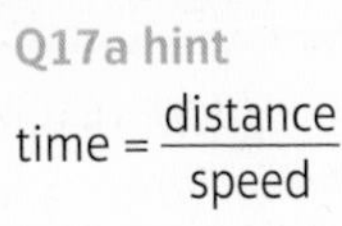
>

2.4 Formulae

Objectives

- Substitute numbers into formulae.
- Rearrange formulae.
- Distinguish between expressions, equations, formulae and identities.

Why learn this?

You can use a formula to work out the acceleration of a Formula 1 racing car.

Fluency

Use the formula $A = lw$ to calculate the area of a rectangle of length 3 m and width 2 m.

1 Work out

a $6 - (3 - 1)$ b $25 - 3 \times 4$ c 2×4^2 d $2 \times 4 - 3 \times 5 + 4 \times 6$

2 a Write 75 million in standard form.

b Write 3×10^8 as an ordinary number.

3 Use a calculator to work out 1.05^4. Round your answer to 2 decimal places.

Warm up

Key point 7

An **expression** contains letter and/or number terms but no equal signs, e.g. $2ab$, $2ab + 3a^2b$, $2ab - 7$

An **equation** has an equals sign, letter terms and numbers. You can solve it to find the value of the letter, e.g. $2x - 4 = 9x + 1$

An **identity** is true for all values of the letters, e.g. $\frac{4x}{2} \equiv 2x$, $x(x + y) \equiv x^2 + xy$

A **formula** has an equals sign and letters to represent different quantities, e.g. $A = \pi r^2$
The letters are **variables** as their values can vary.

4 Write whether each of these is an expression, an equation, an identity or a formula.

a $E = mc^2$ b $4x + 7 = 2x$ c $2v$ d $2(x + y) = 2x + 2y$
e $2p^2q^3$ f $C = 2\pi r$ g πd h $2\pi r = 7$
i $(uv^2)^4 = u^4v^8$ j $\frac{2x}{5} = 9$

Reflect Write your own examples of an expression, an equation, an identity and a formula. Beside each one, write how you know what it is.

5 Use the formula $Q = 2P^3$ to work out the value of Q when
a $P = 10$ b $P = -1$

Q5 hint Use the priority of operations.

6 Use the formula $D = 2X^2 + Y$ to work out the value of D when
a $X = 10$ and $Y = 150$ b $X = -2$ and $Y = 0$

7 **Real / Reasoning** The instructions describe how to cook a joint of beef.

Cook for 30 minutes at 220 °C, followed by 40 minutes per kilogram at 160 °C.

a Work out the total time taken to cook a 2.5 kg joint of beef.
b Write a formula for the total time, T (minutes) to cook m kg of beef.

8 **Problem-solving** a Write a formula, in terms of b and h, for the area, A, of the triangle.

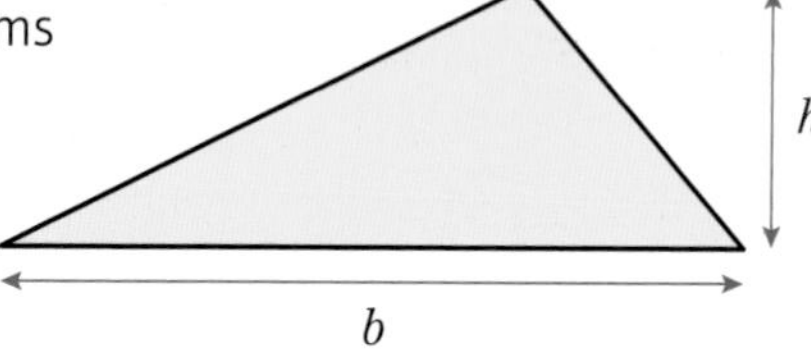

b Use the formula to work out the value of
i A when $b = 6$ and $h = 3$
ii b when $A = 20$ and $h = 4$

9 **Finance** An amount £P is put into a bank account offering r% interest. After n years the value of the savings, S, is given by the formula

$$S = P\left(1 + \frac{r}{100}\right)^n$$

Joe invests £10 000 in this account in January 2015. The interest rate is 4.6%.
How much will his investment be worth in January 2020?
Give your answer to the nearest penny.

10 **STEM / Problem-solving** A car, initially travelling at a speed of u m/s, accelerates at a constant rate of a m/s². The distance, s, travelled in t seconds is given by the formula $s = ut + \frac{1}{2}at^2$.

a A car joins a motorway travelling at 10 m/s and has a constant acceleration of 0.6 m/s².
Work out the distance travelled by the car in 20 s.
b Work out the acceleration of a Formula 1 car which starts from rest and travels 70 m in 2.5 s.

Q10a strategy hint Write down the information you are given, u = □, a = □, t = □. Then substitute into the formula.

Q10b hint At rest, $u = 0$. Substitute all the information into the formula. Then find a.

Key point 8

The **subject** of a formula is the letter on its own, on one side of the equals sign.

Example 2

a Make a the subject of the formula $v^2 = u^2 + 2as$

b Make x the subject of the formula $y = \frac{ax + b}{c}$

a $v^2 = u^2 + 2as$

$v^2 - u^2 = 2as$ — Subtract u^2 from both sides.

$\frac{v^2 - u^2}{2s} = a$ — Divide both sides by $2s$.

$a = \frac{v^2 - u^2}{2s}$ — Re-write in the form $a = \ldots$

b $y = \frac{ax + b}{c}$

$cy = ax + b$ — Multiply both sides by c.

$cy - b = ax$ — Subtract b from both sides.

$\frac{cy - b}{a} = x$ — Divide both sides by a.

$x = \frac{cy - b}{a}$ — Re-write in the form $x = \ldots$

11 Change the subject of each formula to the letter given in the brackets.

a $v = u + at$ [a]

b $E = m - 2n$ [n]

c $W = \frac{3G}{H}$ [G]

d $R = \frac{Q}{7} + C$ [Q]

e $T = \frac{V - W}{3}$ [V]

f $s = ut + \frac{1}{2}at^2$ [a]

12 STEM The formula, $F = \frac{9C}{5} + 32$ is used to convert temperatures from degrees Celsius to degrees Fahrenheit.

a Convert 28 °C into degrees Fahrenheit.

b Make C the subject of the formula.

c Convert 104 °F into degrees Celsius.

13 STEM a Make T the subject of the formula $S = \frac{D}{T}$

b Sometimes the distance between the Earth and Mars is about 57.6 million kilometres. The speed of light is approximately 3×10^8 m/s. Estimate the time taken for light to travel from Mars to the Earth.

14 Exam-style question

a Make a the subject of the formula $c = 3(2ab + 3)$ **(3 marks)**

b Find a when $b = 1.5$ and $c = 63$. **(1 mark)**

Exam hint

Check the value of **a** by substituting the values of all three letters into the original formula.

15 STEM / Reasoning The formula $d = \sqrt{2Rh}$, where $R \approx 6.37 \times 10^6$ metres is the radius of the Earth, gives the approximate distance to the horizon of someone whose eyes are h metres above sea level. Use this formula to estimate the distance (to the nearest metre) to the horizon of someone who stands

a at sea level and is 1.7 m tall

b on the summit of Mount Taranaki, New Zealand, which is 2518 m above sea level.

2.5 Linear sequences

Objectives

- Find a general formula for the nth term of an arithmetic sequence.
- Determine whether a particular number is a term of a given arithmetic sequence.

Why learn this?

Patterns linking data are often used to recognise trends in the data.

Fluency

- What is the next term in the sequence 3.7, 4.1, 4.5, 4.9, 5.3, …?
- What is the value of $6n + 1$ when $n = 1$? … $n = 2$? … $n = 0$?

Warm up

1 Work out the outputs when each of these numbers is used as an input to this function machine.

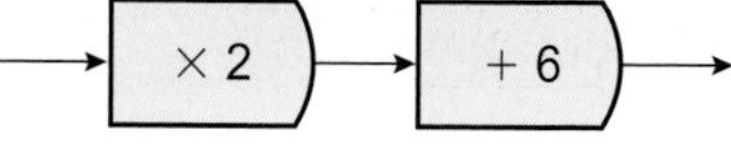

a 0 b 5 c 10

2 Write down the previous term and the next term in this sequence.

☐, 7, 10, 13, 16, 19, 22, ☐, …

3 Write down the first five terms of the sequence with nth term

a $2n$ b $3n + 1$ c $-4n$ d $-2n + 3$

Key point 9

u_n denotes the nth term of a sequence. u_1 is the first term, u_2 is the second term and so on.

4 Work out the 1st, 2nd, 3rd, 10th and 100th terms of the sequence with nth term

a $u_n = 7 + 3n$ b $u_n = 100 - 2n$ c $u_n = 6$

Key point 10

In an **arithmetic sequence**, the terms increase (or decrease) by a fixed number called the **common difference**.

5 For each arithmetic sequence, work out the common difference and hence find the 3rd term.

a 0.63, 0.65, … b $\frac{1}{4}$, $\frac{3}{4}$, …
c 2, −3, … d 0.569, 1.569, …

> **Q5 communication hint**
> 'Hence' means 'use what you have just found to help you'.

Key point 11

The nth term of an arithmetic sequence = common difference × n + zero term

Example 3

a Work out the nth term of the sequence 3, 7, 11, 15, … b Is 45 a term of the sequence?

a $4n$ 4, 8, 12, 16, …) −1
3, 7, 11, 15, …
+4 +4

The nth term is $4n - 1$.

> The common difference is 4. Write out the first five terms of the sequence for $4n$, the multiples of 4. Work out how to get from each term in $4n$ to the term in the sequence.

b $45 = 4n - 1$
$46 = 4n$
$11.5 = n$
45 cannot be in the sequence because 11.5 is not an integer.

> Write an equation using the nth term and solve it.

ActiveLearn Homework, practice and support: Higher 2.5

6 **Reasoning**

a Find the common difference for each sequence you wrote in **Q2** and **Q3**.

b Where does the common difference appear in the nth term?

c Predict the common difference for each sequence.

i nth term $5n - 2$ ii $u_n = -3n + 4$

d Work out the first three terms of each sequence to check your predictions.

7 Write down, in terms of n, expressions for the nth term of these arithmetic sequences.

a 3, 5, 7, 9, 11, … b 14, 18, 22, 26, 30, … c 2, 12, 22, 32, 42, ….

d 13, 10, 7, 4, 1, … e 5, 10, 15, 20, 25, 30, …

8 **Reasoning** a Show that 596 is a term of the arithmetic sequence 5, 8, 11, 14, …

b Show that 139 cannot be a term of the arithmetic sequence 4, 11, 18, 25, …

Reflect How did the worked example help you to answer this question?

9 **Exam-style question**

Here are the first five terms of an arithmetic sequence.

3, 9, 15, 21, 27

a Find an expression, in terms of n, for the nth term of this sequence. **(2 marks)**

b Ben says that 150 is in the sequence.

Is Ben right? Explain your answer. **(1 mark)**

Exam hint

Explain means show your working, then answer the question with either:

Yes, Ben is correct because …

No, Ben is not correct because …

10 **Reasoning** The nth term of the sequence 5, 13, 21, 29, 37, … is $8n - 3$.

a Solve $8n - 3 = 1000$

b Use your answer to part **a** to find the first term in the sequence that is greater than 1000.

Q10b hint What is the next integer value of n? Substitute this into the nth term.

11 **Reasoning** a Find the first term in the arithmetic sequence 2, 11, 20, 29, 38, … that is greater than 4000.

b Find the first term in the arithmetic sequence 400, 387, 374, 361, … that is less than 51.

Q11a hint Begin by finding a formula for the nth term, u_n, and then solve the inequality $u_n > 4000$.

12 **Real / Modelling** Frank weighs 100 kg and goes on a diet losing 0.4 kg a week.

a How much does he weigh after

i 1 week ii 2 weeks iii 3 weeks?

b After how many weeks will Frank weigh less than 89 kg?

Q12b hint Find the nth term of the sequence. Write and solve an inequality.

13 **Real / Modelling** Martina trains for a marathon. In her first week of training she runs 5 miles. Each week after that she increases her run by 0.8 miles. How many weeks of training will it take before she runs more than 26 miles?

Q13 hint Begin by writing the first few terms of the sequence.

14 **Reasoning** The nth term of an arithmetic sequence is $u_n = 7n + 3$.

a Write down the values of the first four terms, u_1, u_2, u_3, u_4.

b Write down the value of the common difference, d.

c By substituting $n = 0$, work out the value of the zero term, u_0.

Discussion What do you notice about your answers to parts **b** and **c**, and the numbers that appear in the formula, $u_n = 7n + 3$? What can you say about the zero terms of the sequences in **Q4**?

15 a Find the outputs when the terms in each of these arithmetic sequences are used as inputs to the function machine.

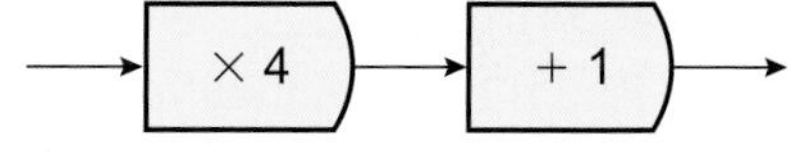

i 2, 5, 8, 11, 14, … ii 10, 20, 30, 40, 50, …

b Compare the common differences for each input sequence with the common difference for the output sequence.
How are these related to the operations used in the function machine?

Key point 12

When an arithmetic sequence with common difference d is input into this function machine, the output sequence has common difference $p \times d$.

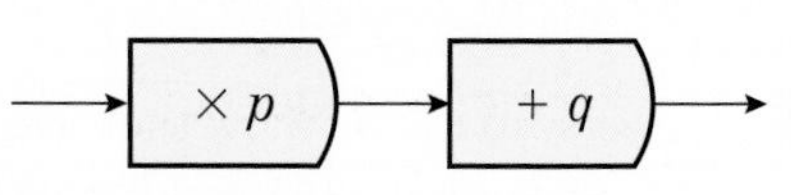

16 Reasoning When 3 is input into this function machine, the output is 10.
When 7 is input into the function machine, the output is 18.

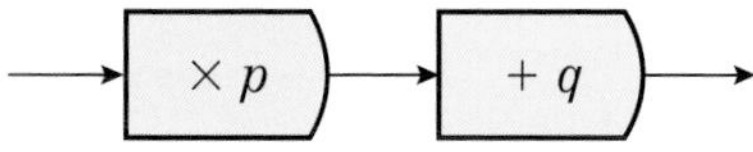

a Work out the difference between the two inputs.
b Work out the difference between the two outputs.
c Use your answers to parts **a** and **b** to find the value of p in the function machine.
d Work out the value of q.

Q16c hint $p = \frac{\text{difference between outputs}}{\text{difference between inputs}}$

Q16d hint Put either of the inputs 3 and 7 into the machine. As a check, they should both work.

17 Reasoning Find the values of p and q in this function machine when the inputs 2 and 7 produce outputs of 20 and 55, respectively.

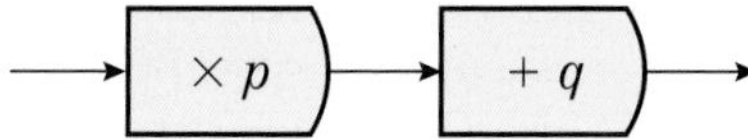

2.6 Non-linear sequences

Objectives

- Solve problems using geometric sequences.
- Work out terms in Fibonacci-like sequences.
- Find the nth term of a quadratic sequence.

Why learn this?

The amount of money you have in a savings account increases using a geometric sequence.

Fluency

- What is the next term of each sequence?
  1, 2, 4, 7, 11, 16, 22, …
  0.25, 1, 4, 16, …
- Are these sequences arithmetic or geometric? Why?

Warm up

1 a Increase £1200 by 4%.
b Decrease £180 by 15%.

2 Find the term-to-term rule and work out the missing numbers in these geometric sequences.
a 3, 6, 12, 24, ☐, ☐, … b 81, ☐, 9, 3, 1, ☐, … c 2, −6, ☐, −54, 162, −486, …

ActiveLearn Homework, practice and support: Higher 2.6

Key point 13

In a Fibonacci type sequence the next number is found by adding the previous two numbers together. e.g. 1, 1, 2, 3, 5, 8, 13, 21, … is a Fibonacci type sequence because $1 + 1 = 2$, $1 + 2 = 3$, $2 + 3 = 5$ and so on.

3 Find the next three terms in each of these Fibonacci-like type sequences.

a 2, 3, □, □, □, … b 1, 4, □, □, □, … c −2, 1, □, □, □, …

Key point 14

In an **geometric sequence** the terms increase (or decrease) by a **constant multiplier**.

4 Write down the first four terms of each sequence.

a $u_n = \frac{1}{n}$ b $u_n = 2^n$ c $u_n = 0.3^n$

Discussion Which of these are geometric sequences?

Q4 hint Substitute $n = 1, 2$ etc. into these formulae.

5 Write down the first five terms of these geometric sequences.

a first term = $\sqrt{2}$; term-to-term rule is 'multiply by $\sqrt{2}$'

b first term = 3; term-to-term rule is 'multiply by $2\sqrt{3}$'

6 **Finance / Problem-solving** Ian is a millionaire. He promises to donate £10 to charity one month, £20 the next month, £40 the next month and so on. Predict how many months until he is donating over £1000.

Q6 Communication hint **Predict** means finding a good guess about what might happen. Check your guess and improve it if you need to.

7 **Finance / Modelling** John invests £8000 in a bank account at 5% interest.

a How much money does John have after 1 year?

b He leaves the interest in the account each year. How much money does he have after

i 2 years ii 3 years?

c How long will it be before his investment exceeds £10 000?

8 **Finance / Modelling** Sarah gets pocket money every week from the age of 5 until her 21st birthday and is given a choice of two options.

Option 1: Get the same number of pounds each week as her age.

Option 2: Get £5 a week aged 5 and increasing by 15% a year.

Which option should Sarah choose?

Give reasons for your answer.

Q8 hint Work out the total amount for each option separately and see which is larger.

Key point 15

A **quadratic sequence** has n^2 and no higher power of n in its nth term.

9 **Reasoning** a Write down the first six terms of the sequence $u_n = n^2$.

b Work out a formula for the nth term of each sequence.

i 2, 5, 10, 17, 26, 37, …

ii 0, 3, 8, 15, 24, 35, …

iii 4, 9, 16, 25, 36, 49, …

Q9b hint Compare with the sequence for n^2. What do you need to add or subtract?

10 Copy and complete this diagram to work out the next term in the sequence 0, 1, 8, 21, …

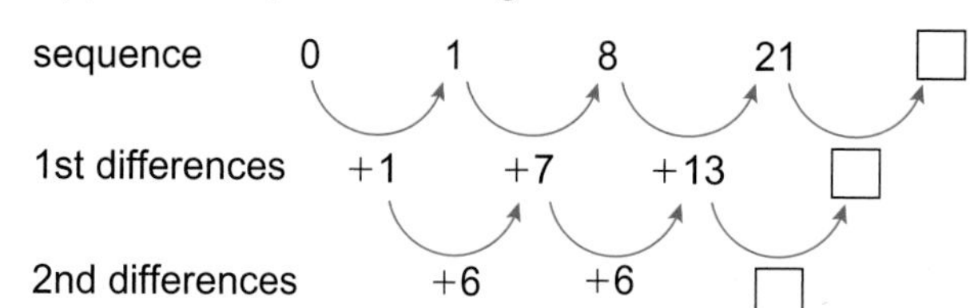

Q10 hint Begin with the second difference box, then the first difference box and finally the sequence box.

11 Work out the next term of each sequence.

a 6, 21, 46, 81, … b 2, 7, 16, 29, … c 0, 1, 3, 6, …

Q11 hint Work out the first and second differences.

12 a Copy and complete to work out first and second differences for the sequence $u_n = n^2 + 7$.

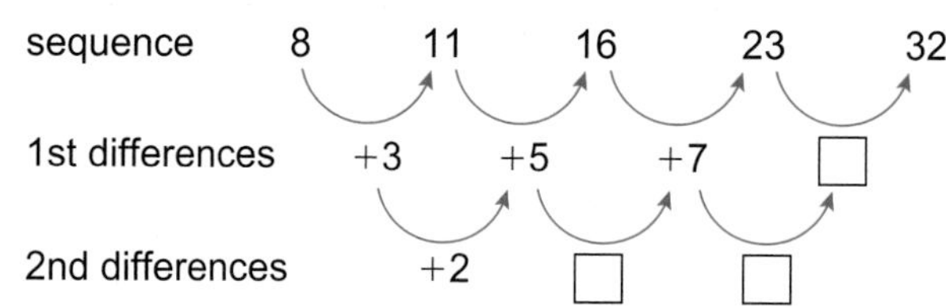

b Copy and complete for the sequence $v_n = 3n^2 - n - 2$

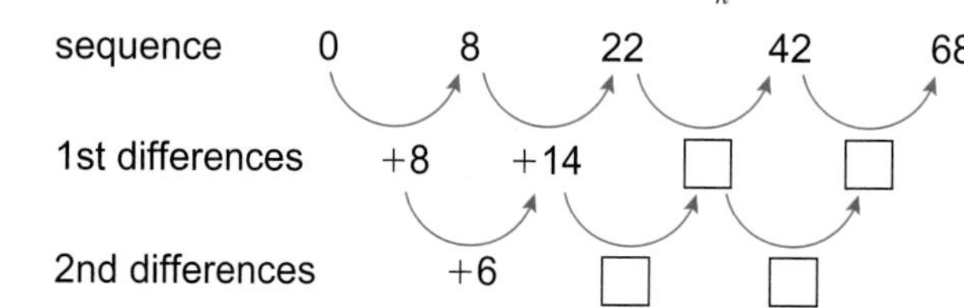

Discussion Are the second differences increasing, decreasing or constant? What is the connection between the formula for the nth term and the second differences?

Key point 16

The second differences of a quadratic sequence, $u_n = an^2 + bn + c$, are constant and equal to $2a$.

Example 4

Find a formula for the nth term of the sequence 8, 23, 48, 83, 128, …

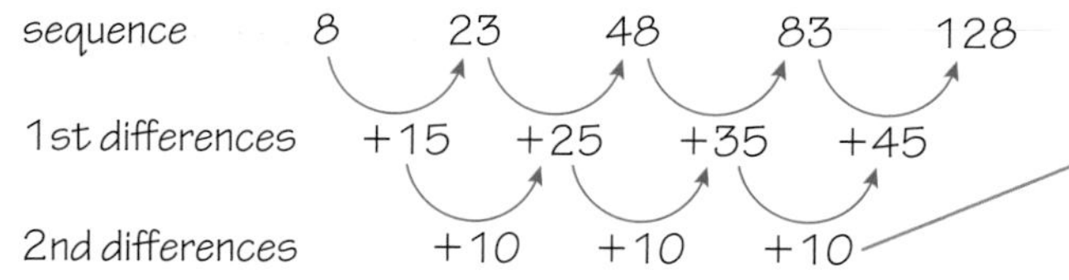

Work out the second differences.

So $a = 10 \div 2 = 5$

The formula has a $5n^2$ term in it.

Halve the second difference to find the coefficient of n^2.

$5n^2$	5	20	45	80	125
Sequence	8	23	48	83	128

Compare the given sequence with $5n^2$.

The nth term is $5n^2 + 3$

The numbers in the second row are 3 more than those in the first row.

13 Reasoning Find a formula for the nth term of each of these quadratic sequences.

a 3, 9, 19, 33, 51, …

b −2, 7, 22, 43, 70, …

c 4.5, 6, 8.5, 12, 16.5, …

14 **Reasoning** Each number in Pascal's triangle is found by adding the pair of numbers immediately above it.

Row 0 1

Row 1 1 1

Row 2 1 2 1

Row 3 1 3 3 1

Row 4 1 4 6 4 1

a Work out the numbers in the next row.

b Copy and complete the table for the sum of the numbers in each row.

Row, n	0	1	2	3	4	5
Sum	1	2				

c Work out a formula for the sum of the numbers in row n.

15 The sequence 2, 7, 14, 23, 34, … has nth term in the form $u_n = an^2 + bn + c$

a Find the second differences and show that $a = 1$.

b Subtract the sequence n^2 from the given sequence.

	2	7	14	23	34
−	1	4	9	16	25
	1	□	□	□	9

c Find the nth term of this linear sequence.

d Write the nth term of 2, 7, 14, 23, 34, …

$n^2 + \square n - \square$

Key point 17

The nth term of a quadratic sequence can be worked out in three steps.

Step 1 Work out the second differences.

Step 2 Halve the second difference to get the an^2 term.

Step 3 Subtract the sequence an^2. You may need to add a constant, or find the nth term of the remaining terms.

16 Find the nth term of each sequence.

a 4, 10, 18, 28, 40, …

b 0, 1, 4, 9, 16, …

c 5, 12, 23, 38, 57, …

d 3, 11, 25, 45, 71, …

Q16 hint Use the method in **Q15**.

17 **Communication** The nth term of a sequence is $u_n = 10^n$. Show that the product of u_5 and u_8 is u_{13}.

Q17 hint $x^m \times x^n = x^{m+n}$

18 **Exam-style question**

a Write down the first four terms in the sequence with nth term $u_n = 2^n$. **(2 marks)**

b State the term-to-term rule. **(1 mark)**

c Use algebra to show that the product of any two terms in the sequence is also a term in the sequence. **(2 marks)**

Exam hint
The question refers to *any* two terms so no credit is given for just checking it out for particular numbers.

2.7 More expanding and factorising

Objectives

- Expand the product of two brackets.
- Use the difference of two squares.
- Factorise quadratics of the form $x^2 + bx + c$

Did you know?

Expanding two brackets is a skill needed for graphing and analysing quadratic functions.

Fluency

What is the square root of 64?
What are the factor pairs of **i** 12 **ii** −6?

Warm up

1 Find a pair of numbers whose
 a product is 6 and sum is 5 b product is 4 and sum is −5.

2 Simplify
 a $(2x)^2$ b $(5y)^2$

3 a Copy and complete this expression for the area of the whole rectangle. $(x + \square)(\square + 1)$
 b Write an expression for the sum of the areas of the smaller rectangles.
 Collect like terms and simplify.

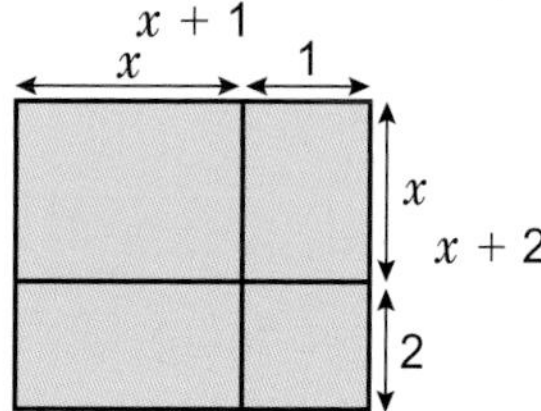

Key point 18

To expand double brackets, multiply each term in one bracket by each term in the other bracket.

Example 5

Expand and simplify $(x + 3)(x + 5)$

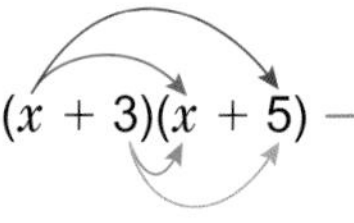

$(x + 3)(x + 5)$ — Multiply each term in the 2nd bracket by each term in the 1st bracket. FOIL: Firsts, Outers, Inners, Lasts

$= x^2 + 5x + 3x + 15$ — Collect like terms.

$= x^2 + 8x + 15$

4 Expand and simplify
 a $(x + 6)(x + 10)$ b $(x + 6)(x - 3)$ c $(x - 4)(x + 10)$ d $(x - 3)(x - 4)$

5 **Problem-solving** Find the missing terms in these quadratic expressions.
 a $(x + 2)(x + \square) = x^2 + \square x + 6$ b $(x - \square)(x + 8) = x^2 + 5x - \square$

Key point 19

To square a single bracket, multiply it by itself, then expand and simplify.
$(x + 1)^2 = (x + 1)(x + 1) = x^2 + 2x + 1$

6 Expand and simplify
 a $(x + 2)^2$ b $(x - 3)^2$ c $(x + 5)^2$ d $(x - 4)^2$

7 a Copy and complete to evaluate $51^2 - 49^2$ without a calculator.
$(51 + 49)(51 - 49) = 2 \times \square = \square$

b Without using a calculator work out
i $101^2 - 99^2$ ii $1.03^2 - 0.97^2$

8 Expand and simplify
a $(x - 4)(x + 4)$ b $(x - 2)(x + 2)$

Discussion Why can your answers be called 'difference of two squares'?

9 Factorise
a $x^2 - 25$ b $y^2 - 49$ c $t^2 - 81$

> **Q9a hint** Factorising is the inverse of expanding.
> $x^2 - 25 = (x \quad)(x \quad)$

Example 6

Factorise $x^2 + 5x + 6$

$x^2 + 5x + 6$

$(x \quad)(x \quad)$ — Write a pair of brackets with x in each one. This gives the x^2 term when multiplied.

$1 \times 6 \qquad 2 \times 3$ — Work out all the factor pairs of 6, the number term.

$1 + 6 = 7 \qquad 2 + 3 = 5$ — Work out which factor pair will **add** to give 5, the number in the x term.

$(x + 2)(x + 3)$ — Then write each number in each of the brackets with x.

Check: $(x + 2)(x + 3) = x^2 + 5x + 6$ — The expression is now factorised. Expand the brackets to check it is correct.

10 Factorise

a $x^2 + 8x + 7$ b $x^2 + 7x + 12$
c $x^2 + 8x + 15$ d $x^2 + 2x - 3$
e $x^2 - 2x - 3$ f $x^2 - 6x + 8$
g $x^2 - 6x - 7$ h $x^2 - 7x + 12$
i $x^2 - 4x + 4$ j $x^2 - 14x + 24$
k $x^2 - 6x - 16$ l $x^2 + 2x + 1$

> **Q10a hint** Find two numbers with product 7 and sum 8.

> **Q10d hint** For a product of −3, one number must be positive and one number is negative.

> **Q10f hint** For a positive product but a negative sum both numbers must be negative.

11 **Problem-solving** A rectangular piece of paper has length $(x + 5)$ cm and width $(x + 2)$ cm.
A square with sides of length, x cm is removed.

a Write an expression for the area of the rectangle before the square is cut out.
Expand the brackets.

b Write an expression for the shaded area

c Find x if the shaded area is 31 cm².

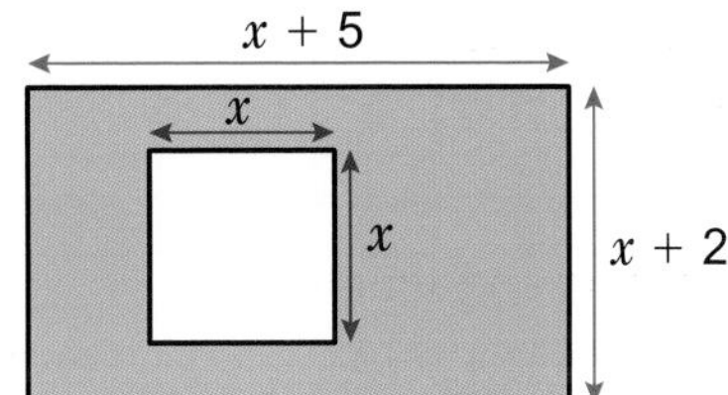

12 **Problem-solving / Reasoning** The two rectangles shown have the same area. Find x.

13 Copy and complete these factorisations.

a $4x^2 - 9 = (2x)^2 - \square^2 = (2x - \square)(2x + \square)$

b $16y^2 - 1 = (\square y)^2 - \square^2 = (\square y - \square)(\square y + \square)$

14 Factorise

a $9m^2 - 25$ b $25c^2 - 81$ c $x^2 - 49y^2$

15 **Exam-style question**

a Factorise $x^2 + 11x + 30$ **(2 marks)**

b Expand $(3u - 4v)^2$ **(3 marks)**

Exam hint
In part **a**, check your factorisation by expanding the brackets.

2 Problem-solving

Objective
- Use smaller numbers to help you solve problems.

Example 8

A factory makes boxes of Christmas crackers. Each box contains 12 crackers.
The factory has 13 machines. Each day, each machine makes 1638 boxes of Christmas crackers.
The same number of boxes is loaded on to each of 18 lorries.

a How many crackers are on one lorry?

b Write an expression for the number of crackers on a lorry when
- each box contains c Christmas crackers
- they are made in a factory that has m machines
- each machine makes b boxes of Christmas crackers per day
- the same number of boxes is loaded on to each of n lorries.

a Using smaller numbers:

Each box of Christmas crackers contains 2 crackers.

3 machines

Each machine makes 5 boxes of crackers.

4 lorries

Replace the numbers in the question with smaller numbers.

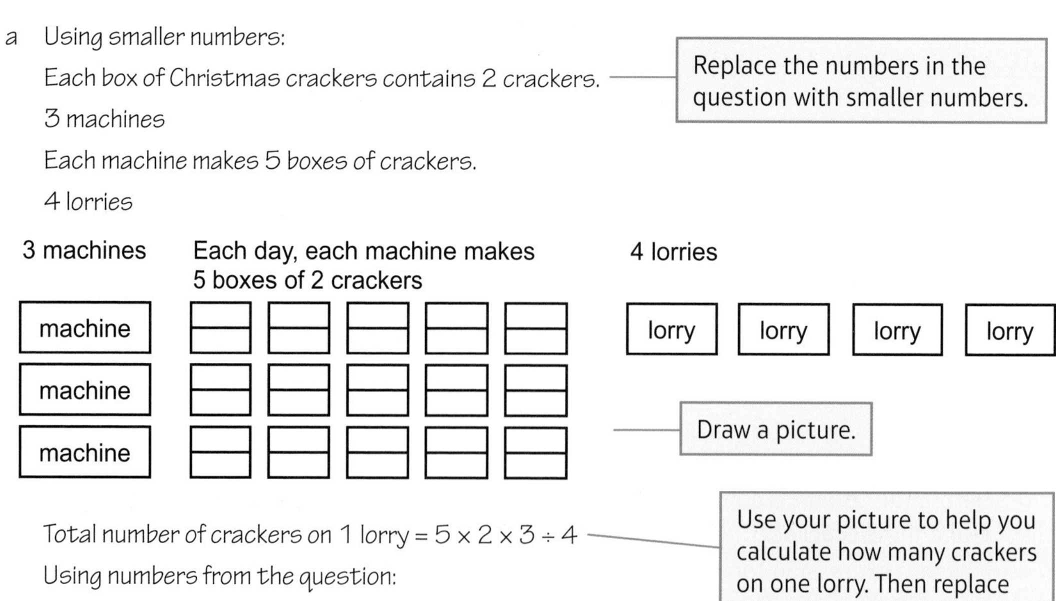

Draw a picture.

Total number of crackers on 1 lorry = $5 \times 2 \times 3 \div 4$

Using numbers from the question:

Total number of crackers on 1 lorry = $1638 \times 12 \times 13 \div 18$
$= 14196$

Use your picture to help you calculate how many crackers on one lorry. Then replace the smaller numbers with the corresponding numbers in the question.

b Total number of crackers on a lorry = $\frac{b \times c \times m}{n}$
$= \frac{bcm}{n}$

Replace the numbers in your calculation with the corresponding letters to write an expression.

1 Luke is revising for a Spanish exam. Every day he reads 11 pages of his Spanish vocabulary book. There are 15 words on every page. After three weeks, he has only 35 words left.

a How many words are in his Spanish vocabulary book?

b Write an expression for the number of words in a vocabulary book when someone reads x pages per day, there are y words on every page, and after z weeks, the person has only m words left.

Q1a hint Replace each number in the question with a smaller number and draw a picture too. There is no 'correct' picture. There is also no 'correct' smaller number.

2 A farmer grows strawberries. The farmer employs 28 fruit pickers.
Each day, each fruit picker picks 36 kg of strawberries.
All the strawberries are packaged into plastic tubs.
Each plastic tub contains 0.25 kg of strawberries.
Then the same number of plastic tubs is put into each of 63 boxes.

a How many plastic tubs are put into each box?

b Write an expression for the number of plastic tubs in a box when p fruit pickers each pick q kg of strawberries, these are packaged into plastic tubs, each containing s kg of strawberries, and the same number of tubs is loaded into each of t boxes.

Q2a hint Use a whole number instead of 0.25 kg.

3 A vending machine has 24 different products. It stores 15 of each of these products.
On average, people buy 35 products from the machine every day.

a How many products remain in the vending machine at the end of 7 days?

b Write an expression for the number of products remaining in the vending machine at the end of n days when a vending machine has x different products, it stores y of each of these products, and on average, people buy m products from the machine every day.

4 22 office workers send an email to each other.

a How many emails are sent altogether?

b Write an expression for the number of emails sent by x workers.

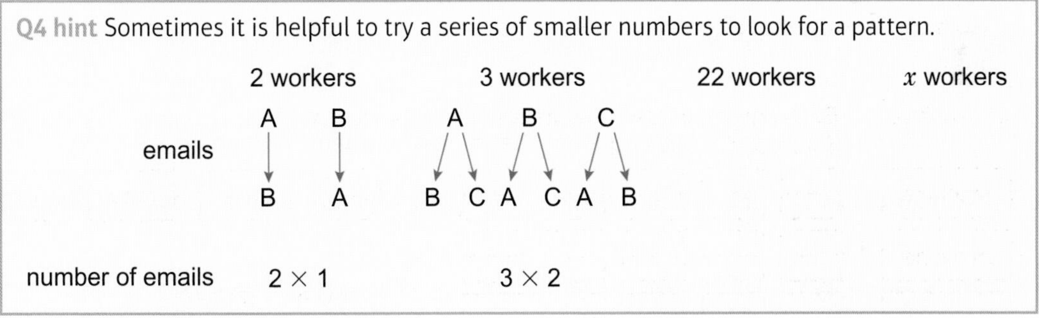

5 a Write down the 137th odd number.

b Write an expression for the nth odd number.

Q5 hint Write down the 1st, 2nd, 3rd odd numbers. How do you find the 137th odd number?

6 How many times do two or more odd digits appear in a number when counting from 0 to 999?

Q6 hint How many times do two or more odd digits appear in numbers
0 – 9 10 – 19 20 – 29 30 – 39…
100 – 109 110 – 119 120 – 129 130 – 139…

7 **Reflect** Did using smaller numbers help you?
Is this a strategy you would use again to solve problems?
What other strategy or strategies helped you to solve these problems?

2 Check up

Log how you did on your Student Progression Chart.

Simplifying, expanding and factorising

1 Simplify

a $4p \times 5p^3$ b $15x^4 \div 3x^2$ c $(b^2)^{-3}$

2 Expand and simplify $3(2p + q) - 2(3p - q)$

3 Factorise

a $2xy - 6y$ b $3ab - 6a^2$

4 Expand and simplify

a $(x + 4)(x - 6)$ b $(x + 5)^2$

5 Simplify

a $2x^{-2}$ b $4x^0$

c $(9c^2)^{\frac{1}{2}}$ d $\dfrac{16p^{-2}}{4p^3}$

6 Expand and simplify $(2s - r)(s + 3r)$

7 Factorise

a $x^2 - 81$ b $x^2 - 9x + 14$

Equations and formulae

8 Write whether each of these is an expression, an equation, an identity or a formula.

a $v = u + at$ b $a^2 - b^2 = (a - b)(a + b)$

c mv d $4a = 5$

9 Solve $4x - 3 = 2x + 6$

10 Solve $2(3x + 1) = 5(x - 3)$

11 Use the formula $z = f^2 - 2fg$ to work out the value of z when $f = 10$ and $g = 3$.

12 **Communication** Show that the equation $x^3 + 4x = 6$ has a solution between 1.1 and 1.2

13 An electrician charges a £25 call-out fee, plus £36 per hour.
Write a formula for his total charge £C for n hours' of work.

14 a Make y the subject of the fomrula $2x + 3y = 4$
b Make b the subject of the formula $S = 6ab + 4a^2$

15 Solve $\dfrac{x}{3} - \dfrac{x}{4} = \dfrac{5}{6}$

Sequences

16 Write down the next two terms in the Fibonacci sequence 3, 4, 7, 11, ….

17 a Find the nth term of the arithmetic sequence 2, 11, 20, 29, …
b Show that 167 cannot be a term in this sequence.
c Find the first number in the sequence that is greater than 167.

18 Find the nth term of the sequence 10, 19, 34, 55, …

ActiveLearn Homework, practice and support: Higher 2 Check up

Reflect

19 How sure are you of your answers? Were you mostly

Just guessing ☹ Feeling doubtful 😐 Confident ☺

What next? Use your results to decide whether to strengthen or extend your learning.

* Challenge

20 a Multiply together the four pairs of connected terms and expand your answers.

b Add together your answers and simplify the result. Would the result have been the same if you had expanded in a different order?

c Factorise your simplified expression.

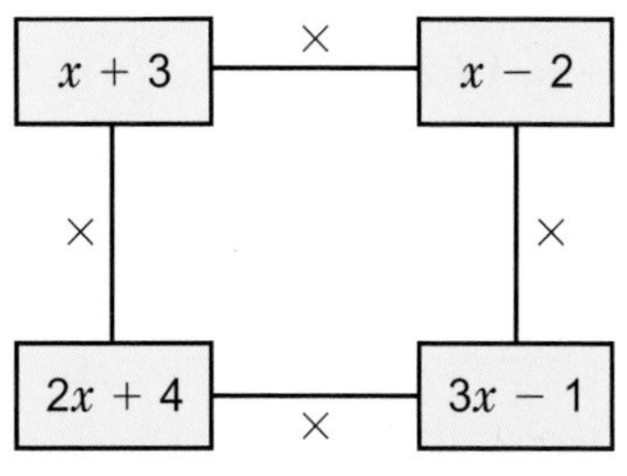

2 Strengthen

Simplifying, expanding and factorising

1 Simplify

a $t^3 \times t^2$ b $t^4 \times t^3$

c $t \times t^3$ d $t^{-2} \times t^4$

e $t^{-6} \times t^{-1}$ f $t^{\frac{1}{2}} \times t^{\frac{3}{2}}$

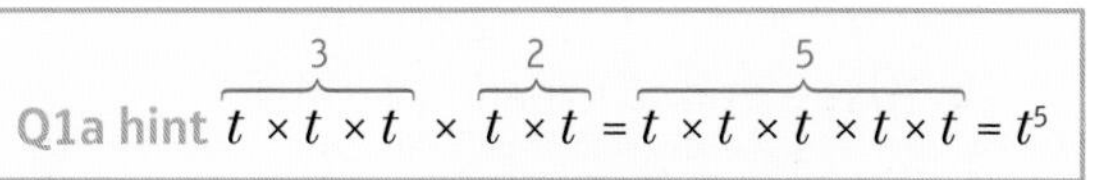

Q1a hint $\overbrace{t \times t \times t}^{3} \times \overbrace{t \times t}^{2} = \overbrace{t \times t \times t \times t \times t}^{5} = t^5$

Q1d hint Add the indices.

2 Simplify

a $3p^2 \times 6p^3$ b $8z \times 9z^4$

c $7b^3 \times 2b^5$ d $2r^5 \times 4r^{-2}$

e $2x^{\frac{2}{3}} \times 3x^{\frac{4}{3}}$ f $5s^{-2} \times 2s^{-4}$

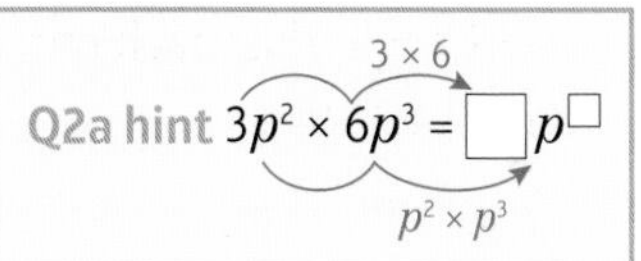

Q2a hint $3p^2 \times 6p^3 = \square p^{\square}$
3 × 6
$p^2 \times p^3$

3 Copy and complete

a $t^6 \div t^2 = t^{\square}$ b $t^5 \div t^2 = t^{\square}$ c $t^3 \div t^3 = t^{\square} = \square$

Q3a hint What do you multiply t^2 by to get t^6?

4 Simplify

a $20p^6 \div 4p^2$ b $\dfrac{12a^7}{4a^2}$

c $\dfrac{9y^{-1}}{3y^2}$ d $\dfrac{6p^{\frac{1}{2}}}{3p^{-\frac{1}{2}}}$

Q4a hint $20 \div 4 \quad p^6 \div p^2$
$= \square \quad p^{\square}$
$= \square p^{\square}$

Q4c hint $\square y^{-1-2} = \square y^{\square}$

5 Copy and complete

a $(x^2)^2 = \square^{\square} \times \square^{\square} = \square^{\square}$

b $(x^2)^3 = \square^{\square} \times \square^{\square} \times \square^{\square} = \square^{\square}$

c $(x^2)^4 = \square^{\square} \times \square^{\square} \times \square^{\square} \times \square^{\square} = \square^{\square}$

d What do you notice about powers and brackets? What is the rule?

6 Simplify

a $(a^4)^{\frac{1}{2}}$ b $(r^2)^{-1}$ c $(2g^{\frac{1}{3}})^3$

Q6a hint Use the rule you noticed in **Q5d**.

7 a Expand $3(2x + y)$

b Expand $2(3x - 4y)$

c Expand and simplify $3(2x + y) + 2(3x - 4y)$

Q7a hint Use a multiplication grid.

×	$2x$	y
3		

8 Expand and simplify

a $2(4c + 5d) + 3(c - 3d)$

b $6(3m + n) - 4(m - n)$

Q8 hint Use grids to help you.

9 Copy and complete the factorisations.

a $3ab^2 - 2ab = ab(\square \ldots \square)$

b $8xy + 6x = 2\square(\square \ldots \square)$

c $3st^2 - 6st = \square(\square \ldots \square)$

d $14ab^2 + 21b = \square(\square \ldots \square)$

Q9a hint Find the common factors.
$3ab^2 = 3 \times a \times b \times b$
$-2ab = -2 \times a \times b$
Write the common factors first.
$3ab^2 = ab \times 3b$
$-2ab = ab \times -2$

Q9b hint Start with
$8xy = 2 \times 4 \times \square \times \square$
$6x = 2 \times 3 \times \square$
Now follow the method for part **a**.

10 a Copy and complete the multiplication grid.

×	x	+5
x	x^2	$5x$
+4		+20

b Use your answer to part **a** to expand

$(x + 4)(x + 5) = x^2 + 5x + \square + 20$
$= x^2 + \square x + 20$

11 a Use this grid to expand $(x - 6)^2$

×	x	−6
x	x^2	
−6		

b Use a grid to expand $(x - 4)(x + 4)$

12 Use the grids to expand and simplify

a $(x + 8)(3x + 2)$

×	$3x$	+2
x		
+8		

b $(2x + 1)(5x + 3)$

×	$2x$	+1
$5x$		
+3		

c $(3x - 7)(x + 4)$

×	$3x$	−7
x		
+4		

13 Match the expresions to their factorisations.

$(x - 3)^2$ | $x^2 + 6x + 5$ | $(x + 1)(x + 5)$ | $x^2 - 9$

$x^2 - 6x + 9$ | $(x + 3)(x + 2)$ | $x^2 + 5x + 6$ | $x^2 - x - 6$

$(x + 2)(x - 3)$ | $(x - 3)(x + 3)$

Q13 hint Expand the brackets to check.

14 a There are three pairs of positive integers whose product is 12. One pair is 1 and 12.
Write down the other two pairs.

b Which pair of numbers in part **a** add up to 8?

c Use your answer to part **b** to factorise
$x^2 + 8x + 12 = (x + \square)(x + \square)$

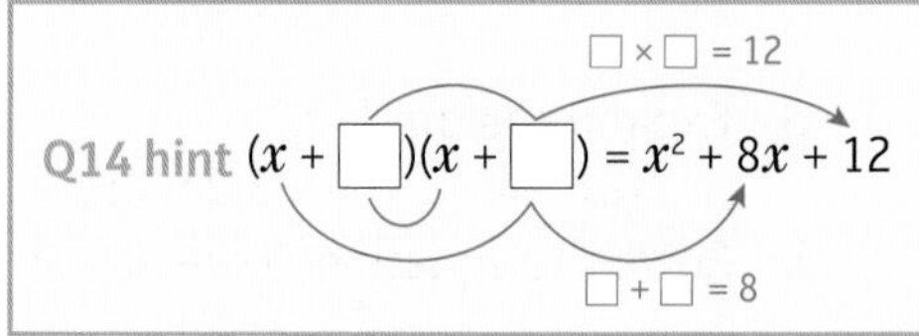

15 Factorise

a $x^2 + 13x + 12$ b $x^2 + 7x + 12$

16 a There are four pairs of integers whose product is −10. One pair is −2 and +5.
Write down the other three pairs.

b Use your answers to part **a** to factorise

i $x^2 - 9x - 10$ ii $x^2 + 9x - 10$

iii $x^2 + 3x - 10$ iv $x^2 - 3x - 10$

17 a There are four pairs of integers that multiply to 24 and add up to a negative number. One pair is −8 and −3. Write down the other three pairs.

b Use your answer to part **a** to write down the factorisation of

i $x^2 - 25x + 24$

ii $x^2 - 14x + 24$

iii $x^2 - 10x + 24$

iv $x^2 - 11x + 24$

Equations and formulae

1 Write whether each of these is an identity, a formula, an expression or an equation.

a $2x$

b $x + 2x = 3x$

c $y = 2x$

d $2x = 1$

Q1 hint When there is no = sign it is … When the two sides are always equal it is … When you can solve it to find the value of the letter it is …

2 When $U = 5$ and $V = 3$, work out

a V^2 b $4V^2$ c $4V^2 + U$

3 Use the formula $m = 2x^2 + g$ to work out m when $x = 3$ and $g = 5$.

Q3 hint $m = 2x^2 + g = 2 \times \square^2 + \square$

4 Use the formula $t = r^2 - 3rs$ to work out t when $r = 5$ and $s = 2$.

5 Make x the subject of the formula $y = 2x - 4$

Q5 hint Use inverse operations.

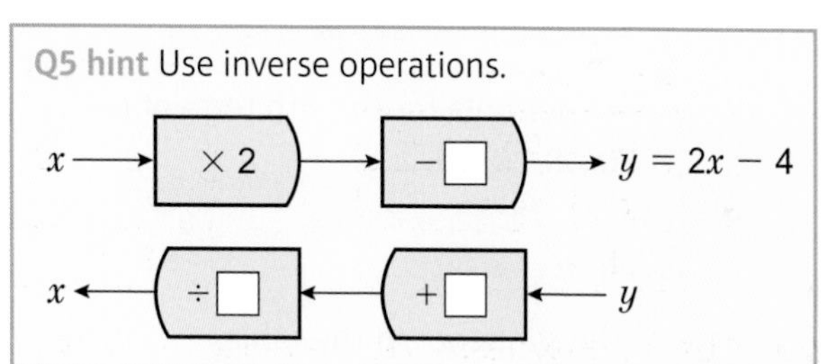

6 Make Q the subject of the formula $P = \frac{Q}{a} + b$

Q6 hint Use function machines to help you.

7 a Make b the subject of $c = \frac{3b}{4}$

b Make s the subject of $v^2 = u^2 + 2as$

8 Solve the equation $5x - 1 = 3x + 7$

Q8 hint You need to get only unknowns (x) on one side of the equals, and only numbers on the other side.

9 a Expand the brackets.

i $7(2x - 4)$ ii $2(3x + 5)$

b Solve the equation $7(2x - 4) = 2(3x + 5)$

Q9b hint Rewrite the equation using your expressions from part **a**.

10 a Expand and simplify $7(2x + 1) - 3(4x + 3)$

b Solve $7(2x + 1) - 3(4x + 3) = 5$

11 Simplify

a $\frac{7x}{7}$ b $\frac{4x}{2}$ c $\frac{20x}{5}$

12 Solve these equations. Start by multiplying both sides of the equation by 5.

a $\frac{x}{5} = 4$ b $\frac{3x}{5} = 2$

13 Solve these equations

a $\frac{x}{6} = 3$ b $\frac{4x}{7} = 1$

c How do you decide what to multiply by?

14 Solve these equations

a $\frac{x}{4} - \frac{x}{5} = 3$ b $\frac{x+1}{4} = \frac{x}{2}$

Q14a hint Multiply by $4 \times 5 = 20$

Sequences

1 Write down the next two terms in each of these Fibonacci sequences.

a 1, 1, 2, 3, 5, 8, …

b 5, 7, 12, 19, 31, …

c 2, 4, 6, 10, 16, 26, …

Q1 hint The rule is 'add two terms to get the next'.

2 Work out the first three terms of the sequence with nth term

a $2n + 3$ b $50 - 2n$

c $n^2 + 1$ d $10n^2$

Q2 hint Substitute $n = 1, n = 2, n = 3$ into each formula.

3 The first five terms of an arithmetic sequence are 3, 6, 9, 12, 15.

a These are multiples of □.

b What is the 12th term?

c Copy and complete this statement.
The general term is □n

Q3a hint Which times tables are these numbers in?

4 Work out a formula for the nth term of each of these arithmetic sequences.

a 10, 20, 30, 40, 50, …

b 7, 14, 21, 28, 35, …

c 12, 24, 36, 48, 60, …

Q4 hint Use the method from **Q3**.

5 Look at the sequence in the table.

Term number	1	2	3	4	5	6
Term	7	8	9	10	11	12

a What number do you add to each number in the top row of the table to get the number in the bottom row?

b Copy and complete this statement.
The nth term is n + □

Q5b hint Check your answer by substituting $n = 1, n = 2, n = 3$.

6 Write down a formula for the nth term of each of these arithmetic sequences.

a 3, 4, 5, 6, 7, …

b 13, 14, 15, 16, 17, …

c −3, −2, −1, 0, 1 …

Q6a hint Use the method from **Q5**.

Q6c hint Try n − □

7 These two sequences have the same common difference.

Sequence A: 4, 8, 12, 16, 20, …

Sequence B: 7, 11, 15, 19, 23, …

a Work out the nth term of sequence A.

b What do you add to each term in sequence A to get the terms in sequence B?

c Write the nth term of sequence B.

Q7a hint The numbers in sequence A are multiples of □ so the nth term is □n.

Q7c hint Use your answers from parts **a** and **b**.

8 a Write down the next two terms in each of these arithmetic sequences.

  i 6, 12, 18, 24, …
  ii 1, 3, 5, 7, …
  iii 4, 7, 10, 13, …
  iv 25, 20, 15, 10, …

b Find the nth term of each sequence in part **a**.

> **Q8b hint** Use the method from **Q7**.

9 a Write down the first five terms of the sequence with nth term $u_n = 10 + 4n$.

b Explain why 351 cannot be a term of this sequence.

c Which term of the sequence is 102?

> **Q9b hint** Can this sequence have odd numbers in it?

> **Q9c hint** Solve $10 + 4n = 102$

10 The nth term of an arithmetic sequence is $5n + 7$.

a Which term of the sequence is 107?

b Find the first term in the sequence which is bigger than 108.

c Find the first term in the sequence which is bigger than 150.

> **Q10a hint** What equation do you need to solve?

> **Q10c hint** Solve $5n + 7 = 150$. Use the next integer value of n.

11 a Copy and complete the first and second differences for this sequence and work out the next term.

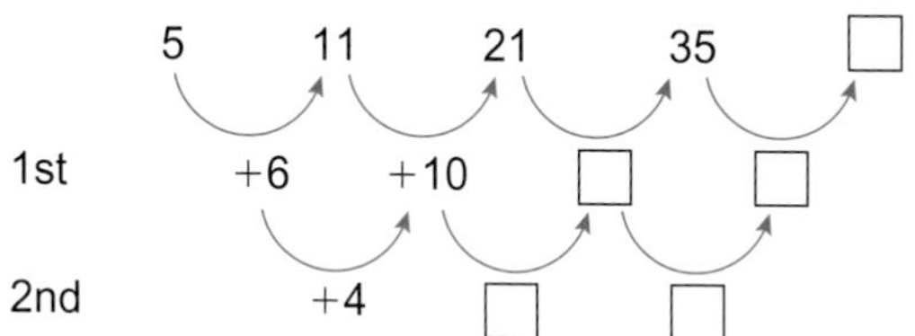

b Find a formula for the nth term.

> **Q11b hint** The formula is $an^2 + b$ where a is half the second difference. For the first term $n = 1$ and the term is 5. Substitute $n = 1$ and your value of a into $an^2 + b = 5$ to find b.

12 Find a formula for the nth term of each of these quadratic sequences.

a 9, 21, 41, 69, … b −9, −6, −1, 6, …

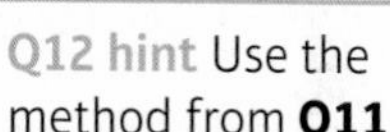

> **Q12 hint** Use the method from **Q11**.

2 Extend

1 **Reasoning** a Write down the next three terms in each sequence.

  i $u_1 = 5, \dots u_{n+1} = u_n + 1$
  ii $u_1 = 40, \dots u_{n+1} = \frac{1}{2}u_n$
  iii $u_1 = 7, \dots u_{n+1} = u_n - 4$
  iv $u_1 = 1, \dots u_{n+1} = -3u_n$

b Which of these sequences are arithmetic and which are geometric?

> **Q1a i hint** $u_2 = u_1 + 1$

2 **Reasoning** a The 1st term of an arithmetic sequence is 0.341 and the 2nd term is 0.407 Work out the 3rd term.

b The 1st term of an arithmetic sequence is 9 and the 3rd term is 14. Work out the 2nd term.

c The 1st term of an arithmetic sequence is 4 and the 5th term is 16. Work out the 4th term.

d The 1st term of an arithmetic sequence is 5.8 and the 2nd term is 5.9. Work out the 100th term.

3 **Modelling** A clothing store monitors sales in-store and online.
Sales for the last few years are shown in the table.

Year	2010	2011	2012	2013	2014	2015
In-store	31 250	25 000	20 000	16 000		
Online	640	960	1440	2160		

Assuming both types of sales form a geometric sequence

a work out the sales of each type for the next two years

b work out the year when online sales are predicted to exceed in-store sales.

4 **Finance** The formula gives the monthly repayments, £M, needed to pay off a mortgage over n years when the amount borrowed is £P and the interest rate is r%.

$$M = \frac{Pr(1 + \frac{1}{100}r)^n}{1200[(1 + \frac{1}{100}r)^n - 1]}$$

Calculate the monthly repayments when the amount borrowed is £250 000 over 25 years and the interest rate is 5%.

5 **Problem-solving** Raj attempts a multiple choice test with 20 questions. He scores 5 marks for a correct answer but loses 2 marks if it is incorrect. Raj attempts all 20 questions and gets a total score of 51. How many answers did he get right?

Q5 strategy hint
Let x be the number of correct answers.

6 **Real** The deposit, D, needed when booking a skiing holiday is in two parts:

- a non-returnable booking fee, B
- one-tenth of the total cost of the holiday, which is worked out by multiplying the price per person, P, by the number of people, N, in the party.

$$D = B + \frac{NP}{10}$$

a Find the deposit needed to book a holiday for four people when the cost per person is £2000 and the booking fee is £150.

b Make P the subject of the formula.

c What is the price per person when D = £500, B = £150 and $N = 5$?

7 Change the subject to the letter given in the brackets.

a $v^2 = u^2 + 2as$ $[a]$

b $V = \frac{1}{3}\pi r^2 h$ $[h]$

c $S = \dfrac{a(r^n - 1)}{r - 1}$ $[a]$

d $a^2x - b^2y = c$ $[y]$

8 Simplify

a $\dfrac{4c^2d}{2c^{-2}d^{-3}}$

b $4x^{\frac{1}{2}}y^{-2} \times 3x^{\frac{3}{2}}y^3$

c $\left(2m^{-\frac{1}{4}}n^{\frac{3}{4}}\right)^4$

d $\sqrt[3]{8p^{-3}q^{12}}$

9 **Exam-style question**

Work out a simplified expression for the area of this shape.
All the angles are right angles.

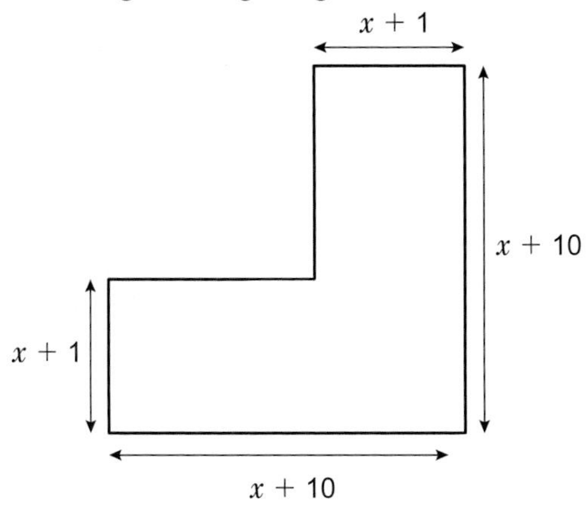

(4 marks)

> **Exam hint**
> Mark any lengths you find on the diagram.

10 **Exam-style question**

a Expand and simplify $(p + 9)(p - 4)$ **(2 marks)**

b Solve $\frac{5w - 8}{3} = 4w + 2$ **(3 marks)**

c Factorise $x^2 - 9$ **(1 mark)**

d Simplify $(9x^8y^3)^{\frac{1}{2}}$ **(2 marks)**

June 2012, Q14, 1MA0/2H

> **Exam hint**
> Check your solution to part **b** by substituting back into the equation.

11 **Exam-style question**

You can change temperatures from °F to °C by using the formula

$$C = \frac{5(F - 32)}{9}$$

F is the temperature in °F.
C is the temperature in °C.
The minimum temperature in an elderly person's home should be 20 °C.
Mrs Smith is an elderly person.
The temperature in Mrs Smith's home is 77 °F.

a Decide whether or not the temperature in Mrs Smith's home is lower than the minimum temperature should be. **(3 marks)**

b Make F the subject of the formula $C = \frac{5(F - 32)}{9}$ **(3 marks)**

June 2014, Q12, 1MA0/1H

> **Exam hint**
> You need to show calculations to support your decision in part **a**.

12 **Communication**

a Explain why $2n + 1$ is an odd number for any integer n.

b Show that the product of two odd numbers is always odd.

> **Q12b hint** How could you write the product of two different odd numbers, algebraically?

13 Factorise completely

a $x^2 - 12x + 32$ b $x^2 - 12x + 36$ c $x^2 - x - 2$ d $\frac{x^2}{25} - \frac{y^2}{49}$

14 Solve

a $3(2x - 1) - 4(3x - 2) = 10$

b $\frac{2}{3}(x + 4) = \frac{4}{5}(x - 1)$

c $\frac{x}{6} - \frac{3x}{8} = 1$

d $\frac{5x + 7}{14} = \frac{1 - 2x}{21}$

15 Communication Show that the difference between consecutive square numbers is always an odd number.

16 Find the nth term of each sequence.

a $1, -5, -15, -29, -47, \ldots$ b $0, -1, -4, -9, -16, \ldots$

Q16 hint The second differences are negative so $-\square n^2 + bn + c$

2 Knowledge check

- $x^m \times x^n = x^{m+n}$ $x^m \div x^n = x^{m-n}$ $(x^m)^n = x^{mn}$
  $x^0 = 1$ $x^{-m} = \frac{1}{x^m}$ $x^{\frac{1}{n}} = \sqrt[n]{x}$ .... *Mastery lesson 2.1*
- When the two sides of a relation such as $2(x + 5) = 2x + 10$ are equal for all values of x it is called an **identity** and we write $2(x + 5) \equiv 2x + 10$ using the '$\equiv$' symbol. .... *Mastery lesson 2.2*
- An **equation**, such as $2x = 6$, is only true for certain values of x (in this case $x = 3$). .... *Mastery lesson 2.2*
- To expand a bracket, multiply each term inside the brackets by the term outside the brackets. $x(y + z) \equiv xy + xz$ .... *Mastery lesson 2.2*
- Unless a question asks for a decimal answer, give non-integer solutions to an equation as exact fractions. .... *Mastery lesson 2.3*
- To solve an equation involving fractions, multiply each term on both sides by the LCM of the denominator. .... *Mastery lesson 2.3*
- An **expression** contains letter and number terms but no equals sign, e.g. $2ab$, $2ab + 3a^2b$, $2ab - 7$ .... *Mastery lesson 2.4*
- An **equation** has an equals sign, terms in one letter and numbers, e.g. $2x - 4 = 9x + 1$
  You can solve it to find the value of the letter. .... *Mastery lesson 2.4*
- An **identity** has an equals sign and is true for all values of the letters, e.g. $\frac{4x}{2} = 2x$, $x(x + y) \equiv x^2 + xy$ .... *Mastery lesson 2.4*
- A **formula** has an equals sign and letters to represent different quantities, e.g. $A = \pi r^2$
  The letters are **variables** as their values can vary. .... *Mastery lesson 2.4*
- The **subject** of a formula is the letter on its own, on one side of the equals sign. .... *Mastery lesson 2.4*
- In an **arithmetic sequence** the terms increase (or decrease) by a fixed number called the **common difference**. .... *Mastery lesson 2.5*
- When an arithmetic sequence with common difference d is input into this function machine, the output sequence has common difference $p \times d$. .... *Mastery lesson 2.5*
- In a Fibonacci-like sequence the next number is found by adding the previous two numbers together. .... *Mastery lesson 2.6*
- In a **geometric sequence** the terms increase (or decrease) by a **constant multiplier**. The nth term is ar^n. .... *Mastery lesson 2.6*
- A **quadratic sequence** has n^2 and no higher power of n in its nth term. *Mastery lesson 2.6*
- The second differences of a quadratic sequence, $u_n = an^2 + bn + c$ are constant and equal to $2a$. .... *Mastery lesson 2.6*

- The nth term of a quadratic sequence can be worked out in three steps.
  Step 1 Work out the second differences.
  Step 2 Halve the second difference to get the an^2 term.
  Step 3 Subtract the sequence an^2. You may need to add a constant, or find the nth term of the remaining terms. .... *Mastery lesson 2.6*
- To expand **double brackets**, multiply each term in one bracket by each term in the other bracket. .... *Mastery lesson 2.7*
- To **square** a single bracket, multiply it by itself, then expand and simplify, e.g. $(x + 1)^2 = (x + 1)(x + 1) = x^2 + 2x + 1$ .... *Mastery lesson 2.7*
- A **quadratic expression** has a squared term (and no higher power), e.g. $x^2 + 8x + 10$ .... *Mastery lesson 2.7*

Reflect

Choose A B or C to complete each statement about algebra.

In this unit, I did...	**A** well	**B** OK	**C** not very well
I think algebra is...	**A** easy	**B** OK	**C** hard
When I think about doing a algebra, I feel	**A** confident	**B** OK	**C** unsure

Did you answer mostly As and Bs? Are you surprised by how you feel about algebra? Why?
Did you answer mostly Cs? Find the three questions in this unit that you found the hardest. Ask someone to explain them to you. Then complete the statements above again.

2 Unit test

Log how you did on your Student Progression Chart.

1 Work out the next two terms of the Fibonacci sequence, 4, 7, 11, 18, 29... *(2 marks)*

2 Write whether each of these is an expression, a formula, an equation or an identity.
 a $4(3x + 1) = 5x - 6$ *(1 mark)*
 b $4(3x + 1)$ *(1 mark)*
 c $4(3x + 1) = 12x + 4$ *(1 mark)*
 d $y = 4(3x + 1)$ *(1 mark)*

3 Solve
 a $4(5x - 2) = 32$ *(3 marks)*
 b $7x + 3 = 2x - 12$ *(3 marks)*

4 Simplify
 a $7q^2 \times 9q^3$ *(2 marks)*
 b $\frac{25y^4}{5y}$ *(2 marks)*
 c $(c^4)^2$ *(1 mark)*

5 Expand
 a $3x(4x + y)$ *(2 marks)*
 b $(x + 4)(x - 3)$ *(2 marks)*
 c $(x - 7)^2$ *(2 marks)*

6 Find the first three terms of the sequence with nth term $u_n = 81 \times \left(\frac{1}{3}\right)^n$ *(3 marks)*

7 a Find the nth term of the arithmetic sequence 4, 10, 16, 22, 28, ... *(2 marks)*
 b Show that 231 is not in the sequence. *(1 mark)*
 c Find the smallest number in this sequence which is greater than 234. *(3 marks)*

ActiveLearn Homework, practice and support: Higher 2 Unit test

8 **Reasoning** The value of a car goes down by 10% a year. A car costs £40 000 when new.

- **a** How much is it worth after
  - **i** 1 year **ii** 2 years? *(2 marks)*
- **b** After how many years is it worth less than half of its original price? *(2 marks)*
- **c** Does the answer to part **b** increase, decrease or stay the same when
  - **i** the cost of the new car is changed to £12 000 *(1 mark)*
  - **ii** the rate of decrease is changed to 20%? *(1 mark)*

9 Make x the subject of the formula $y = 2x + 5$ *(2 marks)*

10 **Reasoning** Find the nth term of the sequence 2, 11, 26, 47, … *(6 marks)*

11 **Reasoning** The diagram shows an isosceles triangle. All lengths are in centimetres.

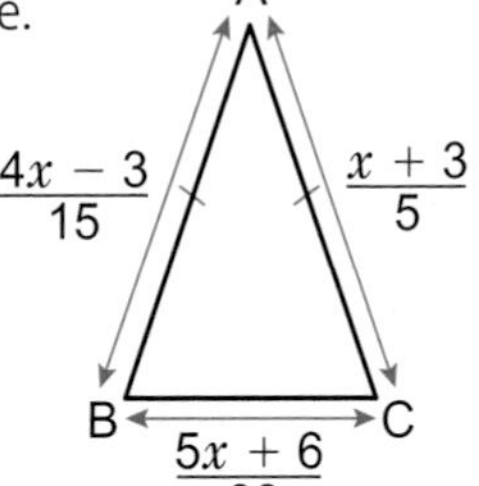

- **a** Write down an equation for x. *(1 mark)*
- **b** Solve the equation. *(2 marks)*
- **c** Work out the length of BC. *(2 marks)*

12 $E = \frac{1}{2}mv^2$

Find E when $m = 6 \times 10^{-8}$ and $v = 3 \times 10^8$. *(1 mark)*

13 Factorise completely

- **a** $x^2 - 16$ *(1 mark)*
- **b** $6y^2 - 9xy$ *(2 marks)*
- **c** $x^2 + 3x - 10$ *(2 marks)*

Sample student answer

- **a** How does drawing the 3D shapes help?
- **b** Where can you get help with the formulae in an exam?
- **c** How does the student's layout of the answer help ensure no mistakes are made?
- **d** Why has the student used a capital R for the radius of the cone?

Exam-style question

A sphere of metal, radius 5 cm, is melted down and made into a cone of the same volume.
The perpendicular height of the cone needs to be 5 cm.
What will the base radius of the cone be? **(3 marks)**

Student answer

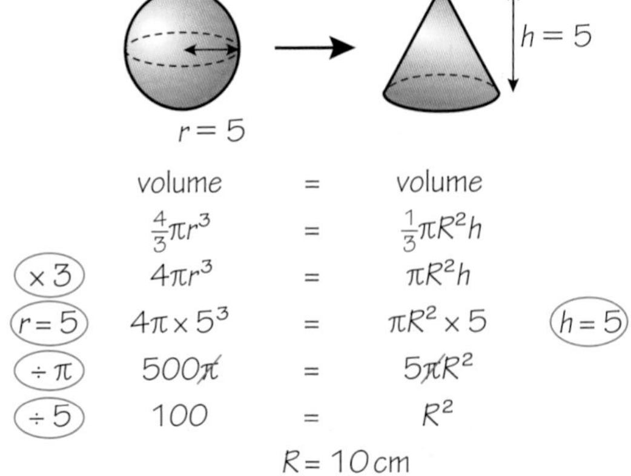

	volume	=	volume	
	$\frac{4}{3}\pi r^3$	=	$\frac{1}{3}\pi R^2 h$	
×3	$4\pi r^3$	=	$\pi R^2 h$	
r = 5	$4\pi \times 5^3$	=	$\pi R^2 \times 5$	h = 5
÷π	500π	=	$5\pi R^2$	
÷5	100	=	R^2	

$R = 10$ cm

Master p.63 | Problem-solve p.81 | Check p.82 | Strengthen p.84 | Extend p.89 | Test p.94

3 INTERPRETING AND REPRESENTING DATA

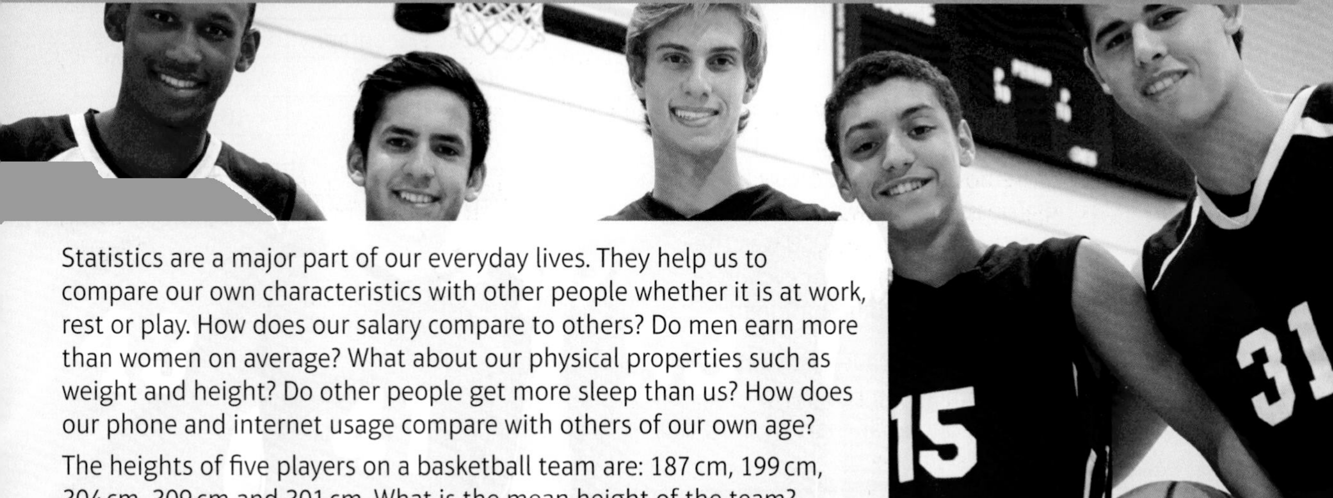

Statistics are a major part of our everyday lives. They help us to compare our own characteristics with other people whether it is at work, rest or play. How does our salary compare to others? Do men earn more than women on average? What about our physical properties such as weight and height? Do other people get more sleep than us? How does our phone and internet usage compare with others of our own age?

The heights of five players on a basketball team are: 187 cm, 199 cm, 204 cm, 209 cm and 201 cm. What is the mean height of the team?

3 Prior knowledge check

Numerical fluency

1 Work out

a $0 \times 4 + 1 \times 2 + 2 \times 5 + 3 \times 2 + 4 \times 1$

b $\frac{1}{2}(11 + 16)$

c $\frac{8 \times 2 + 7 \times 3 + 3 \times 4 + 2 \times 5}{8 + 7 + 3 + 2}$

2 Write the number that is halfway between

a 10 and 12 b 11 and 17 c 1 and 6

Fluency with data

3 The table shows the times of trains from York to London.

Train	A	B	C	D
Depart	09:30	10:01	10:59	14:06
Arrive	11:42	12:23	12:51	16:11

a How long does the 09:30 train take to get to London? Give your answer in hours and minutes.

b What time does the 14:06 train arrive in London? Give your answer as an am/pm time.

c Mary arrives at York station at 9:20 am. Which train does she catch to London?

4 Find the mean, median, mode and range of these data sets.

a 0, 1, 1, 3, 5

b 1, 3, 3, 3, 5, 7, 8, 13

c 5, 0, 1, 6, 5, 4, 1

5 In 2005, a group of office workers were asked how they travelled to work. The survey was repeated in 2010 and 2015. The compound bar chart shows the results.

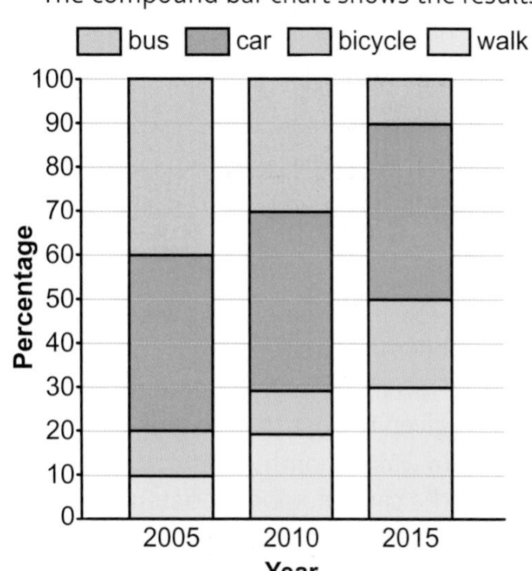

a What percentage of workers walked to work in 2005?

b What percentage of workers cycled to work in 2015?

c For each statement write T if it is definitely true, F if it is definitely false or CT if you cannot tell because there is not enough information.

i The number of workers walking to work increased in every survey between 2005 and 2015.

ii The proportion of workers travelling by bus decreased in every survey between 2005 and 2015.

iii The number of workers travelling to work by car in 2010 was the same as the number cycling or travelling by bus combined.

6 The table shows the age distribution of men and women in a company. Draw a compound bar chart to show this information.

Age	18–24	25–39	40–69
Number of men	10	40	50
Number of women	20	50	30

7 The dual bar chart shows the number of school detentions given to boys and girls during the year.

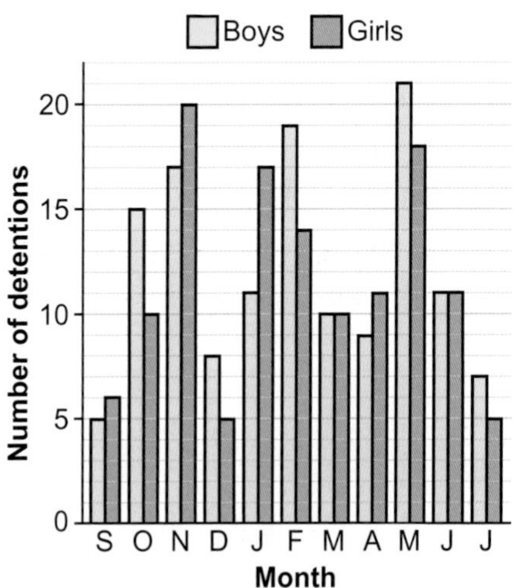

a How many detentions were given to boys in March?

b How many more detentions were given to boys than to girls in October?

c In which months did boys and girls get the same number of detentions?

d In which month did girls get 6 more detentions than boys?

e In which month did the school give the largest number of detentions?

f Compare the total number of detentions given to boys and girls.

8 The table shows the marks that two students obtained in their end of year exam. Draw a dual bar chart to show this information.

Subject	Geography	Biology	Spanish
Kyle	9	10	5
Adil	3	10	2

9 The pie chart shows the GCSE grades awarded to 720 students.

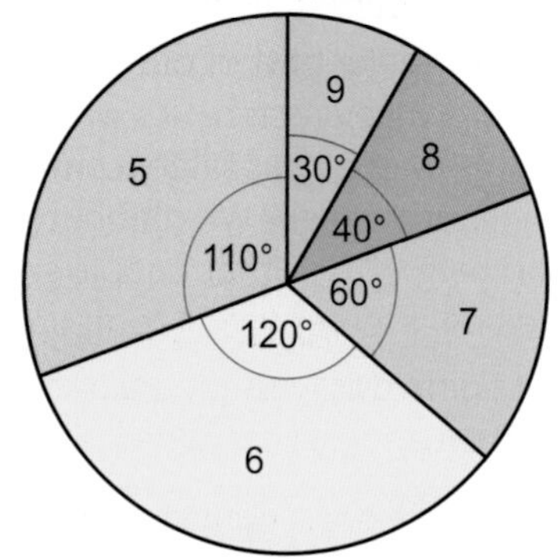

a What is the modal grade?

b How many students were awarded grade 8?

10 The table shows the holiday destinations of 120 people. Draw a pie chart for this data.

Destination	UK	France	USA	Spain
Number of people	55	15	20	30

11 The bar chart shows the scores obtained when a dice is rolled.

a How many times did the dice land on a 2?

b Work out the total number of rolls.

c What is the modal score?

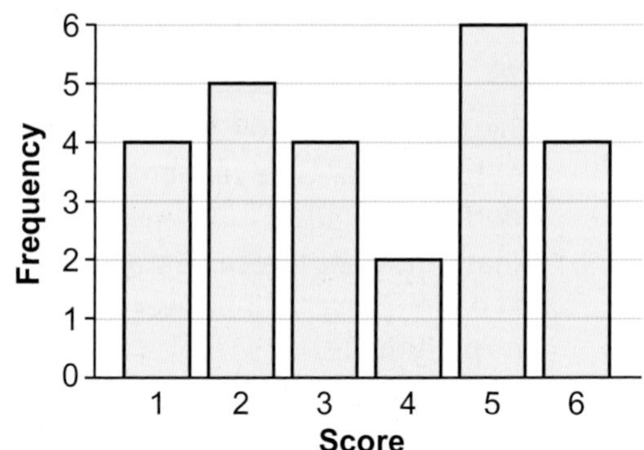

12 A social scientist is studying urban life in a developing country. She uses a data collection sheet to record information about the size of families living in a city.

Number of children	Tally
0	||||
1	~~||||~~ ~~||||~~
2	~~||||~~ |||
3	~~||||~~ |
4	~~||||~~
5	||

a Work out the range of the number of children per family in the city.

b Work out the total number of families in the survey.

c Calculate the mean. Round your answer to 1 decimal place (1 d.p.).

d The mean number of children per family in a rural community is 3.1. Compare this with the mean for city families.

13 The data set shows points awarded in the 2013/14 season of the Premier League.

69, 86, 36, 33, 40, 38, 84, 37, 72, 45, 49, 50, 56, 64, 30, 42, 79, 82, 38, 32

a Display the data on a stem and leaf diagram.

3	
4	
5	
6	
7	
8	

Key
3 | 0 means 30

b How many teams are in the Premier League?

c How many of these teams scored over 72 points?

14 Vinay conducts a survey to find out the types of TV programmes people like to watch.

He asks people whether they like to watch sport, drama, news or documentaries.

Design a suitable data collection sheet Vinay could use to collect the information.

15 The heights, in centimetres, of a dozen students in Year 10 are

162, 154, 174, 165, 175, 149, 160, 167, 171, 159, 170, 163

a Is the data discrete or continuous?

b Work out the missing numbers, x and y, in the grouped frequency table.

Height h (cm)	Frequency
$140 \leqslant h < 150$	1
$150 \leqslant h < 160$	2
$160 \leqslant h < 170$	x
$170 \leqslant h < 180$	y

c Draw a frequency diagram.

* Challenge

16 Find a set of five positive whole numbers with

- range 10
- mode 4
- median 6
- mean 7.

Is there more than one possible set?

Repeat for a set of six numbers. Find as many possible answers as you can.

3.1 Statistical diagrams 1

Objectives

- Construct and use back-to-back stem and leaf diagrams.
- Construct and use frequency polygons and pie charts.

Why learn this?

Diagrams provide a quick way of comparing the salaries of men and women.

Fluency

What are the mode, median and range of 2, 2, 4, 7?

Warm up

1 The table shows the median and range of scores obtained by Sophie and Celia after playing many rounds of golf.

	Median	Range
Sophie	71	13
Celia	93	25

Q1 strategy hint Compare the median scores and range of scores. In golf a lower score is better.

Write two sentences comparing the performances of Sophie and Celia.

Questions in this unit are targeted at the steps indicated.

2 **Real / Communication** The pie charts show the ages of people attending an open air theatre and a music festival.
1500 attended the theatre and 20 000 attended the festival.

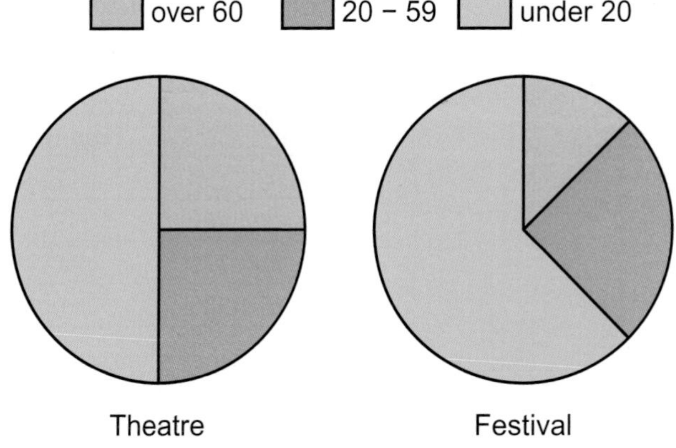

a How many people under 20 attended the theatre?
b Which pie chart has the larger sector for over-60s?
c Show that there are more over-60s at the festival than at the theatre.

Q2b hint What fraction is the over-60 sector?

Discussion Does a larger sector always represent a larger number?

3 **Reasoning** The stem and leaf diagram shows the masses of a group of people in a lift.

5	4			
6	3	4	7	
7	0	2	6	8
8	3	9		

Key
5 | 4 means 54 kg

a How many people are in the lift?
b What is the mass of the heaviest person in the lift?
c What is the range?
d What is the median?
e A safety notice in the lift reads, 'Maximum 12 persons, total weight 800 kg'.
Explain whether this group of people can travel in the lift safely.
f Calculate the mean mass of the people in the lift.

Q3d hint The median is the $\frac{n+1}{2}$th value, where n is the total number of values.

Q3e strategy hint Mention both of the facts on the safety notice in your answer.

Key point 1

A **back-to-back stem and leaf diagram** compares two sets of results.

Example 1

The annual salaries of employees working in an ICT company are displayed in the back-to-back stem and leaf diagram.

Key Male — 8 | 1 represents a salary of £18 000; Female — 1 | 9 represents a salary of £19 000

Male					Female			
			8	1	9	9		
9	5	2	0	2	1	2	6	7
8	7	3	0	3	0	4	4	
				4	5	6		
				5	4	8		

Compare the distribution of salaries of the male and female employees.

Male range: 38 000 − 18 000 = £20 000

Female range: 58 000 − 19 000 = £39 000

There are 9 males, so median male salary is: $\frac{9+1}{2}$ = 5th value = £29 000

There are 13 females so median female salary is: $\frac{13+1}{2}$ = 7th value = £30 000

Female employees' salaries have a larger range but the median salaries of men and women are similar.

Write a sentence comparing ranges and medians.

4 **Real / Problem-solving** A group of students take maths and English exams. The back-to-back stem and leaf diagram shows their results.

Compare the distribution of marks obtained by the students for the two exams.

Q4 hint Refer to the context (exam marks) in your comparison.

Key

4 | 3 represents 34 marks on the Maths exam

4 | 1 represents 41 marks on the English exam

Maths				English			
	5	4	3				
			4	1	5		
9	4	0	5	3	4	8	8
	3	1	6	0	2	9	
	6	6	7	8			
		8	8				

Discussion What does the shape of a back-to-back stem and leaf diagram show you?

5 **Real / Problem-solving** The heights (in cm, measured to the nearest cm) of two types of tulips are recorded.

Type A: 24, 37, 52, 26, 29, 46, 47, 29, 30, 36, 48, 55, 59

Type B: 16, 23, 34, 37, 31, 13, 64, 52, 53, 37, 43, 39, 38, 42, 42, 37

a Draw a back-to-back stem and leaf diagram for this data.

b Use the shape of your diagram to compare the distribution of heights of the two types of tulip.

Q5b hint A stem and leaf diagram is similar to a bar chart turned on its side. Compare the outlines of the two charts.

6 **Exam-style question**

Jeevan counted the number of letters in each sentence of a newspaper article. He showed his results in a stem and leaf diagram.

0	8	8	9					
1	1	2	3	4	4	8	9	
2	0	3	5	5	7	7	7	
3	2	2	3	3	6	6	8	8
4	1	2	3	3	5			

Key

4 | 1 represents 41 letters

a Write down the number of sentences with 36 letters.

b Work out the range.

c What is the modal number of letters?

d Work out the median. **(4 marks)**

Exam hint
Find the total number of values to help you find the median value.

Key point 2

To draw a **frequency polygon** you can join the midpoints of the tops of the bars in a frequency diagram with straight lines.

7 The table shows the heights of 100 students.

a Copy and complete the frequency diagram.

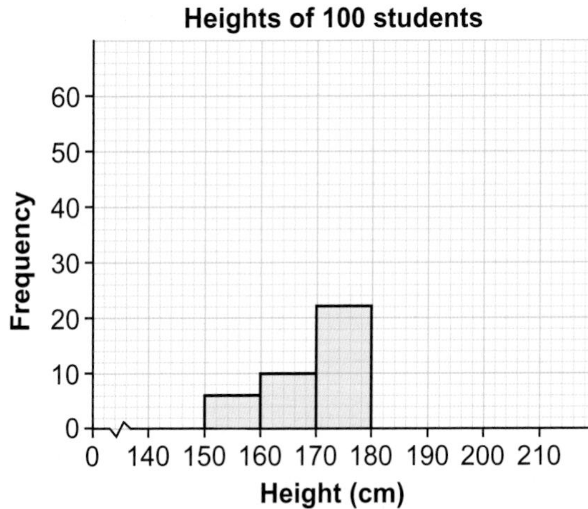

Height (h cm)	Frequency
$140 \leqslant h < 150$	0
$150 \leqslant h < 160$	6
$160 \leqslant h < 170$	10
$170 \leqslant h < 180$	22
$180 \leqslant h < 190$	52
$190 \leqslant h < 200$	10
$200 \leqslant h < 210$	0

Q7a hint Data is continuous, so no gaps between bars.

b Draw a frequency polygon on the same diagram.

Discussion Is there a quicker way to draw a frequency polygon without drawing a frequency diagram first?

Q7b hint Draw straight lines to connect the midpoints of the tops of the bars.

Key point 3

To draw a frequency polygon, plot the frequency against the midpoints for each group.

8 **Real** The frequency table shows the time taken for competitors to complete a charity fun run.

a How many runners took part?

b Work out the percentage of runners who took more than 100 minutes.

c Estimate the range.

d Copy and complete the frequency polygon.

Time (t mins)	Frequency
$20 \leqslant t < 40$	4
$40 \leqslant t < 60$	5
$60 \leqslant t < 80$	12
$80 \leqslant t < 100$	9
$100 \leqslant t < 120$	7
$120 \leqslant t < 140$	3

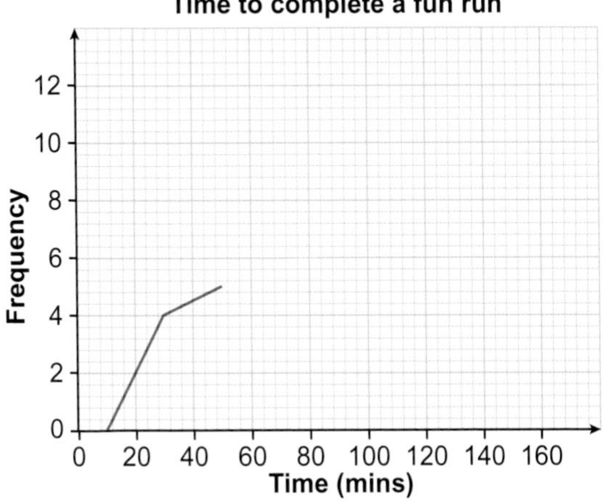

Q8c hint Do we know the actual times for the 4 runners in the $20 \leqslant t < 40$ group? What is the shortest time taken?

Q8d hint Work out the midpoint for each group.

Discussion Why is your answer to part **c** only an estimate? What assumptions did you make?

9 **Reasoning** A group of Year 10 students are each asked to guess the price of a can of cola and a small bunch of bananas.
The results are displayed on the frequency polygons.

a Which data set has the greater range?

b Would you expect the median of data set A to be greater than, less than or about the same as the median of data set B?

c Which data set do you think gives the prices for cola?

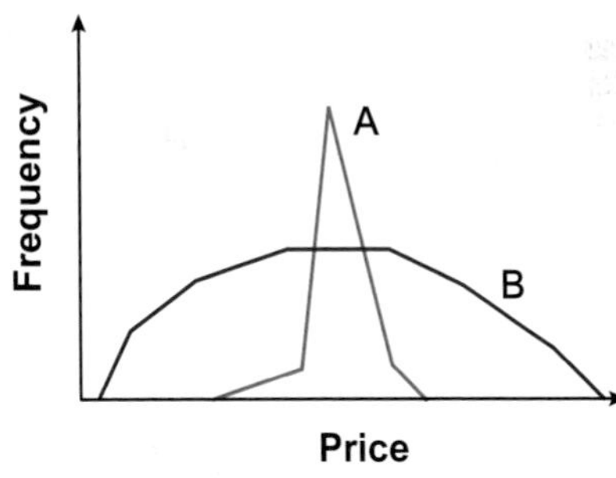

3.2 Time series

Objectives

- Plot and interpret time series graphs.
- Use trends to predict what might happen in the future.

Why learn this?

Scientists can use trends in weather patterns to investigate climate change.

Fluency

Is this sequence increasing or decreasing?
1, 5, 9, 13, 17, 21, 25, 29

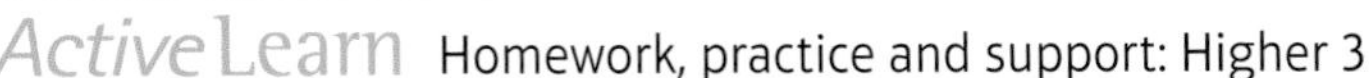

Warm up

1 Match each sequence with the correct description.

a 3, 5, 7, 9, 11, 13, 15, 17, 19, 21
b 0, 26, 46, 60, 69, 76, 81, 84, 85, 85
c 4, 4, 4, 4, 4, 4, 4, 4, 4, 4
d 12, 11, 10, 9, 8, 7, 6, 5, 4, 3
e 1, 3, 7, 2, 0, 4, 8, 6, 2, 5

i The values are constant.
ii The values increase at a constant rate.
iii The values decrease at a constant rate.
iv The values increase at a decreasing rate.
v The values fluctuate up and down.

2 Predict the next two terms in these sequences.

a 4, 6, 8, 4, 6, 8, 4, 6, 8, …
b 3, 4, 6, 9, 13, 18, 24, 31, …

3 The price of a book rises from £16 to £20. Work out the percentage increase.

Key point 4

A **time series** graph is a line graph with time plotted on the horizontal axis.

4 **Real / Reasoning** The table shows the temperature of a hospital patient recorded on the hour every 2 hours during a 24-hour period.

Time	00	02	04	06	08	10	12	14	16	18	20	22
Temperature (°C)	36.7	36.8	37.1	37.4	37.8	38.3	38.0	38.2	37.4	37.3	37.2	37.1

a What is the patient's temperature at 6 pm?
b What is the patient's maximum temperature during this period? At what time did it occur?
c Work out the mean temperature. Give your answer to 1 decimal place (1 d.p.).

Discussion Why is the mean temperature an estimate?

d Represent this time series on a line graph. Comment on the variation of temperature.

Q4d hint Use a vertical axis from 36 to 39.

5 Each week of the autumn term, a teacher records the number of pieces of late homework.

Week	1	2	3	4	5	6	7	8	9	10	11	12
Number of late homeworks	38	34	26	14	8	7	18	10	7	15	25	40

Draw a time series graph for this data. Comment on how late homework varies during the course of the term.

6 **Real / Reasoning** The time series graph shows the prices of two magazines over the last 6 years.

a What was the price of Magazine A in 2012?
b Suzy says the price of Magazine A has risen more than the price of Magazine B during this period.
Is she correct? Give a reason for your answer.
c Suzy also says that the rate of increase in the price of Magazine A is slowing down.
Is she correct? Give a reason for your answer.
d Predict the prices of each magazine in 2018.

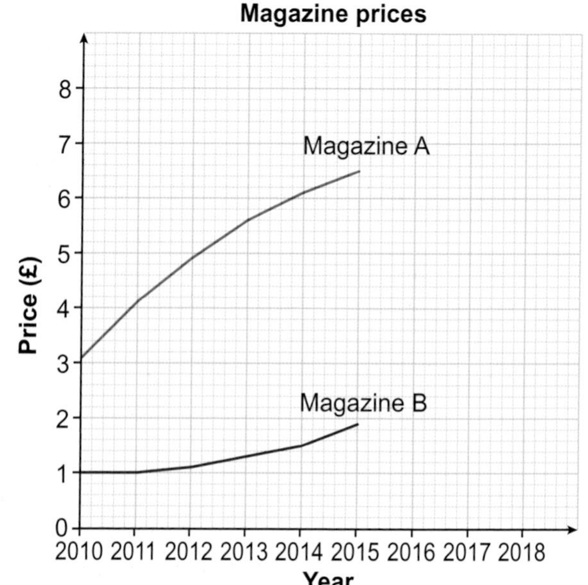

Example 2

The table shows the quarterly price of a tonne of wheat (in dollars) during the last three years.

2012				2013				2014			
Q1	Q2	Q3	Q4	Q1	Q2	Q3	Q4	Q1	Q2	Q3	Q4
250	279	101	157	348	371	230	264	451	477	322	347

Communication hint Prices are recorded every 3 months so the first quarter covers January, February and March.

a What is the price in the third quarter (Q3) of 2013?
b In which quarter is the price the lowest?
c Draw a time series graph of the data.
d Describe the variation in prices during this period and comment on the overall trend.

a $230
b The lowest price is $101 which occurs in the third quarter of 2012.
c

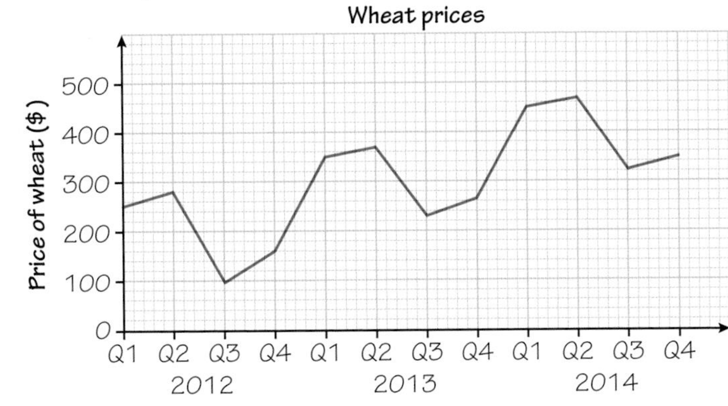

d The price of wheat fluctuates up and down during the course of each year. However the overall trend shows a general increase in prices.

7 **Real / Problem-solving** The table shows the quarterly sales (in thousands) of umbrellas during the last three years.

2012				2013				2014			
Q1	Q2	Q3	Q4	Q1	Q2	Q3	Q4	Q1	Q2	Q3	Q4
89	75	24	85	80	66	19	76	75	62	17	73

a What are the sales in the second quarter of 2013?
b In which quarter are sales the highest?
c Draw a time series graph for this data.
d Describe the variation in umbrella sales during this period and comment on the overall trend.
e **Reflect** What do you think the term 'trend' means? Write a definition in your own words.

Q7a hint Sales are in thousands.

8 **Real / Problem-solving** The table shows the profit (in millions of pounds) of an ICT company over the past 10 years. The profit for 2010 is not known.

Year	2005	2006	2007	2008	2009	2010	2011	2012	2013	2014
Profit	0.5	0.7	0.8	1.1	1.4	?	2.4	3.1	4.1	5.3

a Draw a line graph for this time series.
b Describe the overall trend.
c Estimate what the profit might have been in 2010.
d Predict what the profit might be in 2015.

Discussion How reliable are your values in parts **c** and **d**?

Q8a hint Draw a horizontal axis from 2005 to 2015 and a vertical axis up to 7, to help with part **d**.

Q8d hint Continue the 'curved' shape of the graph for two more years. Profit is in millions of pounds.

9 **Exam-style question**

The table shows the height of sea water in a harbour between midnight and noon.

Time	Midnight	1 am	2 am	3 am	4 am	5 am	6 am	7 am	8 am	9 am	10 am	11 am	Noon
Height (m)	10	12.5	14.3	15	14.3	12.5	10	7.5	5.7	5	5.7	7.5	10

a Draw a time series graph to show how water height varies with time.

b At what time does high tide occur in the morning?

c Predict the height at 3 pm.

d It is only safe for a ship to enter the harbour when the water height exceeds 7.5 m.
During which times of the day is it not safe for ships to enter the harbour? **(7 marks)**

Exam hint
Use both the table and the graph to help you spot the pattern.

3.3 Scatter graphs

Objectives

- Plot and interpret scatter graphs.
- Determine whether or not there is a linear relationship between two variables.

Why learn this?

Scatter graphs help us to see whether there is a connection between two sets of data. For example, it would be useful for a shop to know if there is a link between sales of bottled water and temperature.

Fluency

State whether each line has a positive, negative or zero gradient.

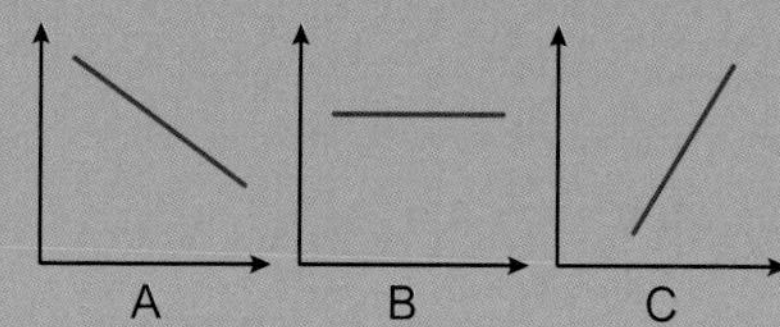

Warm up

1 Draw x and y-axes on graph paper from 0 to 8. Plot five points with coordinates A(2, 4.75), B(5, 3.5), C(4, 3.75), D(7, 3) and E(3, 4).
Four of the points lie on a line. Which point does not lie on the line?

Key point 5

Bivariate data has two variables. Plotting these on a **scatter graph** can show whether there is a relationship between them.

2 Eight students took a maths test and a science test. Their marks are displayed in the table.

Student	A	B	C	D	E	F	G	H
Maths mark	3	9	7	3	6	10	5	1
Science mark	4	8	4	2	5	7	3	1

a Copy and complete the scatter graph.
Student A scored 3 in maths and 4 in science so draw a cross at (3, 4).
For student B, draw a cross at (9, 8).

b Use the scatter graph to copy and complete the sentence.
In general, students with higher maths scores got .... science scores and students with lower maths scores got .... science scores.

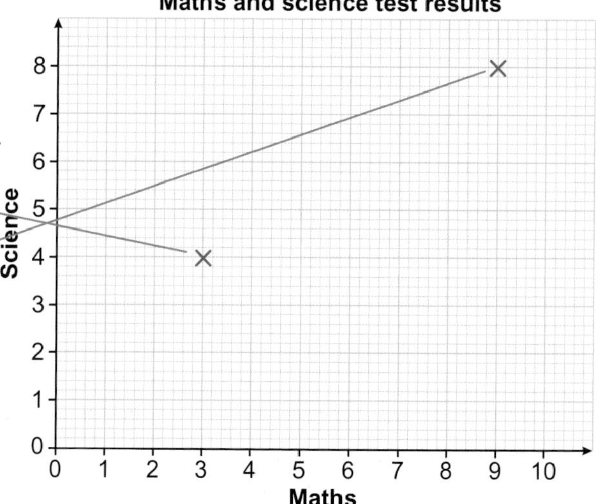

Key point 6

A scatter graph shows a relationship or correlation between variables.

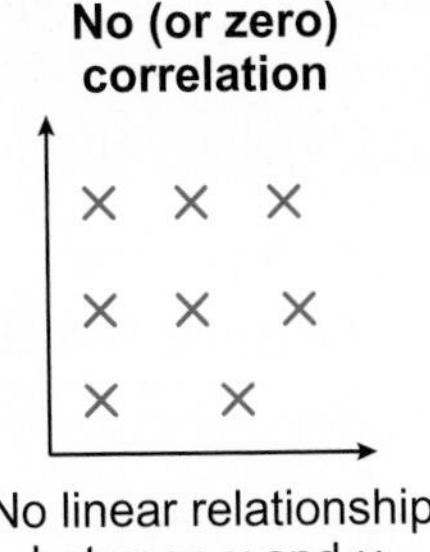

No linear relationship between x and y.

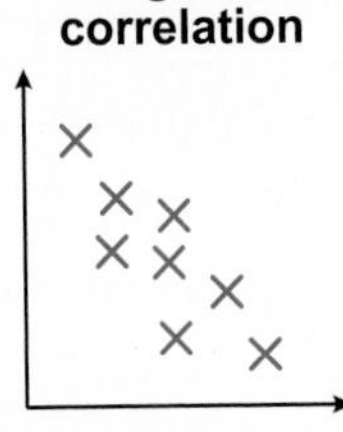

Points lie close to a downward-sloping straight line. As x increases y decreases.

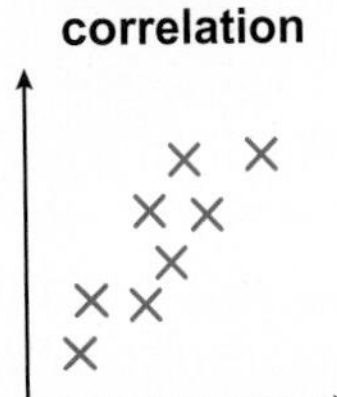

Points lie close to an upward-sloping straight line. As x increases y increases.

3 The daily sales and price of ice cream are recorded together with the maximum outside temperature. Three scatter graphs are plotted from the data.

a

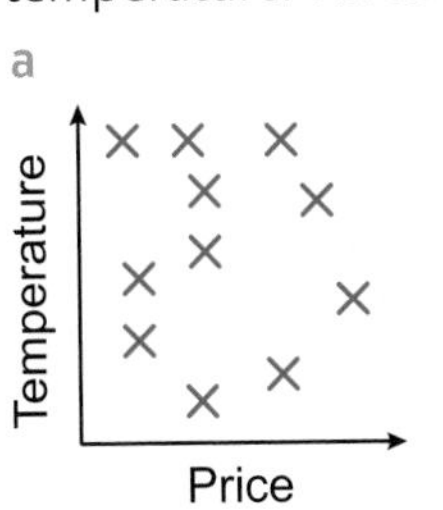

b

c

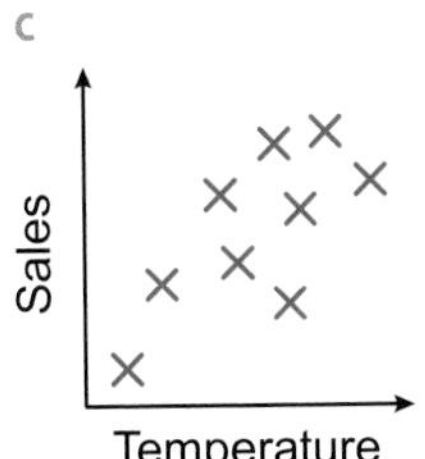

Q3 hint To describe what the correlation means in words, you could say, 'As the price of ice creams increase, sales ... .'

For each graph state whether there is positive, negative or no correlation and describe in words what this means.

4 **Real** A car dealer notes the engine size of seven models of car and the distance they travel on a litre of petrol.

Engine size (litres)	1	1.4	1.6	2	3	3.5	4
Distance (km)	16	14.2	13.5	11.7	9.2	8.4	7.1

a Draw a scatter graph for this data.

b Describe any relationship between these two variables.

Q4b hint State the type of correlation and then write a sentence beginning, 'The larger the engine size, the ... .'

Q4a hint Put engine size on the horizontal axis from 0 to 5 and distance on the vertical axis from 6 to 17.

5 An auction house asks an art dealer to award six paintings marks out of 10, without disclosing the names of the artists.

Score	2	7	5	3	8	4
Market value	3.5	1.8	5.6	4.3	8.4	2.5

The market value (in £100 000s) and dealer's score for each painting are in the table.
Draw a scatter graph and describe any relationship between the score and the value of the paintings.

6 **Reasoning** A survey of seven British towns records the number of serious road accidents in a week, together with the number of takeaway restaurants.

Number of restaurants	85	15	10	52	71	25	90
Number of accidents	27	9	4	19	17	12	19

a Draw a scatter graph and comment on any relationship between the two variables.

b A local councillor notices that there has been a sharp increase in the number of road accidents in recent years. She puts the blame on an increase in the number of takeaway restaurants. Does the scatter graph provide statistical evidence to support the councillor's view?

Discussion Why do you think there is correlation in this data set?

7 **Real** What sort of correlation would you expect to find between:

a height above sea level and air temperature

b adults' weekly calorie intake and their weight

c a student's shoe size and marks on a French exam?

Discussion Give *two* other practical examples of data sets that illustrate each of the three types of correlation.

8 **STEM / Reasoning** In a chemistry experiment, the mass of chemical produced, y, and temperature, x, are recorded.

x (in °C)	100	110	120	130	140	150	160	170	180	190	200
y (in mg)	34	39	41	45	48	47	41	35	26	15	3

a Plot these points on a scatter graph.

b State the type of correlation between mass and temperature.

c Describe in words what happens to the mass of chemical produced as the temperature increases from 100 to 200 °C.

d Estimate the maximum mass and the temperature required to achieve this.

9 **Exam-style question**

A manufacturer of mp3 players monitors the cost of quality control (in £10 000s) and the percentages of faulty items.

The results are shown on the scatter graph.

Effect of quality control

Percentage of faulty items (0–5) against Cost of quality control (£10 000s) (0–9)

Exam hint
This graph is drawn on mm-squared paper, so make sure you read off the values accurately. Be aware of the units when you give your answers.

a State the type of correlation between these variables and interpret your answer.

b What was the highest percentage of faulty items in the data set?

c Find the range of the cost of quality control. **(4 marks)**

3.4 Line of best fit

Objectives

- Draw a line of best fit on a scatter graph.
- Use the line of best fit to predict values.

Did you know?

You can use a line of best fit to predict someone's height from their shoe size.

Fluency

Copy and complete this sentence.

When the correlation between x and y is negative, larger values of x are associated with … values of y.

1 Read off the value of
   a y when $x = 6$
   b x when $y = 2.5$

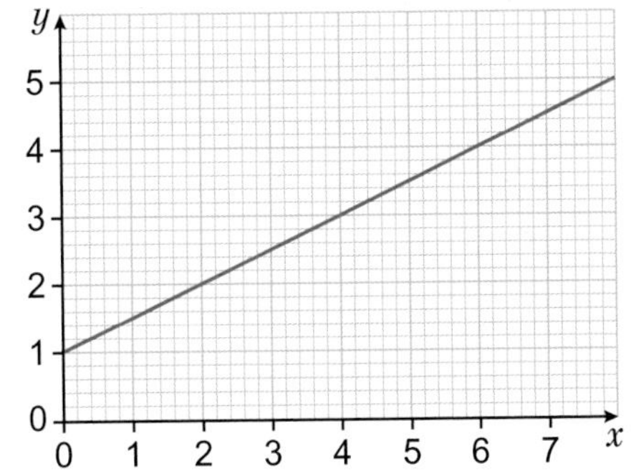

Key point 7

A **line of best fit** is the line that passes as close as possible to the points on a scatter graph.

2 Which line, A, B or C, is the best line of best fit for the data points on the scatter graph?

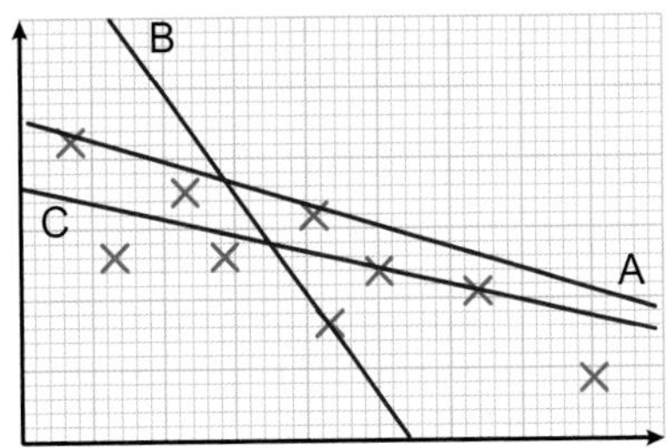

Example 3

The scatter graph shows the GDP per capita (in $1000s) and life expectancy (in years) for eight countries.

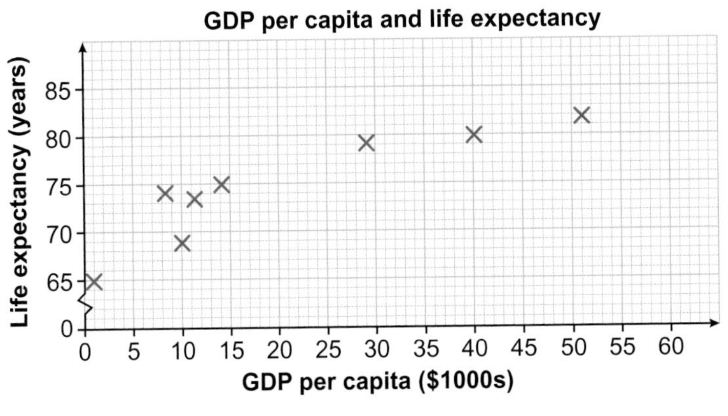

Communication hint
The gross domestic product (GDP) measures the value of goods and services produced by a country. The GDP per capita is the GDP divided by the number of people in that country.

a Draw a line of best fit.

b The GDP per capita in the UK is $36 000. Estimate the life expectancy of a baby born in the UK.

a

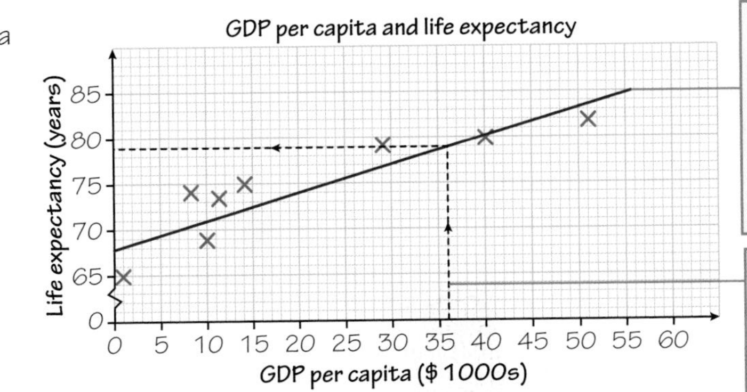

Position a transparent ruler over your scatter graph so it follows the overall trend. Move it slightly so you have roughly the same number of points above and below the line.

To estimate life expectancy, start at $36 000 on the horizontal axis, go up to the line of best fit and read off the answer on the vertical axis.

b Estimated life expectancy in the UK is 79 years.

3 The table shows the height and weight of eight athletes.

Height (cm)	155	166	170	175	178	192	193	198
Weight (kg)	50	65	64	77	67	85	115	95

a Draw a scatter graph for this data.
b Draw a line of best fit on your graph.
c Use your line of best fit to estimate the weight of an athlete who is 185 cm tall.
d Estimate the height of an athlete who weighs 60 kg.

4 **STEM** A chemical engineer heats gas inside a sealed tank and measures the temperature (in degrees Kelvin, °K) and pressure (in atmospheres, atm).

Temperature (°K)	300	303	304	312	325	339	343	351
Pressure (atm)	1.4	1.5	1.7	2.0	2.2	2.3	2.5	3.0

Draw a line of best fit on a scatter graph and use the line to
a estimate the temperature required to create a pressure of 2.8 atm
b estimate the pressure when the temperature is 308 °K.

Q4 hint Draw lines on your scatter graph to show how you obtained your estimates.

5 **Real / Reasoning** The table shows the height and shoe size of a group of male college students:

Height (cm)	Shoe size
158	4
168	5
164	6
167	8.5
174	8.5
178	10
173	10.5
185	12

a Draw a line of best fit on a scatter graph and use it to estimate
i the shoe size of someone who is 175 cm tall
ii the height of someone with shoe size 7
iii the height of someone with shoe size 13.5.
b Which of these estimates do you think is the least reliable? Give a reason for your answer.

Discussion As a general rule, is it better to use a line of best fit to make predictions about values inside or outside the existing range of data points?

Key point 8

Using a line of best fit to predict data values within the range of the data given is called **interpolation** and is usually reasonably accurate.

Using a line of best fit to predict data values outside the range of the data given is called **extrapolation** and many not be accurate.

6 **Real / Reasoning** Jack and Joe perform an identical experiment in a science lesson. Their results are shown on the scatter graphs.

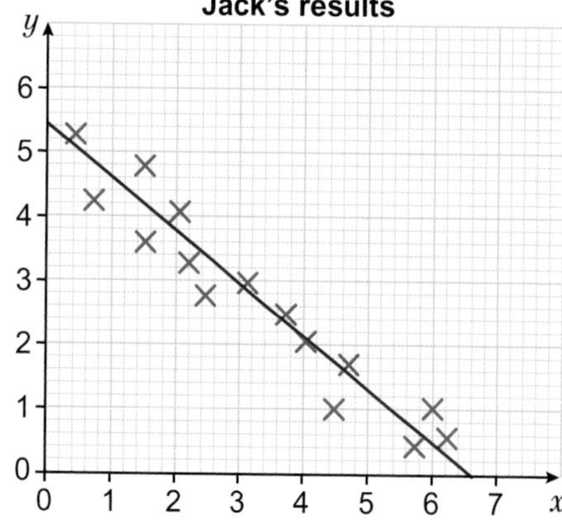

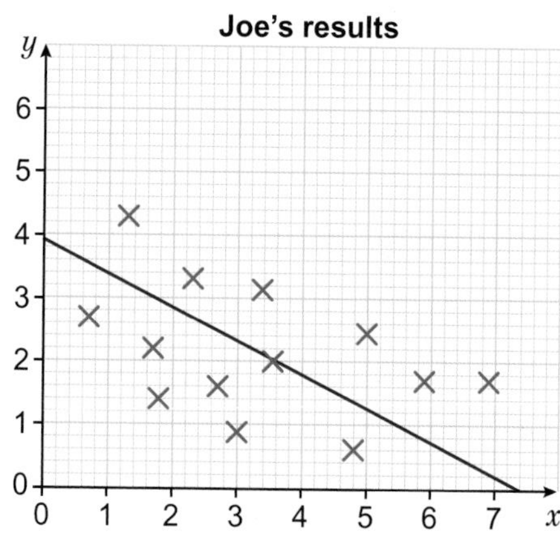

a Use the given lines of best fit to work out two estimates for the value of y at $x = 3$.
b Which of the estimates is likely to be more reliable? Give *two* reasons for your answer.

Q6b hint The points on one diagram are very close to the line of best fit. On the other diagram the points are more scattered.

Key point 9

Individual points which are outside the overall pattern of a scatter graph are called **outliers**. If they are likely to be from incorrect readings you can ignore them.

7 **STEM / Reasoning** An elastic rope is suspended from the ceiling and stretched vertically by hanging weights on the end. The table shows the weight, W (in newtons), and length, L (in cm), of the elastic.

W (N)	1	2	4	4.5	5	6.5	8.5	9	10
L (cm)	12.4	14.8	18.6	20.4	21.2	24.5	29.4	30.4	40

Q7e hint When the elastic is not stretched, no weight is suspended.

a Draw a scatter graph for this data.
b Why is the last point classified as an outlier? Suggest a possible reason for this.
c Draw a line of best fit passing close to the remaining eight points.
d Use the line to estimate the length of the elastic when a mass of 7 N is suspended.
e Estimate the length of the elastic when it is not stretched.

Reflect A lot of the questions in this lesson ask you to estimate. Write a definition of the word 'estimate' in your own words.

8 **Modelling** The table shows the age, x, and mass, y, of a sample of 11 boys.

x (years)	2	4	6	8	10	12	14	16	18	20	22
y (kg)	13	16	21	25	33	40	51	60	67	71	72

a Draw a scatter graph of this data.
b Assuming that weight can be modelled using a line of best fit:
i estimate the weight of a 15-year-old ii estimate the weight of a 24-year-old.
c Which of the answers in part **b** is likely to be the more reliable?
d By drawing a smooth curve close to the data points, make new estimates of the weights in part **b**.
e Which of the two models is the more accurate? Give a reason for your answer.

9 **Exam-style question**

The table shows the distance, d, of ten apartments from a city centre and the monthly rent, M, for each.

Distance, d (km)	0.4	0.8	0.9	1.4	1.8	2.3	2.3	3.2	3.4	4
Rent, M (£)	510	470	430	340	400	290	320	140	100	120

a Plot the points on a scatter graph.
b Describe the relationship between the distance from the city centre and monthly rent.
c Estimate the rent of an apartment that is 2.7 km from the city centre.

(5 marks)

Exam hint
Always draw lines on your diagram for any readings from your graph. If you get the answer wrong, you may still get marks for using the correct method.

3.5 Averages and range

Objectives

- Decide which average is best for a set of data.
- Estimate the mean and range from a grouped frequency table.
- Find the modal class and the group containing the median.

Did you know?

You can compare aspects of your lifestyle with averages.

Fluency

Work out the range of 2, 6, 4, 9, 1 and 14.

ActiveLearn Homework, practice and support: Higher 3.5

Warm up

1 The table shows the scores when a dice is rolled.

Score	1	2	3	4	5	6
Frequency	4	1	3	3	2	1

a How many times is the dice rolled in total?
b What is the modal score?

2 a Write down the number which is halfway between 11 and 20.
b What is the middle value in the interval $20 \leqslant x < 40$?

3 **Real / Finance** The annual salaries of staff who work in a cake shop are
£12 000, £12 000, £15 000, £18 000, £40 000

a Work out the mean, median and mode of staff salaries.
b The company wishes to quote one of the averages in an advertisement for new staff.
Which of the averages would be the most appropriate? Give reasons for your answer.

Q3b hint The number quoted in the advert must represent a typical salary that you could reasonably expect to earn.

4 **Reasoning** The sizes of shoes sold in a shop during a morning are
5, 5.5, 5.5, 6, 7, 7, 7, 7, 8.5, 9, 9, 10, 11, 11.5, 12, 13

a Work out the mean, median and mode of these shoe sizes.
b The shop manager wishes to buy more stock but is only allowed to buy shoes of one size.
Which one of these averages would be the most appropriate to use?
Give reasons for your answer.

Reflect What do you think the differences are between mean, median and mode? Write notes in your own words and include an example of when each would be appropriate to use.

5 **Real / Finance** The monthly costs of heating a shop in the winter months are shown in the table.

Month	Heating cost
Nov	£180
Dec	£190
Jan	£270
Feb	£240
Mar	£180

a Work out the mean, median and mode of heating costs.
b The shop must provide a report of expenses and overheads to its accountant.
Which of the averages is the most appropriate to provide in the report? Give reasons for your answer.

6 **Reasoning** State whether it is better to use the mean, median or mode for these data sets. Give reasons for your answers.

a Time taken for five people to perform a task (in seconds):
6, 25, 26, 30, 30.
b Car colour: red, red, grey, black, black, black, blue.

7 Identify the outliers of the data sets and find the range of each.

a The masses of six members of a local wrestling team:
7 kg, 76 kg, 82 kg, 89 kg, 96 kg, 101 kg
b The salaries of the six people who work in a small restaurant:
£14 000, £15 000, £15 000, £17 500, £19 000, £38 000

Q7a hint Is it possible for someone in the team to weigh 7 kg?

Q7 strategy hint Think about whether you should include the outlier in your calculations.

Q7b hint Who might be earning £38 000 in a restaurant? It is likely that this figure is correct.

8 **STEM / Finance** Identify the outliers of the data sets. Calculate a sensible value of the range. Give a reason why the outlier has been included or excluded in your calculation.

a Nine temperature readings (in °C) recorded during a science experiment:
34, 44, 30, 27, 500, 30, 40, 45, 2.9

b The profit or loss made by a firm during the last six years:
£100 000, −£250 000, £50 000, £75 000, £150 000, −£25 000

Q8b hint The negative numbers indicate a loss.

Example 4

The table shows the times, T, taken for 100 people to queue for a rollercoaster at a theme park.

a Estimate the mean waiting time.

b Explain why the mean is only an estimate.

The third column gives an estimate of the waiting time in each class.

a

Time, T (mins)	Frequency, f	Class midpoint, x	xf
$0 \leqslant T < 20$	14	10	$10 \times 14 = 140$
$20 \leqslant T < 40$	55	30	$30 \times 55 = 1650$
$40 \leqslant T < 60$	31	50	$50 \times 31 = 1550$
Total	100		3340

The fourth column gives an estimate of the total waiting time in each class.

$$\text{Mean} = \frac{\text{sum of waiting times}}{\text{total number of people}} = \frac{3340}{100}$$

$= 33.4$ minutes

b The mean is an estimate because we don't know the exact times taken.

Discussion What assumptions have been made about the data?

9 **Real** The grouped frequency table shows the length of Kate's phone calls during the last month.

Time, T (mins)	Frequency, f	Midpoint, x	xf
$0 \leqslant T < 4$	27	2	$2 \times 27 = 54$
$4 \leqslant T < 10$	34		
$10 \leqslant T < 20$	15		
$20 \leqslant T < 60$	4		
Total			

a Copy and complete the table to estimate the mean length of phone calls.

b A call costs 0.05 p/min. Estimate the total cost of these calls.

Key point 10

If the total frequency in a grouped frequency table is n, then the median lies in the class containing the $\frac{n+1}{2}$th item of data.

The **modal class** has the highest frequency.

10 **Reasoning** The times taken for students to do their maths homework are shown in the table.

t (mins)	Frequency
$0 \leqslant t < 10$	3
$10 \leqslant t < 20$	5
$20 \leqslant t < 30$	8
$30 \leqslant t < 40$	5

a How many students took less than 10 minutes to do their homework?

b How many students altogether took less than 20 minutes to do their homework?

c How many students altogether took less than 30 minutes to do their homework?

d State the modal class.

e Explain why the median is the 11th data value.

f Use your answers to parts **a–d** to work out which class interval contains the median.

11 **Real / Problem-solving** The table shows the distances jumped by two athletes training for a long jump event.

Distance (d m)	Ben's frequency	Jamie's frequency
$6.5 \leqslant d < 7.0$	3	8
$7.0 \leqslant d < 7.5$	7	18
$7.5 \leqslant d < 8.0$	25	21
$8.0 \leqslant d < 8.5$	1	3
$8.5 \leqslant d < 9.0$	0	1

a How many jumps did Ben do in training?
b Explain why Ben's median distance is halfway between the 18th and 19th items in the data set.
c In which class interval is Ben's median?
d Work out which class interval contains Jamie's median distance.
e On average, which athlete jumps the furthest in training?
f State the modal class for Ben and Jamie.
g At the long jump event, both athletes must compete against the current champion, who jumped 8.31 m.
Who stands the better chance of beating him? Explain your answer.

12 **Exam-style question**

A rail company monitors delays on its peak time weekday service.

The results for the last month are shown in the frequency polygon.

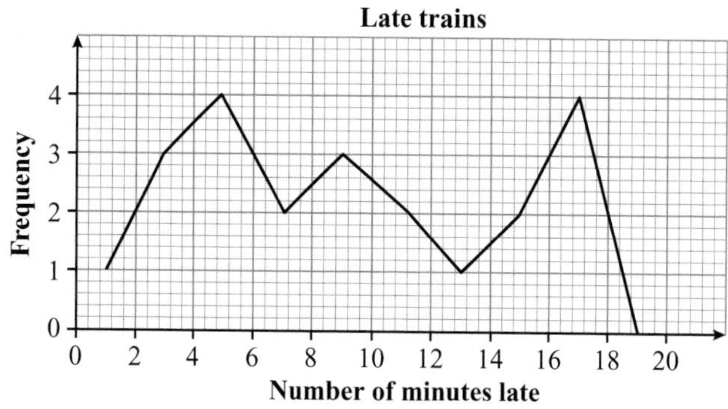

Exam hint
Make sure you understand what the diagram represents before you begin.

a How many trains were more than 14 minutes late last month?

b The company offers compensation to its monthly season ticket holders if the mean delay on its peak weekday trains exceeds 10 minutes.
Should the company offer compensation for last month? **(5 marks)**

3.6 Statistical diagrams 2

Objectives

- Construct and use two-way tables.
- Choose appropriate diagrams to display data.
- Recognise misleading graphs.

Why learn this?

Being able to recognise misleading graphs can help you to avoid being conned by misleading adverts.

Fluency

There are 670 boys in a school of 1200. How many girls are there?

Warm up

1 The table shows the drink choices of a group of 40 people.
Draw a pie chart to represent this data.

Drink	Tea	Coffee	Cola	Water
Frequency	8	20	5	7

2 Draw a bar chart for the data in **Q1**.

ActiveLearn Homework, practice and support: Higher 3.6

3 A group of 180 students are asked whether they did their maths homework last night. The table shows some information about their responses.

	Yes	No	Total
Boys		30	
Girls	25		100
Total			180

Q3 hint Begin by using the total in the second row to work out the number of girls who did not do their homework.

Copy and complete the table.

4 **Reasoning / STEM** A clinical trial is carried out to compare the effect of two drugs for the treatment of hay fever.

One hundred hay fever sufferers were given *either* Drug A or Drug B. After a week the patients were asked to choose one of three responses: no change, improved or much improved.

	No change	Improved	Much improved	Total
Drug A	10			60
Drug B			13	
Total	17	65		100

a Copy and complete the table.

b What fraction of these patients were given Drug B?

c Which drug performed best in this trial? Give reasons for your answer.

Q4b hint Divide the number of patients given Drug B by the total number of patients.

5 **Reasoning** Students were asked whether they were in favour of having more lockers in the school changing rooms.

In Year 10, 110 of the 180 students were in favour. In Year 11, 100 of the 210 students were against the idea.

a Display this information in a two-way table.

b The school will only buy new lockers if at least 60% of Year 10 and 11 students are in favour. Explain whether the school will buy the lockers.

6 The bar chart shows the level of support for a new high speed rail link.

The government claims that this provides convincing evidence that people are in favour of the plans.

a Explain why this bar chart is misleading.

Q6a hint Look carefully at the vertical scale.

b Draw a correct version. Comment on what information this provides about the level of support for the new rail link.

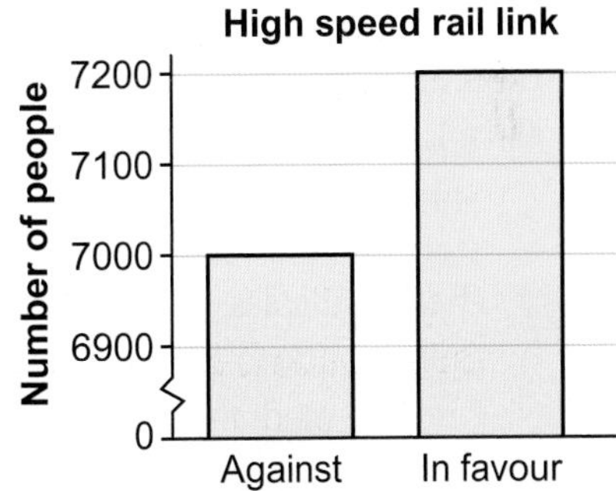

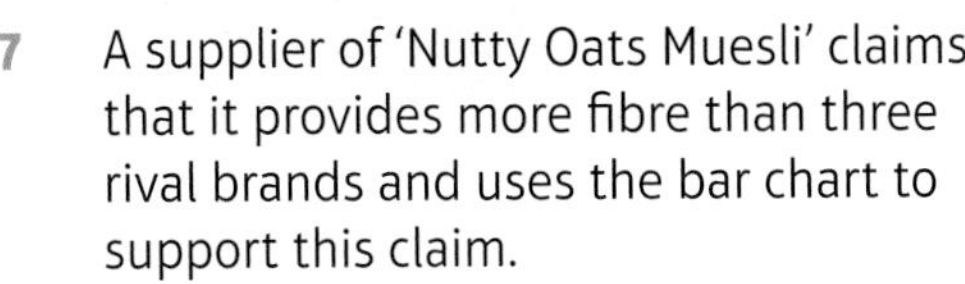

7 A supplier of 'Nutty Oats Muesli' claims that it provides more fibre than three rival brands and uses the bar chart to support this claim.

Give *two* reasons why this diagram is misleading.

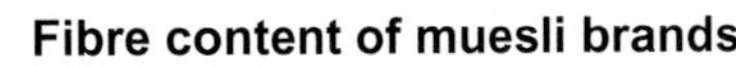

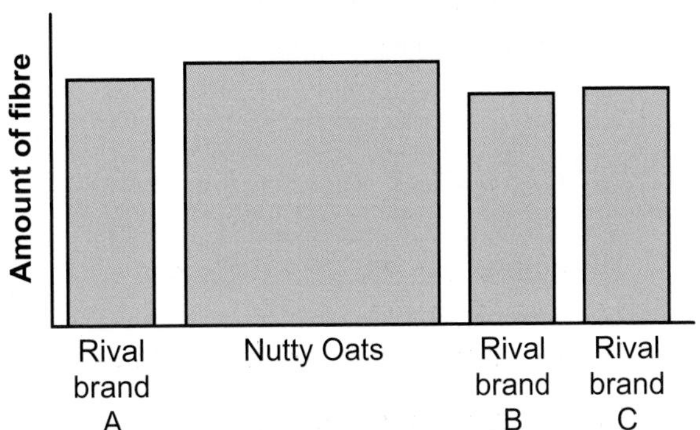

8 **Finance / Reasoning** The line graph shows the share price of an ICT company on the first of each month.

a Find the share price on 1 May.

b Amelia bought 250 shares on 1 February. She sold them on 1 October. How much profit did she make?

c In which months should Amelia have bought and then sold her shares to make the highest profit?

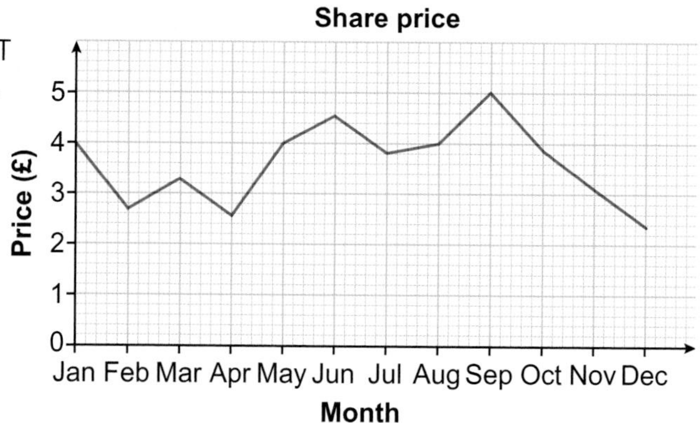

Key point 11

Line graphs are useful for tracking changes over time. Pie charts are good when comparing parts of a whole. Bar charts are used to compare the frequencies of two data sets.

9 **Reasoning** A tuck shop sells four types of crisps: ready salted (RS), cheese and onion (CO), salt and vinegar (SV), and smoky bacon (SB).

On one day, the shop sells these flavours to boys and girls:

Boys: CO, RS, CO, SV, SV, RS, RS, SB, CO, CO, RS, SB, RS, RS, CO, CO, RS, SV, SV, RS

Girls: RS, SB, SV, SV, CO, CO, RS, RS, RS, SV, RS, CO, SB, SV, SB, SV, SB, CO, SB, CO

a Explain why it is not possible to display this data on a frequency polygon.

b The shop manager decides to display the data on either a pie chart or a bar chart. Which should he use if he is most interested in:

i the proportion of each flavour bought by the boys or girls combined

ii comparing the number of each flavour bought by boys and girls?

c The shop manager orders 720 packets of crisps from a supplier. How many packets of cheese and onion crisps should he order?

10 **Communication** A teacher records the marks awarded to boys and girls on a test.

Boys: 23, 7, 10, 34, 10, 5, 3, 39, 31, 6, 7, 15, 21

Girls: 1, 15, 25, 39, 17, 24, 11, 28, 6, 39, 20, 16

> **Q10a hint** What information do you lose when you draw a bar chart?

> **Q10c hint** Use $\frac{(n+1)}{2}$

a State one advantage of using a back-to-back stem and leaf diagram instead of a dual bar chart to display this information.

b Draw a back-to-back stem and leaf diagram for this data.

c Find the median marks to compare the performance of boys and girls on this test.

11 **Exam-style question**

A researcher wants to compare the waiting times at several hospital accident and emergency departments.

The table shows the waiting times at one of the hospitals over a morning.

> **Exam hint**
> Imagine trying to draw each diagram. You should also think about what the diagram is going to be used for.

Waiting time, T (mins)	$0 \leq T < 30$	$30 \leq T < 60$	$60 \leq T < 90$	$90 \leq T < 120$	$120 \leq T < 150$	$150 \leq T < 180$
Frequency	20	35	30	24	15	12

a Which *one* of these statistical diagrams would be the best diagram to use to display this data?

stem and leaf pie chart frequency polygon scatter graph

b Give reasons for your choice. **(4 marks)**

3 Problem-solving: Pollution particulates

Objectives

- Be able to estimate the mean from a frequency polygon.
- Be able to construct a statistical argument and identify limitations arising from estimating the mean.

Air pollution limits

Companies are not allowed to cause too much air pollution. This includes the number of airborne particulates they produce.

> **Communication hint** Particulates are very small particles. For example, dust and soot are both particulates.

The legal limits for particulates are a yearly mean of

- 40 mg in every cubic metre of air for PM10 (coarse) particles
- 25 mg in every cubic metre of air for PM2.5 (fine) particles.

Fact

PM2.5 particles have a diameter of less than 2.5 micrometres.
PM10 particles have diameters between 2.5 and 10 micrometres.
1 micrometre = 10^{-6} metres

Case study

A national newspaper recently accused a company of being over the legal air pollution limits. The company tried to persuade the newspaper to withdraw the story by publishing two frequency polygons. According to the company, these frequency polygons showed that their yearly means for both types of particulates were below the legal limits.

The newspaper refused to withdraw the story. It claimed that the company's frequency polygons did not fully prove that levels of particulates fell below the legal limits.

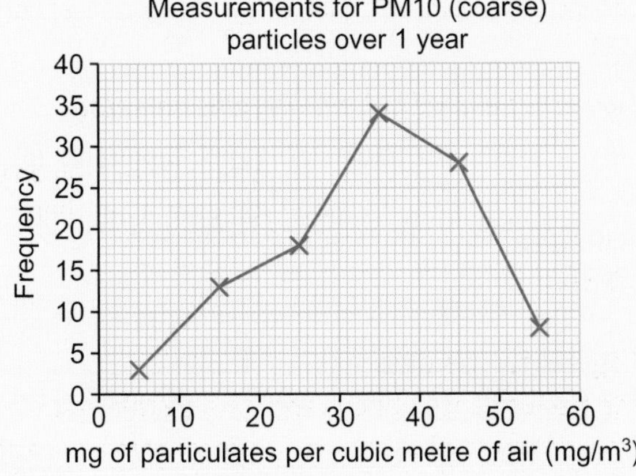

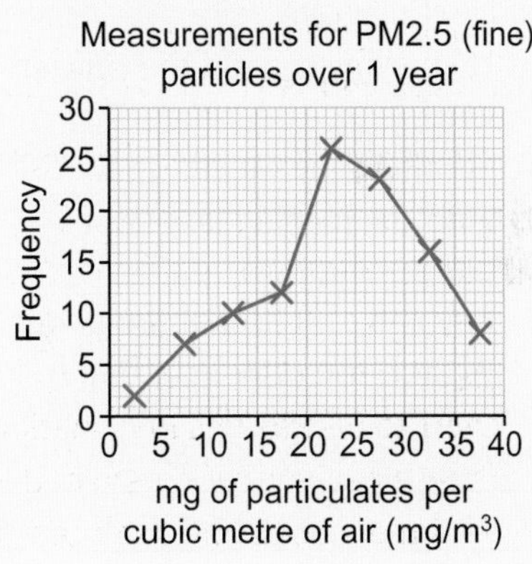

1 Are the means in the case study above or below the legal limits for particulates?

> **Q1 hint** Use the frequency polygons to produce two grouped frequency tables. From these tables you can estimate the mean for each type of particulate.

2 Construct a statistical argument supporting the claims made by the newspaper. Is there evidence to support the claim that one of the yearly averages could be over the legal limit?

> **Q2 hint** Are your answers to **Q1** exact means? What assumption are you making, and how does this work to the company's advantage? What would happen if the actual values were typically higher than the midpoint in some or all of the groups? Could you estimate a maximum value for the mean?

3 Check up

Log how you did on your Student Progression Chart.

Statistical diagrams

1 Copy and complete the two-way table, which shows the voting intentions of a group of men and women.

a How many women are in the group?

b How many women intend to vote for Party A?

	Party A	Party B	Total
Men	120		200
Women			
Total		130	380

2 A group of 48 adults are asked what their favourite subject was at school. They can choose from maths, English and science.

A group of 32 school children are asked the same question.

The pie charts show the choices for the adults and children separately.

a How many adults chose English?

b Reeshma says that an equal number of adults and children chose maths.

Is she correct? Give a reason for your answer.

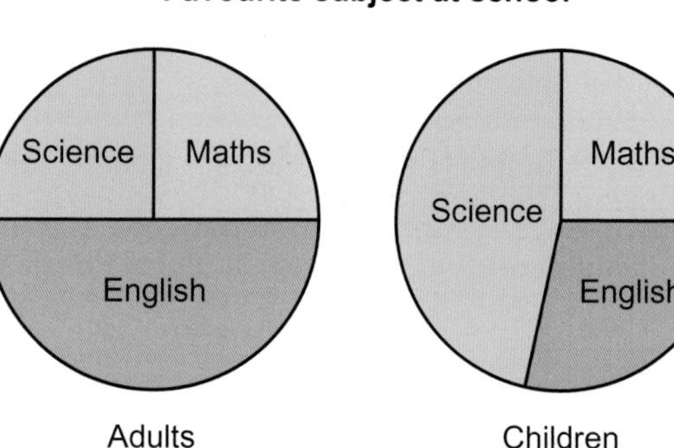

3 The lengths of cod and plaice caught by a fishing trawler are shown in the back-to-back stem and leaf diagram.

a How many cod did the trawler catch?

b Compare the ranges.

c Find the median length of plaice caught.

c Any fish caught that are under 22 cm in length must be returned to the sea. What is the total length of fish returned?

Cod		Plaice
7 4 1 0	1	7 8 8 9 9
9 7 4 1	2	7
8 7 3 0	3	0
9 4 3	4	2 5 5 5 5

Key Cod 0 | 1 represents 10 cm   Plaice 1 | 5 represents 15 cm

4 Draw a frequency polygon for this data.

Time	$0 \leqslant t < 20$	$20 \leqslant t < 40$	$40 \leqslant t < 60$	$60 \leqslant t < 80$	$80 \leqslant t < 100$
Frequency	6	5	1	2	5

Averages and range

5 Identify the outliers in the data sets and work out the range of each, including the outliers.

a Price of TVs: £450, £450, £640, £690, £4800

b Heights of plants in cm: 42, 47, 39, 51, 22, 24

6 For the grouped frequency table:

a estimate the mean

b state the group containing the median

c estimate the range

d state the modal class.

Class	Frequency
$0 \leqslant x < 5$	7
$5 \leqslant x < 10$	15
$10 \leqslant x < 15$	13
$15 \leqslant x < 20$	4
$20 \leqslant x < 25$	1

ActiveLearn Homework, practice and support: Higher 3 Check up

Scatter graphs and time series

7 **Reasoning** The time series graph shows the sales (in 1000s) of bottles of sun cream during the last three years.

a How many bottles were sold in the third quarter (Q3) of 2012?

b Give a possible reason why the sales fluctuate up and down.

c Describe the overall trend in sales over this three-year period.

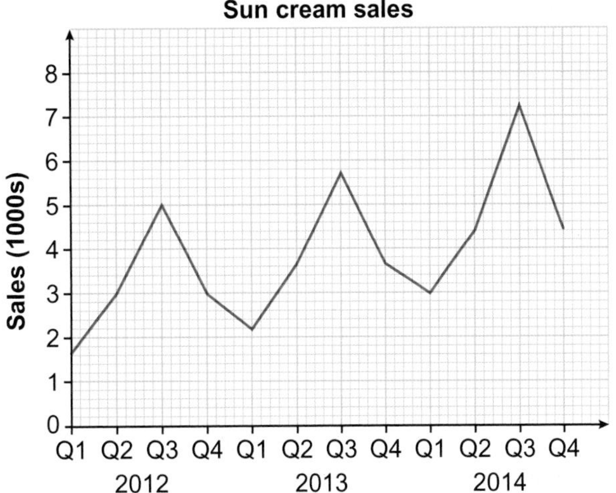

8 **Reasoning** The table shows the marks awarded to seven candidates taking two maths exams. The information is displayed in the scatter diagram together with a line of best fit.

Candidate	A	B	C	D	E	F	G
Marks for Paper 1	57	68	42	24	28	71	34
Marks for Paper 2	45	61	28	14	22	64	24

a State whether this diagram shows positive, negative or no correlation.

b Use the line of best fit to estimate the mark that someone might get on:

i Paper 2 if they get 50 on Paper 1

ii Paper 1 if they get 90 on Paper 2.

c Which of your answers to part **b** would you expect to be more reliable?
Give reasons for your answer.

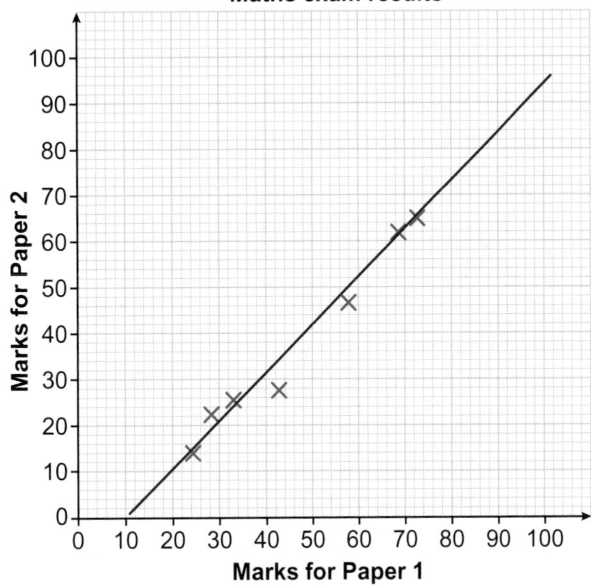

9 How sure are you of your answers? Were you mostly

Just guessing ☹ Feeling doubtful  Confident ☺

What next? Use your results to decide whether to strengthen or extend your learning.

Reflect

★ Challenge

10 a Write down any four numbers and calculate the mean.

b Add 3 to each number in your list and calculate the new mean.

c What do you notice about your answers to parts **a** and **b**?

d Use algebra to show that this works for any set of four numbers.

e What happens to the mean when each number is multiplied by c?
Use algebra to show you are correct.

Q10 hint Let the four numbers be w, x, y and z.

3 Strengthen

Statistical diagrams

1 A group of 40 boys and 72 girls are asked to choose their favourite Olympic sport from archery, badminton, cycling and diving.

The pie charts show the choices for the boys and girls separately.

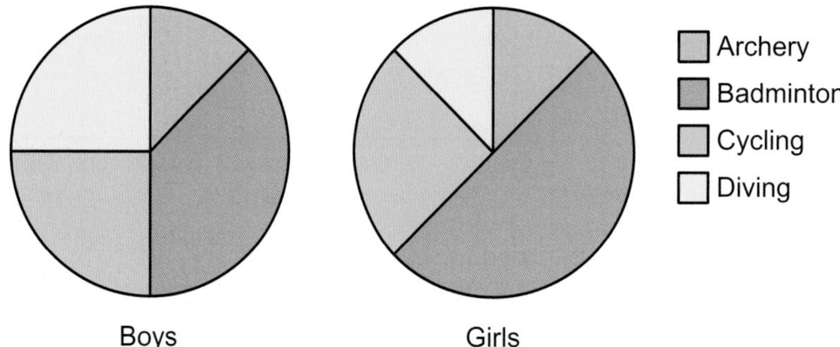

a How many girls chose badminton?

b How many boys chose cycling?

c Sam says that equal numbers of boys and girls chose archery. Is he correct?
Give a reason for your answer.

Q1a hint What fraction of the girls pie chart represents badminton?

Q1c hint Work out the numbers for each pie chart and compare.

2 **Real** A group of students are asked whether they study a science or arts subject at college.

The information is shown in the two-way table.

	Science	Arts	Total
Men	25	15	40
Women	20	20	40
Total	45	35	80

a How many of the students study an arts subject?

b How many of the students are women studying science?

c What fraction of the group are men?

Q2a hint The arts students are in the second column.

Q2b hint Go along the row labelled 'Women' and down the column labelled 'Science'.

Q2c hint The last column shows that there are 40 men and 80 students altogether.

3 A group of 20 children are asked if they have a pet. The information is shown in the two-way table.

	Yes	No	Total
Boys		4	
Girls	3		
Total	5		20

a Work out the number of boys who have a pet.

b Work out the total number of boys.

c Work out the total number of girls.

d Fill in the remaining numbers in the table.

Q3a hint Fill in the number in the top left-hand corner. The first column should add up to 5.

Q3b hint Add up the two numbers in the top row.

Q3c hint Fill in the middle number in the last column. The column should add up to 20.

4 The times taken (in minutes) to complete a task by a group of boys and girls are shown in the back-to-back stem and leaf diagram.

Boys					Girls			
			4	1	5	9	9	
9	8	0	0	2	0	2	7	7
8	7	3	0	3	0	2	4	
			3	4	5	6		

Key Boys: 4 | 1 represents 14 mins. Girls: 1 | 5 represents 15 mins

a How many boys are there in the group?

b What is the shortest time for the boys?

c How many girls took longer than 40 minutes to complete the task?

d What is the longest time overall? Is this achieved by a boy or a girl?

Q4a hint Try writing out the complete list for the boys. The 10s digit is in the middle column, so the longest time for the boys is 43 (not 34).

Q4c hint The girls are on the right, so you read their times forwards. For example, 15, 19, 19, … Write out the complete list for the girls.

5 a Copy and complete the back-to-back stem and leaf diagram for the data sets.

A: 20, 27, 30, 30, 32, 38, 49

B: 26, 28, 28, 32, 33, 40

Set A			Set B		
7	0	2	6	8	8
		3			
		4			

Key Set A: 7 | 2 represents 27. Set B: 2 | 6 represents 26

Q5a hint Begin with Set A. The first two numbers are 20 and 27. They have been written backwards on the first row. Do the same for the second row with the next four numbers.

b For each data set find

i the median

ii the range.

Q5b i hint The median is the middle value. If there are two middle values, the median is halfway between them.

Q5b ii hint The range is the difference between the largest and smallest values.

c Compare the two data sets.

Q5c hint Write about the ranges and medians in your answer.

6 **Real** Twenty people record the time, t hours, they spend on the internet during a day.

Time (t hours)	$0 \leqslant t < 2$	$2 \leqslant t < 4$	$4 \leqslant t < 6$	$6 \leqslant t < 8$	$8 \leqslant t < 10$
Frequency	1	3	9	5	2
Midpoint	1	3			

a Copy and complete the table to show the midpoints.

b Copy and complete the frequency polygon. Plot the midpoints on the horizontal axis and frequency on the vertical axis.

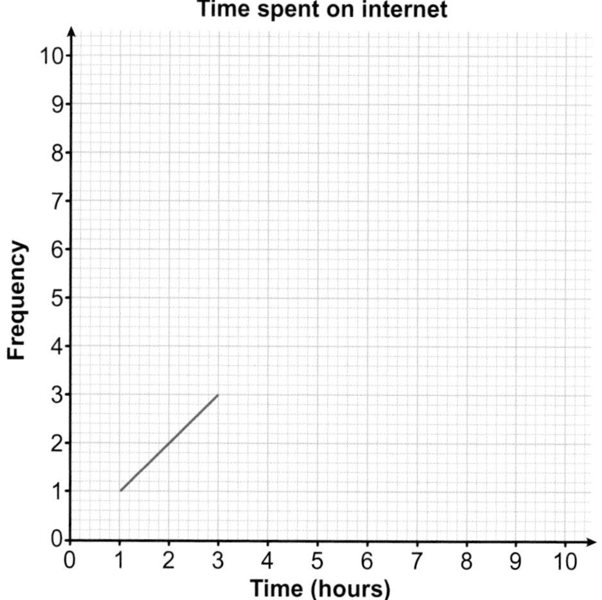

7 The frequency polygon shows the length (in minutes) of 15 songs.

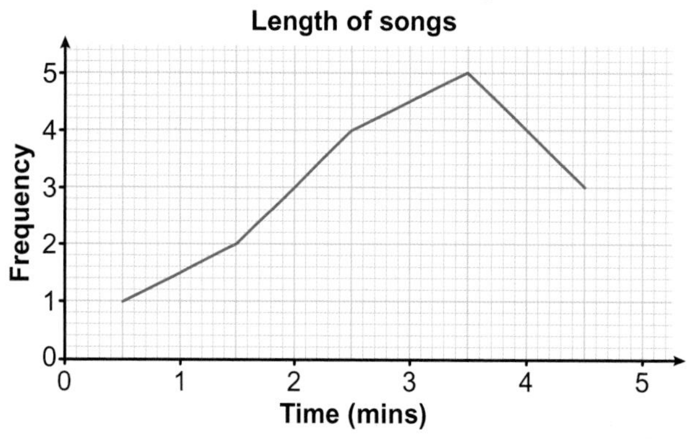

Q7 hint The midpoints are the values shown on the horizontal axis on the graph. The frequencies are the values shown on the vertical axis on the graph.

Q7 hint Check that the frequencies add up to 15.

Copy and complete the table.

Midpoint	0.5	1.5			
Interval	$0 \leqslant t < 1$	$1 \leqslant t < 2$			
Frequency	1	2			

Averages and range

1 Jason records the following lengths during a physics experiment.

3.4 cm, 5.8 cm, 2.9 cm, 4.8 cm, 46 cm, 5.8 cm

a Which one of these values is an outlier?

b Explain why it is appropriate to remove this outlier from the data set.

c Calculate the range.

Q1a hint Look at the list and find a value that is much bigger that the rest.

Q1b hint Outliers can be removed if you think they are clear errors. What do you think Jason might have done here?

Q1c hint Do not include the outlier value in your calculation.

2 **STEM** A zoologist measures the lengths of 50 snakes (to the nearest cm).

Lengths (L cm)	Frequency	Midpoint	Frequency × midpoint
$0 \leqslant L < 10$	7	5	$7 \times 5 = 35$
$10 \leqslant L < 20$	12	15	$12 \times 15 = 180$
$20 \leqslant L < 30$	20		
$30 \leqslant L < 40$	8		
$40 \leqslant L < 50$	3		
Total	50		

On a copy of the table

a complete the third column to show the midpoints of each class

b complete the fourth column to find the total length of all the snakes

c use your answer to part **b** to work out the mean length.

> **Q2a hint** To find the midpoint of $20 \leqslant L < 30$, add the endpoints and divide by 2:
> $20 + 30 = 50$
> $50 \div 2 = 25$

> **Q2c hint** Divide the total length by the total number of snakes.

3 **Real** A speed camera records the speeds of passing cars during the morning rush hour.

Speed (x mph)	Frequency	Midpoint	Frequency × midpoint
$0 \leqslant x < 20$	8	10	$8 \times 10 = 80$
$20 \leqslant x < 25$	90		
$25 \leqslant x < 30$	184		
$30 \leqslant x < 40$	18		
Total			

a How many cars were there altogether?

b What percentage of cars broke the speed limit of 30 mph?

c Copy and complete the table to work out the mean speed.

> **Q3a hint** Find the total of the frequency column.

> **Q3b hint** Divide the number of cars in the $30 \leqslant x < 40$ class by the total number of cars and multiply by 100.

> **Q3c hint** Follow the method you used in **Q2**.

4 Write down the modal class for the grouped frequency table in

a **Q2**

b **Q3**.

> **Q4 hint** The modal class is the one with the highest frequency.

5 Work out the class containing the median in

a **Q2**

b **Q3**.

> **Q5a hint** There are 50 numbers in the grouped frequency table.
> The median is the $\frac{50 + 1}{2} = 25.5$th number, so is halfway between the 25th and 26th numbers.
> There are 7 numbers in the first group, 12 in the second (making 19 so far) and 20 in the third (making 39). The 25th and 26th numbers must be in the … group.

Scatter graphs and time series

1 State whether each scatter graph indicates positive, negative or zero correlation.

Correlation graphs

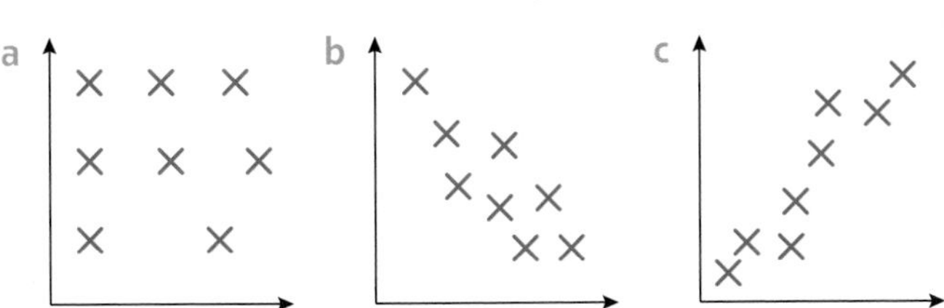

> **Q1 hint** Do the points lie close to a line? If not, there is zero correlation. An uphill line shows positive correlation and a downhill line shows negative correlation.

2 **STEM** In a science experiment a ball is dropped onto the ground five times to see if there is a connection between the height of the drop and the height of the bounce.

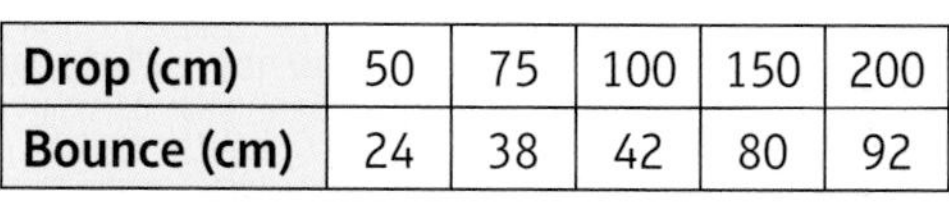

Drop (cm)	50	75	100	150	200
Bounce (cm)	24	38	42	80	92

a Copy and complete the scatter graph.

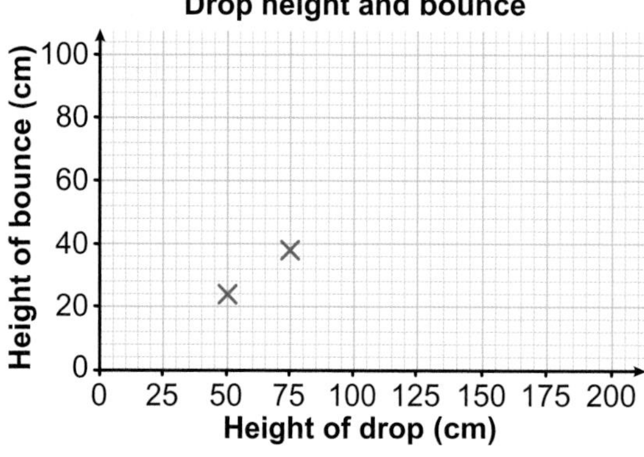

Q2a hint Height of drop is plotted on the horizontal axis and height of bounce is plotted on the vertical axis. The vertical axis goes up by 4 units for each small square.

Q2c hint As you look at the diagram from left to right, do the points slope upwards, downwards or neither?

Q2d hint Use a transparent ruler and try to draw the line so you have two or three points above the line and a similar number below.

b Copy and complete the sentence.
As the height of the drop increases, the height of the bounce … .

c Does the scatter graph show positive, negative or zero correlation?

d Draw a line on your diagram that passes as close as possible to the five points.

e Use the line of best fit to predict the height of bounce when the ball is dropped from a height of 125 cm.

Q2e hint Start at 125 cm on the horizontal axis. Draw a vertical line up until you hit the line of best fit. Now draw a horizontal line from that point across to the vertical axis and read off the answer.

3 **Real** A company reviews its pricing policy by monitoring monthly sales (in thousands of items) and prices.

Price (£)	32	30	25	15	12	8	5
Sales (1000s)	12	60	16	30	37	41	44

Q3a hint Each small square on the horizontal axis is worth one unit.

Q3c hint State the correlation and write a sentence that begins, 'As price increases, sales …'.

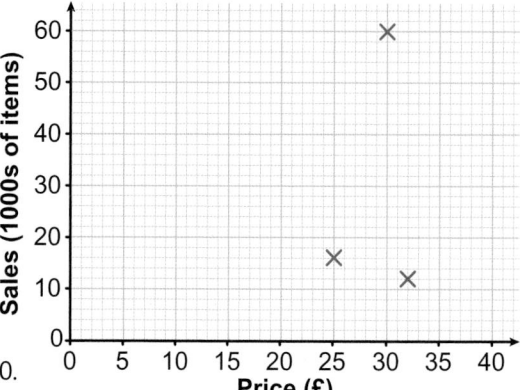

a Copy and complete the scatter graph.

b One of the points on the graph is an outlier and should be ignored. Which point is it?

c Describe the relationship between sales and price.

d Draw a line of best fit on your diagram.

e Use the line of best fit to estimate monthly sales when the price is
 i £20 ii £36

f Which of your answers is the most reliable in part **e**?
Give a reason for your answer.

g Use the line of best fit to find the price needed to get monthly sales figures of 20 000.

Monthly sales

Sales (1000s of items): 60, 50, 40, 30, 20, 10, 0

0 5 10 15 20 25 30 35 40
Price (£)

Q3d hint Ignore the outlier when drawing your line.

Q3e hint Follow the method you used in **Q2**.

Q3f hint Interpolation (predicting values within the range of points on your graph) is reliable but extrapolation (predicting values outside the range of points on your graph) is unreliable.

Q3g hint Start with the sales on the vertical axis, go along to the line and read off the price on the horizontal axis. Always draw the lines on your diagram.

4 **Reasoning** The table shows the number of ice lollies sold by a kiosk at a seaside resort last year.

Month	Jan	Feb	Mar	Apr	May	Jun	Jul	Aug	Sep	Oct	Nov	Dec
Number	10	80	150	220	290	360	360	290	220	150		

a Copy and complete the time series graph.

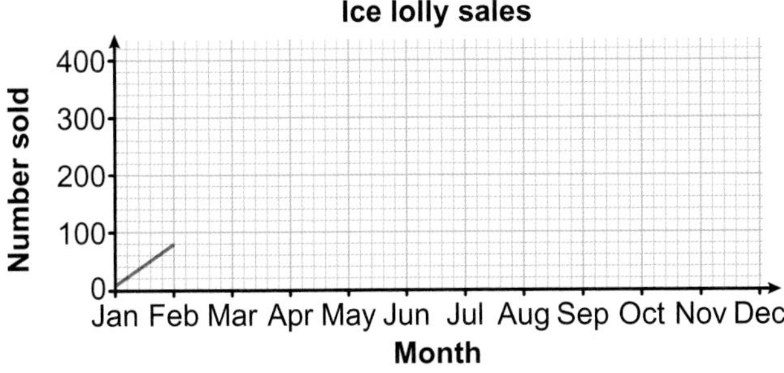

b Describe in words how ice lolly sales vary during the year.

c Assuming that the pattern of sales continues, estimate the sales for November and December.

Q4c hint Read the numbers in the table from left to right to spot the pattern.

3 Extend

1 **Reasoning** Each point on the scatter graph shows the mass of a bag of fertiliser and its price.

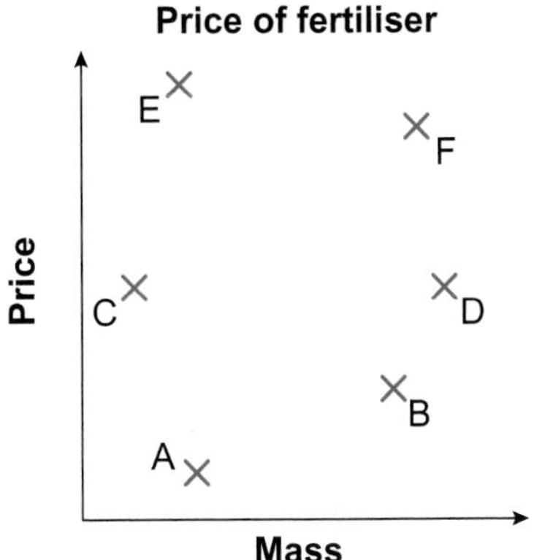

a Which two bags are the same price?

b Which of B and D gives the better value for money?

c Which two bags give the same value for money?

d Which bag gives the worst value for money?

e Describe the correlation.

Give reasons for your answers.

2 **Reasoning** There is positive correlation between variables x and y, negative correlation between y and z and negative correlation between z and w.

State what type of correlation you would expect between

a x and z b y and w c x and w.

Q2 strategy hint You could draw diagrams to help you.

3 **Reasoning** Find the value of the missing number, x, in this list:

2, 4, 5, 5, 6, 7, 7, x

a if the mode is 7 b if the mean is 6.

4 **STEM / Problem-solving** Two dozen tomato plants are grown in a greenhouse and the total weight of fruit produced by each plant is recorded. The same number of tomato plants of this variety is grown outdoors.

The information is displayed in the frequency polygons.

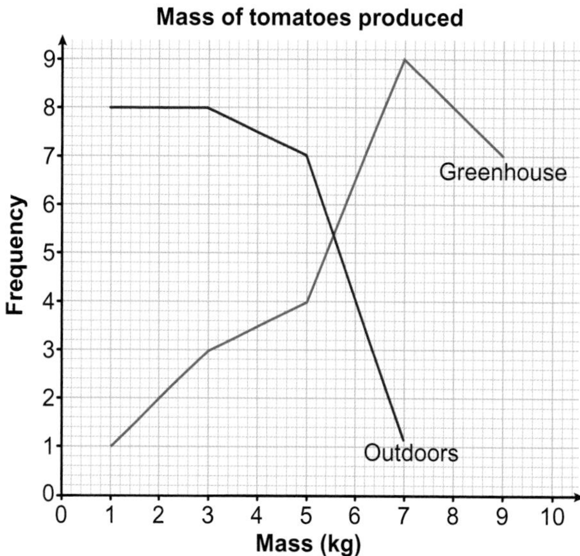

a Estimate the greatest mass of tomatoes grown by a plant outdoors.

b How many plants grown in the greenhouse produce between 4 kg and 6 kg of fruit?

c The grower claims that the average yield of plants grown in the greenhouse exceeds the average yield of outdoor plants by more than 3 kg. Does this data support the claim?

> **Q4c communication hint**
> Yield means the amount of fruit produced.

5 **Reasoning** A group of 50 students take a maths test. Their marks out of 40 are shown in the table.

Marks	20–22	23–25	26–30	31–40
Frequency	1	15	22	12

a Estimate the mean mark and explain why this is only an estimate.

b Explain which class contains the median.

c Estimate the range.

d The student whose mark was between 20 and 22 has her paper remarked. She is awarded a new mark between 23 and 25.

Without doing any calculations, state whether the following will increase, decrease or stay the same.

i Range   ii Mean

6 **Reasoning** Every student in Year 10 completes a questionnaire in which they are asked to choose their favourite takeaway food from pizza, burger and curry.

The results are shown on the pie chart.

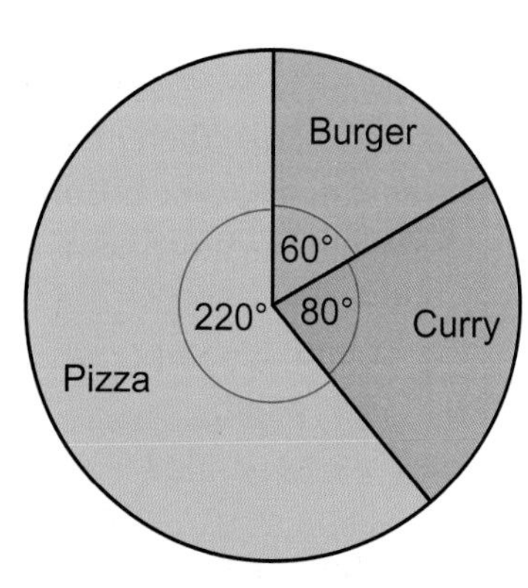

a If 132 students chose pizza, how many chose burger?

b On a similar pie chart for Year 11, the angle of the sector for pizza is 200°.

Explain why this does not necessarily mean that fewer students in Year 11 chose pizza as their favourite.

7 **Reasoning** The stem and leaf diagram shows the ages in complete years of a sample of 24 men in a tennis club.

a Find the median age.

b The ages, in complete years, of a sample of 14 women from the club are:
82, 58, 53, 9, 23, 81, 45, 48, 31, 77, 16, 23, 64, 62
Draw a back-to-back stem and leaf diagram for the men's and women's ages.

c Without doing any further calculations, make one comparison between the ages of men and women at the club.

0	8				
1	2	8			
2	1	6	7		
3	1	4	7	9	
4	0	0	0	2	5
5	0	7	7	8	
6	2	3	4		
7	0				
8	3				

Key
1 | 2 means 12 years

Q7c hint Look at the outline shapes of the two sides of the diagram.

8 **Exam-style question**

Suzy did an experiment to study the times, in minutes, it took 1 cm ice cubes to melt at different temperatures.

Some information about her results is given in the scatter graph.

The table shows information from two more experiments.

Temperature (°C)	15	55
Time (minutes)	22	15

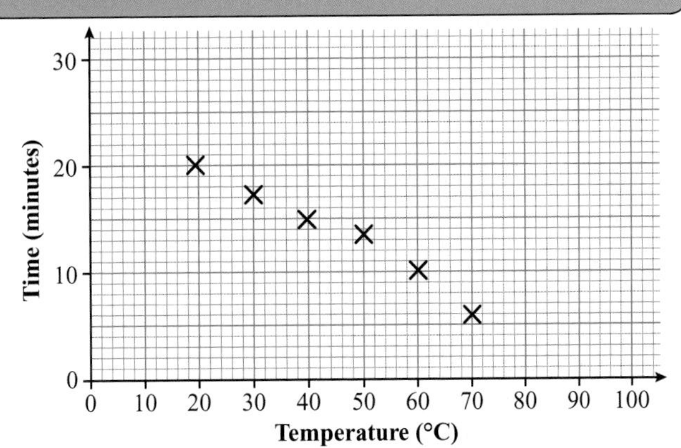

a On the scatter graph, plot the information from the table. **(1 mark)**

b Describe the relationship between the temperature and the time it takes a 1 cm ice cube to melt. **(1 mark)**

c Find an estimate for the time it takes a 1 cm ice cube to melt when the temperature is 25 °C. **(2 marks)**

Exam hint
Draw a line of best fit on the graph to help you answer part **c**.

Suzy's data cannot be used to predict how long it will take a 1 cm ice cube to melt when the temperature is 100 °C.

d Explain why. **(1 mark)**

Nov 2011, Q11, 1380/3H

Key point 12

A line of best fit passes through the mean point, $(\bar{x}, \bar{y})$, where $\bar{x}$ is the mean of the x coordinates and $\bar{y}$ is the mean of the y coordinates.

9 **Real / Problem-solving** A car showroom has five cars, all of the same model but of different ages. The table shows the age, x (in years), and the price, y (in £1000s) of each car.

Age (years)	1	2	3	6	7
Price (£1000s)	25	18	15	9	6

a Plot the points on a scatter graph and state the type of correlation.

b Find the mean, $\bar{x}$, of the ages of the cars.

c Find the mean, $\bar{y}$, of the values of the cars.

d Plot the point $(\bar{x}, \bar{y})$ on your scatter diagram and draw a line of best fit passing through this point.

e Use your line to estimate the cost of a car that is 5 years old.

f Explain why it would not be sensible to use the line to estimate the value of a new car of this model.

10 **Modelling** The table shows the age and mass of a group of 10 people.

Age	5	8	12	13	15	30	40	45	52	55
Mass (kg)	18	30	45	49	57	76	79	81	83	84

a Plot these points on a scatter graph.

b Draw a line of best fit on the diagram passing through the mean point and with roughly the same number of points above and below the line.

c Use your line to estimate the mass of someone who is 23 years old.

d A better model is to use two different straight lines, one up to 20 years old and the other for over 20. Draw these on your diagram and use them to obtain a more reliable answer to part **c**.

11 **Real / Problem-solving** The table shows the age distribution of male and female teachers in a school.

a Draw frequency polygons for these two sets of data on the same diagram.

b By calculating the mean of each data set, compare the age distributions of male and female teachers.

c What feature of your diagram confirms that your comparison is correct?

Age (x years)	Male	Female
$20 \leqslant x < 25$	1	0
$25 \leqslant x < 30$	2	9
$30 \leqslant x < 35$	3	10
$35 \leqslant x < 40$	7	12
$40 \leqslant x < 45$	10	8
$45 \leqslant x < 50$	10	7
$50 \leqslant x < 55$	12	4
$55 \leqslant x < 60$	4	0
$60 \leqslant x < 65$	1	0

12 **Finance** The table shows the annual salaries of 200 employees of a company.

	Under £30 000	At least £30 000	Total
Men	60		90
Women		50	
Total			200

a Copy and complete the table.

b What percentage of employees are women?

c What percentage of men earn under £30 000?

d What percentage of employees earning at least £30 000 are women?

13 **Reasoning** Every student in Year 7 must attend a lunchtime club. They can choose from music, drama or sport.

Out of the 185 students in Year 7, 65 choose music and 49 choose drama. Of the 95 boys in the year, 35 choose music. 32 girls choose sport.

Copy and complete the two-way table for this information.

	Music	Drama	Sport	Total
Boys				
Girls				
Total				

14 **Reasoning** One hundred students studying music at school are asked to choose their preference from rap, jazz and classical.

Of the 29 who choose rap, 13 are girls. Of the 21 who choose jazz, 10 are girls.

There are 54 boys altogether.

a Draw a two-way table and use it to work out the percentage of students who

i prefer classical music   ii are girls who prefer classical music.

b What percentage of boys prefer rap?

15 **Problem-solving** A car manufacturer compares colour preferences of men and women. Potential customers are asked to pick their favourite colour from a list of four. The results are shown in the table.

Colour	creamy white	platinum grey	midnight blue	charcoal black
Men	2	13	25	20
Women	4	8	6	2

Which one of these statistical diagrams could be used?

scatter graph   frequency polygon
back-to-back stem and leaf diagram   pie chart

Explain your choice.

Reflect For each of the statistical diagrams in this question, write down an example of when it would be the appropriate diagram to use.

Key point 13

Means of time series data from several consecutive periods are called **moving averages**. These smooth out the ups and downs, enabling you to spot the trend.

16 **Real / Reasoning** A retail company recorded the number of customer complaints each month.

Month	1	2	3	4	5	6	7
Complaints	45	35	20	40	30	15	30

a Draw a time series graph for this data.

b Show that the mean number of complaints during the first 3 months is 33.3 (1 d.p.).

c Show that the mean number of complaints for months 2, 3 and 4 is 31.7 (1 d.p.).

d Copy and complete the table of 3-month moving averages.

Months	1–3	2–4	3–5	4–6	5–7
Moving average (1 d.p.)	33.3	31.7			

e What do these averages tell you about the overall trend in customer complaints?

Q16d hint To find the moving average for months 3–5, add the values for months 3, 4 and 5 and divide by 3.

Q16e hint You can plot the moving averages on your time series graph. Make sure you plot each value in the middle of the 3 months it covers.

17 **Exam-style question**

A garage sells used cars.

The table shows the number of used cars it sold from July to December.

July	August	September	October	November	December
28	25	34	46	28	40

a Work out the 3-point moving averages for the information in the table. The first two have been worked out for you.

29   35   ....   **(2 marks)**

b Comment on the trend shown by the 3-point moving averages. **(1 mark)**

Nov 2011, Q15, 1380/4H

Exam hint
Make sure your comment is in context by relating it to the number of cars sold.

18 A biologist measures the lengths of 40 fish and records the results in a grouped frequency table.

Length (L cm)	$0 \leqslant L < 10$	$10 \leqslant L < x$	$x \leqslant L < 3x$
Frequency	6	24	10

Find x if the mean length is 15.75 cm.

3 Knowledge check

- A **back-to-back stem and leaf diagram** compares two sets of results. On the left-hand side the numbers are read backwards. .... *Mastery lesson 3.1*
- A **frequency polygon** is a graph made by joining the midpoints of the tops of the bars in a bar chart with straight lines. .... *Mastery lesson 3.1*
- A quicker way of drawing a frequency polygon is to plot the frequency against midpoints of each group. .... *Mastery lesson 3.2*
- The **modal class** (or modal group) has the highest frequency. .... *Mastery lesson 3.5*
- To estimate a mean from a grouped frequency table, add together the products of class midpoints and their frequencies, and divide by the total frequency. .... *Mastery lesson 3.5*
- If the total frequency in a grouped frequency table is n, then the median lies in the group containing the $\frac{n+1}{2}$th item of data. .... *Mastery lesson 3.5*
- A **time series** graph is a line graph with time plotted on the horizontal axis. *Mastery lesson 3.3*
- **Bivariate data** is data that has two variables. Points can be plotted on a **scatter diagram** to see if there is a link between them. .... *Mastery lesson 3.4*
- Data displays positive correlation if the points on a scatter diagram lie close to an upward-sloping straight line. Data displays negative correlation if the points on a scatter diagram lie close to a downward-sloping straight line. .... *Mastery lesson 3.3*
- **A line of best fit** is the line that passes as close as possible to the points on a scatter graph. .... *Mastery lesson 3.4*
- Using a line of best fit to predict data values within the range of the data given is called **interpolation** and is usually reasonably accurate. .... *Mastery lesson 3.4*
- Using a line of best fit to predict data values outside the range of the data given is called **extrapolation** and may not be accurate. .... *Mastery lesson 3.4*
- Individual points which are outside the overall pattern of a scatter diagram are called **outliers**. They can be removed from a data set provided a reason for their removal is given. .... *Mastery lesson 3.5*
- The line of best fit passes through the mean point, $(\overline{x}, \overline{y})$. .... *Extend 3*
- Means of time series data from several consecutive periods are called moving averages. .... *Extend 3*

Reflect

Look back at the questions you answered in this test.

a Which one are you most confident that you have answered correctly? What makes you feel confident?

b Which one are you least confident that you have answered correctly? What makes you feel least confident?

c Discuss the question you feel least confident about with a classmate. How does discussing it make you feel?

3 Unit test

Log how you did on your Student Progression Chart.

1 **Reasoning** To get a grade 9 in a GCSE exam, Bhavik must have a mean of at least 93%, averaged over three exam papers.
His mean score on the first two papers is 89%. Is it still possible for him to achieve a top grade overall? *(3 marks)*

ActiveLearn Homework, practice and support: Higher 3 Unit test

2 **Reasoning** The time series graph shows the rate of unemployment in a country over a 3-year period.

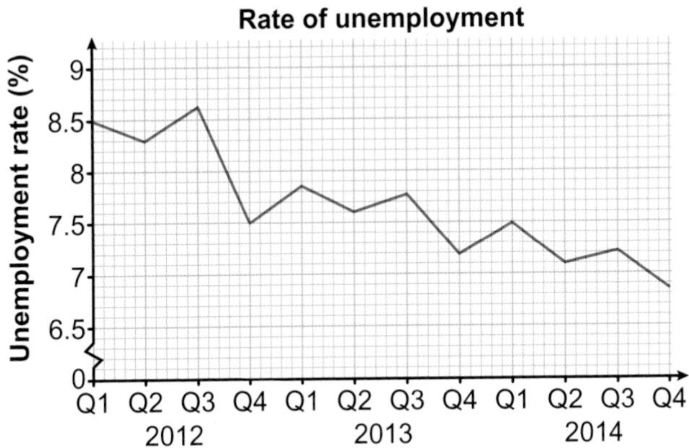

a What was the rate of unemployment in the second quarter of 2014? Give your answer correct to 1 d.p. *(1 mark)*

b Which quarter experienced the greatest fall in the rate of unemployment? *(1 mark)*

c Describe what this time series suggests about the trend in unemployment over the 3 years. *(1 mark)*

3 The back-to-back stem and leaf diagram shows the number of apples growing on two types of apple tree in an orchard.

Autumn Gold		Ruby Red
4 5 1	2	0 7
4 3 0	3	2 2 4 6
7 4	4	1 3 7 8
2 5 0	5	4 8
4	6	

Key Autumn Gold: 1 | 2 represents 21 apples   Ruby Red: 2 | 0 represents 20 apples

a How many 'Autumn Gold' trees are there? *(1 mark)*

b Find the range of the number of apples growing on 'Ruby Red' trees. *(1 mark)*

c Find the median number of apples growing on 'Autumn Gold' trees. *(1 mark)*

4 **Reasoning** The table shows the time that 80 customers spent queuing in a bank.

Time (t minutes)	Frequency
$0 \leqslant t < 2$	18
$2 \leqslant t < 4$	24
$4 \leqslant t < 6$	20
$6 \leqslant t < 8$	10
$8 \leqslant t < 10$	6
$10 \leqslant t < 12$	2

a Estimate the mean queuing time. *(3 marks)*

b In which group is the median time? *(1 mark)*

c It is discovered that one of the customers whose time was originally recorded as 9 minutes only queued for 3 minutes.

Without doing any calculations, state whether your answers to parts **a** and **b** will increase, decrease or stay the same. *(2 marks)*

5 The table shows the number of hours six students spent revising for a maths test and their mark.

Time	5	2	8	1	6	4
Mark	80%	50%	90%	40%	75%	60%

a Plot this data on a scatter diagram. *(3 marks)*

b Describe the correlation and explain what this means in this context. *(2 marks)*

c Draw a line of best fit on your diagram and use it to estimate the mark of someone who revises for 3 hours. *(2 marks)*

d Is it sensible to use the graph to estimate someone's mark if they have done no revision at all? Give a reason for your answer. *(1 mark)*

6 **Reasoning** An estate agent keeps a record of the types of property sold in the last month.

semi-detached	terraced	terraced	flat	detached
detached	flat	terraced	flat	flat
semi-detached	flat	detached	terraced	detached
detached	terraced	flat	flat	terraced

a Explain why it is not possible to present this information using a frequency polygon. *(1 mark)*

b Draw a pie chart for this data. *(3 marks)*

c Next month's sales are also displayed on a pie chart. The angle for the sector representing flats is 140°. Explain whether this shows that she has sold more flats than last month. *(1 mark)*

7 **Reasoning** 100 people each bought an electronic tablet with a choice of 16 GB, 32 GB or 64 GB of memory.

53 of the customers are women. 12 of the women bought a 16 GB tablet.

15 of the men bought a 32 GB tablet. 20 of the 40 customers who bought a 64 GB tablet are men.

a Draw a two-way table for this data. *(3 marks)*

b What fraction of customers bought a 32 GB tablet? *(1 mark)*

8 The frequency polygon shows the distribution of times taken to clean a car.

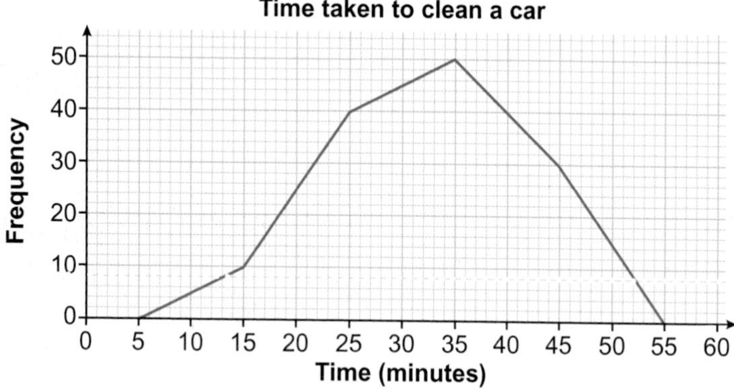

a How many cars are cleaned in total? *(1 mark)*

b How many cars took between 40 and 50 minutes to be cleaned? *(1 mark)*

c State the modal class. *(1 mark)*

Sample student answers

a Who has got the answer correct? b Who gets the most marks? Explain why.

Exam-style question

Ed has 4 cards.

There is a number on each card.

The mean of the 4 numbers on Ed's cards is 10.

12	6	15	?

Work out the number on the 4th card. **(3 marks)**

June 2013, Q6, 1MA0/1H

Student A

7

Student B

$$\frac{12 + 6 + 15 + ?}{4} = 10$$

$$33 + ? = 40$$

4th card = 6

4 FRACTIONS, RATIO AND PERCENTAGES

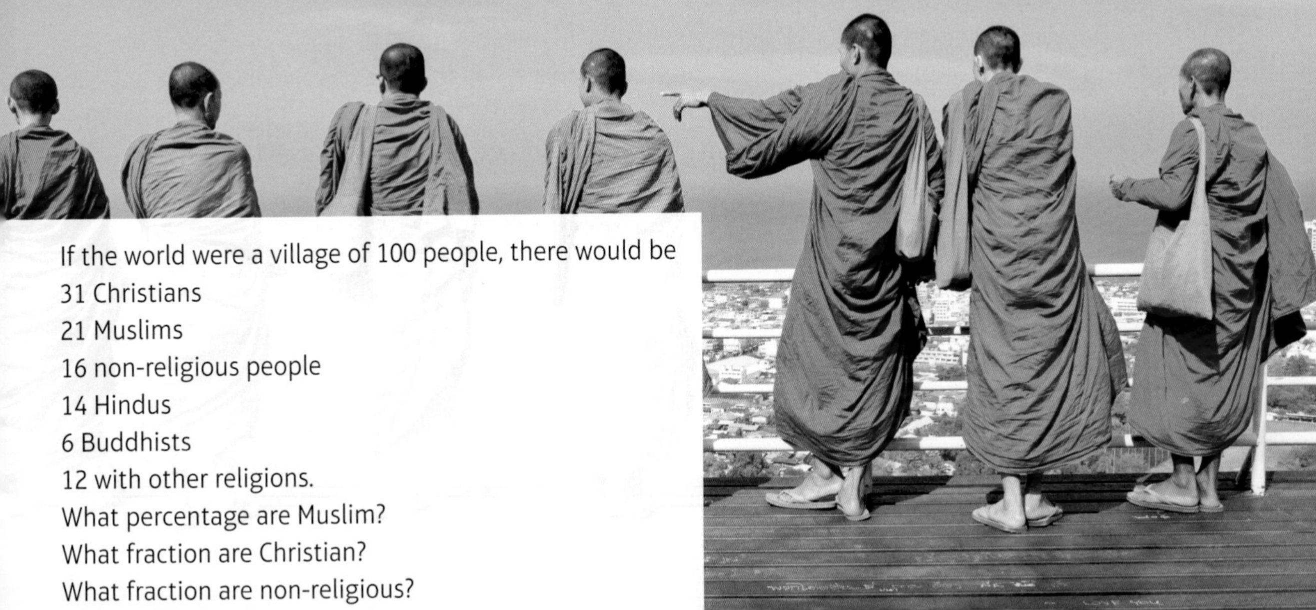

If the world were a village of 100 people, there would be
31 Christians
21 Muslims
16 non-religious people
14 Hindus
6 Buddhists
12 with other religions.
What percentage are Muslim?
What fraction are Christian?
What fraction are non-religious?
What percentage have a religion?

4 Prior knowledge check

Numerical fluency

1 A selection box contains four types of sweets: 6 caramels, 7 chocolates, 4 bonbons and 3 nougats.
  a What fraction are caramels?
  b What fraction are not caramels?

2 Work out
  a $\frac{1}{6}$ of 18 kg b $\frac{3}{10}$ of £25
  c $\frac{5}{9}$ of 45 litres d $\frac{7}{8}$ of 64 m

3 Work out
  a $\frac{3}{4} \times \frac{1}{5}$ b $\frac{2}{3} \times \frac{9}{10}$

4 Work out
  a 25% of £12 b 30% of 150 g

5 Work out
  a $3 \times \frac{1}{4}$ b $5 \times \frac{9}{20}$

6 Work out
  a $\frac{2}{5} \div 3$ b $\frac{7}{10} \div \frac{2}{3}$ c $\frac{3}{5} \div \frac{6}{7}$

7 Work out
  a $\frac{1}{4} + \frac{3}{8}$ b $\frac{1}{5} + \frac{1}{6}$ c $\frac{3}{5} - \frac{4}{9}$

8 Giving your answers as mixed numbers, work out
  a $2 \times 1\frac{1}{5}$ b $3 \times 2\frac{5}{8}$

9 Giving your answers as mixed numbers where appropriate, work out
  a $3 \div \frac{1}{4}$ b $7 \div \frac{3}{4}$

10 Write each ratio in its simplest form.
  a 10:25 b 63:7

11 There are 40 girls and 25 boys at a holiday camp.
What is the ratio of girls to boys?
Give your answer in its simplest form.

12 A loom band bracelet uses green and blue rubber bands in the ratio 3:1.
What fraction of the rubber bands are blue?

13 Here are the ingredients needed to make 8 scones

275 g self-raising flour
25 g sugar 50 g butter 1 egg

a How many eggs would you need to make 24 scones?
b How much butter would you need to make 12 scones?

14 Use a multiplier to calculate these percentages.
a 20% of £78 b 70% of 52 kg
c 45% of 340 ml d 8% of 510 m

15 Write these test scores as percentages.
a 15 out of 25 b 49 out of 60
c Which was the best score?

16 The price of a sofa is £480. Lucy pays a deposit of 15% of the price.
Work out the amount she must pay as a deposit.

17 Write as a percentage of £50
a £37 b £75

18 Write these fractions as both decimals and percentages. Where necessary, round your answers to 3 significant figures.
a $\frac{13}{20}$ b $\frac{3}{7}$ c $\frac{11}{8}$ d $2\frac{3}{4}$

19 Write down the single (decimal) number you can multiply by to work out an increase of
a 17% b 38% c 6% d 210%

20 Increase £28 by 12%.

21 Decrease 5 m by 8%.

22 Write $3.\dot{2}\dot{4}$ to 6 decimal places.

23 Convert to a decimal
a $\frac{1}{3}$ b $\frac{2}{9}$

* Challenge

24 Which of these amounts would you prefer to win?

55% of £150	$\frac{3}{4}$ of £120
0.8 of £90	300% of £28

4.1 Fractions

Objectives

- Add, subtract, multiply and divide fractions and mixed numbers.
- Find the reciprocal of an integer, decimal or fraction.

Why learn this?

You can use reciprocals to work out the gradients of perpendicular graphs, as well as to simplify calculations.

Fluency

Which of these numbers are
a unit fractions **b** improper fractions **c** mixed numbers?
i $2\frac{1}{5}$ **ii** $\frac{3}{2}$ **iii** $\frac{15}{6}$ **iv** $\frac{1}{7}$ **v** $3\frac{7}{8}$ **vi** $\frac{1}{9}$

Warm up

1 Write
a $3\frac{3}{8}$ as an improper fraction
b $\frac{17}{6}$ as a mixed number.

2 Work out
a $35 \times \frac{2}{7}$ b $24 \times \frac{3}{8}$
c $8 \times \frac{5}{12}$ d $6 \times \frac{11}{15}$

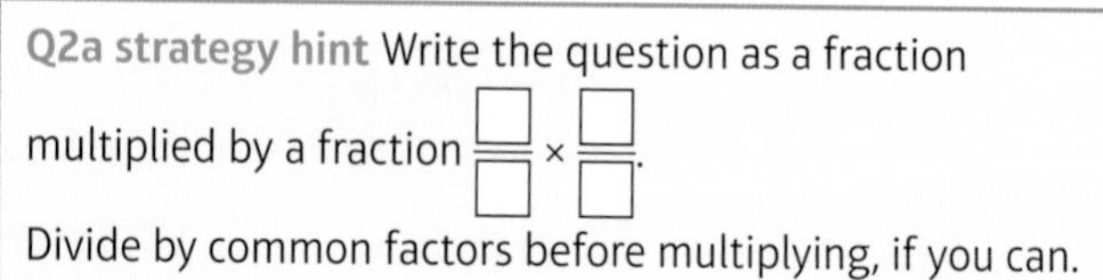
Q2a strategy hint Write the question as a fraction multiplied by a fraction $\frac{\square}{\square} \times \frac{\square}{\square}$.
Divide by common factors before multiplying, if you can.

3 Work out
a $21 \times 36 \div 14$
b $32 \times 45 \div 36$
c $9 \times 24 \div 8$

Questions in this unit are targeted at the steps indicated.

4 Tonia and Trinny are twins. Their friends give them identical cakes for their birthday.
Tonia eats $\frac{1}{8}$ of her cake and Trinny eats $\frac{1}{6}$ of her cake.
How much cake is left? Give your answer as a mixed number.

Key point 1

The **reciprocal** of the number n is $\frac{1}{n}$. You can also write this as n^{-1}.

5 Find the reciprocal of each number.

a 8 b 0.145 c 4.8 d $\frac{2}{3}$

Use a calculator to check your answers.

Reflect What method did you use to work out the reciprocals of the decimal numbers?

Discussion What happens when you multiply a number by its reciprocal?

Q5d hint To find the reciprocal of a fraction, swap the numerator and the denominator.
For example, the reciprocal of $\frac{3}{4}$ is $\frac{4}{3}$.

6 Find the reciprocal of

a $\frac{1}{2}$ b $\frac{2}{5}$

c $\frac{15}{4}$ d $5\frac{2}{3}$

Discussion Is it possible to find the reciprocal of zero? Explain your answer.

Q6d hint To find the reciprocal of a mixed number, first convert it into an improper fraction.

Key point 2

It is often easier to write mixed numbers as improper fractions before doing a calculation.

7 Giving your answer as a mixed number where appropriate, work out

a $1\frac{1}{4} \times \frac{1}{6}$ b $1\frac{2}{5} \times \frac{2}{3}$ c $3\frac{3}{8} \times 1\frac{2}{9}$ d $2\frac{4}{5} \times 2\frac{1}{7}$

8 Work out

a $2\frac{1}{2} \div 5$ b $3 \div \frac{9}{10}$

9 Giving your answers as mixed numbers where appropriate, work out

a $1\frac{4}{7} \div \frac{2}{3}$ b $2\frac{3}{5} \div 1\frac{7}{9}$ c $2\frac{7}{10} \div \frac{9}{25}$ d $3\frac{1}{7} \div 2\frac{3}{4}$

10 **Reflect** Katherine says, 'Dividing by a fraction is the same as multiplying by the reciprocal of that fraction.'
Is she correct? Show some working to explain your answer.

11 Copy and complete the calculation.

$4\frac{7}{10} + 3\frac{1}{2}$

$= \square\frac{7}{10} + \frac{1}{2}$

$= \square\frac{7}{10} + \frac{\square}{10} = \square\frac{\square}{10}$

$= \square\frac{\square}{\square}$

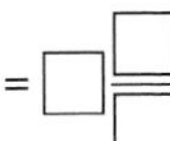

12 Work out

a $1\frac{9}{10} + 2\frac{3}{5}$ b $2\frac{7}{8} + 3\frac{1}{4}$

c $4\frac{4}{5} + 6\frac{5}{6}$ d $3\frac{4}{9} + 5\frac{3}{4}$

Q12c hint Sometimes both denominators must be changed to add fractions.

13 **Exam-style question**

A part has broken on a machine and needs to be replaced.
The replacement must be between $7\frac{1}{18}$ cm and $7\frac{3}{18}$ cm long in order to fit the machine.
The diagram shows the replacement part.

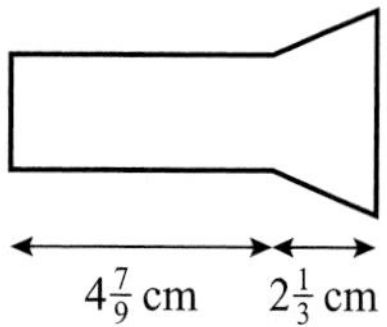

Will this part fit the machine?
You must explain your answer. **(5 marks)**

Exam hint
Explain your answer by showing your calculations. Write a sentence, 'The part will/will not fit the machine because …'

Example 1

Work out $4\frac{1}{2} - 1\frac{4}{5}$

$4\frac{1}{2} - 1\frac{4}{5} = \frac{9}{2} - \frac{9}{5}$ — Write both numbers as improper fractions.

$= \frac{45}{10} - \frac{18}{10}$ — Write both fractions with a common denominator.

$= \frac{27}{10}$

$= 2\frac{7}{10}$ — Write the answer as a mixed number.

14 Work out these subtractions.

a $6\frac{9}{10} - 2\frac{2}{5}$ b $5\frac{3}{5} - 1\frac{7}{8}$ c $3\frac{1}{6} - 4\frac{1}{9}$ d $4\frac{1}{4} - 5\frac{3}{8}$

Discussion Do you always need to change mixed numbers to improper fractions to subtract?

15 **Problem-solving** In an engineering factory, the production line takes up $\frac{2}{3}$ of the floor area.
Out of the remaining floor area, a total of $\frac{3}{5}$ is taken up by office space and the canteen.
The rest is warehouse space.
The warehouse space occupies 2000 m^2.
Work out the floor area of the production line.

16 **Problem-solving** Alice watched two films at the cinema.
The first film was $1\frac{5}{6}$ hours long and the second was $2\frac{1}{4}$ hours long.

a Work out the total length of the two films.

Alice drove to the cinema.
She arrived 10 minutes before the first film began and had to wait for half an hour between the two films.
She left immediately after the second film finished.
Car park tickets can be bought in multiples of 1 hour.

b How many hours of parking did Alice need to buy?

4.2 Ratios

Objectives

- Write ratios in the form $1:n$ or $n:1$.
- Compare ratios.
- Find quantities using ratios.
- Solve problems involving ratios.

Did you know?

Hairdressers use ratios to mix different dyes together to get the correct hair colour.

Fluency

Find the missing numbers.

a $\frac{1}{5} \times \square = 1$ **b** $\frac{2}{3} \times \square = 1$ **c** $\square \times \frac{5}{9} = 1$

Warm up

1 Simplify these ratios, giving your answer without units.

a 3:6 b 15:25 c 24:42

d 3 cm:15 mm e 450 g:1.8 kg f 1.8 litres:240 ml

> **Q1 hint** Give answers for **d** to **f** without units.

2 Write each ratio in the form $1:n$

a 4:20

b 28:14

c $\frac{1}{3}:2$

d $\frac{3}{7}:\frac{2}{8}$

> **Q2a hint** The question tells you to make the left side of the ratio equal to 1.
> Divide both sides of the ratio by the number on the left.
> $\div 4$ (4 : 20) $\div 4$
> 1 : $\square$
> The number on the right may not be a whole number.

Key point 3

You can compare ratios by writing them as **unit ratios**.
In a unit ratio, one of the numbers is 1.

3 Write each ratio in the form $n:1$

a 12:4 b 30:45

c $3:\frac{1}{5}$ d $\frac{3}{4}:\frac{9}{10}$

> **Q3d hint** Make the right-hand side of the ratio equal to 1.

4 Write these ratios in the form $1:n$

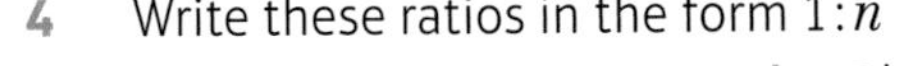

a £3:60p b 5 kg:80 g

c 2 hours:45 minutes d 20 cm:7.3 m

> **Q4 hint** If the answer is not an integer, you can use fractions or decimals. Choose whichever is most accurate.

5 **Reasoning** In a school there are 52 teachers and 598 students.

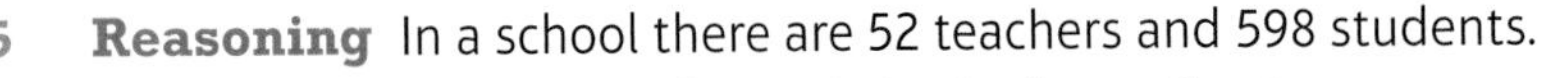

a Write the student:teacher ratio in the form of $n:1$.

Another school has 85 teachers and 1020 students.

b Which school has larger number of teachers per student?

6 **Problem-solving** Julie and Hammad each make a glass of orange squash. Julie uses 42 ml of squash and 210 ml of water. Hammad uses 30 ml of squash and 170 ml of water.
Who has made their drink stronger?

> **Q6 hint** Write the ratios in the form $1:n$

7 **Reasoning** Archie and Ben share some money in the ratio 7:11.
Ben gets £132. How much money does Archie get?

> **Q7 hint**

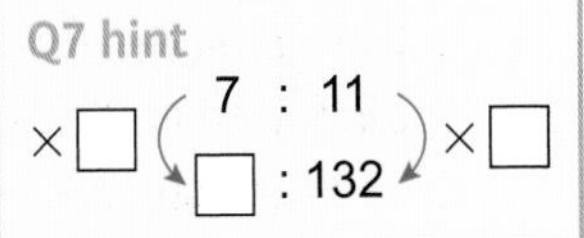

8 To make a tough adhesive, Paul mixes 5 parts of resin with 2 parts of hardener.

a Write down the ratio of resin to hardener.

b To fix a birdbath, Paul uses 9 g of hardener. How many grams of resin does he use?

c On another project, Paul used 12 g of resin. How much hardener did he use?

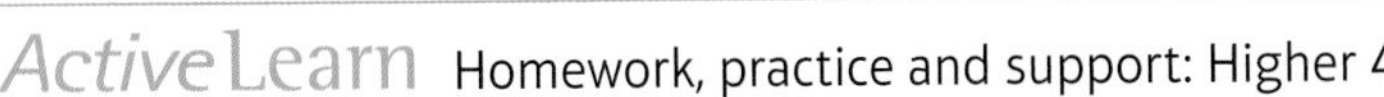

Homework, practice and support: Higher 4.2

9 A scale model of Tower Bridge in London is 22 cm high. The real bridge is 66 m high.

a Work out the scale of the model. Write it as the ratio of real height to model height.

The bridge is 243 m long in real life.

b How long is the model?

10 In a school, the ratio of the number of students to the number of computers is $1:\frac{3}{5}$.
There are 210 computers in the school. How many students are there?

11 Sally and David divide £35 in the ratio 3 : 2.

a What fraction does Sally get?

b What fraction does David get?

c How much money does each person get?

Example 2

Share £126 between Lu and Katie in the ratio 2 : 5.

2 + 5 = 7 parts — Find out how many parts there are in total.

1 part = £126 ÷ 7 = £18 — Find out how much one part is worth.

Lu: 2 × £18 = £36

Katie: 5 × £18 = £90 — Find 2 parts and 5 parts.

Check: £36 + £90 = £126 ✓

12 Share 465 building blocks between Benji and Freddie in the ratio 7 : 8.
How many blocks does each person get?

Discussion Which is easier, working out fractions first (like in **Q11**) or using the method in the worked example? Why?

13 James and Freya share a piece of fabric 20.4 m long in the ratio 3 : 2.
What length of fabric does Freya get?

14 Share each quantity in the given ratio.

a £374 in the ratio 2 : 4 : 5

b £46.70 in the ratio 1 : 3 : 4

c 87m in the ratio 3 : 1 : 6

d 774 kg in the ratio 2 : 7 : 3

Discussion How should you round your answer when working with money?
What about with kg?

15 **Exam-style question**

Talil is going to make some concrete mix.
He needs to mix cement, sand and gravel in the ratio 1 : 3 : 5 by weight.
Talil wants to make 180 kg of concrete mix.
Talil has

15 kg of cement
85 kg of sand
100 kg of gravel.

Does Talil have enough cement, sand and gravel to make the concrete mix? **(4 marks)**

Nov 2012, Q13, 1MA0/1H

Exam hint
Work out how much of each ingredient is needed for 180 kg of concrete mix and comment on each ingredient to say if there is enough.

16 Write each ratio as a whole number ratio in its simplest form.

a 20:36.5

b 71:120.5

c 20.1:46.9

d 90.3:6.02

Q16a hint Multiply first by powers of ten to make both sides of the ratio whole numbers, then simplify.

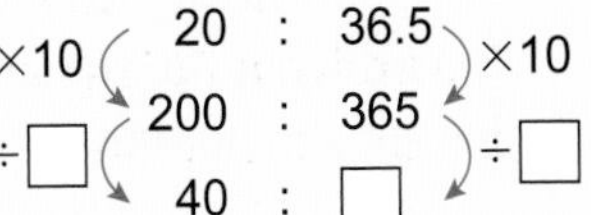

17 **Real / Reasoning** Ben wants to make some turquoise paint. He is going to mix blue, green and yellow paint in the ratio 2.4:1.5:0.1.
Copy and complete the table to show how much of each colour Ben needs to make the paint quantities shown.

Size	Blue	Green	Yellow
1 litre			
2.5 litres			
5.5 litres			

Q17 hint Write the ratio in whole numbers first, then share the amount of paint in the new ratio.

4.3 Ratio and proportion

Objectives

- Convert between currencies and measures.
- Recognise and use direct proportion.
- Solve problems involving ratios and proportion.

Why learn this?

When you are on holiday, it is useful to be able to convert between currencies, to work out the price you would pay for an item back home.

Fluency

A wildlife sanctuary has 7 adult tigers and 2 tiger cubs.
What proportion of the tigers are cubs?

1 Which of these ratios are equivalent?

1 : 3   4 : 7   2 : 5   6 : 15   5 : 7.5

2 The exchange rate between pounds and Australian dollars (AUD) is £1 = $1.80.

a Convert £200 to dollars.

b Convert $756 to pounds.

3 **Problem-solving** Kirsty buys a pair of jeans in England for £52.
On holiday in Hong Kong, she sees the same jeans on sale for HK$620.
The exchange rate is £1 = HK$12.40.
Where are the jeans cheaper?

4 **Reasoning** Ned and Adrian both go out for a bicycle ride one day. Ned rides for 23.5 miles.
Adrian rides for 41 km. 5 miles = 8 km.

a Write the ratio of miles to kilometres in the form $1:n$.

b Work out who has ridden further and by how much.

5 **Reasoning** Craig is painting his room orange.
He buys a tin of paint with red and yellow in the ratio 5:4.
Another tin of paint has yellow and red in the ratio 16:20.
Are the two tins of paint the same shade of orange?
Explain your answer.

Q5 hint Are the proportions of red and yellow paint the same in both tins?

6 **Problem-solving** Joe is paid £63 for 12 hours' work in a supermarket.

a What fraction of this is Joe paid for 7 hours' work?

b Work out how much is he paid for 7 hours' work.

Joe is paid more than the minimum wage for his age.

c How old is he? How specific can you be?

Age	21 and over	18 to 20	Under 18	Apprentice
Current minimum wage (2014)	£6.50	£5.13	£3.79	£2.73

7 In a cake, the ratio of butter, b, to sugar, s, is 3 : 4.

Copy and complete.

$s = b \times \square = \square b$

$b = s \times \square = \square s$

Q7 hint

$b:s$	$s:b$
3 : 4	4 : 3
1 : $\square$	1 : $\square$

8 **Reasoning** Caroline makes spicy beetroot chutney.

For every 500 g of beetroot, she uses 2 hot chillies.

a Write a formula for c, the number of chillies used with n grams of beetroot.

b Caroline has 2.75 kg of beetroot. How many chillies does she need?

Caroline wants to make the chutney much spicier, so she doubles the number of chillies.

c Write a formula for the new recipe.

Key point 4

When two quantities are in **direct proportion**, as one is multiplied by a number, n, so is the other.

9 Are these pairs of quantities in direct proportion?

a 10 bread rolls cost £1.60, 15 rolls cost £2.24

b 3 bitcoins cost £10.80, 7 bitcoins cost £25.20

c 5 people weigh 391 kg, 9 people weigh 767 kg.

10 **STEM / Modelling** In a science experiment, Kishan measures how far a spring extends when he adds different weights to it. The table shows his results.

Are the weight and extension in direct proportion?

Weight (w)	1 N	2 N	3 N	4 N	5 N
Extension (e)	12 mm	24 mm	36 mm	48 mm	60 mm

11 The table gives readings P and Q in a science experiment.

a Are P and Q in direct proportion? Explain.

b Write a formula for Q in terms of P.

c Write the ratio P : Q in its simplest form.

P	5	10	14
Q	7.5	15	21

12 The values of A and B are in direct proportion.

Work out the missing values P, Q, R and S.

Value of A	32	P	Q	20	72
Value of B	20	30	35	R	S

13 The cost of ribbon is directly proportional to its length. A 3.5 m piece of ribbon costs £2.38.

Work out the cost of 8 m of this ribbon.

14 **Problem-solving** The length of the shadow of an object is directly proportional to the height of the object.
A lamp post 4.8 m tall has a shadow 2.1 m long.
Work out the height of a nearby bus stop with a shadow 1.05 m long.

15 **Exam-style question**

Margaret is in Switzerland.
The local supermarket sells boxes of Reblochon cheese.
Each box of Reblochon cheese costs 3.10 Swiss francs. It weighs 160 g.

3.10 Swiss francs
160 g

In England, a box of Reblochon cheese costs £13.55 per kg.
The exchange rate is £1 = 1.65 Swiss francs.
Work out whether Reblochon cheese is better value for money in Switzerland or England. **(4 marks)**

Nov 2010, Q5, 5MB1H/01

Exam hint
You only need to convert one of the prices into the other currency, not both, before you look at the weight.

4.4 Percentages

Objectives

- Work out percentage increases and decreases.
- Solve real-life problems involving percentages.

Why learn this?

Percentage change calculations help us to compare the cost of living, to see if we are spending more or less of our money on basic necessities from one year to the next.

Fluency

Find these percentages of £50.

a 10% **b** 20% **c** 5% **d** 120%

1 Write down the single number you can multiply by to work out an increase of
a 15% b 30% c 5%

2 Write down the single number you can multiply by to work out a decrease of
a 25% b 10% c 6%

3 Karen gets a gas bill. The cost of the gas before the VAT was added was £361.20.
VAT is charged at 5% on domestic fuel bills.
What was the cost of the gas bill, including VAT?

Q3 communication hint **Value Added Tax (VAT)** is charged at 20% on most goods and services. Domestic fuel bills have a lower VAT rate of 5%.

4 **Exam-style question**

Petra booked a family holiday.
The total cost of the holiday was £3500 *plus* VAT at 20%.
Petra paid £900 of the total cost when she booked her holiday.
She paid the rest of the total cost in 6 equal monthly payments.
Work out the amount of each monthly payment. **(5 marks)**

June, 2013 Q7, 5MB3H/01

Exam hint
Read one sentence at a time and decide what calculation you need to do.

ActiveLearn Homework, practice and support: Higher 4.4

5 A holiday costing £875 in the brochure is reduced by 12%.
How much does the holiday cost now?

6 **Reasoning** Curtis buys a car for £9600. The value of the car depreciates by 20% each year.
Work out the value of the car after
a 1 year
b 2 years.

Q6 communication hint
Depreciates means loses value.

Q6b hint The value at the end of year 1 depreciates another 20% in year 2.

Key point 5

Simple interest is the interest calculated only on the original amount invested.
It is the same each year.

7 **Finance** a Work out the amount of simple interest earned in one year for each of these investments.
i £1500 at 2% per year. ii £700 at 8% per year.
b Martina invests £14 500 for 3 years at 6.75% simple interest.
How much is the investment worth at the end of the 3 years?

Q7b hint Work out the amount of interest she earns each year and multiply by 3.

8 **Finance / Problem-solving** Income tax is paid on any money you earn over your personal tax allowance.
The personal tax allowance is currently set at £10 000.
Above this amount, tax is paid at different rates, depending on how much you earn.
The table shows the rates for 2014/15.

Q8 communication hint
Your **income** means the amount of money you earn or are paid, and 'per annum' (abbreviated to p.a.) means each year.

Tax rate	Taxable income above your personal allowance
Basic rate 20%	£0 to £31 865
Higher rate 40%	£31 866 to £150 000

Work out the amount of income tax each of these people paid in the 2014/15 tax year.
a Ella earns £26 500 per annum.
b Sammy earns £28 760 p.a.
c Antony earns £47 000 p.a.
d Pippa earns £73 850 p.a.

Q8 hint Subtract the personal tax allowance before working out the tax owed.

Key point 6

You can calculate a percentage change using the formula

$$\text{percentage change} = \frac{\text{actual change}}{\text{original amount}} \times 100$$

9 **Finance** Inder invests £3200. When her investment matures, she receives £3328.
a What was the actual increase?
b Work out the percentage increase in her investment.

10 In 2014, the Croftshire County Council raised £18.64 million in council tax. In 2004, they raised £17.18 million. What was the percentage increase over the decade?

11 Reena bought a jacket for £45. Six months later, she sold it for £34.65.
What was her percentage loss?

Key point 7

$$\text{Percentage loss (or profit)} = \frac{\text{actual loss (or profit)}}{\text{original amount}} \times 100$$

12 Guy spent £11.40 buying ingredients to make cupcakes.
He sold all the cakes for a total of £39.90.
What percentage profit did Gary make?

Q12 strategy hint When you are working out profits, remember to subtract any costs first.

13 **Reasoning** The price of a magazine costing £1.20 increased by 150% over 2 years.
Jo says the magazine is now $1\frac{1}{2}$ times more expensive.
Eric says it is $2\frac{1}{2}$ times more expensive.
Who is correct?

Key point 8

You can use inverse operations to find the original amount after a percentage increase or decrease.

Example 3

In one year, the value of a car dropped by 12% to £9240.
How much was the car worth at the start of the year?

100% − 12% = 88% = 0.88

Original number → ×0.88 → 9240

9240 → ÷0.88 → 10500 — Draw a function machine

The car was worth £10 500 at the start of the year.

14 Stuart pays £52.56 for his office stationery order. This price includes VAT at 20%.
What was the cost of the stationery before VAT was added?

15 The cost of living increased by 2% one year.
The next year it increased by 3%.
Copy and complete the calculation to work out the total percentage increase over these two years.

$£x$ → ×1.02 → ×1.03 → □

→ ×□ →

16 **Problem-solving** Manjit bought a house. The value of her house went up by 5% in the first year. In the second year, the value went up by 2%. At the end of the two years, her house was worth £171 360.

a What was the total percentage increase? Do not round your answer.

b Work out the amount Manjit paid for her house.

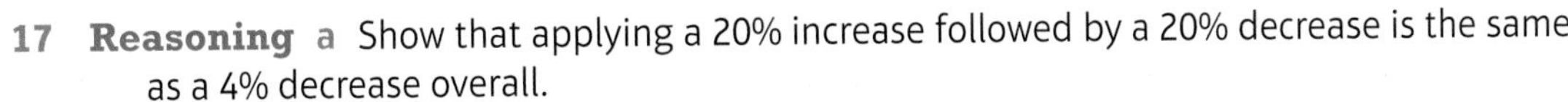

17 **Reasoning** a Show that applying a 20% increase followed by a 20% decrease is the same as a 4% decrease overall.

b Will the final amount be the same or different if you apply the decrease first, then the increase?

4.5 Fractions, decimals and percentages

Objectives

- Calculate using fractions, decimals and percentages.
- Convert a recurring decimal to a fraction.

Why learn this?

Converting fractions, decimals and percentages can make calculations simpler.

Fluency

What are the decimal and percentage equivalents for

a $\frac{1}{2}$ **b** $\frac{3}{4}$ **c** $\frac{1}{5}$ **d** $\frac{3}{10}$

Warm up

1 Solve these equations

a $9m = 3$ b $2 - n = 6$ c $\frac{p}{4} = 8$

2 Copy and complete this table.

Fraction	Decimal	Percentage
$\frac{1}{8}$		
	0.45	
$\frac{2}{3}$		
		80%
	1.5	

3 Work out

a $\frac{3}{8}$ of 10 b 0.25 of 40

c $\frac{4}{5}$ of 16 d 0.48 of 350

e $12\frac{1}{2}$ of 64 f 250% of £19

Q3a hint Change $\frac{3}{8}$ to a decimal.

Q3b hint Change 0.25 to a fraction.

4 **Reasoning** A restaurant manager bought a case of 12 bottles of sparkling water.
He paid 90p per bottle.
He sold $\frac{1}{4}$ of the bottles for £2.10 per bottle and the rest of the case for £2.40 per bottle.

a How much profit did he make?

b Express this profit as a percentage of the total cost price.

5 **Exam-style question**

Mr Mason asks 240 Year 11 students what they want to do next year.
15% of the students want to go to college.
$\frac{3}{4}$ of the students want to stay at school.
The rest of the students do not know.
Work out the number of students who do not know.

(4 marks)

June 2013, Q2, 1MA0/1H

Q5 strategy hint You could draw a diagram.

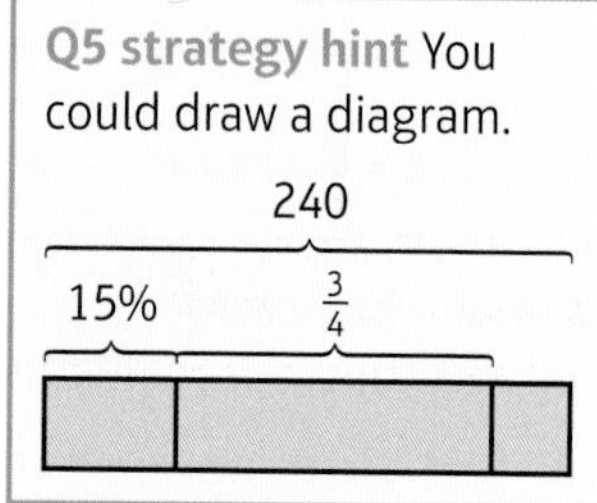

6 **Exam-style question**

A farmer uses 1.8 out of every 5 acres of land to grow crops.
He grows corn on $\frac{5}{6}$ of the land he uses for crops.
What percentage of the total area of his land does he use to grow corn? **(3 marks)**

ActiveLearn Homework, practice and support: Higher 4.5

7 **Exam-style question**

Which is closer to 30%: $\frac{1}{3}$ or $\frac{2}{7}$?

You must show your working. **(3 marks)**

8 **Problem-solving** The table shows the number of days, absence for Year 9 students in each school term over 2 years. Write three sentences comparing the absences in the 2 years. Use fractions, decimals, percentages, ratio or proportion.

	Term 1	Term 2	Term 3
Year 9 (2012/2013)	46	76	24
Year 9 (2013/2014)	28	64	36

Q8 hint Choose calculations that will help you to compare.

9 **Problem-solving** Work out $\frac{1}{3}$ of 0.25 of 48% of £340. Show all your working out.

10 **Reasoning** Write the sum of the sequence $\frac{6}{10} + \frac{6}{100} + \frac{6}{1000} + \ldots$ as a fraction (where … indicates that the sequence goes on forever.) Explain your answer.

Q10 hint Write the fractions as decimals.

11 **Reasoning** Two variables s and t are connected by the formula

$$s = 4t$$

a Are s and t in direct proportion? Explain.

b Write t as a fraction of s.

c Write the ratio $s:t$.

Example 4

Write $0.\dot{3}$ as a fraction.

$0.\dot{3} = 0.333333333\ldots = n$ — Call the recurring decimal n.

so $10n = 3.333333333\ldots$ — Multiply the recurring decimal by 10 to shift the sequence one place left.

$10n - n = 3.333333333\ldots - 0.333333333\ldots$

$= 3.000000000\ldots$ — Subtract the value of n from the value of $10n$. This makes all the numbers after the decimal point 0.

$9n = 3$

$n = \frac{3}{9}$ — Solve the equation.

$n = \frac{1}{3}$ — Simplify the fraction if possible.

Key point 9

All recurring decimals can be written as exact fractions.

12 Write these recurring decimals as exact fractions. Write each fraction in its simplest form.

a $0.\dot{6}$ b $0.\dot{1}$

c $0.\dot{5}\dot{2}$ d $0.181818\ldots$

e $0.\dot{7}4\dot{3}$ f $0.\dot{2}6\dot{1}$

Q12 strategy hint Multiply by a power of ten. If 1 decimal place recurs, multiply by 10. If 2 decimal places recur, multiply by 100. If 3 decimal places recur, multiply by 1000.

13 Which of these fractions are equivalent to recurring decimals. Show your working out.

a $\frac{7}{25}$ b $\frac{11}{42}$ c $\frac{29}{80}$ d $\frac{4}{15}$

Discussion How can you tell whether a fraction will give a recurring or terminating decimal?

4 Problem-solving

Objective

- Use bar models to help you solve problems.

Example 5

Sophie spent $\frac{1}{8}$ of her clothes allowance. She spent £11. Crista spent $\frac{1}{6}$ of her clothes allowance. Now Crista has £2 less than Sophie. How much is Crista's clothes allowance?

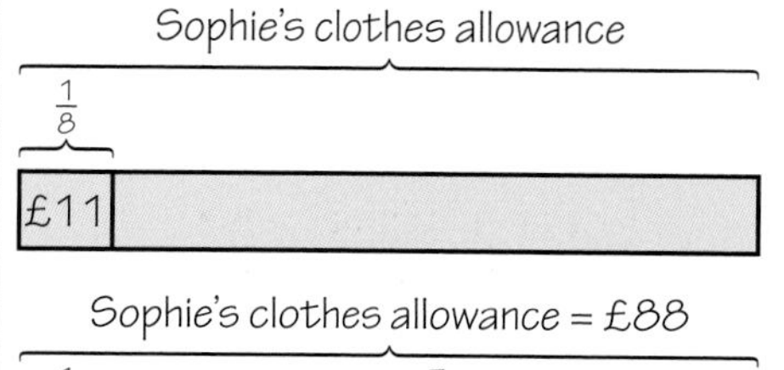

Draw a bar to represent Sophie's clothes allowance. Add the information from the question.

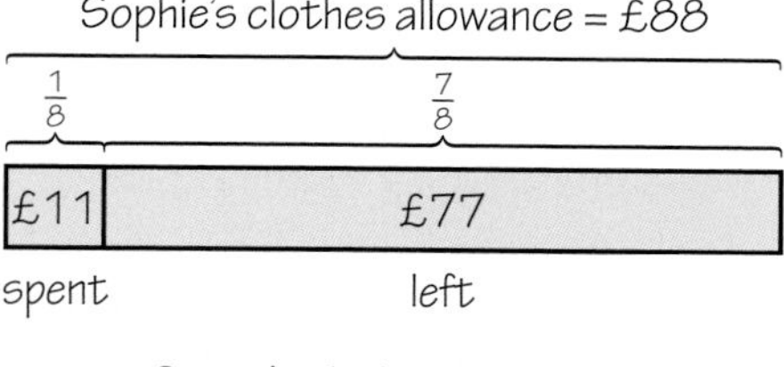

$\frac{1}{8}$ of Sophie's clothes allowance = £11.
So, Sophie's clothes allowance = £11 × 8 = £88
Sophie has $\frac{7}{8}$ of £88 left = £77 left

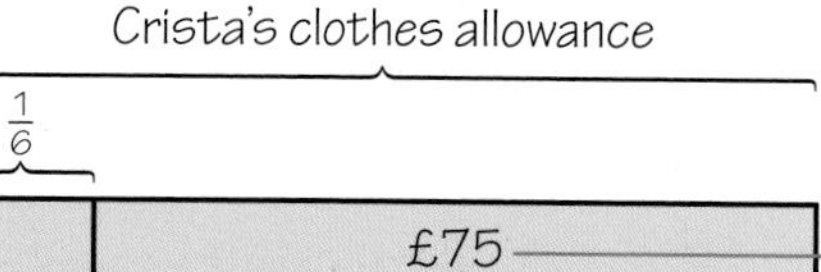

Draw a bar to represent Crista's clothes allowance. Add the information from the question.

Crista has £2 less than Sophie. £77 – £2 = £75

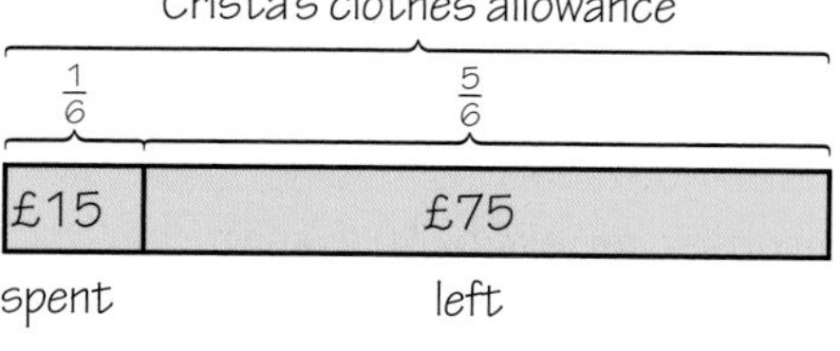

$\frac{5}{6}$ of Crista's clothes allowance = £75. So, Crista spent $\frac{1}{6}$ of her clothes allowance = £75 ÷ 5 = £15

Crista's clothes allowance = £15 + £75 = £90

Use your bar model to answer the question.

Check:

Check your answer works.

Sophie spent $\frac{1}{8}$ of her clothes allowance = $\frac{1}{8}$ of £88 = £11
Amount left = £88 – £11 = £77

Crista spent $\frac{1}{6}$ of her clothes allowance = $\frac{1}{6}$ of £90 = £15

Amount left = £90 – £15 = £75

Now Crista has £2 less than Sophie. ✓

1 This Christmas, Mr Smith spent $2\frac{1}{4}$ times his budget for presents. He spent £405. Mrs Smith spent $1\frac{3}{5}$ times her budget for presents. She spent £5 less than Mr Smith spent.

a How much was Mr and Mrs Smith's total budget for presents?

b How much did they overspend?

Q1 hint

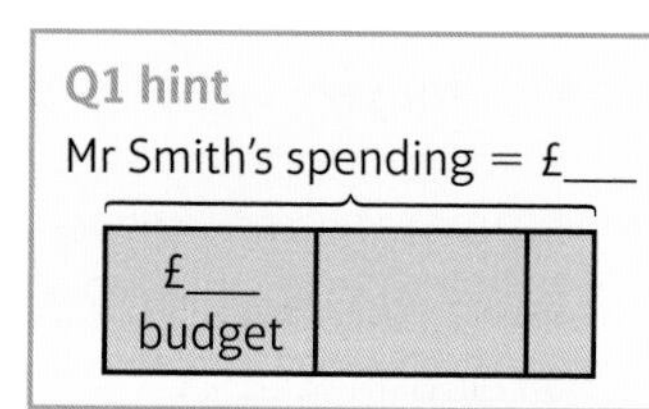

2 Caroline and Naomi share a flat with a monthly rent of £1025. Caroline's bedroom is $1\frac{1}{2}$ times the size of Naomi's, so she agrees to pay $1\frac{1}{2}$ times the rent of Naomi. How much do they each pay?

Q2 hint

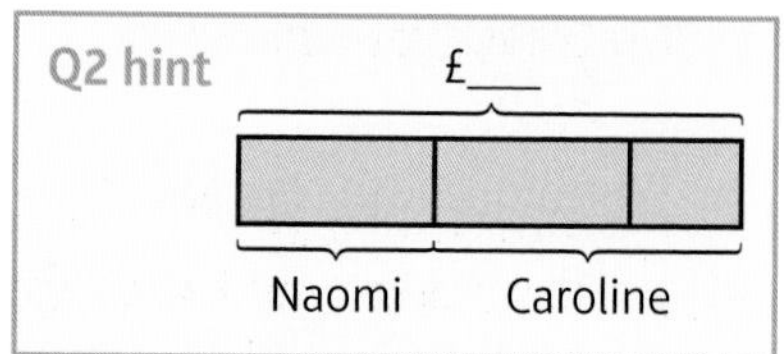

3 A petting zoo has rabbits, goats and llamas in the ratio 6 : 3 : 2. The zoo has 8 more rabbits than llamas. How many goats does it have?

Q3 hint Draw a bar model showing the ratio 6 : 3 : 2. Compare rabbits and llamas. How many sections represent 8 rabbits?

4 Amateur boxers can only fight other boxers in the same weight class.
The table shows three of the weight classes.
Two amateur boxers have weights in the ratio 2.5 : 3. Their total weight is 165 kg.
Can the boxers fight each other?
Explain.

Weight class	Boxer's weight (kg)
Heavyweight	81–91
Light heavyweight	75–81
Middleweight	69–75

Q4 hint Draw a bar to represent the total weight. Split the bar into 0.5 sections.
One section = ___ kg

5 Jamie invests some money. In the first year it increases to 110% of its original value. He spends 20% of the profit on a cricket bat and $\frac{1}{8}$ of the remainder on a cricket jumper. He is left with £140 profit. How much did Jamie invest?

6 Flu is passed around an accounts department. The clerk has $2\frac{1}{2}$ times the days off sick than the accountant. The accountant has $\frac{2}{3}$ the time off sick than the book-keeper. In total they all take 10 sick days. How many sick days do they each take?

Q6 hint Draw the accountant's section of a bar first. Label it A. Next draw the clerk's section. Label it C. Be very careful when drawing the book-keeper's section, B.

7 8 adults, 6 children and 2 seniors swim lengths at a swimming pool session. The mean number of lengths swum by the adults is 40, the mean swum by the children is 7 and the mean swum by the seniors is 35. Work out the mean number of lengths swum by everyone in the session.

Q7 hint Draw a bar model to represent all the swimmers. 8 adults swum a mean of 40 lengths. How many lengths did they swim altogether?

8 **Reflect** Look back at the exam-style questions in lessons **4.4** and **4.5**. How could you answer these questions using bar models?
Is drawing bar models a strategy you would use again to solve problems?

4 Check up

Log how you did on your Student Progression Chart.

Fractions

1 Work out

a $1\frac{3}{5} + 2\frac{5}{8}$ b $3\frac{1}{6} - 2\frac{7}{9}$

2 Work out

a $2\frac{1}{3} \times 1\frac{1}{4}$ b $2\frac{4}{5} \div \frac{7}{10}$

Ratio and proportion

3 Write each ratio as simply as possible, without units.

a 350 ml : 2 litres b 0.7 kg : 3.2 kg

4 Write each ratio in the form 1 : n

a 4 : 28 b $\frac{5}{6}$: 3

5 The euro exchange rate was £1 = €1.27. Work out

a how many euros I would get for £45 b the price in pounds of a sofa priced at €488.95

6 **Reasoning** Ellis makes some biscuits. For every 200 g of flour he uses, he needs 75 g of butter.
  a Write a ratio for the amount of flour to the amount of butter.
  b Write a formula for f, the amount of flour, in terms of the amount of butter, b.
  Ellis makes 24 biscuits using 300 g of flour.
  c How many biscuits can he make with 375 g of butter?

7 Share £132 in the ratio 3 : 2 : 1

Fractions, decimals and percentages

8 Work out the final amount when
  a £450 is increased by 7.5%
  b 877.2 kg is decreased by 3.2%

9 Simon scores 68 marks in his second maths test. In his first maths test he scored 85 marks. What is the percentage decrease in Simon's score?

10 The price of a laptop increases by 35%.
The new price is £972.
What was the original price?

11 Barbara invests £14 000.
In the first year, she earns 5.9% interest.
In the second year, she earns 3.2% interest.
  a What was the total percentage increase over the 2 years?
  b How much money does she have after 2 years?

Reflect

12 How sure are you of your answers? Were you mostly
Just guessing Feeling doubtful Confident 
What next? Use your results to decide whether to strengthen or extend your learning.

* Challenge

13 Find the cube root of the reciprocal of the square root of the reciprocal of 64.
Write a problem similar to this.
Make sure you know the answer.

4 Strengthen

Fractions

1 Work out
  a $\frac{1}{5} \div \frac{1}{4}$
  b $\frac{1}{3} \div \frac{3}{4}$
  c $\frac{4}{5} \div \frac{3}{10}$
  d $\frac{3}{8} \div \frac{1}{5}$
  e $\frac{9}{10} \div \frac{2}{5}$
  f $\frac{7}{12} \div \frac{5}{6}$

Q1a hint $\frac{1}{5} \div \frac{1}{4} = \frac{1}{5} \times \square$

2 Giving your answers as mixed numbers, work out
  a $3 + \frac{13}{12}$
  b $\frac{7}{3} + 4$
  c $\frac{21}{9} + 5$
  d $2 + \frac{48}{15}$

Q2a hint Change the improper fraction to a mixed number:
$3 + \frac{13}{12} = 3 + 1\frac{\square}{12} = 4\frac{\square}{12}$

ActiveLearn Homework, practice and support: Higher 4 Strengthen

3 Work out

a $2\frac{1}{3} + 3\frac{3}{4} = 5 + \frac{1}{3} + \frac{3}{4} = 5 + \frac{\square}{\square}$

b $1\frac{2}{3} + 5\frac{1}{8}$ c $2\frac{4}{5} + 1\frac{2}{9}$ d $6\frac{1}{7} + 4\frac{7}{10}$

4 Work out these subtractions.
Give your answers as mixed numbers.

Q4a hint Write both numbers as halves.

a $6 - 2\frac{1}{2} = \frac{12}{2} - \frac{\square}{2}$

b $5\frac{1}{4} - 2\frac{3}{8}$ c $3\frac{2}{7} - 1\frac{3}{5}$ d $4\frac{3}{8} - \frac{5}{6}$ e $2\frac{1}{10} - 1\frac{7}{8}$

5 Work out

a $1\frac{4}{5} \div \frac{1}{4}$ b $2\frac{1}{3} \div \frac{5}{8}$

c $2\frac{4}{5} \div \frac{7}{10}$ d $3\frac{1}{8} \div \frac{5}{12}$

Q5a hint Change the mixed number to an improper fraction. Then use the method in **Q1**.

Ratio and proportion

1 Which of these ratios are in the form

a $1:n$? b $n:1$?

A 3:1 **B** 5:2 **C** 2:1 **D** 1:7 **E** 8:9

2 Write each ratio in the form

i $1:n$ ii $n:1$

The first one has been started for you.

a 3:15

$1 : n$ — $3 : 15$, ÷3 → $1 : \square$

$n : 1$ — $3 : 15$, ÷15 → $\square : 1$

b 8:2 c 7:56 d 42:24

3 Simplify each of these ratios.

a 6.5:4 b 5:8.5 c 2.8:4 d 5.6:8.8

Q3 strategy hint Choose a number to multiply by that will give you a whole number on both sides of the ratio.
×2 (6.5 : 4) ×2 → 13 : 8

4 In 2007 the exchange rate from pounds to US dollars was £1 = \$2.

a How many dollars could you buy for
 i £4 ii £5?
 iii What calculation did you use to work it out?

b How many pounds could you buy for
 i \$6 ii \$10?
 iii What calculation did you use to work it out?

Q4 hint Work out the multiplier to change £ to \$. Use the inverse operation to change \$ to £.
£1 ×□ → = \$2 ÷□ → £1

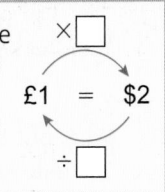

5 **Reasoning** A and B are in direct proportion.
Fill in the missing values for X and Y in this table.

A	4	8	Y
B	9	X	45

Q5 hint

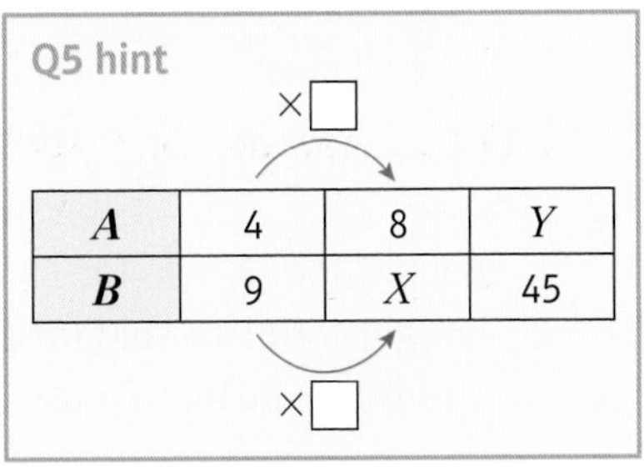

6 **Reasoning** P and Q are in direct proportion.
Find the missing value in the table.
Show your working.

P	Q
3	5
	9

7 Kiran, Lewis, Stephen and Jane are paid in the ratio 2:5:4:7, according to the number of hours they have worked.
Lewis is paid £35.50.
Work out how much money Kiran, Stephen and Jane receive.

Q7 hint
K : L : S : J
2 : 5 : 4 : 7
□ : £35.50 : □ : □

Fractions, decimals and percentages

1 Convert these percentages to decimals.
- a 104%
- b 126.5%
- c 98.3%

2 The price of a theatre ticket increases by 3.5% from £45
- a What percentage of £45 will the new price be?
- b Write your answer to part **a** as a decimal.
- c Work out the new price.

Q2a hint 100% + □% = □%

Q2c hint □ × 45 = □

3 The price of a pedicure decreases by 4.2% from £36 on promotion.
- a What percentage of £36 will the new price be?
- b Write your answer to part **a** as a decimal.
- c Work out the new price.

4 Betsy invests £3000. When her investment matures, she receives £3135.
- a Copy and complete the working to calculate the percentage increase of Betsy's investment.

  original amount = 3000
  actual change = 3135 − 3000 = 135

  $$\text{percentage change} = \frac{\text{actual change}}{\text{original amount}} \times 100 = \frac{135}{3000} \times 100 = \square\%$$

- b Check your answer by increasing £3000 by the percentage you calculated. Do you get £3135?

Q4 hint Draw this information as a bar model.

£3000 £□
£3135

5 Work out the percentage loss made on each of these items. For each part copy and complete the following working. Check your answers.

original amount = □
actual change = □

$$\text{percentage change} = \frac{\text{actual change}}{\text{original amount}} \times 100 = \frac{\square}{\square} \times 100 = \square\%$$

- a Bought for £8, sold for £5.75
- b Bought for £145, sold for £120
- c Bought for £615, sold for £500

6 Work out

a 0.75×150 g b $\frac{3}{8}$ of £10

c $33.\dot{3}\%$ of £36

> **Q6 hint** Use an equivalent fraction, decimal or percentage to make the calculation easier.

7 The price of a computer game after a 28% increase is £13.44

a What decimal number do you multiply by to increase a value by 28%?

b Draw a function machine for this calculation.

c Work backwards through the function machine to find the original price.

> **Q7b, c hint**
> Original price: → × □ → £13.44
> Work backwards: □ ← ÷ □ ← £13.44

8 Find the original price of

a a sofa that costs £585 after a 25% discount

b a house priced at £192 030 after its value rose by 3.8%.

> **Q8 hint** Use the method from **Q7**.

9 Carol put £5000 in a savings account for 2 years.
The first year she earned 2.5% interest.
The second year she earned 3.1% interest.

a Write a calculation to find the amount of money Carol had at the end of the first year.

b Multiply your calculation from part **a** by 1.031 to find the amount in the account after 2 years.

4 Extend

1 **Problem-solving** Amanda, Chen and Mark shared some money in the ratio 2:4:9.
Mark got £84 more than Amanda.
How much money did Chen get?

2 Work out 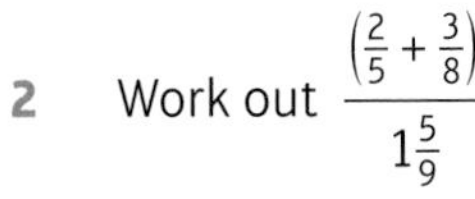 $\dfrac{\left(\frac{2}{5}+\frac{3}{8}\right)}{1\frac{5}{9}}$

3 **Reasoning** The diagram shows a rectangle ABCD.
AB is twice the length of BC.
E is the midpoint of AB. F is the midpoint of BC.
Work out the area of each of these triangles
Give your answer as a fraction of the rectangle.

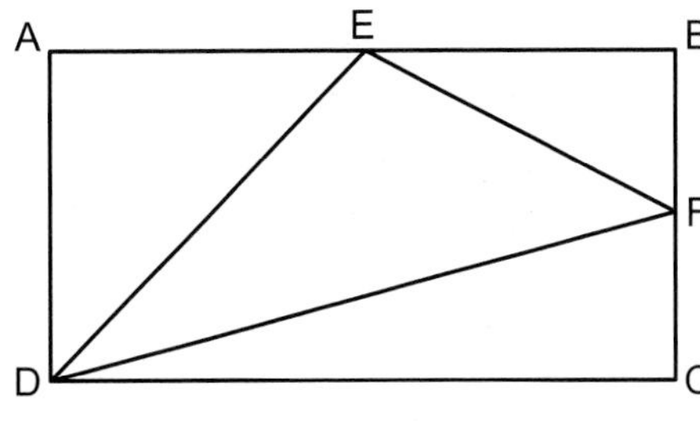

a ADE b BEF

c CDF d DEF

Discussion Would these fractions change if length AB was 3 times the length of BC?

4 A photocopier increases the sides of a square in the ratio 2:3. What percentage increase is this?

5 **Problem-solving** The diagram shows three identical shapes, A, B and C.
$\frac{4}{7}$ of shape A is shaded.
$\frac{5}{6}$ of shape C is shaded.
What fraction of shape B is shaded?

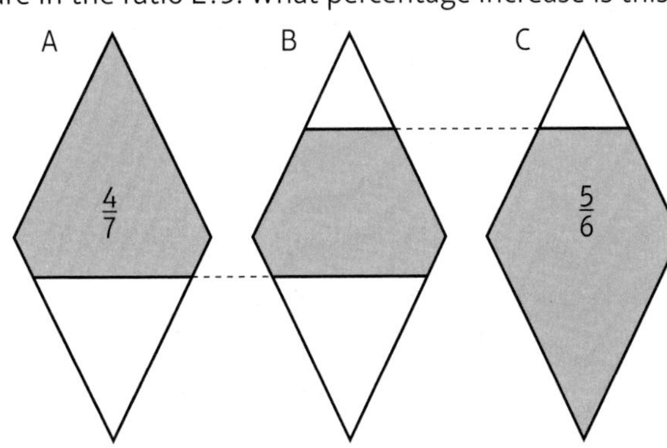

6 **Finance / Reasoning** Gareth sells cupcakes.
He adds 40% profit to the cost price.
He sells the cupcakes for £1.68 each.
He wants to increase his profit to 60% of the cost price.
How much should he sell each cupcake for?

7 **Problem-solving** In a company, 65% of the workers are female.
40% of the women drive to work.
50% of the men drive to work.
What percentage of the company's employees drive to work?

8 Here is some information about a class.

	Boys	Girls
Left-handed	4	3
Right-handed	8	9

a Write down the ratio of right-handed boys to left-handed boys. Give your answer in its simplest form.
b What percentage of the girls are left-handed?

9 **Exam-style question**

VWXY is a rectangle with length 20 cm and width 12 cm.

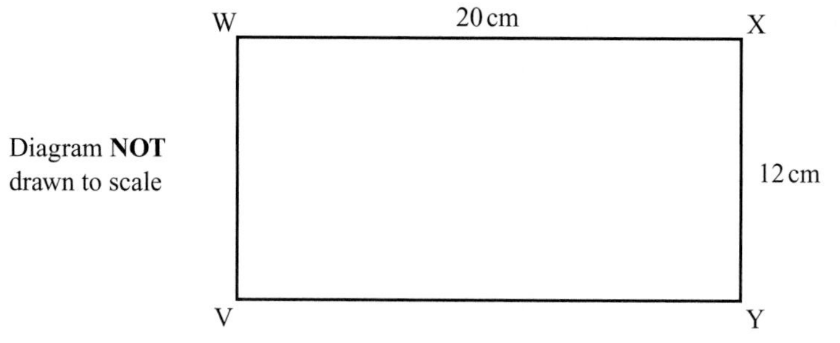

The length of the rectangle is increased by 30%.
The width of the rectangle is increased by 10%.
Find the percentage increase in the area of the rectangle. **(5 marks)**

Q9 strategy hint
What single number would you multiply each length by to find the new area? What happens if you multiply these two numbers together?

10 Work out
a $2^{-1} \div \frac{1}{2}$
b $173^{-1} \div \frac{1}{173}$
c $3^{-4} + 3^{-2}$

11 **Real / Reasoning** Sian has some sheep.
The sheep produce an average of 15.8 litres of milk per day for 146 days.
Sian sells the milk in $\frac{1}{4}$ litre bottles.
Work out an estimate for the total number of bottles that Sian will be able to fill with the milk.
Show clearly how you worked out your estimate.

12 **Exam-style question**

Each day a company posts some small letters and some large letters.
The company posts all the letters by first class post.
The tables show information about the cost of sending a small letter by first class post and the cost of sending a large letter by first class post.

Small Letter

Weight	First class post
0–100 g	60p

Large Letter

Weight	First class post
0–100 g	£1.00
101–250 g	£1.50
251–500 g	£1.70
501–750 g	£2.50

One day the company wants to post 200 letters.
The ratio of the number of small letters to the number of large letters is 3 : 2.
70% of the large letters weigh 0–100 g.
The rest of the large letters weigh 101–250 g.
Work out the total cost of posting the 200 letters by first class post. **(5 marks)**

Nov 2013, Q11, 1MA0/1H

Exam hint
Show a separate calculation for each of the last four lines of the question to reach your final answer.

13 **Exam-style question**

Mr Layton needs to buy some oil for his central heating.
He can put up to 2500 litres of oil in his oil tank.
There are already 750 litres of oil in the tank.
Mr Layton is going to fill the tank with oil.
The price of oil is 58.4 p per litre.
Mr Layton gets 6% off the price of the oil.
How much does Mr Layton pay for the oil he needs to buy? **(4 marks)**

14 **Exam-style question**

Boris, Carla and Dean share some money.
Boris gets $\frac{1}{10}$ of the money.
Carla and Dean share the rest of the money in the ratio 4 : 5.
What percentage of the money does Dean get? **(2 marks)**

15 **Exam-style question**

Linda is going on holiday to the Czech Republic.
She needs to change some money into koruna.
She can only change her money into 100 koruna notes.
Linda only wants to change up to £200 into koruna.
She wants as many 100 koruna notes as possible.
The exchange rate is £1 = 25.82 koruna.
How many 100 koruna notes should she get? **(3 marks)**

June 2012, Q9, 1MA0/2H

Exam hint
Start with the values you are given and write down each step in your reasoning.

4 Knowledge check

- It is often easier to write mixed numbers as improper fractions before doing a calculation. *Mastery lesson 4.1*
- You should divide by common factors before multiplying, if you can. *Mastery lesson 4.1*
- The **reciprocal** of the number n is $\frac{1}{n}$. You can also write this as n^{-1}. *Mastery lesson 4.1*
- To find the reciprocal of a fraction, swap the numerator and the denominator. For example, the reciprocal of $\frac{3}{4}$ is $\frac{4}{3}$. *Mastery lesson 4.1*
- To find the reciprocal of a mixed number, first convert it into an improper fraction. *Mastery lesson 4.1*
- Sometimes both denominators must be changed to add fractions. *Mastery lesson 4.1*
- You can compare ratios by writing them as **unit ratios**. In a unit ratio, one of the numbers is 1. The other number may or may not be a whole number. *Mastery lesson 4.2*
- To share a quantity in a given ratio you could work out what fraction of the total amount each person receives, and then multiply each fraction by the total amount. Another method is to work out how much one part is worth, and then multiply by the number of parts each person receives. *Mastery lesson 4.2*
- When two quantities are in **direct proportion**, as one is multiplied by a number, n, so is the other. Their ratio also stays the same as they increase or decrease. *Mastery lesson 4.3*
- **Simple interest** is the interest calculated only on the original amount invested. It is the same each year. *Mastery lesson 4.4*
- You can calculate a percentage change using the formula
  $$\text{percentage change} = \frac{\text{actual change}}{\text{original amount}} \times 100$$
  Mastery lesson 4.4
- $$\text{Percentage loss (or profit)} = \frac{\text{actual loss (or profit)}}{\text{original amount}} \times 100$$
  You can use inverse operations to find the original amount after a percentage increase or decrease. *Mastery lesson 4.4*
- **Value Added Tax (VAT)** is charged at 20% on most goods and services. Domestic fuel bills have a lower VAT rate of 5%. On some things no VAT is charged. *Mastery lesson 4.4*
- **Depreciates** means loses value. *Mastery lesson 4.4*
- Your income means the amount of money you earn or are paid, and 'per annum' (abbreviated to p.a.) means each year. *Mastery lesson 4.4*
- When you are working out profits, remember to subtract any costs first. *Mastery lesson 4.4*
- All recurring decimals can be written as exact fractions. *Mastery lesson 4.5*
- If 1 decimal place recurs, multiply by 10.
  If 2 decimal places recur, multiply by 100.
  If 3 decimal places recur, multiply by 1000. *Mastery lesson 4.5*

Reflect

For this unit, copy and complete these sentences.

- I showed I am good at ____
- I found ___ hard
- I got better at ____ by ___
- I was surprised by _____
- I was happy that ____
- I still need help with ____

4 Unit test

Log how you did on your Student Progression Chart.

1 Work out a $2\frac{4}{5} \div 1\frac{3}{7}$ b $5\frac{2}{3} - 2\frac{7}{8}$ *(6 marks)*

2 **Reasoning** Alice has $8\frac{1}{4}$ acres of orchards.
Alice grows apple trees in $4\frac{5}{6}$ acres of the orchards.
She grows pear trees in the rest.
How many acres of pear trees does Alice have? *(3 marks)*

3 **Reasoning** Selika gives her garden a makeover.
She spends money on plants, materials and labour in the ratio 1 : 5 : 12.
She spends £848.75 on materials.
Work out
a how much money she spends on labour costs *(2 marks)*
b how much she spends in total. *(1 mark)*

4 On a hospital ward, there are 16 nurses and 68 patients.
a Write the nurse : patient ratio in the form of $1 : n$. *(1 mark)*
Another ward has 18 nurses and 81 patients.
b Which hospital has the best nurse : patient ratio?
Explain your answer. *(2 marks)*

5 **Problem-solving** Cameron is going on holiday to Spain.
He needs to change some money into euros.
He can only change his money into €20 or €50 notes.
Cameron has up to £540 to change.
He wants to take as many euros as possible.
The exchange rate is £1 = €1.27.
How many euros will Cameron get? *(3 marks)*

6 **Reasoning** *J* is directly proportional to *K*. Work out the missing values, **a** and **b**, in the table. *(4 marks)*

J	*K*
52	36
39	a
b	22.5

c Write a formula for *J* in terms of *K*.

7 Nathan makes fudge and sells it at a Christmas fayre.
He spends £2.14 on ingredients.
Nathan sells all the fudge and has £9.63 in the cash box at the end of the sale.
What percentage profit does Nathan make on the fudge? *(2 marks)*

8 a A tourist attraction experienced a 3.75% fall in visitor numbers in June, compared to the previous month, due to exceptionally bad weather.
There were 121 660 visitors in June.
How many visitors were there in May? *(3 marks)*
b In July there were 4.5% more visitors than in May.
What was the percentage increase in visitor numbers from June to July? *(2 marks)*

ActiveLearn Homework, practice and support: Higher 4 Unit test

9 **Exam-style question**

The diagram shows rectangle EFGH.

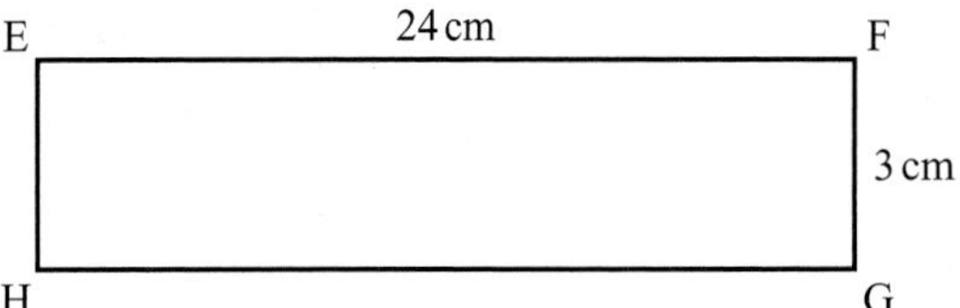

Length EF is 24 cm. Width FG is 3 cm.

The length of the rectangle decreases by 40% and the width increases by 30%.

What is the overall percentage change to the area of the rectangle?

State clearly if this is an increase or decrease. **(6 marks)**

Sample student answer

Why will the student only get 2 marks for this answer?

Exam-style question

Stacey bought a handbag in Paris.

The handbag cost €64.80

In Manchester, the same type of handbag costs £52.50

The exchange rate was £1 = €1.20

Compare the cost of the handbag in Paris with the cost of the handbag in Manchester. **(3 marks)**

June 2012, Q2b, 5MB1/01

Student answer

£1 = €1.20

£52.50 × 1.20 = €63

5 ANGLES AND TRIGONOMETRY

In 1936 work began on building more than 11 000 'trig pillars' across the United Kingdom. By measuring angles and using trigonometry, surveyors could work out the distances between pillars. This enabled them to make a map of the whole country, accurate to a few metres.

The diagram shows the angles between two trig points. Estimate the distance AC. Explain your reasoning.

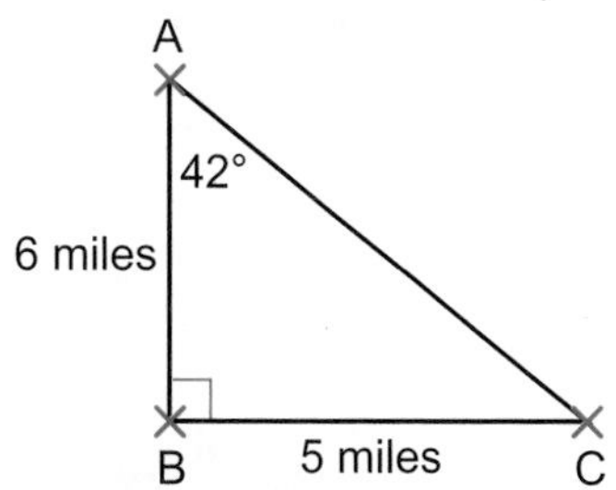

5 Prior knowledge check

Numerical fluency

1 Work out
- a $3^2 + 4^2$
- b $\sqrt{3^2 + 4^2}$
- c $\sqrt{10^2 - 6^2}$
- d $\sqrt{6^2 + 8^2}$

2 Work out
- a $\sqrt{6^2 - 5^2}$
- b $\sqrt{4^2 + 6^2}$
- i Give your answers as a decimal correct to 2 decimal places.
- ii Leave your answers in surd form.

Geometrical fluency

3 List ALL the quadrilaterals which have
- a exactly ONE pair of parallel sides
- b four right angles
- c two or more lines of symmetry
- d no pairs of parallel sides
- e no right angles.

4 What is the name given to a regular
- a triangle
- b quadrilateral?

5 Which of these shapes are regular?

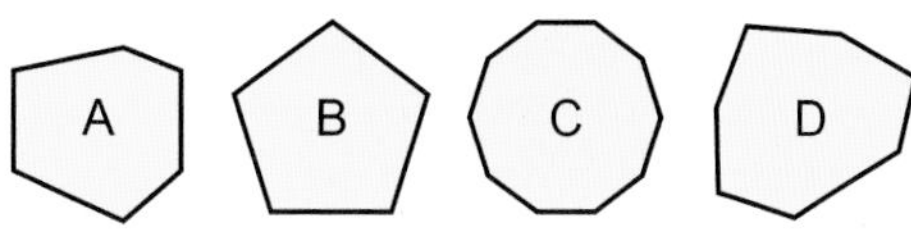

6 Find the size of the angles marked with letters.
Give reasons for your answers.

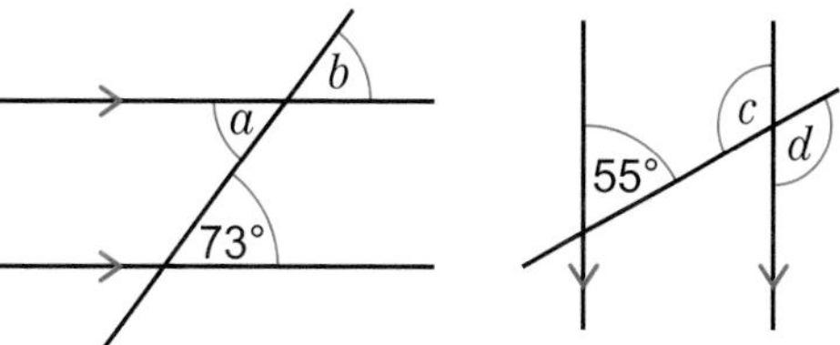

7
- a Sketch an isosceles triangle PQR and draw on any lines of symmetry.
- b Use your diagram to show that two angles are equal.

8 Work out the size of

a angle DAB b $\angle AZY$ c $M\hat{N}K$

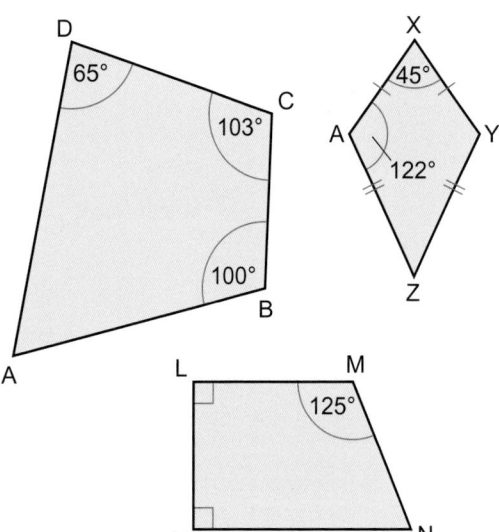

Algebraic fluency

9 $a = 3$ and $b = 11$. Work out

a $x = a^2 + b^2$ b $x = b^2 - a^2$

10 $x = 8$ and $y = 24$. Work out $\frac{x}{y}$.

Give your answer as a fraction in its simplest form.

11 Make y the subject of each formula.

a $2y = x$ b $\frac{y}{3} = 2x$ c $\frac{x}{y} = 4$

12 Write and solve an equation to calculate the size of each angle in degrees.

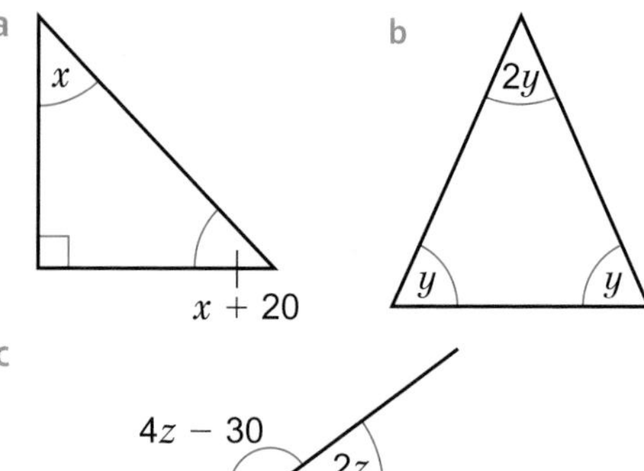

c

$4z - 30$

$2z$

* Challenge

13 In an isosceles triangle, one angle is twice the size of one of the other angles. Work out the value of each angle.

5.1 Angle properties of triangles and quadrilaterals

Objectives

- Derive and use the sum of angles in a triangle and in a quadrilateral.
- Derive and use the fact that the exterior angle of a triangle is equal to the sum of the two opposite interior angles.

Did you know?

The angle at which you hit a tennis ball affects its trajectory.

Fluency

Name these shapes.

a 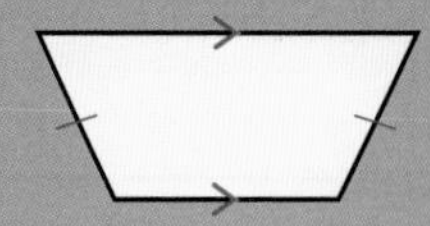

b 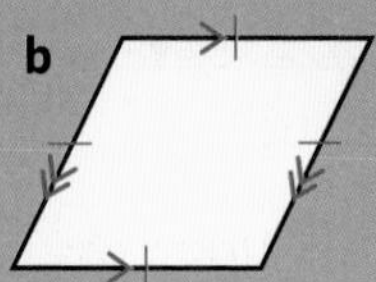

c 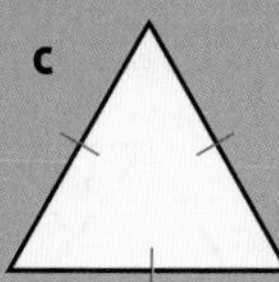

d 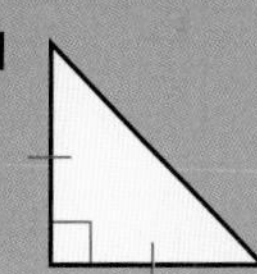

Warm up

1 What is the size of any angle in an equilateral triangle?

2 An isosceles triangle has one angle of 130°.
What are the sizes of the other two angles?

3 Work out the size of angle a. Give reasons for your answer.

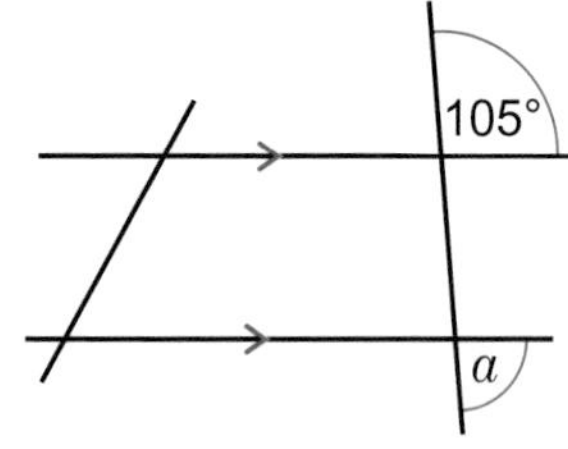

Questions in this unit are targeted at the steps indicated.

4 ABC and CDE are straight lines. AE is parallel to BD.
Work out the size of

a $A\hat{B}D$ b $B\hat{D}C$
c $A\hat{E}C$ d $A\hat{C}E$

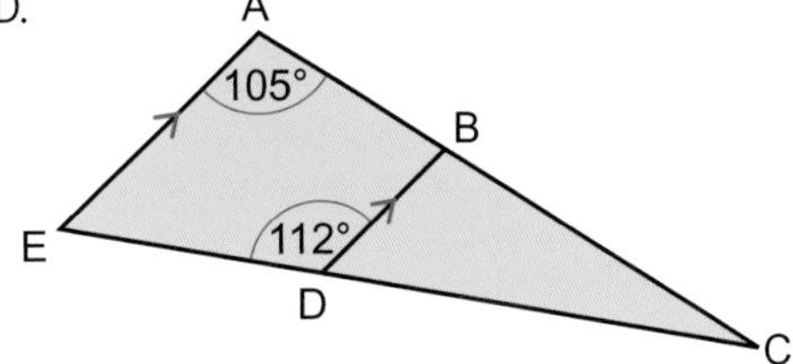

5 **Reasoning** ABCD is a parallelogram.

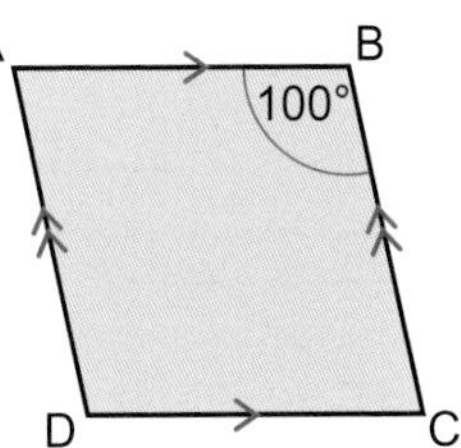

Q5b hint Line BA is parallel to CD.
Line CB is parallel to DA.

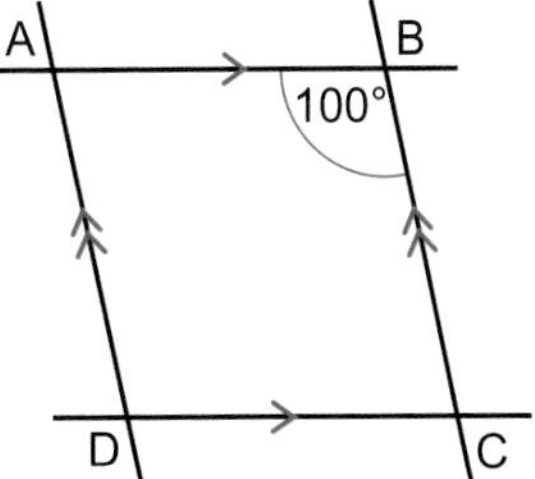

a Copy the parallelogram and extend each side.
b Work out the other angles in the parallelogram.
c What do you notice about the opposite angles?
d Repeat with different parallelograms.
Is your observation in part **c** still true?
e **Reflect** What property of parallelograms have you shown?

6 Triangle ABC is shown. DE is parallel to AB.

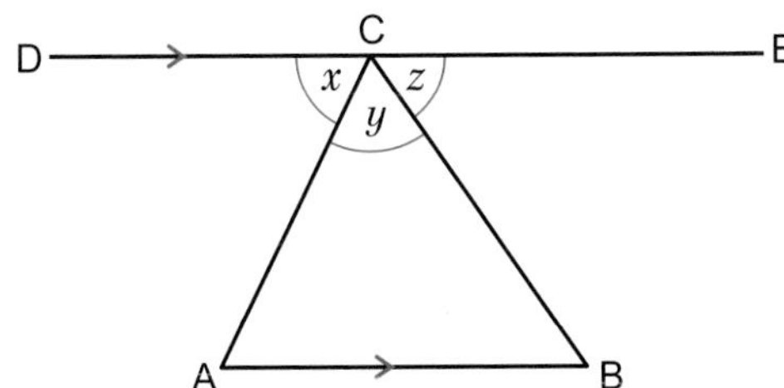

a What is the value of $x + y + z$? Give a reason for your answer.
b Copy the diagram. Mark on the size of each of these angles in terms of x, y and z.
i angle CAB ii angle ABC
Give reasons for your answers.
c Use your answer to part **a** to derive the sum of angles in a triangle.

Q6c hint
$x + y + z =$ ____°
The angle sum of a triangle is ____°

7 **Exam-style question**

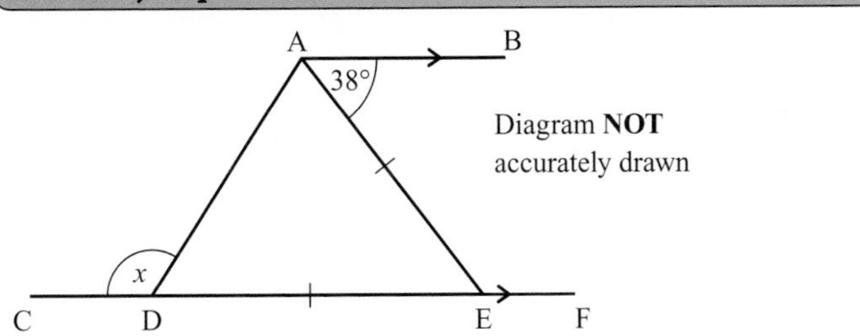

CDEF is a straight line.
AB is parallel to CF. DE = AE.
Calculate the size of the angle marked x.
You must give reasons for your answer. **(4 marks)**

Exam hint
Mark on the diagram the size of any angles that you know.
Start by using the rules of parallel lines to find another angle on the diagram. Then find which two angles are equal in the isosceles triangle by looking to see which two sides are equal.

8 **Communication / Reasoning**
In this diagram a diagonal divides the quadrilateral into two triangles.
Use the diagram to prove that the angle sum of a quadrilateral is 360°.

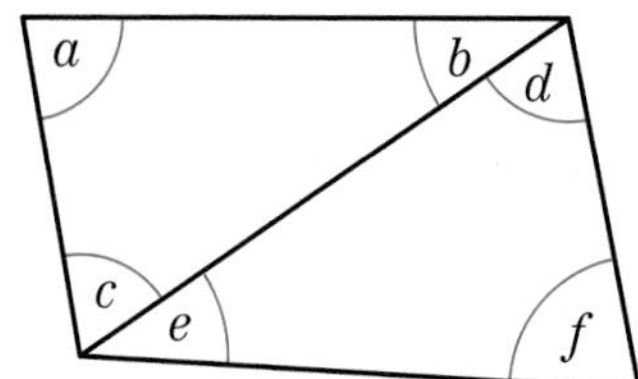

Q8 hint Begin
$a + b + c =$ _____°

Key point 1

When one side of a triangle is extended at the vertex:
- the angle marked x is called the **interior angle**.
- the angle marked y is called the **exterior angle**.

$x + y = 180°$ (angles on a straight line)

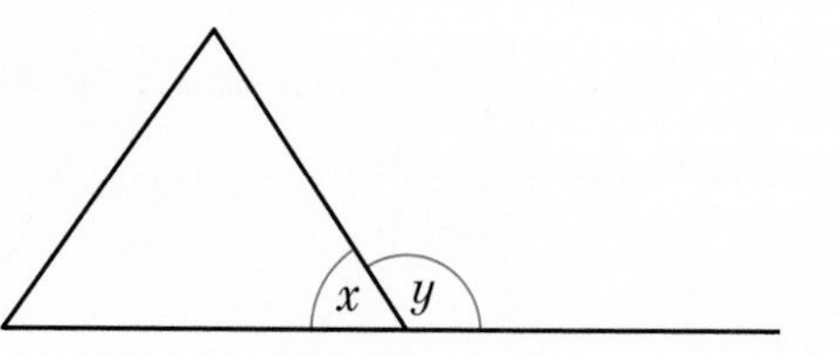

9 Work out the size of each angle marked with a letter.

a
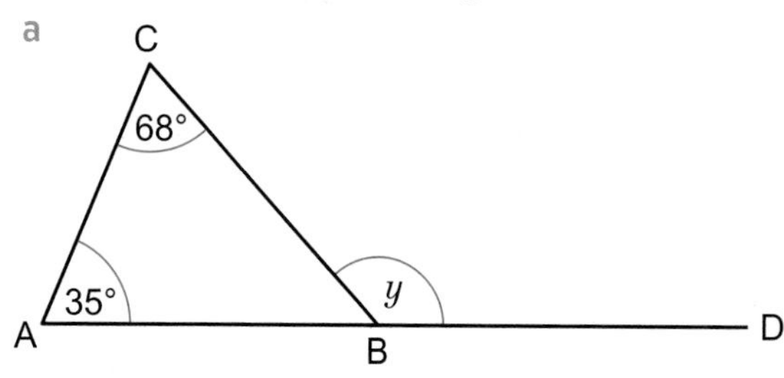

b
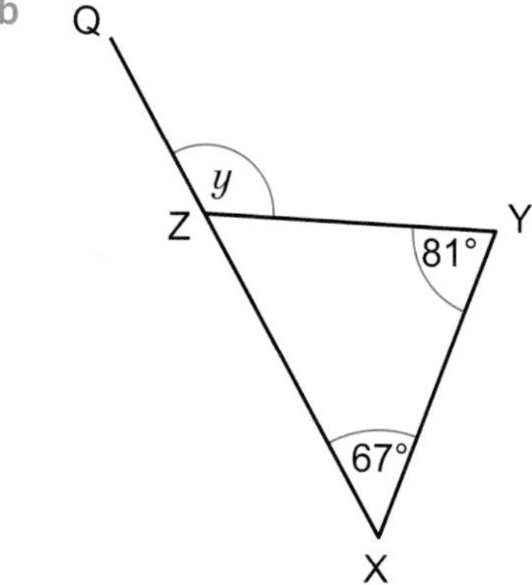

c
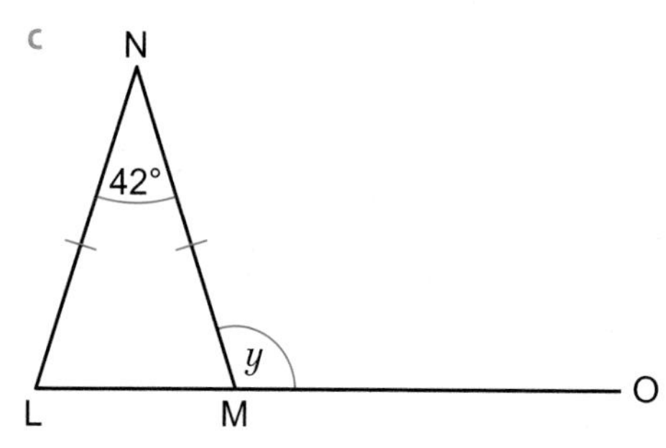

d
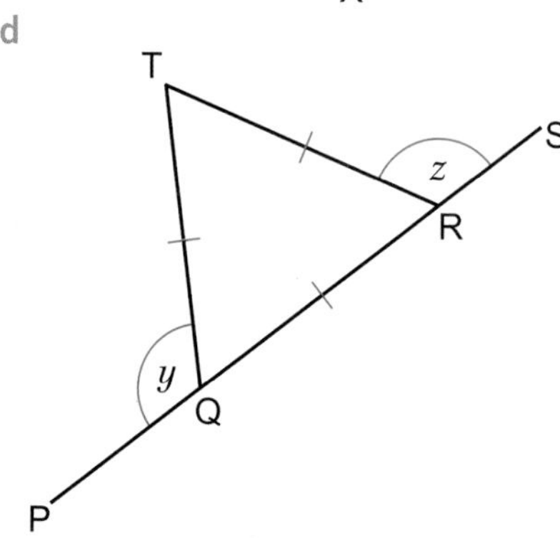

Discussion What do you notice about the relationship between the exterior angle of a triangle and the interior angles at the other two vertices?

Key point 2

The exterior angle of a triangle is equal to the sum of the interior angles at the other two vertices.

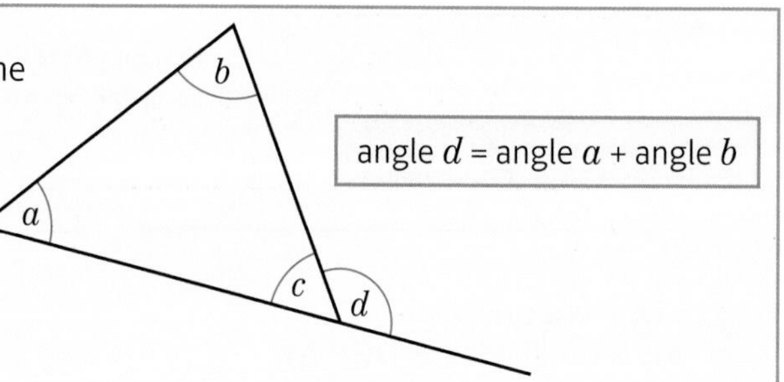

angle d = angle a + angle b

Example 1

Work out the size of angle ABC. Give reasons for your answer.

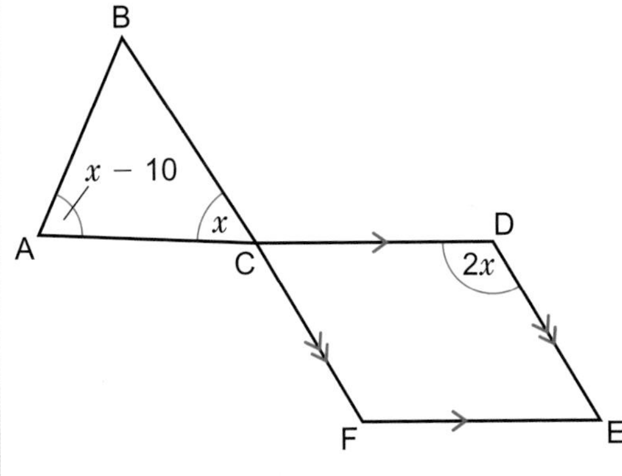

$\angle DCF = x$ (vertically opposite angles are equal)

Write a reason every time you write an angle.

$\angle CDE + \angle DCF = 180°$ (co-interior angles add up to 180°)

$x + 2x = 180°$ — Form an equation using the angle property.

$3x = 180°$ — Collect like terms.

$x = 60°$ — Solve the equation to find x.

$\angle BCA = 60°$

$\angle BAC = x - 10 = 60 - 10 = 50°$ — Substitute the value of x to find angle BAC.

$\angle ABC = 180° - (60 + 50)°$ (angles in a triangle add up to 180°)

$= 180° - 110°$

$= 70°$

10 Reasoning
ABCD is an isosceles trapezium.
BCE is an isosceles triangle.
DCEF is a straight line.
Angle BEF = 132°.
Work out the size of angle DAB.
Give reasons for your working.

Q10 hint Start with:
angle BCE + angle CBE = ___°
(exterior angle is equal to the sum of the____)
angle BCE = angle CBE (_____)
Write any angles you work out on a sketch of the diagram.

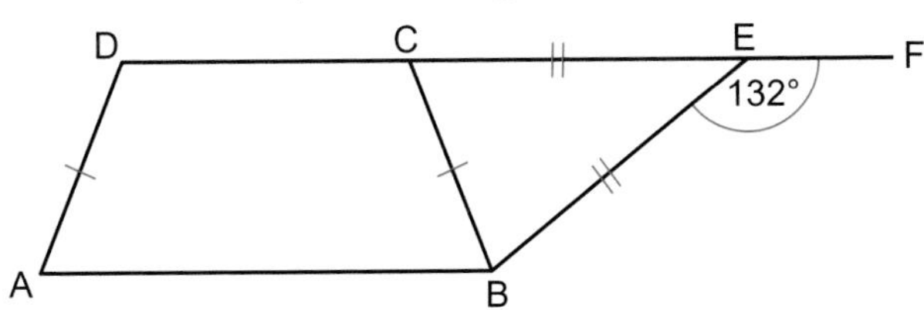

Q10 communication hint
To 'give reasons for your working' write the angle property you use for every angle you find.

11 Reasoning Work out the size of angle ACB. Give reasons for your working.

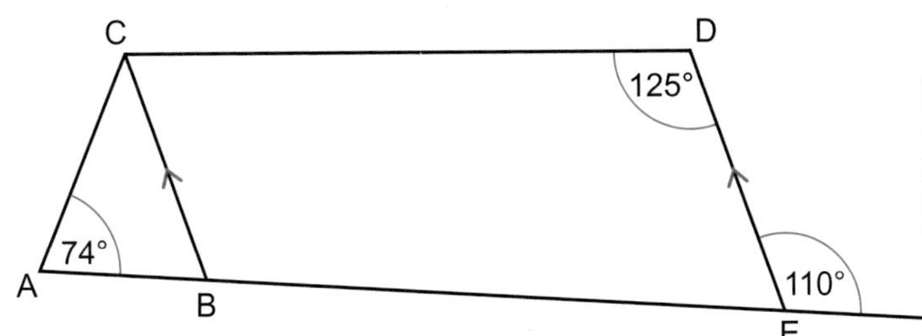

Q11 hint Use the properties on the diagram:
BC is parallel to DE.

12 Work out the size of angle CBD.
Give reasons for your working.

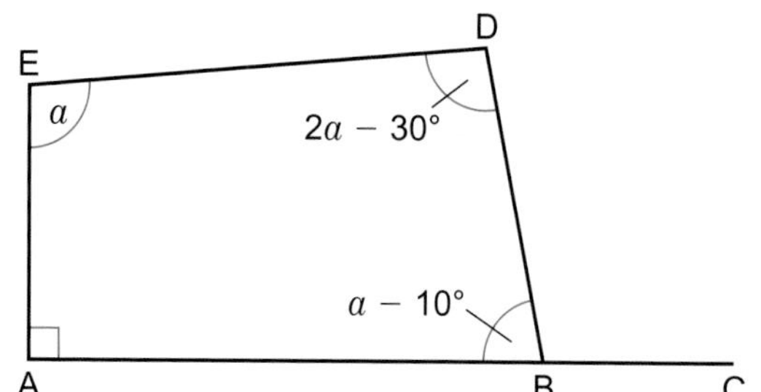

Q12 hint Use the fact that angles in a quadrilateral add up to 360° to write an equation.

5.2 Interior angles of a polygon

Objectives

- Calculate the sum of the interior angles of a polygon.
- Use the interior angles of polygons to solve problems.

Why learn this?

Polygons are used in the construction of buildings and bridges due to their strength and beauty.

Fluency

- Name these polygons.

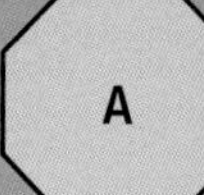

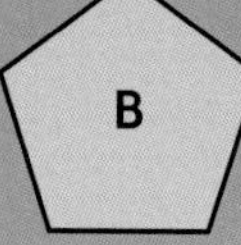

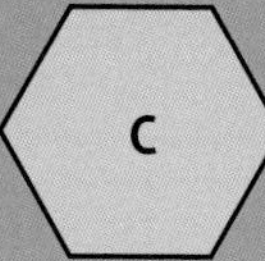

- What can you say about the sides and angles in a regular polygon?

Warm up

1 For each value of n, work out $(n - 2) \times 180$

a $n = 3$ b $n = 5$ c $n = 7$ d $n = 8$

2 Work out the size of each angle marked with a letter.

a

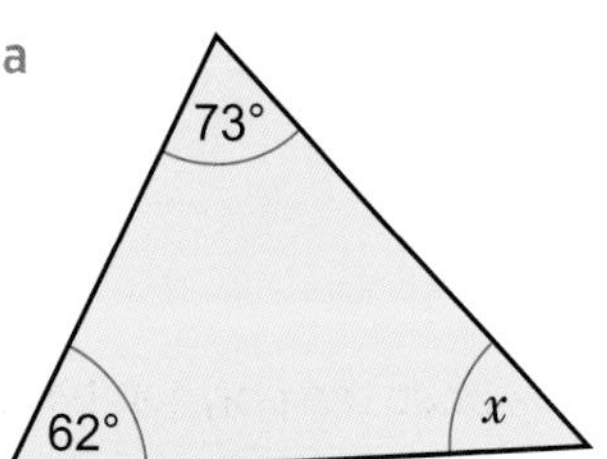

b

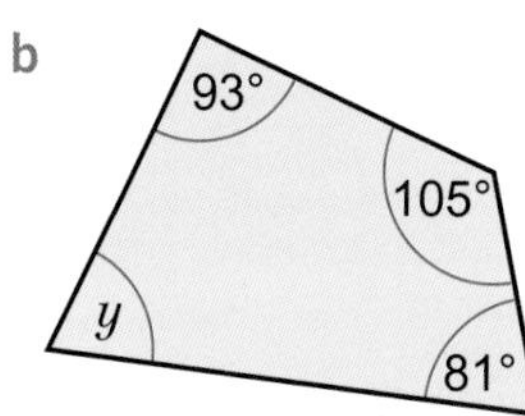

c

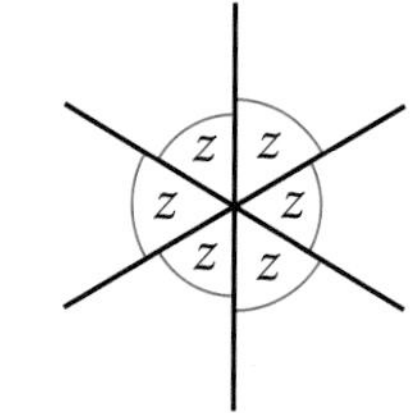

Example 2

Work out the sum of the interior angles of a pentagon.

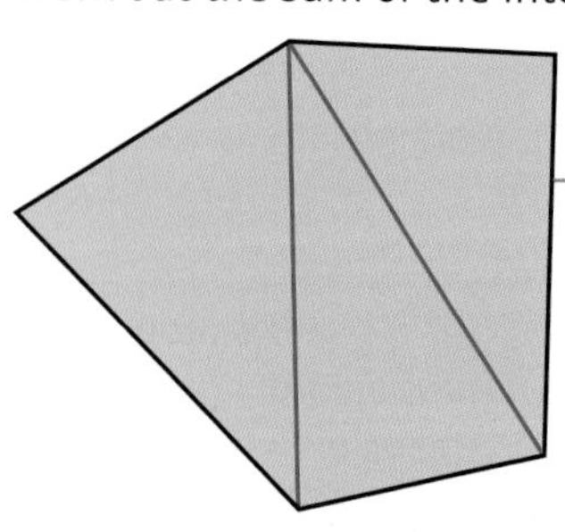

A pentagon has five sides.
Sketch a pentagon.
Draw in the diagonals from one vertex to all the other vertices.

The pentagon has been divided into 3 triangles.
The angle sum of each triangle is 180°.

Sum of the interior angles of a pentagon = 3 × 180° = 540°

ActiveLearn Homework, practice and support: Higher 5.2

3 Work out the sum of the interior angles of a hexagon.
Discussion Does it matter if the hexagon is regular or irregular?

> **Q3 hint** Use the same method as Example 2.

4 **Reasoning** Copy and complete the table.

Polygon	Number of sides (n)	Number of triangles formed	Sum of interior angles
Triangle	3	1	180°
Quadrilateral	4		
Pentagon	5	3	540°
Hexagon	6		
Heptagon	7		

Discussion What do you think the sum of the angles in a 12-sided polygon (dodecagon) is?

Key point 3

The sum of the interior angles of a polygon with n sides = $(n - 2) \times 180°$

5 A regular polygon has 20 sides.
a Work out the sum of the interior angles of the polygon.
b Work out the size of the interior angle.

> **Q5a hint** Substitute into $(n - 2) \times 180°$.

6 **Reasoning** Work out the size of each interior angle of
a a regular pentagon
b a regular octagon
c a regular heptagon
d a regular polygon with 15 sides.

7 **Reasoning** Work out the size of each unknown interior angle.

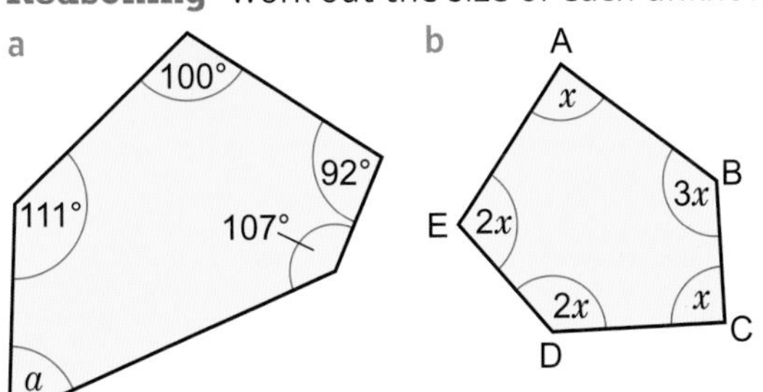

> **Q7a hint** First work out the sum of the interior angles for a 5-sided polygon.

> **Q7b hint** To find x solve
> $x + 3x + x + 2x + 2x =$ ____°
> Then work out *each* interior angle.

Example 3

The sum of the interior angles of a polygon is 1620°. How many sides does the polygon have?

$(n-2) \times 180° = 1620°$ — Form an equation using the sum of interior angles.

$n - 2 = \frac{1620}{180}$ — Divide both sides by 180.

$n - 2 = 9$

$n = 11$ — Add 2 to both sides.

8 The sum of the interior angles of a polygon is 3060°.
How many sides does the polygon have?

9 **Problem-solving** A regular pentagon is divided into 5 isosceles triangles.
Work out the size of
a angle x
b angle y
c angle z.

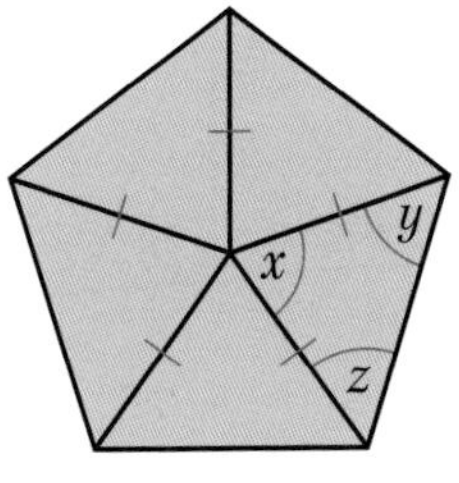

10 **Reasoning** **Q9** shows a pentagon made from isosceles triangles.
What polygon can you make from equilateral triangles?

Reflect Besides triangles, which other regular polygons can fit together like this to create a pattern without leaving any gaps? Explain.

11 **Exam-style question**

The diagram shows a regular hexagon and a regular octagon.

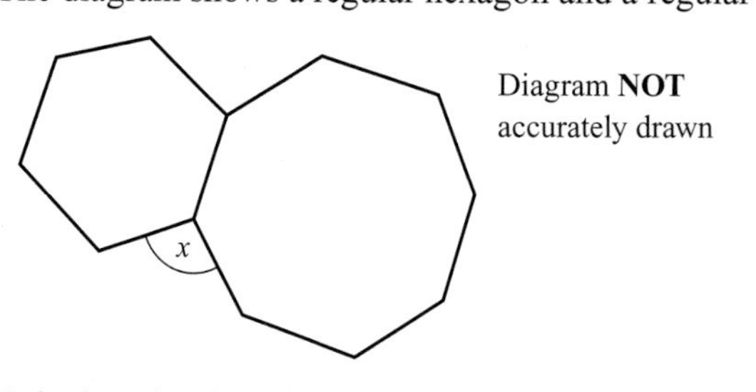

Calculate the size of the angle marked x.
You must show all your working. **(4 marks)**

June 2012, Q13, 1MA0/1H

Exam hint
Mark the angles on the diagram as you work them out. Write the angle property you use for each one.

5.3 Exterior angles of a polygon

Objectives

- Know the sum of the exterior angles of a polygon.
- Use the angles of polygons to solve problems.

Did you know?

Polygons have been used for thousands of years to create decorative patterns called mosaics.

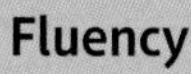

Fluency

Work out the size of the unknown angles.

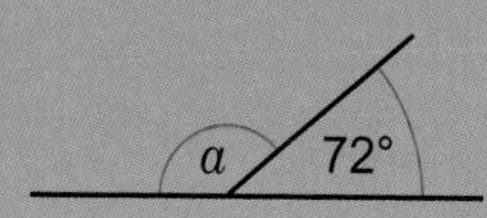

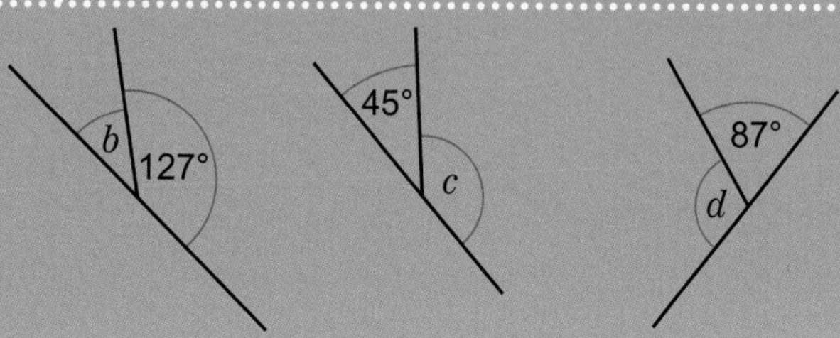

Warm up

1 Work out the sum of the interior angles of
a a heptagon b a pentagon c a decagon.

2 For each triangle work out
a the sizes of angles a, b and c
b the value of $a + b + c$.

i
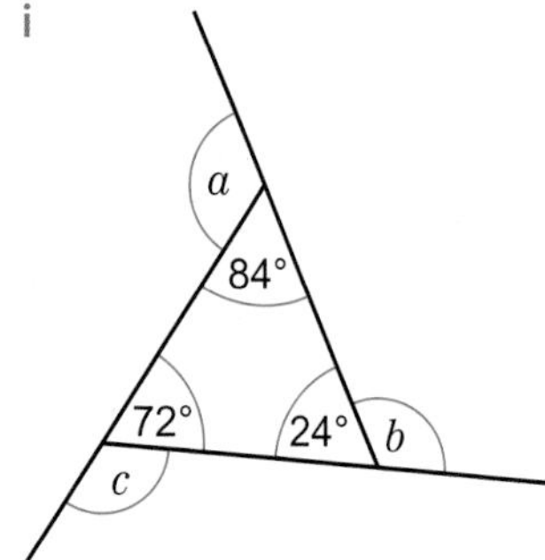

ii
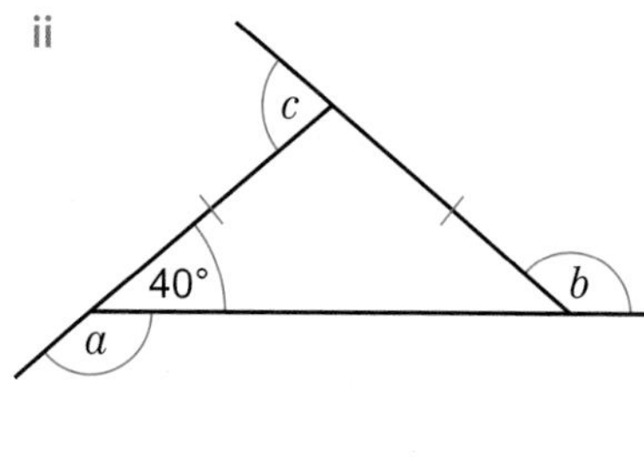

iii
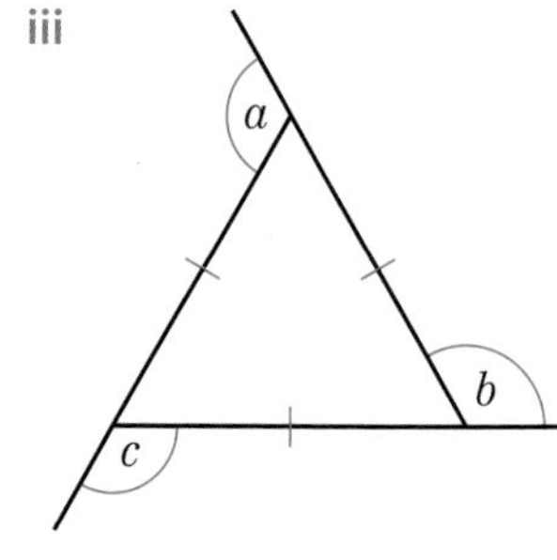

3 For each quadrilateral work out

a the sizes of angles a, b, c and d

b the value of $a + b + c + d$.

i

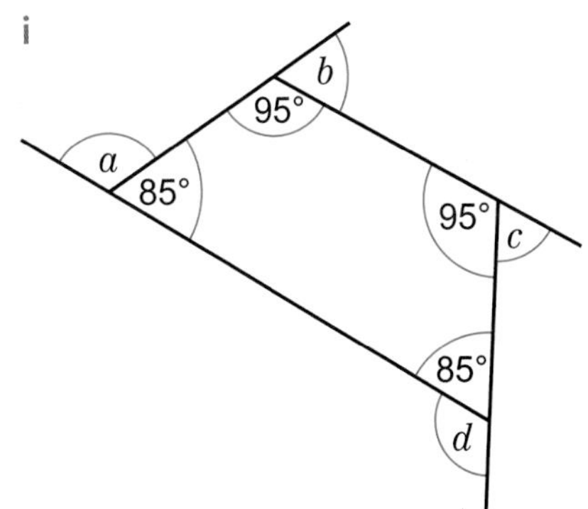

ii

Key point 4

When one side of a polygon is extended at a vertex:

- angle x is the interior angle
- angle y is the exterior angle.

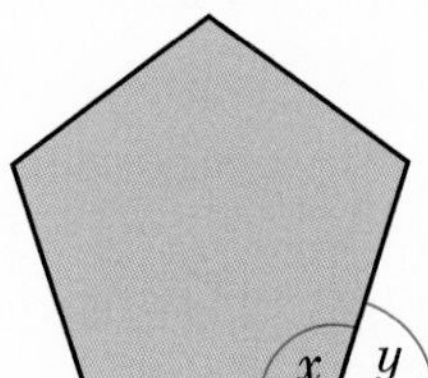

interior angle + exterior angle = 180°
(angles on a straight line add up to 180°)

4 **Reasoning** A pentagon and a hexagon are shown.

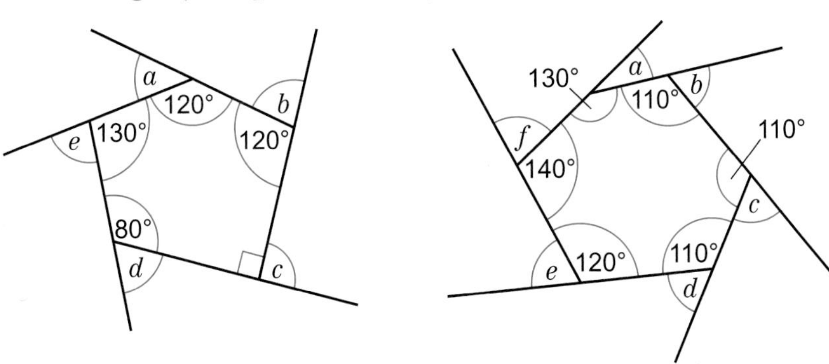

a Work out the sizes of the angles marked with letters.

b Work out the sum of the exterior angles.

c What do you notice about the sum of the exterior angles?

Discussion Does it matter if the polygon is regular or irregular?

Q4a hint For the pentagon work out the value of $a + b + c + d + e$.

Key point 5

The sum of the exterior angles of a polygon is always 360°.

In a regular polygon all the angles are the same size, so exterior angle = $\frac{360°}{\text{number of sides}}$

5 Work out the sizes of an exterior angle of a regular hexagon.

6 Work out the sizes of the angles marked with letters.
The first one has been started for you.

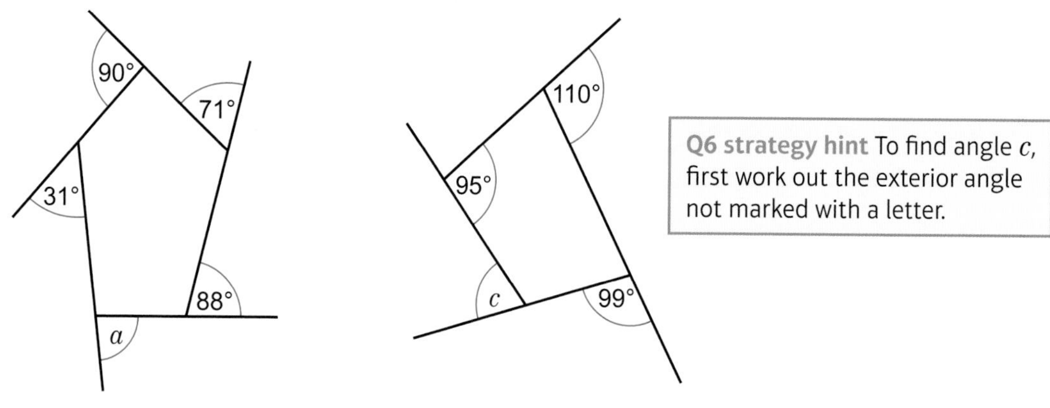

Q6 strategy hint To find angle c, first work out the exterior angle not marked with a letter.

$a + 31° + 90° + 71° + 88° = 360°$

7 **Reasoning** The sizes of seven of the exterior angles of an octagon are 42°, 110°, 13°, 67°, 55°, 11° and 53°.
Work out the size of each interior angle.

Q7 hint Work out all eight interior angles.

8 **Reasoning** Work out the sizes of each unknown exterior angle in this polygon.

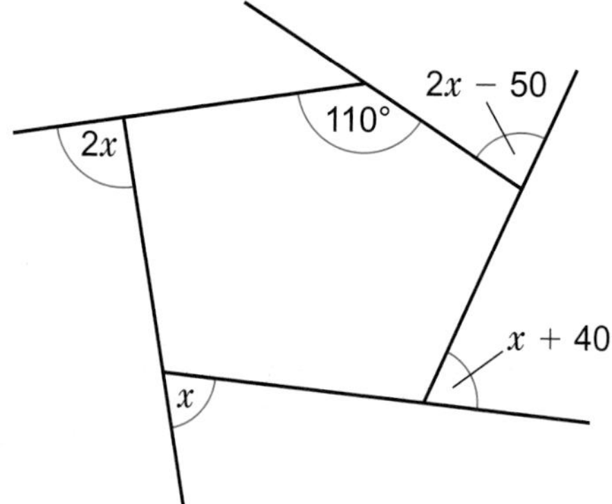

Example 4

Each interior angle of a regular polygon is 140°.
How many sides does the polygon have?

Exterior angle = 180° − 140° = 40°

Use interior angle + exterior angle = 180° to work out the size of an exterior angle.

Number of sides = $\frac{360°}{40°} = 9$

For a regular polygon, exterior angle = $\frac{360°}{\text{number of sides}}$
so number of sides = $\frac{360°}{\text{exterior angle}}$

9 How many sides does a regular polygon have if its exterior angle is
a 10° b 72° c 20°?

10 How many sides does a regular polygon have if its interior angle is
a 120° b 150° c 140°?

11 **Reflect** Can the exterior angle of a regular polygon be 70°?
Explain.

12 **Problem-solving** One side of a regular hexagon ABCDEF forms the side of a regular polygon with n sides.

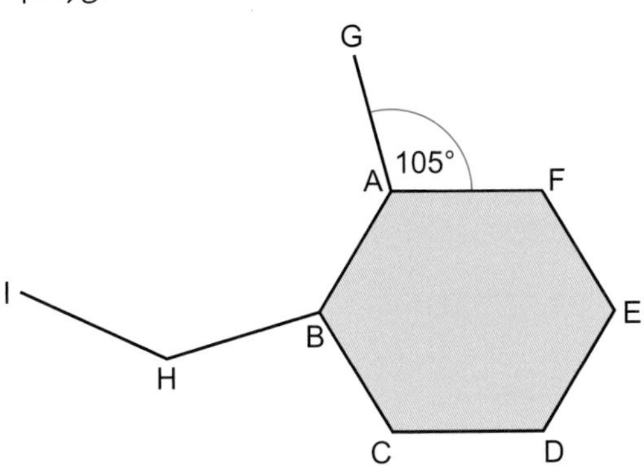

Angle GAF = 105°.
Work out the value of n.

13 **Problem-solving** The exterior angle of a regular polygon is half the size of its interior angle.
How many sides does the polygon have?

Q13 hint Work out the size of the interior angle first.

5.4 Pythagoras' theorem 1

Objectives

- Calculate the length of the hypotenuse in a right-angled triangle.
- Solve problems using Pythagoras' theorem.

Did you know?

Mesopotamian, Chinese and Indian mathematicians all independently discovered Pythagoras' theorem. However, the Greek Pythagoras ended up getting all the credit.

Fluency

- Calculate **a** 9^2 **b** 12^2 **c** 20^2
- Work out the area of a square of side length 11 cm.

Warm up

1 Work out
a $\sqrt{100}$
b $\sqrt{25}$
c $\sqrt{9}$
d $\sqrt{49}$

2 $a = 4.5$ and $b = 6.2$
Work out
a $a^2 + b^2$
b $\sqrt{a^2 + b^2}$
Give your answers correct to 1 decimal place.

3 Find the positive solution of each equation.
Give your answers correct to 3 significant figures.
a $x^2 = 12$
b $x^2 = 12^2 + 8^2$

Key point 6

In a right-angled triangle the longest side called the hypotenuse.
Pythagoras' theorem states that, in a right-angled triangle, the square of the hypotenuse is equal to the sum of the squares of the other two sides.

$c^2 = a^2 + b^2$

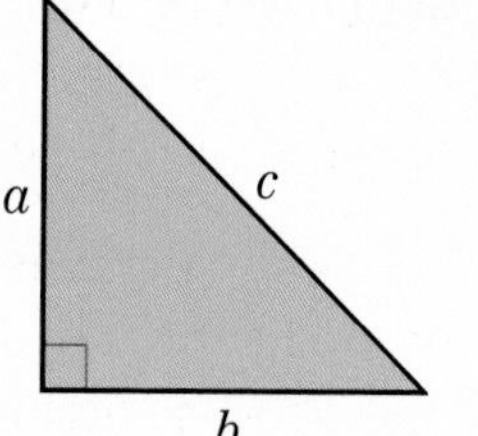

Example 5

Calculate the length of the hypotenuse.
Give your answer correct to 2 significant figures.

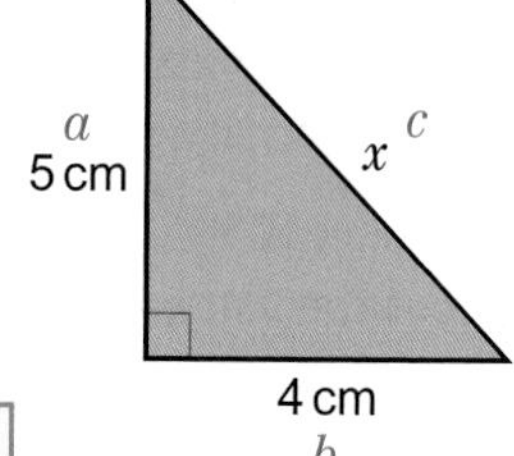

$a = 5, b = 4, c = x$

$c^2 = a^2 + b^2$

$x^2 = 5^2 + 4^2$ — Sketch the triangle. Label the hypotenuse c and the other two sides a and b. Substitute the values of a, b and c into the formula for Pythagoras' theorem.

$x^2 = 25 + 16$

$x^2 = 41$

$x = \sqrt{41}$ — Use a calculator to find the square root.

$x = 6.4031...$

$x = 6.4$ cm (to 2 s.f.) — Round your answer to 2 significant figures and put the units in your answer.

Discussion Does it matter which side is a and which is b?

4 **Reflect** Dawn and Eleri are answering the same question.
Parts of their working are shown.

Dawn's working	Eleri's working
$x^2 = 33.846$	$x^2 = 33.846$
$x = \sqrt{33.8}$	$x = \sqrt{33.846}$
$x =$ _____	$x =$ _____

a Which working is the more accurate? Why?
b Will the accuracy of the working affect the answer?

5 Calculate the length of the hypotenuse in each triangle.
Give your answers correct to 2 significant figures.

a
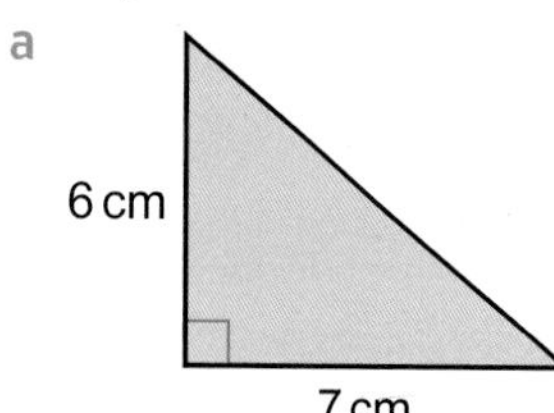

b
9 cm
4 cm

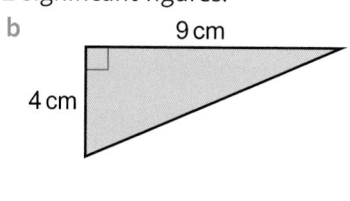

c
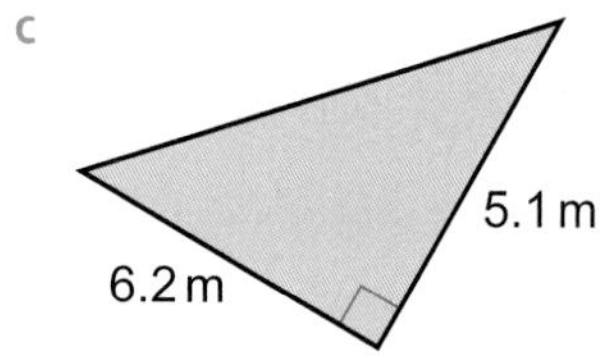

Q5 hint Do not round *before* taking the square root. Use all the figures on your calculator display.

6 ABC is a right-angled triangle.
Calculate the length of AC.
Give your answer correct to 3 significant figures.

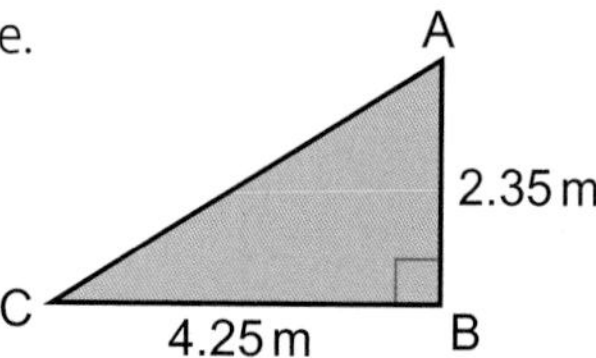

7 Calculate the length of the diagonal of a rectangle measuring 5 cm by 3.5 cm.
Discussion What is a sensible rounding for the answer to this question? Why?

> **Q7 strategy hint** Sketch a right-angled triangle and label it. State the degree of accuracy after your answer e.g. 2 s.f. or 1 d.p.

8 **Real** A zip wire runs from a vertical height of 20 feet. The total horizontal distance travelled is 32 feet. What is the length of the zip wire?

9 **Problem-solving** A ship sails 5 miles North and then 8.1 miles East.
It then returns directly to its starting point.
What is the total distance the ship travels?

> **Q9 hint** The question is asking for the *total* distance.

10 **Real / Problem-solving** A roof truss is made of wood.
The vertical support bisects the horizontal span.
Work out the total length of wood needed to make the truss.

> **Q10 communication hint** Bisect means divide in half.

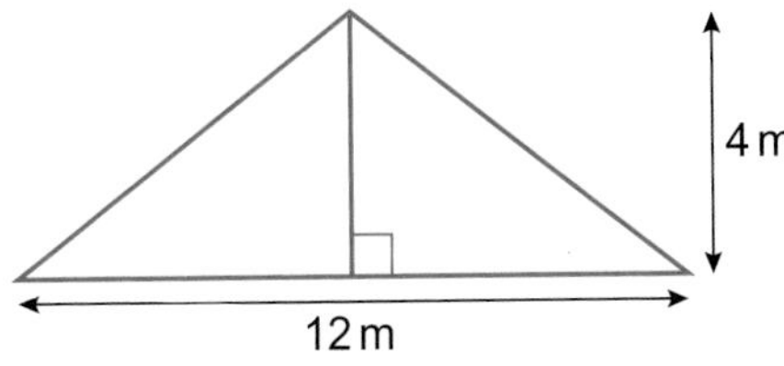

11 **Exam-style question**

ABCD is a trapezium.
AB = 6 cm
BC = 8 cm
AD = 12 cm

Calculate the perimeter of ABCD.
Give your answer correct to 1 decimal place. **(3 marks)**

> **Exam hint**
> Divide the shape into a rectangle and a right-angled triangle and then fill in the measurements of the sides of the triangle.

Key point 7

A triangle with sides a, b and c, where c is the longest side, is right-angled *only* if $c^2 = a^2 + b^2$.

12 **Reasoning** Can a right-angled triangle have sides of length
a 4 cm, 5 cm, 8 cm
b 9 cm, 12 cm, 15 cm
c 5 cm, 12 cm, 13 cm?
Explain your answers.

> **Q12a hint** If the triangle is right-angled, the longest side will be the hypotenuse.

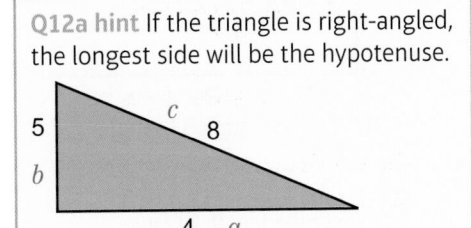

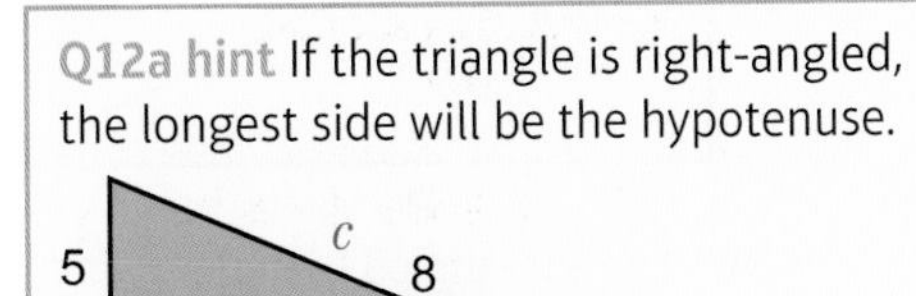

5.5 Pythagoras' theorem 2

Objectives

- Calculate the length of a shorter side in a right-angled triangle.
- Solve problems using Pythagoras' theorem.

Why learn this?

Pythagoras' theorem is used to calculate the distances travelled by aircraft.

Fluency

- Find **a** $\sqrt{16}$ **b** $\sqrt{100}$ **c** $\sqrt{25}$
- Which of these have an exact answer? $\sqrt{9}$, $\sqrt{5}$, $\sqrt{64}$, $\sqrt{37}$

Warm up

1 Calculate the length of the hypotenuse in this right-angled triangle. Give your answer correct to 1 decimal place.

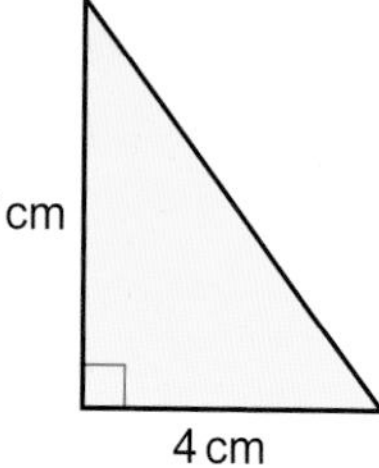

2 Solve these equations.

a $5^2 = a^2 + 4^2$ b $10^2 = 6^2 + b^2$ c $5^2 + c^2 = 13^2$

Example 6

Calculate the length m in this right-angled triangle. Give your answer correct to 3 significant figures.

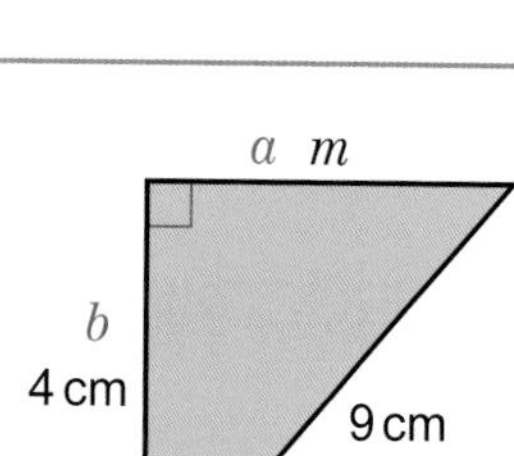

$c^2 = a^2 + b^2$

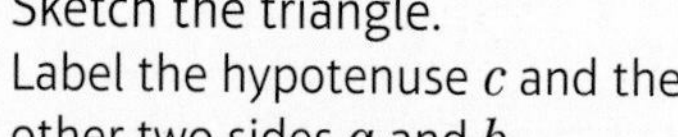

Sketch the triangle. Label the hypotenuse c and the other two sides a and b.

$9^2 = m^2 + 4^2$ — Substitute the values of a, b and c into Pythagoras' theorem.

$81 = m^2 + 16$

$m^2 = 81 - 16 = 65$ — Solve the equation.

$m = \sqrt{65}$ — Use a calculator to find the square root.

$m = 8.0622...$

$m = 8.06$ cm (to 3 s.f.) — Give your answer correct to 3 s.f. and include the units.

3 **Exam-style question**

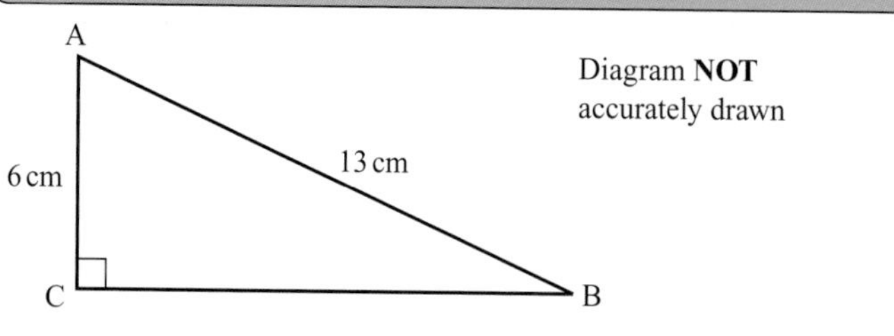

ABC is a right-angled triangle.
AC = 6 cm
AB = 13 cm
Calculate the length of BC.
Give your answer correct to 3 significant figures. **(3 marks)**

March 2013, Q13, 1MA0/2H

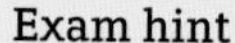

Exam hint

Label a, b and c on the diagram. Write down the theorem you use before you substitute in.

4 **Modelling / Real** A ladder of length 5 m leans against a vertical wall.
The foot of the ladder is 4.2 m from the base of the wall.
How far is the top of the ladder from the ground?

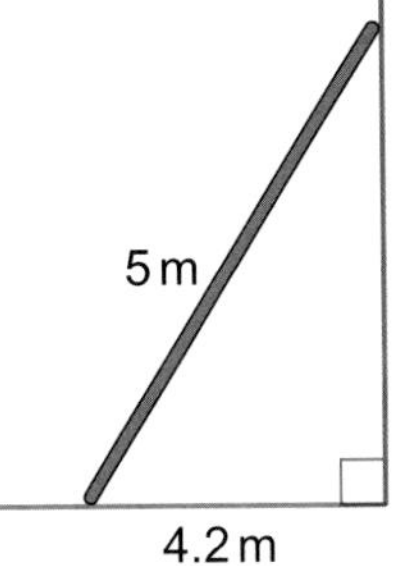

Q4 hint State the degree of accuracy after your answer.

5 **Modelling / Real** A ramp is to be used to go up one step.

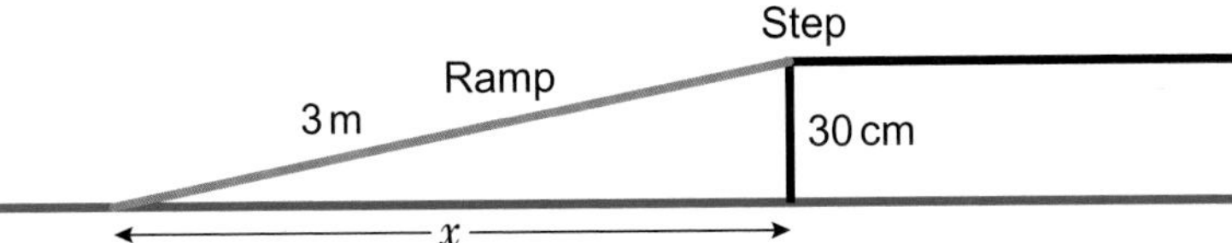

The ramp is 3 m long.
The step is 30 cm high.
How far away from the step (x) does the ramp start?
Give your answer in metres, to the nearest centimetre.

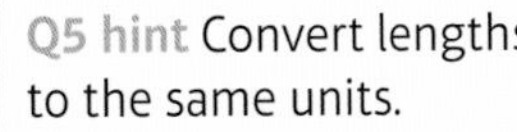

Q5 hint Convert lengths to the same units.

6 Calculate the vertical height of trapezium ABCD.
Give your answer in centimetres, to the nearest millimetre.

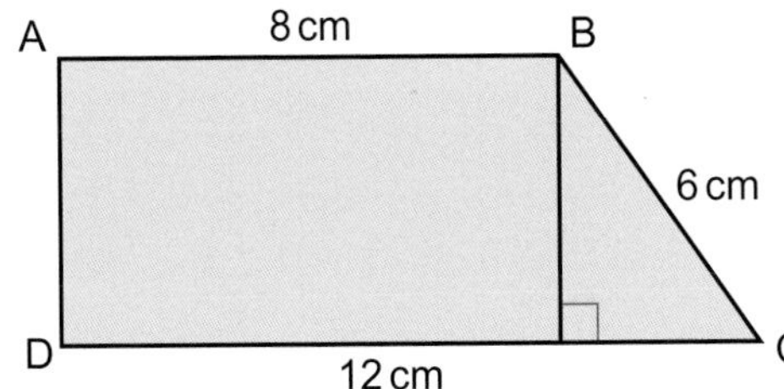

7 **Problem-solving** a Calculate the length of the side of the largest square that *fits inside* a 12 cm diameter circle.
b Work out the length of the side of the smallest square that *surrounds* a 12 cm diameter circle.

8 Work out the length of the unknown side in each right-angled triangle.
Give your answers in surd form.

a
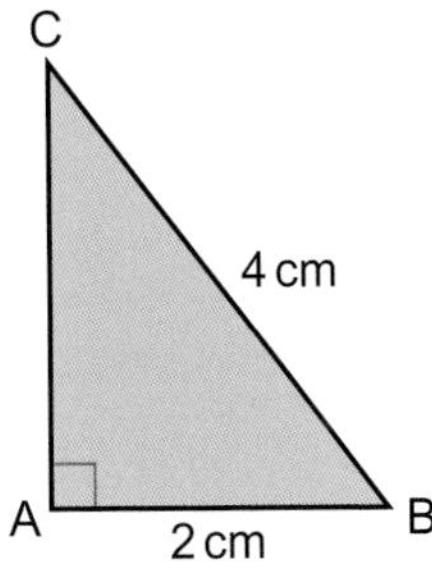

b
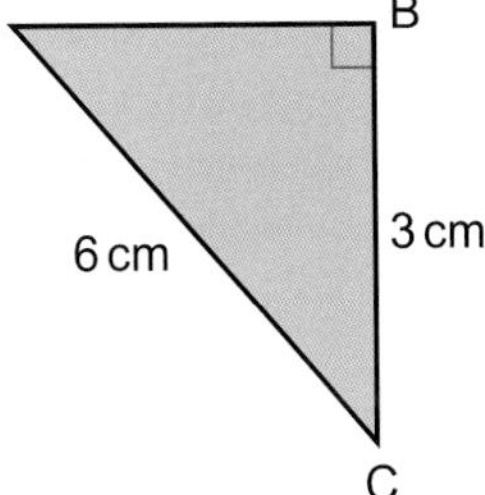

Q8a hint Simplify the surd so your answer looks like this:
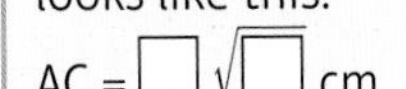
AC = $\square\sqrt{\square}$ cm

c
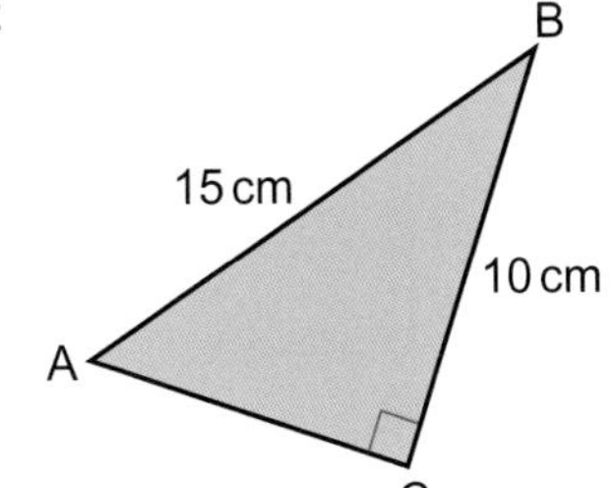

Q8 communication hint Giving an answer in 'surd form' means *don't* work out the square root.

9 **Problem-solving** ABC is an equilateral triangle.
D is the midpoint of BC.
Work out the height of the triangle.
Give your answer in surd form.

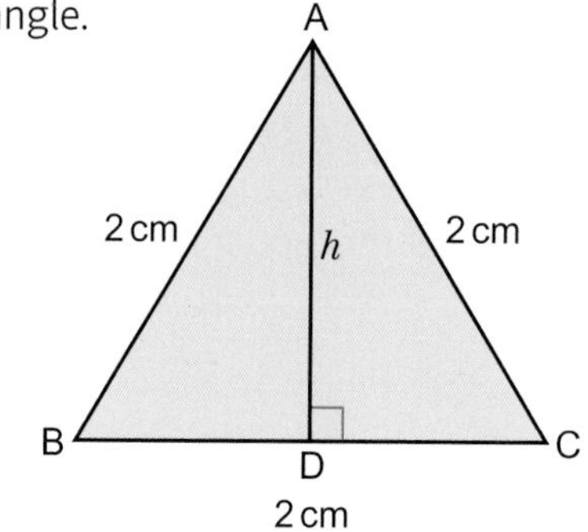

10 **Problem-solving**
Work out
a the length of AD
b the length of CD
c the perimeter of the triangle.
Give all your answers to 3 s.f.

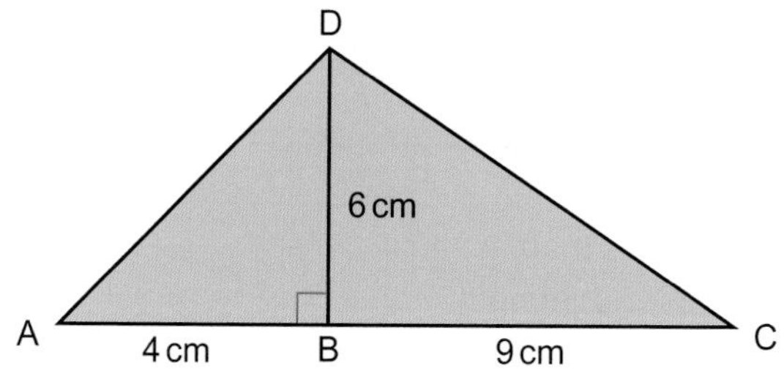

5.6 Trigonometry 1

Objectives

- Use trigonometric ratios to find lengths in a right-angled triangle.
- Use trigonometric ratios to solve problems.

Did you know?

Trigonometry was used to map the British Isles.

Fluency

- Convert each fraction to a decimal. $\frac{1}{2}, \frac{1}{4}, \frac{1}{5}, \frac{1}{10}, \frac{2}{5}$
- Name the hypotenuse in each triangle.

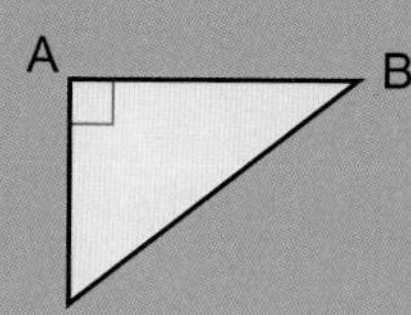

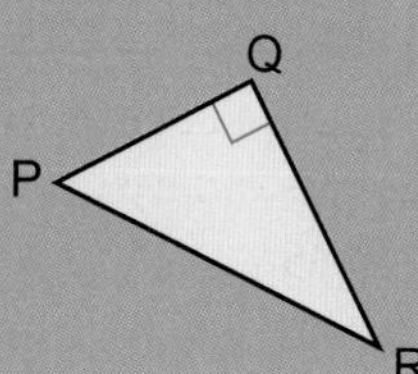

Warm up

1 Work out the size of each unknown angle.

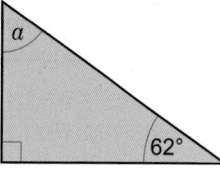

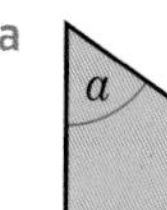

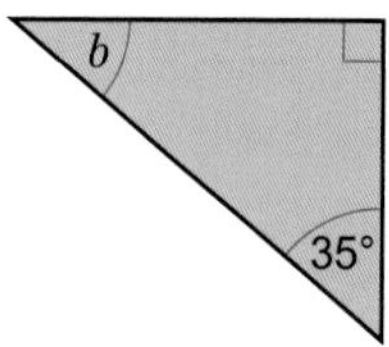

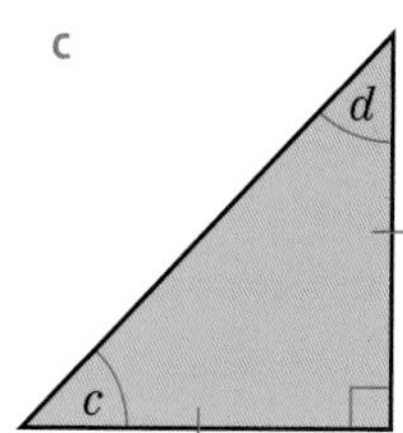

2 Solve these equations, correct to 2 decimal places where necessary.

a $\frac{x}{5} = 4$ b $\frac{10}{x} = 5$ c $3.5 = \frac{x}{2.1}$ d $9.5 = \frac{10}{x}$

Key point 8

The side opposite the right angle is called the **hypotenuse**.
The side opposite the angle θ is called the **opposite**.
The side next to the angle θ is called the **adjacent**.

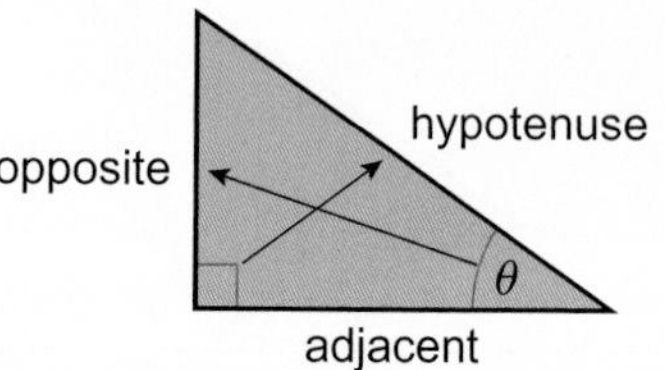

3 **Reasoning** Draw triangle ABC accurately using a ruler and protractor.
Angle A = 90°, angle B = 30° and AB = 5 cm.

a Label the **hyp**otenuse (**hyp**), **opp**osite side (**opp**) and **adj**acent side (**adj**).

b Measure each unknown side to the nearest millimetre.

c Write the fraction

i $\frac{\text{opposite}}{\text{hypotenuse}}$ ii $\frac{\text{adjacent}}{\text{hypotenuse}}$ iii $\frac{\text{opposite}}{\text{adjacent}}$

Convert each fraction to a decimal.
Give your answer correct to 1 decimal place.

d Repeat parts **a** to **c** for triangle ABC with

i angle A = 90°, angle B = 30° and AB = 7 cm

ii angle A = 90°, angle B = 30° and AB = 8 cm

Discussion What do you notice about the ratios of sides in a triangle with angles 30°, 60° and 90°?

Key point 9

In a right-angled triangle:

The **sine** of angle θ is the ratio of the opposite side to the hypotenuse, $\sin\theta = \frac{\text{opp}}{\text{hyp}}$

The **cosine** of angle θ is the ratio of the adjacent side to the hypotenuse, $\cos\theta = \frac{\text{adj}}{\text{hyp}}$

The **tangent** of angle θ is the ratio of the opposite side to the adjacent side, $\tan\theta = \frac{\text{opp}}{\text{adj}}$

You can find the sine, cosine and tangent of an angle using the sin, cos, tan keys on your calculator.

4 Use your calculator to find, correct to 1 decimal place where necessary

a sin 35° b cos 17°

c tan 82° d cos 73°

e tan 12° f tan 49°

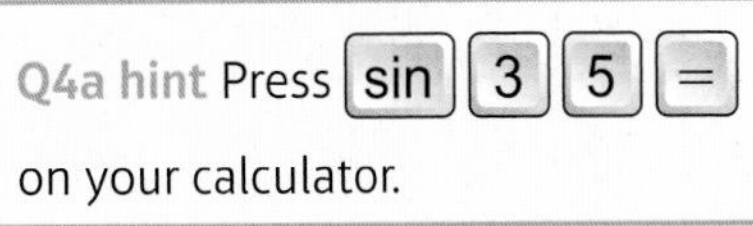

Example 7

Calculate the length of the side marked x.
Give your answer correct to 3 significant figures.

opp x hyp 10 cm 32° adj

$\sin\theta = \frac{\text{opp}}{\text{hyp}}$

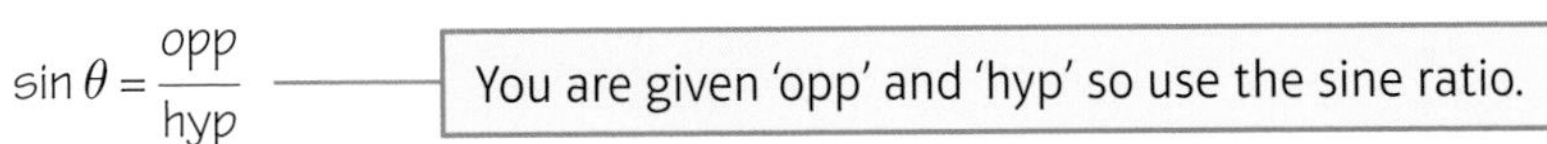

You are given 'opp' and 'hyp' so use the sine ratio.

$\sin 32° = \frac{x}{10}$

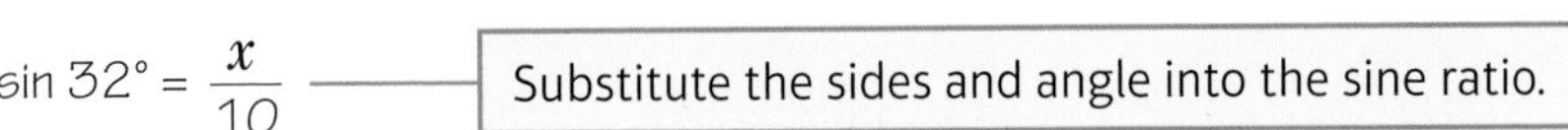

Substitute the sides and angle into the sine ratio.

$x = 10 \times \sin 32°$

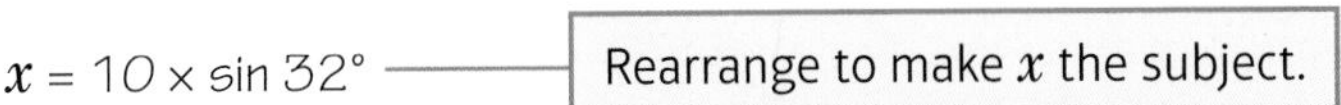

Rearrange to make x the subject.

$x = 5.2991\ldots$

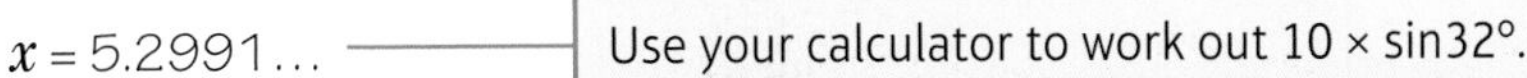

Use your calculator to work out 10 × sin32°.

$x = 5.30$ cm (to 3 s.f.) Round your answer to 3 s.f. and put in the units.

5 Calculate the length of the side marked x in each triangle. Give your answers correct to 3 significant figures.

Q5a hint Use $\sin\theta = \frac{\text{opp}}{\text{hyp}}$

Q5b hint Use $\tan\theta = \frac{\text{opp}}{\text{adj}}$

a

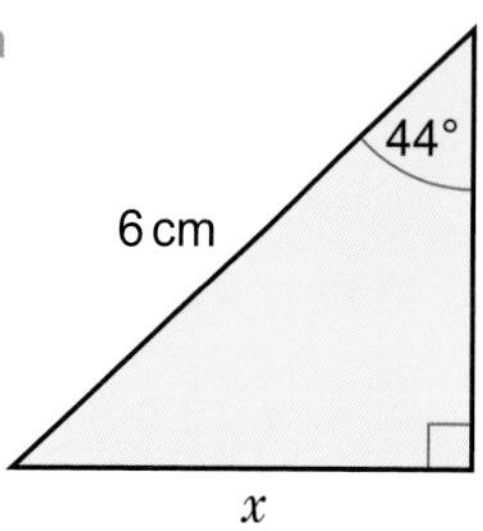

b

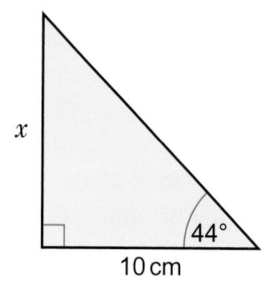

x
44°
10 cm

c

12 cm
81°
x

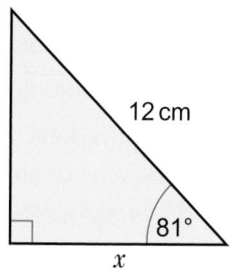

6 **Reflect** The mnemonic SOHCAHTOA can be used to remember the sine, cosine and tangent ratios.
Does it help you? Can you devise a mnemonic of your own?

7 Calculate the length of the side marked x. Give your answer correct to 1 decimal place.

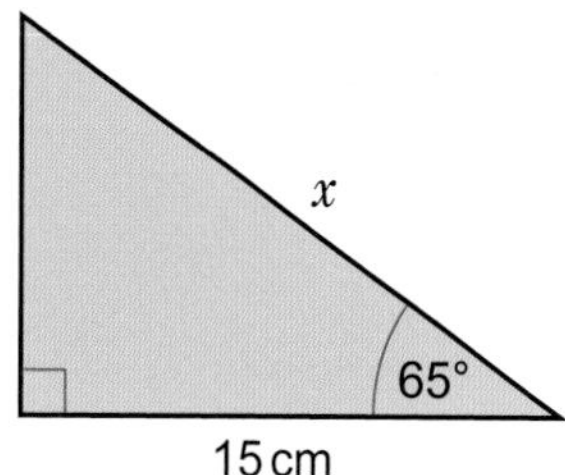

Q7 hint You are given adj and hyp so use the cosine ratio.

8 Calculate the length of the side marked x in each triangle.
Give your answers correct to 1 decimal place.

a

31°
x
5.1 cm

b

52°
7.2 m
x

c

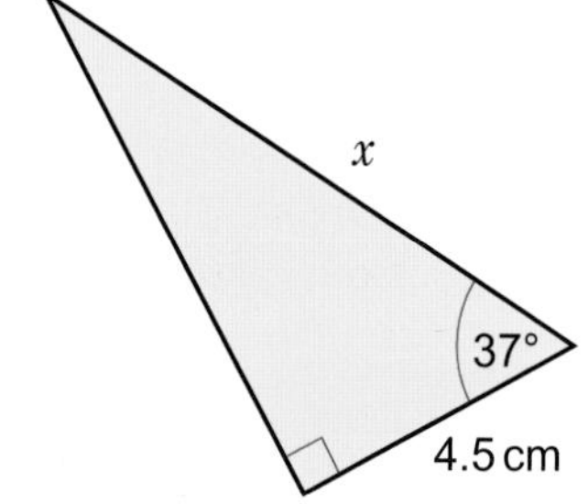

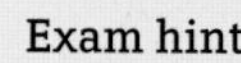

9 **Exam-style question**

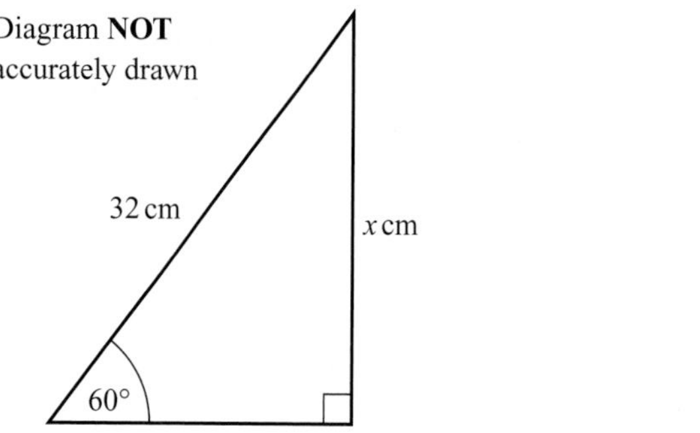

Calculate the value of x.
Give your answer correct to 3 significant figures. **(3 marks)**

Nov 2012, Q17, 1MA0/2H

Exam hint
Use SOHCAHTOA to write an equation involving x. Show your unrounded answer to at least 5 digits before rounding to 3 s.f.

10 Problem-solving / Reasoning A shed roof makes an angle of 41° with the horizontal.

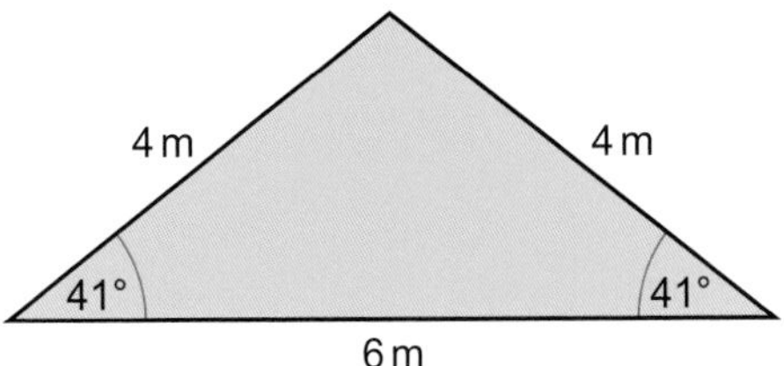

The width of the shed is 6 m.
The length of each slope is 4 m.
Calculate the height of the roof.

Key point 10

The **angle of elevation** (e) is the angle measured upwards from the horizontal.
The **angle of depression** (d) is the angle measured downwards from the horizontal.

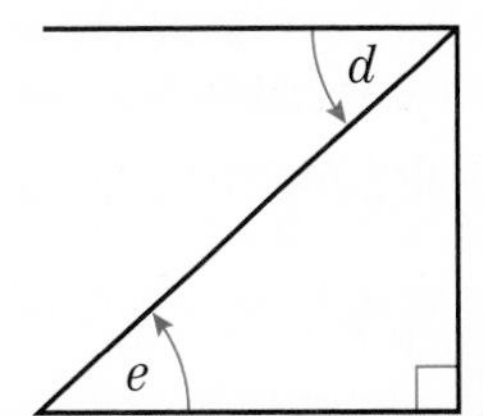

11 Real / Modelling A ladder 7 m long is leaning against a wall.
The angle of elevation is 72°.
What height does the ladder reach?

Q11 strategy hint Use a sketch.

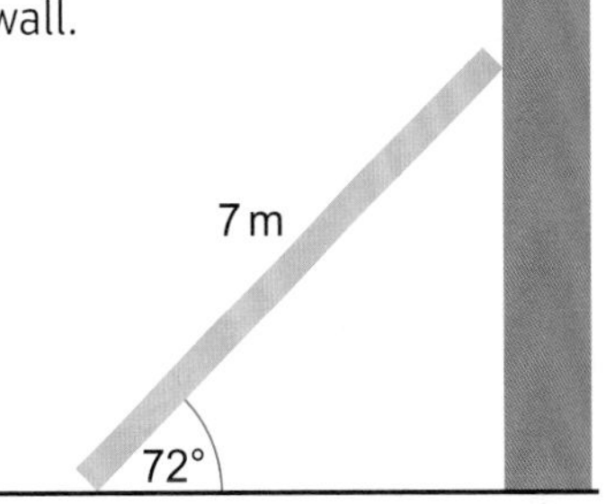

5.7 Trigonometry 2

Objectives

- Use trigonometric ratios to calculate an angle in a right-angled triangle.
- Find angles of elevation and angles of depression.
- Use trigonometric ratios to solve problems.
- Know the exact values of the sine, cosine and tangent of some angles.

Did you know?

A sextant is a navigation tool used at sea. It uses trigonometry to calculate the angle between two fixed points.

Fluency

Name the opposite and adjacent sides in these triangles.

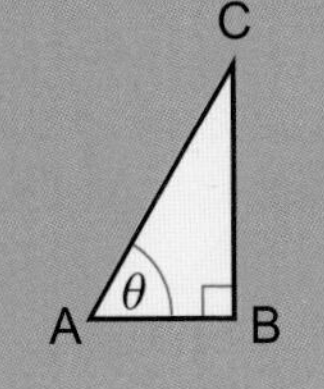

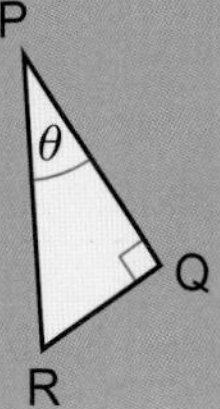

Warm up

1 Use your calculator to find, correct to 2 decimal places

a tan 49° b cos 16° c sin 75°

2 ABC is a right-angled triangle.
Calculate the length of AB, correct to 2 decimal places.

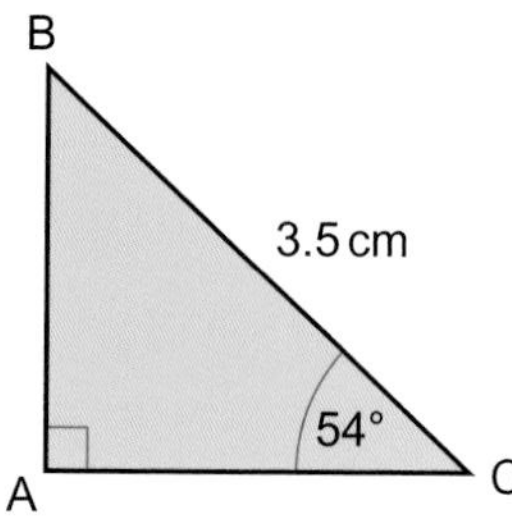

Key point 11

If the lengths of two sides of a right-angled triangle are given, you can find a missing angle using the **inverse trigonometric functions**:

$\sin^{-1}$ $\cos^{-1}$ $\tan^{-1}$

Make sure you know how to use $\sin^{-1}$, $\cos^{-1}$ and $\tan^{-1}$ on your calculator.

3 Use the inverse function on your calculator to find the value of θ correct to 0.1°.

a $\sin\theta = 0.562$ b $\cos\theta = 0.805$

c $\tan\theta = 0.246$ d $\sin\theta = \frac{4}{5}$

e $\cos\theta = \frac{11}{14}$ f $\tan\theta = \frac{8.5}{11.5}$

Q3 communication hint Correct to 0.1° means give your answer to 1 d.p.

Q3d hint Enter $\frac{4}{5}$ as a fraction.

Example 8

Calculate the size of angle x.

angle = x

opposite = 5 cm

hypotenuse = 9 cm

Identify the information given: angle, opposite and hypotenuse.

$\sin\theta = \frac{\text{opp}}{\text{hyp}}$ — You are given 'opp' and 'hyp' so use the sine ratio.

$\sin x = \frac{5}{9}$ — Substitute the sides and angle into the sine ratio.

$x = \sin^{-1}\left(\frac{5}{9}\right)$ — Use $\sin^{-1}$ to find the angle.

$x = 33.7489\ldots$

$x = 33.7°$ (to 1 d.p.) — Round your answer to 1 d.p.

5 cm
9 cm
x

4 Calculate the size of angle x in each triangle.
Give your answers correct to 1 decimal place.

a

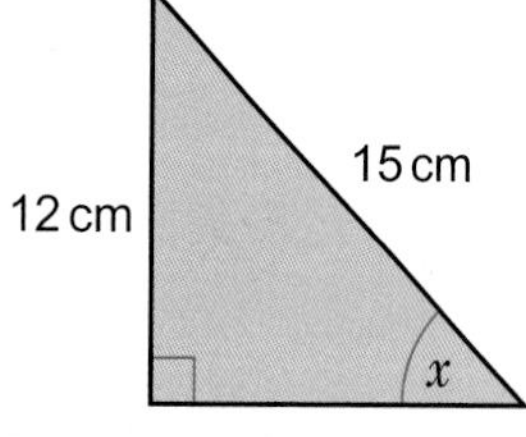

b

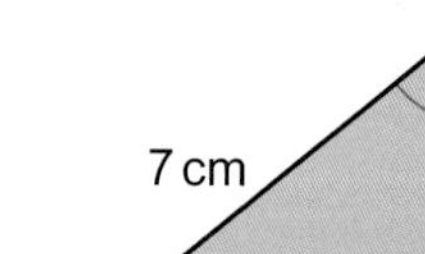

c

4.5 cm
x
7.2 cm

Q4a hint Use $\sin x = \frac{\text{opp}}{\text{hyp}}$

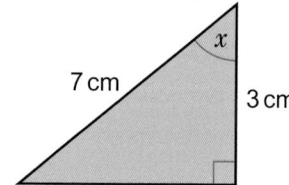

Q4b hint Use $\cos x = \frac{\text{adj}}{\text{hyp}}$

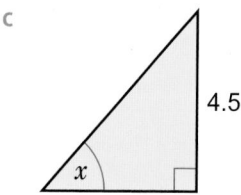

Q4c hint Use $\tan x = \frac{\text{opp}}{\text{adj}}$

5 **Exam-style question**

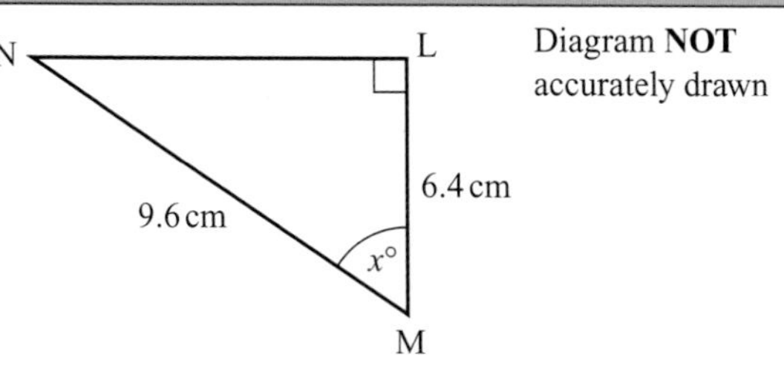

LMN is a right-angled triangle.
MN = 9.6 cm.
LM = 6.4 cm.
Calculate the size of the angle marked $x°$.
Give your answer correct to 1 decimal place. **(3 marks)**

June 2012, Q16, 1MA0/2H

Exam hint
Do not round until the very end of your calculation.

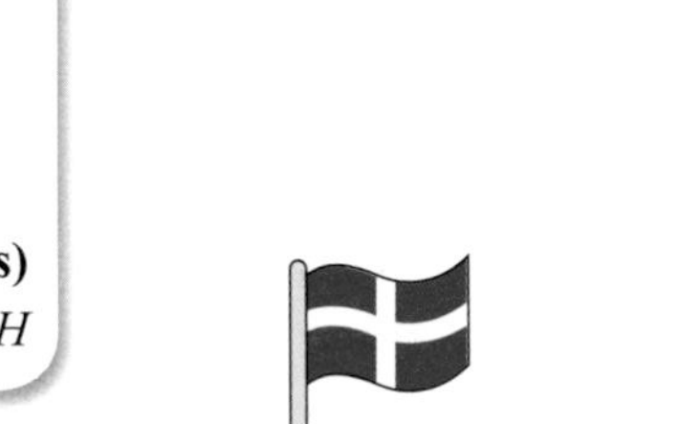

6 **Real / Problem-solving** A flagpole is secured to the ground by wires. Each wire is 4 m long.
The wires attach to the flagpole at a height of 3 m.
What is the size of the angle (x) the wire makes with the ground?

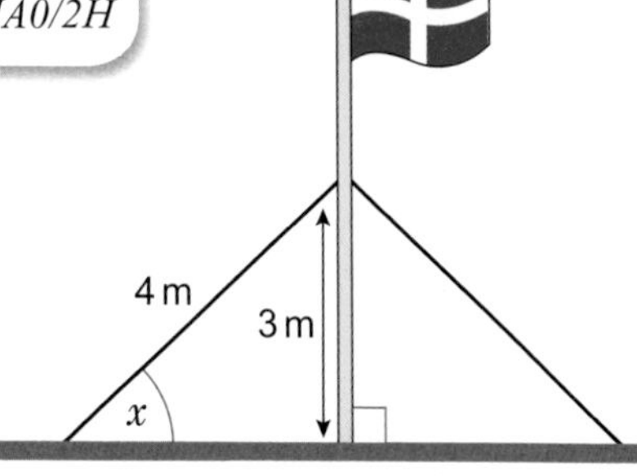

7 **Real / Problem-solving** A tree 20 m in height stands on horizontal ground.
Work out the angle of elevation of the top of the tree from point A.

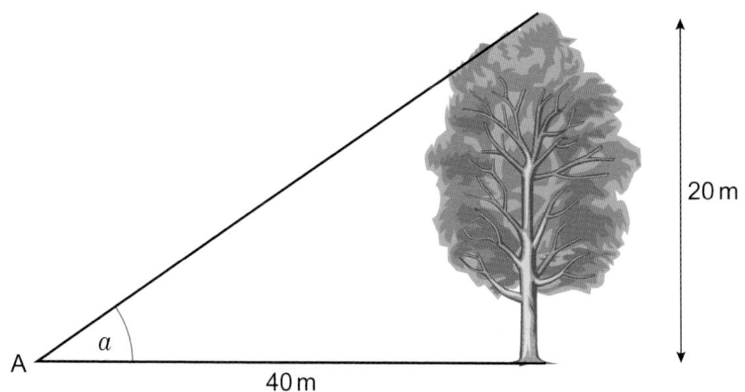

Discussion How can you work out the angle of depression of point A from the top of the tree?

8 **Real / Problem-solving** From P, a ship sails 3 km East and then 5 km North to its destination.
A helicopter flies from P directly to the ship.

a How far does the helicopter fly?

b On what angle (x) from North should the helicopter fly?

Give your answers correct to 1 decimal place.

Q8a hint Use Pythagoras' theorem.

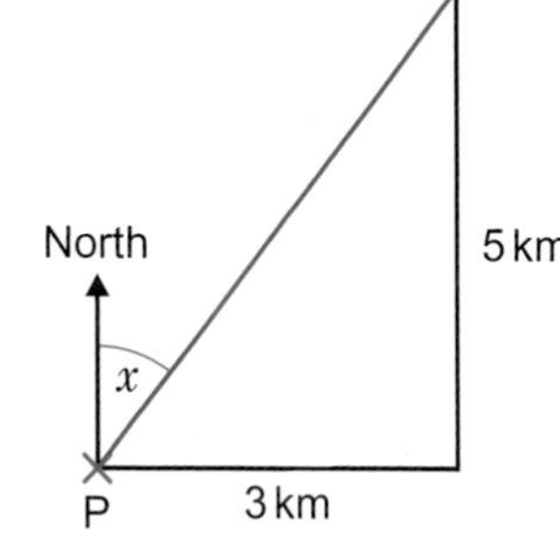

9 **Real / Problem-solving** From the top of a vertical cliff, 65 m high, a lifeguard can see a boat out at sea.
The boat is 42 m from the base of the cliff.
What is the angle of depression of the boat from the top of the cliff?

Q9 hint Sketch and label a right-angled triangle to show this information.

10 **Problem-solving** Work out the area of this isosceles triangle.

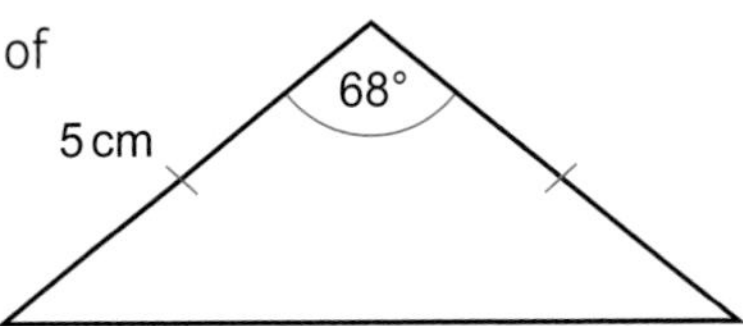

11 **Problem-solving** Calculate the size of angle θ in this diagram.

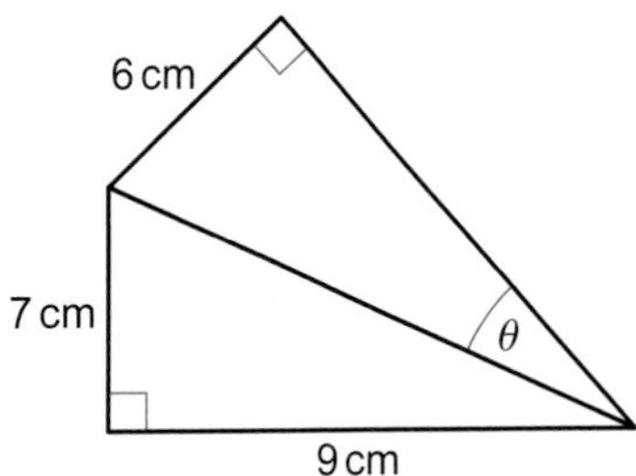

12 ABC is an isosceles triangle.

a Use the diagram to write the value of tan 45°.

b Use Pythagoras' theorem to find the length of BC. Leave your answer in surd form.

c Write these ratios as exact values using surds.

i sin 45° ii cos 45°

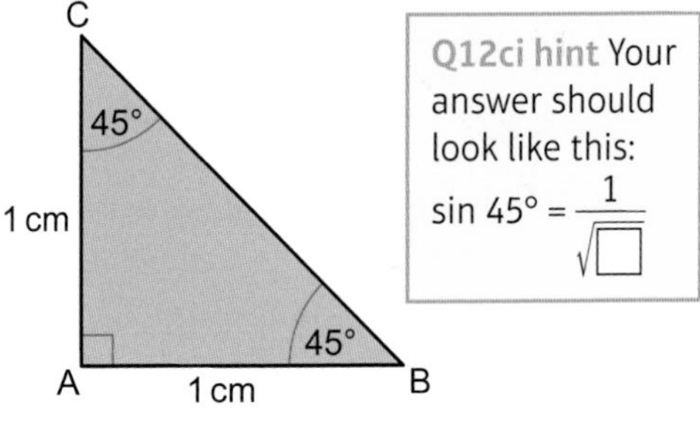

Q12ci hint Your answer should look like this:

$\sin 45° = \frac{1}{\sqrt{\square}}$

13 ABC is an equilateral triangle. D is the midpoint of AC.

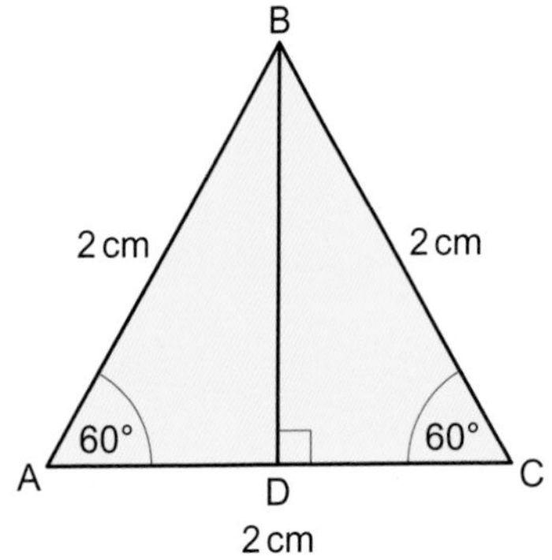

Q13a hint
Sketch right-angled triangle ABD. Add length AD and angle ABD to your diagram.

a Use the diagram to write these ratios as fractions.

i cos 60° ii sin 30°

b Work out the length of BD. Leave your answer in surd form.

c Write these ratios as exact values using surds.

i sin 60° ii tan 60° iii cos 30° iv tan 30°

Q13ci hint
Your answer should look like this:

$\sin 60° = \frac{\sqrt{\square}}{\square}$

Key point 12

The sine, cosine and tangent of some angles may be written exactly.

	30°	**45°**	**60°**	**0**	**90°**
sin	$\frac{1}{2}$	$\frac{\sqrt{2}}{2}$	$\frac{\sqrt{3}}{2}$	0	1
cos	$\frac{\sqrt{3}}{2}$	$\frac{\sqrt{2}}{2}$	$\frac{1}{2}$	1	0
tan	$\frac{\sqrt{3}}{3}$	1	$\sqrt{3}$	0	

14 Find the exact value of x in these triangles.

a

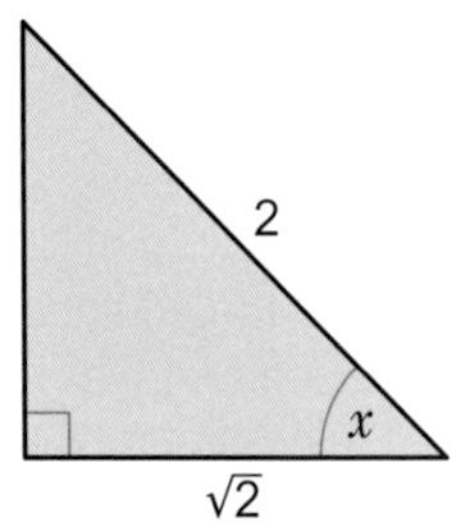

b

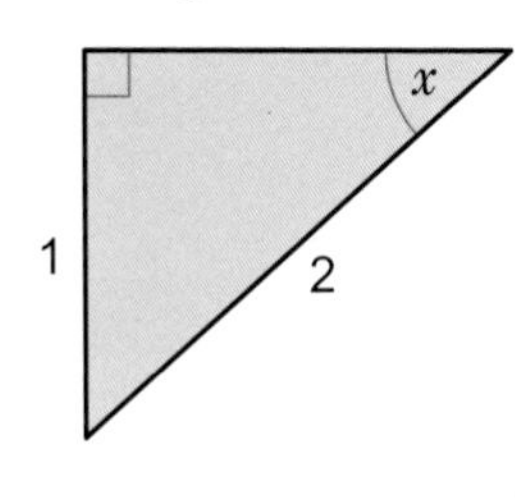

c

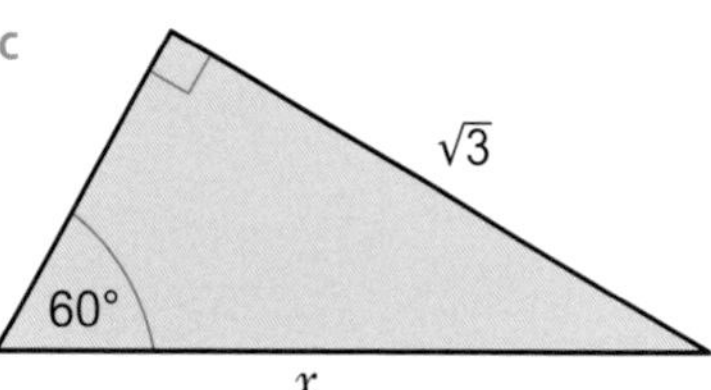

d

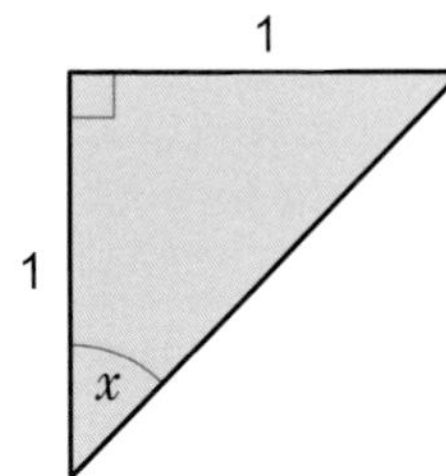

e

30°, 3, x

Q14 strategy hint Sketch the triangle. Label the hyp, opp and adj. Decide on the ratio to use by looking at Key point 12. Substitute the values you are given.

5 Problem-solving

Objective

- Use x for the unknown to help you solve problems.

Example 9

In this quadrilateral, angles PQR and QRS are equal.
Angle PSR is $\frac{4}{7}$ angle QRS. Angle QPS is $\frac{6}{7}$ angle QRS.

a Find angle PSR.

b Show that angle QPS is a right angle.

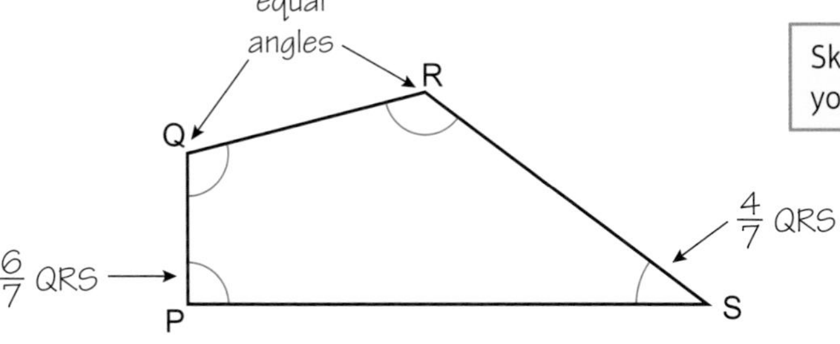

Sketch the diagram and write your findings on it as you go.

Let angle QRS be x.

Angle QRS is used to define a lot of other angles. So call this 'unknown' angle $x°$.

Angle PSR $= \frac{4}{7}x$

Angle QPS $= \frac{6}{7}x$

Angles in the quadrilateral PQRS $= x + x + \frac{4}{7}x + \frac{6}{7}x = \frac{24}{7}x = 360°$

Angles in a quadrilateral sum to 360°.

Therefore $x = \frac{7}{24} \times 360 = 105°$

Write and solve an equation.

a Angle PSR $= \frac{4}{7}x = \frac{4}{7} \times 105 = 60°$

Use the fact that PSR $= \frac{4}{7}x$

b Angle QPS $= \frac{6}{7}x = \frac{6}{7} \times 105 = 90°$

Therefore angle QPS is a right angle.

Show that angle QPS is a right angle, i.e. 90°.

1 In this triangle, angle BAC is 25% of angle ABC.
The ratio angle ACB : angle ABC is 1 : 4.

a What is the size of angle ABC? b What kind of triangle is this?

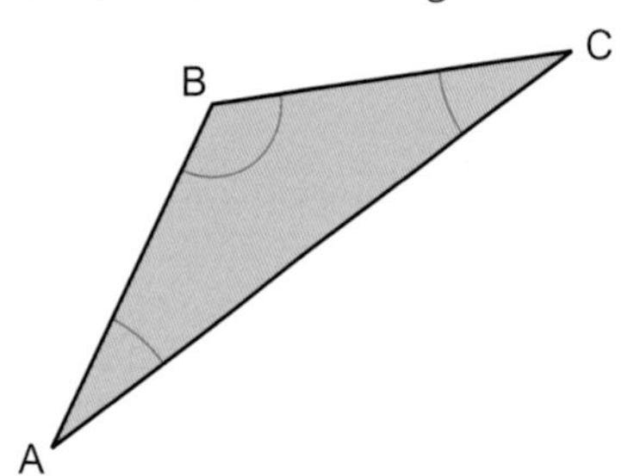

Q1a hint Copy the diagram. Call angle ABC x. Rewrite the percentage and ratio as fractions, then label the other angles in terms of x.

2 Antony thinks of three numbers.

- The second number is 4 times the first.
- The third number is 4 less than the second.
- The first number multiplied by the second number is 25.

Find the three numbers.

Q2 hint Let the first number be x.

3 A rectangle has a length twice its width. Its diagonal is $\sqrt{45}$ cm. What are the length and width of the rectangle?

Q3 hint Draw a rectangle with a diagonal. Label the shorter sides x.

4 A right-angled triangle has a hypotenuse that is 1.6 times the length of the base. What are its angles?

5 This brand logo is a heptagon with a vertical line of symmetry. Angle A is three times angle C. Angle B is four times angle C. Angle D is 260°. Find angles E, F and G.

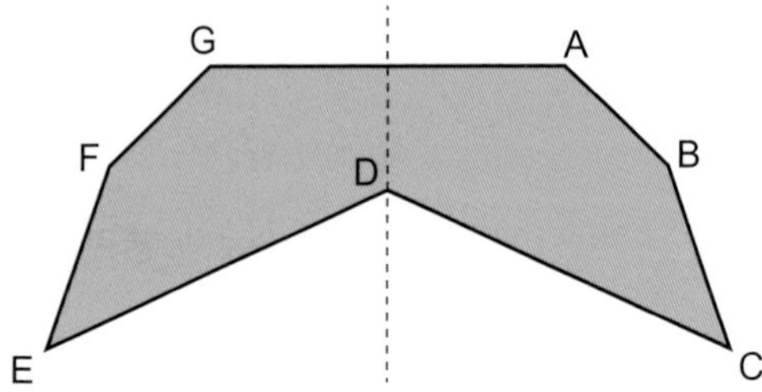

Q5 hint Sketch the diagram. Angle C seems important (it is mentioned twice) so call it x.
Label the other angles in terms of x, then use these to form an equation.

6 **Reflect** Choose A, B or C:
Solving problems by using x for the unknown is:

A always easy B sometimes easy, sometimes hard C always hard

Discuss with a classmate or your teacher what you did find easy or hard.

5 Check up

Log how you did on your Student Progression Chart.

Angles and polygons

1 a What is the size of each interior angle of a regular decagon?
b What is the size of an exterior angle of a regular pentagon?

2 Part of a regular polygon is shown.

How many sides does the polygon have?

3 Work out the size of angle x.

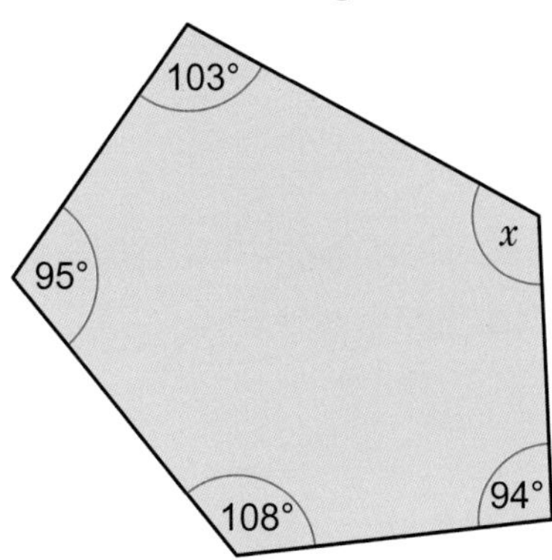

4 DEA is a straight line.
ABEF is a rectangle.
Angle ACD = 90° and angle EAB = 36°.
Work out the size of angle DEB.
Give reasons for your working.

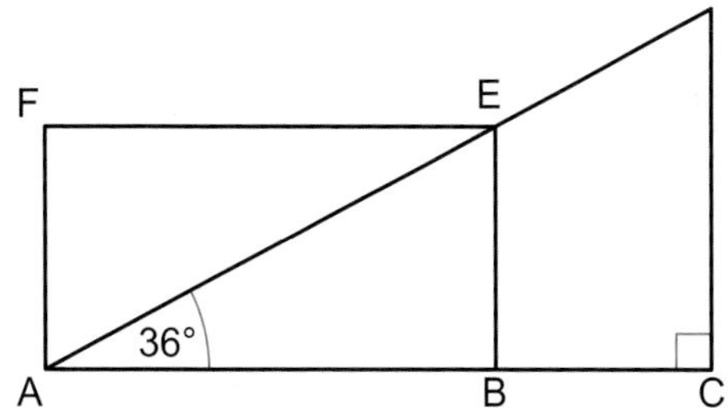

5 Work out the size of angle ABE.
Give reasons for your working.

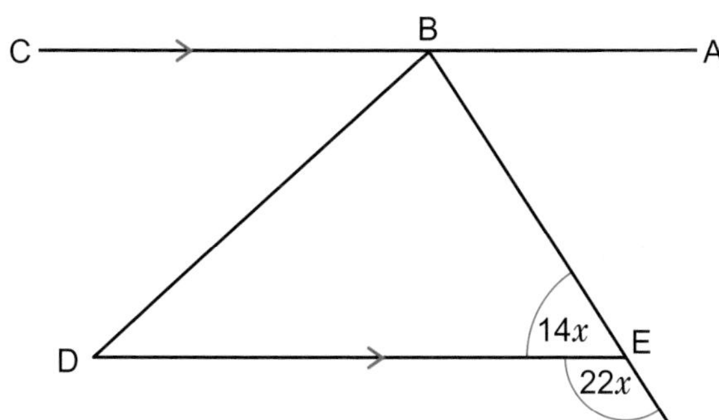

6 **Communication** Show that for any quadrilateral
$a + b + c + d = 360°$

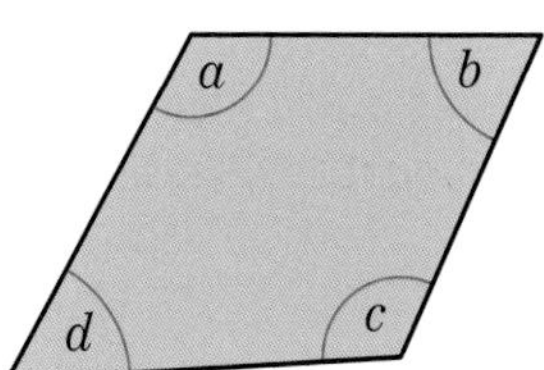

7 **Reasoning** BCD is an isosceles triangle.
AC is parallel to ED.
AE is parallel to BD.
Angle BAE = 62°.
Work out the size of the angle marked x.
Give reasons for your working.

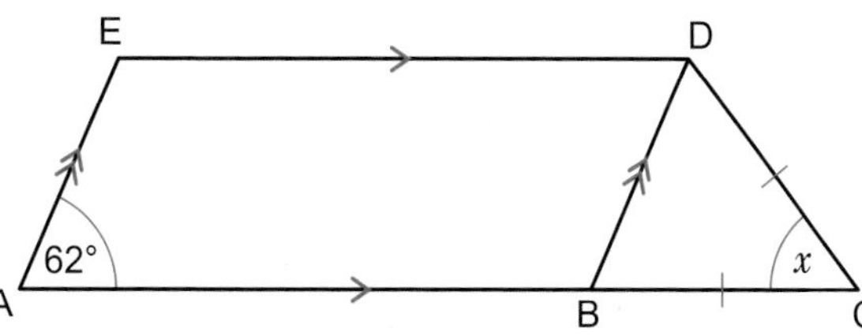

Pythagoras' theorem

8 Calculate the length of x in each right-angled triangle.
Give your answers correct to 2 significant figures.

a
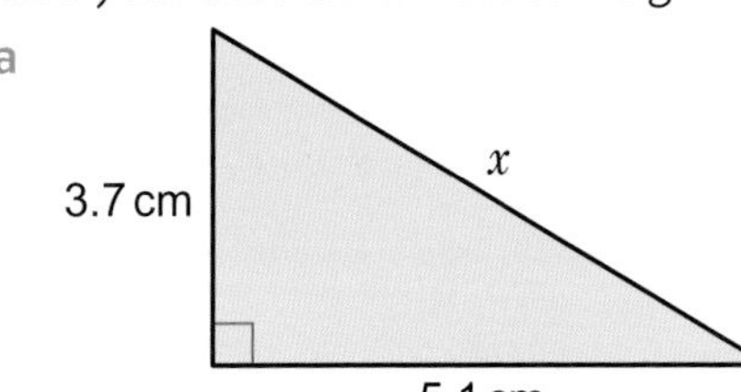

b
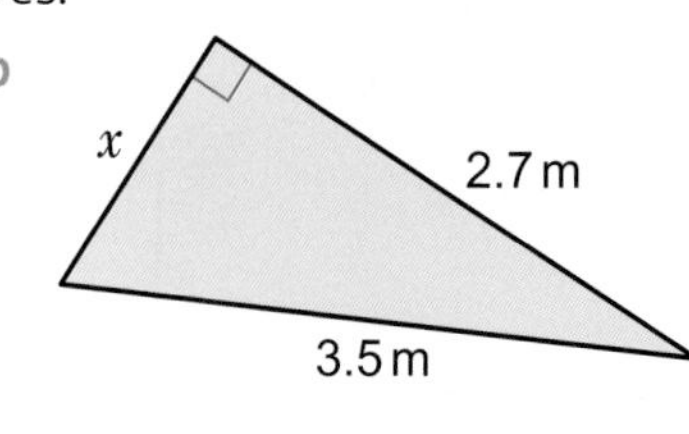

9 **Reasoning** A triangle has sides of length 3 cm, 6 cm and 7 cm.
Is the triangle a right-angled triangle? Explain your answer.

10 Work out the length of AC in this right-angled triangle.
Give your answer in surd form.

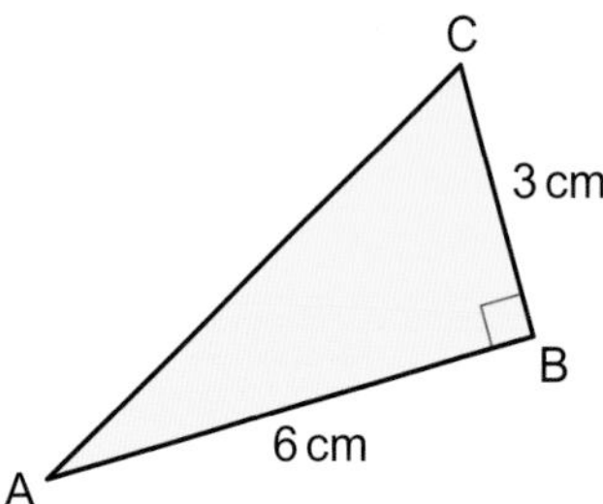

Trigonometry

11 Calculate the length of the side marked x in each triangle.
Give your answers correct to 3 significant figures.

a
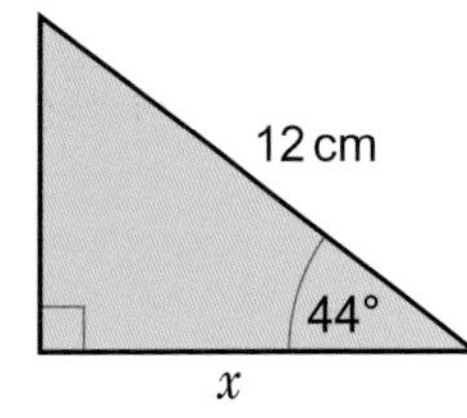

b
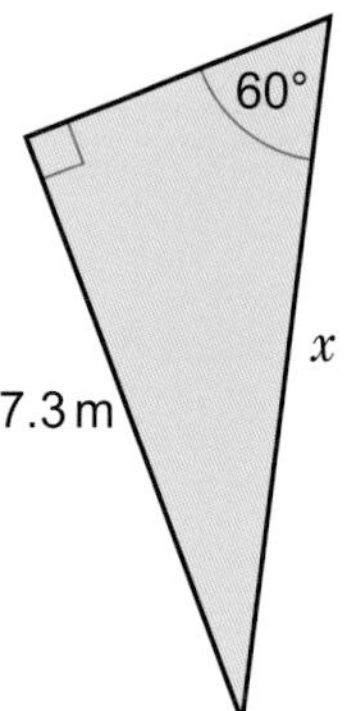

12 Calculate the size of angle y in this triangle.
Give your answer correct to 3 significant figures.

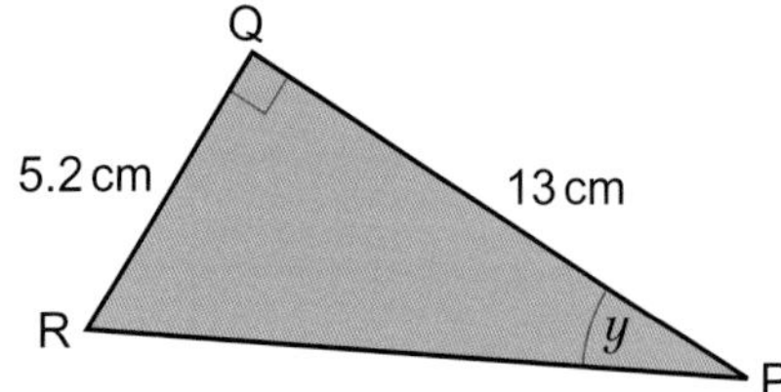

13 A kite is flying at a height of 11.7 m.
The string of the kite is 14 m long.
What is the angle of elevation of the kite? Give your answer correct to 1 decimal place.

14 Write down the value of

a tan 45° b sin 30° c cos 60°

15 How sure are you of your answers? Were you mostly

Just guessing Feeling doubtful  Confident 

What next? Use your results to decide whether to strengthen or extend your learning.

Reflect

★ Challenge

16 The square and the isosceles triangle have the same area.
Find tan θ.

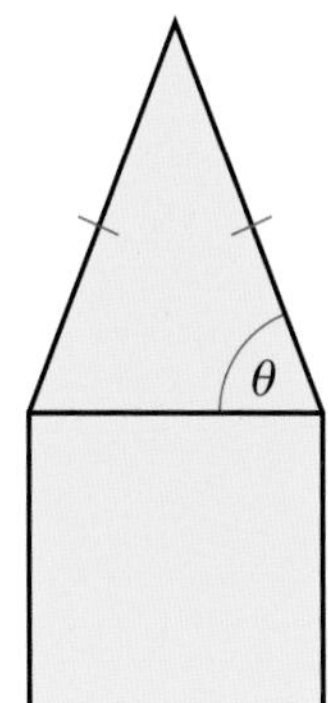

Not to scale

5 Strengthen

Angles and polygons

1 **Reasoning** Mario divides some shapes into triangles to work out the sum of the interior angles.

a Copy and complete his table.

Polygon	Quadrilateral	Pentagon	Hexagon	Heptagon
Number of sides (n)	4			
Number of triangles	2			
Sum of interior angles	2 × 180° = 360°			

b Copy and complete Mario's working to find an expression for the sum of the interior angles of *any* polygon.

Number of sides = n

Number of triangles = n – __

Sum of interior angles = (n – __) × __°

c Work out the sum of the interior angles of a decagon.

Q1c hint Use your answer to part **b**.

2 Work out the size of each interior angle of these shapes. The first one has been started for you.

a a regular nonagon

Sum of interior angles = $(n - \_\_) \times \_\_°$

$= (9 - \_\_) \times \_\_°$

$= \_\_\_°$

Interior angle $= \_\_\_° \div 9$

$= \_\_°$

b a regular polygon with 12 sides

c a regular polygon with 20 sides.

Q2 communication hint In a **regular polygon** all the sides and all the angles are equal.

3 The sum of the exterior angles of any polygon is 360°.
What is the size of an exterior angle of a

a regular quadrilateral

b regular decagon

c regular polygon with 18 sides?

Q3a hint Divide 360° by the number of sides.

4 a Rearrange the formula to make n, number of sides, the subject.

exterior angle $= \frac{360°}{n}$

b How many sides does a regular polygon have if the exterior angle is

i 90° ii 60° iii 30° iv 12°?

Q4a hint $n = \frac{\square}{\square}$

Q4b hint Use your answer to part **a**.

5 Part of a regular polygon is shown.

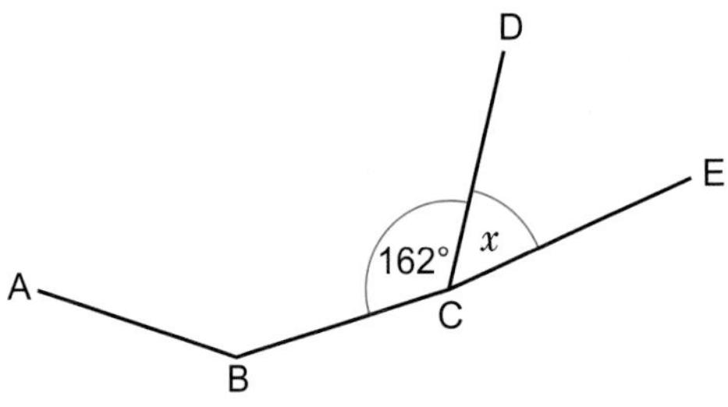

Q5b hint Use the fact that angles on a straight line add up to 180°.

a Identify the interior and exterior angle.

b Work out the size of x.

c How many sides does the polygon have?

6 a What is the sum of the interior angles of a hexagon?

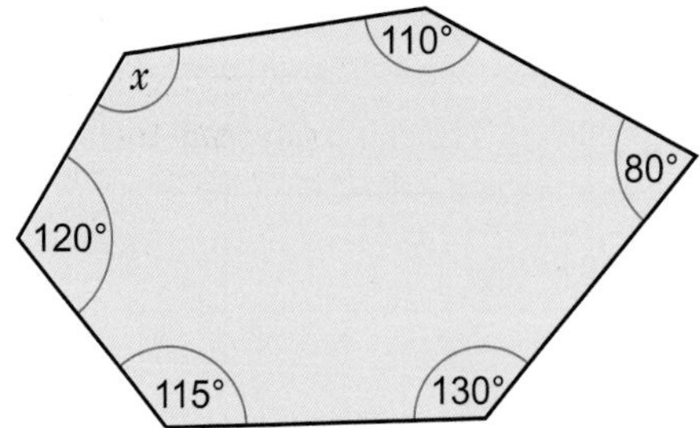

Q6b hint Use your answer to part **a** to form an equation.
$x + 110° + 80° + 130° + 115° + 120° = \_\_\_°$
Solve it for x.

b Work out the size of x.

7 **Communication** Work out the size of angle ABC. Give reasons for your working. The working has been started for you.

$x + 3x + x + 5 + x + 7 = \_\_°$

(angles in a quadrilateral __________)

$6x + \_\_ = \_\_°$

$6x = \_\_°$

$x = \_\_°$

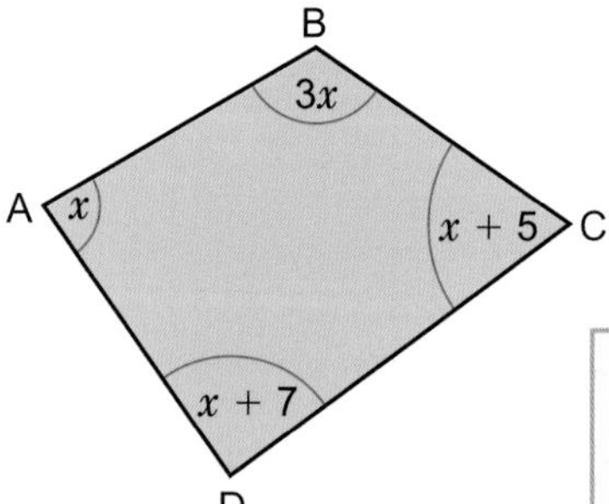

Q7 hint Have you answered the question asked?

8 **Reasoning / Communication** Work out the value of x and y. Give reasons for your working.

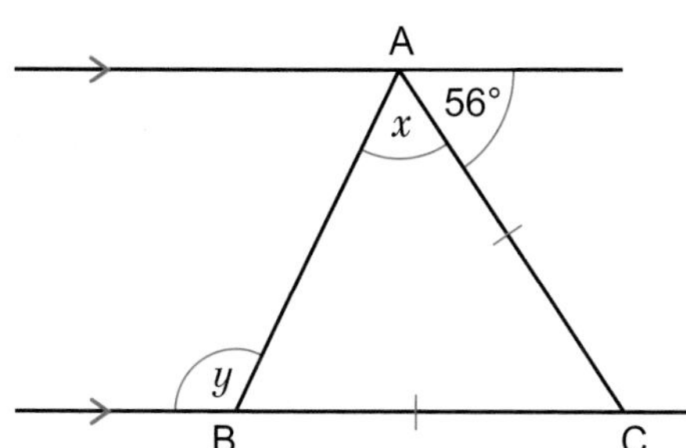

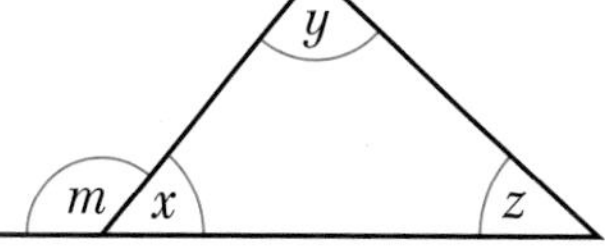

Q8 hint Use these reasons:
Alternate angles are equal.
Angles in a triangle sum to 180°.
Base angles of an isosceles triangle are equal.
Angles on a straight line add up to 180°.

9 **Reasoning / Communication**
Are these statements true or false? Explain your answers.

a $m = x$ b $z = x + y$ c $m = y + z$
d $m = 180° - x$ e $x = 180° - (y + z)$

10 **Reasoning** Copy each diagram.

i

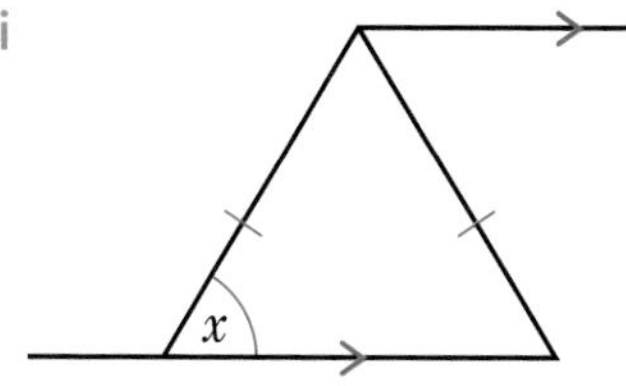

ii

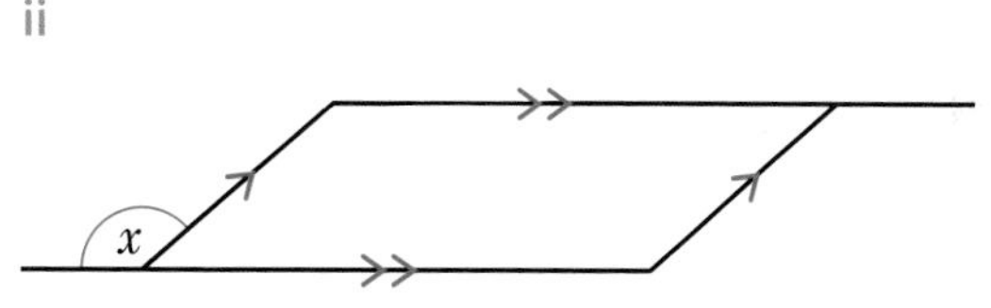

a Write x in all the angles equal to x in each diagram.
b Write y in all the angles equal to $180 - x$ in each diagram.
c Look at your diagram for part **ii**. What properties of parallelograms are shown?

Pythagoras' theorem

1 Name the hypotenuse in each triangle.

a

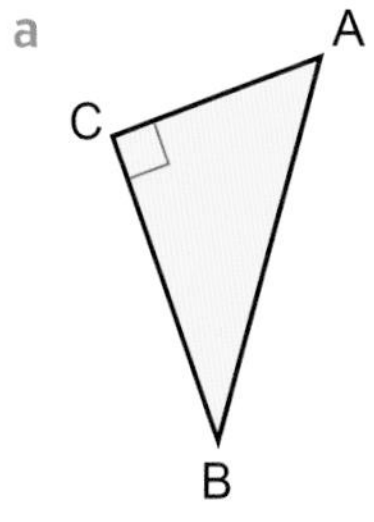

b

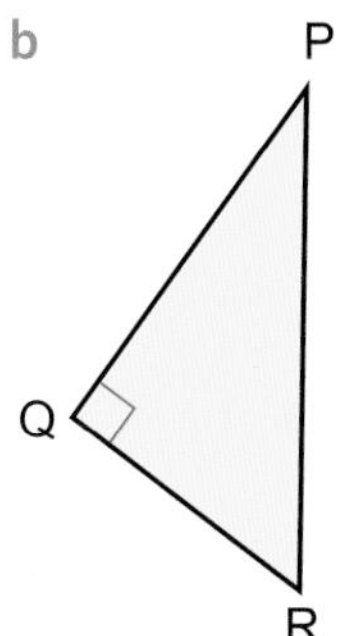

c

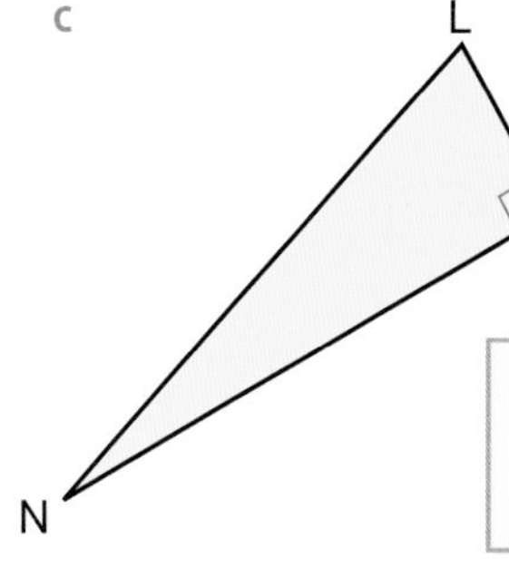

Q1 hint The hypotenuse is the longest side and is opposite the right angle.

2 The hypotenuse of this right-angled triangle is labelled c.
Copy and complete these steps to find the value of c.

$c^2 = 8^2 + \_\_^2$

$c^2 = \_\_$

$c = \sqrt{\_\_}$

$c = \_\_$ cm

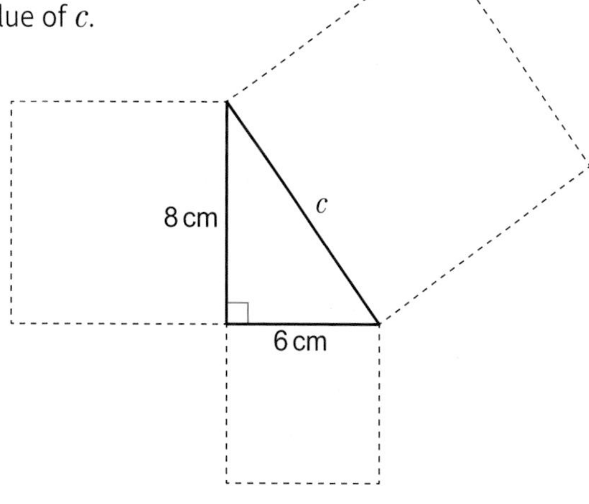

Q2 hint In a right-angled triangle, the square of the hypotenuse is equal to the sum of the squares of the other two sides. This is Pythagoras' theorem.

3 Calculate the length of c in these right-angled triangles.
Round your answer to 1 decimal place where necessary.

a

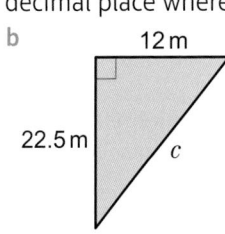

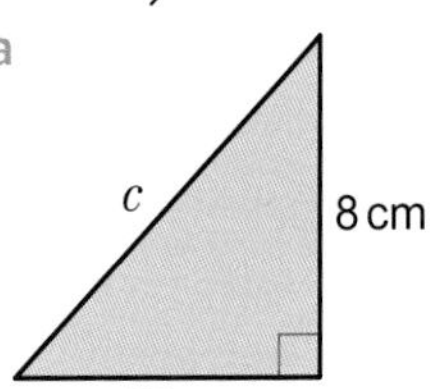

b

12 m
22.5 m
c

> **Q3 hint** Use the same method as in **Q2**.

> **Q3b hint** Do not round before taking the square root. You should find the square root of 650.25.

4 One of the shorter sides of this right-angled triangle is labelled b.
Copy and complete these steps to find the value of b.

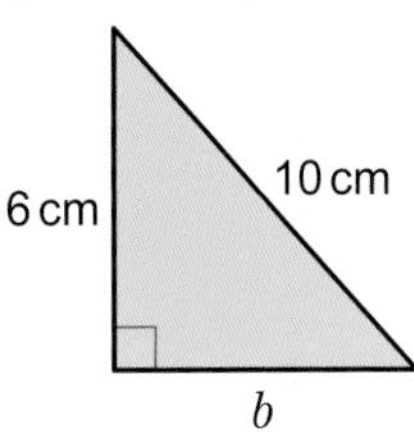

> **Q4 hint** Pythagoras' theorem: $c^2 = a^2 + b^2$

$c^2 = a^2 + b^2$

$10^2 = \_\_^2 + b^2$

$100 = \_\_ + b^2$

$b^2 = 100 - \_\_$

$b = \sqrt{\_\_}$

$b = \_\_$ cm

5 Calculate the length x in each right-angled triangle, correct to 2 d.p. where necessary.

a

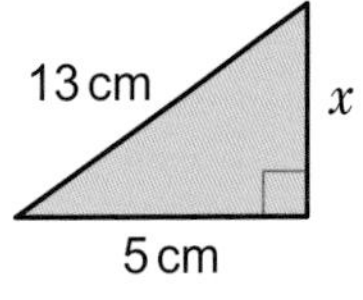

b

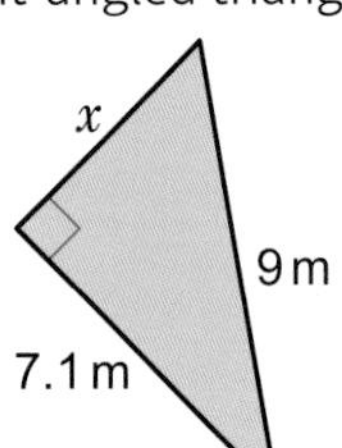

> **Q5a hint** Check your answer.

6 **Reasoning** Aaron says, 'Triangle ABC is right-angled'. Is Aaron correct? Explain.

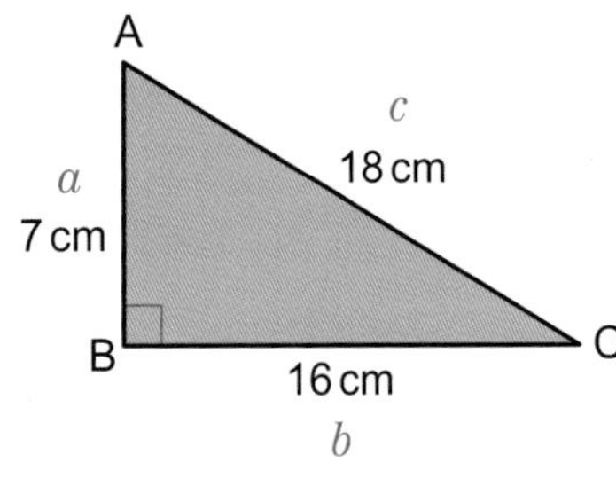

> **Q6 hint** A triangle is only right-angled if $c^2 = a^2 + b^2$, where c is the longest side.
> Start:
> $c^2 = 18^2 = \ldots$
> $a^2 + b^2 = \ldots$

7 Work out the value of x in each right-angled triangle.
Give your answers in surd form.

> **Q7 hint** Make sure any surds are in their simplest form.

a

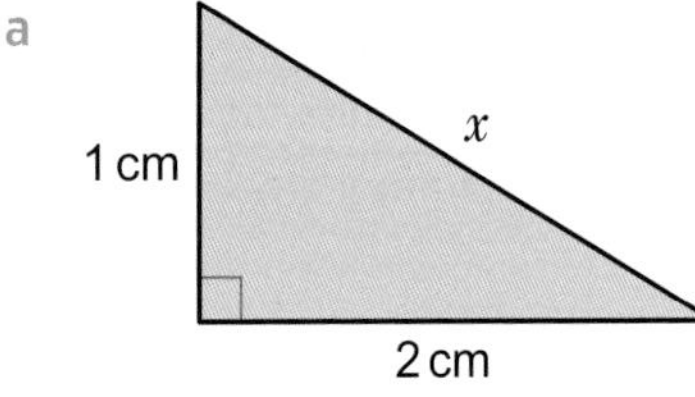

b

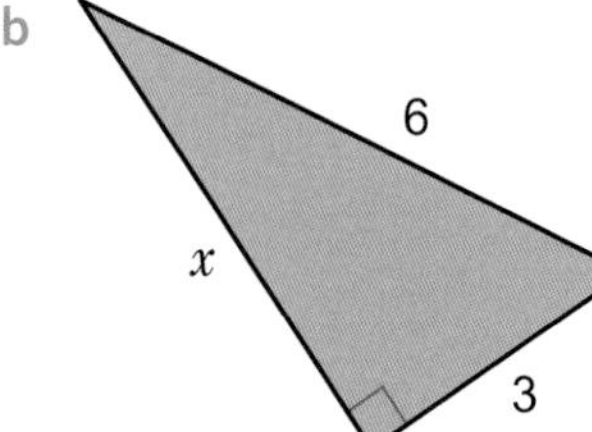

c

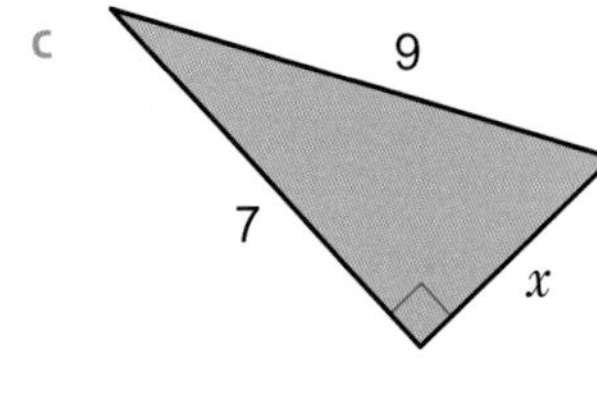

Trigonometry

1 Sketch these triangles and label the hypotenuse, opposite and adjacent sides.

a
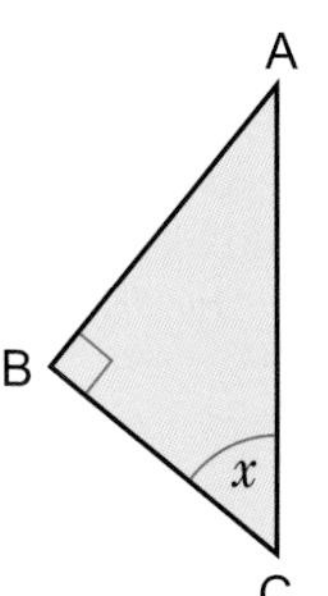

b
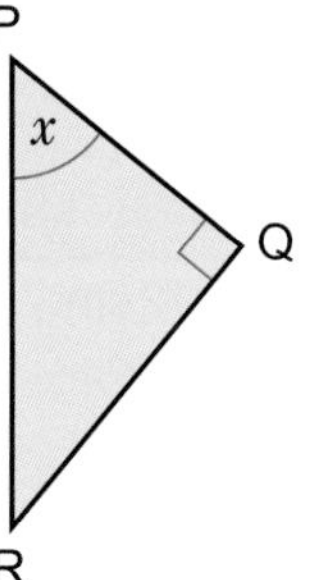

c
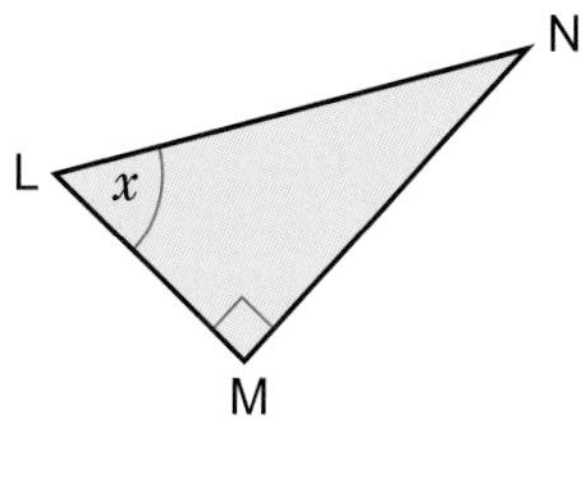

2 Write $\sin x$, $\cos x$ and $\tan x$ as fractions for this triangle.
The first one has been started for you.

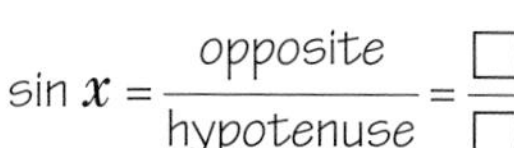

$\sin x = \frac{\text{opposite}}{\text{hypotenuse}} = \frac{\square}{\square}$

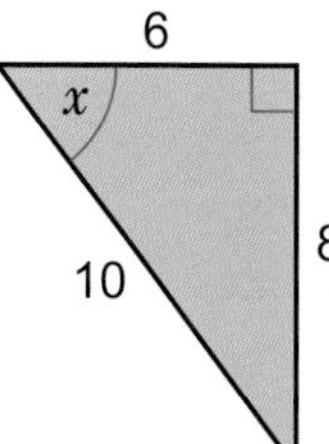

Q2 hint Use SOH CAH TOA.

3 Use your calculator to find, correct to 2 decimal places

a $\sin 22°$ b $\tan 36°$ c $\cos 70°$ d $\tan 58°$

4 Copy and complete. Use your calculator to find each angle to 1 decimal place.

a $\tan x = 0.345$ $x = \tan^{-1}(\square)$

b $\sin \theta = 0.806$ $\theta = \square^{-1}(0.806)$

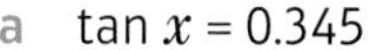

c $\cos y = 0.7625$ $y = \cos^{-1}(\square)$

d $\sin a = \frac{2}{3}$ $a = \square^{-1}\left(\frac{2}{3}\right)$

e $\cos b = \frac{4.8}{5.1}$ $b = \cos^{-1}(\square)$

5 Sophie is calculating the length of the side marked x in this triangle.
Copy and complete her working.

SOH CAH (TOA)

$\tan = \frac{\text{opp}}{\text{adj}}$

$\tan 36° = \frac{x}{5}$

$x = 5 \times$ ______

$x =$ ______

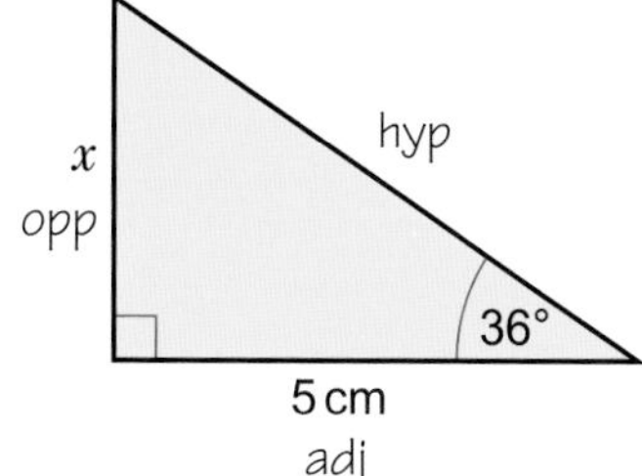

Q5 hint Round your answer to 1 d.p. and put in the units.

6 Calculate the length of the side marked x in each triangle.
Give your answers correct to 1 decimal place.

a
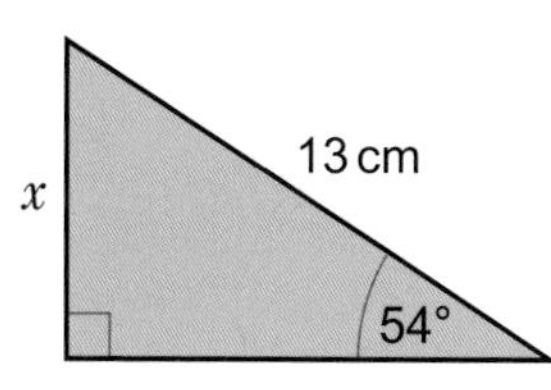

b
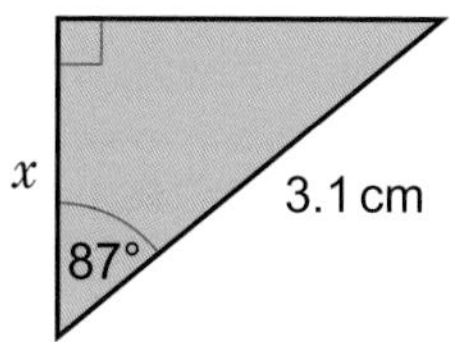

Q6 hint Use the same method as in **Q5**.
Write SOH CAH TOA.
Underline the information you are given.
Use the ratio with two underlines.

7 Calculate the size of angle x.
The working has been started for you.

SOH(CAH)TOA

$$\cos = \frac{\text{adj}}{\square}$$

$$\cos x = \frac{\square}{\square}$$

$$x = \cos^{-1}\left(\frac{\square}{\square}\right)$$

$$x = \_\_°$$

Q7 hint Round your answer to 1 d.p.

8 Calculate the size of angle ABC in this triangle.
Give your answer correct to 0.1°.

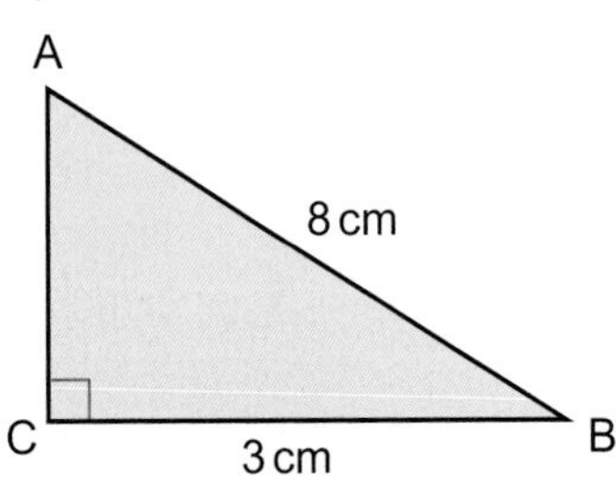

Q8 hint Sketch triangle ABC.
Label angle ABC.
Then use the same method as in Q8.

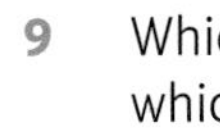

9 Which angle is an angle of elevation and which is an angle of depression?

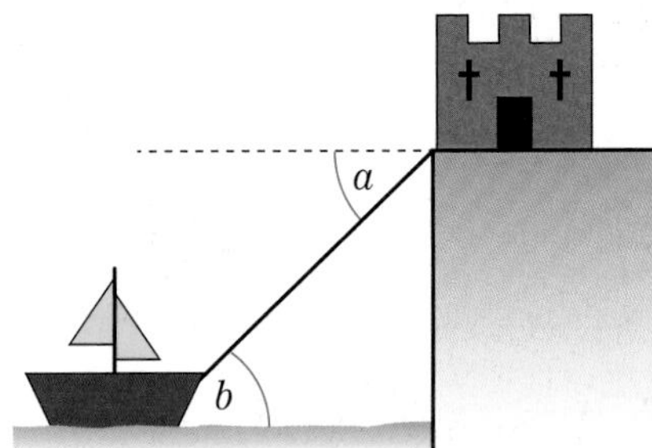

10 **Real / Problem-solving** From the top of a vertical cliff, C, a boat, B, can be seen out at sea. The cliff is 90 m high. The boat is 110 m from the base of the cliff.

a Copy and complete the diagram to show the information you are given.

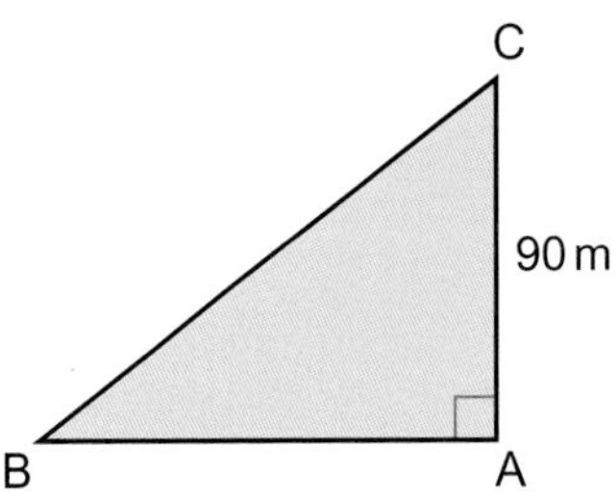

b What is the angle of elevation of the cliff top from the boat?
Give your answer correct to 1 decimal place.

Q10b hint Find the angle at B.

11 Copy and complete this table. Leave your answers in surd form.

	30°	**45°**	**60°**	**0**	**90°**
sin					
cos					
tan					

Q11 hint Look back at Key point 12.

5 Extend

1 **Problem-solving** A rectangle BCEF is constructed inside a regular hexagon ABCDEF. Work out the size of angle DEC.

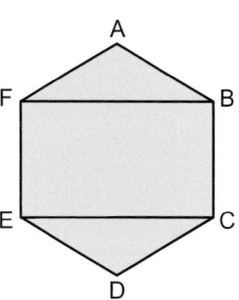

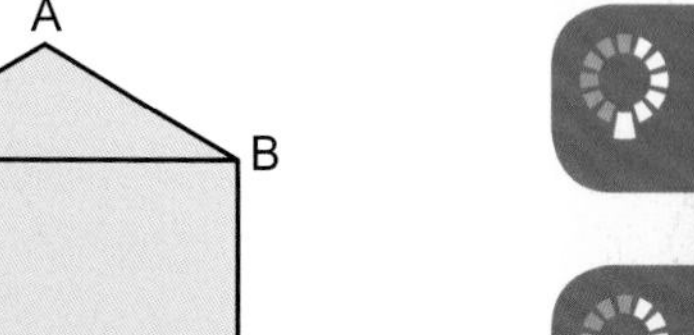

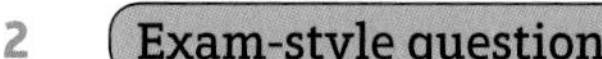

2 **Exam-style question**

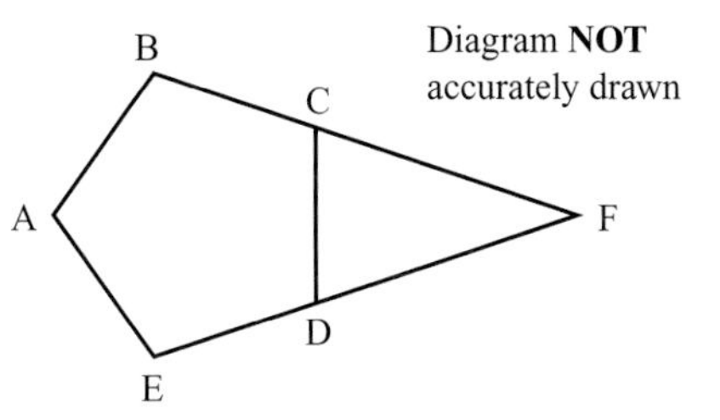

ABCDE is a regular pentagon.
BCF and EDF are straight lines.
Work out the size of angle CFD.
You must show how you got your answer. **(3 marks)**

June 2014, Q11, 1MA0/1H

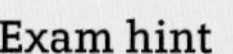

Exam hint
You could start by working out the size of the exterior angles of the pentagon.

3 **Problem-solving** Work out the size of angle QSP.
Give reasons for your working.

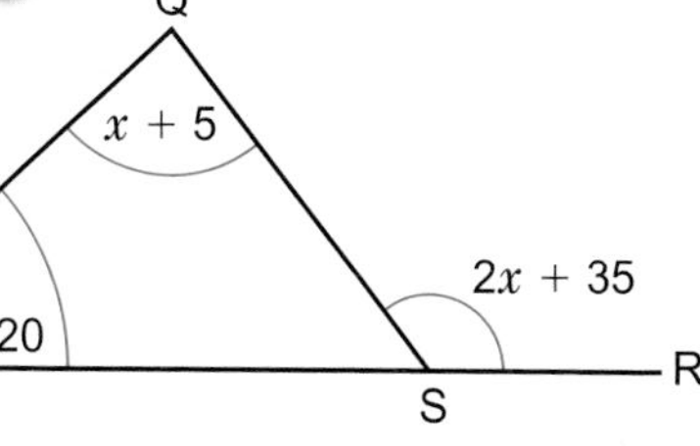

4 **Exam-style question**

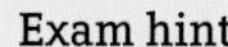

ABCD is a trapezium.

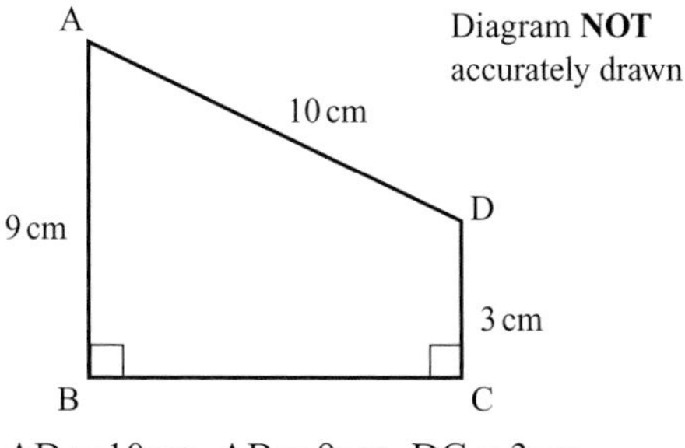

AD = 10 cm, AB = 9 cm, DC = 3 cm
Angle ABC = angle BCD = 90°
Calculate the length of AC.
Give your answer correct to 3 significant figures. **(5 marks)**

Nov 2012, Q15, 1MA0/2H

Exam hint
Divide the trapezium into a rectangle and a triangle and put on the sizes of the three sides of the triangle.
Calculate the length BC first.

5 **Problem-solving** A rectangular garden measures 25 m by 20 m.
A path is laid diagonally across the garden and along the whole of its perimeter.
What is the total length of the path?

6 **Problem-solving** The ratio of the interior angles of a pentagon is 1:2:3:4:5.
Work out the sizes of all five interior angles.

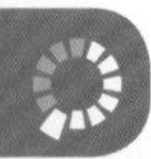

7 **Problem-solving** The exterior and interior angles of a regular polygon are in the ratio 1 : 2. How many sides does the polygon have?

8 **Problem-solving** The area of triangle ABC is 42 cm².

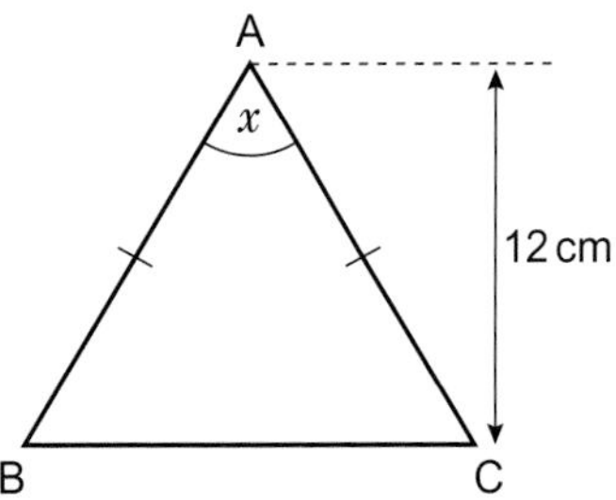

Calculate the size of angle x.
Give your answer correct to 1 decimal place.

9 **Reasoning** Sarah sees an aeroplane.
She estimates it is flying at a height of 56 000 feet.
The angle of elevation to the aeroplane is 49°.
What is the horizontal distance between Sarah and the plane? Give your answer correct to 3 sf.

10

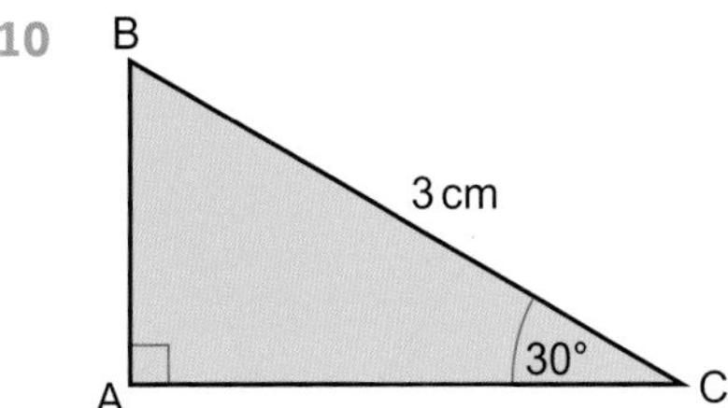

a Find the length (to 2 d.p.) of
   i AB   ii AC.
b Work out the perimeter of the triangle.

11 **Problem-solving** In a right-angled triangle the shortest side is 4 cm and the longest side is 8 cm.
Work out the exact perimeter of the triangle.

12 **Problem-solving** The length of the diagonal of a square is 10 cm.
Work out the length of a side of the square.
Give your answer correct to 1 decimal place.

13 **Problem-solving** The diagram shows a circle centre O. Angle OBA is 90°.

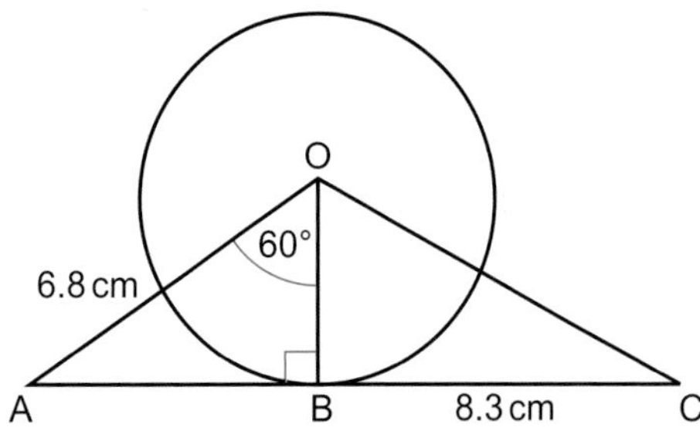

a Work out the radius of the circle.
b Work out angle OCB.

5 Knowledge check

- The angle marked x is called the **interior angle**. The angle marked y is called the **exterior angle**.

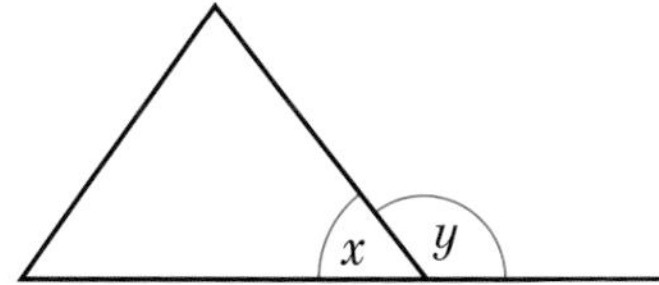

$x + y = 180°$ (angles on a straight line add up to 180°) .... *Mastery lesson 5.1*

- For any polygon, interior angle + exterior angle = 180°. .... *Mastery lesson 5.3*
- The exterior angle of a triangle is equal to the sum of the interior angles at the other two vertices.

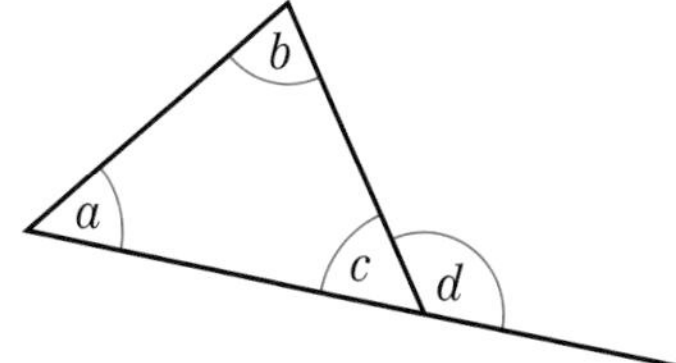

angle d = angle a + angle b .... *Mastery lesson 5.1*

- The sum of the interior angles of a polygon with n sides = $(n - 2) \times 180°$. *Mastery lesson 5.2*
- The sum of the exterior angles of a polygon is always 360°. .... *Mastery lesson 5.3*
- The exterior angle of a regular n-sided polygon is .... *Mastery lesson 5.3*
- In a right-angled triangle the longest side is called the **hypotenuse** and is opposite the right angle. .... *Mastery lesson 5.4*
- Pythagoras' theorem states that in a right-angled triangle, the square of the hypotenuse is equal to the sum of the squares of the other two sides.

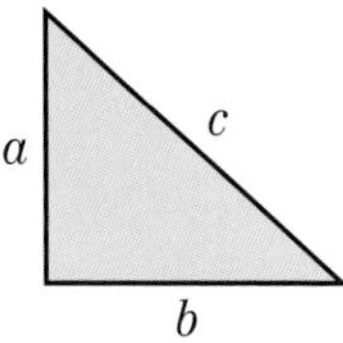

$c^2 = a^2 + b^2$ .... *Mastery lesson 5.4*

- A triangle with sides a, b and c, where c is the longest side, is right-angled *only* if $c^2 = a^2 + b^2$. .... *Mastery lesson 5.4*
- In a right-angled triangle, the side opposite the angle θ is called the **opposite**. The side next to the angle θ is called the **adjacent**. .... *Mastery lesson 5.6*

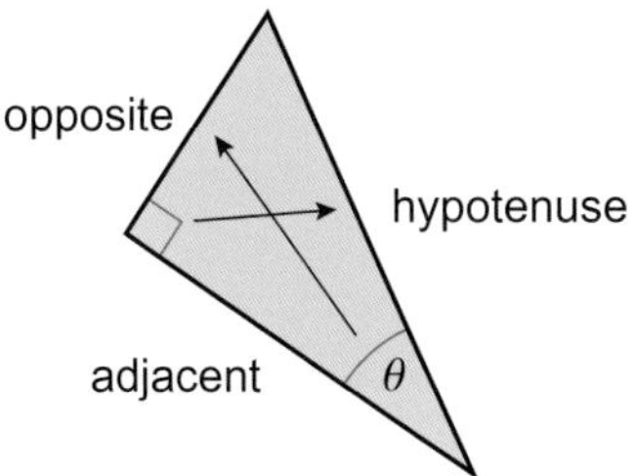

- The **sine** of angle θ is the ratio of the opposite side to the hypotenuse, $\sin\theta = \frac{\text{opp}}{\text{hyp}}$. .... *Mastery lesson 5.6*
- The **cosine** of angle θ is the ratio of the adjacent side to the hypotenuse, $\cos\theta = \frac{\text{adj}}{\text{hyp}}$. .... *Mastery lesson 5.6*
- The **tangent** of angle θ is the ratio of the opposite side to the adjacent side, $\tan\theta = \frac{\text{opp}}{\text{adj}}$. .... *Mastery lesson 5.6*
- You can use $\sin^{-1}$, $\cos^{-1}$ or $\tan^{-1}$ on your calculator to find an angle when you know its sin, cos or tan. .... *Mastery lesson 5.7*
- The **angle of elevation** (e) is the angle measured upwards from the horizontal. The **angle of depression** (d) is the angle measured downwards from the horizontal. .... *Mastery lesson 5.6*

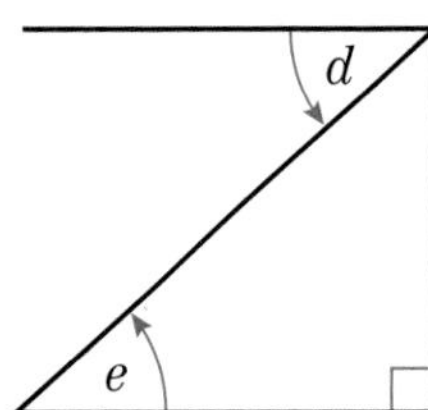

- The sine, cosine and tangent of some angles may be written exactly. ... *Mastery lesson 5.7*

	30°	**45°**	**60°**	**0**	**90°**
sin	$\frac{1}{2}$	$\frac{\sqrt{2}}{2}$	$\frac{\sqrt{3}}{2}$	0	1
cos	$\frac{\sqrt{3}}{2}$	$\frac{\sqrt{2}}{2}$	$\frac{1}{2}$	1	0
tan	$\frac{\sqrt{3}}{3}$	1	$\sqrt{3}$	0	

Reflect

'Notation' means symbols. Mathematics uses a lot of notations.

For example:

= means is equal to   ° means degrees   ∟ means a right angle

Look back at this unit. Write a list of all the maths notation used.

Why do you think this notation is important?

Could you have anwsered the questions in this lession without understanding the maths notation?

5 Unit test

Log how you did on your Student Progression Chart.

1 **Reasoning** ABCDEFGHI is a regular nonagon.

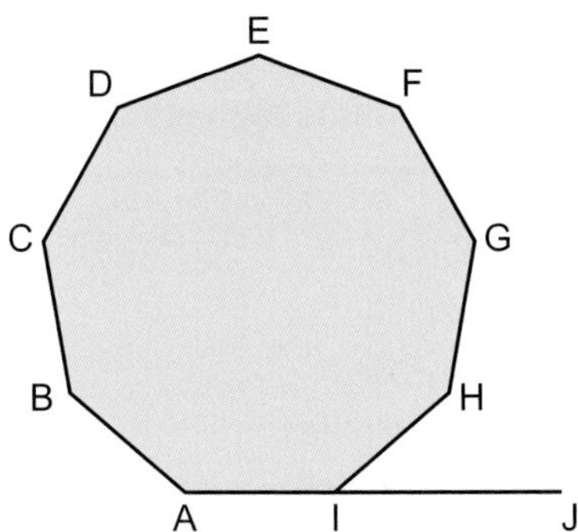

a What is the sum of the interior angles?

b Work out the size of angle HIJ. *(3 marks)*

2 PQR is a right-angled triangle.

PR = 12.3 cm

RQ = 6.4 cm

Calculate the length of PQ.

Give your answer correct to 2 decimal places.

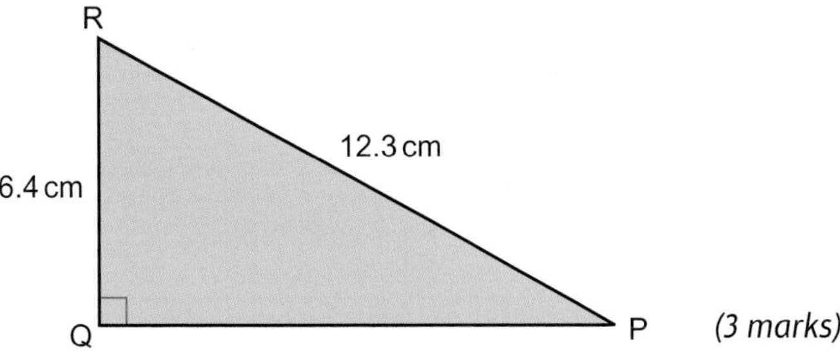

(3 marks)

3 **Communication**

ABC and DEF are straight lines.

AC is parallel to DF.

BE = CE

Work out the value of x.

Give reasons for your answer.

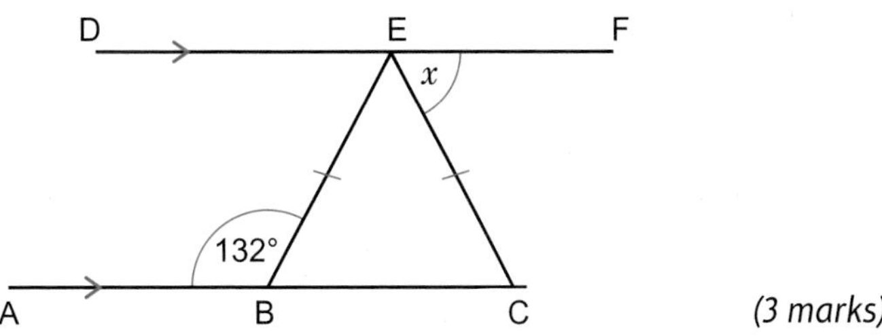

(3 marks)

4 **Communication**

ABC and DEFG are straight lines.

AC is parallel to DG.

Prove that the angle sum of any triangle is 180°.

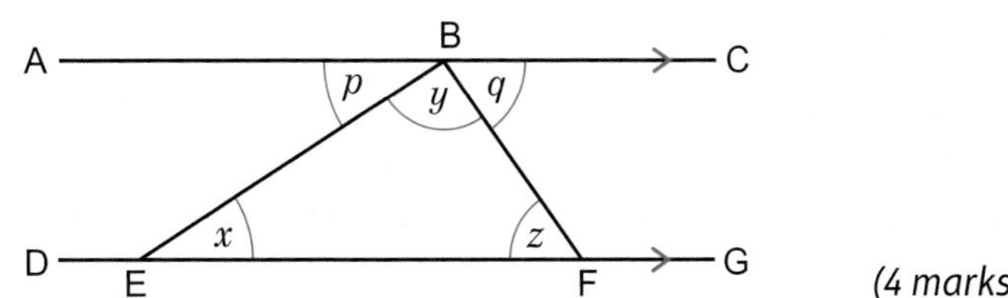

(4 marks)

5 Write down the value of

a tan 0°

b sin 90°

c cos 0°

d cos 45°

e sin 60° *(5 marks)*

6 XYZ is a right-angled triangle.
YZ = 5.4 m
XZ = 7.6 m
Calculate the size of the angle marked x.
Give your answer correct to 1 decimal place. *(3 marks)*

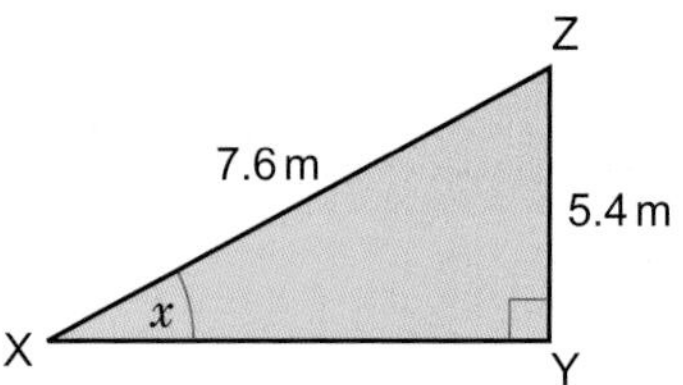

7 **Reasoning** Kari builds a skate ramp with 2 metres of wood. She wants the vertical height of the ramp to be 1 metre.
What does the angle of elevation need to be? *(3 marks)*

8 **Reasoning** A ship is sighted from the top of a lighthouse.
The angle of depression from the lighthouse to the ship is 45°.
The distance from the top of the lighthouse directly to the ship is 4 miles.
Calculate the horizontal distance of the ship from the bottom of the lighthouse.
Give your answer correct to 2 decimal places. *(3 marks)*

9 **Problem-solving** A rectangular lawn has a diagonal path running across it.
The lawn is 10 m wide and 15 m long.
Work out the length of the path.
Give your answer in surd form. *(3 marks)*

Sample student answers

Which student gives the best answer? Explain.

Exam-style question

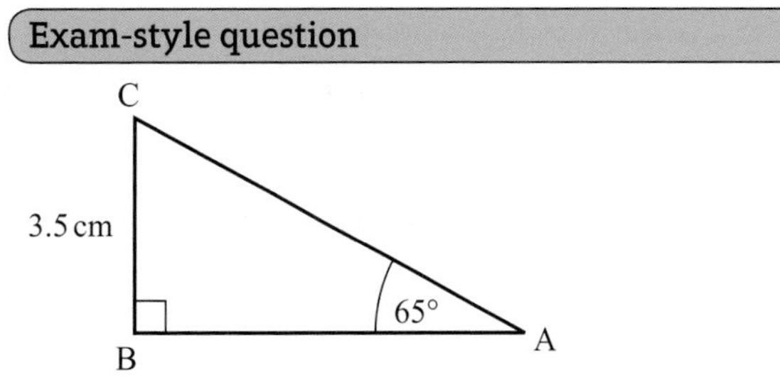

ABC is a right-angled triangle.
BC = 3.5 cm
Angle ABC = 90° and angle BAC = 65°
Calculate the length of AC.
Give your answer correct to 2 decimal places. **(3 marks)**

Student A

$\sin 65° = \frac{\text{opp}}{\text{hyp}}$

$0.9 = \frac{3.5}{AC}$

$AC = \frac{3.5}{0.9}$

$AC = 3.89$ cm

Student B

$\sin 65° = \frac{3.5}{\text{hyp}}$

$\text{hyp} = 3.5 \sin 65°$

$= 3.172077255$

$= 3.17$ cm

Student C

$\sin 65° = \frac{3.5}{x}$

$x = \frac{3.5}{\sin 65°}$

$x = 3.86$ cm

6 GRAPHS

Top athletes and sports teams use graphs to track their progress, particularly when new training methods are introduced, so that they know how much of an improvement they have made.

Two groups of athletes followed two different training programmes. They recorded the soreness in their muscles using a scale of 0 to 5. The graph shows their results.

Giving your answers as a scale reading,

a how much difference was there in the first readings for the two groups?

b how much difference was there in the last two readings?

c Which group had the better training session?

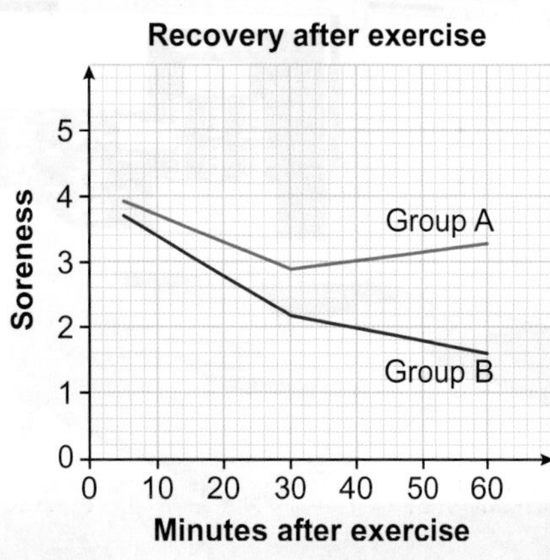

6 Prior knowledge check

Numerical fluency

1 Work out

a $4^2 + 2 \times 4 - 3$ b $(-2)^3 + 7 \times -2$

2 Write down the reciprocal of

a 7 b $\frac{1}{4}$ c -3 d $-\frac{2}{5}$

Algebraic fluency

3 Miguel walks 7.5 km in 1.5 hours. What is his speed?

> **Q3 hint**
> $\text{Speed} = \frac{\text{distance}}{\text{time}}$

4 When $x = 5$, work out

a $x + 4$ b $2x - 3$ c $\frac{1}{x}$

d x^2 e x^3

5 Solve

a $3x + 5 = 9$ b $3x - 2 = -x + 10$

Graphical fluency

6 Write down the equation for each line.

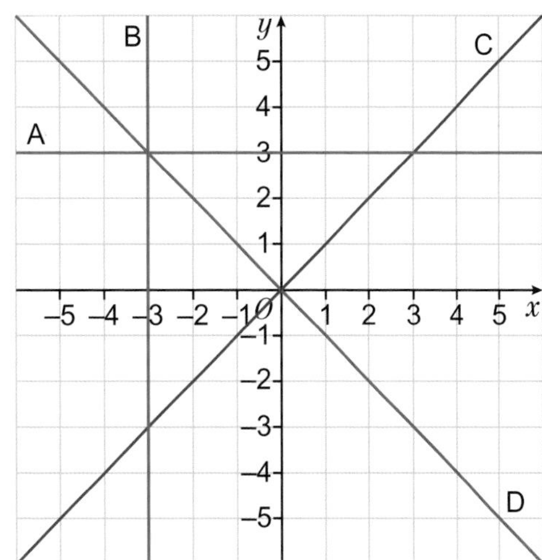

7 a Copy and complete the table of values for $y = 2x + 1$.

x	−3	−2	−1	0	1	2	3
y							

b Plot the graph of $y = 2x + 1$.

8 For each of these graphs, work out

i the gradient

ii the y-intercept.

a

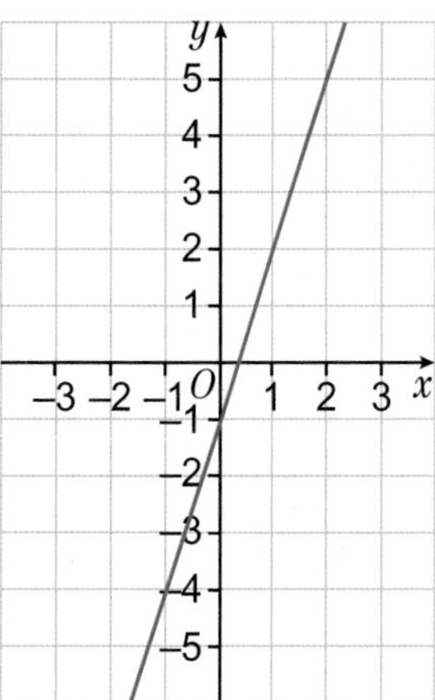

b

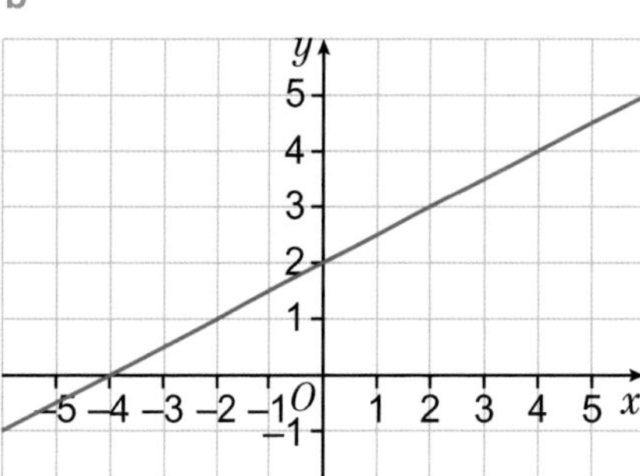

c

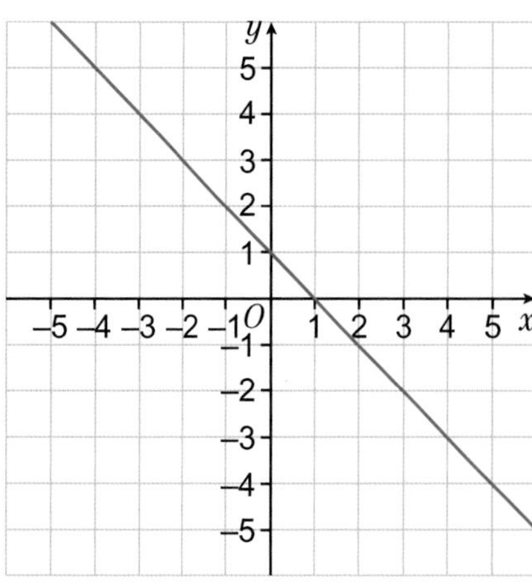

9 **Real** The graph shows the amount an electrician charges his customers.

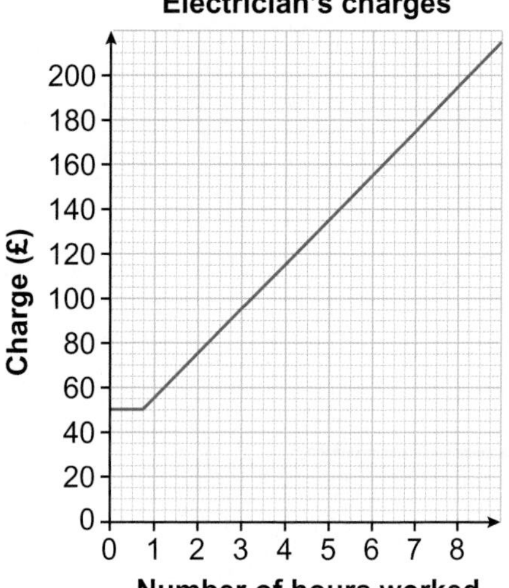

a How much does the electrician charge

i for 1 hour's work

ii for $5\frac{1}{4}$ hours' work?

b The electrician charges a call-out fee.

i How much is the call-out fee?

ii How many minutes of work does the call-out fee include?

iii How much does the electrician charge per hour after the initial call-out fee?

★ Challenge

10 The graphs of $y = x^2 + 2$ and $y = 2x + 5$ are plotted on the grid.

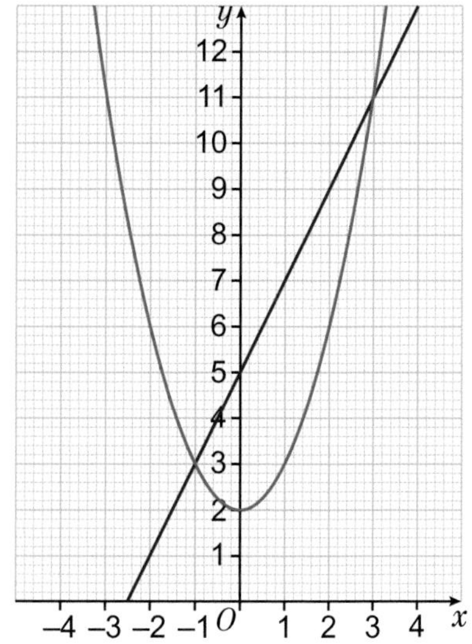

Give the coordinates of the two points of intersection of these graphs.

6.1 Linear graphs

Objectives

- Find the gradient and y-intercept from a linear equation.
- Rearrange an equation into the form $y = mx + c$.
- Compare two graphs from their equations.
- Plot graphs with equations $ax + by = c$.

Why learn this?

You can use linear graphs to show how two values are related, like converting money from pounds to dollars.

Fluency

Which graph has positive gradient?
Which has negative?
What are the x- and y-intercepts of each graph?

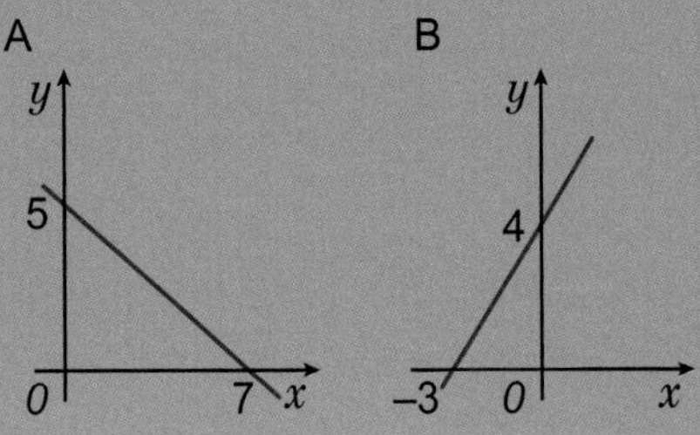

1 Rearrange $2x - y = 5$ to make y the subject.

Questions in this unit are targeted at the steps indicated.

2 On squared paper, draw a line with gradient

a 5 b $\frac{1}{2}$ c -3

3 Copy and complete this table for the graphs on the grid.

Equation of line	Gradient	y-intercept
$y = 2x + 4$		
$y = 2x$		
$y = 2x - 3$		

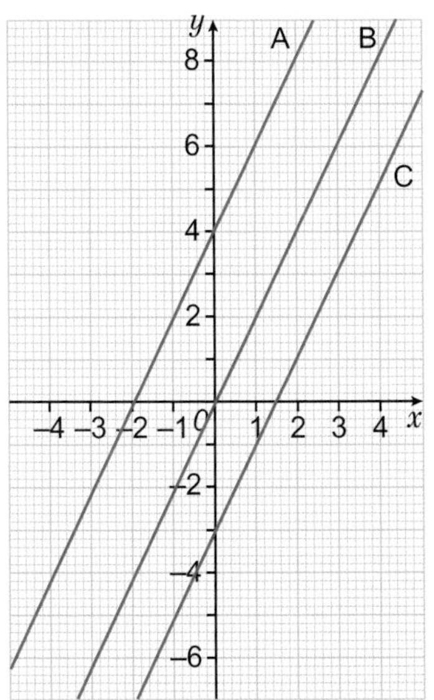

Discussion How can you find the gradient and the y-intercept from the equation of a line?

Key point 1

A **linear equation** generates a straight-line (linear) graph.

The equation for a straight-line graph can be written as $y = mx + c$ where m is the gradient and c is the y-intercept.

Example 1

Write the equation of

a line A

b line B.

a $y = mx + c$ — Write down the formula.

gradient $m = 2$ — Work out the gradient from points on the line. Find the y-intercept.

y-intercept is (0, −8), so $c = -8$

Equation of line A is $y = 2x - 8$ — Substitute the values into the formula.

b $y = mx + c$

gradient, $m = -3$

y-intercept is (0, 2), so $c = 2$

Equation of line B is $y = -3x + 2$

4 a Match each line to an equation.

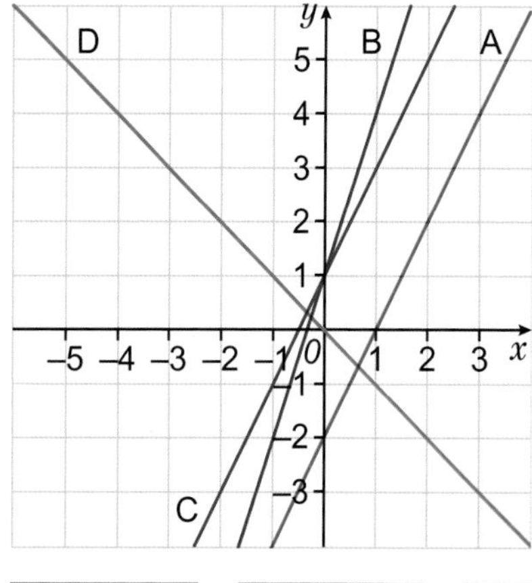

$y = -x$	$y = 3x + 1$	$y = 2x + 1$	$y = 2x - 2$	$y = 2x - 2$

b Which line passes through the origin?

c Which line is the steepest?

d Which lines have the same intercept?

e Which lines are parallel?

Q4b communication hint The origin is the point (0, 0).

Q4e hint Parallel lines have the same gradient.

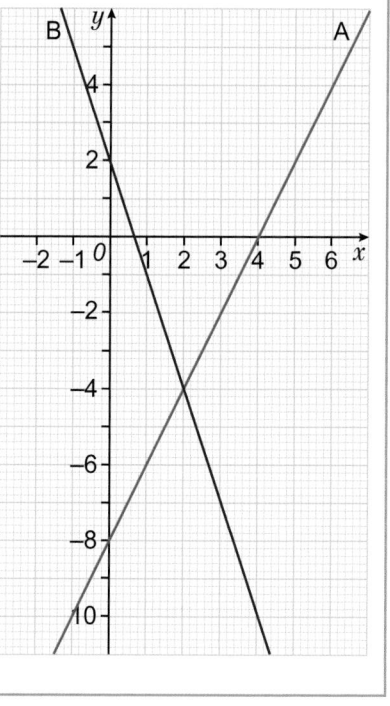

5 Write the equations of these lines.

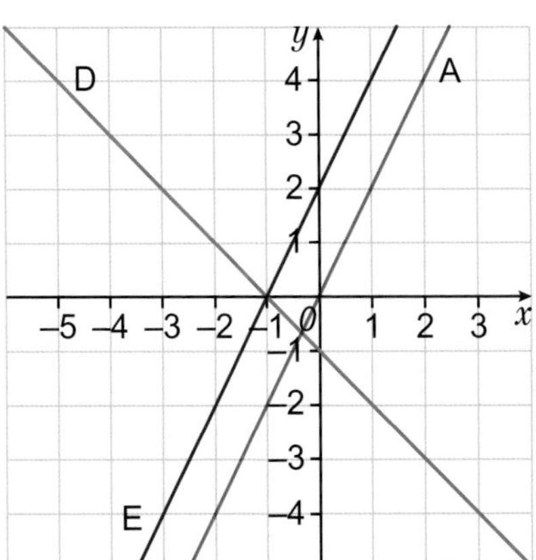

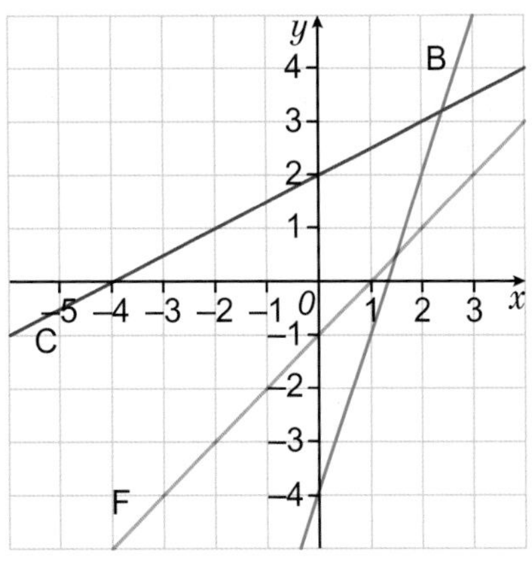

Q5 hint Read the scale of both axes carefully.

6 Here are the equations of some linear graphs. Which of these graphs

a cross the y-axis at the same point

b are parallel?

i $y = 2x - 3$ ii $y = 3x + 1$ iii $y = x - 1$ iv $y = 2x + 1$ v $y = -x$

Key point 2

To find the y-intercept of a graph, find the y-coordinate where $x = 0$.
To find the x-intercept of a graph, find the x-coordinate where $y = 0$.

7 a For the equation $2y - x = 3$

i copy and complete the table of values

x	0	
y		0

ii plot the graph on suitable axes.

Q7a i hint When $x = 0$, what is the value of y?

b Repeat part **a** for the lines with equation

i $x + y = 4$ ii $x + y = 7$

Discussion Where do you think the graph of $x + y = 3$ will cross the axes? Where will $x + y = -1$?

8 In **Q7** you drew the graphs of $2y - x = 3$, $x + y = 4$ and $x + y = 7$.

a Rearrange each equation to make y the subject.

b Read the gradients and y-intercepts from each.

c Look back at your graphs in **Q7** to check the gradients and y-intercepts are correct.

Key point 3

To compare the gradients and y-intercepts of two straight lines, make sure their equations are in the form $y = mx + c$.

9 **Reasoning** Which is the steepest line?

a $y = \frac{1}{3}x - 2$ b $2y + 5x = 7$ c $3x + \frac{1}{2}y = 2$

d $y = 1 - 4x$ e $6x - 2y = 9$

Q9 hint Rearrange to $y = mx + c$ if necessary.

10 **Communication / Problem-solving** Which of these lines pass through (0, 3)? Show how you worked it out.

A $y = 3x - 3$ **B** $4y - 8x = 12$ **C** $5y = 3x - 15$ **D** $2x - y = 3$ **E** $3x + y = 3$

6.2 More linear graphs

Objectives

- Sketch graphs using the gradient and intercepts.
- Find the equation of a line, given its gradient and one point on the line.
- Find the gradient of a line through two points.

Did you know?

You can plot a straight-line graph using just the gradient and intercept – you don't have to work out a table of values.

Fluency

Which lines are parallel? Which have the same y-intercept?

- $y = 3x + 1$
- $y = -x + 1$
- $y = x + 2$
- $y = 5 - x$

Warm up

1 Write the equation of each line.

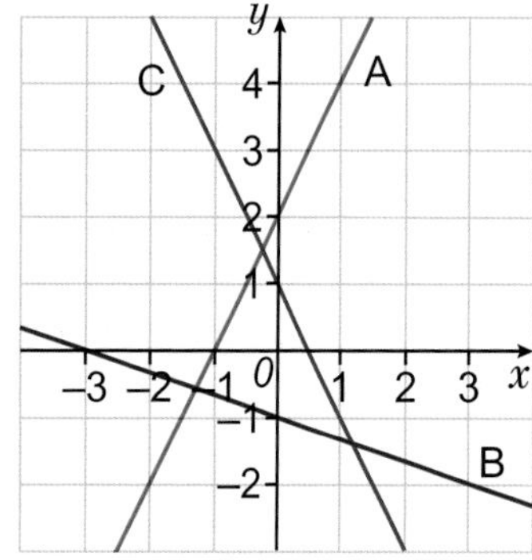

2 The equation of a line is $y = 3x + c$. Find the value of c when $x = 4$ and $y = 15$.

Example 2

On the same grid, draw these graphs from their equations.

a $y = 2x - 1$

b $y = -x + 4$

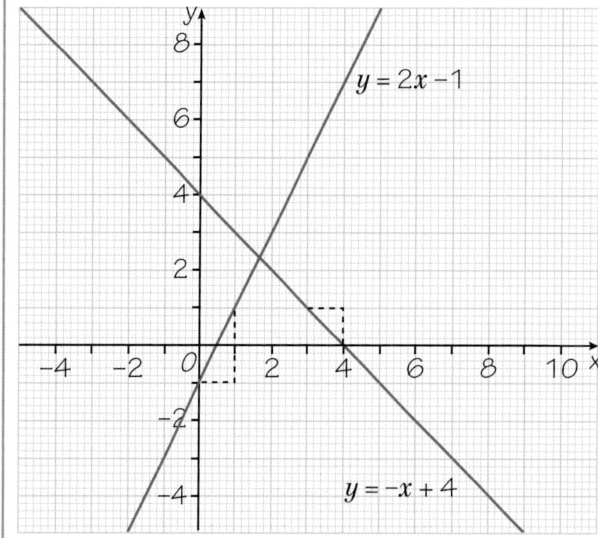

Plot the y-intercept.

Decide if the gradient is positive or negative.

Draw a line with this gradient, starting from the y-intercept.

Extend your line right across the grid.

Label the line with its equation.

ActiveLearn Homework, practice and support: Higher 6.2

3 Draw these graphs from their equations.
Use a coordinate grid from −10 to +10 on both axes.

a $y = 2x + 4$ b $y = 2x - 3$ c $y = 3x$
d $y = \frac{1}{2}x + 2$ e $y = -2x + 1$ f $y = -3x + 2$

4 **Reasoning** Match each equation to one of these sketch graphs.

$y = 5x + 1$	$y = 2x + 3$	$y = -x + 4$	$y = -3x$	$y = \frac{1}{2}x + 3$

A
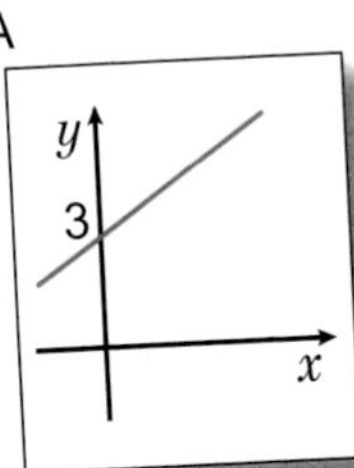

B
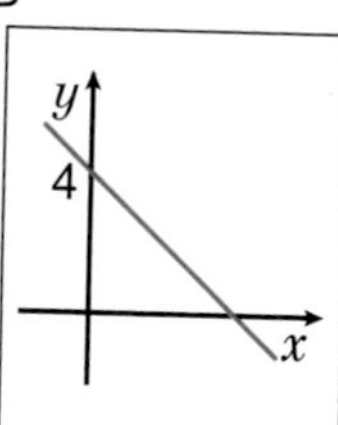

C
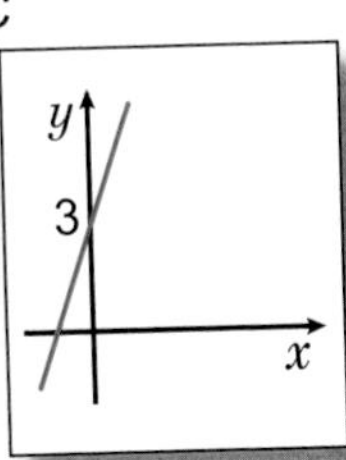

D
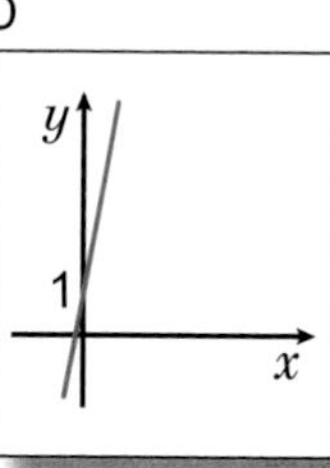

E
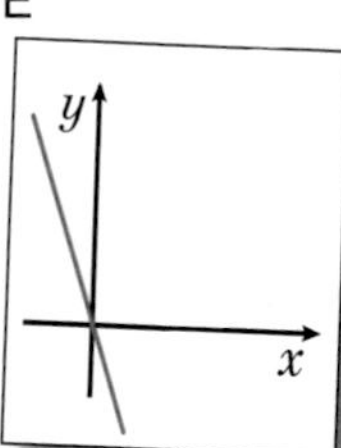

Reflect What does it mean in maths to sketch a graph? What information do you include on a sketch? How is this different from plotting a graph?

5 Sketch the graphs of

a $y = 2x$ b $y = 3x + 1$ c $x + y = 5$

Q5 hint Find the x- and y-intercepts. Join them with a straight line.

6 a Find the x-intercept and y-intercept of the graph with equation

i $x + y = 3$
ii $3x + y = -6$
iii $y - x = 2$
iv $y - 2x = 4$

b Sketch the graphs.

Q6 hint Mark the x- and y-intercept and join with a straight line.

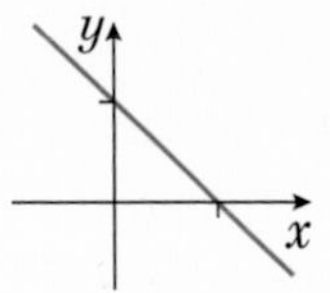

Key point 4

A linear function has a graph that is a straight line.

7 **Reasoning** Which of these are linear functions?

a $y = -3x$ b $y = \frac{x}{4}$ c $y = 2x + 1$
d $3x + 2y = 5$ e $y = x^2 + 4$ f $y = \frac{4}{x}$

Q7 hint Can you write them as $y = mx + c$?

8 **Reflect** $y = mx + c$ is a linear equation.
In your own words, how would you describe what 'linear' means?

9 **Reasoning**

a Does the point (3, 6) lie on the line $y = \frac{1}{2}x$?
b Does the point (2, 9) lie on the line $y = 2x + 5$?
c Does the point (−2, −7) lie on the line $y = -4x - 1$?

Q9 hint Substitute the values of x and y into the equation of the line.
Do both sides of the equation have the same value? What does it mean if they do? What does it mean if they do not?

10 **Problem-solving** A straight line has gradient 2. The point (4, 5) lies on the line. Find the equation of the line.

Q10 strategy hint Substitute the gradient (m) into the equation $y = mx + c$.
Then substitute the given values of x and y (the coordinates of the point) and solve to find c.

11 Problem-solving Work out the equations of these straight-line graphs.

a The line with gradient 3 that passes through the point (0, 5)

b The line with gradient −1 that passes through the point (3, 0)

c The line with gradient $\frac{1}{2}$ that passes through the point (6, 1)

d The line with gradient −2 that passes through the point (5, −4)

12 Find the gradient of the line joining points A (−3, −2) and B (5, 4)

a by drawing the graph and using the formula

$$\text{gradient} = \frac{\text{difference in } y\text{-coordinates}}{\text{difference in } x\text{-coordinates}}$$

b using the formula $m = \frac{y_2 - y_1}{x_2 - x_1}$ where

$A = (x_1, y_1)$ and $B = (x_2, y_2)$
(−3, −2) (5, 4)

Q12a hint

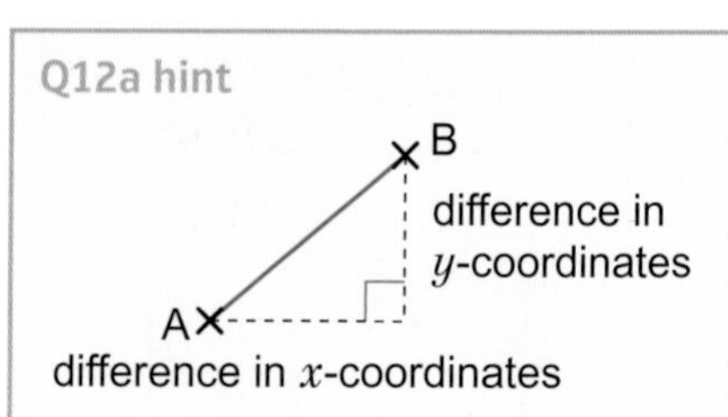

Discussion Which method do you prefer? If you didn't draw a graph, could you still use method **a**?

13 Reasoning P is the point (−2, 6). Q is the point (10, 0).

a Find the gradient of line PQ.

b Write $y = mx + c$ using your gradient from part **a**. Substitute the coordinates of Q into this equation. Solve to find c.

c Write the equation of the line PQ.

14 To find the coordinates of the point where these graphs intersect

$y = 4x - 3$ $y = -x + 12$

a write the two equations equal to each other

b solve to find x

c substitute x into one of the first equations to find y

d write the coordinates (x, y)

Q14a hint $4x - 3 = -x + 12$

15 Find the coordinates of the point where these graphs intersect.

$y = -x + 2$ $3x + 2y = 5$

Q15 hint Write both as $y = mx + c$.

6.3 Graphing rates of change

Objectives

- Draw and interpret distance–time graphs.
- Calculate average speed from a distance–time graph.
- Understand velocity–time graphs.
- Find acceleration and distance from velocity–time graphs.

Why learn this?

A rate of change tells us how fast something changes in a given time period. Your speed measures how fast your position changes over time.

Fluency

A car travels 17 miles in $\frac{1}{2}$ hour. What is its speed?

Warm up

1 Find the area of each shape.

a

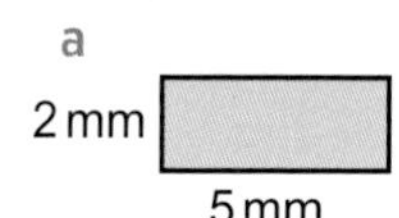

b

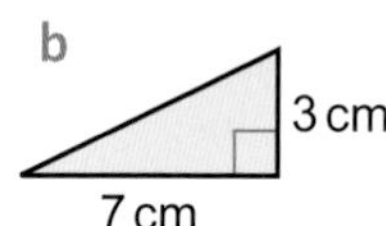

Q1 hint Remember to give units for your answer.

ActiveLearn Homework, practice and support: Higher 6.3

Key point 5

A **distance–time graph** represents a journey.
The vertical axis represents the *distance* from the starting point.
The horizontal axis represents the *time* taken.

2 **Real** Sophie drives from her house to a cinema.
The distance–time graph shows her journey.

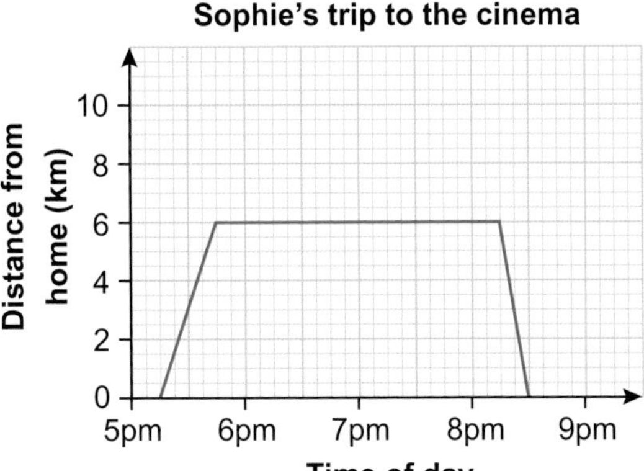

a How far is Sophie's house from the cinema?
b What time does Sophie arrive at the cinema?
c How long does she take to drive to the cinema?
d How long is she at the cinema?
e What was her speed on the way to the cinema?
f Work out the gradient for her drive to the cinema. What do you notice?

Discussion What does a horizontal line mean on a distance–time graph? What does the gradient mean?

Key point 6

On a distance–time graph, the gradient is the speed.

3 **Real / Modelling** Amal drives to her friend's house.
She drives 150 km in 2.5 hours. Then she stops for a half-hour break.
She then drives 70 km in 1 hour and arrives at her friend's house.

a On graph paper draw a horizontal axis from 0 to 4 hours and a vertical axis from 0 to 220 km.
Draw a distance–time graph to show Amal's journey.
b Work out her speed for the first part of the journey.

4 Kirsty is practising speed skating.
She covers the 1200 m straight course in 75 seconds.
She rests for 1 minute then skates back to the start line at 10 m/s.
a Draw a distance–time graph to show Kirsty's skating practice.
b Work out the fastest speed she travelled.

Q4a hint Work out how far Kirsty travels in 1 second, or in 10 seconds. Plot this as a point.

Q4b hint What units do you need to use?

5 **Exam-style question**

Simon went for a cycle ride.
He left home at 2pm.
The travel graph represents part of Simon's cycle ride.
At 3pm Simon stopped for a rest.

a How many minutes did he rest? **(1 mark)**

b How far was Simon from home at 5pm? **(1 mark)**

At 5pm Simon stopped for 30 minutes.
Then he cycled home at a steady speed.
It took him 1 hour 30 minutes to get home.

c Complete the travel graph. **(2 marks)**

March 2013, Q3, 1MA0/2H

Exam hint
In an examination, graphs are marked online so make sure the examiner can see your pencil drawings. Use a pencil that is easy to see over the grid lines.

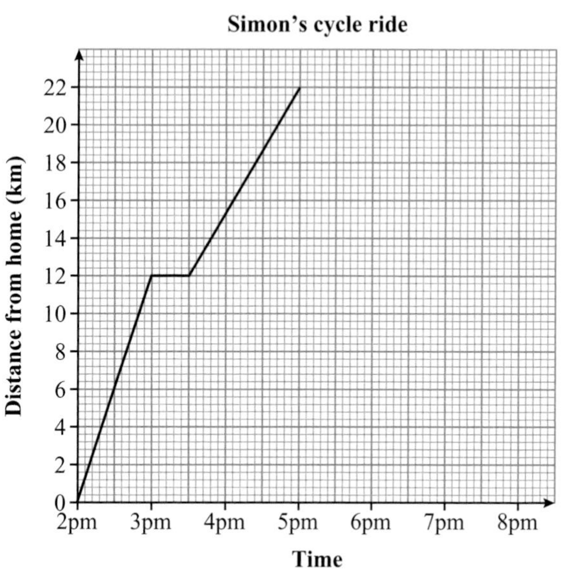

Q5 communication hint
Steady speed means travelling the same distance each minute.

Key point 7

Average speed = $\frac{\text{total distance}}{\text{total time}}$

Make sure your units match.

6 **Real / Modelling** The table shows a train journey from Birmingham to Shrewsbury. The train stops at Wolverhampton and Telford on the way.

Station	Time
Birmingham New Street (departing)	1432
Wolverhampton (arriving)	1442
Telford (arriving)	1459
Shrewsbury (arriving)	1519

When a train arrives at a station, it stays for 3 minutes before leaving for the next station. There are 16 miles between each pair of stations.

a Draw a distance–time graph for this journey.

b Work out the speed of the train between Birmingham and Wolverhampton.

c Work out the speed of the train between Telford and Shrewsbury.

d What was the average speed for the whole journey?

Q6d hint Do you think the average speed for the whole journey will be faster or slower than the speed over each part of the journey? Why?

7 Look at the graph you drew for **Q3**. What was Amal's average speed for the whole journey?

8 **Real / Modelling** Train A travels from Manchester to London.

Train B travels from London to Manchester.

a Use the graph to estimate how far they are from London when they pass each other.

b Work out the speed for each part of the journey for Train A.

c When was Train B travelling fastest? How can you tell this from the graph?

d Which train travelled faster on average?

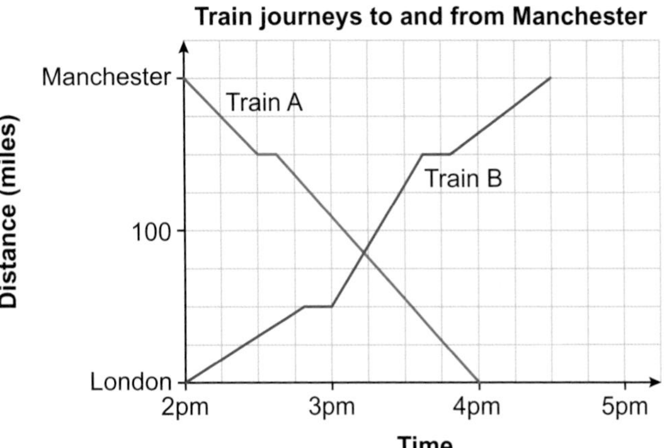

Discussion Are these distance–time graphs good models for train journeys? What assumptions have been made?

> **Q8a hint** Look at the units on the graph.

Key point 8

The gradient of a straight line graph is the rate of change.

9 **Reasoning** Josh runs water into these three containers at a constant rate.

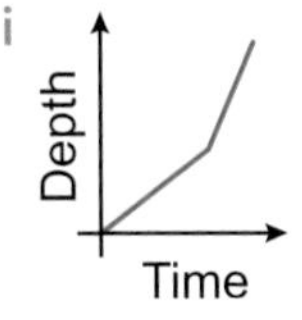

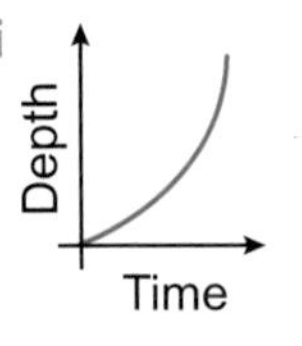

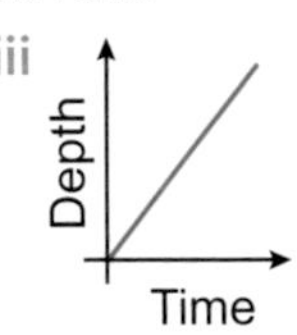

a In which container does the depth of water increase by the same amount every second?

b Which graph shows the depth of water increasing steadily?

c Match each graph to one container.

Discussion Why is graph **ii** curved?

> **Q9 communication hint** **Constant rate** means the same amount flows in every second.

10 **Reasoning** Here are three vases.
They are all cylinders and all the same height.
Skye fills the vases with water at the same rate.

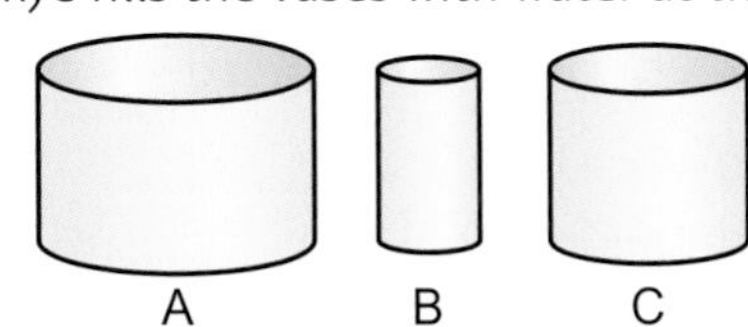

On the same axes, sketch three graphs showing the rate at which water fills the vases.

> **Q10 hint** Which vase will be full first? Which will be full last?

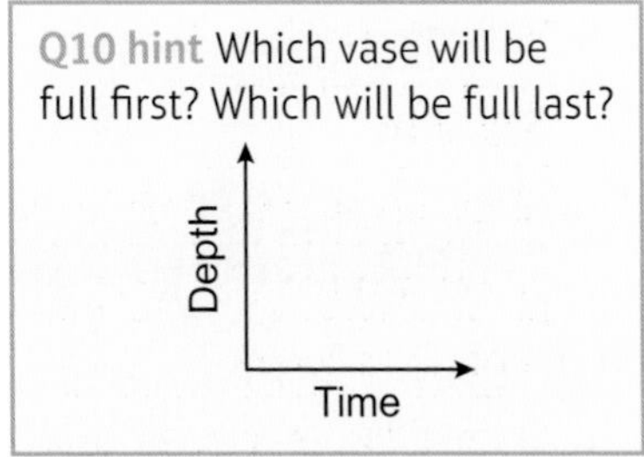

Key point 9

A **velocity–time graph** has time on the x-axis and velocity on the y-axis.
The gradient is the rate of change of velocity, or acceleration.
A positive gradient means an object is speeding up.

$$\text{Acceleration} = \frac{\text{change in velocity}}{\text{time}}$$

The area under a velocity–time graph is the distance travelled.

11 **Real** Gavin goes for a run.
The graph shows his journey.

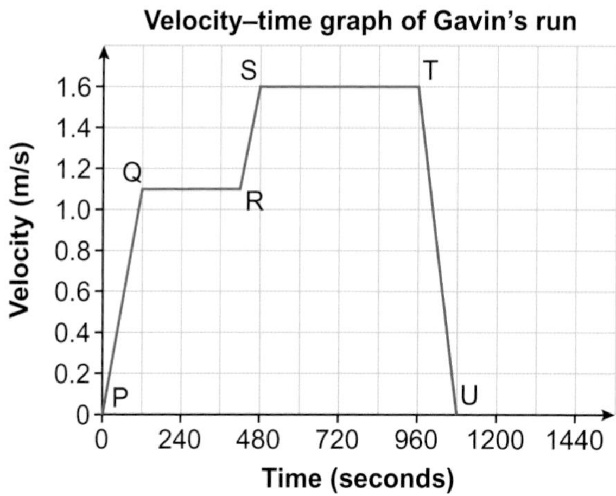

Q11b hint Read seconds from the graph and change to minutes.

Q11c hint Gradient of the line segment PQ.

acceleration (m/s²)

$$= \frac{\text{change in velocity (m/s}^2\text{)}}{\text{time (s)}}$$

Q11d hint Find the area of the triangle under the line segment TU. Read the height from the velocity axis and the base from the time axis.

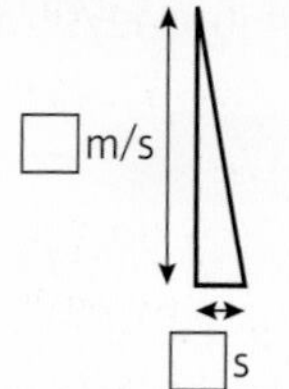

Work out

a Gavin's maximum velocity

b how many minutes he ran at 1.1 m/s

c his acceleration for the first part of the journey

d the distance Gavin ran during the last 120 seconds.

e Copy and complete this description of Gavin's run.

He accelerated at □ m/s² for the first □ minutes,

then ran at a constant velocity of □ m/s for □ minutes.

Next …

Discussion How do you show constant speed, constant acceleration and constant deceleration on a velocity–time graph?

Communication hint **Deceleration** is negative acceleration. It means that an object is slowing down.

12 **Reflect** Why do we call the graphs in this lesson 'rate of change' graphs?
What kind of rates of change might you find in everyday life?

Q12 hint What can you think of that changes?

6.4 Real-life graphs

Objectives

- Draw and interpret real-life linear graphs.
- Recognise direct proportion.
- Draw and use a line of best fit.

Why learn this?

Engineers use graphs showing the performance of car engines to work out the most efficient speeds to save fuel.

Fluency

Here is a sketch of a graph with a gradient of $-\frac{1}{2}$.
What is its equation?

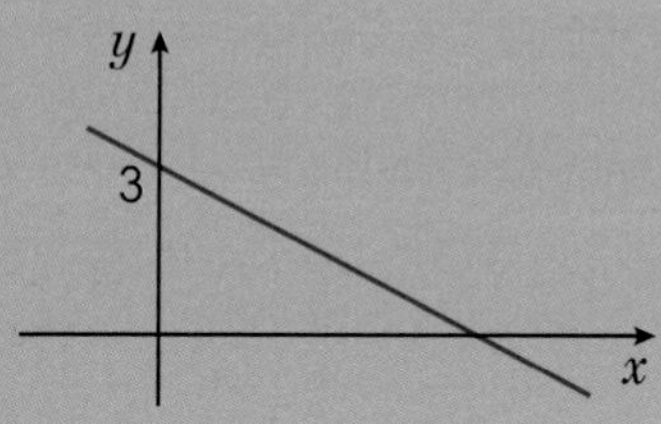

1 The table shows the charge for using different numbers of units of electricity.

Units	0	200	500	700	900	1000
Charge (£)	12	40	82	110	138	152

a Plot these points on a grid.

b i Use your graph to find the charge for using 800 units of electricity.

ii Declan receives a bill for £60. How many units of electricity has he used?

Key point 10

Graph axes do not have to start at zero.
A zigzag line —⁄\/— shows that values have been missed out.

2 **Real / Problem-solving** Gurpreet is buying some pens to give away at an exhibition.

The graph shows the price per pen depending on how many pens are ordered.

a How much would a single pen cost?

b Gurpreet buys 60 pens. How much does he spend altogether?

c For another event, Gurpreet is given a budget of £75. How many pens can he afford to buy?

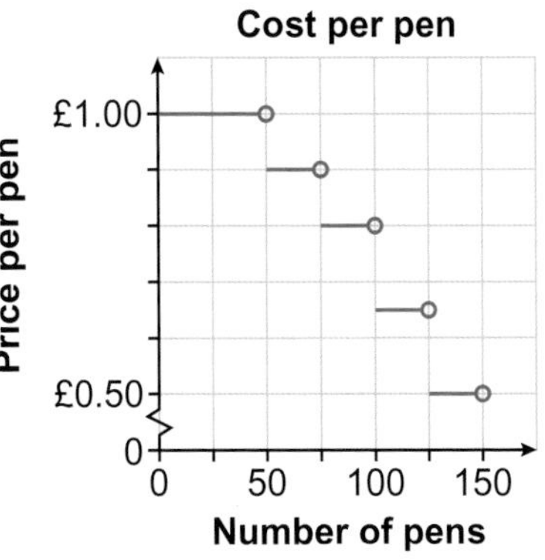

Q2 hint The open circles show that the upper limit of each bar is not included at that price.

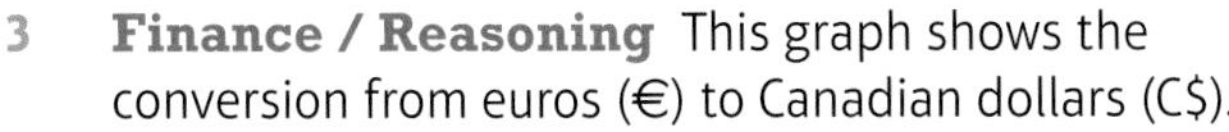

3 **Finance / Reasoning** This graph shows the conversion from euros (€) to Canadian dollars (C$).

a How many dollars do you get for €10?

b How many euros do you get for C$1?

c Work out the gradient of the graph.

Discussion What does the gradient tell you?

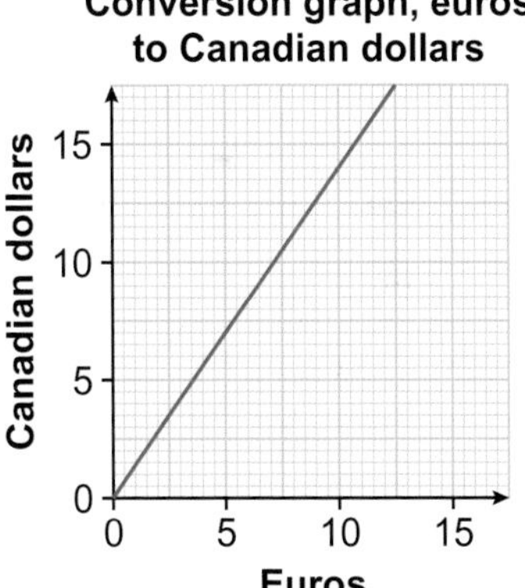

4 **Real / Reasoning** This graph shows the charge to hire a van for a number of days.

a Calculate the gradient of the line.

b What is the initial charge before you add on the daily hire charge?

c Write down the equation of the line.

Discussion What does each part of the equation represent?

d Alice has £450. For how many days can she hire the van?

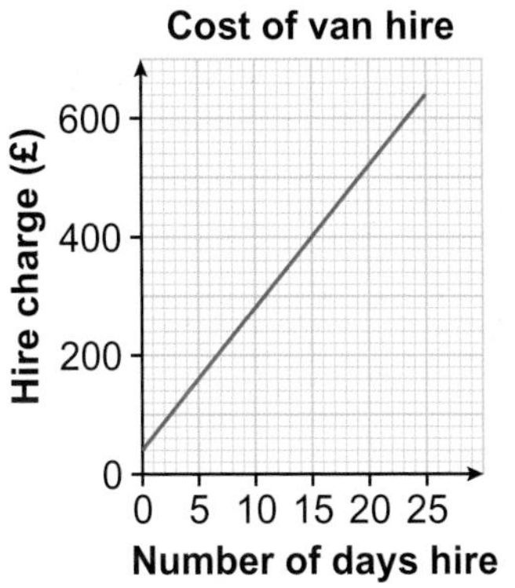

Key point 11

When two quantities are in **direct proportion**

- their graph is a straight line through the origin
- when one variable is multiplied by n, so is the other.

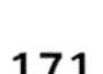

5 **Modelling** Which of these graphs show one variable in direct proportion to another?

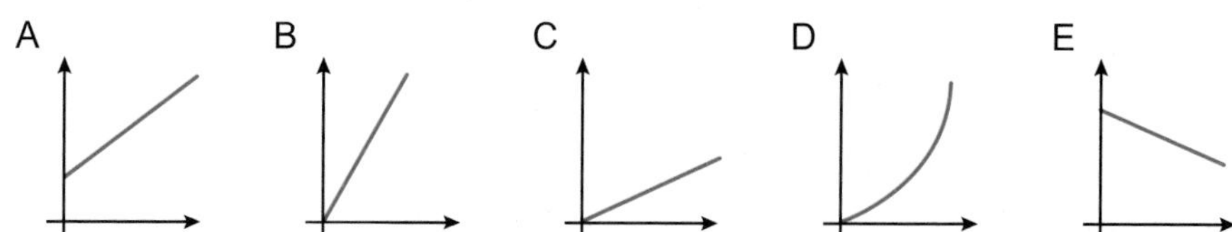

6 **Modelling** Look at the graph you drew for **Q1**, and the graphs in **Q3** and **Q4**. Which show direct proportion?

7 **Real / Reasoning** A recipe uses a spice mix including chilli powder and cumin in the ratio 2 : 5.

a Copy and complete this table.

Chilli powder (grams)	1	4	10
Cumin (grams)			

b Draw a graph showing grams of cumin (y) against grams of chilli (x).

c Write the equation linking x and y.

d How much chilli would you need for a recipe using 85 g of cumin?

Discussion Does extending the graph give accurate values?

Q7c hint You could write the equation of the line.

Q7d hint How can you use the values in the table to help you?

8 **Reasoning / Modelling** Zadie has a new freezer delivered to her house. Zadie turns on the freezer and a sensor records the temperature inside the freezer. The graph gives information about the temperature, T °C, of the air inside the freezer.

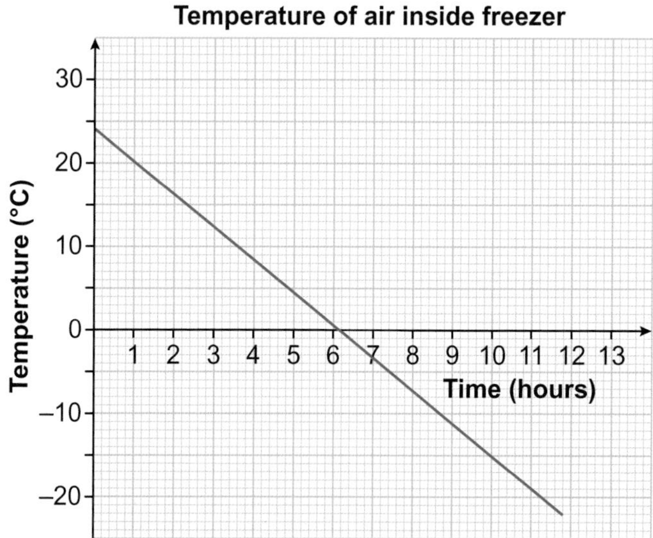

a What does the y-intercept tell you? What does the x-intercept tell you?

b Use the graph to estimate the temperature 3.5 hours after Zadie turns on the freezer.

c How much does the temperature fall over the first 5 hours?

d Is the rate of decrease of temperature constant? How can you tell from the graph?

Discussion Can you predict the temperature when $x = 14$? When $x = 36$?

9 **Modelling / Reasoning** The table shows the largest quantity of a sugar, k grams, which will dissolve in a cup of coffee at temperature t °C

t (°C)	44	50	62	70	78	85
k (grams)	265	300	360	400	440	475

a On a suitable grid, plot the points and draw a graph to illustrate this information.

b Use your graph to find

i the lowest temperature at which 120 g of sugar will dissolve in the coffee

ii the largest amount of sugar that will dissolve in the coffee at 81 °C.

The equation of the graph is in the form $k = at + b$.

c Use your graph to estimate the values of the constants a and b.

d Will 4 teaspoons of sugar dissolve in the coffee at 90 °C? Use the equation to decide. Justify your answer.

Q9d hint
1 teaspoon of sugar = 5 grams

10 **Reasoning / Finance** The graph shows two different Pay As You Go mobile phone tariffs, Plan A and Plan B.

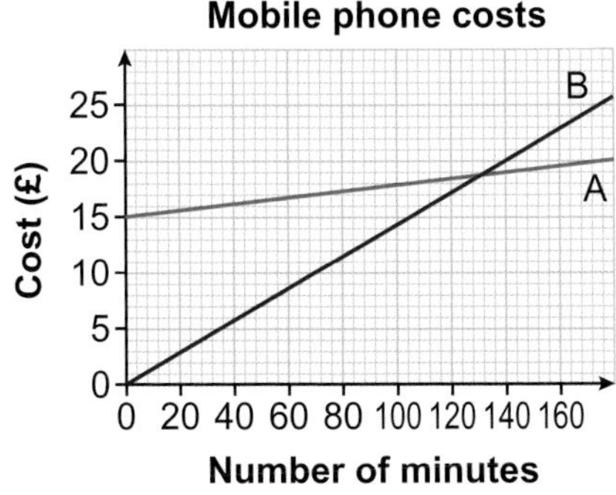

a How much does 100 minutes cost on

i Plan A ii Plan B?

What is the practical meaning of

b the y-intercept value on Plan A

c the point where the two graphs intersect?

d Another tariff, Plan C, is introduced. On Plan C you will pay £18.50 per month for unlimited minutes.
Which plan should each person choose?
Molly: Average 150 minutes of calls per month.
Theo: Average 100 minutes of calls per month.

11 **Finance / Real / Reasoning** Beth wants to sell her car.
She has tracked the online sale price of the same model of car for a month.
Here are her results.

Car age (years)	1.1	3	2	5	4.2	1.7	5.5	2.5
Price (£)	11 800	9000	10 250	4900	6000	10 700	4500	9800

a Plot a scatter graph of Beth's results.

b What type of correlation does this graph show?

c Draw in a line of best fit.

d Write the equation of your line of best fit.

e Beth's car is $3\frac{1}{2}$ years old.
Use your equation to work out how much she should sell it for.

Q11a hint Plot years against price in £1000s.

Discussion Can you use your equation to predict the price of a brand new car?

12 **Exam-style question**

The table shows life expectancy (in years) for females born in the UK from 2000 to 2013.

a From this data, work out the life expectancy of a girl born in

i 2020

ii 2050. **(4 marks)**

b Which answer is more reliable? Why? **(2 marks)**

Year of birth	Life expectancy (years)
2000	80.2
2001	80.4
2002	80.5
2003	80.5
2004	81.1
2005	81.2
2006	81.5
2007	81.7
2008	81.7
2009	82.3
2010	82.4
2011	82.6
2012	82.8
2013	83.0

Source: ONS

Q12 strategy hint You could draw a graph to show this information and extend it.

6.5 Line segments

Objectives

- Find the coordinates of the midpoint of a line segment.
- Find the gradient and length of a line segment.
- Find the equations of lines parallel or perpendicular to a given line.

Why learn this?

Parallel and perpendicular lines are useful for drawing constructions and working out angle questions.

Fluency

- Which pairs of lines are parallel and which are perpendicular?
- What can you say about graphs of parallel lines?

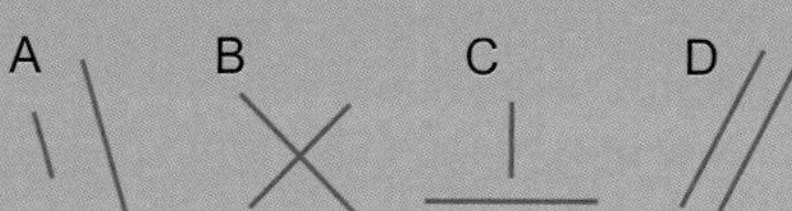

Warm up

1 What is the value half way between

a 5 and 9 b −2 and 4 c 3 and 8 d −5 and −2?

2 Here is a right-angled triangle.
What is the length of side AB?

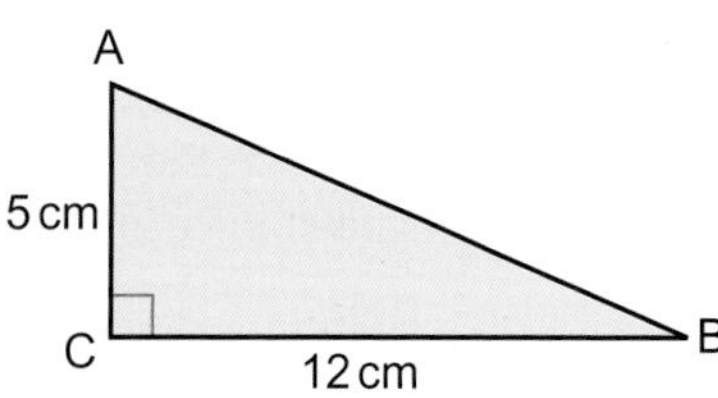

Q2 hint Use Pythagoras' theorem.

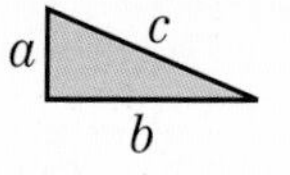

3 Write down the gradient and y-intercept of the line $y = 2x - 3$.

Q3 hint Look back at lesson **6.1** if you are stuck.

4 Work out the midpoint of a line segment AB, where

a A is (0, 3) and B is (4, 7)

b A is (2, 9) and B is (9, 2)

c A is (3, 8) and B is (−1, 6)

d A is (−4, −1) and B is (0, 0).

Discussion How can you find the midpoints of line segments without drawing?

Q4 hint Draw the lines on a grid with axes from −10 to 10.

Q4 communication hint A **line segment** is a part of a straight line.

Q4 discussion hint What value is half way between the two x-coordinates? And between the two y-coordinates?

Key point 12

The coordinates of the **midpoint** of a line segment are $\left(\frac{x_1 + x_2}{2}, \frac{y_1 + y_2}{2}\right)$

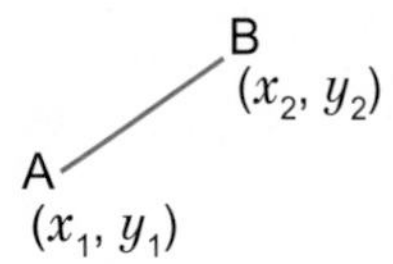

5 Work out the midpoint of a line segment PQ, where

a P is (0, 1) and Q is (3, 10)

b P is (2, 3) and Q is (6, −5)

c P is (−3, 3) and Q is (7, −2)

d P is (−7, −4) and Q is (5, 0).

Q5 hint Work out these midpoints without drawing the graphs. You can use a quick sketch to check your answer *after* you have worked it out.

6 Work out the gradient of each line segment in **Q5**.

Q6 hint Use the formula

$$\text{gradient} = \frac{\text{change in } y}{\text{change in } x} \quad \text{or} \quad m = \frac{y_2 - y_1}{x_2 - x_1}$$

7 What is the length of the line segment with end points

a E (−3, 2) and F (0, −2)

b G (−4, −1) and H (2, 7)

c J (−5, 3) and K (8, −1)?

Q7a hint Sketch a right-angled triangle and use Pythagoras' theorem to work out the length of the hypotenuse.

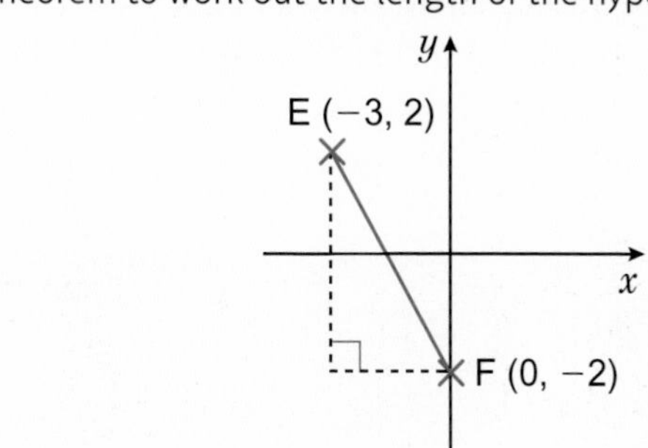

8 **Reasoning** A line is parallel to the line $y = 2x - 7$ and passes through the point (2, −5).

a Substitute the value of m for this line into $y = mx + c$.

b Substitute the coordinates of the known point to work out the equation of the line.

9 **Reasoning / Finance** The graph shows the profits of two companies who sell garden furniture.
Write the equation for the profit of company B.

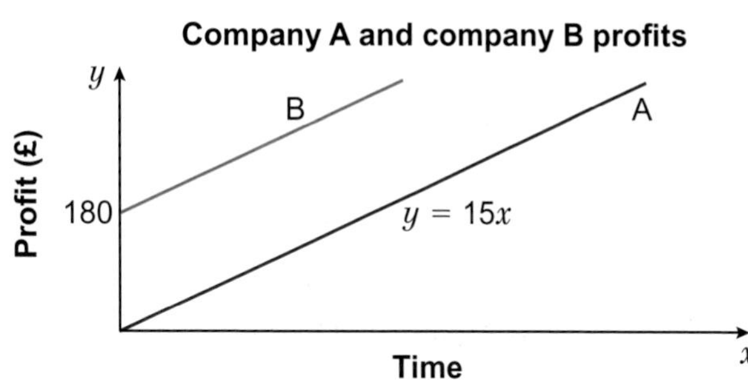

10 **Problem-solving** Write the equation of a line parallel to $y = \frac{1}{3}x + 2$, which passes through the point (9, −2).

11 **Problem-solving** Find the equation of a line that passes through the point (−2, −2) and is parallel to the line with equation $y - 3x = 7$.

12 Here are three pairs of perpendicular lines.

a Write down the gradient of each line.

b Multiply the gradients in each pair together. What do you notice?

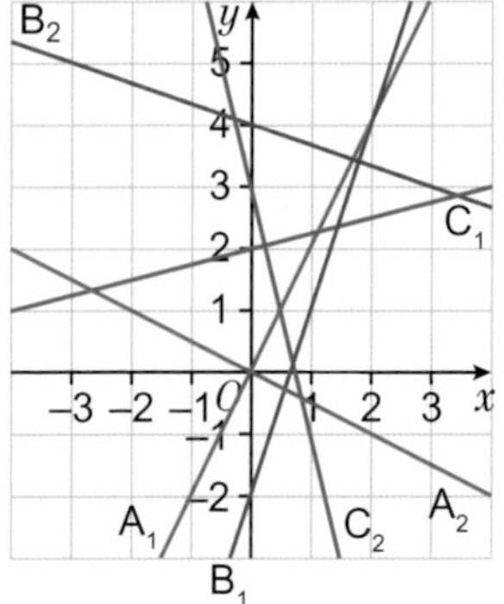

Key point 13

When two lines are **perpendicular**, the product of the gradients is −1.

When a graph has gradient m, a graph perpendicular to it has gradient $\frac{-1}{m}$.

13 Write down the gradient of a line perpendicular to

a $y = 3x - 1$ b $y = -\frac{1}{4}x + 2$ c $y = \frac{2}{5}x + 3$

14 **Exam-style question**

Find the equation of a line

a that is perpendicular to the line with equation $y = \frac{1}{2}x$ and passes through the point (−2, 9) **(2 marks)**

b that is perpendicular to the line with equation $x + y = 6$ and passes through the point (−3, −7). **(2 marks)**

6.6 Quadratic graphs

Objectives

- Draw quadratic graphs.
- Solve quadratic equations using graphs.
- Identify the line of symmetry of a quadratic graph.
- Interpret quadratic graphs relating to real-life situations.

Did you know?

Quadratic graphs help us work out the path followed by projectiles as they move through the air, like footballs or juggling balls.

Fluency

Which of these are quadratic expressions?

- $x^3 + x^2$
- $4x + 2$
- $1 - x^2$
- $5x^2 - 6x + 1$

ActiveLearn Homework, practice and support: Higher 6.6

1 Write down the equation of each line.

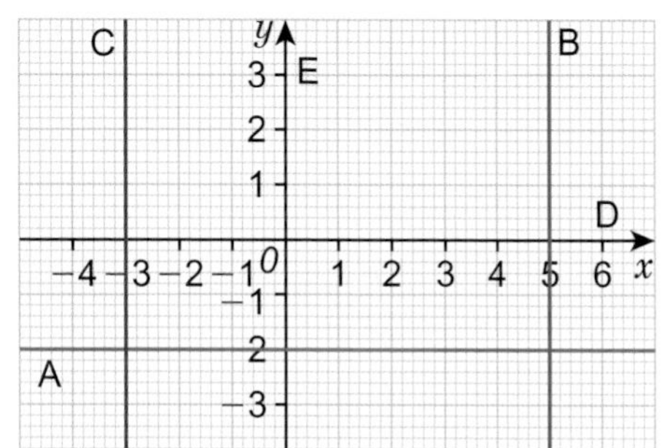

2 Copy and complete the table of values for $y = x^2$.

x	−4	−3	−2	−1	0	1	2	3	4
y									

3 Plot the graph of $y = x^2$ using your table of values from **Q2**.
Draw an x-axis from −5 to +5 and a y-axis from 0 to +20.
Plot the coordinates from your table of values.
Join the points with a smooth curve.
Label your graph $y = x^2$.

Q3 strategy hint It is easier to draw a curve with your hand 'inside it' and moving outwards. Turn your paper round so you can draw the curve comfortably.

Key point 14

A **quadratic equation** contains a term in x^2 but no higher power of x.
The graph of a quadratic equation is a curved shape called a **parabola**.

4 a Copy and complete this table of values for $y = x^2 - 3$.

x	−3	−2	−1	0	1	2	3
x^2							
−3	−3	−3	−3	−3	−3	−3	−3
y							

Q4a hint For quadratic functions with more than one step, you can include a row for each step in the table.

b Plot the graph of $y = x^2 - 3$.

Discussion What do you think the graph of $y = x^2 + 2$ will look like?

5 **Exam-style question**

Draw the graph of $y = -x^2$ for $-3 \leqslant x \leqslant 3$.

(4 marks)

Q5 strategy hint Work out the value of y for all the integer values of x from −3 to 3.
You could draw a table for these values.

Exam hint
Here are some common mistakes people make when drawing graphs.

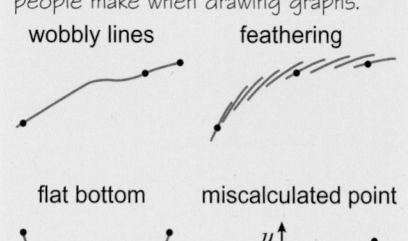

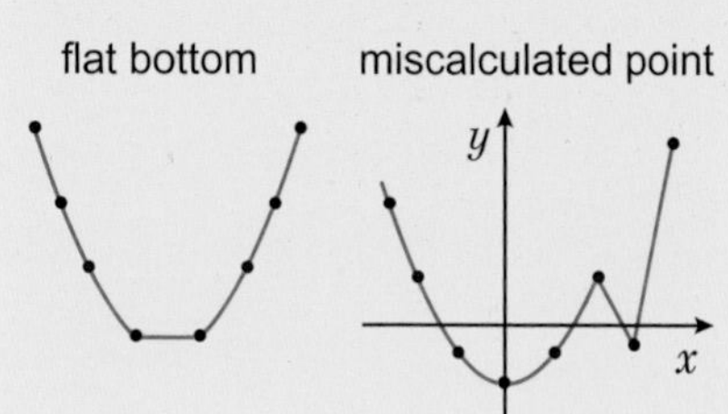

6 a Copy and complete this table of values for $y = 3x^2$.

x	−2	−1	0	1	2
y					

b Plot the graph of $y = 3x^2$.

Key point 15

A quadratic graph has either a **minimum point** or a **maximum point** where the graph turns.

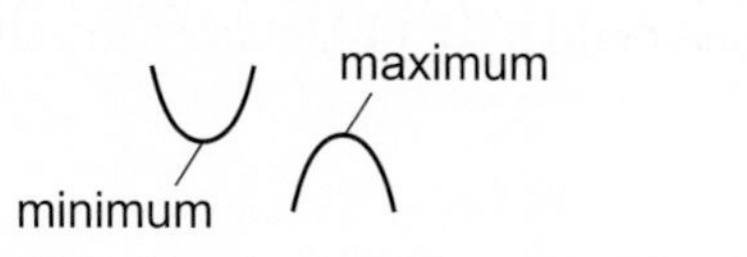

7 **Reasoning** Compare your graphs from **Q3**, **Q4**, **Q5** and **Q6**.

a What is the same about these graphs?

b Which ones have a minimum point? Which ones have a maximum point?

c Find the coordinates of the minimum/maximum point for each graph.

d Describe the symmetry of each graph by giving the equation of its mirror line.

8 **Modelling / STEM** Some maths students are investigating the effects of gravity on bottle rockets.
The students measure the rocket's height until it falls back to the ground.
The graph shows the rocket's height, h metres, at time t seconds after take-off.

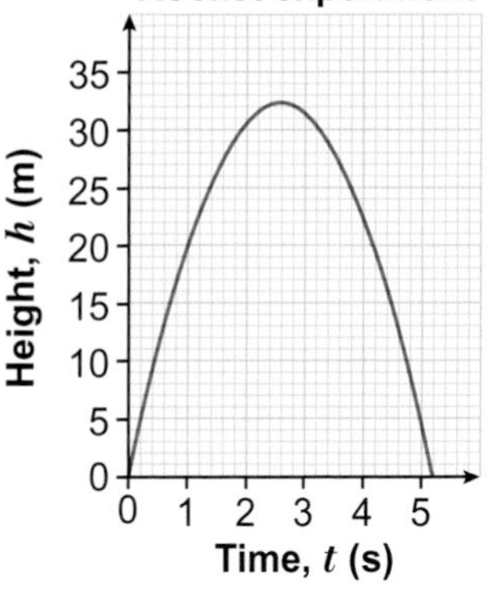

a What type of graph is this?

b When is the rocket travelling fastest?

c When is the rocket's speed zero?

> **Q8c hint** Faster speed = steeper gradient

d What is the maximum height that the rocket reaches?

e How long is the rocket in the air?

9 Here is the graph of $y = x^2 - 1$.
Use the graph to solve the equation $x^2 - 1 = 0$.
Discussion How could you use the graph to solve $x^2 - 1 = 2$?

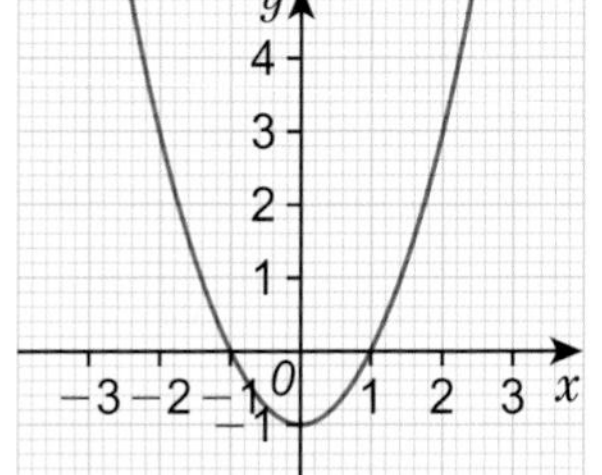

> **Key point 16**
>
> A quadratic equation can have 0, 1 or 2 solutions.

10 Here are four graphs.
Use these graphs to solve the equations

a $2x - x^2 = 0$

b $x^2 - 2x + 1 = 0$

c $2x^2 + 5x - 3 = 0$

d $6 - x^2 - x = 0$

> **Q10b hint**
> Only one solution.

a $y = 2x - x^2$

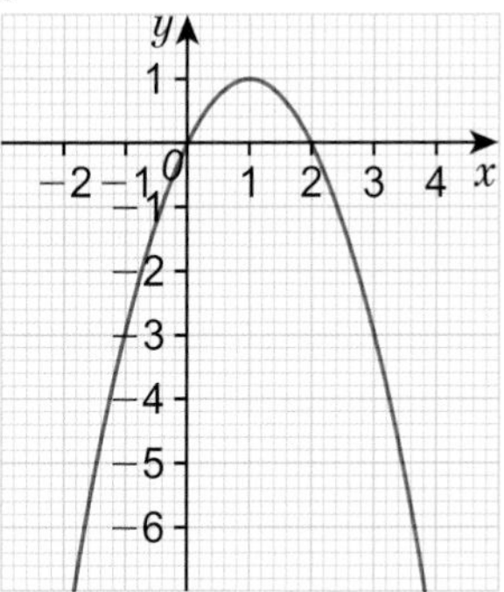

b $y = x^2 - 2x + 1$

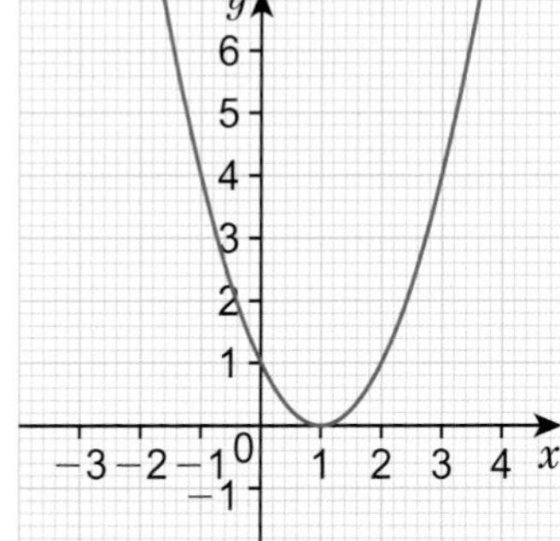

c $y = 2x^2 + 5x - 3$

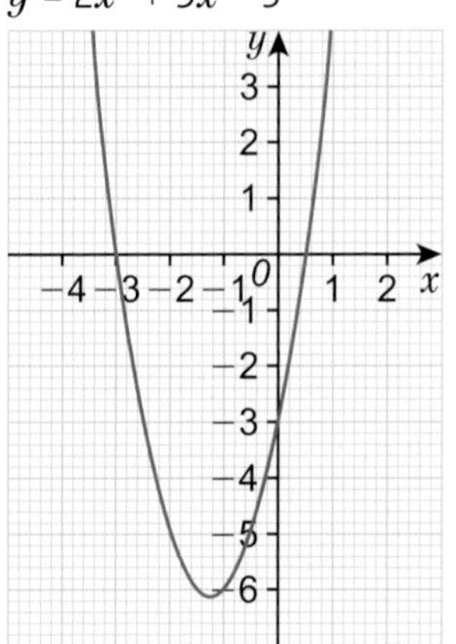

d $y = 6 - x^2 - x$

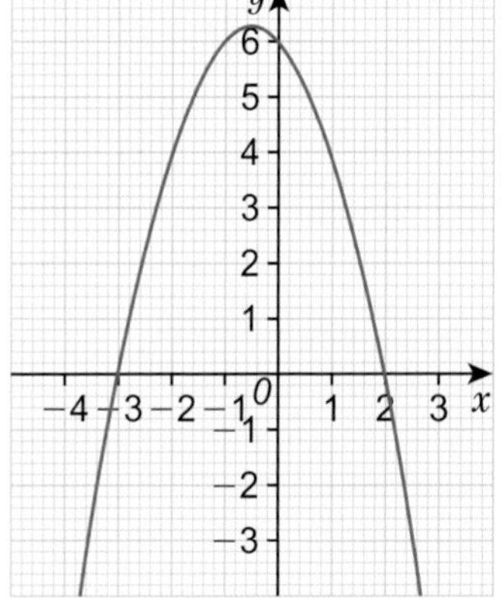

Example 3

Here is the graph of $y = x^2 - 3x - 2$.
Use the graph to solve the equation $x^2 - 3x - 8 = 0$.
Give your answers correct to 1 decimal place.

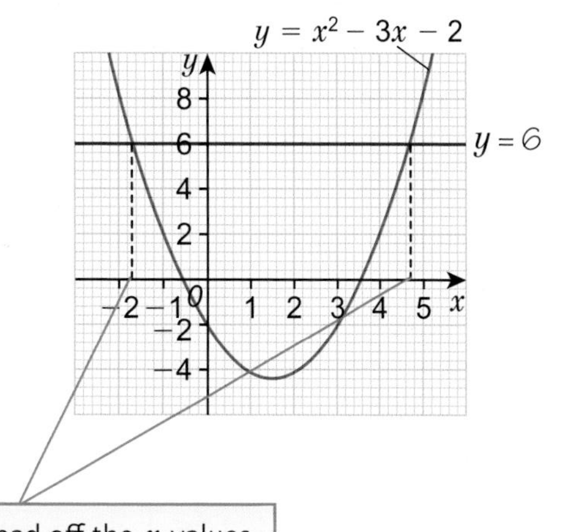

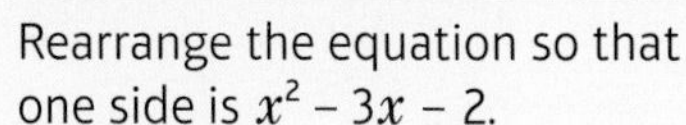

Rearrange the equation so that one side is $x^2 - 3x - 2$.

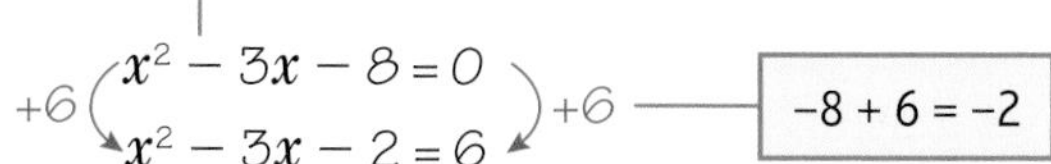

−8 + 6 = −2

Find where $y = x^2 - 3x - 2$ intersects $y = 6$.

Read off the x-values.

$x = -1.7$
$x = 4.7$

11 **Reasoning** Use the graphs in **Q10** to solve the equations

a $2x - x^2 + 2 = 0$
b $x^2 - 2x - 3 = 0$
c $2x^2 + 4x - 3 = 0$
d $1 - x^2 - 3x = 0$
e Explain why $2x - x^2 = 3$ has no solutions.

Q11c hint

$+\square$ ($2x^2 + 4x - 3 = 0$
$2x^2 + 5x - 3 = \square$) $+\square$

12 **Exam-style question**

a Complete the table for $y = 2x^2 - 3x - 4$.

x	−2	−1	0	1	2	3	4
y		1			−2	5	

(2 marks)

b Draw the graph of $y = 2x^2 - 3x - 4$. **(2 marks)**

c By drawing a suitable line on your graph, solve the equation $2x^2 + x - 20 = 0$. **(2 marks)**

Q12 strategy hint

$+\square$ ($2x^2 - 3x - 4 = 0$
$2x^2 + x - 20 = \square$) $+\square$

13 **Modelling / Real** Carla throws a rounders ball.
This table gives data for the height, h metres, of the rounders ball at time, t seconds, after Carla has thrown it.

Time, t (seconds)	0	1	2	3	4
Height, h (metres)	1.2	3.7	4.7	4.2	2.2

Q13 communication hint
The **trajectory** of an object is the path it follows.

a Use this data to draw a graph showing the trajectory of the rounders ball.
b Continue the graph to predict when the rounders ball will land.

6.7 Cubic and reciprocal graphs

Objectives

- Draw graphs of cubic functions.
- Solve cubic equations using graphs.
- Draw graphs of reciprocal functions.
- Recognise a graph from its shape.

Why learn this?

You may see reciprocal graphs in science, when you do experiments on volume and pressure.

Fluency

What shape is
- a linear graph?
- a quadratic graph?

Warm up

1 Copy and complete this table of values for $y = x^3$.

x	−3	−2	−1	0	1	2	3
y							

2 Work out the value of $\frac{1}{x}$ when

a $x = -4$

b $x = \frac{1}{3}$

Give your answers as fractions and as decimals to 2 d.p.

Key point 17

A **cubic function** contains a term in x^3 but no higher power of x.
It can also have terms in x^2 and x and number terms.

3 Using your table of values from **Q1**, draw the graph of $y = x^3$ for $-3 \leqslant x \leqslant 3$.

Q3 hint What values do you need to include on the x-axis? And on the y-axis?

4 **Reasoning** Use your graph from **Q3** to estimate

a 1.7^3

b $\sqrt[3]{-11}$

Use a calculator to work out

c 1.7^3

d $\sqrt[3]{-11}$

Discussion Which of your answers are most accurate? Explain.

5 Draw the graph of $y = -x^3$ for $-3 \leqslant x \leqslant 3$.

Reflect What is the same and what is different about this graph and the one you drew in **Q3**?

Q5 hint Make a table of values like the one in **Q1**.

6 a Plot graphs of $y = x^3 + 1$ and $y = x^3 - 2$ for $-3 \leqslant x \leqslant 3$.

b Compare these two graphs and the graphs from **Q3** and **Q5**. What similarities can you see? What are the differences?

Discussion What do you think the graph of $y = x^3 + 5$ would look like? What about $y = x^3 - \frac{1}{2}$?

Q6a hint Draw both graphs on the same axes.

Key point 18

A reciprocal function is in the form $\frac{k}{x}$ where k is a number.

ActiveLearn Homework, practice and support: Higher 6.7

Example 4

Draw the graph of $y = \frac{1}{x}$, where $x \neq 0$, for $-3 \leqslant x \leqslant 3$.

x	-3	-2	-1	$-\frac{1}{2}$	$-\frac{1}{4}$	$\frac{1}{4}$	$\frac{1}{2}$	1	2	3
y	$-\frac{1}{3}$	$-\frac{1}{2}$	-1	-2	-4	4	2	1	$\frac{1}{2}$	$\frac{1}{3}$

Make a table with x-values from −3 to 3. Do not include 0.

Work out the y-values and complete the table.

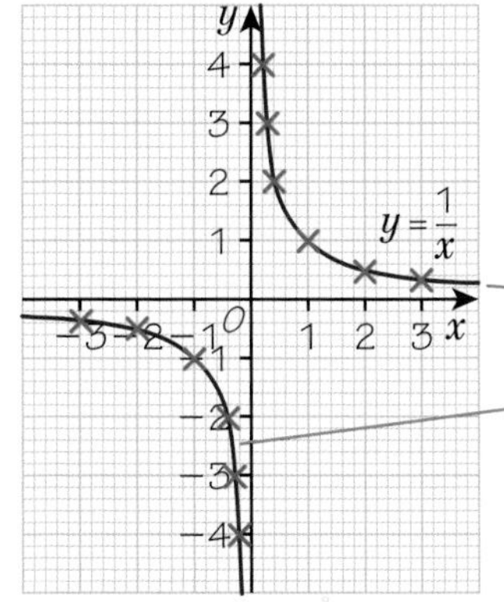

Plot the points. Join the two parts with smooth curves.

Key point 19

The x and y axes are **asymptotes** to the curve. An asymptote is a line that the graph gets very close to, but never actually touches.

Discussion Why can't you read the value of y when $x = 0$ from this graph?

7 **Reasoning** a Draw a table of values for $y = -\frac{1}{x}$, where $x \neq 0$, for $-3 \leqslant x \leqslant 3$.

b Draw the graph of $y = -\frac{1}{x}$.

c What is same and what is different about $y = \frac{1}{x}$ and $y = -\frac{1}{x}$?

8 a Draw the graph of $y = \frac{3}{x}$, where $x \neq 0$, for $-4 \leqslant x \leqslant 4$.

b Use your graph to find the value of y when

i $x = 3$ ii $x = -1$ iii $x = -2.5$

9 **Reasoning** Match each equation to a graph.

a $y = x^2 - 1$ b $y = x^3 - 2$ c $y = 2x$

d $y = \frac{1}{x}$ e $y = -x^3$ f $y = -3x$

A
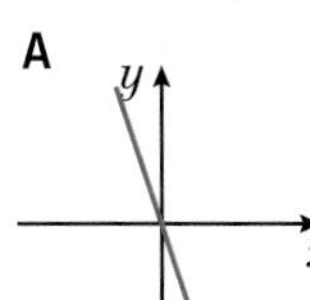

B
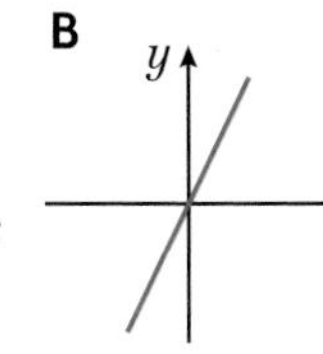

C
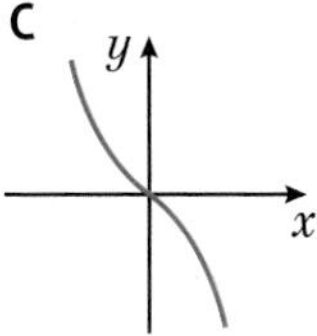

D
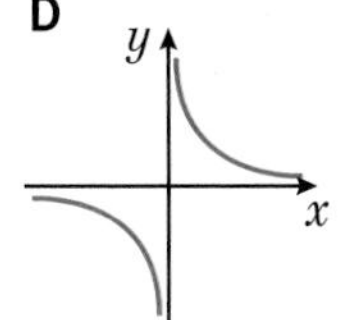

E
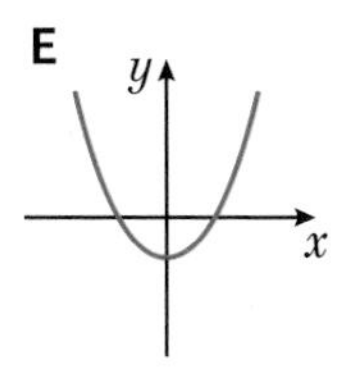

F
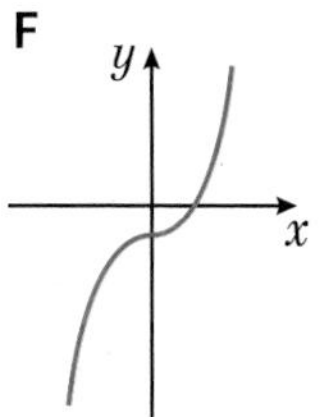

Key point 20

A cubic equation can have 1, 2 or 3 solutions.

10 **Reflect** Write a hint on how to remember the shapes of different types of graphs. Include sketches in your hints.

11 Use the graphs you drew in **Q6** to solve the equations.
a $x^3 - 2 = 0$ b $x^3 + 1 = 0$ c $x^3 - 2 = -3$

Q11c hint Read off the x-values where the curve crosses $y = -3$.

12 **Exam-style question**

a Complete the table of values for $y = x^3 - 5x$. **(2 marks)**

x	−3	−2	−1	0	1	2	3
y			4	0			12

b Draw the graph of $y = x^3 - 5x$ from $x = -3$ to $x = 3$. **(2 marks)**

c Hence or otherwise, solve $x^3 - 5x = 2$. **(2 marks)**

Nov 2013, Q17, 1MA0/2H

Exam hint
Think about what shape your graph should be. About where will it cross the axes?
In part **c**, 'Hence or otherwise' means that it will be easier to answer this question using parts **a** and **b** (the graph you have drawn).

13 This is the graph of $y = 11x + 2x^2 - 5x^3$.
By drawing suitable lines on the graph
a solve the equation $11x + 2x^2 - 5x^3 = 0$
b solve the equation $11x + 2x^2 - 5x^3 = 8$.

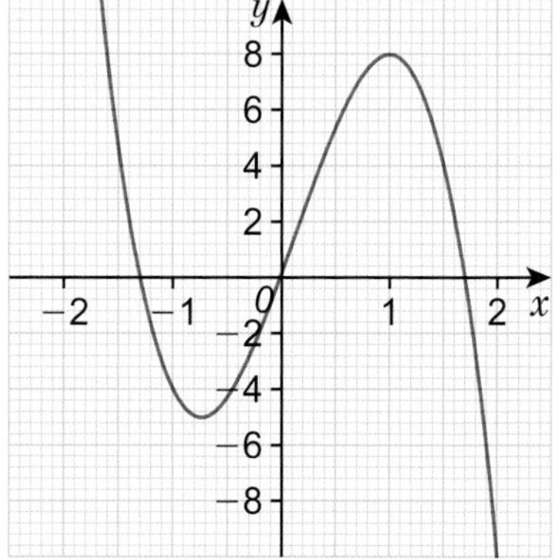

6.8 More graphs

Objectives

- Interpret linear and non-linear real-life graphs.
- Draw the graph of a circle.

Why learn this?

You can apply all the things you have learned about graphs to many interesting and practical contexts that you often come across in daily life.

Fluency

These graphs show how the depth of water in two containers changes over time, when you pour water in at a steady rate.
Which container fills with water faster?

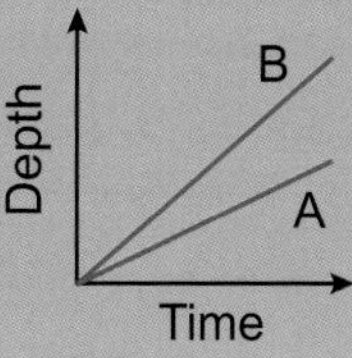

Warm up

1 From the graph, find
a the value of x when $y = 3$
b the value of y when $x = 1$.

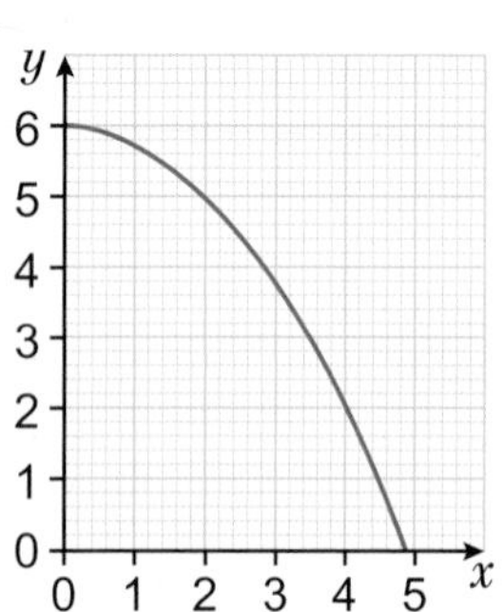

2 Construct a circle of radius 5 cm.

ActiveLearn Homework, practice and support: Higher 6.8

3 **Reasoning / Modelling** The distance–time graphs represent the journeys made by a bus and a car starting in Exeter, travelling to Cheltenham and returning to Exeter.

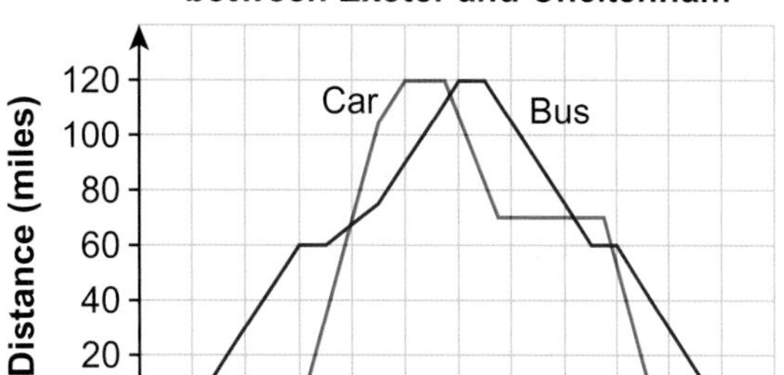

a How far is it from Exeter to Cheltenham?

b Including stops, how much longer than the car did the bus take to complete the journey from Exeter to Cheltenham?

c Work out the greatest speed of the car during the journey.

The bus stopped at Weston-super-Mare on its journey.

d On the return journey, at what time did the bus reach Weston-super-Mare?

e Work out the average speed of the car over the whole journey.

> **Q3e hint** Include the stops.

f What does the change in gradient on the bus's journey from Weston-super-Mare to Cheltenham show?

4 **Modelling / STEM** A skydiving instructor jumps from an aircraft flying at 3000 m. The graph models her motion as she falls. At what height does she open her parachute? Explain how you know.

> **Q4 hint** When does she start descending at a constant speed?

Key point 21

No correlation or weak correlation shows that there is *no* linear relationship between two quantities, because their graph is not close to a straight line.
When the points follow a curve, there may be a non-linear relationship between the quantities.

5 **Reasoning** Here are two sets of data.

> **Q5 hint** For the graph of data set B, make sure your y-axis extends to 40.

Data set A

x	3	4	5	6	6	7	7	8	10
y	7	8	10	13	14	14	16	16	21

Data set B

x	3	3.6	4	4	4.5	4.7	5.1	5.3	5.6	5.7
y	9	13	15	14	21	23	26	27	31	33

a Plot each set of data on a scatter graph.

b Describe the correlation for each set.

c Draw a line of best fit for the graph for data set A. What does this show?

d Draw the graph of $y = x^2$ on the same grid as the graph for data set B. Copy and complete this table of values to help you.

x	3.0	3.2	3.5	3.7	4.0	4.5	4.8	5.0	5.5	6.0
y										

e What do you notice from your graph? What do you think the relationship is between x and y in data set B?

6 **Real / Reasoning** The petrol consumption of a car, in kilometres per litre (km/l), depends on the speed of the car.

The table gives some information about the petrol consumption of a car at different speeds.

Speed (km/h)	62	68	76	86	93	99	103
Petrol consumption (km/l)	12.6	13.9	14.7	15	14.6	13.7	12.2

a Draw axes on graph paper, using 5 cm to represent 20 km/h on the horizontal axis and 4 cm to represent 1 km/l on the vertical axis.

Start the horizontal axis at 60 and the vertical axis at 12. Show the discontinuities clearly on the axes.

Plot the values from the table and join them with a smooth curve.

From your graph, estimate

b the petrol consumption at 75 km/h

c the speeds which give a petrol consumption of 13.5 km/l.

7 The graph shows the numbers of rats recorded in a colony.

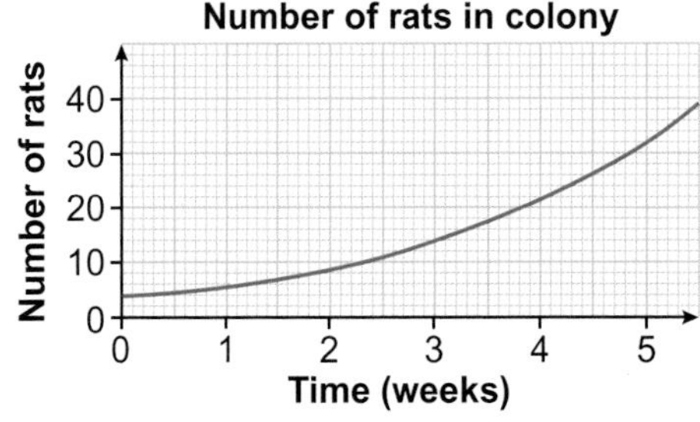

a How many rats were there at the start of the study?

b Explain how you found your answer to part **a**.

c Estimate the number of rats at

i 3 weeks ii 5 weeks.

d Describe the change in the number of rats from week 3 to week 5.

> Q7d hint Is it an increase or a decrease? By how much? Write a sentence beginning, 'The number of rats …'

8 **STEM / Reasoning** The graph shows the count rate against time for magnesium-27, which is a radioactive material.

The count rate is the number of radioactive emissions per second.

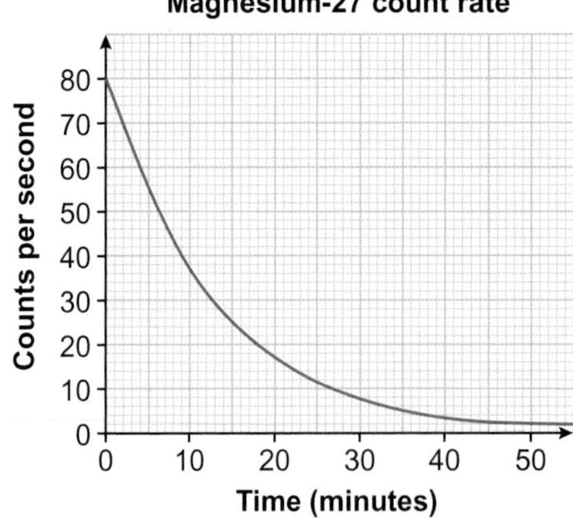

> Q8 communication hint
> The **half-life** is the time it takes for the count rate to halve.

a Estimate the count rate after 20 minutes.

b After how many minutes is the count rate 30?

c Estimate the half-life of magnesium-27.

d Does the count rate ever reach zero?

9 **Finance / Problem-solving** The graph shows the value of an investment over a 5-year period.

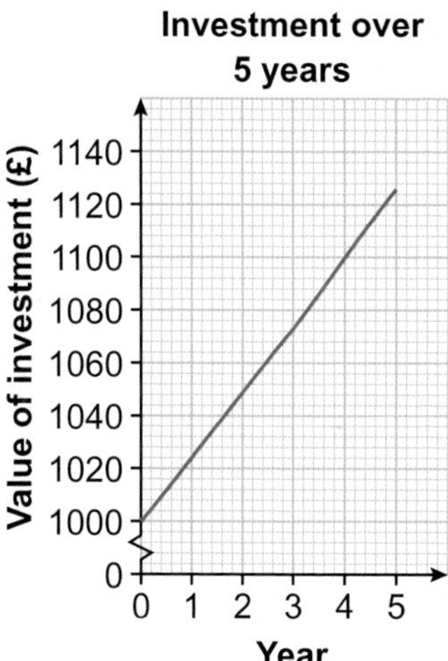

a What was the initial value of the investment?

b Estimate the value of the investment after 5 years.

c How much did the value increase in the first year?

d The rate of interest remained the same for the 5 years. Work out the percentage interest rate.

> **Q9d hint**
> $\frac{\text{actual change}}{\text{original amount}} \times 100$

Key point 22

The equation of a circle with centre (0, 0) and radius r is $x^2 + y^2 = r^2$.

Example 5

Construct the graph of $x^2 + y^2 = 36$.

$r = \sqrt{36} = 6$ — Compare $x^2 + y^2 = 36$ with $x^2 + y^2 = r^2$.

$x^2 + y^2 = 36$

Using compasses set to 6 units, draw a circle, centre *0*.

10 On graph paper, draw the graphs of

a $x^2 + y^2 = 1$ b $x^2 + y^2 = 16$ c $x^2 + y^2 = 49$ d $x^2 + y^2 = 81$

6 Problem-solving: Profit parabolas

Objective • Use quadratic functions to model real-life situations.

Tom sells trainers online. He has hired you to work out the best (most profitable) price for his trainers.

T is the price at which Tom sells each pair of trainers, Q is the quantity (or number) of pairs of trainers that Tom sells per week, and P is Tom's weekly profit.

1 **Finance** Tom buys each pair of trainers from the supplier for £25.
Explain why his weekly profit can be modelled using the formula $P = Q(T - 25)$.

> **Q1 hint** Start by writing an expression for the profit Tom makes on one pair of trainers.

2 Tom knows (from experience) that if he sells the trainers at £40 nobody buys them, as they are too expensive.
However, every time he lowers his price by £1, he sells 10 more pairs a week on average.
Show that this can be modelled by the formula:
$Q = 10(40 - T)$

> **Q2 hint** Show that when the sale price is £40, no trainers are sold. What happens when the sale price is £39, £38 and so on?

3 These formulae can be combined as $P = 10(40 - T)(T - 25)$.
Expand the brackets to write this new formula in full.

4 **Finance** Starting with your new formula, choose an appropriate method to help you find
- a the selling price that will maximise profits
- b the most profit Tom can make in a week.

> **Q4 hint** You might choose to plot a graph of profit against selling price. Create a table of values first to help you choose the scales for your axes.

5 **Finance** Tom's supplier raises their price to £26.
How will this affect your result?
Make a prediction and then test it by adjusting the formula for P.

> **Q5 hint** You will need to change one of the numbers in the formula given in **Q3**.

6 Check up

Log how you did on your Student Progression Chart.

Linear graphs

1 **Reasoning** A line has equation $2x + 3y = 7$.
Write down the gradient and y-intercept of the line.

2 Write down the equations of these four lines.

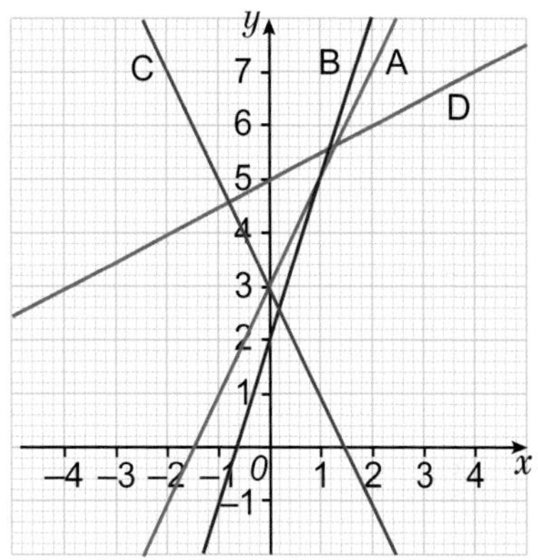

3 Draw a graph of the equation $y = -3x - 1$.
Do not use a table of values.

4 Without drawing a graph, find the gradient of the line through each pair of points.
a G (−4, 5) and H (4, 1) b P (1, −4) and Q (4, 5)

5 **Reasoning** Hamzah goes for a bike ride with his friends. The graph shows the five stages of his journey.
a What is the gradient for the first stage of the bike ride? What does this represent?
b Work out Hamzah's average speed for the whole journey, including any stops.
c On which stage of the journey was he travelling fastest?
d Work out his speed for that stage.

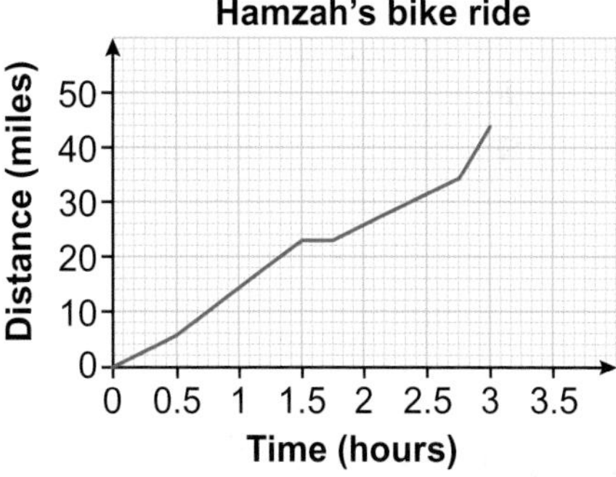

6 **Reasoning** Annie buys some cakes from the Kupkake Faktory.
There is a minimum order of three cupcakes.
Then there is a fixed price for each cupcake.
The graph shows the price structure.
a What does the gradient tell you?
b What does the y-intercept tell you?
c Are x and y in direct proportion?
Explain your answer.

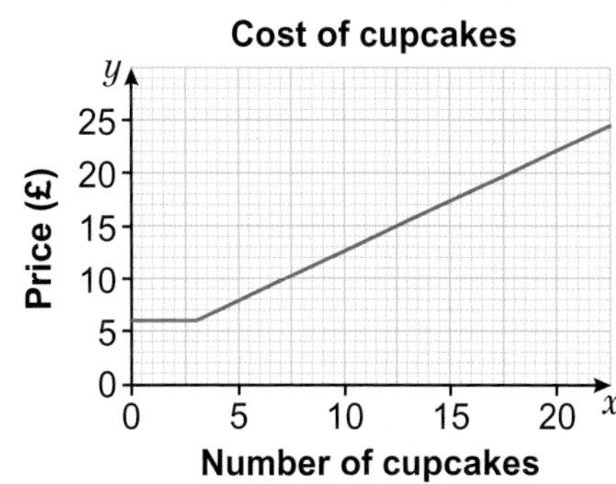

7 **Reasoning** J is the point (2, −5) and K is the point (−3, −1). Work out
a the midpoint of the line segment JK b the length of the line segment.

8 **Reasoning** The equation of a line is $y = 3x + 1$. Work out
a the equation of a line parallel to $y = 3x + 1$ which goes through the point (2, −7).
b the equation of any line perpendicular to $y = 3x + 1$ that does not share its y-intercept.

ActiveLearn Homework, practice and support: Higher 6 Check up

Non-linear graphs

9 **Reasoning** Match each equation to one of the graphs below.

a $y = x^2$ b $y = \frac{1}{x}$ c $y = -x^2$ d $x^2 + y^2 = 9$

e $y = x^3$ f $y = -\frac{1}{x}$ g $y = -x^3$

A
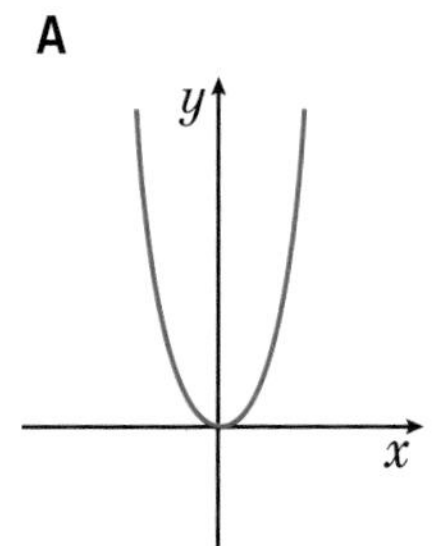

B
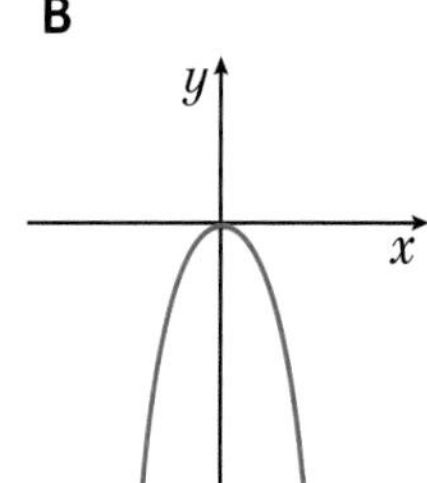

C
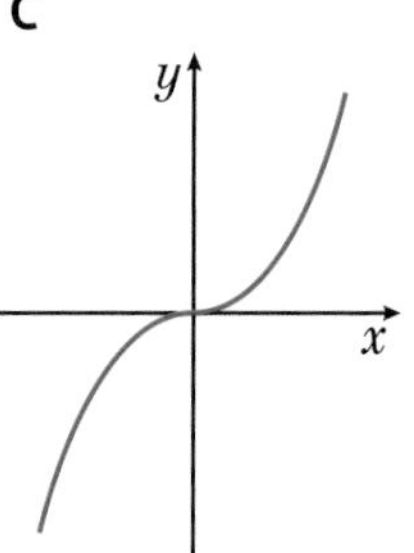

D
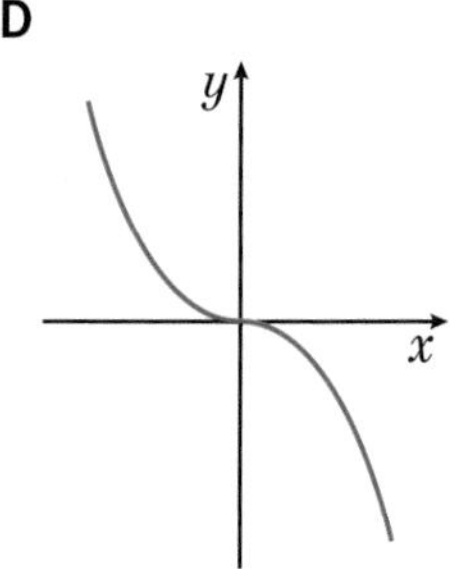

E
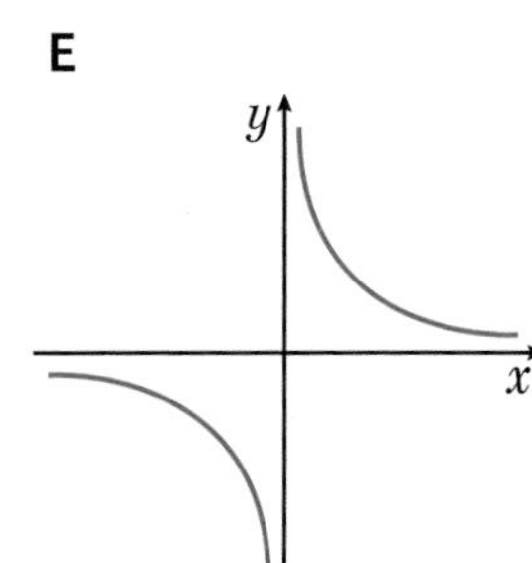

F
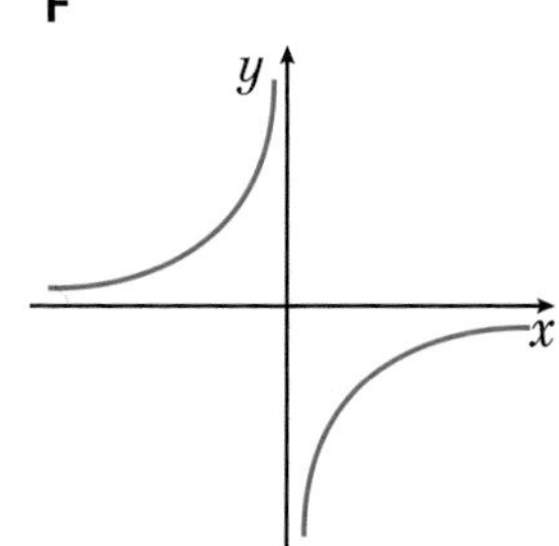

G
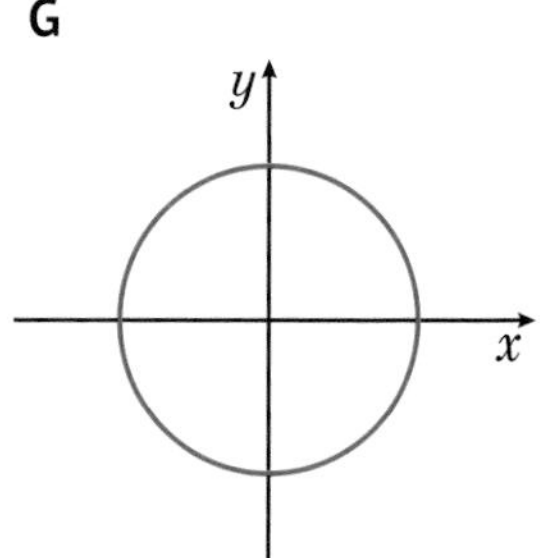

10 The equation $-x^3 + 3x - 1 = 0$ has three solutions.
Use the graph of $y = -x^3 + 3x - 1$ to estimate all three solutions.

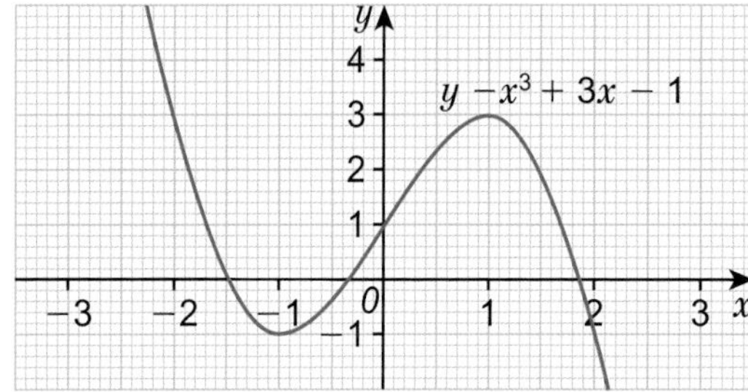

Real-life graphs

11 **Real / Modelling** Hannah is watering her garden. The water coming out of the hosepipe forms a smooth curve.

This graph models the curve.

a Give the coordinates of the maximum point of this graph.

b What was the maximum height that the water reached?

c How long did the water take to hit the ground after leaving the hosepipe?

d What is the practical meaning of the start point of the graph (the y-intercept)?

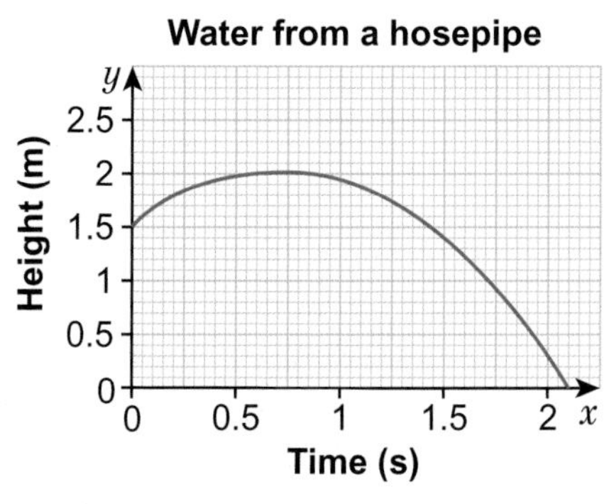

12 **Reasoning** Mike fills these containers with water at a constant rate.
Match each container to a graph.

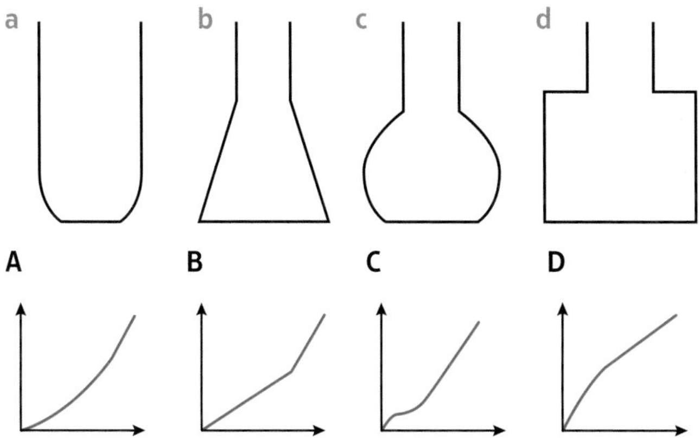

13 How sure are you of your answers? Were you mostly

Just guessing   Feeling doubtful    Confident 

What next? Use your results to decide whether to strengthen or extend your learning.

14 **★ Challenge**

Design a game where you have to get from the start to the finish, collecting all the gold coins and avoiding all the obstacles.

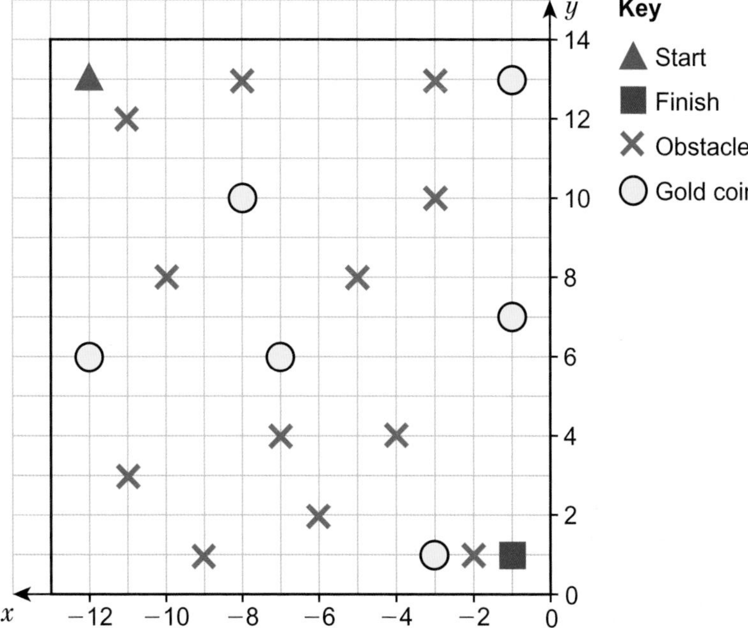

Design your game within a border.

There must be at least 10 obstacles and 5 gold coins. You cannot touch the borders of your grid to escape your obstacles. Record your moves.

Here are two examples of a possible first move for this game:
Walk 7 units south, or walk 7 units down the line $x = -12$.

After everyone has finished Check up, challenge a friend to complete your game in fewer moves.

6 Strengthen

Linear graphs

1 a Which of these equations of lines are in the form $y = mx + c$?

i $y = 3x + 1$ ii $2y = 4x + 6$ iii $x + y = 4$

iv $y = 2x - 1$ v $10x + 2y = 1$

b Rearrange the other equations in the form $y = mx + c$.

2 The graph shows four straight-line graphs.

a Which lines have y-intercept (0, 1)?

b Which graphs have a positive gradient?

c Which graphs have a negative gradient?

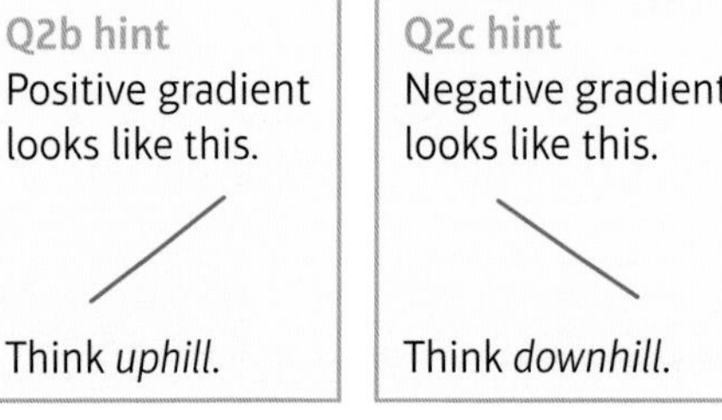

Q2b hint Positive gradient looks like this.

Think *uphill*.

Q2c hint Negative gradient looks like this.

Think *downhill*.

d Which equation matches which line?

i $y = x - 3$ ii $y = -3x + 1$

iii $y = 2x + 1$ iv $y = -x + 2$

3 Match each graph to an equation.

a $y = -2x$

b $y = 3x - 2$

c $y = -x + 2$

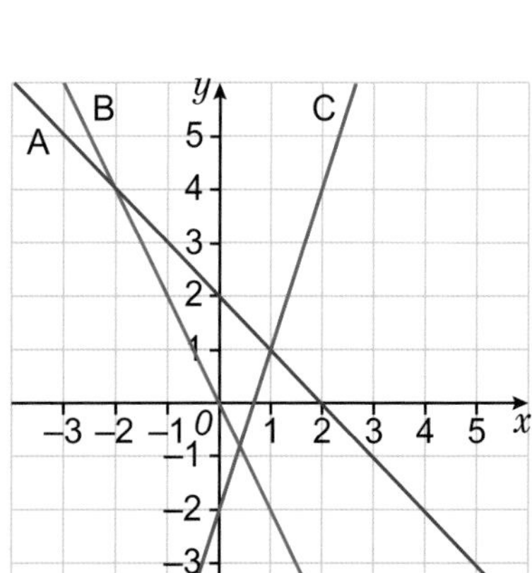

4 On squared paper, draw lines with these gradients.

a 4 b −1 c $\frac{1}{2}$

5 Draw lines for these equations.

a $y = 4x - 3$

b $y = -x + 2$

c $y = \frac{1}{2}x$

Q5a hint Identify the intercept from the equation. Mark this on the y-axis first. Then draw the gradient. Use **Q4** to help you.

Q5b hint $-x$ means '−1 lot of x'.

Q5c hint What does it mean if there is not a number on the end of the equation?

*Active*Learn Homework, practice and support: Higher 6 Strengthen

6 The sketch graph shows the points A (–3, 1) and B (2, 6). From the sketch, work out

a the change in x

b the change in y

c the gradient of this line segment, using $\frac{\text{change in } y}{\text{change in } x}$.

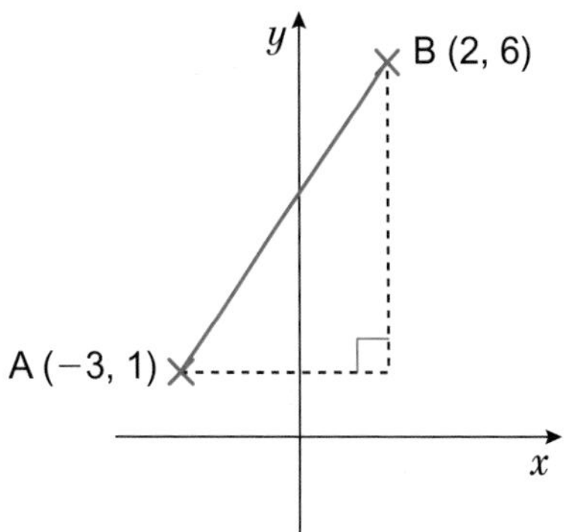

7 Work out the gradient of the line segment between each pair of points.

a C (–2, 3) and D (1, 5)

b E (–4, –1) and F (0, 4)

Q7 hint Use the same method as **Q6**. Draw a sketch.

8 For the same pairs of points as **Q7**, find the length of each line segment. Leave your answers in surd form.

Q8 hint Sketch the line segment. Use Pythagoras' theorem.

9 For the same pairs of points as **Q7**, find the midpoint of each line segment.

	(x	, y)
C	(–2	, 3)
D	(1	, 5)
M	(□	, □)
	$\frac{(-2+1)}{2}$	$\frac{(3+5)}{2}$

Q9 hint Copy and complete the calculation to work out the coordinates of M, the midpoint.

10 a Choosing from these equations, which pairs of lines are parallel?

A $y = 2x - 3$	**B** $y = -x - 1$	**C** $y = -2x$
D $y = 4 - x$	**E** $y = x + 5$	**F** $y = 2x + 1$

b Write the equation of another line parallel to each pair.

Q10a hint Which lines have the same gradient? Look for the same value of m in $y = mx + c$.

Q10b hint Your line must have the same gradient, but you can use any *different* value of c.

11 Find the negative reciprocal of

a 4

b $\frac{1}{3}$

c –10

d $-\frac{3}{5}$

Q11 hint Find the reciprocal. Then change the sign.

12 Choosing from these equations, which pairs of lines are perpendicular?

A $y = -\frac{1}{2}x - 3$	**B** $y = 2x - 1$	**C** $y = -2x + 5$
D $y = 1 - x$	**E** $y = x + 6$	**F** $y = \frac{1}{2}x + 1$

Q12 hint Which lines have gradients which are negative reciprocals? Look for one whole number and one fraction (with the same denominator as the whole number). One gradient needs to be positive and the other needs to be negative.

13 Reasoning This is the graph of $y = \frac{1}{2}x + 4$.

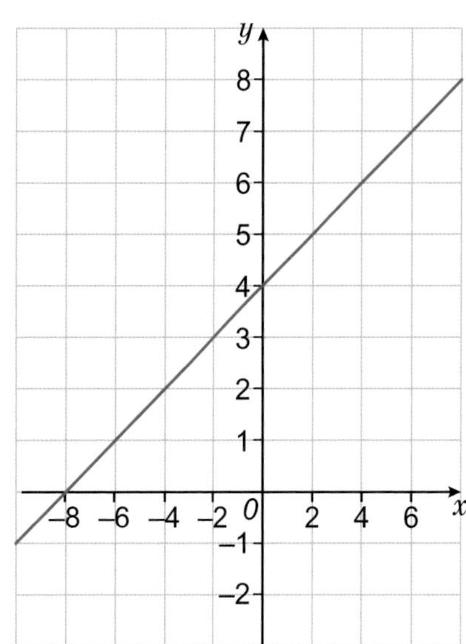

Q13 communication hint To 'satisfy the equation' means 'make the equation true'. Substitute the x and y values into the equation – are both sides equal?

a Which of these points are on this line?
L (0, 4) M (3, 6) N (–5, –10) P (–4, 2)

b Do the points L, M, N and P satisfy the equation?

14 Reasoning a What is the gradient of any line parallel to $y = 3x - 3$?

b Write the equation of a line parallel to $y = 3x - 3$ that goes through (–1, 7).

Q14b hint Substitute the given values of x and y, and the value of m you found in part **a** into $y = mx + c$. Solve to find c.

Q14a hint In the equation $y = mx + c$, m is the gradient.

Non-linear graphs

1 Reasoning Enzo makes a table of values and plots the graph of $y = x^2 + 2$.
From the graph, which points do you think are incorrect?

Q1 hint Which points do not fit the shape of an x^2 graph? Check if the points fit the equation.

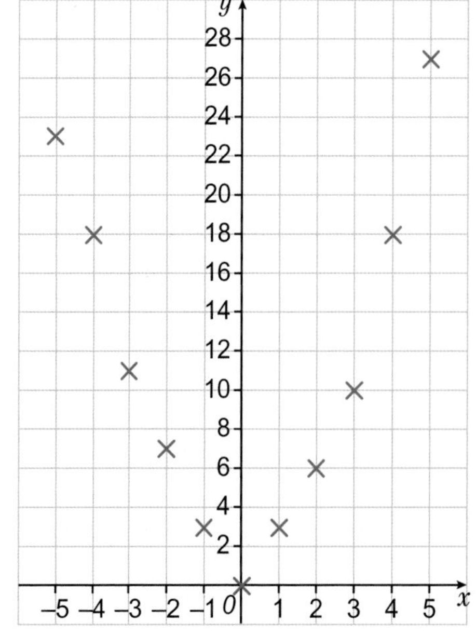

2 **Reasoning** Shona makes a table of values and plots the graph of $y = x^3 - 1$.

a From the graph, which points do you think are incorrect?

> **Q2a hint** Which points do not fit the shape of an x^3 graph?

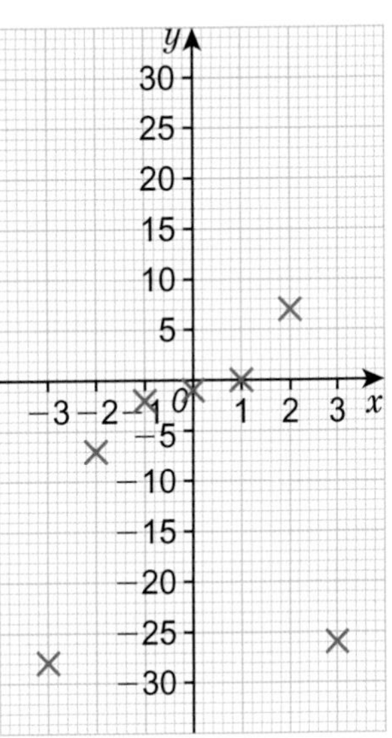

b Now find the incorrect points in Shona's table of values.

x	−3	−2	−1	0	1	2	3
y	−28	−7	−2	−1	0	7	−26

c Work out the correct values.

3 **Reasoning** Match the words to the graphs and the equations.

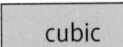

a 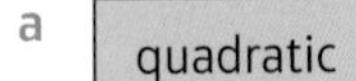

b cubic
c reciprocal
d linear

> **Q3 hint** Some words match more than one equation or graph. Which two graphs are quadratic? Which is x^2? Which is $-x^2$? Look back at lessons **6.6** and **6.7** to help you. Do the same for cubic and reciprocal graphs.

A

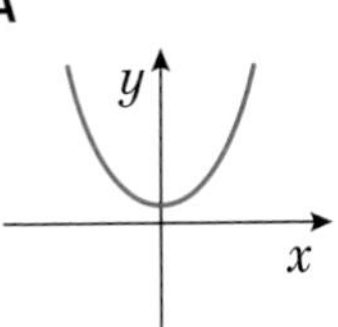

B

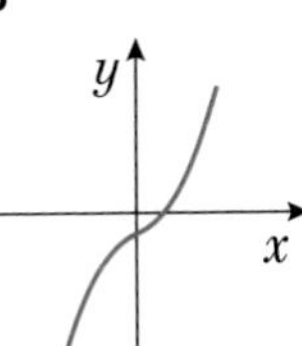

C

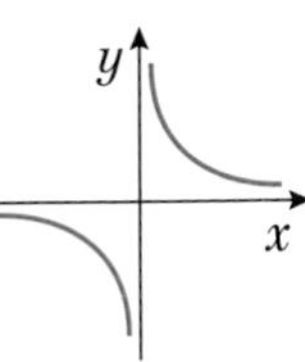

D

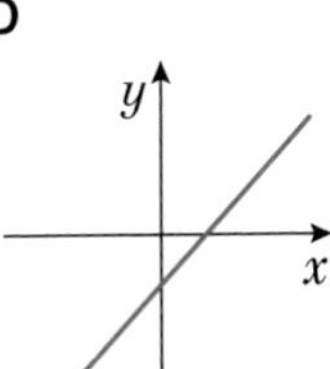

E

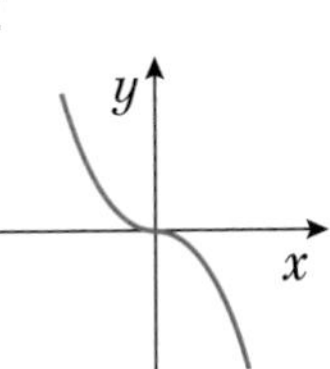

F

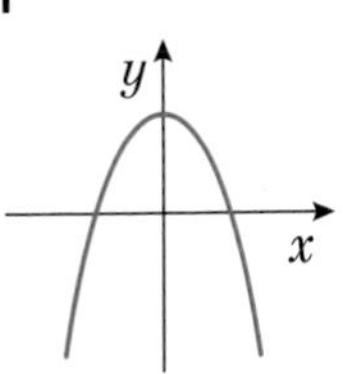

$y = x - 2$ | $y = -x^2 + 4$ | $y = -x^3$ | 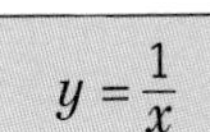

| $y = x^2 + 1$ | 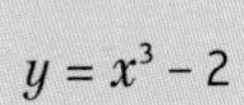

Real-life graphs

1 **Reasoning** Frankie kicks a rugby ball for a conversion after a try.
He kicks the ball from 6 m in front of the goal posts.
The graph shows the path followed by the rugby ball.

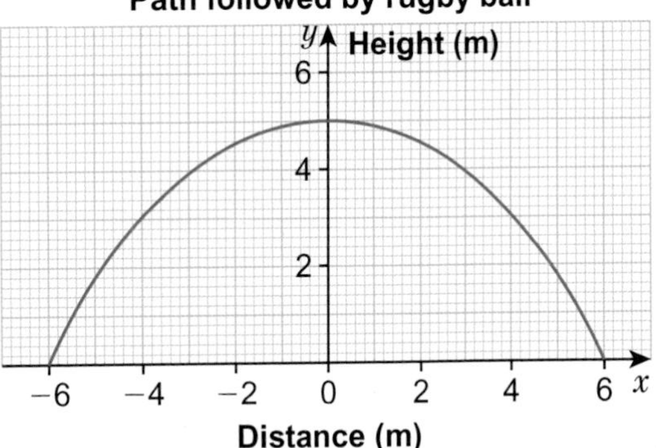

a What type of graph is this?
Explain your answer.

> **Q1a hint** Think about the shape of the graph.

b What are the coordinates of the maximum point?

> **Q1b hint** Look carefully at the axes before you read the values.

c What is the value of y when $x = 2$?

d What is the value of x when $y = 3.5$?

e What do the negative values of x mean in this context?

f Find the height of the ball as it goes past the posts.

g The bar on a rugby goal post is set at 3 m.
Assuming he kicked the ball straight at the goal, has the ball gome over the bar?

> **Q1e hint** Read the question again for a reminder.

2 Reasoning Here are two vases.

Vase 1

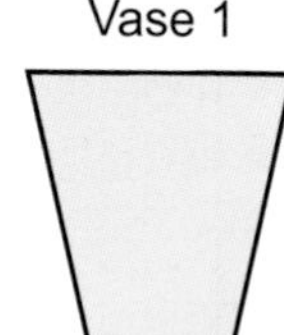

Vase 2

a Use the words 'faster' or 'slower' to write sentences about how the water level in these vases changes as they fill up.

> **Q2a hint** You could write a sentence like 'The water level in this vase rises faster first, then slower.'

b The graphs show the depth of water against time when water pours into the vases at a constant rate. Which graph matches which vase?

> **Q2b hint** The steeper the graph, the faster the rise in water level.

Graph A

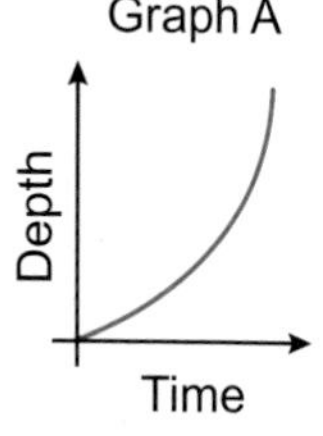

Graph B

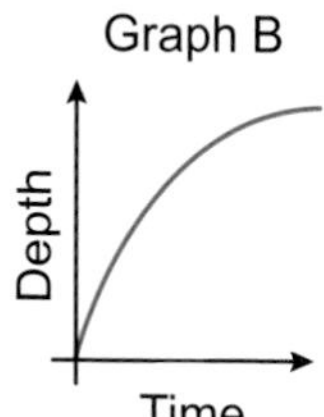

3 Finance The graph shows the profit made on the sales of light bulbs by the Like Bulbs company. The line goes through the origin.

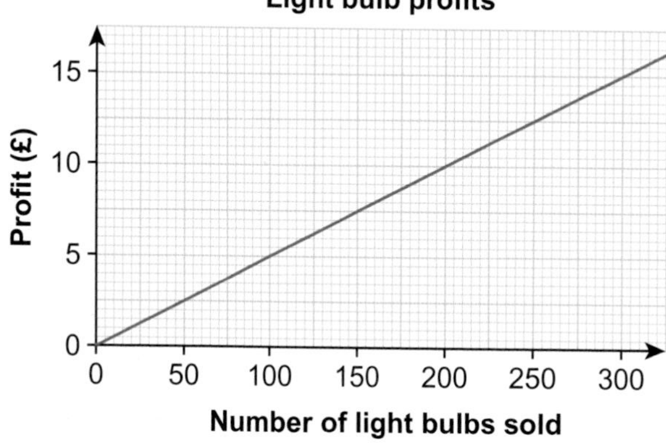

a Copy and complete

i 25 bulbs = £☐ profit

ii £13 profit = ☐ bulbs

iii £☐ profit = 175 bulbs

iv ☐ bulbs = £4.80 profit

b Describe the relationship between profit and the number of light bulbs sold.

6 Extend

1 **Reasoning** Without plotting the graphs, work out which of these functions

a have the same y-intercept b have the same gradient.

A $y = \frac{1}{2}x + 4$ **B** $y + 4x = 8$ **C** 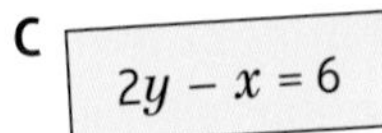

 D $x + y = 3$

2 a A is the point (−1, 2). B is the point (7, 5).
Find the coordinates of the midpoint of AB.

b P is the point (−4, 1).
Q is the point (1, −3).
Find the gradient of PQ.

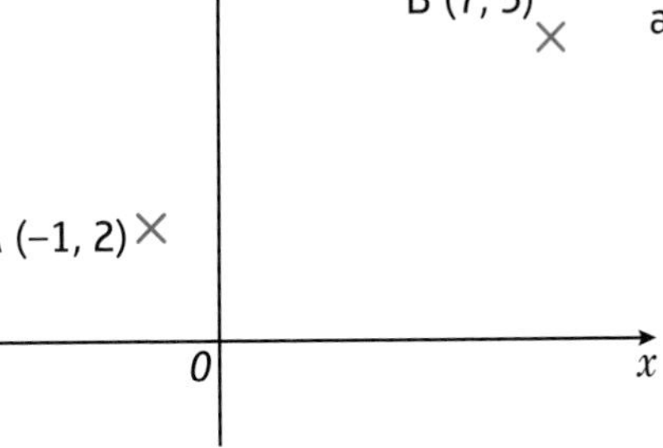

Diagram NOT accurately drawn

3 **Problem-solving** Point L has coordinates (−1, −3). Point M has coordinates (5, 5).
Point M is the midpoint of the line LN.

a What are the coordinates of point N?

b What is the gradient of the line segment LN?

Discussion Do you need to work out the equation of the line?

4 **STEM / Real** A biologist conducts a study into plant diversity in a nature reserve.
For each plant, he records the average size of the seeds and the number of seeds on the plant.
His results are shown in the graph.
The line of best fit has been drawn in for you.
Write a formula that models the link between seed size and number of seeds in this sample.

Q4 hint Find the equation of the line of best fit.

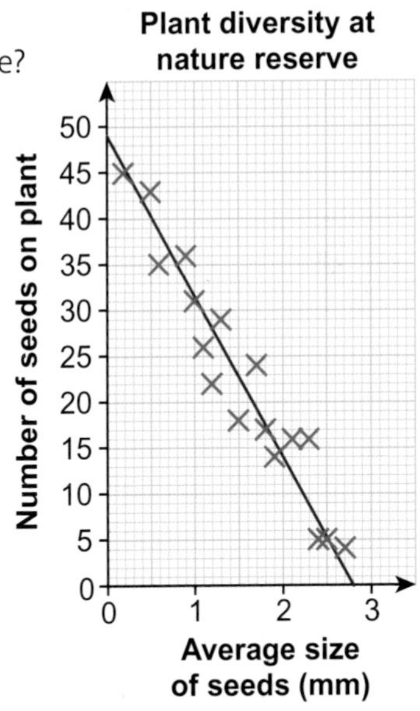

5 **Problem-solving** A recipe for ratatouille uses aubergines and tomatoes in the ratio 2 : 5.

a Write an equation showing the relationship between aubergines, a, and tomatoes, t.

b Plot the graph of the equation from $a = 0$ to $a = 10$.

c Find the gradient of the graph.

Discussion What is the meaning of the gradient in the context of the question?

d Hence or otherwise, work out the quantity of aubergines needed in a recipe that uses 600 g of tomatoes.

Q5a hint Check your equation by substituting the values from the ratio. Both sides of the equation must have the same value.

Q5d communication hint 'Hence or otherwise' tells you to use the information you've already worked out in the question, or any other method you can think of.

6 **Reasoning** Here are the cross-sections of three different concrete-transporter lorries.

a

b

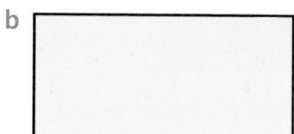

c

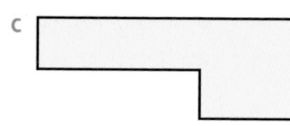

A builder empties the lorries by pumping out the concrete from the bottom at a steady rate. Here are three sketch graphs showing the relationship between the depth of the concrete left in the lorry and the number of minutes since the pump was switched on.

A

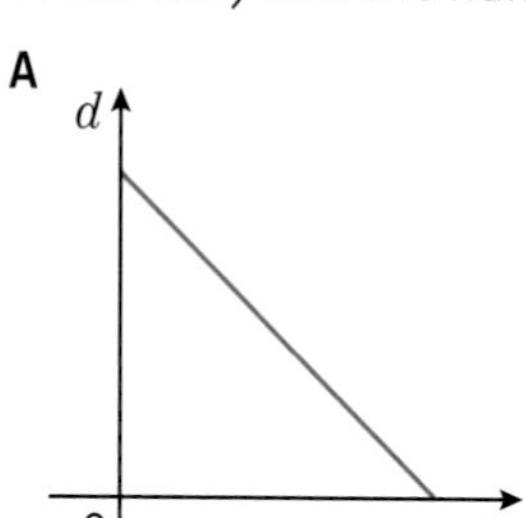

B

d

0

t

C

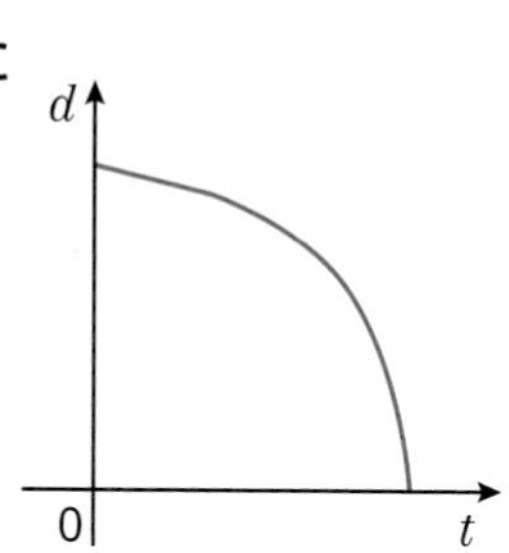

Match each lorry with one graph.

7 a **Communication** Show that the equation $x^2 - 3x - 1 = 2x - 1$ can be rewritten as $x^2 - 5x = 0$.

b Find the equation of the straight line shown.

c Solve the equation $x^2 - 5x = 0$.

> **Q7c hint** You can solve the equation $x^2 - 5x = 0$ by finding the intersection of the graph of $y = x^2 - 3x - 1$ with the graph of a different straight line.

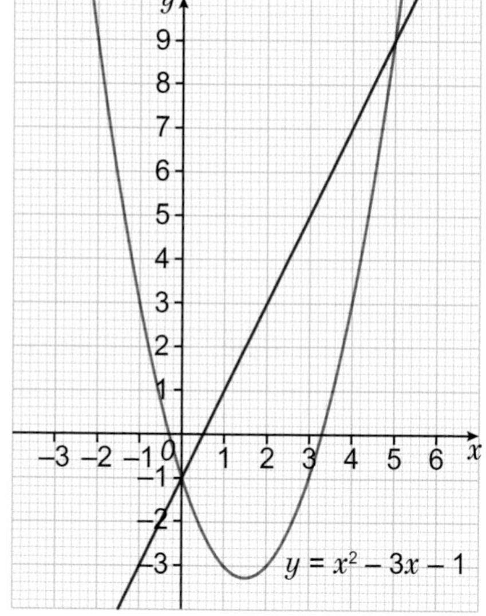

8 a Copy and complete this table of values for the equation $y = 1 - \frac{2}{x}$, $x \neq 0$.

x	−3	−2	−1	−0.5	−0.1	0.1	0.5	1	2	3
y	1.7		3	5		−19				0.3

b Draw the graph of $y = 1 - \frac{2}{x}$ for $-3 \leqslant x \leqslant 3$.

c Write the equations of the two asymptotes for this graph.

9 **Modelling** A mobility scooter accelerates from rest at a constant rate of 1.2 m/s^2. The distance, s, covered by the scooter is given by the formula $s = \frac{1}{2}at^2$, where a is the acceleration in m/s^2 and t is the time in seconds.

a Draw a graph of the distance covered by the scooter for values of t from 0 to 10 seconds.

b What is the distance covered after 4.5 seconds?

c How many seconds does the scooter take to cover 40 m?

10 Write the equation of a line

a parallel to the line $3x + 5y = -10$

b parallel to the line $2y - 4x = 5$, and which goes through the point (3, 7)

c perpendicular to the line $8x + 6y = -1$, and which goes through the point (−2, 1).

11 **Problem-solving** A rectangle is made using four straight lines on centimetre squared paper. Three of these lines are shown on the grid.
The point (−4, 0) lies on the missing side.

a Work out the equation of the missing side.
Label the two missing corners C (in quadrant 4) and D (in quadrant 3).

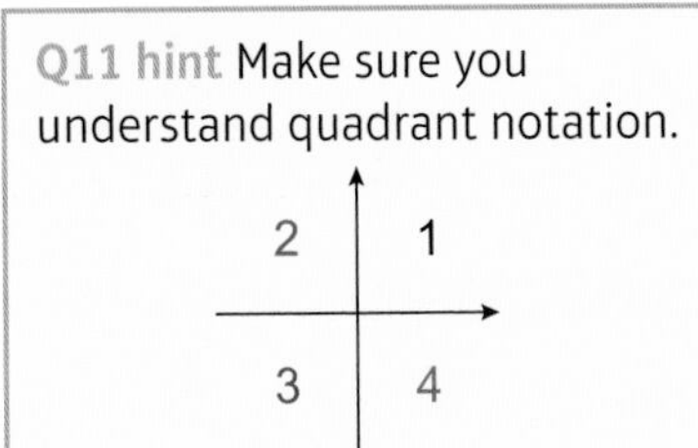

Q11 hint Make sure you understand quadrant notation.

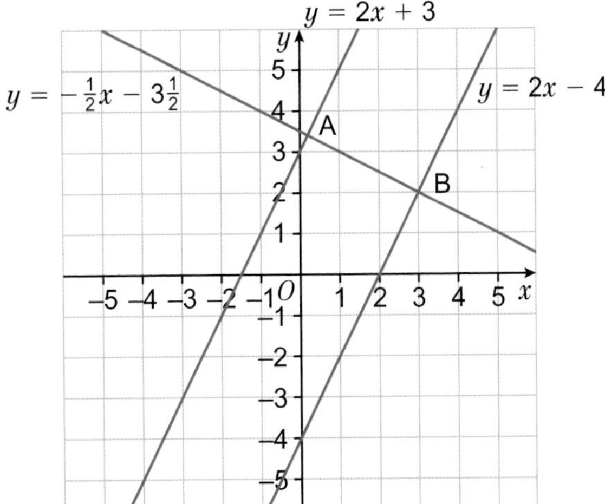

b Work out the coordinates of corner D.

c Work out the length of the diagonal BD.
Give your answer to 1 d.p.

12 **Communication** The point D (4, k) lies on the line $y = 3x - 5$.
Show that the point D also lies on the line $y = 2x - 1$.

13 **Exam-style question**

ABCD is a square.
P and D are points on the y-axis.
A is a point on the x-axis.
PAB is a straight line.
The equation of the line that passes through the points A and D is $y = -2x + 6$.
Find the length of PD.

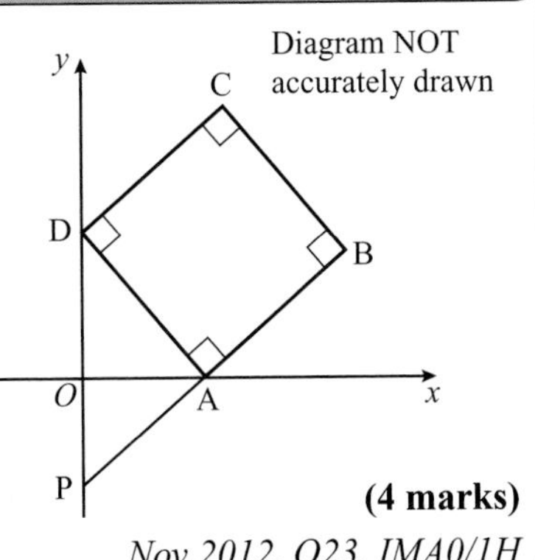

(4 marks)

Nov 2012, Q23, IMA0/1H

Q13 strategy hint You can start by using the equation $y = -2x + 6$ to work out where A and D cut the axes.

14 **Reasoning**

a Match each equation with its graph.

i $y = x^2 - 2$ ii $y = \frac{1}{x}$ iii $y = 2x^2 + 5$ iv $y = x^3 - 2$

A

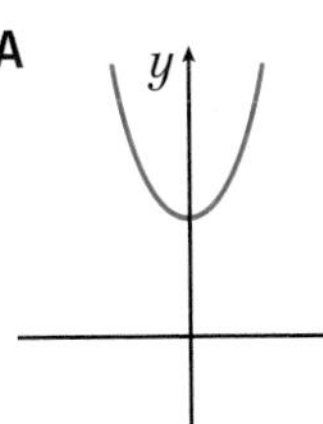

B

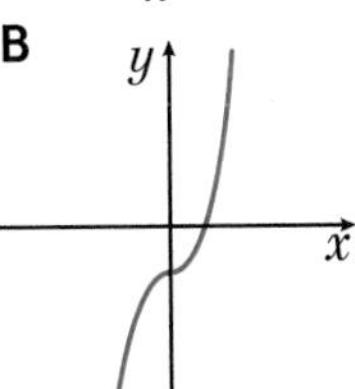

C

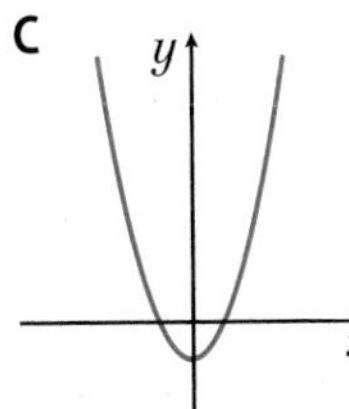

D

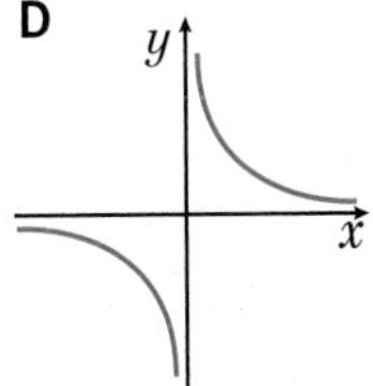

b Find the equation of the line of symmetry for each graph.

15 Problem-solving Match each equation with its graph.

a $y = x^3 - 4x^2 + 5$ b $y = -x^3 + 2x^2 - 3$

c $y = 3x^3 + 2x + 5$ d $y = -2x^3 + 3x^2$

Q15 hint To find the y-intercept, substitute $x = 0$.

A

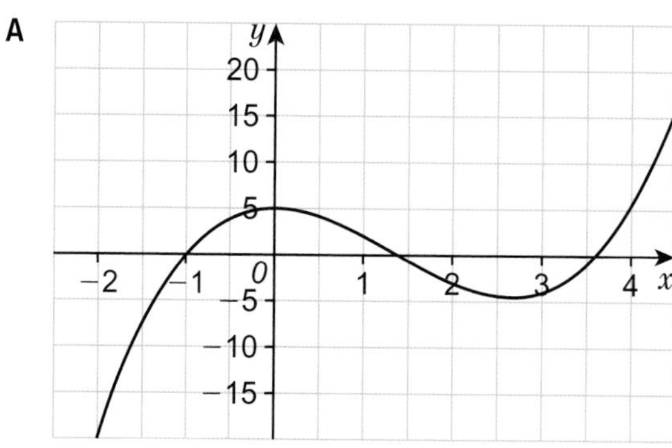

B

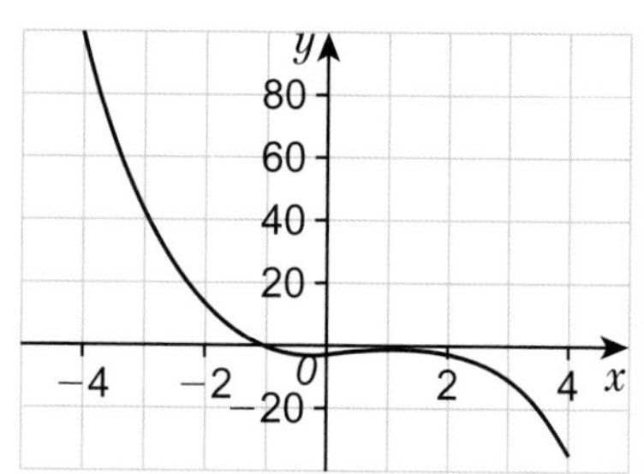

C

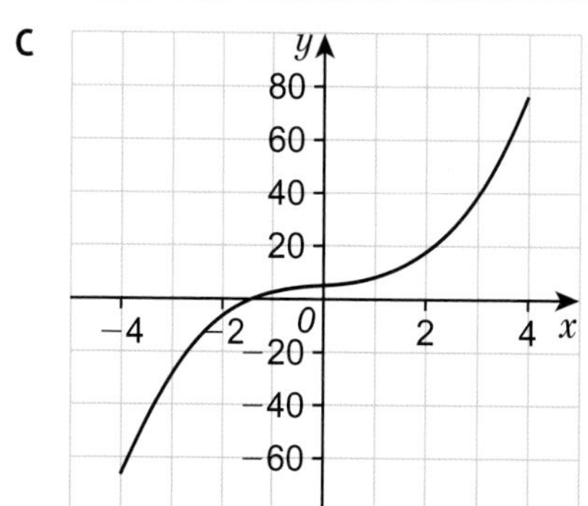

D

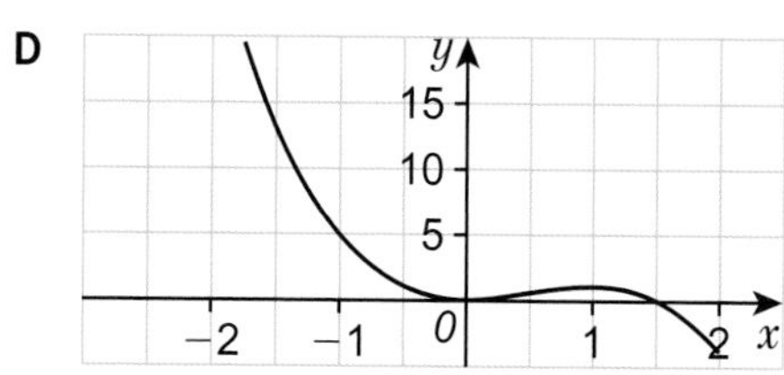

6 Knowledge check

- A **linear equation** generates a straight-line (linear) graph. .... *Mastery lesson 6.1*
- The equation for a straight-line graph can be written as $y = mx + c$ where m is the gradient and c is the y-intercept. .... *Mastery lesson 6.1*
- **Parallel lines** have the same gradient. .... *Mastery lesson 6.1*
- To find the y-intercept of a graph, find the y-coordinate where $x = 0$. To find the x-intercept of a graph, find the x-coordinate where $y = 0$. .... *Mastery lesson 6.1*
- To compare the gradients and y-intercepts of two straight lines, make sure their equations are in the form $y = mx + c$. .... *Mastery lesson 6.1*
- A linear function has a graph that is a straight line. .... *Mastery lesson 6.2*
- A **distance–time graph** represents a journey.
  - Straight lines mean constant speed
  - horizontal lines mean no movement
  - the gradient is the speed, since average speed = $\frac{\text{total distance}}{\text{total time}}$
  - Average speed = $\frac{\text{total distance}}{\text{total time}}$

    Make sure your units match. .... *Mastery lesson 6.3*
- The gradient of a straight-line graph is the rate of change. .... *Mastery lesson 6.3*
- On a **velocity-time graph**
  - straight lines mean constant acceleration
  - horizontal lines mean no change in velocity (i.e. travelling at a constant velocity)
  - the gradient is the acceleration, since acceleration = $\frac{\text{change in velocity}}{\text{time}}$
  - the area under a velocity–time graph is the distance travelled.

- Graph axes do not have to start at zero.
  A zigzag line shows that values have been missed out. .... *Mastery lesson 6.4*

- When two quantities are in **direct proportion**
  - the graph is a straight line through the origin
  - when one variable is multiplied by n, so is the other. .... *Mastery lesson 6.4*
- The coordinates of the **midpoint** of a line segment are $\left(\frac{x_1 + x_2}{2}, \frac{y_1 + y_2}{2}\right)$ .... *Mastery lesson 6.5*

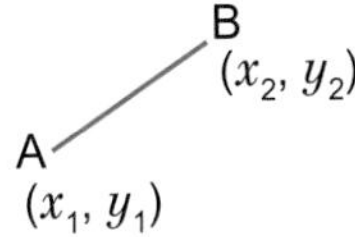

- When two lines are perpendicular, the product of the gradients is −1.
  When a graph has gradient m, a graph perpendicular to it has gradient $\frac{-1}{m}$. ... *Mastery lesson 6.5*
- **Reciprocal functions** are in the form $\frac{k}{x}$ where k is a number. .... *Mastery lesson 6.7*

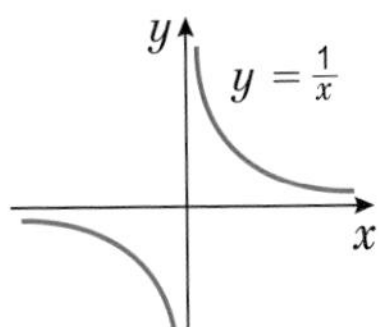

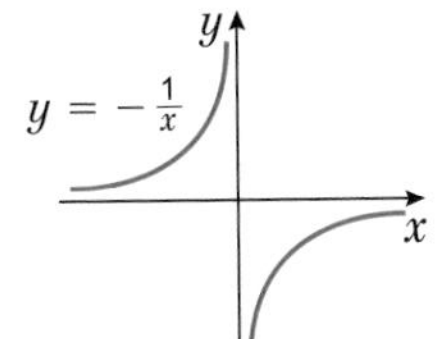

- A **quadratic equation** contains a term in x^2 but no higher power of x.
  The graph of a quadratic equation is a curved shape called a **parabola**. ...... *Mastery lesson 6.6*

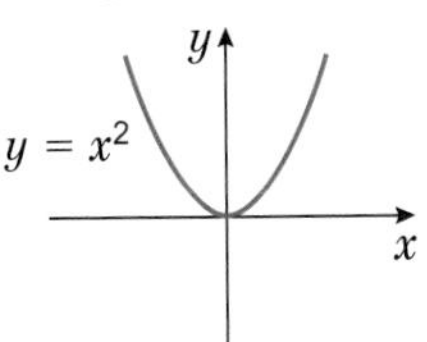

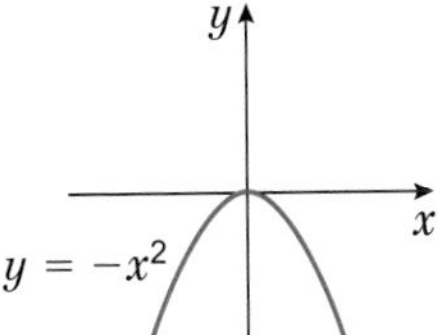

- A quadratic graph has either a **minimum point** or a **maximum point** where the graph turns. .... *Mastery lesson 6.6*

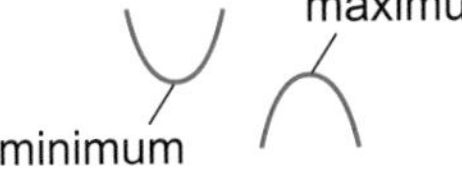

- A quadratic equation can have 0, 1 or 2 solutions. .... *Mastery lesson 6.6*
- A **cubic function** contains a term in x^3 but no higher power of x.
  It can also have terms in x^2 and x and number terms. .... *Mastery lesson 6.7*

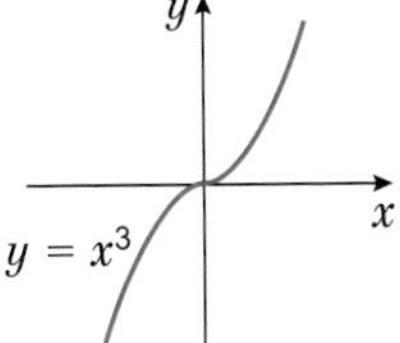

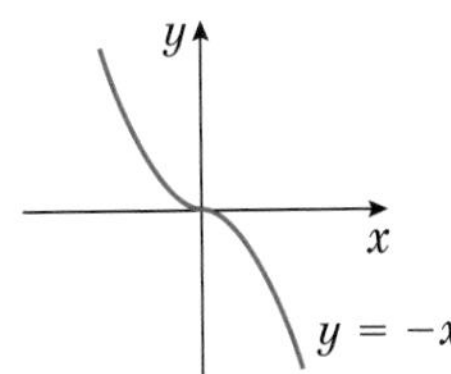

- A cubic function can have 1, 2 or 3 solutions. .... *Mastery lesson 6.7*
- No correlation or weak correlation shows that there is *no* linear relationship between two quantities, because their graph is not close to a straight line.
  When the points follow a curve, there may be a non-linear relationship between the quantities. .... *Mastery lesson 6.8*
- The equation of a circle with centre (0, 0) and radius r is $x^2 + y^2 = r^2$. .... *Mastery lesson 6.8*

Reflect

In this unit, which was easier:

a **plotting and drawing** graphs or b **reading and interpreting** graphs?

Copy and complete this sentence to explain why:

I find __________ graphs easier, because ____________________

6 Unit test

Log how you did on your Student Progression Chart.

1 On squared paper, draw the graph of $y = 2x - 1$ for $-2 \leqslant x \leqslant 5$.
Do not make a table of values. *(4 marks)*

2 Line segment GH is drawn on centimetre squared paper.
G is the point (−1, 7). H is the point (2, −2).

a Find the gradient of the line segment GH.

b Find the midpoint of the line segment GH.

c Find the length of the line segment GH. Give your answer correct to 1 d.p. *(3 marks)*

3 Here is a sketch of a scatter graph.

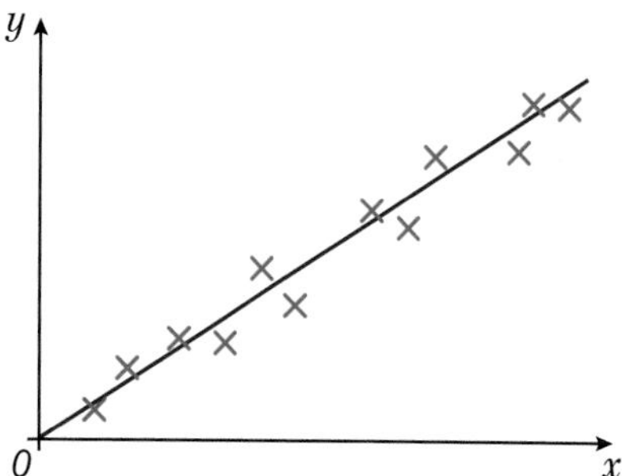

Describe a the correlation and b the relationship between x and y. *(2 marks)*

4 **Exam-style question**

Elliot went for a cycle ride. He left home at 10am.

The travel graph represents part of Elliot's cycle ride.

Elliot's cycle ride

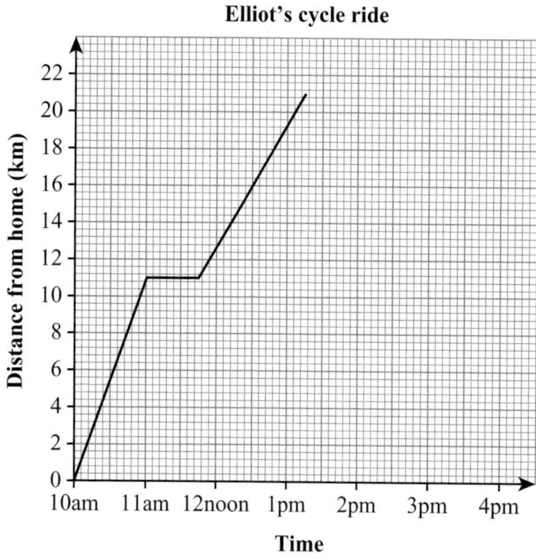

At 11am Elliot stopped for a rest.

a How many minutes did he rest?

b How far from home was Elliot at 12:30pm?

At 1:15pm Elliot stopped for 30 minutes.

Then he cycled home at a steady speed. It took him 1 hour 45 minutes to get home.

c Complete the travel graph. **(4 marks)**

ActiveLearn Homework, practice and support: Higher 6 Unit test

5 A ball bearing rolls down a ramp onto a table.
The graph shows its velocity for the first 20 seconds.

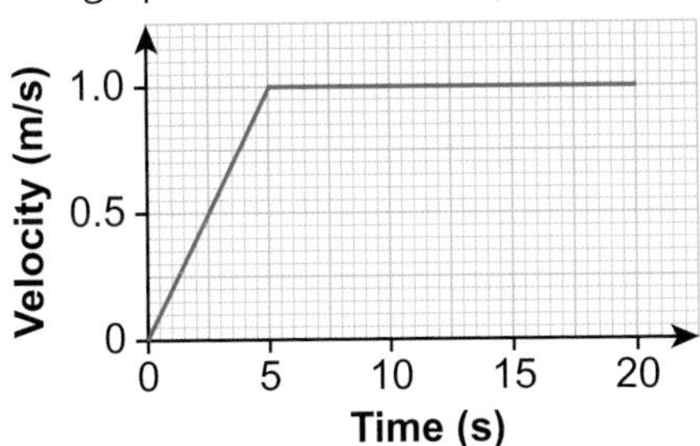

a What was its acceleration in the first 5 seconds?

b How long is the ramp? *(4 marks)*

6 Here is a scatter graph with the line of best fit drawn on.

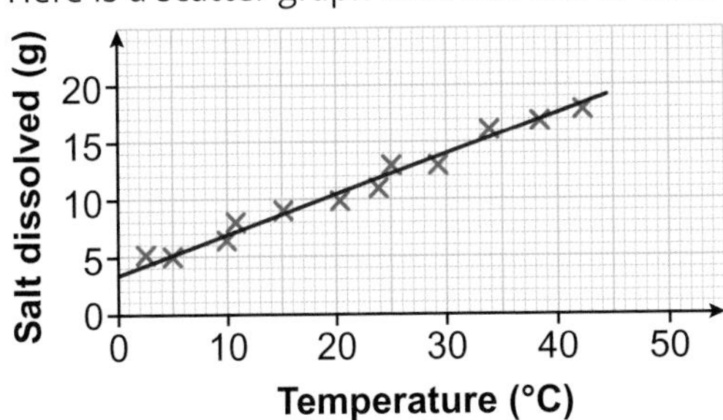

Work out the equation of the line of best fit. *(2 marks)*

7 **Reasoning** The formula KE = $\frac{1}{2}mv^2$ gives the kinetic energy (in joules) of an object with mass m kg and velocity v m/s.

a Work out the kinetic energy of a ball with mass 1.5 kg, travelling at 3 m/s.

b Draw a graph of KE against v for values of v between 0 and 4 m/s.

c What is the velocity of the ball if its kinetic energy is 9 joules? *(5 marks)*

8 Write the equation of a line which is

a parallel to $y = 3x + 2$

b parallel to $2y + 5x = 1$

c perpendicular to $y = -2x + 1$ and passes through the point (4, −3). *(4 marks)*

9 **Reasoning** Here is the graph of a quadratic function.

a What are the coordinates of the minimum point?

b What is the equation of the line of symmetry of the graph?

c Identify the equation of the graph from the options below. Justify your answer.

i $y = x^2 + 2$

ii $y = 2x^2 + 2x$

iii $y = 2x^2 - 4x$

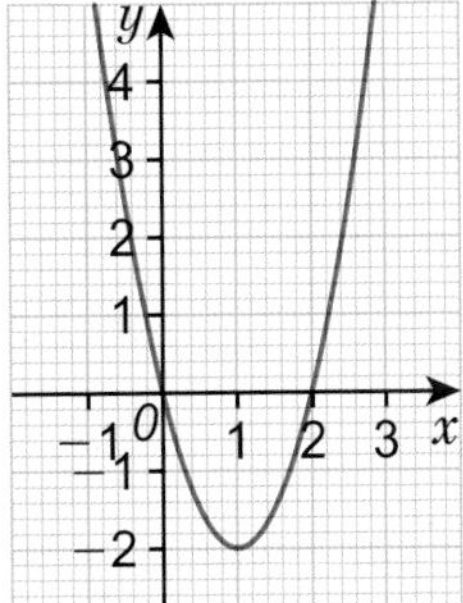

(3 marks)

10 **Problem-solving** Using the graph in **Q9**

a solve the equation $2x^2 - 4x = 2$

b find the coordinates of the point where the line $y = x + 1$ intersects the given graph. *(4 marks)*

Sample student answers

Whose method do you prefer? Why?

Exam-style question

Debbie drove from Junction 12 to Junction 13 on a motorway.

The travel graph shows Debbie's journey.

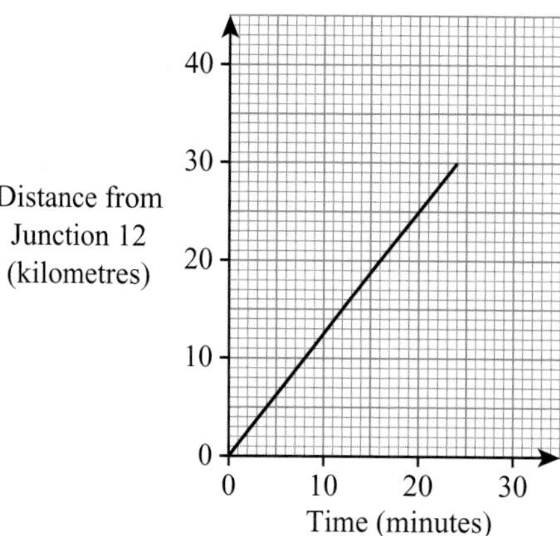

Ian also drove from Junction 12 to Junction 13 on the same motorway.
He drove at an average speed of 66 km/hour.
Who had the faster average speed, Debbie or Ian?
You must explain your answer. **(4 marks)**

June 2013, Q11, 1MA0/1H

Student A

Ian: 66 km/h = 33 km in 30 minutes

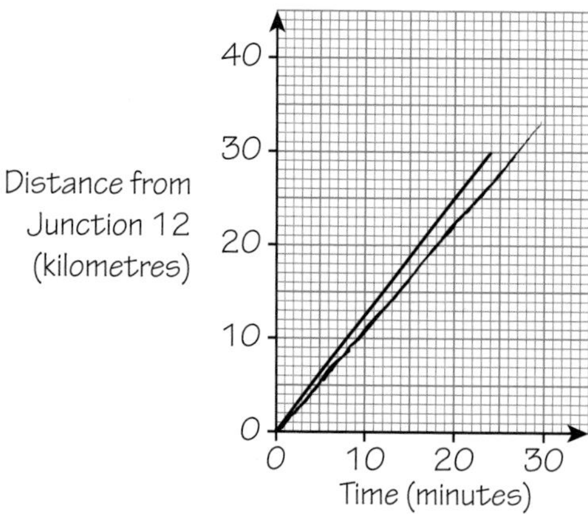

Debbie had the faster average speed as her graph is steeper.

Student B

$S = \frac{D}{T}$

$= 30 \div \frac{2}{5}$

$= 30 \times 5 \div 2$

$= 75$ km/h

Debbie had the faster average speed as she has 75 km/h and Ian has 66 km/h.

7 AREA AND VOLUME

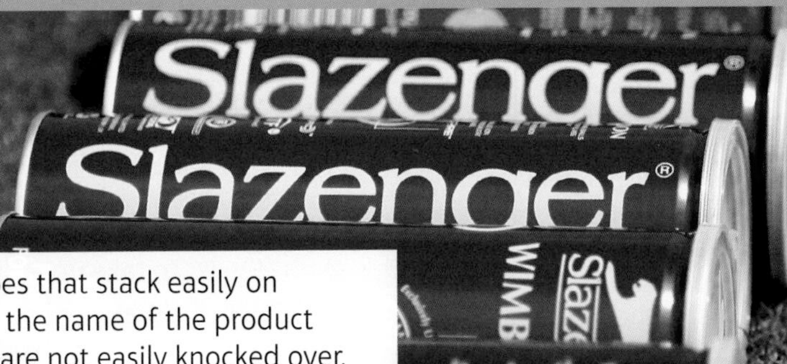

Goods are packaged in shapes that stack easily on shelves, have space to show the name of the product and attractive pictures, and are not easily knocked over. Packaging also needs to be cheap to produce, and should not leave too much empty space around the item inside.

A manufacturer can choose either a cylindrical box or a cuboid to package 4 tennis balls. Each tennis ball has a diameter of 6.5 cm. Work out the possible dimensions for each box.

7 Prior knowledge check

Numerical fluency

1 Calculate an estimate for $5.2 \times 4.4 \times 18.9$

2 Round to 1 decimal place (1 d.p.).
 a 3.57 b 2.06 c 4.99

3 Round to 2 d.p.
 a 9.402 b 13.9834

Fluency with measures

4 Match each object to the amount of liquid it can hold.

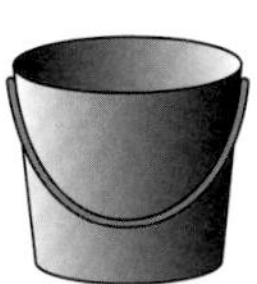

5 litres

5 Copy and complete.
 a 5.2 m = □ cm b 24 cm = □ mm
 c 1 m = □ mm d 3.41 km = □ m
 e 0.327 litres = □ ml
 f 2400 ml = □ litres

Algebraic fluency

6 When $a = 3$, $b = 2$ and $c = 4$, work out
 a $2a^2$ b $4ac + 2b$
 c ab^2c d $\frac{1}{3}ab$

7 Make x the subject of
 a $y = cx$ b $a = bxz$
 c $m = \frac{1}{2}xy$

Geometrical fluency

8 Sketch a circle. Draw and label its centre, radius and diameter.

9 a Sketch the net of this triangular prism. Label the lengths on your net.

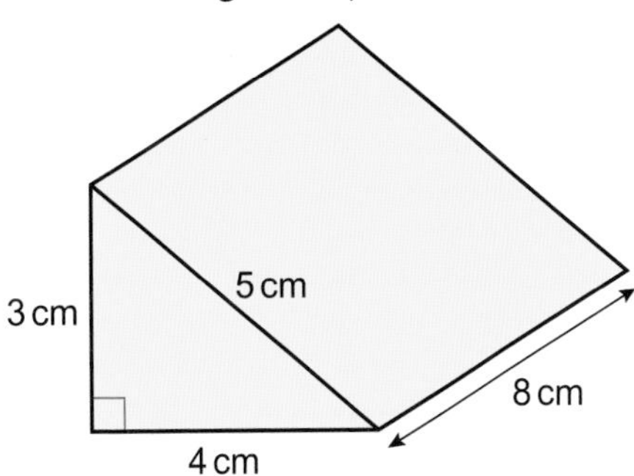

 b Work out the surface area of the prism.

10 a Sketch the solid formed by this net.

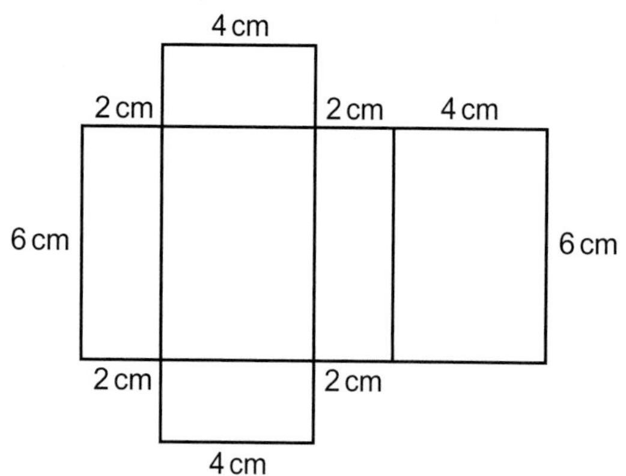

b Draw sketches to show its planes of symmetry.

c Calculate the volume of the solid.

11 Work out the area of this parallelogram.

12 Work out the volume of this 3D solid. Write your answer to 2 significant figures (2 s.f.).

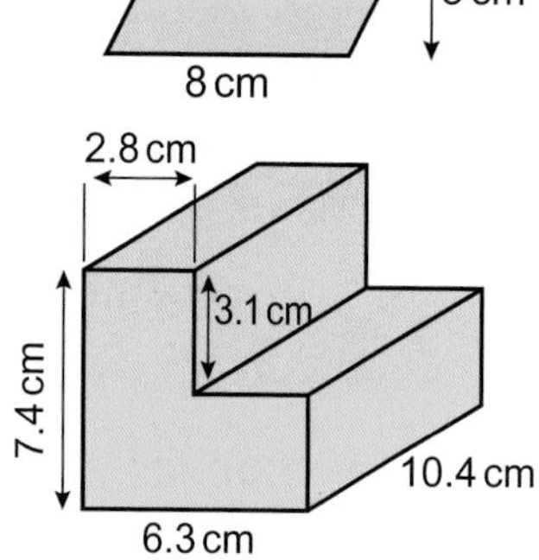

*** Challenge**

13 These chocolate bars are packed into cuboid-shaped boxes for transport.

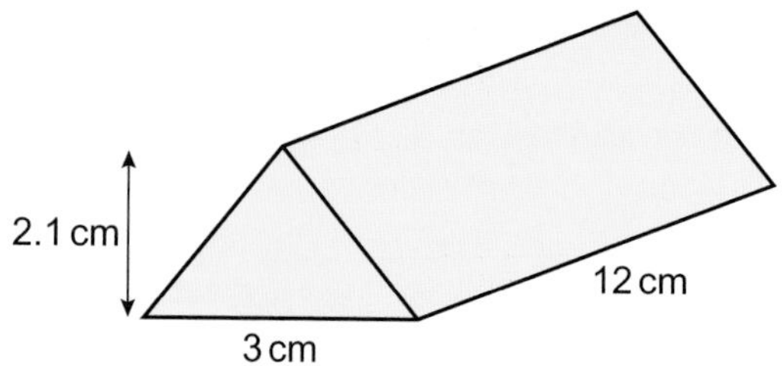

Sketch how you could pack 6 of these bars to take up as little room as possible.

What size box do you need for 48 chocolate bars?

What other shapes of box could you use to transport them? Which shape box would be most practical?

7.1 Perimeter and area

Objectives

- Find the area and perimeter of compound shapes.
- Recall and use the formula for area of a trapezium.

Did you know?

The name trapezium comes from the Greek word, trapeza, meaning table.

Fluency

Which of these measurements are areas and which are perimeters?

3 cm² 5 km 32 mm 7 m² 46 mm² 10 cm

Warm up

1 Work out the area and perimeter of each shape.

a

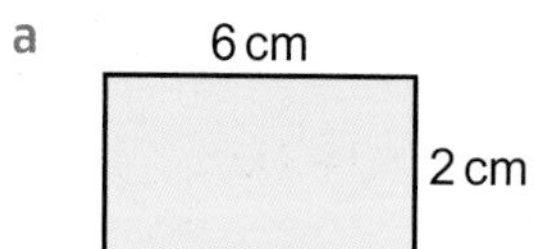

b

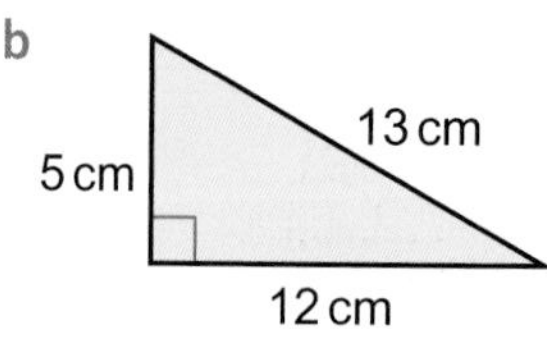

c

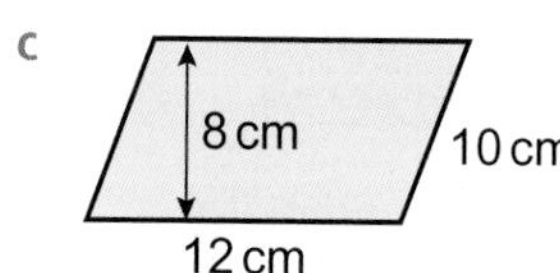

2 Solve

a $3x = 12$ b $\frac{1}{2}x = 6$ c $4(x + 2) = 28$

ActiveLearn Homework, practice and support: Higher 7.1

Questions in this unit are targeted at the steps indicated.

3 a Work out the area and perimeter.

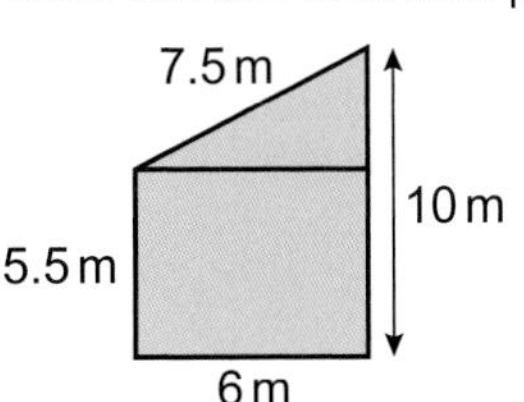

b Work out the area.

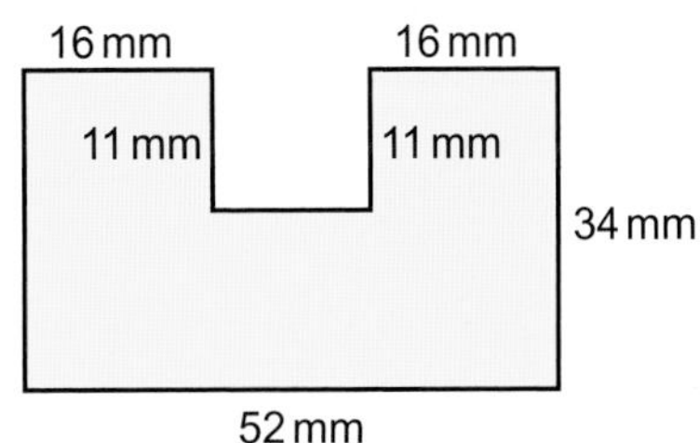

c Work out the shaded area. Give your answer to the nearest square mm.

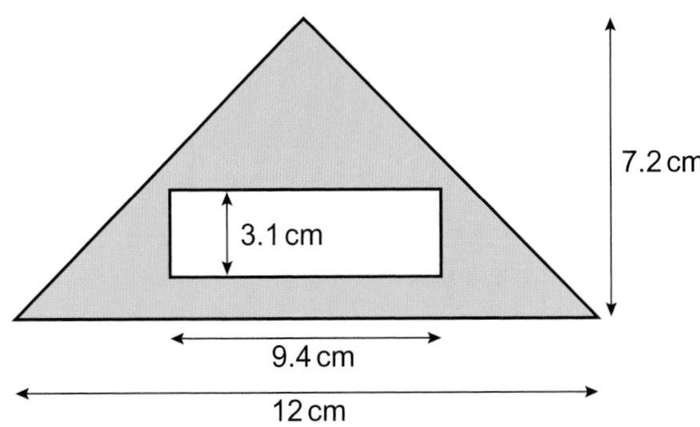

Discussion How did you work out the area in part **b**? How else could you have done it? Which way is most efficient?

Key point 1

This trapezium has parallel sides, a and b, and perpendicular height, h.

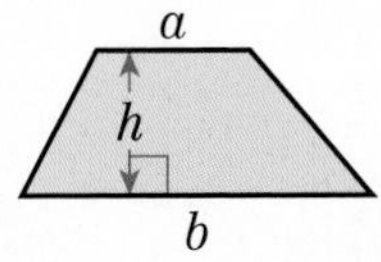

Two trapezia put together make a parallelogram, with base $(a + b)$ and perpendicular height, h.

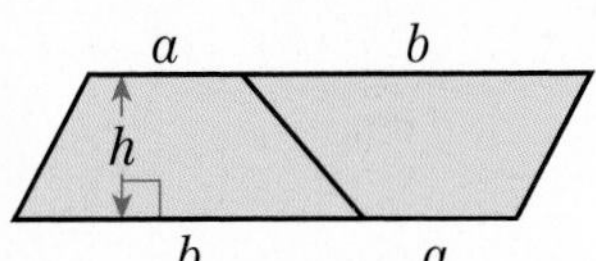

Communication hint Trapezia is the plural of trapezium.

Area of 2 trapezia = base × perpendicular height = $(a + b)h$

Area of a trapezium = $\frac{1}{2}(a + b)h$

4 Calculate the areas of these trapezia. Round your answers to 1 decimal place (1 d.p.) where necessary.

Q4a hint Use the formula: area = $\frac{1}{2}(a + b)h$

a

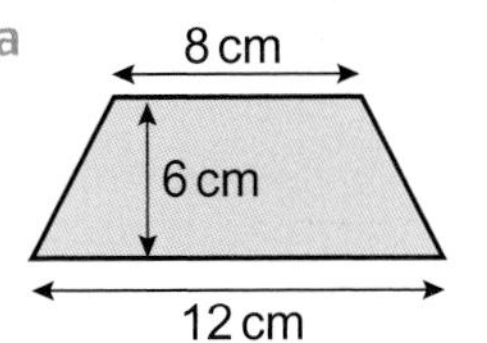

b

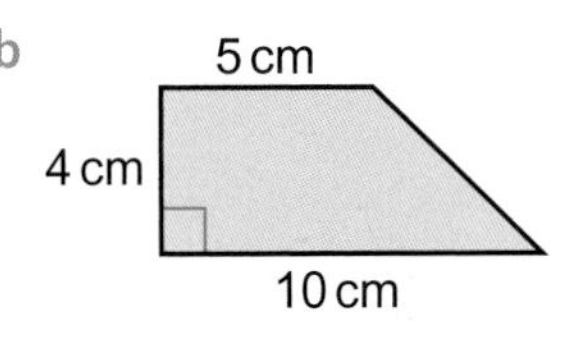

c d

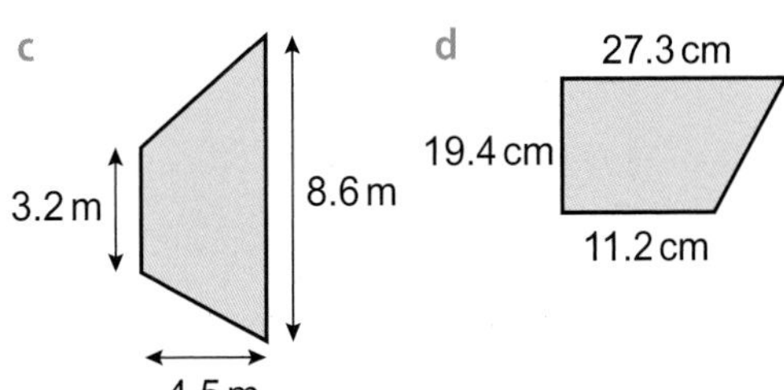

5 Calculate the area and perimeter of this isosceles trapezium.

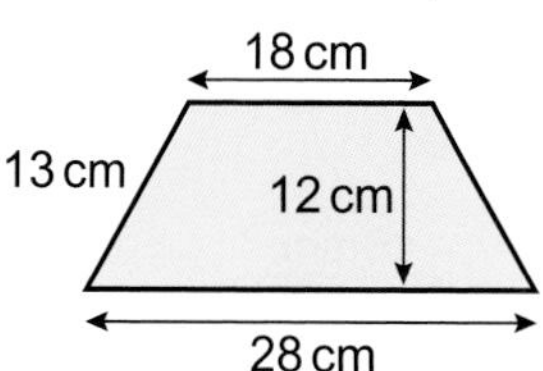

Q5 communication hint An **isosceles trapezium** has one line of symmetry. Its two sloping sides are equal.

Reflect How can you remember how to find the area of a trapezium?

6 **Exam-style question**

Here is a diagram of Jim's garden.

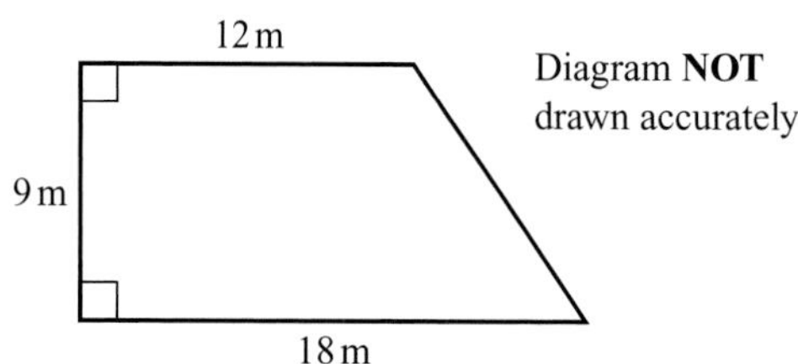

Jim wants to cover his garden with grass seed to make a lawn.
Grass seed is sold in bags.
There is enough grass seed in each bag to cover 20 m^2 of garden.
Each bag of grass seed costs £4.99.
Work out the least cost of putting grass seed on Jim's garden. **(4 marks)**

Nov 2012, Q25, 1MA0/1F

Exam hint
Show your working by writing all the calculations you do on your calculator.

7 **Problem-solving**
Here is the plan of a play area.
Work out its area.

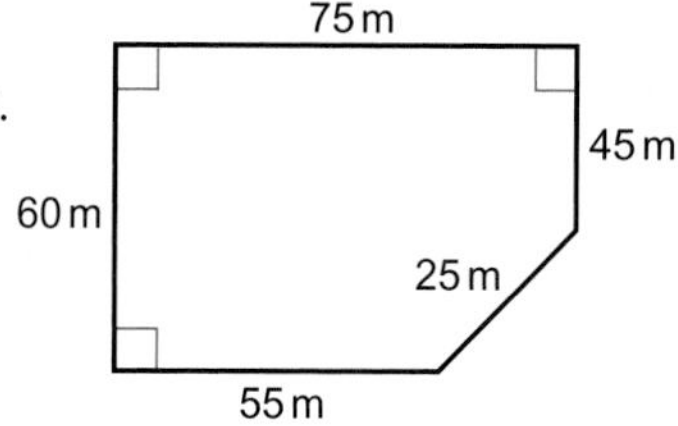

8 The area of this trapezium is 96 cm^2.

a Substitute the values of a, b and A into the formula $A = \frac{1}{2}(a + b)h$

b Simplify to get an equation.

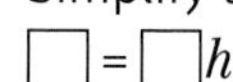

c Solve to find h.

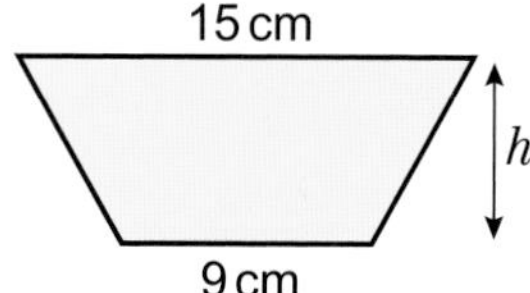

9 **Problem-solving** A trapezium has area 32 cm^2, and parallel sides 5.5 cm and 10.5 cm. Work out its height.

Q9 strategy hint Sketch and label the trapezium. Use the method from **Q8**.

Example 1

This trapezium has area 70 m^2.
Find the length of the shorter parallel side.

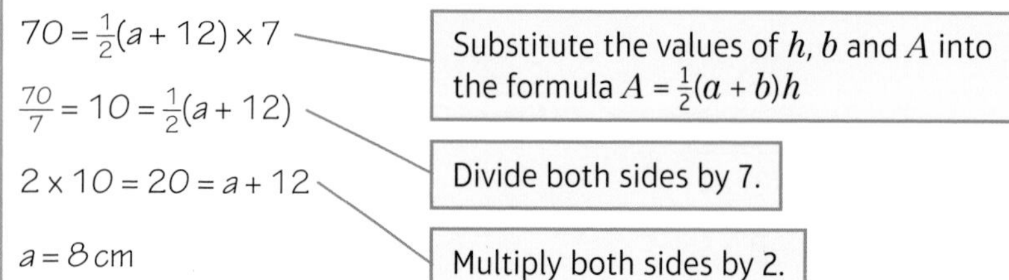

$70 = \frac{1}{2}(a + 12) \times 7$ — Substitute the values of h, b and A into the formula $A = \frac{1}{2}(a + b)h$

$\frac{70}{7} = 10 = \frac{1}{2}(a + 12)$ — Divide both sides by 7.

$2 \times 10 = 20 = a + 12$ — Multiply both sides by 2.

$a = 8$ cm

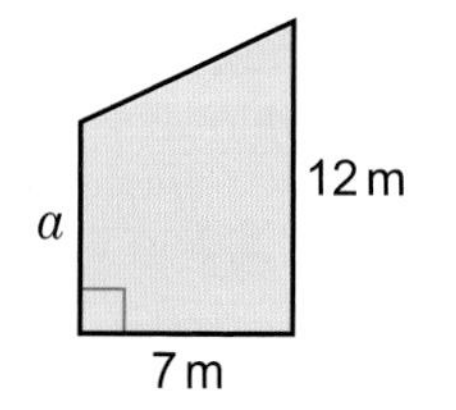

10 **Reasoning** Find the missing lengths.

a b

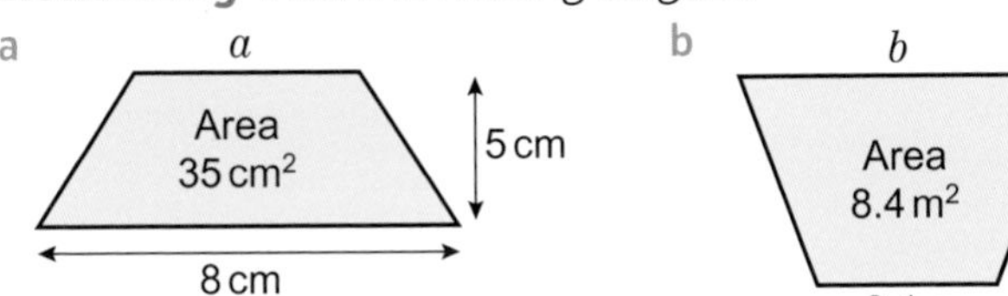

11 **Problem-solving** One corner of a rectangular piece of paper is folded up to make this trapezium.

Work out the area of the trapezium.

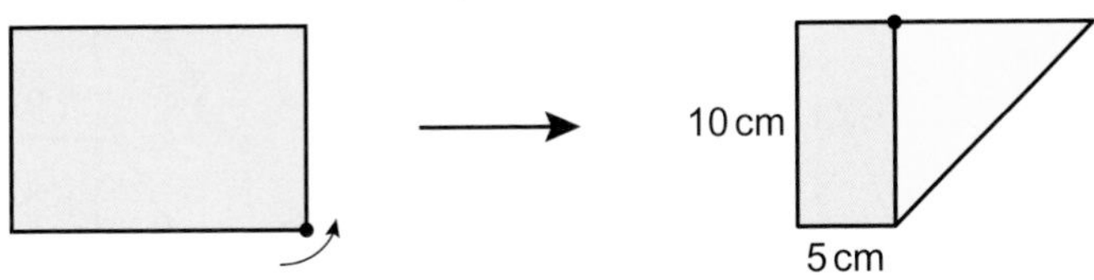

7.2 Units and accuracy

Objectives

- Convert between metric units of area.
- Calculate the maximum and minimum possible values of a measurement.

Did you know?

The accuracy of a measurement depends on the instrument you measure it with.

Fluency

What is the formula for the area of a square, triangle, rectangle, parallelogram and trapezium?
What are the possible values of x when $3 \leqslant x < 6$ and x is an integer?

1 Round each number to the level of accuracy given.

a 3.567 (1 d.p.) b 320.6 (2 s.f.) c 8.495 (2 d.p.) d 15.721 (3 s.f.)

2 Work out

a i 10% of 25 kg ii 10% less than 25 kg iii 10% more than 25 kg

Q2aii hint Subtract 10% of 25 kg from 25 kg.

b i 5% of 40 m ii 5% less than 40 m iii 5% more than 40 m

3 a Explain why these two squares have the same area.

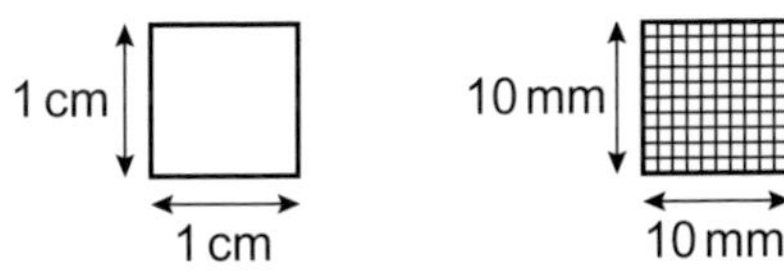

b Work out the area of each square.

c Copy and complete.

$1\text{ cm}^2 = \square\text{ mm}^2$

Key point 2

To convert from cm^2 to mm^2, multiply by 100.
To convert from mm^2 to cm^2, divide by 100.

Warm up

4 Reasoning

a Sketch a square with side length 1 m and a square with side length 100 cm.

b Copy and complete.

$1\text{ m}^2 = \square\text{ cm}^2$

c How do you convert from cm^2 to m^2?

5 Convert

a 250 mm^2 to cm^2
b 5.2 m^2 to cm^2
c 7000 cm^2 to m^2
d 3.4 cm^2 to mm^2
e 8.85 m^2 to cm^2
f 1246 mm^2 to cm^2
g 0.37 m^2 to mm^2
h $2\,800\,000\text{ mm}^2$ to m^2

> **Q5g hint**
> Convert m^2 to cm^2 then to mm^2.

6 Calculate these areas.

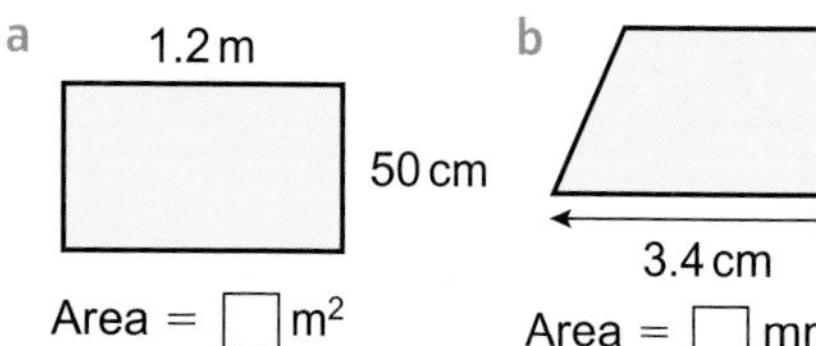

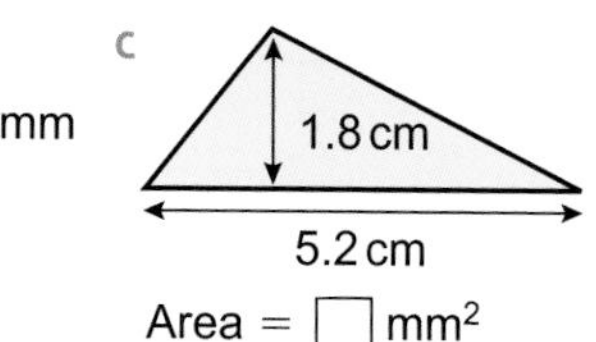

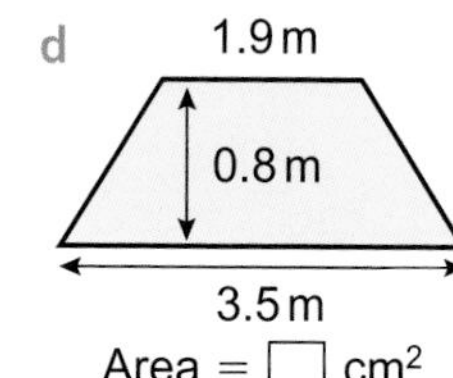

e **Reflect** In part **c**, which was the easiest way to find the area in mm^2?

- Find the area in cm^2, then convert to mm^2.
- Convert the lengths to mm, then find the area.

f **Reflect** Which was the easiest way to find the area of part **d** in cm^2?

> **Key point 3**
>
> 1 hectare (ha) is the area of a square 100 m by 100 m.
> $1\text{ ha} = 100\text{ m} \times 100\text{ m} = 10\,000\text{ m}^2$
> Areas of land are measured in hectares.
>
> 100 m
> 100 m

7 Here is the plan of a playing field.

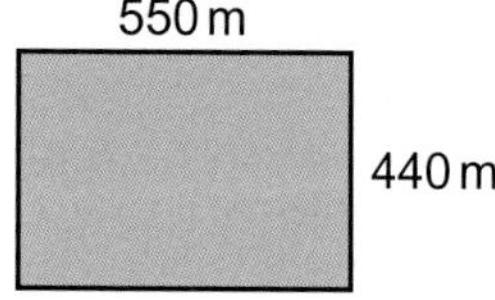

Work out the area of the field in hectares.

8 STEM / Problem-solving A farmer counts 2 wild oat plants in a 50 cm by 50 cm square of a field. The whole field has area 20 ha.
Estimate the number of wild oat plants in the field.

> **Q8 strategy hint** Work out the area of the square in m^2. How many of these would fit in the field?

> **Key point 4**
>
> A 10% error interval means that a measurement could be up to 10% larger or smaller than the one given.

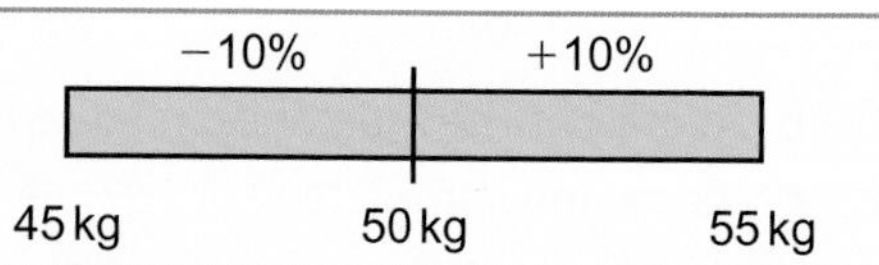

9 A factory makes bolts 30 mm long, with a 10% error interval.

a Work out the largest and smallest possible lengths of the bolts.

b Write the possible lengths as an inequality.

$\square\text{ mm} \leqslant \text{length} \leqslant \square\text{ mm}$

10 Sweets are packed in 20 g bags, with a 5% error interval.
Work out the possible masses of the bags of sweets.

Q10 hint □ g $\leqslant$ mass $\leqslant$ □ g

11 Reasoning a Each measurement has been rounded to the nearest cm.
Write its smallest possible value.

i 36 cm   ii 112 cm

b Each measurement has been rounded to 1 d.p. Write its smallest possible value.

i 2.5 cm   ii 6.7 kg

Discussion What is the largest possible value that rounds down to 36 cm?

Key point 5

Measurements rounded to the nearest unit could be up to half a unit smaller or larger than the rounded value. The possible values of x that round to 3.4 to 1 d.p. are $3.35 \leqslant x < 3.45$

12 Each measurement has been rounded to the accuracy given.
Write an inequality to show its smallest and largest possible values. Use x for the measurement.

a 18 m (to the nearest metre)   b 24.5 kg (to 1 d.p.)
c 1.4 m (to 1 d.p.)   d 5.26 km (to 2 d.p.)

Key point 6

The upper bound is half a unit greater than the rounded measurement.
The lower bound is half a unit less than the rounded measurement.

$12.5 \leqslant x < 13.5$
lower bound   upper bound

13 Write i the lower bound ii the upper bound of each measurement.

a 8 cm (to the nearest cm)   b 5.3 kg (to the nearest tenth of a kg)   c 11.4 m (to 1 d.p.)
d 2.25 litres (to 2 d.p.)   e 5000 m (to 1 s.f.)   f 32 mm (to 2 s.f.)   g 1.53 kg (to 3 s.f.)

Example 2

The length of the side of a square is 5.34 cm to 2 d.p.
Work out the upper and lower bounds for the perimeter.
Give the perimeter to a suitable degree of accuracy.

	Lower bound	Upper bound
Side length	5.335 cm	5.345 cm
Perimeter	5.335 x 4 = 21.34 cm	5.345 x 4 = 21.38 cm

Use the upper and lower bounds of the side length to calculate the upper and lower bounds of the perimeter.

21.34 = 21.3 to 1 decimal place   21.38 = 21.4 to 1 decimal place

Round the upper and lower bounds to 1 decimal place. Do they give the same value?

21.34 = 21 (nearest cm)   21.38 = 21 (nearest cm)

Perimeter = 21 cm

Round to nearest cm. They both give the same value so we can be sure that to the nearest cm, the perimeter is 21 cm.

14 Reasoning A rectangle measures 15 cm by 28 cm to the nearest cm.

a Work out the upper and lower bounds for the length and width.

b Calculate the upper and lower bounds for the perimeter of the rectangle. Give the perimeter to a suitable degree of accuracy.

Q14 strategy hint Sketch a diagram. Which two possible lengths will give the largest possible perimeter?

15 Reasoning A parallelogram has base length 9.4 m and height 8.5 m. Both measurements are given to 1 d.p. Work out the upper and lower bounds for its area.

16 Reasoning A parallelogram has area 24 cm^2 to the nearest whole number. Its height is 6.2 cm (to 1 d.p.).

a Write the upper and lower bounds for the area and the height.

b Work out

i $\frac{\text{upper bound for area}}{\text{upper bound for height}}$ ii $\frac{\text{upper bound for area}}{\text{lower bound for height}}$

c What is the upper bound for the base of the parallelogram?

Q16c hint Which calculation in part **b** gives the higher value?

7.3 Prisms

Objectives

- Convert between metric units of volume.
- Calculate volumes and surface areas of prisms.

Did you know?

The shapes in a child's shape sorter toy are prisms.

Fluency

Calculate the volume of this cuboid.
How could you work out its surface area?

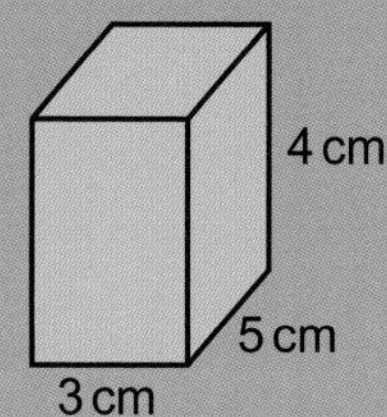

Warm up

1 Work out the volume of this 3D solid made from two cuboids.

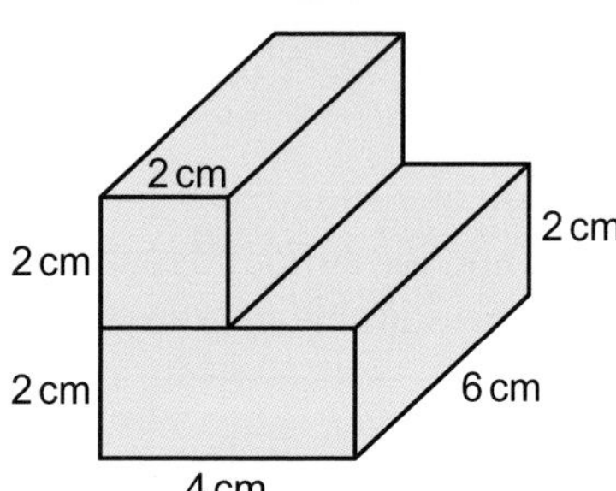

2 Solve

a $9b = 72$ b $\frac{25}{2}h = 50$

Key point 7

The **surface area** of a 3D solid is the total area of all its faces.

3 Reasoning

a Sketch a cuboid 4 cm by 5 cm by 7 cm.

b Work out the area of the top of your cuboid.
Which other face of the cuboid is identical to this one?

c Work out the area of the front and side of your cuboid.
Which other faces of the cuboid are identical to them?

d Work out the total surface area of your cuboid.

Discussion How can you calculate the surface area of a cuboid without drawing its net?

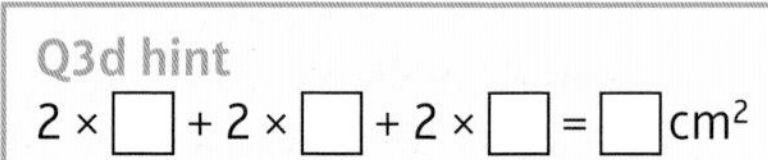

4 Calculate the surface area of a cuboid 3 cm × 2 cm × 6 cm.

ActiveLearn Homework, practice and support: Higher 7.3

Key point 8

A **prism** is a 3D solid that has the same cross-section all through its length.

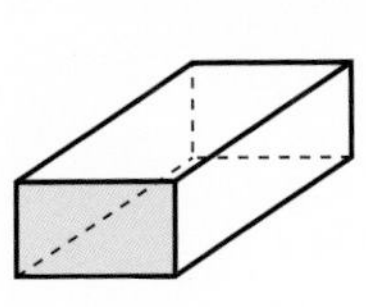
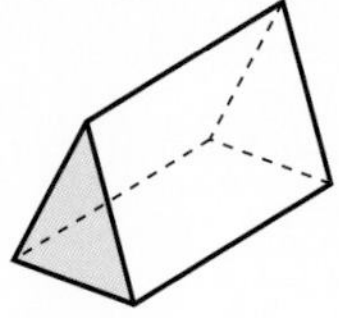
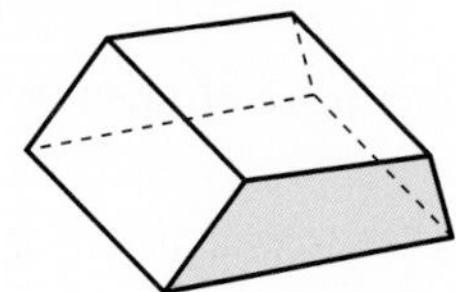

5 **Communication** a Is the 3D solid in **Q1** a prism? Explain your answer.

b Work out the area of the cross-section of the solid in **Q1**.

c Multiply the area of the cross-section by the length of the solid. What do you notice?

Discussion For a cuboid, why is multiplying area of cross-section by the length the same as multiplying length × width × height?

Key point 9

Volume of a prism = area of cross-section × length

6 Work out the volume of each prism.

a

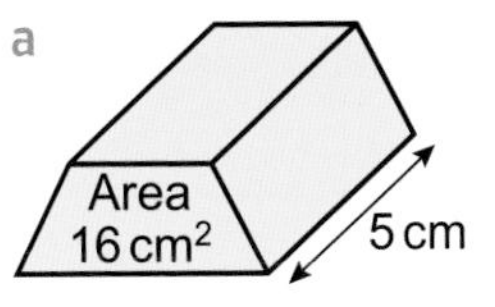

b

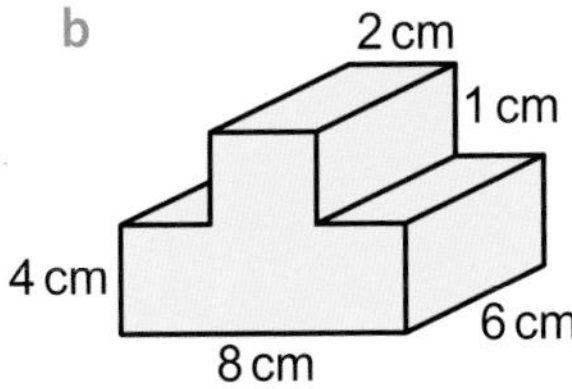

c

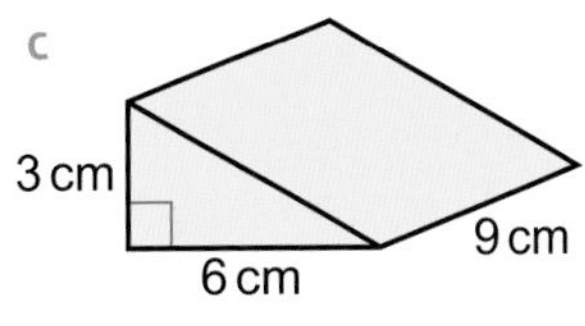

7 **Reasoning** This triangular prism has volume 48 cm³.

a Work out its height.

b Sketch its net and work out its surface area.

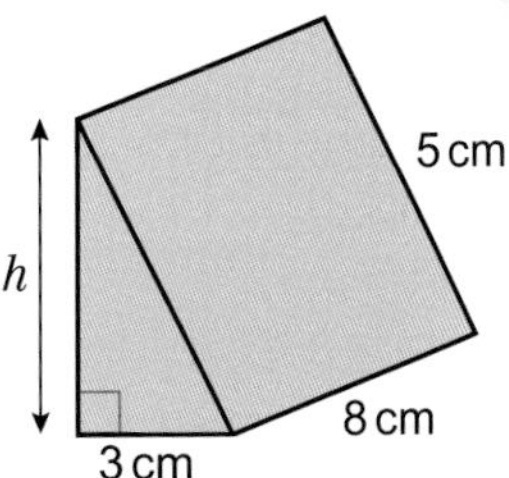

Q7a hint Write and solve an equation:
volume 48 = $\frac{1}{2}$ × □ × h × □

8 **Exam-style question**

Jane has a carton of orange juice.
The carton is in the shape of a cuboid.

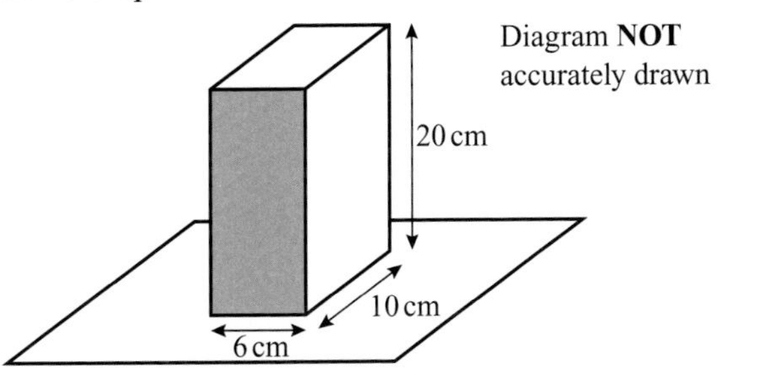

The depth of the orange juice in the carton is 8 cm.
Jane closes the carton.
Then she turns the carton over so that it stands on the shaded face.
Work out the depth, in cm, of the orange juice now. **(3 marks)**

June 2012, Q12, 1MA0/1H

Exam hint
Add the level of the juice to the diagram. Sketch the carton when it stands on the shaded face.

Key point 10

Volume is measured in mm³, cm³ or m³.

9 **Reasoning**

a Sketch a cube with side length 1 cm and a cube with side length 10 mm.

b Copy and complete.

$1\text{ cm}^3 = \square\text{ mm}^3$

c How do you convert from mm³ to cm³?

10 **Reasoning**

a Work out the volumes of a cube with side length 1 m and a cube with side length 100 cm.

b How do you convert from m³ to cm³?

Key point 11

Capacity is measured in ml and litres.

$1\text{ cm}^3 = 1\text{ ml}$

$1000\text{ cm}^3 = 1\text{ litre}$

11 Convert

a 4.5 m^3 into cm³

b 52 cm^3 into mm³

c $9\,500\,000\text{ cm}^3$ into m³

d 3421 mm^3 into cm³

e 5200 cm^3 into litres

f 0.7 litres into cm³

g 175 ml into cm³

h 3 m^3 into litres.

Q11h hint Convert 3 m^3 to cm³ first.

12 **Problem-solving** A water tank is a cuboid 140 cm tall, 80 cm wide and 2 m long. Kate paints all the faces except the base.

a Work out the total area she paints, in square metres.

b 1 tin of paint covers 4 m^2. How many tins of paint does Kate need?

Q12a hint Sketch the tank.

13 **Problem-solving / Reasoning**

A cube has surface area 507.8 cm^2 to 1 d.p.

What is the length of 1 side of the cube?

Give your answer to 1 d.p.

Q13 hint First work out the area of 1 face.

14 **STEM / Modelling** A scientist collects a sample of leaf mould 20 cm deep from a 0.25 m^2 area in a wood.

a By modelling the sample as a prism, calculate the volume of leaf mould she collects.

b In the leaf mould sample she counts 12 worms. Estimate the number of worms in the top 20 cm of leaf mould in 2 hectares of the wood.

15 **Communication** Show that the volume of this prism is $20x^3$.

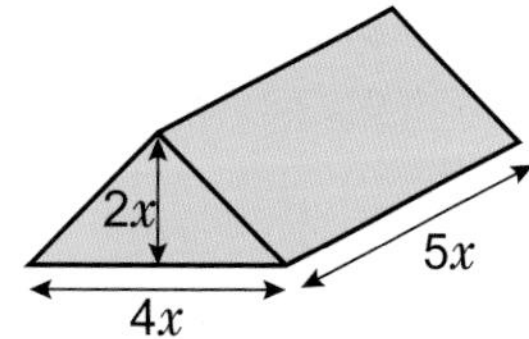

Q15 hint Work out the volume using the measurements in x. Simplify your answer.

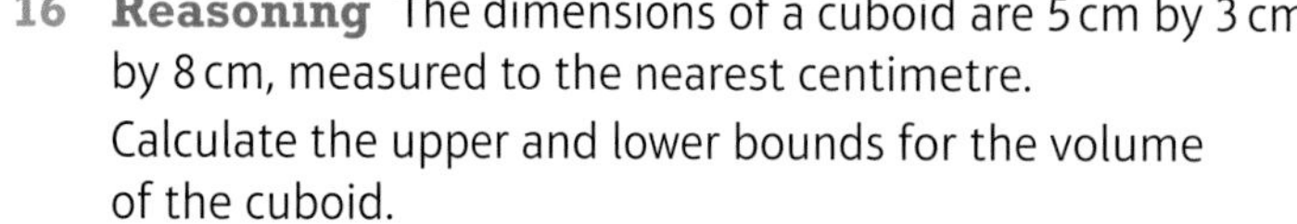

16 **Reasoning** The dimensions of a cuboid are 5 cm by 3 cm by 8 cm, measured to the nearest centimetre.

Calculate the upper and lower bounds for the volume of the cuboid.

Q16 strategy hint First calculate the upper and lower bound of each measurement.

7.4 Circles

Objectives

- Calculate the area and circumference of a circle.
- Calculate area and circumference in terms of π.

Did you know?

Speedometers record the number of revolutions of the wheel and the time taken. They then use the circumference of the wheel to work out the distance travelled in that time, and then the speed.

Fluency

What is the radius of each circle?
What is the diameter?

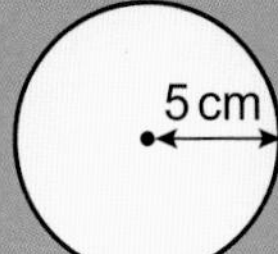

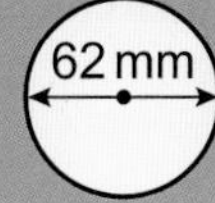

Warm up

1 Solve

a $35 = 7r$

b $75 = 3r^2$

2 Make x the subject of

a $y = mx$

b $t^2 = x^2$

c $p = x^2$

Key point 12

The **circumference** of a circle is its perimeter.

3 **Reasoning** The table gives the diameter and circumference of some circles.

Diameter	Circumference
10 cm	31.4 cm
54 m	169.6 m
36 mm	113 mm

a Work out the ratio $\frac{\text{circumference}}{\text{diameter}}$ for each one. What do you notice?

b The ratio $\frac{\text{circumference}}{\text{diameter}}$ of a circle is represented by the Greek letter π (pi).

Find the π key on your calculator.

Write the value of π to 8 d.p.

Discussion How can you work out the circumference of a circle if you know its diameter? What if you know its radius?

Key point 13

For any circle
circumference = π × diameter
$C = \pi d$ or $C = 2\pi r$

4 Work out the circumference of each circle. Give your answers to 1 d.p. and the units of measurement.

a
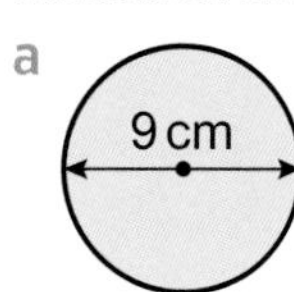

b
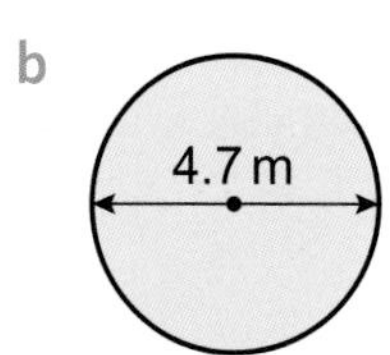

c 12 mm

d 3.4 cm

Reflect Do you need to remember both formulae for the circumference of a circle, or just one?

Key point 14

The formula for the area, A, of a circle with radius r is $A = \pi r^2$.

5 Find the area of each circle.

a
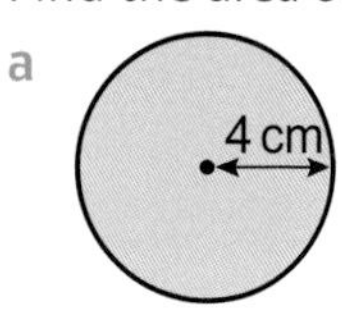

b
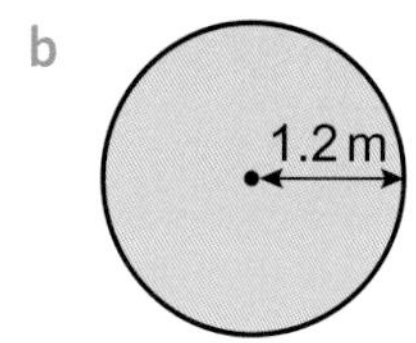

c
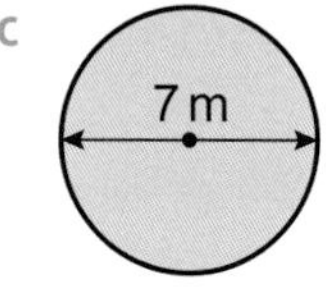

d
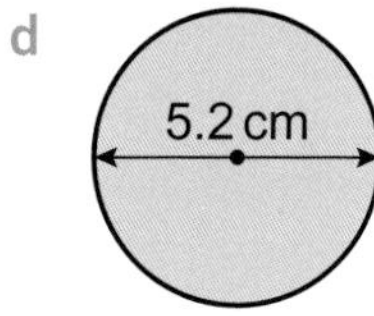

6 **Reflect** One of these expressions is for the area of a circle and one is for the circumference.

$2\pi r$ $\qquad$ πr^2

How can you remember which is which?

Q6 hint Think about the units of area and circumference.

7 **Exam-style question**

Mr Weaver's garden is in the shape of a rectangle.
In the garden there is a patio in the shape of a rectangle and two ponds in the shape of circles with diameter 3.8 m.
The rest of the garden is grass.

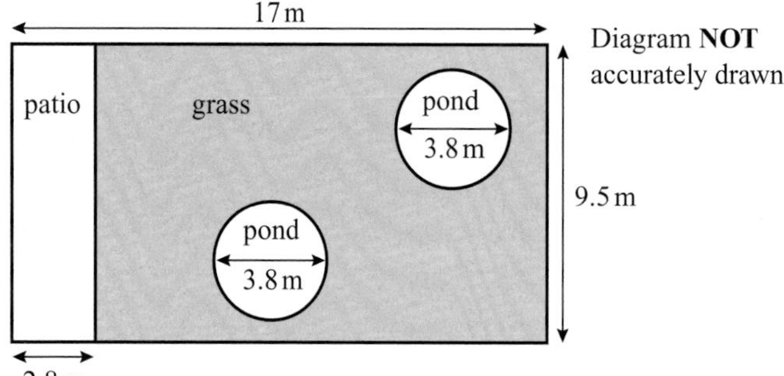

Mr Weaver is going to spread fertiliser over all the grass.
One box of fertiliser will cover 25 m^2 of grass.
How many boxes of fertiliser does Mr Weaver need?
You must show your working. **(5 marks)**

June 2012, Q5, 1MA0/2H

Exam hint

Show clearly which areas you are working out. You could use:
Area □ =
Area ○ =

8 **STEM / Modelling** A biologist needs to estimate the area of this mould sample.

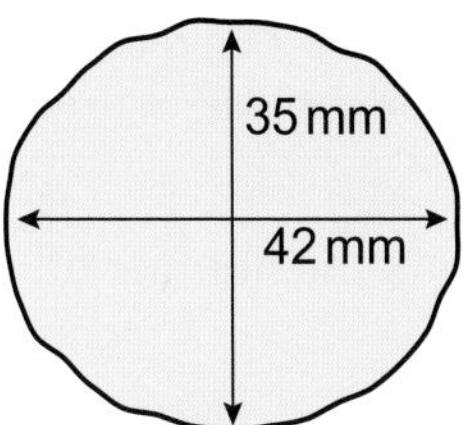

Q8b hint Use your mean from part **a** as an estimate for the diameter.

She measures the distance across it in two directions.

a Work out the mean distance across the sample.

b By modelling the sample as a circle, calculate an estimate for its area to the nearest square millimetre.

9 **Real / Problem-solving** A bicycle wheel has radius 32 cm.
How many complete revolutions does the wheel make in 1 km?

10 **Reasoning** The areas and circumferences of these circles are given in terms of π.
Match each circle to its area and circumference.

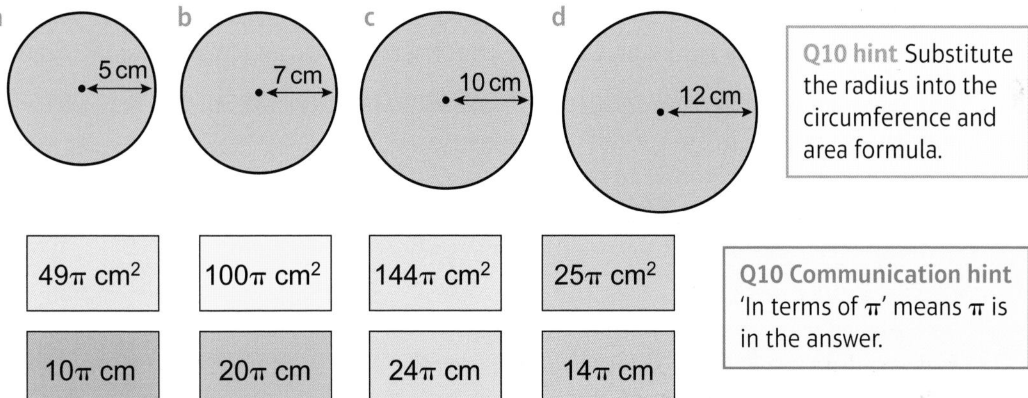

Q10 hint Substitute the radius into the circumference and area formula.

Q10 Communication hint 'In terms of π' means π is in the answer.

11 **Reasoning** a Work out the area and circumference of a circle with radius 6 cm

i in terms of π

ii to 2 s.f.

b Which values for the area and circumference are the most accurate?

12 The circumference of a circle is 104 cm.

a Substitute the value for C into the formula $C = \pi d$.

b Solve the equation to find the diameter to 1 d.p.

Q12 hint $C = \pi d$ so $\frac{C}{\square} = d$

13 Find the radius of a circle with circumference 24 cm.

Q13 hint Find the diameter and halve it.

Example 3

A circle has area 50 m². Find its radius, to the nearest cm.

$50 = \pi r^2$ — Substitute A = 50 into the area formula.

$\frac{50}{\pi} = r^2$ — Rearrange to make r^2 the subject.

$\sqrt{\frac{50}{\pi}} = r$ — Square root both sides to find r.

$r = 3.99$ m $= 399$ cm

14 a Find the radius of a circle with area 520 m^2.
Give your answer to the nearest cm.
b Find the diameter of a circle with area 630 cm^2.
Give your answer to the nearest mm.

Q14b hint Find the radius and double it.

15 Tim is using circles in a scale diagram. The area of each circle represents the number of people in a group.

Circle	Number of people	Area of circle
X	40	40 cm^2
Y	25	25 cm^2
Z	70	70 cm^2

a Copy and complete to make r the subject of the formula for area of a circle.

$A = \pi r^2$

$\frac{A}{\square} = r^2$

$\sqrt{\frac{A}{\square}} = \square$

Q15b communication hint Radii is the plural of radius.

b Use your formula for r from part **a** to work out the radii of circles X, Y and Z to the nearest mm.

16 **Real / Reasoning** In a bakery, pastry is rolled into rectangles 0.5 m by 4 m.
Circles of diameter 6 cm are cut out of the pastry.
The remaining pastry is thrown away.
a Calculate the area of pastry thrown away from each rectangle.
b What percentage of the pastry is thrown away?

Q16a hint How many circles fit across and along the rectangle of pastry?

7.5 Sectors of circles

Objectives

- Calculate the perimeter and area of semicircles and quarter circles.
- Calculate arc lengths, angles and areas of sectors of circles.

Did you know?

Half a circle is called a semicircle, but half a sphere is called a hemisphere. Semi is from the Latin word for half, and hemi is from the Greek word for half.

Fluency

What fraction of the whole circle is each sector?

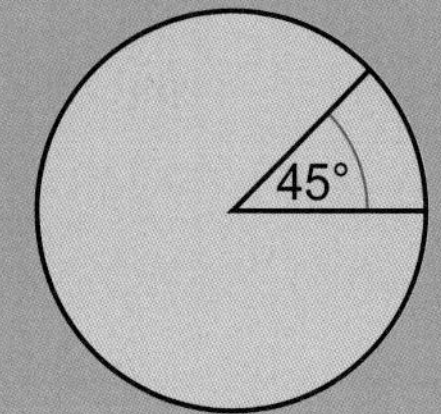

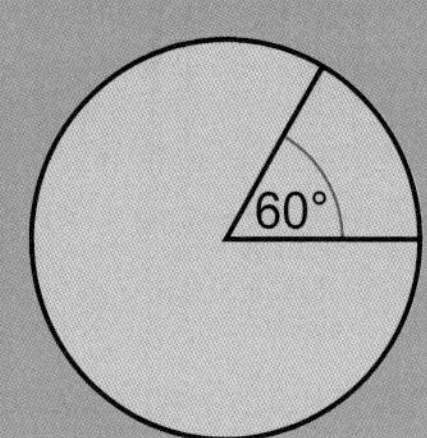

Warm up

1 Find the circumference and area of this circle
a in terms of π
b correct to 3 significant figures.

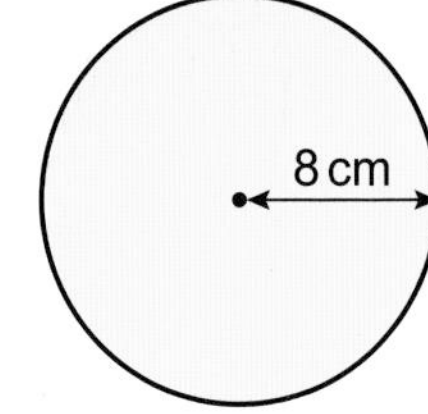

2 Simplify
a $2\pi + 2\pi$
b $2\pi + 6 + 4.2$
c $5 + 3\pi + 2$

Q2b hint Collect like terms, $\square\pi + \square$

ActiveLearn Homework, practice and support: Higher 7.5

3 Work out the area of

a the semicircle b the quarter circle

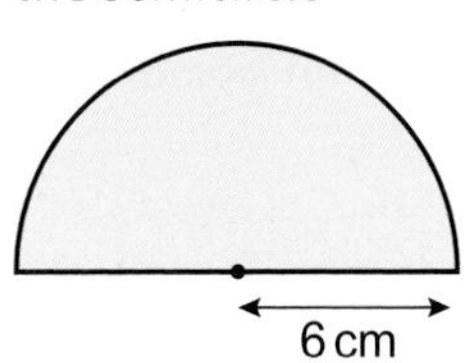

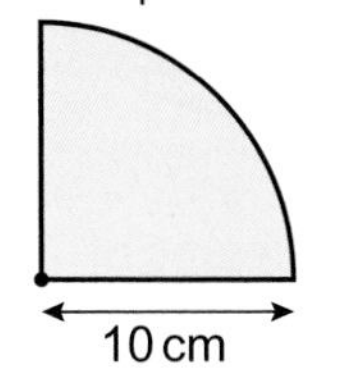

i in terms of π ii correct to 3 s.f.

Q3 strategy hint Find the area of the whole circle first.

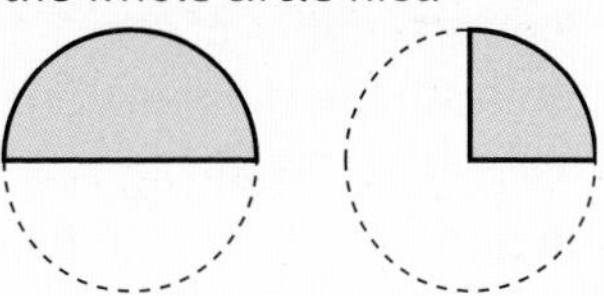

4 Work out the perimeter of each semicircle

i in terms of π

ii to 1 d.p.

a

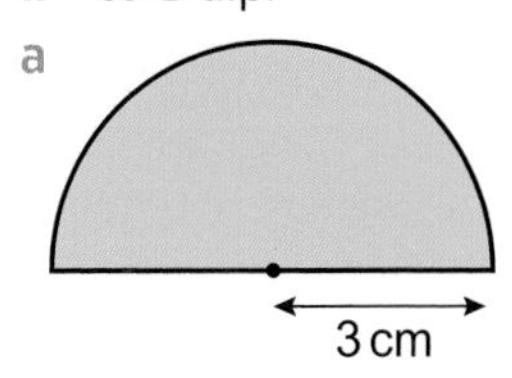

b

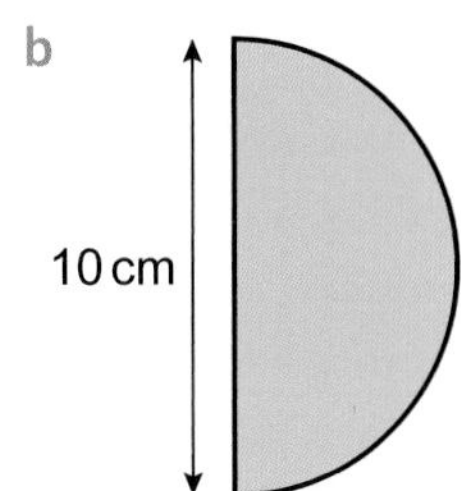

Q4 hint Find half the circumference of the whole circle. Add the diameter to it.

5 Work out the perimeter of this quarter circle

a in terms of π

b to 1 d.p.

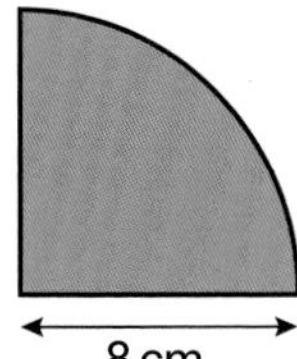

Q5 hint Find one quarter of the circumference of the whole circle. Add two radii to it.

6 **Problem-solving** A window is made from a rectangle and a semicircle.

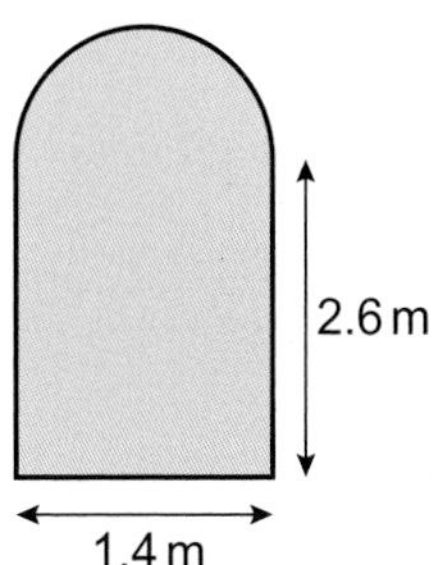

Q6 strategy hint What is the diameter of the semicircle?

Work out to 1 d.p.

a the area

b the perimeter of the window.

7 **Problem-solving** Four quarter circles are cut from a 10 cm square like this.

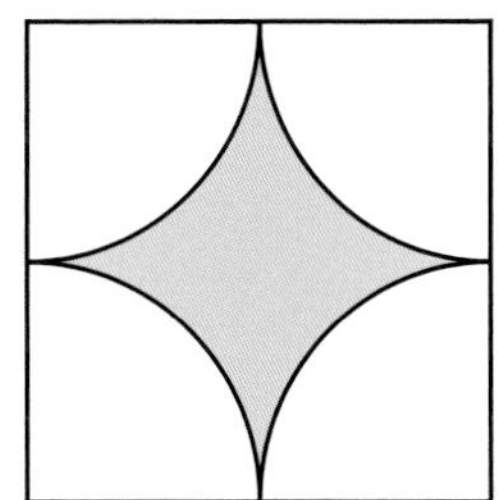

Work out the shaded area.

8 The area of this semicircle is $15\,cm^2$ to the nearest whole number.

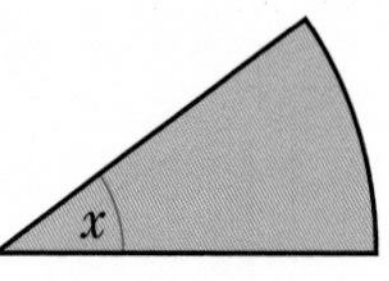

a Find the radius of the semicircle. Write all the numbers on your calculator display.

b Round your answer to part **a** to a suitable degree of accuracy.

Discussion What is a suitable level of accuracy?

Q8a strategy hint Calculate the area of the whole circle, then find its radius.

Q8b hint How many decimal places could you measure to with a ruler?

Key point 15

For a sector with angle $x°$ of a circle with radius r

Arc length $= \frac{x}{360} \times 2\pi r$

Area of sector $= \frac{x}{360} \times \pi r^2$

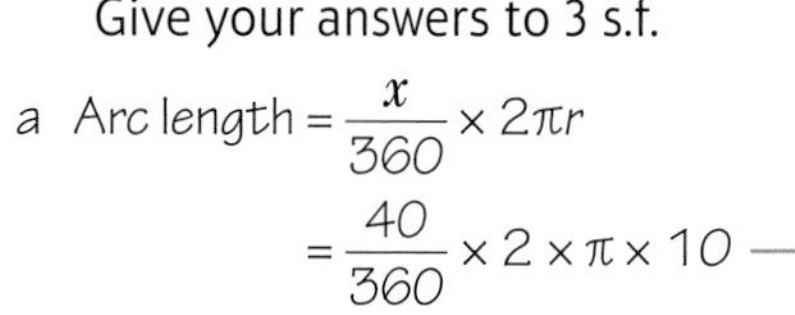

Communication hint An **arc** is part of a circle.

Example 4

Work out

a the arc length

b the perimeter

c the area of this sector.

Give your answers to 3 s.f.

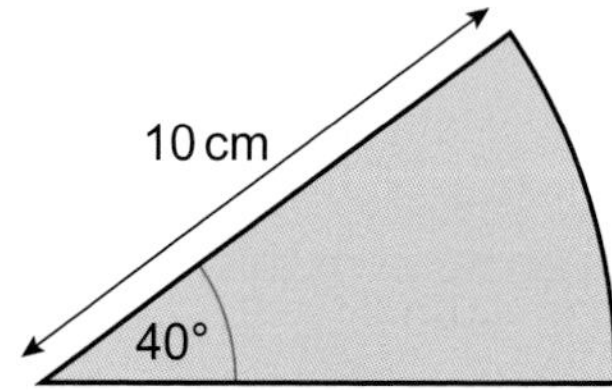

a Arc length $= \frac{x}{360} \times 2\pi r$

$= \frac{40}{360} \times 2 \times \pi \times 10$ — Write the formula, substitute the angle x and radius.

$= 6.98$ cm (3 s.f.)

b Perimeter $= 6.98 + 10 + 10$ — Perimeter = arc length + 2 radii

$= 27.0$ cm (3 s.f.)

c Area $= \frac{x}{360} \times \pi r^2$

$= \frac{40}{360} \times \pi \times 100$ — Write the formula, substitute the angle x and radius.

$= 34.9\,cm^2$ (3 s.f.)

9 **Exam-style question**

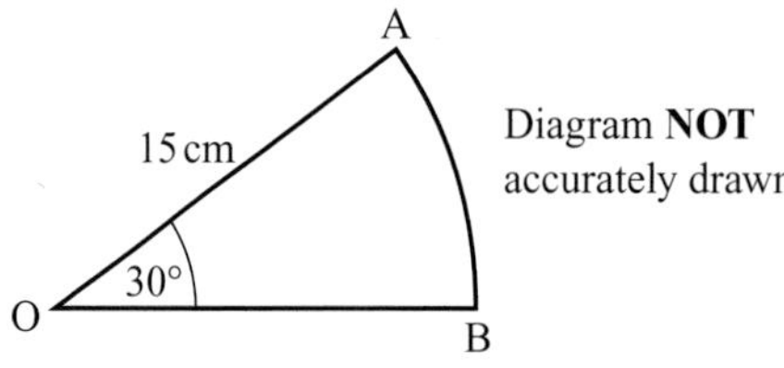

OAB is a sector of a circle, centre *O*.
The radius of the circle is 15 cm.
The angle of the sector is 30°.
Calculate the area of sector *OAB*.
Give your answer correct to 3 significant figures. **(2 marks)**

March 2013, Q19, 1MA0/2H

Exam hint
Write the formula you are using and show how you will substitute the given numbers into it. Make sure you write down your unrounded answer on your calculator before rounding.

10 Work out the arc length and perimeter of the sector in **Q9**.
Give your answers to 3 s.f.

11 **Problem-solving** a Work out the arc length and area of this sector.

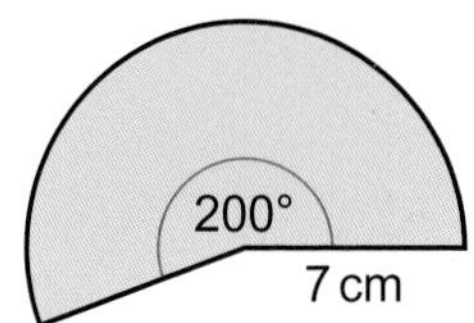

b The radius of the circle was measured to the nearest cm.
Work out the upper and lower bounds for the area of the sector.

12 The area of this sector is 10 cm².

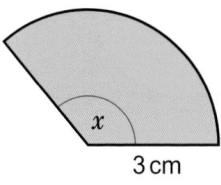

a Substitute $r = 3$ and area = 10 into the formula
Area = $\frac{x}{360} \times \pi r^2$

b Solve your equation from part **a** to find the angle of the sector, to the nearest degree.

13 **Problem-solving** Find the angle of this sector.

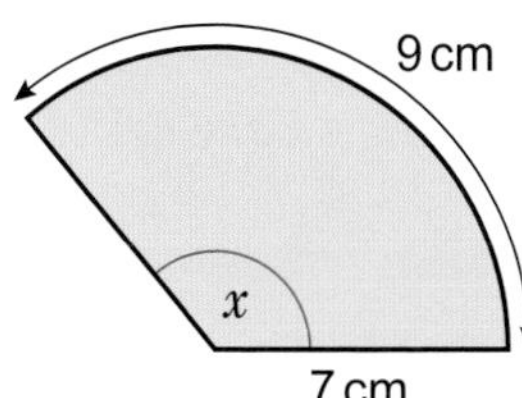

14 **Problem-solving** This sector has area 16 m².
Find the radius.
Give your answer to a suitable degree of accuracy.

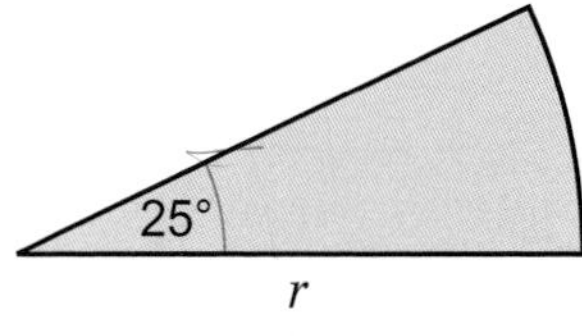

15 **Problem-solving** Angle AOB is 45°.
The sector AOB has area 5π cm².
Find the length of the arc AB.

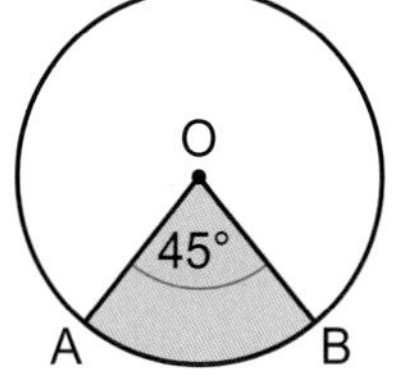

Q15 strategy hint
First find the radius.

16 **Problem-solving** Calculate the area of the shaded region.
Give your answer in terms of π.

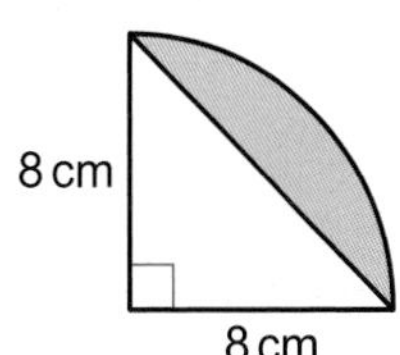

7.6 Cylinders and spheres

Objectives

- Calculate volume and surface area of a cylinder and a sphere.
- Solve problems involving volumes and surface areas.

Why learn this?

The volume of drink in a can, or the volume of water in a pipe, can be modelled as a cylinder.

Fluency

What is the area and circumference of this circle in terms of π?

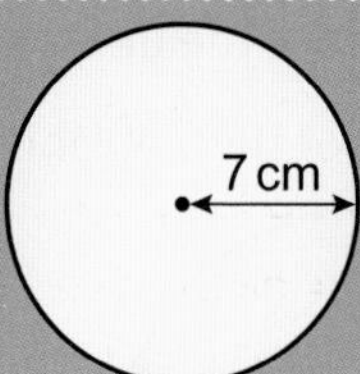

Warm up

1 Sketch the net of a cylinder.

2 Find the value of r. Give answers to 1 decimal place where appropriate.

a $72 = 2r^2$ b $54 = \frac{2}{3}r^3$ c $32 = 4\pi r^2$ d $30\pi = 3\pi r^3$

3 **Reasoning** A cylinder is a prism.

a Write an expression for the area of its cross-section.

b Write a formula for its volume.

Q3b hint $V = \square \times h$

Key point 16

The volume of a cylinder of radius r and height h is $V = \pi r^2 h$

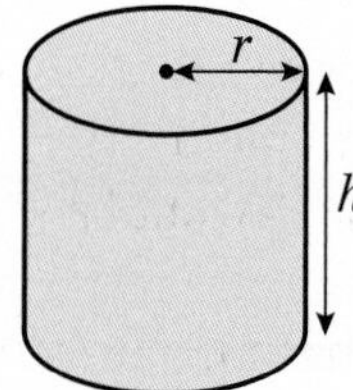

4 Work out the volume of each cylinder. Give your answers to 1 d.p.

a

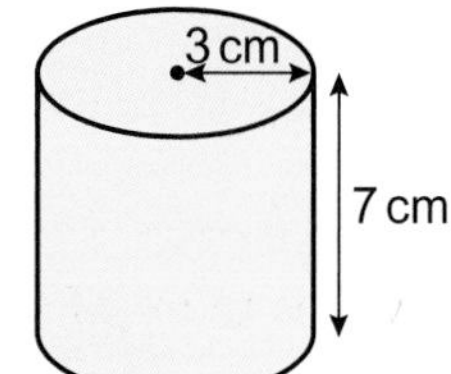

b

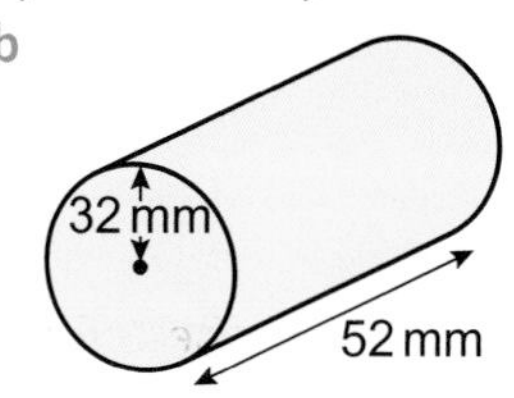

c

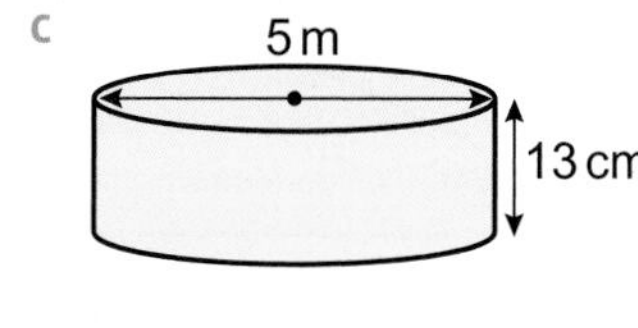

5 **Modelling / STEM** A scientist takes a circular section of ice, with diameter 1 m. The mean thickness of the ice is 34 cm. Estimate the volume of ice in the sample. Give your answer in m³.

Q5 strategy hint Draw a diagram.

6 A cylinder has radius 3.5 cm and volume 125 cm³. Work out its height.

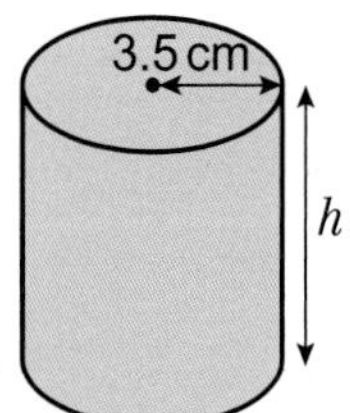

Q6 strategy hint Substitute the values into the volume formula and solve to find h.

ActiveLearn Homework, practice and support: Higher 7.6

Example 5

Calculate the total surface area of this cylinder. Give your answer to 1 d.p.

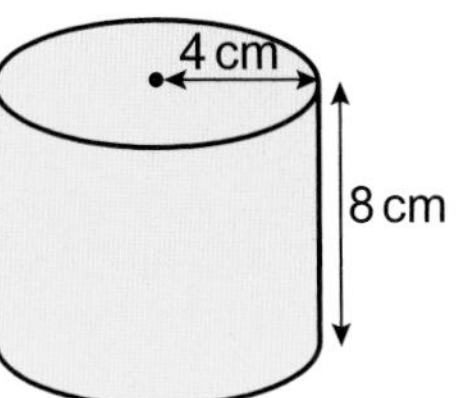

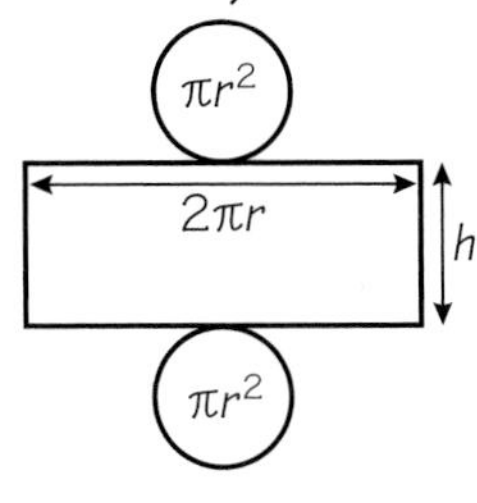

Sketch a net.
Each circle has area πr^2.
The length of the rectangle is the circumference of the circle, $2\pi r$.
The width of the rectangle is the height of the cylinder, h.

Area of each circle = $\pi \times 4^2 = 16\pi$

Area of rectangle = $2\pi rh = 2 \times \pi \times 4 \times 8 = 64\pi$

Surface area = $2 \times 16\pi + 64\pi$ — Two circles plus rectangle.

$= 32\pi + 64\pi$

$= 96\pi$

$= 301.6\,\text{cm}^2$

Key point 17

The total surface area of a cylinder of radius r and height h is $2\pi r^2 + 2\pi rh$

7 Calculate the total surface area of each cylinder in **Q4**.

8 **Problem-solving** A cylinder has total surface area $3900\,\text{mm}^2$ and radius 15 mm.
Work out its height, to the nearest millimetre.

Q8 strategy hint Substitute the values into the surface area formula and solve.

9 **Problem-solving** When $120\,\text{cm}^3$ of water is poured into a cylinder, it reaches a height of 8 cm.
More water is poured into the cylinder, until it reaches a height of 20 cm.
How much water is in the cylinder now?

Q9 hint Work out the radius of the cylinder.

Key point 18

For a sphere of radius r

Surface area = $4\pi r^2$

Volume = $\frac{4}{3}\pi r^3$

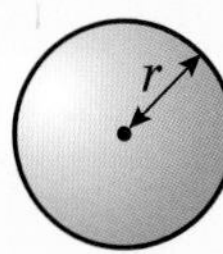

10 Calculate the surface area and volume of each sphere. Give your answers in terms of π.

a

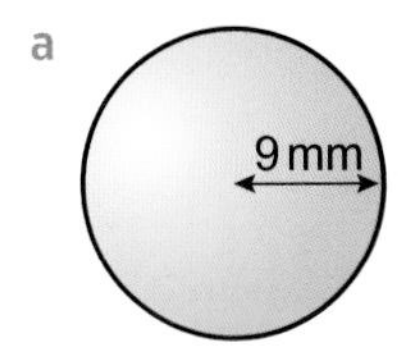

b

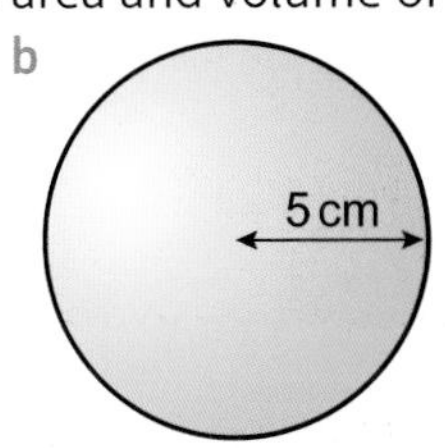

11 **Real** In kitchen cupboards, plastic hemispheres prevent the doors banging when they are closed.
A factory produces these hemispheres with diameter 1 cm with a 10% error interval.
Work out the possible volumes of plastic used for each hemisphere, to the nearest cubic millimetre.

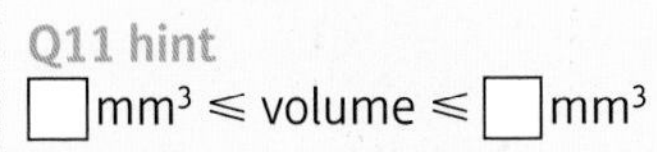

Q11 hint
$\square\,\text{mm}^3 \leqslant \text{volume} \leqslant \square\,\text{mm}^3$

12 **Exam-style question**

The diagram shows a solid glass paperweight, made from a hemisphere on top of a cylinder.
The height of the cylinder is 20 mm.
The radius of the cylinder is 4 mm.

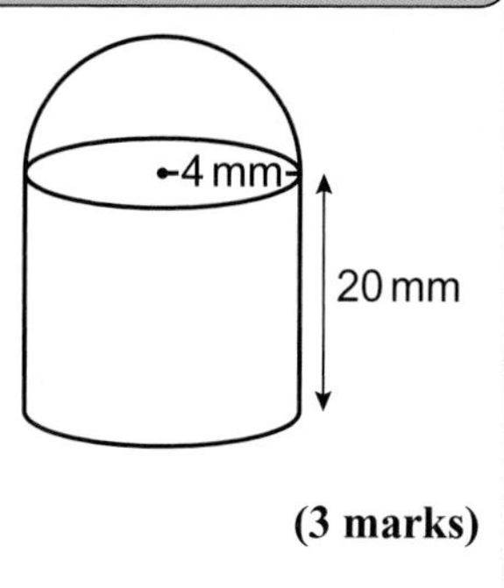

a Calculate the total volume of the paperweight. Give your answer correct to 3 significant figures. **(3 marks)**

b Calculate the surface area of the paperweight. Give your answer correct to 3 significant figures. **(3 marks)**

Exam hint
Work on each part of the solid separately. Set out your working clearly.

Discussion How do you calculate the curved surface area of a hemisphere? Is this the same as the surface area of a hemisphere?

13 **Problem-solving** A spherical ball bearing is made from 20 ml of molten steel. Work out its radius, to the nearest millimetre.

14 **Problem-solving** What is the radius of a sphere with surface area 500 m^2?

Q14 hint Give your answer to an appropriate level of accuracy.

15 **Problem-solving / STEM** A 10 m long cylinder of brass with radius 2 cm is melted down to make a sphere. Work out the radius of the sphere.

Q15 hint The two solids have the same volume. Write an equation in terms of π and solve it.

16 **Reflect** In this lesson you have used four formulae, for volume and surface area of cylinders and of spheres.
How can you remember which is which? Do you need to remember the cylinder formulae or can you work them out using what you know about prisms and circles?

7.7 Pyramids and cones

Objectives

- Calculate volume and surface area of pyramids and cones.
- Solve problems involving pyramids and cones.

Did you know?

A cone is a pyramid with a circular base.

Fluency

What is the volume of a cube of side 5 mm?
A cube has volume 64 cm^3. How long is one of its sides?

Warm up

1 Work out the area of each triangle.

a

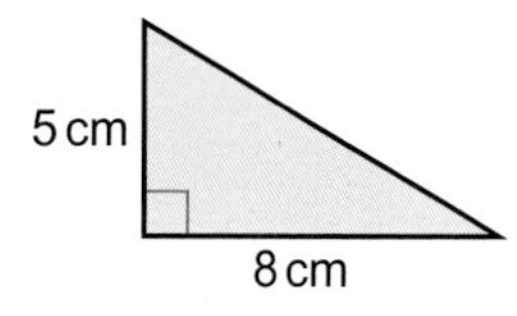

b

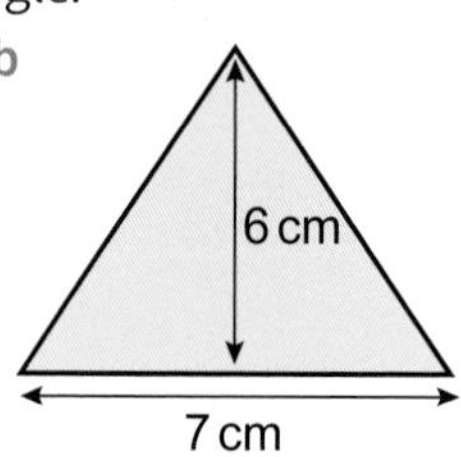

ActiveLearn Homework, practice and support: Higher 7.7

2 Find the length of the sloping side in the triangle in **Q1a**.
Give your answer to 1 d.p.

3 Here is a square-based pyramid.

a Sketch a net of this pyramid.

b Work out the area of each face.

c Calculate the surface area of the pyramid.

Discussion Do you need to sketch the net to work out the surface area of a pyramid?

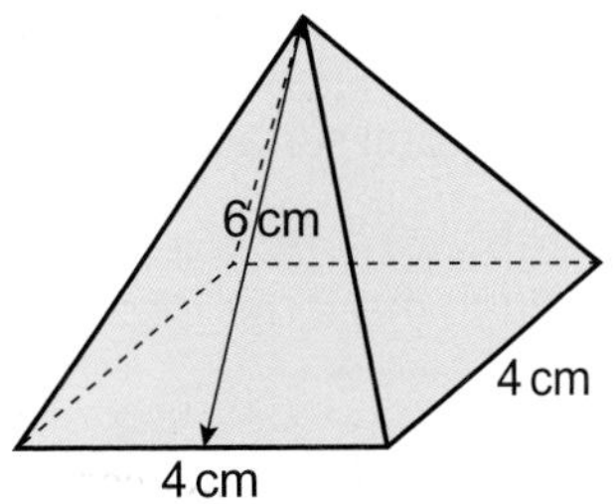

> **Q3a hint** On a square-based pyramid the triangular faces are identical.

Key point 19

Volume of pyramid = $\frac{1}{3}$ area of base × vertical height

Volume of cone = $\frac{1}{3}$ area of base × vertical height

$= \frac{1}{3}\pi r^2 h$

4 This pyramid has a square base of side 8 cm, and vertical height 10 cm.
Calculate its volume to 3 s.f.

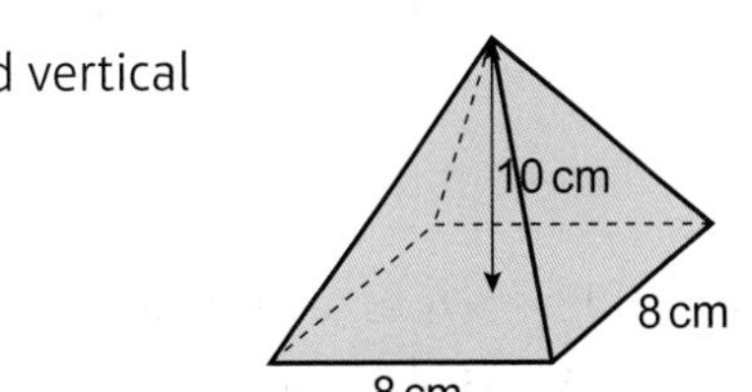

5 **Exam-style question**

This solid is made from a square-based pyramid and a cube.
The square-based pyramid is 12 cm high, and has volume 300 cm^3.

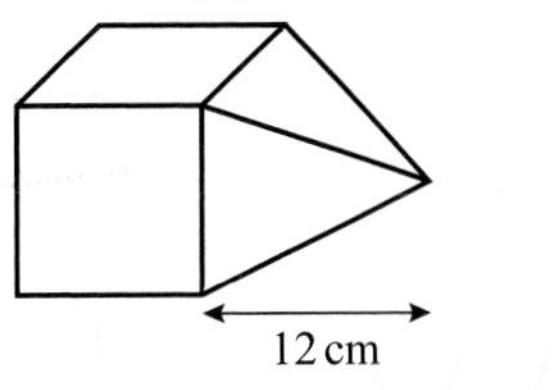

Find

a the length of one side of the square base

b the total volume of the solid. **(6 marks)**

> **Exam hint**
> Give your answers to a suitable degree of accuracy.

6 A cone has base radius 6 cm and height 8 cm.
Calculate its volume

a in terms of π

b to 3 s.f.

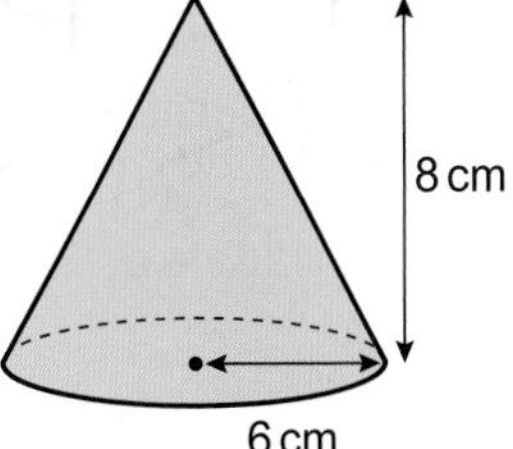

Key point 20

Curved surface area of a cone = $\pi r l$, where r is the radius and l is the slant height.

Total surface area of a cone = $\pi r l + \pi r^2$

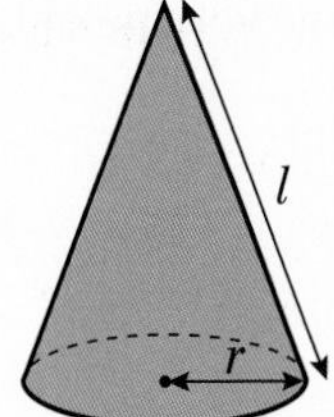

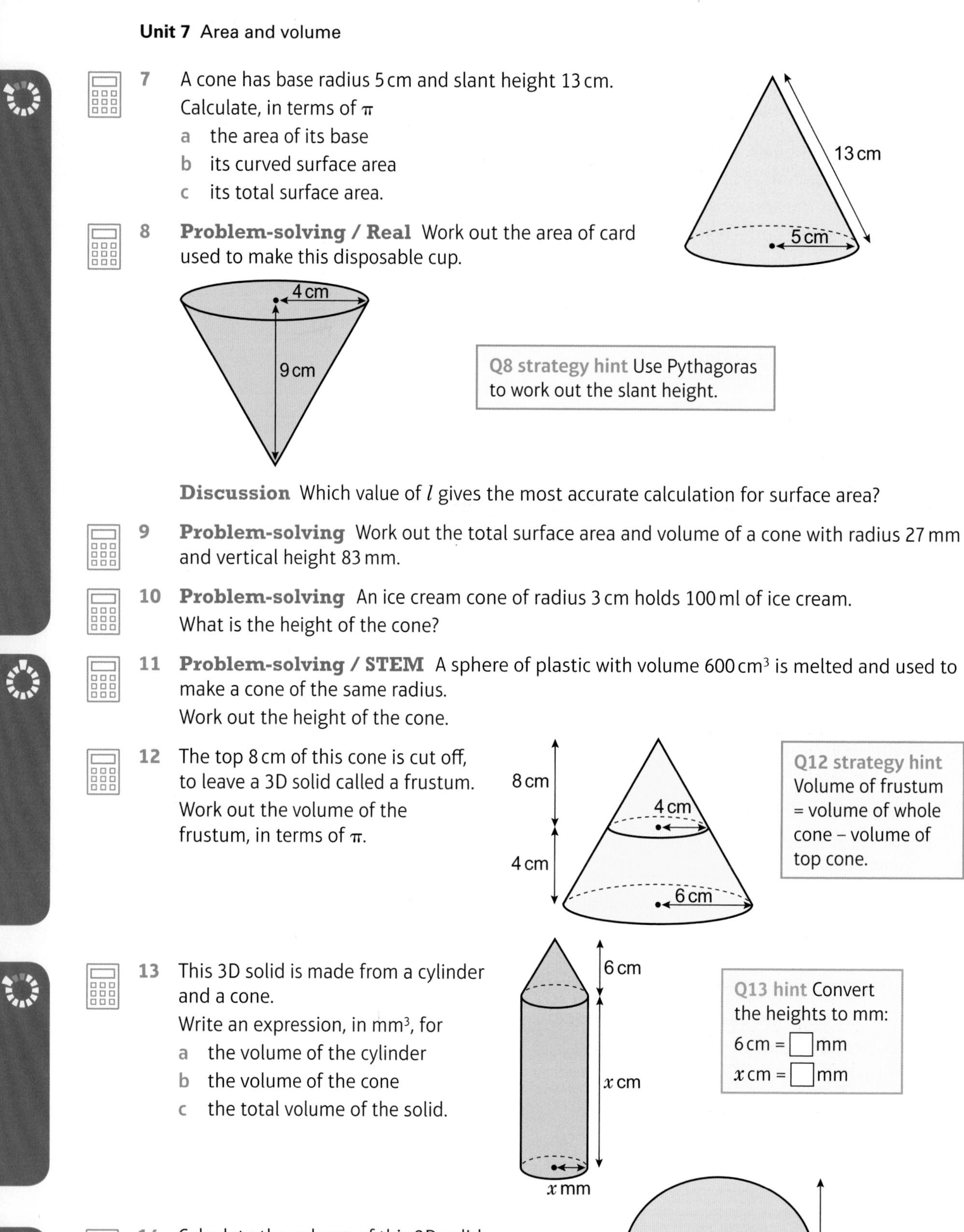

7 A cone has base radius 5 cm and slant height 13 cm.
Calculate, in terms of π

a the area of its base

b its curved surface area

c its total surface area.

8 **Problem-solving / Real** Work out the area of card used to make this disposable cup.

Q8 strategy hint Use Pythagoras to work out the slant height.

Discussion Which value of l gives the most accurate calculation for surface area?

9 **Problem-solving** Work out the total surface area and volume of a cone with radius 27 mm and vertical height 83 mm.

10 **Problem-solving** An ice cream cone of radius 3 cm holds 100 ml of ice cream.
What is the height of the cone?

11 **Problem-solving / STEM** A sphere of plastic with volume 600 cm^3 is melted and used to make a cone of the same radius.
Work out the height of the cone.

12 The top 8 cm of this cone is cut off, to leave a 3D solid called a frustum.
Work out the volume of the frustum, in terms of π.

Q12 strategy hint
Volume of frustum = volume of whole cone − volume of top cone.

13 This 3D solid is made from a cylinder and a cone.
Write an expression, in mm^3, for

a the volume of the cylinder

b the volume of the cone

c the total volume of the solid.

Q13 hint Convert the heights to mm:
6 cm = ☐ mm
x cm = ☐ mm

14 Calculate the volume of this 3D solid.

7 Problem-solving

Objective • Use a flow diagram to help you solve problems.

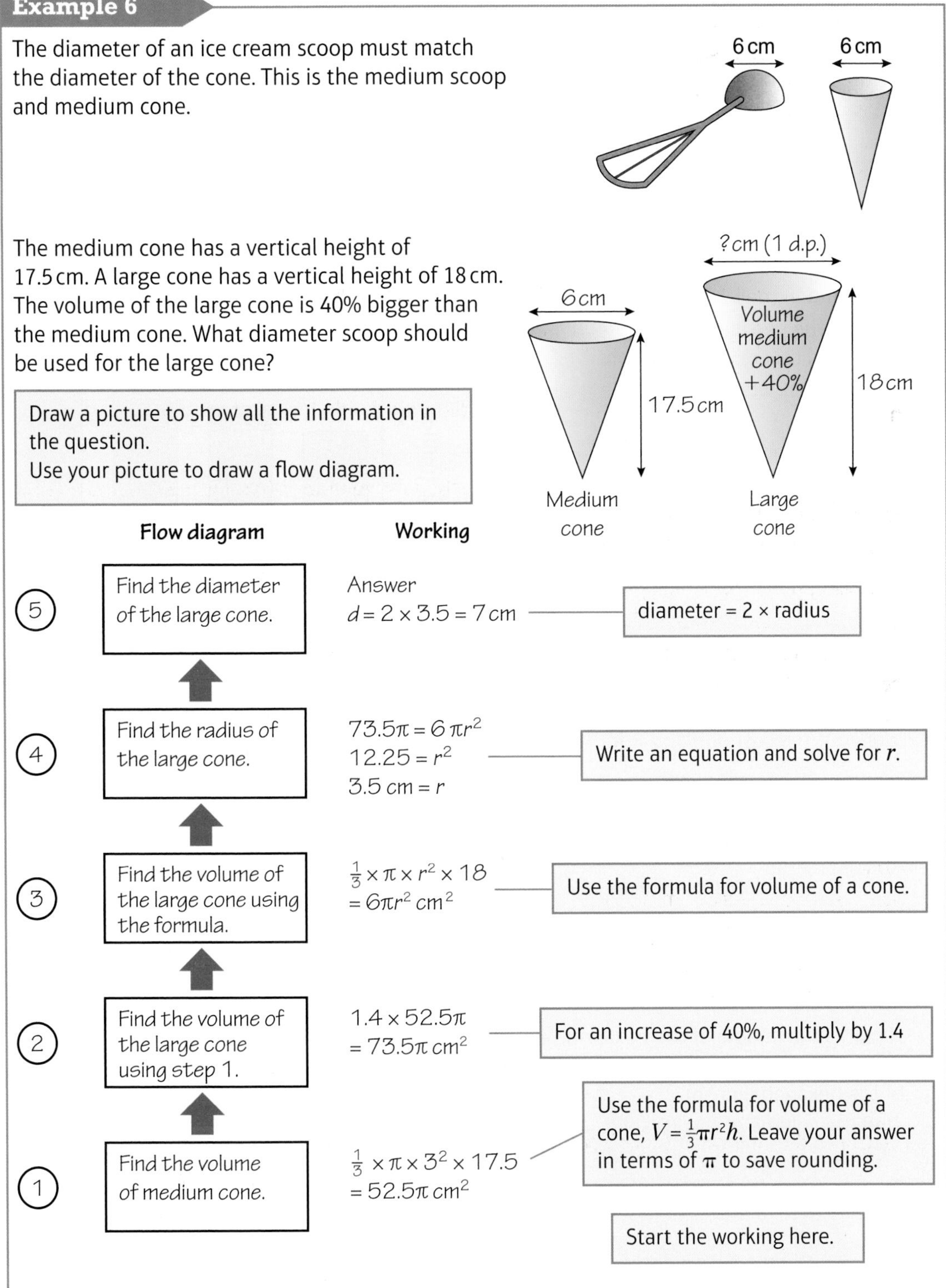

Example 6
The diameter of an ice cream scoop must match the diameter of the cone. This is the medium scoop and medium cone.
6 cm
6 cm
The medium cone has a vertical height of 17.5 cm. A large cone has a vertical height of 18 cm. The volume of the large cone is 40% bigger than the medium cone. What diameter scoop should be used for the large cone?
Draw a picture to show all the information in the question.
Use your picture to draw a flow diagram.
?cm (1 d.p.)
6 cm
Volume medium cone +40%
17.5 cm
18 cm
Medium cone
Large cone
Flow diagram
Working
5 Find the diameter of the large cone.
Answer
d = 2 × 3.5 = 7 cm
diameter = 2 × radius
4 Find the radius of the large cone.
73.5π = 6πr²
12.25 = r²
3.5 cm = r
Write an equation and solve for r.
3 Find the volume of the large cone using the formula.
1/3 × π × r² × 18
= 6πr² cm²
Use the formula for volume of a cone.
2 Find the volume of the large cone using step 1.
1.4 × 52.5π
= 73.5π cm²
For an increase of 40%, multiply by 1.4
Use the formula for volume of a cone, V = 1/3πr²h. Leave your answer in terms of π to save rounding.
1 Find the volume of medium cone.
1/3 × π × 3² × 17.5
= 52.5π cm²
Start the working here.

1 A football has a diameter of 14.2 cm. The volume of a tennis ball is 10% of the volume of the football. What is the diameter of the tennis ball? Give your answer to 1 decimal place.

2 A cake recipe requires a round tin with diameter 18 cm and height 5 cm. The recipe says that the mixture will exactly fill the tin. Renee makes the mixture. Then she realises she only has this tin available:

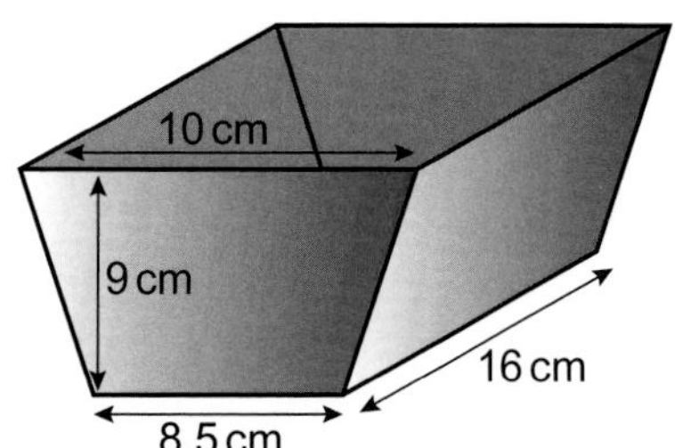

What height does the mixture come to in this tin? Give your answer to 1 decimal place.

3 The bar chart shows the area taken up by woodland in the UK from 1965 to 2012.

What is the percentage increase in area of woodland covering the UK from 1965 to 2012?

Give your answer to 3 significant figures.

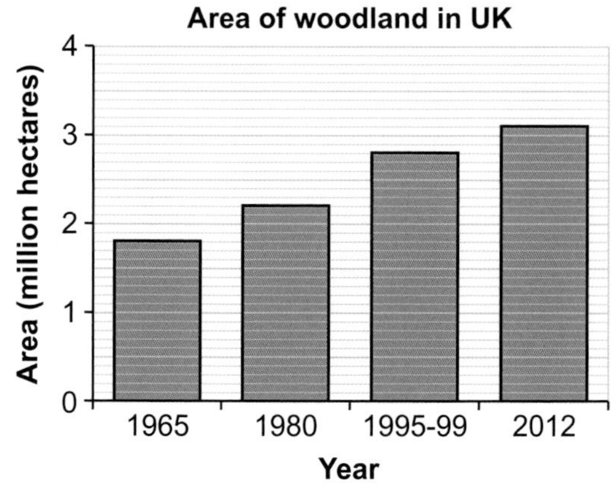

Q3 hint Draw a flow diagram. Check when you increase the area of the woodland in 1965 by your answer, you get the area of woodland in 2012.

4 **Finance** The diagram shows a tank for storing central heating oil.

At the end of the summer, the tank is one quarter full.

Work out the cost of filling it for the winter.

Central heating oil

52p per litre + 5% VAT

Delivery charge: up to 1000 litres **£25**

1000 litres or more **£35**

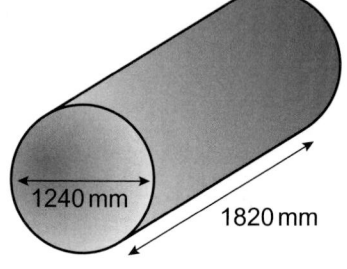

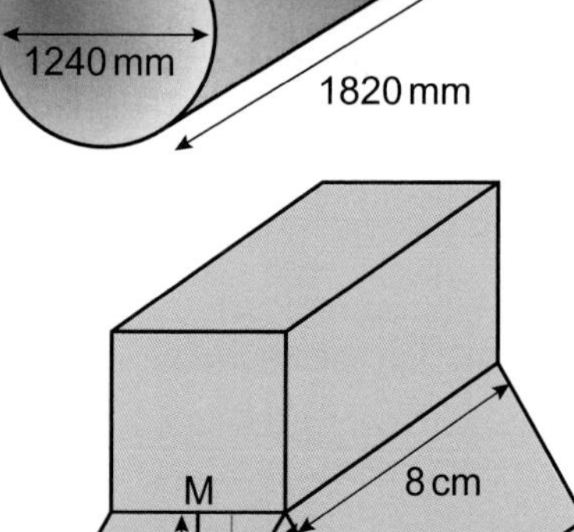

5 A cuboid with a square cross-section sits exactly on top of a regular hexagonal prism to make a 3D solid like this:

H marks the middle of the hexagon. M is the midpoint of the side of the square and the side of the hexagonal prism.

Find the volume of the 3D solid.

M
8 cm
4 cm
5 cm
H

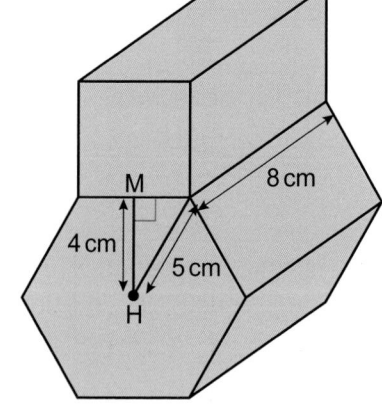

6 **Reflect** Did the flow diagrams help you?

Is this a strategy you would use again to solve problems?

What other strategies did you use to solve these problems?

7 Check up

Log how you did on your Student Progression Chart.

2D shapes

1 Calculate

a the area

b the perimeter of this isosceles trapezium.

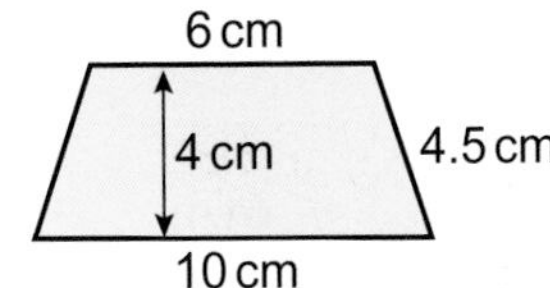

2 **Reasoning** The area of this trapezium is 45 cm². Find x.

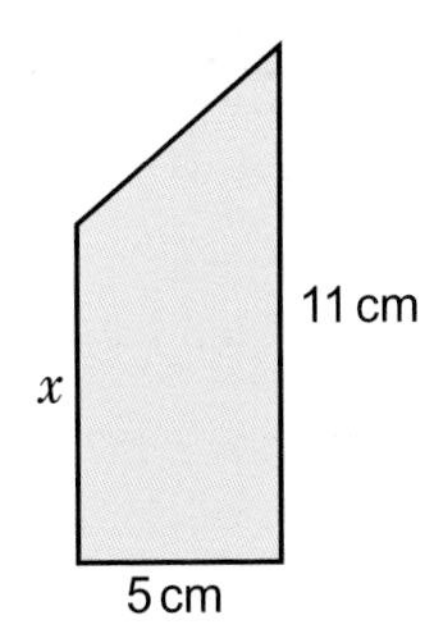

3 a Calculate the circumference of this circle.

b Find the area of the circle in terms of π.

Give your answers to 1 d.p.

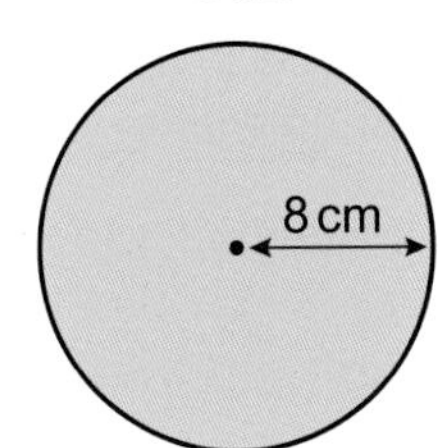

4 Work out the perimeter of this semicircle.
Give your answer correct to 1 d.p.

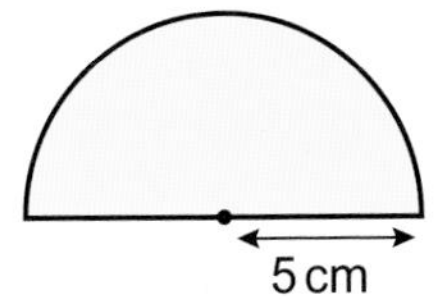

5 Calculate

a the area

b the arc length of this sector.

Give your answers correct to 1 d.p.

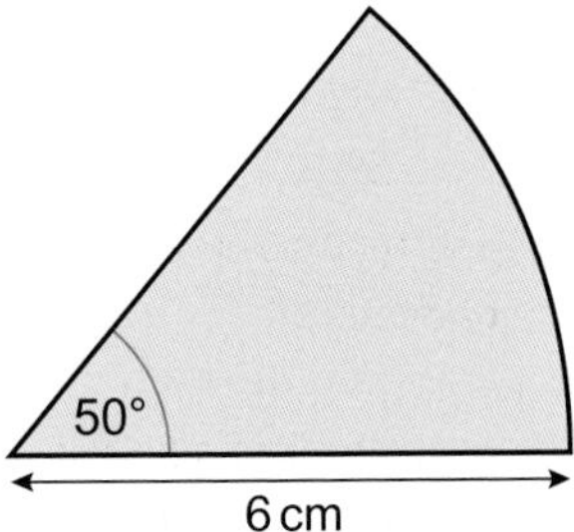

Accuracy and measures

6 Copy and complete.

a $4\text{ m}^2 = \square\text{ cm}^2$

b $5600\text{ cm}^2 = \square\text{ m}^2$

c $9.5\text{ million cm}^3 = \square\text{ m}^3$

d $3\text{ litres} = \square\text{ ml} = \square\text{ cm}^3$

7 Ball bearings are manufactured with volume 10 cm³, with an error interval of 5%.
Write an inequality to show the possible values.

8 Write an inequality to show the upper and lower bounds of each measurement.

a 36 m rounded to the nearest metre.

b 9.2 cm rounded to the nearest mm.

c 23.6 km rounded to 1 d.p.

ActiveLearn Homework, practice and support: Higher 7 Check up

3D solids

9 Work out the volume of this triangular prism.

3 cm 5 cm 6 cm 4 cm

10 Calculate the surface area of this cylinder, with radius 5.3 cm and height 9.5 cm.

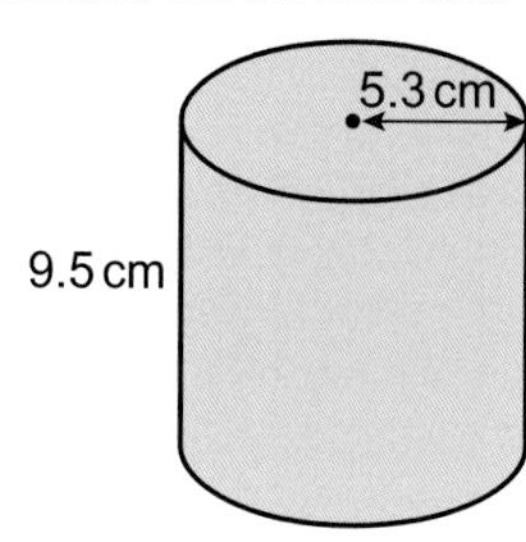

11 A sphere has radius 3 cm.
Work out its volume in terms of π.

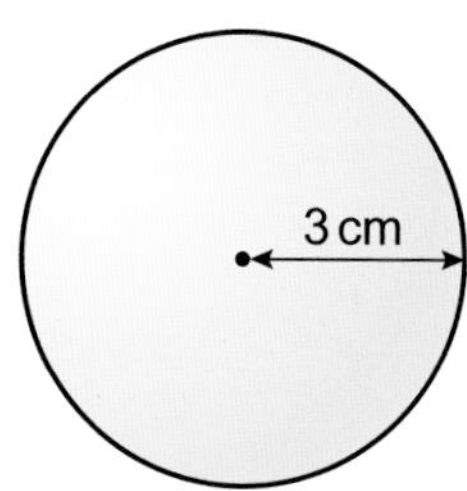

12 Calculate the volume of this cone.

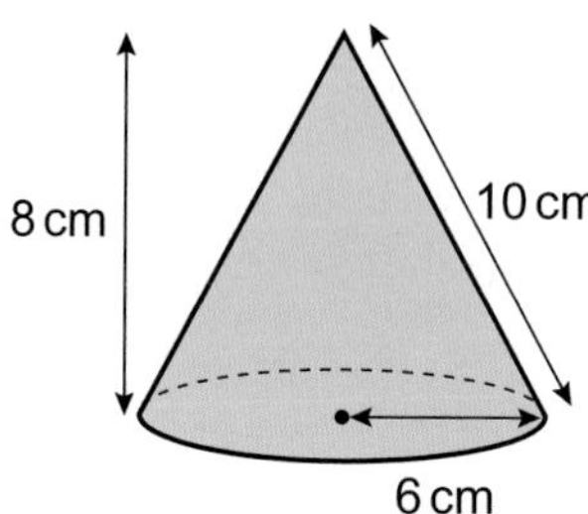

13 How sure are you of your answers? Were you mostly

Just guessing ☹ Feeling doubtful 😐 Confident 

What next? Use your results to decide whether to strengthen or extend your learning.

* Challenge

14 These two cardboard boxes each hold 4 tennis balls, with diameter 6.5 cm.

Q14 hint You could think about:
- stacking and transport
- eye catching design
- amount of card needed to make the box
- amount of empty space inside, around the tennis balls.

Which shape would you recommend to the manufacturer and why?

7 Strengthen

2D shapes

1 Find

a the vertical height

b the perimeter of this trapezium.

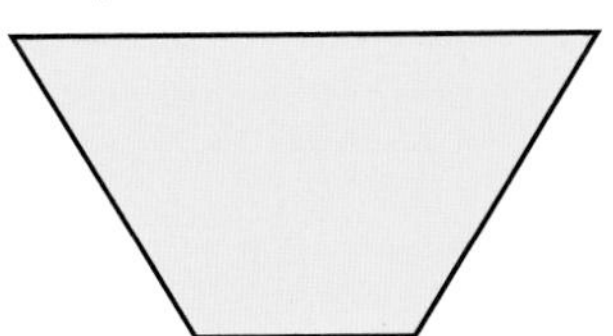

Q1a hint Use a ruler to measure the lengths you need.

2 **Reasoning** This trapezium has area 55 cm².

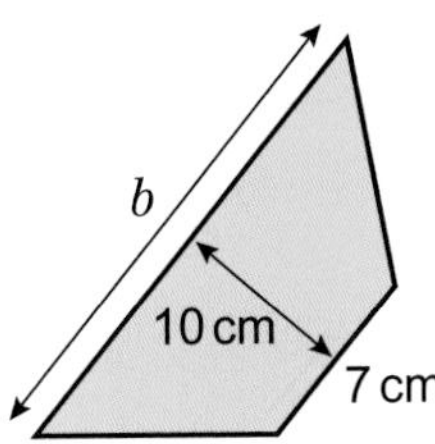

Q2a hint Copy the diagram and label a, b and h.

a Substitute the values for A, a and h into the formula $A = \frac{1}{2}(a + b)h$

b Simplify the right hand side of your equation from part **a**. Multiply out the brackets.

c Solve the equation to find b.

3 Work out the circumference of each circle

a

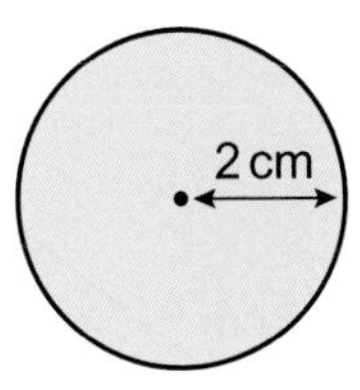

b

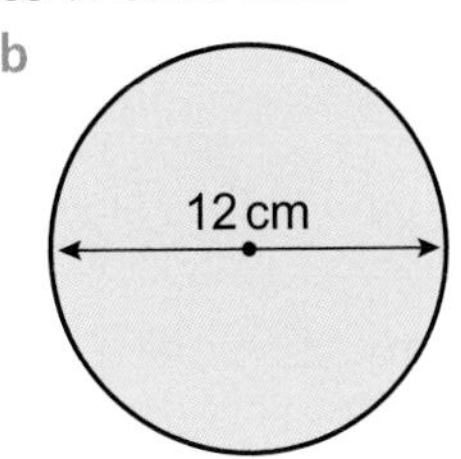

Q3 hint $C = \pi d$ and $C = 2\pi r$
If you know the diameter, choose the formula with d in it.
If you know the radius, choose the formula with r in it.

i in terms of π

ii correct to 1 d.p.

4 Here are two expressions used with circles.

$2\pi r$ πr^2

Q4 hint Circumference is in cm.
Area is in cm².
πr^2 $2\pi r$

a Match each expression to the correct measurement.

area = 34 cm² circumference = 19.5 cm

b Write the formula for area of a circle.

c Write the formula for circumference of a circle.

5 Work out the area of each circle in **Q3**

i in terms of π

ii correct to 1 d.p.

Q5i hint Substitute the radius into the formula for area. Work out r^2 but leave π as it is.

6 Here is a semicircle with radius 7 cm.

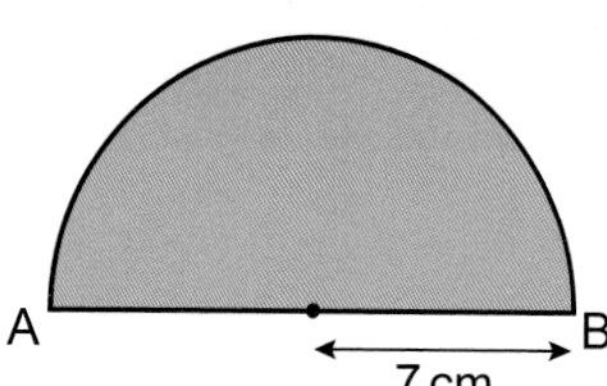

Work out, giving your answers to 1 d.p.

- a the area of a full circle with radius 7 cm
- b the area of this semicircle with radius 7 cm
- c the circumference of a full circle with radius 7 cm
- d the arc length *AB*
- e the diameter *AB*
- f the perimeter of the semicircle.

> **Q6b hint** What fraction of a circle is it?

7 What fraction of each circle is shaded?

a

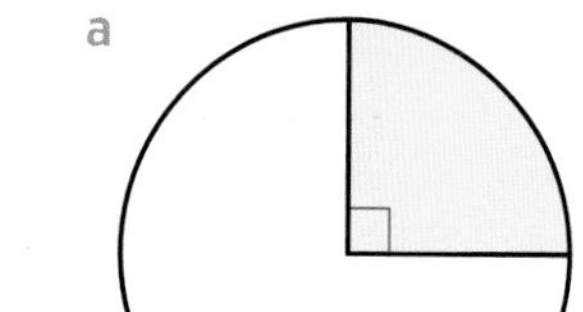

b

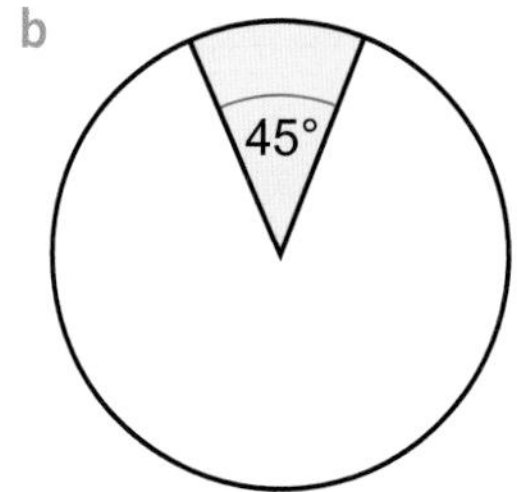

c

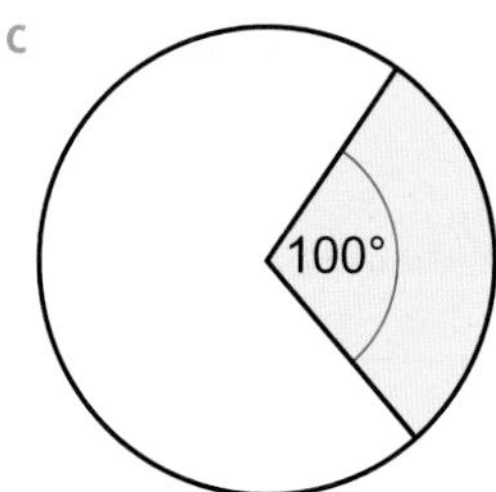

> **Q7 hint** $\frac{\square}{360} = \frac{\square}{\square}$

8 Here is a sector of a circle with radius 8 cm.

- a What fraction of a circle is this sector?
- b Work out, giving your answers to 1 d.p.
  - i the area of a full circle with radius 8 cm
  - ii the area of this sector with radius 8 cm
  - iii the circumference of a full circle with radius 8 cm
  - iv the arc length *AB*
  - v the total length from *A* to *O* to *B*
  - vi the perimeter of the sector.

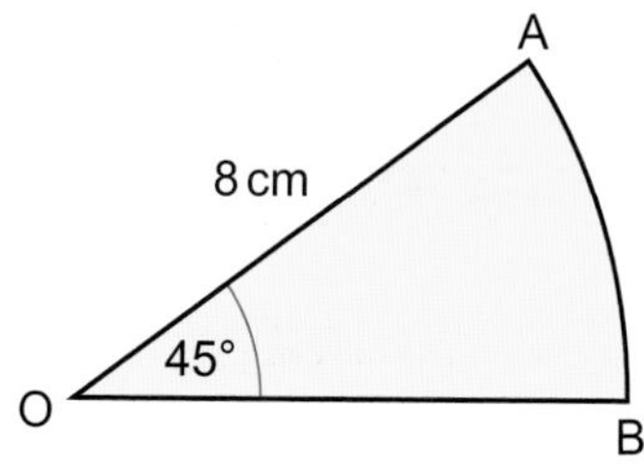

Accuracy and measures

1 a Work out the area of each shape in cm^2.

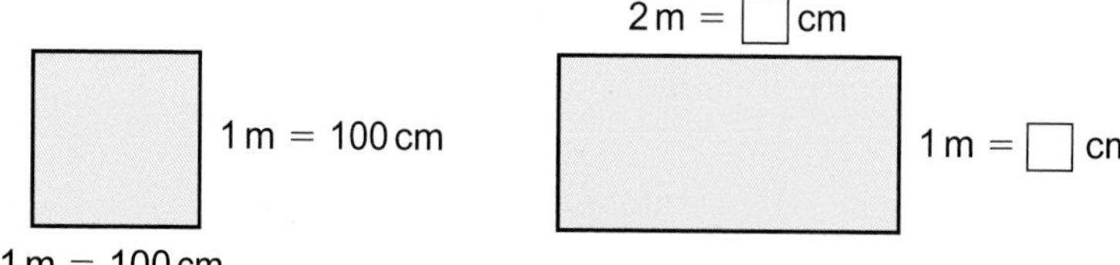

b Copy and complete this double number line for cm^2 and m^2.

> **Q1b hint** Use the areas of the rectangles from part **a**. Follow the pattern.

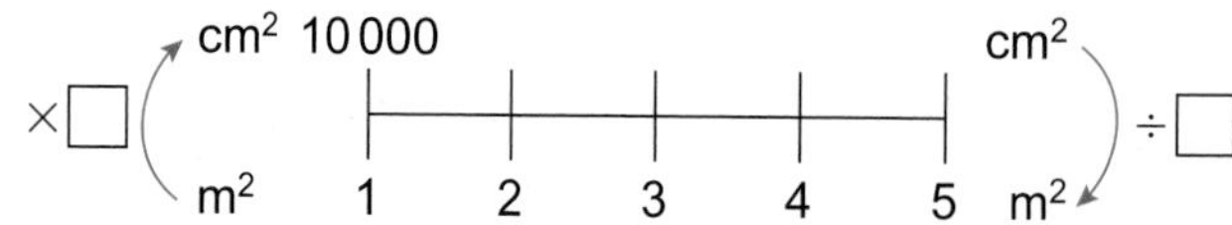

2 a Work out the volume of each cuboid in cm³.

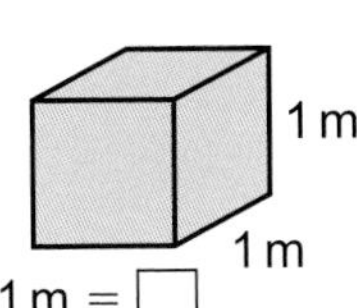

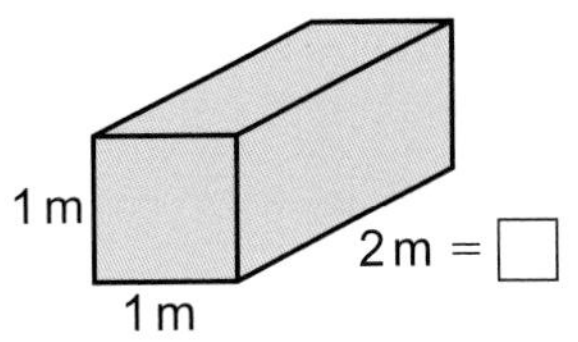

b Copy and complete this double number line for cm³ and m³.

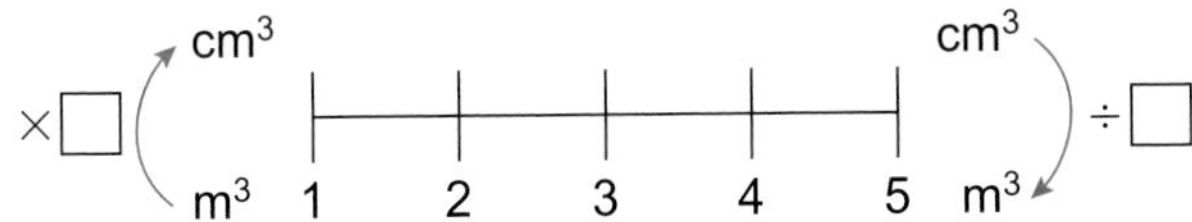

3 Pencils are made 25 cm long, with an error interval of 10%.

a Work out 10% of 25.

b Work out 25 + 10% and 25 − 10%.

c Write an inequality to show the possible values.

Q3c hint

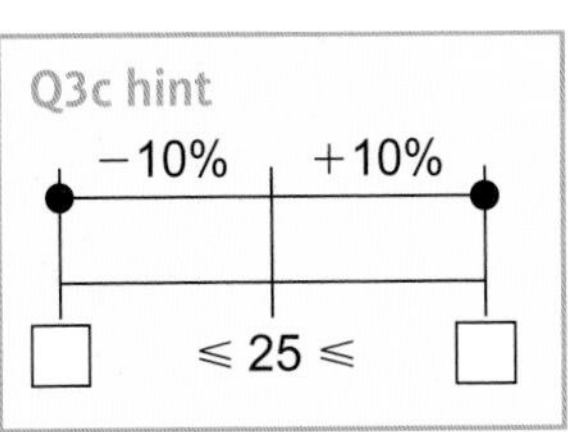

4 a A pen is 23 cm long, to the nearest cm.
The diagram shows the measurements that round to 23 cm, to the nearest cm.
Write an inequality to show the upper and lower bounds of this measurement. Use l for the length of the pen.

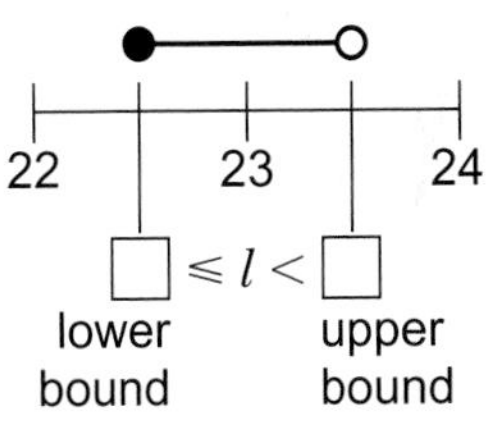

b A pencil sharpener is 32 mm long, to the nearest mm.
Write an inequality to show the upper and lower bounds of this measurement. Use l for the length of the pencil sharpener.

Q4b hint Draw a diagram, like the one in part **a**.

3D solids

1 Work out the volume of these prisms by

- working out the area of the front face
- multiplying the area by the length of the prism.

Q1 hint Sketch the front face to help you work out its area.

a

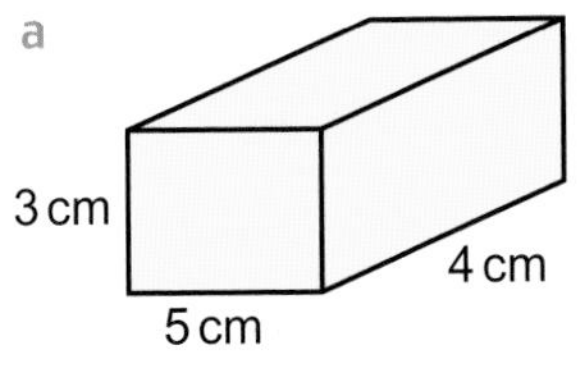

b

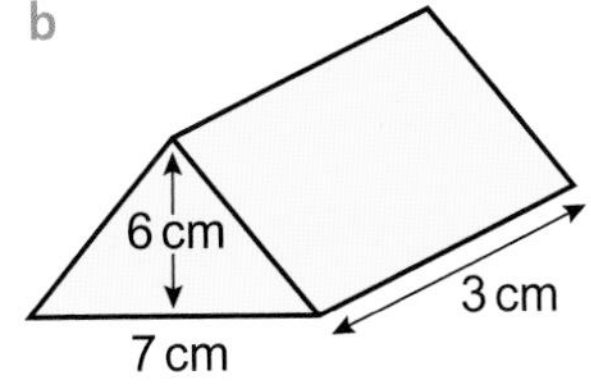

c

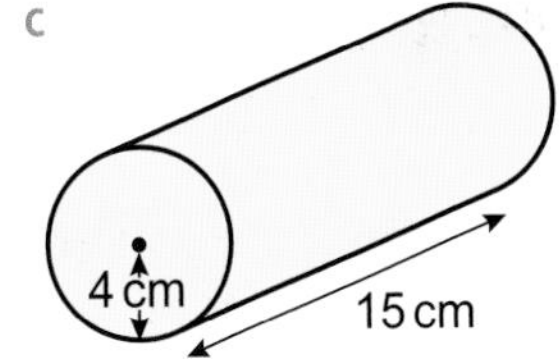

2 a Sketch the net of this cylinder.

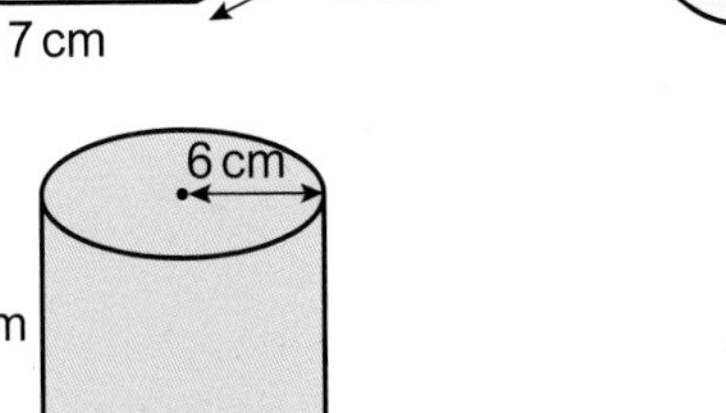
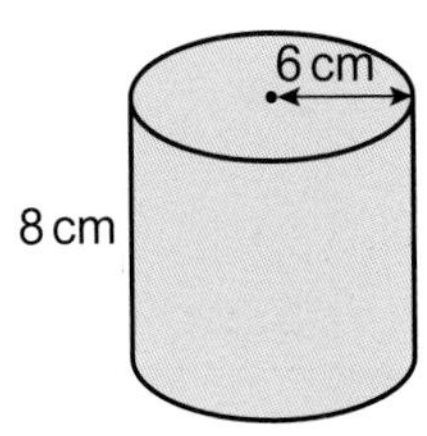
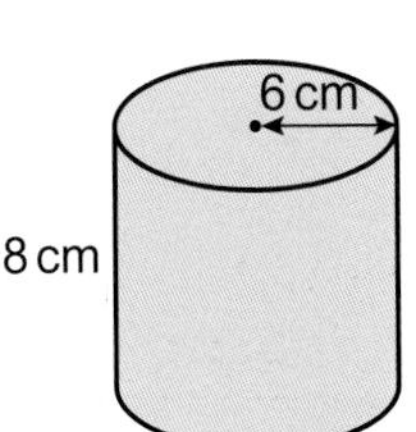

b Work out (giving your answers to 1 d.p.)

i the area of the circle

ii the length of the rectangle

iii the area of the rectangle

iv the total surface area of the cylinder.

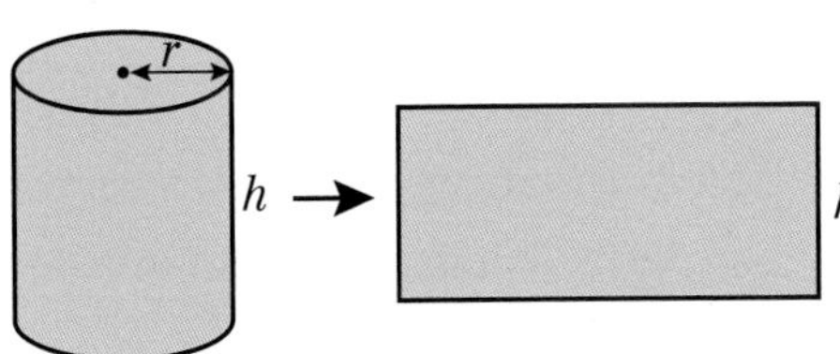

Q2bii hint The rectangular face wraps right round the circle. Circumference = $2\pi r$

Q2biv hint

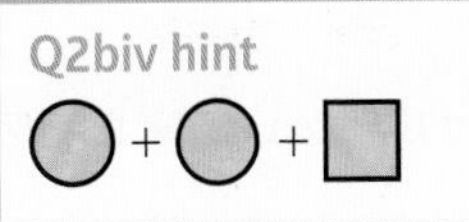

3 Here are two expressions used with spheres.

a Match each expression to the correct measurement.

volume = 340 cm^3 surface area = 746 cm^2

Q3a hint Area is in cm$^{\boxed{2}}$. Volume is in cm$^{\boxed{3}}$.
$4\pi r^{\boxed{2}}$ $\frac{4}{3}\pi r^{\boxed{3}}$

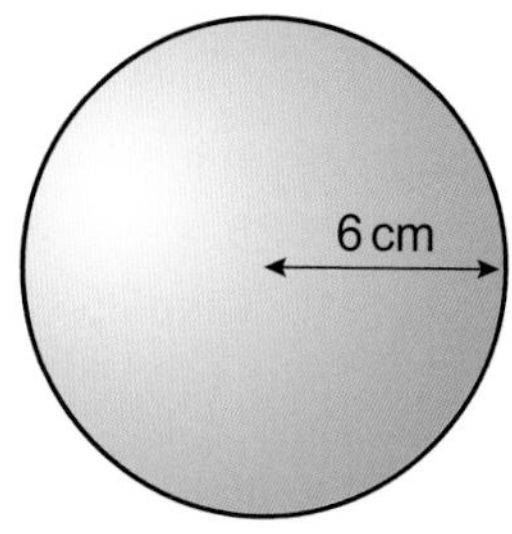

b Write the formula for surface area of a sphere.
c Write the formula for volume of a sphere.
d For this sphere, work out
 i the surface area
 ii the volume.
Give your answers to 2 d.p.

4 Here is a cone.

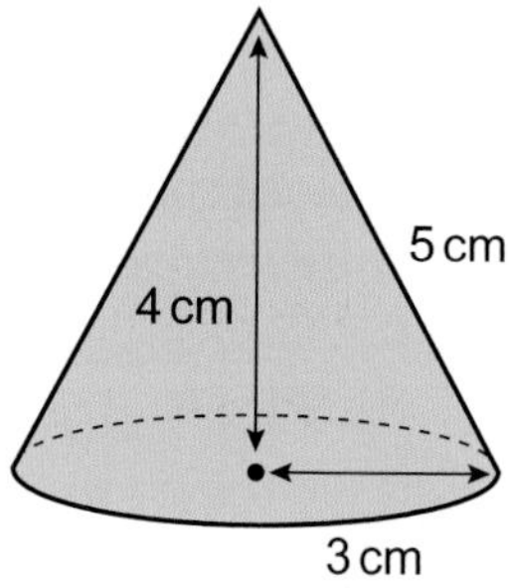

a What is its vertical height?
b What is its slant height?
c Use the formula to work out the volume of the cone.
$V = \frac{1}{3}\pi r^2 h$, where h is vertical height
d Use the formula to work out the surface area of the cone.
Surface area = $\pi r^2 + \pi r l$, where l is the slant height.
Give your answer correct to 1 d.p.

7 Extend

1 **Exam-style question**

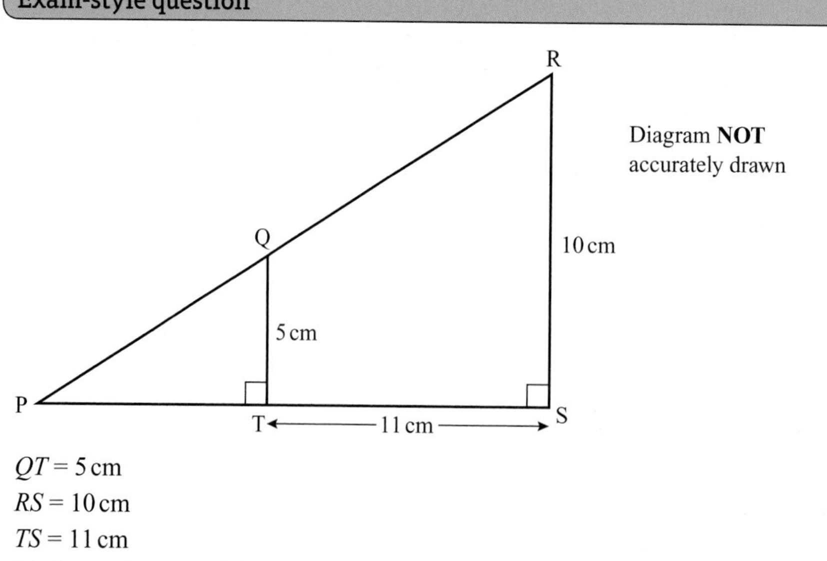

$QT = 5$ cm
$RS = 10$ cm
$TS = 11$ cm
Work out the area of the trapezium $TQRS$. **(2 marks)**

Exam hint
Write down the formula for finding its area before substituting in the measurements.

2 Write an expression for
   a the area of the cross section of this prism in mm^2
   b the volume of the prism in mm^3.

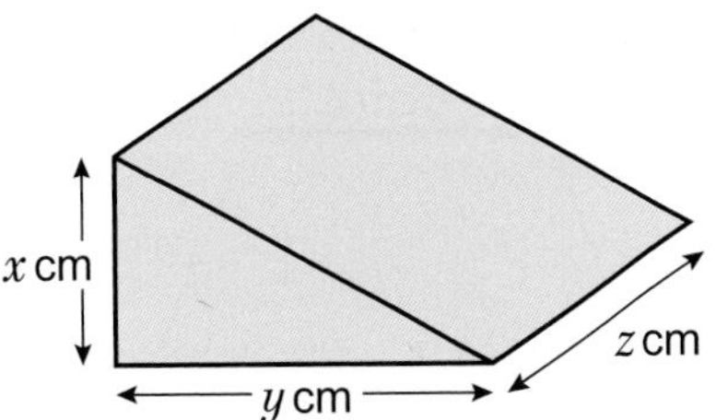

3 **Problem-solving** The diagram shows a lawn with a wall along one side.

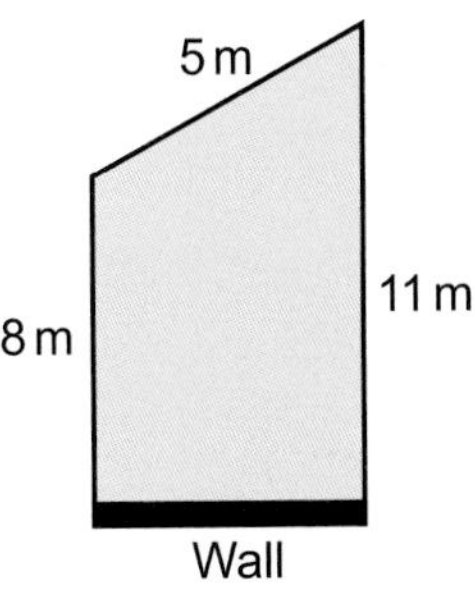

   a Show clearly that the wall is 4 m long.
   b A 500 ml bottle of lawn feed treats 20 m^2 of lawn. How many bottles are needed to treat this lawn?

Q3a hint Split the shape into a rectangle and a right-angled triangle. Use Pythagoras.

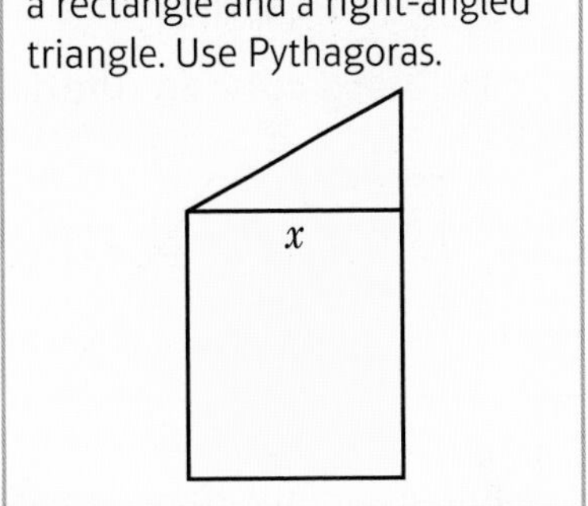

4 **STEM** The graph shows the velocity of a remote-controlled car during a test. Work out the distance the car travelled during the test.

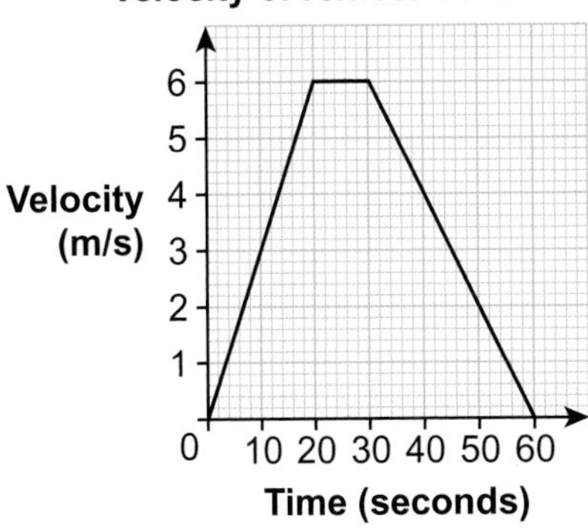

Q4 hint Distance = area under velocity–time graph.

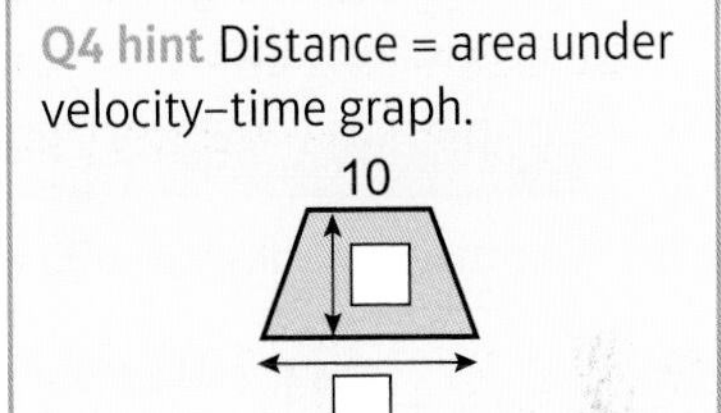

5 **Problem-solving / Real** A cylindrical water tank has height 1.2 m and radius 50 cm. Water flows into the tank at a rate of 300 ml per second. How long will it take for the tank to fill? Give your answer to the nearest minute.

6 Sketch three copies of this net for a triangular prism.

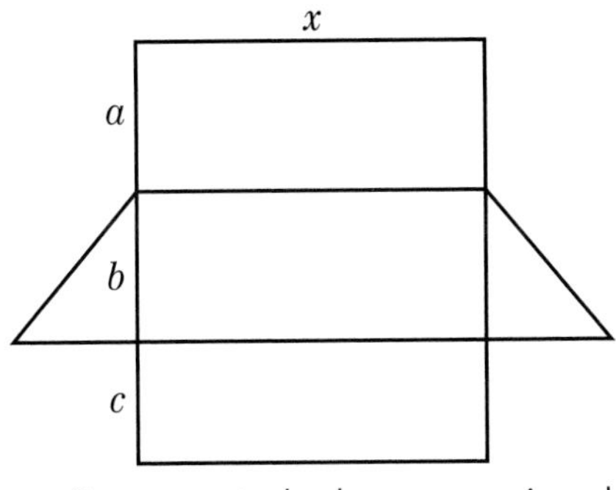

Q6b hint Expand the brackets.

   a On one net, shade an area given by the expression ax.
   b On the second net, shade an area given by the expression $(a + b)x$.
   c On the third net, shade an area given by the expression $\frac{1}{2}bc$.

7 **Problem-solving** The diagram shows an isosceles trapezium.

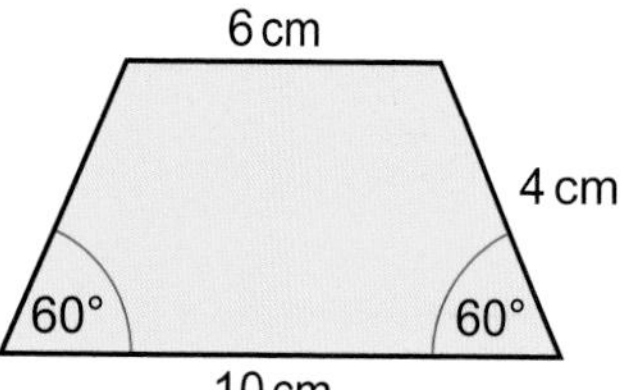

Q7a hint Find x using $\sin 60°$.

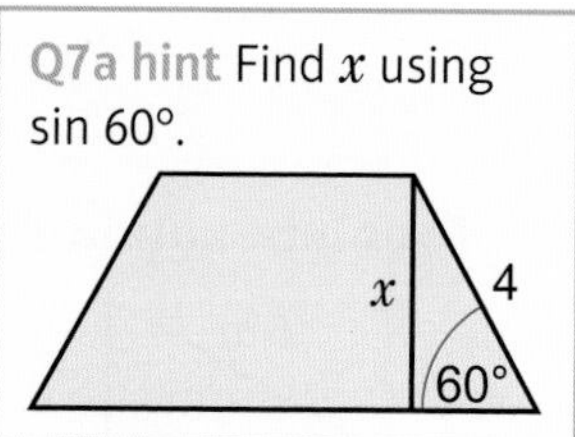

Calculate

a the height

b the area of the trapezium.

8 **Problem-solving** The diagram shows a trapezium. All the lengths are in centimetres.

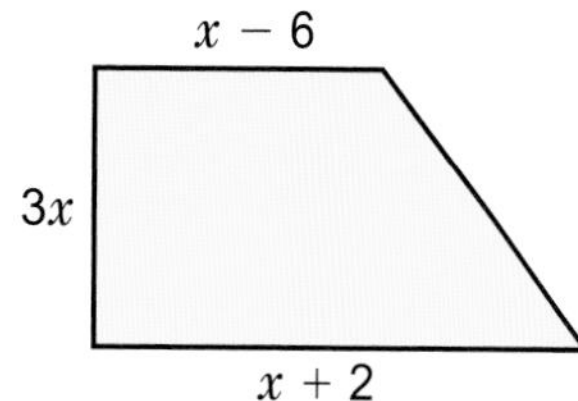

Q8 hint Substitute the measurements given into the formula for area of a trapezium.

The area of the trapezium is 144 cm^2.
Show that $3x^2 - 6x = 144$

9 **Problem-solving** A paintbrush is 30 cm long.
It just fits diagonally inside a cylindrical paint tin.
The tin is 24 cm high. What is the capacity of the tin, to the nearest litre?

Q9 hint Sketch the tin. Find the diameter, then the radius.

10 **Exam-style question**

The diagram shows a solid metal cylinder.

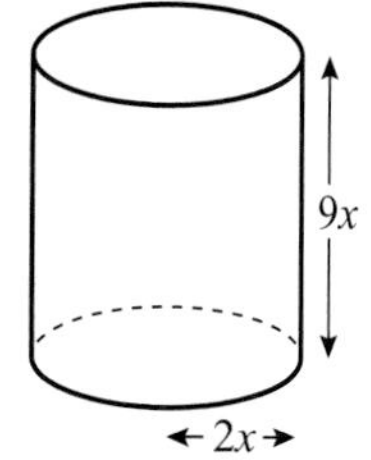

Diagram **NOT** accurately drawn

The cylinder has base radius $2x$ and height $9x$.
The cylinder is melted down and made into a sphere of radius r.
Find an expression for r in terms of x. **(3 marks)**

June 2012, Q25, 1MA0/1H

Exam hint Write down the volume formula before you substitute the given lengths into it.

11 $x = 15.6$ $y = 4.2$ $z = 5.8$
All values have been rounded to 1 d.p.
Work out the upper bounds of

a $x + y$ b yz c xyz d $\frac{x}{y}$

12 $d = 520$ (2 s.f.) $e = 13.8$ (3 s.f.)
Work out the lower bounds of

a $d + e$ b d^2 c $\frac{d}{e}$

13 **STEM** The diameter of the moon is 3.5×10^6 metres.
Calculate its surface area. Assume the moon is a sphere.

Q13 hint Give your answer in standard form.

7 Knowledge check

- Area of a trapezium $= \frac{1}{2}(a + b)h$

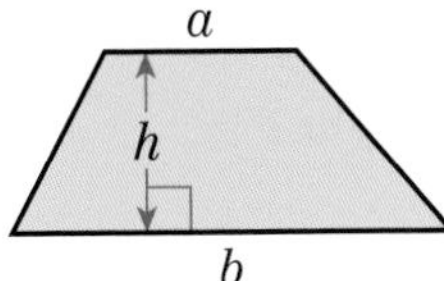

.... *Mastery lesson 7.2*

- To convert from cm^2 to mm^2, multiply by 100. To convert from mm^2 to cm^2, divide by 100. .... *Mastery lesson 7.2*
- 1 hectare (ha) is the area of a square 100 m by 100 m.
  1 ha = 100 m × 100 m = 10 000 m^2 .... *Mastery lesson 7.2*
- A 10% error interval means that a measurement could be up to 10% larger or smaller than the one given.

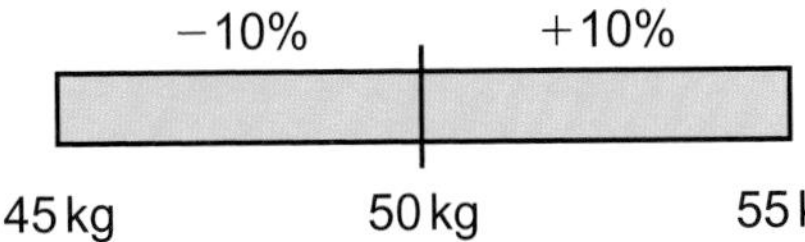

.... *Mastery lesson 7.2*

- Measurements rounded to the nearest unit could be up to half a unit smaller or larger than the rounded value. The possible values of x that round to 3.4 to 1 d.p. are $3.35 \leqslant x < 3.45$ .... *Mastery lesson 7.2*
- The upper bound is half a unit greater than the rounded measurement. The lower bound is half a unit less than the rounded measurement. .... *Mastery lesson 7.2*

  $12.5 \leqslant x < 13.5$

  lower bound upper bound
- When giving the answer to a calculation to an appropriate degree of accuracy, round the upper and lower bounds by the same amount. If the upper and lower bound give the same value when rounded, then the answer is to an appropriate degree of accuracy. .... *Mastery lesson 7.2*
- **Volume** is measured in mm^3, cm^3 or m^3. .... *Mastery lesson 7.3*
- **Capacity** is measured in ml and litres.
- 1 cm^3 = 1 ml, 1000 cm^3 = 1 litre .... *Mastery lesson 7.3*
- The **surface area** of a 3D solid is the total area of all its faces. .... *Mastery lesson 7.3*
- A **prism** is a 3D solid that has the same cross-section all through its length. .... *Mastery lesson 7.3*
- Volume of a prism
  = area of cross-section × length.

.... *Mastery lesson 7.3*

- The **circumference** of a circle is its perimeter. For any circle
  circumference = π × diameter
  $C = \pi d$ or $C = 2\pi r$ .... *Mastery lesson 7.4*

- The formula for the area, A, of a circle with radius r is $A = \pi r^2$

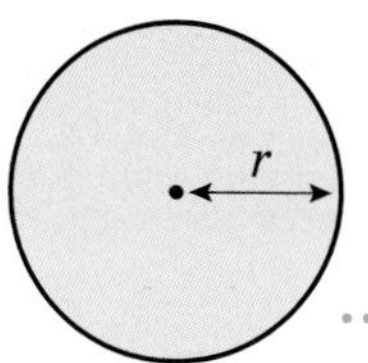

Mastery lesson 7.4

- The volume of a cylinder of radius r and height h is $V = \pi r^2 h$ .... *Mastery lesson 7.6*

- The surface area of a cylinder of radius r and height h is $2\pi r^2 + 2\pi rh$

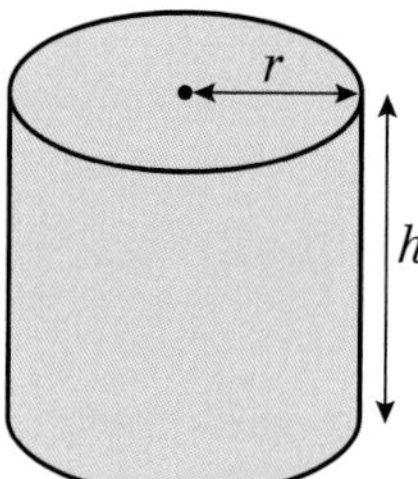

Mastery lesson 7.6

- For a sphere of radius r
surface area = $4\pi r^2$
volume = $\frac{4}{3}\pi r^3$

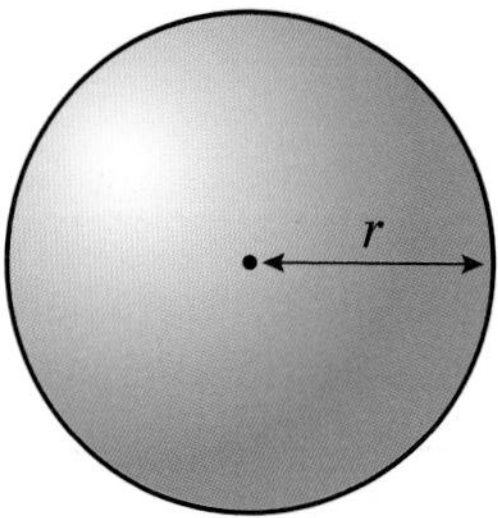

Mastery lesson 7.6

- Volume of pyramid = $\frac{1}{3}$ area of base × vertical height .... *Mastery lesson 7.7*
- Volume of cone = $\frac{1}{3}$ area of base × vertical height = $\frac{1}{3}\pi r^2 h$ .... *Mastery lesson 7.7*

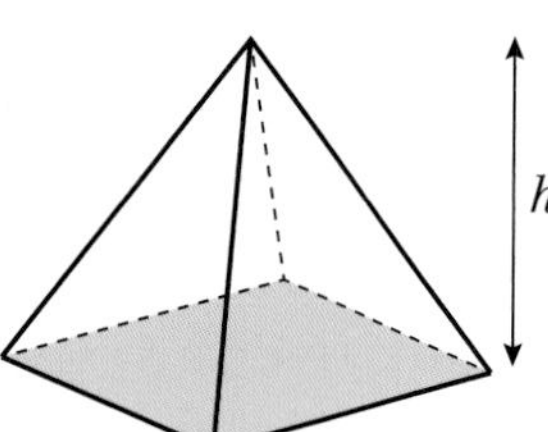

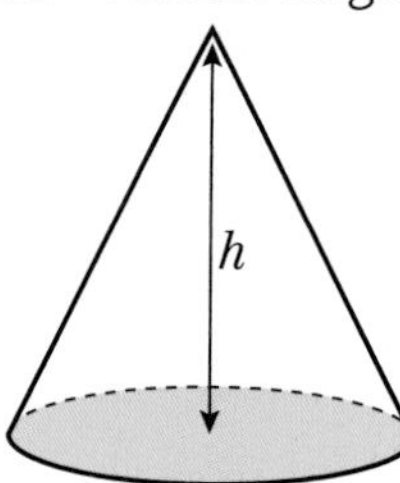

Mastery lesson 7.7

- Curved surface area of a cone = πrl, where r is the radius and l is the slant height. .... *Mastery lesson 7.7*
- Total surface area of a cone = $\pi rl + \pi r^2$.

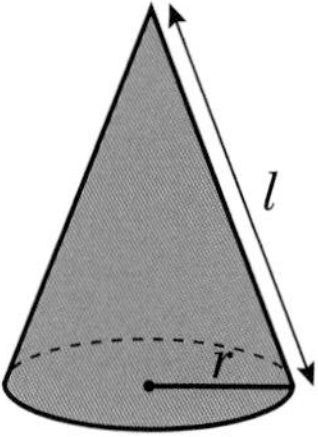

Mastery lesson 7.7

Reflect

Look back at this unit.
Which lesson made you think the hardest? Write a sentence to explain why.
Begin your sentence with, 'Lesson ____ made me think the hardest because ____'

7 Unit test

Log how you did on your Student Progression Chart.

1 **Exam-style question**

A circle has a diameter of 140 cm.
Work out the circumference of the circle.
Give your answer correct to 3 significant figures. **(2 marks)**

Nov 2013, Q12, 1MA0/2H

2 **Reasoning** Work out

a the area

b the perimeter of this trapezium. *(4 marks)*

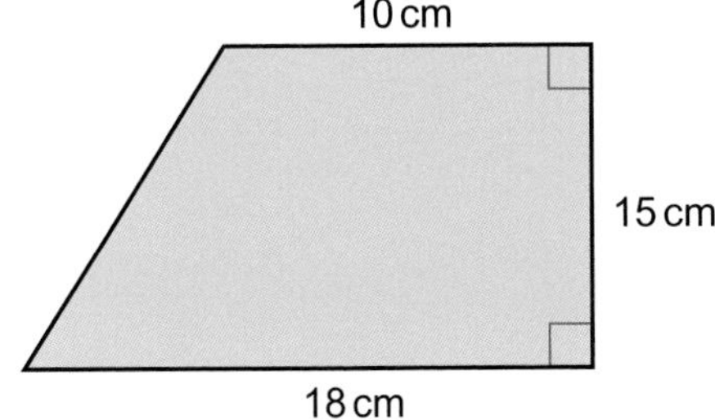

3 **Reasoning** A trapezium of area 60 cm² has parallel sides of length 12 cm and 8 cm.
What is the distance between the two parallel sides? *(3 marks)*

4 **Reasoning** A cylindrical water tank holds 26 litres when full. The water tank is 36 cm tall.
Work out the radius of the tank to 2 d.p. *(4 marks)*

5 **Reasoning** This triangular prism has volume 6720 cm³.

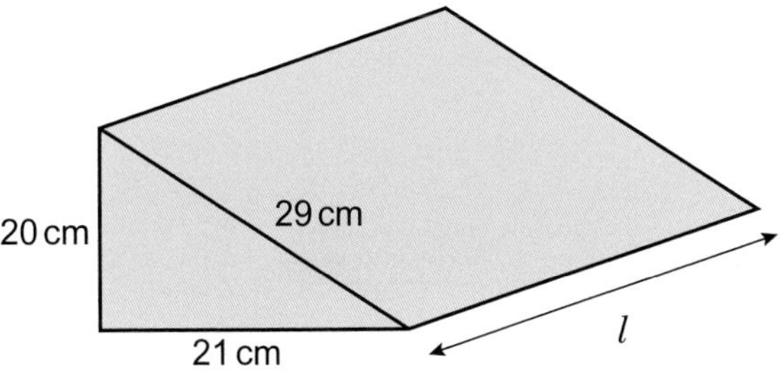

Work out

a the length of the prism

b its surface area. *(5 marks)*

6 **Reasoning** a What is the area of a circle with radius 8 cm? Give your answer in terms of π.

b Dan draws a circle with double the area of the circle in part **a**.
What radius does he use? Give your answer to a suitable degree of accuracy. *(3 marks)*

7 **Reasoning** Calculate the perimeter of this shape.
Give your answer to 3 s.f.

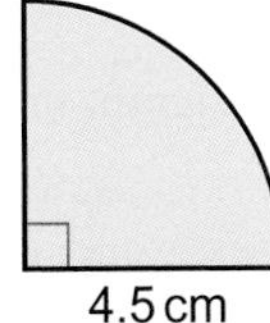

(2 marks)

8 Calculate the surface area of a sphere of radius 5.6 cm. Give your answer to 3 s.f. *(2 marks)*

9 This solid is made from a cone and a cylinder of radius 4cm.

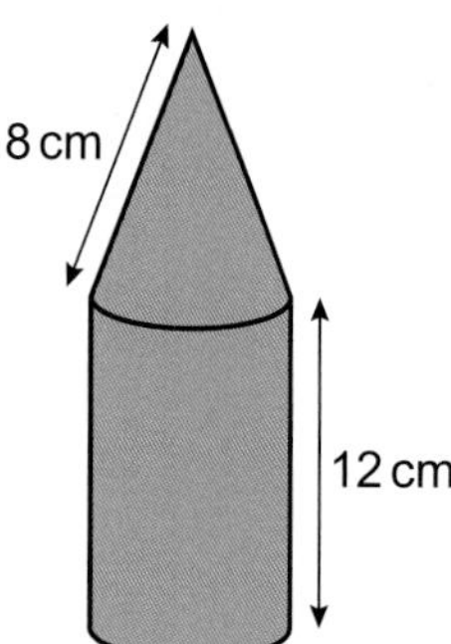

Find the total surface area of the solid, including the base, in terms of π. *(4 marks)*

10 Find the angle of this sector. *(3 marks)*

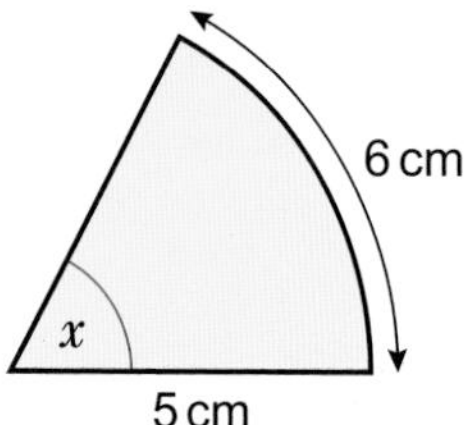

11 The lengths of this square-based pyramid are given to 1 d.p.
Calculate the upper and lower bounds for its volume. *(3 marks)*

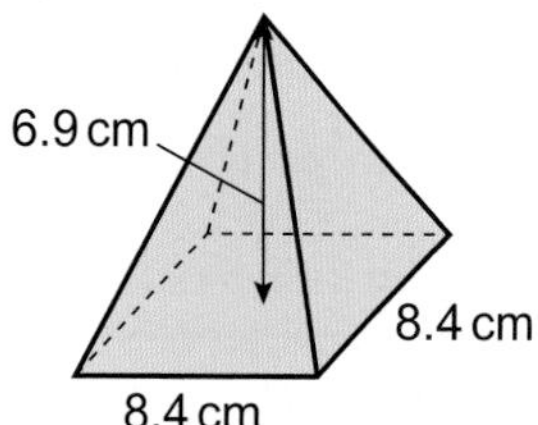

Sample student answer

Why will the student only get 1 mark for this answer?

Exam-style question

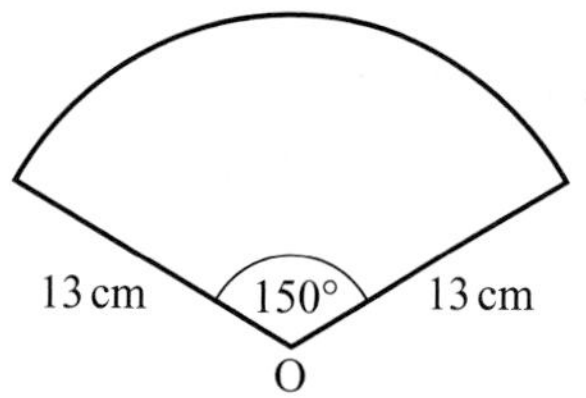

Diagram **NOT** accurately drawn

The diagram shows a sector of a circle, centre O.
The radius of the circle is 13 cm.
The angle of the sector is 150°.
Calculate the area of the sector.
Give your answer correct to 3 significant figures. **(2 marks)**

June 2008, Q19, 5540/4H

Student answer

$\frac{150}{360} \times \pi \times 13^2 = 221.2\,\text{cm}^2$

8 TRANSFORMATIONS AND CONSTRUCTIONS

Scale drawings and bearings are used in navigation. Ships and planes use them to guide their direction of travel.

What is your bearing when travelling

a south
b west
c south-east
d north-west
e south-west?

8 Prior knowledge check

Numerical fluency

1 Copy and complete

a $62 \times 4 = \square$
b $8 \times \square = 40$
c $270 \div 9 = \square$
d $128 \div \square = 32$

2 Work out

a $\frac{1}{2} \times 30$
b $\frac{1}{3} \times 63$
c $\frac{2}{3} \times 24$
d $\frac{3}{4} \times 84$

3 Copy and complete

a $3.6 \times 10 = \square$
b $4.82 \times \square = 482$
c $23 \div 100 = \square$
d $532 \div \square = 5.32$

Fluency with measures

4 Copy and complete

1 m = $\square$ cm
1 km = $\square$ m
1 km = $\square$ cm

5 How many

a cm in 4 m
b m in 6.2 km?

Geometrical fluency

6 a Which of these shapes are congruent?
b Which of these shapes are similar?

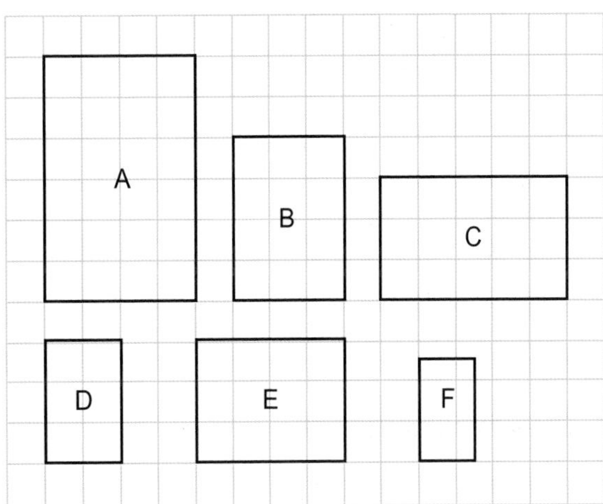

7 a Copy this diagram and draw the reflection of the triangle in
i the y-axis ii $y = 1$ iii $x = -2$

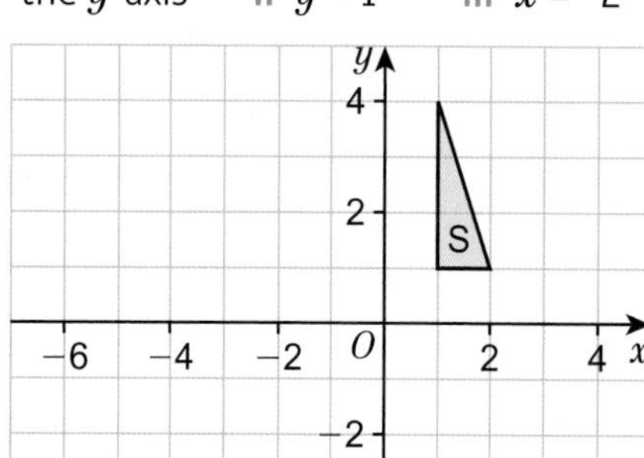

b Translate shape S
3 squares left and 2 squares down.

c Rotate shape S
90° anticlockwise about the origin.

8 Copy and rotate this shape 180° about the centre of rotation.

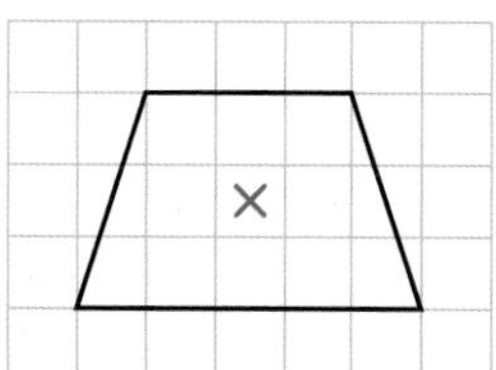

9 Copy this shape.
Enlarge it by scale factor 2.

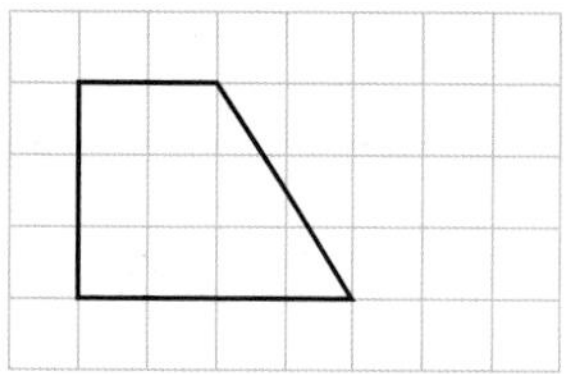

10 Is the enlarged shape in **Q9** similar or congruent to the original?

11 Draw this triangle accurately.

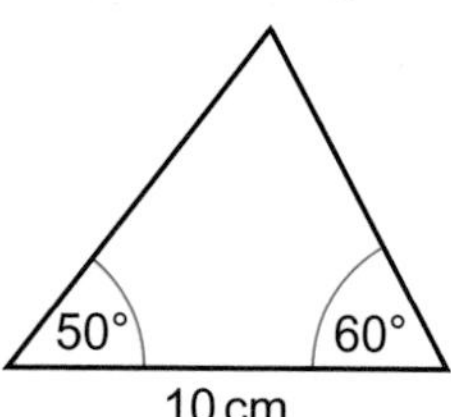

★ Challenge

12 Draw a grid with x- and y-axes from −5 to 5.

a Work out the coordinates of points needed to make
i a kite
ii a trapezium with one line of symmetry.

b Draw the line of symmetry on each shape.

c The line of symmetry cuts each shape in half. Write down the coordinates of the vertices of
i one half of the kite
ii one half of the trapezium.

d Give the coordinates from part **c** and the equation of the line of symmetry to one of your classmates. Can they use these to reproduce your original shapes?

8.1 3D solids

Objective

- Draw plans and elevations of 3D solids.

Why learn this?

Architectural drawings show plans and elevations of buildings.

Fluency

What are the dimensions of the top, side and front of this cuboid?

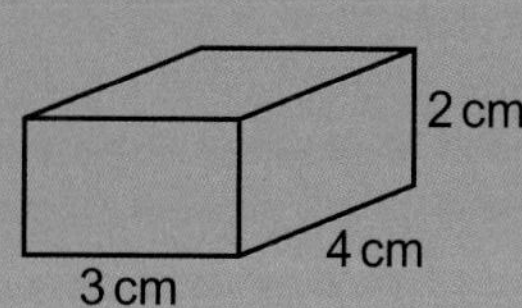

Warm up

1 On an isometric grid, draw
a a cube b a cuboid c a triangular prism.

ActiveLearn Homework, practice and support: Higher 8.1

Key point 1

The **plan** is the view from above the solid.
The **front elevation** is the view of the front of the solid.
The **side elevation** is the view of the side of the solid.

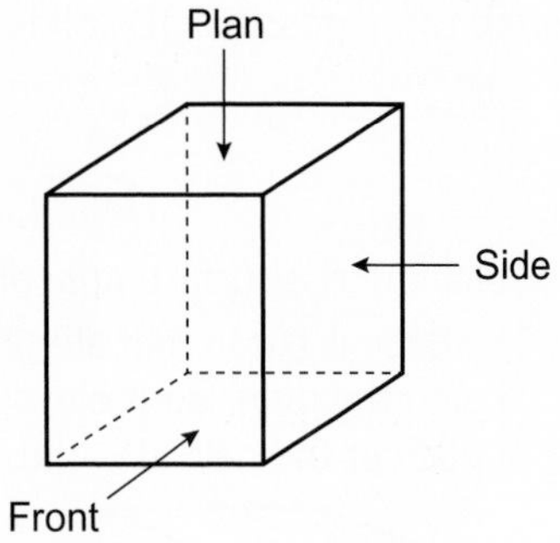

Plan
Side
Front

Example 1

Draw the **plan, front elevation** and **side elevation** of this solid on squared paper.

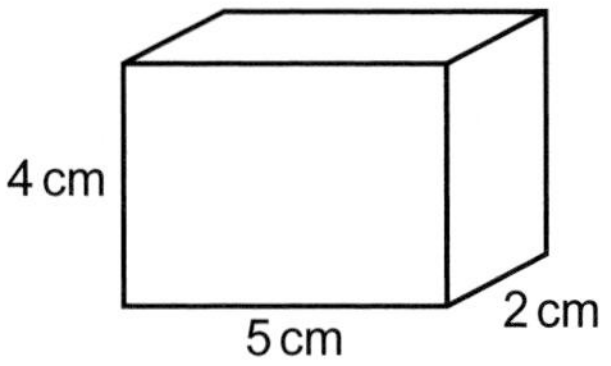

4 cm
5 cm
2 cm

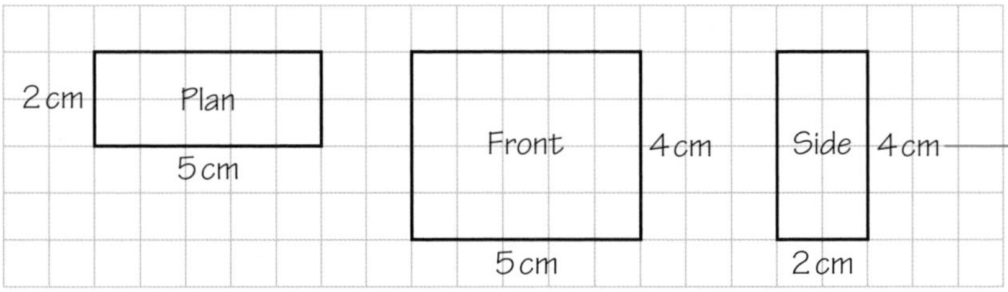

2 cm Plan 5 cm
Front 4 cm 5 cm
Side 4 cm 2 cm

Use a ruler.
Measure accurately.
Label the lengths.

Questions in this unit are targeted at the steps indicated.

2 On squared paper, draw and label the plan, front elevation and side elevation of these solids.

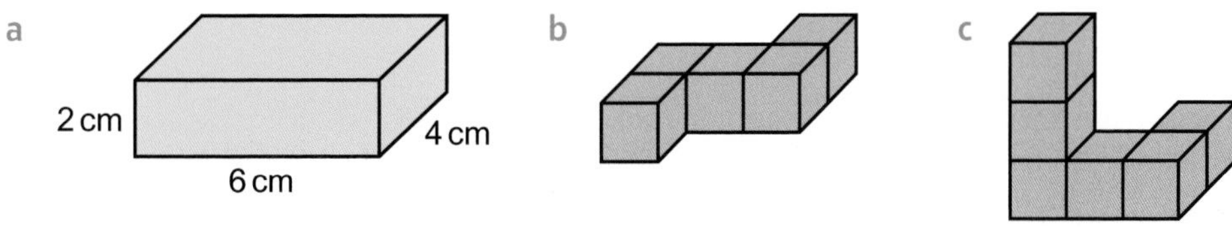

3 **Reasoning** Sketch the solids represented by these plans and elevations.

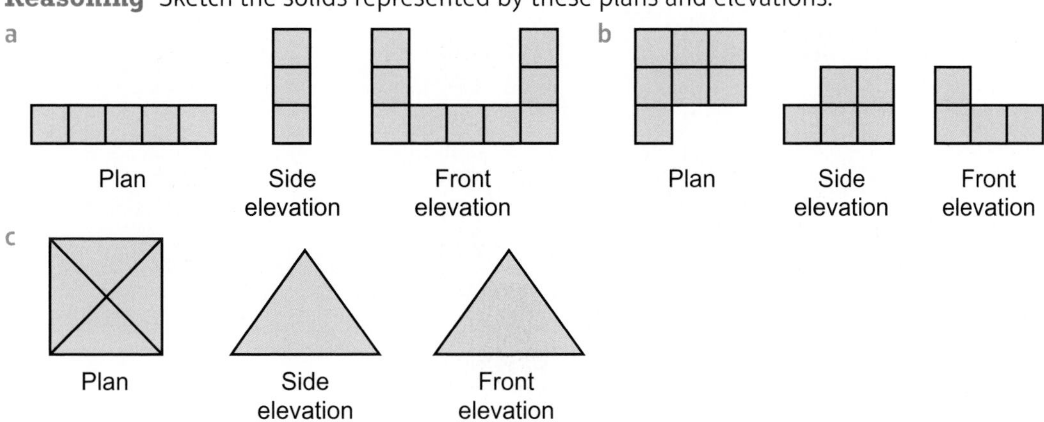

4 **Problem-solving** Here is the side elevation of a 3D solid. Sketch three possible 3D solids it could belong to.

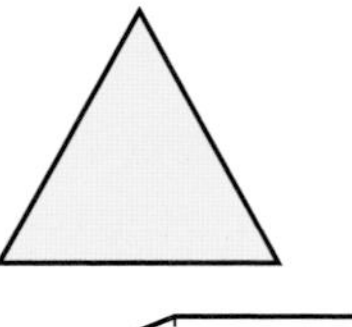

5 **Problem-solving / Reasoning** Here is a cube.

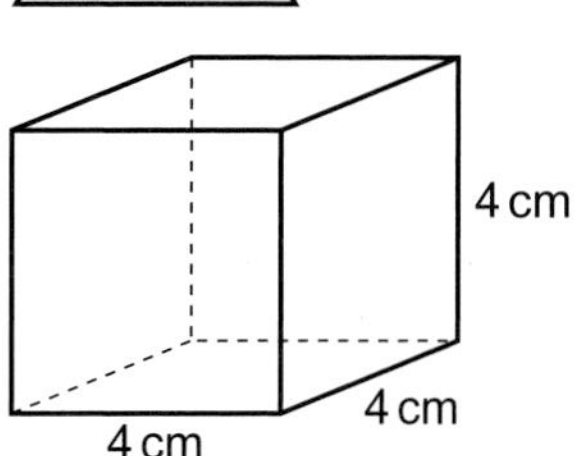

a Calculate the surface area of the cube.

b The cube is cut in half along the red plane. Sketch the plan, front elevation and side elevation of each of the new 3D solids.

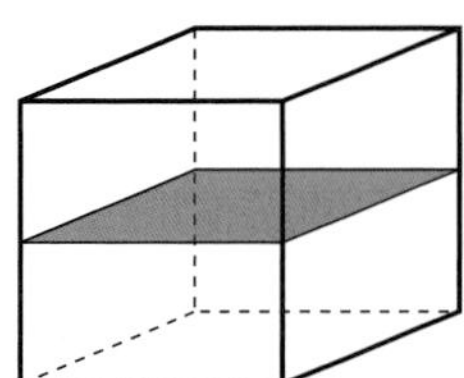

c Calculate the surface area of each of the new solids.

d Repeat parts **b** and **c** for the cube cut along this red plane.

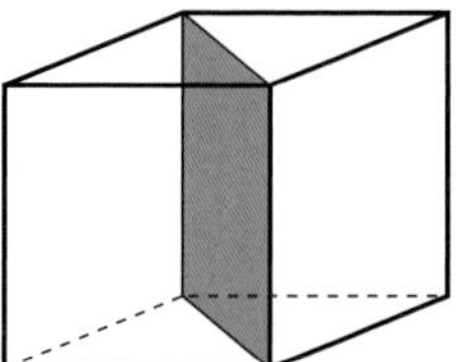

Discussion Why does the surface area of the two parts not equal the surface area of the original cube?

6 **Exam-style question**

Here is a solid prism.

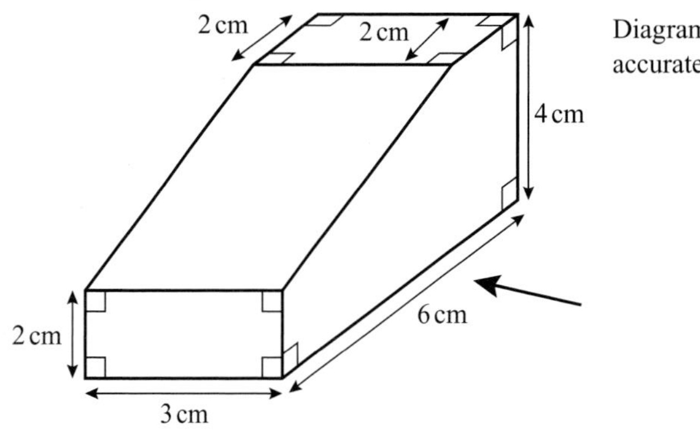

Diagram **NOT** accurately drawn

On a centimetre-square grid, draw an accurate side elevation of the solid prism from the direction of the arrow. **(2 marks)**

March 2013, Q10, 5MB2H/01

Exam hint
Use a pencil and make sure the lines are dark enough so that an examiner can see them.

8.2 Reflection and rotation

Objectives

- Reflect a 2D shape in a mirror line.
- Rotate a 2D shape about a centre of rotation.
- Describe reflections and rotations.

Why learn this?

Car mechanics and engineers need to know how far a part has rotated when checking engine parts.

Fluency

When you reflect a shape, are the object and image congruent? What about a rotation?

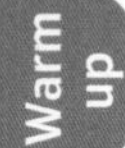

1 Draw a coordinate grid from −5 to +5 on both axes. Draw these straight lines.

a $y = -3$ b $x = 4$ c $y = x$ d $y = -x$

Key point 2

Reflections and rotations are types of transformation.
Transformations move a shape to a different position.
To describe a reflection, you need to give the equation of the mirror line.

Key point 3

An original shape is called an **object**. When the object is transformed, the resulting shape is called an **image**.

2 Draw a coordinate grid from −5 to +5 on both axes.
- a Draw rectangle Q with vertices at coordinates (1, 1), (1, 3), (5, 3), (5, 1).
- b Reflect rectangle Q in the x-axis. Label the image R.
- c Reflect rectangle R in $x = 1$. Label the image S.
- d Reflect rectangle S in the x-axis. Label the image T.
- e Describe the single reflection that maps rectangle T onto rectangle Q.

Q2 hint Use tracing paper to help.

3 **Reasoning** Draw a coordinate grid from −3 to +3 on both axes.
- a Draw triangle A with coordinates (−1, −1), (−1, 2), (2, −1).
- b Reflect triangle A in the line $y = 1$. Label the image B.
- c Reflect triangle A in the line $y = x$. Label the image C.

Discussion What is special for triangle A about the line $y = x$?
Explain how you could use this in part **c**.

4 Describe the reflection that maps
- a P onto Q
- b P onto R
- c P onto S
- d P onto T.

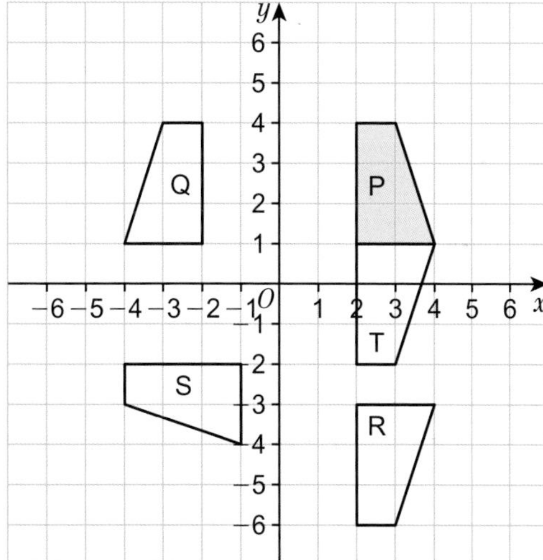

5 Describe the reflection that maps
- a A to B
- b A to C.

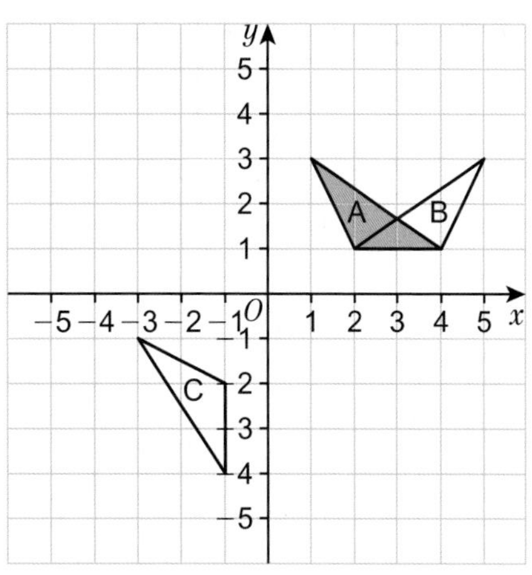

6 Draw a coordinate grid from −5 to +5 on both axes.

a Draw shape A with vertices at coordinates (−1, 2), (−1, 4), (1, 4), (1, 2).

b Reflect shape A in the line $y = x$. Label the image B.

c Reflect shape A in the line $y = -x$. Label the image C.

d Reflect shape A in the x-axis. Label the image D.

e Describe the reflection that maps shape D to shape B.

7 **Reasoning** Draw a coordinate grid from −8 to +8 on both axes.

> **Q7 hint** Use tracing paper to help.

a Draw triangle A with coordinates (1, 1), (1, 5), (4, 5).

b Rotate triangle A

i 90° clockwise about (1, 1)

ii 180° about (1, 0)

iii 90° anticlockwise about (0, 1)

iv 180° about (−2, 1)

v 90° clockwise about (2, 5)

vi 180° about (2, 3).

Label your results i, ii etc.

Discussion Why don't you need to give the direction for a rotation of 180°?

Key point 4

To describe a rotation you need to give
- the direction of turn (clockwise or anticlockwise)
- the angle of turn
- the **centre of rotation**.

Example 2

Describe the rotation that takes shape A onto shape B.

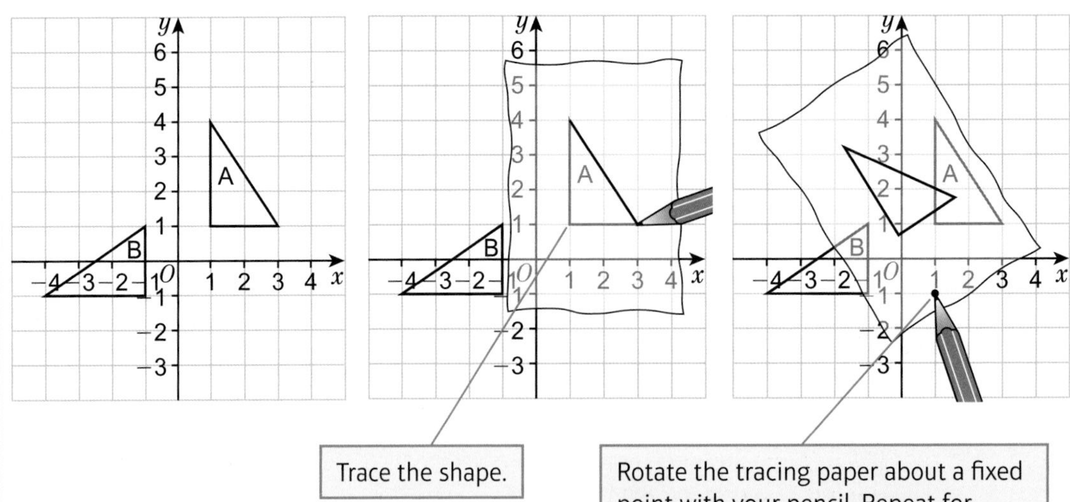

Trace the shape.
Rotate the tracing paper about a fixed point with your pencil. Repeat for different positions until your tracing ends up on top of the image.

Rotation anticlockwise 90° about (1, −1) — Give the direction, angle and centre of rotation.

8 Describe the rotation that takes each shape to its image.

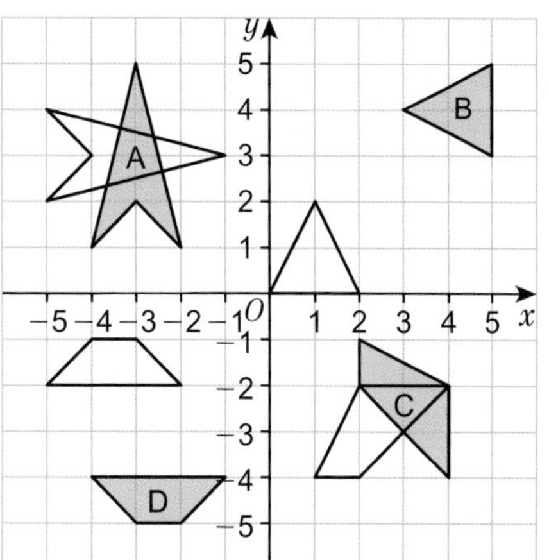

9 **Problem-solving** Draw a coordinate grid from –5 to +5 on both axes.
 a Draw shape A with vertices at coordinates (1, 1), (1, 3), (3, 3), (4, 1).
 b Reflect shape A in the line $y = x$. Label the image B.
 c Reflect shape B in the y-axis. Label the image C.
 d Describe the transformation that takes shape A to shape C.

10 **Reasoning** 'A reflection in one axis followed by a reflection in the other axis is the same as a rotation.'
Decide whether this statement is: sometimes true, always true or never true.

11 **Exam-style question**

Describe fully the single transformation that maps triangle A onto triangle B.

(3 marks)

June 2012, Q9, 1MA0/1H

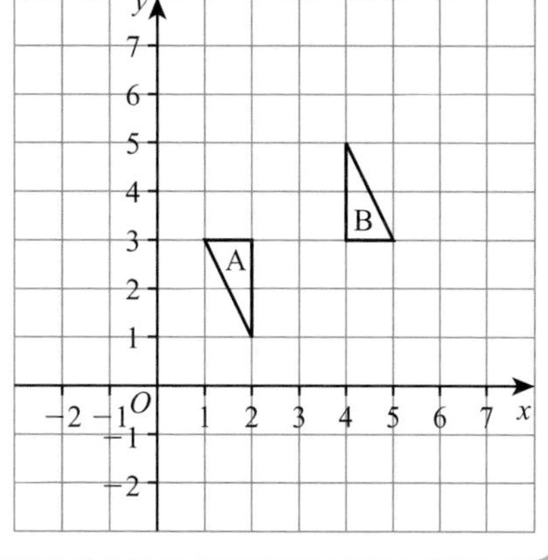

Exam hint
The question is worth 3 marks which means you need to give 3 pieces of information about the transformation.

8.3 Enlargement

Objective

- Enlarge shapes by fractional and negative scale factors about a centre of enlargement.

Why learn this?

Special effects artists use enlarged shapes when designing images for a background scene.

Fluency

What is the scale factor of the enlargement of shape A to shape B?
Which vertex on shape B corresponds to F on shape A?

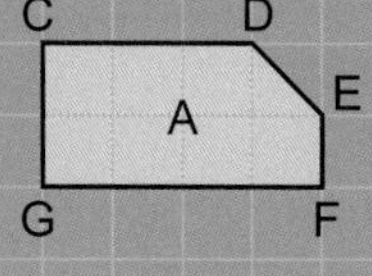

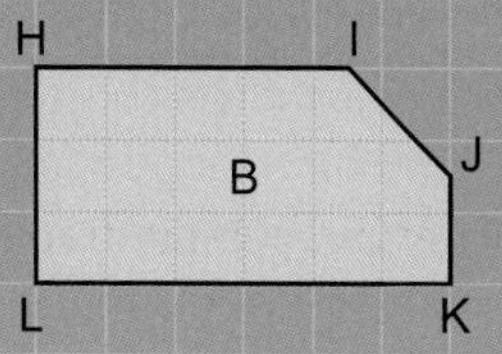

Warm up

1 Copy this diagram and draw
- a an enlargement with scale factor 3, centre (0, 0)
- b an enlargement with scale factor 1.5, centre (1, 1).

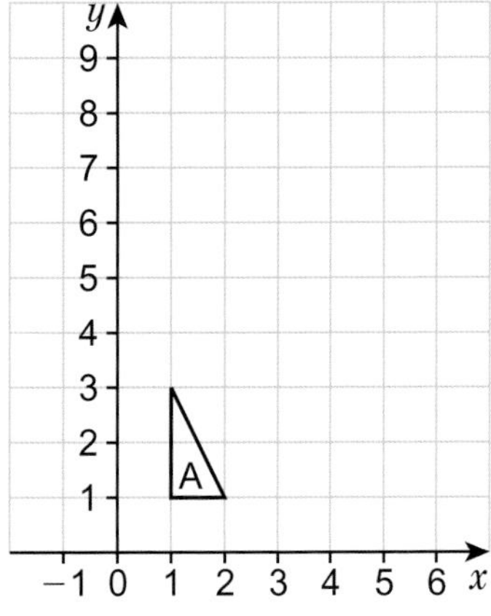

Key point 5

An enlargement is a transformation where all the side lengths of a shape are multiplied by the same **scale factor**.

2 Copy the diagram.
Enlarge the triangle by scale factor 2, with these centres of enlargement.
- a (3, 5)
- b (4, 3)
- c (2, 2)

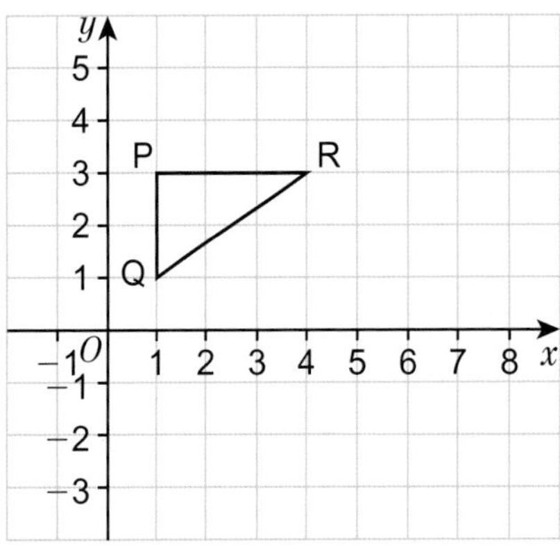

Key point 6

To describe an enlargement you need to give the centre of enlargement and the scale factor.

3 Triangle ABC has been enlarged to give triangle PQR.
- a What is the scale factor of the enlargement?
- b Copy the diagram.
  Join corresponding vertices on the object and the image with straight lines.
  Extend the lines until they meet at the centre of enlargement.
- c Write down the coordinates of the centre of enlargement.
- d Copy and complete to describe the enlargement from A to B.
  Enlargement by scale factor _____, centre (_____, _____).

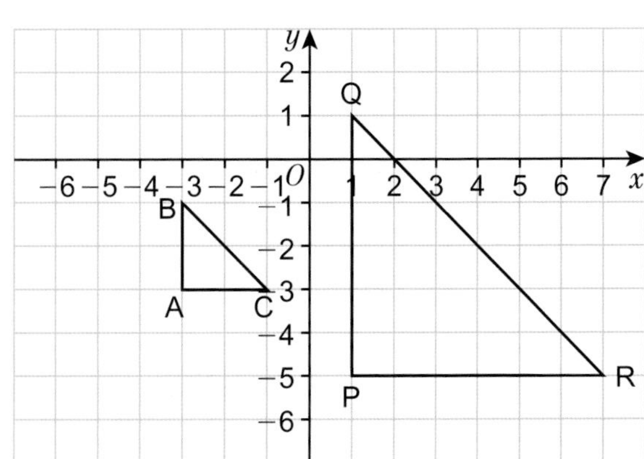

4 **Problem-solving** Draw a rectangle A, with base 3 cm and height 2 cm.

a Work out the area of the rectangle.

b Shape A is enlarged by scale factor 2 to make shape B. Work out the area of shape B.

c Shape A is enlarged by scale factor 3 to make shape C. Work out the area of shape C.

d Shape A is enlarged by scale factor 4 to make shape D. Work out the area of shape D.

e Copy and complete this table.

Shape	Scale factor	$\frac{\text{Area of enlarged shape}}{\text{Area of shape A}}$
B	2	
C	3	
D	4	

Discussion When a shape is enlarged by scale factor k, what happens to its area?

Key point 7

When a shape is enlarged by scale factor k, the area is enlarged by scale factor k^2.

5 Copy these diagrams. Enlarge each shape by the scale factor given.

a scale factor $\frac{1}{2}$

Q5a hint Multiply all the lengths by $\frac{1}{2}$.

b scale factor $\frac{1}{3}$

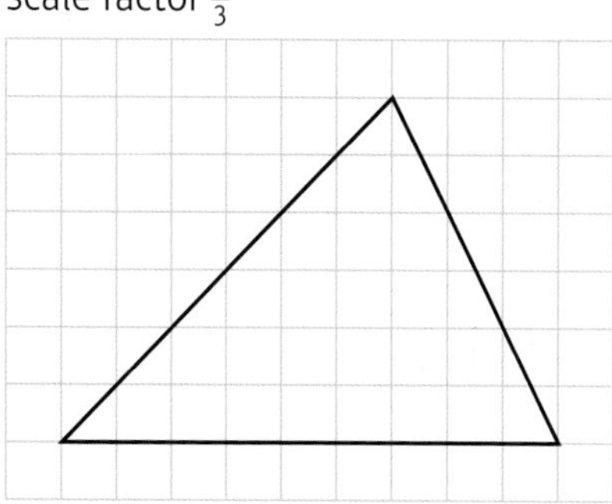

Q5b hint Divide the base and vertical height by 3.

Key point 8

To enlarge a shape by a fractional scale factor, multiply all the side lengths by the scale factor. When a centre of enlargement is given, multiply the distance from the centre to each point on the shape by the scale factor.

6 a Copy and enlarge each shape by the given scale factor about the centre of enlargement shown.

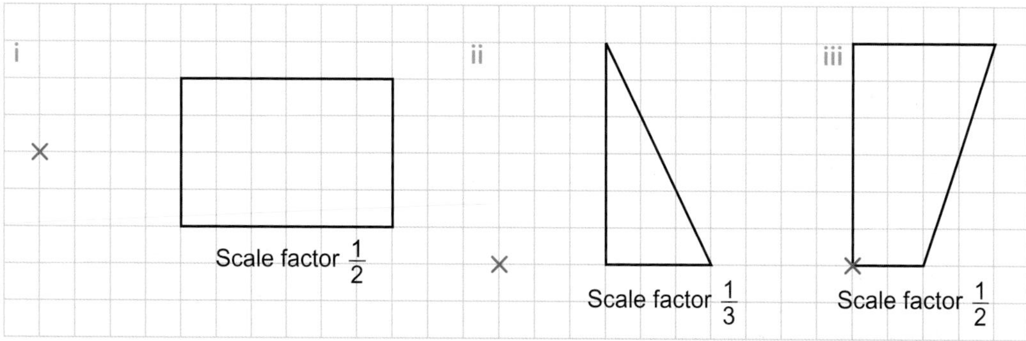

b **Reasoning** When a shape is enlarged by scale factor $\frac{1}{2}$, is its area enlarged by scale factor $\left(\frac{1}{2}\right)^2$? Explain.

7 Problem-solving

a Describe the enlargement that maps shape A onto shape P.

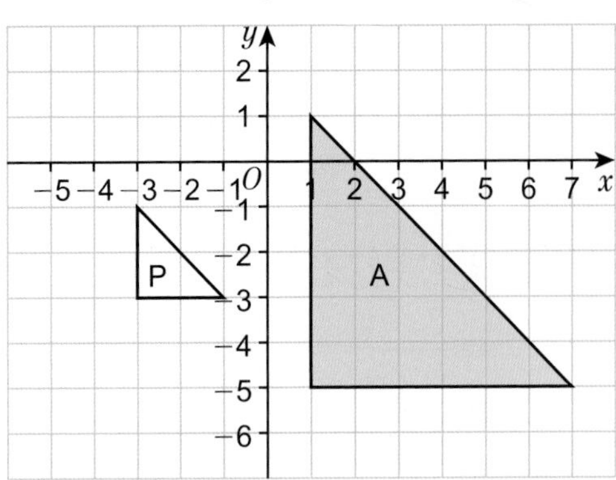

b Describe the enlargement that maps shape C onto shape R.

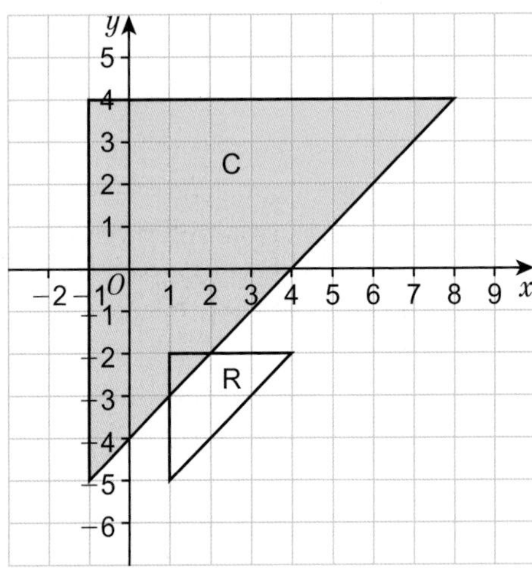

Key point 9

A **negative scale factor** takes the image to the opposite side of the centre of enlargement.

Example 3

Enlarge triangle A by scale factor –2 about centre (–1, 2).

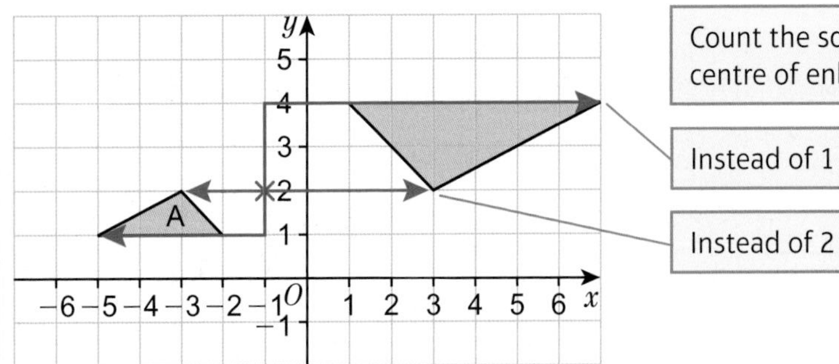

Count the squares from the centre of enlargement.

Instead of 1 down, 4 left, go 2 up, 8 right.

Instead of 2 left, go 4 right.

8 Draw a coordinate grid from –12 to +12 on both axes. Join the points (1, 2), (4, 4) and (4, 1) to make a triangle. Enlarge the triangle

a by scale factor –2, centre of enlargement (–1, 0)

b by scale factor –2, centre of enlargement (–1, 3)

c by scale factor –2, centre of enlargement (4, 4).

9 **Problem-solving** Describe fully the single transformation that maps shape S onto shape T.

> **Q9 strategy hint** Draw lines to find the centre.

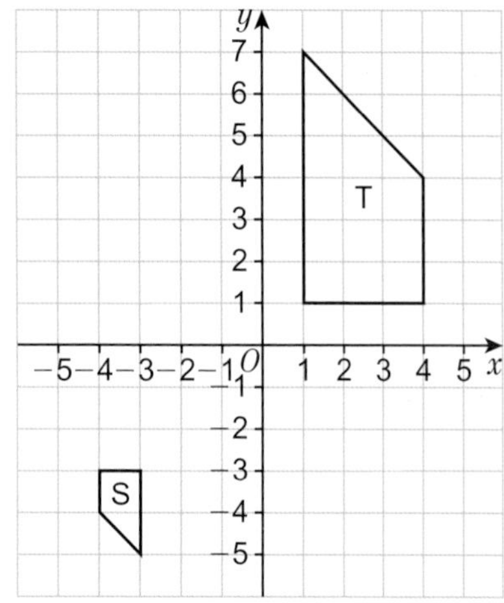

10 **Reflect** Jamie said, 'An enlargement always makes a shape bigger than the original shape.' Is Jamie correct? Explain your answer.

11 **Exam-style question**

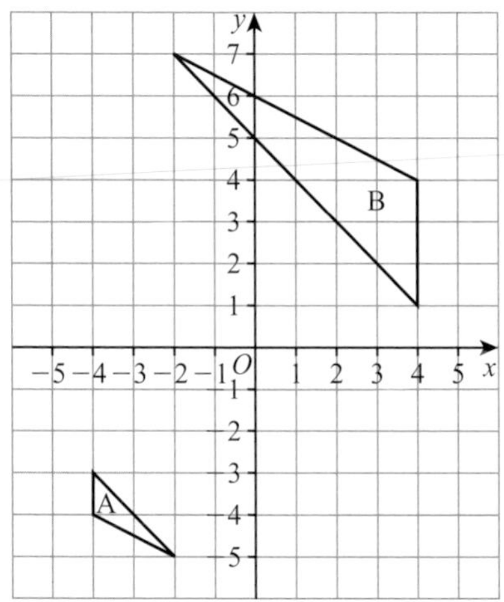

Describe fully the single transformation that maps shape A onto shape B. **(3 marks)**

> **Exam hint**
> For 3 marks, give 3 pieces of information about the transformation.

8.4 Translations and combinations of transformations

Objectives

- Translate a shape using a vector.
- Carry out and describe combinations of transformations.

Why learn this?

Interior designers draw a plan of a room and translate and transform furniture so that it fits well.

Fluency

Describe two possible transformations that could take shape A to shape B.

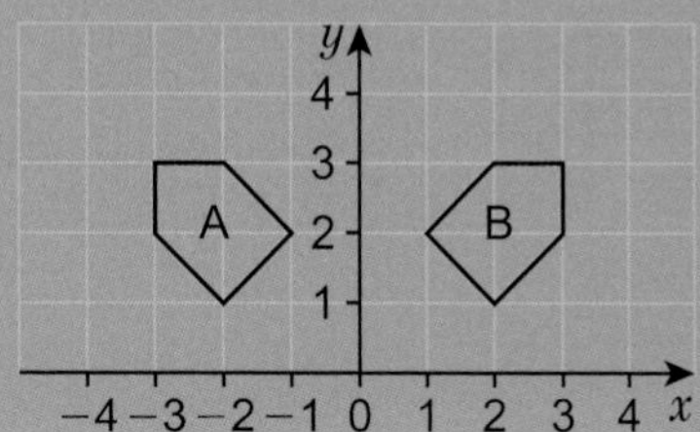

Warm up

1 Describe the translation that moves each shape to its image.

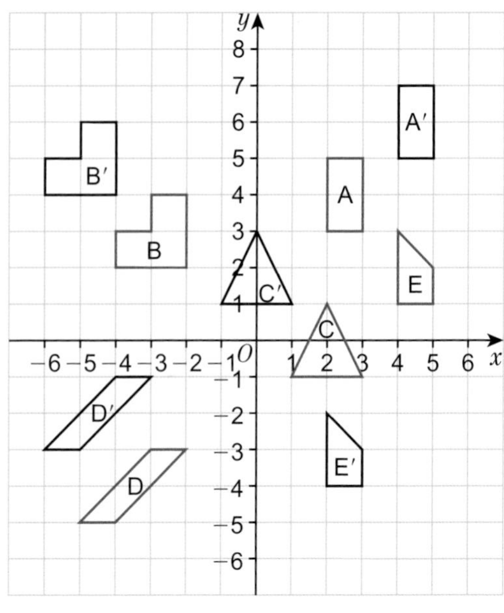

Q1 hint ☐ squares left, ☐ squares right.

Key point 10

In a translation, all the points on the shape move the same distance in the same direction.

Key point 11

You can describe a translation by using a **column vector**.

The column vector for a translation 2 squares right and 3 squares down is $\begin{pmatrix} 2 \\ -3 \end{pmatrix}$.

The top number gives the movement parallel to the x-axis.

The bottom number gives the movement parallel to the y-axis.

Example 4

Translate triangle A by the vector $\begin{pmatrix} -3 \\ 2 \end{pmatrix}$.

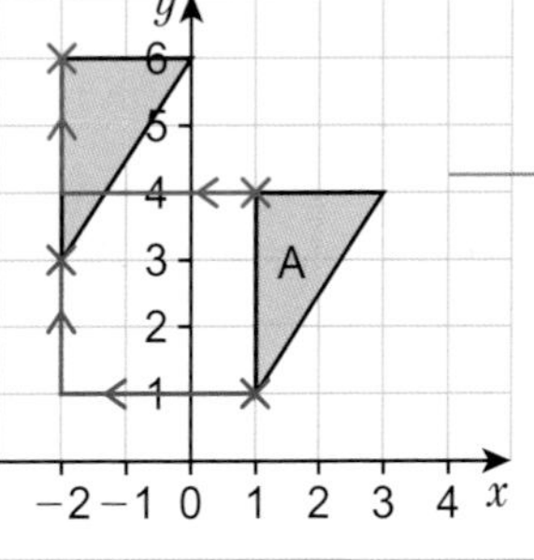

Move each point on the original shape 3 squares left and 2 squares up.

2 Copy this diagram. Translate shape A by the vectors

a $\begin{pmatrix} 2 \\ 3 \end{pmatrix}$ to B

b $\begin{pmatrix} 3 \\ -4 \end{pmatrix}$ to C

c $\begin{pmatrix} -2 \\ 0 \end{pmatrix}$ to D

d $\begin{pmatrix} 0 \\ 3 \end{pmatrix}$ to E

e $\begin{pmatrix} -1 \\ -4 \end{pmatrix}$ to F

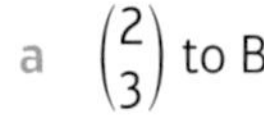

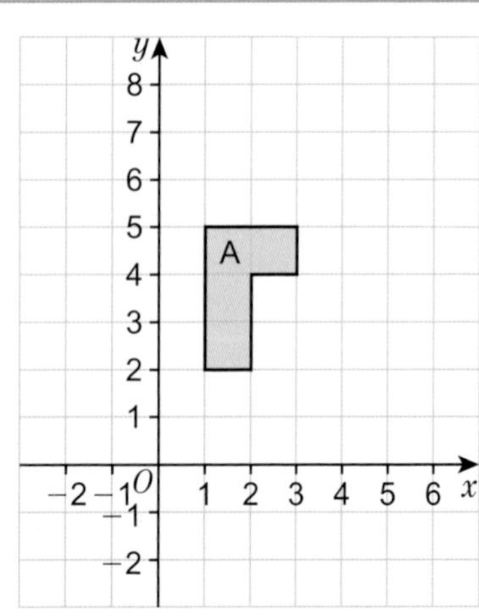

3 Describe these translations using column vectors.

a B to A

b A to C

c B to E

d D to E

e E to D

Discussion How can you use your answer to part **d** to help you find the answer to part **e**?

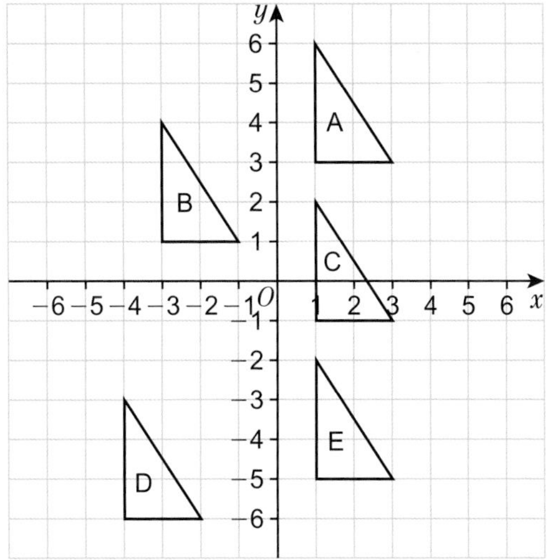

4 **Reasoning** A shape is translated by vector $\begin{pmatrix} a \\ b \end{pmatrix}$.

What vector would translate the shape back to its original position? Explain your answer.

5 Draw a coordinate grid from −6 to +6 on both axes.

a Plot a triangle with vertices at (1, 1), (3, 1) and (1, −2). Label the triangle P.

b i Translate triangle P by vector $\begin{pmatrix} 1 \\ 4 \end{pmatrix}$. Label the image Q.

ii Translate triangle Q by vector $\begin{pmatrix} -2 \\ 1 \end{pmatrix}$. Label the image R.

c Describe the translation of triangle P to triangle R, using a single vector.

Discussion What do you notice about the vectors in parts **b** and **c**?

Key point 12

The **resultant vector** is the vector that moves the original shape to its final position after a number of translations.

6 a **Reasoning** A shape is translated by vector $\begin{pmatrix} 3 \\ 4 \end{pmatrix}$ followed by a translation by vector $\begin{pmatrix} 1 \\ -3 \end{pmatrix}$.

What is the **resultant vector**?

b What is the resultant vector for a translation of $\begin{pmatrix} a \\ b \end{pmatrix}$ followed by a translation of $\begin{pmatrix} c \\ d \end{pmatrix}$? Explain your answer.

7 Copy this diagram and shape A only on a coordinate grid from −6 to +6 on both axes.

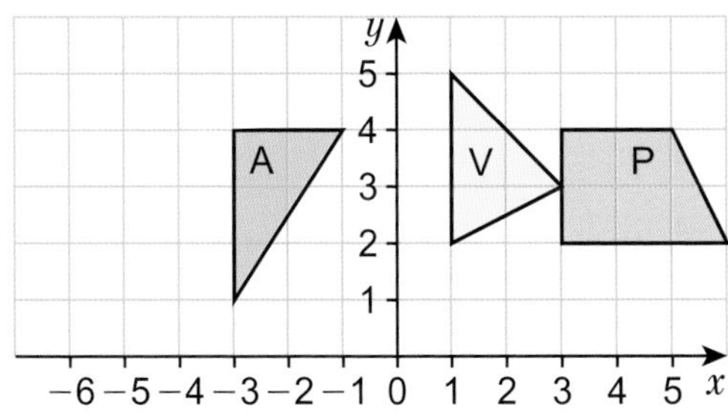

a Translate shape A by vector $\begin{pmatrix} 3 \\ -2 \end{pmatrix}$. Label the image B.

b Reflect shape B in the line $y = -2$. Label the image C.

8 Copy the diagram from **Q7** and shape P only on a coordinate grid from −6 to +6 on both axes.

a Reflect shape P in the line $y = 1$. Label the image Q.

b Rotate shape Q through 180°, about point (1, −1). Label the image R.

c Translate shape R by vector $\begin{pmatrix} 2 \\ 4 \end{pmatrix}$. Label the image S.

d Describe the reflection that maps shape P onto shape S.

9 Copy the diagram from **Q7** and shape V only on a coordinate grid from −6 to +6 on both axes.

a Reflect triangle V in the line $y = x$. Label the image W.

b Translate triangle W by vector $\begin{pmatrix} -4 \\ -2 \end{pmatrix}$. Label the image X.

c Rotate triangle X through 90° anticlockwise about point (−2, 2). Label the image Y.

d Describe the single transformation that maps triangle V onto triangle Y.

10 **Problem-solving / Reasoning** A company has based its logo on a triangle.
Draw a coordinate grid from −6 to +6 on both axes.

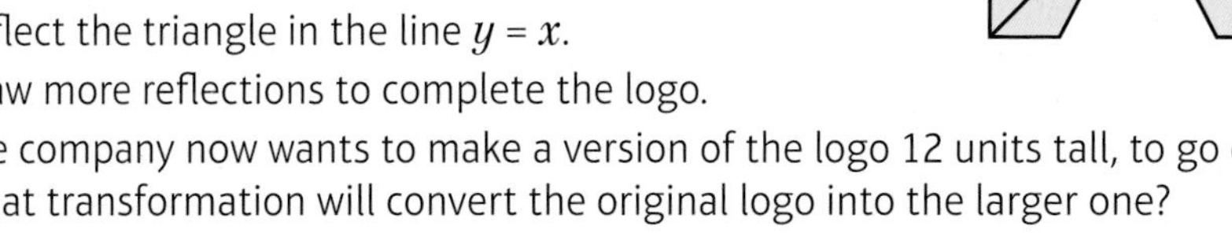

a Plot the points (0, 0), (1, 2) and (2, 2) and join them to make a triangle.

b Reflect the triangle in the line $y = x$.

c Draw more reflections to complete the logo.

d The company now wants to make a version of the logo 12 units tall, to go on a desk sign. What transformation will convert the original logo into the larger one?

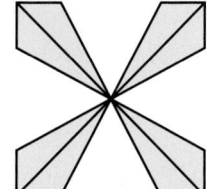

11 **Exam-style question**

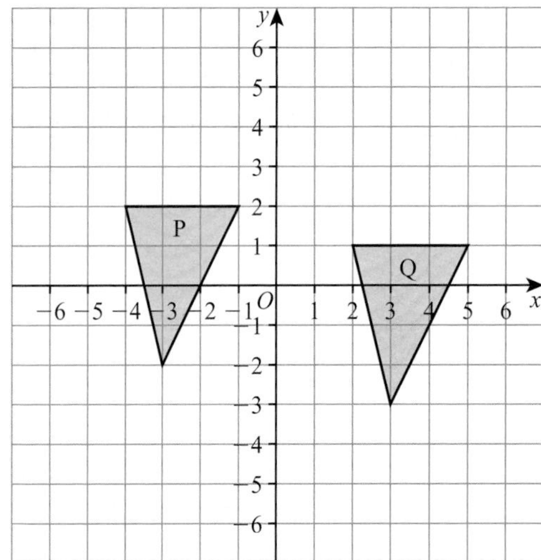

a Describe fully the single transformation that maps triangle P onto triangle Q. **(2 marks)**

b Reflect triangle P in the x-axis and label the image S.
Then translate shape S by vector $\begin{pmatrix} 6 \\ 3 \end{pmatrix}$ and label the image R. **(3 marks)**

c Describe the single transformation that maps triangle Q onto triangle R. **(2 marks)**

Exam hint
Make sure you label each triangle with the correct letter.

12 **Reflect** Adam says, 'A shape and its transformed image are always congruent.' Do you agree with this statement?
If not, give a counter example and explain your answer.

Q12 communication hint
A counter example is an example where the statement is not true.

8.5 Bearings and scale drawings

Objectives

- Draw and use scales on maps and scale drawings.
- Solve problems involving bearings.

Why learn this?

Bearings are used in plane and boat navigation as the north line is fixed.

Fluency

Convert
- 50 000 cm to metres
- 5000 m to kilometres

Warm up

1 a On a scale drawing, 1 cm represents 2 m. What does 10 cm on the drawing represent?
b On a map, 1 cm represents 10 km. What is the length on the map for a real-life distance of 25 km?

2 a Make an accurate scale drawing of this triangular garden. Use a scale of 1 cm to 1.5 m.
b What is the perimeter of the real-life garden?

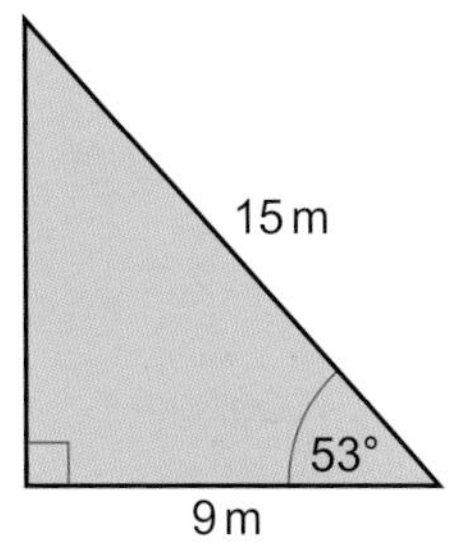

Q2 hint Use a ruler and a protractor.

3 Describe the bearing of B from A.

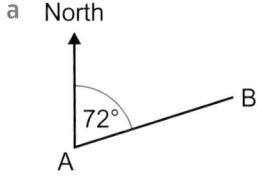

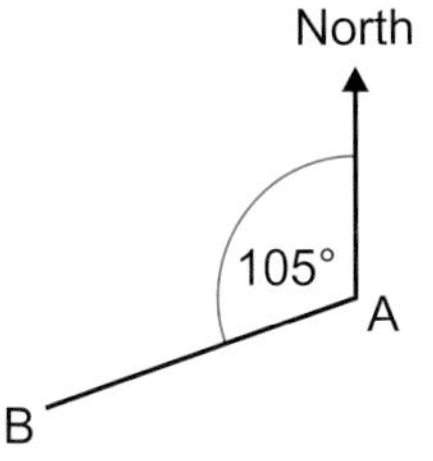

Q3 hint A bearing always has three digits, for example 090°.

4 Plot a point A. Plot a point B on a bearing of 285° from A.

5 **Real / Problem-solving** Here is a map of a town.

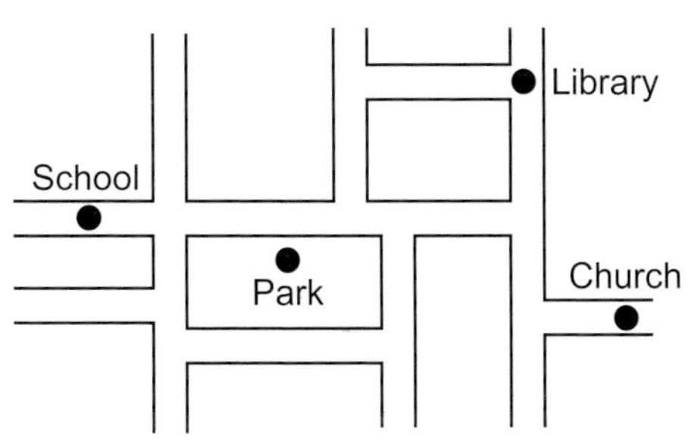

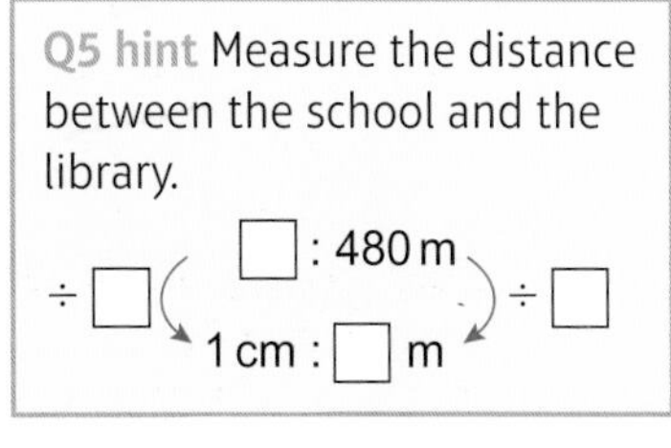

The real-life distance between the school and the library 'as the crow flies' is 480 m.
a What scale has been used on the map?
b From the map, estimate the distance as the crow flies between
i the church and the park ii the church and the school.
c John can walk 100 m in 40 seconds.
How long will it take him to walk from the library to the school?
Write your answer in minutes.

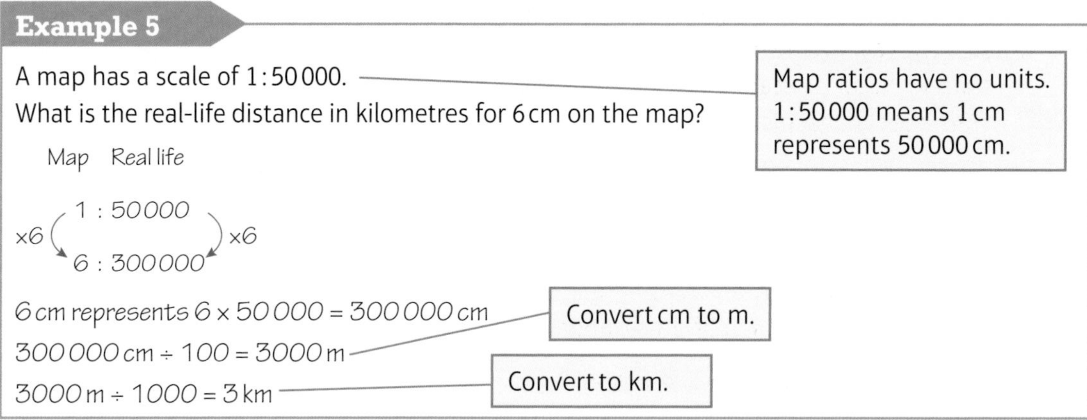

Example 5

A map has a scale of 1:50 000.
What is the real-life distance in kilometres for 6 cm on the map?

Map ratios have no units. 1:50 000 means 1 cm represents 50 000 cm.

Map   Real life

$\times 6$ 1 : 50000 $\times 6$
6 : 300000

6 cm represents $6 \times 50\,000 = 300\,000$ cm

$300\,000 \text{ cm} \div 100 = 3000$ m — Convert cm to m.

$3000 \text{ m} \div 1000 = 3$ km — Convert to km.

6 Paul is using a map with a scale of 1:50 000.
He measures these distances.
What are the distances in real life? Write your answers in kilometres.

a 10 cm b 6 cm c 2.5 cm d 0.5 cm

7 The scale on a map is 1:200 000.
What is the distance on the map, in cm, for a real distance of

a 20 km b 10 km c 8 km

Q7 hint
Map Real life
1 cm : 200 000 cm = ☐ m = ☐ km

8 **Real** The scale on a map is 1:25 000.

a On the map, the distance between two schools is 10 cm. Work out the real distance between the schools. Give your answer in km.

b The real distance between two farms is 4 km. Work out the distance between the farms on the map. Give your answer in cm.

9 **Real**

a Calculate the distance in km between
i Galway and Sligo ii Dublin and Belfast

b Which place is 180 km from Dublin and 150 km from Belfast?

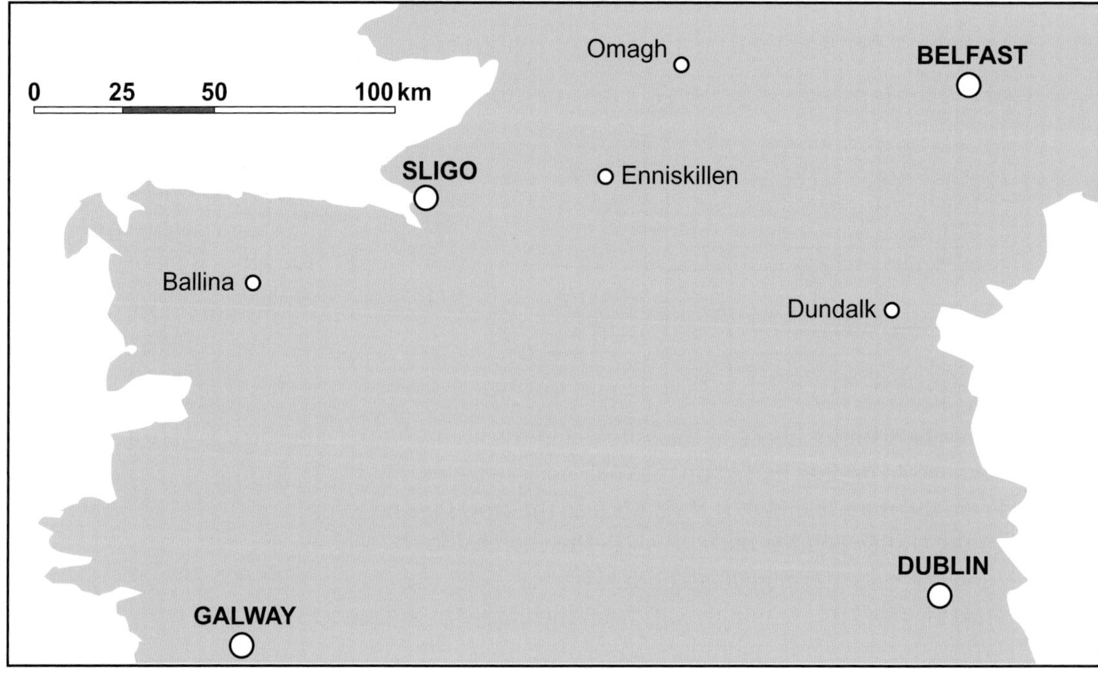

10 **Modelling** The diagram shows two satellites A and B detecting an aeroplane (C).

a Make an accurate scale drawing using a scale of 1 : 2 000 000.

b Work out the real distances AC and CB.

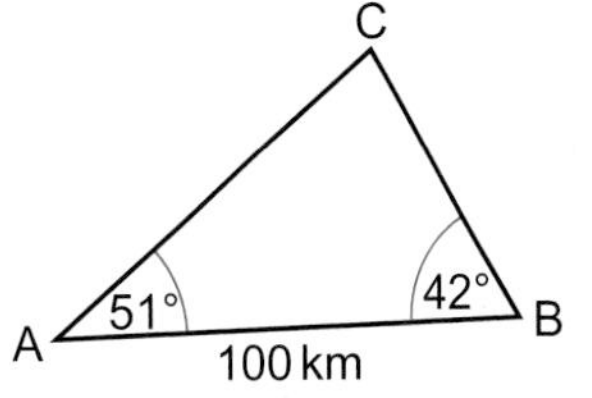

11 **Real / Problem-solving** The distance between Manchester Airport and Luton Airport is 215 km.
The bearing of Luton Airport from Manchester Airport is 135°.
Make an accurate scale map of the locations of the two airports, using a scale of 1 cm to 40 km.

> **Q11 strategy hint** Draw a sketch first.

12 **Problem-solving** A plane is 80 km west of an airport.
The plane then flies on a bearing of 050° for 120 km.

a Make an accurate scale drawing. Use a scale of 1 cm to 20 km.

b What is the bearing of the airport from the plane?

13 **Problem-solving** A ship sails for 24 km on a bearing of 060°.
It then turns and sails for 18 km on a bearing of 160°.

a Use a scale of 10 cm to 30 km to draw an accurate scale drawing of the journey of the ship.

b How far is the ship from its starting point to the nearest kilometre?

c What is the bearing the ship should sail to return to its starting point?

14 **Real / Problem-solving** The bearing of Palermo Airport from Paris Airport is 143°.
Calculate the bearing of Paris Airport from Palermo Airport.

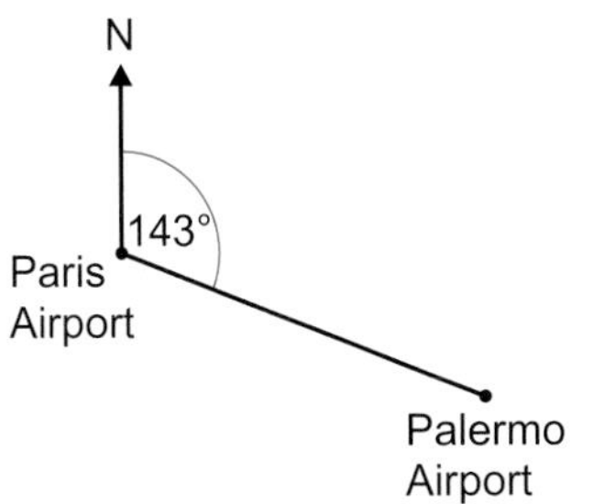

Q14 strategy hint

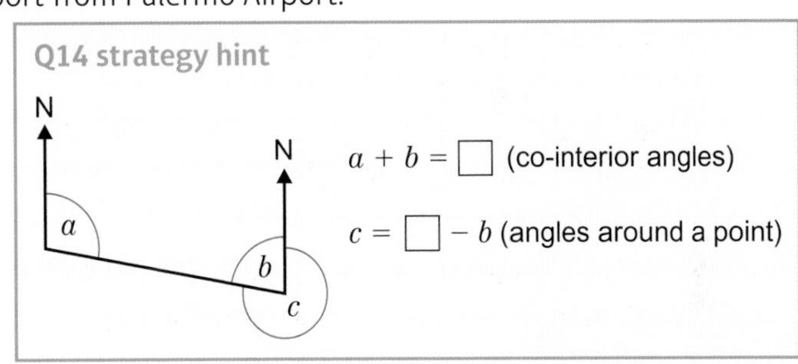

N
N
a
b
c
$a + b = \square$ (co-interior angles)
$c = \square - b$ (angles around a point)

15 **Problem-solving** a The bearing of B from A is 080°. Work out the bearing of A from B.

b The bearing of C from D is 230°. Work out the bearing of D from C.

> **Q15 strategy hint** Draw a diagram.

16 **Exam-style question**

The diagram shows the position of two boats, B and C.

Boat T is on a bearing of 060° from boat B.
Boat T is on a bearing of 285° from boat C.
Draw an accurate diagram to show the position of boat T.
Mark the position of boat T with a cross (×).
Label it T. **(3 marks)**

June 2013, Q6, 5MB3H/0

> **Q16 strategy hint** Draw each bearing from the North line clockwise. Make sure the bearing lines are long enough so that they meet.

8.6 Constructions 1

Objectives

- Construct triangles using a ruler and compasses.
- Construct the perpendicular bisector of a line.
- Construct the shortest distance from a point to a line using a ruler and compasses.

Why learn this?

Traditional architects use compasses and rulers to draw accurate scale drawings.

Fluency

What do these words mean: perpendicular, bisect, arc?

Warm up

1 Make an accurate drawing of this triangle.

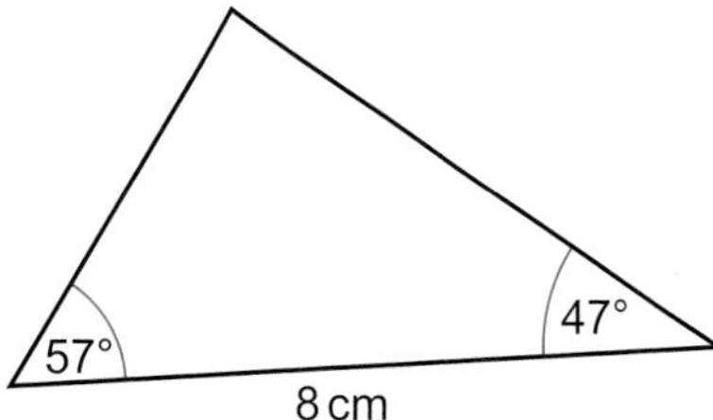

2 **Reasoning** Make an accurate drawing of a triangle with these three angles.

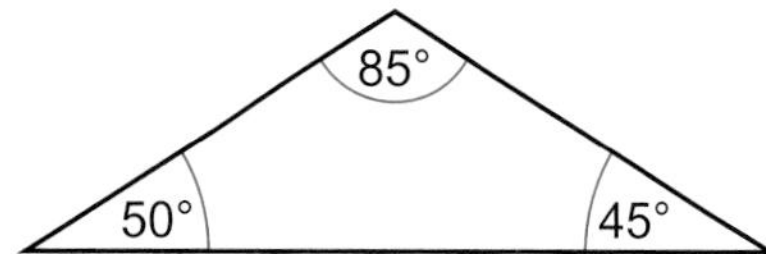

Discussion Can you draw a different triangle with the same angles?

Key point 13

To **construct** means to draw accurately using a ruler and compasses.

Example 6

Construct a triangle with sides 11 cm, 8 cm and 6 cm.

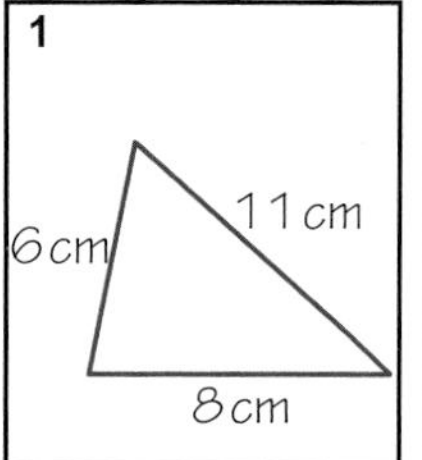

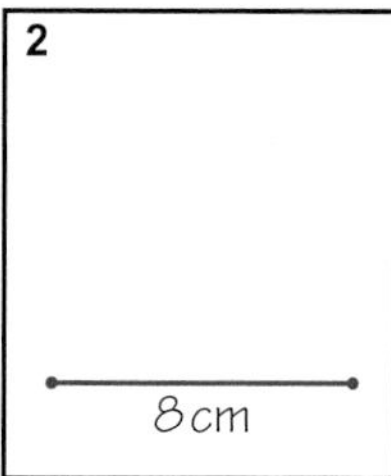

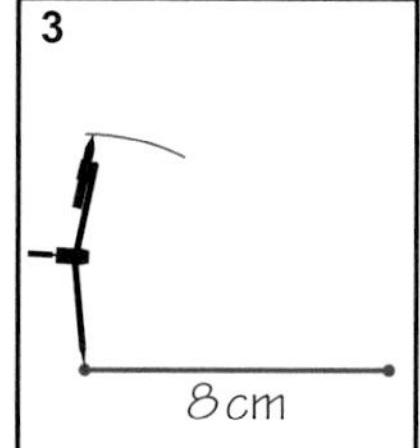

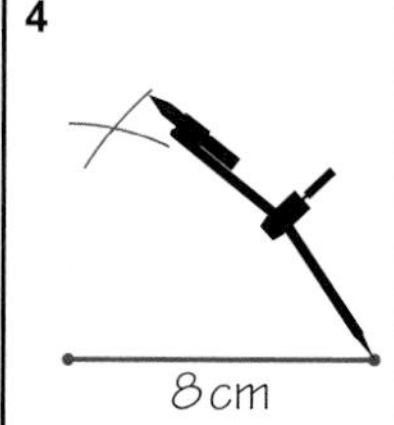

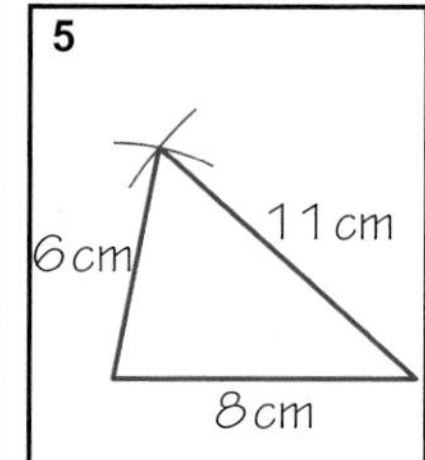

1 Sketch the triangle first.
2 Draw the 8 cm line.
3 Open your compasses to 6 cm. Place the point at one end of the 8 cm line. Draw an arc.
4 Open your compasses to 11 cm. Draw another arc from the other end of the 8 cm line. Make sure your arcs are long enough to intersect.
5 Join the intersection of the arcs to each end of the 8 cm line. Don't rub out your construction marks.

3 Construct an accurate drawing of this triangle.

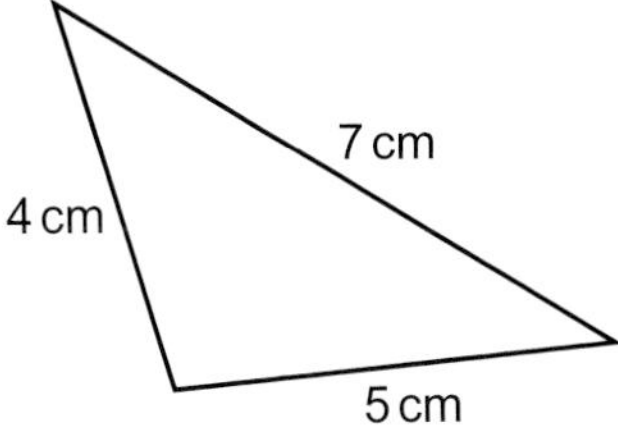

4 Construct each triangle ABC.

a AB = 5 cm, BC = 6 cm, AC = 7 cm

b AB = 10 cm, AC = 5 cm, CB = 6 cm

c AB = 8.5 cm, BC = 4 cm, AC = 7.5 cm

Q4 strategy hint Sketch each triangle first and label the lengths.

5 Construct an equilateral triangle with sides 6.5 cm.
Check the angles using a protractor.

6 **Reasoning** Explain why it is impossible to construct a triangle with sides 6 cm, 4.5 cm, 11 cm.

Q6 hint Try the construction.

7 **Real** Construct an accurate scale drawing of this skateboard ramp.
Use a scale of 1 cm to 20 cm.

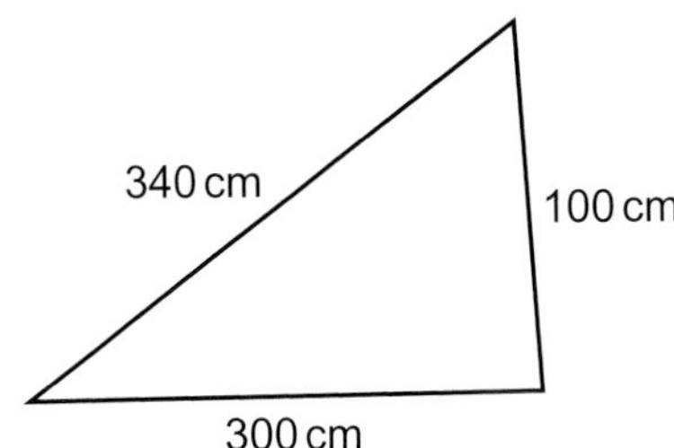

8 **Real** The diagram shows the end elevation of a house roof.
Using a scale of 1 cm to 2 m, construct an accurate scale drawing of this elevation.

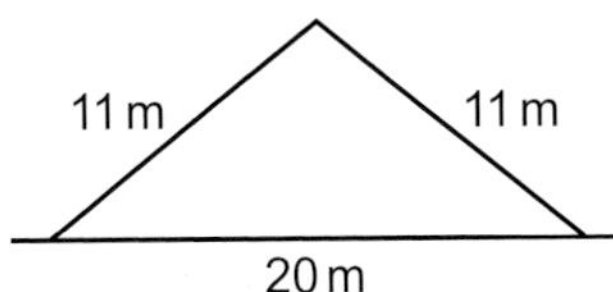

9 **Real / Problem-solving** This chocolate box is in the shape of a tetrahedron.
Each face is an equilateral triangle with side length 24 cm.
Construct an accurate net for the box.
Use a scale of 1 cm to 4 cm.

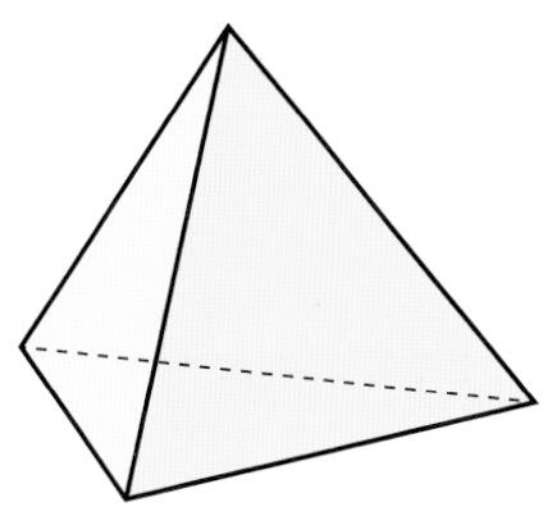

Key point 14

A **perpendicular bisector** cuts a line in half at right angles.

Example 7

Draw a line 9 cm long. Construct its **perpendicular bisector**.

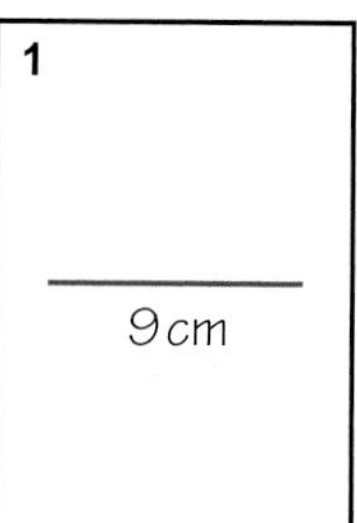

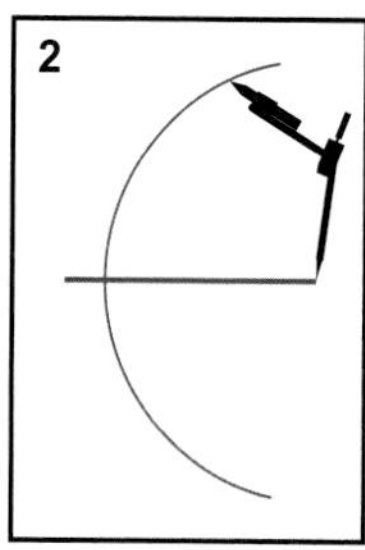

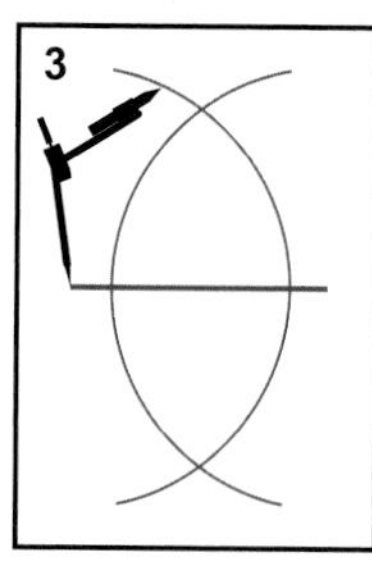

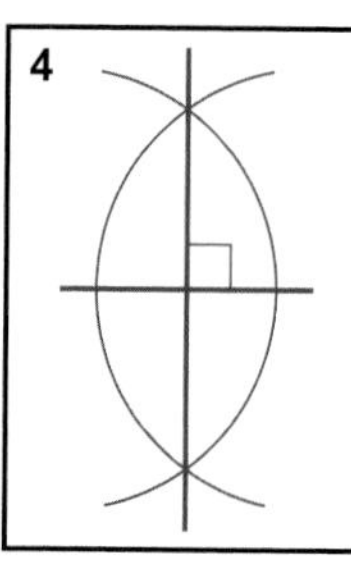

1 Use a ruler to draw the line.
2 Open your compasses to more than half the length of the line.
Place the point on one end of the line and draw an arc above and below.
3 Keeping the compasses open to the same distance, move the point of the compasses to the other end of the line and draw a similar arc.
4 Join the points where the arcs intersect.
Don't rub out your construction marks.
This vertical line is the perpendicular bisector.

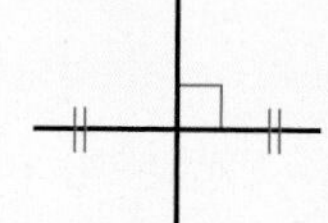

10 a Draw a line segment AB 7 cm long. Construct the perpendicular bisector of AB.
b Use a ruler and protractor to check that it bisects your line at right angles.
c Mark any point P on your perpendicular bisector. Measure its distance from A and from B.
Discussion How can you find a point the same distance from A as from B?

11 **Problem-solving** Two ships, S and T, are 50 m apart.
a Using a scale of 1 cm to 5 m, draw an accurate scale drawing of the ships.
b A lifeboat is equidistant from both ships.
Construct a line to show where the lifeboat could be.

> **Q11 communication hint**
> 'Equidistant' means 'at equal distance from'.

12 Follow these instructions to draw the perpendicular from point P not on the line to the line AB.
a Draw a line segment AB and point P not on the line.
b Open your compasses and draw an arc with centre P. Label the two points where it intersects the line AB S and T.
c Construct the perpendicular bisector of the line ST.
Discussion What is the shortest distance from P to AB?

P ×

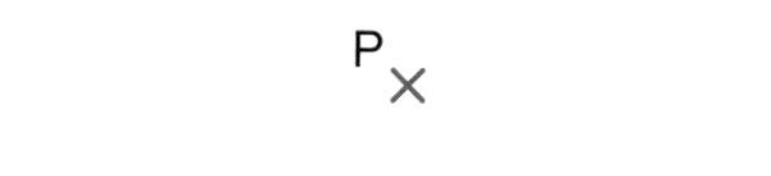

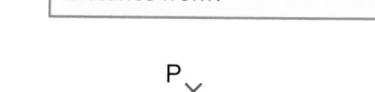

13 Follow these instructions to construct the perpendicular at point P on a line.
a Draw a line segment and point P on the line.
b Open your compasses. Put the point on P and draw arcs on the line on either side of point P. Label the points where they intersect the line X and Y.
c Construct the perpendicular bisector.

Key point 15

The shortest path from a point to a line is perpendicular to the line.

14 **Problem-solving** A swimmer wants to swim the shortest distance to the edge of a swimming pool. The scale is 1 cm to 5 m.

a Trace the diagram and construct the shortest path for the swimmer to swim to each side of the swimming pool.

b Work out the difference in the distances.

c The swimmer swims 2 m every second. How long would the shortest distance take?

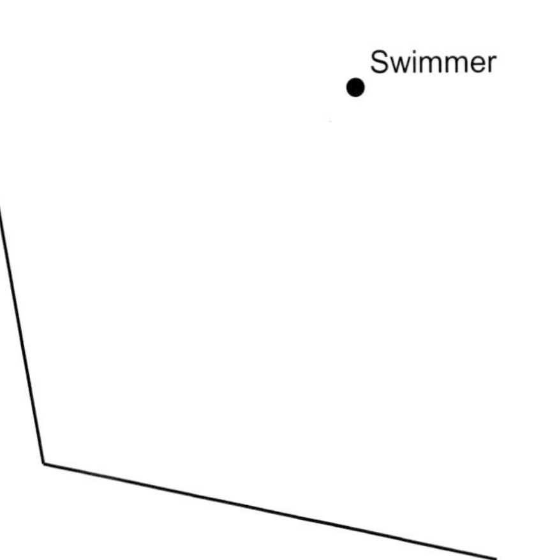

8.7 Constructions 2

Objectives

- Bisect an angle using a ruler and compasses.
- Construct angles using a ruler and compasses.
- Construct shapes made from triangles using a ruler and compasses.

Why learn this?

Constructing shapes accurately reduces errors, which can be costly and even dangerous.

Fluency

Use a protractor to draw an angle of 45°.

Warm up

1 a Construct a triangle with sides 10 cm, 8 cm, 6 cm using a ruler and compasses.
b What type of triangle have you drawn?

2 a Construct an equilateral triangle with side 5 cm using a ruler and compasses.
b What is the size of each interior angle in your triangle?

Key point 16

An **angle bisector** cuts an angle exactly in half.

Example 8

Draw an angle of 80°.
Construct the **angle bisector**.

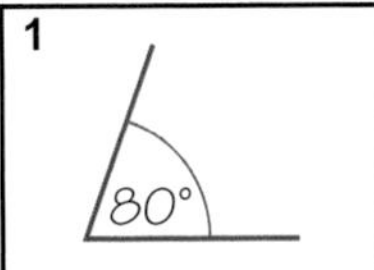

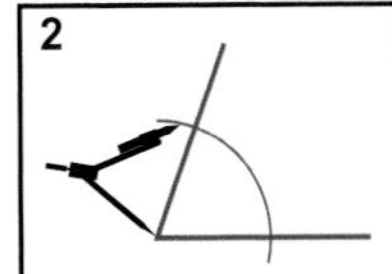

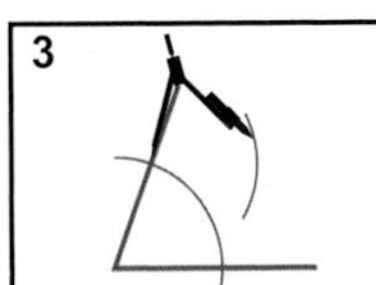

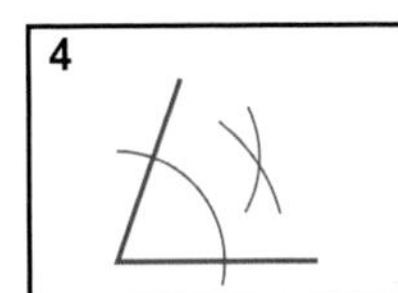

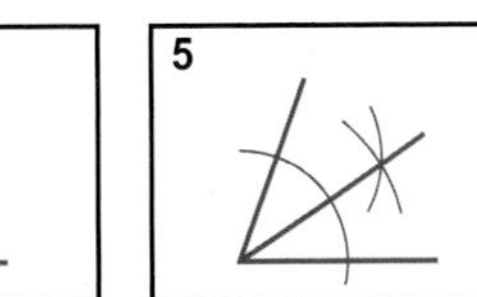

1 Draw an angle of 80° using a protractor.
2 Open your compasses and place the point at the vertex of the angle. Draw an arc that crosses both arms of the angle.
3 Keep the compasses open to the same distance. Move them to one of the points where the arc crosses an arm. Make an arc in the middle of the angle.
4 Do the same for where the arc crosses the other arm.
5 Join the vertex of the angle to the point where the two small arcs intersect. Don't rub out your construction marks. This line is the angle bisector.

3 For each angle
 i trace the angle
 ii construct the angle bisector using a ruler and compasses
 iii check your two smaller angles using a protractor.

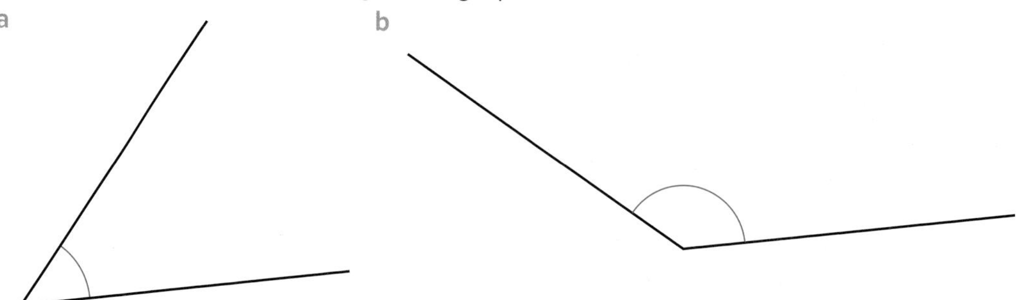

4 **Problem-solving** Use a ruler and compasses to construct these angles.
 a 90° b 45°

Q4a hint What angle will you get when you bisect a straight line?

5 **Problem-solving** Use a ruler and compasses to construct these angles.
 a 60° b 30°

6 **Problem-solving** Use a ruler and compasses to construct a 120° angle.

7 **Real / Problem-solving** A gardener wants to divide a rectangular garden into two sections. The triangular section will be decking and the rest of the garden will be grass.

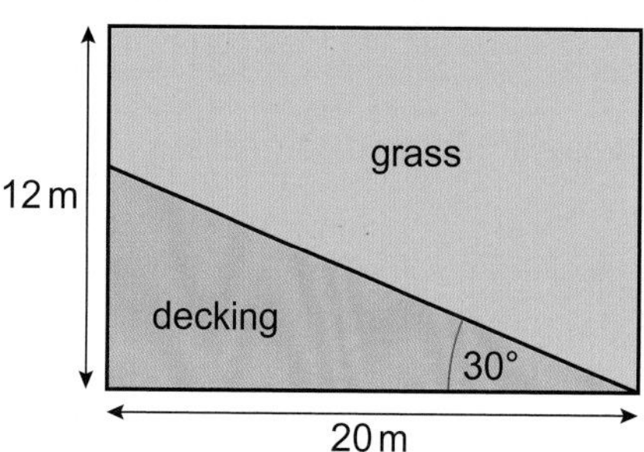

Q7a hint To draw a right angle at a vertex, you need to extend the line beyond the vertex. Then mark two points on the line an equal distance either side of the point and construct the perpendicular bisector.

 a Make a scale drawing of the rectangular garden. Use a scale of 1 cm to 4 m.
 b Use a ruler and compasses to construct an angle of 30°.
 c Calculate the area of the decking.

8 **Real / Problem-solving** Four pairs of wires connect a flagpole to the ground. The lower wire BD bisects angle ABC.

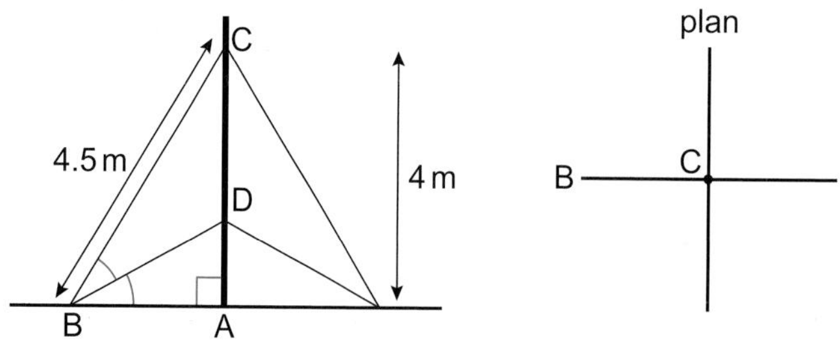

 a Construct a scale drawing. Use a scale of 1 cm to 1 m.
 b Measure the length of the wire BD.
 c How much wire is used in total?

9 Problem-solving

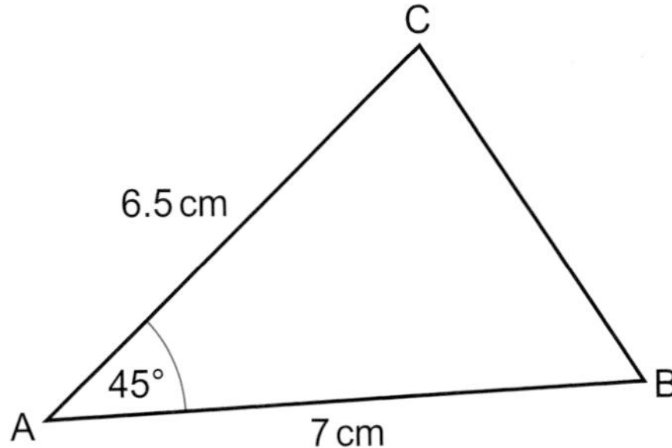

a Use a ruler, protractor and compasses to construct the triangle ABC.
b Construct a line that is perpendicular to AB and passes through C.
c Calculate the area of the triangle to the nearest cm.

10 Problem-solving Construct this trapezium made from equilateral triangles using a ruler and compasses.

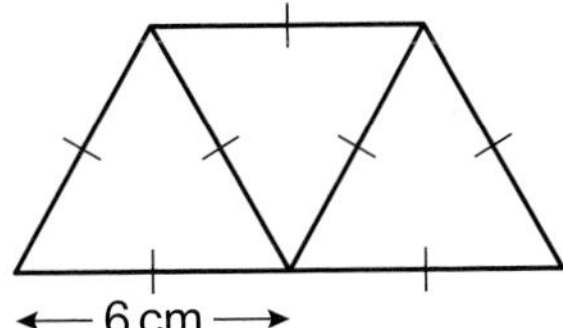

11 a Construct a triangle with sides 5 cm, 8 cm and 10 cm.

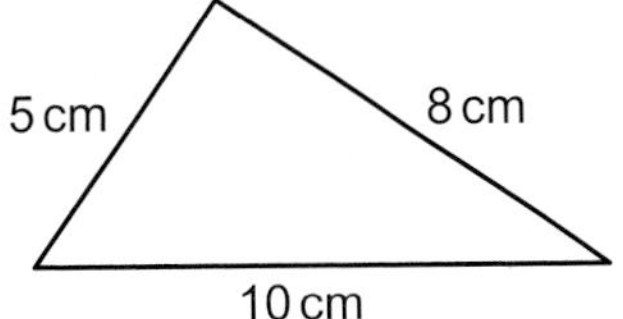

b Construct the bisector of each angle.
c The angle bisectors cross at the same point. Label this point O.
d Construct the perpendicular to one of the sides from the point you found in part **c**. Label the point where the perpendicular meets the side A.
e Draw a circle with radius OA.

What do you notice about your circle?

12 Reasoning

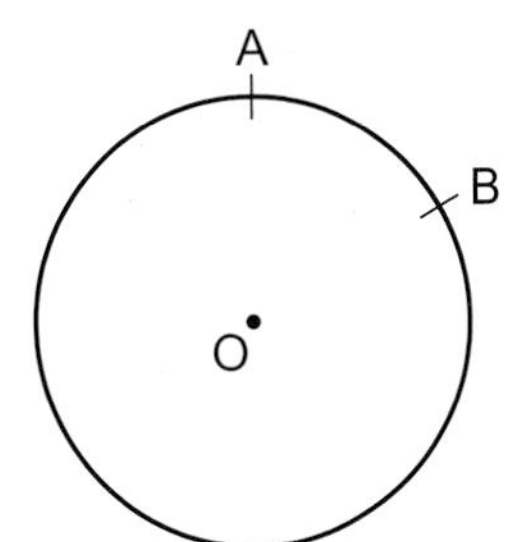

a Draw a circle with centre O and radius 5 cm. Mark a point A on its circumference.
b Keep the compasses the same size as the radius and draw an arc from point A. Label the point where the arc cuts the circle B.
c Keeping the compasses the same, repeat from point B. Repeat until you have six points on the circumference.
d Join the points and name the shape that you have drawn.

Discussion What is the size of angle AOB? Explain why.

13 Problem-solving Draw a regular octagon in a circle of radius 5 cm.

> **Q13 hint** What angle will you need to construct from the centre to two consecutive points on the circumference?

8.8 Loci

Objectives

- Draw a locus.
- Use loci to solve problems.

Why learn this?

Telephone companies use loci to plan where they will put their telephone masts.

Fluency

Are all the points on the dotted line the same distance from the solid line?

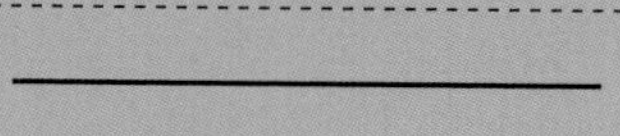

Warm up

1 Draw a small cross. Mark ten points which are 4 cm from it.
What shape do they make?

Key point 17

A **locus** is the set of all points that obey a certain rule. Often a locus is a continuous path.
A circle is the locus of a point that moves so that it is always a fixed distance from a fixed point.

2 **Real** A teacher asks some children to sit 5 m from her while she reads a story.
Sketch the locus of where the children are sitting.

3 **Problem-solving** Draw a line 6 cm long.
Draw the locus of all points which are 3 cm from the line.

Q3 hint First use compasses to draw points 3 cm from each end.

4 **Real / Problem-solving** The diagram shows a fenced area in a park.
a Draw a plan of the fenced area using a scale of 1 cm to 5 m.

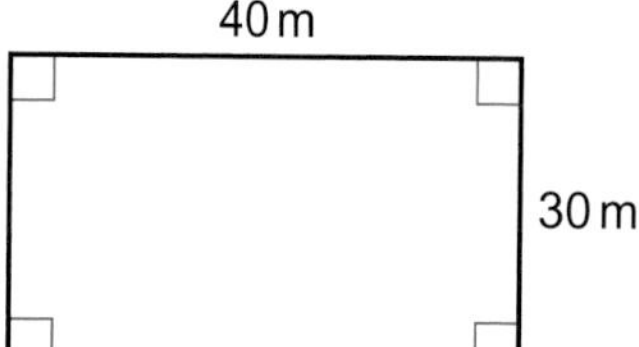

Q4 hint Think carefully about what happens at the corners.

A runner runs round the fenced area, staying exactly 10 m from the fence.
b Construct the locus of his path.

5 **Reasoning / Problem-solving**
a Draw two points 10 cm apart and label them A and B.
b i Mark a point which is 5 cm from A and 5 cm from B.
ii Mark two points which are 6 cm from A and 6 cm from B.
iii Mark two points which are 7 cm from A and 7 cm from B.
c Join the points with a straight line.
Discussion Can you use this line to show *all* the points that are equidistant from A and B?

Q5 communication hint Points that are the same distance from points A and B are **equidistant** from A and B.

Key point 18

Points equidistant from two points lie on the perpendicular bisector of the line joining the two points.

6 **Problem-solving** A library is to be built equidistant from two towns, Arton and Borham.
The towns are 2 km apart.
Using a scale of 1 cm to 250 m, construct the locus of the places where the library can be built.

Key point 19

Points equidistant from two lines lie on the angle bisector.

7 Clare wants to place a lamp in her living room so that it is equidistant from the two marked walls.

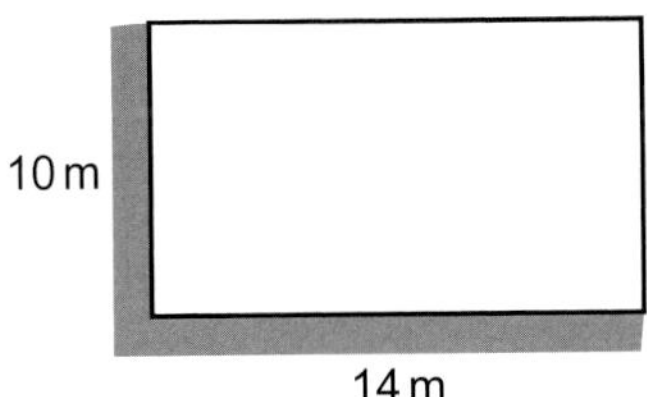

a Copy the diagram using a scale of 1 cm to 2 m.

b Construct the locus of the places where she can position the lamp.

8 **Problem-solving** This rectangle is rotated about D. Copy the diagram.

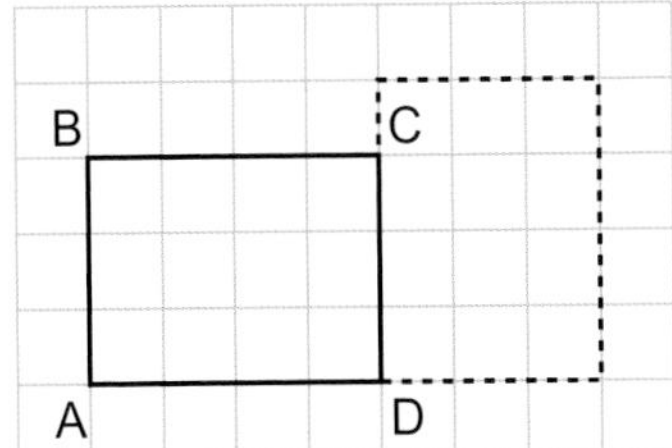

The rectangle is then rotated 90° clockwise about C. Add this to the diagram.

a Draw the locus of vertex A.

b Draw the locus of vertex B.

Example 9

A and B are two points 4 cm apart.

Shade the points that are less than 3 cm from A and less than 2 cm from B.

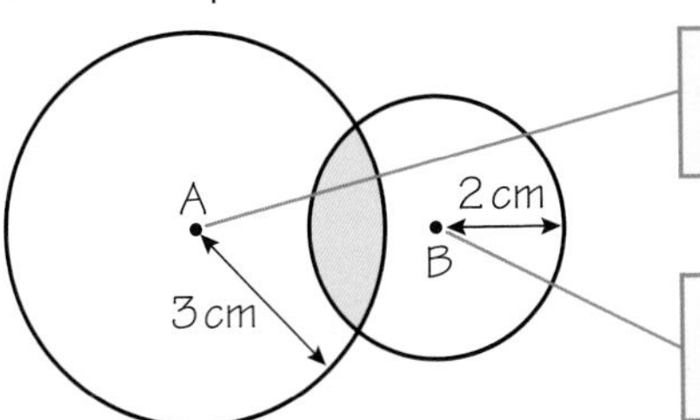

Draw a circle at A with radius 3 cm. All the points inside this circle are less than 3 cm from A.

Draw a circle at B with radius 2 cm. All the points inside this circle are less than 2 cm from B

Shade the region which satisfies both rules.

9 **Real / Problem-solving** Radio masts A and B are 120 km apart.

The bearing of radio mast B from radio mast A is 120°.

The radio masts each transmit a signal over a distance of 80 km.

Draw an accurate scale drawing of the radio masts using a scale of 1 cm to 20 km.

Shade the region which can receive signals from both radio masts.

10 **Exam-style question**

Here is a map.

The map shows two towns, Burford and Hightown.

A company is going to build a warehouse.

The warehouse will be less than 30 km from Burford **and** less than 50 km from Hightown.

Shade the region on the map where the company can build the warehouse. **(3 marks)**

Nov 2012, Q10, 1MA0/1H

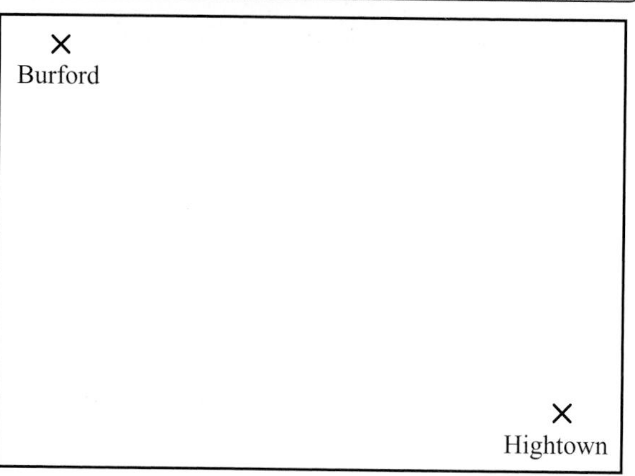

Scale: 1 cm represents 10 km

Exam hint Use a pair of compasses and make sure the pencil lines are dark enough for an examiner to see.

11 **Real / Problem-solving** Make an accurate scale drawing of this garden. Use a scale of 1 cm to 4 m.

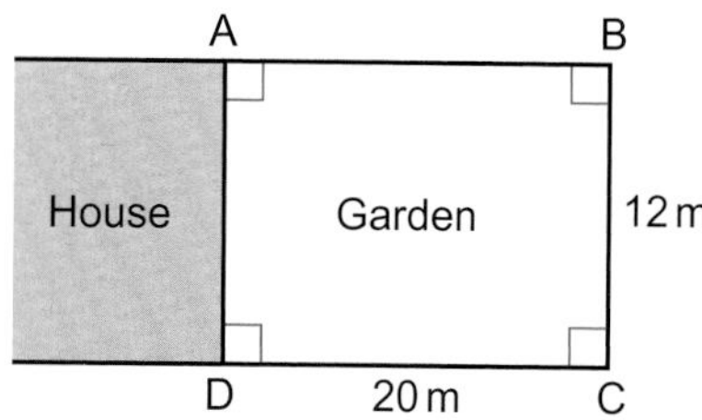

A tree can be planted between 10 m and 4 m from corner C.

It must be planted at least 14 m from the house.

Accurately shade the region where the tree could be planted.

12 **Problem-solving** ABCD is a square of side 8 cm.

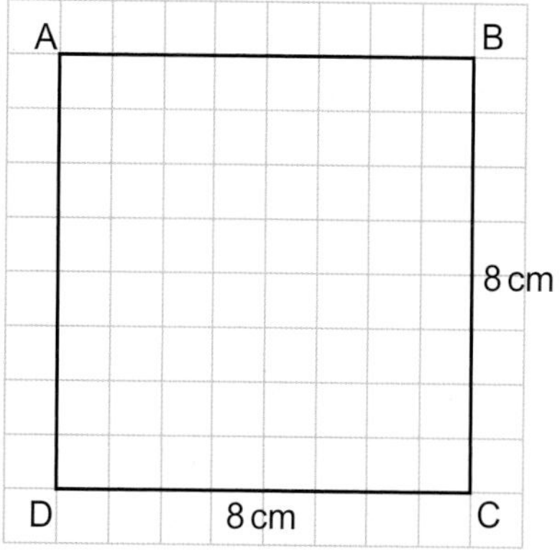

Q11 hint Use the angle bisector.

Copy the diagram. Shade the region that is less than 5 cm from A and closer to side BC than to side CD.

13 A graph $x^2 + y^2 = 16$ shows the boundary of the region covered by a fire engine, where x and y are in km.

a Draw a coordinate grid from −5 to +5 on both axes. Plot the graph.

b What area does the fire engine cover?

8 Problem-solving: Under construction

Objective • Use a ruler and compasses to construct given figures.

Engineers, architects and designers can use geometric constructions when manually drafting diagrams.

1 By combining the constructions you have met in this chapter, construct a protractor on a piece of plain paper. Your labelled angles should go up in multiples of 15 degrees and range from 0 to 180 degrees.

> **Q1 hint** Start by constructing a perpendicular bisector.

2 In lesson 8.7 **Q12** you learned how to construct a regular hexagon inside a circle.
Extend this method to construct a regular dodecagon (12 sides).

> **Q2 hint** You will need to construct three perpendicular bisectors.

3 Follow these instructions to construct a square.
- Draw one side of your square, AB. Extend this line segment through B to help you construct a perpendicular bisector at B.
- Set the width of your compasses to the length of AB. With the point of your compasses at B, draw an arc through the perpendicular bisector to find vertex C.
- Finally, set the point of your compasses at A and then C, drawing an arc each time to locate the final vertex of the square. Label the intersection of these arcs D and draw in the remaining lines.

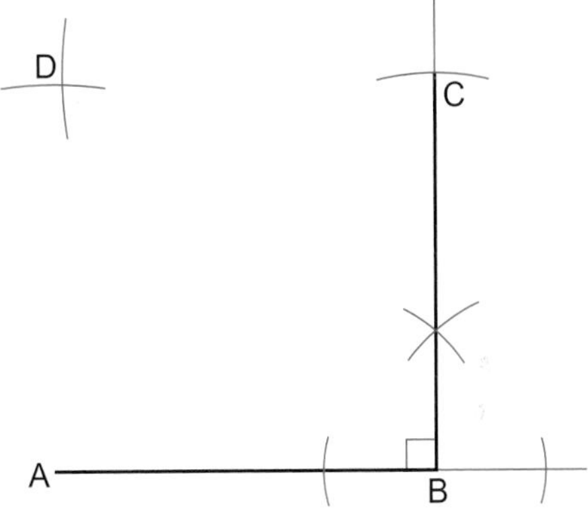

4 How could you use your square and adapt the method you used to answer **Q2** to construct a regular octagon?

> **Q4 hint** You might start by constructing a square with all of its vertices on a circle.

5 Which other regular polygons could you construct by adapting this method?

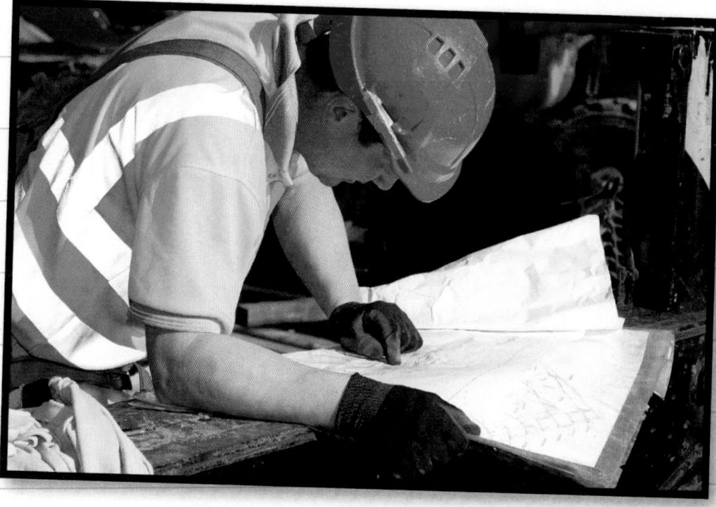

8 Check up

Log how you did on your Student Progression Chart.

Transformations

1 Enlarge this shape by scale factor $\frac{1}{3}$ with centre of enlargement S.

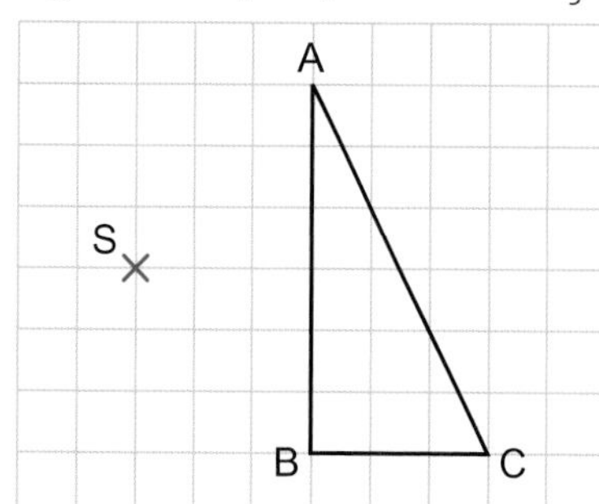

2 a Describe the reflection that maps shape P onto shape Q.

b Describe the rotation that maps shape Q onto shape R.

c Use a vector to describe the translation that maps shape R onto S.

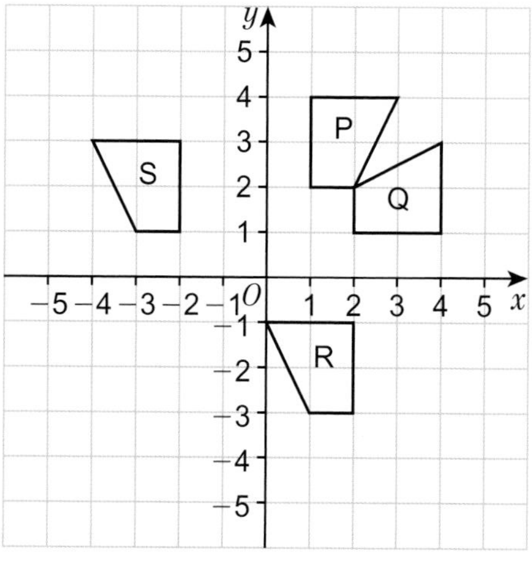

3 Describe fully the transformation that maps

a shape A onto shape B

b shape A onto shape C

c shape A onto shape D.

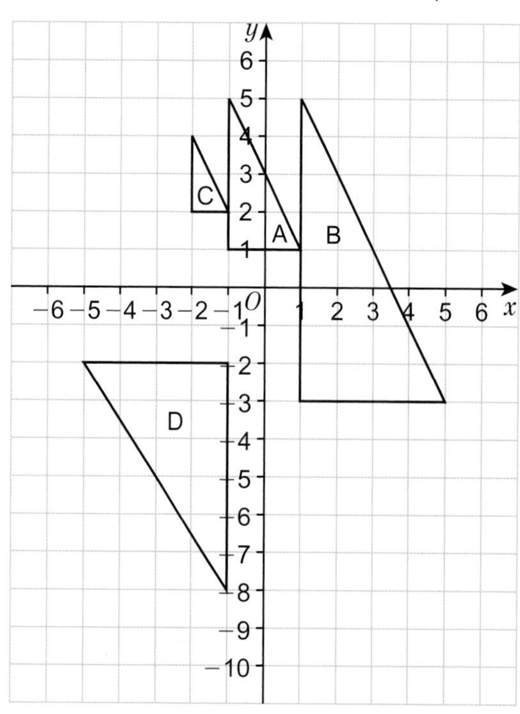

4 a Copy the diagram on a coordinate grid from −6 to +6 on both axes. Reflect shape A in the line $y = -1$. Label the image B.

b Rotate shape B by 180° about point (2, 0). Label the image C.

c Translate shape C by vector $\begin{pmatrix} -6 \\ -2 \end{pmatrix}$. Label the image D.

d Describe fully the single transformation that maps shape A onto shape D.

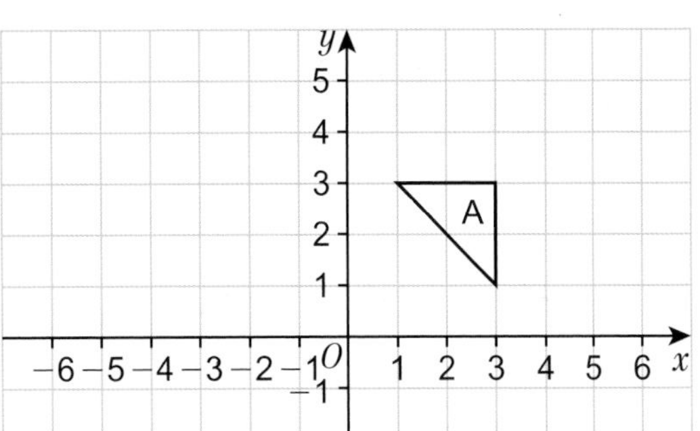

Drawings and bearings

5 Draw the plan, front elevation and side elevation of this shape.

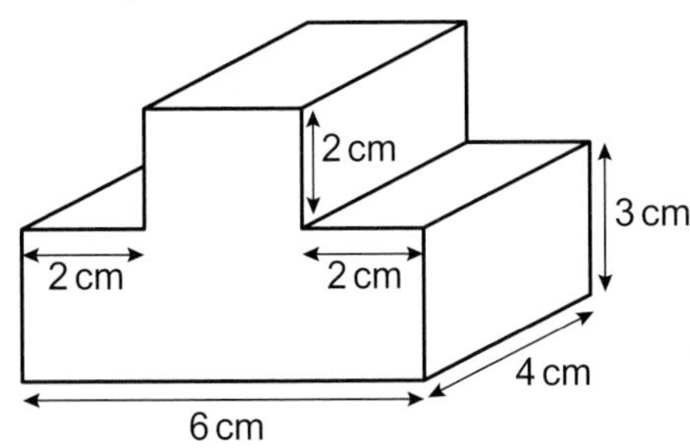

6 A map has a scale of 1:100 000. Find in cm the distance on the map for a real distance of 4 km.

7 The bearing of a ship from a lighthouse is 110°. What is the bearing of the lighthouse from the ship?

8 A plane flies 250 km from an airport on a bearing of 130°. The plane then turns and travels for 200 km on a bearing of 050°.

a Using a scale of 1 cm to 50 km, draw an accurate scale drawing of the flight of the plane.

b Find the bearing that the plane must travel on to return to the airport.

Constructions and loci

9 Draw a line 10 cm long. Construct the perpendicular bisector using a ruler and compasses.

10 Trace this angle. Bisect the angle using a ruler and compasses.

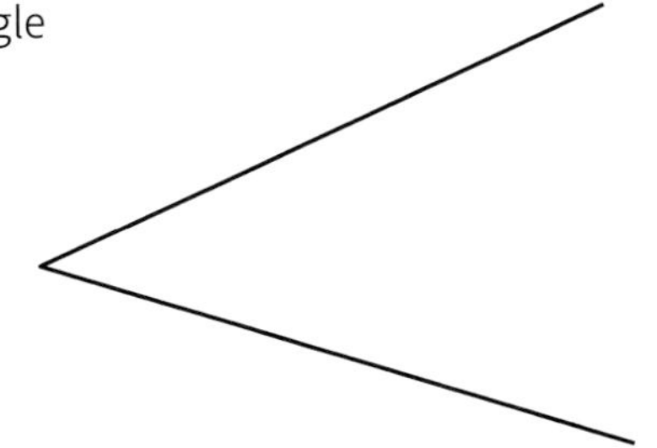

11 Three radio stations can transmit signals up to 100 km.

a Using a scale of 1 cm to 20 km, construct a triangle with the radio stations at the corners of the triangle.

b Shade the region where someone could hear all three radio stations.

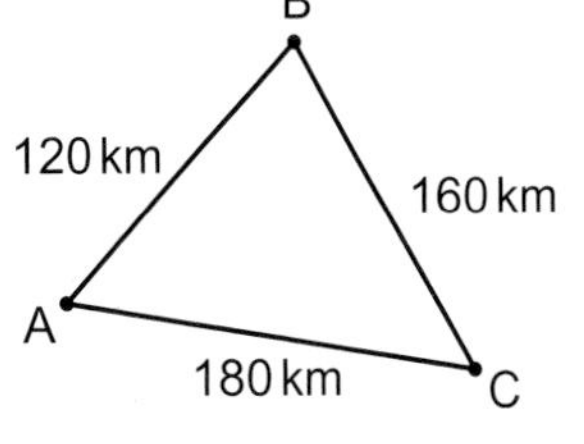

Reflect

12 How sure are you of your answers? Were you mostly

Just guessing ☹ Feeling doubtful 😐 Confident ?

What next? Use your results to decide whether to strengthen or extend your learning.

✱ Challenge

13 What regular polygons can you construct using only ruler and compasses?

8 Strengthen

Transformations

1 Joanne has started to reflect the triangle in the line $y = x$.

a Copy the diagram.

b Turn the page so the mirror line is vertical and continue the reflection.

c Trace your completed diagram. Fold your diagram along the line $y = x$. What happens to the image and the object?

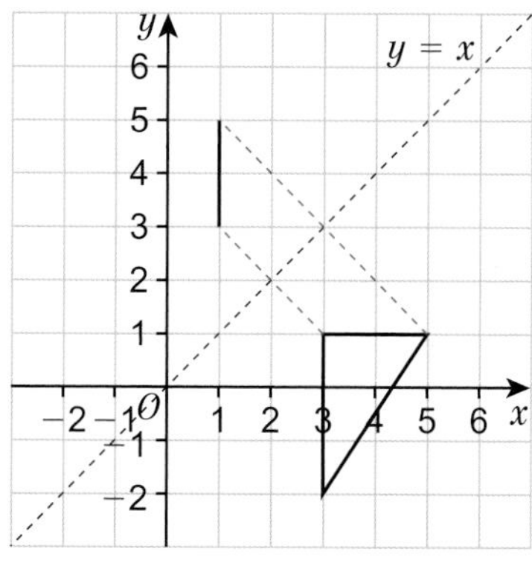

Q1 hint Reflect each vertex in the mirror line.

2 Draw a coordinate grid from −4 to +4 on both axes.

a Plot and join the points (−1, 1), (−1, 4), (1, 4), (1, 1).

b Draw the line $y = -x$.

c Reflect the shape from part **a** in the line $y = -x$.

3 Shape A is reflected to give image B.

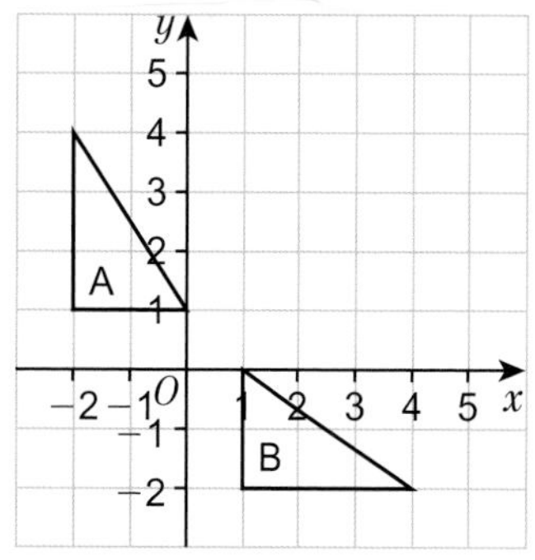

Q3 hint Join corresponding vertices on the object and the image and find the midpoints. Join the midpoints to find the mirror line.

Copy and complete this sentence.

Shape A is reflected in the line $y =$ ______ to give image B.

4 Write as a column vector

a 3 right, 2 up

b 5 right, 1 up

c 2 right, 1 down

d 3 left, 2 up

e 6 left, 3 down

f 3 right, 4 down

Q4 hint The vector is $\begin{pmatrix} \text{horizontal movement} \\ \text{vertical movement} \end{pmatrix}$.
Left ← and *down* ↓ are negative.

5 Describe the translation that takes
   a A to B
   b A to C
   c A to D
   d A to E

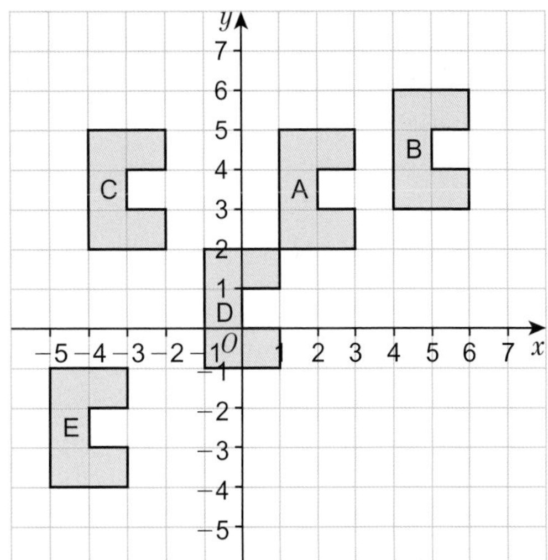

6 This triangle has been rotated through 90°.

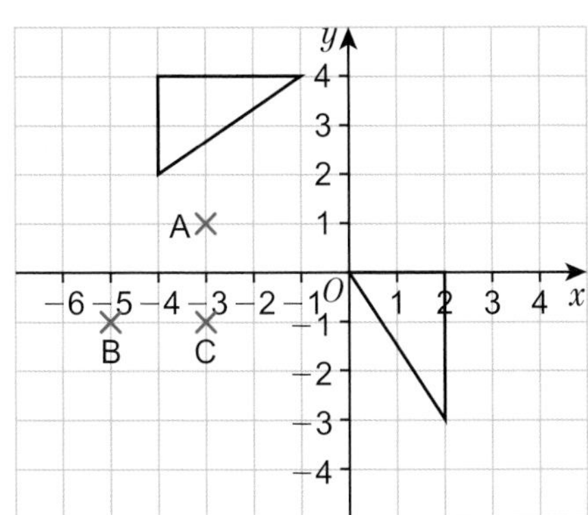

Q6 hint Use tracing paper to help.

Is A, B or C the centre of rotation?

7 Describe the rotation that takes shape A to
   a shape B
   Rotation of ____° ____________ about (____, ____)
   b shape C
   c shape D

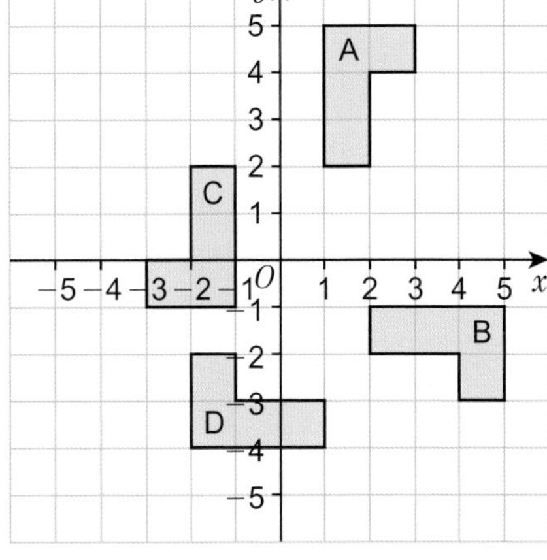

Q7 hint You need to give angle and direction of rotations and the centre of rotation. No direction is needed for rotations of 180°.

8 Fernando has started to enlarge this rectangle by scale factor $\frac{1}{2}$ about the centre of enlargement P.

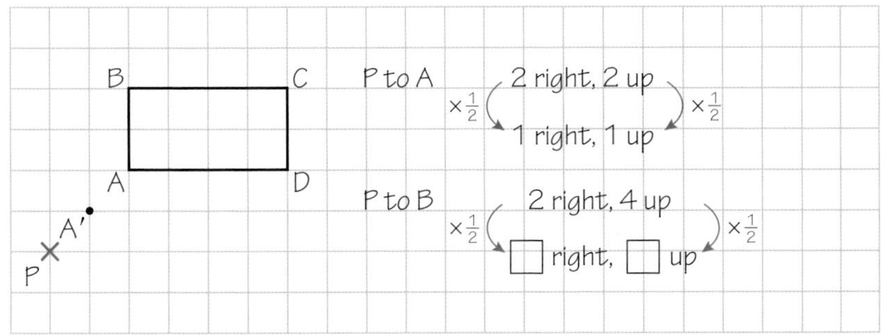

Q8 hint A′ is the position of A after the enlargement.

   a Copy the diagram.
   Work out the horizontal and vertical distances from P to points A and B.
   Halve them and mark the new points.
   b Repeat for points C and D.
   c Plot the new points and join them up.

Q8 hint Check that the lengths on the enlargement are half as long as the original.

9 Copy this diagram.
Complete the workings to enlarge the shape by scale factor −2 about centre of enlargement Y.

Q9 hint The negative sign tells you to change direction.

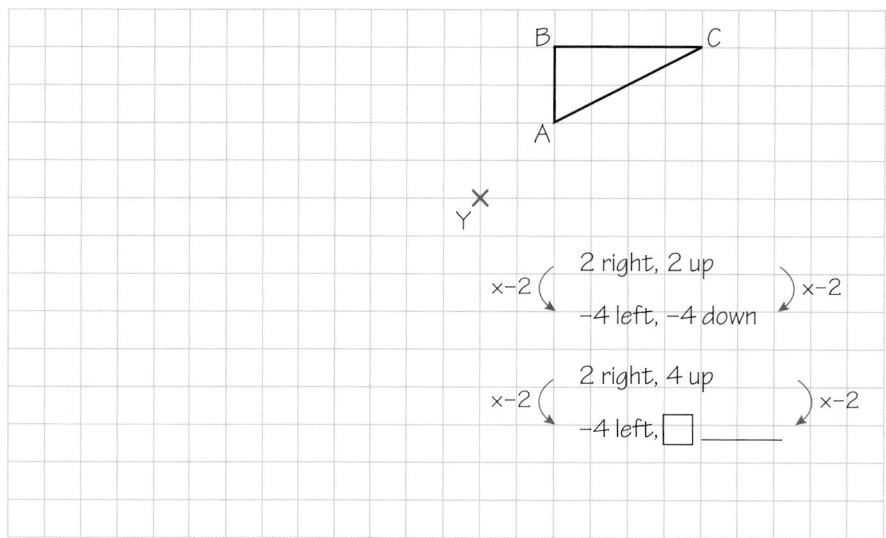

10 Copy this diagram in the top right of a 25 × 20 square grid.
Enlarge the shape by scale factor −3 about centre of enlargement O.

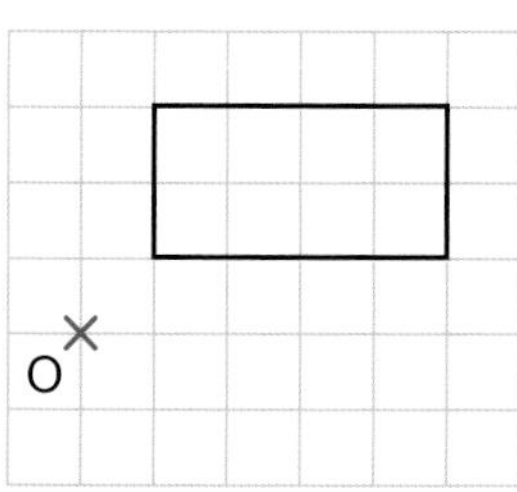

11 Copy and complete the list of information needed to describe:

Reflection line of reflection

Rotation _______, _________ and ________

Translation _______ and _________ (or _________)

Enlargement _______ and ________

Drawings and bearings

1 This solid is made from cubes.

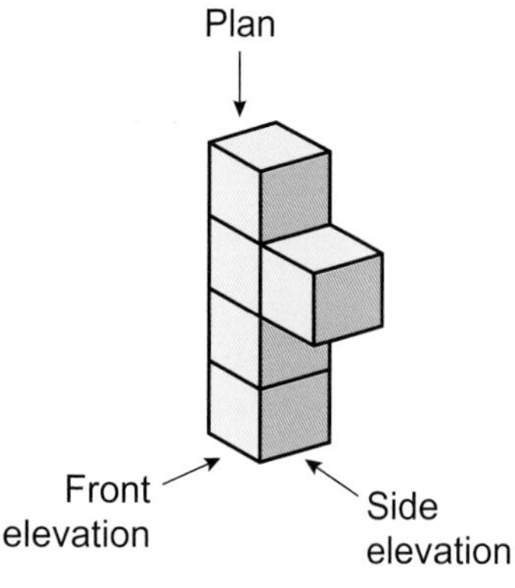

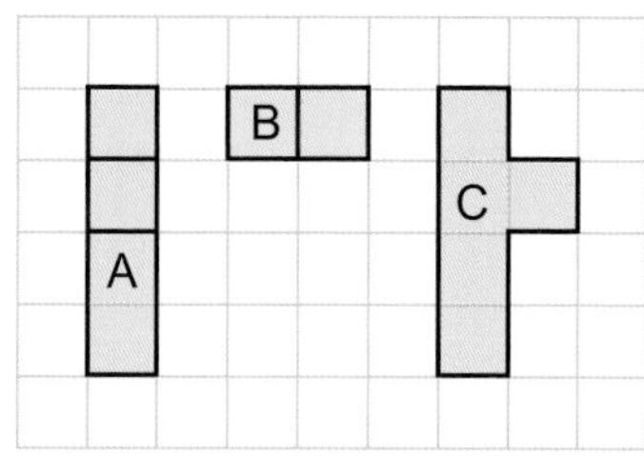

a Which is the plan?

b Which is the side elevation?

c Which is the front elevation?

2 Simeon started to draw the plan and elevations of this solid.
Copy and complete Simeon's drawings.

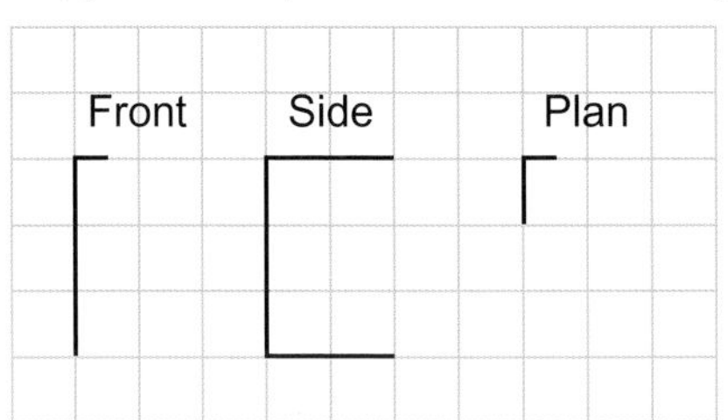

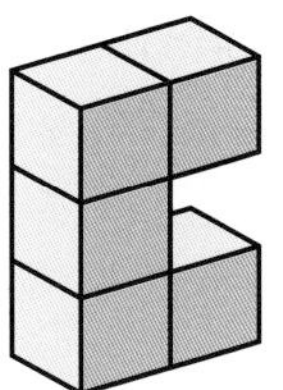

Q2 hint In the question the side elevation has been shaded.

3 A map has a scale of 1:50 000.
What real distances represent a map distance of

a 2 cm
b 5 cm
c 12 cm
d 8.5 cm?

Q3 hint
1 cm represents 50 000 cm = 0.5 km.

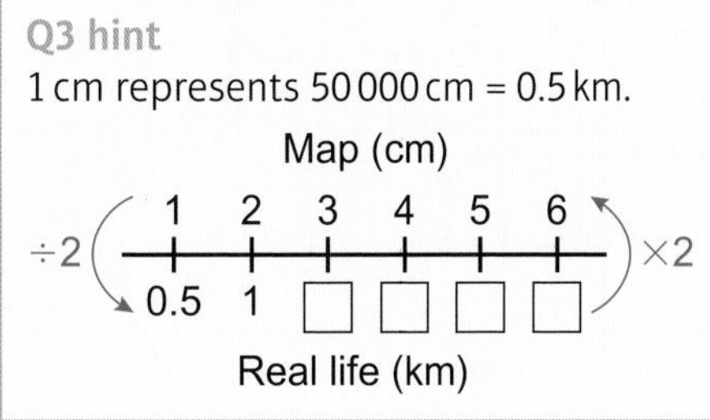

4 A map has a scale of 1:300 000.
What distances on the map represent a real-life distance of

a 9 km
b 21 km
c 30 km
d 22.5 km?

Q4 hint 1 cm represents 300 000 cm = 3 km.
Draw a number line like the one in **Q3**.

5 **Reasoning** The bearing of B from A is 115°.
Copy and complete the working to find the bearing of A from B. Give the reason for each step.

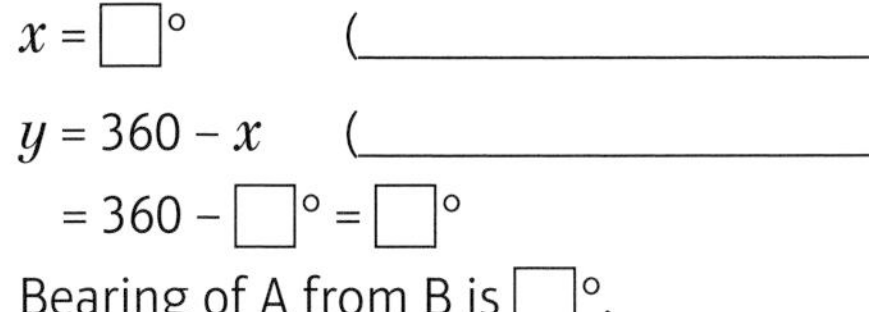

$x = \square°$ (____________________)

$y = 360 - x$ (____________________)

$= 360 - \square° = \square°$

Bearing of A from B is $\square°$.

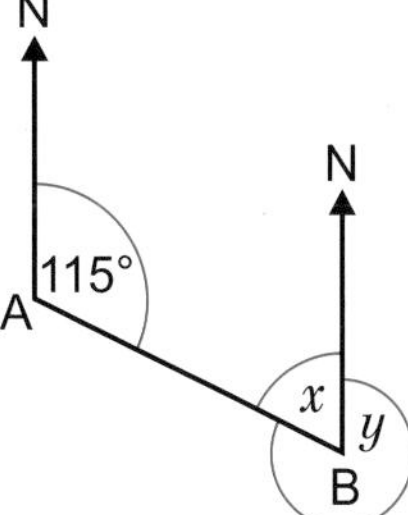

6 **Problem-solving** A ship sails 20 km from port on a bearing of 050°.
It then turns and sails for 30 km on a bearing of 160°.

a Copy and complete the sketch of the ship's journey.

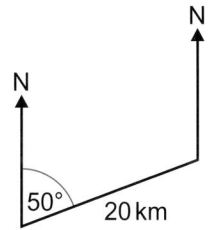

Q6a communication hint
A sketch is not an accurate drawing.

N
N
50°
20 km

b Make an accurate scale drawing using a scale of 1 cm to 5 km.
c Use your diagram to work out
  i how far the ship is from port to the nearest km
  ii the bearing the ship needs to sail on to get back to port.

Q6c hint Complete the triangle you drew in part **b**. Use the third side to find the distance and bearing.

Constructions and loci

> **Q1 hint** When you are asked to construct, use only a ruler and compasses.

1 Follow these instructions to accurately construct a triangle with sides 6 cm, 7 cm and 10 cm.

a Use a ruler to draw the 10 cm side accurately.

b The 6 cm side starts at the left-hand end of this line. Open your compasses to exactly 6 cm and draw an arc from the left-hand end of the line.

c Open your compasses to exactly 7 cm and draw an arc from the other end.

d Use the point where the arcs cross to create the finished triangle.

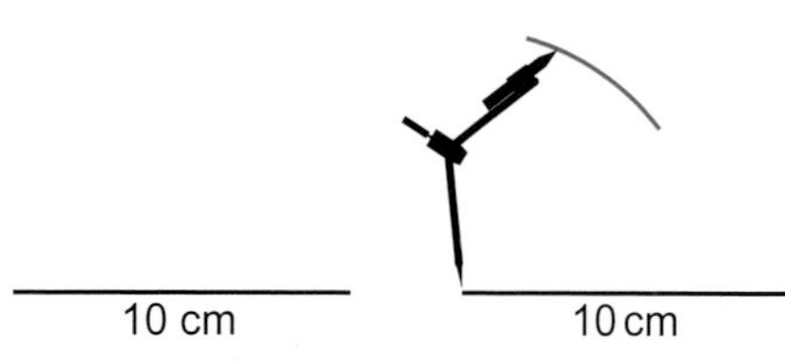

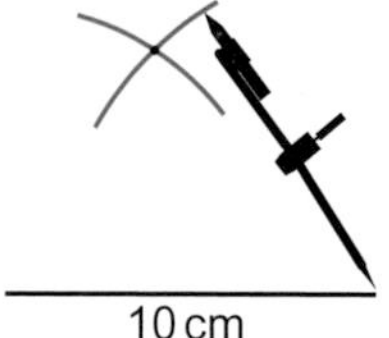

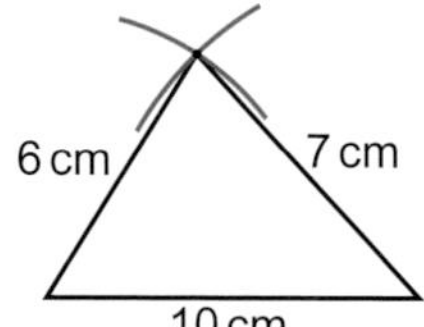

2 Construct this triangle.

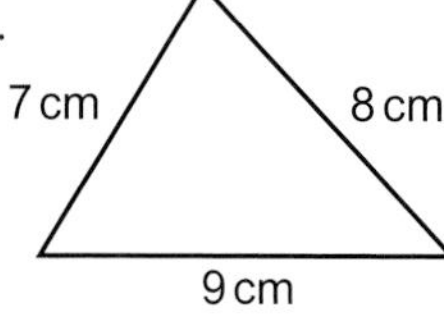

3 Draw a line 12 cm long. Follow these instructions to construct the perpendicular bisector.

a Draw the line. Open your compasses to more than half the length of the line.

12 cm

b Draw the first arc.

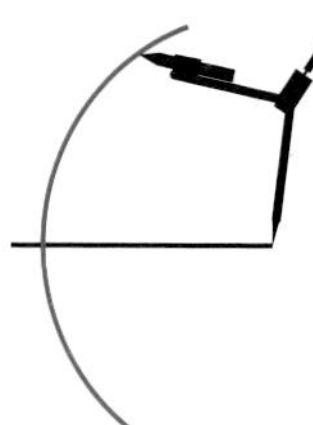

> **Q3 strategy hint** Remember this diagram.
>
> 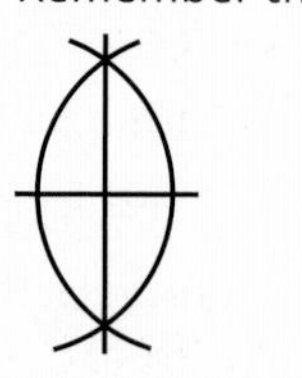

c Draw the second arc.

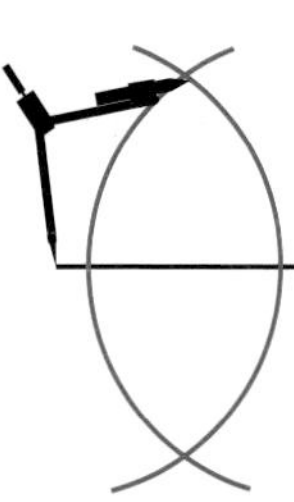

d Draw the perpendicular bisector.

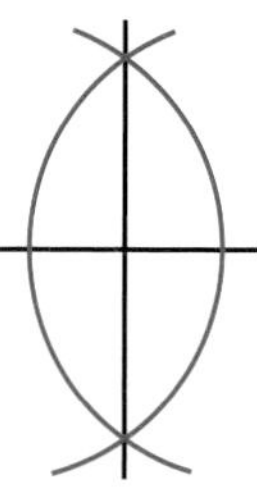

> **Q3 hint** Check by measuring that the angle is 90° and that the line is cut in half.

4 Draw a line 7 cm long. Construct the perpendicular bisector.

5 Use a protractor to draw an angle of 70°.
Follow these instructions to construct the angle bisector.

a Draw the angle.

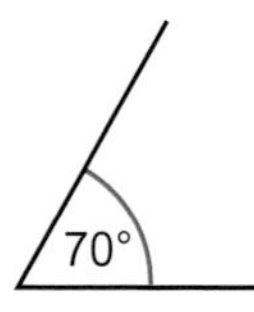

b Draw an arc from the vertex of the angle.

c Draw another arc between the two sides of the angle.

d Draw a second arc.

e Draw the angle bisector.

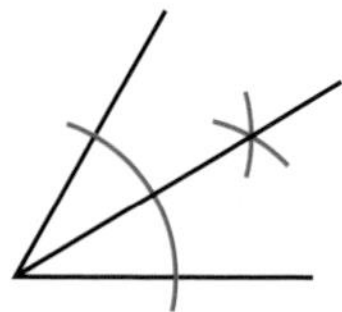

6 Use a protractor to draw an angle of 100°. Construct the angle bisector using a ruler and compasses.

Q6 strategy hint Remember this diagram. 

7 a Draw a dot in the middle of a blank piece of paper.
Draw as many dots as you can exactly 4 cm from your dot.
b What shape have you created?
c Copy and complete this sentence.
Points that are equidistant from a centre make a ________.

8 Draw two crosses 6 cm apart. Label them A and B.
a Mark all the points which are 4 cm from cross A.
Shade lightly the region that is less than 4 cm from A.
b Mark all the points which are 4 cm from cross B.
Shade lightly the region that is less than 4 cm from B.
c Shade darkly the region which is less than 4 cm from cross A and less than 4 cm from cross B.

Q8c hint Shade darkly the region that has been lightly shaded from both A and B.

8 Extend

1 **Problem-solving** Draw a coordinate grid from −5 to +5 on both axes.
Join the points (1, 1), (1, 3) and (4, 1) to make triangle M.
a Enlarge triangle M by scale factor −1 about the origin. Label the new triangle N.
b What rotation also maps triangle M onto triangle N?
c Does this always work? Try other shapes.

2 **Problem-solving** Triangle P with vertices at (−1, 0), (−1, 3), (1, 0) is transformed to give triangle Q with vertices at (−1, 0), (−1, −6), (−5, 0).
Describe fully the single transformation that maps triangle P onto triangle Q.

3 **Problem-solving** Draw a coordinate grid from −4 to +4 on both axes.
a Plot the points A (1, 2), B (3, 4), C (2, −1).
b Reflect points A, B and C in the x-axis.
c What do you notice about the coordinates of A, B and C when they are reflected in the x-axis?
d Repeat parts **a** and **b**, reflecting in the y-axis.
e What would be the coordinates of the point (p, q) if it was reflected in
i the x-axis ii the y-axis iii the x-axis, then the y-axis?

4 **Problem-solving** Draw a coordinate grid from −4 to +4 on both axes.
a Plot the points A (1, 3), B (−4, −2), C (−1, 3), D (1, −2).
b Reflect points A, B, C and D in the line $y = x$.
c Reflect points A, B, C and D in the line $y = -x$.
d What would be the coordinates of the point (p, q) if it was reflected in
i the line $y = x$ ii the line $y = -x$ iii the line $y = x$, then the line $y = -x$?

Discussion Compare your answer to part **d iii** with your answer to **Q3e iii**. What do you notice?

5 **Problem-solving** Quadrilateral A with vertices at (2, 1), (4, 1), (3, 5), (5, 5) is reflected in the x-axis to give image B.
Quadrilateral B is reflected in the y-axis to give image C.
Without drawing, work out the vertices of image C.

Q5 hint Look back at **Q3** to help you.

6 **Problem-solving** The point S(4, 3) is reflected to give point T. Point T is reflected to give point U. The coordinates of U are (–3, 4). Without drawing, find two combinations of reflections that could map point S onto point U.

7 **Problem-solving** Draw a coordinate grid from –4 to +4 on both axes.

a Plot the points A(–1,3) and B(3,–1).

b Work out the equation of line AB.

c What is the gradient of a line perpendicular to AB?

d Construct the perpendicular bisector of line AB. Check its gradient matches your answer to part **c**.

8 **Reasoning / Problem-solving** The bearing of a ship from port is a°, where $0 < a < 180$.

a Show that the bearing of the port from the ship is $(180 + a)$°.

b Work out the bearing of the port from the ship when $180 < a < 360$.

Q8 hint Draw a diagram with North arrows marked from both port and ship.

9 The diagram shows three sides of a regular hexagon. ABP is a straight line. The length of each side is 4 cm.

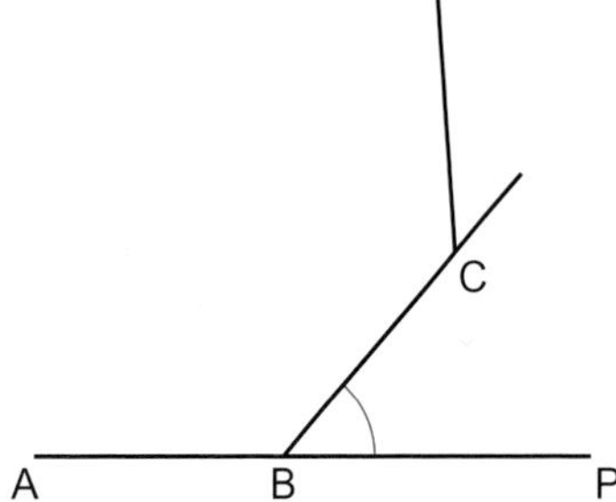

a Work out the marked exterior angle.

b Draw accurately sides AB and BC.

c Continue in the same way to draw the hexagon.

10 **Problem-solving** Here is a rectangle.

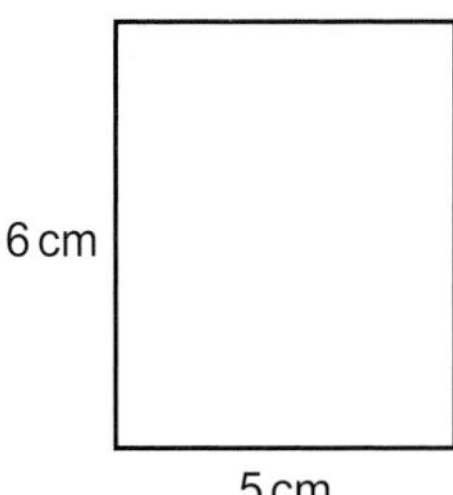

Using a ruler and compasses, construct a triangle with the same area as this rectangle.

11 **STEM**

a A firework rocket reaches a height of 100 m before exploding in all directions to a distance of 20 m. What 3D shape does the burst form?

b What is the 2D shape of the burst as seen from the ground?

12 **Exam-style question**

a Draw an accurate plan, front elevation and side elevation of this prism. **(3 marks)**

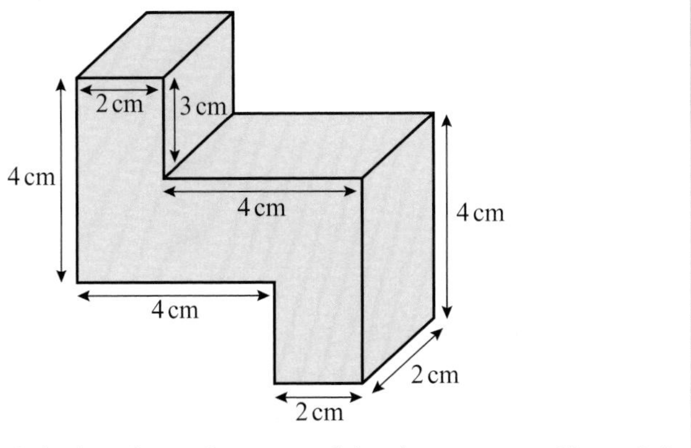

b Calculate the surface area of the shape. **(4 marks)**

Exam hint
An exam question like this will usually give you centimetre squared paper to accurately draw your answers to part **a**. Be sure to label each diagram. You can use the answers to part **a** to work out the total surface area of the shape.

13 **Problem-solving** Describe a combination of two transformations that map triangle A onto triangle B.

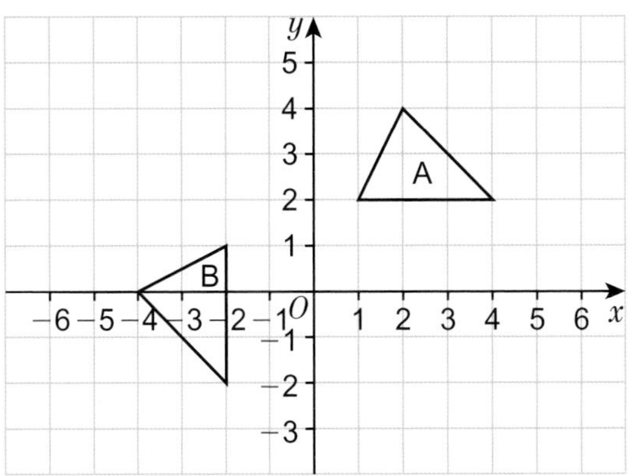

Q14 hint Is the shape congruent to the original? This will help you decide which transformations to use.

14 **Problem-solving** Draw a coordinate grid from −8 to +8 on both axes.

a Plot the points A(−5, 6) and B(3, −2).

b Find where the perpendicular bisector of AB intersects the graph $x^2 + y^2 = 25$.

8 Knowledge check

- The **plan** is the view from above an object. The **front elevation** is the view of the front of the object. The **side elevation** is the view of the side of the object. .... *Mastery lesson 8.1*
- A **transformation** moves a shape to a different position. **Reflections**, **rotations**, **translations** and **enlargements** are all types of transformation. .... *Mastery lessons 8.2 and 8.3*
- An original shape is called an **object**. When the object is reflected, rotated, translated or enlarged, the resulting shape is called an **image**. .... *Mastery lessons 8.2 and 8.3*

- To describe a **rotation** you need to give the direction of turn (clockwise or anticlockwise), the angle of turn and the **centre of rotation**. .... *Mastery lesson 8.2*
- An **enlargement** is a transformation where all the side lengths of a shape are multiplied by the same **scale factor**. .... *Mastery lesson 8.3*
- To describe an enlargement you need to give the **centre of enlargement** and the scale factor. To find the centre of enlargement, join corresponding points of the object and the image. .... *Mastery lesson 8.3*
- To enlarge a shape by a fractional scale factor, multiply the distance from the centre to each point on the shape by the scale factor. .... *Mastery lesson 8.3*
- A negative scale factor takes the image to the opposite side of the centre of enlargement. .... *Mastery lesson 8.3*
- When a shape is enlarged the area increases by (scale factor)^2. .... *Mastery lesson 8.3*
- You can describe a translation using a **column vector**. The column vector for a translation 2 squares right and 3 squares down is $\begin{pmatrix} 2 \\ -3 \end{pmatrix}$.
  The top number in the column vector gives the movement parallel to the x-axis and the bottom number gives the movement parallel to the y-axis. .... *Mastery lesson 8.4*
- The **resultant vector** is the vector that moves the original shape to its final position after a number of translations or other transformations. .... *Mastery lesson 8.4*
- In reflections, rotations and translations, the object and the image are **congruent**, as the lengths of the sides and the angles do not change. *Mastery lesson 8.4*
- In an enlargement, the object and the image are **similar**. .... *Mastery lesson 8.4*
- A **bearing** is an angle in degrees, clockwise from north. A bearing is always written using three digits. .... *Mastery lesson 8.5*
- To **construct** means to draw accurately using a ruler and compasses. ... *Mastery lesson 8.6*
- A **perpendicular bisector** cuts a line in half at right angles. .... *Mastery lesson 8.6*
- The shortest distance from a point to a line is perpendicular to the line. *Mastery lesson 8.6*
- An **angle bisector** cuts an angle exactly in half. .... *Mastery lesson 8.7*
- A **locus** is the set of all points that obey a certain rule. Often the locus is a continuous path. .... *Mastery lesson 8.8*
- The locus of a point that moves so it is always a fixed distance from a fixed point is a circle. .... *Mastery lesson 8.8*
- Points equidistant from two points lie on the perpendicular bisector of the line joining the two points. .... *Mastery lesson 8.8*
- Points equidistant from two lines lie on the angle bisector. .... *Mastery lesson 8.8*

Reflect

In this unit, you have done a lot of drawing. Write down at least three things to remember when doing drawings in mathematics. Compare your list with a classmate. What else can you add to your list?

8 Unit test

Log how you did on your Student Progression Chart.

1 Sketch the shape represented by these plans and elevations.

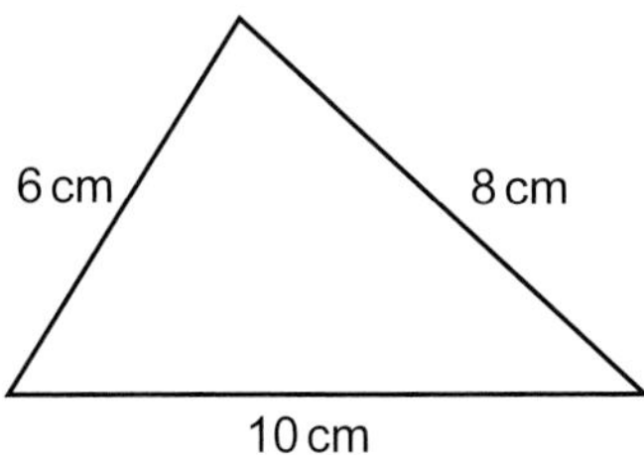

(3 marks)

2 Construct this triangle using a ruler and compasses.

6 cm 8 cm 10 cm

(3 marks)

3 Describe the transformation that maps

a A onto B *(2 marks)*

b B onto C *(3 marks)*

c C onto D *(2 marks)*

d D onto A *(2 marks)*

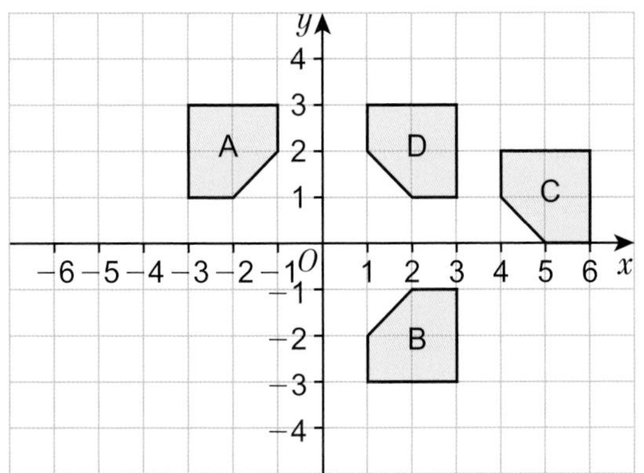

4 Copy this diagram.

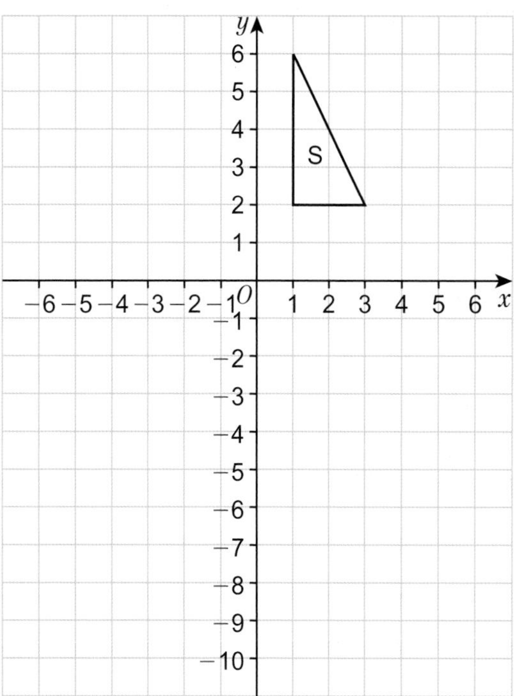

a Enlarge triangle S by scale factor $\frac{1}{2}$ about point (1, 0). Label the image T. *(2 marks)*

b Enlarge triangle S by scale factor –2 about point (0, 1). Label the image U. *(2 marks)*

5 Describe the enlargement from shape A to shape B.

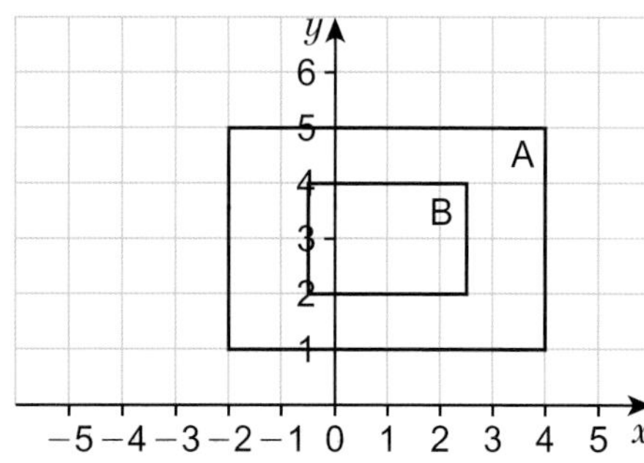

(3 marks)

6 Alistair is going build a fence around the outside of his garden.

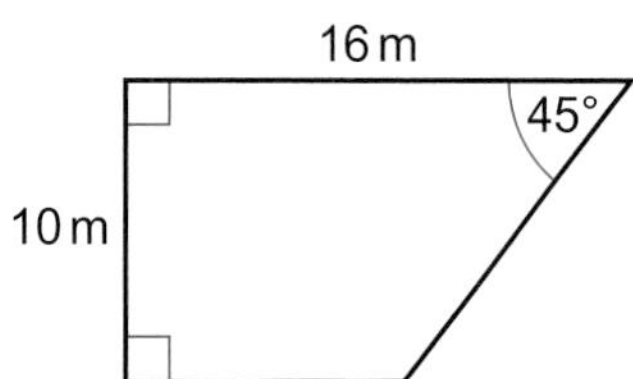

a Draw an accurate scale drawing of the garden using a ruler and compasses. Use a scale of 1 cm to 2 m. *(3 marks)*

b How long will the new fence be? *(1 mark)*

7 Copy this diagram.
Shade the region that is less than 7 cm from D and closer to BC than to AD.

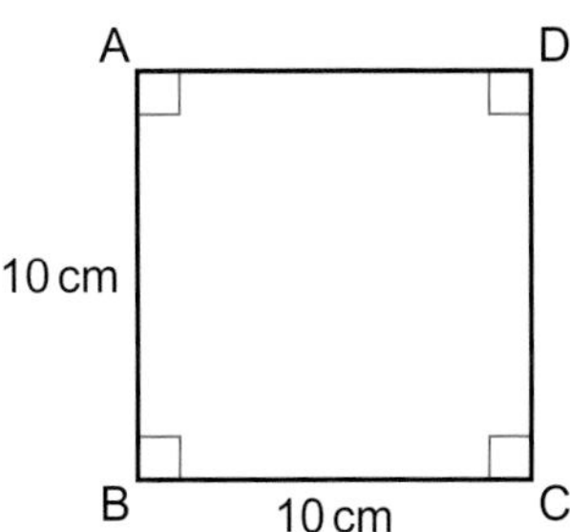

(3 marks)

8 A map has a scale of 1: 200 000.
The length on the map is 4 cm. What is the real distance in km? *(1 mark)*

9 A ship sails 120 km from the port on a bearing of 200°.
The ship turns and travels for 160 km on a bearing of 040°.

a Using a scale of 1 cm to 20 km, draw an accurate scale drawing of the path of the ship. *(2 marks)*

b What is the bearing that the ship must travel on to return to the port. *(1 mark)*

10 Point (2, 3) is reflected in the x-axis and then in the line $y = x$.
Without drawing, write the coordinates of the image. *(2 marks)*

Sample student answers

Which student gives the best answer and why?

Exam-style question

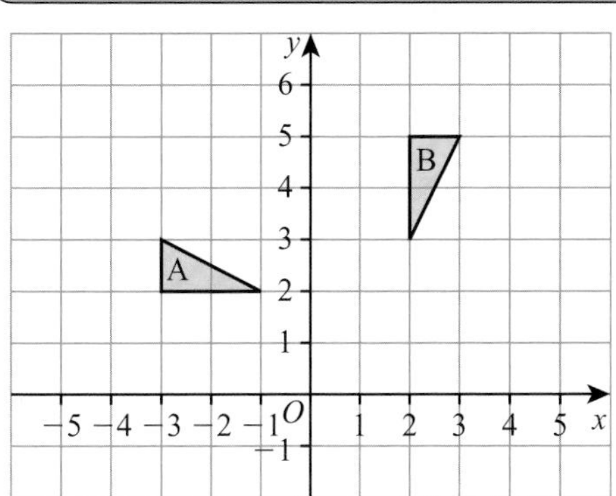

Describe fully the single transformation which maps triangle A onto triangle B. **(3 marks)**

June 2012, Q6, 5MB3H/01

Student A

Rotation, 90°

Student B

Rotation, 90° about centre of rotation (1, 1), anticlockwise

Student C

Rotation, 90° clockwise , centre (−1, 2), and then translated by the vector $\begin{pmatrix} 3 \\ 1 \end{pmatrix}$

Master p.281 | Problem-solve p.296 | Check p.297 | Strengthen p.298 | Extend p.300 | Test p.303

9 EQUATIONS AND INEQUALITIES

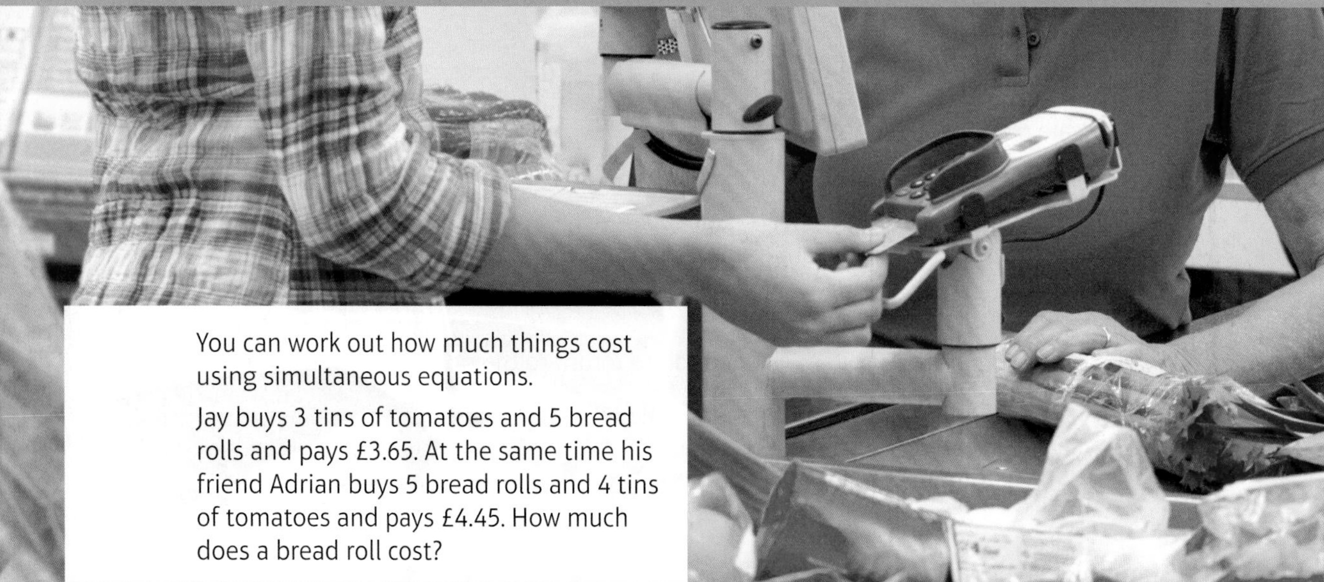

You can work out how much things cost using simultaneous equations.

Jay buys 3 tins of tomatoes and 5 bread rolls and pays £3.65. At the same time his friend Adrian buys 5 bread rolls and 4 tins of tomatoes and pays £4.45. How much does a bread roll cost?

9 Prior knowledge check

Numerical fluency

1 Which of these values of x satisfy $x > 2$?
a −3 b 4 c 0 d 2

2 Which of these values of y satisfy $y \leqslant -4$?
a 6 b 4 c 0 d −4 e −6

3 What is the value of 6^2?

4 Write two solutions to $x^2 = 144$

5 Simplify
a $\sqrt{12}$ b $\sqrt{20}$

Algebraic fluency

6 Find the value of $x^2 + 5x + 6$ when $x = -3$

7 Expand and simplify $(x - 4)(x + 3)$

8 Solve to find x
a $8x - 6 = 10$ b $9 - 3x = -6$
c $4x + 12 = 0$

9 Factorise these expressions.
a $x^2 + 8x$ b $x^2 + 4x + 3$
c $y^2 - 3y - 10$ d $x^2 - 25$ e $4 - y^2$

10 It costs £53 for 2 adults and 3 children to go ice skating.
Write an equation to show this.
Use x for adult price and y for child price.

Graphical fluency

11 This is the graph of $y = x^2 - 3x - 4$
Use the graph to solve the equation $x^2 - 3x - 4 = 0$

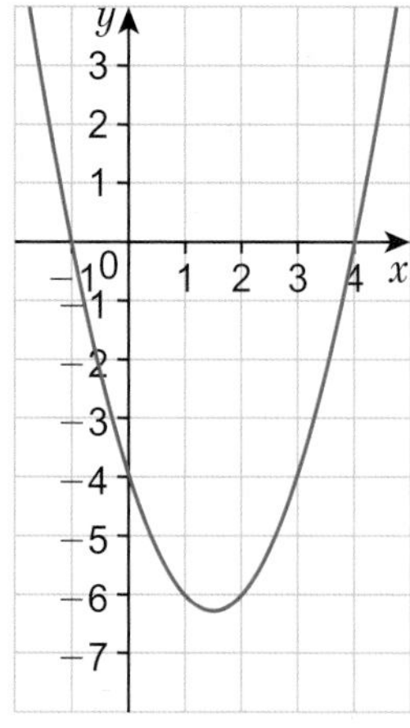

★ Challenge

12 Write down two numbers that have a product of 16 and a sum of −10

9.1 Solving quadratic equations 1

Objectives

- Find the roots of quadratic functions.
- Rearrange and solve simple quadratic equations.

Did you know?

Quadratic equations can have 0, 1 or 2 possible solutions.

Fluency

- Give two possible values of **a** $\sqrt{100}$ **b** $\sqrt{144}$ **c** $\sqrt{49}$
- What are the factors of 15?

Warm up

1 Which factor pairs of −12 have a difference of +8?

2 Factorise
a $x^2 - 5x$ b $y^2 - 4$ c $x^2 + 3x - 10$

3 Solve to find the value of z.
a $3z^2 = 108$ b $2z^2 + 1 = 33$ c $4z^2 - 100 = 0$

Questions in this unit are targeted at the steps indicated.

Key point 1

Solving a quadratic equation means finding values for the unknown that fit.

4 Find the solutions to these quadratic equations.
a $4x^2 = 64$
b $2x^2 + 3 = 101$
c $7x^2 - 175 = 0$

Q4 hint Rearrange to make x^2 the subject. Square root both sides to find two possible values of x.

Example 1

Solve $x^2 + 2x - 8 = 0$

$(x + 4)(x - 2) = 0$ — Factorise

So either $x + 4 = 0$ or $x - 2 = 0$ — The product of the factors is 0 so one or both factors equals 0

$x = -4$ or $x = 2$ — Solve the linear equations.

5 Solve
a $x^2 - 10x + 24 = 0$ b $x^2 + x - 30 = 0$
c $y^2 + 3y + 2 = 0$ d $b^2 - 3b - 10 = 0$

Q5 hint Factorise first.

Key point 2

The **roots** of a quadratic function are its solutions when it is equal to zero.

6 Find the roots of these functions.
a $x^2 - 2x$ b $x^2 - 16$ c $4 - y^2$

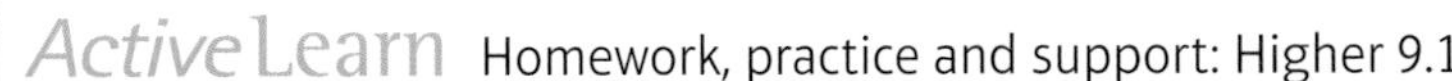

7 a Use this graph to solve the equation $x^2 + x - 6 = 0$

b Factorise $x^2 + x - 6 = 0$ to show that you get the same solution.

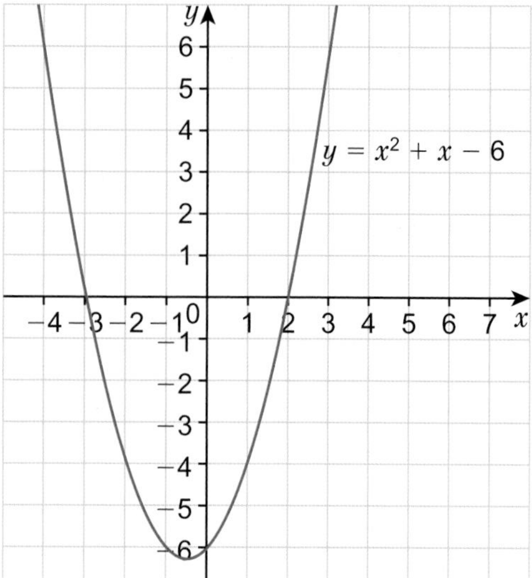

8 Solve

a $x^2 + 7x = -6$ b $x^2 + x = 12$

c $x^2 + 8 = 6x$ d $x^2 = 7x$

Q8a hint Rearrange into the form $x^2 + \square x + \square = 0$

9 **Problem-solving** Write any function that will give the roots $x = 4$ and $x = -6$.

Discussion Compare your function with other people's.
What other functions are possible?
What do you notice about them all?

9.2 Solving quadratic equations 2

Objectives

- Solve more complex quadratic equations.
- Use the quadratic formula to solve a quadratic equation.

Did you know?

The word 'quadratic' comes from the Latin 'quad' meaning square or four sided.

Fluency

- A quadratic equation must contain a squared term – true or false?
- What two values of x satisfy **a** $x^2 = 25$ **b** $x^2 = 64$ **c** $x^2 = 225$?

Warm up

1 Solve

a $x^2 + 4x + 3 = 0$ b $x^2 + 5x + 4 = 0$ c $x^2 - x - 6 = 0$

2 Expand and simplify

a $(2x + 1)(x + 3)$ b $(3x - 1)(x + 2)$ c $(2x + 2)(x - 4)$ d $(4x - 3)(x + 4)$

3 Use your calculator to evaluate

a $\frac{-3 + \sqrt{28}}{6}$ b $\frac{5 - \sqrt{12}}{10}$

4 Simplify

a $\sqrt{24}$ b $\sqrt{28}$ c $\sqrt{40}$

d $\frac{-3 + 3\sqrt{2}}{3}$ e $\frac{-4 + 2\sqrt{3}}{4}$

Q4d hint $\frac{-3}{3} + \frac{3\sqrt{2}}{3} = \square + \sqrt{2}$

5 Write and solve an equation to find x.
Discussion Why is only one of the solutions a value for x?

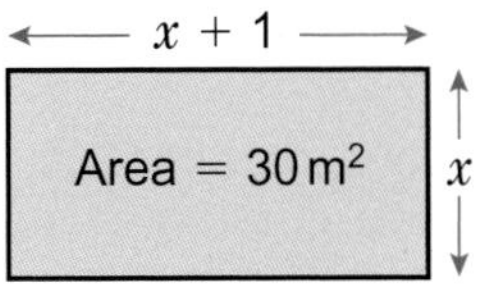

6 **Problem-solving / Modelling** Rugs come in several shapes and sizes.
A small rug has dimensions $a \times a$.
A large one has dimensions $2a \times (a + 1)$.
The area of the large rug is $12\,\text{m}^2$.
What are the dimensions of the small rug?

> **Q6 hint** Draw a diagram. Write an equation for the large rug. Solve to find a.

7 Copy and complete to factorise the expression.
$3x^2 + 5x - 2 = (3x ______)(x ______)$

> **Q7 hint** Write the factor pairs of −2
> Which factor pair fits $3 \times \square + 1 \times \square = 5$?
> Put the factors in the brackets. Then expand and simplify to check you get $3x^2 + 5x - 2$

8 Factorise these expressions.

a $5x^2 + 15x + 10$
b $2x^2 + 3x - 5$
c $4x^2 - 6x - 4$
d $3x^2 + 5x - 12$
e $2x^2 - 7x - 15$

> **Q8a hint** First look for any common factors.

9 Solve

a $(2a + 5)(a - 8) = 0$
b $(2x - 9)(3x + 12) = 0$
c $(3y - 4)(2y + 5) = 0$
d $(4b - 3)(3b - 8) = 0$

> **Q9a hint** Either $2a + 5 = 0$ or $a - 8 = 0$

10 Solve

a $2x^2 - 3x - 5 = 0$
b $3x^2 + 5x - 12 = 0$
c $4x^2 - 6x - 18 = 0$
d $6x^2 + 9x - 15 = 0$

> **Q10a hint** Factorise first by removing the common factor.

11 **Problem-solving** Gerri has a patio that is $4\,\text{m} \times 5\,\text{m}$. She has $10\,\text{m}^2$ of turf and wants to use it to make a border round the patio that is the same width all round.

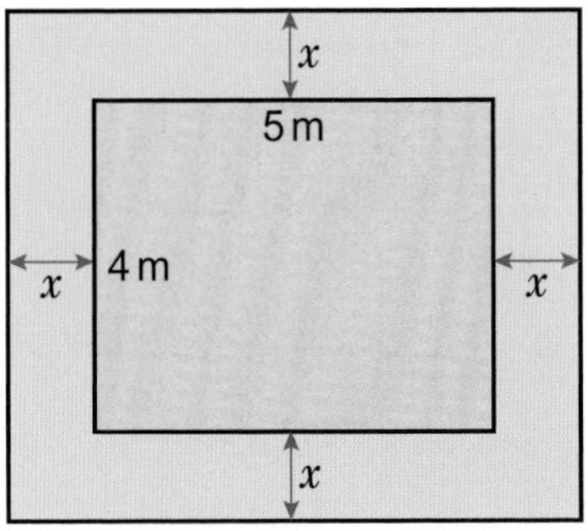

a Write an equation for the area of the grass border.
b Gerri uses all her turf. Solve to find x.

12 **Exam-style question**

Solve, by factorising, the equation $8x^2 - 2x - 21 = 0$
(3 marks)

> **Exam hint**
> Expand the brackets in your answer to check you get back to the original equation.

Key point 3

You can use the **quadratic formula**

$$x = \frac{-b \pm \sqrt{b^2 - 4ac}}{2a}$$

to find the solutions to a quadratic equation $ax^2 + bx + c = 0$

Example 2

Solve $x^2 + 4x + 2 = 0$. Give your solutions in surd form.

$a = 1, b = 4, c = 2$ — Compare with $ax^2 + bx + c$. Write the values of a, b and c.

$x = \frac{-4 \pm \sqrt{4^2 - 4 \times 1 \times 2}}{2 \times 1}$ — Substitute a, b and c into the quadratic formula.

$= \frac{-4 \pm \sqrt{16 - 8}}{2}$

$= \frac{-4 \pm \sqrt{8}}{2}$ — You are asked to give your solutions in surd form, so simplify the surds.

$= \frac{-4 \pm \sqrt{4}\sqrt{2}}{2}$

$= \frac{-4 \pm 2\sqrt{2}}{2} = -2 \pm 2\sqrt{2}$ — ± means 'plus or minus'. + gives one solution and – gives the other.

The solutions are $x = -2 + \sqrt{2}$ and $x = -2 - \sqrt{2}$

13 Solve, giving your solutions in surd form

a $x^2 + 5x + 5 = 0$ b $x^2 + 7x + 2 = 0$ c $x^2 + 2x - 2 = 0$

d $x^2 + 2x - 6 = 0$ e $3x^2 + 9x + 5 = 0$

14 Solve, giving your solutions to 2 decimal places

a $x^2 + 6x - 10 = 0$ b $2x^2 - 5x - 6 = 0$

c $3x^2 + 2x - 2 = 0$ d $2x^2 + 3x - 8 = 0$

Q14 hint Use the quadratic formula and a calculator to find the solutions. Round to 2 decimal places.

15 Solve $2x^2 - 7x - 15$

a by finding factors of –15 b by using the quadratic formula.

Discussion Does it matter which method you use? When is it better to use the formula?

16 **Exam-style question**

Solve $5x^2 + 6x - 2 = 0$

Give your solutions correct to 2 decimal places. **(3 marks)**

June 2013, Q18, 5MB3H/01

Q16 strategy hint The instruction to give your solutions correct to 2 decimal places is a hint that you cannot factorise the equation.

Exam hint

You must know the quadratic formula in the exam. Always make sure you give your answer exactly as the exam question asks.

9.3 Completing the square

Objectives

- Complete the square for a quadratic expression.
- Solve quadratic equations by completing the square.

Why learn this?

Completing the square can help you find the maximum or minimum point of a quadratic curve.

Fluency

Which of these expand and simplify to give an x term of $2x$?

$(x + 4)^2$ $(x + 2)^2$ $(x + 1)^2$ $(x - 2)^2$

Warm up

1 Expand and simplify

a $(x + 4)^2$ b $(x - 3)^2$ c $(2x + 3)^2$

d $(x + 2)^2 + 4$ e $(x + 1)^2 + 7$

2 Simplify these surds.

a $\sqrt{45}$ b $\sqrt{32}$ c $\sqrt{48}$ d $\sqrt{90}$

3 Solve, giving your answers in surd form

a $x - 1 = \pm\sqrt{3}$ b $x + 2 = \pm\sqrt{2}$

c $x - 7 = \pm\sqrt{5}$ d $x - 4 = \pm\sqrt{3} + 1$

Q3a hint $x = \square + \sqrt{3}$, $x = \square - \sqrt{3}$

4 Write a quadratic expression for the area of the large square.

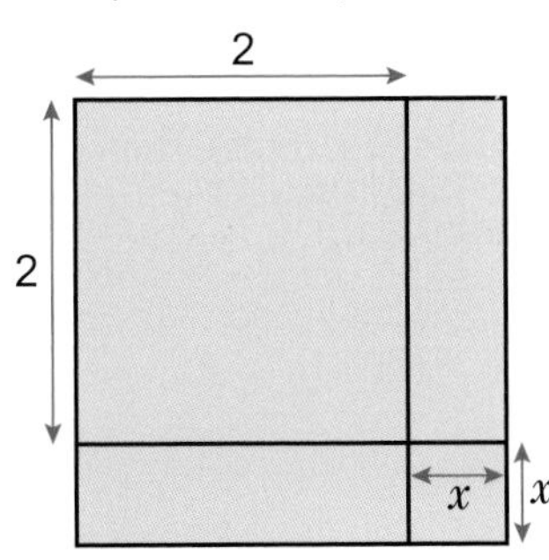

Q4 hint Write an expression with two brackets. Expand to get a quadratic expression.

Key point 4

Expressions such as $(x + 2)^2$, $(x - 1)^2$ and $(x + \frac{1}{2})^2$ are called **perfect squares**.

5 Write these expressions in the form $(x + 2)^2 + \square$ or $(x + 2)^2 - \square$

a $x^2 + 4x + 5$ b $x^2 + 4x + 6$ c $x^2 + 4x - 1$

Q5a hint Compare the expression with your answer to **Q4**.

6 Write these as perfect squares.

a $x^2 + 6x + 9$ b $x^2 + 8x + 16$ c $x^2 + 10x + 25$ d $x^2 + 12x + 36$

Example 3

Write $x^2 + 2x + 7$ in the form $(x + p)^2 + q$

$[x^2 + 2x] + 7$ — Separate the x terms from the constant.

$x^2 + 2x \equiv (x + 1)^2 - 1$ — Find the perfect square which will give the correct x^2 and x terms, then subtract the constant to make the identity true.

So $[x^2 + 2x] + 7 = [(x + 1)^2 - 1] + 7$ — Substitute the identity into the original expression.

$= (x + 1)^2 + 6$ — Simplify the expression.

So $p = 1$ and $q = 6$ — Compare $(x + 1)^2 + 6$ with $(x + p)^2 + q$ and write down the values.

Key point 5

$x^2 + bx + c$ can be written in the form $\left(x + \frac{b}{2}\right)^2 - \left(\frac{b}{2}\right)^2 + c$.

This is called **completing the square**.

7 Write these in the form $(x + p)^2 + q$

a $x^2 + 2x - 1$ b $x^2 + 8x$ c $x^2 + 12x$ d $x^2 + 6x + 11$ e $x^2 - 4x + 6$

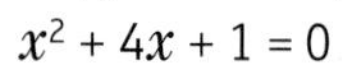

8 Copy and complete to solve the quadratic equation, giving your answer in surd form.

$x^2 + 4x + 1 = 0$

$(x + \square)^2 - \square + 1 = 0$

$(x + \square)^2 = \square$

$(x + \square) = \pm\sqrt{\square}$

$x = \square - \sqrt{\square}$ or $x = \square + \sqrt{\square}$

Q8 hint Complete the square first.

9 Solve these quadratic equations, giving your answer in surd form.

a $x^2 + 6x + 7 = 0$

b $x^2 + 2x - 5 = 0$

c $x^2 + 8x + 9 = 0$

Q9 hint Follow the method used in **Q8**.

10 Copy and complete to write the expression $3x^2 - 12x - 1$ in the form $p(x + q)^2 + r$

$3x^2 - 12x - 1 = 3(\square - \square) - 1$

$= 3[(x - \square)^2 - \square] - 1$

$= 3(x - \square)^2 - 12 - 1$

$= 3(x - \square)^2 - \square$

Q10 hint Factorise the x^2 and x terms. Then complete the square for the expression inside the brackets. Simplify so that you have $p(x + q)^2 + r$

Key point 6

$ax^2 + bx + c$ can be written as $a\left(x^2 + \frac{b}{a}x\right) + c$ before completing the square for the expression inside the brackets.

11 Write these in the form $a(x + p)^2 + q$

a $2x^2 + 12x + 2$ b $3x^2 - 6x + 5$

c $5x^2 + 10x + 25$ d $4x^2 + 12x - 7$

Q11 hint Follow the method used in **Q10**.

12 Solve these equations by completing the square. Give your answer in surd form.

a $2x^2 - 12x + 2 = 0$ b $3x^2 + 12x - 3 = 0$

Q12 hint Complete the square first. Then solve.

13 Copy and complete to solve $4x^2 - 8x - 12 = 0$. Give your answer correct to 2 decimal places.

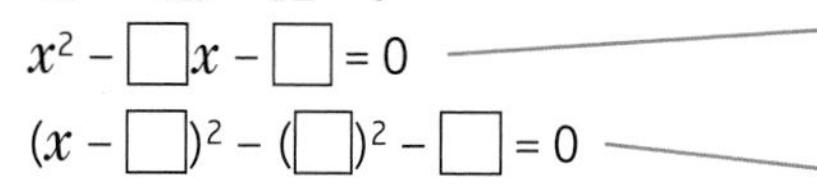

$4x^2 - 8x - 12 = 0$

$x^2 - \square x - \square = 0$ — Divide every term by the coefficient of x^2, 4

$(x - \square)^2 - (\square)^2 - \square = 0$ — Complete the square for the first two terms.

$(x - \square)^2 = (\square)^2 + \square$ — Rearrange terms

$(x - \square) = \pm\sqrt{\square}$ — Take the square root of both sides.

$x = \square + \sqrt{\square}$ or $x = \square - \sqrt{\square}$

$x = \square$ or $x = \square$

14 Solve these quadratic equations, giving your answer correct to 2 decimal places.

a $2x^2 + 4x - 8 = 0$ b $6x^2 - 3x - 2 = 0$ c $3x^2 + 6x - 10 = 0$

d $5x^2 - 15x - 4 = 0$ e $4x^2 + 6x - 5 = 0$

Q14 hint Follow the method used in **Q13**.

Discussion Which is more accurate, giving an answer as a surd or to 2 decimal places?

15 **Exam-style question**

Solve $6x^2 + 3x - 13 = 0$ by completing the square.
Give your answer correct to 2 decimal places. **(3 marks)**

Exam hint
When you are told the method to use, make sure you use this method.

9.4 Solving simple simultaneous equations

Objectives

- Solve simple simultaneous equations.
- Solve simultaneous equations for real-life situations.

Did you know?

Simultaneous means 'at the same time'. The only way to solve equations in two variables is to have two equations that both variables satisfy.

Fluency

When $y = 3$ what is **a** $2y$ **b** $-3y$ **c** $5y$?

Warm up

1 Rearrange these equations to make b the subject.
a $b - 12 = 2a$ b $2b + 6c = 10$ c $5a - 3b = 5$

2 Write an equation for each of these.
a The sum of x and y is 12.
b The difference between x and y is 4.

3 Which of these have the value 0 for *any* value of the variable?
a $-3y + (-3y)$ b $2y - (-2y)$ c $-4x + 4x$ d $-3z - (-3z)$

Key point 7

When there are two unknowns, you need two equations to find their values. These are called **simultaneous equations**.

4 Solve the simultaneous equations

a $y = 3$
$2x + y = 11$

b $y = 5$
$3x - y = 4$

c $y = -6$
$3x + 2y = 30$

d $y = 4x$
$3x + 2y = 11$

e $y = x + 2$
$x + y = 20$

f $y = x + 3$
$x + 3y = 17$

g $y - 3x = 0$
$2x + 2y = 24$

h $2x - y = 0$
$5x + 4y = 26$

Q4a hint Substitute the value of y into the second equation: $2x + \square = 11$. Now solve.

Q4c hint When $y = -6$, $2y = \square$. Substitute this value into the second equation.

Q4e hint Substitute the value of y into the second equation: $x + x + 2 = 20$

Q4g hint Rearrange the first equation to make y the subject.

5 **Problem-solving** Two meals and a bottle of wine cost £36. The bottle of wine costs £3 more than a meal.
How much is one meal?
How much is a bottle of wine?

Q5 strategy hint Let the cost of a meal be x and the wine be y. Write an equation for the first sentence. Write an equation for the second sentence.

6 **Problem-solving / Modelling** Jake buys 2 lamb chops and 2 sausages and pays £7 for them. At the same time Jamie buys 3 lamb chops and 4 sausages and pays £11.
How much does a sausage cost?

Example 4

Solve the simultaneous equations

$x + y = 6$

$3x - y = 10$

① $x + y = 6$

② $3x - y = 10$

①+② $4x + 0 = 16$

$x = 4$

$4 + y = 6$

$y = 2$

Check: $3 \times 4 - 2 = 10$ ✓

The terms in y have opposite signs, so add the equations to eliminate the terms in y.

Divide both sides by 4

Substitute $x = 4$ into equation ①

Now check that your solutions work in equation ②

7 Solve these simultaneous equations.

a $2x - y = 4$, $3x + y = 11$

b $5x + y = 15$, $2x + y = 3$

c $4x - 3y = 10$, $5x - 3y = 14$

Q7b hint Subtract the equations.

8 Antony solves simultaneous equations in this way.

$3x - y = 8$

$x + y = 4$

$y = 4 - x$ — Write the second equation with y as the subject.

$3x - (4 - x) = 8$ — Substitute $y = 4 - x$ into the first equation.

$3x - 4 + x = 8$

$3x + x - 4 = 8$

$4x = 12$ — Collect like terms to find x.

$x = 3$

$3 + y = 4$ — Now, substitute $x = 3$ into the second equation.

So $y = 1$

$(3 \times 3) - 1 = 8$ ✓ — Check that the solutions work by substituting them into the first equation.

Solve the simultaneous equations in **Q7a** using Antony's method.

Reflect Which method do you prefer?

9 Copy and complete to solve the simultaneous equations.

a

$4x - 2y = 16$ ①

$3x + y = 17$ ②

×2 $\square x + 2y = \square$ ×2 ③

Q9 hint Multiply every term in equation ② by 2

b Add equations ① and ③. Find x.

$4x - 2y = 16$ ①

$\square x + 2y = \square$ ③

$\square x + 0 = \square$

$x = \square$

Q9 hint Add equations ① and ③.

c Substitute your value of x into equation ① to find y.

10 Solve these simultaneous equations.

a $x + 5y = 22$, $4x - y = 4$

b $3y - 2x = 16$, $x + 4y = 14$

c $5x + 3y = 21$, $3x + y = 11$

d $3y - 2x = -7$, $5y - x = 7$

Discussion How do you decide what number to multiply by?

Q10 hint Follow the method used in **Q9**.

11 **Exam-style question**

Solve the simultaneous equations

$4x + 3y = 5$

$2x + y = 3$ **(3 marks)**

Exam hint
Substitute your values of x and y back into each equation to check they fit.

12 Solve the simultaneous equations

$5x + y = 0$

$x + 5y = 0$

Discussion Does it matter which equation you multiply?

Q12 hint First solve by multiplying the first equation by 5. Then solve by multiplying the second equation by 5.

13 **Problem-solving** The sum of two numbers is 23 and their difference is 5. Let the two numbers be x and y. Write two equations and solve them to find the two numbers.

14 **Finance / Problem-solving** A telephone company charges £x per month for a basic line rental and then £y per 100 minutes. Justin pays £18 for 200 minutes. Teresa pays £21 for 300 minutes.

a Work out the cost of the monthly rental.

b How much would Caron pay for 400 minutes?

15 **Exam-style question**

The diagram shows a rectangle. All sides are measured in centimetres.

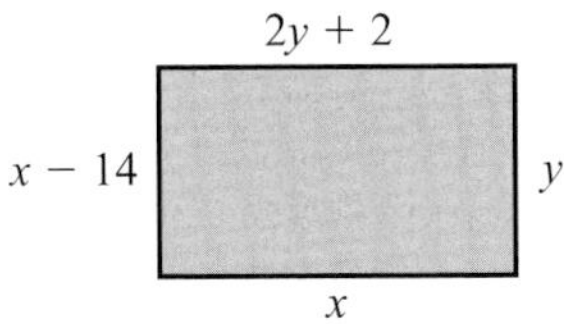

Jean says the perimeter of the rectangle is more than 75 cm.
Show that Jean is correct. **(5 marks)**

Q15 strategy hint Set up two simultaneous equations and solve them.

9.5 More simultaneous equations

Objectives

- Use simultaneous equations to find the equation of a straight line.
- Solve linear simultaneous equations where both equations are multiplied.
- Interpret real-life situations involving two unknowns and solve them.

Did you know?

Sometimes you need to multiply both equations to eliminate a term.

Fluency

What is the equation of a straight line?

Warm up

1 Multiply each equation by 4

a $2x + 3y = 6$ b $x - 6y = 7$

2 Solve these simultaneous equations.

a $y + 2x = 8$
$2y - 2x = 4$

b $3y + 2x = 14$
$5y + x = 14$

3 a Write the equation of a line through (2, 5).
b Write the equation of a line through (3, 8).
c Solve your simultaneous equations from parts **a** and **b** to find m and c.
d Write the equation of the line through the points (2, 5) and (3, 8).

Q3a hint Substitute $x = 2$, $y = 5$ into $y = mx + c$

Q3d hint Substitute your values of m and c from part **c** into $y = mx + c$

4 Find the equation of the line through the points (6, −3) and (−2, 5).

Example 5

Solve the simultaneous equations

$5x + 2y = 16$

$4x - 3y = -1$

① $5x + 2y = 16$ ①×3: $15x + 6y = 48$ ③

② $4x - 3y = -1$ ②×2: $8x - 6y = -2$ ④

Multiply equation ① by 3 and equation ② by 2 to make the coefficients of y equal.

③+④ $23x = 46$

Add these equations to eliminate y.

$x = 2$

$10 + 2y = 16$

Substitute $x = 2$ into equation ①

$2y = 6$

$y = 3$

Check: $4 \times 2 - 3 \times 3 = 8 - 9 = -1$ ✓

Check your answers by substituting into equation ②

5 Solve these simultaneous equations.

a $5x + 3y = 13$
$4x - 2y = 6$

b $3x + 2y = 13$
$4x + 4y = 20$

c $4x + 7y = 1$
$3x + 10y = 15$

d $2x + 5y = 24$
$3x + 7y = 34$

e $4x + 1.5y = 7$
$5x + 2y = 8$

6 **Problem-solving / Modelling** A coffee shop menu shows deals of the day as one coffee and two scones for £3.80, or two coffees and two scones for £5.80.
Work out the cost of a coffee and the cost of a scone.

7 **Problem-solving / Modelling** Daniel pays for two children and himself to go into a sports centre. His friend pays for an adult and three children.
Daniel pays £3.80 for the tickets and his friend pays £4.80.
How much is an adult ticket and how much is a child's ticket?

8 **Problem-solving / Modelling** Amy buys two bananas and three pears in a shop and pays £1.95.
At the same time Jacob buys three bananas and five pears and pays £3.05.
What is the cost of a pear? What is the cost of a banana?

9 **Exam-style question**

Solve the simultaneous equations.

$3x - 4y = 8$

$9x + 5y = -1.5$

(3 marks)

June 2012, Q16b, 5MB3H/01

Exam hint
Number the equations and show any multiplying, adding or subtracting clearly.

10 **Problem-solving** A van can carry 300 kg. Two possible maximum loads are 4 bags of cement and 7 bags of sand, or 6 bags of cement and 3 bags of sand.
What is the mass of a bag of sand? What is the mass of a bag of cement?

11 **Problem-solving** Hire charges for a mini bus consist of £x fixed charge + y pence for each mile of the journey.
A hire for 20 miles costs £25.
A hire for 30 miles costs £30.
How much would a hire for 50 miles cost?

12 The diagram shows a rectangle with all measurements in metres.

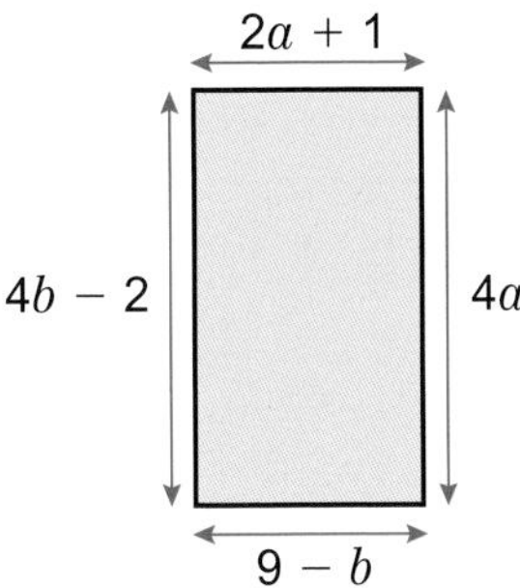

a Write down a pair of simultaneous equations in a and b.
b Solve the equations.
c Give the dimensions of the rectangle.

Q12a hint Rearrange your equations so that they are both $\square a + \square b = \square$

9.6 Solving linear and quadratic simultaneous equations

Objectives

- Solve simultaneous equations with one quadratic equation.
- Use real-life situations to construct quadratic and linear equations and solve them.

Did you know?

A quadratic expression is a way of describing the area of a rectangular shape.

Fluency

Which of these equations is **a** linear **b** quadratic **c** a circle?
$x^2 - 4x + 2 = 0$ $y = 4 - 3x$ $x^2 + y^2 = 16$ $y + x = 4$

Warm up

1 Solve these quadratic equations.
a $x^2 + 3x - 4 = 0$
b $2x^2 - x - 3 = 0$
c $6x^2 + 10x - 4 = 0$

2 Use the quadratic formula to solve these equations, giving your answer to 2 d.p.
a $x^2 + 3x - 5 = 0$
b $3x^2 - x - 3 = 0$
c $2x^2 + 5x - 3 = 0$

3 Find the length of the line joining A(2, 9) to B(−4, 1).

Example 6

Solve these simultaneous equations.

① $2x + y = 3$

② $x^2 + y = 6$

$y = 3 - 2x$ — Rearrange equation ① to make y the subject.

$x^2 + (3 - 2x) = 6$ — Substitute $y = 3 - 2x$ into equation ②

$x^2 - 2x + 3 = 6$

$x^2 - 2x - 3 = 0$ — Expand the bracket and rearrange so the right-hand side is 0.

$(x + 1)(x - 3) = 0$ — Solve the quadratic equation.

So either $(x + 1) = 0$ or $(x - 3) = 0$

$x = -1$ or $x = 3$

$2 \times (-1) + y = 3$ — Substitute $x = -1$ into equation ① to find one value of y.

$-2 + y = 3$

$y = 5$

$2 \times 3 + y = 3$ — Substitute $x = 3$ into equation ① to find the second value of y.

$6 + y = 3$

$y = -3$

So the solutions are $x = -1$, $y = 5$ and $x = 3$, $y = -3$

Key point 8

A pair of quadratic and linear simultaneous equations can have two possible solutions.

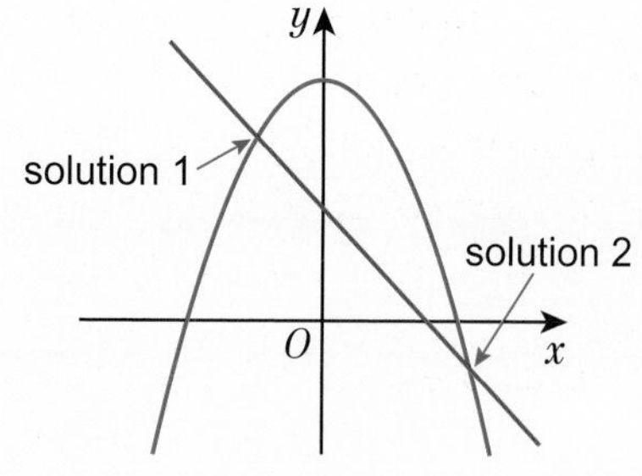

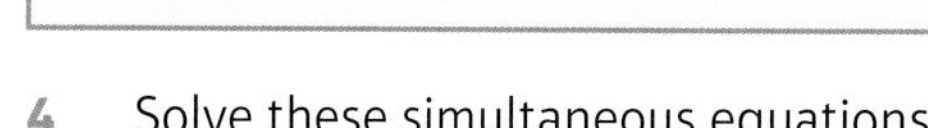

4 Solve these simultaneous equations.

a $y = x$
$x^2 + y = 12$

b $2x - y = 7$
$x^2 - 15 = y$

c $y - 4x = 6$
$y = 2x^2 + 3x + 5$

d $y = 5x - 3$
$y = 3x^2 + 6x - 7$

e $x^2 + y^2 = 4$
$3x + 5 = y$

5 Solve these simultaneous equations. Give your answers correct to 2 decimal places where appropriate.

a $y + 3x = 8$
$y = x^2 + 2x + 4$

b $2y - 4x = 6$
$y = x^2 + x - 5$

Key point 9

To find the coordinates where two graphs intersect, solve their equations simultaneously.

6 **Reasoning** The diagram shows a sketch of the curve $y = 4(x^2 - x)$.
The curve crosses the straight line with equation $y = 4 - 4x$ at two points.
Find the coordinates of the points where they intersect.

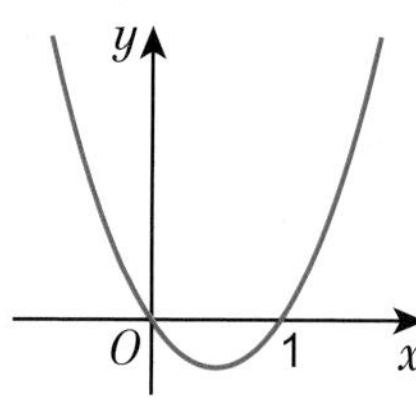

7 Solve these simultaneous equations.

a $y = 2x^2 + x - 2$
$x + y = 2$

b $y = 4x^2 - x - 6$
$y = 2 - x$

c $y = x^2 + 6x + 5$
$y = 5 + x$

d $y = 3x^2 - 4x - 2$
$y = 2x - 3$

e $y = 2x - 1$
$y^2 = 4x + 13$

f $3y = x + 6$
$y^2 = 2x + 7$

8 **Reasoning** The diagram shows a rug laid on a wooden floor. Its width is 2 m less than the width of the room.

a Write an equation to represent the width of the room (y).

b Write an equation to represent the area of the rug, which is 3 m².

c Use these two equations to find the value of x and hence the width of the room (y).

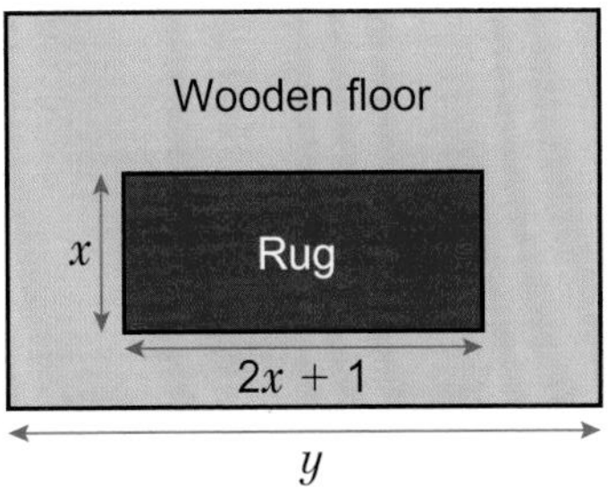

9 A curve with equation $y = x^2 - 4x - 1$ crosses a straight line with equation $y = 2x - 1$ in two places. Find the coordinates of the points where they intersect.

10 **Exam-style question**

C is the curve with equation $y = x^2 - 6x + 6$
L is the straight line with equation $y = 2x - 9$
L intersects C at two points, A and B.
Calculate the exact length of AB. **(6 marks)**

Exam hint
There are 2 marks each for finding A and B, and 2 marks for finding the length AB.

11 **Reasoning** The diagram shows a circle of diameter 4 cm with centre at the origin.

a Write the equation for the circle.

b Use an algebraic method to find the points where the line $y = 2x - 1$ crosses the circle.

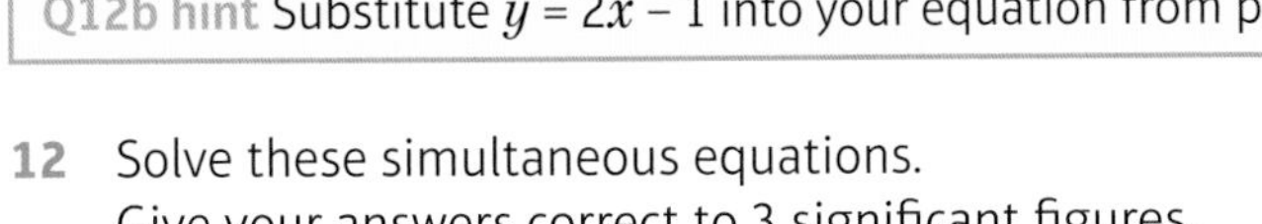

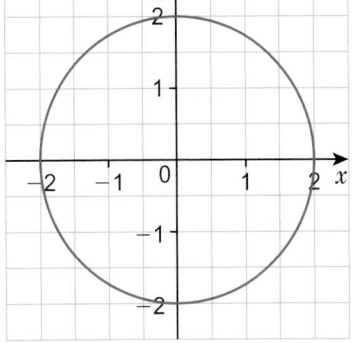

Q12b hint Substitute $y = 2x - 1$ into your equation from part **a**.

12 Solve these simultaneous equations.
Give your answers correct to 3 significant figures.

a $x^2 + y^2 = 28$
$y = x + 3$

b $x^2 + y^2 = 35$
$y = 2 + 5x$

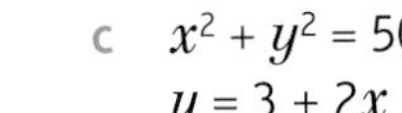

c $x^2 + y^2 = 50$
$y = 3 + 2x$

9.7 Solving linear inequalities

Objective

- Solve inequalities and show the solution on a number line and using set notation.

Why learn this?

Anything you can do with an equation you can also do with an inequality. It helps us to consider a wider range of potential answers to problems.

Fluency

Which is true for $-2 < x < +2$?

- x lies between −2 and +2 but is equal to neither.
- x is bigger than +2 and smaller than −2.
- x is greater than or equal to −2 and less than or equal to +2.
- x is equal to either −2 or +2.

Warm up

1 Solve to find x

a $5 - 4x = 3$ b $4 - 2x = 6 - 3x$ c $4x = 2(2 - x)$ d $9 - 5x = 21 + x$

2 For each inequality, choose three possible integer values for x from the cloud.

a $x \geqslant 5$ b $x < 4$

c $x < 10$ d $-2 \leqslant x < 4$

0, −2, 3, 12, 5, 10

3 Write down four integers that satisfy each inequality.

a $x + 2 > 3$ b $x - 1 < 0$ c $3x + 1 > 4$ d $5x - 3 > 2$

e $-5 < x \leqslant 3$ f $0 < x \leqslant 4$ g $3 > x \geqslant -3$

Key point 10

You can show **inequalities** on a number line.

An empty circle O shows that the value is not included.

A filled circle ● shows that the value is included.

An arrow O⟶ shows that the solution continues towards infinity.

You can rearrange an inequality in the same way as you rearrange an equation.

4 Write the inequalities that these number lines represent for x.

a
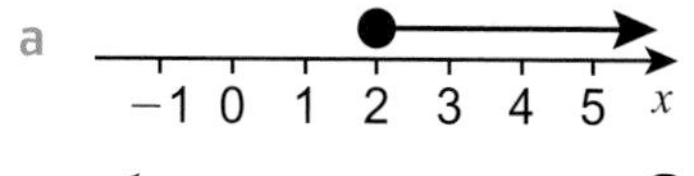

b
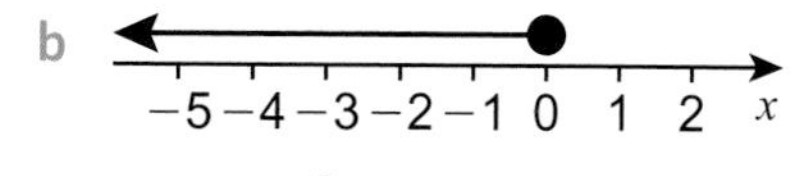

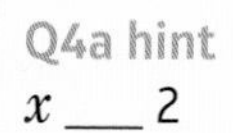

c
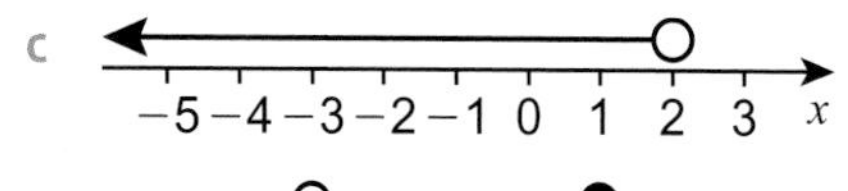

d

−3 −2 −1 0 1 2 3 4 5 x

e

−1 0 1 2 3 4 5 6 x

f
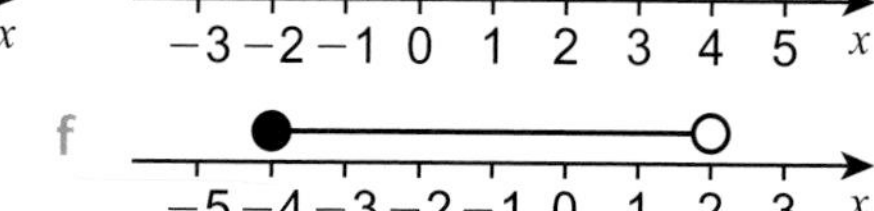

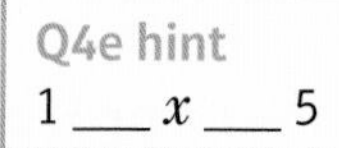

5 **Exam-style question**

$-4 < n \leqslant 1$ where n is an integer.

a Write down all the possible values of n. **(2 marks)**

b Write down the inequality represented on the number line.

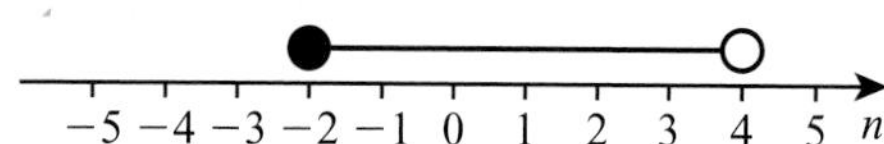

(2 marks)

March 2013, Q4, 5MB3H/01

Q5b strategy hint Write your inequality in terms of x.

6 Draw number lines to show these inequalities.

a $x \geqslant 1$ b $1 \leqslant x \leqslant 5$ c $-1 < x < 3$ d $0 > x \geqslant -2$

Discussion Are number lines a good way to represent inequalities?

Key point 11

You can write the solution to an inequality using **set notation**.

$\{x : x > 2\}$

the set of x such that

7 $\{x : x > 3\}$ means the set of all x values such that x is greater than 3.

Write the meaning of these sets.

a $\{x : x < 2\}$ b $\{x : x \leqslant -2\}$ c $\{x : x \geqslant 0\}$

d $\{x : x \leqslant 0\}$ e $\{x : x > -1)$

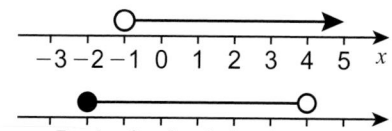

8 Show the sets in **Q7** on number lines.

Example 7

Solve $3x - 2 > 6 - x$. Show your answer on a number line and write the solution set using set notation.

$3x > 6 - x + 2$ — Add 2 to both sides.

$4x > 8$ — Add x to both sides.

$x > 2$ — Divide both sides by 4.

In set notation: $\{x : x > 2\}$ — This tells us that there is a set of values of x, not just one value.

9 Solve each inequality and show your answer on a number line.
Write the solution set using set notation.

a $4x > 12$ b $5x < 10$ c $3x - 5 > 4$ d $2x + 3 \leqslant 7$

10 **Exam-style question**

$3x + 5 > 16$, where x is an integer.
Find the smallest value of x. **(3 marks)**

Nov 2012, Q1, 5MB3H/01

Exam hint
Use the inequality symbol on each line of your working.

11 Solve these inequalities and write the solution using set notation.

a $3(x - 2) > 6$ b $2(x - 1) < 5x + 7$
c $2(3x + 4) > 4x - 3$ d $3(4 - 2x) < 2(2x - 3)$

Q11a hint Expand the brackets first.

Key point 12

When inequalities have a lower limit and an upper limit, solve the two sides separately.

12 Solve

a $-7 < 2x + 1 \leqslant 5$ b $-5 < 2x + 1 \leqslant 9$
c $-2 \leqslant \frac{2x}{3} \leqslant 6$ d $-1 < \frac{3x - 1}{4} \leqslant 2$

Q12a hint Write the two inequalities separately: $-7 < 2x + 1$ and $2x + 1 \leqslant 5$. Solve each one separately.

13 a Multiply both sides of the inequality $5 > 3$ by -1.
Is the inequality still true?

b Divide both sides of the inequality $8 < 16$ by -2.
Is the inequality still true?

Discussion What happens when you multiply or divide an inequality by a negative number?

Key point 13

When you multiply or divide an inequality by a negative number, reverse the inequality signs.

14 Find the possible integer values of x in these inequalities.

a $-8 < -x < -2$ b $-4 < -2x < 10$
c $-8 < 4 - 3x \leqslant 10$ d $-6 \leqslant 4 - 2x \leqslant 8$

15 Solve these inequalities and write the solution using set notation.

a $3(x + 2) \geqslant 2x + 3$ b $2 - x < 2x + 5$
c $-3 < 2x + 1 \leqslant 9$ d $-2 > 4(1 - x) \geqslant -8$

9 Problem-solving: Overtaking

Objectives

- Be able to substitute into formulae.
- Be able to factorise quadratic expressions.
- Be able to solve linear and quadratic simultaneous equations.

If a vehicle is accelerating at a constant rate we can use the formula below to find the distance it has travelled after a certain time.

$$s = ut + \frac{1}{2}at^2$$

s = distance (m)
u = initial velocity (m/s)
a = acceleration (m/s^2)
t = time (s)

1 Find the distance travelled by a car in 10 seconds if its acceleration is 8 m/s^2 and it starts at a speed of 5 m/s.

2 A car is travelling at a constant speed of 3 m s^{-1}. At time $t = 0$ it starts to accelerate at 4 m/s^2. At what time will it have travelled 20 m?

> **Q2 hint** Substitute into the formula and factorise to solve the quadratic equation. Which solution would be the correct answer?

3 A car leaves a set of traffic lights from stationary and accelerates at a constant rate of 6 m/s^2. A lorry passes the traffic lights just as they go green. It is travelling at a constant speed of 18 m/s.
How many seconds after it leaves the traffic lights does the car overtake the lorry?

> **Q3 hint** Use $s = ut + \frac{1}{2}at^2$ to find the equation that models the car leaving the set of traffic lights. You could answer this problem by drawing distance–time graphs for the car and the lorry on the same set of axes (time from 0 s to 18 s).

4 At the next set of traffic lights the car stops and waits for the lights to go green.
When the lights change, it accelerates at a constant rate and is overtaken by a motorcycle travelling at a constant speed.

a Find equations for both vehicles so that the car overtakes the motorcycle after 5 seconds.

b Can you find more than one solution to this problem?

9 Check up

Log how you did on your Student Progression Chart.

Quadratic equations

1 Solve
   a $x^2 + 3x = 0$
   b $2x^2 + 2x - 12 = 0$

2 Factorise $3x^2 - 4x - 4 = 0$

3 Write and solve an equation to find x.

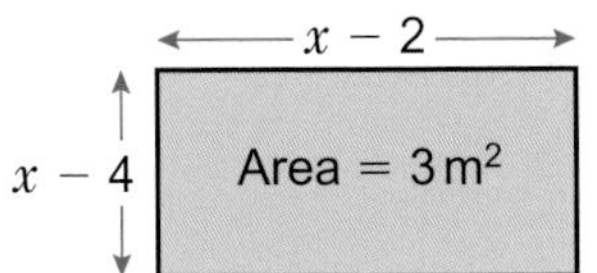

4 Use the quadratic formula to solve $x^2 - 2x - 6 = 0$
Give your answer in surd form.

5 Write $x^2 + 6x + 3$ in the form $(x + p)^2 + q$

6 Solve $x^2 + 6x - 3 = 0$ by completing the square, giving your answer in surd form.

Simultaneous equations

7 Solve these simultaneous equations.
   a $y = x + 3$
   $x - 2y = -1$
   b $2x + y = 4$
   $3x + 2y = 6$

8 **Problem-solving** Find the equation of the line through the points (4, 5) and (0, −3).

9 Solve these simultaneous equations.
$2x + y = 5$
$x^2 + 2x = y$

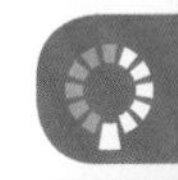

Inequalities

10 Find the possible integer values for x in $-10 < -5x \leqslant 25$

11 a Solve the inequality $-3 < x - 3 \leqslant 6$
   b Show your answer on a number line.
   c Write the solution using set notation.

12 How sure are you of your answers? Were you mostly
Just guessing  Feeling doubtful  Confident 
What next? Use your results to decide whether to strengthen or extend your learning.

Reflect

★ Challenge

13 **a** My daughter's age in 3 years' time will be the square of her age 3 years ago.
How old is she now?
   b Write a problem like this for a friend.
Check that it works and that you can find the answer.

9 Strengthen

Quadratic equations

1 Solve $x(x + 7) = 0$

> **Q1 hint** $x \times (x + 7) = 0$
> Either $x = 0$ or $x + 7 = 0$

2 a Factorise $x^2 + 5x$
b Solve $x^2 + 5x = 0$

> **Q2a hint** $x(\square + \square)$

> **Q2b hint** Use your answer to part **a** to help.
> $x(\square + \square) = 0$
> Either $x = 0$ or $(\square + \square) = 0$

3 a Find two numbers whose product is −12 and whose sum is −1
b Factorise $x^2 - x - 12$
c Solve $x^2 - x - 12 = 0$

> **Q3a hint** Which two factors of −12 have a difference of 1? The **−**12 means the signs are different.

> **Q3b hint** Use your answer to part **a**.
> $(x$ + one factor$)(x$ − other factor$)$

4 Solve
a $x^2 + 3x - 18 = 0$
b $x^2 - 7x + 12 = 0$
c $x^2 + 2x - 15 = 0$

> **Q4a hint** Follow the method used in **Q3**. Begin by finding two numbers whose product is −18 and whose sum is +3

5 a Write down the factor pairs of −6.
b Try each factor pair in these brackets.
$(2x + \square)(x + \square)$
Which pair gives $2x^2 + x - 6$ when you expand?
c Solve $2x^2 + x - 6 = 0$

> **Q5c hint** Use your answer to part **b**.
> $(2x + \square)(x + \square) = 0$

6 Solve
a $3x^2 + 5x - 12 = 0$ b $3x^2 - 2x - 8 = 0$ c $4x^2 - 14x + 12 = 0$

> **Q6a hint** Follow the method used in **Q5**.

7 a A rectangle is x cm long and $(x - 2)$ cm wide. Write an expression for its area.
b A square has side $(x + 4)$. Write an expression for its area.

8 The diagram shows a rectangle.

$x - 2$ (width), $x + 1$ (height)

a Write an expression for the area of the rectangle.
b The area of the rectangle is 4 m².
Write an equation for this.
c Rearrange your equation so you have 0 on the right-hand side.
d Solve this equation to find x

> **Q8b hint** Use your answer to part **a**.

> **Q8d hint** Follow the method used in **Q3**.

9 What are the values of a, b and c in each of these equations?
a $2x^2 + 3x + 1 = 0$
b $2x^2 - 4x - 6 = 0$
c $3x^2 + 4x - 1 = 0$

> **Q9 hint** Write each equation below the general equation.
> $ax^2 + bx + c$
> $2x^2 + 3x + 1 = 0$

10 What is the value of $b^2 - 4ac$ for each equation in **Q9**?

> **Q10 hint** Use the values of a, b and c you found in **Q9**. Substitute the values into the expression.

11 Use the quadratic formula $x = \dfrac{-b \pm \sqrt{b^2 - 4ac}}{2a}$ to solve the equations in **Q9**.

> **Q11 hint** Use the values of $b^2 - 4ac$ you found in **Q10**.

12 a Expand i $(x + 3)^2$ ii $(x - 5)^2$

b Predict what the x term will be when you expand $(x + 6)^2$

c Expand to check your prediction.

13 a Copy and complete

$(x + \square)^2 = x^2 + 4x + \square$

$(x + \square)^2 - \square = x^2 + 4x$

b Solve $x^2 + 4x = 8$ by completing the square.

Q13b hint Use your answer from part **a**. Rearrange so you have $(\quad)^2 = \square$
Take square roots of both sides.

14 a Copy and complete

$(x - \square)^2 = x^2 - 6x + \square$

$(x - \square)^2 - \square = x^2 - 6x$

b Solve $x^2 - 6x = 9$ by completing the square.

Q14b hint Follow the method used in **Q13**.

Simultaneous equations

1 Solve $3x + y = 10$ when

a $y = 3$

b $y = 2x$

Q1a hint Substitute 3 in place of y and rearrange to find x

Q1b hint Substitute $2x$ in place of y and rearrange to find x

2 Solve the simultaneous equations

$4x + y = 12$ and $y = 2x$

Q2 hint Follow the method used in **Q1**.

3 a Check this solution to a pair of simultaneous equations.

① $2x + 3y = -5$

② $3x + 4y = -6$

Rearrange ① to make y the subject: $y = \dfrac{-2x - 5}{3}$ ③

Substitute ③ into ②: $3x + 4\left(\dfrac{-2x - 5}{3}\right) = -6$

Simplify and solve: $9x + 4(-2x - 5) = -18$

$9x - 8x - 20 = -18$

$x - 20 = -18$

$x = 2$

Substitute $x = 2$ into ①: $2 \times 2 + 3y = -5$

Solve for y: $4 + 3y = -5$

$3y = -9$

$y = -3$

b Check the answers by substituting them into $3x + 4y = -6$.

4 Solve the simultaneous equations

$3x + 2y = -8$

$3y - 2x = 14$

Q4 hint Write them with x and y terms above each other. Follow the method used in **Q3**.

5 Here are two simultaneous equations.

$x^2 - 2x = 6 + y$

$5x + y = 4$

a Rearrange $5x + y = 4$ to make y the subject.

b Substitute your answer to part **a** in place of y in the first equation.

c Simplify the equation by collecting like terms.

d Solve the quadratic equation.

e Substitute for x to find y.

6 Solve the simultaneous equations
$x^2 + 4x = 4 + y$
$y + x = 2$

Q6 hint Follow the method used in **Q5**.

Inequalities

1 Write the inequalities for these number lines.

a 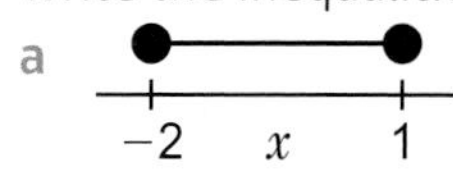

b 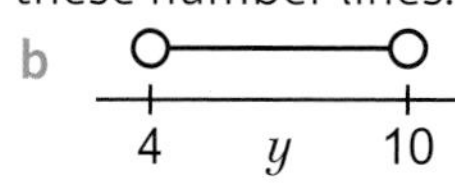

c 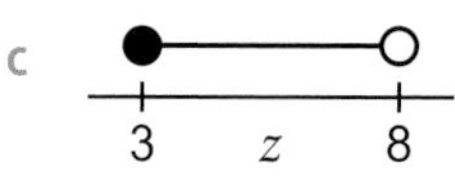

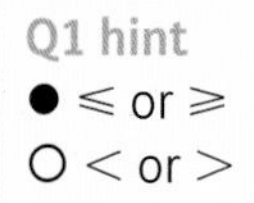

2 a Show these inequalities on number lines.
i $x > 3$ ii $x \leqslant 7$ iii $3 \leqslant x < 7$
b Write two integer values that satisfy each inequality.

3 Draw a number line to show each inequality.
a $x \geqslant 0$ b $x < 0$ or $x \geqslant 1$ c $-2 < x \leqslant 4$ d $-4 \leqslant x \leqslant 4$
e Write all the possible integer values for **c** and **d**.

4 Solve these inequalities.
a $x - 3 > 5$ b $7 - x < 4$
c $-2 < x + 1 < 5$ d $0 \leqslant x - 4 \leqslant 8$

Q4 hint Solve using the same method as for an equation. Remember:
$-x < -3$ (× −1) → $x > \square$ (× −1)

5 a Solve $2 < 3x - 1$
b Solve $3x - 1 < 11$
c Find the possible integer values for $2 < 3x - 1 < 11$

Q5c hint Use your answers from **a** and **b** to write $\square < x < \square$

6 Find the possible integer values for $4 < 2x + 5 \leqslant 9$

Q6 hint Follow the method used in **Q5**.

7 Show the solutions to **Q5** and **Q6** on number lines.

8 In set notation, $x \leqslant 3$ is $\{x : x \leqslant 3\}$.
Write these inequalities in set notation.
a $x > 4$ b $x < 7$ c $x \geqslant 5$ d $0 \leqslant x < 3$

Q8 hint Write $\{x : \ \}$ around the inequality.

9 Extend

1 The product of two consecutive numbers is 30.
a Write this using algebra.
b Solve your equation to find the two numbers.

Q1a hint Use x for the first number. The next consecutive number is $x + \square$

2 A lawn is 4 m longer than it is wide. The total area of the lawn is 30 m^2.
What is its perimeter? Give your answer to 2 decimal places.

3 Write $3x^2 + 2.4x$ in the form $a(x + p)^2 + q$. State the values of a, p and q.

4 **Reasoning** The football pitch in the diagram has area 7140 m^2.
What are the dimensions of the pitch?

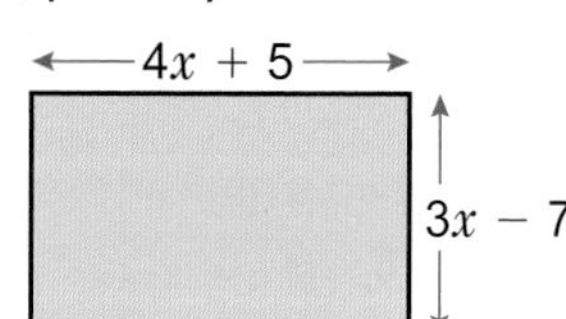

5 The diagram shows a 6-sided shape.
All the corners are right angles and all the measurements are in centimetres.
The area of the shape is 75 cm^2.
a Show that $2x^2 + 5x - 75 = 0$
b Solve the equation $2x^2 + 5x - 75 = 0$

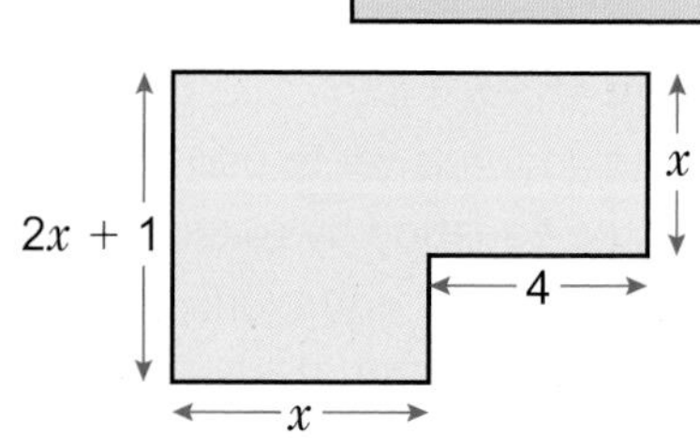

6 **Problem-solving** Gary buys three mobile phone cases and two charging leads and pays £29.50. At the same time his friend buys two phone cases but takes a charging lead back, gets a full refund on it and pays £11.50.
What is the cost of a mobile phone case? What is the cost of a charging lead?

7 **Reasoning** This trapezium has area 20 m². The area of a trapezium is given by the formula $A = \frac{1}{2}(a + b)h$
Find the value of x.

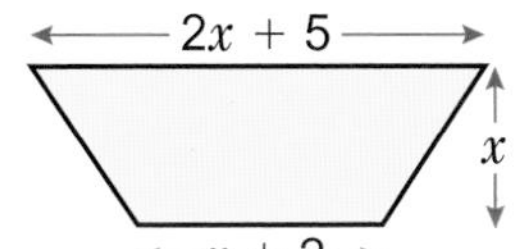

8 The diagram shows a sketch of the curve $y = 2x^2 + 4x - 2$
The line $y = 6 - x$ crosses the curve at points A and B.
Use an algebraic method to find the coordinates of A and B.
Give your answers to 1 decimal place.

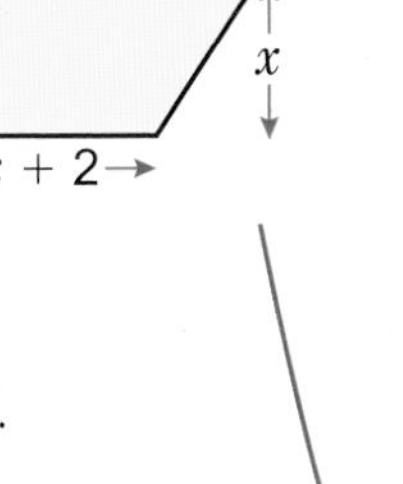

9 **Reasoning** Jodie and Kate play netball for the same team.
In a month they score fewer than 15 goals. Kate scored 4 more goals than Jodie. What is the maximum number of goals Jodie could have scored?

Q9 hint Write and solve an inequality.

10 A ball thrown in the air travels at speed $s = (20 - 4t)$ metres per second, where t is the time after being thrown. For which values of t is the speed between 5 m/s and 15 m/s?

11 **Problem-solving** A pond is 8 m wide by 10 m long.
It has a path around it that is exactly the same width all round.
The area of the path is 80% of the area of the pond.
How wide is the path?

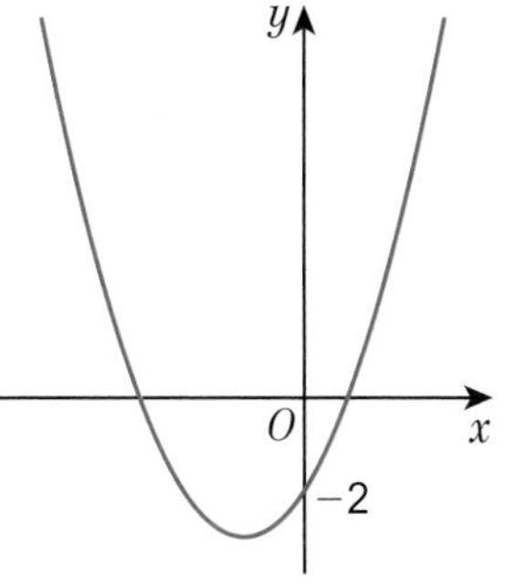

12 **Exam-style question**

Solve the simultaneous equations
$x^2 + y^2 = 25$
$y = 2x + 5$ **(6 marks)**

Nov 2012, Q21, 5MB3H/01

Exam hint
Write your solutions in two pairs:
$x = \square$, $y = \square$
and
$x = \square$, $y = \square$

13 Solve $x - 4 < 3x + 6 \leqslant 9$.
Show the solutions on a number line and write as a solution set.

14 **Communication** a Show that $(x + p)^2 + q = x^2 + 2px + p^2 + q$

b Jake uses this method to write $x^2 + 4x + 5$ in the form $(x + p)^2 + q$
$(x + p)^2 + q = x^2 + 2px + p^2 + q = x^2 + 4x + 5$
Comparing the x terms: $2px = 4x$ so $2p = 4$, $p = 2$
Comparing number terms: $p^2 + q = 5$
$4 + q = 5$
$q = 1$
So $(x + p)^2 + q = (x + 2)^2 + 1$
Use Jake's method to write these expressions in the form $(x + p)^2 + q$
i $x^2 + 6x + 15$ ii $x^2 + 8x - 3$ iii $x^2 - 4x + 2$ iv $x^2 + 3x + 7$

Q14a hint Expand and simplify the LHS.

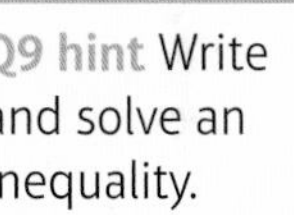

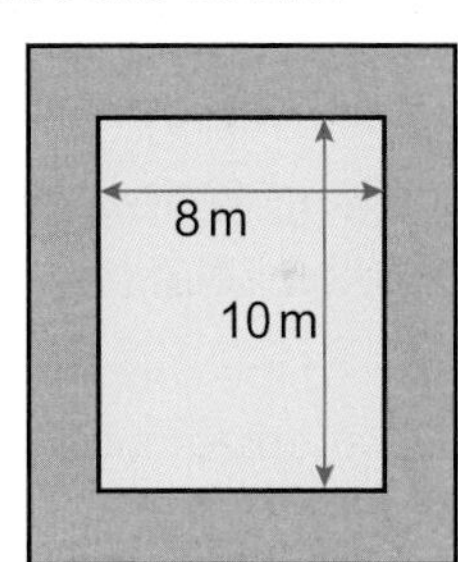

9 Knowledge check

- Solving a quadratic equation means finding values for the unknown that fit. .... *Mastery lesson 9.1*
- The **roots** of a quadratic function are its solutions when it is equal to zero. .... *Mastery lesson 9.1*
- You can solve equations of the form $ax^2 + bx + c = 0$ by fractorising. .... *Mastery lesson 9.2*
- You can use the quadratic formula to find the solutions to the **quadratic equation** $ax^2 + bx + c = 0$

  $x = \dfrac{-b \pm \sqrt{b^2 - 4ac}}{2a}$ .... *Mastery lesson 9.2*
- Expressions such as $(x + 2)^2$, $(x - 1)^2$ and $(x + \frac{1}{2})^2$ are called **perfect squares**. .... *Mastery lesson 9.3*
- $x^2 + bx + c$ can be written in the form $\left(x + \dfrac{b}{2}\right)^2 - \left(\dfrac{b}{2}\right)^2 + c$

  This is called **completing the square**. .... *Mastery lesson 9.3*
- $ax^2 + bx + c$ can be written as $a\left(x + \dfrac{b}{a}x\right)^2 + c$

  before completing the square for the expression inside the brackets. .... *Mastery lesson 9.3*
- When there are two unknowns, you need two equations to find their values. These are called **simultaneous equations**. .... *Mastery lesson 9.4*
- A pair of quadratic and linear simultaneous equations can have two possible solutions. .... *Mastery lesson 9.6*
- To find the coordinates where two graphs intersect, solve their equations simultaneously.

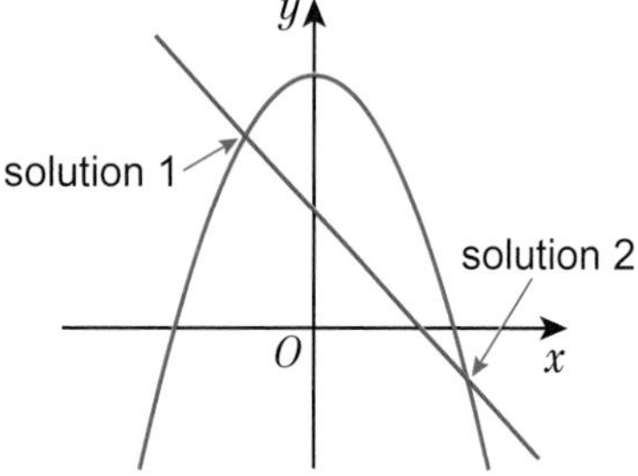

.... *Mastery lesson 9.6*

- You can show **inequalities** on a number line. An empty circle ○ shows that the value is not included. A filled circle ● shows that the value is included. An arrow ○⟶ shows that the solution continues towards infinity. .... *Mastery lesson 9.7*
- You can rearrange an inequality in the same way as you rearrange an equation. .... *Mastery lesson 9.7*
- You can write the solution to an inequality using **set notation**.

  $\{x : x > 2\}$ .... *Mastery lesson 9.7*

  the set of ↗ ↖ x such that
- When inequalities have a lower limit and an upper limit, solve the two sides separately. .... *Mastery lesson 9.7*

- When you multiply or divide an inequality by a negative number, reverse the inequality signs. .... *Mastery lesson 9.7*

Reflect

For this unit, copy and complete these sentences.

I showed I am good at ________

I found ________ hard

I got better at ________ by ________

I was surprised by ________

I was happy that ________

I still need help with ________

9 Unit test

Log how you did on your Student Progression Chart.

1 Solve the quadratic equation $15 = 2x^2 - 7x$ *(3 marks)*

2 Factorise $9x^2 - 3x - 2$ *(2 marks)*

3 **Reasoning** a One side of a rectangular room is 2 m longer than the other side. Write the area of the room as an expression in x (the length of the shorter side). *(1 mark)*
b The area of the room is 15 m². What is the value of x? *(3 marks)*

4 Solve the equation $3x^2 - 2x - 2 = 0$.
Give your solutions to 3 significant figures. *(3 marks)*

5 **Problem-solving** A picture frame is designed to take a 6 cm × 8 cm picture.
The border is exactly the same width all round and its area is 32 cm².
How wide is the border?

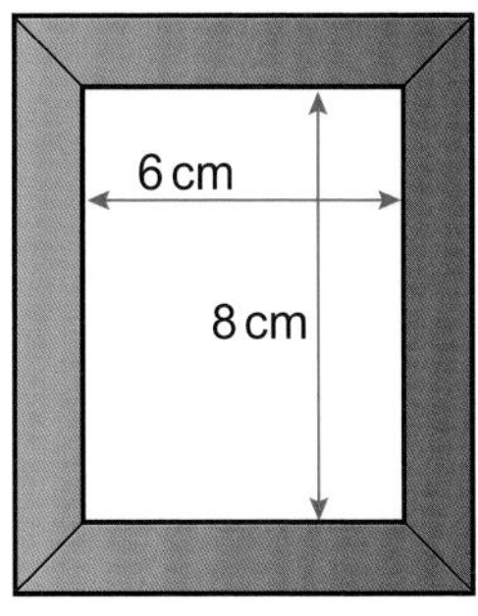

(4 marks)

6 Solve these simultaneous equations.

a $3x - 2y = 5$
$4x + 2y = 16$ *(3 marks)*

b $5x + 2y = 18$
$2x + 3y = 5$ *(4 marks)*

7 **Problem-solving** Jay buys 2 apples and 3 pears in a supermarket and pays £1.90.
He later returns and buys 4 pears and 5 apples and pays £3.35.
What is the cost of one apple? What is the cost of one pear? *(5 marks)*

8 Solve the inequalities

a $2y + 6 < 4y + 8$ *(3 marks)*

b $-3 < 2(x - 4) \leqslant 6 - x$ *(3 marks)*

ActiveLearn Homework, practice and support: Higher 9 Unit test

9 Find the integer x that satisfies both the inequalities $x + 4 > 5$ and $3x - 4 < 5$ *(3 marks)*

10 **Reasoning** This trapezium has area 3 m^2.

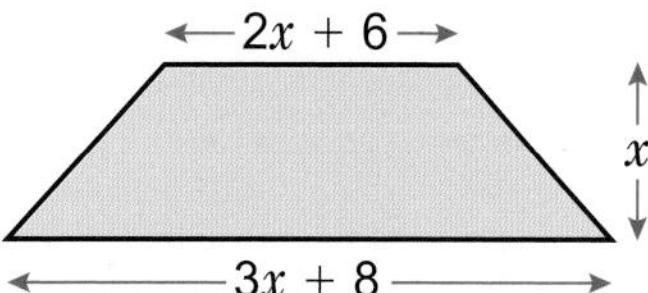

The area of a trapezium is given by the formula $A = \frac{1}{2}(a + b)h$
Find the value of x. *(4 marks)*

11 A curve with equation $y = x^2 - 6x - 2$ crosses a straight line with equation $y = 3x - 2$ in two places. Find the coordinates of the points where they intersect. *(4 marks)*

Sample student answer

a Why is it a good idea to label each equation with a number?

b What could be the problem with the algebra letters that the student has chosen (s and l)?

c Why will the student only get 4 out of the 5 marks?

Exam-style question

Paper clips are sold in small boxes and in large boxes.

There is a total of 1115 paper clips in 4 small boxes and 5 large boxes.

There is a total of 530 paper clips in 3 small boxes and 2 large boxes.

Work out the number of paper clips in each small box and in each large box. **(5 marks)**

June 2014, Q17, 5MB3H/01

Student answer

① $4s + 5l = 1115$ ①× 3: $12s + 15l = 3345$ ③

② $3s + 2l = 530$ ②× 4: $12s + 8l = 2120$ ④

③ – ④: $7l = 1225$

$1225 \div 7 = 175$

Put into ②:

$3s + 350 = 530$

$3s = 180$

$s = 60$

10 PROBABILITY

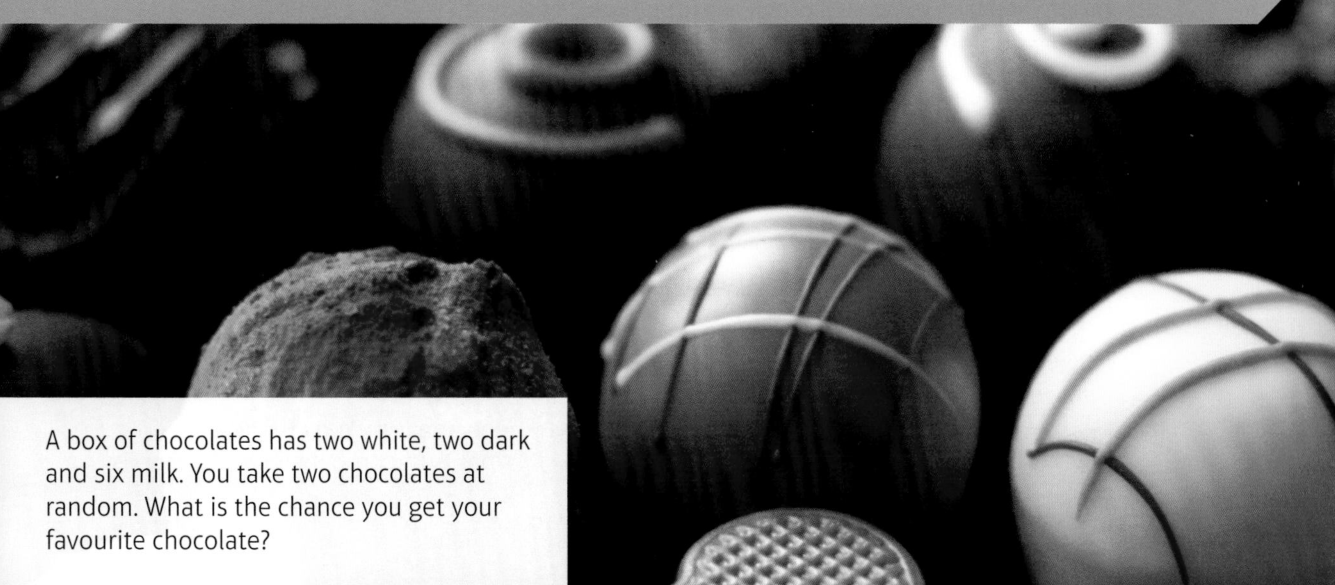

A box of chocolates has two white, two dark and six milk. You take two chocolates at random. What is the chance you get your favourite chocolate?

10 Prior knowledge check

Numerical fluency

1 Work out
 a $0.45 + 0.52$ b $0.6 + 0.25$
 c $0.52 - 0.17$ d $0.7 - 0.32$

2 Work out
 a $1 - 0.22$ b $1 - 0.76$
 c $1 - \frac{3}{5}$ d $1 - \frac{5}{12}$
 e $100\% - 27\%$ f $100\% - 68\%$

3 Write three equivalent fractions for each fraction.
 a $\frac{1}{4}$ b $\frac{2}{5}$ c $\frac{5}{6}$ d $\frac{7}{10}$

4 Write these fractions in ascending order.
 $\frac{3}{8}$ $\frac{5}{12}$ $\frac{2}{7}$ $\frac{1}{3}$ $\frac{2}{5}$

5 Work out
 a 0.25×160 b 0.2×120
 c 0.48×55 d $\frac{1}{3} \times 150$
 e $\frac{5}{8} \times 480$ f $\frac{7}{10} \times 180$

6 Convert these fractions to decimals and percentages.
 a $\frac{1}{4}$ b $\frac{3}{10}$ c $\frac{3}{5}$
 d $\frac{3}{8}$ e $\frac{17}{20}$ f $\frac{37}{80}$

7 Work out
 a 20% of 340
 b 35% of 234
 c 82% of 250
 d 7% of 60

8 Work out
 a $\frac{2}{3} + \frac{1}{4}$ b $\frac{3}{5} + \frac{1}{6}$
 c $\frac{5}{6} - \frac{1}{4}$ d $\frac{7}{12} - \frac{3}{7}$

Fluency with probability

9 a Copy the probability scale.

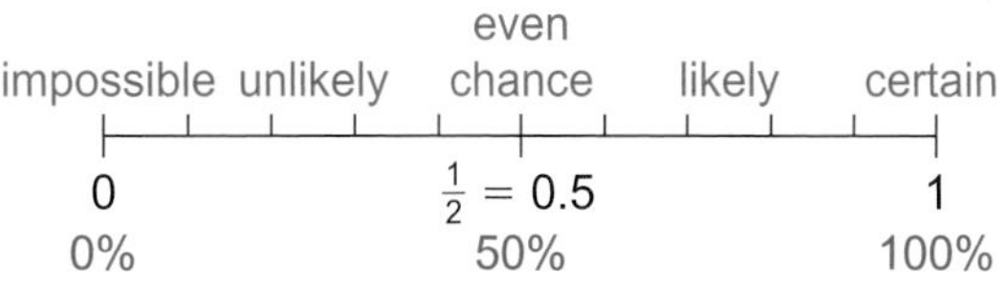

 b Write the capital letter of each event in the correct place on your scale.
  A Rolling an even number on a fair six-sided dice.
  B The probability of landing a 5 on a biased spinner is 30%.
  C The probability that two people will share their birthday from a class of 30 students is 0.7.

D The probability of an 18–24-year-old having a car accident within two years of passing their driving test is $\frac{1}{4}$.
E The probability of a 60-year-old male not surviving to his next birthday is 1%.

10 The frequency table shows the outcomes in a study of a new weight-loss regime to tackle obesity.
a Copy the table.

Outcome	Frequency	Relative frequency
'Healthy' category	78	
'Overweight' category	17	
'Obese' category	5	
Total frequency		

b Work out the total frequency.
c Calculate the relative frequencies.
d A further 300 people sign up to the weight-loss regime.
How many of them would you expect to fall into the 'healthy' category?

11 **Reasoning** An ordinary six-sided dice is rolled once.
a List all the possible outcomes.
b What is the probability of each outcome?
c What is the sum of all their probabilities?

12 **Reasoning** In a game, players spin the coloured spinner to move around the board.

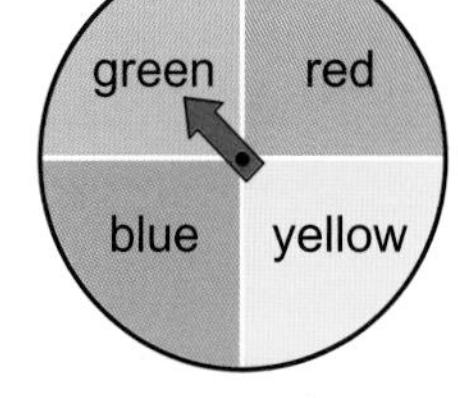

a What is the theoretical probability of landing on green if the spinner is fair?
b How many times would you expect to land on each colour in 100 spins?

In a game there are 100 spins. Here are the results:

Colour	red	yellow	blue	green
Frequency	25	28	24	23

c What is the experimental probability of getting yellow?

13 The table gives the numbers of boys and girls in a group who wear glasses.

	Glasses	No glasses	Total
Boys		10	
Girls	6		18
Total			32

a Copy and complete the table.
A person is picked at random.
What is the probability the person is
b a boy without glasses
c a girl?

14 A bag contains red, blue and yellow counters. The probability of choosing a red counter is 0.2 and the probability of choosing a blue counter is 0.35.
Work out the probability of choosing a red or blue counter.

* Challenge

15 Use the numbers 1, 2 and 3 as many times as you like on the spinner to make these statements true.
- P(odd number) > P(even number)
- P(1) = 2P(3)

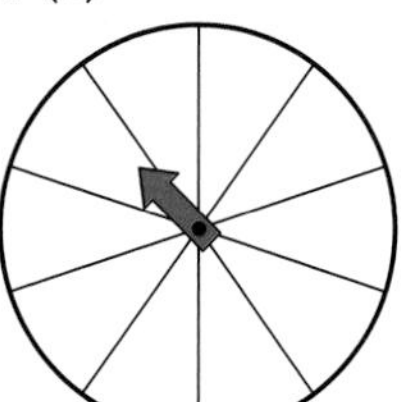

Find a different solution.

10.1 Combined events

Objectives

- Use the product rule for finding the number of outcomes for two or more events.
- List all the possible outcomes of two events in a sample space diagram.

Why learn this?

The word probability comes from the Latin word 'probabilitas' which can have different meanings. In Europe it is a measure of the 'authority' of a witness in legal cases.

Fluency

List all the possible outcomes when rolling a 6-sided dice.

Warm up

1 Jess is buying a car. She has a choice of four colours, red, blue, silver and white, and a choice of two models, an estate and saloon.

a How many possible combinations of colour and model are there?

b How many combinations are there for 5 colours and 2 models?

c How many combinations are there for 6 colours and 2 models?

d How many combinations are there for 6 colours and 3 models?

e How many combinations are there for m colours and n models?

Questions in this unit are targeted at the steps indicated.

Key point 1

$$\text{Probability} = \frac{\text{number of successful outcomes}}{\text{total number of possible outcomes}}$$

2 a From a set menu of 5 main courses and 3 desserts, write all the combinations of main courses and desserts.

b How many possible combinations are there?

c What is the probability that the combination chosen from the menu is lasagne and profiteroles?

Main courses	Desserts
Beef stew	Apple pie with custard
Lasagne	Profiteroles with cream
Chicken chasseur	Strawberry cheesecake with cream
Haddock	
Mushroom and broccoli bake	

Q2c hint Lasagne and profiteroles are 1 combination out of a total of □ combinations so

P(lasagne and profiteroles) = $\frac{1}{\square}$.

d What is the probability that the combination chosen is haddock and a dessert served with cream?

3 Highfield Technology School has two students from each year group on the school council. Amy, Beth, Callum, Dan and Ellie would like to represent Year 11.

a How many combinations of two students are there?

b What is the probability that Amy will represent Year 11?

c What is the probability that Ellie will represent Year 11?

d What is the probability that Callum and Dan will represent Year 11?

4 **Modelling** Two fair coins are flipped at the same time.

a Write a list of all the possible outcomes.

b How many outcomes are there altogether?

c Work out

i P(two tails) ii P(head and tail).

Q4c i communication hint P(two tails) means the probability of getting two tails.

Key point 2

A **sample space diagram** shows all the possible outcomes of two events.

Example 1

Two fair five-sided spinners are spun and the results are added together.

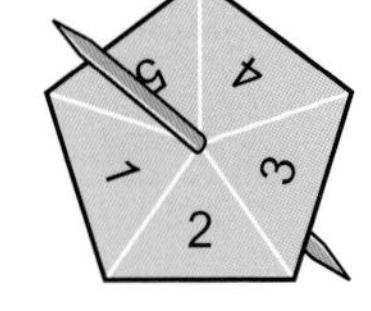

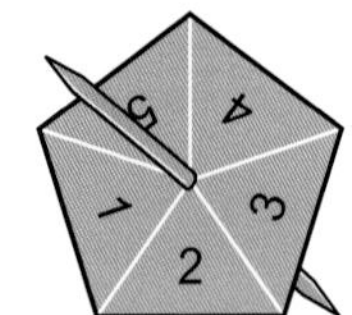

a Draw the sample space diagram to show all the possible outcomes.

b Work out the probability of getting a total of 2.

c Work out the probability of getting a total of 6.

d Work out the probability of getting a total that is a prime number.

a

		Red spinner				
		1	2	3	4	5
Blue spinner	1	(2)	3	4	5	6
	2	3	4	5	6	7
	3	4	5	6	7	8
	4	5	6	7	8	9
	5	6	7	8	9	10

Add the number on the red spinner to the number on the blue spinner.

b $P(2) = \frac{1}{25}$

$\frac{\text{number of ways of scoring 2}}{\text{total number of scores}}$

c $P(6) = \frac{5}{25} = \frac{1}{5}$

d $P(\text{prime}) = \frac{11}{25}$

The outcomes that are prime numbers are 2, 3, 5 and 7.

5 **Modelling** Jake rolls a fair six-sided dice and spins a fair four-sided spinner and adds the results together.

		Dice					
		1	**2**	**3**	**4**	**5**	**6**
Spinner	**1**	2	3				
	2	3					
	3						
	4						

a Copy and complete the sample space diagram to show all the possible outcomes.

b Work out the probability of getting

i a total of 4

ii a total that is a square number

iii a total of 12.

6 **Exam-style question**

Louise spins a four-sided spinner and a five sided spinner.

The four-sided spinner is labelled 2, 4, 6, 8.

The five-sided spinner is labelled 1, 3, 5, 7, 9.

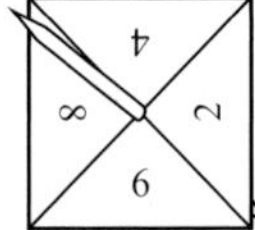

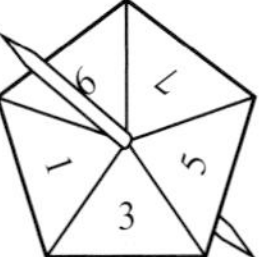

Louise adds the score on the four-sided spinner to the score on the five-sided spinner.

She records the possible total scores in a table.

4-sided spinner

5-sided spinner

+	2	4	6	8
1	3	5	7	9
3	5	7	9	11
5	7	9	11	13
7	9	11		
9	11	13		

Q6 strategy hint Use your completed table to answer parts **b** and **c**.

a Complete the table of possible scores. **(1 mark)**

b Write down all the ways in which Louise can get a total score of 11
One way has been done for you.
(2, 9) **(2 marks)**

Both spinners are fair.

c Find the probability that Louise's total score is less than 6 **(2 marks)**

Nov 2010, Q3, 1380/3H

7 **Modelling**

a Sasha rolls two fair six-sided dice. Draw a sample space diagram to show all of the possible outcomes.

b How many possible outcomes are there altogether?

c Work out the probability of getting a total of

i 3

ii an even number

iii greater than 8.

d Which total are you most likely to get when rolling two six-sided dice?

8 Chloe has two bags of sweets, A and B.
In bag A she has a strawberry flavour, an orange flavour, a lime flavour and a blackcurrant flavour sweet.
In bag B she has a strawberry flavour, a blackcurrant flavour and a lime flavour sweet.
Chloe takes a sweet at random from each bag.

a Draw a sample space diagram to show all the possible outcomes.

b Work out the probability that the sweets will be

i both strawberry flavour

ii the same flavour

iii different flavours.

9 Harry flips a fair coin three times. Work out the probability that the coin lands heads up for all three flips.

Q9 hint List all possible outcomes.

10 Jade, Kyle, Laura, Max and Nicole take part in a chess tournament.
Each player in the tournament plays every other player.
There are 10 matches altogether.
Two players are picked at random to play the first game.
Work out the probability that the first game will be played by a male player and a female player.

10.2 Mutually exclusive events

Objectives

- Identify mutually exclusive outcomes and events.
- Find the probabilities of mutually exclusive outcomes and events.
- Find the probability of an event not happening.

Did you know?

The oldest known dice ever excavated is 5000 years old. Dice used to be called 'bones' because they were made from a bone in the ankle of hoofed animals.

Fluency

Work out

a $0.53 + 0.2$ **b** $0.4 + 0.35$ **c** $1 - 0.72$ **d** $1 - 0.65$ **e** $1 - \frac{2}{5}$ **f** $1 - \frac{5}{9}$

Warm up

1 Boys and girls in a class are in the ratio of 1 : 2. What fraction of the class are boys?

2 A fair 6-sided dice is rolled once. What is the probability of rolling

a a square number b a multiple of 2 c a multiple of 3?

3 Here are some lettered tiles.

One of these tiles is selected at random.
Work out the probability of getting an E or an S.

Key point 3

Two events are **mutually exclusive** if they cannot happen at the same time. For example, when you roll an ordinary dice, you cannot get a 3 and an even number at the same time.

4 Which two events from **Q2** are mutually exclusive?

5 A fair 6-sided dice is rolled once. Work out the probability of rolling

a a square number or a multiple of 5
b a prime number or an even number
c a square number or a multiple of 2.

Key point 4

When events are mutually exclusive you can add their probabilities.
For mutually exclusive events P(A or B) = P(A) + P(B)

6 **Modelling** A standard pack of cards is shuffled and a card is chosen at random. Find the probability of choosing

a a diamond or spade
b a black ace or a heart.

Q6 hint A standard pack of 52 cards is equally split into four suits; hearts, diamonds, clubs and spades. For each suit there is an Ace, 2, 3, 4, 5, 6, 7, 8, 9, 10, Jack, Queen and King.

7 The chance of it raining on a day in July in London is 71%. There is a 6% chance of a dry day with cloud coverage. What is the probability that on a visit to London in July, you will have a clear blue sunny sky with no rain?

Q7 hint The sum of the probabilities of all possible outcomes is □%.

8 The numbered cards are shuffled. A card is chosen at random.

1 2 3 4 5 6 7 8 9 10

Work out the probability of choosing a square number, a prime number or a multiple of 6.

Key point 5

For 3 or more mutually exclusive events, P(A or B or C or…) = P(A) + P(B) + P(C) + …
The probabilities of an exhaustive set of mutually exclusive events sum to 1.

9 **Problem-solving** The table gives the probability of getting each of 1, 2, 3 and 4 on a biased 4-sided spinner.

Number	1	2	3	4
Probability	$4x$	$3x$	$2x$	x

Work out the probability of getting

a 2 or 4 b 1 or 2 or 3.

Key point 6

For mutually exclusive events A and not A, P(not A) = 1 – P(A). A and not A are always mutually exclusive.

Example 2

A bag contains 20 counters. 7 of the counters are red. A counter is taken at random from the bag. Work out the probability that the counter will be

a red b not red.

a $P(\text{red}) = \frac{7}{20}$

b $P(\text{not red}) = 1 - P(\text{red})$

$= 1 - \frac{7}{20}$

$= \frac{13}{20}$

$P(A) = \dfrac{\text{number of successful outcomes}}{\text{total number of possible outcomes}}$

P(not A) = 1 – P(A)

10 **Modelling** A fair 6-sided dice is rolled. Work out the probability of rolling

a 1 b not 1.

11 **Modelling** A standard pack of cards is shuffled and a card is chosen at random. Find the probability of choosing

a a heart b not a heart.

12 The probability that it will rain tomorrow is 0.88.
Work out the probability that it will not rain tomorrow.

13 **Exam-style question**

Josie rolls a biased dice.

The probability that the dice will land on 1 or 2 or 3 or 4 or 5 is given in the table.

Score	1	2	3	4	5	6
Probability	0.15	0.25	0.20	0.10	0.15	

Work out the probability that the dice will land on 6. **(2 marks)**

Nov 2011, Q2, 2381/6B

14 **Reasoning** Anna has a box of chocolates. In the box there are milk, plain and white chocolates in the ratio 4:3:2. Anna doesn't like white chocolates. Work out the probability that she will pick a chocolate that is not white.

> **Q14 hint** How many parts are there in total for the ratio 4:3:2?
> P(not white) = $1 - \frac{\square}{\square}$

15 A and B are two mutually exclusive events.
P(A) = 0.25 and P(A or B) = 0.6
Work out the value of P(B).

> **Q15 hint** For two mutually exclusive events, P(A or B) = P(A) + P(B).

16 **Problem-solving** C and D are two mutually exclusive events.
P(D) = 0.4 and P(C or D) = 0.78
Work out P(not C).

10.3 Experimental probability

Objectives

- Work out the expected results for experimental and theoretical probabilities.
- Compare real results with theoretical expected values to decide if a game is fair.

Did you know?

Scientists use results from their experiments to work out the chance of a drug or treatment being successful.

Fluency

Simplify

a $\frac{9}{12}$ **b** $\frac{15}{45}$ **c** $\frac{24}{36}$ **d** $\frac{27}{45}$

Warm up

1 Work out

a 50×0.3 b 200×0.7 c $210 \times \frac{1}{3}$ d $150 \times \frac{4}{5}$

2 Write the correct sign, < or >, between each pair of fractions.

a $\frac{1}{3}$ and $\frac{2}{5}$ b $\frac{5}{9}$ and $\frac{7}{11}$ c $\frac{3}{10}$ and $\frac{4}{15}$ d $\frac{5}{12}$ and $\frac{3}{8}$

3 Ella dropped a drawing pin on the table lots of times. It landed either point up or point down. She recorded her results in a frequency table.

Position	Frequency
Point up	43
Point down	7

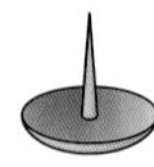

a Work out the total frequency.

b Work out the experimental probability of the drawing pin landing

i point up ii point down.

c She drops the drawing pin 100 times. How many times do you expect it to land point up?

Discussion When you repeat an experiment, will you get exactly the same results? Why is experimental probability only an estimate? How can you improve the accuracy of the estimate?

Key point 7

In a probability experiment a trial is repeated many times and the outcomes recorded.
The relative frequency of an outcome is called the **experimental probability**.

$$\text{Experimental probability of an outcome} = \frac{\text{frequency of outcome}}{\text{total number of trials}}$$

ActiveLearn Homework, practice and support: Higher 10.3

4 **Exam-style question**

Abraham and Betty have a biased dice.
They each want to find an estimate for the probability that the dice will land on a six.
Abraham is going to roll the dice 60 times.
He will record the number of sixes he gets.
Betty is going to roll the dice 600 times.
She will record the number of sixes she gets.
Who is more likely to get the better estimate?
Give a reason for your answer. **(1 mark)**

March 2012, Q4, 2381/6B

Exam hint
You only get the mark for the name and a reason. Just writing the correct name scores 0 marks.

Key point 8

Theoretical probability is calculated without doing an experiment.

Example 3

Josh uses this spinner for a game.

a What is the theoretical probability that the spinner will land on the letter B?

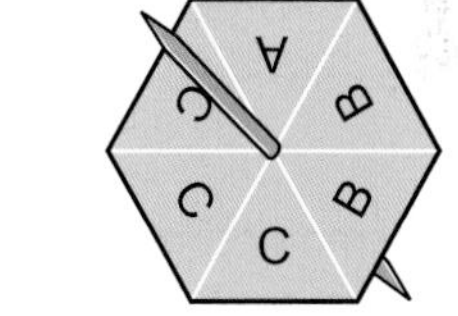

Josh is going to spin this spinner 300 times.

b Estimate how many times the spinner will land on the letter B.

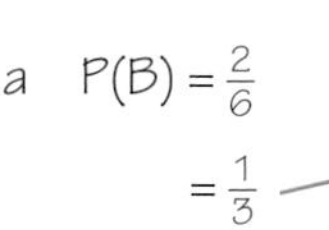

a $P(B) = \frac{2}{6}$

$= \frac{1}{3}$ — Simplify fractions where possible.

b $\frac{1}{3} \times 300$ — Expected number of outcomes = number of trials × probability

= 100 times

5 **Modelling** Mia makes this 8-sided spinner for an experiment.

a What is the theoretical probability that the spinner will land on blue?

Mia is going to spin this spinner 400 times.

b Estimate how many times the spinner will land on blue.

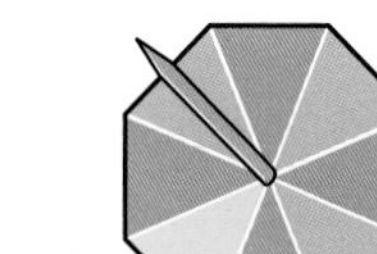

6 **Modelling** There are red, blue and yellow counters in a bag in the ratio of 5:6:1.

a What is the probability of choosing a red counter?

A counter is picked at random from the bag and then replaced.
This is done 180 times.

b How many times would you expect a red counter to be picked?

7 The probability of England losing their next football match is 0.28
The probability of England drawing their next football match is 0.36
Work out an estimate for the number of times England will win over their next 50 football matches.

Key point 9

As the number of experiments increases, the experimental probability gets closer and closer to the theorectical probability.

8 **Modelling** The table shows the results of rolling a six-sided dice.

a Copy and complete the table, calculating the relative frequency for each outcome.

b What is the experimental probability of rolling a 6?

c When the dice is rolled 500 times, how many times would you expect to get a 6?

d Is the dice fair? Give a reason for your answer.

Number	Frequency	Relative frequency
1	23	
2	22	
3	21	
4	18	
5	9	
6	7	

9 **Modelling** The table shows the results of spinning a five-sided spinner.

Number	1	2	3	4	5
Frequency	46	39	37	40	38

Is the spinner fair? Give a reason for your answer.

10 **Modelling** Holly flips two coins and records the result. She does this 160 times. One possible outcome is (tail, tail). Estimate the number of times she will get two tails.

11 **Modelling** The probability of winning a prize in a raffle is $\frac{1}{200}$. Sarah says that if she buys 200 tickets she will win a prize. Is she right? Give a reason for your answer.

12 **Modelling** Ben rolls two dice 180 times.
How many times would you expect to get a total

a of 12 b of 7 c that is a prime number?

13 **Modelling** A dentist estimates that the probability a patient will come to see him needing a filling is 0.235.
Of the next 160 patients who come to see him, 25 need a filling.
How good is the dentist's estimate of this probability? Explain your answer.

10.4 Independent events and tree diagrams

Objectives

- Draw and use frequency trees.
- Calculate probabilities of repeated events.
- Draw and use probability tree diagrams.

Why learn this?

When you use a probability tree diagram to find the probability of two or more events you will avoid missing any combinations.

Fluency

Work out

a $\frac{2}{5} + \frac{1}{5}$ **b** $\frac{4}{9} + \frac{3}{9}$ **c** $\frac{1}{2} + \frac{1}{4}$ **d** $\frac{3}{8} + \frac{1}{4}$

Warm up

1 Work out these calculations; give your answers in their simplest form.

a $\frac{7}{10} \times \frac{2}{9}$ b $\frac{5}{9} \times \frac{3}{4}$ c $\frac{5}{7} \times \frac{13}{20}$

d 0.4×0.2 e 0.6×0.7 f 0.55×0.8

Key point 10

A **frequency tree** shows two or more events and the number of times they occurred.

ActiveLearn Homework, practice and support: Higher 10.4

2 In a football tournament 20 matches were played. In 8 of the matches Team A scored the first goal. In 5 of these matches they also scored the second goal. Team B scored the first two goals in 3 of the matches. Copy and complete the frequency tree for the 20 matches played.

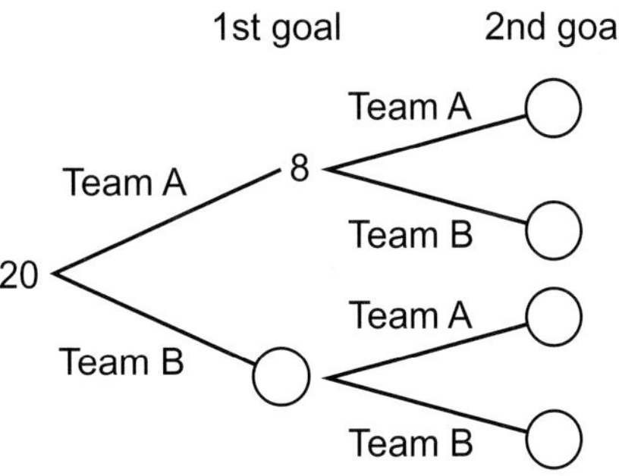

3 A garage records the MOT test results on 40 cars.
Of the 40 cars tested, 15 of them are less than 5 years old.
11 of the cars under 5 years old passed.
28 cars passed altogether.

a Copy and complete the frequency tree.
b Work out the probability that a car fails its MOT.

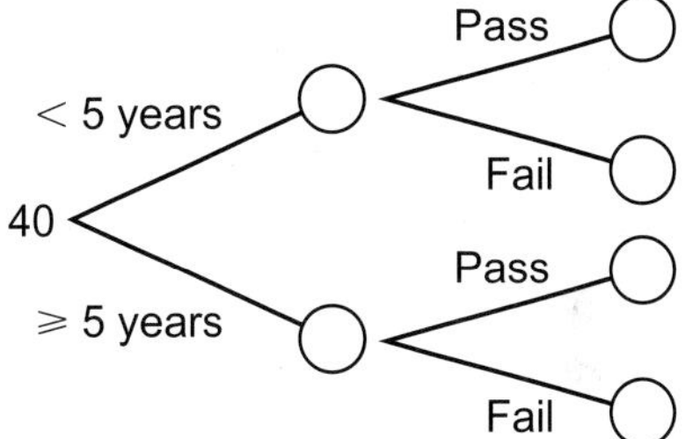

4 **STEM** 80 people with similar symptoms were tested for a virus using a new trial medical test.
19 of the people tested showed a positive result.
The virus only developed in 11 of the people who tested positive.
A total of 67 people did not develop the virus at all.

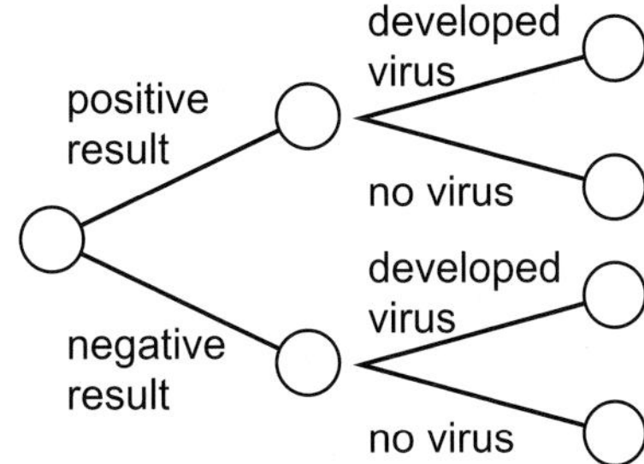

a Copy and complete the frequency tree.
b Work out the probability that a person develops the virus.

Key point 11

Two events are **independent** if one event does not affect the probability of the other.
For example, flipping heads with a coin has no effect on rolling an even number with a dice, so they are independent events.
To find the probability of two independent events, multiply their probabilities.
P(A and B) = P(A) × P(B)

5 Connor and Ryan compete against each other over a 100 metre sprint and over a 100 metre swim.
The probability that Connor will win the sprint is 0.3
The probability that Connor will win the swimming is 0.8
Assuming that the two events are independent, work out the probability that Connor will win both races.

6 **Reasoning** There are two sets of traffic lights on Matthew's car journey to school.
The probability that he has to stop at the first set of traffic lights is 0.45.
The probability that he has to stop at the second set of traffic lights is 0.35.
Work out the probability that he will
- a not stop at the first set of traffic lights
- b not stop at the second set of traffic lights
- c not stop at either set of traffic lights.

Discussion What assumption did you make?

7 A card is taken at random from each of two ordinary packs of cards, pack A and pack B.
Work out the probability of getting
- a a black card from pack A and a black card from pack B
- b a heart from pack A and a spade from pack B
- c a queen from pack A and an even-numbered card from pack B
- d an ace from pack A and an ace of clubs from pack B
- e a queen of hearts from each pack.

8 Joe plays a spin-the-wheel game at the fair.
The probability that he wins is $\frac{3}{5}$.
Calculate the probability that he wins three successive games.

Key point 12

A **tree diagram** shows two or more events and their probabilities.

Example 4

This fair five-sided spinner is spun twice.
- a Draw a tree diagram to show the probabilities.
- b What is the probability of both spins landing on red?
- c What is the probability of landing on one red and one blue?

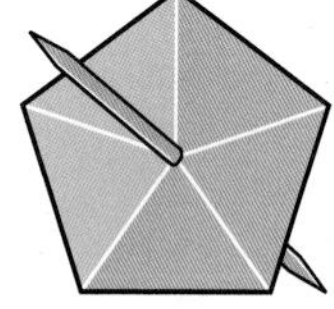

a

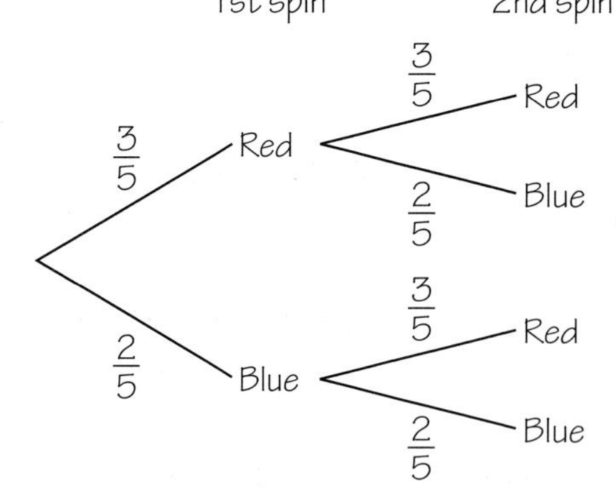

Write the probability on each branch of the diagram.

b $P(R, R) = \frac{3}{5} \times \frac{3}{5} = \frac{9}{25}$

Go along the branches for Red, Red. The 1st and 2nd spins are independent, so multiply the probabilities.

c $P(R, B) = \frac{3}{5} \times \frac{2}{5} = \frac{6}{25}$

$P(B, R) = \frac{2}{5} \times \frac{3}{5} = \frac{6}{25}$

Go along the branches for Red, Blue and Blue, Red.

$P(R, B \text{ or } B, R) = \frac{6}{25} + \frac{6}{25} = \frac{12}{25}$

The outcomes Red, Blue and Blue, Red are mutually exclusive, so add the probabilities of their outcomes.

9 On a hook-a-duck game at a fundraising event you win a prize if you pick a duck with an 'X' on its base. Aaron picks a duck at random, replaces it and then picks another one.

      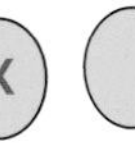  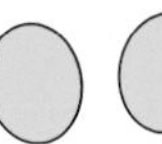

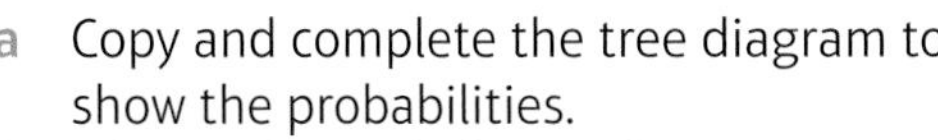

a Copy and complete the tree diagram to show the probabilities.

b What is the probability of
   i winning two prizes
   ii winning nothing
   iii winning one prize
   iv winning at least one prize?

> **Q9b iv hint** Winning 'at least one' means winning one or more.

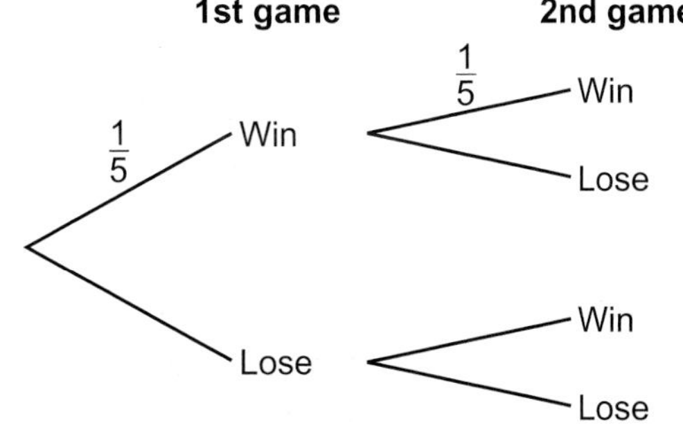

10 Megan has two bags of counters, labelled A and B.
In bag A there are 3 red and 5 green counters.
In bag B there are 1 red and 5 green counters.
A counter is chosen at random from each bag.

a Copy and complete the tree diagram to show the probabilities.

b Work out the probability of choosing
   i two counters the same colour
   ii one red and one green counter
   iii no red counters
   iv at least one red counter.

Bag A — Bag B: Red (Red, Green); Green (Red, Green)

Discussion How did you calculate P(at least one red)? Is there another way?

11 Rachel and Max each have a box of chocolates.
Rachel has 5 milk and 2 white chocolates in her box.
Max has 7 milk and 3 white chocolates in his box.
They each choose a chocolate from their own box at random.

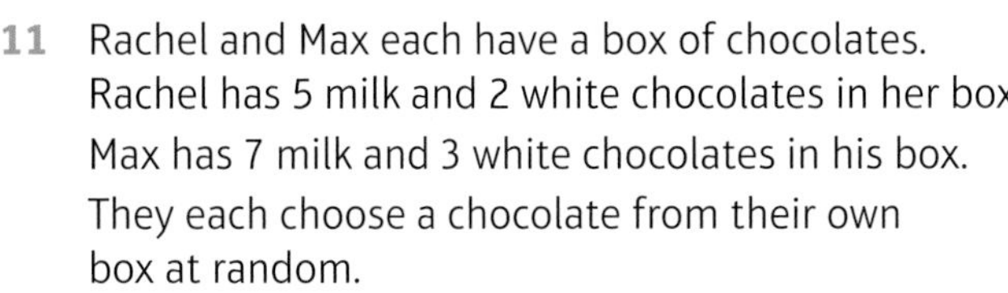

a Copy and complete the tree diagram to show the probabilities.

b Work out the probability that
   i they both choose a milk chocolate
   ii one of them chooses a milk chocolate and one of them chooses a white chocolate.

Rachel — Max: Milk (Milk, White); White (Milk, White)

Discussion What assumptions have you made when calculating the probabilities in part **b**?

12 **Exam-style question**

There are only red and yellow marbles in a bag.
There are 5 red marbles and 3 yellow marbles.
Ethan takes at random a marble from the bag, notes the colour and then puts the marble back in the bag.
Ethan then repeats this process.

a Complete the probability tree diagram. **(2 marks)**

b Work out the probability that Ethan takes marbles of different colours. **(3 marks)**

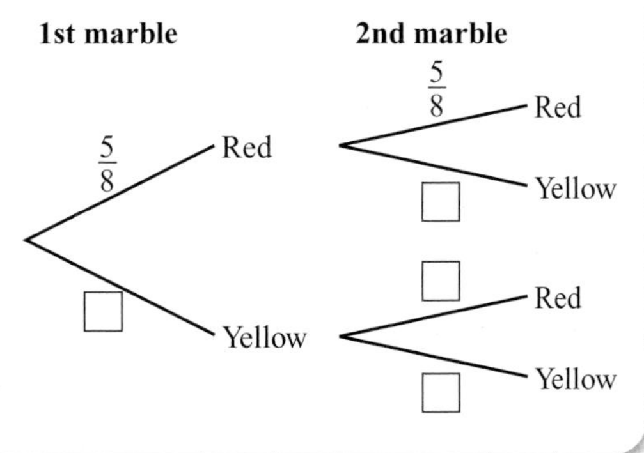

13 **Reasoning** Caitlin spins two spinners.
On spinner 1, P(pink) = 0.2
On spinner 2, P(pink) = 0.35

a Draw a tree diagram to show all the possible outcomes.
b What is the probability of only one spinner landing on pink?
c What is the probability of both spinners landing on pink?
d If each spinner was spun 500 times, how many times would you expect them both to land on pink?

10.5 Conditional probability

Objectives

- Decide if two events are independent.
- Draw and use tree diagrams to calculate conditional probability.
- Draw and use tree diagrams without replacement.
- Use two-way tables to calculate conditional probability.

Why learn this?

Conditional probability is used in most type of statistics.

Fluency

There are 4 red counters in a bag of 10 counters. What is
a P(red) **b** P(not red)?

Warm up

1 This spinner is spun twice.

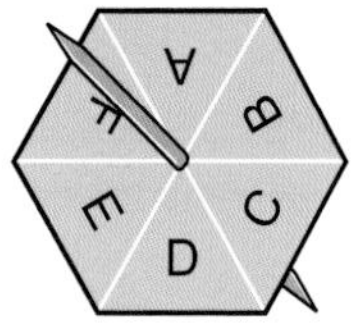

a Copy and complete the tree diagram to show all the probabilities.
b Work out the probability of the spinner landing on a vowel for both spins.

1st spin
2nd spin
$\frac{1}{3}$ vowel
vowel
consonant
consonant
vowel
consonant

2 The two-way table shows the subjects students like best.
Work out the probability that a student picked at random
a likes English best
b does not like science best.
c Work out the probability that a male student picked at random likes maths best.

	English	Maths	Science	Total
Male	15	23	22	60
Female	32	17	21	70
Total	47	40	43	130

Key point 13

If one event depends upon the outcome of another event, the two events are **dependent events**. For example, removing a red card from a pack of playing cards reduces the chance of choosing another red card.
A tree diagram can be used to solve problems involving dependent events.

ActiveLearn Homework, practice and support: Higher 10.5

3 For each of the events, state if the events are independent or dependent.

a Randomly choosing a chocolate from a box, eating it, and then choosing another.

b Rolling two six-sided dice.

c Flipping a coin three times.

d Choosing two socks from a drawer.

e Randomly choosing a counter from a bag, replacing it and then choosing another counter.

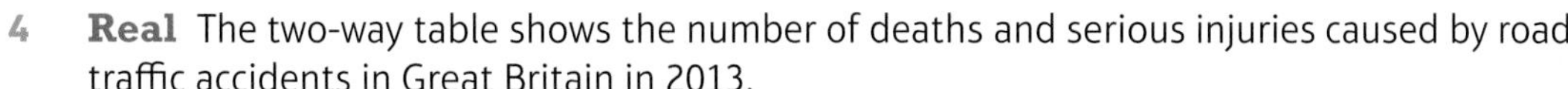

Key point 14

A **conditional probability** is the probability of a dependent event. The probability of the second outcome depends on what has already happened in the first outcome.

4 **Real** The two-way table shows the number of deaths and serious injuries caused by road traffic accidents in Great Britain in 2013.

		Speed limit			
		20 mph	**30 mph**	**40 mph**	**Total**
Type of injury	**Fatal**	6	520	155	681
	Serious	420	11 582	1662	13 664
	Total	426	12 102	1817	14 345

Work out an estimate for the probability

a that the accident is fatal given that the speed limit is 30 mph

b that the accident happens at 20 mph given that the accident is serious

c that the accident is serious given that the speed limit is 40 mph.
Give your answers to 2 decimal places.

Q4a hint

$\dfrac{\text{number of fatal accidents at 30 mph}}{\text{total number of accidents at 30 mph}}$

Example 5

There are 10 pencils in Toby's pencil case.
Seven of the pencils are HB pencils.
Toby takes two pencils out of his pencil case.

a Draw a tree diagram to show all the possible outcomes.

b Work out the probability that he picks out at least one HB pencil.

a 1st pencil 2nd pencil

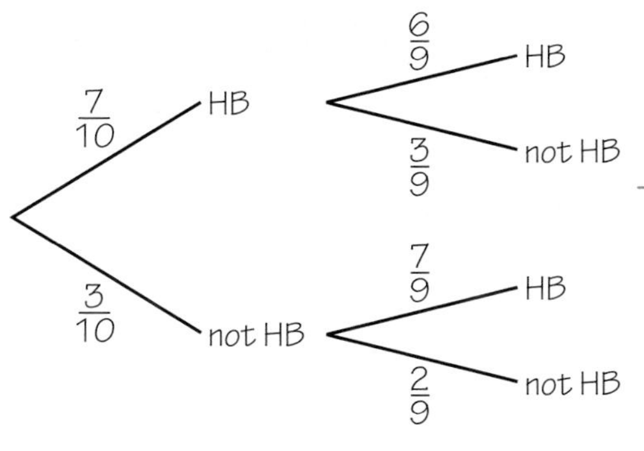

Taking two pencils from the pencil case at the same time is the same as taking one pencil, then another (without replacement).

b P(at least 1 HB) = 1 − P(no HB)

P(not HB, not HB) = $\frac{3}{10} \times \frac{2}{9} = \frac{6}{90} = \frac{1}{15}$

P(at least 1 HB) = $1 - \frac{1}{15} = \frac{14}{15}$

You don't need to simplify probability fractions, but sometimes it makes calculations easier.

5 Chris has a bag containing 5 red and 3 orange sweets. He chooses a sweet at random and eats it. He then chooses another sweet at random.

a Copy and complete the tree diagram to show all the probabilities.

b Work out the probability that the sweets will be

i the same colour

ii one of each colour

iii not orange.

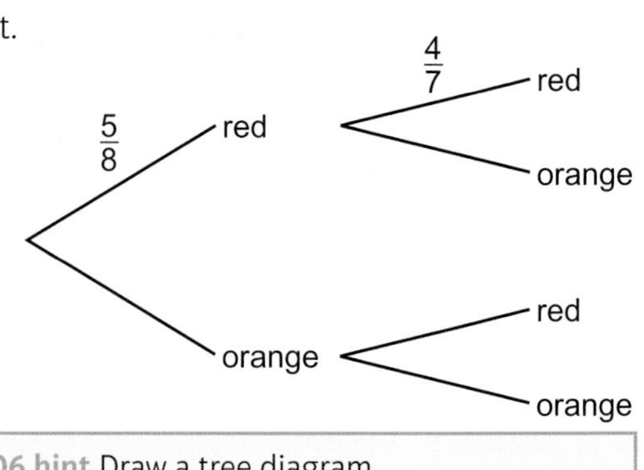

6 **Problem-solving** In a group of students, 55% are girls. 30% of these girls prefer to play electronic games on a hand-held gaming device.

75% of the boys prefer to play electronic games on a games console.

One student is chosen at random.

Find the probability that this is a boy who prefers to play games on a hand-held gaming device.

Q6 hint Draw a tree diagram.

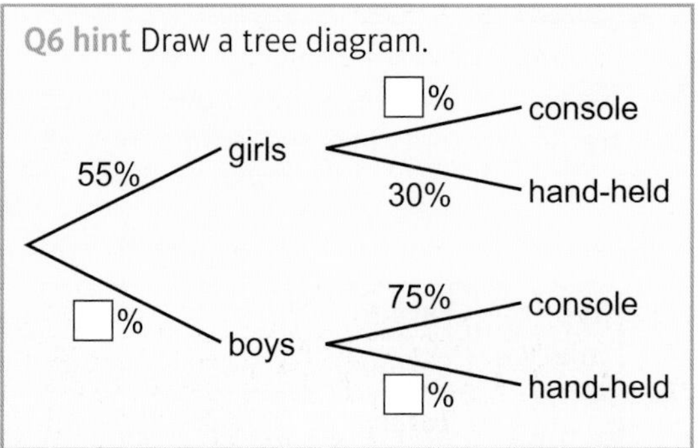

7 **Problem-solving** Callum either walks to school or travels by car.

The probability that he walks to school is 0.65. If he walks to school, the probability that he will be late is 0.3

If he travels to school by car, the probability that he will be late is 0.05

Work out the probability that he will not be late.

8 Emily and Zoe have 5 chocolate biscuits, 3 shortbread biscuits and 2 ginger biscuits in a tub. Emily picks a biscuit, at random, from the tub. Then Zoe picks a biscuit.

a Copy and complete the diagram to show all the probabilities.

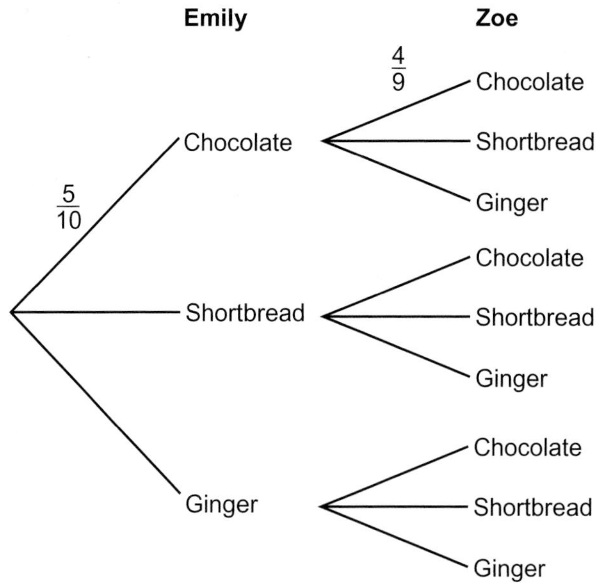

b What is the probability that they both pick the same type of biscuit?

9 **Problem-solving** There are 5 boxes of cornflakes and 7 boxes of puffed wheat.

Mike and Reece both choose a box at random.

Work out the probability that they do not choose the same type.

10 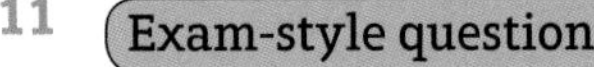 A bag contains 4 red, 3 blue and 2 green marbles.
Jamie chooses two counters, at random, from the bag.
Work out the probability that his two counters are the same colour.

11 **Exam-style question**

There are 3 orange sweets, 2 red sweets and 5 yellow sweets in a bag.

Sarah takes a sweet at random.

She eats the sweet.

She then takes another sweet at random.

Work out the probability that both sweets are the same colour. **(4 marks)**

June 2010, Q26, 1380/3H

Q11 strategy hint
Once a sweet has been taken, and eaten, remember the total number of sweets in the bag is now one less.

10.6 Venn diagrams and set notation

Objectives

- Use Venn diagrams to calculate conditional probability.
- Use set notation.

Why learn this?

John Venn, born in Kingston upon Hull, first introduced Venn diagrams in 1880.

Fluency

What are the integer values in each set?

a $\{x: 0 < x \leqslant 5\}$ **b** $\{x: -3 \leqslant x < 2\}$

Warm up

1 Amber surveyed Year 10 students to see how many owned a smartphone and tablet.
The Venn diagram shows her results.

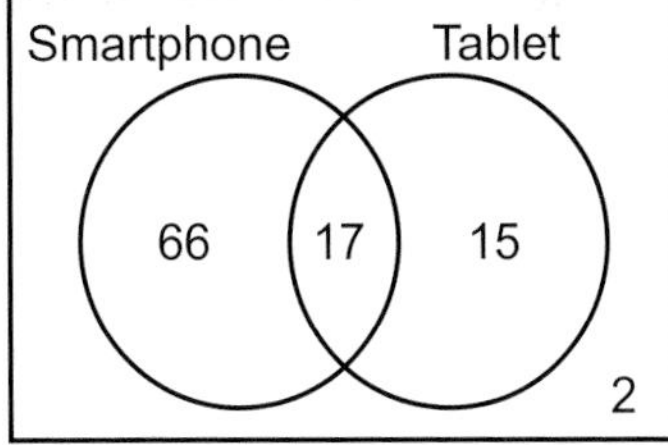

- **a** How many students did not own either a smartphone or tablet?
- **b** How many students owned a tablet?
- **c** How many students owned both a smartphone and a tablet?
- **d** How many students took part in Amber's survey?

Key point 15

Curly brackets { } show a set of values.
∈ means 'is an element of'.

Communication hint
An element is a 'member' of a set.

2 A = {positive even numbers < 10}
B = {prime numbers < 10}

- **a** List the numbers in each set.
  A = {2, …} B = {…}
- **b** Write 'true' or 'false' for each statement.
  - i 6 ∈ A
  - ii 1 ∈ B
  - iii 5 ∈ B

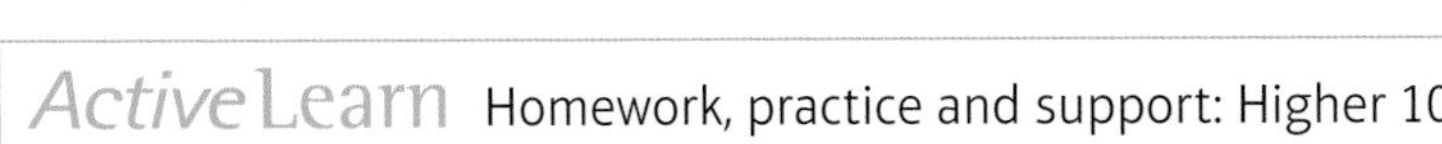

3 The Venn diagram shows two sets, P and Q, and the set of all numbers being considered, ξ.
Write all the elements of each set inside curly brackets { }.

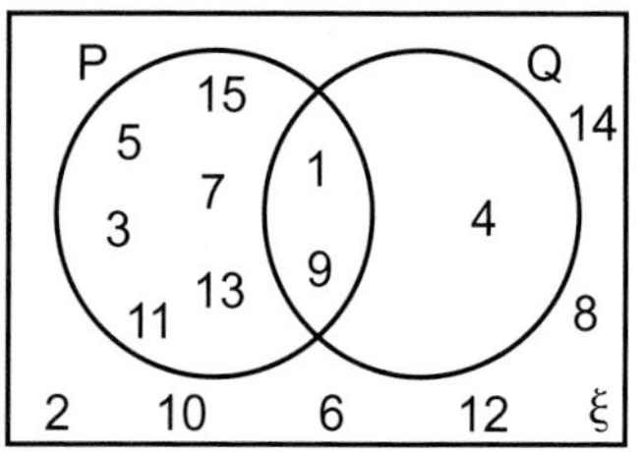

a P
b Q
c ξ
d Which set is {square numbers < 16}?
e Write descriptions of the other two sets.

Key point 16

A ∩ B means 'A intersection B'. This is all the elements that are in A *and* in B.

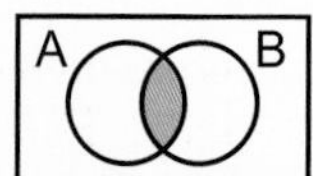

A ∪ B means 'A union B'. This is all the elements that are in A *or* B *or* both.

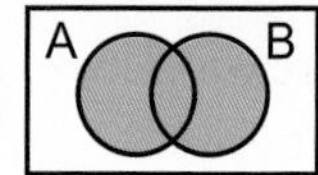

A′ means the elements *not* in A.

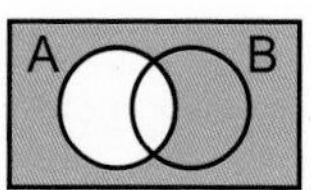

ξ means the universal set – all elements being considered.

4 For the Venn diagram in **Q3**, write these sets.

a P ∪ Q
b P ∩ Q
c P′
d Q′
e P′ ∩ Q
f Q′ ∩ P

Key point 17

You can calculate probabilities from Venn diagrams.

Example 6

The Venn diagram shows the number of students studying German (G) and Mandarin (M).
A student is picked at random. Work out

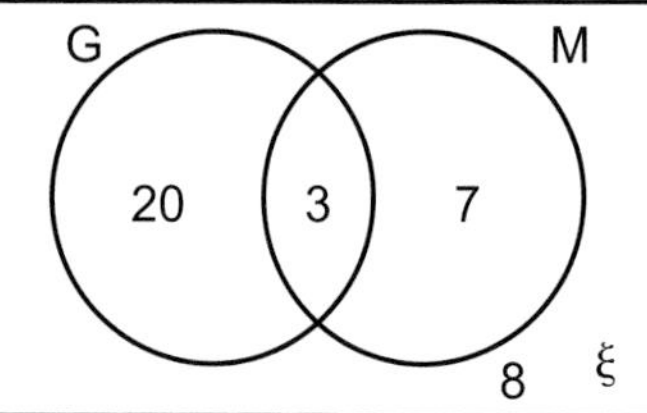

a P(G ∩ M)
b P(G′)
c P(G ∪ M)

a $20 + 3 + 7 + 8 = 38$ — Work out the total number of students.

$P(G \cap M) = \frac{3}{38}$ — $\frac{\text{Number of students in } G \cap M}{\text{total number of students}}$

b $P(G') = \frac{7 + 8}{38} = \frac{15}{38}$ — $\frac{\text{Number of students in } G'}{\text{total number of students}}$

c $P(G \cup M) = \frac{20 + 3 + 7}{38} = \frac{30}{38}$

5 The Venn diagram shows two events when a 12-sided dice is rolled: prime numbers and multiples of 2.

X = {number is prime}

Y = {number is a multiple of 2}

Work out

a P(X) b P(Y) c P(X ∩ Y)

d P(X ∪ Y) e P(X′) f P(Y′)

g P(X ∩ Y′) h P(X′ ∪ Y)

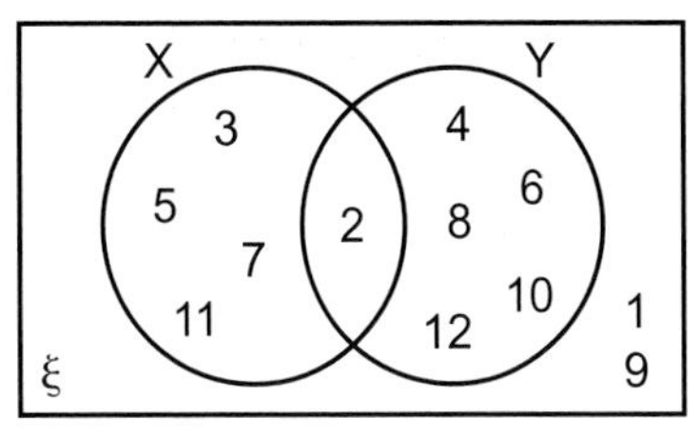

6 **Reasoning** Charlie asks the 30 students in his class if they passed their English (E) and maths (M) tests.

21 students passed both their English and maths tests.

2 students didn't pass either test.

25 students passed their maths test.

a Draw a Venn diagram to show Charlie's data.

b Work out

i P(E) ii P(E ∩ M) iii P(E ∪ M) iv P(E′ ∩ M)

Discussion What does P(E′ ∩ M) mean in the context of the question?

7 Lucy carried out a survey of 150 students to find out how many students play an instrument (I) and how many play for a school sports team (S).

63 students play on a school sports team.

27 students play an instrument and play on a school sports team.

72 students do not either play an instrument or play on a school sports team.

a Draw a Venn diagram to show Lucy's data.

b Work out the probability that a student plays an instrument.

c Work out the probability that a student plays an instrument given that they play on a school sports team.

Q6c hint

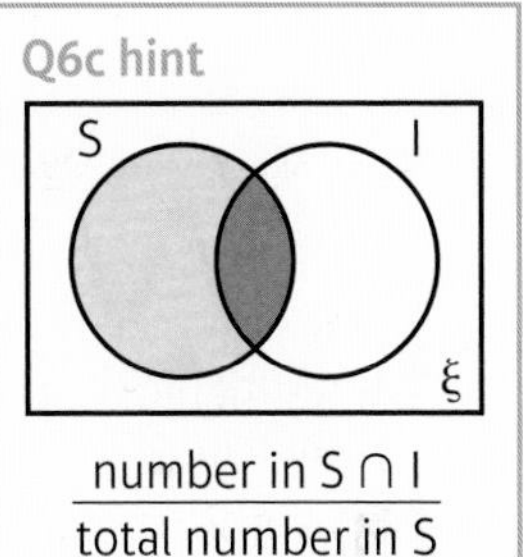

$\frac{\text{number in S} \cap \text{I}}{\text{total number in S}}$

Key point 18

A ∩ B ∩ C means the **intersection** of A, B and C.

A ∪ B ∪ C means the **union** of A, B and C.

P(A ∩ B | B) means the probability of A and B given B.

8 Caitlin did a survey of pet owners owning cats (C), dogs (D) and fish (F). The Venn diagram shows her results.

a How many people took part in the survey?

One of the pet owners is chosen at random.

b Work out

i P(C)

ii P(C ∩ D | D)

iii P(D ∩ F | F).

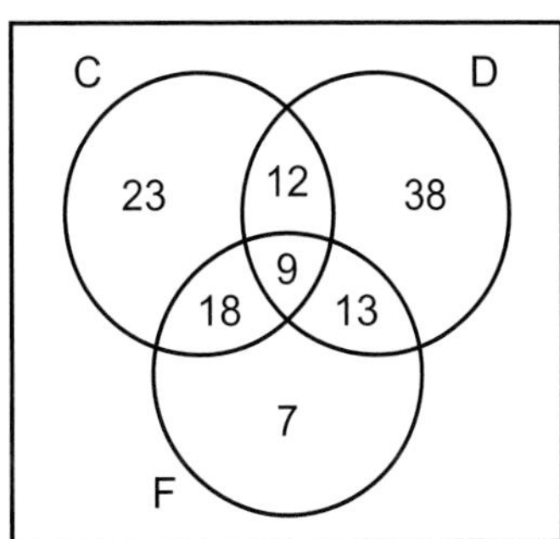

Q8b ii hint P(C ∩ D | D) means the probability of a cat and dog owner given that pet owner owns a dog.

9 The Venn diagram shows people's choice of chocolate (C), strawberry (S) and vanilla (V) ice cream flavours for the 'three scoops' dessert.

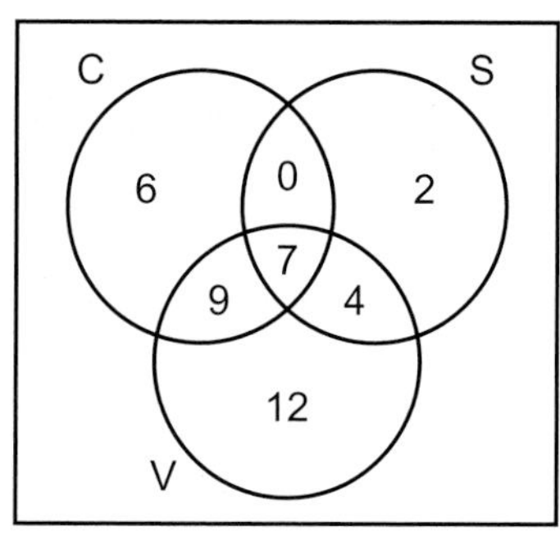

a How many people had all three flavours?

b How many people chose ice cream for dessert?

c Work out
i $P(C \cap S \cap V)$
ii $P(S \cap V)$
iii $P(S \cap V \mid S)$.

10 **Exam-style question**

There are 80 students at a language school.

All 80 students speak at least one language from French, Italian and Spanish.

7 of the students speak French, Italian and Spanish.

15 of the students speak French and Italian.

26 of the students speak French and Spanish.

17 of the students speak Italian and Spanish.

41 of the students speak French.

52 of the students speak Spanish.

> **Q10 strategy hint** Once you have put the number 7 in the overlap of all three circles, you can work out another number to fill in on your diagram for each sentence in the question.

a Draw a Venn diagram to show this information. **(3 marks)**

One of the students is chosen at random.

b Work out the probability that this student speaks French but not Italian. **(2 marks)**

Given that the student speaks Spanish,

c work out the probability that this student also speaks French. **(2 marks)**

10 Problem-solving: Drug testing

Objectives

- Be able to construct and use a tree diagram to find probabilities.
- Be able to calculate a cost and compare two different scenarios.

Drug testing is widely used in sport, to help ensure fair play. These tests aim to identify any athletes guilty of taking performance-enhancing drugs. However, they are never 100% accurate and different methods have different success rates.

The table below shows the two different types of drug tests offered by one company.

It is suspected that 5% of athletes at an international athletics event are taking performance-enhancing drugs.

1 Complete the tree diagram below for test A by filling in the probabilities. What percentage of people taking the test will test positive, even though they haven't taken any drugs?

> **Q1 communication hint** A 'positive' test result means the test shows that drugs have been taken. A 'negative' test result means the test shows that drugs have not been taken.

2 Draw a tree diagram for test B. Using your tree diagram, explain why half the people who tested positive are actually innocent.

> **Q2 hint** Compare the probability of a false positive (a result that incorrectly shows the athlete has taken drugs) and a genuine positive result.

The organisers of the event must choose to use test A or test B. They need to consider the overall cost and the reliability of the test results. (Reputations would be damaged if they accused innocent competitors of using drugs, or if they let guilty competitors get away with drug taking.)

At the event a random sample of 600 participants are tested. To improve accuracy, all positive results are tested again at the same cost. If the retest shows a positive result the competitor will be found guilty of using performance-enhancing drugs.

3 By comparing the costs and the overall reliability of the tests, decide which of the two tests you would recommend to the organisers. Make sure you include calculations to create a convincing argument.

> **Q3 hint** Start by working out how many tests will need to be taken. This needs to include any retests.

Drug tests

Test	Accuracy	Cost per test
A	98%	£52
B	95%	£40

Test A results

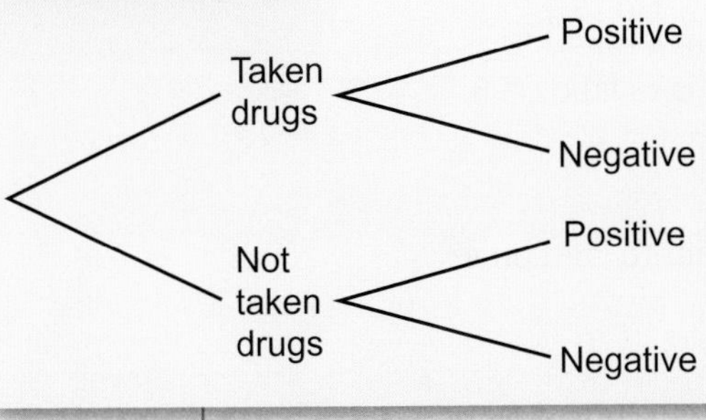

10 Check up

Log how you did on your Student Progression Chart.

Calculating probability

1 The probability that it will rain today is 0.3.
The probability that it will rain tomorrow is 0.25.
The two probabilities are independent.
 a Work out the probability that it will not rain tomorrow.
 b Work out the probability that it will rain today and tomorrow.

2 A bag contains toy animals. Emma takes an animal from the bag at random.
 a The probability of choosing a sheep is $\frac{1}{5}$.
 What is the probability of choosing an animal that is not a sheep?
 b The probability of choosing a pig is $\frac{1}{2}$.
 What is the probability of choosing a sheep or a pig?

3 Ewan spins the spinner and rolls the 6-sided dice.
He finds the total of the outcomes.

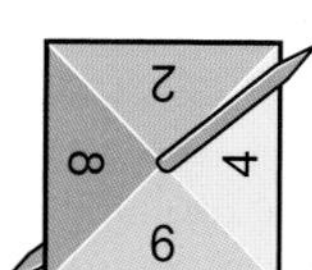

 a Draw a sample space diagram to show all the possible outcomes.
 b Work out the probability of scoring a total of
  i 8
  ii a multiple of 3
  iii more than 9.

4 Jane records the hair colour of students in her class and whether they wear glasses.
The two-way table shows her results.

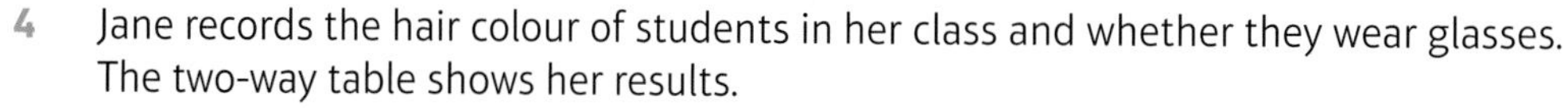

A student is chosen at random. Given that a student has dark hair, what is the probability that they wear glasses?

	Hair colour			
	Fair	**Dark**	**Ginger**	**Total**
Wears glasses	0	2	1	3
Does not wear glasses	13	15	1	29
Total	13	17	2	32

5 A and B are mutually exclusive. P(B) = 0.45, P(A or B) = 0.6. Work out P(A).

Experimental probability

6 The table shows the probability of each number on a six-sided spinner.

Number	1	2	3	4	5	6
Probability	0.2	0.3		0.15		0.15

The spinner is equally as likely to land on 3 as it is to land on 5.
 a Copy and complete the table.

The spinner is spun 300 times.
 b How many times would you expect the spinner to land on 4?
 c Is the spinner fair? Explain your answer.

Tree diagrams and Venn diagrams

7 Grace spins two spinners, A and B.
The probability of getting an even number on spinner A is 0.4.
The probability of getting an even number on spinner B is 0.65.

a Copy and complete the tree diagram to show all the possible probabilities.

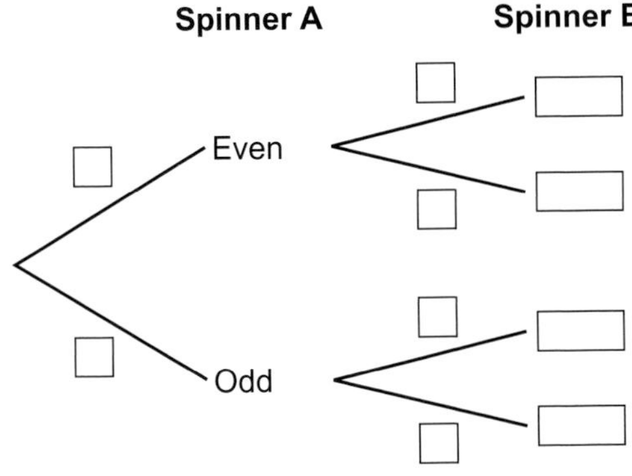

b Work out the probability of getting an even number on
i neither spinner ii only one spinner.

8 **Reasoning** Lewis surveyed 140 students in his year group to find out if they sent text messages or emails last week.
79 students sent text messages and emails.
126 students sent text messages.

a Draw a Venn diagram to show Lewis's data.

A student is chosen at random.

b Work out the probability that the student sent an email last week.

c Given that the student sent a text message, work out the probability that they sent an email.

9 The Venn diagram shows two sets, R and S.
Copy and complete

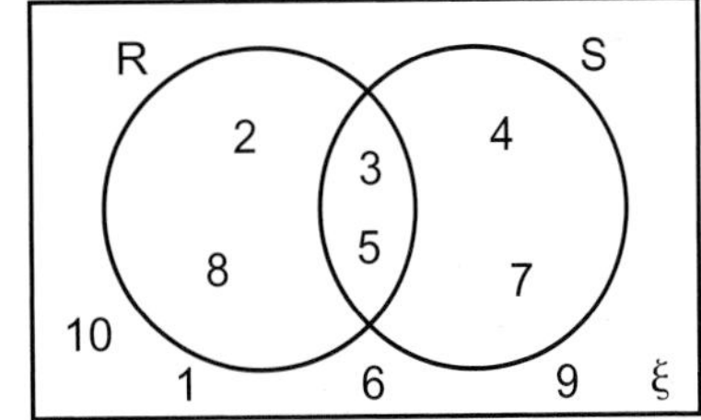

a $R = \{ \quad \}$

b $R' = \{ \quad \}$

c $R \cap S = \{ \quad \}$

d $\xi = \{ \quad \}$

e Is $4 \in R \cup S$ true?

10 How sure are you of your answers? Were you mostly

Just guessing  Feeling doubtful  Confident 

What next? Use your results to decide whether to strengthen or extend your learning.

Reflect

* Challenge

11 **a** Use the numbers 1 to 6 to fill in the sectors of these two spinners so that when the results are added together the probability of getting a total of 7 is $\frac{3}{20}$. Draw a sample space diagram to help you.

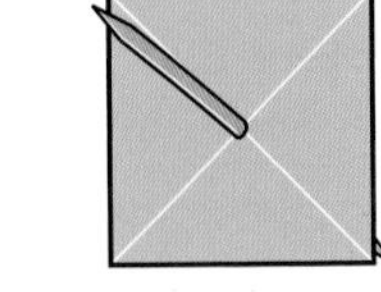

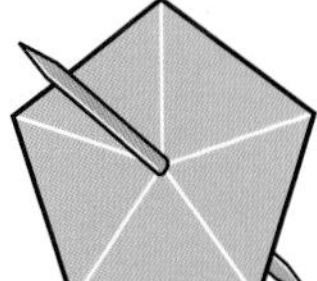

b Both of your spinners are spun once and the results are added together. This is then repeated. Draw a tree diagram to show the probability of getting a total of 7 or not a total of 7 on each spin.

c Use your tree diagram to work out the probability of spinning two 7s.

10 Strengthen

Calculating probability

1 Below are some lettered tiles.

One of these tiles is selected at random. Work out

a P(I) b 1 − P(I) c P(not I).

d What do you notice about your answers to parts **b** and **c**?

e Work out

i P(T) ii P(not T).

Q1c hint List all the letters not I.

Q1e hint Use your answer to part **d** to help you.

2 Using the tiles in Q1, work out

a P(A) b P(S) c P(A or S).

d What do you notice about your answers to parts **a**, **b** and **c**?

e Work out P(I or T).

Q2c hint How many letters are A or S?

Q2e hint Use your answer to part **d** to help you.

3 The probability of getting a letter 'S' on a tile in a word game is 0.05. The probability of getting tiles 'S' or 'E' is 0.2. Work out the probability of getting the tile 'E'.

Q3 hint P(S or E) = P(S) + P(E)

4 Leah is playing in two netball matches tomorrow. Her team can win, draw or lose their matches.

a Copy and complete the sample space diagram to show all the possible outcomes for both matches.

		1st match		
		Win	**Draw**	**Lose**
2nd match	**Win**	W, W	D, W	
	Draw			
	Lose	D, L		

b How many possible outcomes are there?

c What is the probability that Leah's team draw both of their matches?

d Write down the outcomes that will give Leah's team at least one win.

e What is the probability of the team winning at least once?

Q4e hint 'Winning at least once' means one or more than one win.

5 Brad spins these two fair spinners.

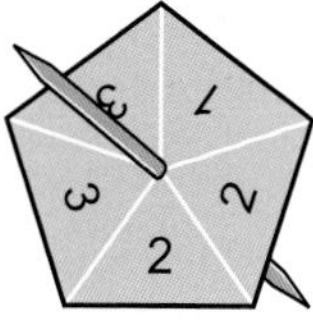

Spinner A Spinner B

Q5a hint Put spinner A on the horizontal axis and spinner B on the vertical axis.

a Draw a sample space diagram to show all the possible outcomes. How many are there?

b Work out the probability of

i a 3 ii one number being double the other.

c Which is more likely: two even numbers or two odd numbers?

d Brad spins the two spinners and adds the two numbers together. Draw a new sample space diagram to show the scores.

e Which score is most likely?

f What is the probability of scoring at least 6?

Q5d hint Use the same axes as in part **a**. For the result 1, 2, the score is 1 + 2 = 3.

6 Mohammed surveyed the students in his year group to see who had school dinners and who had a packed lunch. The two-way table shows his results.

	School dinner	Packed lunch	Total
Male	27	46	73
Female	36	41	77
Total	63	87	150

a How many female students are there?
b A student is chosen at random. Given that the student is female, work out the probability that the student has a packed lunch.

Q6b hint What fraction of females have packed lunches?

7 A pack of cards is numbered from 1 to 20.
For each of the questions, state if you need to add or multiply the probabilities.

a Two cards are chosen at random. Work out the probability of choosing an even number and an odd number.
b A card is chosen at random. Work out the probability of choosing a multiple of 5 or a multiple of 9.
c Two cards are chosen at random. Work out the probability of choosing two prime numbers.
d Two cards are chosen at random. Work out the probability of choosing two multiples of 3.

Q7 hint P(A or B) = P(A) + P(B)
P(A and B) = P(A) × P(B)

Experimental probability

1 A machine used to pack crisps had 200 bags tested for weight. 10 of the bags were underweight.

a Estimate the probability that the next bag of crisps from the machine is underweight.
b The machine packs 500 bags.
How many would you expect to be underweight?
c The machine packs 720 bags. How many would you expect to be underweight?

Q1a hint It is an 'estimate' because it is not theoretical probability. 'Estimate' does not mean 'take a guess', it means use the information given.

Q1b hint Use your answer to part **a** to help you.

2 Dylan makes a six-sided spinner. He spins it 180 times and gets 15 threes.

a How many threes would you expect in 180 spins of a fair spinner?
b Do you think Dylan's spinner is fair? Explain.

Q2 hint Is the expected number close to the actual number?

3 Megan spun a spinner 80 times. The table shows her results.

Number	Frequency
1	16
2	12
3	14
4	17
5	21

Q3a hint Copy the table and include an extra column to work out the relative frequency.

a Estimate the probabilities for each of the five outcomes.
b Megan spun the spinner another 150 times. How many times would you expect it to land on 1?

Tree diagrams and Venn diagrams

1 A pack of cards is numbered from 1 to 20.
For each question, state if it is with or without replacement.

a Two cards are chosen at random. Work out the probability of choosing an even number and an odd number.

b A card is chosen at random and then shuffled in the pack before another card is chosen. Work out the probability of choosing a multiple of 5 and a multiple of 9.

c Two cards are chosen at random. Work out the probability of choosing two prime numbers.

2 A pack of cards is numbered from 1 to 20.
For which questions would you draw a tree diagram?

a Two cards are chosen at random. Work out the probability of choosing an even number and an odd number.

b Two cards are chosen at random. Work out the probability of choosing two prime numbers.

c A card is chosen at random. Work out the probability of not getting a prime number.

3 The tree diagram shows the probabilities of picking a counter from a bag, replacing the counter and then choosing another counter.

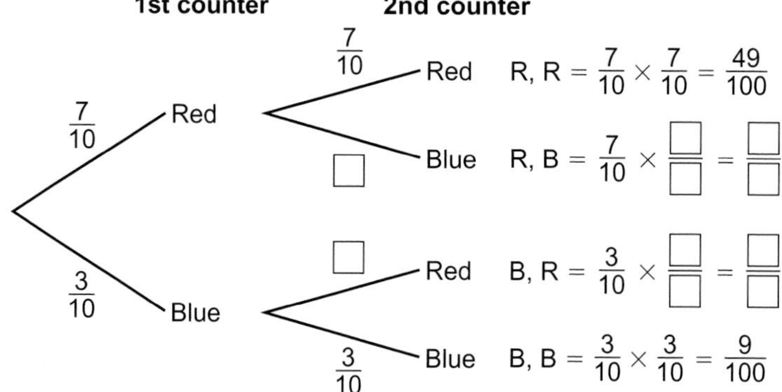

a Copy and complete the diagram and calculations to show all the probabilities.

b Work out the probability of getting one red counter.

Q3b hint P(one Red) = P(R, B or B, R) = P(R, B) + P(B, R)

4 Jane has a box of crayons with 3 blue crayons and 5 green crayons. She chooses a crayon at random, uses it, replaces it and then chooses another crayon at random.

a Copy and complete the tree diagram to show all the probabilities and calculations.

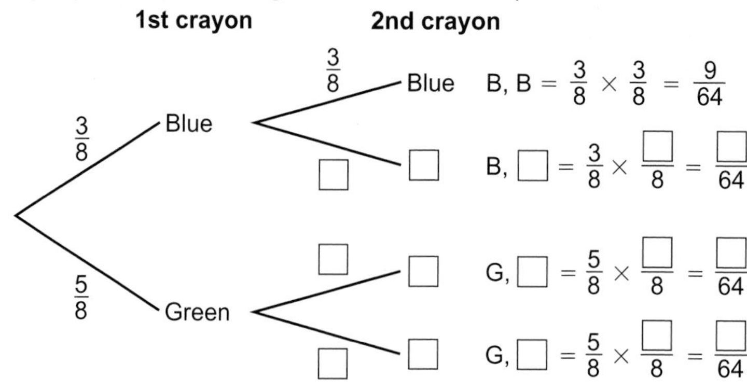

b Work out the probability of Jane choosing

i a blue and green crayon

ii 2 green crayons

iii at least one green crayon.

Q4b iii hint Which outcomes have one or more green?

5 Paige shuffles an ordinary pack of cards. She turns a card over, returns it to the pack for another shuffle and then turns over another card.

a Draw a tree diagram to show all the probabilities and calculations of choosing a heart.

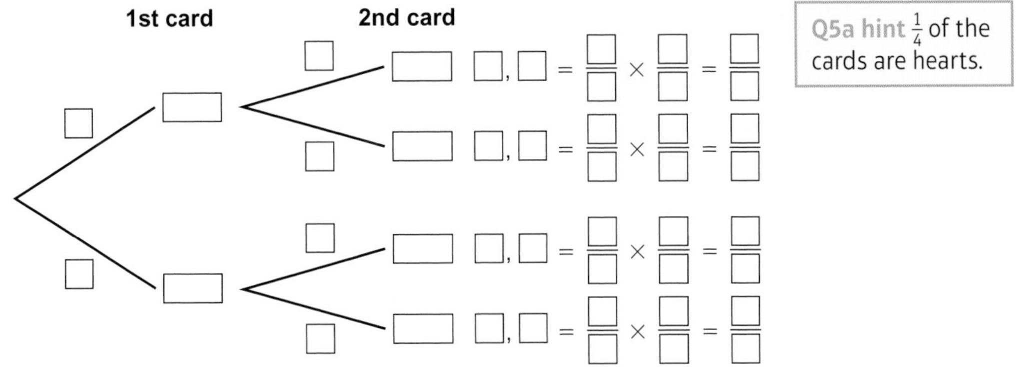

Q5a hint $\frac{1}{4}$ of the cards are hearts.

Work out the probability that

b both cards are hearts

c one card is a heart.

6 Louis has a box of chocolates. 6 of the chocolates are caramel centres and 4 of them are strawberry centres. Louis picks a chocolate at random and eats it. He then picks another chocolate at random.

a What is P(caramel) for the first chocolate?

b Louis picks and eats a caramel. What is P(caramel) for his second chocolate?

c Copy and complete the tree diagram to show all the probabilities and calculations.

Q6b hint

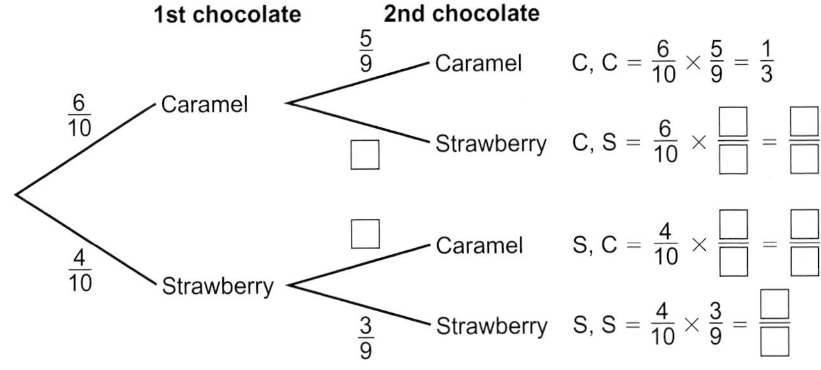

d Work out the probability that Louis picks

i two different chocolates

ii at least one strawberry.

7 Nathan has 8 cartons of juice in the fridge. 5 of the cartons are apple juice and 3 of the cartons are orange juice. Louis chooses two cartons at random.

a Copy and complete the tree diagram to show all the probabilities and calculations.

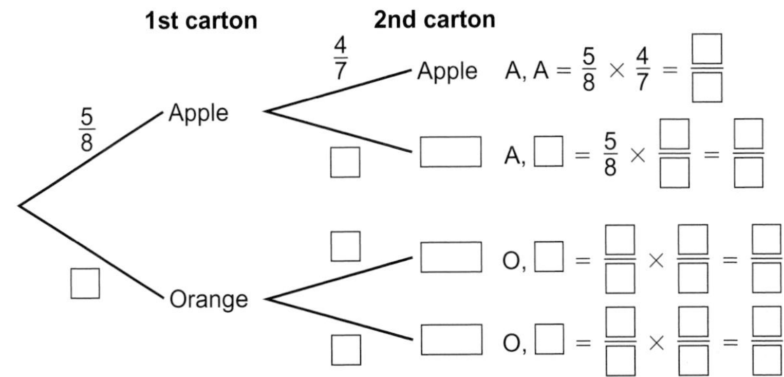

Work out the probability of

b both cartons being orange juice

c one of each flavour.

8 Shannon has a bag of balloons to blow up for a party. In the bag are 7 red balloons and 5 yellow balloons. Shannon takes a balloon at random, blows it up and then takes another balloon at random.

a Draw a tree diagram to show all the probabilities and calculations.

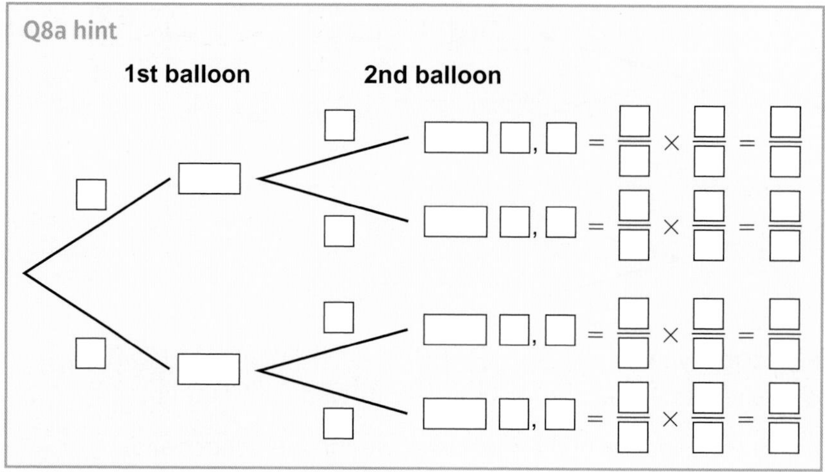

Work out the probability of

b the two balloons being the same colour

c the first balloon being red and the second yellow.

9 **Reasoning** Rachel surveys people swimming at her local swimming pool.
15 people swim front crawl.
12 people swim breaststroke.
7 people swim both front crawl and breaststroke.

a Draw a Venn diagram.

i Write the number for front crawl and breaststroke in the section where the circles overlap.

ii How many people need to go in the rest of the front crawl circle? Write your answer on your Venn diagram.

iii How many people need to go in the rest of the breaststroke circle? Write your answer on your Venn diagram.

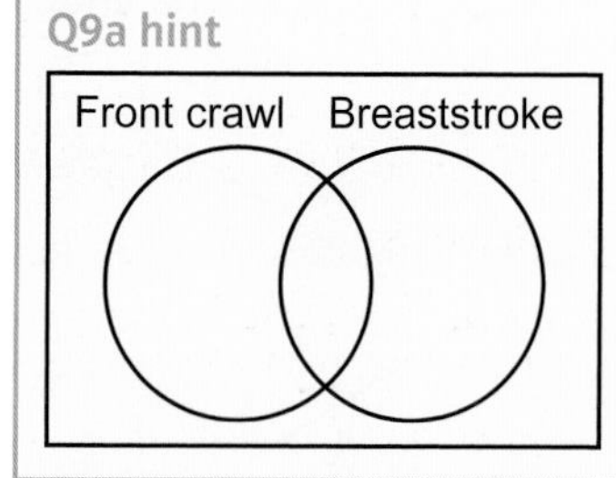

Q9a ii hint The total in the whole front crawl circle needs to be 15.

b Work out the total number of people in the Venn diagram.

c What is the probability that a person chosen at random

i swims front crawl and breaststroke

ii swims front crawl only?

d Given that a person picked at random swims front crawl, what is the probability they also swim breaststroke?

Q9d hint What fraction of 'front crawl' people also swim breaststroke?

10 Georgia carries out a survey of 50 people.
5 people are married and aged under 25.
27 people are married.

a Draw a Venn diagram to show Georgia's survey results.

b Work out the probability that a person chosen at random is married.

c Work out the probability that a person chosen at random is married given that they are under 25.

Q10 hint Shade all those married on your Venn diagram. How many are under 25?

$$\text{P(married given that they are under 25)} = \frac{\text{number of people married and under 25}}{\text{total number of people under 25}}$$

11 Look at the Venn diagram.

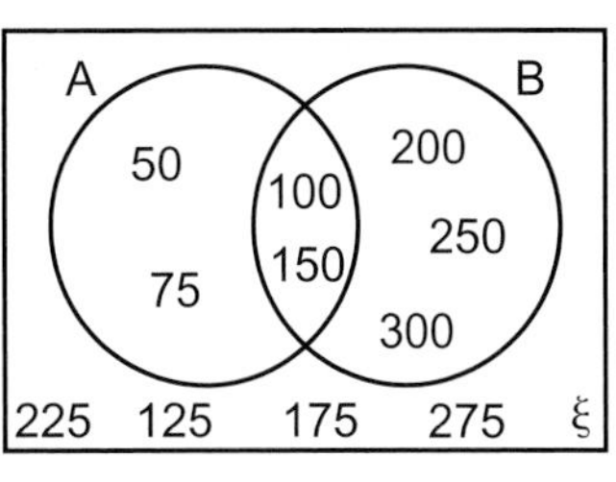

a List the numbers in

i A ii B iii A ∪ B

Q11a iii hint A ∪ B means A or B or both.

iv A ∩ B

Q11a iv hint A ∩ B means A and B.

v ξ

Q11a v hint ξ includes A and B.

b Copy and complete

i 200 ∈ □ ii 175 ∈ □ iii 100 ∈ □ ∩ □

Q11b hint 50 ∈ A means 50 is an element of A.

10 Extend

1 Design a set of counters so that P(red) = P(green) = $\frac{1}{5}$ and P(blue) = $\frac{1}{2}$ and there are half as many yellow counters as red ones.

2 **Problem-solving** Two players are playing a card game with these sets of cards.

Player A

    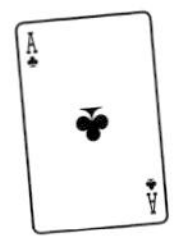     

Player B

Both players shuffle their cars and turn over the top card.
Make up a rule for each player to win so that the game is fair. Use probability to show that the game is fair.

3 **Exam-style question**

Here is a four-sided spinner.
The spinner is biased.
The table shows the probabilities that the spinner will land on a 1 or on a 3.

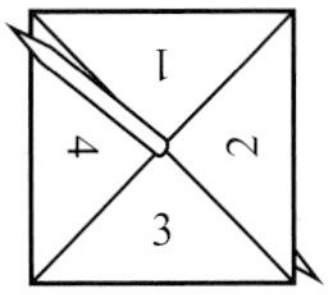

Number	1	2	3	4
Probability	0.2		0.1	

The probability that the spinner will land on 2 is the same as the probability that the spinner will land on 4.

a Work out the probability that the spinner will land on 4. **(3 marks)**

Shunya is going to spin the spinner 200 times.

b Work out an estimate for the number of times the spinner will land on 3. **(2 marks)**

March 2013, Q4, 1MA0/2H

Exam hint
Write down calculations, even if you do them in your head.

4 **Finance** The number of FTSE 100 company share prices that went down from the previous day were recorded for 50 days.

Number of share prices that went down	Frequency
1–20	3
21–40	19
41–60	12
61–80	9
81–100	7

a Estimate the probability that on the next day
  i 41–60 share prices will go down
  ii more than 60 share prices will go down.
 Give your answers as percentages.

b The London stock exchange trades for 253 days in a year. On how many days would you expect fewer than 41 share prices to fall?

c Estimate the probability that fewer than 41 share prices will fall on each of two consecutive days.
 Give your answer as a percentage.

Q4 communication hint The largest 100 companies on the London stock market are called the FTSE 100. Each day, their share prices can go up, down or stay the same.

5 **Exam-style question**

There are 17 girls and 14 boys in Mr Taylor's class.

Mr Taylor is going to choose at random 3 children from his class.

Work out the probability that he will choose exactly 2 girls and 1 boy.

(4 marks)

March 2012, Q4, 2381/6A

Q5 strategy hint Start by writing down the probability of choosing a girl.

6 **Problem-solving** Ali has a bag of red, yellow and blue counters in the ratio 2 : 1 : 3.
Brad has a bag of red, yellow and blue counters in the ratio 4 : 3 : 1.
Ali and Brad have 12 red counters each.
Ali takes a counter out of his bag and puts it into Brad's bag. Brad then takes a counter out of his bag at random.
Work out the probability that they both choose a counter of the same colour.

7 **Reasoning** Tom has a bag with these shapes in.

 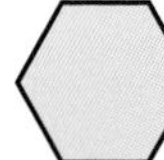  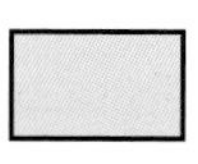 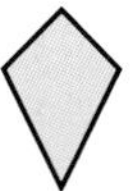 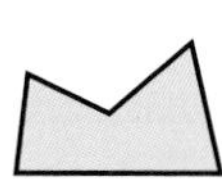 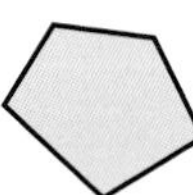 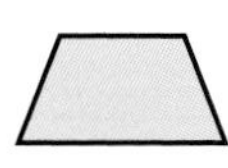

Tom drops the bag and two shapes fall out.

a Work out the probability that the two shapes are not regular polygons.

b Work out the probability that the two shapes have an interior angle sum of 540°.

c Work out the probability that the one of the shapes has an interior angle sum of 360° and the other has an interior angle sum of 540°.

8 **Reasoning** The fair 12-sided dice is rolled and the fair eight-sided spinner is spun.
The numbers rolled by the dice are used for the x-coordinates, and the numbers spun by the spinner are used for the y-coordinates.
Find the probability that the point generated by the two numbers lies on each of the following lines.

a $x = 2$ b $x + y = 9$ c $y = x + 3$ d $y = \frac{1}{2}x - 1$

9 **Problem-solving** There is an 85% chance that a battery will last longer than the advertised life of the battery.

The batteries are sold in packets of two.

A shop has 200 packets of the batteries in stock.

Find an estimate for the number of packets that will have exactly one battery that lasts longer than the advertised life of the battery.

10 **Reasoning** Mike is a stamp collector. The Venn diagram shows information about his stamp collection.

ξ = {Mike's full collection of 720 stamps}

C = {stamps from the 20th century}

B = {British stamps}

A stamp is chosen at random.

It is from the 20th century.

Work out the probability that it is British.

C B
$x - 9$ x $x(x - 2)$
ξ

11 Use set notation to describe the shaded area in each Venn diagram.

a
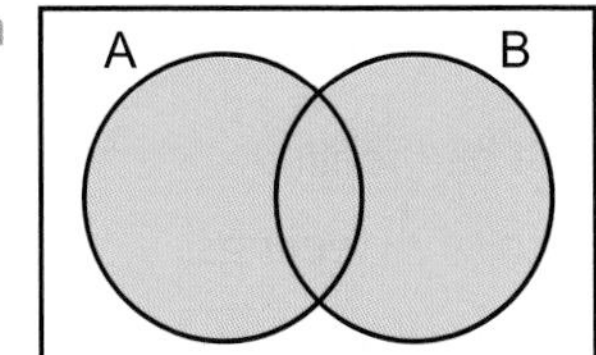

b
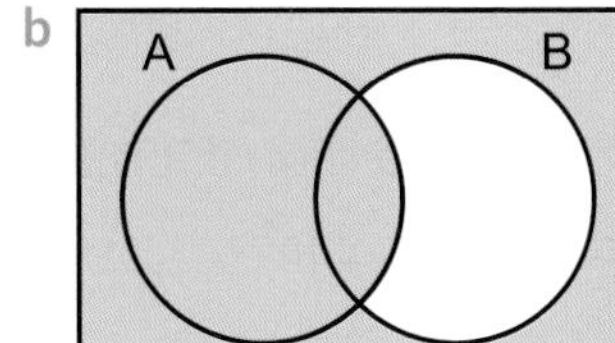

c
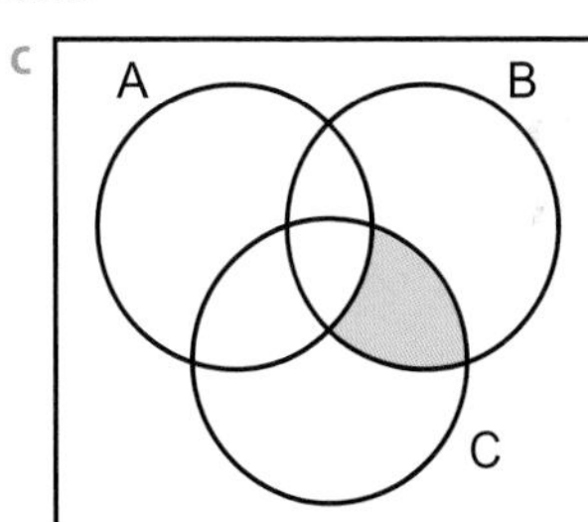

10 Knowledge check

- A **sample space diagram** shows all the possible outcomes of two events. .... *Mastery lesson 10.1*
- Two events are **mutually exclusive** if they cannot happen at the same time. .... *Mastery lesson 10.2*
- When two events are mutually exclusive you can add their probabilities. The probabilities of an exhaustive set of mutually excusive events sum to 1. .... *Mastery lesson 10.2*
- For mutually exclusive events, P(not A) = 1 − P(A) .... *Mastery lesson 10.2*
- If there are m outcomes for one event and n outcomes for another event, the product rule states that the total number of outcomes for the two events is $m \times n$. .... *Mastery lesson 10.3*
- Expected number of outcomes = number of trials × probability. .... *Mastery lesson 10.3*
- Relative frequency = $\frac{\text{frequency}}{\text{total number of trials}}$ .... *Mastery lesson 10.3*
- As the number of experiments increases, the experimental probability gets closer and closer to the theorectical probability. .... *Mastery lesson 10.3*
- A **tree diagram** shows two or more events and their probabilities. .... *Mastery lesson 10.4*
- Two events are **independent** if one happening does not affect the probability of the other. .... *Mastery lesson 10.4*

- To find the probability of two independent events multiply their probabilities, P(A and B) = P(A) × P(B) .... *Mastery lesson 10.4*
- The probability for a repeated independent event is the probability multiplied by itself, P(A and A) = P(A) × P(A), P(A and A and A) = P(A) × P(A) × P(A), etc. .... *Mastery lesson 10.4*
- A **conditional probability** is when one outcome affects another outcome. .... *Mastery lesson 10.5*
- P(A ∩ B) means the probability of the **intersection** of A and B. .... *Mastery lesson 10.6*
- P(A ∪ B) means the probability of the **union** of A and B. .... *Mastery lesson 10.6*
- P(A ∪ B) = P(A) + P(B) − P(A ∩ B) .... *Mastery lesson 10.6*
- P(A ∩ B | B) means the probability of the intersection of A and B given B. .... *Mastery lesson 10.6*

Reflect

Write down a word that describes how you feel

a before a maths test.

b during a maths test

c after a maths test.

Hint Here are some possible words: OK, worried, excited, happy, focused, panicked, calm.

Beside each word, draw a face, ☺ or ☹ to show if it is a good or a bad feeling.

Discuss with a classmate what you could do to change ☹ feelings to ☺ feelings.

10 Unit test

Log how you did on your Student Progression Chart.

1 **Exam-style question**

Riki has a packet of flower seeds.

The table shows each of the probabilities that a seed taken at random will grow into a flower that is pink or red or blue or yellow.

Colour	pink	red	blue	yellow	white
Probability	0.15	0.25	0.20	0.16	

a Work out the probability that a seed taken at random will grow into a white flower. **(2 marks)**

There are 300 seeds in the packet.
All the seeds grow into flowers.

b Work out an estimate for the number of red flowers. **(2 marks)**

March 2012, Q9, 1380/4H

Exam hint
As each part is worth 2 marks, you will need to write down the calculations you do before writing the answer.

2 A company launches a new smartphone.
The phone is made in five different colours with three different storage capacities.

a How many combinations are there? *(2 marks)*

b One of the phones is pink with 16Gb of memory. What is the probability that this combination is bought? *(2 marks)*

3 In a football tournament at group stage there are five football teams in a group, Brazil, England, Scotland, Argentina and France. Each team plays every other team in their group. There are ten matches altogether. Two teams are picked at random to play the first match. Work out the probability that the first game will be played by a European team and a South American team. *(3 marks)*

ActiveLearn Homework, practice and support: Higher 10 Unit test

4 **Problem-solving** A and B are two mutually exclusive events.
P(A) = 0.35 and P(A or B) = 0.8. Work out P(not B). *(2 marks)*

5 State whether each pair of events is independent or dependent.

a Randomly taking two sweets from a bag. *(1 mark)*

b Spinning a five-sided spinner three times. *(1 mark)*

c Randomly taking a marble from a bag and then taking another one. *(1 mark)*

d Randomly taking a marble from a bag, replacing it, and then taking another one. *(1 mark)*

6 **Exam-style question**

Tom plants 3 seeds.
The probability that a seed will germinate is $\frac{4}{5}$.

a Calculate the probability that all 3 seeds will germinate. **(2 marks)**

b Calculate the probability that at least one of the seeds will germinate. **(3 marks)**

March 2010, Q5, 2381/6B

7 **Exam-style question**

There are 11 small jars on a table.
Three of the jars contain honey.
Eight of the jars contain jam.
Rosie takes, at random, two of the jars for her breakfast.
Work out the probability that she takes at least one jar of honey. **(4 marks)**

Nov 2010, Q5, 2381/6B

8 The Venn diagram shows the numbers of people who chose baguettes (B) and soup (S). *(4 marks)*

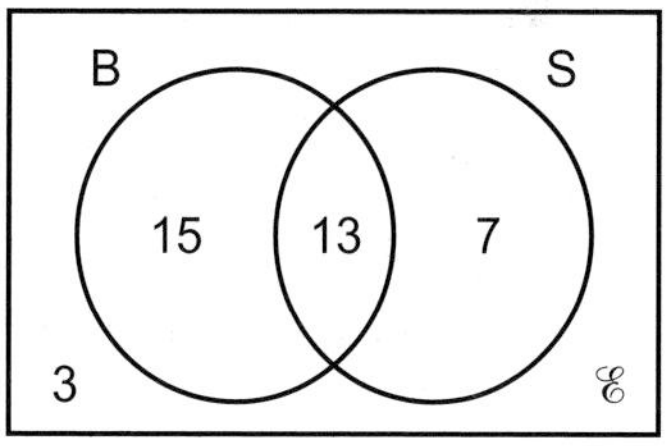

B, S: 15, 13, 7; outside: 3; ℰ

Use the Venn diagram to find

a P(B)

b P(S′)

c $P(B \cap S)$

d $P(B \cup S)$

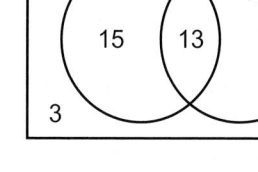

9 The Venn diagram shows customers' choice of sausages (S), bacon (B) and egg (E) fillings for an all-day-breakfast sandwich in a cafe.

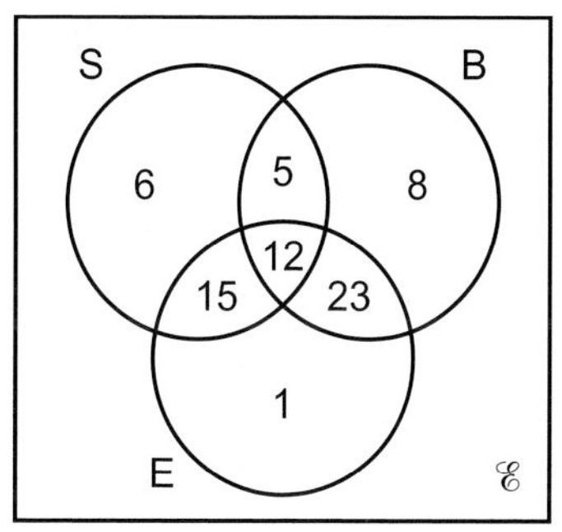

a How many people chose all three fillings? *(1 mark)*

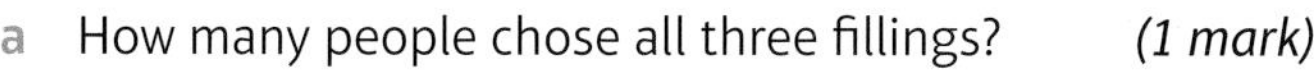

A customer is chosen at random. Work out

b i $P(B \cap E)$ *(1 mark)*

ii $P(S \cap B \cap E)$ *(2 marks)*

iii $P(S \cap E \mid S)$ *(2 marks)*

10 **Problem-solving** A class of 27 students is split between boys and girls in the ratio of 5 : 4.
Work out the probability that two students chosen at random are both boys. *(4 marks)*

Sample student answer

a What is missing from the tree diagram?

b What is missing from the calculations on the right of the tree diagram?

c What else is missing from their response?

Exam-style question

Sally has a bag of 9 sweets.

In the bag, there are

3 orange flavoured sweets

4 strawberry flavoured sweets

and 2 lemon flavoured sweets

Sally takes, at random, two of the sweets.

She eats the sweets.

Work out the probability that the two sweets Sally eats are *not* of the same flavour. **(4 marks)**

Mock paper, Q23, 1MA0/1H

Student answer

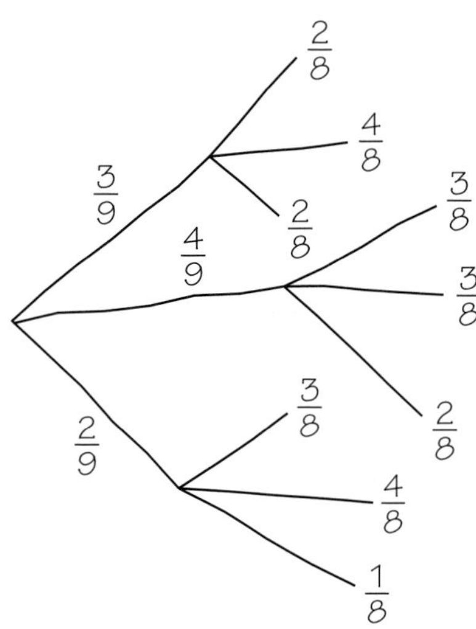

$$\frac{3}{9} \times \frac{4}{8} = \frac{12}{72}$$

$$\frac{3}{9} \times \frac{2}{8} = \frac{6}{72}$$

$$\frac{4}{9} \times \frac{3}{8} = \frac{12}{72}$$

$$\frac{4}{9} \times \frac{2}{8} = \frac{8}{72}$$

$$\frac{2}{9} \times \frac{3}{8} = \frac{6}{72}$$

$$\frac{2}{9} \times \frac{4}{8} = \frac{8}{72}$$

$$\frac{52}{72}$$

Master p.340 | Problem-solve p.351 | Check p.352 | Strengthen p.354 | Extend p.357 | Test p.360

11 MULTIPLICATIVE REASONING

Stiletto heels can damage wooden floors.
The pressure exerted on the floor is worked out using the equation $\text{pressure} = \frac{\text{force}}{\text{area}}$.

Because the stiletto heel reduces the area of shoe in contact with the floor, the pressure on the floor due to the weight of the person increases. This can lead to deep dents being made in the floor surface.

Sophie's weight is 500 N. She wears

a flat shoes with an area of 130 cm^2 in contact with the floor

b stiletto heels with an area of 1.6 cm^2 in contact with the floor.

What is the difference in the pressure Sophie applies to the floor in N/cm^2?

11 Prior knowledge check

Numerical fluency

1 3 metres of material cost £1.56
Work out the cost of 5 metres of the same material.

2 Connor is paid £54 for 8 hours work in a supermarket.
How much is he paid for 10 hours work?

3 An orchestra of 10 people takes 5 minutes to play a song.
How long will it take 20 people?

4 A recipe for 6 people uses 750 g of mince.
How much mince is needed for 16 people?

5 Milk is sold in two sizes of bottle.
A 4-pint bottle of milk costs £0.98.
A 6-pint bottle of milk costs £1.44.
Which bottle of milk is the best value for money?
Show all your working.

6 3 men build a wall in 2 days.
How long will it take

a 1 man

b 2 men?

Fluency with measures

7 The ratio 1 m : 1 cm = 100 : 1.
Copy and complete.

a 1 kg : 1 g = □ : □

b 1 cm : 1 mm = □ : □

c 1 litre : 1 ml = □ : □

d 1 minute : 1 second = □ : □

e 1 hour : 1 minute = □ : □

8 Copy and complete.

a 180 cm = □ m

b 28 000 cm = □ m

c 54 600 m = □ km

9 12 inches = 1 foot, 3 feet = 1 yard.
Use this to work out

a 4 feet in inches

b 5 yards in feet

c 58 inches in feet and inches.

10 20 fluid ounces = 1 pint, 8 pints = 1 gallon.
Use this to work out

a 4 pints in fluid ounces

b 5 gallons in pints

c 20 pints in gallons and pints.

11 5 miles = 8 km. Use this to work out

a 40 miles in km b 48 km in miles.

12 Copy and complete.

a $1\frac{1}{2}$ hours = ☐ minutes

b 50 minutes = ☐ seconds

c 225 minutes = ☐ hours ☐ minutes

13 Jess goes on holiday to New York. The exchange rate of £:US dollars is 1:1.613.

a She changes £500 into US dollars. How many US dollars should she get?

b After her holiday, Jess changes 80 dollars back into pounds. The exchange rate is the same. How much money should she get? Give your answer to the nearest penny.

14 Copy and complete.

1 cm = ☐ mm

1 m = ☐ cm

1 cm^2 = ☐ mm^2

1 m^2 = ☐ cm^2

1 cm^3 = ☐ mm^3

1 m^3 = ☐ cm^3

Algebraic fluency

15 Change the subject of each formula to the letter given in brackets.

a $v = u + at$ (t)

b $D = \frac{M}{V}$ (M)

c $P = \frac{F}{A}$ (A)

16 a $v = u + at$

Work out the value of v when $u = 20$, $a = 10$ and $t = 3$

b $s = ut + \frac{1}{2}at^2$

Work out the value of s when $u = 10$, $a = 8$ and $t = 4$

★ Challenge

17 Do you know on which day of the week you were born?

You can use Zeller's algorithm to work it out from your date of birth.

An algorithm is a sequence of precise instructions to solve a problem.

Example	**15 May 1999**
Let: day number = D	$D = 15$
month number = M	$M = 5$
and year = Y	$Y = 1999$
When M is 1 or 2 add 12 to M	$M = 5$ (no change)
and subtract 1 from Y	$Y = 1999$ (no change)
Let C be the first two digits of Y	$C = 19$
and Y' be the last two digits of Y	$Y' = 99$
Add together the integer parts of:	
$(2.6 \times M - 5.39)$, $(Y' \div 4)$ and $(C \div 4)$,	$7 + 24 + 4$
then add on D and Y' and subtract $2C$.	$+15 + 99 - 38 = 111$
Find the remainder when this quantity is divided by 7. When the remainder is 0 Sun, 1 Mon, 2 Tue, …	$111 \div 7 = 15$ remainder 6 15 May 1999 was a Saturday

Q17 hint The integer part of a number is the whole-number part, e.g. the integer part of 19.75 is 19

11.1 Growth and decay

Objectives

- Find an amount after repeated percentage changes.
- Solve growth and decay problems.

Why learn this?

Repeated proportional change can be used to predict changes in population size over short periods of time.

Fluency

Work out

a $4 \times 4 \times 4 \times 4 = 4^{☐}$ **b** $3 \times 3 \times 3 \times 3 \times 3 = 3^{☐}$

Warm up

1 Work out the multiplier as a decimal for
 a an increase of 30%
 b a decrease of 14%
 c an increase of 7.2%
 d a decrease of 2.5%.

Q1a hint 100% + 30% = 130%
130% = □ as a decimal number

Q1b hint 100% − 14% = □%
□% = □ as a decimal number

Questions in this unit are targeted at the steps indicated.

2 **Real** Wes bought a car for £6500. It lost 35% of its value in the first year. It lost 15% of its value in the second year. Work out
 a the multiplier to find the value of the car at the end of the first year
 b the value of the car at the end of the first year
 c the multiplier to find the value of the car at the end of the second year
 d the value of the car at the end of the second year
 e the decimal multiplier that the original value of the car can be multiplied by to find its value at the end of two years.

Q2a hint Decrease by 35%

Q2c hint Decrease by □%

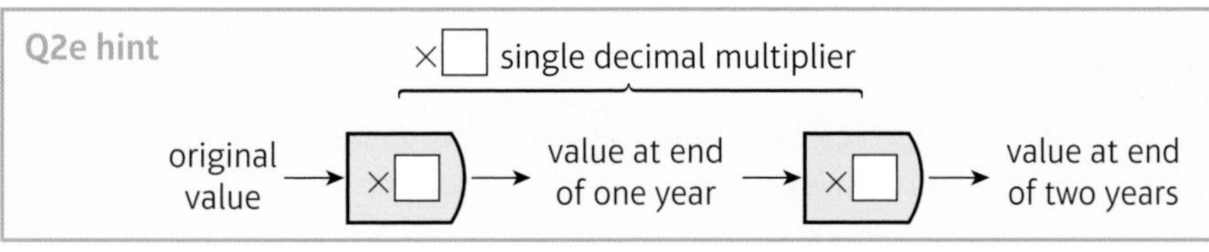

Discussion What do you notice about your answers to parts **a**, **c** and **e**? What is a quick way to find the answer to part **d**, without steps **a** to **c**?

3 Work out the decimal multiplier that represents
 a an increase of 12% for 3 years
 b a decrease of 15% for 4 years.

Q3a hint $\times 1.2^{\square}$

Discussion How could you write these multipliers as a power rather than a decimal number?

4 **Finance** Abdul has a job with an annual salary of £35 000. At the end of the first year he is given an increase of 2%. At the end of the second year he is given an increase of 3.5%. Work out Abdul's salary at the end of two years.

Q4 hint Work out the decimal multiplier that the original salary can be multiplied by to find the salary at the end of two years.

Q4 communication hint **Annual** means 'yearly'.

5 **Reasoning** Becky says an increase of 15% followed by an increase of 22% is the same as an increase of 37%. Is Becky correct? Explain.

Q5 hint Show your working. Write a sentence to explain.

6 Work out the multiplier as a decimal number for
 a an increase of 5% followed by an increase of 3%
 b a decrease of 20% followed by a decrease of 15%
 c an increase of 9% followed by a decrease of 6%.

7 **Finance** Tristan buys a flat for £35 000. In the first year, the value of the flat increases by 12%. In the second year, the value of the flat decreases by 3%. Work out the value of the flat after the 2 years.

8 Abi buys a motor bike for £8200. In the first year the motorbike depreciates by 25%. In the second year it depreciates by 12%. What is the value of the motorbike at the end of the two years?

> **Q8 communication hint** To **depreciate** means to decrease in value.

9 **Finance** £2500 is invested for 2 years at 4.3% per annum compound interest. Work out the total amount in the account after 2 years.

> **Q9 hint** Multiplier = ☐
> Amount in account after 2 years = 2000 × ☐ × ☐

> **Q9 communication hint** In **compound interest** the interest earned each year is added to money in the account and earns interest the next year. Most interest rates are compound interest rates.

10 **Finance / Reasoning** £3500 is invested for 2 years at 4.1% per annum compound interest. Work out the total amount in the account after 2 years.

11 **Reasoning** Anthony says you can work out compound interest using the formula

$$\text{amount after } n \text{ years} = \text{initial amount} \times \left(\frac{100 + \text{interest rate}}{100}\right)^n$$

Show this formula works for **Q9** and **Q10**.

> **Q11 hint** Do you get the same answers if you use the formula for **Q9** and **Q10**?

Key point 1

You can calculate an amount after n years' compound interest using the formula

$$\text{amount} = \text{initial amount} \times \left(\frac{100 + \text{interest rate}}{100}\right)^n$$

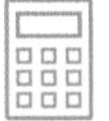

12 **Finance** £3000 is invested for 2 years at 3.8% per annum compound interest.
Work out the **total interest** earned over the 2 years.

> **Q12 hint** Total interest = amount in the account at the end of the investment − amount invested

Example 1

Paul invests £4500 in an account for 2 years. The account pays 3.2% compound interest per annum. Paul has to pay 20% tax on the interest earned each year. The tax is taken from the account at the end of each year.
Paul thinks that at the end of the 2 years he will have at least £4700 in this account.
Is Paul correct? Show all your working.

Year 1

Interest = 0.032 × 4500 = £144 — 3.2% = 0.032

Tax = 0.2 × 144 = £28.80

Amount in account at end of year 1 = 4500 + 144 − 28.80
= £4615.20 — Amount in account at end of year 1: £4500 + interest − 20% of interest

Year 2

Start with £4615.20

Interest = 0.032 × 4615.20 = £147.69

Tax = 0.2 × 147.69 = £29.54

Amount in account at end of year 1 = 4615.20 + 147.69 − 29.54
= £4733.35 — Amount in account at end of year 2: £4615.20 + interest − 20% of interest

Paul is correct. £4733.35 is more than £4700.

13 **Exam-style question**

Katie invests £200 in a savings account for 2 years.
The account pays compound interest at an annual rate of
3.3% for the first year
1.5% for the second year.

a Work out the total amount of money in Katie's account at the end of 2 years. **(3 marks)**

Katie travels to work by train.
The cost of her weekly train ticket increases by 12.5% to £225
Katie's weekly pay increases by 5% to £535.50

b Compare the increase in the amount of money Katie has to pay for her weekly train ticket with the increase in her weekly pay. **(3 marks)**

June 2014, Q18, 1MA0/2H

Exam hint
Make sure you show all your working clearly when doing the comparison.

Q13a strategy hint
What is the multiplier for the 2 years?

14 **Problem-solving / Finance** Laura invests £3600 in a savings account for 2 years. The account pays 3.52% compound interest per annum. Laura has to pay 40% tax on the interest earned each year. The tax is taken from the account at the end of each year. How much is in the account at the end of 2 years?

15 **Reasoning / Finance** Fidel invested £4200 in a savings account. He is paid 3.25% per annum compound interest. How many years before he has £4928.33 in the savings account?

Q15 strategy hint Try different values of n in the formula.

16 **STEM** The level of activity of a radioactive source decreases by 5% per hour. The activity is 1400 counts per second at one point.

a What will it be 2 hours later?
b After how many complete hours will the count be less than 500 counts per second?

17 **Real** In 2014 a fast-food chain has 180 outlets in the UK. The number of outlets is increasing at a rate of 9% each year.

a How many outlets will it have in 2020?
b **Reflect** What is an appropriate degree of accuracy to give for this question? Why?

18 A population of insects increases by 25% per day. At the end of one week there are 2145 insects. How many insects were there at the beginning of the week?

Q18 hint Use x for the number of insects at the start of the week.

11.2 Compound measures

Objectives

- Calculate rates.
- Convert between metric speed measures.
- Use a formula to calculate speed and acceleration.

Why learn this?

Police Accident Investigation Teams use kinematics formulae to work out the speed of cars involved in serious accidents.

Fluency

Find the rates for

a £60 for a 6-hour day £☐ per hour **b** 300 km on 20 litres of petrol ☐km per ☐

Warm up

1 $a = \frac{b}{c}$. Find
   a a when $b = 18$, $c = 3$ b b when $a = 15$, $c = -2$ c c when $a = -4$, $b = 0.5$

2 a Karl cycles 48 km in 3 hours. What was his average speed?
   b Andy cycles with an average speed of 15 km/h for 2 hours. What distance did he travel?
   c Shakil cycles 42 km at an average speed of 14 km/h. How long does it take him?

Q2 hint Speed = $\frac{\text{distance}}{\text{time}}$

3 Convert these times to hours and minutes.
   a 450 minutes   b 6.2 hours

4 **Problem-solving / Finance** George works a 35-hour week and some overtime. He is paid £8.50 an hour for this work. George is paid time and a half for each hour he works on a Saturday and double time for each hour he works on a Sunday.
   a How much is George paid for a week when he works a 35-hour week, plus 4 hours on Saturday and 3 hours on Sunday?
   b In one week George works a 35-hour week and some hours on Saturday. He is paid £335.75 for the week. How many hours did George work on Saturday?

Q4a hint Rate of pay for Saturday = 1.5 × ☐ = ☐
Rate of pay for Sunday = 2 × ☐ = ☐

Q4b hint How much more is George paid than for his 35-hour week?

5 **Real / Problem-solving** Water is leaking from a water butt at a rate of 4.5 litres per hour.
   a Work out how much water leaks from the water butt in
      i 20 mins   ii 50 mins.
   b Initially there are 180 litres of water in the water butt. Work out how long it takes for all the water to leak from the water butt.

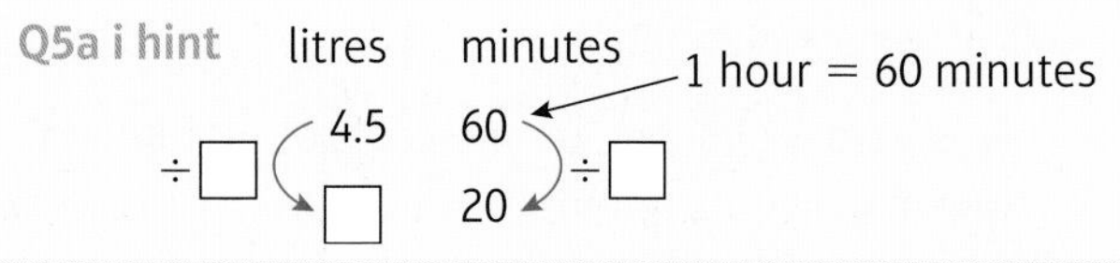

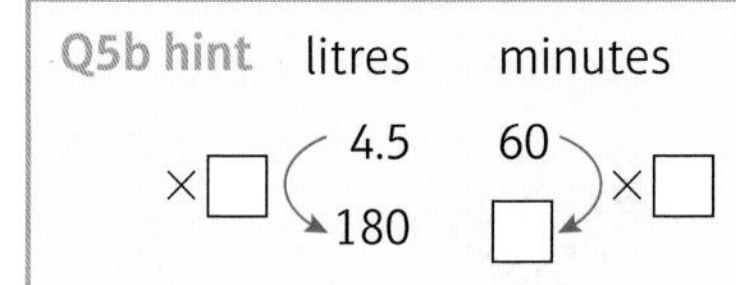

6 **Reasoning / Real** A car travels 320 km and uses 20 litres of petrol.
   a Work out the average rate of petrol usage. State the units with your answer.

   b Estimate the amount of petrol that would be used when the car has travelled 65 km.

Discussion Why does the question ask for 'average rate' rather than 'exact rate'?

Key point 2

Compound measures combine measures of two different quantities. Speed is a measure of distance travelled and time taken. It can be measured in metres per second (m/s), kilometres per hour (km/h) or miles per hour (mph).

Average speed = $\frac{\text{distance}}{\text{time}}$ or $S = \frac{D}{T}$

Example 2

A man walks at an average speed of 5.4 km/h. What is his average speed in m/s?

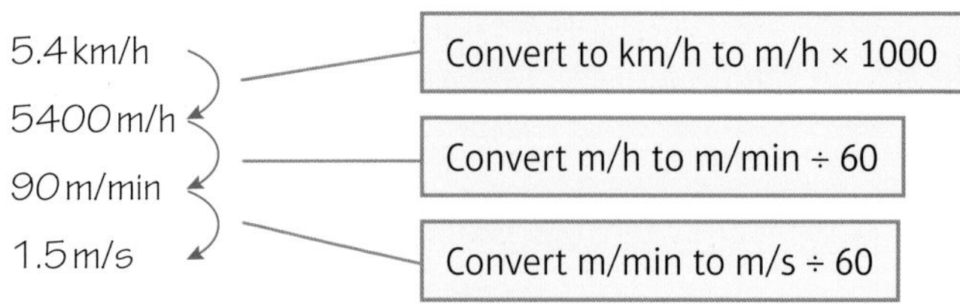

7 Convert these speeds from m/h to km/h.
   a 650 m/h   b 7800 m/h   c 256 000 m/h   d 188 000 m/h

8 **Reasoning** Convert these speeds from metres per second (m/s) to metres per hour (m/h).

a 1 m/s b 12 m/s c 8 m/s d 4.5 m/s

e Would more or less metres be travelled in 1 hour than in 1 second?

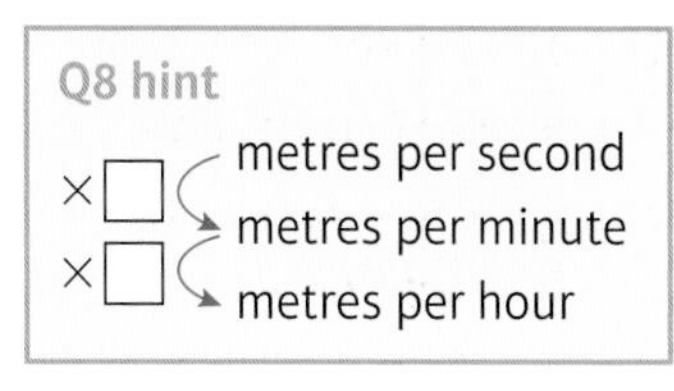

Q8 hint
×□ metres per second
metres per minute
×□ metres per hour

9 Copy and complete the table.

metres per second	kilometres per hour
	54
	72
30	
45	

10 **Real** A commercial aeroplane has a cruising speed of 250 m/s.
What is this speed in km/h?

11 **Problem-solving / Reasoning** A Formula 1 racing car has a top speed of 350 km/h.
A peregrine falcon is the fastest bird with a speed of 108 m/s.
Which is fastest? Explain your answer.

12 a A car travels at x km/h. Write an expression for this speed in m/s.

b A cheetah runs at y m/s. Write an expression for this speed in km/h.

13 **Exam-style question**

Karl travels 35 miles in 45 minutes then 65 km in $1\frac{1}{2}$ hours.

5 miles = 8 kilometres

What is his average speed for the total journey in km/h?

(3 marks)

Q13 strategy hint Average speed for total journey = $\frac{\text{total distance}}{\text{total time}}$

14 **STEM** A swallow flies for 40 minutes at an average speed of 11 m/s. How far does the swallow fly in kilometres?

15 Paul swims 750 metres in 25 minutes.
What is his average speed in km/h?

Q15 strategy hint First convert 750 m to km and 25 minutes to hours but leave each as a fraction before finding the average speed in km/h.

Key point 3

These are kinematics formulae:

$v = u + at$

$s = ut + \frac{1}{2}at^2$

$v^2 = u^2 + 2as$

where a is constant acceleration, u is initial velocity, v is final velocity, s is displacement from the position when $t = 0$ and t is time taken.

In exam questions you will need to decide which equation to use.

Velocity is speed in a given direction, possible units are m/s.

Initial velocity is speed in a given direction at the start of the motion.

Acceleration is the rate of change of velocity, i.e. a measure of how the velocity changes with time, possible units are m/s^2.

16 **STEM** A car starts from rest and accelerates at 5 m/s^2 for 200 m.
Work out the final velocity in m/s.

Q16 strategy hint When the car starts from rest the initial velocity $u = 0$. You are given $a = 5$, $s = 200$ and want to find v. Use the equation that contains u, a, s and v but not t.

17 **STEM** A tram has an initial velocity of 300 m/minute. It travels a distance of 0.5 km in 20 seconds. What is the acceleration of the tram in m/s^2?

18 **STEM** A bus travels with an acceleration of 2 m/s^2 and reaches a speed of 45 km/h in 5 seconds. What was the initial velocity of the bus in m/s?

11.3 More compound measures

Objective

- Solve problems involving compound measures.

Why learn this?

Pressure and density are both examples of compound measures. Water pressure increases with depth and so is an important factor to consider in scuba diving.

Fluency

- ☐ cm^2 = 1 m^2
- ☐ cm^3 = 1 m^3
- What is the formula for the area of a circle?
- What is the formula for the volume of a prism?

Warm up

1 Convert

a 7.5 kg to g   b 62 500 cm^2 to m^2 c 95 000 cm^3 to m^3

2 Solve

a $\frac{m}{5} = 6$ b $\frac{8}{v} = 0.5$

Key point 4

Density is the **mass** of substance in g contained in a certain **volume** in cm^3 and is often measured in grams per cubic centimetre (g/cm^3).

Density = $\frac{\text{mass}}{\text{volume}}$ or $D = \frac{M}{V}$

3 **Real** A sample of brass has a mass of 2 kg and a volume of 240 cm^3. What is its density in g/cm^3?

Q3 hint First convert mass to g.

4 A cubic metre of concrete has a mass of 2400 kg. What is the density of the concrete in g/cm^3?

Q4 hint 1 m^3 = ☐ cm^3

Example 3

The diagram shows a block of wood in the shape of a cuboid.
The density of wood is 0.6 g/cm^3.
Work out the mass of the block of wood.

4 cm, 12 cm, 10 cm

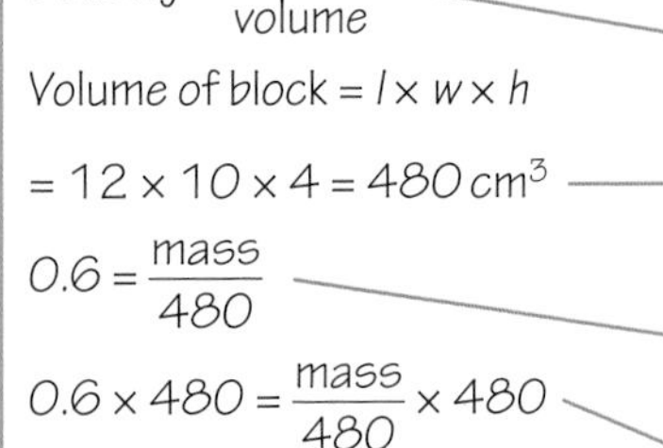

First write down the formula you are going to use.

You are given the density and are asked for the mass. So work out the volume in cm^3.

Substitute values into the formula.

Multiply both sides by 480.

ActiveLearn Homework, practice and support: Higher 11.3

5 **STEM** The area of the cross-section of a plastic prism is 35 cm². Its length is 15 cm. The plastic has a density of 3.9 g/cm³. What is the mass of the prism?

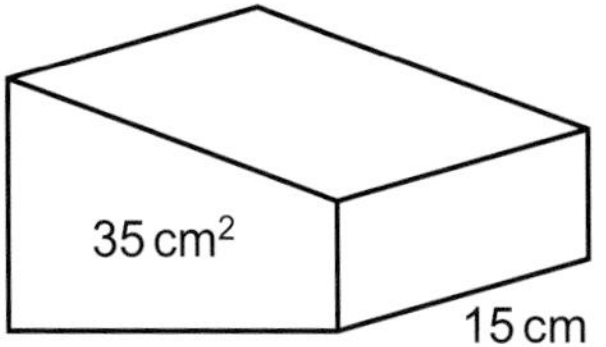

Q6 hint Substitute into the formula.

$\square = \frac{\square}{V}$

Solve to find V.

6 **STEM** Iron has density 8 g/cm³.
The mass of a piece of iron is 5.4 kg.
What is the volume?

Q7 hint

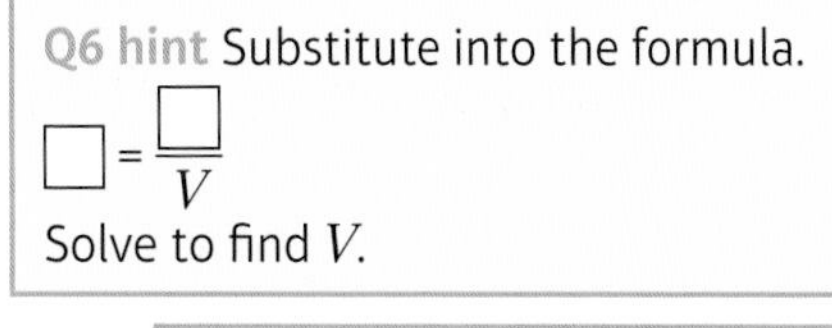

7 **STEM** The density of copper is 8940 kg/m³.
What is the density of copper in g/m³?

Q8 hint

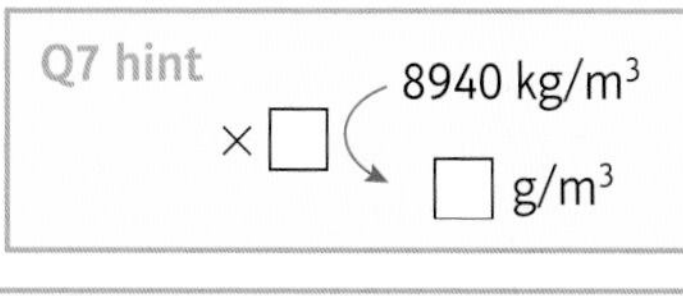

2.70 g/cm³
convert to kg □
□ kg/cm³
convert to m³ □
□ kg/m³

8 **STEM** The density of aluminium is 2.70 g/cm³.
What is the density of aluminium in kg/m³?

9 A metal has density x g/cm³.
Write an expression for its density in kg/m³.

10 **STEM / Reasoning** 1 cm³ of gold has mass 19.32 g.
1 cm³ of platinum has mass 21.45 g.
Which metal is denser? Explain.

11 **Exam-style question**

The density of juice is 1.1 grams per cm³.
The density of water is 1 gram per cm³.
270 cm³ of drink is made by mixing 40 cm³ of juice with 230 cm³ of water.
Work out the density of the drink. **(3 marks)**

Exam hint
Density = $\frac{\text{total mass}}{\text{total volume}}$

Key point 5

Pressure is a compound measure. It is the **force** in newtons applied over an **area** in cm² or m².
It is usually measured in newtons (N) per square metre (N/m²) or per square centimetre (N/cm²).

Pressure = $\frac{\text{force}}{\text{area}}$ or $P = \frac{F}{A}$

Q12 hint First convert the area to m².

12 **STEM** A force of 45 N is applied to an area of 26 000 cm². Work out the pressure in N/m².

Q13 hint Substitute into the formula

$\square = \frac{F}{\square}$

Rearrange to find F.

13 **STEM** A force applied to an area of 4.5 m² produces a pressure of 20 N/m². Work out the force in N.

14 **STEM** Copy and complete the table.
Give your answers to 3 significant figures.

Force	Area	Pressure
60 N	2.6 m²	□ N/m²
□ N	4.8 m²	15.2 N/m²
100 N	□ m²	12 N/m²

15 **STEM** A cylindrical bottle of water has a flat, circular base with a diameter of 0.1 m.
The bottle is on a table and exerts a force of 12 N on the table.
Work out the pressure in N/cm². Give your answer to 3 significant figures.

16 The pressure between a car tyre and the road is 99 960 N/m². The car tyres have a combined area of 0.12 m² in contact with the road.
What is the force exerted by the car on the road? Give your answer to 3 significant figures.

17 **Reasoning** Jamie sits on a chair with four identical legs. Each chair leg has a flat square base measuring 2 cm by 2 cm. Jamie has a mass of 75 kg and the chair has a mass of 5 kg.

Q17 communication hint
Weight is a force on an object due to gravity and is measured in newtons.

a Use $F = mg$ to work out the combined weight of Jamie and the chair, where g is the acceleration due to gravity. Use $g = 9.8\,\text{m/s}^2$.

b Work out the pressure on the floor in N/cm², when only the four chair legs are in contact with the floor.

c The area of Jamie's trainers is 0.04 m². Work out the pressure on the floor when Jamie is standing up. Give your answer in N/m².

d Does Jamie exert a greater pressure on the floor when he is standing up or sitting on the chair?

18 a Convert 50 N/cm² to N/m². b Convert x N/m² to N/cm².

11.4 Ratio and proportion

Objectives

- Use relationships involving ratio.
- Use direct and inverse proportion.

Why learn this?

Speed and time are in inverse proportion. The greater the speed, the shorter the time taken to travel a journey.

Fluency

- When $y = 5x$ then $\frac{y}{x} = \square$
- When $y = \frac{7}{x}$ then $xy = \square$

Warm up

1 Match each graph to an equation.

a $y = \frac{1}{x}$ b $y = x$ c $y = 2x + 3$

A
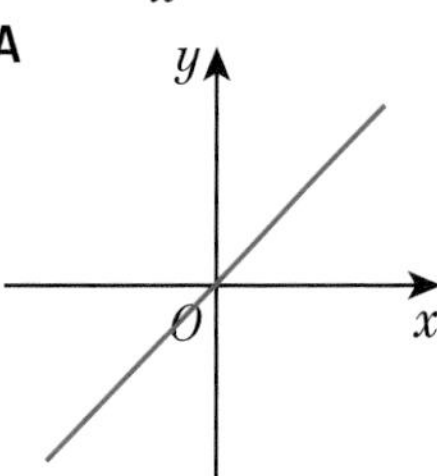

B
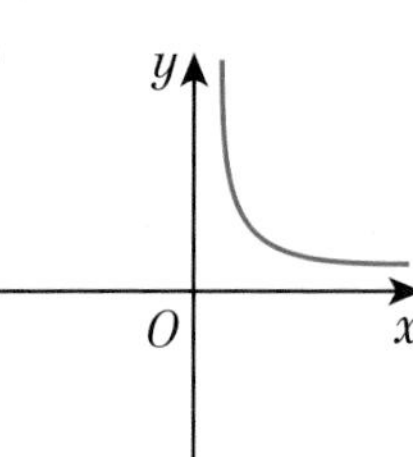

C
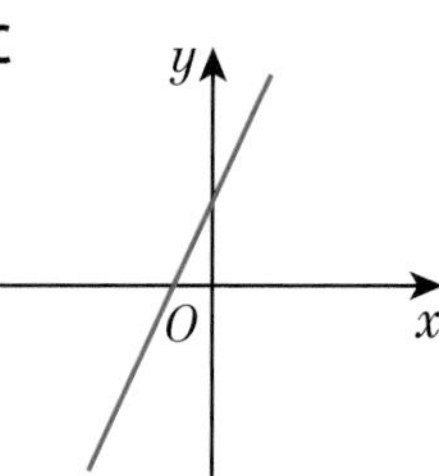

2 What is the gradient of the straight line with equation

a $y = 2x + 5$ b $y = 3x - 7$ c $y = 5x$ d $y = 9x$?

3 a Which of these tables show direct proportion between x and y? Explain your answer.

A

x	2	4	6	8
y	8	16	24	32

B

x	1	2	3	4
y	1	4	9	16

C

x	0	5	10	15
y	2	17	32	47

D

x	2	4	6	8
y	10	20	30	40

b Plot a line graph for the values in the tables which show direct proportion.

c What do you notice about these graphs?

d Work out the equations of the graphs.

 ActiveLearn Homework, practice and support: Higher 11.4

4 Copy and complete these.

a $A:B = 3:5$ so $A = \frac{\square}{\square}B$ b $P:Q = 7:4$, so $P = \frac{\square}{\square}Q$

c $5X = 9Y$ so $X:Y = \square:\square$

Q4a hint $A:B = 3:5$ so $\frac{A}{B} = \frac{3}{5}$ Rearrange to find A.

Q4b hint Divide both sides by 4.

5 **Modelling** The table shows some lengths in both miles and kilometres.

Miles	5	10	15	20
Kilometres	8	16	24	32

a What is the ratio miles : kilometres in the form $1:n$?

b Plot a line graph for these values.

c Are miles and kilometres in direct proportion? Explain your answer.

d What is the gradient of the line?

e Write a formula that shows the relationship between miles and kilometres.

Q5b hint Plot miles on the horizontal axis and kilometres on the vertical axis.

Discussion What is the connection between your answers to parts **a** and **d**? How can you use them to write to formula in part **e**?

6 **Modelling** The table shows the distance (s) in miles travelled by a car over a period of time (t) minutes.

Distance, s (miles)	8	16	24	32	40
Time, t (minutes)	10	20	30	40	50

a Is s in direct proportion to t? Explain.

b What is the relationship between distance (s) and time (t).

c Work out the distance travelled after 25 minutes.

Discussion Did everyone use the same method?

7 **Reflect** You have drawn graphs and calculated ratios to check that ratios are in direct proportion. Which method did you like best? Why?

Key point 6

When x and y are in direct proportion

- $y = kx$, where k is the gradient of the graph of y against x
- $\frac{y}{x} = k$, a constant.

Example 4

A is directly proportional to x. $A = 5$ when $x = 10$.

a Sketch a graph of A against x.

b Use your graph to work out a formula for A in terms of x.

c Use your formula to work out the value of A when $x = 100$

a

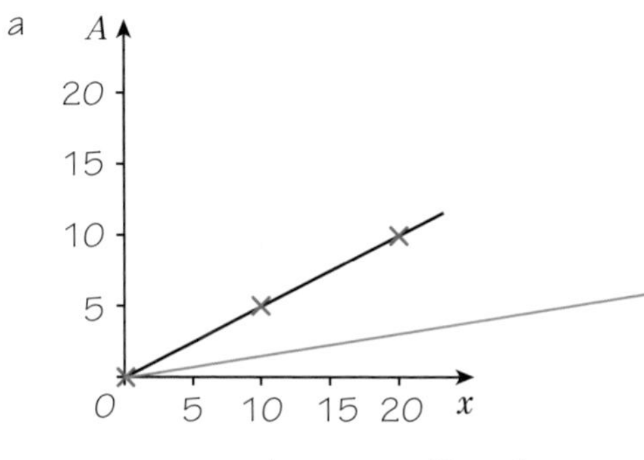

A sketch does not have to be drawn on graph paper. Graph of A against x means A is on the vertical axis. When A and x are in direct proportion, the graph must go through the origin and as A doubles so does x.

b $A = kx$, so $k = \frac{A}{x}$ $k = \frac{5}{10} = \frac{1}{2}$ or 0.5 Substitute $A = 5$ and $x = 10$

$A = 0.5x$

c $A = 0.5 \times 100$ $A = 50$ Substitute $x = 100$ into the formula $A = 0.5x$

8 The cost of buying 20 litres of petrol is £26.

a Show that the cost, £C, of buying the fuel is directly proportional to the amount, x litres, of fuel bought.

b What is the relationship between C and x?

c Work out the cost of 55 litres of fuel.

Q8 hint Draw a table

x	0	5	10	20
C				26

Plot a graph of C against x.

9 **Real** It takes 3 typists 5 hours to type a report.
How long would it take 7 typists?
Give your answer to the nearest minute.

Key point 7

When x and y are in inverse proportion, y is proportional to $\frac{1}{x}$. As one doubles (×2) the other halves (÷2).

10 It takes 5 men 8 hours to build a wall.
How long will it take

a 6 men

b 3 men?

c For both parts **a** and **b**, multiply the exact answer in hours (H) by the number of men (N).
What do you notice?

Q10c hint Write your answer for the time taken in hours in fraction form

Number of hours (H) = $\frac{\square}{\square}$

$H \times N = \frac{\square}{\square} \times \square = \square$

Key point 8

When x and y are in inverse proportion then

- $x \times y$ = a constant
- $xy = k$, so $y = \frac{k}{x}$

11 A and B are in inverse proportion. Work out the values of W, X, Y and Z.

A	10	20	14	Y	Z
B	14	W	X	70	28

12 Do these equations represent direct proportion, inverse proportion or neither?

a $y = 3x$

b $y = \frac{5}{x}$

c $x + y = 9$

d $xy = 10$

e $\frac{y}{x} = 4$

Q12c hint Rearrange the equation to $y = \square$

13 **Problem-solving / STEM** In a circuit, the resistance, R ohms, is inversely proportional to the current, I amps. When the resistance is 12 ohms, the current in the circuit is 8 amps.
Find the current when the resistance in the circuit is 6.4 ohms.

Q13 hint R is proportional to $\frac{1}{I}$ so $R \times I$ = constant

14 r is inversely proportional to t.
$r = 15$ when $t = 0.3$

a Find a formula for r in terms of t.

b Calculate the value of r when $t = 4$

15 a Copy and complete the table for $y = \frac{10}{x}$

x	1	2	5	10
y				

b Use the table to sketch a graph of y against x.

c Work out the value of y when $x = 20$
Does it fit the shape of your graph?

d Work out the value of y when $x = 0.5$. Is your answer consistent with your sketch?

16 **Exam-style question**

The time, T seconds, it takes a water heater to boil some water is directly proportional to the mass of water, m kg, in the water heater.
When $m = 250$, $T = 600$

a Find T when $m = 400$ **(3 marks)**

The time, T seconds, it takes a water heater to boil a constant mass of water is inversely proportional to the power, P watts, of the water heater.
When $P = 1400$, $T = 360$

b Find the value of T when $P = 900$ **(3 marks)**

June 2006, Q16, 5525/05

11 Problem-solving

Objective
- Use arrow diagrams to solve problems.

Example 5

In 2012, visitor numbers to an ice rink increased by 20% compared to the previous year.
In 2013, visitor numbers decreased by 10% compared to the previous year.
In 2013, there were 21 762 visitors. How many visitors were there during 2011?

In 2012, visitor numbers increased by 20%. Draw an arrow and a multiplier of 1.2

In 2013, visitor numbers decreased by 10%. Draw an arrow and a multiplier of 0.9

Year	Number of visitors
2011	?
	×1.2 ↓ ↑ ÷1.2
2012	
	×0.9 ↓ ↑ ÷0.9
2013	21762

Use ? to show that you don't know the number of visitors during 2011.

Draw arrows to work backwards, using the inverse operations: ÷ 0.9 , then ÷ 1.2

Number of visitors in 2011 = 21 762 ÷ 0.9 ÷ 1.2 = 20 150

Use the arrow diagram to calculate the number of visitors in 2011.

Check: 20 150 × 1.2 × 0.9 = 21 762

Check your answer.

1 Judith bought some shares. In the first year they went up in value by 9%. In the second year, they went down in value by 5%. In the third year, they went down in value again by 2.5%. At the end of the third year Judith sold the shares for £2423.07. Did Judith profit or lose on the shares? By how much?

Q1 hint

Judith bought the shares for ?
×□ ↓
At the end of the first year
Continue the arrow diagram.
Make sure you answer the question.

2 The sides of quadrilateral A are 3 cm, 4 cm, 6 cm and 8 cm. A similar quadrilateral B has shortest side 22.5 cm. Find the length of the longest side of the similar quadrilateral B.

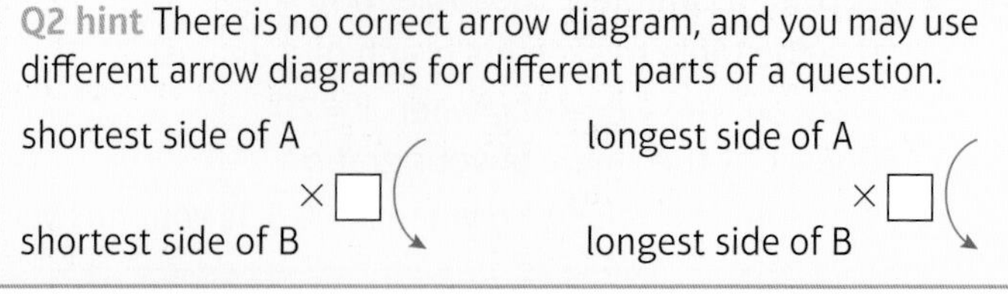

Q2 hint There is no correct arrow diagram, and you may use different arrow diagrams for different parts of a question.
shortest side of A
×□
shortest side of B
longest side of A
×□
longest side of B

3 How many inches per second is 30 miles per hour?

Q3 hint Here are arrow diagrams that may help you:

Distance
miles
× 5280
feet
× 12
inches

Time
hours
÷□
minutes
÷□
seconds

4 **Real / STEM** Concorde flew at an average of 440 metres per second. The distance between London and New York is 5600 km. What was the flight time on Concorde in hours and minutes?

5 **STEM** Vinegar is made of acetic acid and water. A factory fills 200 bottles per minute. Each bottle holds 475 ml of vinegar. There is 4 g of acetic acid per 100 ml of water. How much acetic acid does the factory use in 24 hours? Give your answer in tonnes.

6 **Real / Finance** One day in May 2006, the price of copper hit a high at £4463 per tonne. A 2p coin weighs 7.12 g. Those made before 1992 are 97 per cent copper. How much was a 2p coin, made before 1992, worth in copper on that day in May 2006? Give your answer to the nearest penny.

7 **Reflect** How did the arrow diagrams help you? Is this a strategy you would use again to solve problems?

11 Check up

Log how you did on your Student Progression Chart.

Percentages

1 **Finance** Danny bought a car for £10 000
The value of the car depreciated by 20% in the first year.
Its value depreciated by another 10% in the second year.
Work out the value of Danny's car at the end of two years.

2 **Finance** £3500 is invested for 3 years at 3.4% compound interest.
Work out the total amount in the account after 3 years.

3 **Real** The number of bees in a hive decreases by 3% each year. There are 7500 bees in the hive at the beginning of 2014. How many bees will there be in the hive at the end of 2020?

4 **Reasoning / Finance** Gavin invests £4500 at a compound interest rate of 4.2% per annum. How many years before the investment has grown to £5527.78?

Compound measures

5 **Reasoning / Finance** Scott works a basic 30-hour week. He is paid £8.10 an hour for this work. He is paid time and a quarter for each hour he works on a Saturday and time and a half for each hour he works on a Sunday.
   a Scott works a basic 30-hour week, plus 5 hours on Saturday and 4 hours on Sunday. How much is Scott paid for the week?
   b In one week Scott works a basic 30-hour week and some hours on Saturday. He is paid £303.75 for the week. How many hours did Scott work on Saturday?

6 **Exam-style question**

There are 40 litres of water in a barrel. The water flows out of the barrel at a rate of 125 millilitres per second.

1 litre = 1000 millilitres

Work out the time it takes for the barrel to empty completely. **(3 marks)**

May 2009, Q10, 1380/3H

7 **Reasoning** The mass of this plastic cuboid is 2208 g.
Work out the density of the plastic in grams per cm^3.

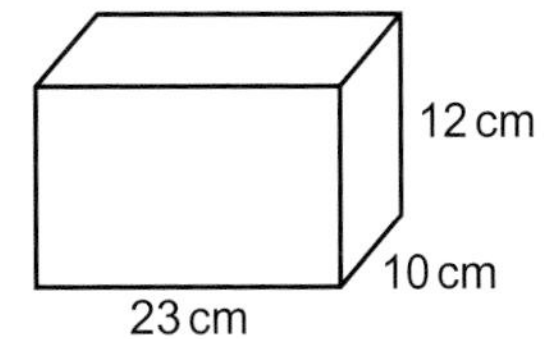

8 **Real** A solid cube of steel has sides of length 5 cm.
The density of steel is 8.05 g/cm^3.

a Convert 8.05 g/cm^3 to kg/m^3.

b Work out the mass of the cube.

9 **Real** A force of 30 N is applied to an area of 3.2 m^2. Work out the pressure in N/m^2.

10 **Communication / Real** The greatest recorded speed of Usain Bolt is 12.3 m/s.
The greatest speed of a great white shark is 40 km/h. Which is faster? Explain your answer.

Ratio and proportion

11 **Modelling** The table shows a comparison of costs in British pounds (P) and euros (E).

Cost in British pounds (P)	1	2	5	10	15
Cost in euros (E)	1.3	2.6	6.5	40	46.5

a Are British pounds and euros in direct proportion? Explain.

b What is the relationship between British pounds (P) and euros (E)?

c Convert £25 to euros.

12 **Communication / Problem-solving** The pressure, P, of water on an object (in bars) is directly proportional to its depth, d (in metres). When the object is at a depth of 8 metres the pressure on the object is 0.8 bars.
A diver's watch has been guaranteed to work at a pressure up to 8.5 bars. A diver takes the watch down to 75 m. Will the watch still work? Give a reason for your answer.

13 **Problem-solving / Real** In a circuit, the resistance, R ohms, is inversely proportional to the current, I amps. When the resistance is 14 ohms, the current in the circuit is 9 amps. Find the current when the resistance is 12 ohms.

14 How sure are you of your answers? Were you mostly

Just guessing  Feeling doubtful  Confident

What next? Use your results to decide whether to strengthen or extend your learning.

Reflect

* Challenge

15 The times, distances and speeds of athletes in a 5 km race and a 10 km race have got mixed up. Sort them into the 5 km and 10 km races and then compile a leader board for each race.

Allia: 18 minutes 38 seconds, 5 km
Billie: 1364 seconds, 13.2 km/h
Chaya: 20 minutes 50 seconds, 4 m/s
Daisy: 10 km, 4.1 m/s
Ellie: 10 km, 13 km/h
Fion: 2105 seconds, 17.1 km/h
Gracie: 45 minutes 3 seconds, 3.7 m/s
Hafsa: 5 km, 3.9 m/s

Q15 strategy hint Use the formula connecting distance, speed and time to work out the missing times and distances for each person.

11 Strengthen

Percentages

1 Write down the multipliers for these percentage increases as a decimal number.

a 20% b 9% c 3.7%

> **Q1a hint** Original = 100%
> After increase = 100% + □% = □%
> Write % as a decimal.

2 Write down the multipliers for these percentage decreases as a decimal number.

a 23% b 6% c 7.5%

> **Q2a hint** Original = 100%
> After decrease = 100% − □% = □%
> Write % as a decimal.

3 Write down the multiplier for

a an increase of 20% followed by an increase of 9%

b an increase of 10% followed by an increase of 15%

c a decrease of 23% followed by a decrease of 6%

d a decrease of 11% followed by a decrease of 9%

e an increase of 12% followed by a decrease of 8%.

> **Q3a hint** Use your answers to **Q1a** and **b**. Multiplier for increase of 20% followed by increase of 9% = □ × □ = □

4 **Finance** Penny buys a new TV for £750. In the first year the value depreciates by 15%. In the second year the value depreciates by 5%. Work out the value of Penny's TV after 2 years.

> **Q4 hint** Write down the multipliers for 15% and 5%.

5 **Finance** Harry invests £400 at 3% **compound interest**. Copy and complete the table.

Year	Amount at start of year	Amount plus interest	Total amount at end of year
1	£400	400 × 1.03	£412
2	£412	$412 \times 1.03 = 400 \times 1.03^2$	£424.36
3	£424.36	$424.36 \times 1.03 = 400 \times 1.03^3$	£437.09
4	£437.09	437.09 × 1.03 =	
5			
6			

6 **Finance / Problem-solving** A company bought a van for £15 000.
Each year the value of the van depreciated by 20%.
Work out the value of the van at the end of four years.

> **Q6 hint** Draw a table.

Year	Value of van (£)
0	1500
1	□

7 **Problem-solving** A population of ants increases at a rate of 30% per day. At the end of one week there are 3500 insects.
How many insects were there at the beginning of the week?

> **Q7 hint** Draw a table.

Day	Number of ants
6	÷□
7	3500

8 **Problem-solving** Molly invests £2000 at a compound interest rate of 2.3% per annum.
After how many years will she have more than £2200?

> **Q8 hint** Draw a table.

Year	Amount (£)
0	2000
1	×□

Compound measures

1 **Reasoning / Finance** Ellie works a basic 35-hour week. Her hourly rate of pay is £6.80. She is paid time and a quarter for each hour she works on a Saturday and time and a half for each hour she works on a Sunday.
How much is Ellie paid for a week when she works a basic 35-hour week, plus 4 hours on Saturday and 3 hours on Sunday?

Q1 hint
Basic pay = £6.80
Pay at time and a quarter = £6.80 × 1.25 = □
Pay at time and a half = £6.80 × □ = □
Total pay = 35 × £6.80 + 4 × □ + 3 × □ = □

2 **Reasoning** A bucket holds 12 litres of water. Water flows out at a rate of 50 ml per second.
Work out

a The amount of water flowing out per minute. Give your answer in litres.

b The time it will take the bucket to empty.

Q2 hint

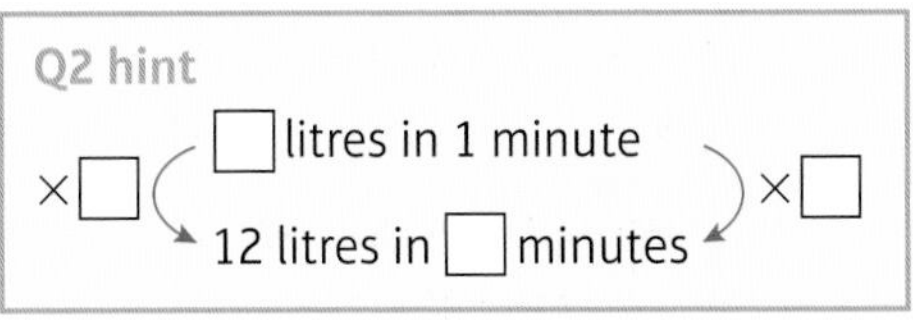

3 **STEM / Modelling** Copy and complete this table of mass, volume and density.
Give your answers to 3 sf.

Metal	Mass (g)	Volume (cm^3)	Density (g/cm^3)
Copper		122	8.96
Lead	450		11.3
Mercury	110	8.15	

Q3 hint Mass = density × volume

Volume = $\frac{\text{mass}}{\text{density}}$ Density = $\frac{\text{mass}}{\text{volume}}$

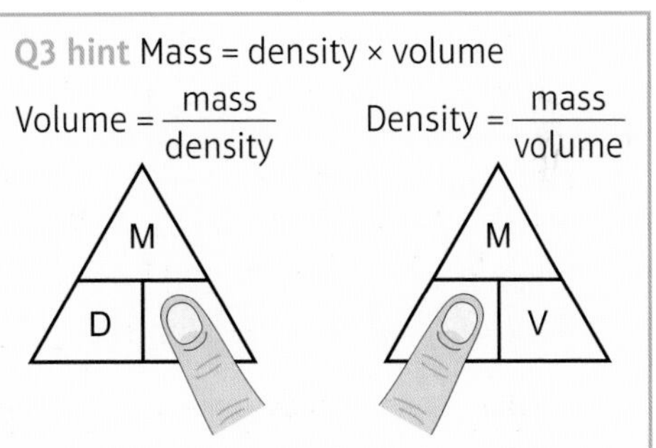

4 **STEM / Modelling** Copy and complete this table of force, area and pressure.

Force (N)	Area (cm^2)	Pressure (N/cm^2)
	13	8
48	12	
65		13

Q4 hint

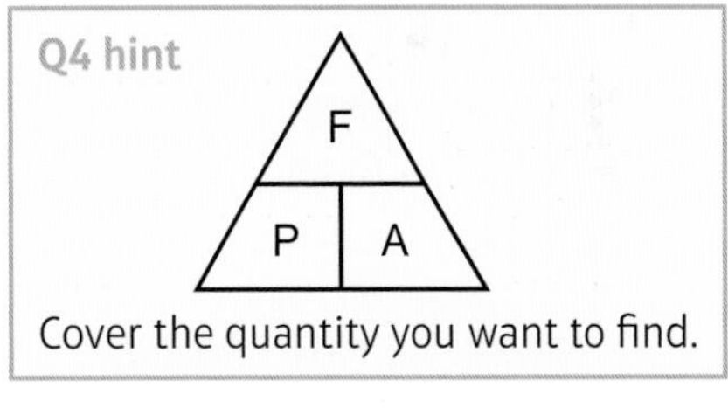

Cover the quantity you want to find.

5 Copy and complete.

a □ g = 1 kg

b □ cm^2 = 1 m^2

c □ cm^3 = 1 m^3

Q5b hint

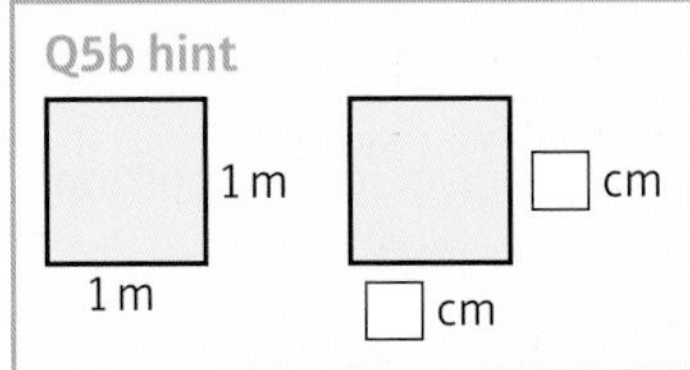

Q5c hint

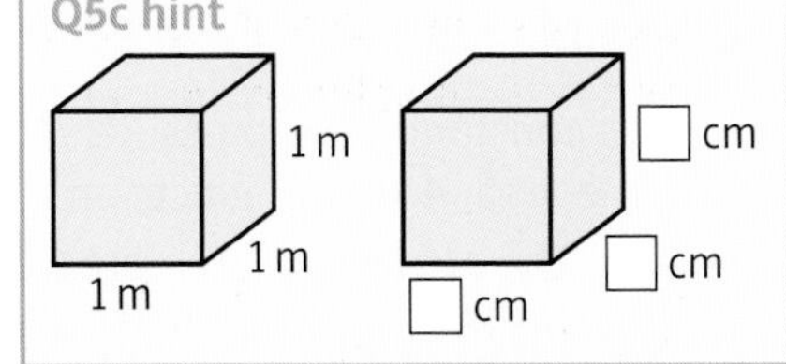

6 a Convert

i 12 000 g to kg

ii 15 kg to g

b Convert

i 270 kg/m^2 to g/cm^2

ii 45 g/cm^2 to kg/m^2

iii 50 g/cm^3 to kg/m^3

iv 20 kg/m^3 to g/cm^3.

Q6a hint Converting smaller to bigger ÷
Converting bigger to smaller ×

Q6b i hint

270 kg per m^2 ×□

□ g per m^2 ×□

□ g per cm^2

Q6b iii hint

50 g per cm^3 ÷□

□ kg per cm^3 ÷□

□ kg per m^3

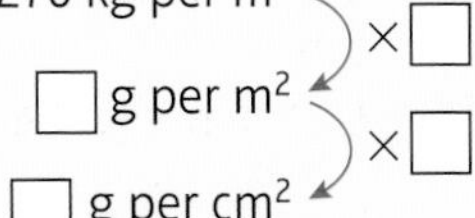

7 Copy and complete this table to convert from km/h to m/s.

km/h	m/h	m/min	m/s
18			
			10
24			
			16

Q7 hint

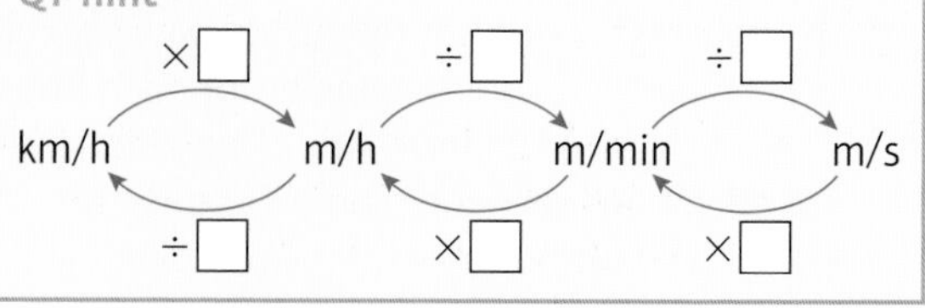

Ratio and proportion

1 A and B are in direct proportion.

A	B
5	10
12	W
X	45
15	Y
Z	36

Find the missing numbers W, X, Y and Z.

Q1 hint
When A and B are in direct proportion, $\frac{A}{B}$ = constant. This means it always has the same value.

A	B	$\frac{A}{B}$
5	10	0.5
12	W	0.5
X	45	0.5

What is W?

What is X?

2 The time, T seconds, it takes a kettle to boil some water is directly proportional to the volume of water, v cm³, in the kettle.
When $v = 300\,\text{cm}^3$, $T = 120$ seconds.
How long will it take to boil 500 cm³ of water?

Q2 hint Draw a table.

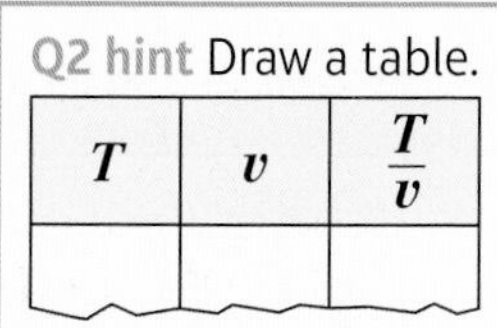

T	v	$\frac{T}{v}$

3 Mike exchanges 100 British pounds (£) for 80 euros (€).

a Show that the number of pounds exchanged, P, is directly proportional to the number of euros received, E, by drawing a graph.

b Find a formula for P in terms of E.

Q3a hint Set up a table for the number of euros for a given number of pounds.

P	10	25	50	100
E				80

Put P on the y-axis and E on the x-axis.
Does the graph pass through (0, 0)?
Do the points lie on a straight line?

Q3b hint When P and E are direct proportion, $\frac{P}{E}$ = constant.
Rearrange $\frac{P}{E} = \square$ to get $P = \square$

4 For a constant force, pressure, P (in N/m²) is inversely proportional to the area, A (in m²) it acts on.
When the area is 2 m², the pressure is 16 N/m².
Work out the pressure when the area is 0.5 m².

Q4 hint Draw a table.

P	A	PA
16		

5 A and B are in inverse proportion.
Work out the values of W, X, Y and Z.

A	B
6	8
8	W
X	16
12	Y
Z	12

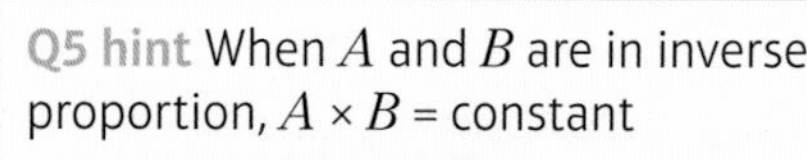

Q5 hint When A and B are in inverse proportion, $A \times B$ = constant

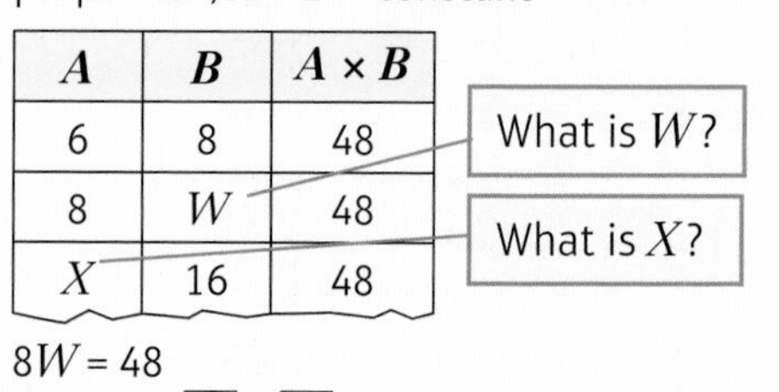

A	B	$A \times B$
6	8	48
8	W	48
X	16	48

What is W?

What is X?

$8W = 48$
$W = 48 \div \square = \square$
$16X = 48$

11 Extend

1 **Real** In a spring, the tension (T newtons) is directly proportional to its extension (x cm). When the tension is 150 newtons, the extension is 6 cm.

a Write a formula for T in terms of x.

b Calculate the tension, in newtons, when the extension is 15 cm.

c Calculate the extension, in cm, when the tension is 600 newtons.

2 **STEM / Problem-solving** The diagram shows a piece of plastic cut into the shape of a trapezium. A force is exerted evenly over the trapezium.
Work out the force required to create a pressure of 20 N/cm^2 on the trapezium.

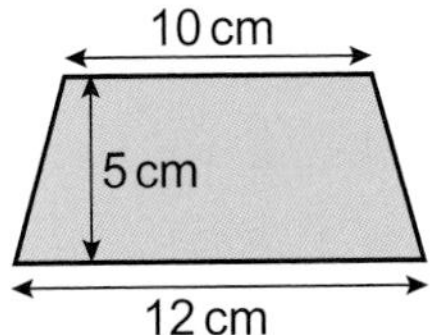

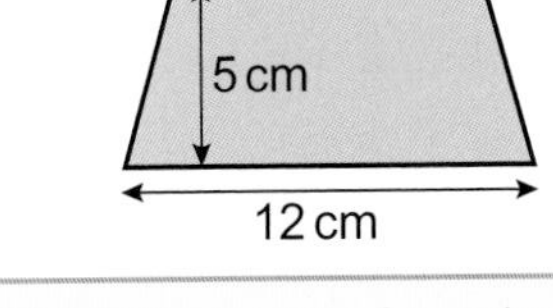

3 Use the kinematics formulae in these questions.

a A train starts from rest and accelerates at 3 m/s^2 for 15 seconds. What is its final velocity in km/h?

b Rafael hits a tennis ball at 20 m/s. It hits the net, 8 m away, at a speed of 12 m/s.
What was its deceleration in m/s^2?

c A ball dropped from the top of a 120 m tower takes 4.9 seconds to reach the ground. Calculate its acceleration.

d A car starts from rest and accelerates to v m/s in 20 seconds.
Find the acceleration in terms of v.

> **Q3 hint** Kinematics formulae:
> $v = u + at$
> $s = ut + \frac{1}{2}at^2$
> $v^2 = u^2 + 2as$

> **Q3b communication hint** Deceleration is negative acceleration.

4 **Finance / Reasoning** George invests £4500 at a compound interest rate of 5% per annum.
At the end of n complete years the investment has grown to £5469.78
Find the value of n.

5 **Finance / Reasoning** Gwen bought a new car.
Each year, the value of her car depreciated by 9%.

a Let £x be the original price of the car.
Write an expression for the value of the car after 1 year.

b After how many years was the car worth less than half of its original price?

> **Q5b hint** When is the value $< 0.5x$?

6 **Finance** The value of a car depreciates by 25% each year.
At the end of 2015 the value of the car was £2560.
Work out the value of the car at the end of 2010.

7 **Problem-solving** Your manager says you can either have a 2.5% pay rise this year and then a 1.5% pay rise next year, or a 3.5% pay rise this year and no pay rise next year.
Which would you prefer?

> **Q7 strategy hint** Decide on a salary figure to work with.

8 **Problem-solving** Plastic block A has a mass of 1.2 kg. Plastic cube B is made from plastic with a density 40% greater than the plastic block. Work out the mass of the plastic cube.
Give your answer in grams to three significant figures.

A

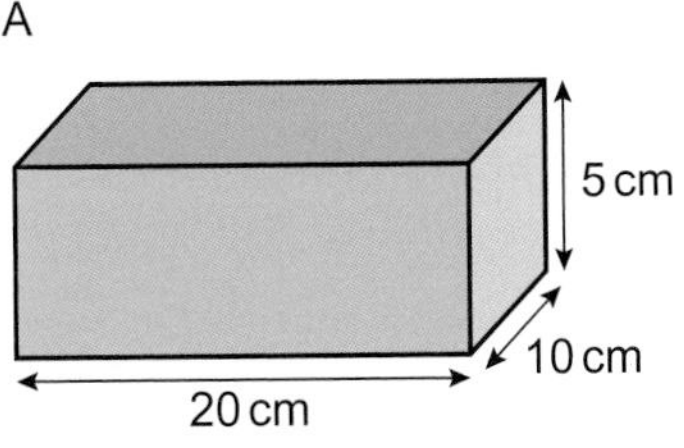

B

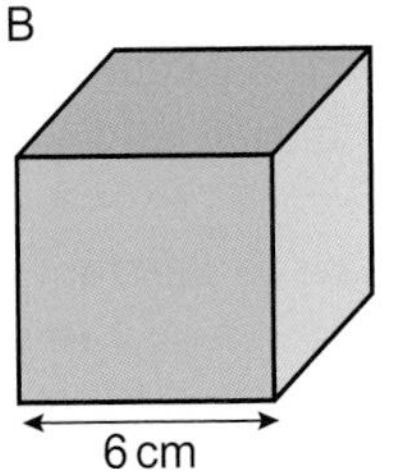

9 **Finance** Chloe invests £3000 at 3.2% per annum compound interest.

a Copy and complete this table of values.

Number of years, n	0	1	2	3	4
Value, y	3000				

Q9a hint Round to the nearest £10.

b Plot the graph of y against n. Join the points.

c Estimate when Chloe will have £3350 in her account.

10 **Problem-solving** Carrie buys a motorbike. The value of the motorbike depreciates by 10% each year. Write down the letter of the graph which best shows how the value of Carrie's motorbike changes with time.

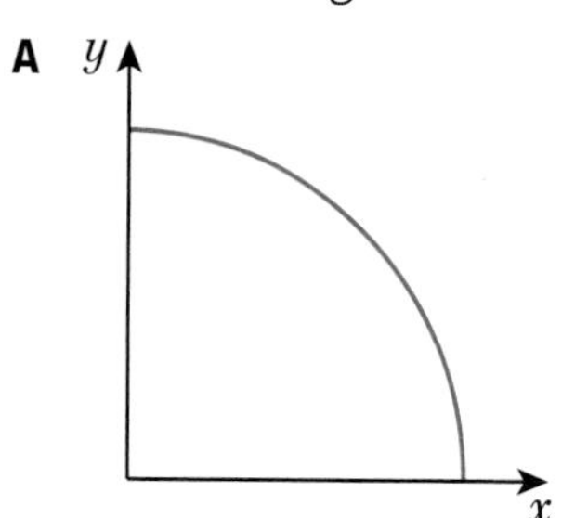

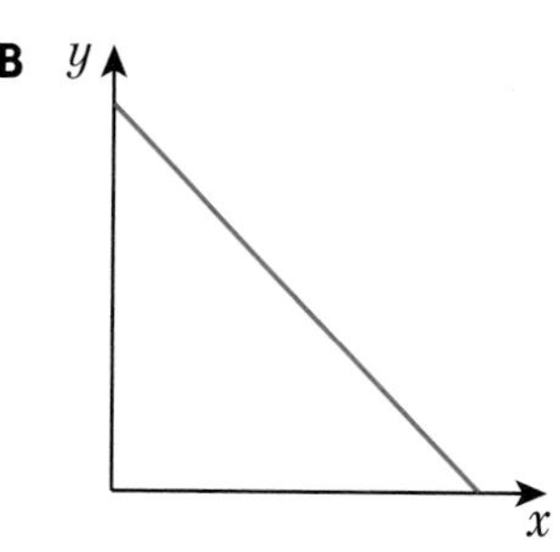

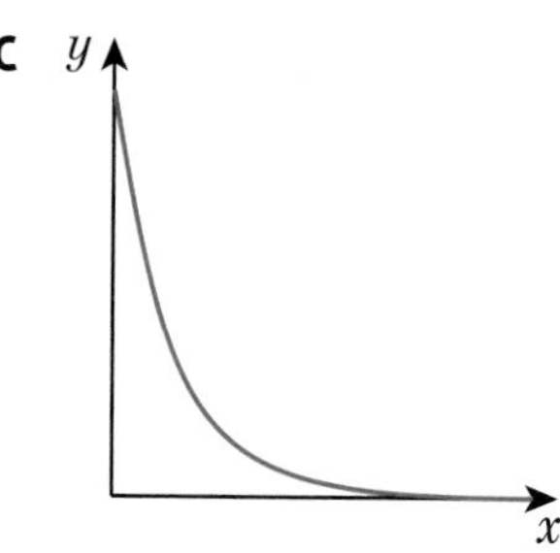

11 **Reasoning** The size of an exterior angle of a regular polygon is inversely proportional to the number of sides of the polygon.

a Sam thinks this means that if one polygon has twice the number of sides of another polygon the size of the exterior angle is half the size of the original exterior angle. Is Sam correct? Explain.

b The size of an exterior angle of a regular hexagon is 60°. What is the size of the exterior angle of a 20-sided polygon?

12 **Problem-solving** The number of outlets of a coffee shop chain in the UK increases at a rate of 9% for 10 years. At the end of the 10 years there are 350 coffee shops. How many were there at the beginning of the 10 years?

13 **Exam-style question**

Sumeet has a pond in the shape of a prism.

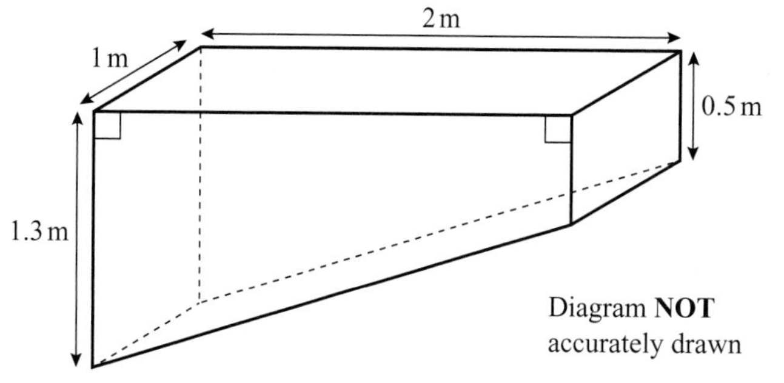

The pond is completely full of water.
Sumeet wants to empty the pond so he can clean it.
Sumeet uses a pump to empty the pond.
The volume of water in the pond decreases at a constant rate.
The level of the water in the pond goes down by 20 cm in the first 30 minutes.
Work out how much more time Sumeet has to wait for the pump to empty the pond completely. **(6 marks)**

June 2013, Q17, 1MA0/1H

Q13 strategy hint First find the volume of the pond, and then find the rate at which the volume of water is emptied in m^3 per hour.

11 Knowledge check

- In **compound interest** the interest earned each year is added to money in the account and earns interest the next year. Most interest rates are compound interest rates. .... *Mastery Section 11.1*
- Total interest = amount in the account at the end of the investment – amount invested .... *Mastery Section 11.1*
- You can calculate an amount after n years' compound interest using the formula
  amount = initial amount × $\left(\frac{100 + \text{interest rate}}{100}\right)^n$ .... *Mastery Section 11.1*
- Compound measures such as speed, density and pressure combine measures of two different quantities. .... *Mastery Section 11.2 and 11.3*
- Speed can be measured in metres per second (m/s), kilometres per hour (km/h) or miles per hour (mph). .... *Mastery Section 11.2*
- Average speed = $\frac{\text{distance}}{\text{time}}$ or $S = \frac{D}{T}$ .... *Mastery Section 11.2*
- These are three kinematics formulae:
  $v = u + at$
  $s = ut + \frac{1}{2}at^2$
  $v^2 = u^2 + 2as$
  where a is constant acceleration, u is initial velocity, v is final velocity, s is displacement from the position when $t = 0$ and t is time taken. ... *Mastery Section 11.2*
- **Velocity** is speed in a given direction, possible units are m/s. .... *Mastery Section 11.2*
- **Initial velocity** is speed in a given direction at the start of the motion. *Mastery Section 11.2*
- **Acceleration** is the rate of change of velocity, i.e. a measure of how the velocity changes with time, possible units are m/s^2. .... *Mastery Section 11.2*
- Density is the **mass** of substance in g contained in a certain **volume** in cm^3 and is often measured in grams per cubic centimetre (g/cm^3).
  Density = $\frac{\text{mass}}{\text{volume}}$ or $D = \frac{M}{V}$ .... *Mastery Section 11.3*
- Pressure is the **force** in newtons applied over an **area**, in cm^2 or m^2. It is usually measured in newtons (N) per square metre (N/m^2) or per square centimetre (N/cm^2).
  Pressure = $\frac{\text{force}}{\text{area}}$ or $P = \frac{F}{A}$ .... *Mastery Section 11.3*
- When x and y are in direct proportion
  $y = kx$, where k is the gradient of the graph of y against x
  $\frac{y}{x} = k$, a constant .... *Mastery Section 11.4*
- When x and y are in inverse proportion, y is proportional to $\frac{1}{x}$.
  As one doubles (×2) the other halves (÷2). .... *Mastery Section 11.4*
- When x and y are in inverse proportion then
  $x \times y$ = a constant
  $xy = k$, so $y = \frac{k}{x}$ .... *Mastery Section 11.4*

This unit is called multiplicative reasoning.
List three ways you have used multiplication or division in this unit.
Why is it good to reason in mathematics?

> **Communication hint** 'Multiplicative' means involving multiplication or division. Reasoning is being able to explain why you have done some maths a certain way.

Reflect

11 Unit test

Log how you did on your Student Progression Chart.

Kinematics formulae

$v = u + at$ $\quad s = ut + \frac{1}{2}at^2$ $\quad v^2 = u^2 + 2as$

1 A car has an initial speed of u m/s.
The car accelerates to a speed of $4u$ m/s in 20 seconds.
Find the acceleration in terms of u. *(3 marks)*

2 A leopard travels 100 metres in 7.19 seconds.
What is its average speed

a in m/s *(2 marks)*

b in km/h? *(2 marks)*

Give your answer to 3 significant figures.

3 **Reasoning** A cylindrical glass has a circular base with radius 4 cm.
The glass exerts a force of 7 N on the table. Work out the pressure in N/m². *(3 marks)*

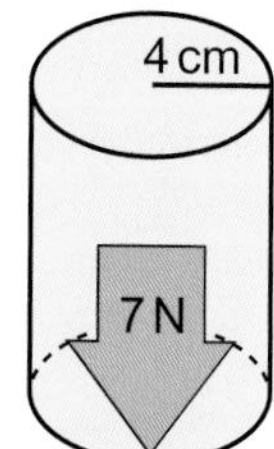

4 **Exam-style question**

The diagram shows a solid prism made from metal.

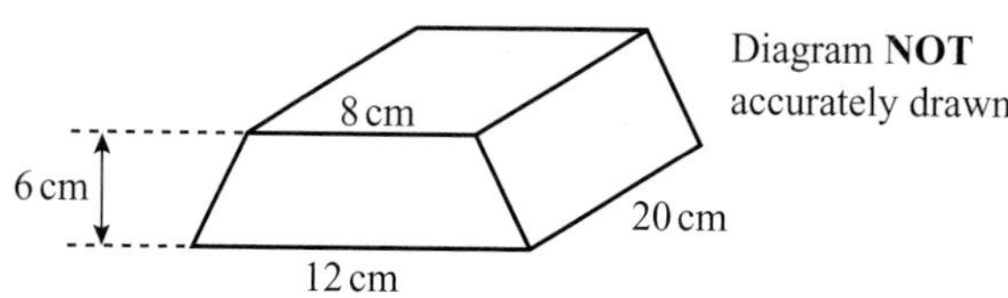

The cross-section of the prism is a trapezium.
The parallel sides of the trapezium are 8 cm and 12 cm.
The height of the trapezium is 6 cm.
The length of the prism is 20 cm.
The density of the metal is 5 g/cm³.
Calculate the mass of the prism.
Give your answer in kilograms. **(5 marks)**

Nov 2011, Q16, 1380/3H

5 When travelling at constant speed the distance, D, travelled by a particle is directly proportional to the time taken, t.
When $t = 20$, $D = 45$

a Find a formula for D in terms of t. *(3 marks)*

b Calculate the value of D when $t = 48$ *(1 mark)*

c Calculate the value of t when $D = 12$
Give your answer correct to 3 significant figures. *(2 marks)*

6 **Exam-style question**

A company bought a van that had a value of £12 000.
Each year the value of the van depreciates by 25%.

a Work out the value of the van at the end of three years. **(3 marks)**

The company bought a new truck.
Each year the value of the truck depreciates by 20%.
The value of the new truck can be multiplied by a single number to find its value at the end of four years.

b Find this single number as a decimal. **(2 marks)**

June 2004, Q12, 5506/06

7 A television loses 4% of its value every month. It was bought for £950 at the beginning of January. How much will it be worth at the end of June? *(3 marks)*

8 When a constant force is applied, the resulting pressure P, is inversely proportional to the area, A. When $A = 8$, $P = 5$

a Find a formula for P in terms of A. *(3 marks)*

b Calculate the value of P when $A = 2$ *(1 mark)*

9 **Exam-style question**

Emma invests £4000 in an account for 2 years. The account pays 3.8% compound interest. Emma has to pay 20% tax on the interest earned each year. The tax is taken from the account at the end of each year.
How much will Emma have in the account at the end of 2 years? **(4 marks)**

10 **Exam-style question**

Nicola invests £8000 for 3 years at 5% per annum **compound** interest.

a Calculate the value of her investment at the end of the 3 years. **(3 marks)**

Jim invests a sum of money for 30 years at 4% per annum **compound** interest.

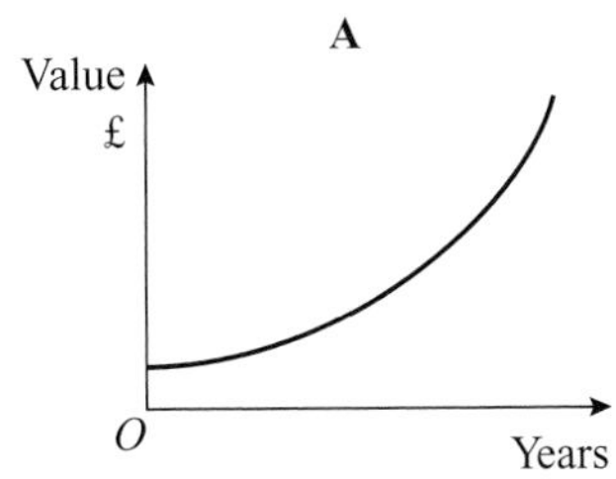

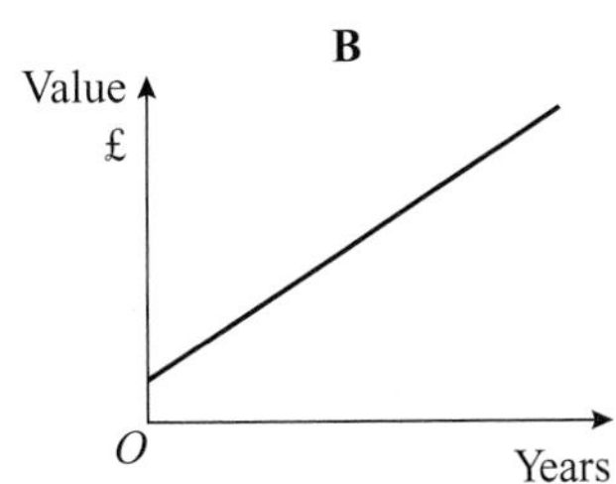

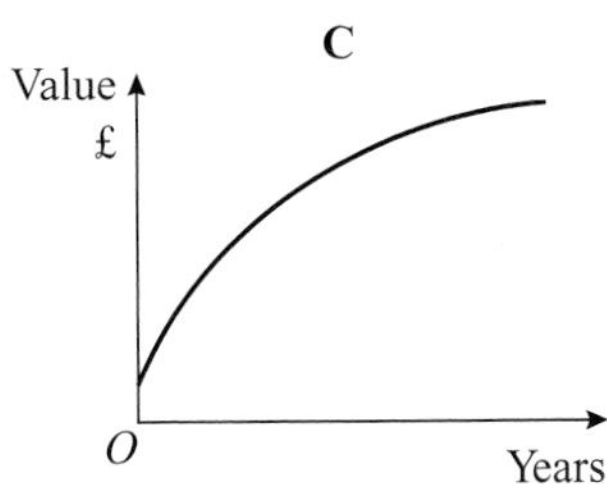

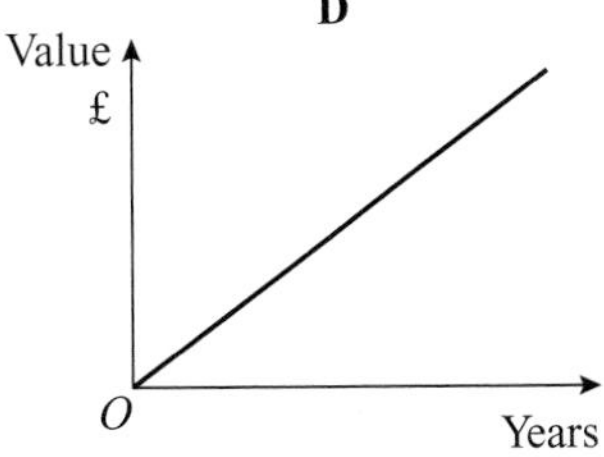

b Write down the letter of the graph which best shows how the value of Jim's investment changes over the 30 years. **(1 mark)**

Hannah invested an amount of money in an account paying 5% per annum **compound** interest. After 1 year the value of her investment was £3885

c Work out the amount of money that Hannah invested. **(3 marks)**

Nov 2004, Q14, 5504/04

Sample student answers

Which student gives the better answer?
Explain.

11 **Exam-style question**

Viv wants to invest £2000 for 2 years in the same bank.

The International Bank	**The Friendly Bank**
Compound interest	**Compound interest**
4% for the first year	5% for the first year
1% for each extra year	0.5% for each extra year

At the end of 2 years, Viv wants to have as much money as possible.
Which bank should she invest her £2000 in? **(4 marks)**

June 2013, Q14, 1MA0/2H

Student A

$2000 \times 1.04 \times 1.01 = 2100.8$

$2000 \times 1.05 \times 1.005 = 2110.5$

Student B

	International Bank	Friendly Bank
Year 1	$2000 \times 1.04 = 2080$	$2000 \times 1.05 = 2100$
Year 2	$2080 \times 1.01 = £2100.80$	$2100 \times 1.005 = £2110.50$ ✓

Viv will get more money if she invests in the Friendly Bank.

Master p.364 | Problem-solve p.382 | Check p.383 | Strengthen p.385 | Extend p.390 | Test p.395

12 SIMILARITY AND CONGRUENCE

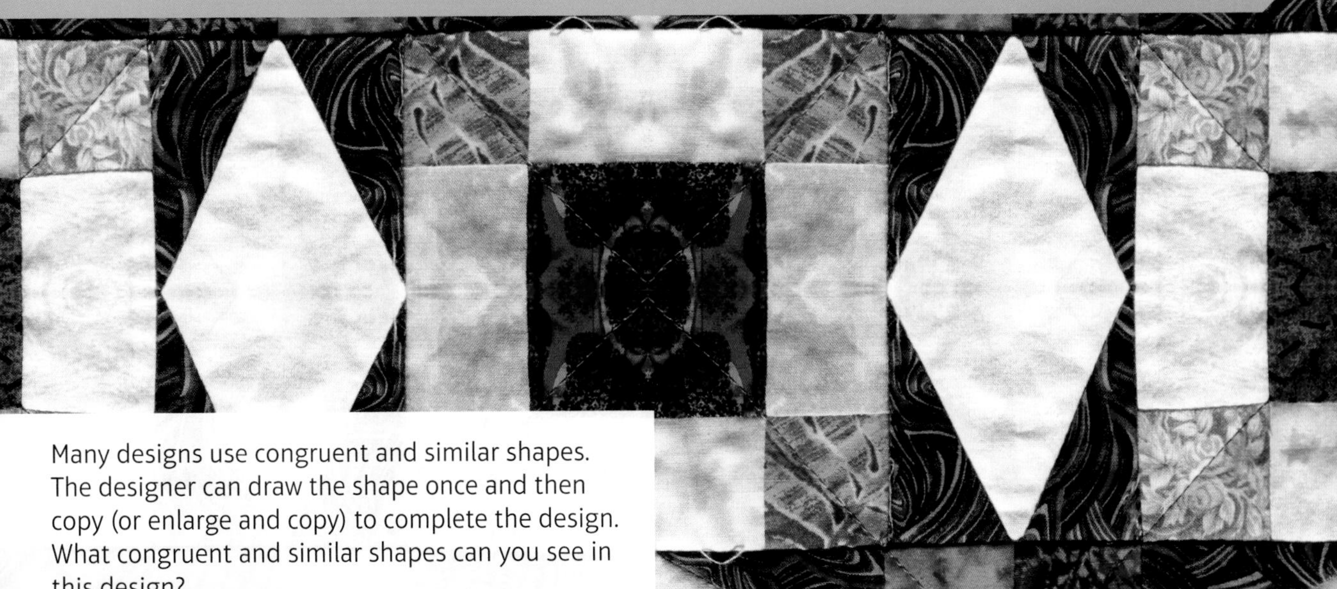

Many designs use congruent and similar shapes. The designer can draw the shape once and then copy (or enlarge and copy) to complete the design. What congruent and similar shapes can you see in this design?

12 Prior knowledge check

Numerical fluency

1 Write as a decimal.
   a $\frac{6}{4}$
   b $\frac{32}{5}$

2 Work out
   a 6^2
   b square root of 9
   c $\sqrt{\frac{16}{49}}$
   d 2^3
   e cube root of 64
   f $\sqrt[3]{\frac{27}{8}}$

3 Simplify these fractions.
   a $\frac{6}{9}$
   b $\frac{4}{20}$

Algebraic fluency

4 Solve these equations.
   a $\frac{x}{5} = \frac{4}{15}$
   b $\frac{3}{4} = \frac{8}{x}$

Geometrical fluency

5 Which shapes are congruent?

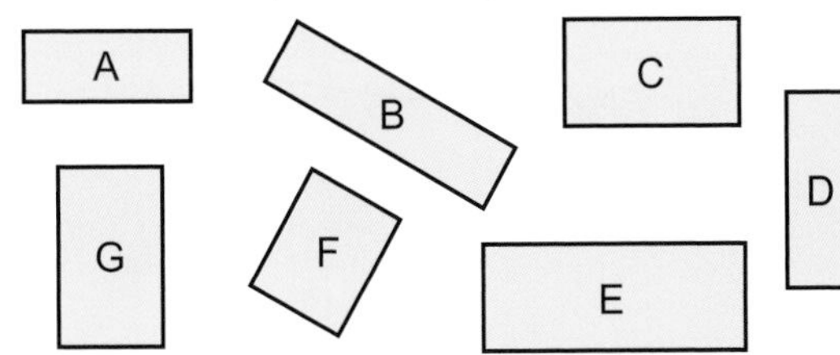

6 What is the scale factor of the enlargement
   a from shape X to shape Y
   b from shape Y to shape X?

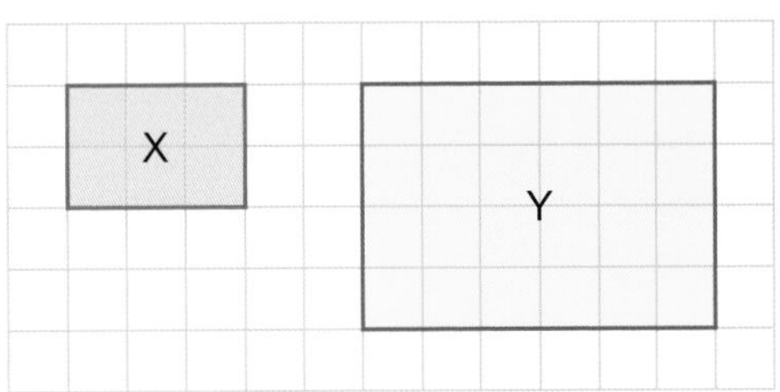

7 This shape has perimeter 12 cm and area 5 cm^2.

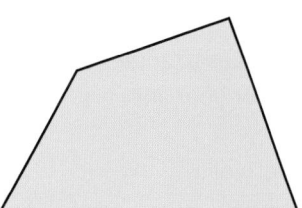

It is enlarged by scale factor 2.
Work out
   a the perimeter of the enlarged shape
   b the area of the enlarged shape.

8 Which angles are equal? Give reasons

a

b

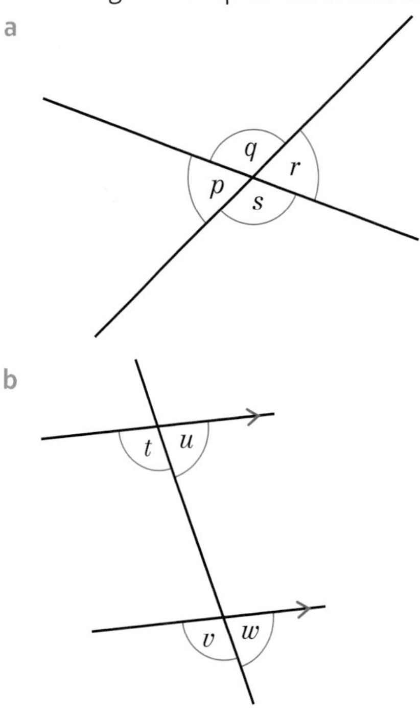

9 Construct this triangle.

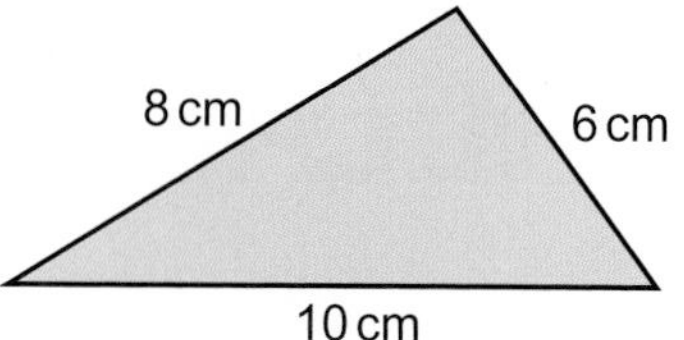

10 Construct each triangle accurately.

a

b

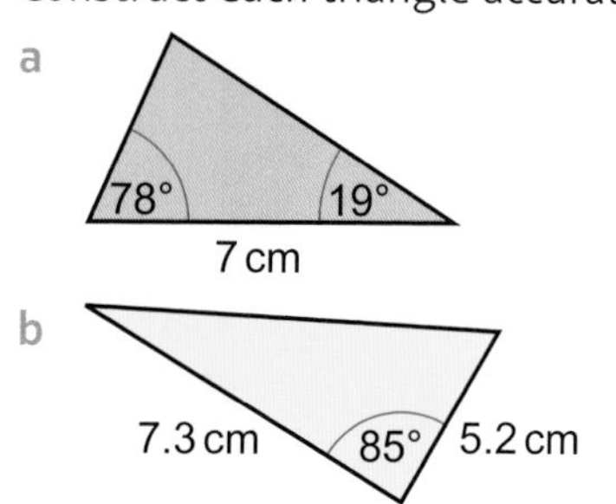

* Challenge

11 There are two possible triangles ABC where $AB = 16$ cm, $BC = 10$ cm and angle $CAB = 40°$. Construct them accurately.

12.1 Congruence

Objectives

- Show that two triangles are congruent.
- Know the conditions of congruence.

Why learn this?

All £1 coins are congruent. This means that coin machines can recognise their value.

Fluency

What do angles in a triangle sum to?

Warm up

1 Which triangles are congruent?

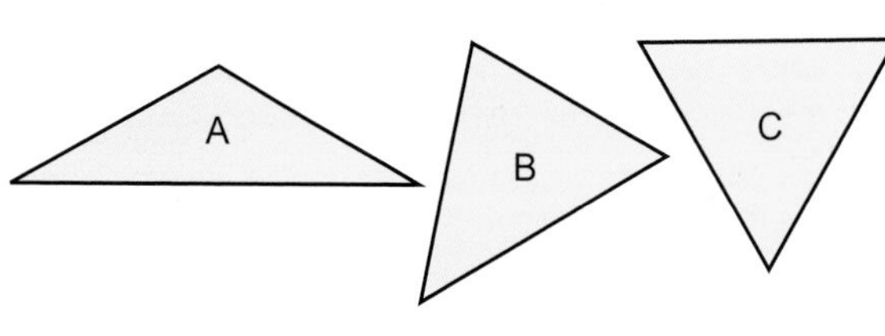

2 Work out the sizes of angles a, b and c.
Give reasons for your answers.

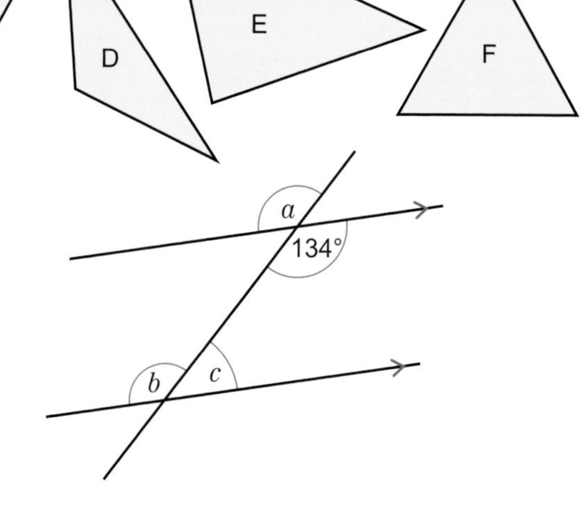

3 Find the length of x.

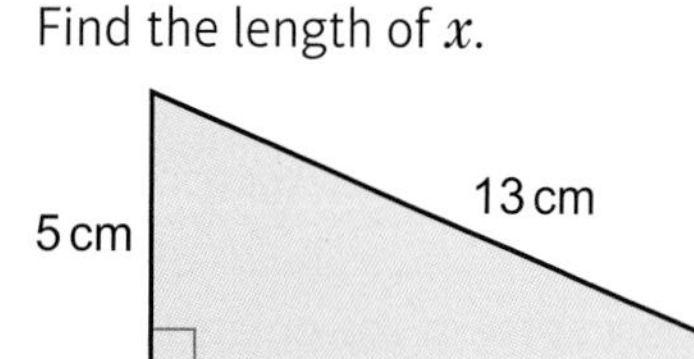

Key point 1

Congruent triangles have exactly the same size and shape. Their angles are the same and corresponding sides are the same length.

Questions in this unit are targeted at the steps indicated.

4 Here is a pair of congruent triangles. Write down

a the size of angle y

b the length of side x.

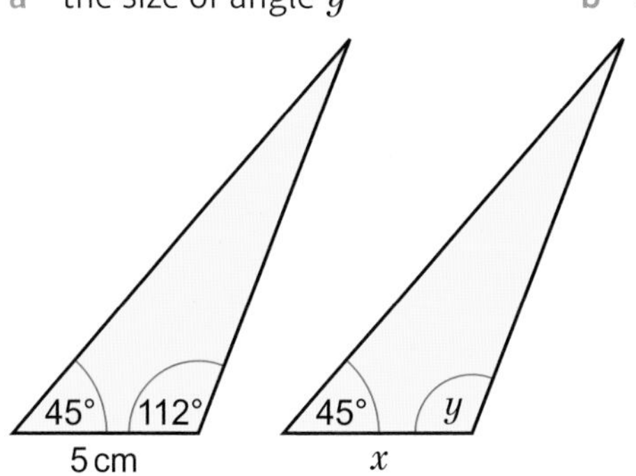

Q4 hint The angles by the 5 cm side will be the same in each triangle.

Key point 2

Two triangles are congruent when one of these conditions of congruence is true.

SSS (all three sides equal)

SAS (two sides and the included angle are equal)

AAS (two angles and a corresponding side are equal)

RHS (right angle, hypotenuse and one other side are equal)

5 Each pair of triangles is congruent. Explain why.

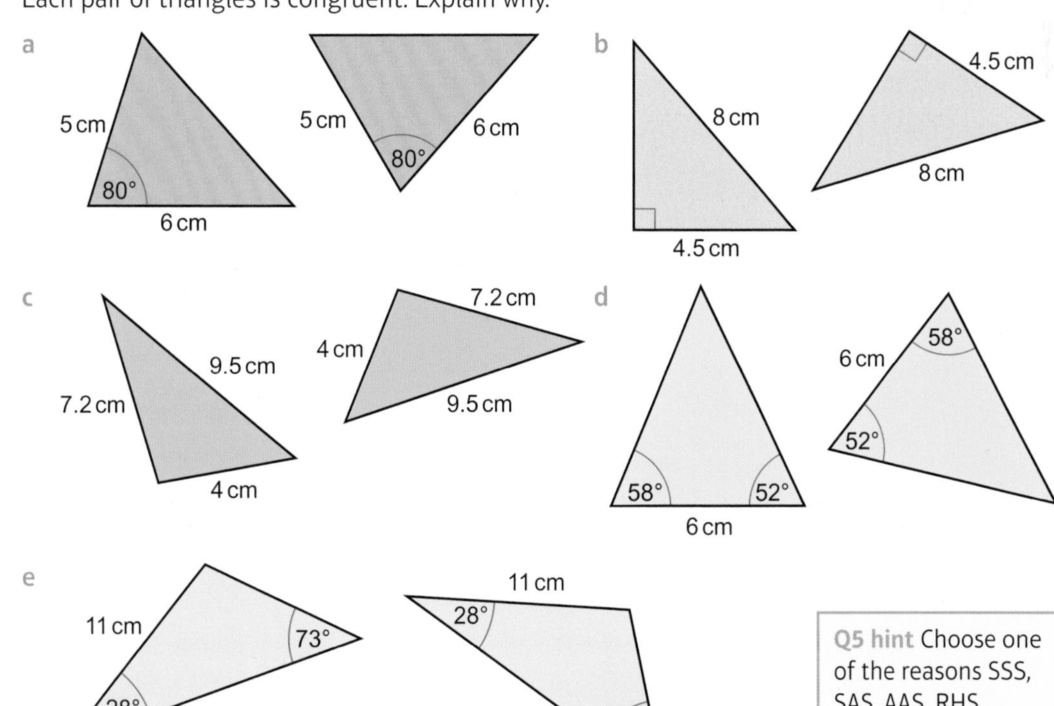

Q5 hint Choose one of the reasons SSS, SAS, AAS, RHS.

6 State whether or not each pair of triangles described below is congruent. If the triangles are congruent, give the reason and write the corresponding vertices in pairs.

a ABC where $AB = 7$ cm, $BC = 5$ cm, angle $B = 42°$
PQR where $PQ = 50$ mm, $QR = 7$ cm, angle $Q = 42°$

b ABC where $AB = 7$ cm, angle $B = 42°$, angle $C = 109°$
PQR where $PQ = 7$ cm, angle $Q = 109°$, angle $R = 42°$

Q6 hint
Sketch each triangle. Angle A corresponds to ☐

7 **Communication** Which of these triangles are congruent to triangle ABC? Give reasons for your answers.

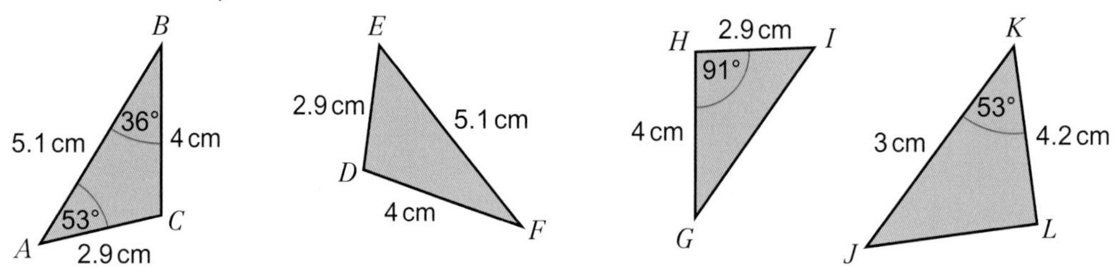

8 **Communication** Are these triangles congruent? Justify your answer.

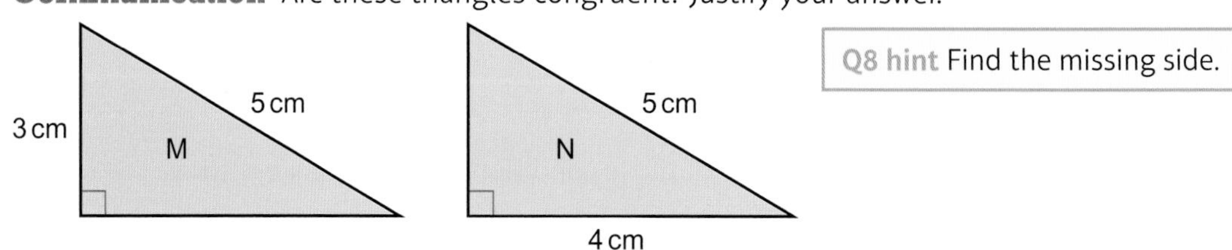

Q8 hint Find the missing side.

9 **Communication** Are all right-angled triangles with one side 5 cm and one side 12 cm congruent? Explain.

Q9 strategy hint
Sketch some triangles.

10 **Communication** AB and CD are parallel lines. $AB = CD$.

a Work out the size of all the angles.

b Show that triangle ABE and triangle CED are congruent.

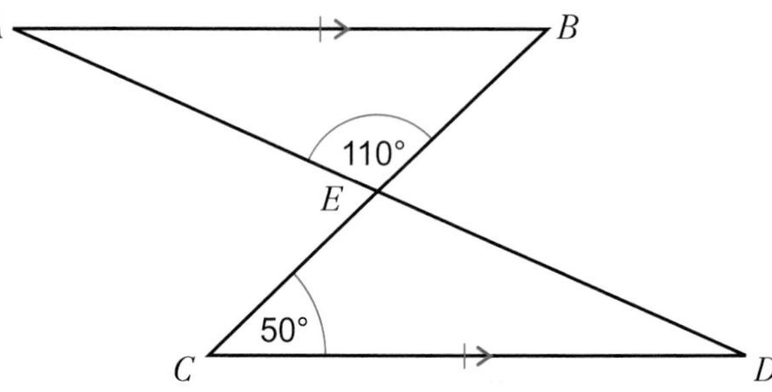

11 **Exam-style question**

In this arrowhead, angle $JKL = 26°$ and angle $KJL = 35°$.

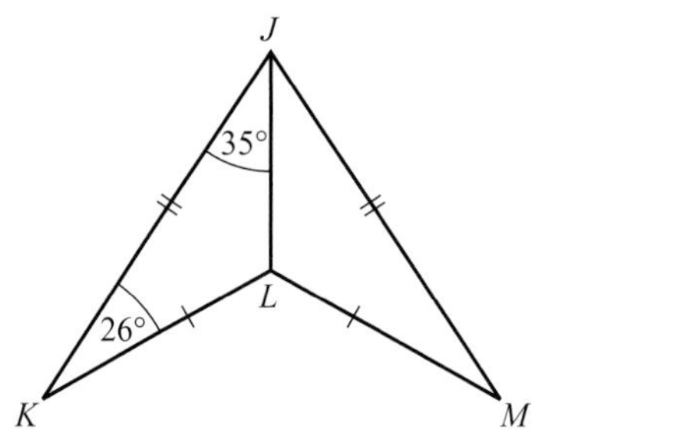

Work out

a angle JLK **(1 mark)**

b angle JLM **(1 mark)**

c angle LJM **(1 mark)**

d Explain why triangles JKL and JLM are congruent. **(2 marks)**

Exam hint
Draw the two triangles separately and label each point.

12.2 Geometric proof and congruence

Objectives

- Prove shapes are congruent.
- Solve problems involving congruence.

Did you know?

A proof is a logical argument that shows something is true. Some mathematicians dedicate their lives to writing proofs.

Fluency

- What are the conditions for congruence in triangles?
- How do you show sides and angles are equal on a shape?

Warm up

1 Sketch each shape. Mark all the equal sides and angles.

rhombus

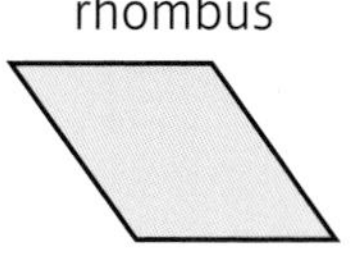

trapezium

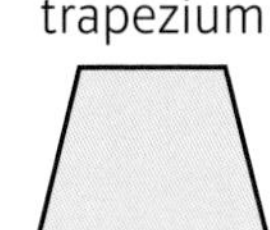

equilateral triangle

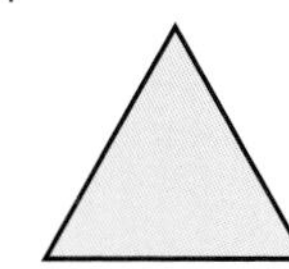

isosceles triangle

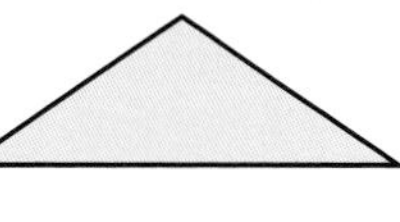

2 M is the midpoint of DE. M is also the midpoint of FG.

a Which length is the same as DM?

b Which length is the same as GM?

c Which angle is the same as angle DMF?

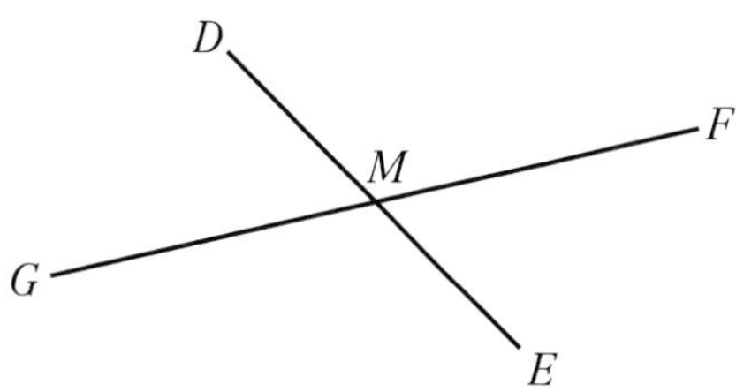

Example 1

$ABCD$ is a parallelogram. Prove triangle ABC is congruent to ADC.

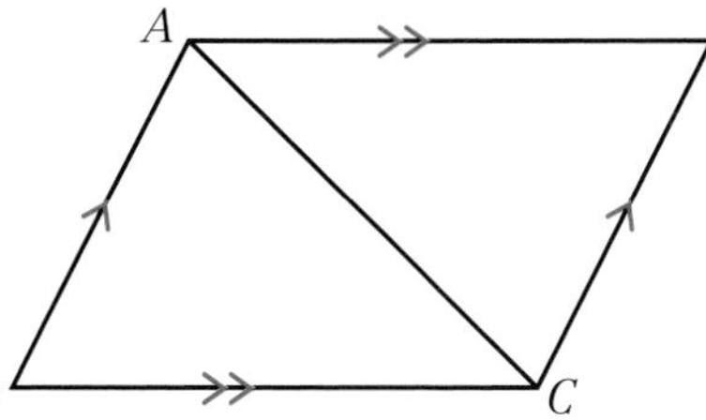

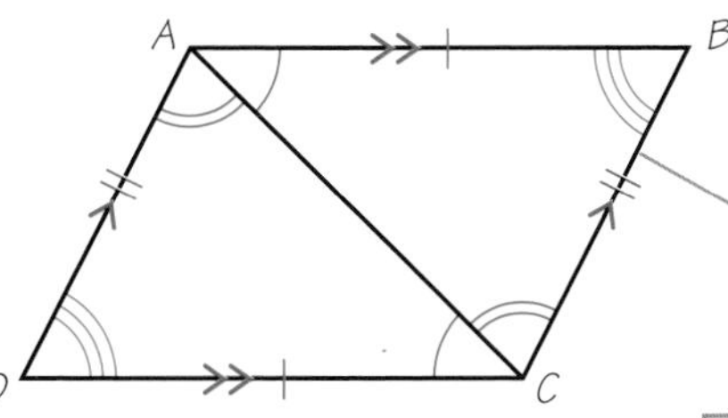

Mark all equal angles and sides.

Length AB = length CD because opposite sides in a parallelogram are equal. — State why $AB = CD$

Length BC = length AD because opposite sides in a parallelogram are equal. — State why $BC = AD$

Length AC is common to both triangles.

So triangle ABC is congruent to triangle ADC (SSS). — State the condition used to prove congruence.

3 **Communication** $WXYZ$ is a rectangle.

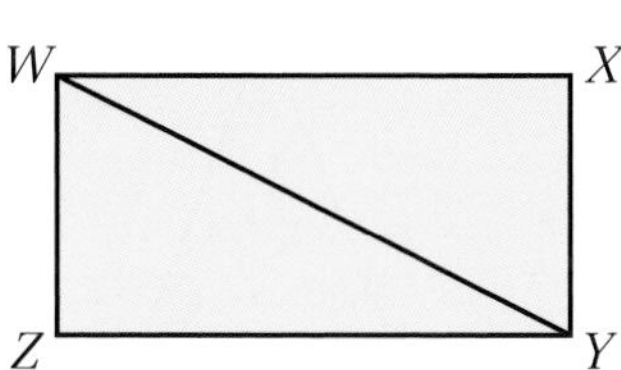

Q3 hint Draw the triangles separately, showing equal sides and angles clearly. Give a reason for congruency.

Q3 communication hint Write each statement of your proof on a new line. Give a reason for every statement you make.

a Prove that triangle WXY is congruent to triangle XYZ.

b Which angle is the same as angle XWY?

4 **Communication** The diagram shows a rhombus $PQRS$.

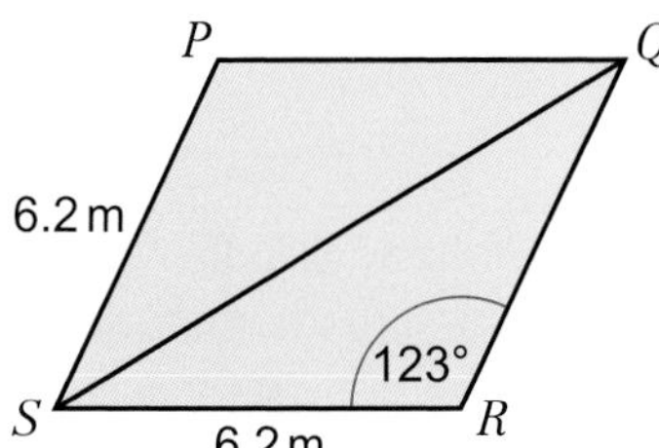

> **Q4 hint** What do you know about opposite sides in a rhombus?

a Prove that triangle PQS and triangle QRS are congruent.

b Find the size of

i angle QPS ii angle RQS.

Reflect Could you use the fact that the angles are the same in the two triangles to prove that they are congruent?

5 **Communication** In the diagram, X is the midpoint of JK. X is also the midpoint of LM. Prove that the triangles JLX and KMX are congruent.

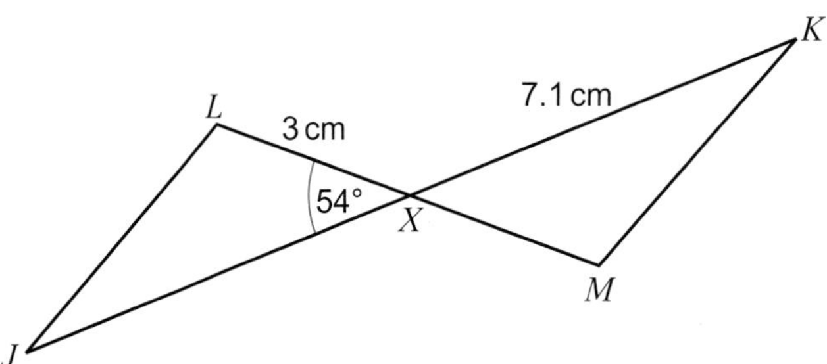

6 **Exam-style question**

In the diagram, $AB = BC = CD = AD$

Prove that triangle ABC is congruent to triangle ACD. **(2 marks)**

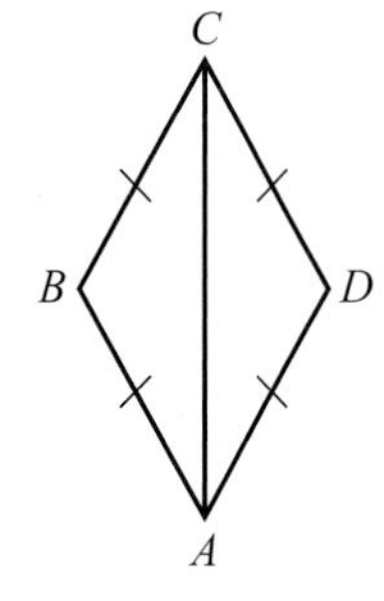

> **Exam hint**
> You need to write a series of logical statements that show the statement is true. You must give a mathematical reason for each statement.

7 **Problem-solving / Communication** FGH is an equilateral triangle. Point E lies on FH. EG is perpendicular to FH.

a Prove that triangle FGE is congruent to triangle GHE.

b Hence, prove that $FE = \frac{1}{2}FH$.

> **Q7 hint** Sketch the triangle.

8 **Communication** RST is an isosceles triangle such that $RS = ST$.
M is the midpoint of line RT.
Use congruent triangles to prove that the line SM which cuts the base of the triangle at right angles also bisects the base.

> **Q8 hint** Show that RM and MT are the same length.

9 **Problem-solving / Communication** Prove that triangles PQR and RST are congruent. Hence, prove that R is the midpoint of PT.

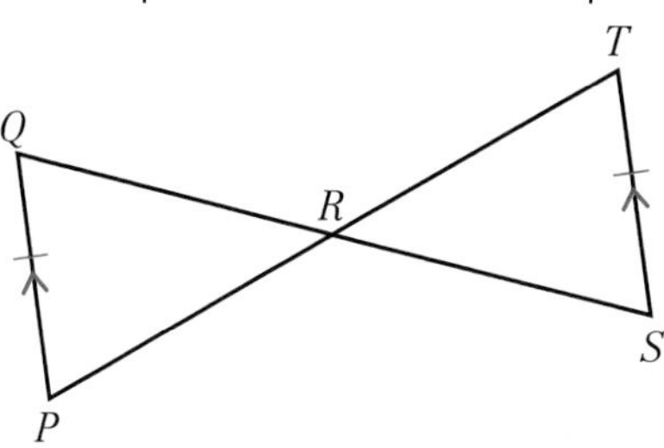

Q9 hint Show that PR and RT are the same length.

10 **Problem-solving / Communication** $CDEF$ is a rhombus. $CD = DE = EF = CF$. CD is parallel to EF. DE is parallel to CF.

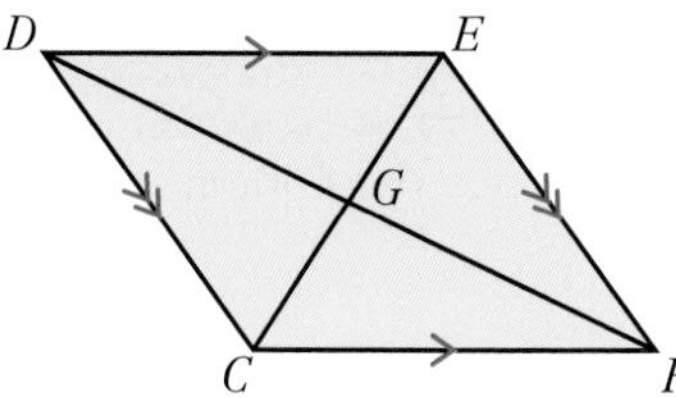

Q10 hint Make sure you only use known facts to justify your argument. Don't assume something if it has not been stated to be true.

Prove that

a triangles DEG and CFG are congruent
b triangles CDG and EFG are congruent
c G is the midpoint of both CE and DF and hence that the diagonals of a rhombus intersect at right angles.

11 **Communication** $ABCD$ is a square. AC and BD are the diagonals of the square, which cross at point E.

a Draw the square showing both diagonals.
b Mark on all equal angles.
c Which triangles are congruent in your diagram?
d What can you say about all the angles at point E?
e Using your answers to parts **b–d**, show that lines AC and BD bisect at point E, at right angles.

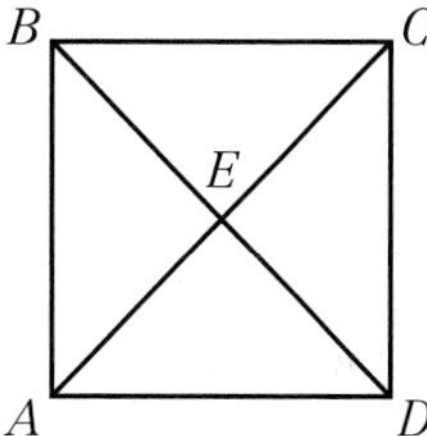

12 **Problem-solving / Communication** $KLMN$ is a parallelogram such that $KN = LM$ and $KL = MN$. The diagonals intersect at O. Prove that the point O is the midpoint of each diagonal in the parallelogram.

Q12 hint Sketch the parallelogram. Use triangles NKO and LMO and show that MO and OK are the same length. Do the same for the other pair of triangles.

13 **Exam-style question**

XYZ is an isosceles triangle with $XY = XZ$.
A and B are points on XY and XZ such that $AX = BX$.
Prove that triangle XAZ is congruent to triangle XBY. **(2 marks)**

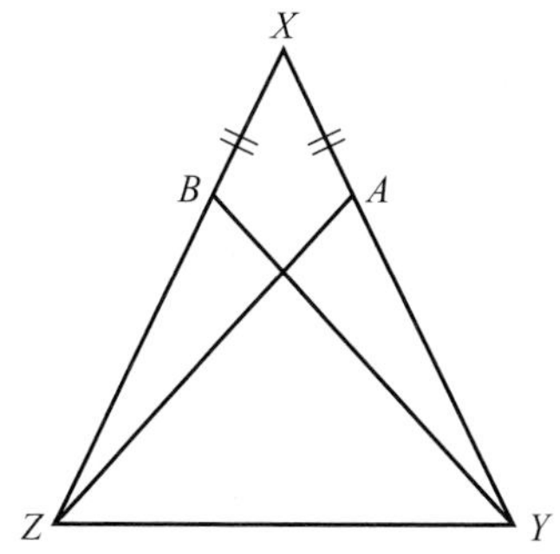

Q13 strategy hint Draw the two triangles separately and label each point. Look for a common angle or side.

12.3 Similarity

Objectives

- Use the ratio of corresponding sides to work out scale factors.
- Find missing lengths on similar shapes.

Why learn this?

We use similarity to draw floor plans to scale.

Fluency

What is the same and what is different in these two regular pentagons?

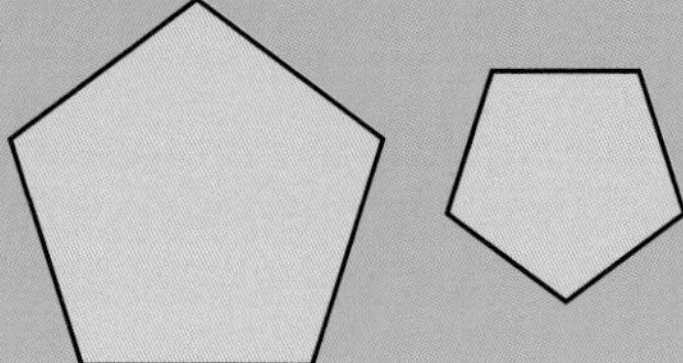

Warm up

1 What is the scale factor of the enlargement from

a A to B b B to A.

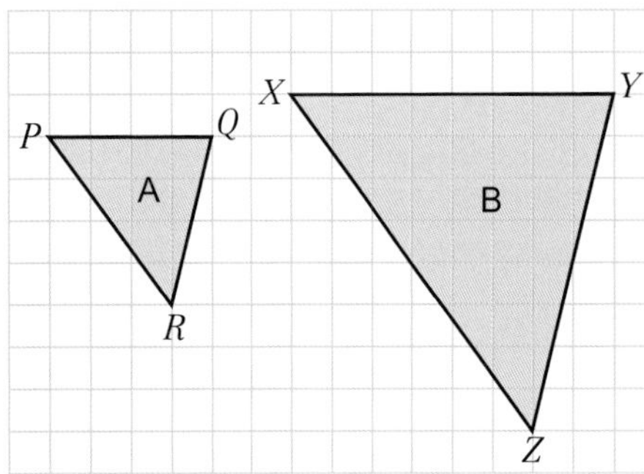

2 Write the pairs of corresponding sides for triangles A and B in **Q1**.

Q2 hint Corresponding sides map on to each other in an enlargement.

3 a Measure the angles and sides in the triangles in **Q1**.

b Write the ratios of corresponding sides as fractions.

Discussion What do you notice about the angles? What do you notice about the ratios?

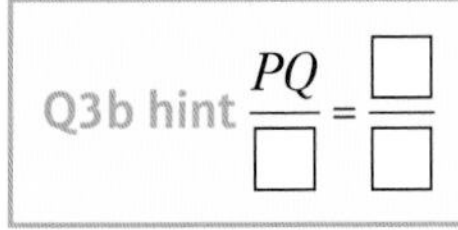

Q3b hint $\frac{PQ}{\square} = \frac{\square}{\square}$

Key point 3

Shapes are similar when one shape is an enlargement of the other. Corresponding angles are equal and corresponding sides are all in the same ratio.

4 **Communication** Here are two parallelograms. Angles SPQ and TUV are the same size.

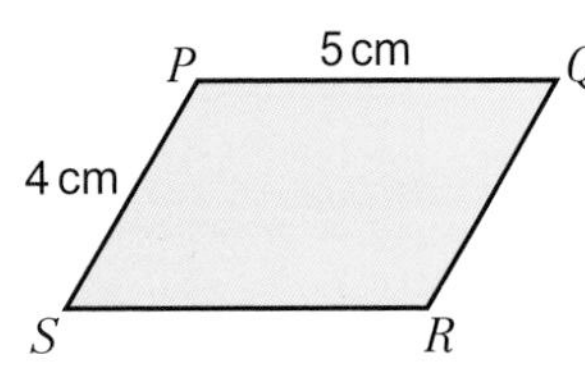

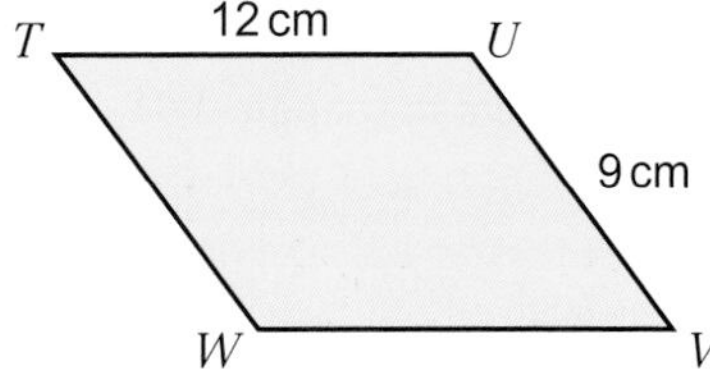

a Which side in $TUVW$ corresponds to

i PQ ii PS?

b Work out the ratio $\frac{PS}{UV}$

c Work out the ratio $\frac{PQ}{TU}$

d Use your answers to parts **a** and **b** to show that the parallograms are not similar.

ActiveLearn Homework, practice and support: Higher 12.3

5 **Communication** The diagram shows parallelograms $ABCD$ and $EFGH$. Angles ABC and FGH are the same size.

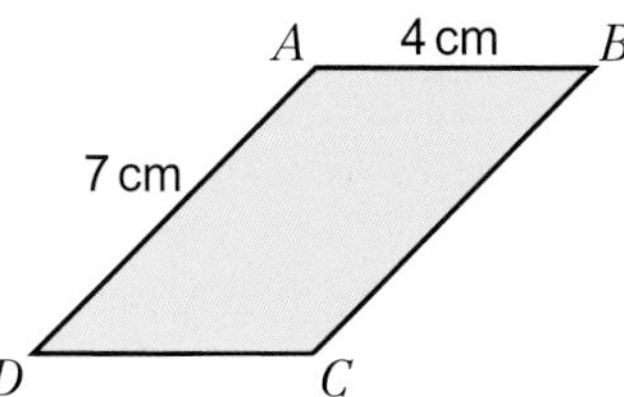

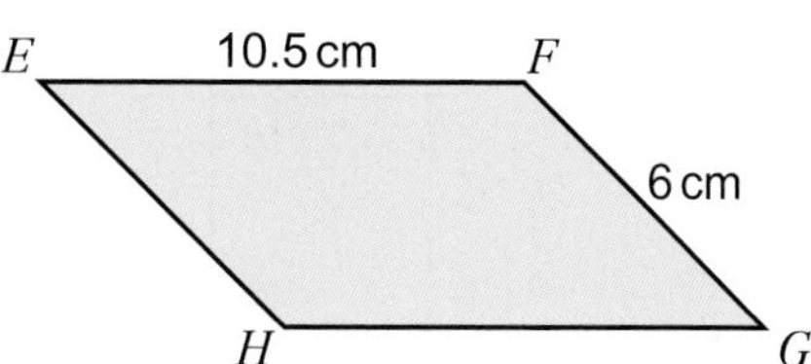

a Write the ratio $\frac{EF}{AD}$

b Write the ratios of the other corresponding sides.

c Are the parallelograms similar?
Explain your answer.

Q5 strategy hint The shapes may not be in the same orientation. Make one ratio of the shorter sides and another ratio of the longer sides.

6 State which of the pairs of shapes are similar.

a

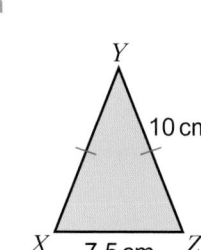

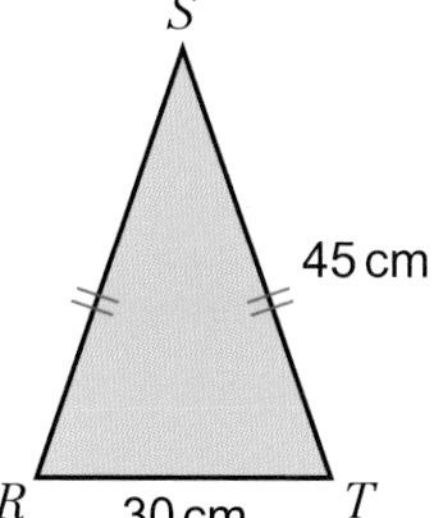

b

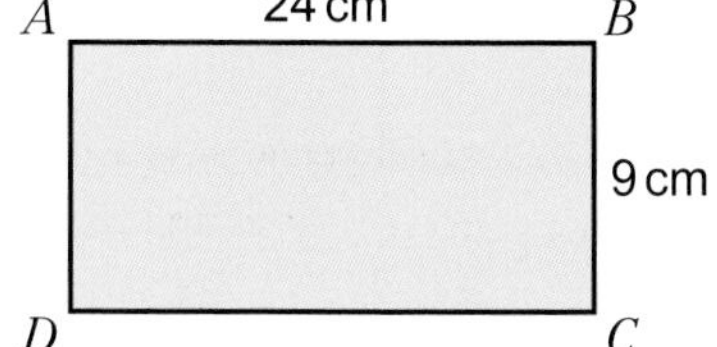

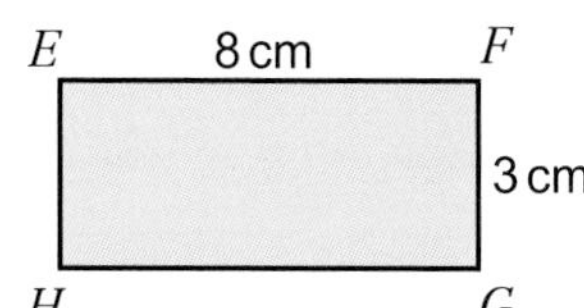

c

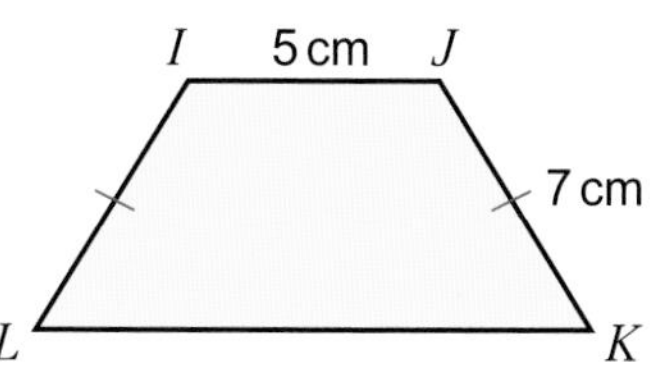

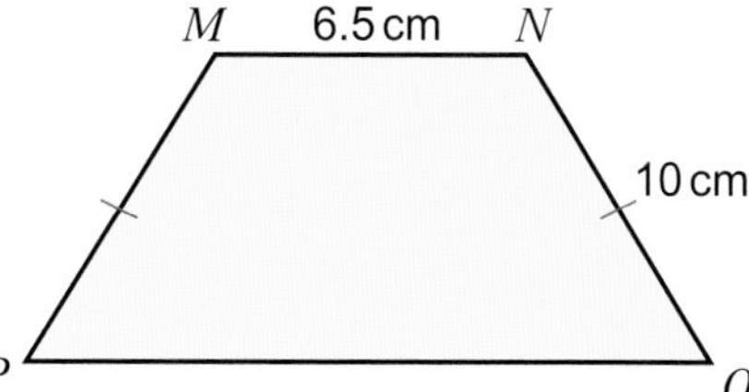

7 Show that pentagon $ABCDE$ is similar to pentagon $VWXYZ$.

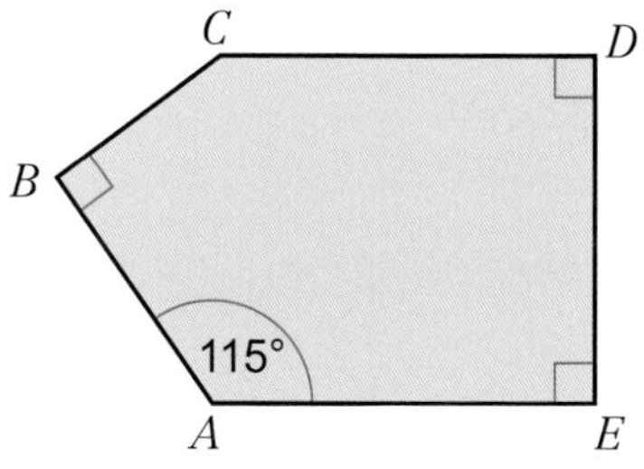

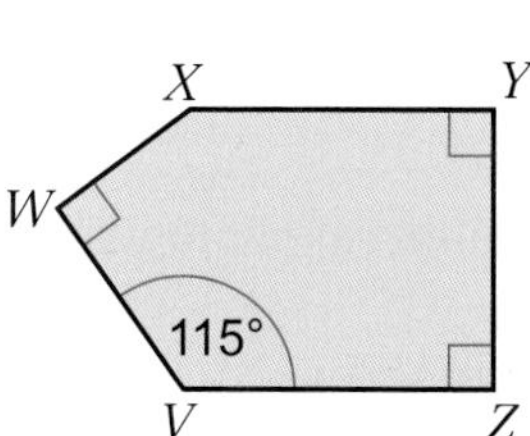

Q7 hint Show that the angles are the same.

Discussion Are similar shapes congruent? Are congruent shapes similar?

Example 2

These two rectangles are similar. Find the missing length x in the smaller rectangle.

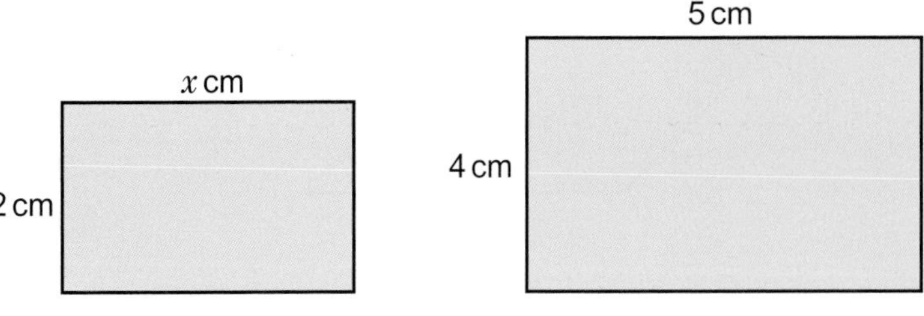

ratio of lengths: $\frac{x}{5}$

ratio of widths: $\frac{2}{4} = \frac{1}{2}$

Write the ratio $\frac{\text{small}}{\text{large}}$ for the lengths and the widths.

$\frac{\text{small}}{\text{large}} = \frac{1}{2} = \frac{x}{5}$

$2x = 5$ — Write an equation to solve for x.

$x = \frac{5}{2} = 2.5\text{ cm}$

8 These two rectangles are similar.
Find the missing side length L in the larger rectangle.

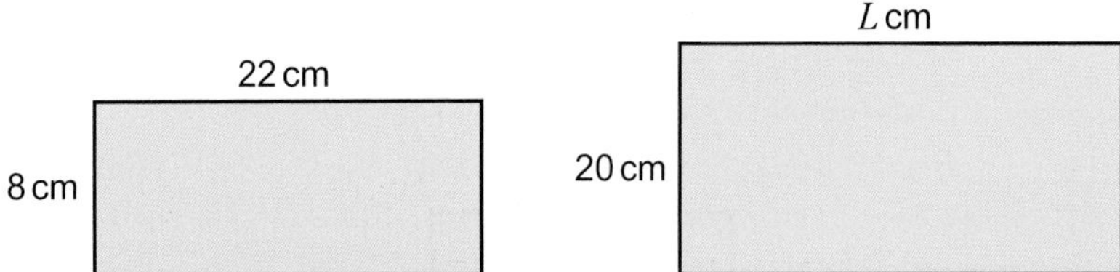

9 **Exam-style question**

A small photograph has a length of 4 cm and a width of 3 cm.
Shez enlarges the small photograph to make a large photograph.
The large photograph has a width of 15 cm.

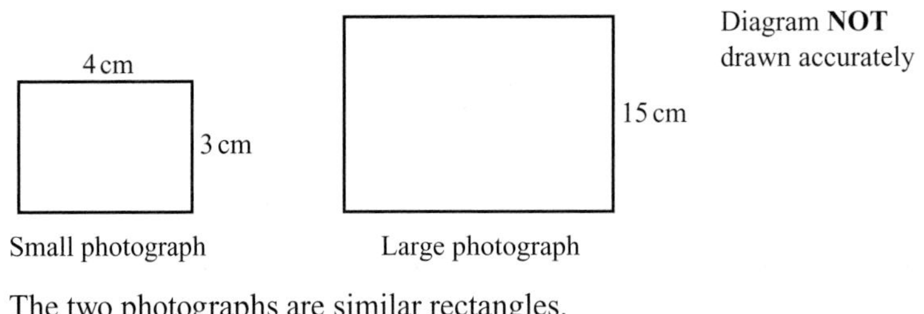

The two photographs are similar rectangles.
Work out the length of the large photograph. **(3 marks)**

June 2012, Q1, 5MB3H/01

Q9 strategy hint Let the length of the large photograph be x. Write the corresponding sides as ratios.

10 **Problem-solving / Real** An aerial photograph shows a campsite with an outdoor swimming pool.
In the photograph, the pool measures 5 cm by 2 cm.
The real pool is 25 m long. How wide is the pool?
Discussion How did you write your ratio?
Does it matter which value is the numerator?

11 a Show that triangles A and B are similar.

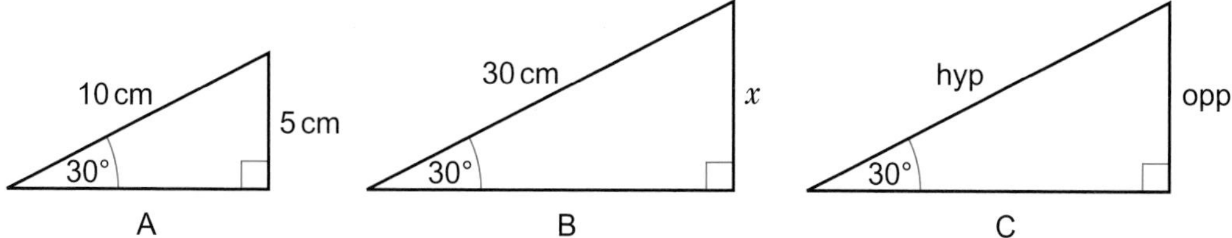

b Work out length x in triangle B.

c Show that triangle C is similar to A and B. Explain.

d Write down the value of $\frac{\text{opposite}}{\text{hypotenuse}}$ for these triangles.

e What is another name for the ratio $\frac{\text{opposite}}{\text{hypotenuse}}$?

12 **Problem-solving / Real** A small can of soup has a height of 10.5 cm and a diameter of 6 cm.
A large can of soup is similar to the small can. It has a diameter of 8 cm.
Find the height of the large can of soup.

> **Q12 hint**
> Sketch the cans.

13 **Problem-solving / Real** The diagram shows one symmetrical panel in a tent. The sizes of the actual tent panel are shown on diagram B. The sizes of the plan are shown in diagram A.

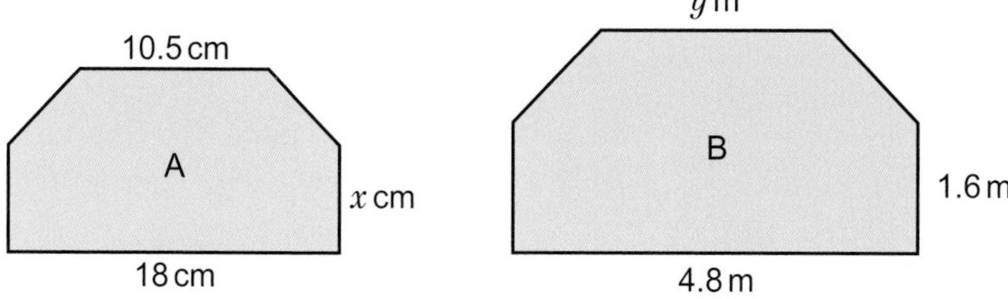

Work out the value of

a x

b y

14 **Communication** Are the triangles in each pair similar? Explain.

a

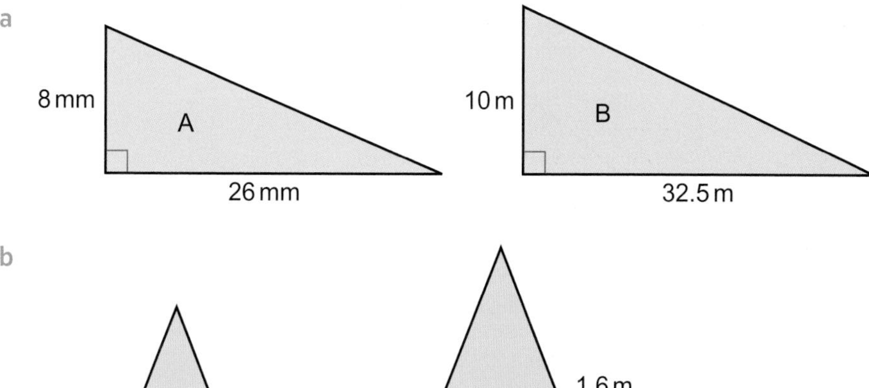

b

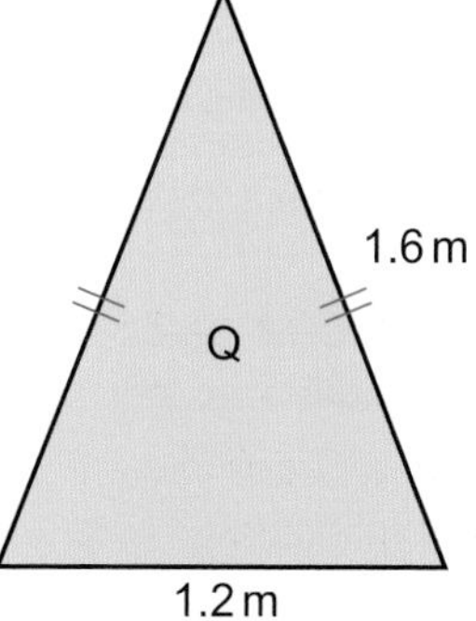

15 Here are two regular hexagons.

a Are they similar? Explain your answer.

b Are all regular hexagons similar?

Discussion Are all circles similar?

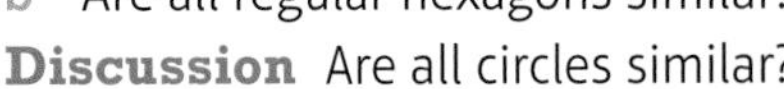

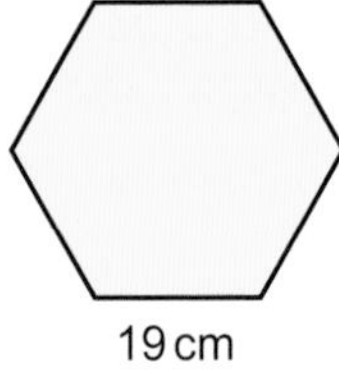

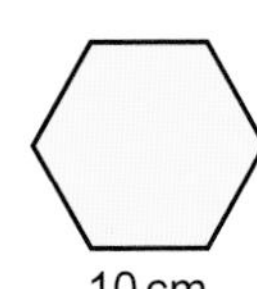

16 Triangle CDE is similar to triangle FGH.

$\angle CDE = \angle FGH$

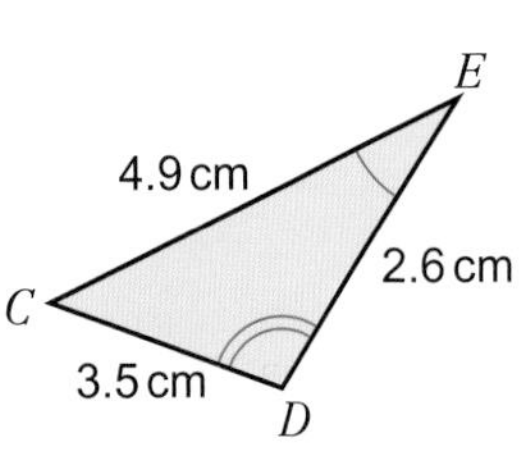

H
9.4 cm
F
G

Calculate the length of

a FG

b FH

17 **Reflect** Write some notes to help you remember how to show that two shapes are similar.

12.4 More similarity

Objectives

- Use similar triangles to work out lengths in real life.
- Use the link between linear scale factor and area scale factor to solve problems.

Why learn this?

You can work out the height of a skyscraper using similar triangles.

Fluency

A shape is enlarged by scale factor 3. What scale factor is its area enlarged by?

Warm up

1 What is the scale factor of the enlargement that maps

a triangle Y onto triangle Z

b triangle Z onto triangle Y.

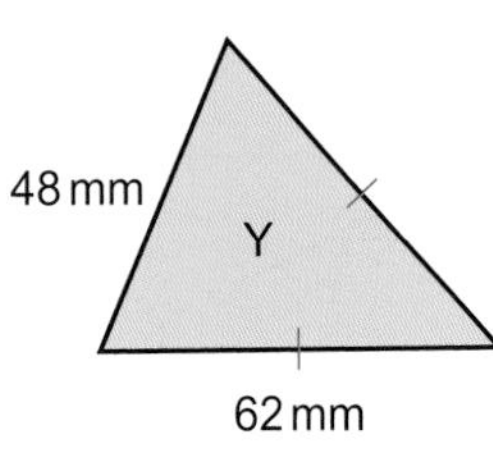

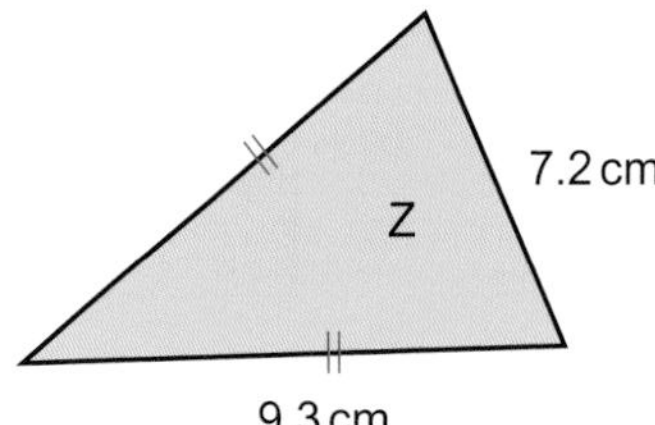

2 **Communication** Are triangles ABE and CDE similar? Explain.

Q2 hint Find the angles in triangle AEB. Give reasons for your answers.

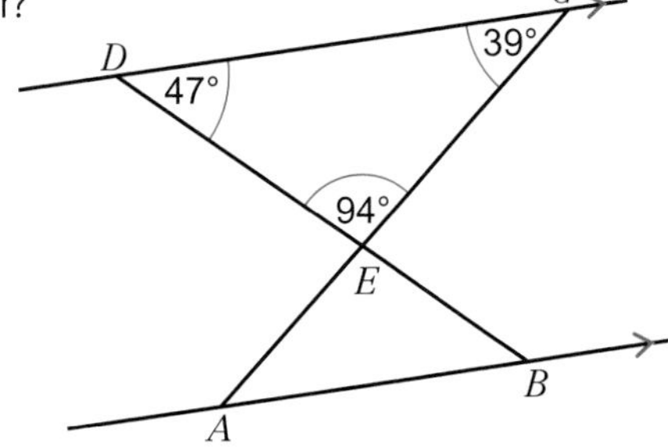

ActiveLearn Homework, practice and support: Higher 12.4

3 **Communication**

a Show that triangles PQR and RST are similar.

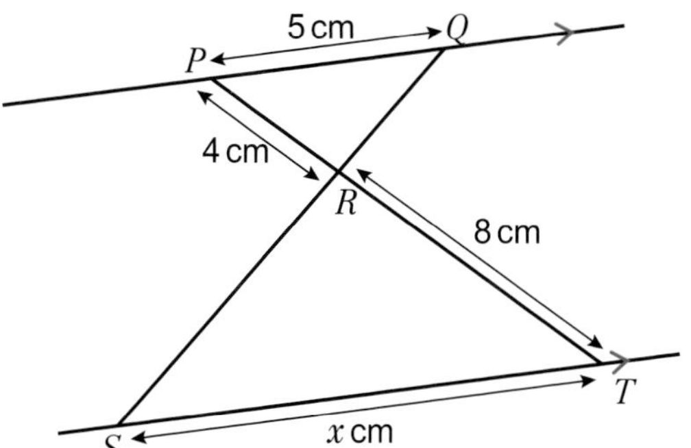

b Find the missing length x.

Q3b hint Mark equal angles.

4 **Communication**

a Explain why triangles FGH and FJK are similar.

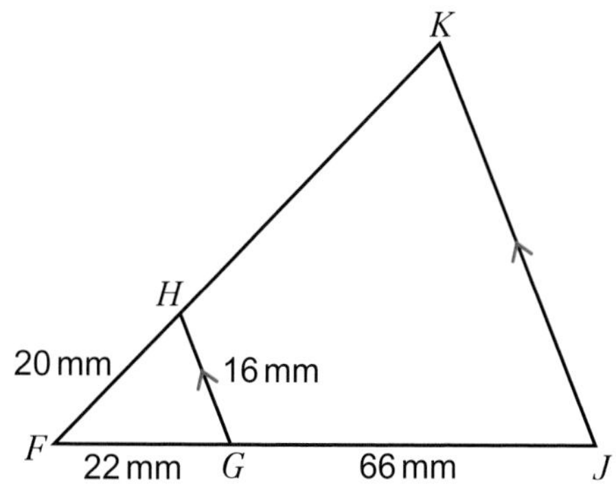

Q4a hint Draw the triangles separately.

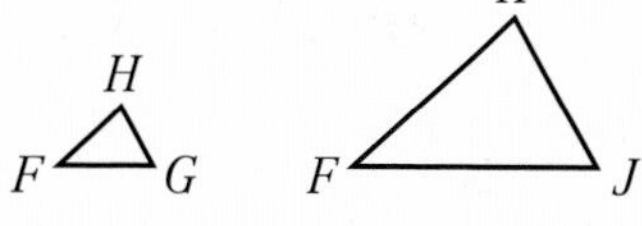

b Calculate the length HK.

c Calculate the length JK.

5 **Communication**

a Find the sizes of angle PQN and angle LMN.

b Explain why triangle LMN is similar to triangle LPQ.

c Find the length of LQ.

d Find the length of NQ.

e Find the length of MP.

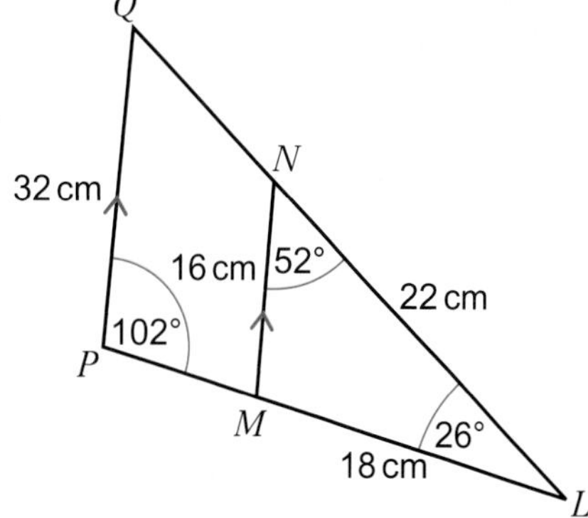

6 **Real** Calculate the height of The Shard using similar triangles.

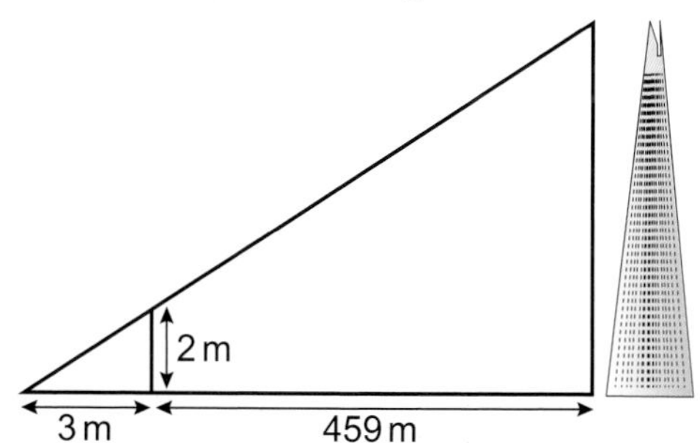

7 The diagram shows two flower beds made in the shape of similar trapezia.

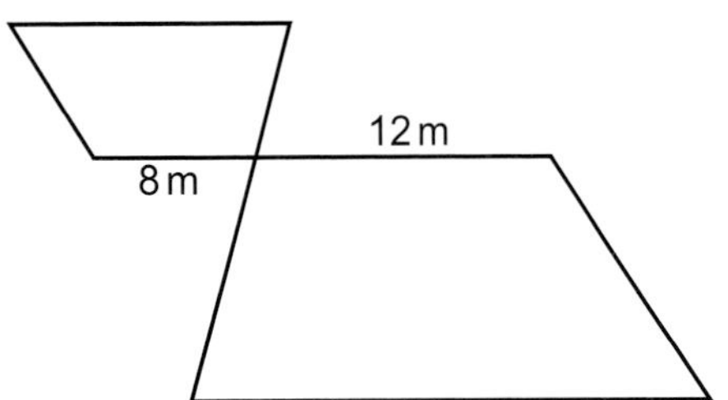

Q7 strategy hint First work out the linear scale factor of the enlargement (k). The area is enlarged by scale factor ☐.

The perimeter of the small flower bed is 36 m.
The area of the small flower bed is 60 m^2.
Work out the perimeter and area of the large flower bed.

8 **Exam-style question**

A company makes teddy bears.

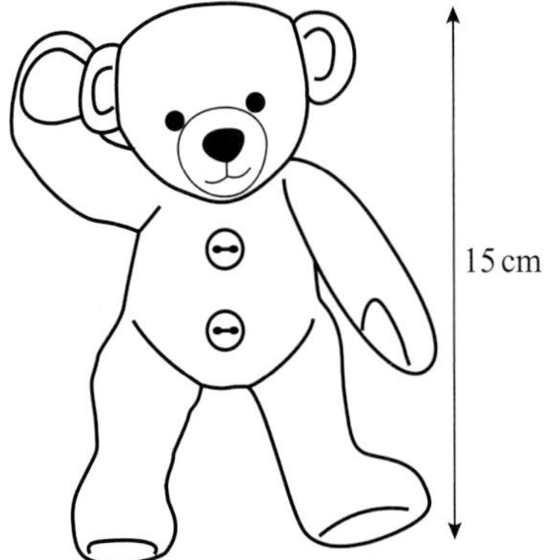

The company makes small bears that are 15 cm tall.
A small bear has a surface area of 200 cm^2.
The same company make giant bears which are 1.8 m tall.
A giant bear is mathematically similar to a small bear.
Work out the surface area of a giant bear. **(3 marks)**

Q8 strategy hint First work out the length ratio. Remember the lengths need to be in the same units.

9 Shape K is similar to shape L.

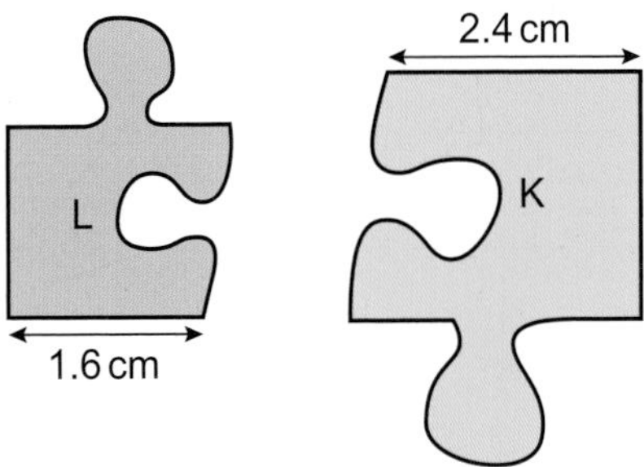

The perimeter of shape K is 7.5 cm and its area is 4.8 cm^2.
Find the perimeter and area of shape L.

Example 3

Shape D is similar to shape E.
Calculate the length of shape E.

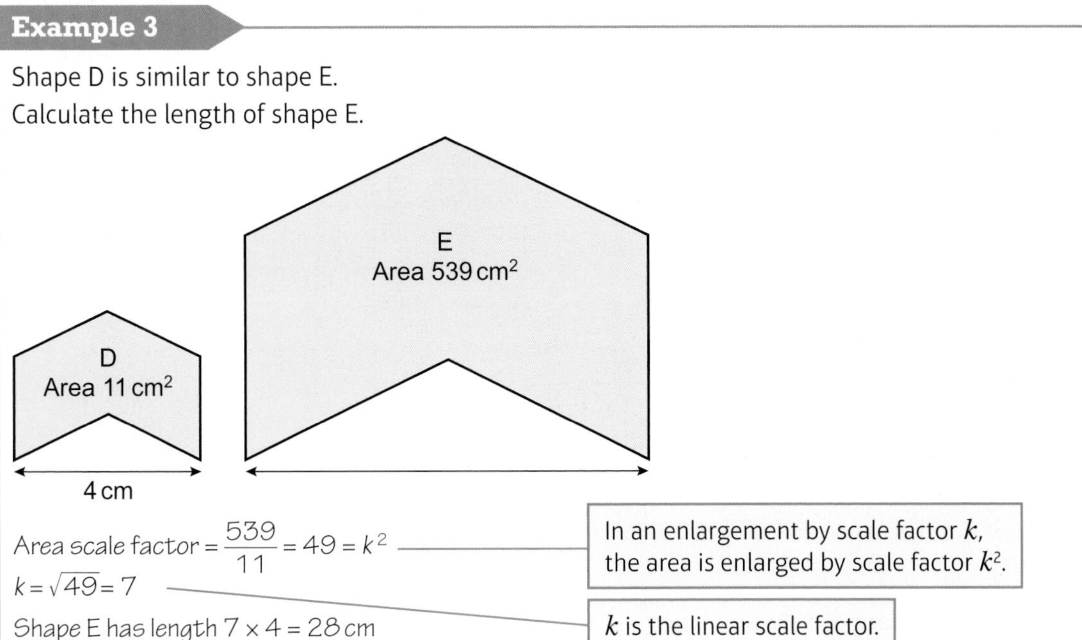

Area scale factor $= \frac{539}{11} = 49 = k^2$

$k = \sqrt{49} = 7$

Shape E has length $7 \times 4 = 28$ cm

In an enlargement by scale factor k, the area is enlarged by scale factor k^2.

k is the linear scale factor.

10 Shape A is similar to shape B.
The area of shape A is $126\,\text{cm}^2$.
The area of shape B is $283.5\,\text{cm}^2$.
Calculate

a length x

b length y.

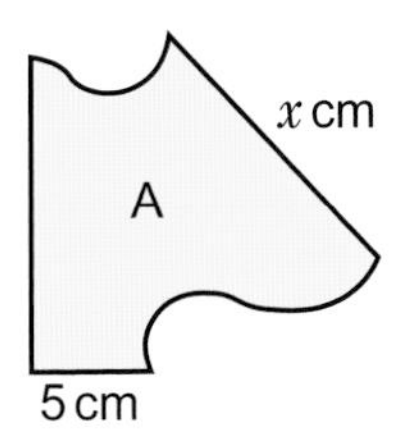

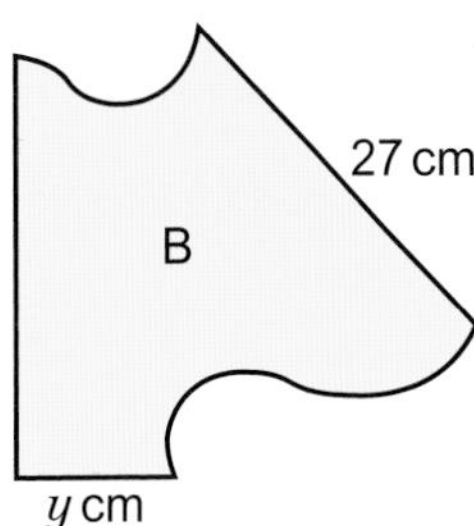

11 Two similar triangles have areas of $36\,\text{cm}^2$ and $100\,\text{cm}^2$ respectively.
The base of the smaller triangle is 3 cm.
Find the length of the base of the larger triangle.

12 A postage stamp has a surface area of $6\,\text{cm}^2$.
What is the area of a similar stamp with lengths that are

a twice the corresponding lengths of the first stamp

b three times the corresponding lengths of the first stamp?

13 A triangle has sides of 4, 5 and 6.4 cm. Its area is $10\,\text{cm}^2$.
How long are the sides of a similar triangle that has an area of $90\,\text{cm}^2$?

14 **Problem-solving** A sheet of A2 paper and a sheet of A4 paper are similar.
The area of a sheet of A2 paper is $2500\,\text{cm}^2$ and the area of a sheet of A4 paper is $625\,\text{cm}^2$.
The width of a sheet of A2 paper is 42 cm.

a Work out the area scale factor.

b Work out the length scale factor.

c Use the length scale factor to work out the width of a sheet of A4 paper.

15 **Problem-solving** The fronts of two cereal packets are similar. The area of front of the larger packet is $1035\,\text{cm}^2$ and the area of the front of the smaller one is $460\,\text{cm}^2$.
The height of the larger packet is 45 cm. Work out the height of the smaller packet.

16 **Reflect** Why are volume scale factors important when going from making a scale model to making the real object?

12.5 Similarity in 3D solids

Objective

- Use the links between scale factors for length, area and volume to solve problems.

Why learn this?

A scale model is similar to the original. Architects use 3D scale models of big projects to give their clients a better understanding of the way a new building will fit into its surroundings.

Fluency

Work out the volume and surface area of this cube.

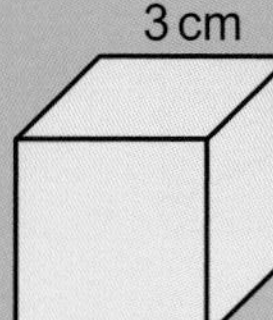

Warm up

1 Work out

a the cube root of 125

b $\sqrt[3]{\frac{64}{27}}$

2 Convert 1.5 litres to cm^3

3 In each pair of diagrams, solid A is enlarged to make solid B.
Copy and complete the table.

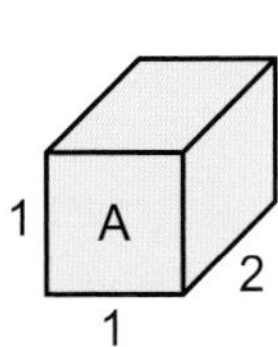

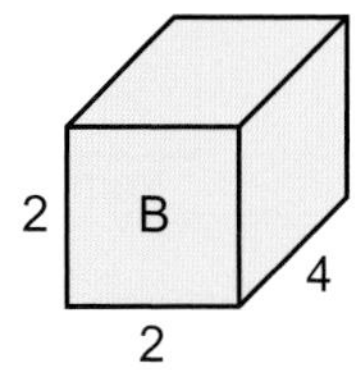

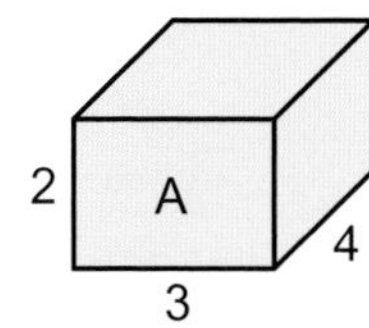

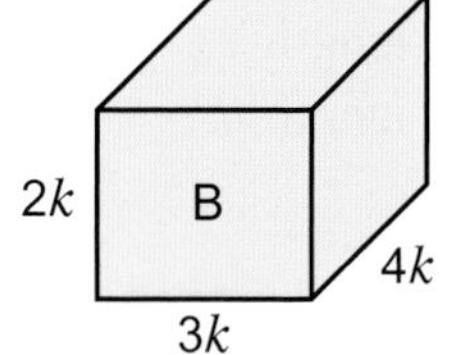

Linear scale factor	Volume A	Volume B	Volume scale factor

Discussion What do you notice about the scale factors of volume?

Key point 4

When a shape is enlarged by linear scale factor k, the volume of the shape is enlarged by scale factor k^3.

4 Prisms A and B are similar.
The volume of prism A is 12 cm^3.
Calculate the volume of prism B.

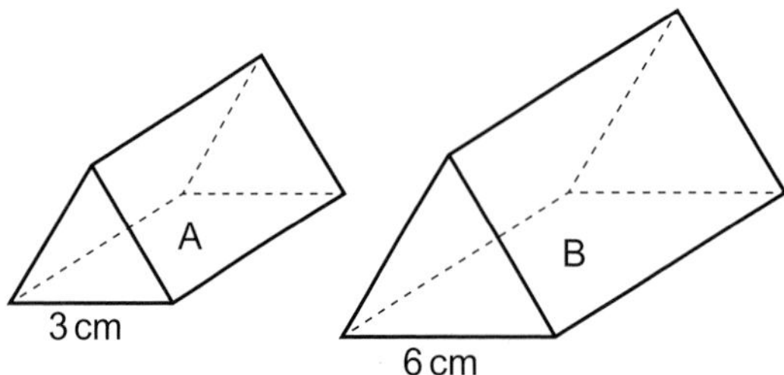

5 Tetrahedrons C and D are similar.
The volume of tetrahedron C is 15 cm³.
Calculate the volume of tetrahedron D.

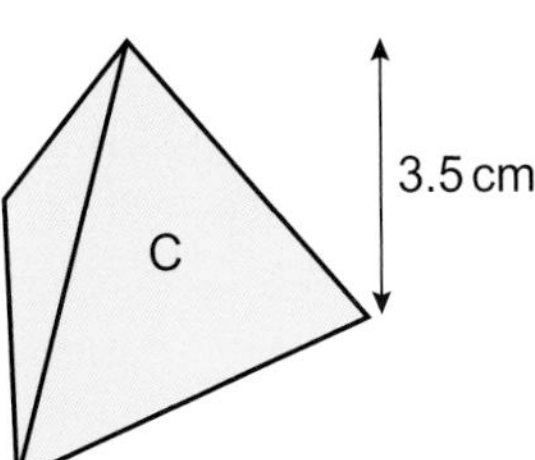

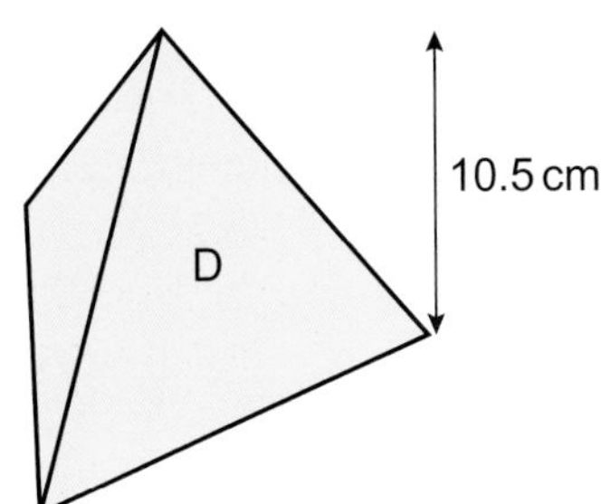

6 Cones E and F are similar.
The volume of cone E is 202.5 cm³.
Calculate the volume of cone F.

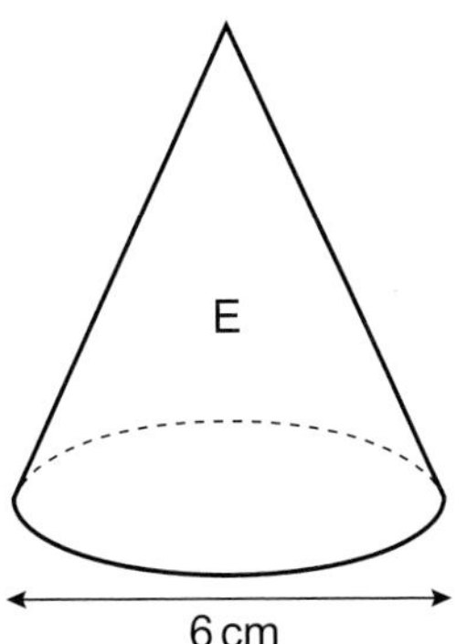

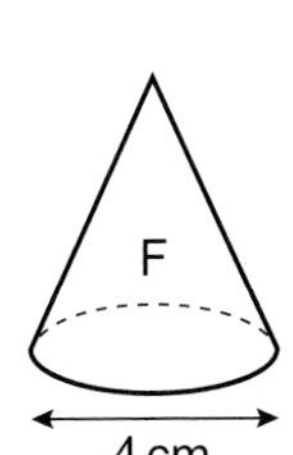

Example 4

Cylinders G and H are similar.
The diameter of G is 6 cm.
The volume of G is 108 cm³. The volume of H is 256 cm³.
Work out the diameter d of cylinder H.

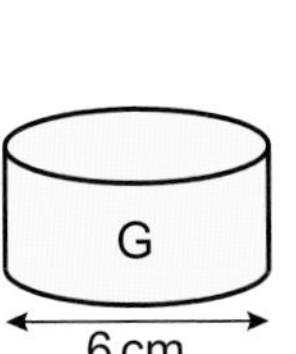

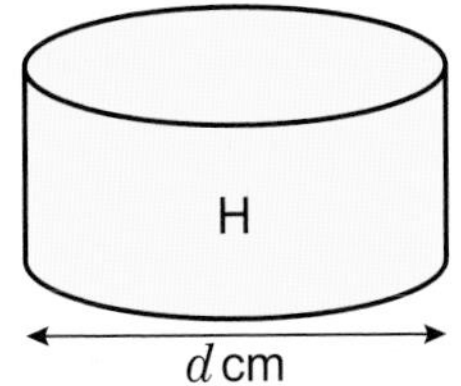

$\text{Volume scale factor} = \frac{\text{large}}{\text{small}} = \frac{256}{108} = \frac{64}{27} = k^3$

In an enlargement be scale factor k, the volume is enlarged by scale factor k^3.

$k = \sqrt[3]{\frac{64}{27}} = \frac{\sqrt[3]{64}}{\sqrt[3]{27}} = \frac{4}{3}$

$d = \frac{4}{3} \times 6 = 8\text{ cm}$

7 Sphere J is similar to sphere K.
The volume of J is 27 times the volume of K.
Work out the diameter of sphere K.

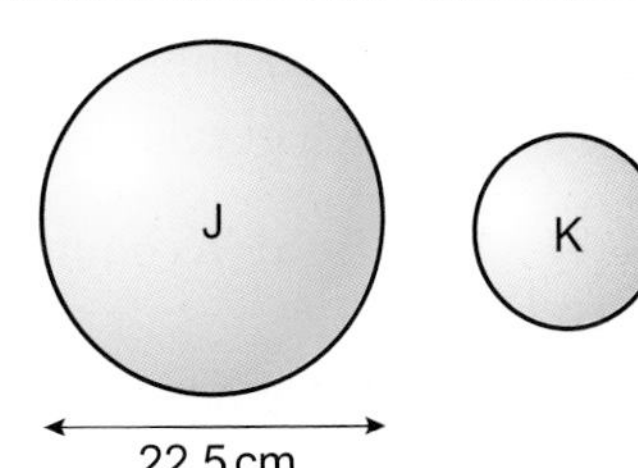

8 Prisms L and M are similar.
The volume of prism M is 343 times the volume of prism L.
Calculate the value of

a length x

b length y.

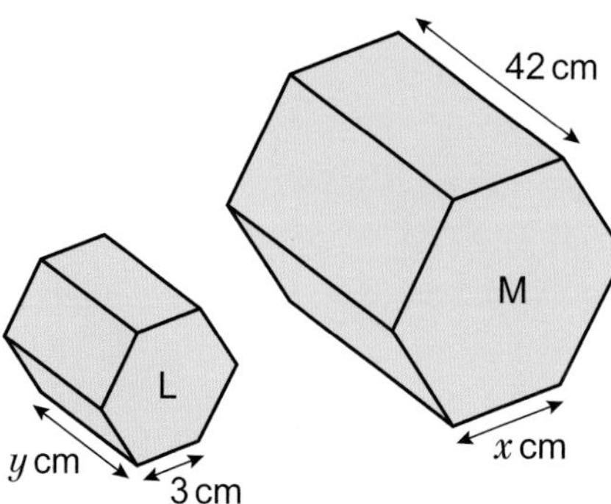

Key point 5

When the linear scale factor is k:

- Lengths are multiplied by k
- Area is multiplied by k^2
- Volume is multiplied by k^3

9 **Reasoning** Pyramid A and pyramid Z are similar.

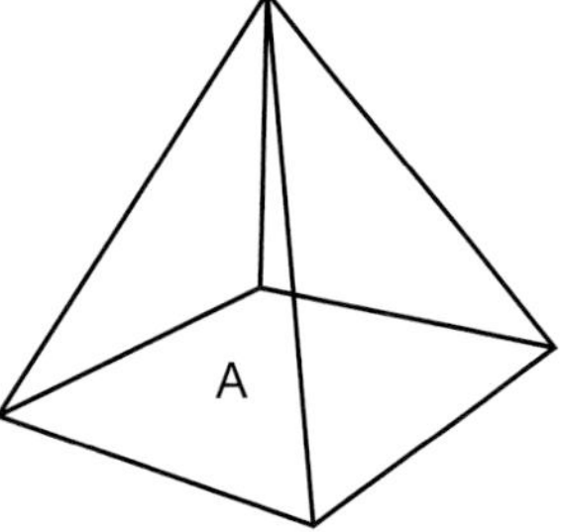

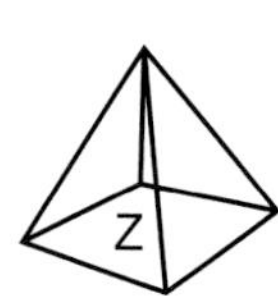

The volume of pyramid A is 125 times the volume of the volume of pyramid Z.
The surface area of pyramid Z is 60 cm^2.

a Write down the volume scale factor, k^3.

b Work out the linear scale factor, k.

c Work out the area scale factor, k^2.

d Calculate the surface area of pyramid A.

For **Q10–15**, give your answers correct to 3 significant figures.

10 **Problem-solving** Cylinders B and C are mathematically similar.
The volume of B is $\frac{1}{64}$ of the volume of C.
The surface area of B is 35.2 cm^2.
Work out the surface area of cylinder C.

Q10 hint Find the linear scale factor first.

11 **Problem-solving / Real** A commercial sink has a surface area of 6300 cm^2 and a capacity of 12.75 litres.
Find the surface area of a similar sink which has a capacity of 7 litres.

Exam hint
Find the linear scale factor first.

12 **Exam-style question**

The bases of two similar cones, A and B, are $207\,\text{cm}^2$ and $92\,\text{cm}^2$ respectively.

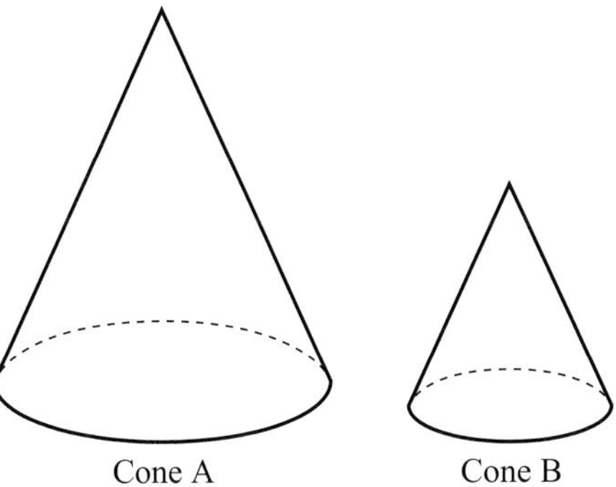

The volume of cone A is $837\,\text{cm}^3$.
Show that the volume of cone B is $248\,\text{cm}^3$. **(5 marks)**

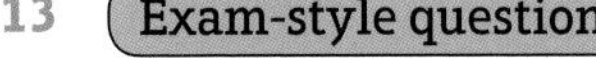

13 **Exam-style question**

The diagram shows two similar solids, A and B.

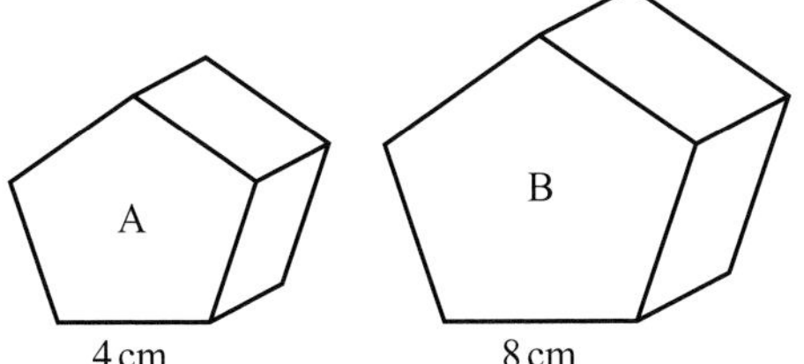

Diagram **NOT** accurately drawn

Solid A has a volume of $80\,\text{cm}^3$.

a Work out the volume of solid B. **(2 marks)**

Solid B has a total surface area of $160\,\text{cm}^2$.

b Work out the total surface area of solid A. **(2 marks)**

Nov 2012, Q25, 1MA0/1H

14 **Real / Reasoning** A can of paint is 18 cm tall and holds 2.5 litres of paint.
A similar can is 1.5 times as tall.
How much paint does it hold?

15 **Reasoning / Communication**
Explain why any two cubes are similar.
Explain why two cuboids are not necessarily similar.

Q15 strategy hint Sketch some cubes and cuboids.

12 Problem-solving

Objective

- Use geometric sketching to help you solve problems.

Example 5

The vertices of triangle ABC are at (0.5, 0), (3.5, 0) and (3.5, 4).
The vertices of triangle DEF are at (−2, −2), (5.5, −2) and (5.5, 8).
Show that the two triangles are similar.

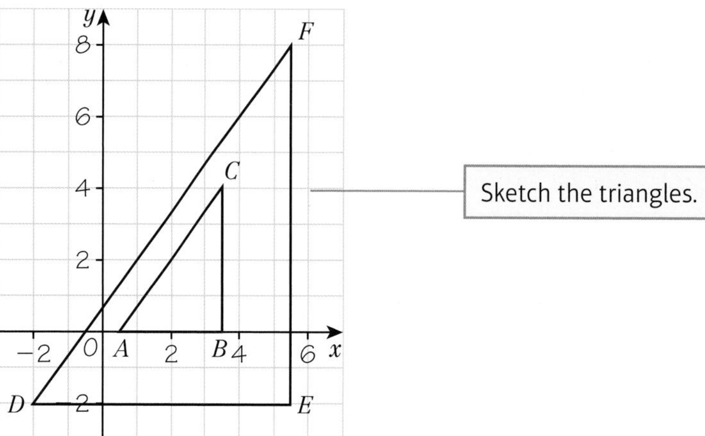

Sketch the triangles.

State how you can show two triangles are similar.

Shapes are similar if corresponding sides are in the same ratio.

Corresponding sides: AB and DE; BC and EF; AC and DF

Use your sketch to identify the corresponding sides in the two triangles.

$AB = 3$ $\quad BC = 4$ $\quad AC^2 = 3^2 + 4^2 = 25$ (Pythagoras' theorem),
so $AC = \sqrt{25} = 5$

Find the lengths of corresponding sides. ABC and DEF are right-angled triangles. You can use Pythagoras' theorem to find the lengths of hypotenuses AC and DF.

$DE = 7.5$ $\quad EF = 10$ $\quad DF^2 = 7.5^2 + 10^2 = 156.25$ (Pythagoras' theorem),
so $DF = \sqrt{156.25} = 12.5$

$\frac{AB}{DE} = \frac{3}{7.5} = 0.4$ $\quad \frac{BC}{EF} = \frac{4}{10} = 0.4$ $\quad \frac{AC}{DF} = \frac{5}{12.5} = 0.4$

Find the ratios of corresponding sides.

All corresponding sides are in the same ratio.
Therefore, triangles ABC and DEF are similar.

Confirm you have shown the triangles are similar.

1 Label the origin on a grid, O. Then draw these line segments:
- $y = 4$ from $x = 0$ to $x = 3$. Label it PQ.
- $y = 2$ from $x = 0$ to $x = 1.5$. Label it ST.
- $y = \frac{4}{3}x$ from $x = 0$ to $x = 3$.

Show that triangles OPQ and OST are similar.

Q1 hint Sketch the line segments. Use the example to help you show that the resulting triangles are similar.

2 A surveyor estimates the height of a tree. He walks 50 paces in a straight line from the bottom of the tree, and puts a 1.2 m pole vertically in the ground. Then he walks another 10 paces on the same straight line. Now when he looks from ground level, the top of the pole and the top of the tree are exactly in line. How tall is the tree?

Q2 hint Sketch a diagram. Use similar triangles to work out the height of the tree.

3 A circular water well has diameter 75 cm. It is surrounded by wooden decking of uniform width. The area of the wooden decking is 5.5 m². What width is the decking?

> **Q3 communication hint** 'Uniform width' means the width of the decking is the same all the way around the well.

> **Q3 hint** Sketch a diagram.

4 A parallelogram $ABCD$ has sides 8 cm and 15 cm. The angles in one pair of angles are twice the size of the angles in the other pair. The diagonal, AC, joins the bigger angles.

a What is the size of angle ABC?

b Prove that triangles ABC and ACD are congruent.

> **Q4a hint** Sketch a diagram. Label all angles in terms of x.

> **Q4b hint** Use the diagram you sketched for part **a** and the conditions for congruence.

5 Show that a triangle with one interior angle of 84° and two exterior angles of 132° is isosceles.

> **Q5 hint** Sketch a diagram. How can you show that a triangle is isosceles?

6 A builder has an 8 m ladder. He needs to reach a gutter that is 7.8 m high. The building site safety rules state that all workers must obey the 4 in 1 ladder rule (for every 4 m in vertical height, the base of the ladder must be 1 m away from the base of the wall). Show that the builder cannot use his ladder.

7 **Reflect** How did geometric sketching help you to solve these problems? For which problems did it help the most? Why?

12 Check up

Log how you did on your Student Progression Chart.

Congruence

1 Which of these triangles are congruent? Give reasons for your answer.

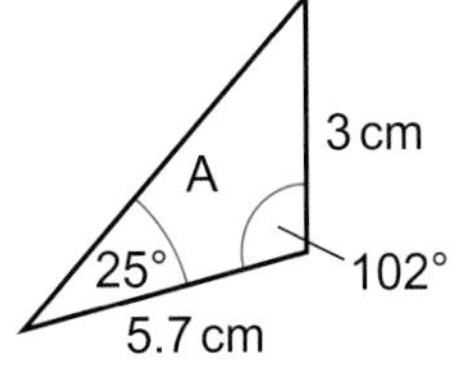

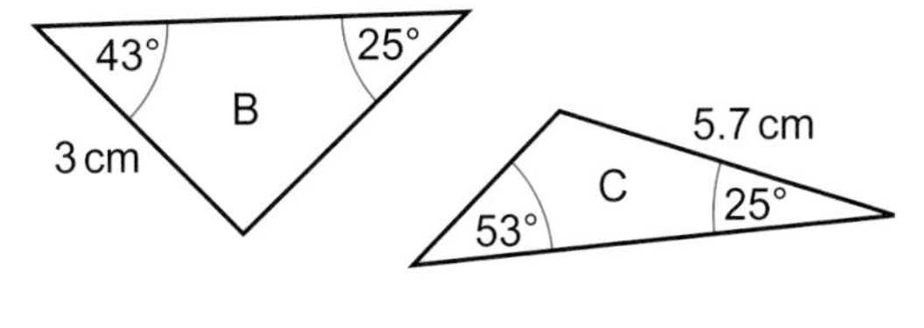

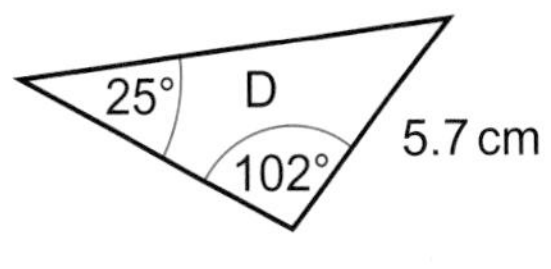

2 $HJKL$ is a rectangle.

a Prove that triangle HJK and triangle HKL are congruent. Explain your answer fully.

b Using a *different* condition for congruence, prove that triangles HJK and HKL are congruent.

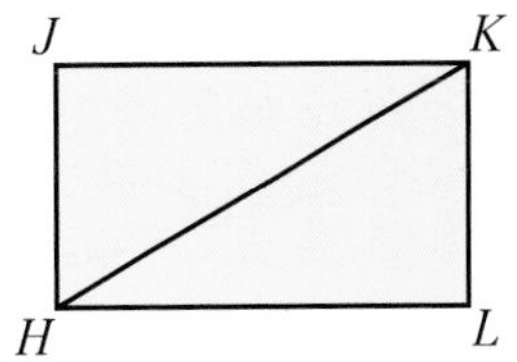

Similarity in 2D shapes

3 Show that each pair of triangles is similar.

a

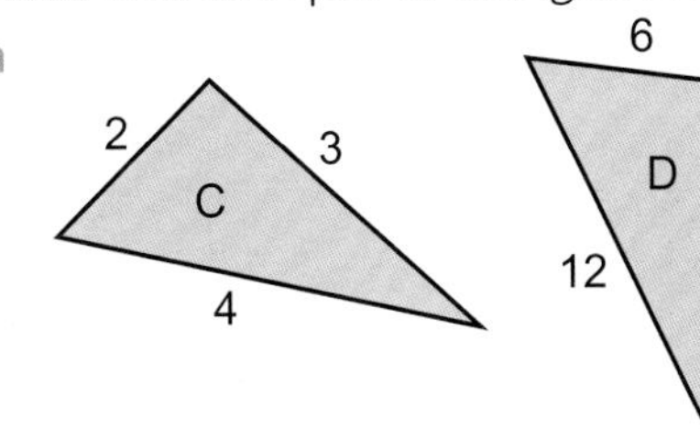

b

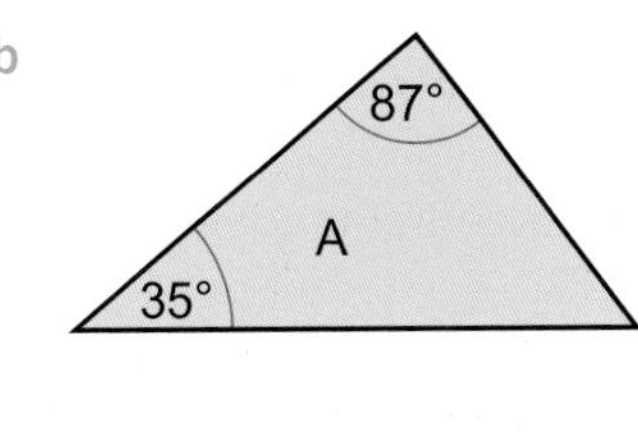

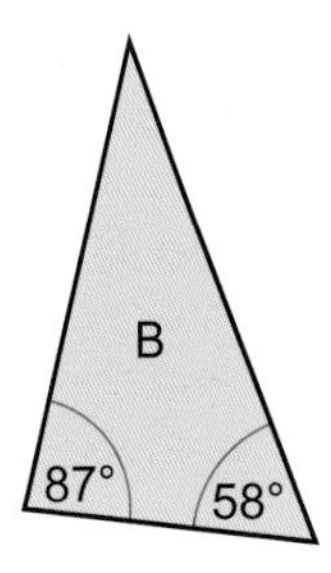

4 Quadrilaterals $ABCD$ and $EFGH$ are similar.

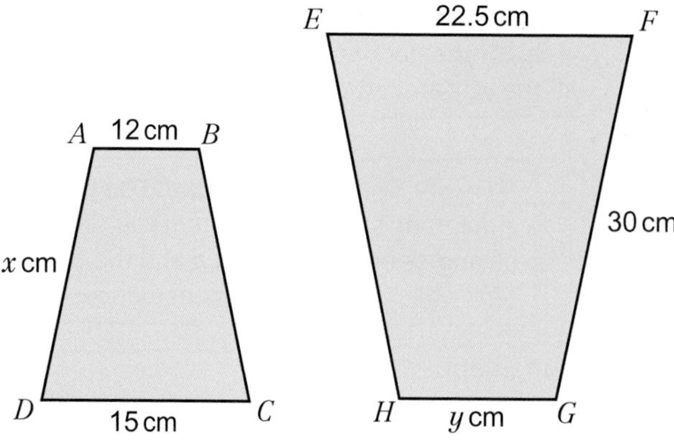

Work out

a length AD

b length GH.

5 **Communication**

a Prove that triangle ABE is similar to triangle ACD.

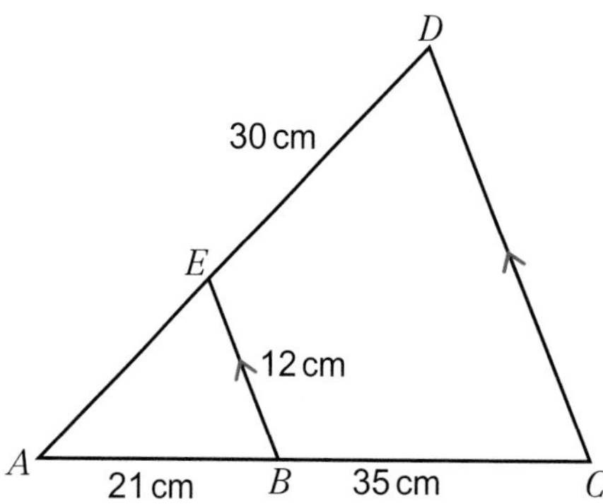

b Work out length CD.

6 a Show that PQR and RST are similar triangles.

b Work out the missing lengths in the diagram, x and y.

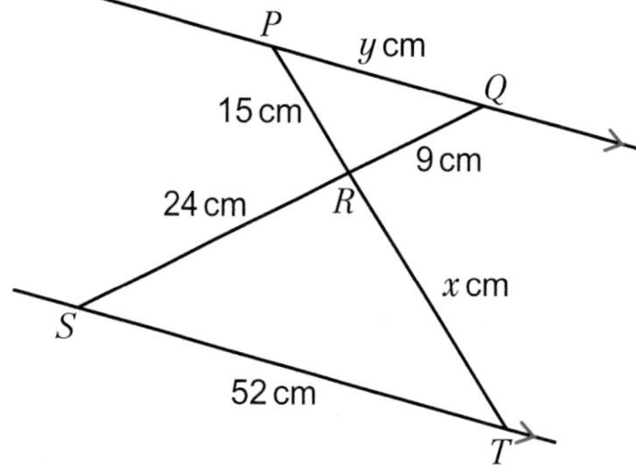

7 Shapes D and E are similar.
Shape D has a perimeter of 41.4 cm and an area of 112.5 cm^2.
Calculate the area and perimeter of shape E.

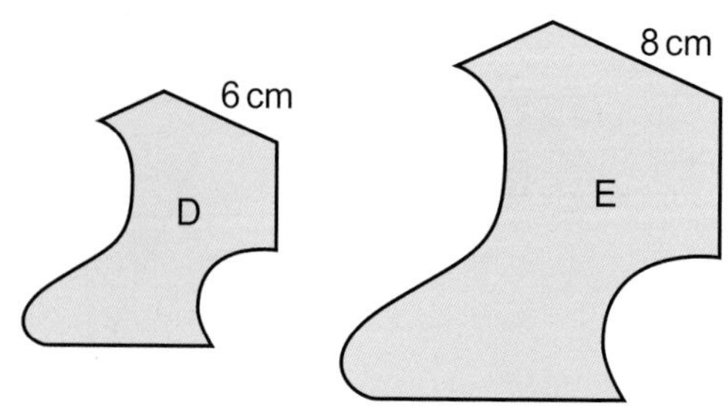

Similarity in 3D solids

8 **Problem-solving** Two whole cheeses are mathematically similar in shape.
The smaller cheese has a radius of 8.4 cm and a volume of 665 cm^3.
The larger cheese has a radius of 13.2 cm.
Work out the volume of the larger cheese.

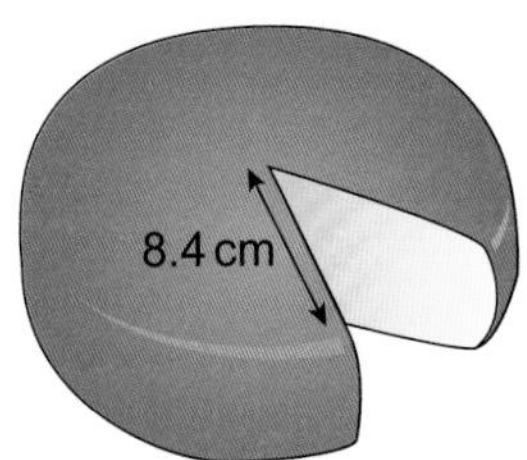

9 **Problem-solving** Pyramids A and B are similar.
The surface area of pyramid A is 1260 cm^2. The surface area of pyramid B is 180 cm^2.
The volume of pyramid A is 9604 cm^3.
Work out the volume of pyramid B.

10 How sure are you of your answers? Were you mostly
Just guessing ☹ Feeling doubtful 😐 Confident ☺

What next? Use your results to decide whether to strengthen or extend your learning.

* Challenge

11 Draw three different right-angled triangles, each with an angle of 45°.
Are they all similar? Explain.
On a calculator, work out the tangent of the 45° angle for each triangle.
What do you notice?
Draw three more right-angled triangles, each with an angle of 60°.
Are they all similar? Explain.
On a calculator, work out the cosine of the 60° angle for each triangle.
What do you notice?

12 Strengthen

Congruence

1 **Reasoning** Which of these triangles is not congruent to triangle A?

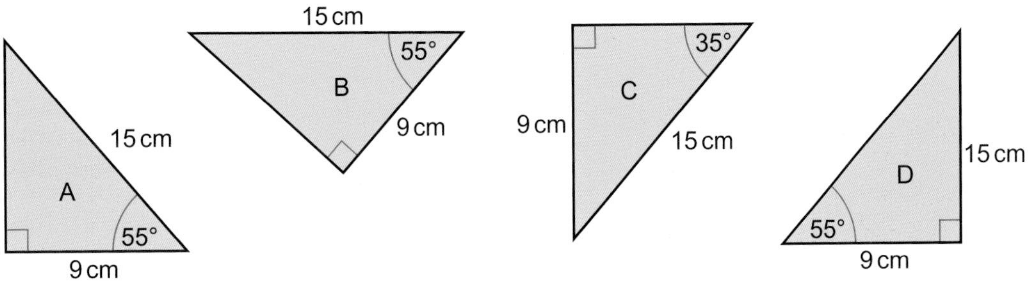

Q1 hint If you rotated triangles B, C and D, which one would not fit exactly on A?

2 Which of these triangles are congruent?

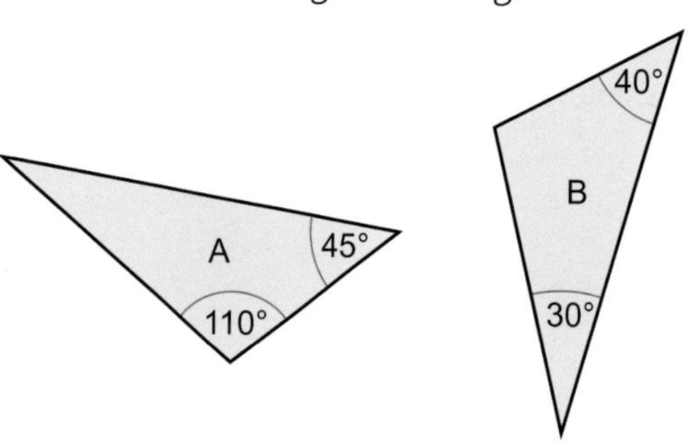

Q2 hint Sketch the triangles and find any missing angles.

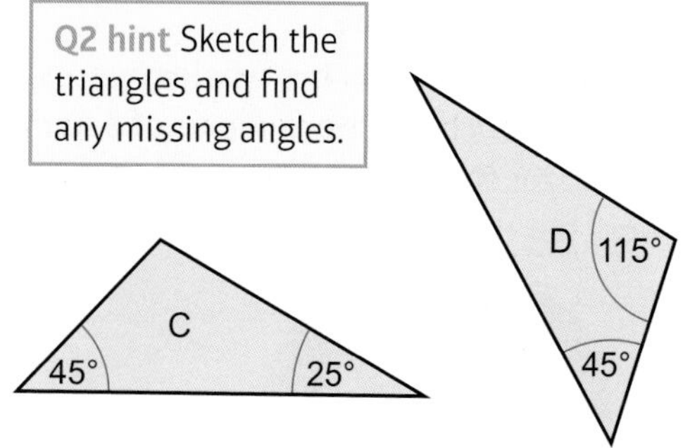

3 Draw two right-angled congruent triangles on paper and cut them out.
What shapes can you make by putting these two triangles together?
You must place the triangles with equal sides touching.

Q3 hint Make sure that sides meet exactly when you match up the triangles.

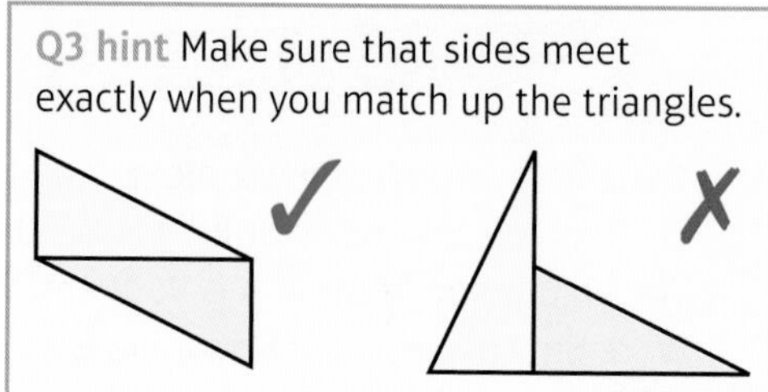

4 **Communication** Why are these pairs of triangles congruent?

Q4 hint Choose from these conditions of congruence: AAS, SAS, SSS, RHS.

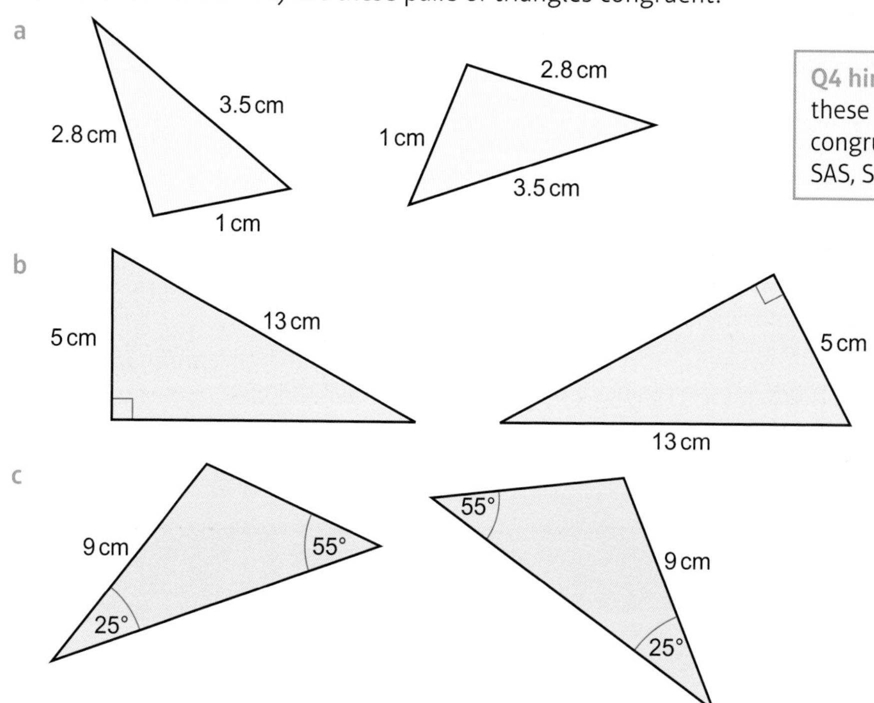

5 **Problem-solving / Communication** *KLMN* is a kite.

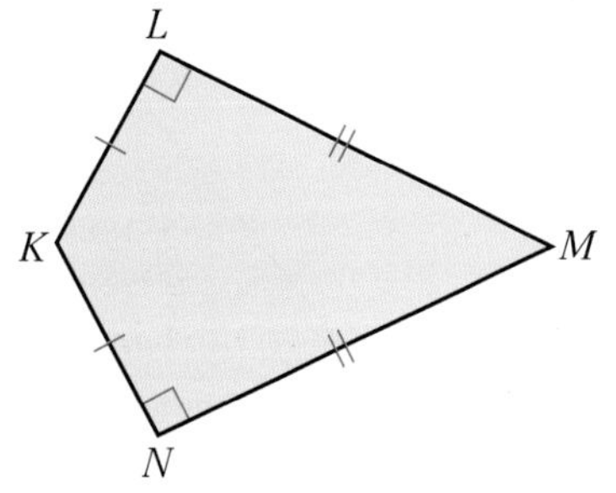

Q5 hint Do the triangles fit exactly on to each other?

a Explain why triangles *KLM* and *KNM* are congruent.
b Explain why triangles *KLN* and *MLN* are not congruent.

Similarity in 2D shapes

1 For each pair of similar triangles:
   i name the three pairs of corresponding sides
   ii state which pairs of angles are equal.

> **Q1 hint** Draw each pair of triangles facing the same way.

a
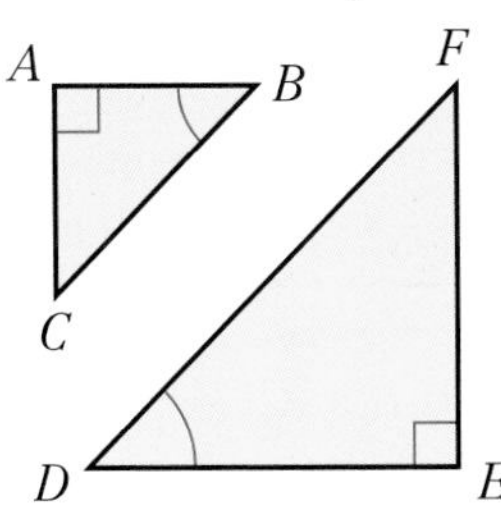

b
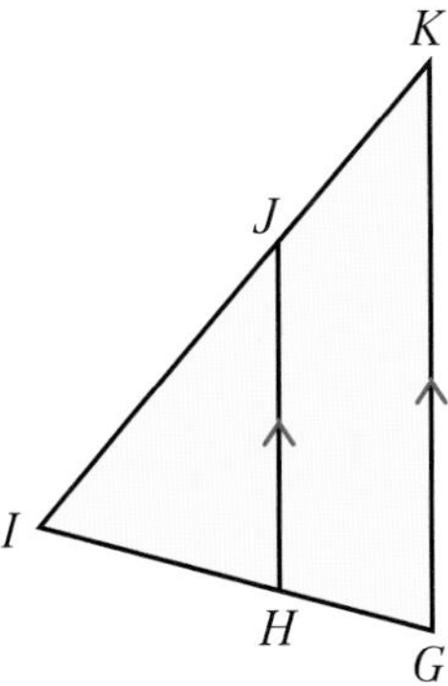

c
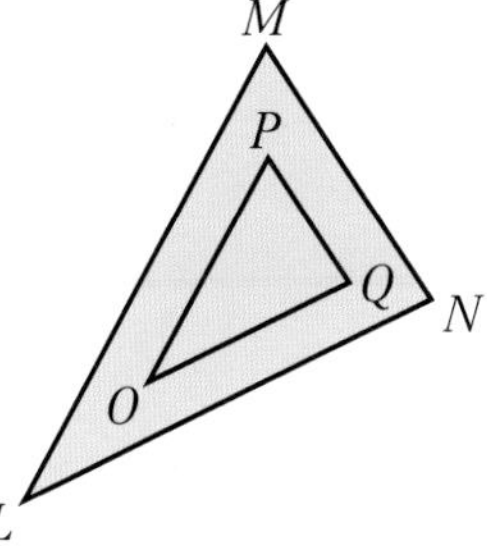

d
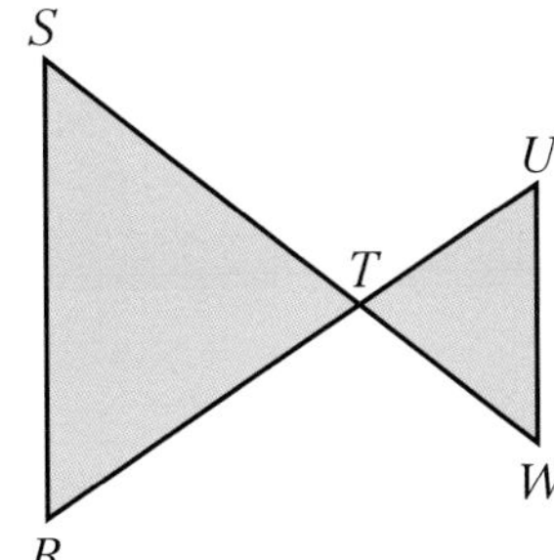

2 Triangles C and D are similar.

a Copy and complete the table showing the pairs of corresponding sides.

C	D	$\frac{C}{D}$
3		
5		
y		

b Use the scale factor to work out x and y.

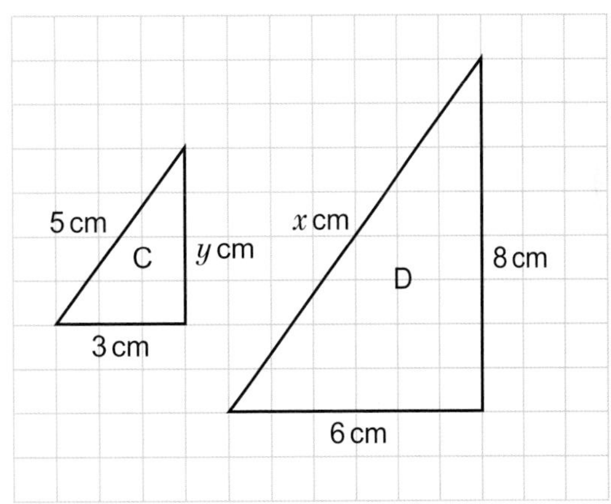

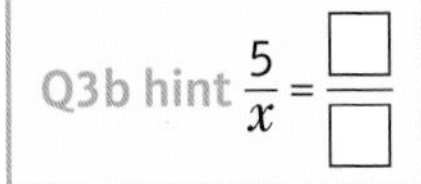

3 **Reasoning** Find the missing length x in these similar shapes.

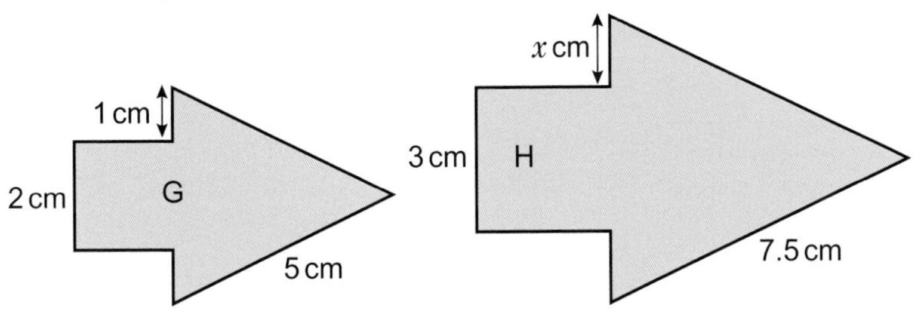

> **Q3 strategy hint** Draw a table for G and H like the one in **Q2**.

4 **Reasoning** All these shapes are similar.
Work out the lengths marked with letters.

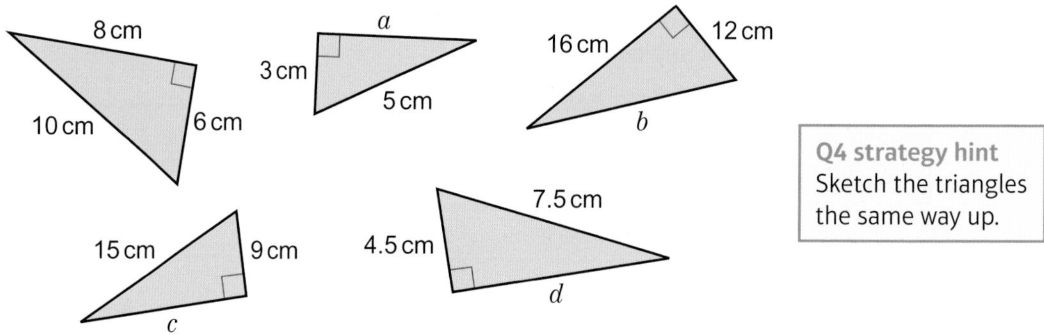

Q4 strategy hint Sketch the triangles the same way up.

5 Which of these triangles are similar to triangle A?

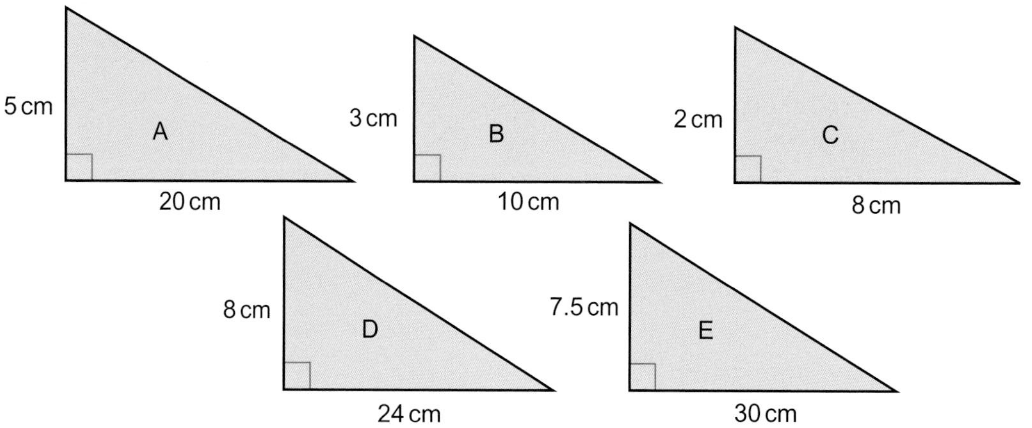

6 **Reasoning / Communication**
The diagram shows two triangles.

a Explain why
   i $a = f$ ii $b = e$ iii $c = d$.

b When three pairs of angles are equal, what does this tell you about the two triangles?

c Trace each angle and compare it with its paired angle.

d Find the missing lengths.

7 a Sketch the diagram.
Mark all the angles that are the same.

Q7a hint Use the parallel lines to find equal angles.

b Prove that triangles ABC and ADE are similar.

Q7b hint Show the triangles have the same angles. For example, angle ACB = angle ☐. Give reasons for your answers.

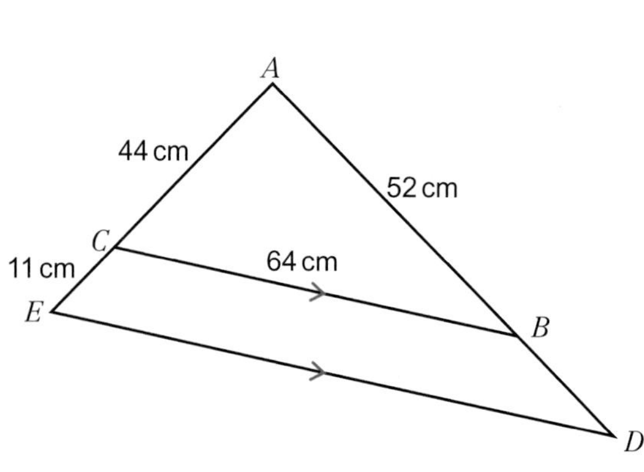

c Trace the small and the large triangle. Sketch them beside each other.
Label the lengths of the sides you know.

d Work out the scale factor and missing lengths.

8 a Draw a rectangle 2 by 5 on squared paper. Label it A.

b Draw an enlargement of A with scale factor 2. Label it B.

c Work out the perimeter of A and the perimeter of B.

d Copy and complete these statements:
lengths on A : lengths on B, scale factor is 2.
perimeter of A : perimeter of B, scale factor is □.

e Enlarge A by scale factor 3. Label it C.

f Predict the perimeter of C.
Now work out the perimeter to check your prediction.

9 Triangle A is similar to triangle B.

a Work out the scale factor of A to B by comparing the given side lengths.

The perimeter of triangle A is 12 cm.

b What is the perimeter of triangle B?

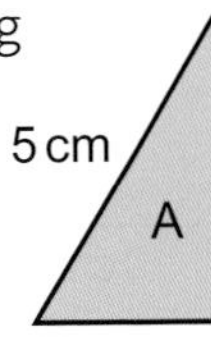

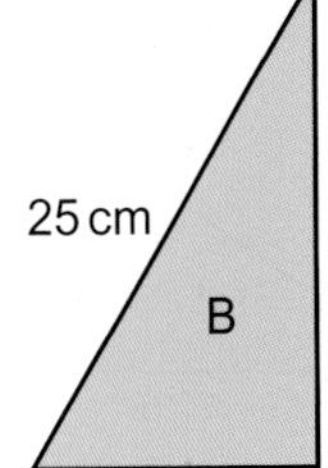

10 Draw a 1 cm by 3 cm rectangle on squared paper. Label it A.
Draw an enlargement of A with scale factor 2. Label the new shape B.
How many times will A fit into B?
Copy and complete
lengths on A : lengths on B, scale factor is 2
area of A : area of B, scale factor is □ = $2^{□}$

11 The area of triangle A in **Q9** is 6 cm². What is the area of triangle B?

Similarity in 3D solids

1 The diagram shows two cubes, with side length 1 cm and side length 2 cm.

a How many times does A fit into B?

b Copy and complete
lengths on A : lengths on B, scale factor is 2
volume of A : volume of B, scale factor is □ = $2^{□}$

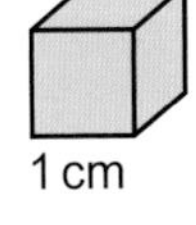

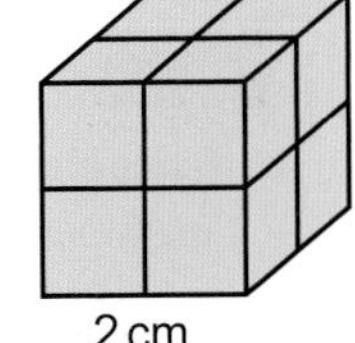

c Cube A is enlarged to make cube C, so the side length is now 3 cm.
What is the scale factor of the lengths for this enlargement?

d Predict: volume of A : volume of C, scale factor is □ = $□^{□}$

e Draw a sketch to check your prediction.

2 Cuboid X and cuboid Y are similar.

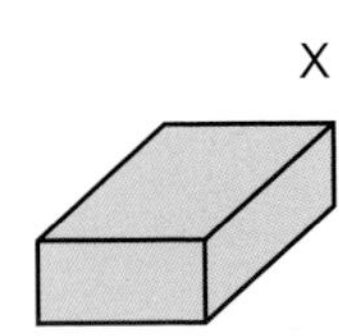

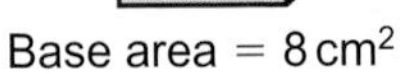

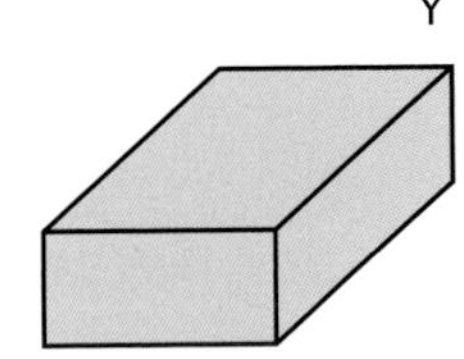

The area of the base of cuboid X is 8 cm².
The area of the base of cuboid Y is 72 cm².

a Work out the area scale factor to change X to Y.

b Find the linear scale factor.

c Find the volume scale factor.

The volume of cuboid X is 12.5 cm³.

d Find the volume of cuboid Y.

Q2b hint Square root the area scale factor.

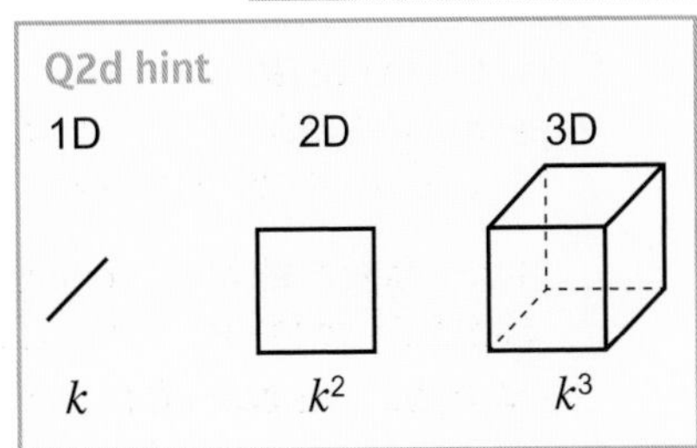

12 Extend

1 Triangle PST is similar to triangle PQR.

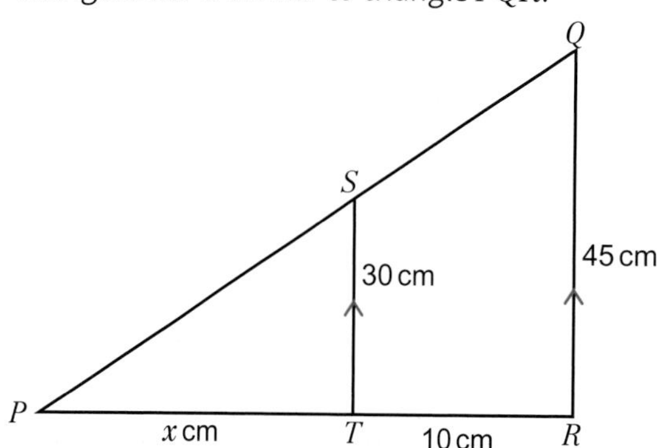

Q1 strategy hint Multiply both sides of the equation by each denominator.

a Draw triangles PQR and PST separately, labelling all sides clearly.

b Write an equation $\frac{PR}{PT} = \frac{QR}{ST}$ and solve to find the value of x.

2 The diagram shows triangle ABC which has a line JK drawn across it.

angle BCA = angle AKJ.

a Prove that triangle AJK is similar to triangle ABC.

b Calculate the length of JK.

c Calculate the length of KC.

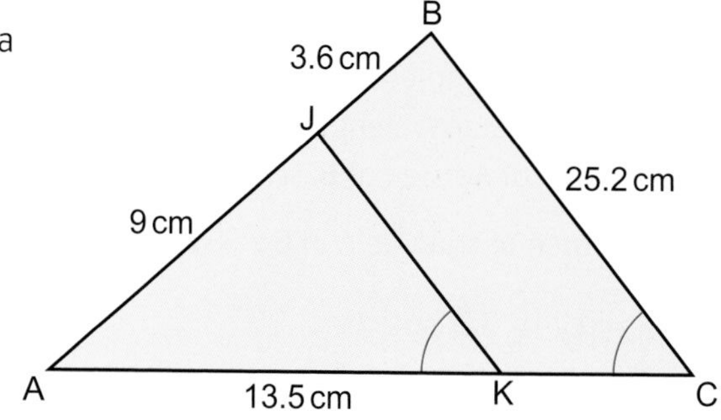

3 **Exam-style question**

Triangle ABC is drawn on a centimetre grid.

A is the point (2, 2).
B is the point (6, 2).
C is the point (5, 5).

Triangle PQR is an enlargement of triangle ABC with scale factor $\frac{1}{2}$ and centre (0, 0).

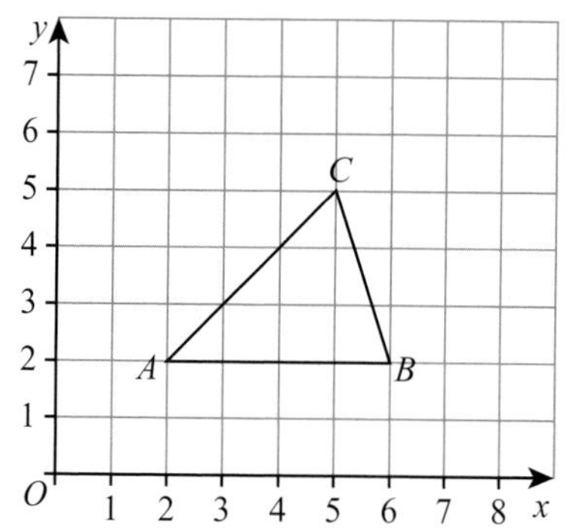

Work out the area of triangle PQR. **(3 marks)**

June 2012, Q18, 1MA0/1H

Exam hint
You could draw the enlargement – but you don't need to.

4 **Problem-solving** ABC is a triangle.
Calculate the perimeter of the trapezium $DBCE$.

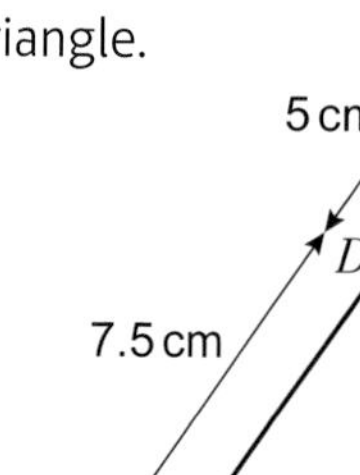

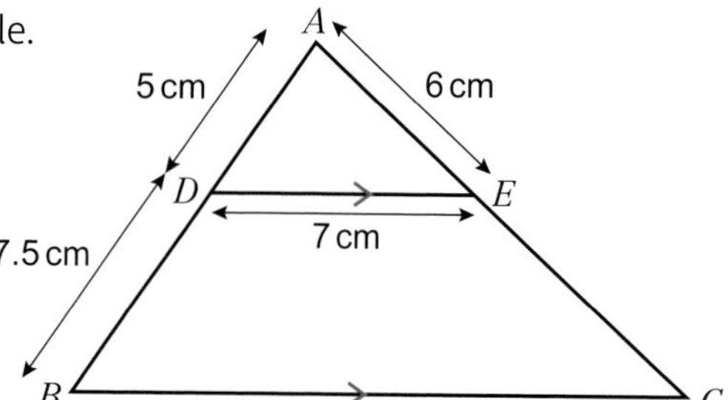

5 The diagram shows a gift box with a surface area of $300\,cm^2$.
A piece of ribbon 46 cm long is tied in a bow around the box.
A larger gift box has a similar bow tied around it, using a piece of ribbon 69 cm long.
What is the surface area of the larger box?

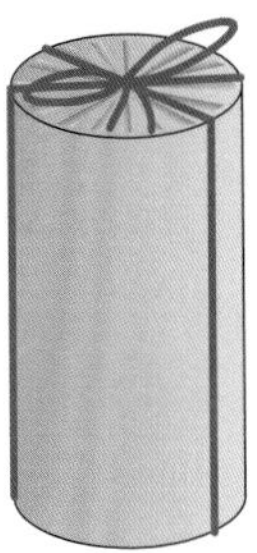

6 **Problem-solving** A statue has a mass of 840 kg.
A similar statue made out of the same material is $\frac{2}{5}$ of the height of the first statue.
What is the mass of the small statue?

Q6 hint mass = volume × density, so mass and volume are in direct proportion.

7 Cylinders Y and Z are similar.
The volume of Y is $6\pi\,cm^3$.
The volume of Z is $93.75\pi\,cm^3$.
Calculate the length of the radius of cylinder Z.

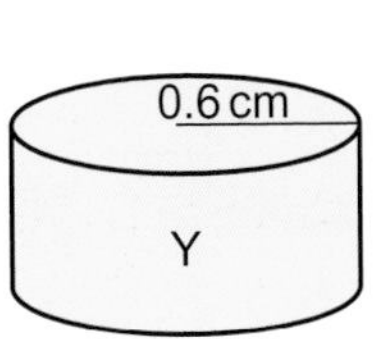

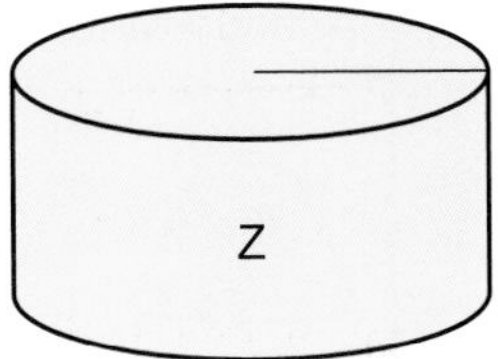

8 **Problem-solving** D and E are two regular pentagonal prisms that are mathematically similar.
Prism D has cross-sectional area $15.5\,cm^2$.
The side length of pentagon D is 3 cm.
The side length of pentagon E is 10.5 cm.
The length of prism D is 20 cm.

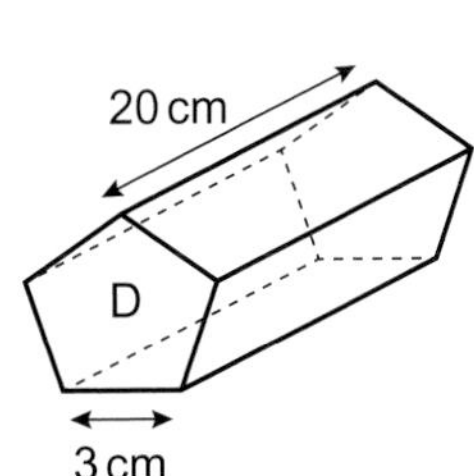

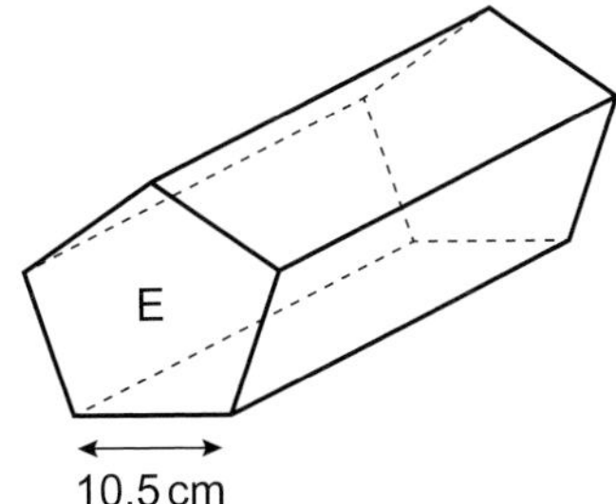

a Work out the volume of prism E.
b How many prisms the same size as prism E could be made from $1\,m^3$ of plastic?

9 In the diagram, $AE = BE$, $AG = HB$, and AB is parallel to DC.

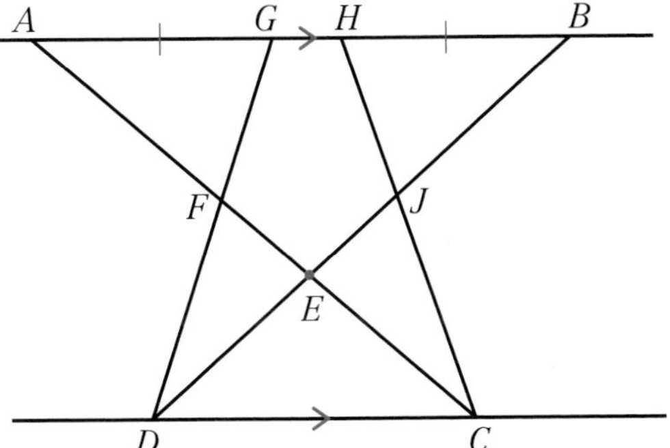

Prove that triangle ACH and triangle BDG are congruent.
Give reasons for your answer.

10 Problem-solving The diagram shows trapezium $ABCD$.
Trapezium $ABCD$ is similar to trapezium $BZXV$.
Lines AD, VW and CB are parallel.
Lines CD and YZ are parallel.
Calculate the perimeter of trapezium $BZXV$.

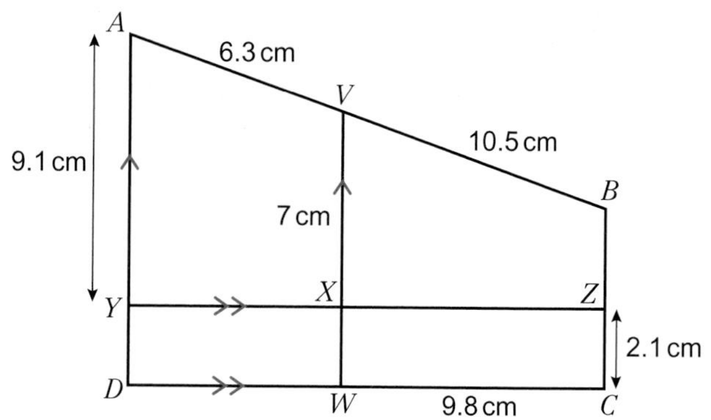

Q11 communication hint A **frustum** is a cone or pyramid with the point cut off, parallel to the base.

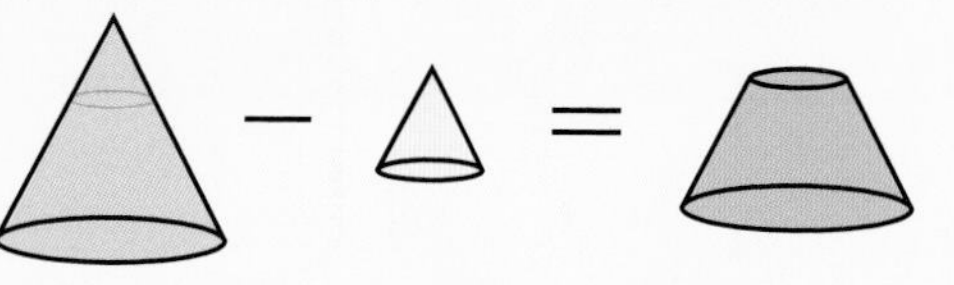

11 Reasoning The diagram shows a frustum.

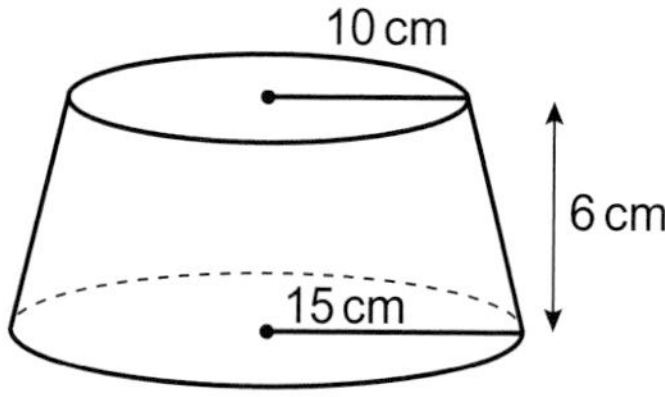

a Find the height of the whole cone.

b Find the volume of the whole cone.

Q11b hint Leave your answer in terms of π.

c Find the volume of the smaller cone (the bit that is missing).

d Hence, find the volume of the frustum.
Give your answer correct to 3 significant figures.

Q11 strategy hint Imagine the whole cone, before the top was cut off to make the frustum. Use similar triangles to work out the heights.

12 Problem-solving This is the frustum of a square-based pyramid.
The length of the base is 18 cm. The length of the top of the frustum is 7.5 cm.
The vertical height of the frustum is 20.4 cm.
Find the volume of the frustum.
Give your answer correct to 3 significant figures.

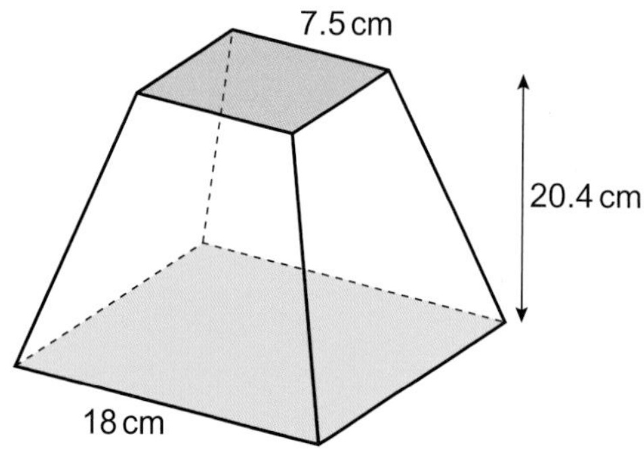

13 **Exam-style question**

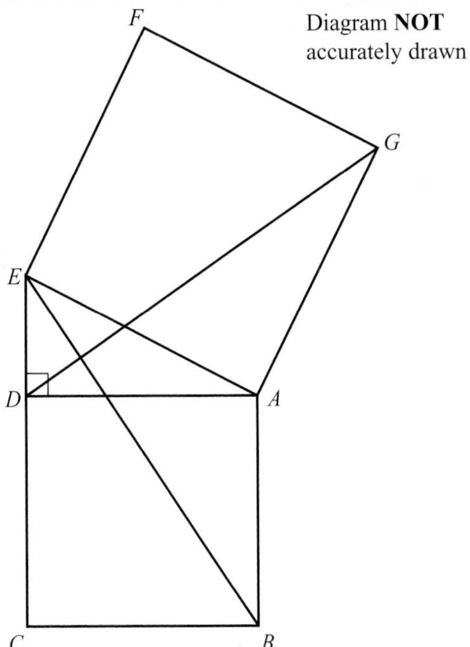

In the diagram,
ADE is a right-angled triangle
$ABCD$ and $AEFG$ are squares.
Prove that triangle ABE is congruent to triangle ADG. **(3 marks)**

March 2013, Q22, 5MB3H/01

14 **Problem-solving / Communication** Triangles STU and VWX are mathematically similar.
The base, SU, of triangle STU has length $2(x - 1)$ cm
The base, VX, of triangle VWX has length $(x^2 - 1)$ cm
The area of triangle STU is 8 cm^2.
The area of triangle VWX is A cm^2.
Prove that $A = 2x^2 + 4x + 2$.

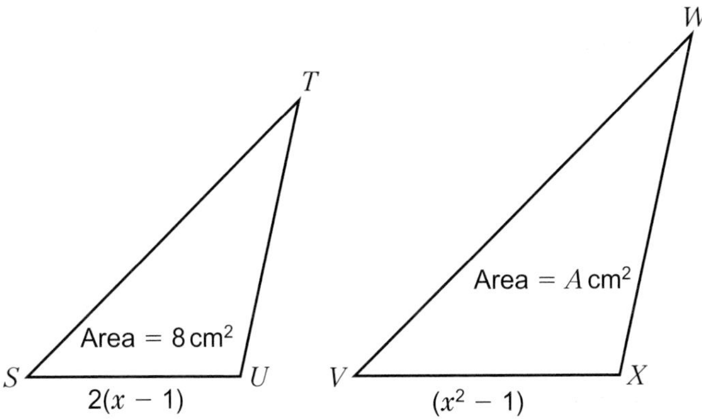

12 Knowledge check

- Congruent triangles have exactly the same size and shape. Their angles are the same and corresponding sides are the same length. .... *Mastery lesson 12.1*
- Two triangles are congruent when one of these conditions of congruence is true.
  SSS (all three sides equal)
  SAS (two sides and the included angle are equal)
  AAS (two angles and a corresponding side are equal)
  RHS (right angle, hypotenuse and one other side are equal) .... *Mastery lesson 12.1*
- You can use congruence to solve problems and prove that shapes are the same. .... *Mastery lesson 12.2*
- To prove something, you write a series of logical statements that show the statement is true. Each statement must be supported by a mathematical reason. .... *Mastery lesson 12.2*
- Shapes are similar when one shape is an enlargement of the other. Corresponding angles are equal and corresponding sides are all in the same ratio. .... *Mastery lesson 12.3*
- When a shape is enlarged by linear scale factor k, the area of the shape is enlarged by scale factor k^2. .... *Mastery lesson 12.4*
- When a shape is enlarged by linear scale factor k, the volume is enlarged by scale factor k^3. .... *Mastery lesson 12.5*
- When the linear scale factor is k:
  Lengths are multiplied by k
  Area is multiplied by k^2
  Volume is multiplied by k^3 .... *Mastery lesson 12.5*

Reflect

For each statement A, B and C, choose a score:

1 – strongly disagree; 2 – disagree; 3 – agree; 4 – strongly agree

A I always try hard in mathematics

B Doing mathematics never makes me worried

C I am good at mathematics

For any statement you scored less than 3, write down two things you could do so that you agree more strongly in the future.

12 Unit test

Log how you did on your Student Progression Chart.

1 JKL and PQR are triangles.
Show that the triangles are not similar. *(4 marks)*

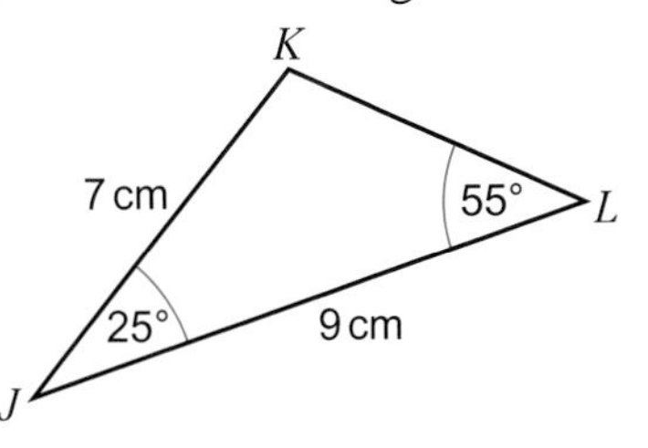

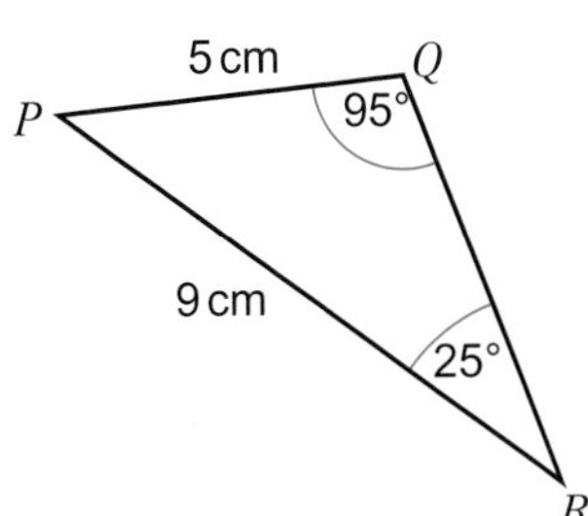

2 Identify two triangles that are congruent and give a reason for your answer. *(2 marks)*

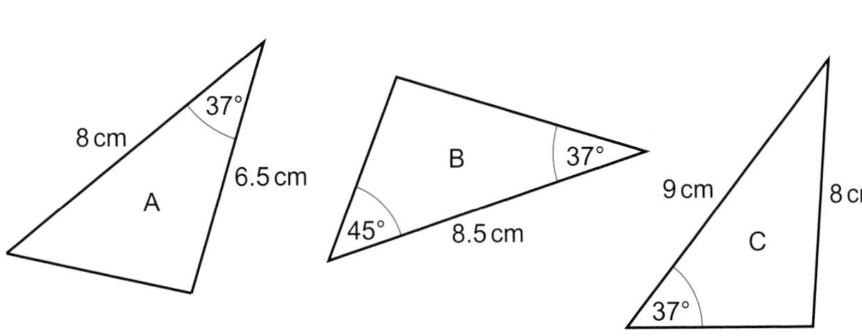

45°
98°
D
8 cm
6.5 cm

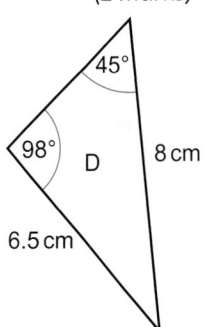

3 M and N are similar shapes.
Work out the missing lengths, x and y. *(3 marks)*

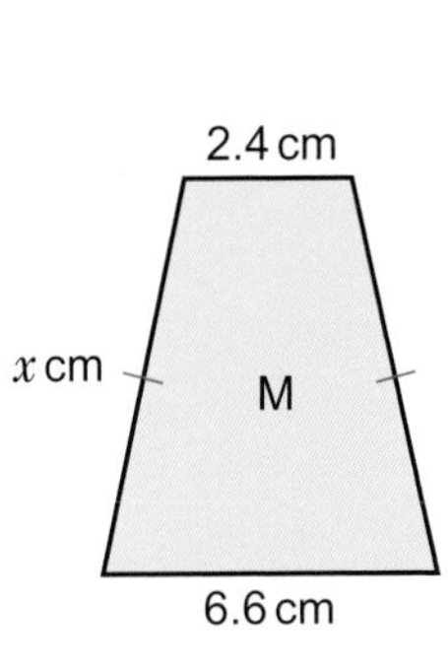

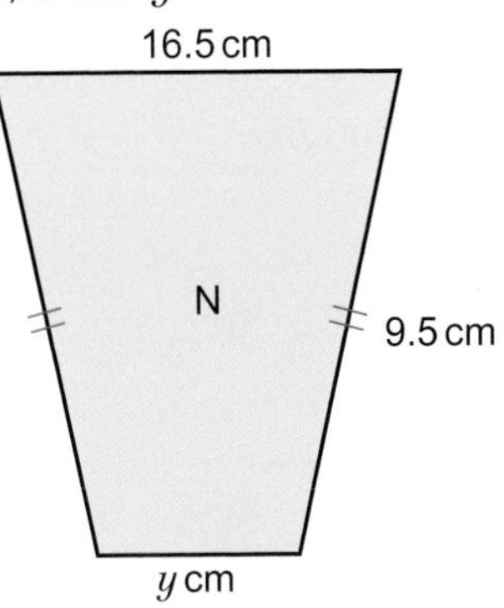

4 **Problem-solving** Work out the lengths x and y.
Explain all the steps in your working. *(4 marks)*

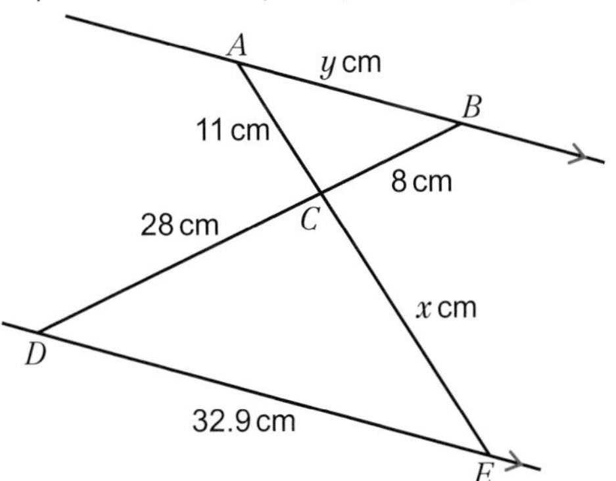

5 BE is parallel to CD. Work out

a the length of BE *(2 marks)*

b the length of BC. *(2 marks)*

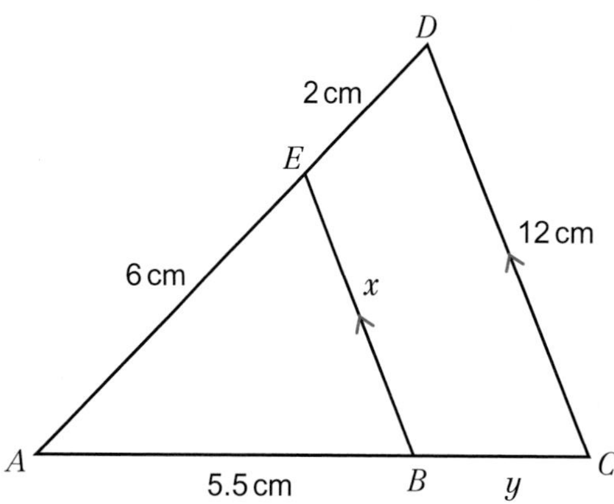

6 $ABCD$ is a rectangle. AC and BD are the diagonals of the rectangle, which cross at point E.

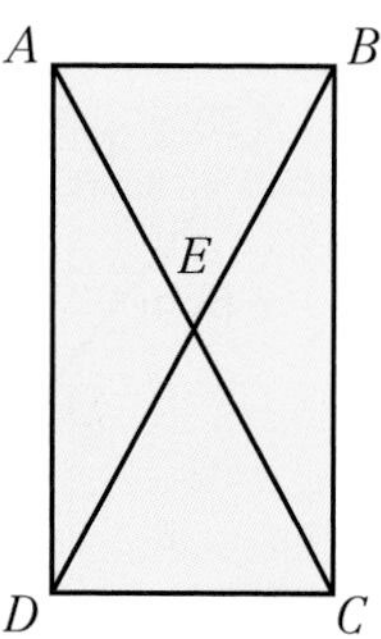

a Prove that both diagonals divide the rectangle into two congruent triangles. *(4 marks)*

b Prove that E is the midpoint of AC and BD. *(2 marks)*

7 Shapes B and C are similar.

The area of B is 18 cm². The area of C is 112.5 cm².

Find the length of side a. *(3 marks)*

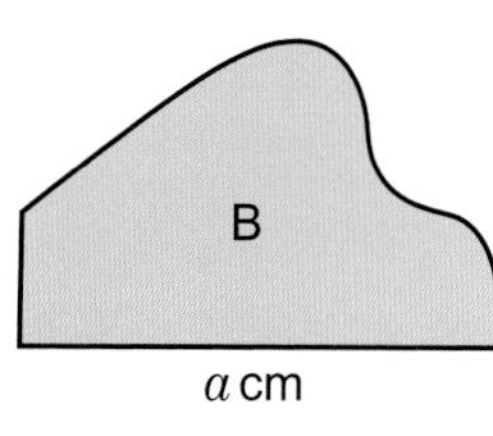

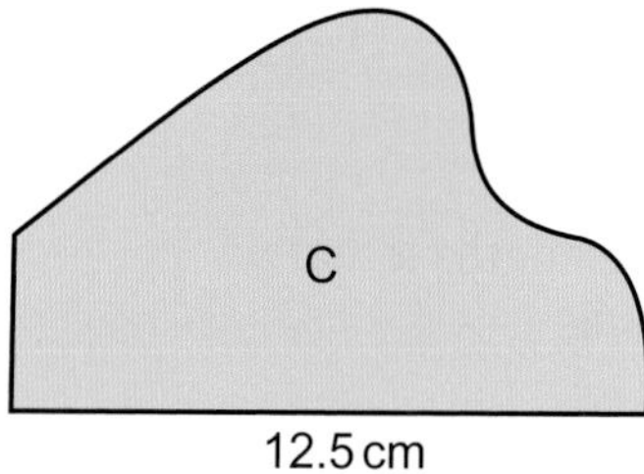

8 **Problem-solving** F and G are mathematically similar greenhouses.

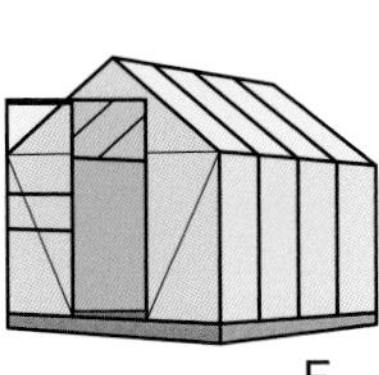

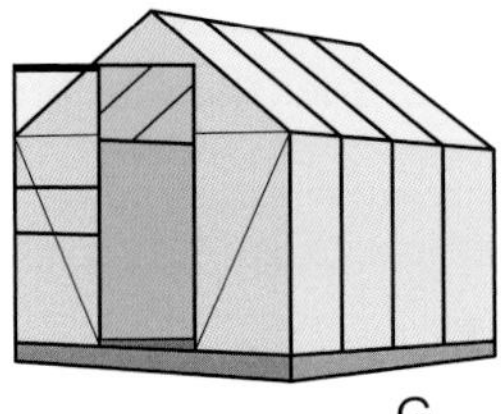

F has a volume of $5.6\,\text{m}^3$ and a surface area of $15\,\text{m}^2$.
The volume of G is $18.9\,\text{m}^3$.
What is the surface area of G? *(4 marks)*

9 **Problem-solving** A cake tin is in the shape of a frustum.

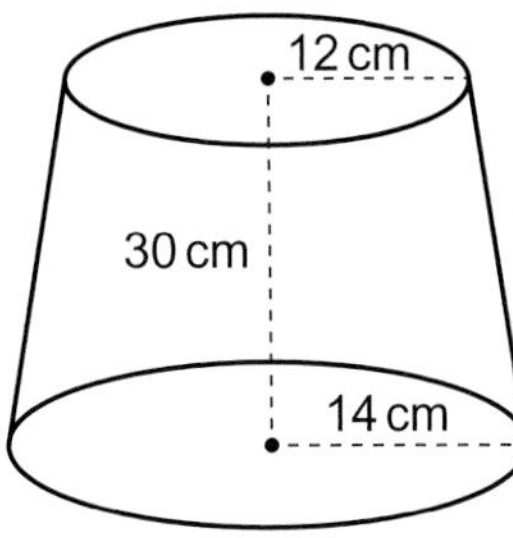

Work out the volume of a cake which exactly fills the tin.
Give your answer correct to 3 significant figures. *(5 marks)*

Sample student answer

a How has the student's diagram helped check similarity?

b How else has the student clearly 'shown' that the triangles are similar?

Exam-style question

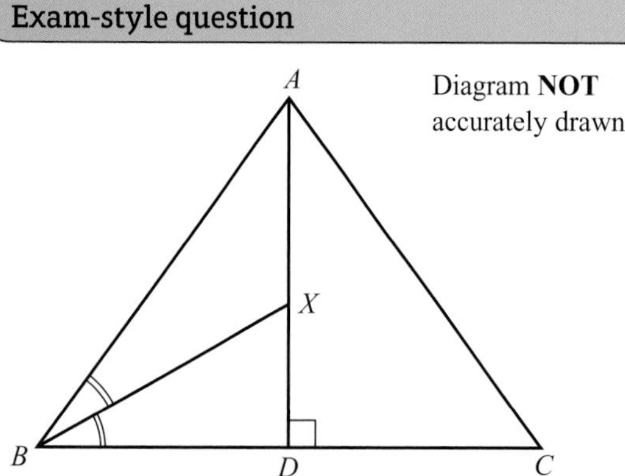

ABC is an equilateral triangle.
AD is the perpendicular bisector of *BC*.
BX is the angle bisector of angle *ABC*.
Show that triangle *BXD* is similar to triangle *ACD*. **(2 marks)**

2010 Practice Paper Set A, Q23a, 1MA0/3H

Student answer

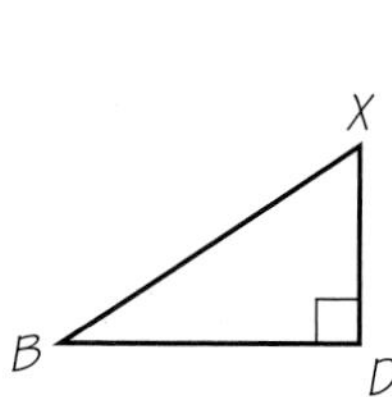

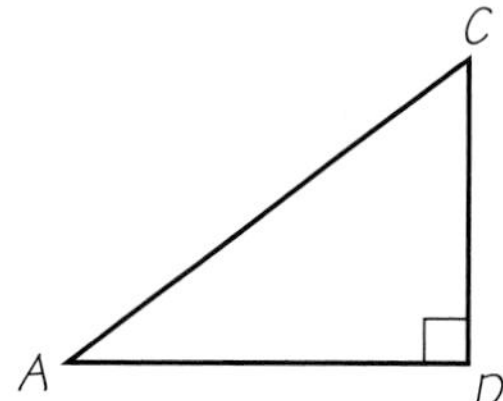

Triangle *ACD*

The line *AD* is perpendicular to *CD* so $\angle CDA$ is a right angle.

Angles in an equilateral triangle are 60°, so $\angle ACD = 60°$

The last angle must be $180 - 90 - 60 = 30°$

Triangle *BXD*

The line *XD* is perpendicular to *BD* so $\angle BDX$ is a right angle.

BX bisects $\angle ABC$, so $\angle DBX$ is 30°

The last angle must be $180 - 90 - 30 = 60°$

Therefore as all the corresponding angles are equal, the triangles must be similar.

Master p.400 | Problem-solve p.426 | Check p.427 | Strengthen p.429 | Extend p.433 | Test p.437

13 MORE TRIGONOMETRY

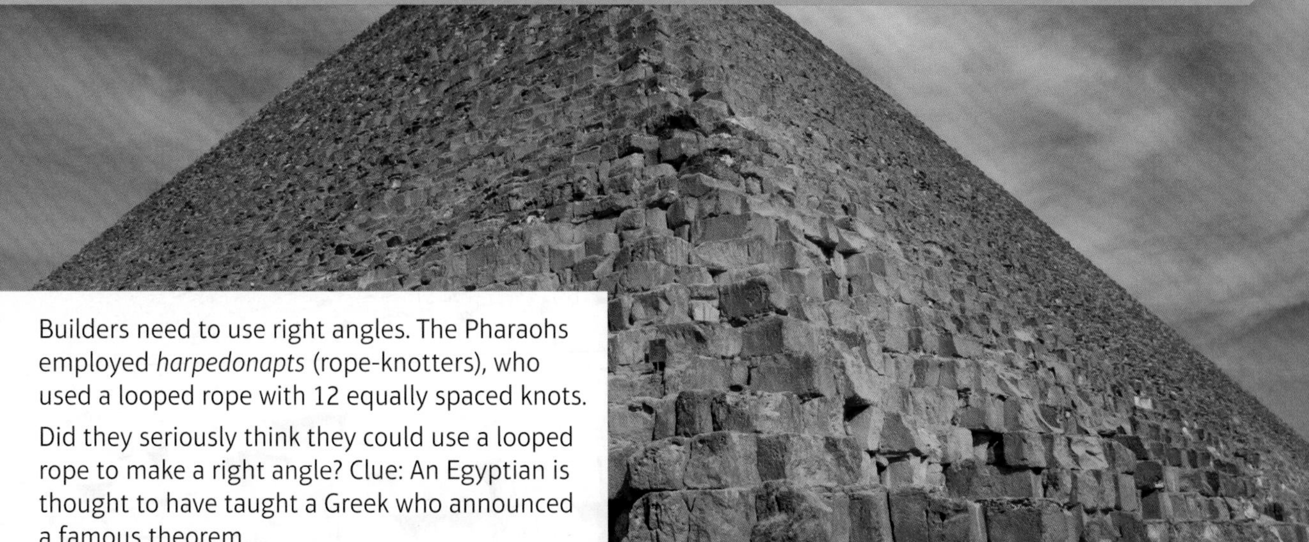

Builders need to use right angles. The Pharaohs employed *harpedonapts* (rope-knotters), who used a looped rope with 12 equally spaced knots.

Did they seriously think they could use a looped rope to make a right angle? Clue: An Egyptian is thought to have taught a Greek who announced a famous theorem…

13 Prior knowledge check

Numerical fluency

1 Work out

a $\frac{0.75}{2} \times 4$ b $3^2 + 4^2 - 2 \times 3 \times 4 \times 0.8$

Algebraic fluency

2 $15^2 = q^2 + 7^2$

Find the value of q to 3 significant figures.

3 $\frac{x}{0.5} = \frac{8.42}{0.749}$

Find the value of x to 3 significant figures.

4 $a = 4$ $b = 6.5$ angle C = 58°

Find the value of $ab \sin C$. Give your answer correct to 3 significant figures.

5 $\cos \theta = \frac{18^2 + 12^2 - 24^2}{2 \times 18 \times 12}$

Find the value of θ correct to 1 decimal place.

Geometrical fluency

6 Find the exact value of x in each triangle.

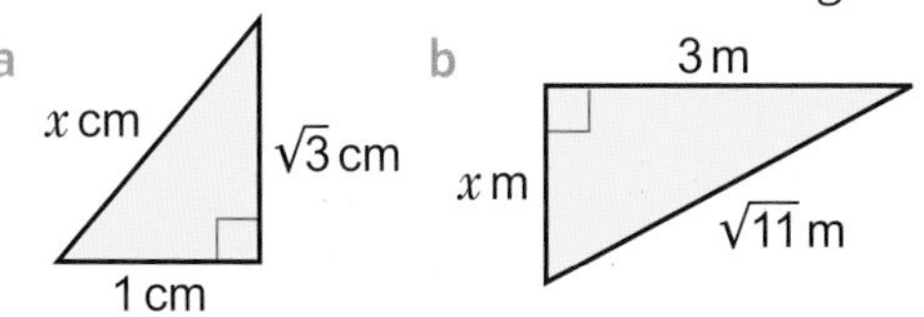

7 Work out the size of each lettered angle. Give each answer correct to 1 decimal place.

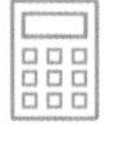

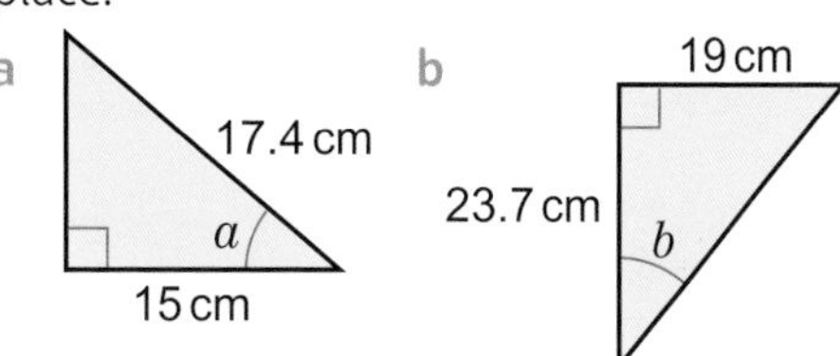

8 Work out the length of each lettered side. Give each answer correct to 3 significant figures.

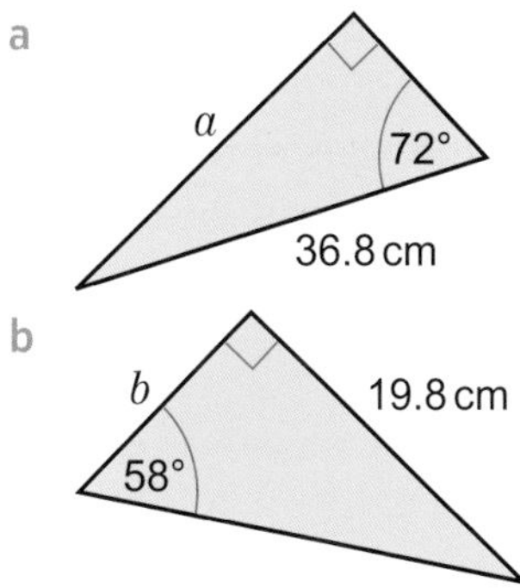

9 Find the area of this triangle.

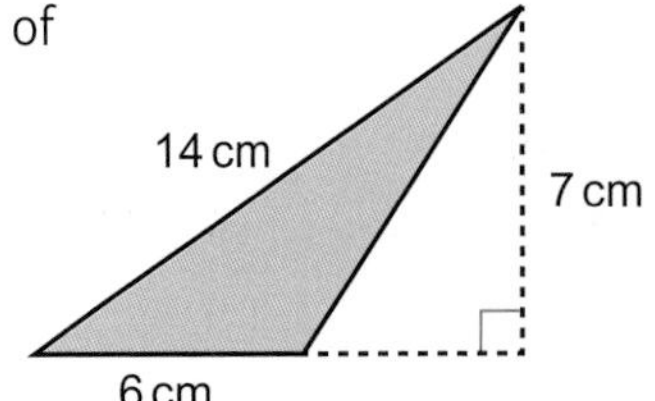

Graphical fluency

10 a Copy and complete the table for $y = x^2 - 6x$.

x	0	1	2	3	4	5	6
y							

b Draw the graph of $y = x^2 - 6x$ for $0 \leqslant x \leqslant 6$.

c Use your graph to solve the equation $x^2 - 6x = -5$.

* Challenge

11 The diagram shows a sequence of isosceles right-angled triangles A, B, C, …

Continue the sequence and find the length of the hypotenuse of triangle H.

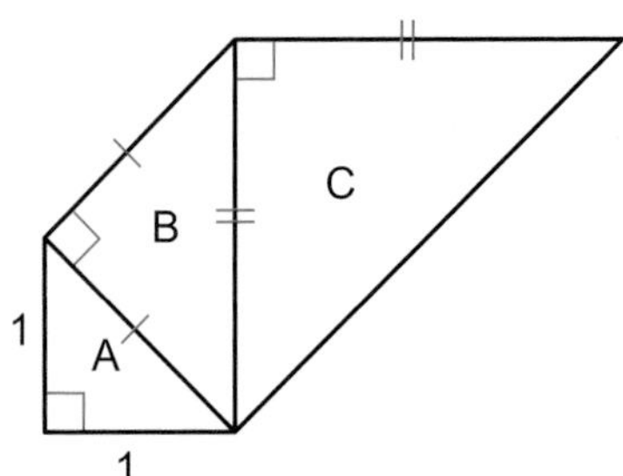

13.1 Accuracy

Objective

- Understand and use upper and lower bounds in calculations involving trigonometry.

Did you know?

The *caesium fountain* atomic clock at the National Physical Laboratory in the UK is the most accurate in the world. In 138 million years it is unlikely to be a second out.

Fluency

The height of a book is measured as 15.4 cm to 1 decimal place. What are the upper and lower bounds of the height of the book?

Warm up

1 $y = 3.6$ to 1 d.p. $z = 9.2$ to 1 d.p. $x = yz$

a Find the upper bound and the lower bound of y.

b Find the upper bound and the lower bound of z.

c Work out the value of the upper bound of x.

d Work out the value of the lower bound of x.

Q1c hint Which bounds of y and z give the largest possible value for yz?

2 $y = 1.2$ to 1 d.p. $z = 0.4$ to 1 d.p. $x = \frac{y}{z}$

a Find the upper bound and the lower bound of y.

b Find the upper bound and the lower bound of z.

c Work out the value of the upper bound of x.

d Work out the value of the lower bound of x.

Q2c hint Check all possible calculations to make sure you have the highest possible value for $\frac{y}{z}$.

Questions in this unit are targeted at the steps indicated.

3 In the diagram, the lengths of AC and BC are given correct to 1 d.p.

a Find the upper bound for the length of
i AC ii BC

b Use your answers to part **a** to work out the upper bound of x.

c Find the lower bound for the length of
i AC ii BC

d Use your answers to part **c** to work out the lower bound of x.

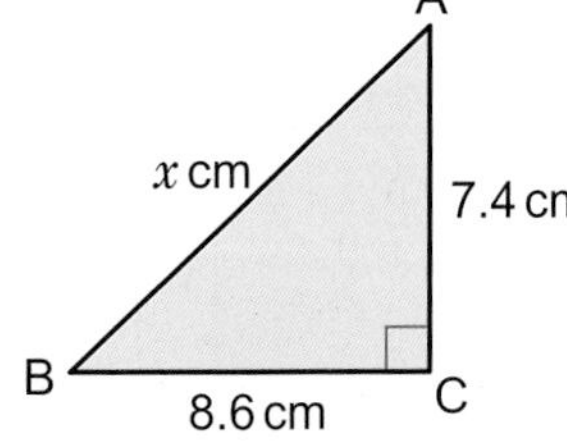

Example 1

In this diagram, the measurements are correct to 3 significant figures.

a Find the upper and lower bounds for the value of x, to 3 decimal places.

b Give the value of x to a suitable level of accuracy.

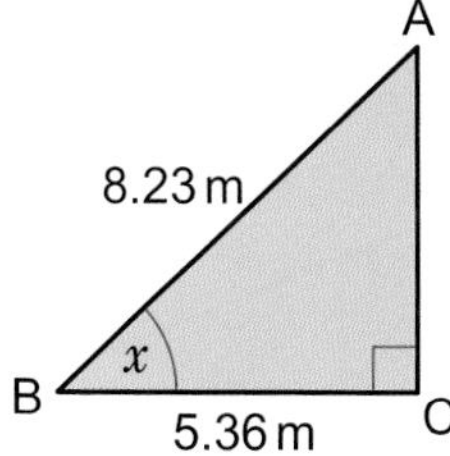

a AB: upper bound = 8.235 m, lower bound = 8.225 m

BC: upper bound = 5.365 m, lower bound = 5.355 m

The upper bound for $\cos x = \frac{5.365}{8.225}$

$= 0.6522796353$

So $x = 49.286°$ (3 d.p.)

The lower bound for $\cos x = \frac{5.355}{8.235}$

$= 0.6502732240$

So $x = 49.438°$ (3 d.p.)

So the upper bound for x is 49.438° and the lower bound is 49.286°

b 49.438° = 49.4 (1 d.p.) 49.286° = 49.3 (1 d.p.)

= 49° (nearest degree) = 49° (nearest degree)

$x = 49°$ (to the nearest degree)

> Find the upper and lower bounds of the lengths of AB and BC.

> The upper bound of a fraction $= \frac{\text{upper bound of the numerator}}{\text{lower bound of the denominator}}$
> Write down all the figures in your calculator display.

> Use $\cos^{-1}$ on your calculator.

> The lower bound of a fraction $= \frac{\text{lower bound of the numerator}}{\text{upper bound of the denominator}}$

> You could write the answer as $49.286° \leqslant x < 49.438°$

> Round the upper and lower bounds to 1 d.p. Do they both give the same value?

> Round to the nearest degree they both give the same value.

4 The upper bound for $\cos x$ is 0.7322834646 and the lower bound is 0.7054263565.

a Find the upper and lower bounds for x, to 3 decimal places.

b What do you notice?

Discussion Why does this happen?
Will it also happen with sine and tangent?

> **Q4b hint** Does the upper bound for $\cos x$ give the upper bound for x?

5 In this diagram, the measurements are correct to 3 significant figures.

a Find the upper and lower bounds for the value of x, to 3 decimal places.

b Write x to a suitable level of accuracy.

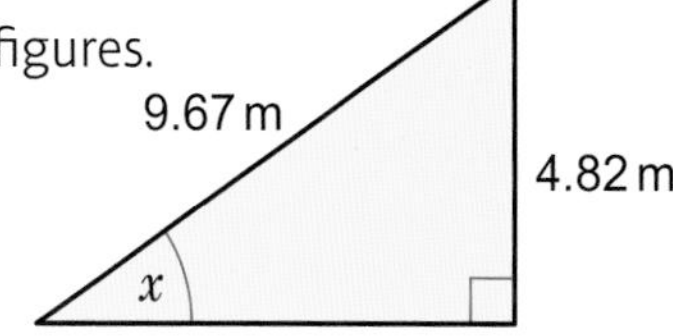

6 In this diagram, the measurements are correct to 2 significant figures.

a Find the upper and lower bounds for the value of x, to 3 decimal places.

b Write x to a suitable level of accuracy.

Discussion Compare your answers with those for **Q5**.
How have the upper and lower bounds for x been affected by reducing the accuracy of the measurements to 2 significant figures?

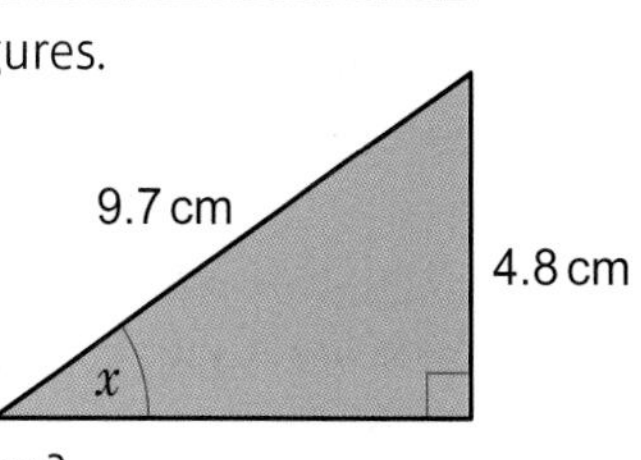

7 In this diagram, the measurements are correct to 2 significant figures.
Find the upper and lower bounds for the value of x.

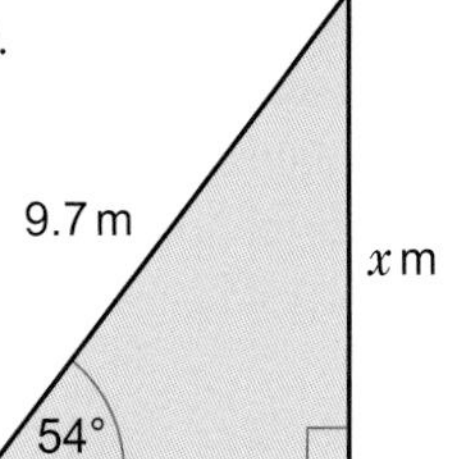

8 In this diagram, the measurements are correct to 2 significant figures.
Find the upper and lower bounds for the value of x.
Discussion Compare your answers with those for **Q7**.
Why is the level of accuracy worse in this question?

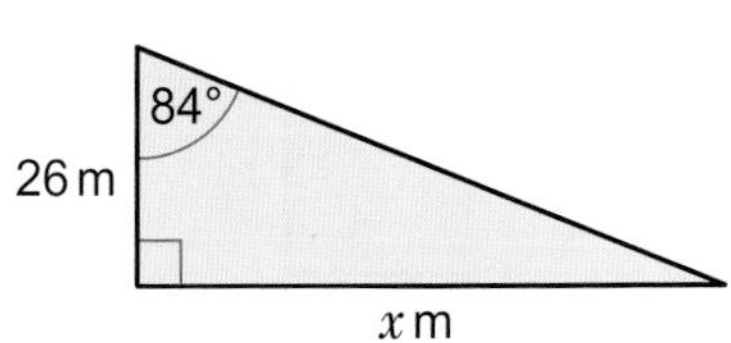

9 **Exam-style question**

Dan does an experiment to find the value of π.
He measures the circumference and the diameter of a circle.
He measures the circumference, C, as 170 mm to the nearest millimetre.
He measures the diameter, d, as 54 mm to the nearest millimetre.
Dan uses $\pi = \frac{C}{d}$ to find the value of π.
Calculate the upper bound and the lower bound for Dan's value of π. **(4 marks)**

June 2013, Q23, 1MA0/2H

Q9 strategy hint
The upper bound of $\frac{C}{d}$ is not $\frac{\text{upper bound of } C}{\text{upper bound of } d}$ and the lower bound of $\frac{C}{d}$ is not $\frac{\text{lower bound of } C}{\text{lower bound of } d}$

13.2 Graph of the sine function

Objectives

- Understand how to find the sine of any angle.
- Know the graph of the sine function and use it to solve equations.

Did you know?

In computer games, the face, body, movement and even the clothing of a character are almost entirely defined by trigonometry.

Fluency

What is the exact value of
sin 30° sin 45° sin 60° sin 90°?

Warm up

1 a Find the value of x in this triangle.
b Find the value of sin θ.

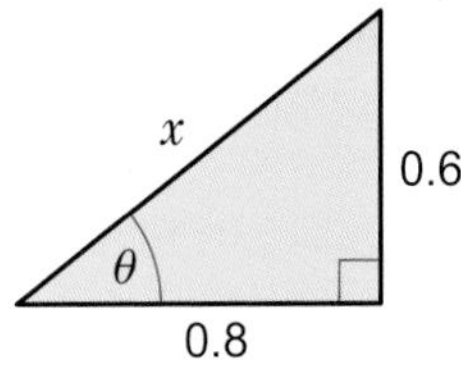

2 Find the value of sin θ in this triangle.

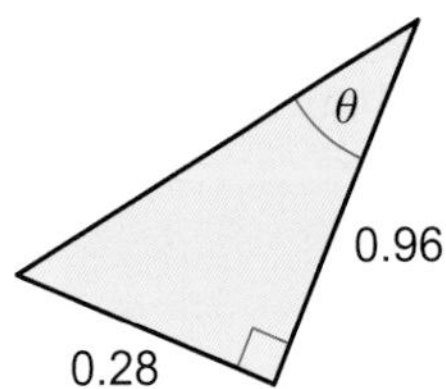

Key point 1

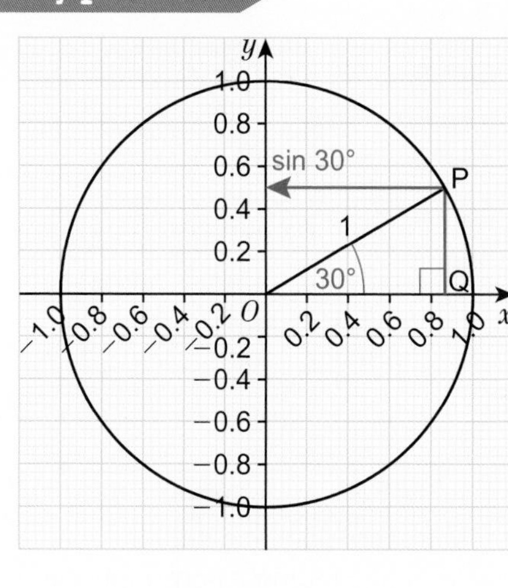

The diagram shows a circle of radius 1 unit with centre at (0, 0).

$\sin 30° = \frac{PQ}{1} = PQ = 0.5$

The length of PQ gives the **sine** of the angle. This is shown on the vertical axis by the position of the arrow.

You can find the sine of any angle using this method.

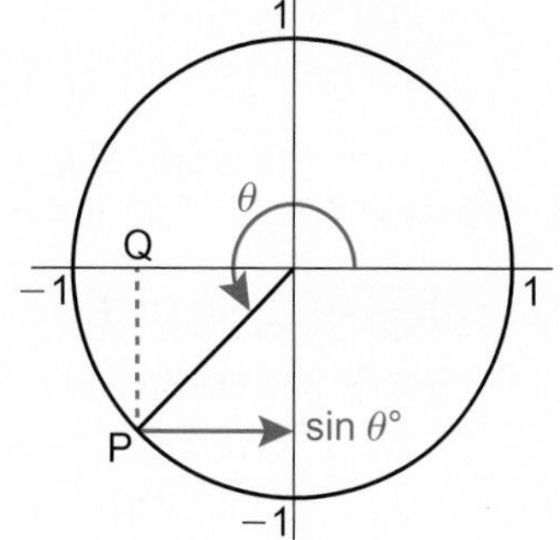

3 Find the value of $\sin\theta$ in each diagram.

a

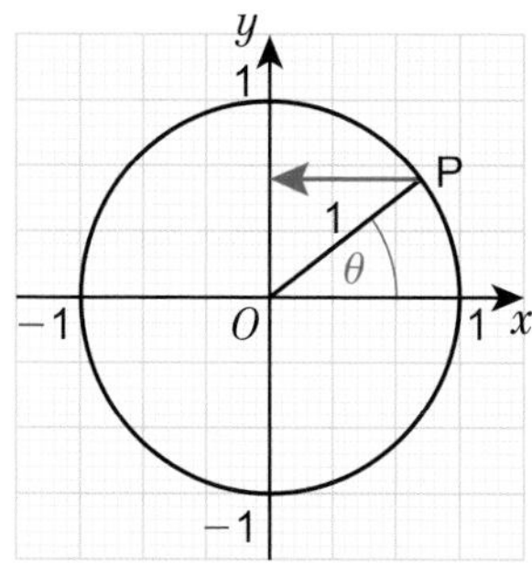

b

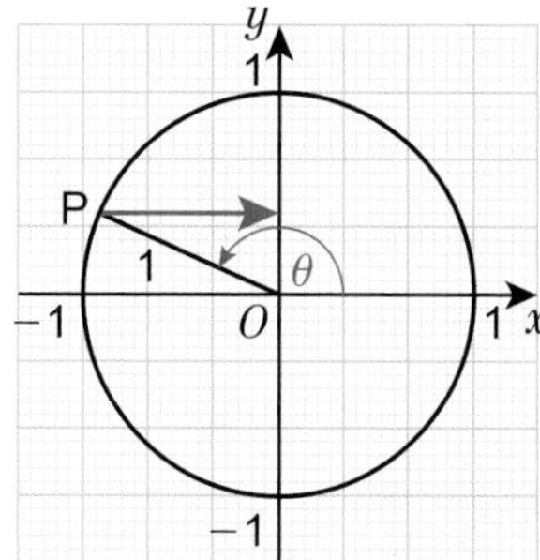

c

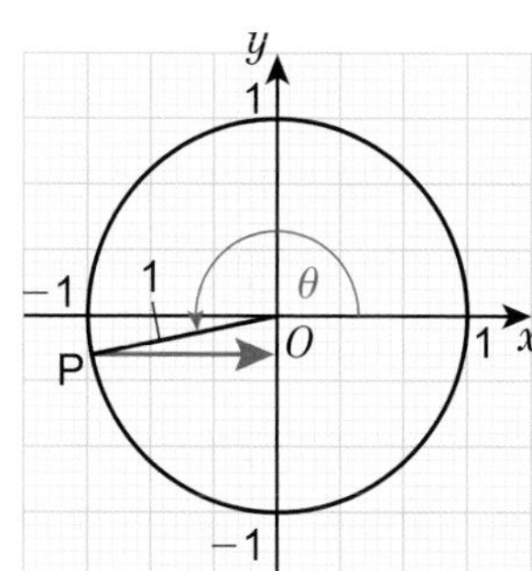

d

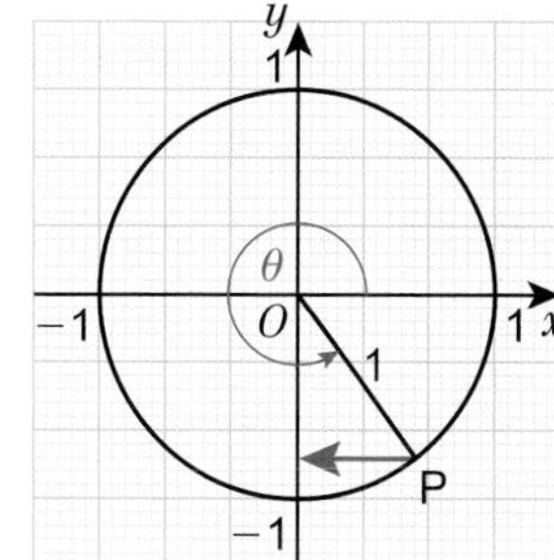

4 a What is the largest value that $\sin\theta$ can take?

b What is the smallest value that $\sin\theta$ can take?

c Find two values of θ so that $\sin\theta = 0$.

Q4 hint Look at the diagrams in **Q3**.

5 $\sin 30° = 0.5$.

Use the diagram to find an obtuse angle θ such that $\sin\theta = 0.5$.

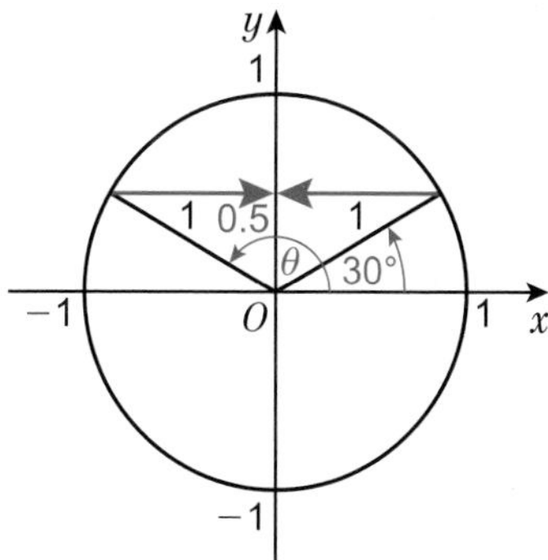

6 As θ increases from 0° to 90°, $\sin\theta$ increases from 0 to 1.
Copy and complete these statements in the same way.

a As θ increases from 90° to 180°, $\sin\theta$ ....

b As θ increases from 180° to 270°, $\sin\theta$ ....

c As θ increases from 270° to 360°, $\sin\theta$ ....

7 Here is the graph of $y = \sin x$ for $0° \leqslant x \leqslant 180°$.

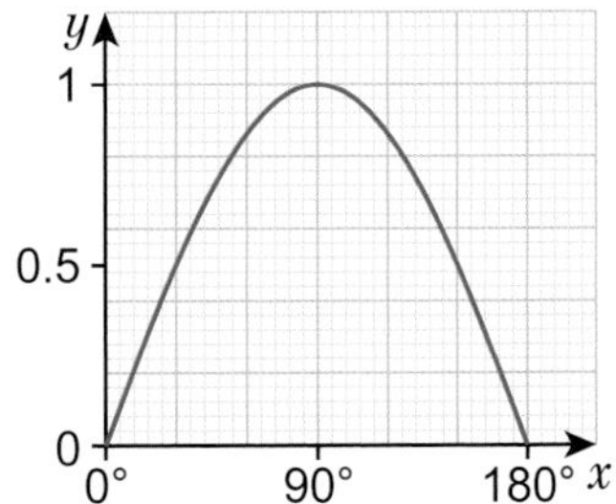

a Use the graph to find
i $\sin 90°$ ii $\sin 75°$.

b Describe the symmetry of the curve.

Q7b hint The graph of $y = \sin x$ for $0° \leqslant x \leqslant 180°$ is symmetrical about $x = \square°$.

c Use the graph to check your answer to **Q6**.

d Copy and complete, inserting numbers greater than 90.
i $\sin 60° = \sin\square°$ ii $\sin 45° = \sin\square°$
iii $\sin 0° = \sin\square°$ iv $\sin 30° = \sin\square°$

Q7d i hint Draw a horizontal line through sin 60° on the graph. Where else does it cross the graph?

e Use the graph to give estimates for the solutions to $\sin x = 0.25$.

Q7e hint Give both values of x.

8 Here is a sketch of the graph of $y = \sin x$ for $0° \leqslant x \leqslant 360°$.

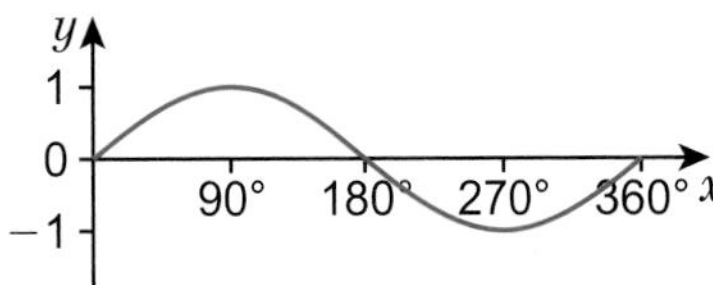

Describe the symmetry of the curve.

9 The graph of $y = \sin x$ repeats every 360° in both directions.

a Sketch the graph of $y = \sin x$ for $0° \leqslant x \leqslant 540°$.

Q9a hint Include x values 0°, 90°, 180°, 270°, 360°, 450°, 540° and y values 1, 0.5, 0, −0.5, −1.

b Use your sketch to find
i $\sin 540°$ ii $\sin 450°$.

c The exact value of $\sin 60°$ is $\frac{\sqrt{3}}{2}$. Write down the exact value of
i $\sin 420°$ ii $\sin 480°$.

d Explain how you worked out your answers to part **c**.

10 a Write down four values of x such that $\sin x = -0.5$.

b Write down four values of x such that $\sin x = -\frac{\sqrt{3}}{2}$.

c Check each of your answers using your calculator.

11 **Exam-style question**

Here is a sketch of $y = \sin x$.

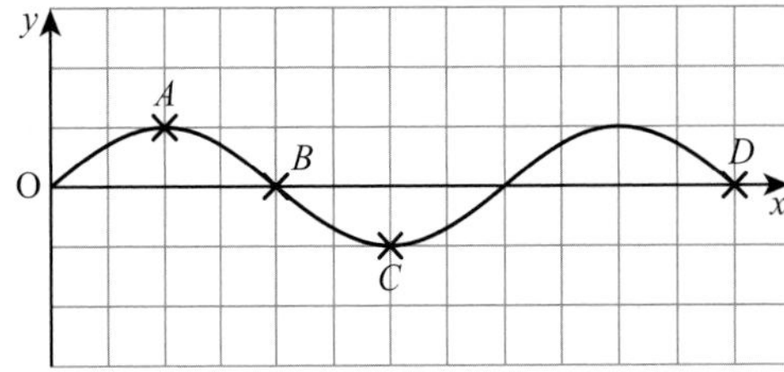

Write down the coordinates of each of the labelled points. **(4 marks)**

Q11 strategy hint Check your answers by seeing if $\sin x = y$

Example 2

Solve the equation $5\sin x = 3$ for values of x in the interval 0° to 540°.

$5\sin x = 3$

$\sin x = \frac{3}{5}$ — Divide both sides of the equation by 5.

$x = \sin^{-1}\left(\frac{3}{5}\right)$

$x = 36.9°$ to 1 d.p. — Use $\sin^{-1}$ to find one value of x from your calculator.

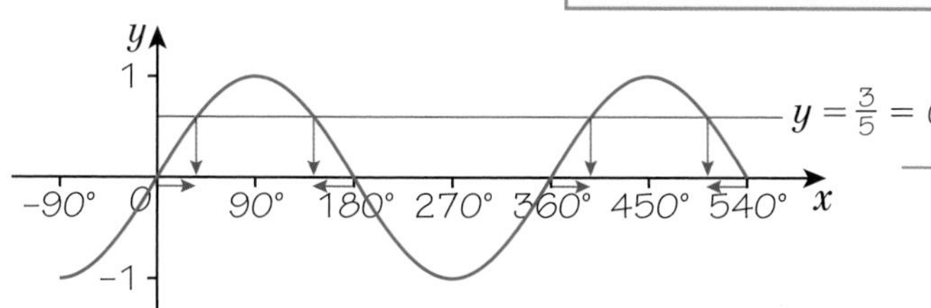

Sketch the graph of $y = \sin x$ for the interval 0° to 540°. Use the graph to find the other values.

From the graph, the other values of x are:

$180° - 36.9° = 143.1°$

$360° + 36.9° = 396.9°$

$540° - 36.9° = 503.1°$

12 Solve the equation $8\sin x = 2.5$ for all values of x in the interval 0° to 720°.

13 Solve the equation $6\sin\theta = 5$ for all values of θ in the interval 0° to 720°.

14 **Reflect** Did you use the worked example to help you answer **Q12** and **Q13**? How does a worked example help you understand and answer questions like this?

13.3 Graph of the cosine function

Objectives

- Understand how to find the cosine of any angle.
- Know the graph of the cosine function and use it to solve equations.

Did you know?

Ultrasound scanners use trigonometry to construct pictures of babies in the womb.

Fluency

What is the exact value of
cos 30°   cos 45°   cos 60°   cos 90°?

1 Find the value of $\cos\theta$ in each triangle.

a

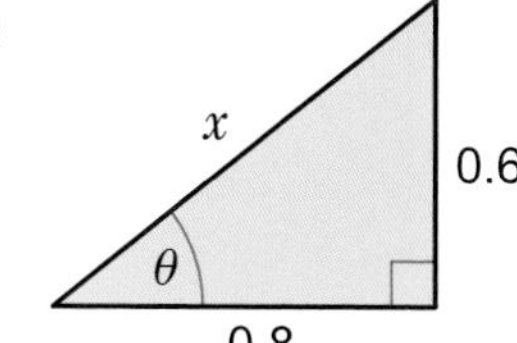

b

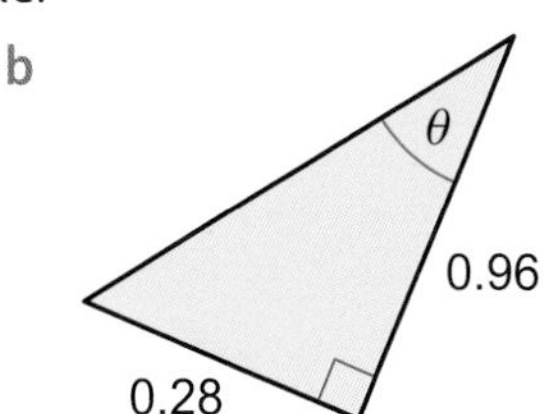

Q1a hint First find the value of x.

Warm up

Key point 2

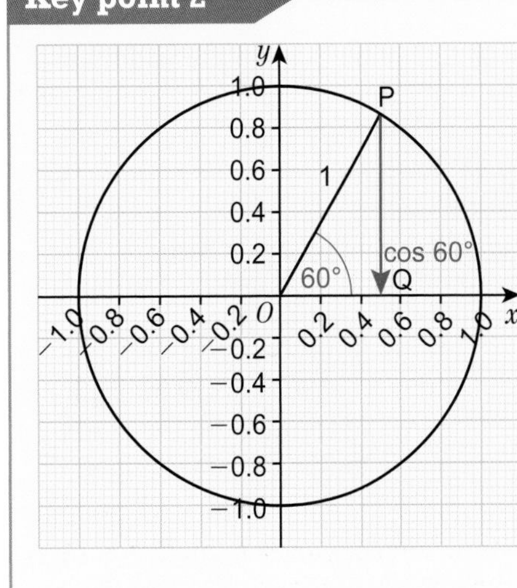

The diagram shows a circle of radius 1 unit with centre at (0, 0).

$\cos 60° = \frac{OQ}{1} = OQ = 0.5$

The length of OQ gives the **cosine** of the angle. This is shown on the horizontal axis by the position of the arrow.

You can find the cosine of any angle using this method.

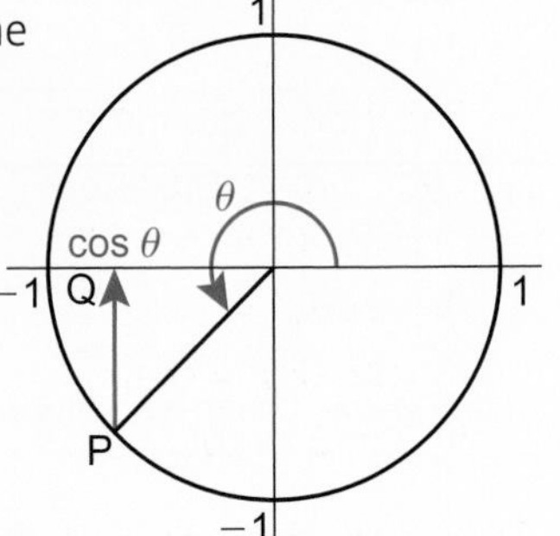

2 Find the value of cos θ in each diagram.

a

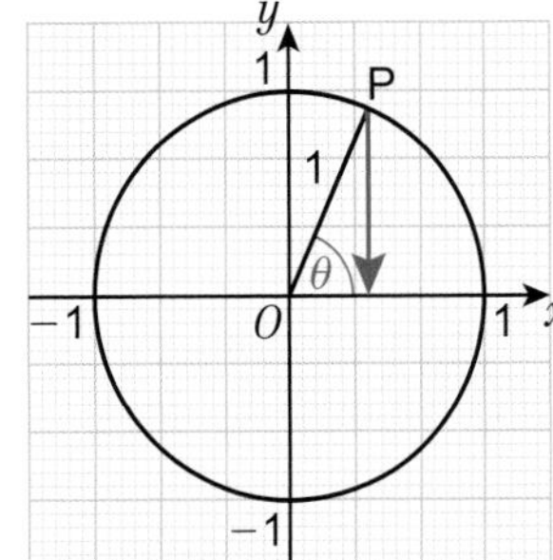

b

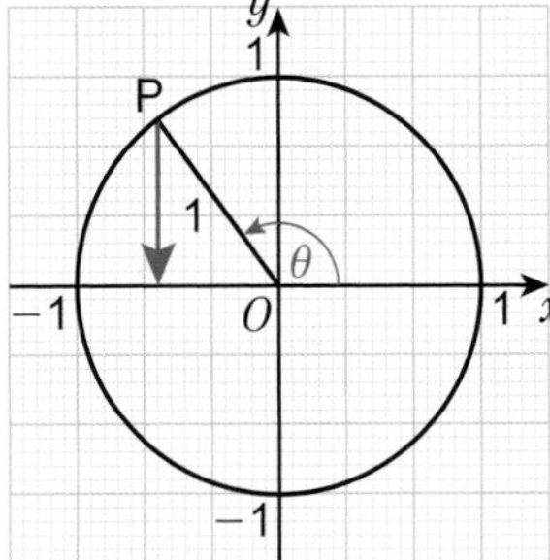

c

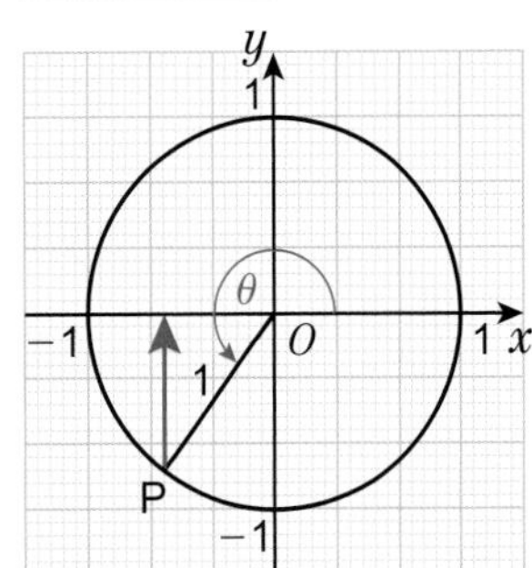

d

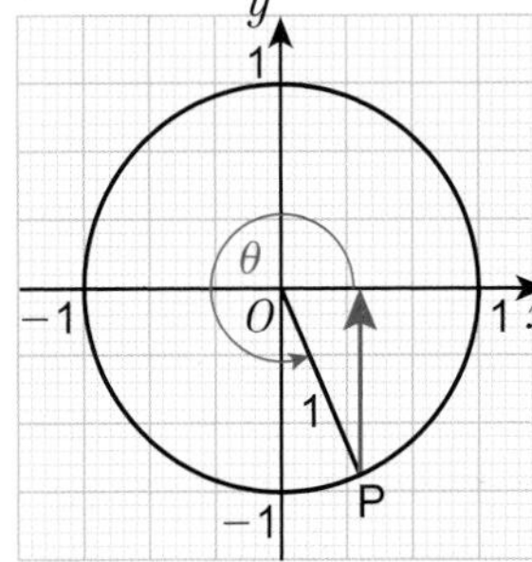

3 cos 60° = 0.5.
Use the diagram to find a reflex angle θ such that $\cos\theta = 0.5$.

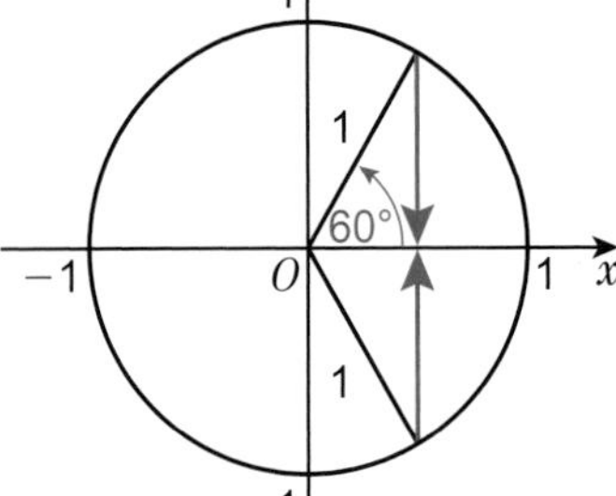

4 $\cos 30° = \frac{\sqrt{3}}{2}$

a Find a reflex angle θ such that $\cos\theta = \frac{\sqrt{3}}{2}$

b Find an obtuse angle such that $\cos\theta = -\frac{\sqrt{3}}{2}$

Q4 hint Draw a diagram like the one in **Q3** and make use of symmetry.

5 As θ increases from 0° to 90°, $\cos\theta$ decreases from 1 to 0.
Copy and complete these statements in the same way.
 a As θ increases from 90° to 180°, $\cos\theta$ ....
 b As θ increases from 180° to 270°, $\cos\theta$ ....
 c As θ increases from 270° to 360°, $\cos\theta$ ....

6 Here is the graph of $y = \cos x$ for $0° \leqslant x \leqslant 360°$.

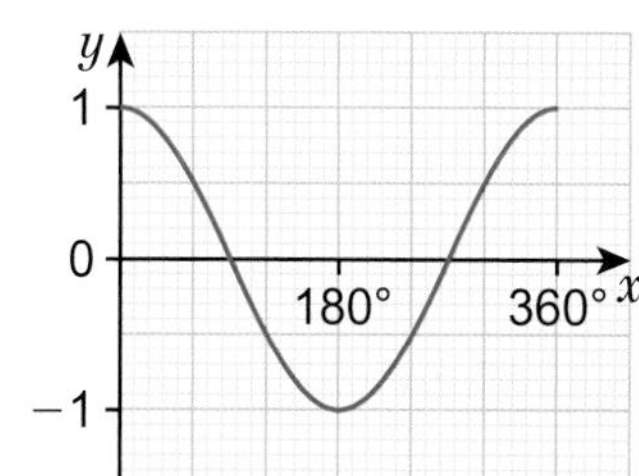

 a Use the graph to find
  i $\cos 120°$
  ii $\cos 180°$.
 b Describe the symmetry of the curve.
 c Copy and complete:
  i $\cos 60° = \cos \square°$
  ii $\cos 90° = \cos \square°$
  iii $\cos 120° = \cos \square°$
  iv $\cos 0° = \cos \square°$

Q6c i hint Draw a horizontal line through $\cos 60°$ on the graph. Where does it cross the graph?

7 The graph of $y = \cos x$ repeats every 360° in both directions.
 a Sketch the graph of $y = \cos x$ for $0° \leqslant x \leqslant 720°$.
 b Use your graph to find
  i $\cos 300°$ ii $\cos 480°$.
 c The exact value of $\cos 30°$ is $\frac{\sqrt{3}}{2}$.
 Write down the exact value of
  i $\cos 390°$ ii $\cos 210°$.

Q7a hint Include x values 0°, 90°, 180°, 270°, 360°, …, 720° and y values 1, 0.5, 0, −0.5, −1.

8 Use your sketch from **Q7** to find four values of x such that
 a $\cos x = 0.5$ b $\cos x = -\frac{\sqrt{3}}{2}$
Check your answers using a calculator.

9 **Exam-style question**

The diagram shows a sketch of the graph $y = \cos x°$

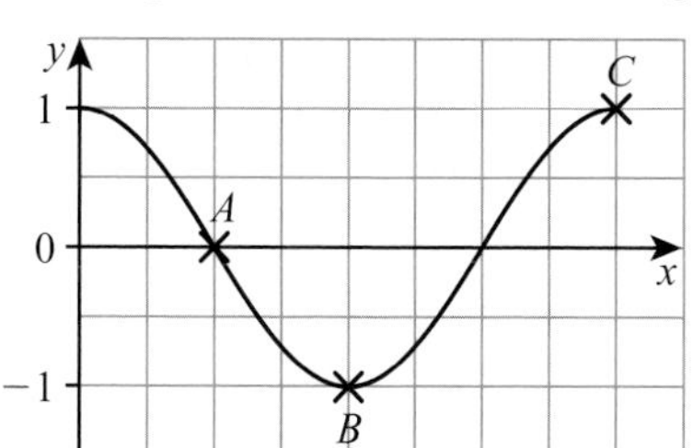

Write down the coordinates of points A, B and C. **(3 marks)**

10 $6\cos x = 4.86$
 a Use $\cos^{-1}$ on your calculator to find one value of x.
 b Sketch the graph of $y = \cos x$ for the interval 0° to 720°.
 c Use your answers to parts **a** and **b** to solve $6\cos x = 4.86$ for values of x in the interval 0° to 720°.

11 Solve the equation $15\cos\theta = -6.8$ for values of θ in the interval 0° to 720°.

13.4 The tangent function

Objectives

- Understand how to find the tangent of any angle.
- Know the graph of the tangent function and use it to solve equations.

Did you know?

Astronomers use trigonometry to predict the positions of comets.

Fluency

What is the exact value of
tan 30°   tan 45°   tan 60°?

Warm up

1 a Find the value of x in this triangle.
b Find the value of $\tan\theta$.

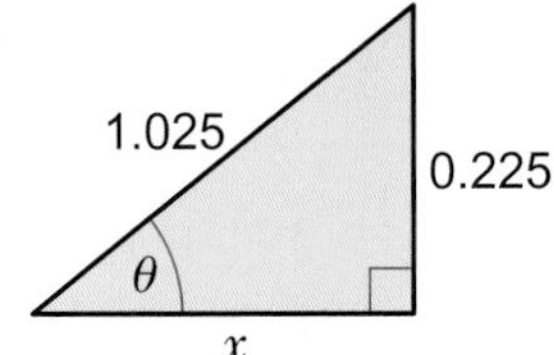

Key point 3

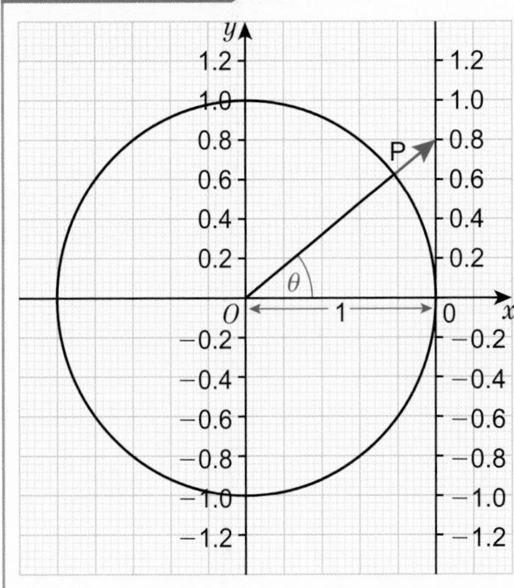

The diagram shows a circle of radius 1 unit with centre at (0, 0).

$\tan\theta = \frac{0.8}{1} = 0.8$

Extending OP to hit the vertical tangent line gives the value of $\tan\theta$.

You can find the **tangent** of any angle using this method except for angles of the form $90° \pm 180n°$

Unlike sine and cosine, the tangent can take *any* value, positive or negative, not just values between −1 and 1.

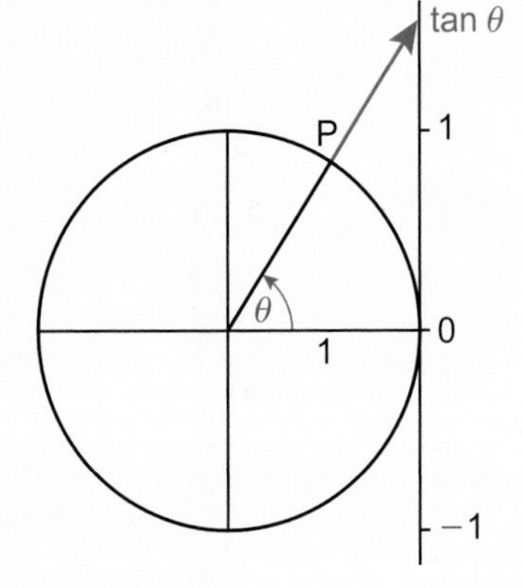

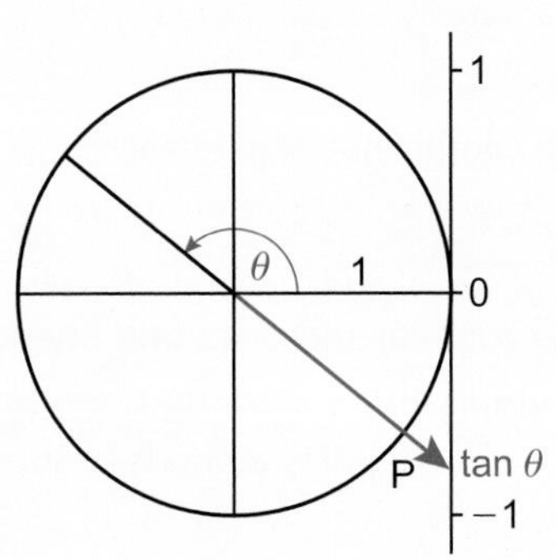

2 Find the value of $\tan\theta$ in each diagram.

a

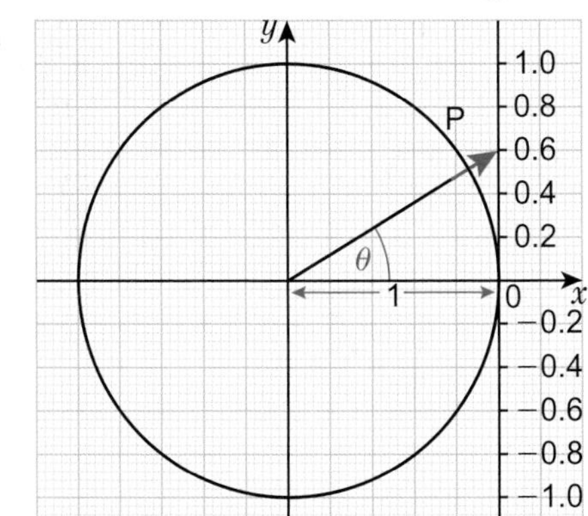

b

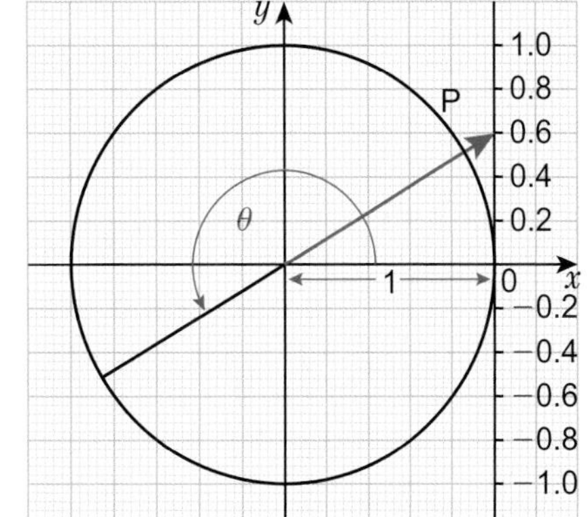

c

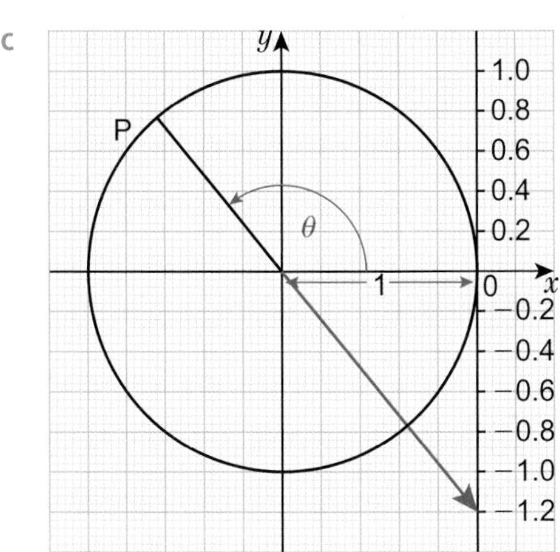

d

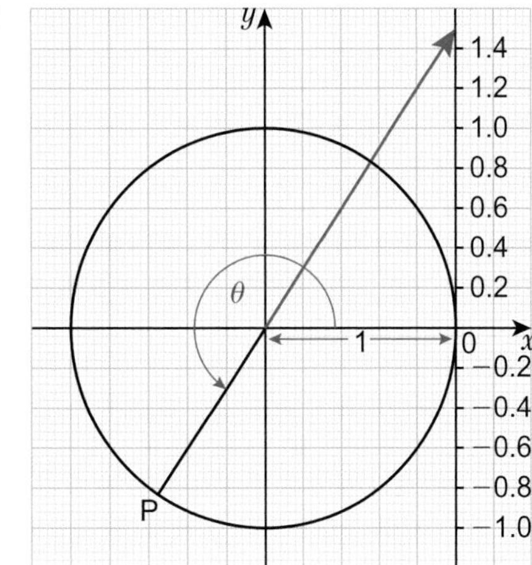

3 $\tan 60° = \sqrt{3}$.
Use the diagram to find a reflex angle θ such that $\tan\theta = \sqrt{3}$.

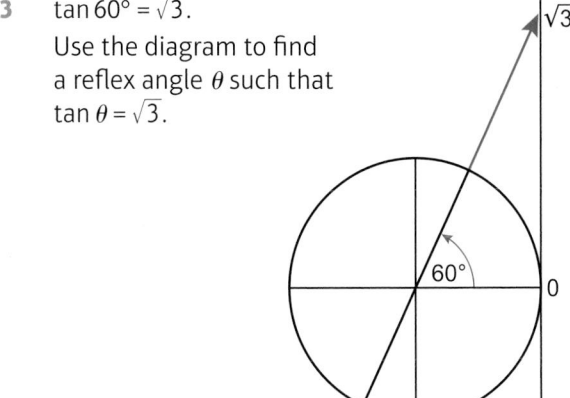

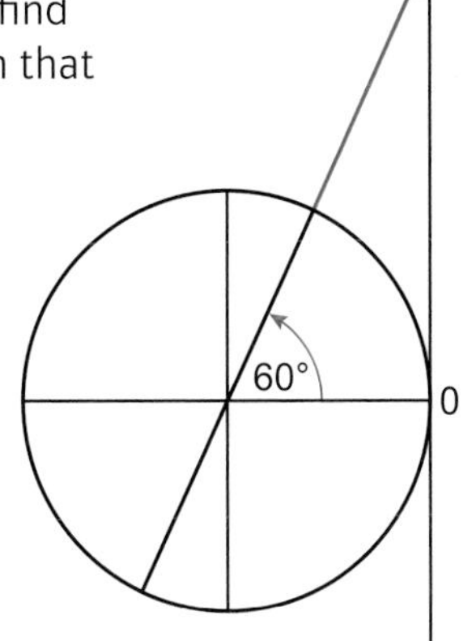

4 $\tan 45° = 1$.

a Find a reflex angle θ such that $\tan\theta = -1$

b Find an obtuse angle such that $\tan\theta = -1$

Reflect The hint suggested drawing a diagram to help you answer this question. Did it help? How?

> **Q4 hint** Draw a diagram like the one in **Q3** and use symmetry.

5 As θ increases from 0° to 90°, $\tan\theta$ increases from 0 to infinity.
Copy and complete these statements in the same way.

a As θ decreases from 180° to 90°, $\tan\theta$ ....

b As θ increases from 180° to 270°, $\tan\theta$ ....

c As θ decreases from 360° to 270°, $\tan\theta$ ....

Discussion $\tan\theta$ is not defined when θ is 90° or 180°, for example. Why do you think this is?

6 Here is the graph of $y = \tan x$ for $0° \leqslant x \leqslant 360°$.

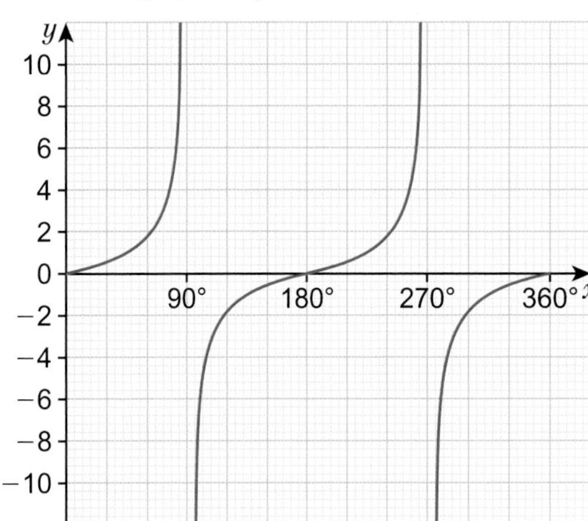

a How often does the graph repeat?

b Use the graph to estimate the value of

i $\tan 60°$ ii $\tan 300°$.

c Describe the symmetry of the curve.

d Copy and complete, inserting numbers greater than 180.

i $\tan 60° = \tan\square°$ ii $\tan 100° = \tan\square°$ iii $\tan 120° = \tan\square°$

7 a Sketch the graph of $y = \tan x$ for $0° \leqslant \theta \leqslant 540°$.

b Use your sketch to find

i $\tan 540°$ ii $\tan 405°$.

c The exact value of $\tan 60°$ is $\sqrt{3}$.
Write down the exact value of

i $\tan 240°$ ii $\tan 120°$.

d Explain how you worked out your answers to part **c**.

8 a Write down four values of x such that $\tan x = 1$.

b Write down four values of x such that $\tan x = -1$.

c Check your answers using a calculator.

9 $3 \tan x = 11$

a Use $\tan^{-1}$ on your calculator to find one value of x.

b Sketch the graph of $y = \tan x$ for the interval 0° to 540°.

c Use your answers to parts **a** and **b** to solve $3 \tan x = 11$ for values of x in the interval 0° to 540°.

10 Solve the equation $4 \tan \theta = 15.7$ for values of θ in the interval $0° \leqslant x \leqslant 720°$.

11 **Exam-style question**

a Sketch the graph of $y = \tan x$ in the interval 0° to 720°.

b Given that $\tan 30° = \frac{1}{\sqrt{3}}$ solve the equation $3 \tan x = \sqrt{3}$ in the interval 0° to 720°. **(4 marks)**

Exam hint
You are expected to use your sketch from part **a** to solve the equation in part **b**.

13.5 Calculating areas and the sine rule

Objectives

- Find the area of a triangle and a segment of a circle.
- Use the sine rule to solve 2D problems.

Did you know?

Sat navs use trigonometry to calculate the position of a vehicle.

Fluency

Calculate the area of each triangle.

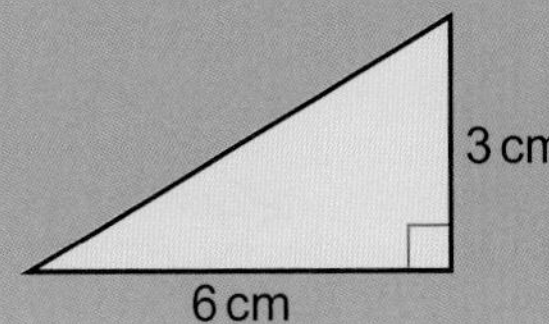

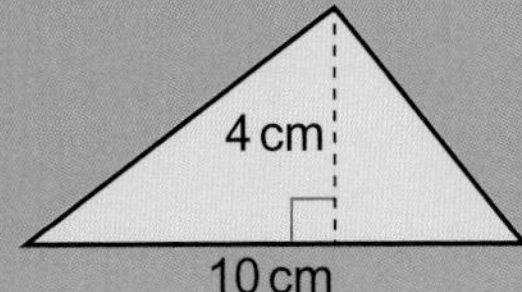

1 Write a formula for calculating the area, A, of each shape.

a 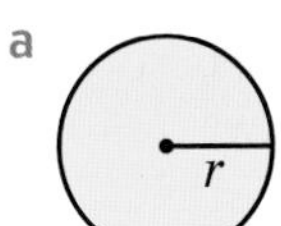

b 

c 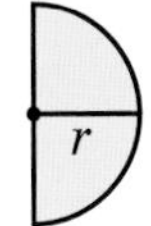

d 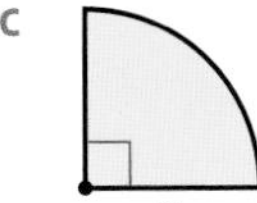

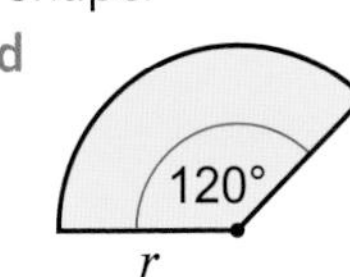

2 Calculate the perpendicular height, h, of this triangle.

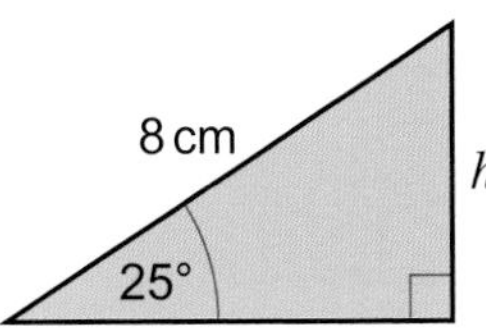

3 a Write h, the perpendicular height of the triangle, in terms of p and θ.

b Write a formula in terms of p and θ to calculate the area of the triangle.

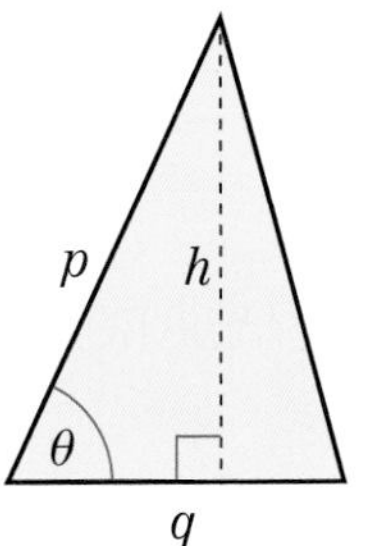

Q3a hint You need to use trigonometry as in **Q2**.

Key point 4

The **area** of this triangle = $\frac{1}{2}ab\sin C$.

a is the side opposite angle A.

b is the side opposite angle B.

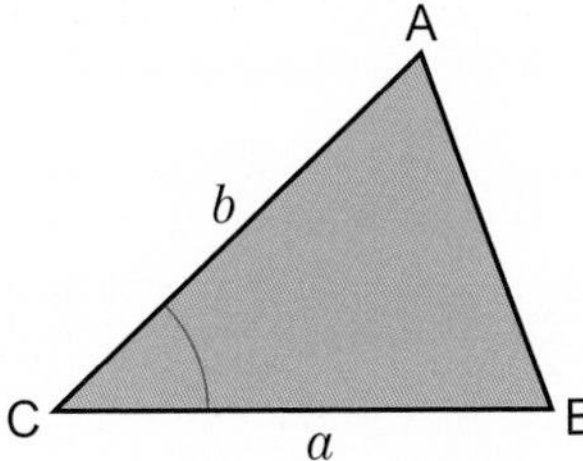

4 Find the area of each triangle.

a 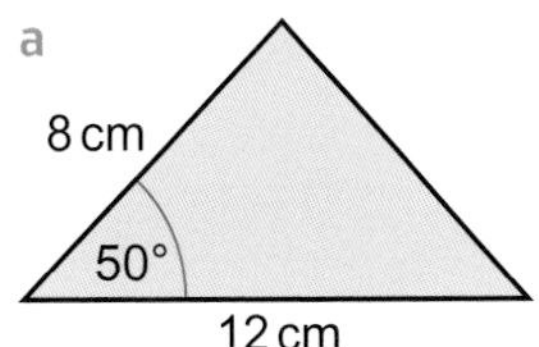

b 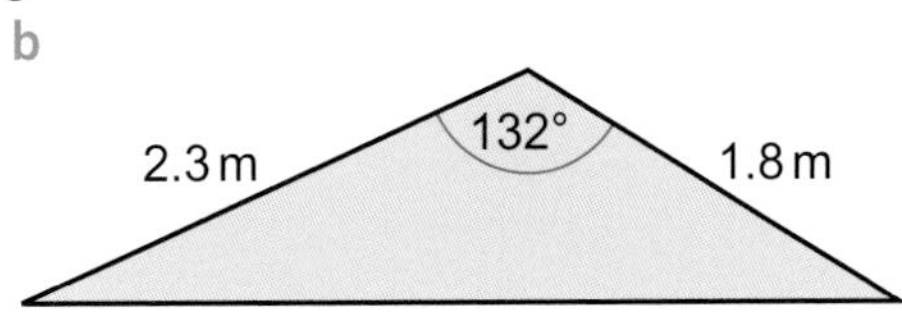

Q4 hint First label vertex C (the given angle). Then label vertices A and B, and their opposite sides a and b.

5 The area of triangle ABC is $38.8\,cm^2$.
Work out the length of AC.

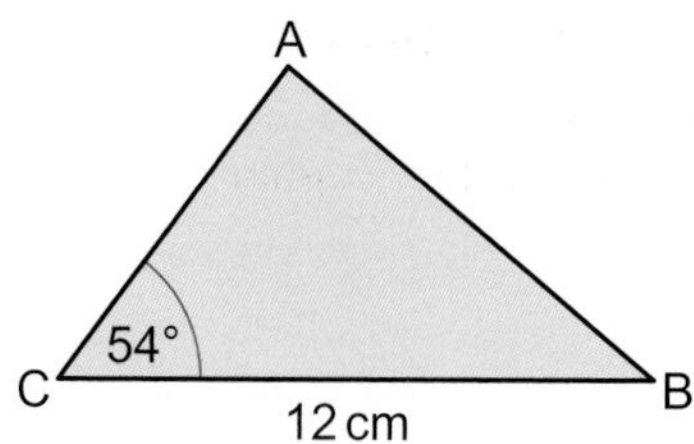

6 a Find the area of triangle AOB in this circle.

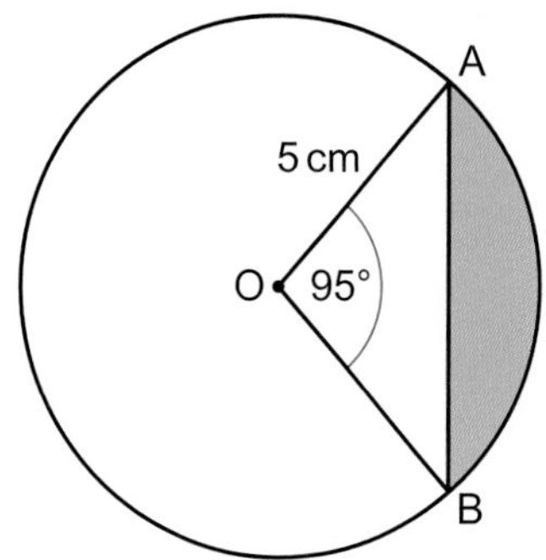

Q6b hint

95°

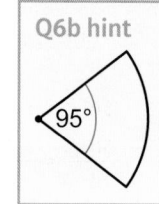

b Find the area of the sector AOB.

c Find the area of the shaded segment of the circle.

Q6c hint How can you use your answers to parts **a** and **b** to find the answer to part **c**?

7 **Exam-style question**

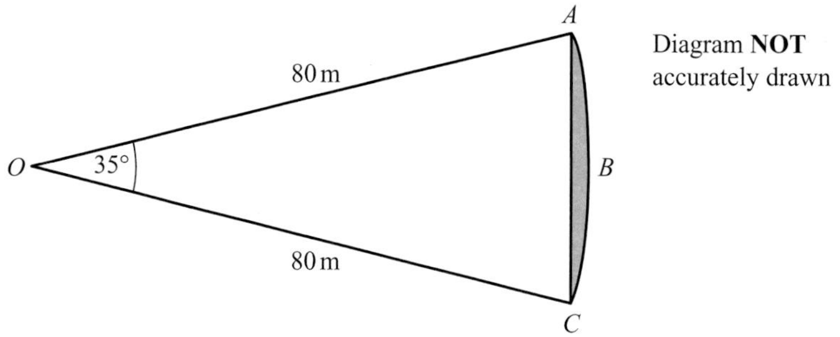

ABC is an arc of a circle centre *O* with radius 80 m.
AC is a chord of the circle.
Angle *AOC* = 35°.
Calculate the area of the shaded region.
Give your answer correct to 3 significant figures. **(5 marks)**

March 2012, Q23, 1380/4H

Q7 strategy hint Find the area of sector *OABC* and the area of triangle *OAC*.

8 **Problem-solving**

a Calculate angle AOB. Give your answer correct to 1 decimal place.

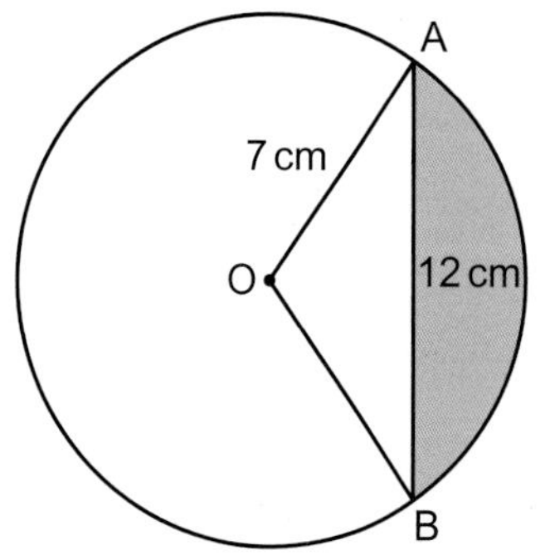

Q8a hint Split the triangle into two right-angled triangles.

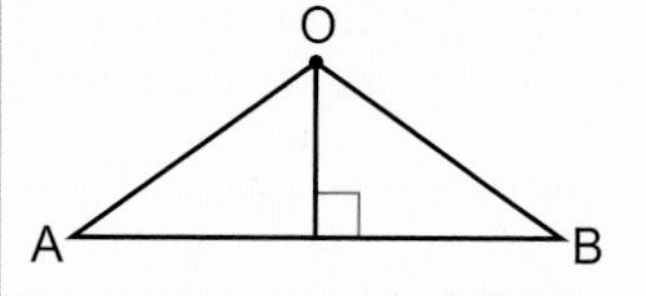

b Work out the area of the shaded segment. Give your answer correct to 3 significant figures.

9 **Problem-solving** In the diagram,
O is the centre of the circle.
Work out the area of the shaded segment.
Give your answer correct to 3 significant figures.

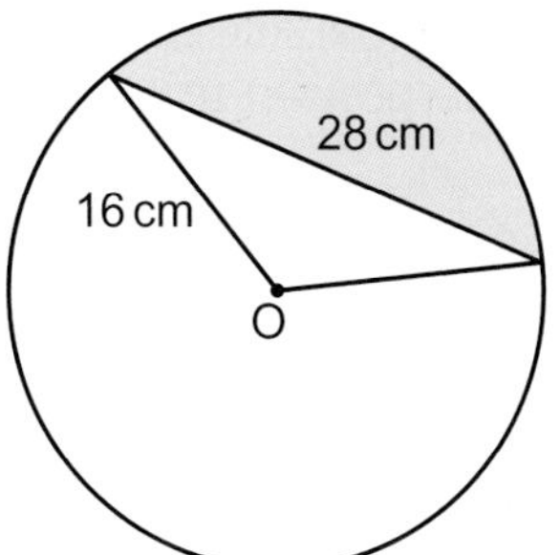

Key point 5

The **sine rule** can be used in any triangle.

- $\frac{a}{\sin A} = \frac{b}{\sin B} = \frac{c}{\sin C}$ Use this to calculate an unknown *side*.
- $\frac{\sin A}{a} = \frac{\sin B}{b} = \frac{\sin C}{c}$ Use this to calculate an unknown *angle*.

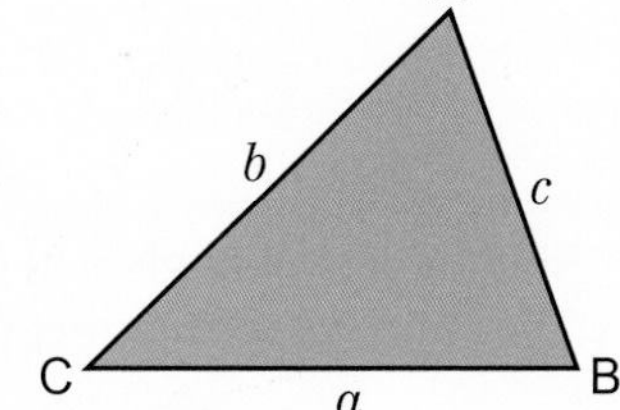

To use the sine rule you need to know one angle and the opposite side. Then:

- if you know another *angle*, you can work out the length of its opposite *side*
- if you know another *side*, you can work out the size of its opposite *angle*.

Example 3

a Find the value of x.
Give your answer to 3 significant figures.

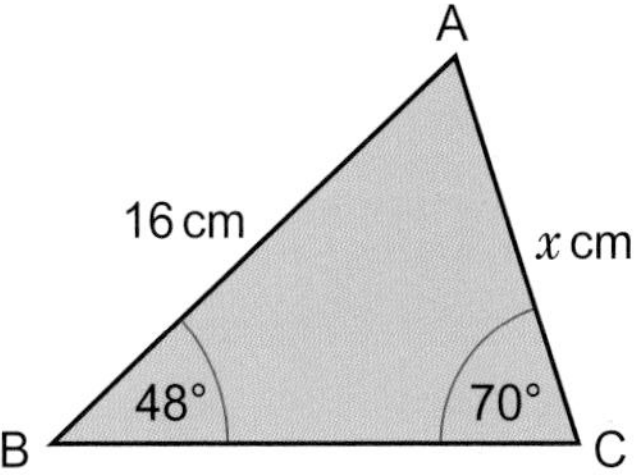

b Find the value of θ.
Give your answer to 1 decimal place.

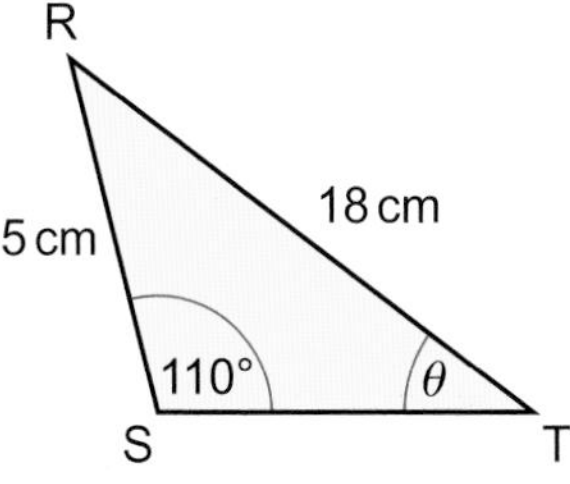

a $\frac{x}{\sin 48°} = \frac{16}{\sin 70°}$ — Use the sine rule $\frac{b}{\sin B} = \frac{c}{\sin C}$

$x = \frac{16 \sin 48°}{\sin 70°} = 12.653...$ — Multiply both sides by sin 48°.

$= 12.7$ cm (3 s.f.)

b $\frac{\sin \theta}{15} = \frac{\sin 110°}{18}$ — Use the sine rule $\frac{\sin T}{t} = \frac{\sin S}{s}$

$\sin \theta = \frac{15 \sin 110°}{18}$ — Multiply both sides by 15.

$\theta = \sin^{-1}\left(\frac{15 \sin 110°}{18}\right)$

$= 51.5°$ (1 d.p.) — Use $\sin^{-1}$ on your calculator.

10 Find the length of the side labelled x in each diagram.
Give your answers correct to 3 significant figures.

a
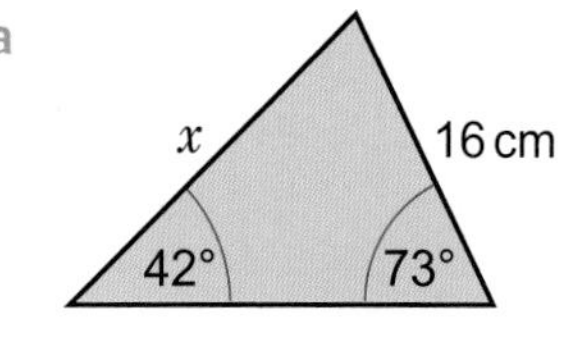

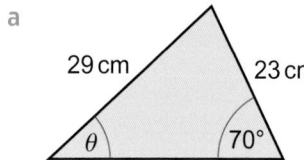

b

c

d
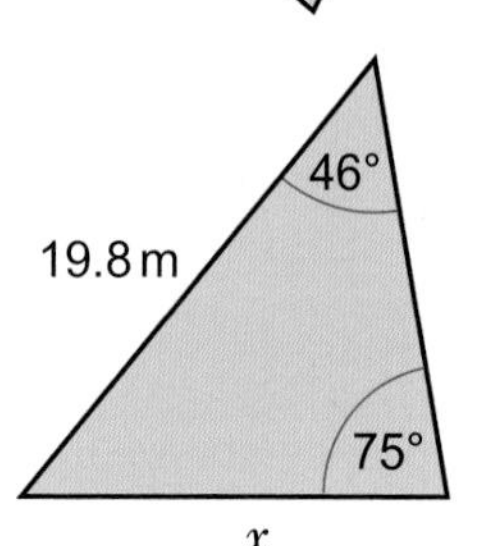

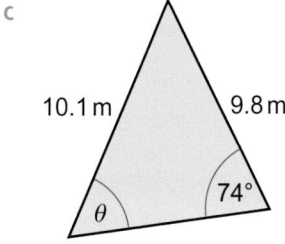

11 Find the size of angle θ in each diagram.
Give your answers correct to 1 decimal place.

a

b
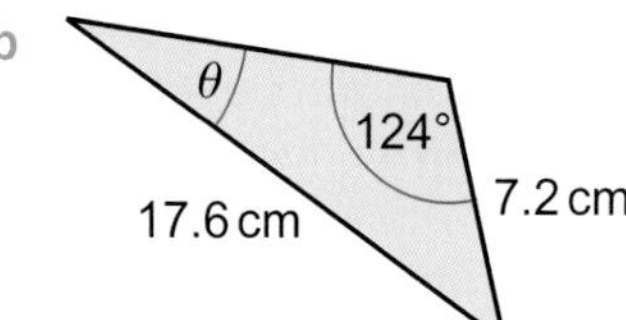

c

d
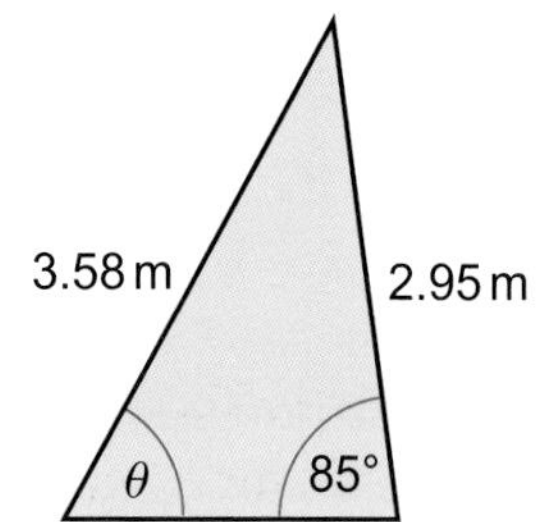

12 a Work out the length of AC.
Give your answer correct to 3 significant figures.
b Work out the size of angle BAC.
Give your answer correct to 1 decimal place.

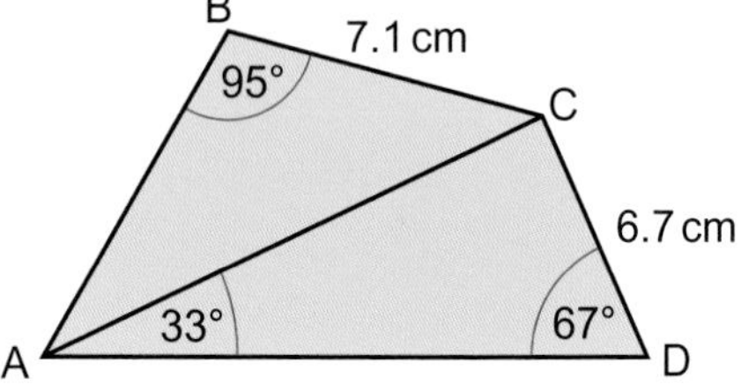

13 In triangle ABC, AB = 8 cm, BC = 6 cm and angle BAC = 40°. Work out the size of angle ACB.

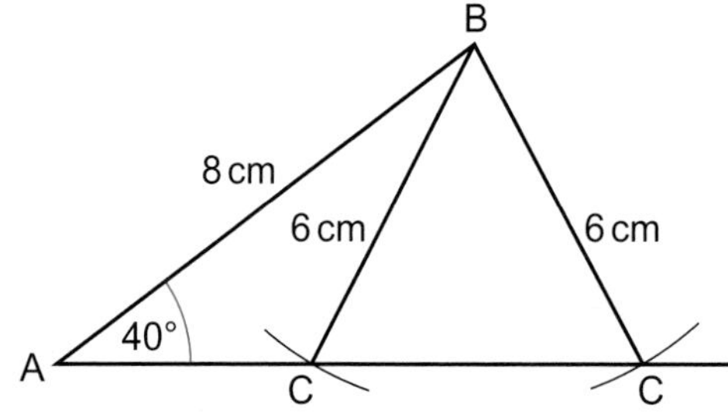

Q13 hint Draw a diagram. Use the sine rule to find the value of sin C. Use the sine graph to find the two possible values.

The diagram shows that there are two possible triangles. Hence there are two possible answers. Give both, correct to 1 decimal place.

14 In triangle XYZ, XY = 12 cm, YZ = 9.5 cm and angle YXZ = 50°. Work out the size of angle XZY.
There are two possible answers. Give each of them correct to 1 decimal place.

13.6 The cosine rule and 2D trigonometric problems

Objectives

- Use the cosine rule to solve 2D problems.
- Solve bearings problems using trigonometry.

Did you know?

Trigonometry was needed to dig the Channel Tunnel. The undersea part is the longest in the world, at 37.9 km. Digging from both sides, the engineers met under the sea and were delighted to find that they were just 2 cm out.

Fluency

Work out the value of

$3 + 4 \times 5$ $\quad$ $9 + 8 - 2 \times 5$

Warm up

1 In the diagram, what is the bearing of

a B from A

b A from B?

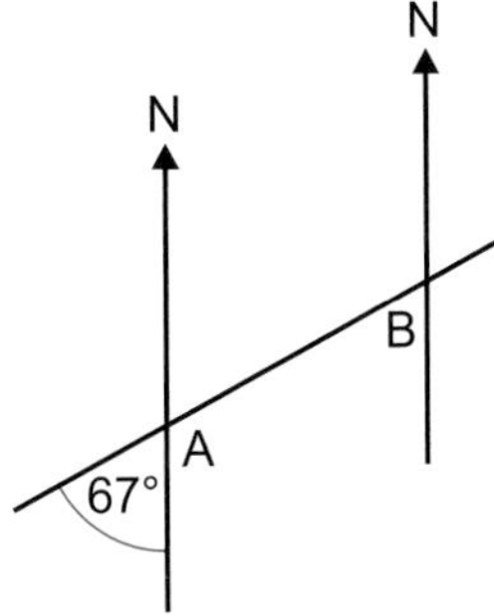

2 Work out the positive value of x when $x^2 = 9^2 + 7^2 - 2 \times 9 \times 7 \times \cos 34°$.
Give your answer correct to 3 significant figures.

3 Find the area of this triangle. Give your answer correct to 3 significant figures.

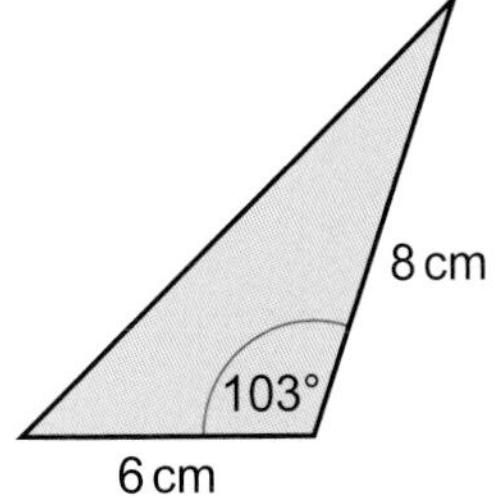

Key point 6

The **cosine rule** can be used in any triangle.

- $a^2 = b^2 + c^2 - 2bc\cos A$ $\quad$ Use this to calculate an unknown *side*.
- $\cos A = \dfrac{b^2 + c^2 - a^2}{2bc}$ $\quad$ Use this to calculate an unknown *angle*.

You can use the cosine rule to find:

- the length of a *side* if you know two sides and the included angle
- an unknown *angle* if you know all three sides.

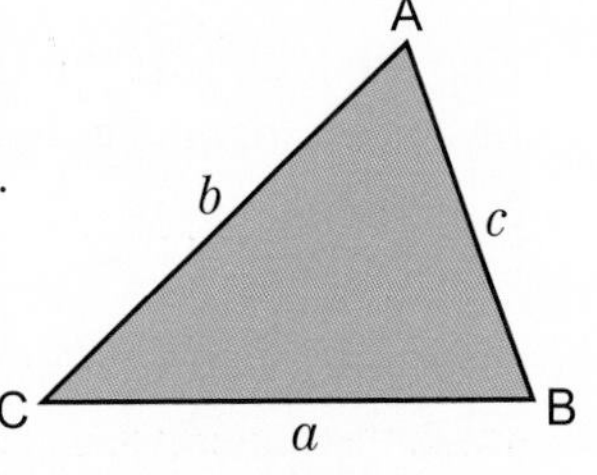

Example 4

a Work out the length of the side labelled x.
Give your answer correct to 3 significant figures.

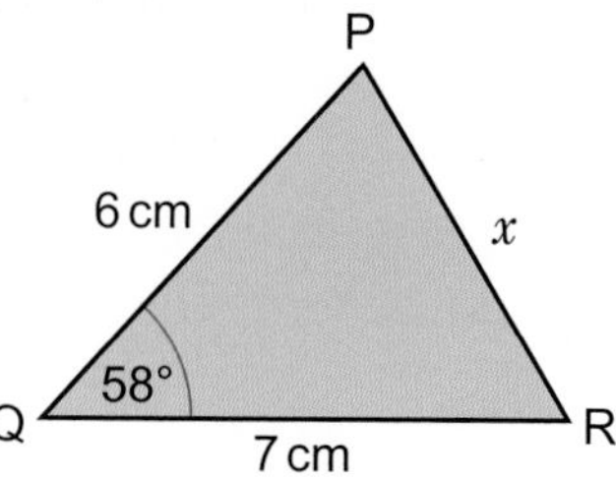

b Work out the size of angle y.
Give your answer correct to 1 decimal place.

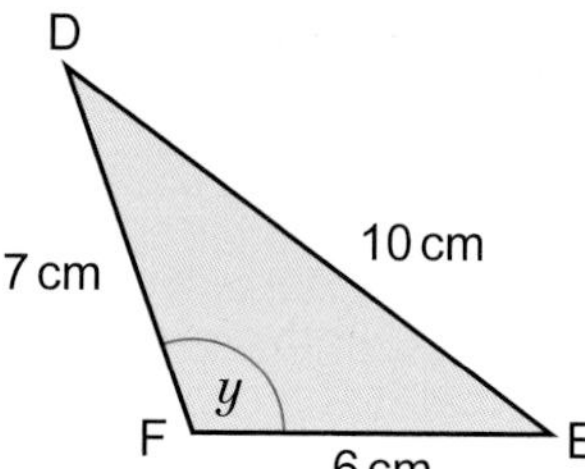

a

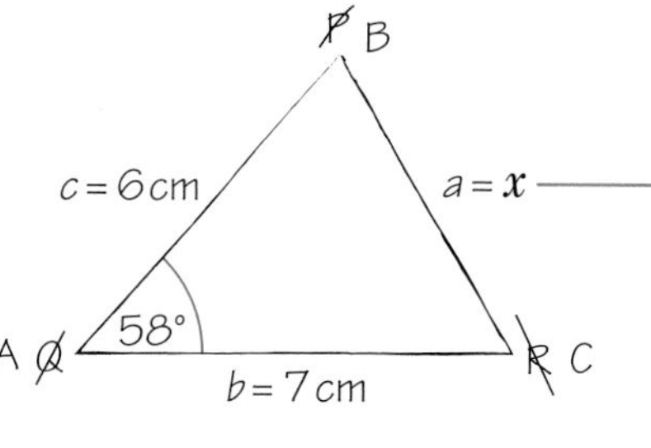

Sketch the triangle. Label the missing side a, and the others b and c.

$a^2 = b^2 + c^2 - 2bc\cos A$

Use the cosine rule to find the side.

$x^2 = 7^2 + 6^2 - 2 \times 7 \times 6 \times \cos 58° = 40.486...$

$x = \sqrt{40.486} = 6.3629... = 6.36\text{ cm (3 s.f.)}$

b

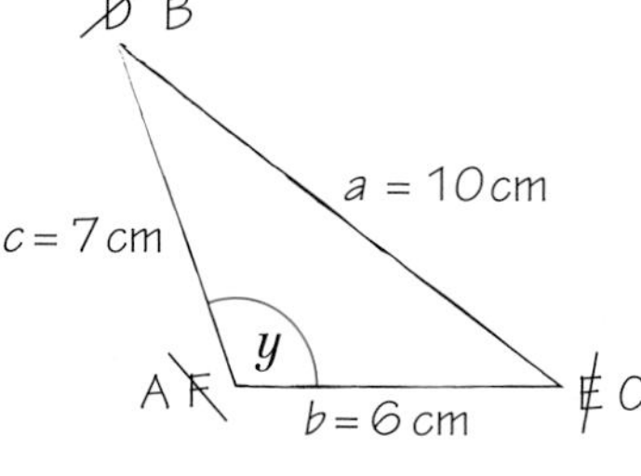

$\cos A = \dfrac{b^2 + c^2 - a^2}{2bc}$

Use the cosine rule to find the angle.

$\cos y = \dfrac{6^2 + 7^2 - 10^2}{2 \times 6 \times 7}$

$y = \cos^{-1}\left(\dfrac{6^2 + 7^2 - 10^2}{2 \times 6 \times 7}\right) = 100.3° \text{ (1 d.p.)}$

4 Find the length of the sides marked with letters in these diagrams.
Give your answers correct to 3 significant figures.

a

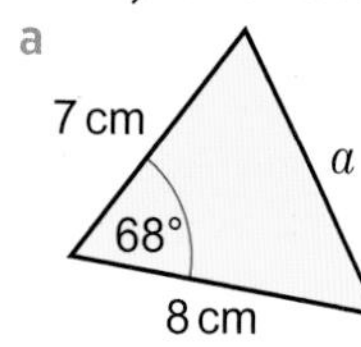

b

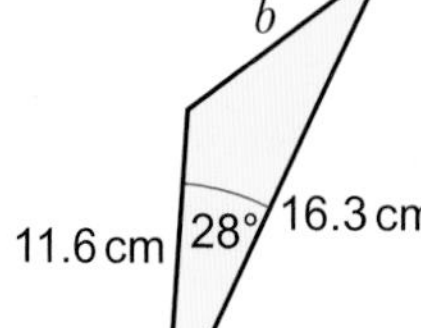

c

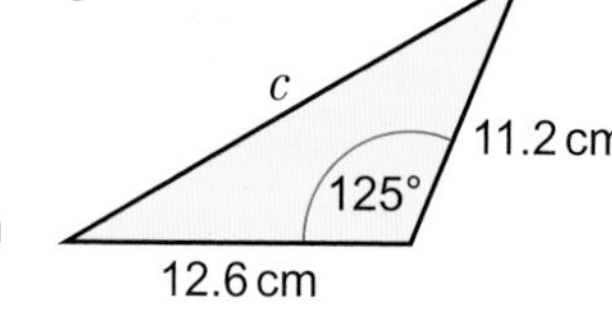

d

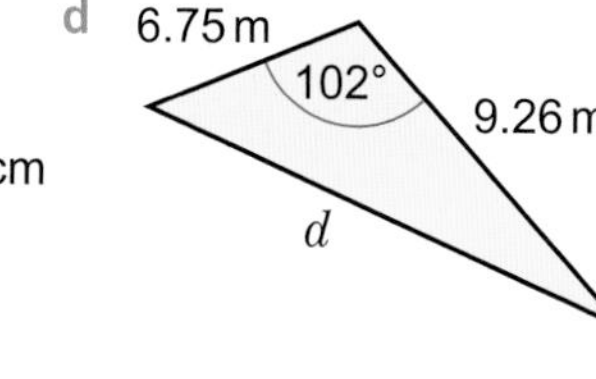

5 Calculate the angles marked with letters in these triangles.
Give your answers correct to 1 decimal place.

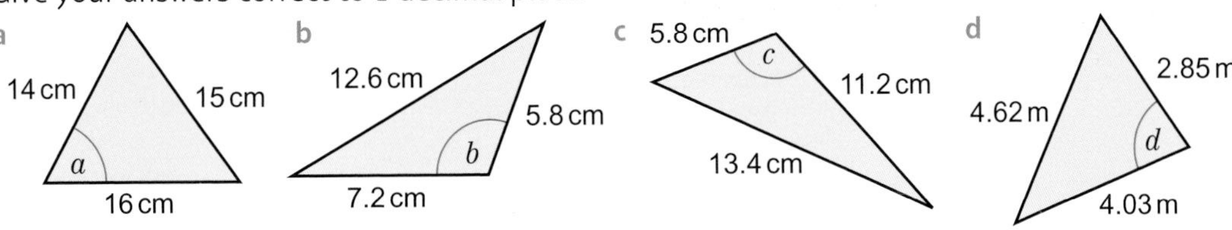

6 In the diagram, O is the centre of the circle of radius 5 cm.
AB is a chord of length 8 cm.
Work out the size of angle AOB.

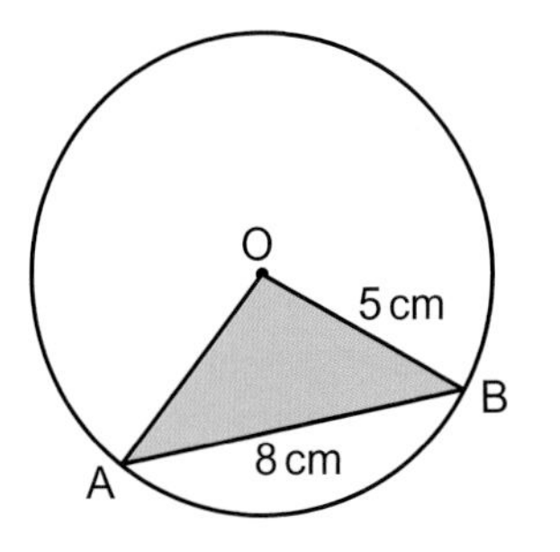

Q6 hint What is the length of OA?

7 **Reasoning**

a Work out the length of PR.
Give your answer correct to 3 significant figures.

b Work out the size of angle QPR.
Give your answer correct to 1 decimal place.

c Work out the area of quadrilateral PQRS.
Give your answer correct to 3 significant figures.

P, 15 cm, Q, 8 cm, 7 cm, 107°, S, 11 cm, R

Reflect In this question you were asked to give answers to 3 significant figures and to 1 decimal place. Explain the difference between giving an answer 'to significant figures' and 'to decimal places'.

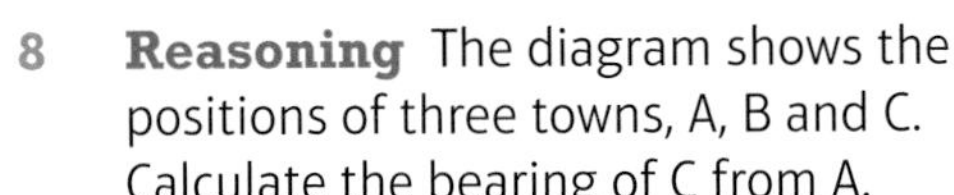

8 **Reasoning** The diagram shows the positions of three towns, A, B and C.
Calculate the bearing of C from A.

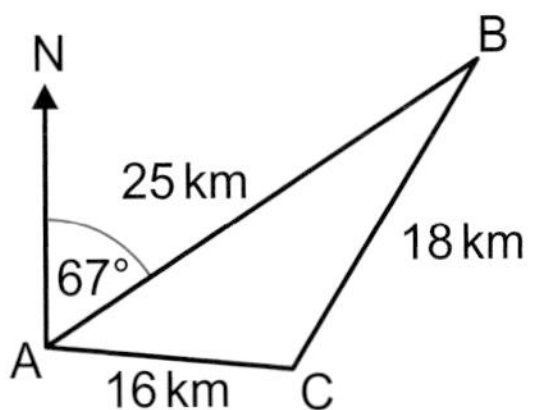

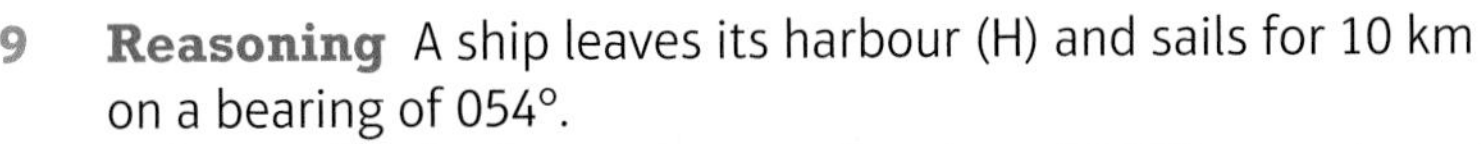

9 **Reasoning** A ship leaves its harbour (H) and sails for 10 km on a bearing of 054°.
It then sails a further 14 km on a bearing of 148° to reach port P.

Q9 hint The north lines are parallel. Use this to find an angle inside the triangle.

a What is the direct distance between H and P?

b What is the bearing of H from P?

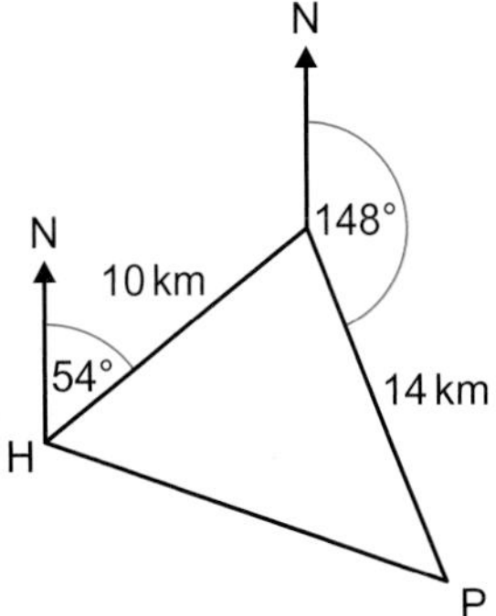

10 Find the size of each lettered angle.

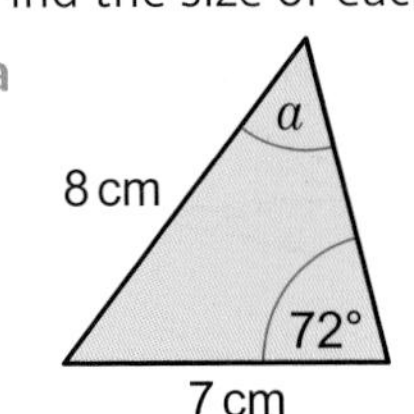

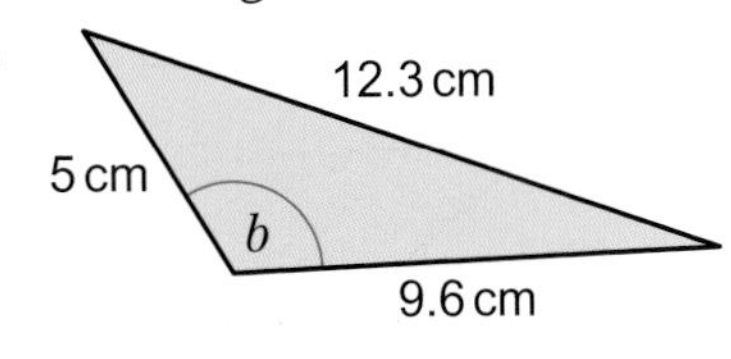

Q10 hint For each one, decide whether to use the sine rule or the cosine rule.

11 **Exam-style question**

ABC is a triangle.

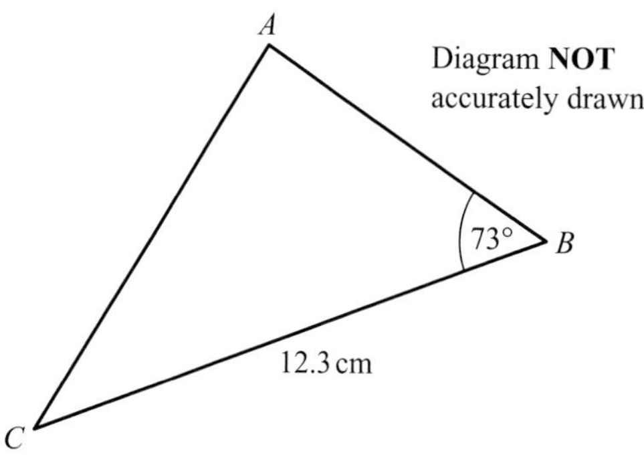

$BC = 12.3\text{ cm}$

Angle $ABC = 73°$

The area of triangle *ABC* is 50 cm^2.

Work out the length of *AC*.

Give your answer correct to 3 significant figures. **(6 marks)**

June 2013, Q19, 5MB3H/01

Exam hint

You need to know the sine rule, the cosine rule and the formula for the area of a triangle. Show any formulae you use in your working.

13.7 Solving problems in 3D

Objectives

- Use Pythagoras' theorem in 3D.
- Use trigonometry in 3D.

Did you know?

Pythagoras' theorem can be extended into three dimensions – and more (if there are any).

Fluency

Which rule would you use to work out x in these diagrams?

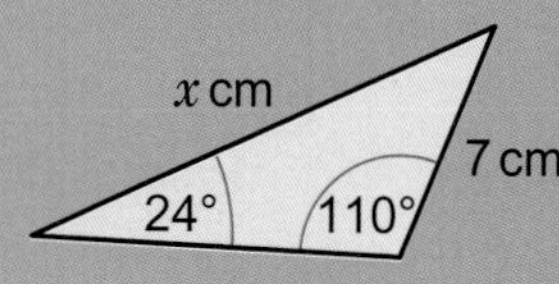

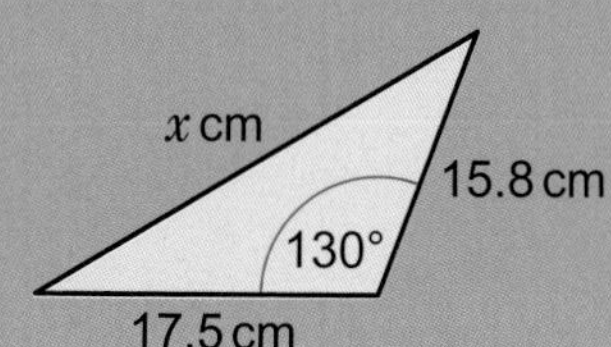

Warm up

1 Find the size of angle θ in each triangle.

a

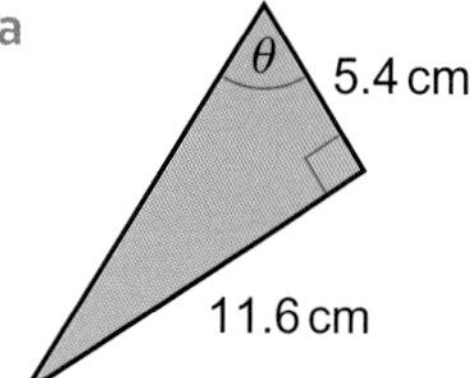

b 9.7 cm, θ, 12.1 cm

c

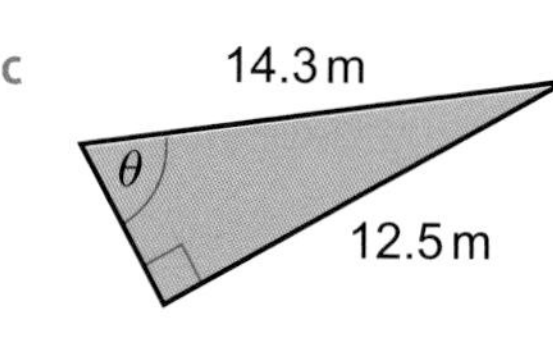

2 Find the value of x in each triangle. Give your answers correct to 3 significant figures.

a

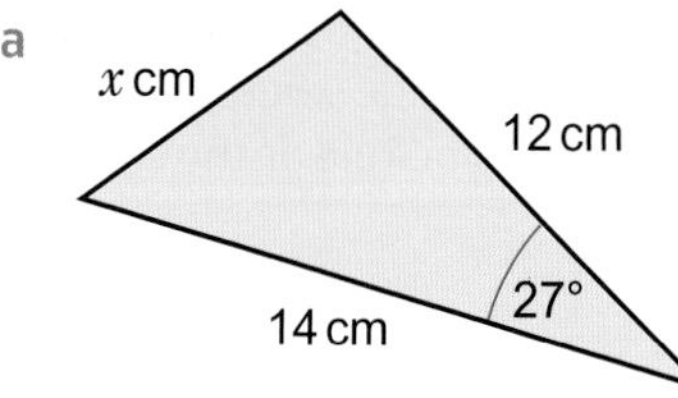

b x cm, 22°, 32.1 cm, 130°

Key point 7

A **plane** is a flat surface. For example, the surface of your desk lies in a horizontal plane; the surface of a wall in your classroom lies in a vertical plane.

In the diagram, BC is perpendicular to the plane WXYZ. Triangle ABC is in a plane perpendicular to the plane WXYZ. θ is the angle between the line AB and the plane WXYZ.

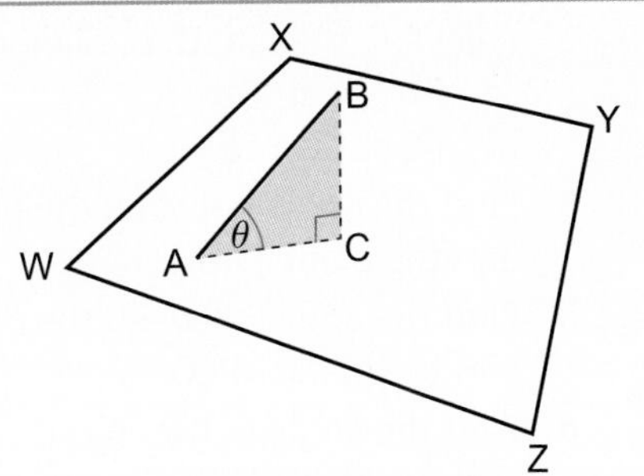

Example 5

a Work out the length of the diagonal, AG, of this cuboid.

Communication hint A **diagonal** is a line joining one vertex, or corner, to another.

b Find the angle that AG makes with the plane EFGH.

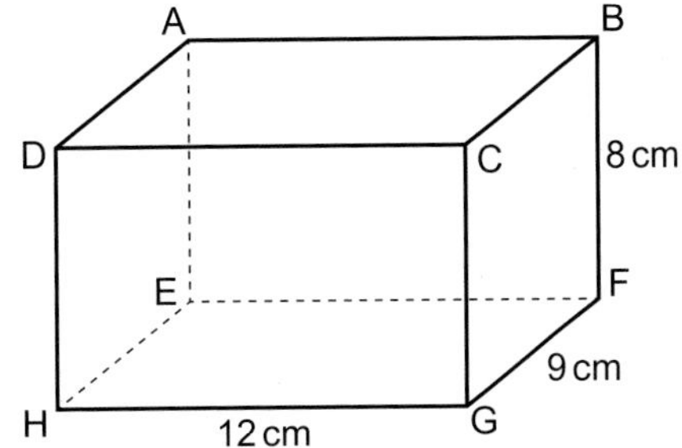

The base EFGH is in a horizontal plane and triangle AEG is in a vertical plane. The length of the diagonal AG is x. The angle that AG makes with EFGH is θ.

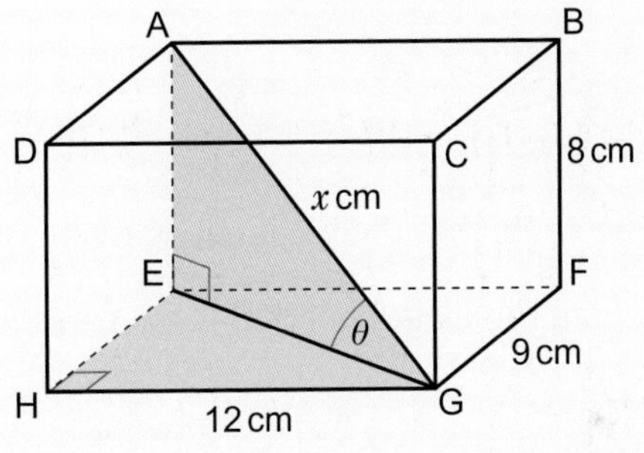

a $EG^2 = 9^2 + 12^2 = 225$

$EG = \sqrt{225} = 15\text{ cm}$

$x^2 = 8^2 + 15^2 = 289$

$x = \sqrt{289} = 17$

The diagonal AG is 17 cm long.

First look at the right-angled triangle EGH. Use Pythagoras' theorem to find length EG.

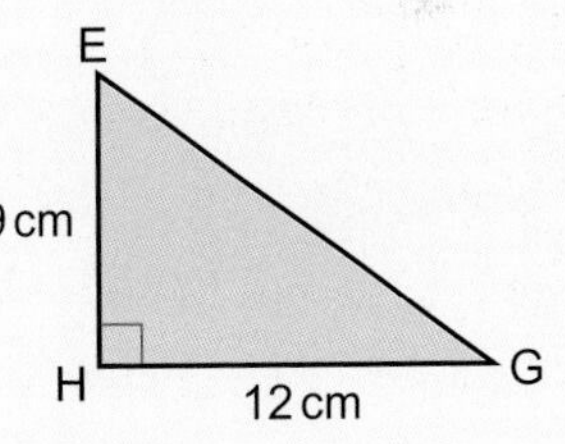

Now look at triangle AEG. Label the length of EG you have just found. Use Pythagoras to work out x.

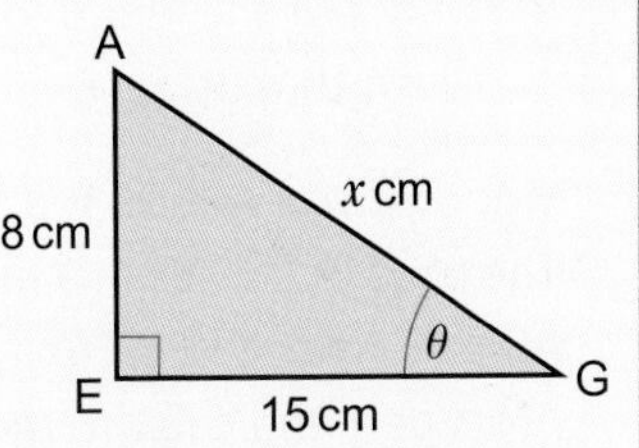

b $\tan\theta = \frac{8}{15}$

$\theta = \tan^{-1}\left(\frac{8}{15}\right)$

$= 28.1°$ (1 d.p.)

The angle that AG makes with the plane EFGH is 28.1° to 1 d.p.

Use the lengths given in the question when you can.

3 **Reasoning** ABCDEFGH is a cuboid.

a Calculate the length of diagonals
   i FH   ii BH
   iii FC   iv CE.

b Find the angle between the diagonal DF and the plane EFGH.

c Find the angle between the diagonal GA and the plane ABCD.

d Find the angle between the diagonal CE and the plane AEHD.

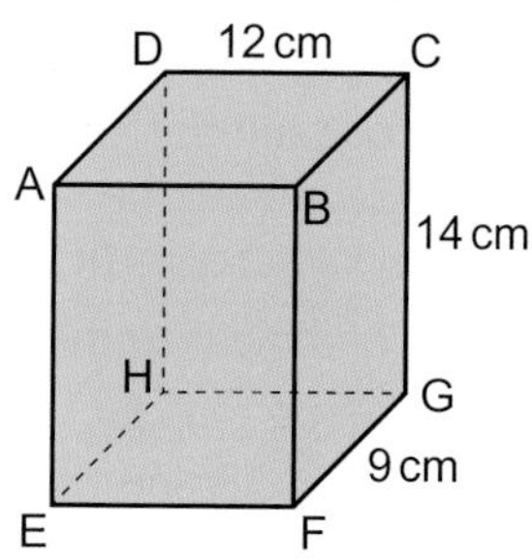

Q3 hint Sketch separate triangles using information from the cuboid.

4 **Problem-solving** In the diagram, ABCDEF is a prism.
The cross-section is a right-angled triangle.
All of the other faces are rectangles.

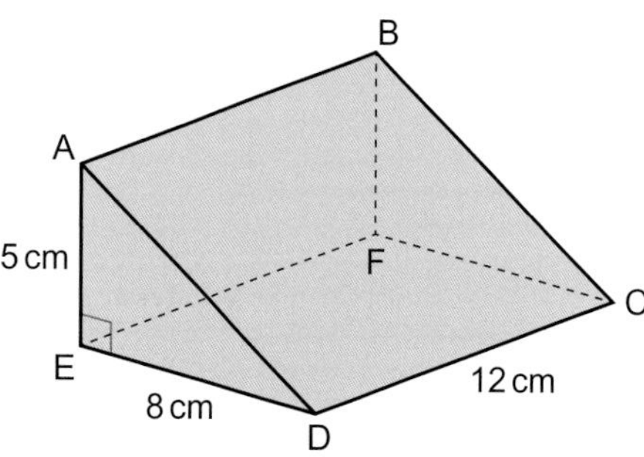

Calculate the angle that the diagonal AC makes with the plane CDEF.

5 **Reasoning** In the diagram, ABCD is a tetrahedron.

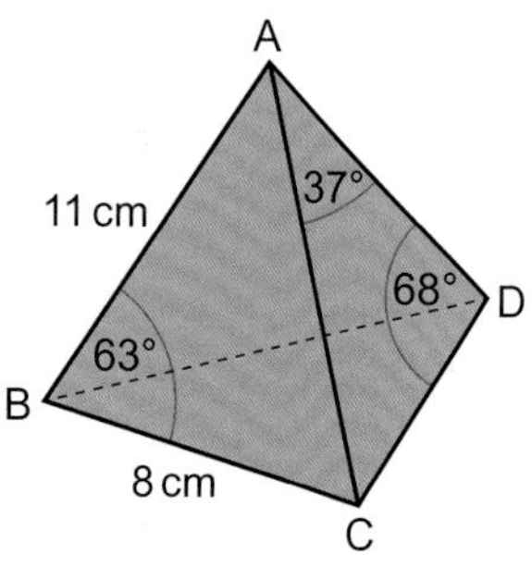

Q5 hint Sketch each triangle separately. Put as many lengths/angles as possible on your sketch to help you answer the questions.

a Work out the length of AC.

b Work out the length of CD.

c Given that BD = 12 cm, calculate angle BCD.

6 **Reasoning** ABCDE is a square-based pyramid.
The base BCDE lies in a horizontal plane.
AB = AC = AD = AE = 24 cm.
AM is perpendicular to the base.

a Calculate the length of
   i CE   ii CM   iii AM.

b Calculate the angle that AB makes with the base, correct to the nearest degree.

c Calculate the angle between AM and the face ACD, correct to the nearest degree.

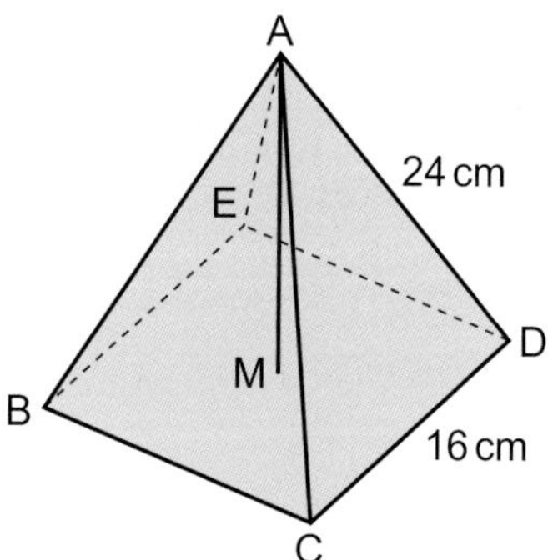

7 **Exam-style question**

The diagram shows a square-based pyramid ABCDE.

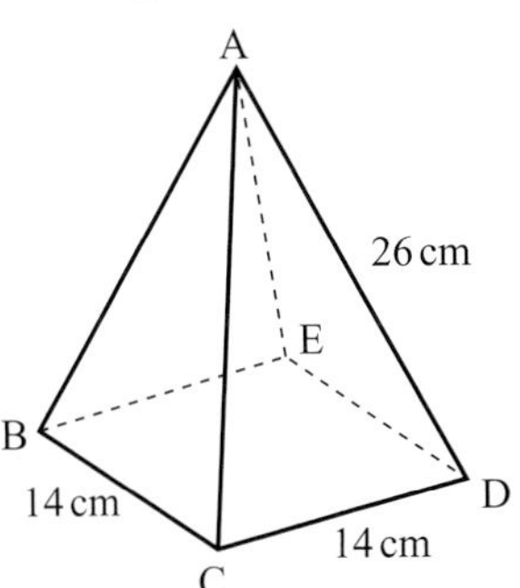

Each triangular face is an isosceles triangle.

a Calculate the length of the diagonal BD.
Give your answer correct to 3 significant figures.

b Calculate the area of triangle ABD.
Give your answer correct to 3 significant figures. **(6 marks)**

Q7 strategy hint For each part, draw a right-angled triangle and label it with the given information.

8 **Problem-solving** ABCDE is a pyramid with a rectangular base.
AB = AC = AD = AE = 28 cm.
Calculate the size of angle BAD correct to the nearest degree.

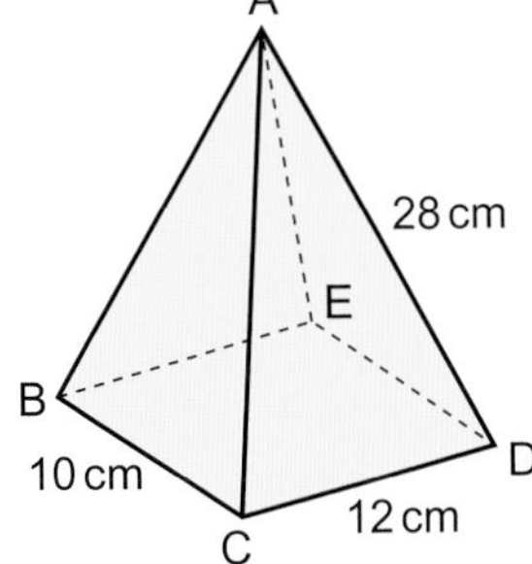

Q8 hint Sketch triangle BAD.

9 **Reflect** When you answered the questions in this lesson, did you make any mistakes? Do you understand where you went wrong? Was there a pattern?

13.8 Transforming trigonometric graphs 1

Objective

- Recognise how changes in a function affect trigonometric graphs.

Did you know?

If you plot the depth of water in a harbour against time, the graph is a simple transformation of a simple sine or cosine curve. The same is true of many quantities that vary over time, including mains voltage and light waves.

Fluency

Find the image of the point (3, 5) under each of these transformations.

a reflection in the x-axis **b** reflection in the y-axis **c** rotation through 180° about (0, 0)

Warm up

1 Write down the exact value of

a $\sin 60°$ b $\tan 30°$ c $\cos 45°$
d $\sin 0°$ e $\cos 90°$ f $\tan 60°$.

2 Sketch these graphs for values of x from 0° to 360°.

a $y = \sin x$ b $y = \cos x$ c $y = \tan x$

3 Here is the graph of $y = \sin x$ for $-180° \leqslant x \leqslant 180°$.

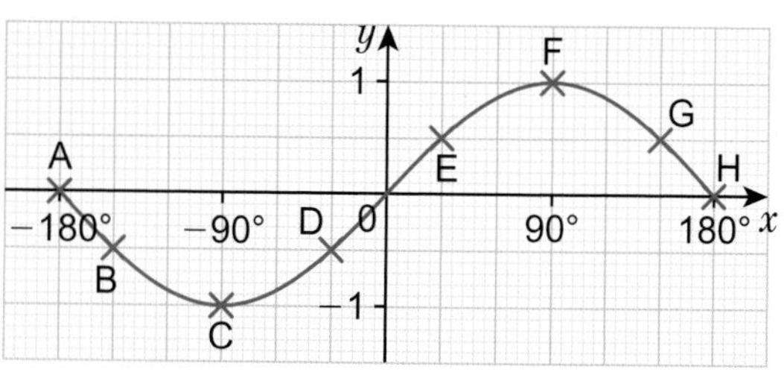

a Copy the table.

i Write in the values of x and $\sin x$ at points A to H on the graph.

ii For each x-value, write in the value of $-\sin x$.

b Sketch the graph of $y = -\sin x$ for $-180° \leqslant x \leqslant 180°$.

c Describe how the graph of $y = \sin x$ is transformed to give the graph of $y = -\sin x$.

	x	$\sin x$	$-\sin x$
A	–180°	0	0
B	–150°	–0.5	0.5
C			

Key point 8

The graph of $\boldsymbol{y = -f(x)}$ is the reflection of the graph of $y = f(x)$ in the x-axis.

4 a Use your table from **Q3**. Add a column for $\sin(-x)$.
Find the sine values from the graph to fill in the $\sin(-x)$ column.

b Sketch the graph of $y = \sin(-x)$ for $-180° \leqslant x \leqslant 180°$.

c Describe the transformation that turns the graph of $y = \sin x$ into the graph of $y = \sin(-x)$.

Key point 9

The graph of $\boldsymbol{y = f(-x)}$ is the reflection of the graph of $y = f(x)$ in the y-axis.

5 Here is the graph of $y = \tan x$ for $-180° \leqslant x \leqslant 180°$.
Sketch the graph of $y = \tan(-x)$ for $-180° \leqslant x \leqslant 180°$.

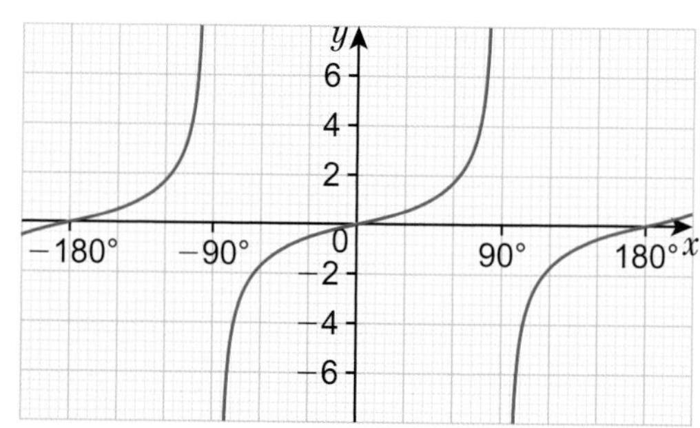

6 a Look at your graph of $y = -\sin x$ for $-180° \leqslant x \leqslant 180°$ in Q3.
What transformations will turn the graph of $y = -\sin x$ into the graph of $y = -\sin(-x)$?

b Sketch the graph of $y = -\sin(-x)$. What do you notice?

Key point 10

The graph of $\boldsymbol{y = -f(-x)}$ is a reflection of the graph of $y = f(x)$ in the x-axis and then the y-axis, or vice versa. These two reflections are equivalent to a rotation of 180° about the origin.

7 **Communication** Explain why the graph of $y = -\sin(-x)$ is the same as the graph of $y = \sin x$.

Q7 hint Sketch the graphs.

8 a Sketch the graph of $y = \cos x$ for $-180° \leqslant x \leqslant 180°$.

b Sketch the graph of $y = -\cos(-x)$.

Q8 hint Rotate $y = \cos x$ by 180° about the origin.

9 a Describe the transformation that maps the graph of $y = \tan x$ to the graph of $y = \tan(-x)$.

b Sketch the graph of $y = \tan x$ and the graph of $y = \tan(-x)$ for the interval 0 to 360°.

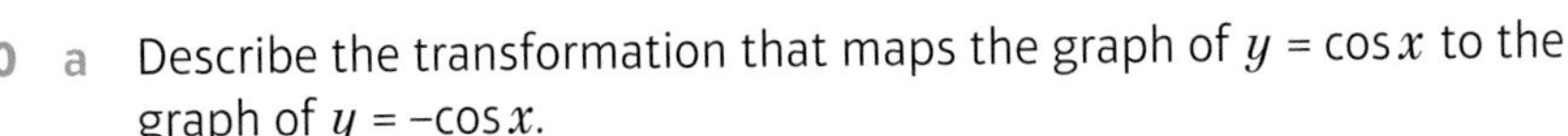

10 a Describe the transformation that maps the graph of $y = \cos x$ to the graph of $y = -\cos x$.

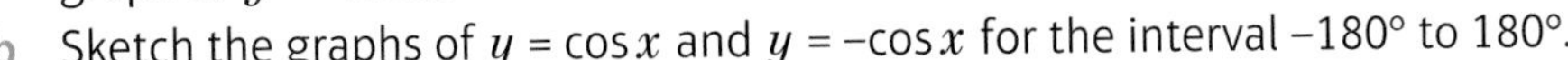

b Sketch the graphs of $y = \cos x$ and $y = -\cos x$ for the interval $-180°$ to $180°$.

11 a Describe the transformation that maps the graph of $y = \tan x$ to the graph of $y = -\tan(-x)$.

b Sketch the graphs of $y = \tan x$ and $y = -\tan(-x)$ for the interval $-180°$ to $180°$.

12 **Exam-style question**

Here is a sketch of the graph of $y = -\sin x$.

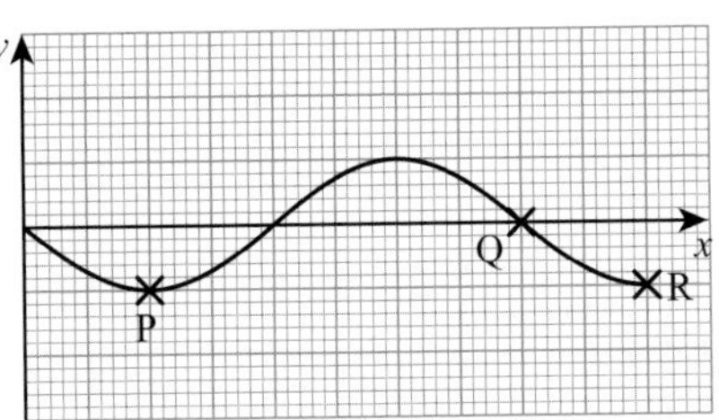

Write down the coordinates of each of the labelled points. **(3 marks)**

Q12 strategy hint This curve is a transformation of the graph of $y = \sin x$.

13.9 Transforming trigonometric graphs 2

Objective

- Recognise how changes in a function affect trigonometric graphs.

Did you know?

The sine and cosine curves are identical in shape but 90° 'out of phase', meaning that you can shift one horizontally to get the other.

Fluency

Is the graph of $y = -\cos(-x)$ the same as the graph of $y = \cos x$?

Warm up

1 a The point (30, 0.5) is translated by the vector $\begin{pmatrix} 0 \\ 2 \end{pmatrix}$.
What are the coordinates of the new point?

b The point (45, 1) is translated by the vector $\begin{pmatrix} -15 \\ 1 \end{pmatrix}$.
What are the coordinates of the new point?

2 a Copy the graph of $y = \sin x$ for $-180° \leqslant x \leqslant 180°$ from **Q3** in lesson 13.8.

b Add 0.5 to the y-coordinate at each of the labelled points.

c Draw the sine graph that passes through the new points. Label it $y = \sin x + 0.5$.

d Describe the transformation from the graph of $y = \sin x$ to this graph.

e Now subtract 0.5 from the y-coordinate at each of the labelled points on the original graph.

f Draw the sine graph that passes through the new points.

g Describe the transformation from the graph of $y = \sin x$ to this graph.

h Write down the equation of the graph.

Key point 11

The graph of $\boldsymbol{y = f(x) + a}$ is the translation of the graph of $y = f(x)$ by $\begin{pmatrix} 0 \\ a \end{pmatrix}$.

3 Write down the equation of each graph.

a
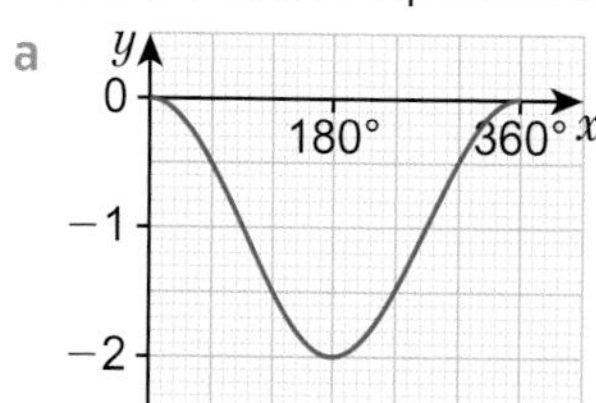

b
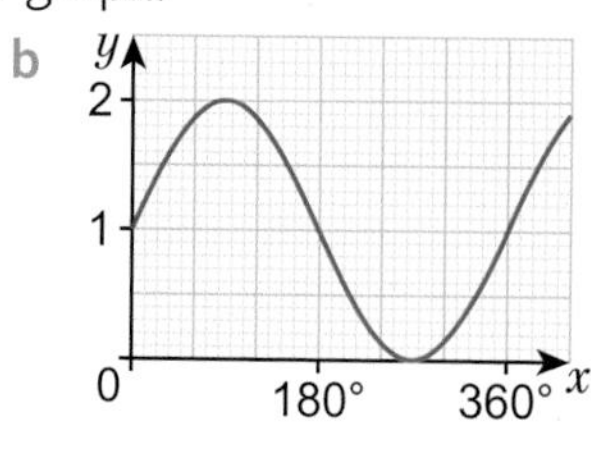

c
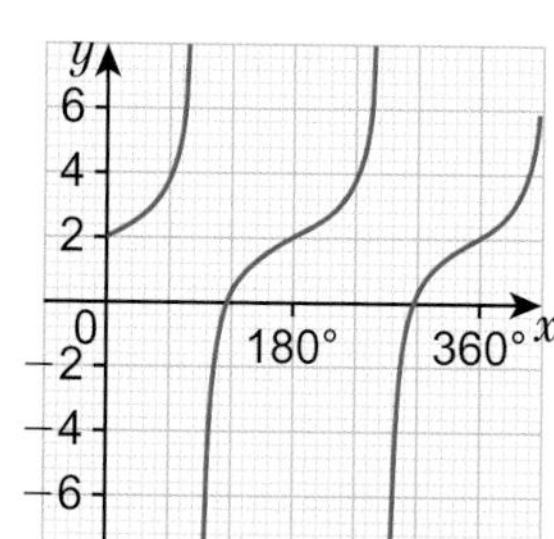

Q3 hint First decide whether it is a sin, cos or tan graph.

4 Here is the graph of $y = \cos x$ for $0° \leqslant x \leqslant 360°$.

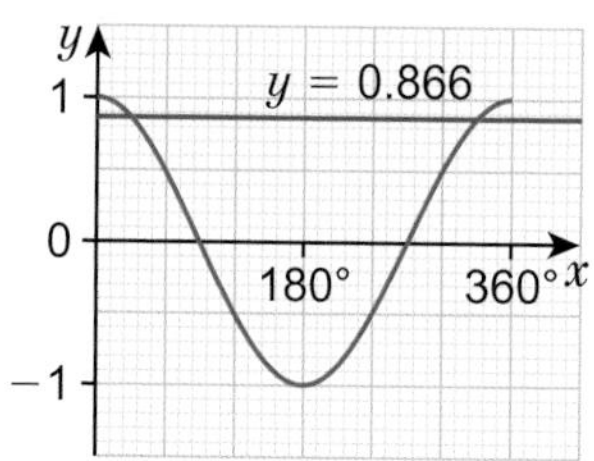

a Copy and complete this table of values for $\cos(x + 30°)$.

x	0°	30°	60°	90°
cos(x + 30°)	cos 30° = □			

b Sketch the graph of $y = \cos(x + 30°)$.

c Describe the transformation that takes the graph of $y = \cos x$ to the graph of $y = \cos(x + 30°)$.

Q4b hint Draw a table of values for x and y.

Key point 12

The graph of $\boldsymbol{y = \mathrm{f}(x + a)}$ is the translation of the graph of $y = \mathrm{f}(x)$ by $\begin{pmatrix} -a \\ 0 \end{pmatrix}$.

5 Describe the transformation of the graph of $y = \cos x$ to make the graph with equation

a $y = \cos(x + 60°)$ b $y = \cos(x + 20°)$ c $y = \cos(x - 30°)$.

6 Describe the transformation of the graph of $y = \tan x$ to make the graph with equation

a $y = \tan(x + 40°)$ b $y = \tan(x + 30°)$ c $y = \tan(x - 60°)$.

7 Match each graph below with one of these equations.

A $y = \tan(x - 30°)$ **B** $y = \sin(x + 45°)$ **C** $y = \cos(x + 45°)$

a
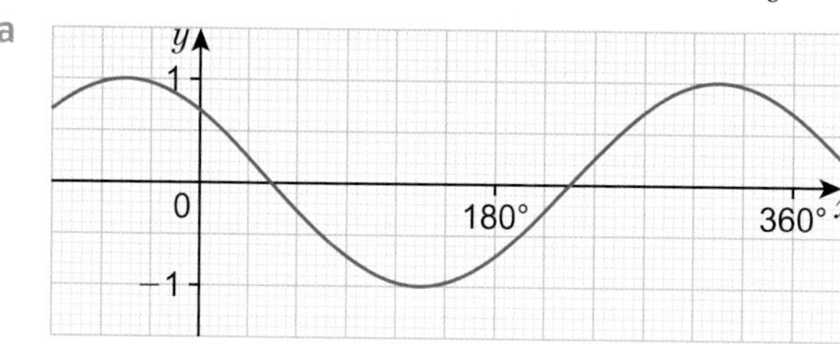

b
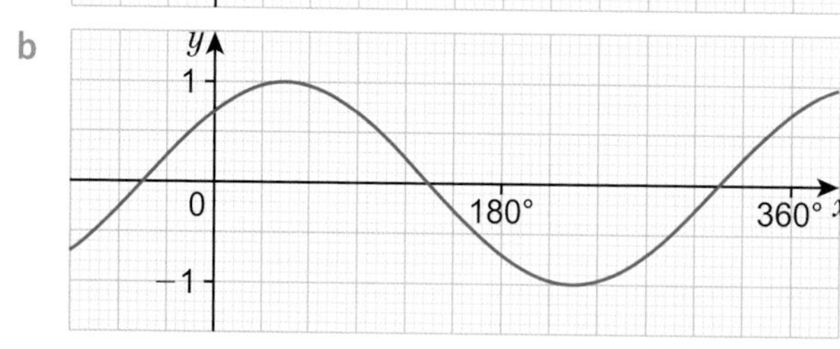

c
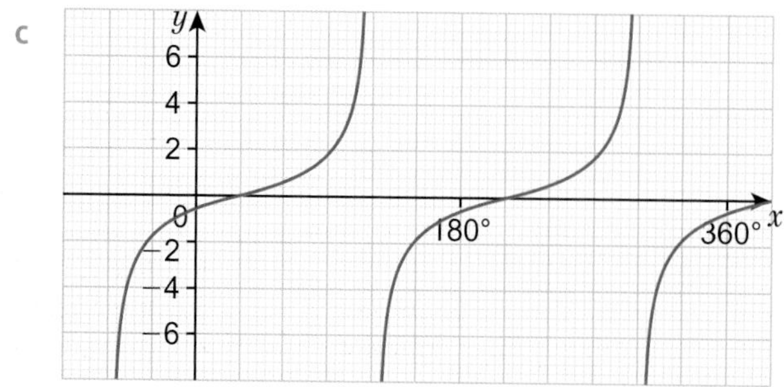

8 a Sketch the graph of $y = \sin x$ for $0° \leqslant x \leqslant 360°$.

b Copy and complete the table of values for $y = 2\sin x$.

x	0°	30°	60°	90°	120°
$\sin x$	0	0.5			
$2\sin x$	0				

c On the same axes, sketch the graph of $y = 2\sin x$.

Key point 13

The graph of $\boldsymbol{y = a\mathrm{f}(x)}$ is a vertical stretch of the graph of $y = \mathrm{f}(x)$, with scale factor a, parallel to the y-axis.

9 Sketch the graphs of these functions for $0° \leqslant x \leqslant 360°$.

a $y = 3\cos x$ b $y = 2\tan x$ c $y = -2\sin x$

10 a Copy your sketch graph of $y = \sin x$ for $0° \leqslant x \leqslant 360°$ from **Q8**.

b Copy and complete the table of values for $y = \sin(2x)$.

x	0°	30°	60°	90°	120°
$\sin(2x)$					

c Sketch the graph of $y = \sin 2x$ on the same axes.

Key point 14

The graph of $\boldsymbol{y = \mathrm{f}(ax)}$ is a horizontal stretch of the graph of $y = \mathrm{f}(x)$, with scale factor $\frac{1}{a}$, parallel to the x-axis.

11 Sketch the graphs of these functions for $0° \leqslant x \leqslant 360°$.

a $y = \sin 3x$ b $y = \cos 2x$ c $y = \tan 2x$

12 **Exam-style question**

The diagram shows part of a sketch of the curve $y = \sin x°$

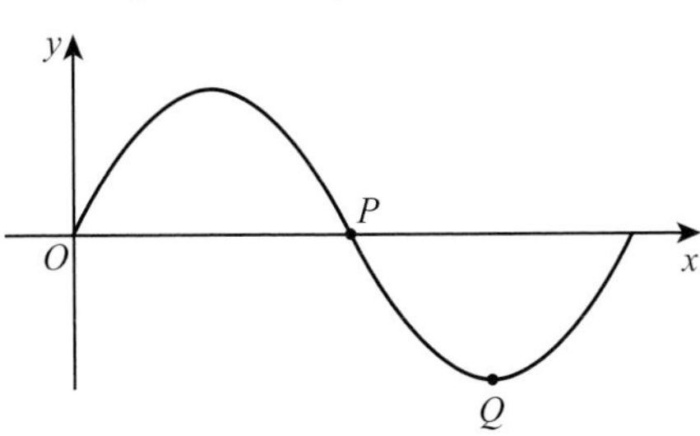

a Write down the coordinates of the point P. **(1 mark)**

b Write down the coordinates of the point Q. **(1 mark)**

Here is a sketch of the curve $y = a\cos bx° + c$, $0 \leqslant x \leqslant 360$

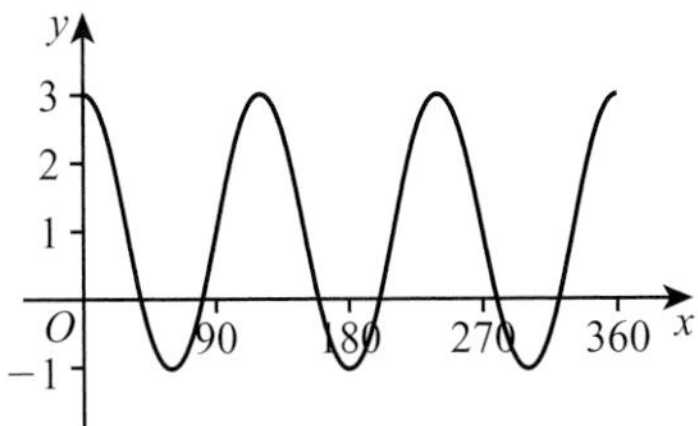

c Find the values of a, b and c. **(3 marks)**

June 2014, Q26, 1MA0/1H

Q12c strategy hint
a is related to the vertical stretch factor, b is related to the horizontal stretch factor and c is related to the vertical translation.

13 Problem-solving: Muddy tracks

Objective

- Use Pythagoras' theorem and the cosine rule to solve problems in 2D.

An expedition needs to cross over the square piece of land X (below) as quickly as possible, moving from A to B. They can walk at 5 km/h in the field, but slow down to a speed of 2 km/h when walking through the wood.

One approach would be to walk in a straight line from A to B.

1 Show that it would take the expedition 9 hours and 54 minutes to travel by this route.

Q1 hint You will need to use the formula, time = $\frac{\text{distance}}{\text{speed}}$

2 Find a route which would be at least 30 minutes quicker to travel.

Land Y (below) has a different layout.

3 Find the quickest possible route over this piece of land. Give your answer to an appropriate degree of accuracy.

Q3 hint Use trigonometry. You might choose to use a trial and improvement approach.

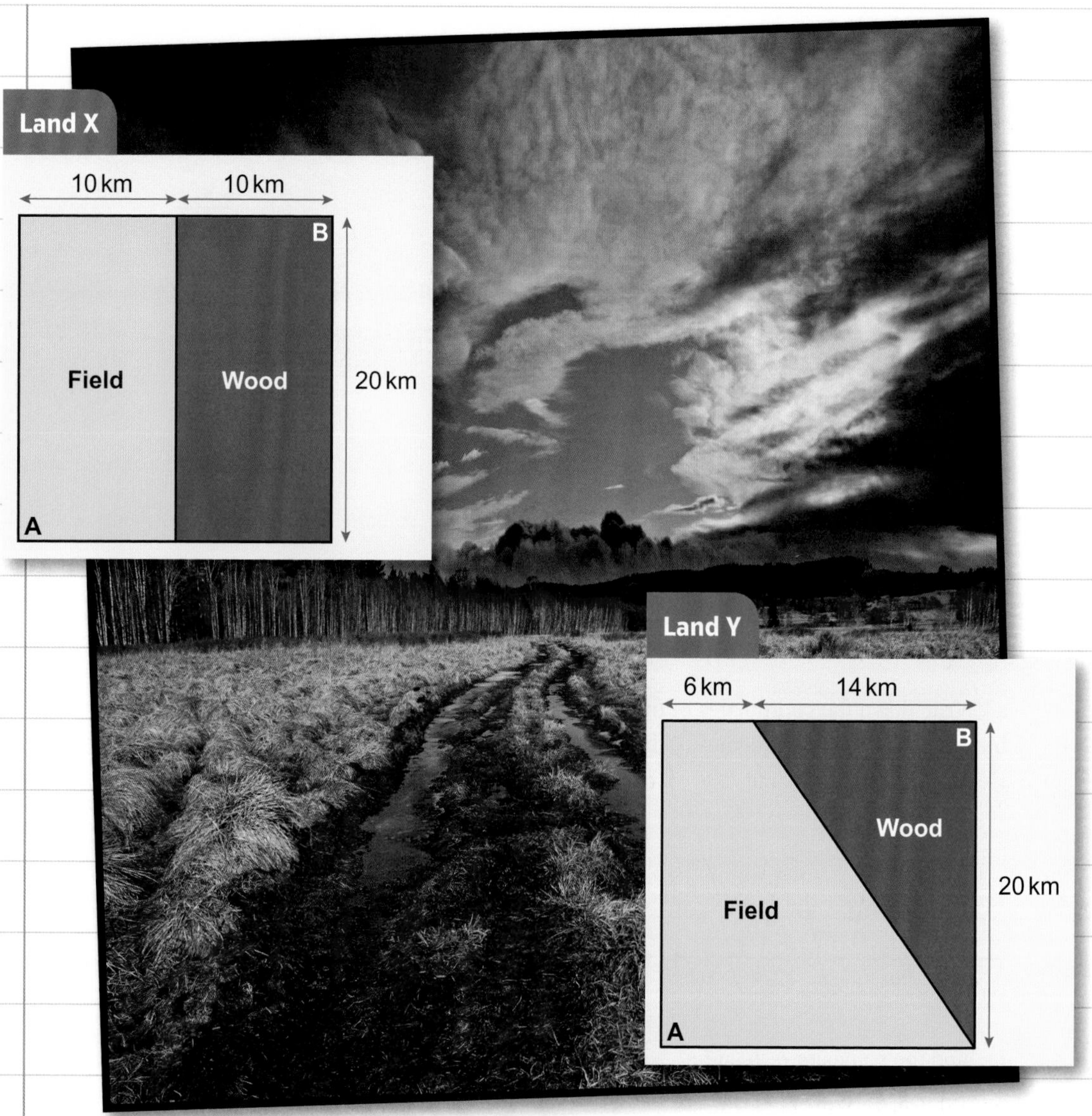

Land X

10 km
10 km
B
Field
Wood
20 km
A

Land Y

6 km
14 km
B
Wood
Field
20 km
A

13 Check up

Log how you did on your Student Progression Chart.

Accuracy and 2D problem-solving

1 Work out the area of this triangle.

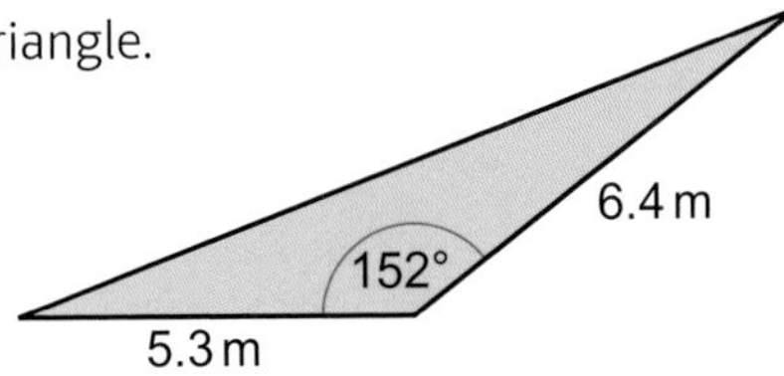

2 Calculate the length of the side labelled x in each diagram.

a

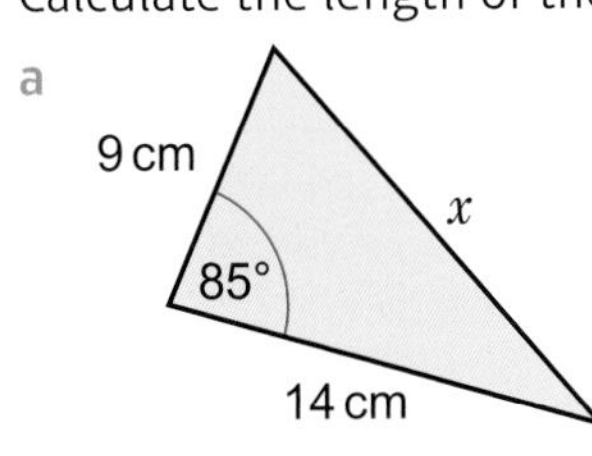

b

98°
25°
x
10.8 cm

3 Find the size of the acute angle θ in these triangles.

a

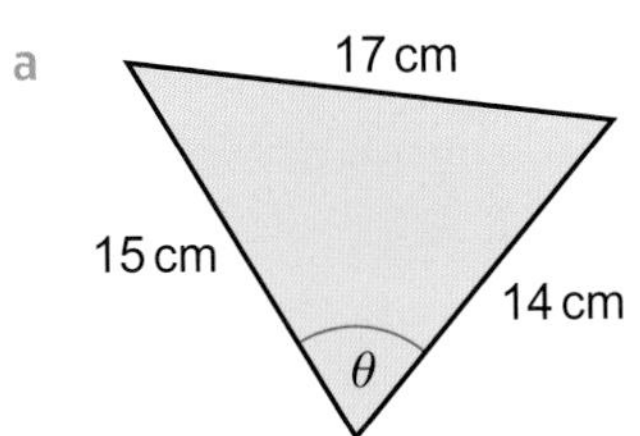

b

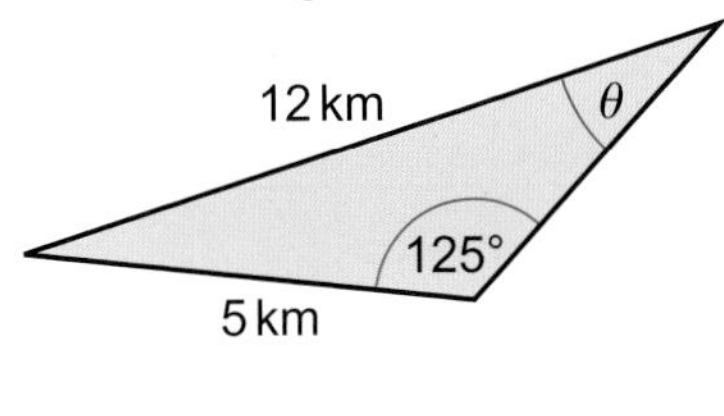

4 Find the upper and lower bounds for the value of x in the diagram. Write x to a suitable level of accuracy.

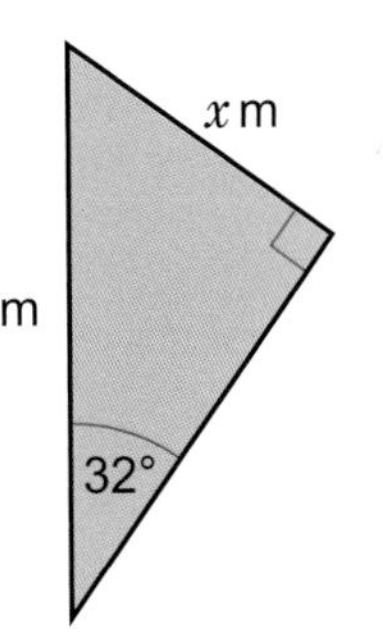

Trigonometric graphs

5 Sketch the graph of $y = \tan\theta$ for $-360° \leqslant \theta \leqslant 360°$.

6 Here is the graph of $y = \cos x$ for $0° \leqslant x \leqslant 360°$.

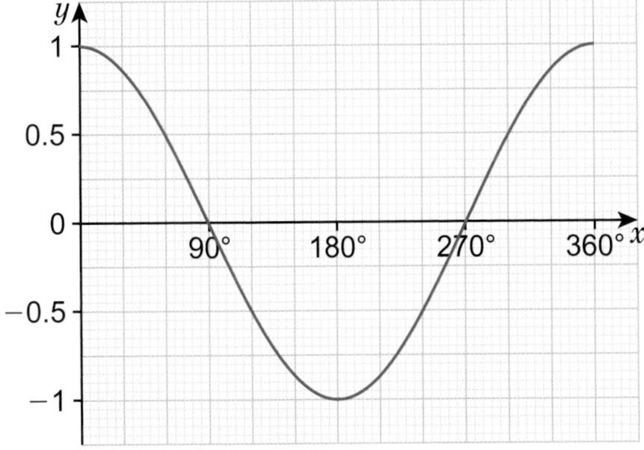

Use the graph to solve $\cos x = 0.4$.

7 Match each graph below with one of these equations.

A $y = 2\sin x + 1$ **B** $y = 3\cos x$ **C** $y = \cos 3x$

a

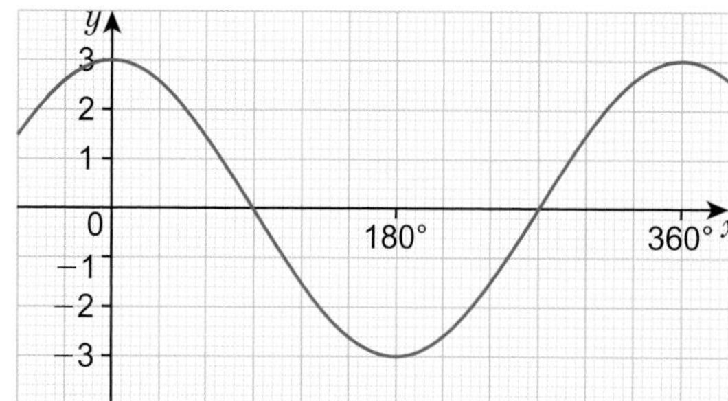

b

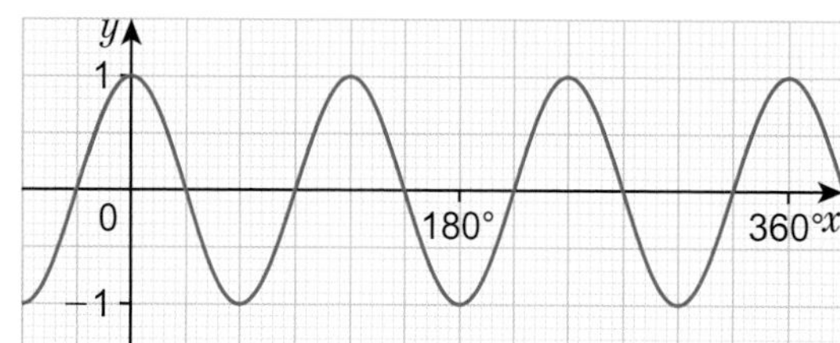

c

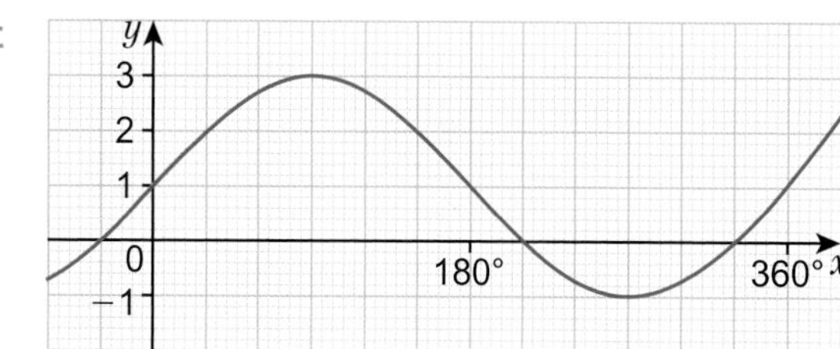

8 Solve the equation $3\sin x = 1$ for $0° \leqslant x \leqslant 720°$.

3D Problem-solving

9 ABCDEFGH is a cuboid.
Calculate the length of the diagonal AG.

D 16 cm C
A B 15 cm
H G
E F 8 cm

10 ABCD is a tetrahedron.
Calculate the length of

a AC

b CD.

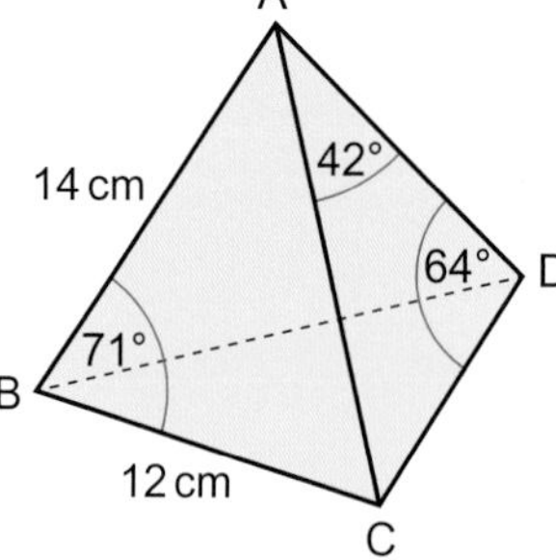

11 How sure are you of your answers? Were you mostly

Just guessing  Feeling doubtful   Confident 

What next? Use your results to decide whether to strengthen or extend your learning.

Reflect

* Challenge

12 The angles that satisfy an equation in the interval $0° \leqslant x \leqslant 720°$ are 30°, 210°, 390° and 570°.
Write down a possible equation.

13 Strengthen

Accuracy and 2D problem-solving

1 a Copy this triangle. Label the vertex at the given angle C, and the other two vertices A and B. Label the sides: a is opposite A and b is opposite B.

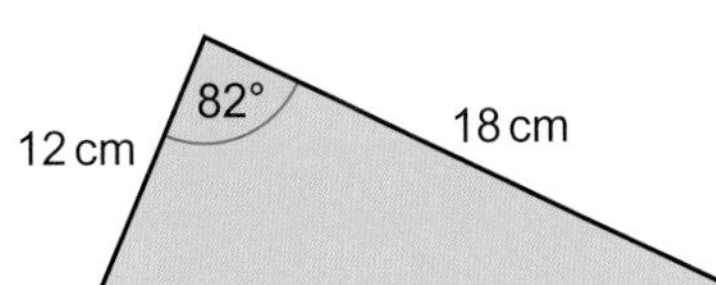

b Use Area = $\frac{1}{2}ab\sin C$ to find the area of the triangle.

2 Find the areas of these triangles.

a

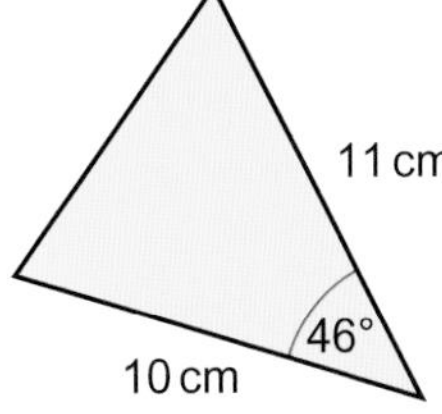

b

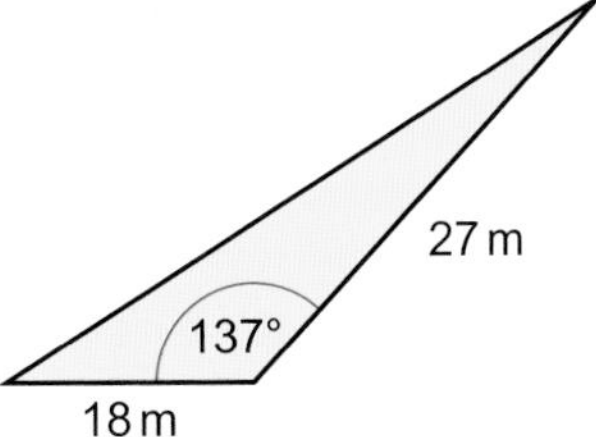

Q2 hint Use the method in **Q1**.

3 a Copy the triangle. Label the vertices A, B and C and the sides a, b and c.

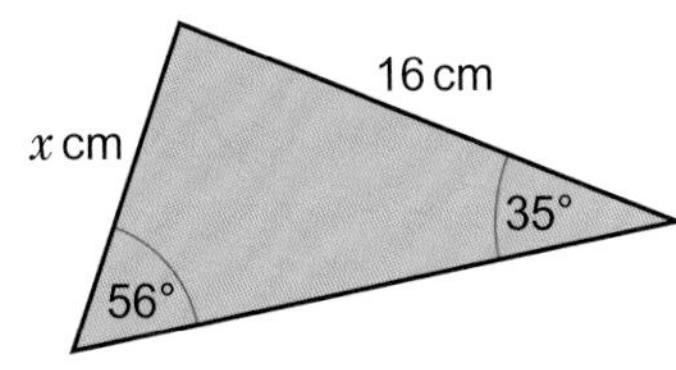

Q3a hint

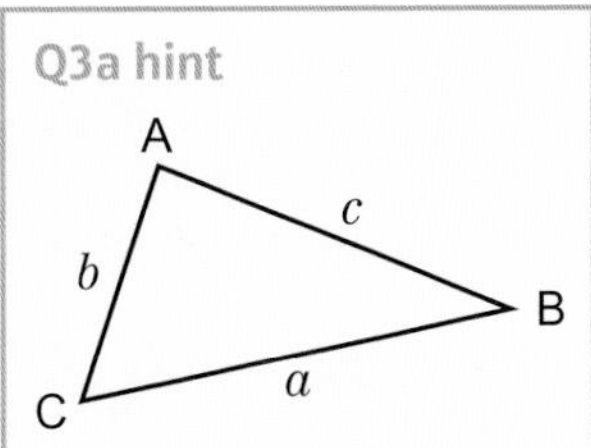

b Substitute the values from the diagram into the sine rule:

$$\frac{a}{\sin A} = \frac{b}{\sin B} = \frac{c}{\sin C}$$

c Use two parts of the sine rule with values in them to find x.
Give your answer correct to 3 significant figures.

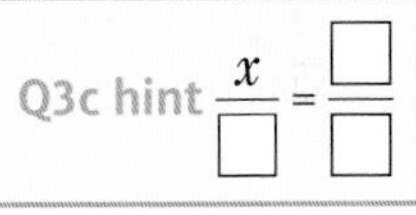

Q3c hint $\frac{x}{\square} = \frac{\square}{\square}$

4 Find the value of x in each triangle. Give your answers correct to 3 significant figures.

a

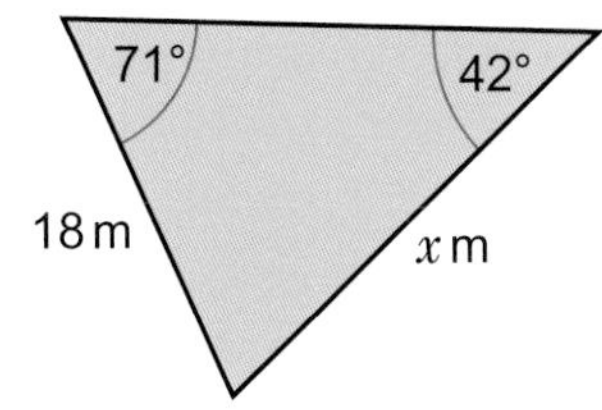

b

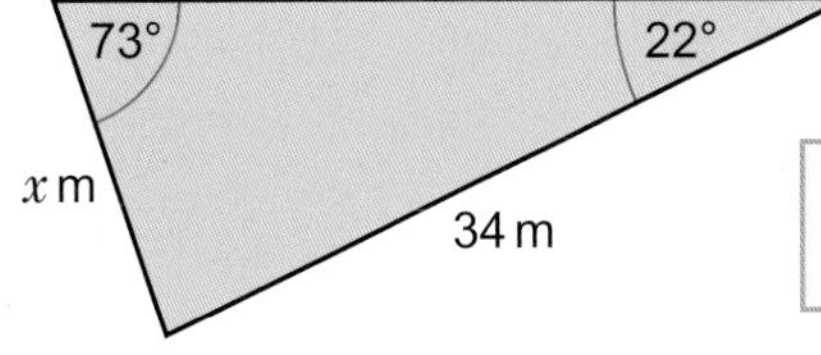

Q4 hint Use the method in **Q3**.

5 a Copy the triangle.
Label the vertices A, B and C and the sides a, b and c.

b Substitute the values from the diagram into the sine rule:

$$\frac{\sin A}{a} = \frac{\sin B}{b} = \frac{\sin C}{c}$$

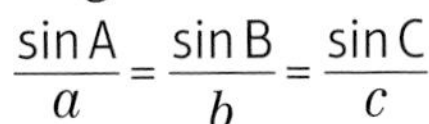

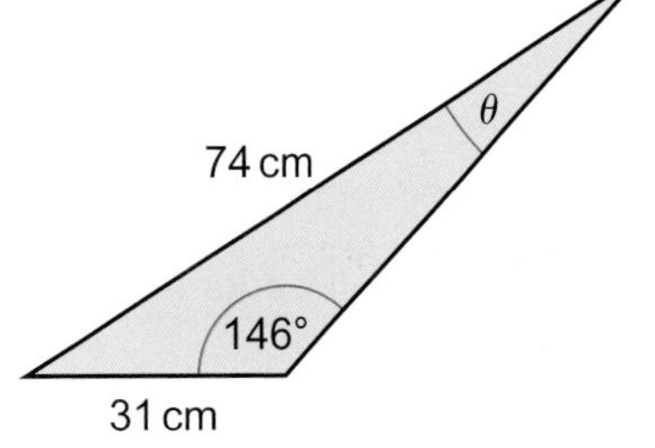

c Use two parts of the sine rule with values in them to find θ.
Give your answer correct to 1 decimal place.

Q5c hint $\frac{\sin\theta}{\square} = \frac{\square}{\square}$

6 Find the size of the acute angle θ in each triangle.
Give your answers correct to 1 decimal place.

a
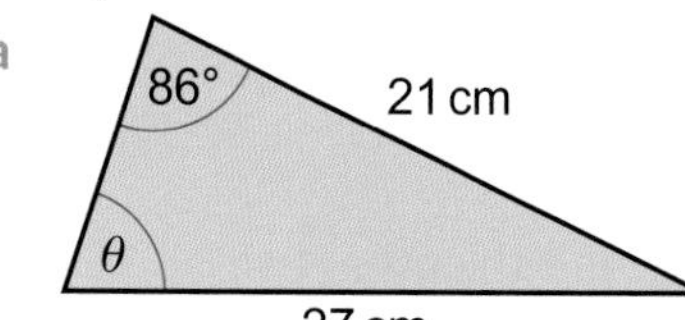

b
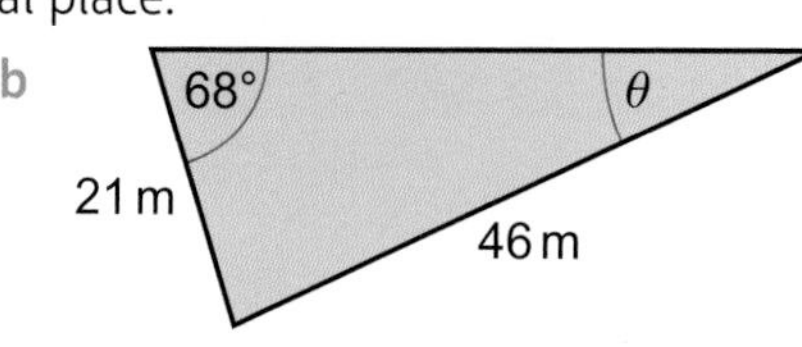

7 a Copy the triangle. Label the x side a, and the others b and c. Label the vertices A, B and C.

b Substitute the values from the triangle into the cosine rule:
$a^2 = b^2 + c^2 - 2bc\cos A$

c Solve the equation to find x, correct to 3 significant figures.

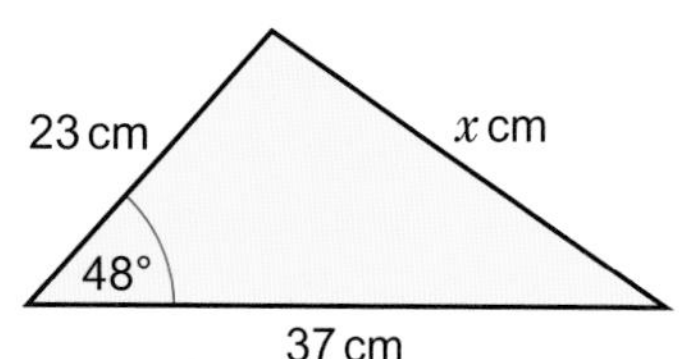

8 Calculate the value of x in each triangle.
Give your answers correct to 3 significant figures.

a
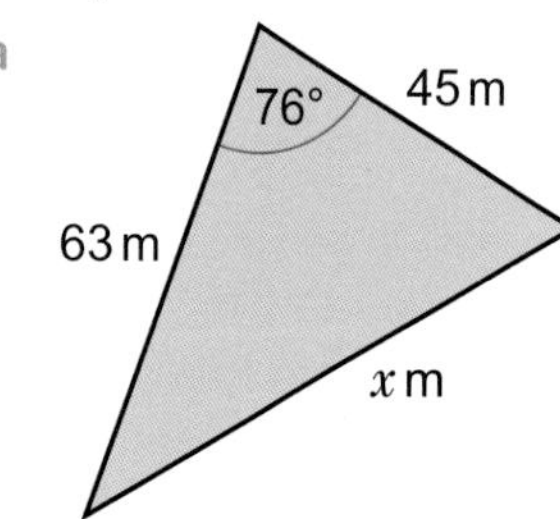

b
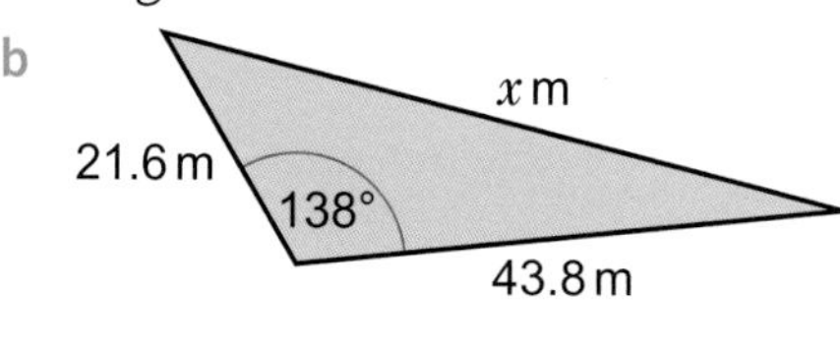

9 a Copy the triangle. Label the vertex and angle θ, A.
Label the other vertices and the other sides.

b Substitute the values from the triangle into the cosine rule:

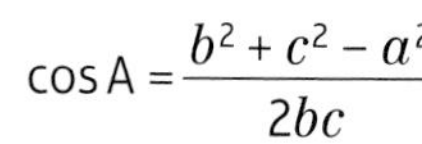

$$\cos A = \frac{b^2 + c^2 - a^2}{2bc}$$

c Calculate the value of θ correct to 1 decimal place.

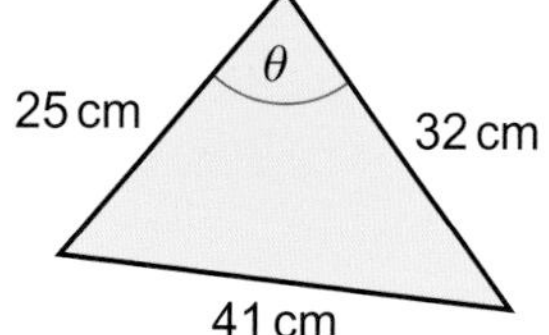

10 Calculate the size of angle θ in each triangle.
Give your answers correct to 1 decimal place.

a

b
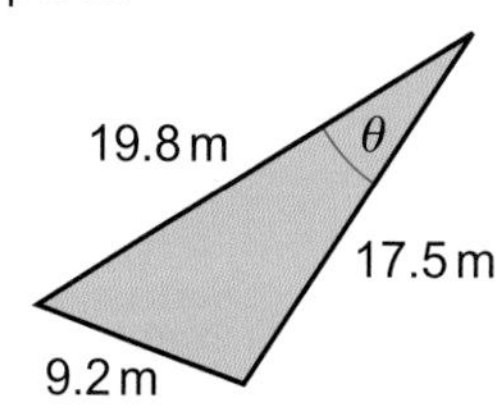

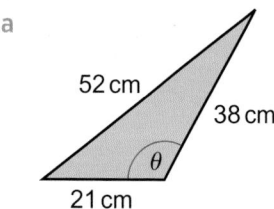

11 All the measures in this equation are given to 2 significant figures.

$$x = \frac{5.7}{\sin 23°}$$

a Copy and complete the table for the upper and lower bounds.

	Upper bound value	Lower bound value
5.7		5.65
23	23.5	

b Find sin(upper bound for 23) and sin(lower bound for 23).

c Which upper and lower bound values give the lower bound for x?

d Which upper and lower bound values give the upper bound for x?

Q11c hint Which values – one from each row of the table – give the smallest possible value of x?

12 All the measures in this equation are given to 2 significant figures.
$x = 14 \tan 36°$
Find the upper and lower bounds for the value of x.

Q12 hint Use the same method as in **Q11**.

Trigonometric graphs

1 a Copy and complete the table for $y = \sin x$.

x	0°	10°	20°	30°	40°	50°	60°	70°	80°	90°
$\sin x$										

b Draw the graph of $y = \sin x$ for $0° \leqslant x \leqslant 90°$.

The sine graph is symmetrical about the line $x = 90°$.

c Sketch the graph of $y = \sin x$ for $0° \leqslant x \leqslant 180°$.

The graph of $y = \sin x$ has rotational symmetry about the point (180°, 0).

d Extend your sketch from part **c** to cover the interval $0° \leqslant x \leqslant 360°$.

2 a Copy and complete the table for $y = \tan x$.

x	0°	10°	20°	30°	40°	50°	60°	70°	80°
$\tan x$									

b Draw the graph of $y = \tan x$ for $0° \leqslant x \leqslant 80°$.

c What happens to the value of $\tan x$ as x increases from 80° towards 90°?

d Sketch the graph of $y = \tan x$ for $0° \leqslant x \leqslant 90°$.

The graph of $y = \tan x$ has rotational symmetry about the point (90°, 0).

e Extend your sketch from part **d** to cover the interval $0° \leqslant x \leqslant 180°$.

The graph of $y = \tan x$ repeats every 180°.

f Extend your sketch from part **e** to cover the interval $0° \leqslant x \leqslant 360°$.

3 Here is the graph of $y = \cos x$ for $0° \leqslant x \leqslant 360°$.

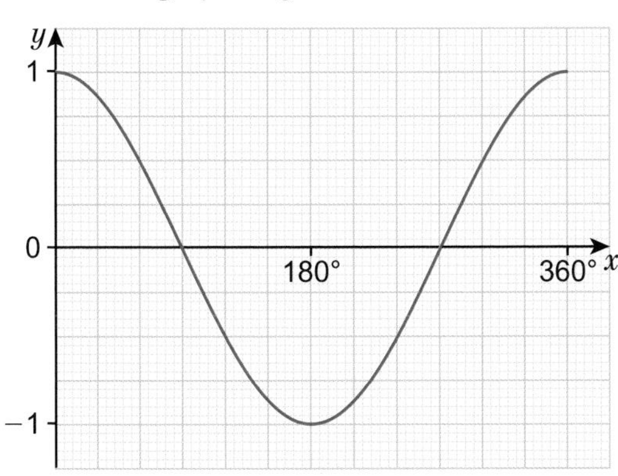

Q3 hint Find –0.6 on the vertical axis and read off the corresponding values from the curve.
There are two values in the interval $0° \leqslant x \leqslant 360°$.

Use the graph to estimate a solution to $\cos x = -0.6$.

4 a Copy and complete this table of values.

x	0°	30°	90°
$\sin x$	0		
$\sin 2x - 1$		$\sin 60° - 1 = \frac{\sqrt{3}}{2} - 1$	
$2\sin x - 1$	$2 \times 0 - 1 = -1$		

b Match each equation to a transformation of the sine or cosine graph.

i $y = 2\sin x - 1$ ii $y = 2\cos x$ iii $y = \sin 2x - 1$

c Match each equation from part **b** to one of these graphs.

A

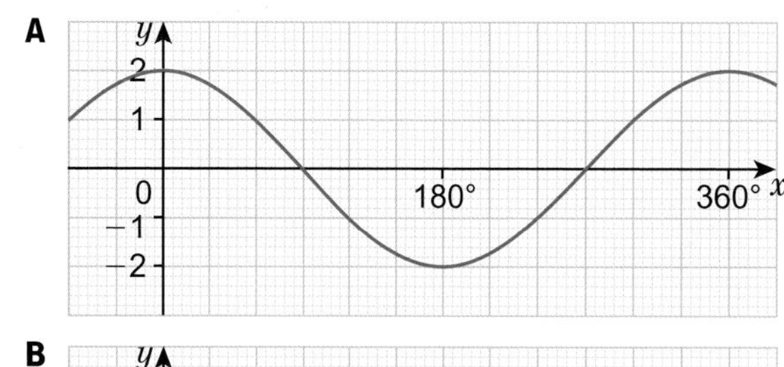

B

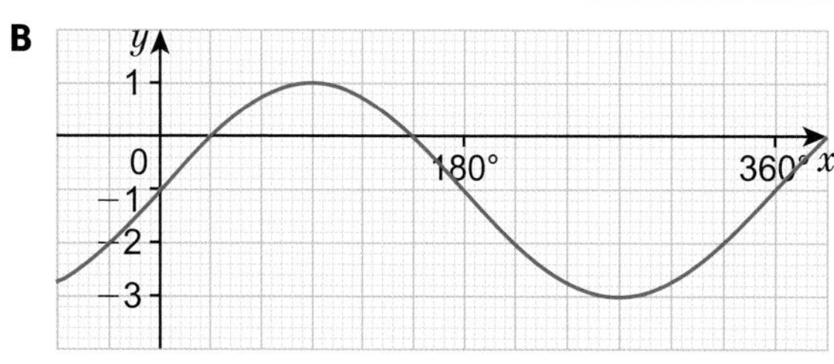

C

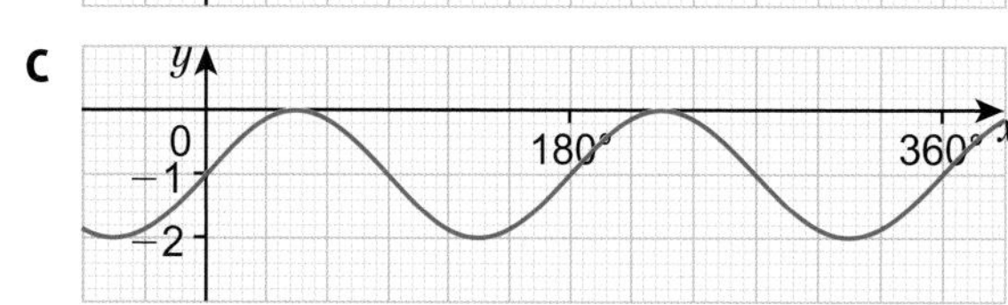

5 a Rearrange the equation $4\cos x = 3$ to make $\cos x$ the subject.

b Sketch the graph of $y = \cos x$ for $0° \leqslant x \leqslant 720°$.

c Use your calculator to find one value of x that satisfies the equation.

d Use your sketch to solve the equation $4\cos x = 3$ for $0° \leqslant x \leqslant 720°$.

3D Problem-solving

1 **Reasoning** The diagram shows a cuboid ABCDEFGH.

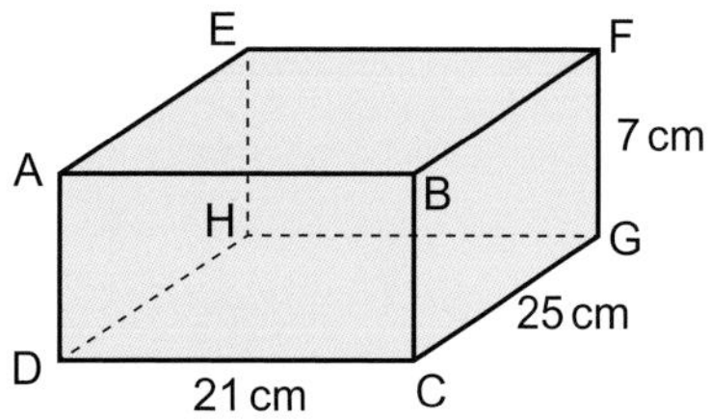

a Sketch triangle CDG.
Label the triangle with the information shown on the diagram.

b Calculate the length of DG.
Give your answer correct to 3 significant figures.

c Sketch triangle DFG.
Label the triangle with the information shown on the diagram and your answer to part **b**.

d Calculate the angle that the diagonal DF makes with the plane DCGH.

> **Q1d hint** The angle between the diagonal DF and the plane DCGH is angle FDG on your sketch for part **c**. Use your **unrounded** answer to part **b**.

2 **Reasoning** In the diagram, ABCD is a tetrahedron.
AB = AC = 11 cm.

a Sketch triangle ABC. Label the triangle with the information shown on the diagram.

b Calculate the length of BC.
Give your answer correct to 3 significant figures.

c Sketch triangle BCD.
Label the triangle with the information shown on the diagram and your answer to part **b**.

d Calculate angle BCD.

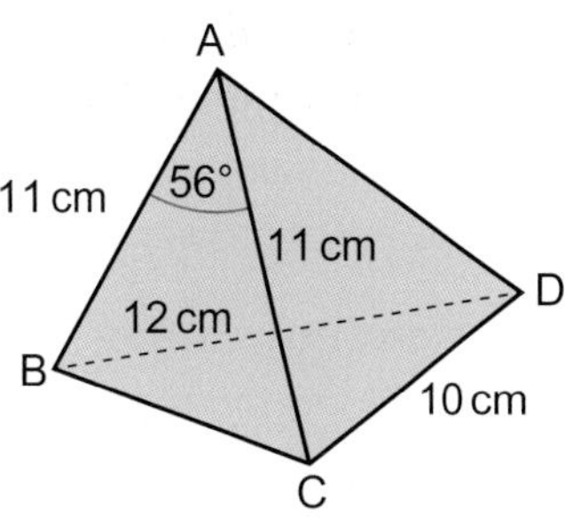

Q2d hint Use your **unrounded** answer to part **b**.

13 Extend

1 a Use the cosine rule to find the length of BC.

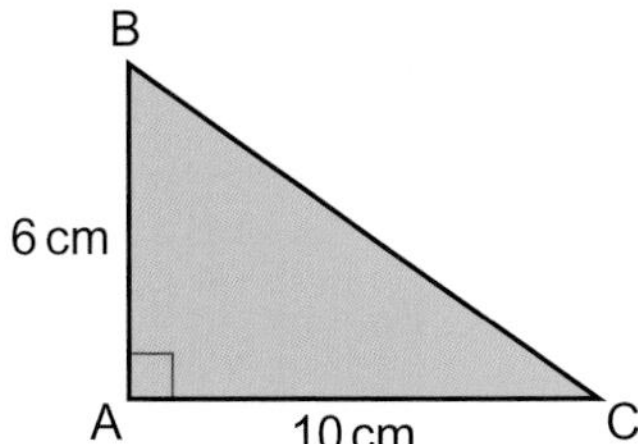

b What do you notice about the cosine rule when A = 90°?

2 a Write an expression for the area of this triangle using angle C.

b Write two more expressions for the area of the triangle using angles B and A.

c Using your answers to parts **a** and **b**, show that

$$\frac{a}{\sin A} = \frac{b}{\sin B} = \frac{c}{\sin C}.$$

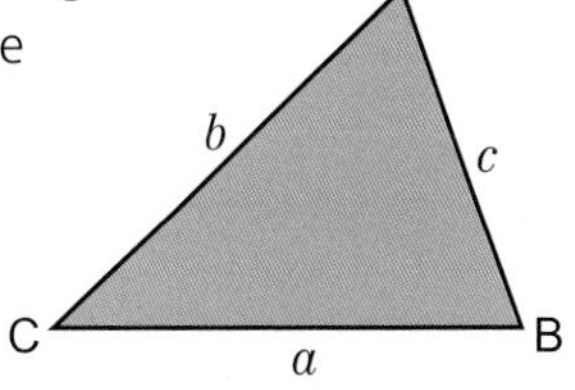

3 The diagram shows a cuboid.
The diagonal of the cuboid has length d.

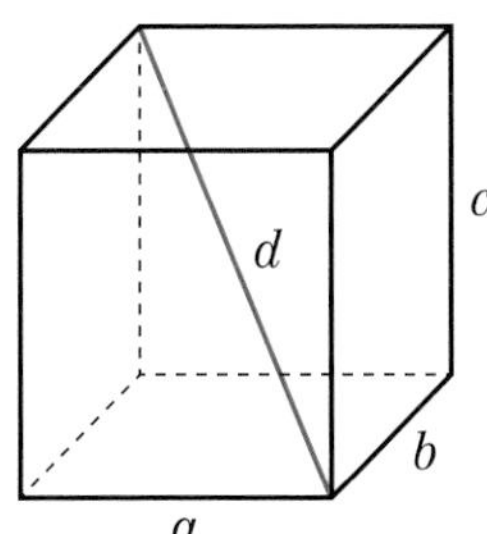

Write d^2 in terms of a, b and c.

4 a In triangle BCD, write a^2 in terms of h, b and x and expand the brackets.

b In triangle ABD, write c^2 in terms of h and x.

c Use your answers to parts **a** and **b** to write a^2 in terms of b, c and x.

d Show that $a^2 = b^2 + c^2 - 2bc\cos A$.

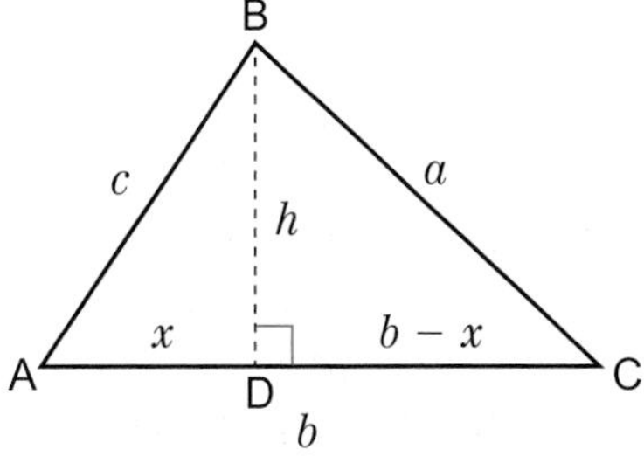

5 **Problem-solving** The area of triangle ABC is 21 cm². Calculate the size of the obtuse angle ABC.
The measurements are rounded to 1 d.p.
Give your answer to a suitable level of accuracy.

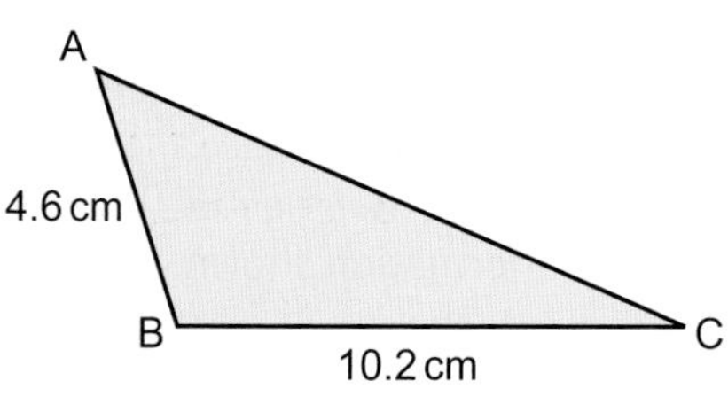

6 **Communication** a Sketch the graph of $y = \sin x$.
b From your graph, find
i $\sin 45°$ ii $\sin(180° - 45°)$.
c What do you notice?
d Explain why $\sin\theta = \sin(180° - \theta)$ for values between 0 and 180°.

Q6d hint Draw lines on your graph.

7 **Problem-solving** Find the area of the shaded segment in terms of r.

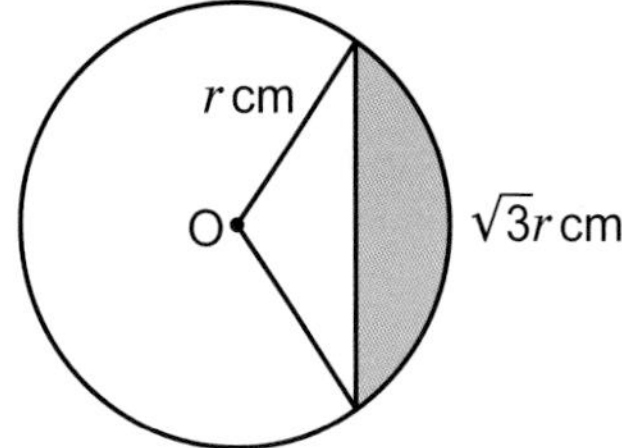

8 **Exam-style question**

Here are two triangles $\mathbf{T_1}$ and $\mathbf{T_2}$.

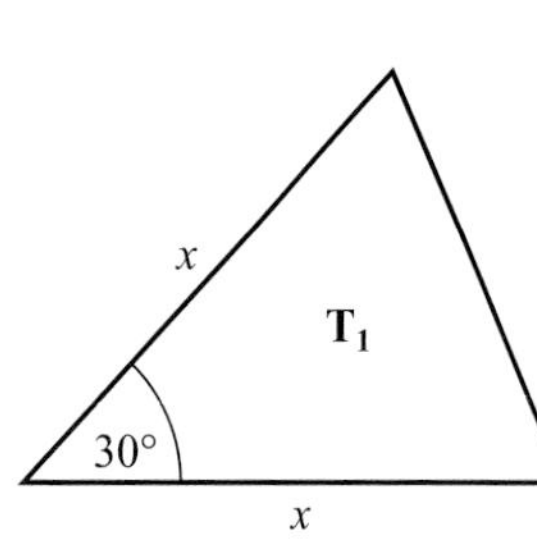

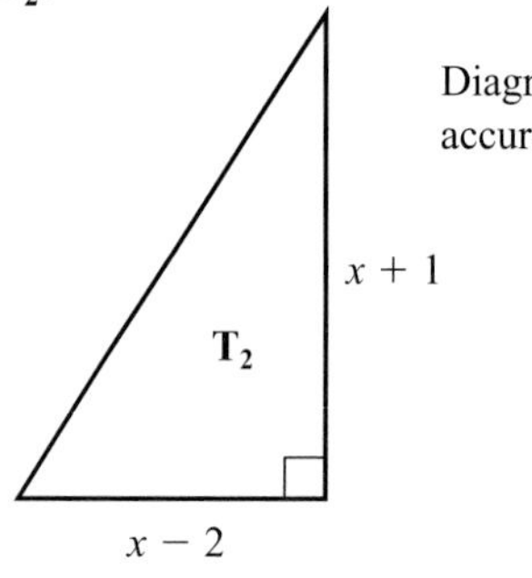

Diagram **NOT** accurately drawn

The lengths of the sides are in centimetres.
The area of triangle $\mathbf{T_1}$ is equal to the area of triangle $\mathbf{T_2}$.
Work out the value of x, giving your answer in the form $a + \sqrt{b}$ where a and b are integers. **(5 marks)**

March 2013, Q25, 1MA0/2H

Q8 strategy hint Write an equation using the area formulae $\frac{1}{2}ab\sin C$ for $\mathbf{T_1}$ and $\frac{1}{2} \times$ base $\times$ height for $\mathbf{T_2}$.

9 a Sketch the graph of $y = \sin x$.
b Copy and complete this table of values for $y = \sin\left(\frac{x}{2}\right)$.

x	0°	60°	90°	120°
$\sin\left(\frac{x}{2}\right)$			$\frac{1}{\sqrt{2}} = 0.7$	

c Sketch the graph of $y = \sin\left(\frac{x}{2}\right)$ on the same axes as your graph of $y = \sin x$.

10 **Modelling** The depth, d metres, of water in a harbour at a time t hours after midnight is $d = 12 + 5\sin(30t)$, where $0 \leqslant t \leqslant 24$.
Sketch the graph of d against t.

11 Problem-solving ABCDE is a square-based pyramid.
Calculate the volume of the pyramid.
Give your answer correct to 3 significant figures.

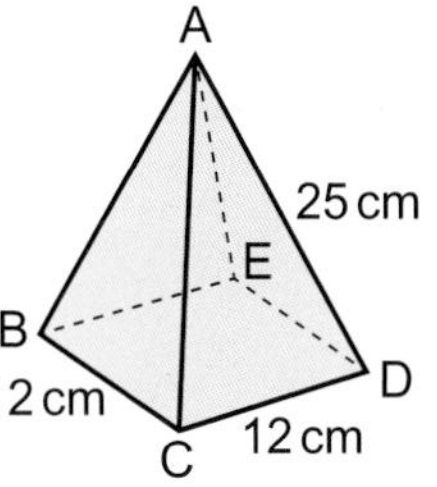

12 Modelling The horizontal distance travelled by a ball in the time, t seconds from when it is kicked, is x metres, where $x = 20t\cos\theta$.
Find the value of θ given that $x = 25$ when $t = 1.5$.

13 Problem-solving

a Find the value of x in triangle ABC.

b The exact value of the area of triangle ABC is $\sqrt{3}\,k\,\text{cm}^2$. Find the value of k.

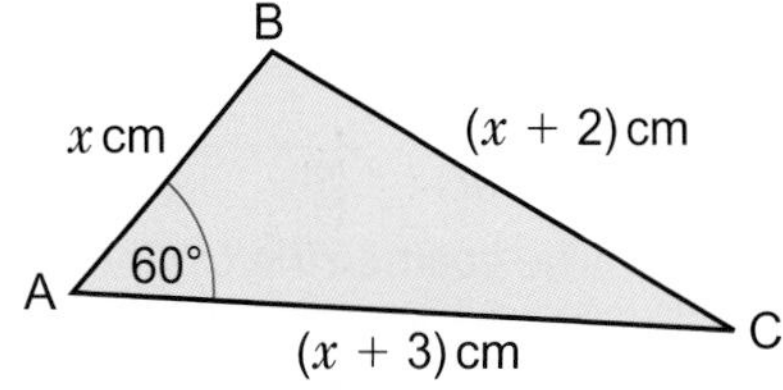

13 Knowledge check

- The **sine** graph repeats every 360° in both directions.

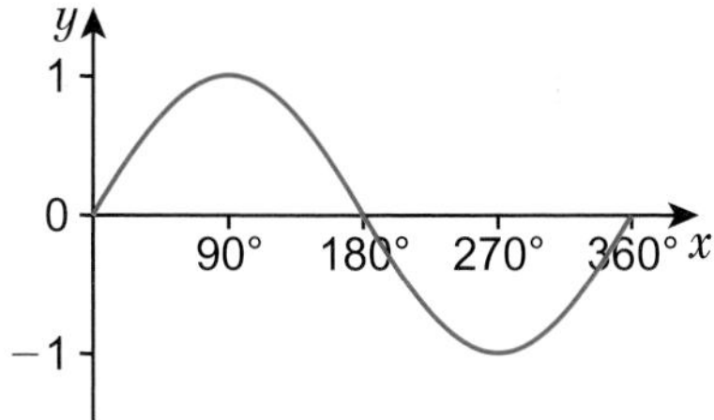

Mastery lesson 13.2

- The **cosine** graph repeats every 360° in both directions.

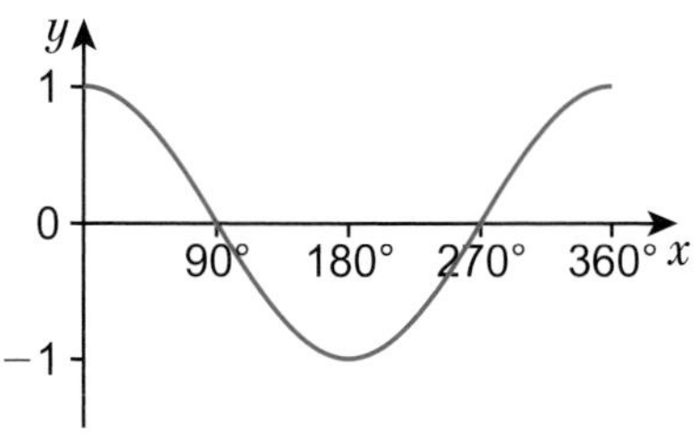

Mastery lesson 13.3

- The **tangent** graph repeats every 180° in both directions.

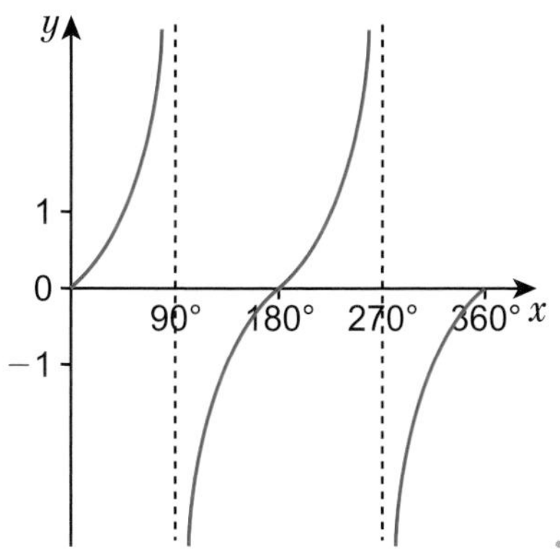

Mastery lesson 13.4

- $\tan x$ is not defined for angles of the form $(90 \pm 180n°)$ Mastery lesson 13.4

- The area of this triangle = $\frac{1}{2}ab\sin C$

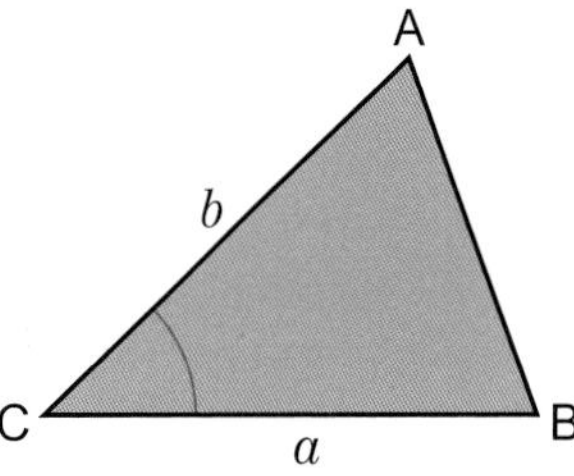

Mastery lesson 13.5

- The **sine rule** can be used in any triangle.
  - $\frac{a}{\sin A} = \frac{b}{\sin B} = \frac{c}{\sin C}$ Use this to calculate an unknown *side*.
  - $\frac{\sin A}{a} = \frac{\sin B}{b} = \frac{\sin C}{c}$ Use this to calculate an unknown *angle*. Mastery lesson 13.5
- The **cosine rule** can be used in any triangle.
  - $a^2 = b^2 + c^2 - 2bc\cos A$ Use this to calculate an unknown *side*.
  - $\cos A = \frac{b^2 + c^2 - a^2}{2bc}$ Use this to calculate an unknown *angle*. Mastery lesson 13.6
- A **plane** is a flat surface. In the diagram
  - BC is perpendicular to the plane WXYZ
  - triangle ABC is in a plane perpendicular to the plane WXYZ
  - θ is the angle between the line AB and the plane WXYZ.

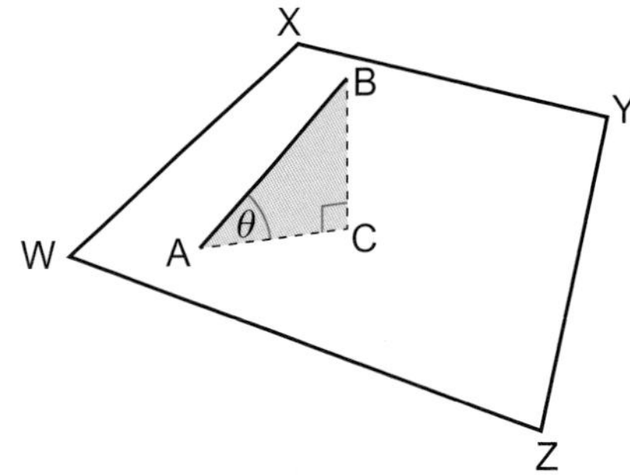

Mastery lesson 13.7

- The graph of $y = -f(x)$ is the reflection of the graph of $y = f(x)$ in the x-axis. Mastery lesson 13.8
- The graph of $y = f(-x)$ is the reflection of the graph of $y = f(x)$ in the y-axis. Mastery lesson 13.8
- The graph of $y = -f(-x)$ is a reflection of the graph of $y = f(x)$ in the x-axis and then the y-axis, or vice versa. These two reflections are equivalent to a rotation of 180° about the origin. Mastery lesson 13.8
- The graph of $y = f(x) + a$ is the translation of the graph of $y = f(x)$ by $\begin{pmatrix} 0 \\ a \end{pmatrix}$. Mastery lesson 13.9
- The graph of $y = f(x + a)$ is the translation of the graph of $y = f(x)$ by $\begin{pmatrix} -a \\ 0 \end{pmatrix}$. Mastery lesson 13.9
- The graph of $y = af(x)$ is a vertical stretch of the graph of $y = f(x)$, with scale factor a, parallel to the y-axis. Mastery lesson 13.9
- The graph of $y = f(ax)$ is a horizontal stretch of the graph of $y = f(x)$, with scale factor $\frac{1}{a}$, parallel to the x-axis. Mastery lesson 13.9

Reflect

For each statement A, B and C, choose a score:

1 – strongly disagree; 2 – disagree; 3 – agree; 4 – strongly agree

A I always try hard in mathematics

B Doing mathematics never makes me worried

C I am good at mathematics

For any statement you scored less than 3, write down two things you could do so that you agree more strongly in the future.

13 Unit test

Log how you did on your Student Progression Chart.

1 **Exam-style question**

The diagram shows the triangle *PQR*.

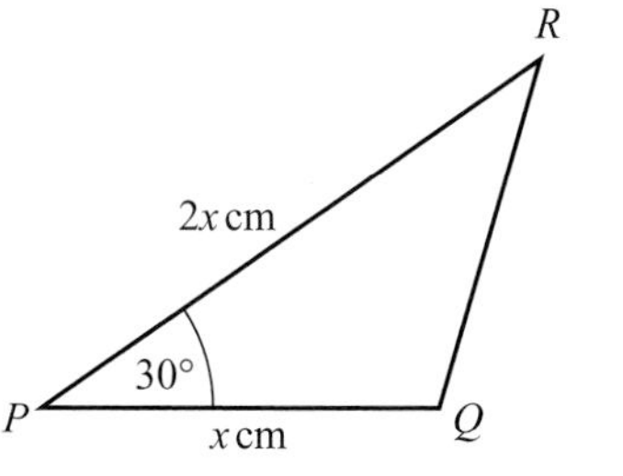

Diagram **NOT** accurately drawn

$PQ = x$ cm

$PR = 2x$ cm

Angle $QPR = 30°$

The area of triangle $PQR = A$ cm^2

Show that $x = \sqrt{2A}$ **(4 marks)**

Nov 2012, Q25, 1MA0/1H

2 **Problem-solving** The diagram shows a circle with centre O and radius 8 cm.
The chord AB has length 14 cm.
Calculate the area of the shaded segment. *(3 marks)*

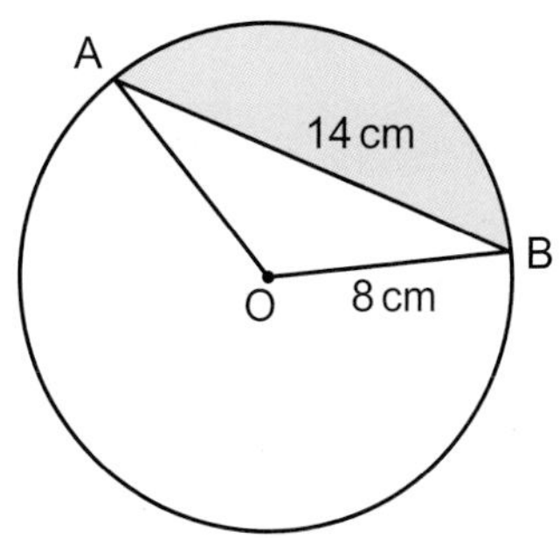

3 **Exam-style question**

PQR is a triangle.

$QR = 7.6$ cm.

Angle $PQR = 47°$.

Angle $PRQ = 64°$.

Calculate the area of triangle *PQR*.

Give your answer correct to 1 decimal place. **(5 marks)**

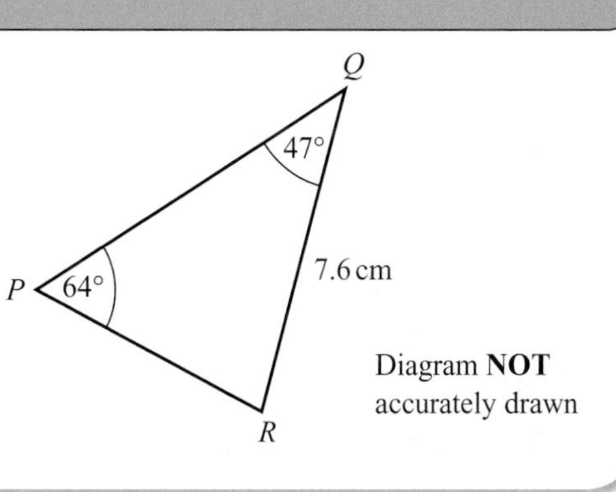

Diagram **NOT** accurately drawn

4 The diagram shows a prism with a triangular cross-section.
The base CDEF lies in a horizontal plane.
The angle between AD and the horizontal is 20°.
The angle between AC and the horizontal is 16°.
Calculate the length of AC.
Give your answer correct to 3 significant figures.

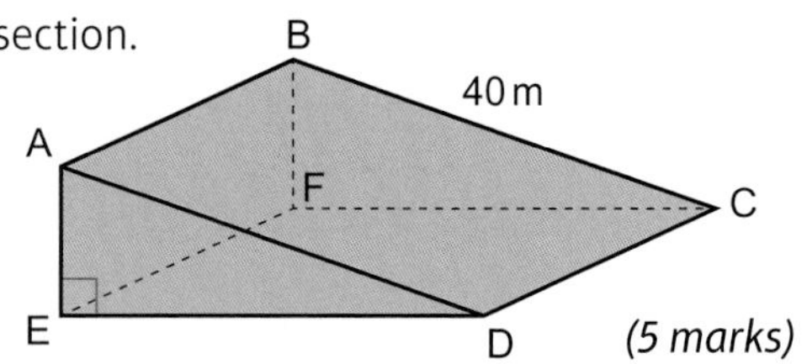

(5 marks)

5 In triangle ABC, AB = 8 cm, BC = 6 cm and angle BAC = 35°.
Find the two possible sizes of angle BCA.
Give your answers correct to 1 decimal place. *(6 marks)*

6 **Reasoning** The diagram shows the positions of towns A, B and C.

a Calculate the direct distance from A to C.

b Calculate the bearing of C from A.

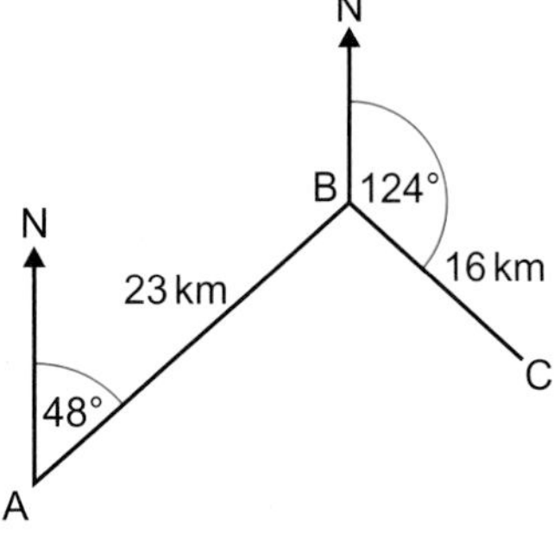

(7 marks)

7 Sketch the graph of $y = -2\sin x$ for $0° \leqslant x \leqslant 360°$. *(3 marks)*

8 a The graph of $y = \cos x$ is reflected in the x-axis.
What is the equation of the new graph?

b The graph of $y = \sin x$ is reflected in the y-axis.
What is the equation of the new graph? *(4 marks)*

9 Sketch the graph of $y = 1 + \cos 3\theta$ for $0° \leqslant \theta \leqslant 360°$. *(3 marks)*

10 a Show that *one* solution to $5\tan\theta = 7$ is 54.5° correct to 1 decimal place.

b Solve the equation $5\tan\theta = 7$ for *all* values of θ in the interval 0° to 720°. *(5 marks)*

Sample student answer

What mistake has this student made?

Exam-style question

ABC is a triangle.
$AB = 8.7$ cm.
Angle $ABC = 49°$.
Angle $ACB = 64°$.
Calculate the area of triangle ABC.
Give your answer correct to 3 significant figures.

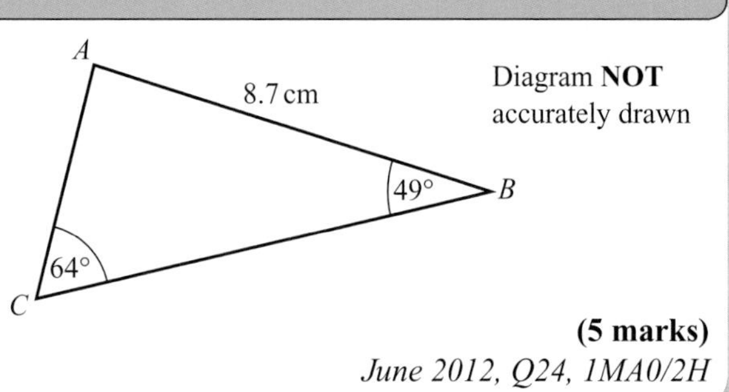

(5 marks)

June 2012, Q24, 1MA0/2H

Student answer

$\tan\theta = \frac{opp}{adj}$

$\tan 49° = \frac{AC}{8.7}$

$AC = 8.7 \times \tan 49°$

$= 10.0082\ldots$

$= 10.0$ cm (1 d.p.)

$\text{Area} = \frac{1}{2}bh$

$= \frac{1}{2} \times 8.7 \times 10.0$

$= 43.5\text{ cm}^2$

14 FURTHER STATISTICS

Statistical diagrams help us to visualise relationships between data.

This table shows the number of vehicles passing through a toll booth in one hour.

Type of vehicle	Car	Van	Lorry	Bus
Frequency	32	15	6	7

Draw a pie chart to represent this data.

14 Prior knowledge check

Numerical fluency

1 a Write 32 as a percentage of 640.
  b Work out 10% of 1120.
  c Work out 20% of 250.

2 Divide 30 people in the ratio 1:5.

3 A group of people is divided in the ratio 2:7. There are 18 people in the smaller group. How many people are there altogether?

Fluency with data

4 Classify each of these as categorical, discrete or continuous.
  a The time taken to run 100 m
  b The number of peas in a pod
  c The colour of cars passing the school gates
  d The masses of chicks in a nest
  e The number of days that it rains in a month

5 Work out the mean, median, mode and range for these data sets.
  a 3, 4, 6, 7, 7
  b 4, 5, 5, 5, 6, 6, 7, 8, 9, 10
  c 2.1, 1.8, 3.2, 4.5, 1.3, 1.8, 5.1, 4.3
  d

Number of eggs	4	5	6	7
Frequency	3	6	5	2

6 The masses of 20 badgers are recorded in this table.

Mass, m (kg)	Frequency
$9 < m \leqslant 10$	5
$10 < m \leqslant 11$	7
$11 < m \leqslant 12$	6
$12 < m \leqslant 13$	2

  a Write down the modal class.
  b Work out an estimate for the mean mass.
  c Write down the interval that contains the median.

* Challenge

7 The mean of six numbers is 7.
The numbers are 3, 5, 5, 10, x and $x - 1$.
Work out
  a the value of x
  b the median value
  c the range.

14.1 Sampling

Objectives

- Understand how to take a simple random sample.
- Understand how to take a stratified sample.

Why learn this?

To understand our behaviour, scientists often need to know our opinions. As they can't ask all 7 billion of us, they need to sample us.

Fluency

- Simplify $\frac{14}{120}$
- Find $\frac{3}{20}$ of 400

Warm up

1 The table shows the members of a leisure centre by age group.

Age	Number of members
10–19	120
20–29	130
30–39	125
40–49	120
50+	125

a How many members are there?

b What fraction of the members are in the 20–29 age group?

c What percentage of the members are in the 40–49 age group? Give your answer correct to 1 d.p.

d Dee asks 10% of the members in the 20–29 age group a question. How many members is this?

e Filipe asks 20 members in the 40–49 age group a question. What fraction of the members is this?

Key point 1

A **population** is the set of items that you are interested in.
A **census** is a survey of the whole population.
A **sample** is a smaller number of items from the population. A sample of at least 10% is considered to be a good-sized sample.

Questions in this unit are targeted at the steps indicated.

2 **Reasoning** John wants to find out about the shopping habits of people in his local town. He wonders whether to ask the first 30 people he sees in the street outside the supermarket.

a Is this sample likely to be representative of the population? Explain.

b Janna suggests that he pick 30 people at random from the Electoral Roll. Is this sample likely to be representative of the population? Explain.

Q2a hint Is every member of the population equally likely to be chosen?

Discussion Why do surveys often use a sample instead of the whole population? What factors must researchers consider when deciding on the size of the sample?

Key point 2

In order to reduce **bias**, a sample must – as far as possible – represent the whole population.

3 **Reasoning** Explain whether each of these samples is biased.

a A medical practice wants to find out patients' opinions of its service. They ask all the people in the waiting room one Friday morning.

b You ask 50 people who use a bottle bank what they think of recycling.

c A head-teacher asks the first 5 people on the register in each tutor group what they think of the new school logo.

d A market research company wants to find out views on a new advertising campaign. It conducts a telephone survey of 2 people in each of 20 towns.

ActiveLearn Homework, practice and support: Higher 14.1

Key point 3

In a **random** sample each item has the same chance of being chosen.
To select a simple random sample, draw names from a hat or use a table of random numbers.

4 Here is a display of random numbers. 86136078784880569909326602317798795097905131992...

a Follow these steps to get seven random numbers between 0 and 50.
Start with 86 and write the digits in pairs: 86, 13, …
Cross out any over 50 and any repeats.
Continue until you have seven numbers.

b Use the display to give six random numbers between 0 and 99.

Example 1

Describe how you could select a random sample of size 15 from a population of 90 people.

List them in alphabetical order of their last names. Number the list from 1 to 90. — Explain how to list the population.

Use a calculator to generate 15 random numbers between 1 and 90. Ignore any repeated numbers or numbers greater than 90. Choose these people from the list. — Explain how to use random numbers to choose the sample from the population list.

5 **Real / Communication** A company with 60 employees needs to try out a new flexi-time scheme. It decides on a random sample of 8 employees.
Explain how it could use this table of random numbers to select a sample. 46126712480699241483783765733947...
Write down the numbers of the employees who will be in the sample.

6 **Reasoning** Jasmine wants to collect some data on the sports enjoyed by members of her local leisure centre. There are 1000 members, 400 of whom are male.

a Jasmine wants a sample of 10% of the members. How many members should she ask?

b Describe how Jasmine could select a simple random sample.
She decides to ask 50 males picked at random and 50 females picked at random.

c What percentage of the males are selected?

d What percentage of the females are selected?

Discussion Is she right to ask the same number of males and females? Explain your answer.

Key point 4

A population may divide into groups such as age range or gender. These groups are called **strata**.
In a **stratified sample**, the number of people taken from each group is proportional to the group size.
Strata is the plural of stratum (meaning 'layer').

7 The table shows the number of students in each school year.

a Alfie chooses a sample of 100 students.
Show that this is 10% of the total number of students.

Q7a hint Work out the total number of students.

b Work out 10% of each year group.

c Show that taking 10% of each year group gives a sample of 100 students in total.

Year	Number of students
7	210
8	190
9	180
10	200
11	220

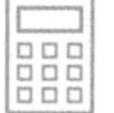

8 **Communication** A sports club manager wants to find out what the members think of the new changing room facilities. There are 350 women and 450 men in the club.

a Explain why a stratified sample should be used.

b The manager wants to survey 10% of the members. How many women and how many men should be asked?

c The club decides to ask 48 members. How many of each gender should it ask?

Q8c hint Work out the sample size as a percentage of the total number of members. $\frac{48}{\square} = \square\%$

9 **Exam-style question**

A school has 450 students.

Each student studies one of Greek or Spanish or German or French.

The table shows the number of students who study each of these languages.

Language	Number of students
Greek	45
Spanish	121
German	98
French	186

An inspector wants to look at the work of a stratified sample of 70 of these students.

Find the number of students studying each of these languages that should be in the sample. **(3 marks)**

Nov 2006, Q15, 5525/06

Exam hint
Show all your calculations clearly. You could add columns to the table.

10 **STEM / Problem-solving** A scientist wishes to find out how many fish are in a lake.

He catches 40 fish and marks them with a small tag.

Two weeks later, he returns to the lake and catches another 40 fish.

Five of the fish he catches have been marked with his tag.

a What fraction of the fish he catches in the second sample are tagged?

b Assume the fraction tagged in the sample is the same as the fraction tagged in the lake. Estimate how many fish are in the lake.

Q10b hint Let f be the number of fish in the lake. Write an expression for the fraction of tagged fish in the lake.

Key point 5

To estimate the size of the population N of an animal species:

- Capture and mark a sample size n.
- Recapture another sample of size M. Count the number marked (m).

$$\frac{n}{N} = \frac{m}{M}$$

So, $N = \frac{n \times M}{m}$

This is the capture–recapture method.

11 **STEM / Problem-solving** A naturalist captures 30 bats in a cave and tags them.
There are approximately 600 bats in the colony.
The naturalist returns a month later and captures 40 bats. How would she expect to find tagged?

14.2 Cumulative frequency

Objectives

- Draw and interpret cumulative frequency tables and diagrams.
- Work out the median, quartiles and interquartile range from a cumulative frequency diagram.

Why learn this?

Having a 'running total' of data helps you work out how many data values are less than or greater than a given number.

Fluency

How many lengths of wood are
- less than or equal to 5 m long
- less than or equal to 10 m long
- more than 5 m long?

Length of wood, l (m)	Frequency
$0 < l \leqslant 5$	4
$5 < l \leqslant 8$	6
$8 < l \leqslant 10$	7

Warm up

1 For 11, 14, 15, 16, 18, 21, 22, 25, 26, 27, 30
 a write down the median
 b which value is
  i $\frac{1}{4}$ of the way into the list ii $\frac{3}{4}$ of the way into the list?

Key point 6

A **cumulative frequency table** shows how many data values are less than or equal to the **upper class boundary** of each data class.
The **upper class boundary** is the highest possible value in each class.

2 The frequency table shows the masses of 50 cats.
Copy and complete the cumulative frequency table.

Q2 hint The cumulative frequency is like a 'running total'.

Mass, m (kg)	Frequency
$3 < m \leqslant 4$	4
$4 < m \leqslant 5$	12
$5 < m \leqslant 6$	17
$6 < m \leqslant 7$	10
$7 < m \leqslant 8$	7

Mass, m (kg)	Cumulative frequency
$3 < m \leqslant 4$	4
$3 < m \leqslant 5$	4 + 12 = ☐
$3 < m \leqslant 6$	
$3 < m \leqslant 7$	
$3 < m \leqslant 8$	

3 **STEM** This frequency table gives the heights of 70 giraffes.
Draw a cumulative frequency table for this data.

Height, h (m)	Frequency
$4.0 < h \leqslant 4.2$	2
$4.2 < h \leqslant 4.4$	3
$4.4 < h \leqslant 4.6$	5
$4.6 < h \leqslant 4.8$	8
$4.8 < h \leqslant 5.0$	12
$5.0 < h \leqslant 5.2$	18
$5.2 < h \leqslant 5.4$	15
$5.4 < h \leqslant 5.6$	7

Q3 hint Start every height group with the shortest height given in the table:
$4.0 < h \leqslant 4.2$, $4.0 < h \leqslant 4.4$, etc.

ActiveLearn Homework, practice and support: Higher 14.2

Key point 7

A **cumulative frequency diagram** has data values on the x-axis and cumulative frequency on the y-axis.

Example 2

The cumulative frequency table shows the amount of pocket money for 40 teenagers.

a Draw a cumulative frequency diagram.

b Use the cumulative frequency diagram to find an estimate for the median amount of pocket money.

c Estimate the range.

Pocket money, x (£)	Cumulative frequency
$0 < x \leqslant 1$	2
$0 < x \leqslant 2$	8
$0 < x \leqslant 3$	19
$0 < x \leqslant 4$	32
$0 < x \leqslant 5$	39
$0 < x \leqslant 6$	40

Plot (0, 0). There are no people with less than £0 pocket money.

a

Pocket money

Cumulative frequency: 0, 10, 20, 30, 40

Pocket money (£): 0, 1, 2, 3, 4, 5, 6

Plot each frequency at the upper class boundary: (1, 2), (2, 8), etc. Draw a smooth curve through the points.

To find the median, draw a line from the halfway cumulative frequency value.

Draw a line down to the x-axis and read off the value.

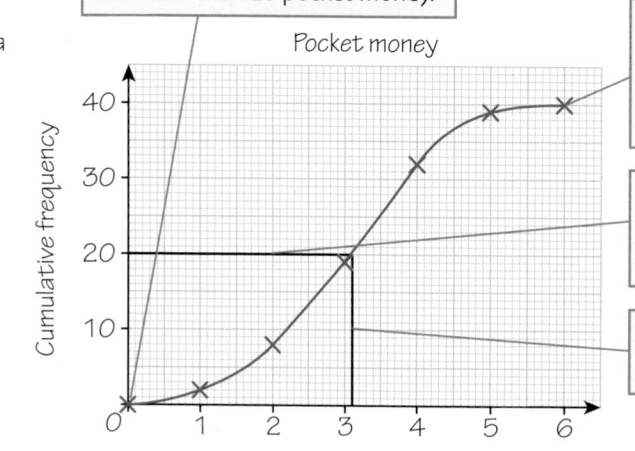

b An estimate for the median is £3.10.

c Lowest possible value = £0

Highest possible value = £6

An estimate for the range = 6 − 0 = £6

The lowest possible value is the lower class boundary of the lowest class.

The highest possible value is the upper class boundary of the highest class.

4 **Exam-style question**

This table gives the times taken by 50 students to solve a maths puzzle.

Time, t (minutes)	Frequency
$0 < t \leqslant 2$	3
$2 < t \leqslant 4$	12
$4 < t \leqslant 6$	19
$6 < t \leqslant 8$	10
$8 < t \leqslant 10$	6

a Draw a cumulative frequency diagram. **(2 marks)**

b Use the diagram to find an estimate for the median time taken. **(1 mark)**

c Estimate the range. **(2 marks)**

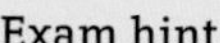

Exam hint

Remember to label both axes and plot your points at the top of the interval.

Discussion Why are your values for the median and range estimates?

5 **STEM**

a Draw a cumulative frequency diagram for the giraffe data in **Q3**.

b Find an estimate for the median height of the giraffes.

Key point 8

For a set of n data values on a cumulative frequency diagram

- the estimate for the **lower quartile** (LQ) is the $\frac{n}{4}$th value
- the estimate for the **upper quartile** (UQ) is the $\frac{3n}{4}$th value
- the **interquartile range** (IQR) = UQ − LQ

6 The time taken for 80 runners to complete a 10-kilometre fun run is shown in the table.

Time, t (minutes)	Frequency
$40 < t \leqslant 45$	3
$45 < t \leqslant 50$	17
$50 < t \leqslant 55$	25
$55 < t \leqslant 60$	26
$60 < t \leqslant 65$	8
$65 < t \leqslant 70$	1

a Draw a cumulative frequency diagram.

b Estimate the median time taken.

c Estimate the lower quartile of the time taken.

Q6c hint Draw a line across to the curve from the cumulative frequency value one quarter of the way up, then down to the x-axis.

d Estimate the upper quartile.

e Use your answers to parts **c** and **d** to work out an estimate for the interquartile range.

7 **STEM / Reasoning** The table shows the masses of 60 hippos.

Mass, m (tonnes)	Frequency
$1.3 < m \leqslant 1.4$	4
$1.4 < m \leqslant 1.5$	7
$1.5 < m \leqslant 1.6$	21
$1.6 < m \leqslant 1.7$	18
$1.7 < m \leqslant 1.8$	10

a Draw a cumulative frequency diagram.

b Estimate the median, quartiles and interquartile range.

c Estimate how many hippos weigh less than 1.55 tonnes.

Q7c hint Draw a line to the curve from 1.55 on the x-axis. Read off the cumulative frequency axis.

d Copy and complete:

40 hippos are estimated to weigh less than ______ tonnes.

Discussion What assumption have you made that could affect your answers to parts **c** and **d**?

8 **Exam-style question**

Charlie drives to work.

The table gives information about the time (t minutes) it took him to get to work on each of 100 days.

Time, t (minutes)	Frequency
$0 < t \leqslant 10$	16
$10 < t \leqslant 20$	34
$20 < t \leqslant 30$	32
$30 < t \leqslant 40$	14
$40 < t \leqslant 50$	4

a Draw a cumulative frequency table. **(1 mark)**

b Draw a cumulative frequency diagram. **(2 marks)**

c Use your diagram to find an estimate of the median time. **(1 mark)**

d Use your diagram to find an estimate for the number of days it took Charlie more than 18 minutes to drive to work. **(2 marks)**

Exam hint

For parts **c** and **d** draw lines on the diagram with a ruler to show how you got your answers.

14.3 Box plots

Objectives

- Find the quartiles and the interquartile range from stem-and-leaf diagrams.
- Draw and interpret box plots.

Why learn this?

Simple diagrams help us to interpret and compare data.

Fluency

- In a set of 50 data values, which value is the median?
- The upper quartile is 26 and the lower quartile is 10. What is the interquartile range?

Warm up

1 The stem-and-leaf diagram gives the ages of the members of a judo club.

1	3 5 8 8
2	0 1 1 2 5 6 6 7 9
3	1 7
4	0 5
5	1

Key: 3 | 7 represents 37 years

a Find the median age. **b** Work out the range.

Key point 9

A **box plot**, sometimes called a **box-and-whisker diagram**, displays a data set to show the median and quartiles.

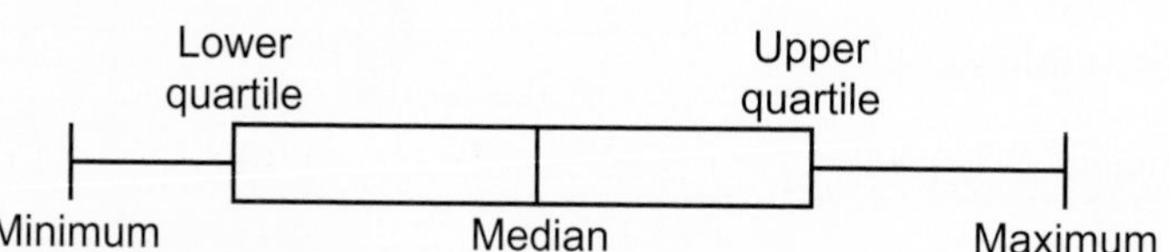

Summary statistics for a set of data are the averages, ranges and quartiles.

Example 3

The table shows **summary statistics** from a data set of the lengths of ladybirds.

Minimum	Lower quartile	Median	Upper quartile	Maximum
3 mm	5 mm	8 mm	9 mm	11 mm

Draw a box plot for the data.

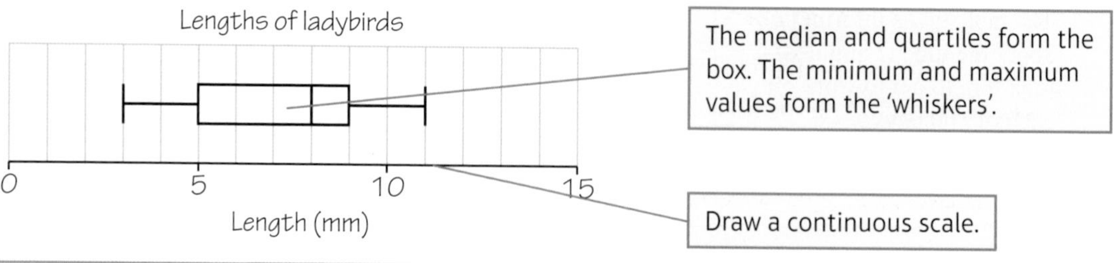

The median and quartiles form the box. The minimum and maximum values form the 'whiskers'.

Draw a continuous scale.

2 Draw a box plot for this data on the masses of tomatoes.

Minimum	LQ	Median	UQ	Maximum
120 g	135 g	140 g	150 g	154 g

Q2 hint LQ = lower quartile
UQ = upper quartile

Key point 10

For a set of n data values

- the lower quartile (LQ) is the $\frac{n+1}{4}$th value
- the upper quartile (UQ) is the $\frac{3(n+1)}{4}$th value.

3 **Reasoning** This data shows the length of time, in minutes, it took 11 students to complete an essay.

15, 18, 19, 21, 22, 25, 26, 26, 28, 30, 31

a Write down the median time taken. b Find the upper and lower quartiles.

c Draw a box plot for the data.

Discussion Why do you use the $\frac{n+1}{4}$th data value for the lower quartile in a data set but the $\frac{n}{4}$th value in a cumulative frequency diagram?

4 This data shows the heights, in metres, of 15 trees.

4.5, 5.3, 11, 4.8, 6.1, 10.2, 5.8, 7.3, 8, 9.6, 6.3, 8.8, 4.9, 6, 8

Draw a box plot for the data.

5 **Reasoning** This stem-and-leaf diagram shows the ages of 37 people on a bus.

a What age was the youngest passenger?

b What is the median age?

c Find the lower and upper quartiles of the ages.

d Work out the interquartile range.

e Draw a box plot.

1	6 7 9 9
2	1 2 2 4 6 7 8
3	3 6 6 7 8 9 9
4	0 0 1 3 4 5 8 8 9 9
5	0 1 2 4 4 5 6 7 9

Key: 3 | 6 represents 36 years

Key point 11

Comparative box plots are box plots for two different sets of data drawn in the same diagram.

6 The ages of children in two different after-school clubs are shown in the comparative box plots.

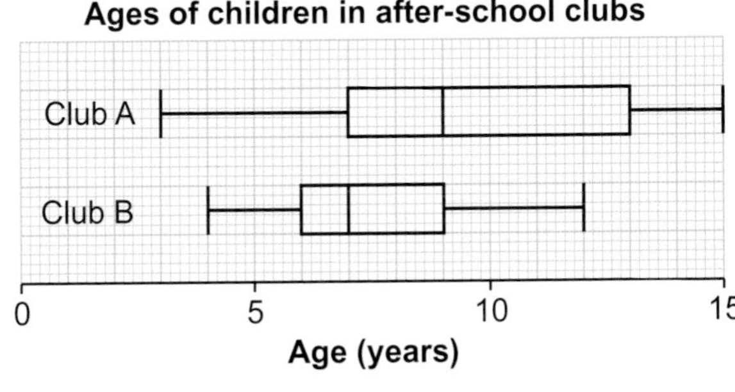

a Which club had the higher median age? b Work out the interquartile range for each club.

c Work out the range for each club.

7 **Reasoning** Summary statistics on the masses, in grams, of two different species of birds are given in this table.

	Minimum	LQ	Median	UQ	Maximum
Species A	45	52	60	65	69
Species B	33	44	65	77	90

a Draw comparative box plots for the two species.

b Compare the two species.

Q7a hint Use the same scale and draw one box plot above the other.

Q7b hint Compare the medians, interquartile ranges and ranges.

8 **STEM / Reasoning** The cumulative frequency graph gives information about the masses, in kilograms, of 36 male and 36 female gibbons.

a Use the graph to find the median and quartiles for each gender.

b Draw comparative box plots for the two genders.

c Compare the two genders.

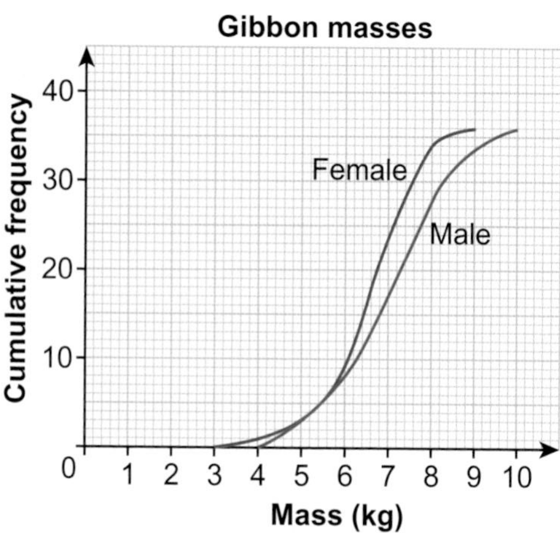

9 **Exam-style question**

40 boys each completed a puzzle.

The cumulative frequency graph gives information about the times it took them to complete the puzzle.

a Use the graph to find an estimate for the median time. **(1 mark)**

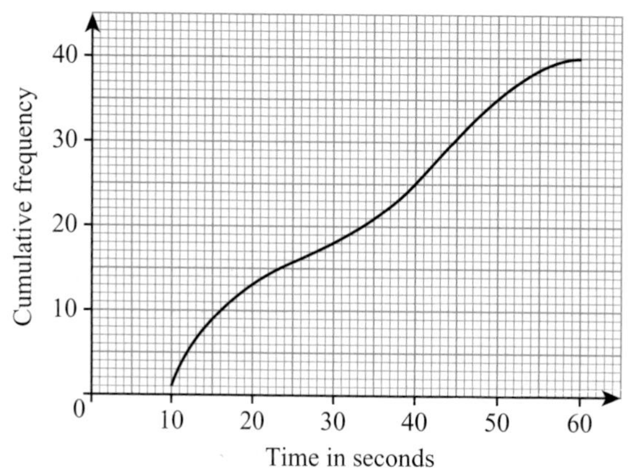

For the boys the minimum time to complete the puzzle was 9 seconds and the maximum time to complete the puzzle was 57 seconds.

b Use this information and the cumulative frequency graph to draw a box plot showing information about the boys' times.
Use a scale like this:

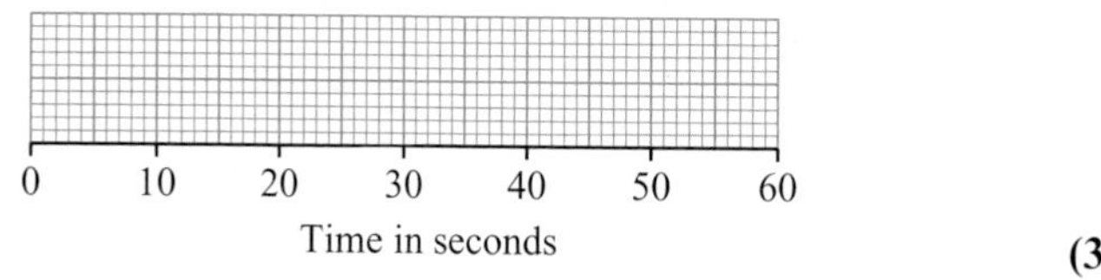

(3 marks)

The box plot below shows information about the times taken by 40 girls to complete the same puzzle.

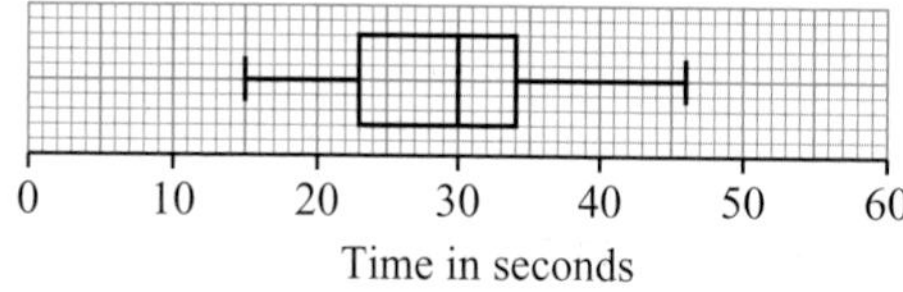

c Make *two* comparisons between the boys' times and the girls' times. **(2 marks)**

June 2004, Q13, 5505/05

Exam hint
Compare an average and a measure of spread.
Be sure to put your comparisons in the context of the question.

14.4 Drawing histograms

Objectives

- Understand frequency density.
- Draw histograms.

Why learn this?

Bar charts and frequency diagrams show data grouped in *equal* class intervals.
For data grouped in *unequal* class intervals, you need a histogram.

Fluency

Work out • 32 ÷ 10 • 158 ÷ 20 • 30 ÷ 0.2 • 8 ÷ 0.05

Warm up

1 The masses of 101 birds are recorded in a table.

Mass, m (grams)	$10 < m \leqslant 12$	$12 < m \leqslant 14$	$14 < m \leqslant 16$	$16 < m \leqslant 18$	$18 < m \leqslant 20$
Frequency	11	23	31	20	16

a Draw a frequency diagram for the data.
b Write down the modal class.
c Calculate an estimate of the mean mass.

Key point 12

In a **histogram** the area of the bar represents the frequency. The height of each bar is the frequency density.

Frequency density = $\dfrac{\text{frequency}}{\text{class width}}$

2 The heights of 70 trees are recorded in this table.

a Work out each class width.
b Work out the frequency density for each class.

Height, h (metres)	**Frequency**	**Class width**	**Frequency density**
$20 < h \leqslant 25$	8	5	$\frac{8}{5} = 1.6$
$25 < h \leqslant 30$	12		
$30 < h \leqslant 40$	35		
$40 < h \leqslant 50$	15		

3 Work out the frequency density for each class in **Q1**.

Example 4

The lengths of 48 worms are recorded in this table.

Length, x (mm)	$15 < x \leqslant 20$	$20 < x \leqslant 30$	$30 < x \leqslant 40$	$40 < x \leqslant 60$
Frequency	6	14	26	2

Draw a histogram to display this data.

$6 \div 5 = 1.2,\ 14 \div 10 = 1.4,\ 26 \div 10 = 2.6,\ 2 \div 20 = 0.1$ — Work out the frequency density for each class

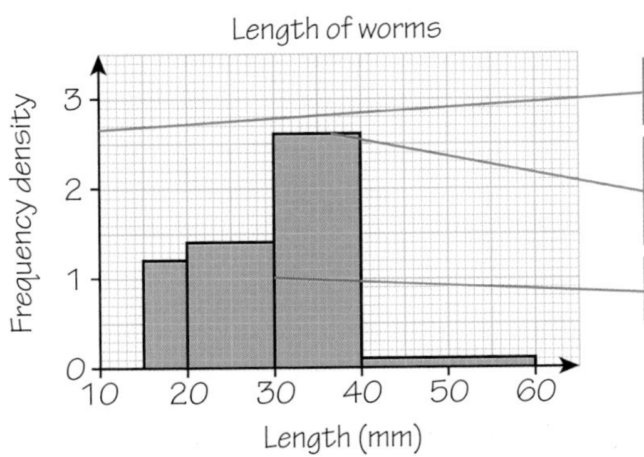

Label the y-axis 'Frequency density'.

The height of each bar is the frequency density for each class.

Draw the bars with no gaps between them.

4 This table shows the times taken for 55 runners to complete a fun run.

Time, t (minutes)	$40 < t \leq 45$	$45 < t \leq 50$	$50 < t \leq 60$	$60 < t \leq 80$
Frequency	4	17	22	12

Draw a histogram for this data.

5 This table contains data on the heights of 76 students.

Height, h (m)	$1.50 < h \leq 1.52$	$1.52 < h \leq 1.55$	$1.55 < h \leq 1.60$	$1.60 < h \leq 1.65$	$1.65 < h \leq 1.80$
Frequency	4	18	25	15	14

Draw a histogram for this data.

6 **Exam-style question**

Fred did a survey on the areas of pictures in a newspaper.
The table gives information about the areas.

Area, A (cm²)	**Frequency**
$0 < A \leq 10$	38
$10 < A \leq 25$	36
$25 < A \leq 40$	30
$40 < A \leq 60$	46

a Work out an estimate for the mean area of a picture. **(4 marks)**

b Draw a histogram for the information given in the table. **(3 marks)**

Nov 2005, Q10, 5525/06

Exam hint
Use the midpoint of each class interval.

14.5 Interpreting histograms

Objective

- Interpret histograms.

Why learn this?

Reading accurately from statistical diagrams is important to draw accurate conclusions.

Fluency

What fraction of a set of data is greater than the median?

Warm up

1 The masses of 80 apples are recorded in a table.

Mass, m (grams)	$100 < m \leq 110$	$110 < m \leq 120$	$120 < m \leq 130$	$130 < m \leq 140$
Frequency	16	22	29	13

a Work out an estimate for the mean mass.

b Find the class interval that contains the median.

2 **STEM / Reasoning** The histogram shows the masses of a number of squirrels.

a How many squirrels weigh between 150 grams and 200 grams?

b How many squirrels weigh between 200 grams and 400 grams?

c How many squirrels are there in total?

Discussion What does the area of the bar on the histogram tell you?

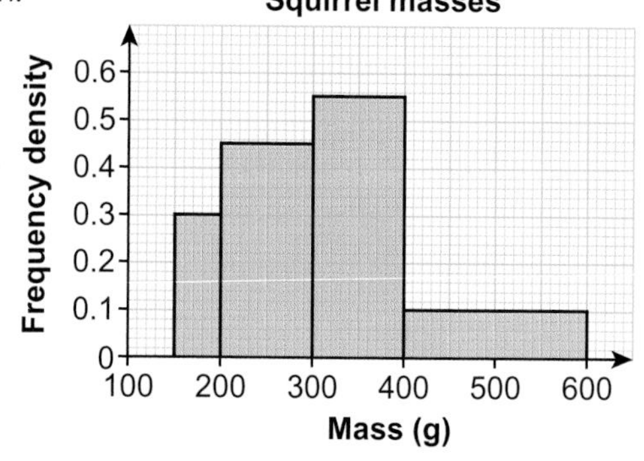

3 **Reasoning** The histogram shows the distance a group of football fans have to travel to a match.

a How many fans travelled less than 5 km?

b Estimate how many fans travelled less than 15 km.

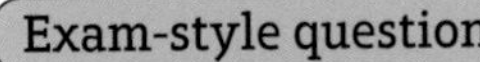

Q3b hint How many fans travelled between 10 and 15 km?

c Estimate how many fans travelled between 25 and 32 km.

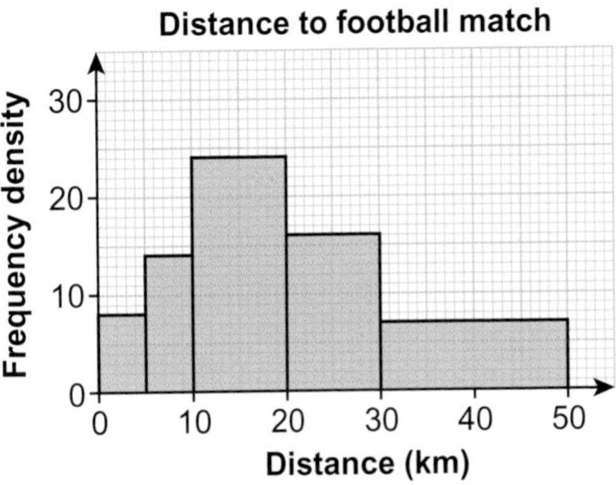

4 **Exam-style question**

The incomplete table and histogram give some information about the ages of the people who live in a village.

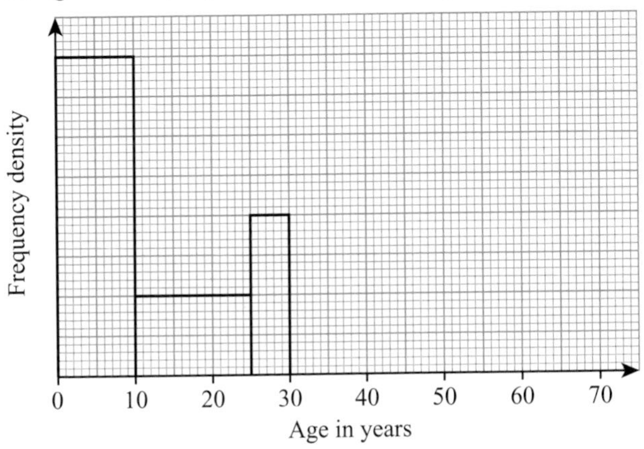

a Use the information in the histogram to complete the frequency table below. **(2 marks)**

Age (x) in years	Frequency
$0 < x \leqslant 10$	160
$10 < x \leqslant 25$	
$25 < x \leqslant 30$	
$30 < x \leqslant 40$	100
$40 < x \leqslant 70$	120

b Complete the histogram. **(2 marks)**

June 2003, Q19, 5505/05

Exam hint
Draw the bars on the histogram neatly with a ruler.
Show your working to calculate the frequencies.

5 **Problem-solving** The histogram shows the times taken for a number of students to complete an arithmetic test.

a Draw a grouped frequency table for the data.

b Work out an estimate for the mean time taken.

c How many students took longer than 19 minutes to complete the test?

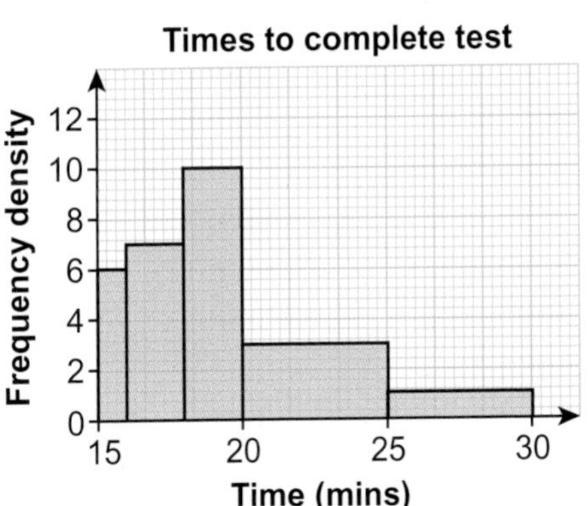

Example 5

The histogram shows the masses of pumpkins in a farm shop.

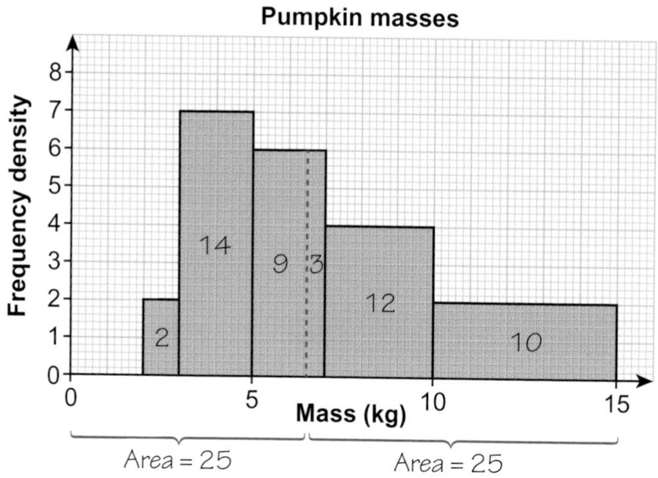

Area = 25 Area = 25

Work out an estimate for the median mass.

Total frequency = $1 \times 2 + 2 \times 7 + 2 \times 6 + 3 \times 4 + 5 \times 2 = 50$

The median is the 25.5th value and lies in the class $5 < m \leqslant 7$.

Frequency = area = 9, frequency density = 6. Class width = $9 \div 6 = 1.5$

An estimate for the median is $5 + 1.5 = 6.5$ kg

Work out the areas of all the bars to find the total frequency.

Work out which class contains the median.

Use frequency density $= \dfrac{\text{frequency}}{\text{class width}}$ to find class width of class from 5 to median.

Add the class width to the lower class boundary.

6 Work out an estimate for the median of the data in **Q4**.

7 **Reasoning** The histogram shows the heights of Year 7 students.

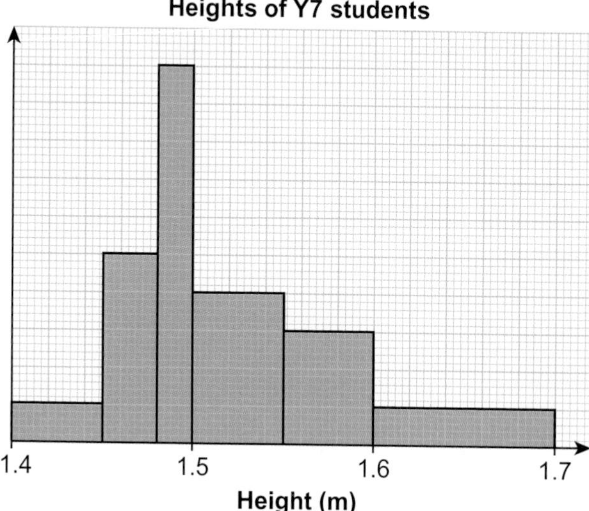

a Five Year 7 students are between 1.4 metres and 1.45 metres tall. Work out the frequency density for that class.

b How many Year 7 students are there in total?

c Work out an estimate for the median height.

d Draw a frequency table for the data in the histogram.

e Work out an estimate for the mean height from your frequency table.

f How many of the students are taller than the mean height?

Q7b hint Label the frequency density scale.

8 **Problem-solving** The histogram shows the masses of some elephants.

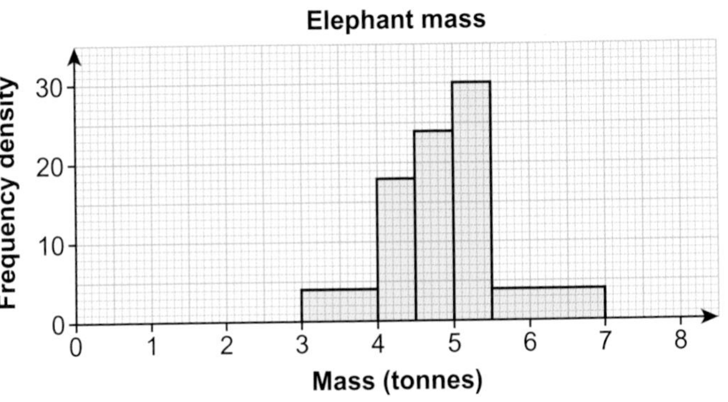

Q8 hint To estimate the median, first find the class containing the median mass.

a How many elephants are there in total?

b Work out an estimate of the median mass.

c Estimate how many elephants weigh more than 5.2 tonnes.

14.6 Comparing and describing populations

Objective

- Compare two sets of data.

Why learn this?

Market research companies analyse our responses to surveys and look for differences between genders and age groups.

Fluency

What are the measures of

- spread
- average?

Warm up

1 Work out the mean, median, mode and range for these sets of data.

a 1.2, 1.3, 1.4, 1.5, 1.5

b 4, 5, 6, 4, 3, 6, 7, 8, 3, 3

2 **STEM** The table gives the masses, in kilograms, of male and female giant tortoises.

Male	280	283	288	290	292	299	300	305	310
Female	260	261	263	265	269	270	271	273	274

a Work out the mean mass for each gender.

b Write down the range of masses for each gender.

Key point 13

The interquartile range measures the spread of the middle 50% of the data.

To describe a data set (or population) give a measure of average and a measure of spread.

To compare data sets, compare a measure of average and a measure of spread.

3 **STEM** Compare the masses of the two genders of tortoise in **Q2**.

4 **STEM** The heights, in centimetres, of female African and Asian elephants are shown in the table.

African	270	275	281	286	290	292	295
Asian	220	221	223	224	226	227	229

a Describe the heights of these two populations of elephants.

b Compare the heights of the two species.

Q4a hint Use the median and the interquartile range.

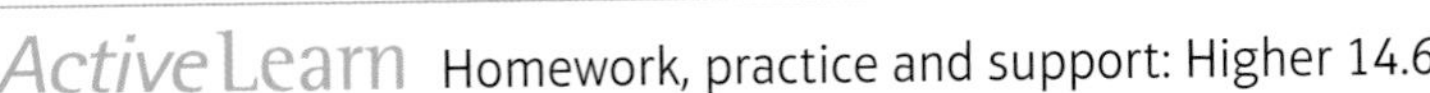

Homework, practice and support: Higher 14.6

5 **Real** The lengths, in minutes, of telephone calls to a helpline are recorded:
10, 11, 13, 15, 17, 18, 18, 19, 21, 22, 95

a Work out the mean call length.

b Work out the median call length.

c Work out the range and interquartile range.

Discussion Which measures of average and spread best represent this data?

Key point 14

The median and interquartile range are not affected by extreme values or **outliers**.

6 **Reasoning** Ten male and ten female cyclists compete in a road race.
The times, in minutes, to complete the course are recorded.

Males: 68, 70, 75, 76, 77, 79, 81, 83, 90, 120

Females: 71, 75, 76, 78, 83, 86, 89, 90, 91, 92

a Explain which of the median and interquartile range or mean and range should be used to describe each set of data.

b Compare the times for males and females.

7 **Real / Problem-solving** The table shows the lengths of delays to trains at Stratfield station.

Length of delay, x (minutes)	Frequency
$0 \leqslant x \leqslant 5$	3
$5 < x \leqslant 10$	7
$10 < x \leqslant 15$	12
$15 < x \leqslant 20$	6
$20 < x \leqslant 25$	2

a Draw a cumulative frequency diagram.

b Find the median and interquartile range.

The box plot shows the delays to trains at Westford station.

Train delays at Westford station

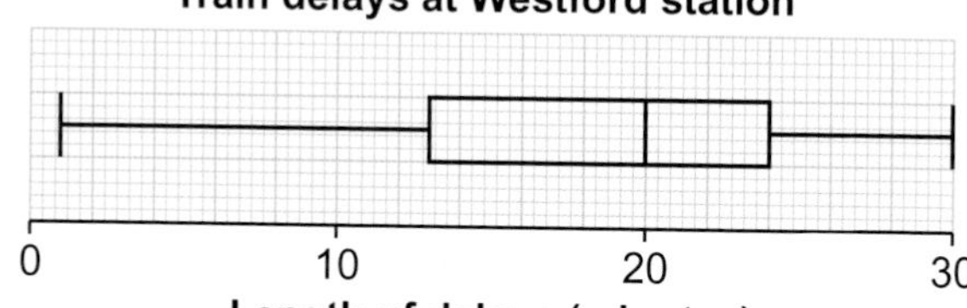

c Compare the lengths of delays at the two stations.

8 **Real / Problem-solving** This back-to-back stem-and-leaf diagram shows the average speeds, in miles per hour, of cars passing two checkpoints.

Checkpoint A		Checkpoint B
1 2 3 8	1	
1 2 4 5	2	3 4 6 8
1 2 2 4 5 6	3	3 1 4 5 6 6 7 9 9
2 3 4 5	4	1 2 2 2 3 4 6 7 8
1 2	5	

Key: 5 | 2 | 3 represents 25 mph and 23 mph

a Describe the speeds at the two check points.

b Compare the speeds at the two checkpoints.

Q8 strategy hint
Compare the medians and interquartile ranges.

9 **Problem-solving** The cumulative frequency graph shows the ages of male and female members of a health club.

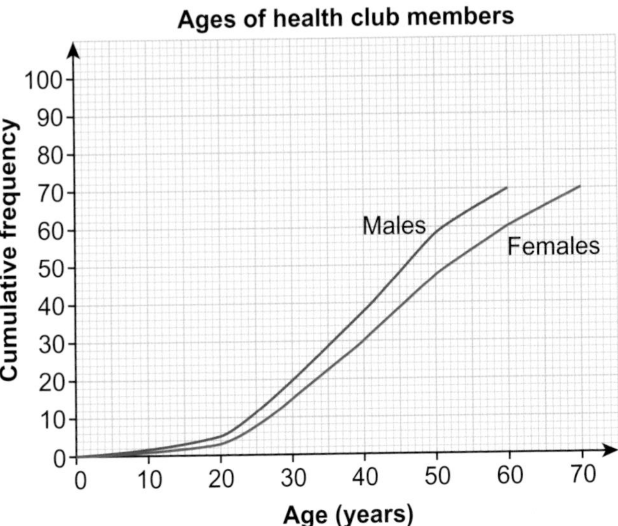

Compare the two sets of data.

10 **Exam-style question**

Chloe did a survey on the amount of money that men spent shopping each week. The cumulative frequency table shows her results.

Amount spent, x (£)	Cumulative frequency
$0 < x \leqslant 10$	9
$0 < x \leqslant 20$	20
$0 < x \leqslant 30$	34
$0 < x \leqslant 40$	51
$0 < x \leqslant 50$	68
$0 < x \leqslant 60$	80

A similar survey of women gave a median of £42 and an interquartile range of £12.

Compare the amounts of money spent by women with the amounts spent by men. **(5 marks)**

Exam hint
Draw a cumulative frequency diagram for the results given in the table.

14 Problem-solving: Brain training

Objectives
- Produce and interpret box plots from lists of data and grouped data.
- Compare distributions and make inferences.

A company wishes to investigate the effectiveness of their brain training software. They test how long it takes a group of teenagers to solve a logical puzzle, allow them to use the software for a week, and then test them again. Below are the results (A), giving the times in seconds.

1 Draw a pair of parallel box plots to display these results.

Q1 hint Parallel box plots are drawn using the same axis.

2 Compare the two distributions. Did the brain training work? Did it have the same effect on everyone?

Q2 hint Look at how the median, the quartiles and the minimum and the maximum have each changed.

The company decides to repeat the investigation with adults, using a larger sample size.
Their results (B) are given in the table below.

3 Use an appropriate method to estimate the maximum, minimum, median and quartiles of these data sets.

Q3 hint You might find it useful to draw another type of graph first to help you estimate the median and quartiles from grouped data.

4 Draw another pair of parallel box plots to display these two new sets of results. Comment on the two 'adult' distributions.

5 Compare and contrast all four sets of data. Does the company's brain training software work?

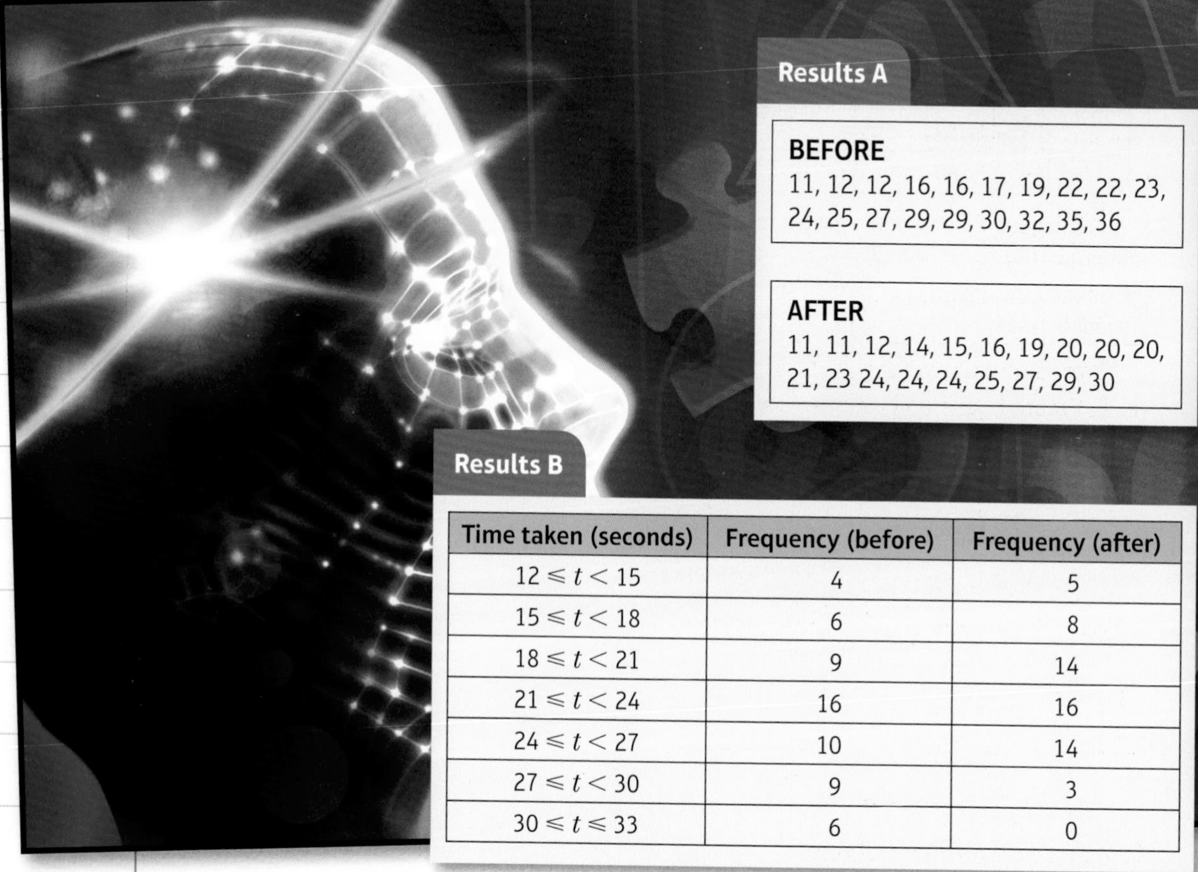

Results A

BEFORE
11, 12, 12, 16, 16, 17, 19, 22, 22, 23, 24, 25, 27, 29, 29, 30, 32, 35, 36

AFTER
11, 11, 12, 14, 15, 16, 19, 20, 20, 20, 21, 23 24, 24, 24, 25, 27, 29, 30

Results B

Time taken (seconds)	Frequency (before)	Frequency (after)
$12 \leqslant t < 15$	4	5
$15 \leqslant t < 18$	6	8
$18 \leqslant t < 21$	9	14
$21 \leqslant t < 24$	16	16
$24 \leqslant t < 27$	10	14
$27 \leqslant t < 30$	9	3
$30 \leqslant t \leqslant 33$	6	0

14 Check up

Log how you did on your Student Progression Chart.

Sampling

1 Jaden wants to take a stratified sample of size 80 from the students in his school.

Year	7	8	9	10	11
Students	320	360	280	340	300

a How many of each year group should he ask?

b He has a numbered list of all the students in Year 7. Use the display of random numbers to write down the numbers of the first five students that he should ask.

1122839815211857534291912552891386004761...

Graphs and charts

2 This table shows the masses of 90 emperor penguins.

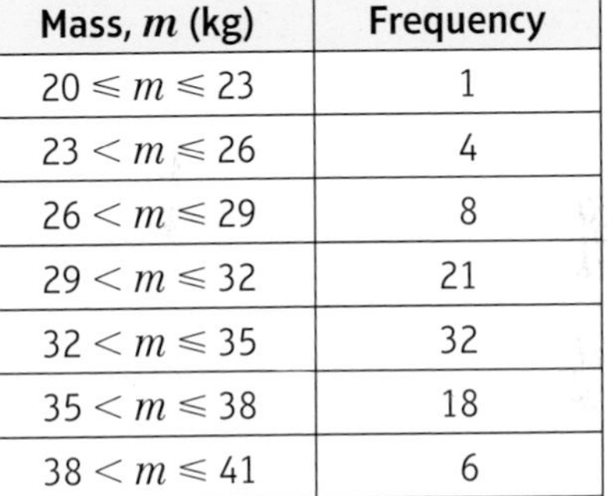

Mass, m (kg)	**Frequency**
$20 \leqslant m \leqslant 23$	1
$23 < m \leqslant 26$	4
$26 < m \leqslant 29$	8
$29 < m \leqslant 32$	21
$32 < m \leqslant 35$	32
$35 < m \leqslant 38$	18
$38 < m \leqslant 41$	6

a Draw a cumulative frequency table.

b Draw a cumulative frequency graph.

c Use your graph to estimate the median mass of the penguins.

d Find the lower quartile, upper quartile and interquartile range.

e Estimate how many penguins weigh
i less than 36 kg ii more than 30 kg.

3 The heights of 100 pine trees are given in this table.

Height, h (m)	**Frequency**
$0 < h \leqslant 10$	3
$10 < h \leqslant 15$	7
$15 < h \leqslant 20$	14
$20 < h \leqslant 22$	31
$22 < h \leqslant 25$	27
$25 < h \leqslant 30$	16
$30 < h \leqslant 40$	2

a Draw a histogram for this data.

b Estimate how many trees are taller than 23 m.

Comparing data

4 The box plots show the times for 15 boys and 15 girls to run 100 m.

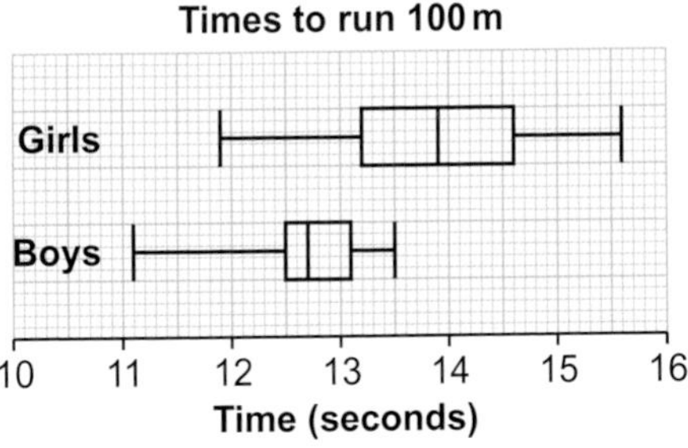

Compare the two distributions.

5 The stem-and-leaf diagram shows the ages of people attending a birthday party at a hotel.

0	3 5 6 8 9
1	2 2 5 8 8 9
2	1 3 6 6 7 8 8 9 9
3	2 4 4 5 6 6 7
4	1 1 4 5 5 8
5	4 7

Key: 1 | 2 represents 12 years

a What is the median age?

b Work out the interquartile range.

c At a second party, the median age was 22 and the interquartile range was 10. Compare the two distributions.

Reflect

6 How sure are you of your answers? Were you mostly

Just guessing ☹ Feeling doubtful 😐 Confident ☺

What next? Use your results to decide whether to strengthen or extend your learning.

* Challenge

7 The histogram shows the masses of a group of porcupines.

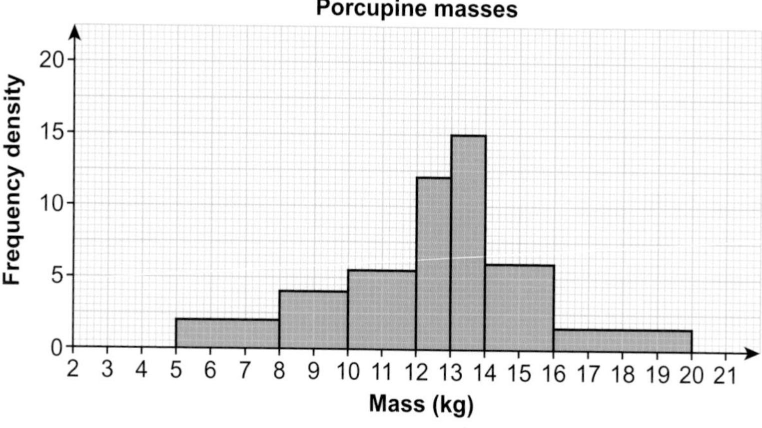

Estimate how many porcupines weigh between 8.5 kg and 16.5 kg.

14 Strengthen

Sampling

1 The membership of a hockey club is 75 men and 40 women.

a The club manager wants to know members' views on the proposed new kit. Should she ask

A the same number of men and women

B more men than women

C more women than men?

> **Q1a hint** Consider the number of men and women in the club.

b She wants to survey 20% of the members. Work out

i 20% of 75   ii 20% of 40.

c There are 115 club members. Work out 20% of 115.

d What do you notice about your answers to parts **b** and **c**?

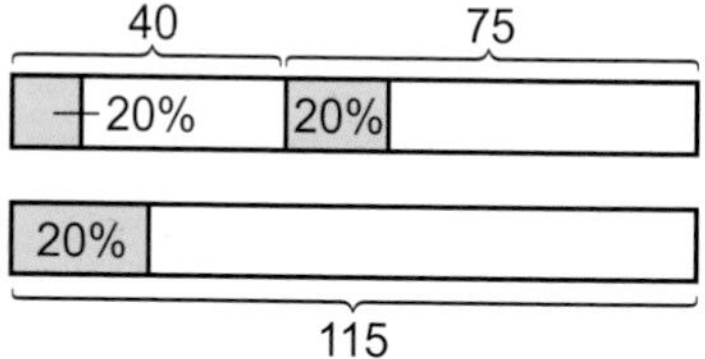

2 Ellen wants to take a stratified sample of the people at her youth club.
The table shows the ages of the members.

Age	13	14	15	16	17
Number of people	16	28	32	24	20

a She wants to ask a sample of 24 people.
What is the total number of people?

b What percentage sample is 24 people?

c Work out this percentage of each age group.

Q2b hint 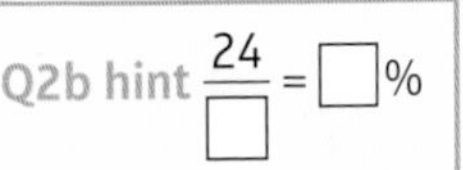

Q2c hint Check that your answers add up to 24.

3 Use this random number display to select five people from a list of 30 people.

02792151210801570187337317701840212420666 2...

a Write the list as a sequence of two-digit numbers.

b Cross out the numbers over 30.

c Cross out any repeats.

d Write down, in order, the numbers of the people who are chosen.

Q3a hint The list starts 02, 79, …

Q3d hint 01 represents the first person on the list.

4 Use the random number table from **Q3** to select a random sample of eight people from a list of 70 people.

Q4 hint Cross out all numbers over 70.

5 Use the random number table from **Q3** to select a random sample of five people from a list of 130 people.

Q5 hint Write the random numbers as 3-digit numbers.

Graphs and charts

1 The table shows the masses of 60 sheep.

Mass, m (kg)	Frequency
$70 \leqslant m \leqslant 75$	1
$75 < m \leqslant 80$	8
$80 < m \leqslant 85$	19
$85 < m \leqslant 90$	15
$90 < m \leqslant 95$	10
$95 < m \leqslant 100$	7

a Copy and complete this cumulative frequency table.

Mass, m (kg)	Cumulative frequency
$70 \leqslant m \leqslant 75$	1
$70 < m \leqslant 80$	1 + 8 = □
$70 < m \leqslant 85$	
$70 < m \leqslant 90$	
$70 < m \leqslant 95$	
$70 < m \leqslant 100$	

Q1a hint For the class $70 < m \leqslant 85$, add the frequencies for all the groups $\leqslant 85$.

b Copy and complete the cumulative frequency graph.

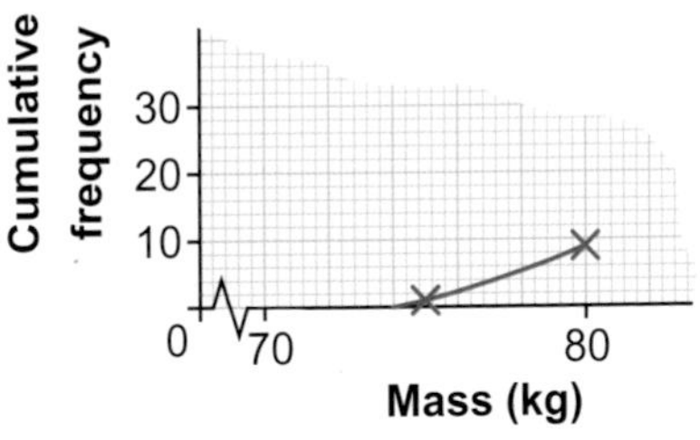

Q1b hint Plot the top value in each class against frequency. Draw a smooth curve through the points.

2 The cumulative frequency graph shows information about the amount of pocket money 40 children receive.

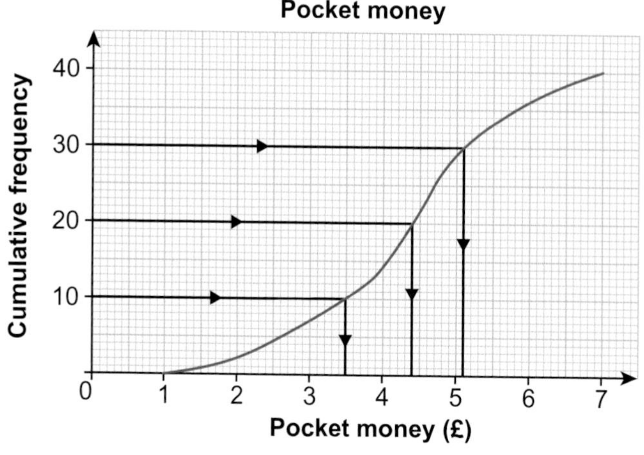

Q2a hint The median is halfway up the cumulative frequency axis. The lower quartile is $\frac{1}{4}$ of the way up. The upper quartile is $\frac{3}{4}$ of the way up.

a Read off the values for the median, lower quartile and upper quartile.

b Work out the interquartile range for the pocket money.

Q2b hint Interquartile range = upper quartile – lower quartile

3 The cumulative frequency graph shows the lengths of sticks gathered to make a camp fire.

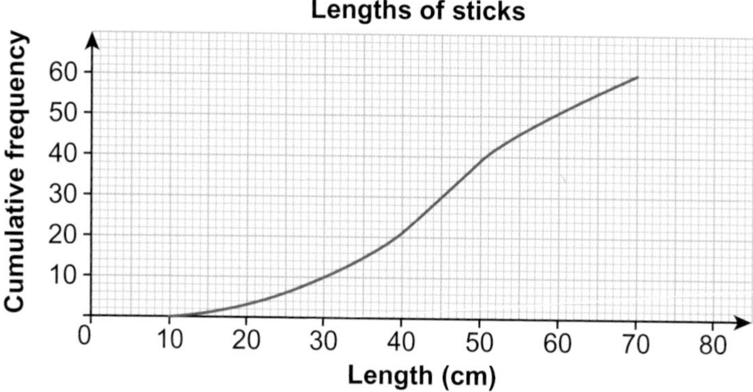

a How many sticks were gathered?

b What value is halfway up the y-axis?

c Work out the median length.

d Work out the lower and upper quartiles.

e Work out the interquartile range.

Q3c hint Draw a line across to the curve from the value you worked out in part **b**. Then go down to the x-axis.

4 The table shows the half-marathon times for 50 elite runners.

Time, t (mins)	Frequency
$65 \leqslant t \leqslant 68$	4
$68 < t \leqslant 71$	7
$71 < t \leqslant 74$	17
$74 < t \leqslant 77$	13
$77 < t \leqslant 80$	9

a Draw a cumulative frequency table.

b Draw a cumulative frequency graph.

c Work out estimates for the median and quartiles.

d Estimate how many runners took less than 72 minutes.

e Estimate how many runners took more than 72 minutes.

Q4d hint Draw a line up to the curve from 72 on the x-axis. Read across to the y-axis.

Q4e hint Total frequency – the number less than 72 minutes

5 Draw a box plot for the data in **Q4**.

a Copy the x-axis scale from **Q4**.

b Mark on the median, lower and upper quartiles with vertical lines.

c Complete the box with ends at the lower and upper quartiles.

d Mark on the lowest possible and highest possible values from the frequency table and join with lines.

Q5 hint

Lower quartile

Upper quartile

Minimum

Median

Maximum

6 The lengths of some caterpillars are shown in the table.

Length, l (mm)	Frequency	Class width	Frequency density
$10 \leqslant l \leqslant 15$	2	$15 - 10 = 5$	$2 \div 5 = 0.4$
$15 < l \leqslant 20$	8	$20 - 15 = \square$	$8 \div \square = \square$
$20 < l \leqslant 30$	15		
$30 < l \leqslant 40$	12		
$40 < l \leqslant 60$	5		

a For the class $15 < l \leqslant 20$, work out

i the class width   ii the frequency density.

b Copy and complete the table.

c Copy and complete the histogram.

Q6b hint Divide the frequency by the class width.

Q6c hint Plot frequency **density** on the y-axis.

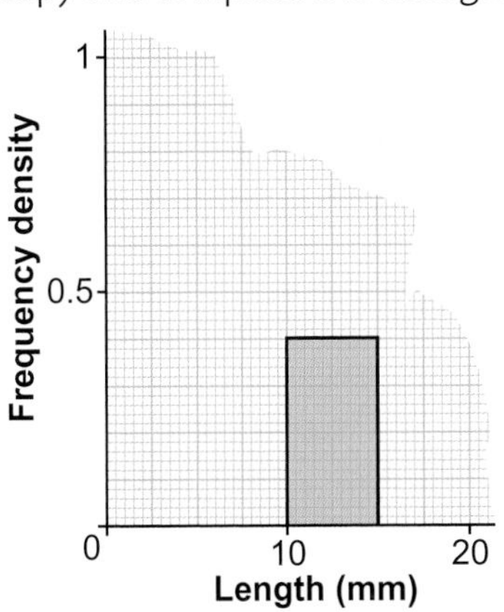

7 This histogram shows how long it took 44 students to complete a test.

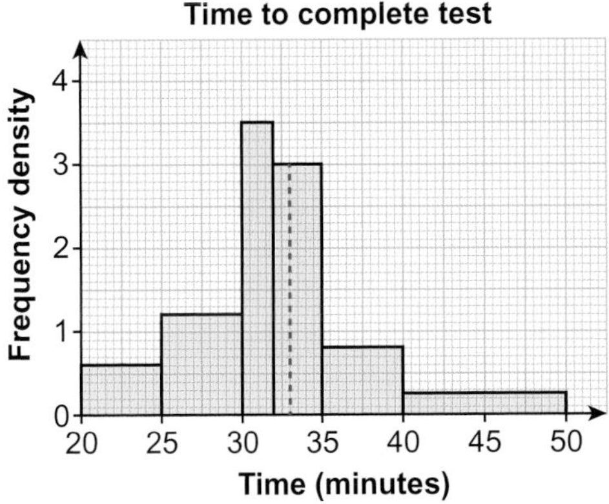

Q7a hint Multiply the class width by the frequency density for that bar.

Q7b hint Add the frequency for the $20 < t \leqslant 25$ bar to your answer to part **a**.

Q7c hint Look at the class $33 < t \leqslant 35$, formed by the red dashed line.

a How many students took between 25 and 30 minutes?

b How many students took less than 30 minutes?

c Estimate how many students took between 33 and 35 minutes.

d Estimate how many students took less than 33 minutes.

Comparing data

1 The box plots show the scores achieved by boys and girls in a test.

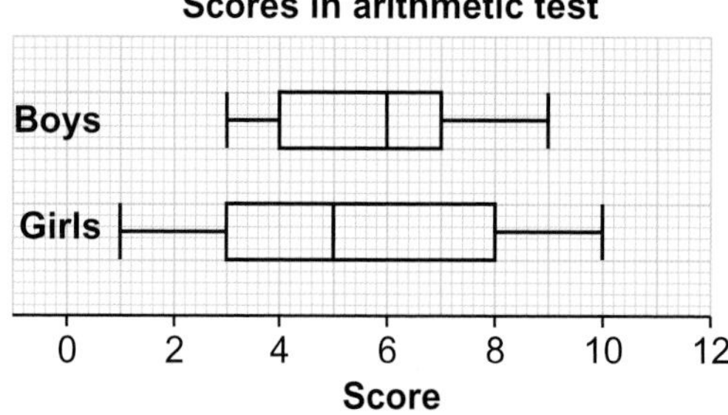

a Copy and complete this table.

	Lower quartile	Median	Upper quartile	Interquartile range
Boys				
Girls				

b Copy and complete

i The median for boys is ____________ than the median for girls.

ii _______ have a smaller interquartile range than _______

2 **STEM** The stem-and-leaf diagram shows the masses, in kilograms, of female wild boars.

5	1 2 2 4 5 9
6	0 3 5 7 7 7 8 9
7	2 3 3 5 6 9
8	0 0 2

Key: 6 | 0 represents 60 kg

a How many boars are recorded in the stem-and-leaf diagram?

b Work out $\frac{n+1}{4}$, $\frac{n+1}{2}$ and $\frac{3(n+1)}{4}$.

c Find these data values by counting from the first one.

d Write down the median value.

e Work out the interquartile range of the masses.

Q2c hint n is the value you worked out in part **a**.

Q2d hint The median value is the $\frac{n+1}{2}$th value.

3 **STEM** The stem-and-leaf diagram shows the masses, in kilograms, of male wild boars.

5	7 8 9 9
6	1 5 6 8 8 8 9
7	2 4 5 7 8 9
8	1 5 7 8

Key: 6 | 0 represents 60 kg

a Find the median mass.

b Find the lower and upper quartiles.

c Find the interquartile range.

d Compare the data in **Q2** and **Q3**.

Q3b hint $n = 21$. The lower quartile is the $\frac{21+1}{4}$th = 5.5th value.

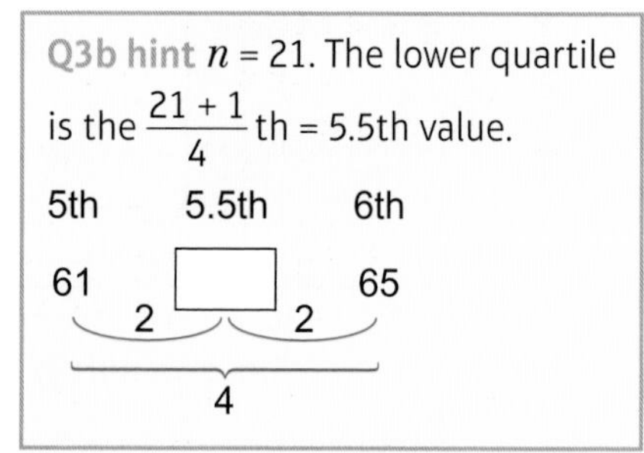

Q3d hint Write sentences like the ones in **Q1b**.

14 Extend

1 **STEM / Reasoning** The speeds of 75 cheetahs are recorded in this table.

a Draw an appropriate graph for the data.

b Find an estimate for

i the median

ii the interquartile range.

Speed, s (mph)	Frequency
$20 \leqslant s \leqslant 25$	3
$25 < s \leqslant 30$	8
$30 < s \leqslant 35$	19
$35 < s \leqslant 40$	21
$40 < s \leqslant 45$	15
$45 < s \leqslant 50$	9

2 The ratio of over-20s to under-20s in a snooker club is 5 : 3. A sample of 120 members is to be used in a survey. How many of each age group should be selected?

Q2 hint Divide 120 in the same ratio as the population.

3 **Reasoning** In a building company there are 450 workers who operate cranes, 620 workers who operate forklift trucks and 130 workers who operate dump trucks. Explain how you would select a sample of 60 workers.

4 **Real / Problem-solving** A construction company owner wants to find out what his employees think of the new car park.

	Builders	Electricians	Plumbers
Male	120	40	20
Female	70	36	34

Explain how he could select a sample of 80 people.

5 **Exam-style question**

A veterinary surgery has 130 clients who have pet rabbits, 150 who have guinea pigs, 75 who have hamsters and 60 who have gerbils. One of the vets carries out a survey using a stratified sample of size 50.

a How many clients should she sample from each group? **(3 marks)**

b Describe how she might choose the clients from within each group. **(2 marks)**

Exam hint Work out the total number of clients.

6 **Real / Problem-solving** The lengths of calls received by a call centre are recorded and shown on this graph.

a Copy and complete:
10% of the calls were less than ☐ minutes.

Q6a hint Draw a line from 10% of the cumulative frequency across to the graph and down to the x-axis.

b Work out the limits between which the middle 80% of the data lies.

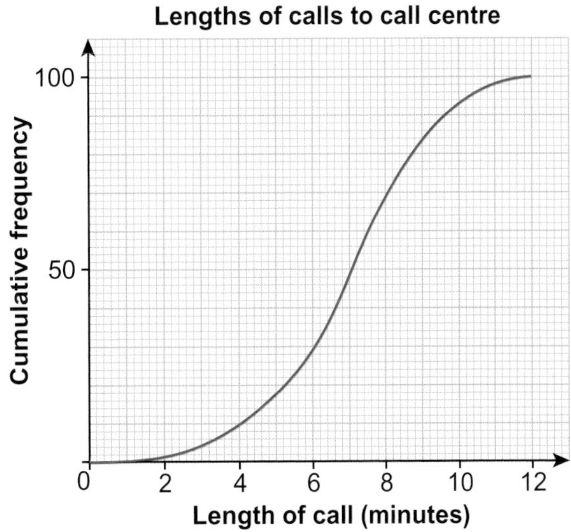

7 **Real / Reasoning** These box plots show the average daily temperature in March and June at a holiday resort.

a Give a reason why a holiday-maker might prefer to go to the resort in March.

b Give a reason why they might prefer to go in June.

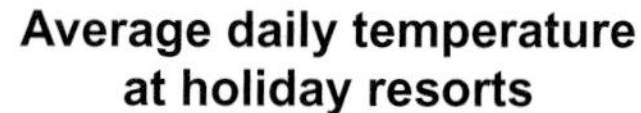

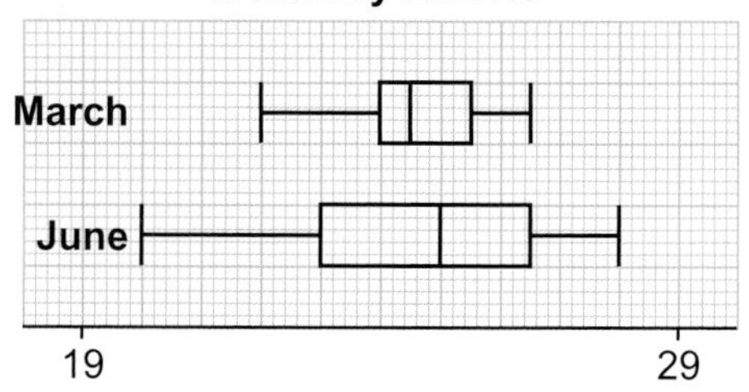

8 **Reasoning** This table records the results of a survey into how many peas there are in a pod.

Peas	Number of pods
3	7
4	11
5	18
6	34
7	20
8	10

a Is the data discrete or continuous?

b Copy and complete this cumulative frequency table.

Peas	Cumulative frequency
$\leqslant 3$	7
$\leqslant 4$	18
$\leqslant 5$	
$\leqslant 6$	
$\leqslant 7$	
$\leqslant 8$	

c Draw a set of axes with x going from 0 to 8 and y going from 0 to 100.

d Plot the coordinate points (peas, cumulative frequency).

e Join the points in a 'staircase' pattern going across and then up from each coordinate point to the next.

f This diagram is called a cumulative frequency step polygon.

Discussion Why is it not appropriate to draw a smooth curve through the points for this data?

g Find the median number of peas in a pod.

Q9e hint

20
10
0 1 2 3 4

9 The number of rabbits born in each of 120 litters is recorded in this table.

Rabbits	Number of litters
2	8
3	15
4	23
5	41
6	25
7	8

a Draw a cumulative frequency table for this data.

b Draw a cumulative frequency step polygon.

c Write down the median number of rabbits in a litter.

d Work out the interquartile range for number of rabbits in a litter.

10 **Reasoning** The time, in seconds, for 15 runners to complete a hurdles race is recorded:
13.5, 14.1, 14.2, 14.3, 14.5, 14.7, 14.7, 14.9, 15.0, 15.1, 15.2, 15.5, 15.7, 15.8, 28.6

a Work out the median time taken.

b Work out the lower quartile, upper quartile and interquartile range.

c Work out 1.5 × interquartile range.

d We can define an outlier as a data value which lies more than 1.5 × the interquartile range below the lower quartile or above the upper quartile.
Use this definition to decide if any of the data points are outliers.

Discussion Assuming all of the runners are of a similar standard, what might account for the extreme value(s)?

11 Reasoning This box plot shows the ages of people at a birthday party.

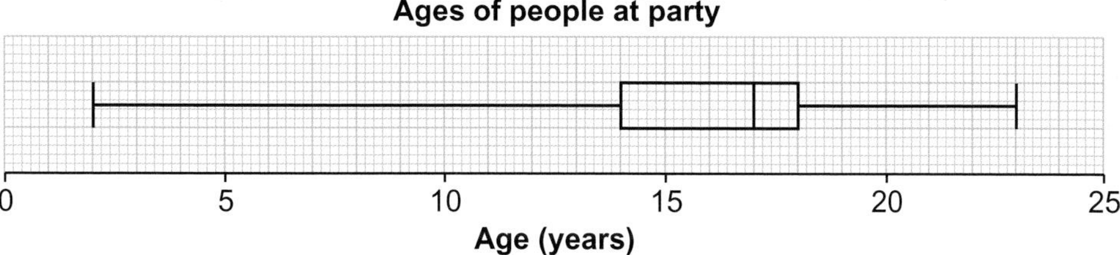

a Write down the interquartile range.

b Using the definition in **Q10d**, work out the range of ages that *would not* be considered outliers.

c The second youngest person at the party was 9 years old and the second oldest person was 21. Write down the values of any outliers.

A box plot can be redrawn to show outliers. Outliers are marked with a cross and the ends of the whiskers are now drawn at the first value *not* considered an outlier.

d Redraw the box plot above to show the outlier(s).

Q11d hint

Minimum LQ Median UQ New max × Outlier

12 Problem-solving The table shows the times taken for players to complete a round of golf.

a Draw a histogram for this data.

b Players who took longer than 4 hours 20 minutes were given a two-shot penalty.
Estimate the number of players who were given a penalty.

Time, t (hours)	Frequency
$3 < t \leqslant 3.5$	7
$3.5 < t \leqslant 3.8$	12
$3.8 < t \leqslant 4.0$	18
$4.0 < t \leqslant 4.5$	24
$5.0 < t \leqslant 6.0$	15

13 STEM / Reasoning The masses of 100 camels are given in this table.

Mass, m (kg)	Frequency
$300 \leqslant m \leqslant 500$	8
$500 < m \leqslant 600$	10
$600 < m \leqslant 700$	15
$700 < m \leqslant 750$	18
$750 < m \leqslant 800$	36
$800 < m \leqslant 900$	11
$900 < m \leqslant 1100$	2

a Draw a histogram for this data.

b Work out an estimate for the mean mass.

c Work out an estimate for the median mass.

d Write down the modal class.

e An estimate for the modal value can be worked out from the histogram using the method shown in this extract from another histogram.
What is the modal value shown in the extract?

f Work out an estimate for the modal mass of the camels.

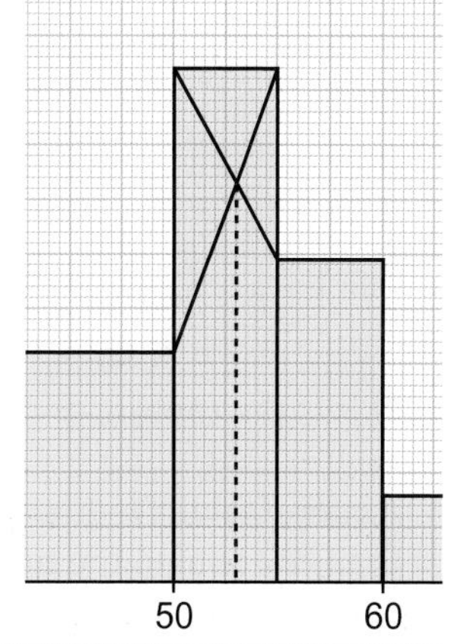

Q13f hint Your value for the mode is only an estimate, as the histogram does not show individual data values.

14 **STEM / Reasoning** The lengths of 50 Barbour's seahorses are recorded in this table.

a Draw a histogram for this data.

b Work out an estimate for the mean length of the seahorses.

c Work out an estimate for the median length of the seahorses.

d Work out an estimate for the mode.

Length, x (cm)	Frequency
$10 \leqslant x \leqslant 12$	3
$12 < x \leqslant 13$	8
$13 < x \leqslant 13.5$	12
$13.5 < x \leqslant 14$	10
$14 < x \leqslant 15$	9
$15 < x \leqslant 17$	8

14 Knowledge check

- A **population** is the set of items that you are interested in. .... *Mastery lesson 14.1*
- A **census** is a survey of the whole population. .... *Mastery lesson 14.1*
- A **sample** is a smaller number of items from the population. A sample of at least 10% is considered to be a good-sized sample. .... *Mastery lesson 14.1*
- In order to reduce **bias**, the sample must represent the whole population. .... *Mastery lesson 14.1*
- In a **random sample** each item has the same chance of being chosen. .... *Mastery lesson 14.1*
- To select a simple random sample draw names from a hat, generate random numbers on a calculator or use a table of random numbers. *Mastery lesson 14.1*
- A population may divide into groups such as age range or gender. These groups are called **strata** (singular **stratum**). .... *Mastery lesson 14.1*
- In a **stratified sample**, the number of people taken from each group is proportional to the group size. .... *Mastery lesson 14.1*
- To estimate the size of the population N of an animal species:
  Capture and mark a sample size n.
  Recapture another sample of size M.
  Count the number marked (m).
  $\frac{n}{N} = \frac{m}{M}$
  So, $N = \frac{n \times M}{m}$
  This is the capture–recapture method. .... *Mastery lesson 14.*
- A **cumulative frequency table** shows how many data values are less than or equal to the **upper class boundary** of each data class. .... *Mastery lesson 14.2*
- A **cumulative frequency diagram** has data values on the x-axis and cumulative frequency on the y-axis. .... *Mastery lesson 14.2*

- The **median** and **quartiles** can be estimated from the cumulative frequency diagram. For a set of n data values
  - the estimate for the **median** is the $\frac{n}{2}$th value
  - the estimate for the **lower quartile** (LQ) is the $\frac{n}{4}$th value
  - the estimate for the **upper quartile** (UQ) is the $\frac{3n}{4}$th value. .... *Mastery lesson 14.2*
  - the **interquartile range** (IQR) = UQ – LQ .... *Mastery lesson 14.2*
- For a set of n data values
  - the lower quartile (LQ) is the $\frac{(n+1)}{4}$th value.
  - the upper quartile (UQ) is the $\frac{3(n+1)}{4}$th value .... *Mastery lesson 14.3*
- A **box plot**, sometimes called a **box-and-whisker diagram**, displays a data set to show the median and quartiles. .... *Mastery lesson 14.3*
- **Comparative box plots** are box plots for two different sets of data drawn on the same scale. .... *Mastery lesson 14.3*
- A **histogram** is similar to a bar chart but is used to represent continuous data. .... *Mastery lesson 14.4*
- In a histogram the area of the bar represents the frequency. The height of each bar is the frequency density. .... *Mastery lesson 14.4*
- Frequency density = $\frac{\text{frequency}}{\text{class width}}$ .... *Mastery lesson 14.4*
- The interquartile range measures the spread of the middle 50% of the data. To describe a data set (or population) give a measure of average and a measure of spread. To compare data sets, compare a measure of average and a measure of spread. .... *Mastery lesson 14.6*
- The median and interquartile range are not affected by extreme values or **outliers**. When there are extreme values, the median and interquartile range should be used rather than the mean and range. *Mastery lesson 14.6*

Choose A, B or C to complete each statement about statistics.

In this unit, I did...	A well	B OK	C not very well
I think _____ is...	A easy	B OK	C hard
When I think about doing _____, I feel...	A confident	B OK	C unsure

Did you answer mostly As and Bs? Are you surprised by how you feel about _____? Why?

Did you answer mostly Cs? Find the three questions in this unit that you found the hardest. Ask someone to explain them to you. Then complete the statements above again.

Reflect

14 Unit test

Log how you did on your Student Progression Chart.

1 **Reasoning** A market research company wants to get the views of members of the public on new chocolate bars.

a Give two reasons why a sample should be taken rather than a census. *(2 marks)*

b They propose to take their sample from the school nearest to the company headquarters. Explain whether this will be a good sample. *(1 mark)*

2 The masses of 100 octopuses are shown in this table.

Mass, m (kg)	Frequency
$3 \leqslant m \leqslant 4$	4
$4 < m \leqslant 5$	11
$5 < m \leqslant 6$	15
$6 < m \leqslant 7$	18
$7 < m \leqslant 8$	23
$8 < m \leqslant 9$	17
$9 < m \leqslant 10$	12

a Draw a cumulative frequency diagram. *(3 marks)*

b Estimate the median mass. *(1 mark)*

c Find the interquartile range. *(2 marks)*

d Estimate how many octopuses weigh more than 6.5 kg. *(2 marks)*

3 **Problem-solving** The times taken by a group of men and a group of women to complete an obstacle course are shown in the box plots.

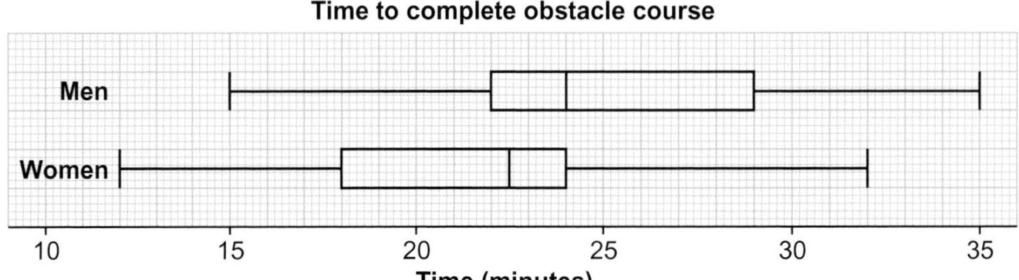

Compare the two data sets. *(3 marks)*

4 The president of a dining society wants to get the opinions of the members on new menu choices. He decides to take a sample of 50 members. The age and gender profile of the society is given in this table.

	18–25	26–35	36–45	45+
Male	19	25	54	30
Female	12	30	45	35

a Explain why a stratified sample should be taken. *(1 mark)*

b How many males under 25 should be in the sample? *(1 mark)*

c How many females aged between 26 and 35 should be in the sample? *(1 mark)*

d The president lists each group of members alphabetically and numbers them He uses a random number table to work out whom to ask.

Use the table below to write down the numbers of the females aged between 26 and 35 that will be surveyed. *(2 marks)*

04897974662531115716402340212556130б...

5 **Reasoning** The ages of 15 people in the cast for a show are
6, 15, 23, 18, 17, 9, 11, 32, 31, 14, 45, 25, 26, 26, 15

a Draw a box plot. *(3 marks)*

b Outliers are defined as being data points at least 1.5 × interquartile range above the upper quartile or below the lower quartile.
Use this definition to work out if the data set contains outliers. *(1 mark)*

c Explain, with a reason, whether these data points should be excluded. *(1 mark)*

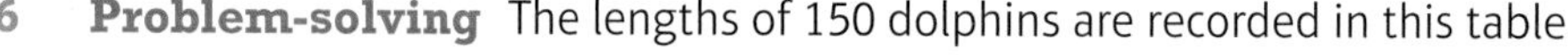

6 **Problem-solving** The lengths of 150 dolphins are recorded in this table.

Length, l (m)	Frequency
$1.5 \leqslant l \leqslant 2.0$	7
$2.0 < l \leqslant 2.5$	19
$2.5 < l \leqslant 2.8$	31
$2.8 < l \leqslant 3.0$	52
$3.0 < l \leqslant 3.5$	34
$3.5 < l \leqslant 5.0$	7

a Draw a histogram. *(3 marks)*

b Estimate the mean length of the dolphins. *(4 marks)*

c Estimate the median length of the dolphins. *(3 marks)*

d Estimate how many dolphins are over 3.25 m long. *(3 marks)*

e Estimate the modal length. *(3 marks)*

Sample student answer

a What is good about the presentation of this graph?

b Which of the values (upper/lower quartile, median) is incorrect, and how has this mistake happened?

Exam-style question

The cumulative frequency graphs give information about the heights of two groups of children, group A and group B.

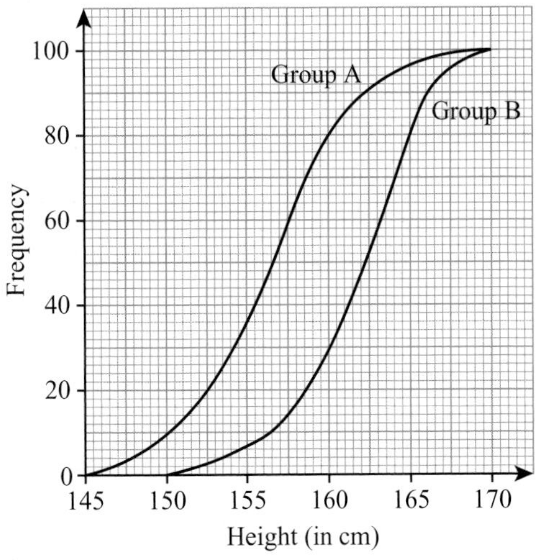

Compare the heights of the children in group A and the heights of the children in group B. **(2 marks)**

Nov 2013, Q11, 5MB1H/01

Student answer

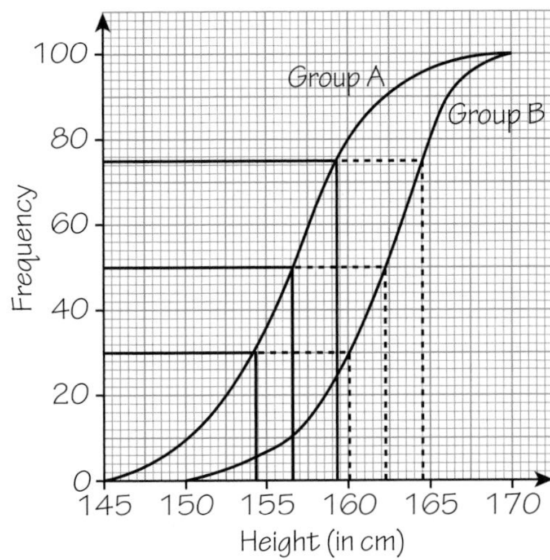

	A	B
Upper	159.5	164.5
Median	156.5	162.5
Lower	154.5	160

On average group B are taller, as shown by their median 162.5 compared to 156.5 for group A.

Group A has a larger interquartile range (5 compared to 4.5 for B) showing that their heights are more spread out.

15 EQUATIONS AND GRAPHS

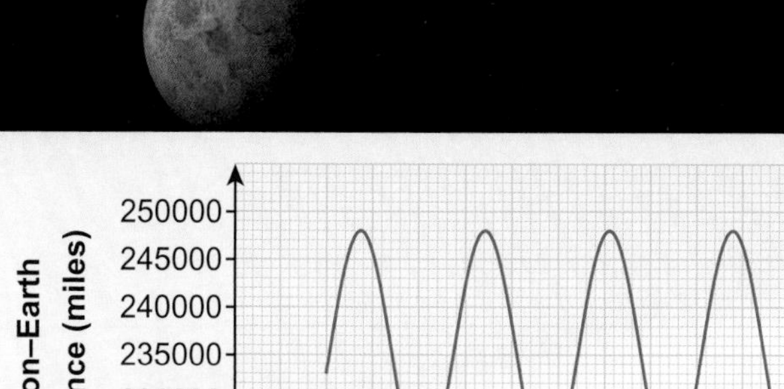

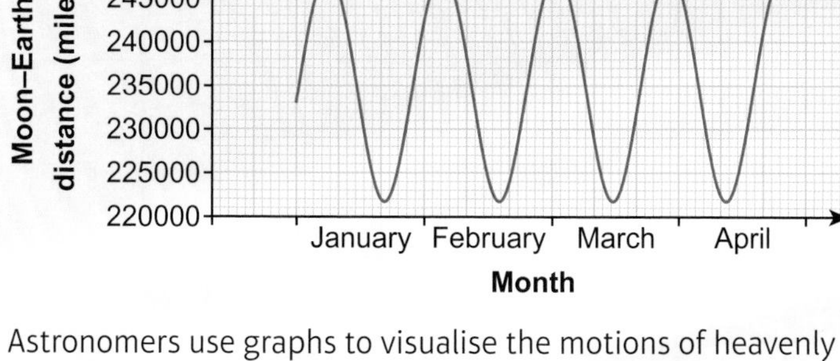

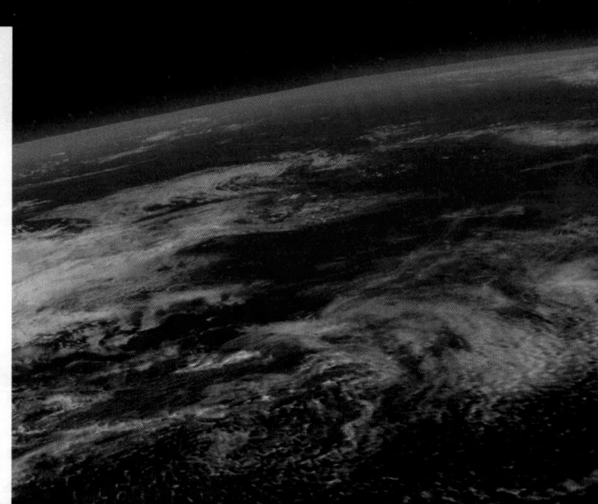

Astronomers use graphs to visualise the motions of heavenly bodies. In the four months shown here, for how long was the Moon less than 230 000 miles away from the Earth?

15 Prior knowledge check

Numerical fluency

1 Simplify these surds.
a $\sqrt{27}$ b $\sqrt{200}$ c $\sqrt{20}$

Algebraic fluency

2 Decide if these functions are
i linear ii quadratic iii cubic.
a $y = 3x^2 + 2x - 5$ b $y = x^3$
c $y = 10 - x^2$ d $y = 2x - 5$
e $2y = 10x - 7$ f $4y = 2x^2 + x^3$
g $0 = y - x$

3 Work out the value of y when $x = 0$ for each of the equations.
a $y = 2x - 7$
b $2y - 3x = 12$
c $y = x^2 - 2x + 7$
d $y = (2x - 3)(4x + 1)$
e $y = 7x^2 - 4x + 12$

4 Solve these inequalities and show the answers on a number line.
a $x + 4 \geqslant 7$ b $12 < 3x + 6$
c $2x - 5 > 9$ d $3x - 2 \leqslant 18 - x$

5 Solve these inequalities and write the solutions using set notation.
a $\frac{x}{3} < -4$ b $3(2x - 7) > -39$
c $\frac{x}{5} \leqslant \frac{x}{3}$ d $10 - 2x \geqslant 4(2x - 1)$

6 Find all the integer values of x which satisfy these inequalities.
a $-3 < x < 4$
b $-10 \leqslant 5x \leqslant 4$
c $-2 < 2(3x + 1) \leqslant 12$
d $-3 \leqslant \frac{10 - x}{4} < -1$

7 Expand and simplify.
a $(x + 3)(x + 7)$ b $(x - 5)(x + 8)$
c $(x - 3)(x - 2)$ d $(x - 4)^2$
e $(2x + 3)(x + 5)$ f $(3x - 5)(x - 2)$
g $(3x + 1)(3x - 1)$

8 Factorise
a $x^2 + 7x + 10$ b $x^2 - 2x - 3$
c $x^2 + 2x - 15$ d $x^2 - 6x - 7$
e $x^2 - 1$ f $3x^2 + 15x + 12$
g $2x^2 + x - 10$

9 Solve the equation by factorising.
$3x^2 + x = 10$

10 Write in the form $a(x + p)^2 + q$
a $x^2 - 6x + 12$ b $x^2 - 8x + 11$
c $3x^2 - 6x + 9$

11 Solve by completing the square. Put your answers in surd form where necessary.
a $x^2 + 4x = 5$ b $x^2 + 2x = 0$
c $x^2 + 3x - 5 = 0$ d $2x^2 + 4x - 6 = 0$

12 Solve the simultaneous equations.
a $x + y = 6$
$y - x = 8$
b $2x - 3y = -9$
$x + y = 8$
c $2x + y = 1$
$x^2 + y = 4$

13 Use the quadratic formula to find the solutions to the equations to one decimal place.
a $0 = x^2 + 5x + 2$ b $6x^2 = 5x + 12$

Graphical fluency

14 a Draw a coordinate grid with −10 to +10 on both axes.
b Plot and label the graphs of these functions by using the gradient and y-intercept.
i $y = 2x + 3$ ii $y - x = 4$
iii $2y = 4x + 7$

15 Which of the graphs shown is
a linear b quadratic c cubic?

i
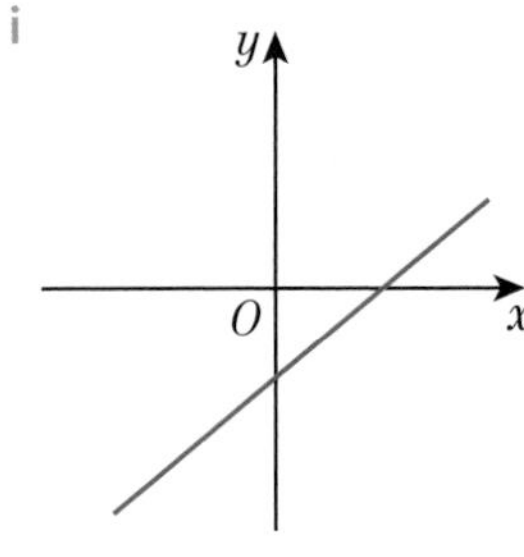

ii
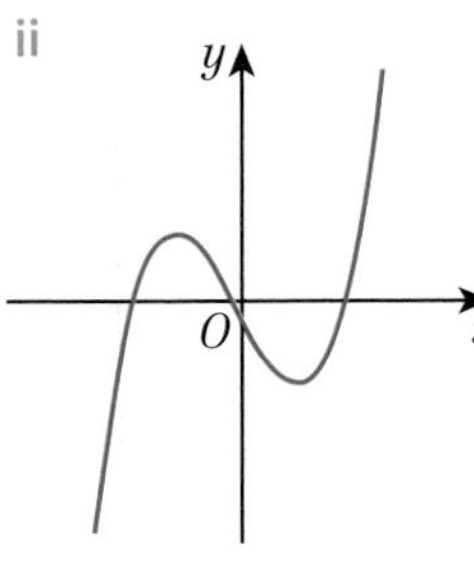

iii

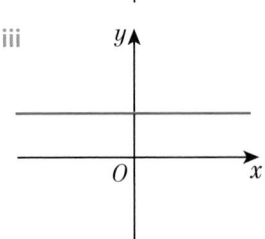

iv
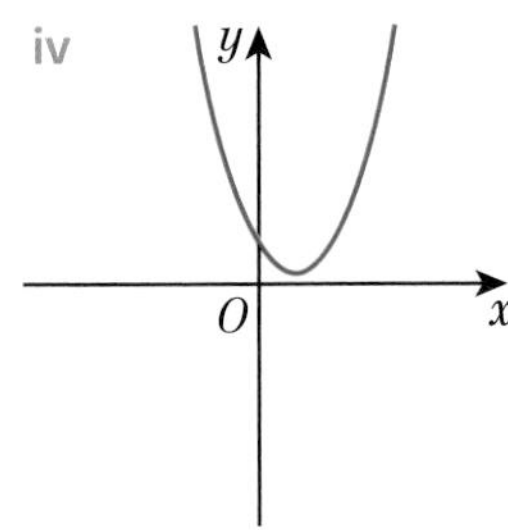

16 a Draw the graph of $y = x^2 + 3x - 4$ for x values from −5 to 2.
b Use your graph to solve
i $x^2 + 3x - 4 = 0$ ii $x^2 + 3x - 4 = 2$

✱ Challenge

17 a The solution to a quadratic equation is $x = -2$ or $x = 7$
Write in the form $ax^2 + bx + c$ three different quadratic equations with this solution.

15.1 Solving simultaneous equations graphically

Objective

- Solve simultaneous equations graphically.

Did you know?

In 200 BC the Chinese discovered a method for solving simultaneous equations.

Fluency

What is the equation of this graph?

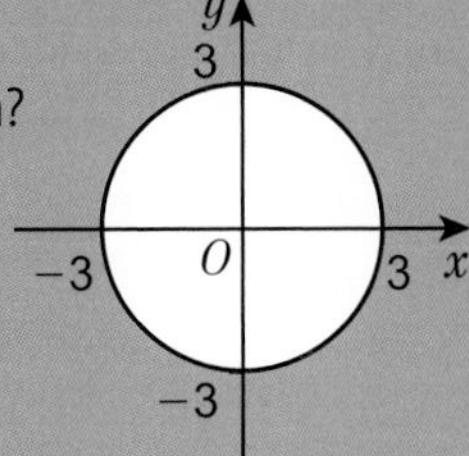

Warm up

1 Draw the graph of $x^2 + y^2 = 25$

Key point 1

You can solve a pair of simultaneous equations by plotting the graph of both equations and finding the point(s) of intersection.

ActiveLearn Homework, practice and support: Higher 15.1

Questions in this unit are targeted at the steps indicated.

2 a Draw a coordinate grid with –10 to 10 on both axes.
Draw the graphs of
i $2y - 4x = 8$ ii $y - x = 6$
b Write down the coordinates of the point of intersection.
c Check your answer to **b** by substituting the x- and y-coordinates into both equations.

3 **Reasoning** a Match the equations to the three lines A, B and C shown.
i $x + y = 5$ ii $y = 2x + 2$ iii $8 = 4y + x$

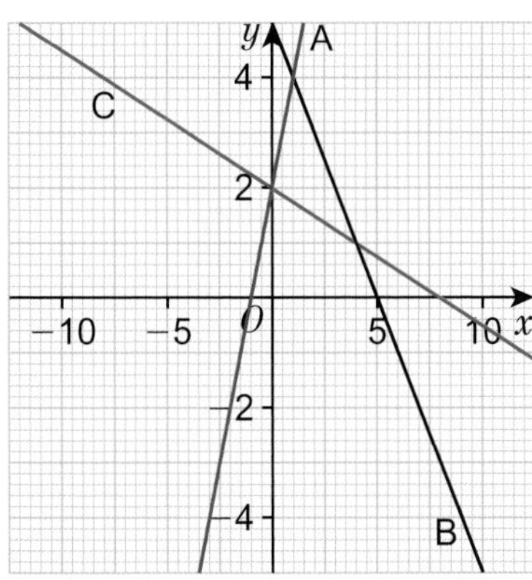

b Hence write down the solutions to these simultaneous equations.
i $x + y = 5$, $y = 2x + 2$ ii $x + y = 5$, $8 = 4y + x$ iii $8 = 4y + x$, $y = 2x + 2$

4 Solve the pairs of simultaneous equations by drawing the graphs.
a $y = 2x + 4$ $y = -2x + 8$
b $2y = 7x - 3$ $x + 2y = 21$
c $0 = y + 2x - 5$ $y = 2x + 9$
d $2x + 3y = -13$ $x + y = -5$
e Show that solving the equations algebraically gives the same solutions to the equations in part **d**.

5 Sara buys 4 bananas and 2 apples for £2.58.
In the same shop, Alex buys 3 bananas and 3 apples for £2.49.
Write a pair of simultaneous equations and solve them graphically to find the cost of
a one apple b one banana.

6 **Real / Reasoning** Two broadband providers offer the following prices:

ONline	**Stream Speed**
No monthly cost	Monthly tariff £20
£2 per GB of data.	£1.50 per GB of data

a For each company form an equation to calculate the monthly cost, with y = total monthly cost and x = GBs of data used.
b Use a graphical method to work out how many GBs of data are used if the cost for both companies is the same.

Discussion When is Stream Speed cheaper than ONline?

Example 1

Solve the simultaneous equations graphically.

$y = x^2 + x - 4$

$y - 2x + 2 = 0$

$y = x^2 + x - 4$

x	−5	−4	−3	−2	−1	0	1	2	3	4
y	16	8	2	−2	−4	−4	−2	2	8	16

Construct a table of values, calculate the points and plot the graph.

$y - 2x + 2 = 0$

$y = 2x - 2$

Rearrange the equation to make y the subject.

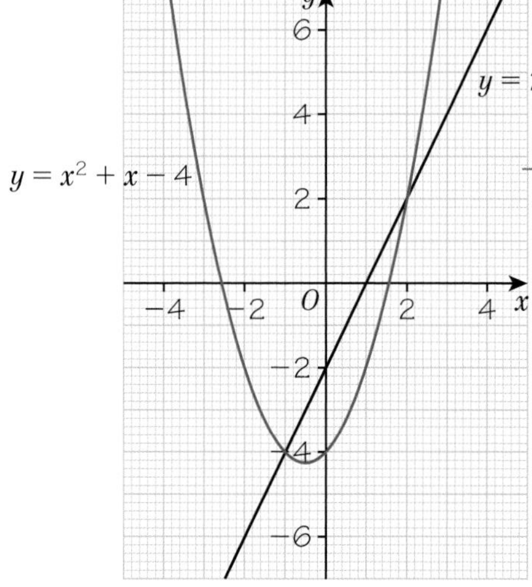

Plot the linear graph on the same grid using the y-intercept and gradient.

The solutions are

$x = 2, y = 2$ and $x = -1, y = -4$

7 Use a graphical method to find an approximate solution to the pair of simultaneous equations

$y + x^2 = 2$

$y + 1 = x$

Q7 hint Start by rearranging the equations, then plot the graphs.

8 **Reasoning**

Solve this pair of simultaneous equations

a graphically

b algebraically to 2 decimal places.

$3x + 2y = 5$

$x^2 + y = 7$

Discussion Which method gives the more accurate solution? Explain.

9 a On a suitable grid draw the graph of $x^2 + y^2 = 25$

b On the same grid draw the graph of $y = x + 1$

c What are the coordinates of the points at which the graphs intersect?

10 Use a graphical method to find an estimate for the solution to the simultaneous equations

$x^2 + y^2 = 9 \qquad x + y = 1$

Discussion How could you check your solution?

11 **Reflect** In **Unit 9** you solved simultaneous equations using algebra. In this lesson, you solved simultaneous equations using graphs. Which method do you prefer? Why?

15.2 Representing inequalities graphically

Objectives

- Represent inequalities on graphs.
- Interpret graphs of inequalities.

Did you know?

In the novel *Animal Farm* by George Orwell, the Pigs declare: 'All animals are equal, but some animals are more equal than others'. Can they be right?

Fluency

Which integer values of x satisfy each inequality?

$x < 12$ $-3 > x$ $3x \geqslant 12$ $21 \leqslant 2x - 1$

1 Solve the inequalities and represent the solutions using **set notation**.

a $2x \leqslant 12$ b $3x - 5 > -11$ c $3(x - 5) \leqslant x - 7$

Warm up

Key point 2

The points that satisfy an inequality such as $y \leqslant 3$ can be represented on a graph.

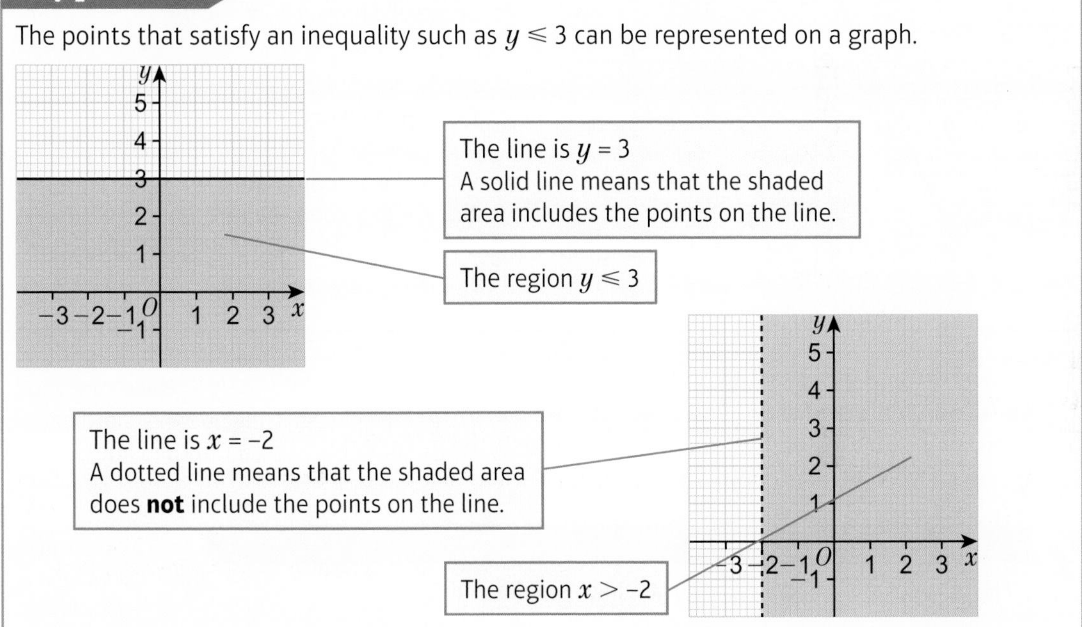

The line is $y = 3$
A solid line means that the shaded area includes the points on the line.
The region $y \leqslant 3$
The line is $x = -2$
A dotted line means that the shaded area does **not** include the points on the line.
The region $x > -2$

2 a Write down the inequalities represented by the shaded regions.

i ii iii

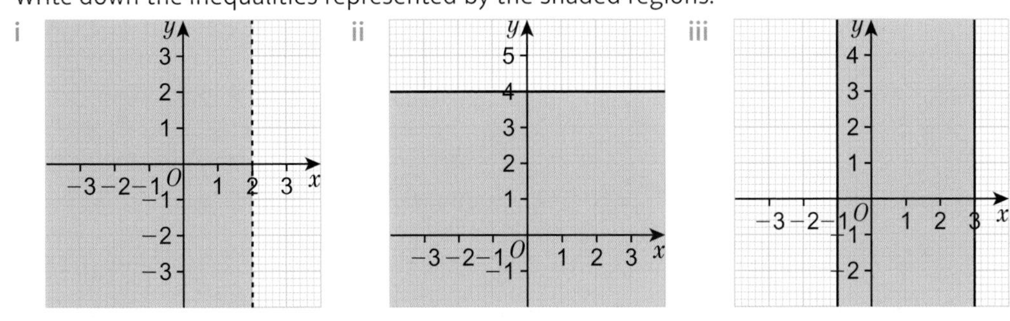

b On a suitable coordinate grid, shade the region of points whose coordinates satisfy

i $x \leqslant -2$ ii $y > 0$

iii $4 > x$ iv $y - 2 < x < 1$

v $2 \leqslant y < 3.5$ vi $-5.5 \leqslant x \leqslant -3$

vii $-4 \leqslant y < -3.5$

Q2b iii hint $4 > x$ is the same as $x < 4$.

3 a Draw a coordinate grid with –5 to 5 on both axes.
  b Draw the graph of $y = 2x + 1$
  c Does the point (2, 3) satisfy the inequality $y < 2x + 1$?
  d Shade the region of points that satisfy $y < 2x + 1$

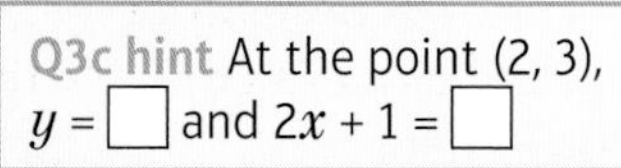

Q3c hint At the point (2, 3),
y = □ and 2x + 1 = □

Q3d hint If (2, 3) satisfies the inequality, then all points in that region will also satisfy the inequality.

4 Draw a coordinate grid with –6 to 6 on both axes.
Shade the region that satisfies each inequality.

a $y > x - 3$ b $y \leqslant 2x - 2$ c $y \geqslant 3x - 3$ d $y < -x + 1$

Q4 hint Test a point to see which region satisfies the inequality.

Example 2

On a coordinate grid, shade the region that satisfies the inequalities
$x < 5$, $y \leqslant 2x + 4$, $y \leqslant 1$ and $y > -2$

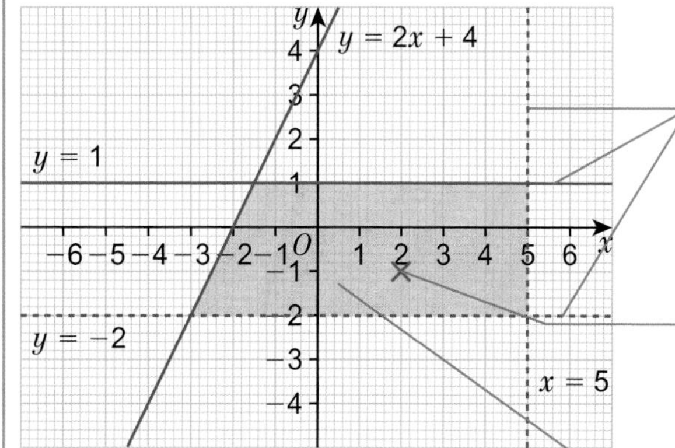

Draw dotted lines $x = 5$ and $y = -2$
Draw solid lines $y = 2x + 4$, $y = 1$

Test a point. For (2, –1)
$y \leqslant 1$ and $y > -2$: the y-coordinate is –1
$x < 5$: the x-coordinate is 2
$2x + 4 = 8$: y-coordinate $\leqslant 8$

Shade the required region.

5 **Reasoning** x and y are integers.
On a coordinate grid with –5 to 5 on both axes, mark on all the coordinates with integer coordinates which satisfy the inequalities
$y > x + 1$ $x \geqslant -3$ $y < 3$

Q5 hint Make sure you include the coordinates on the solid lines.

6 **Exam-style question**

The lines $y = x - 2$ and $x + y = 10$ are drawn on the grid.

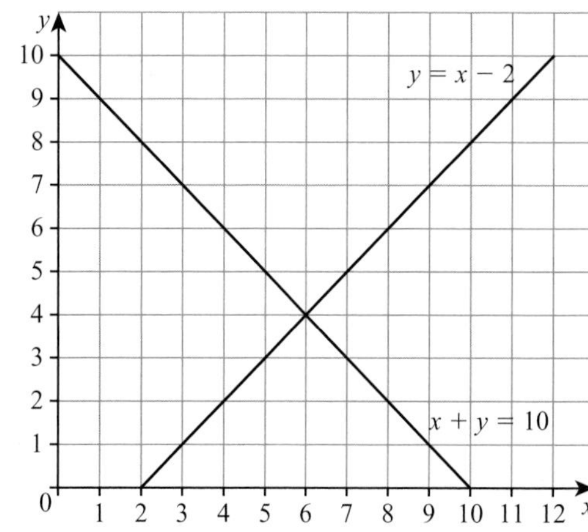

On the grid, mark with a cross (X) each of the points with integer coordinates that are in the region defined by $y > x - 2$, $x + y < 10$ and $x > 3$ **(3 marks)**

Nov 2012, Q17, 1MA0/1H

Exam hint
Substitute your values of x and y in the given equations to check your answers.

7 **Reasoning** The diagrams show a shaded region bounded by three lines. For each diagram
   i write down the equations of the lines
   ii write down the three inequalities satisfied by the coordinates of the points in this region.

a

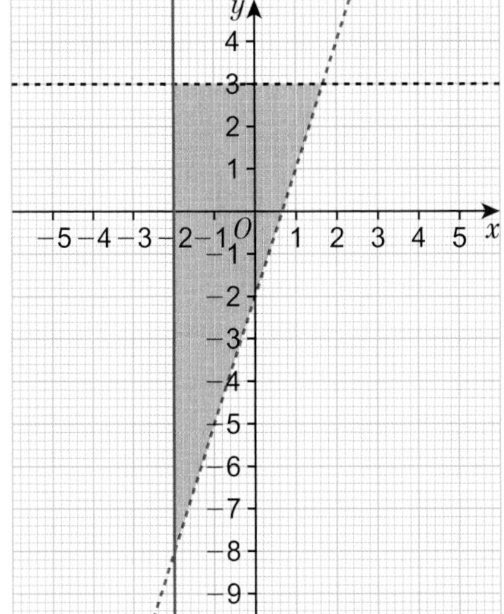

b

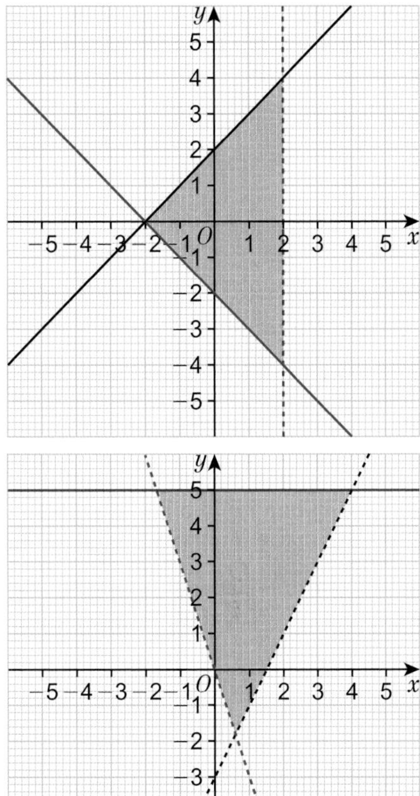

c

8 **Problem-solving** How many points with integer coordinates satisfy these inequalities?

$y > 2x - 3$ $\quad y > -x$ $\quad y < 2$

Q8 hint You could draw a graph.

9 a Draw the graph of $y = x^2$ for values of x from −3 to +3.
   b Draw the line $y = 7$ on the same axes.
   c Shade the region that satisfies $y > x^2$ and $y < 7$

10 This is the graph of $y = 2x^2 - 2x - 4$.

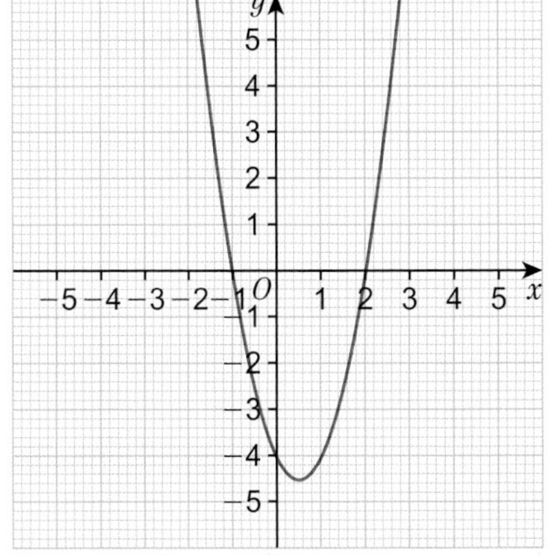

Q10 hint Write your answer as two inequalities: $x <$ ☐ and $x >$ ☐

a For what integer values of x is the graph above the x-axis?
b For what integer values of x is $0 < 2x^2 - 2x - 4$?

Key point 3

You can write solution sets using set notation.
The inequality $x^2 - 9 \geqslant 0$ is satisfied when $-3 \leqslant x \leqslant 3$
This is written: $\{x : -3 \leqslant x \leqslant 3\}$
The inequality $0 < x^2 - 4$ is satisfied when $x < -2$ or $x > 2$
This is written: $\{x : x < -2\} \cup \{x : > 2\}$
The symbol $\cup$ means that the solution includes all the values satisfied by either inequality.

11 **Reasoning** a Sketch the graph of $y = 3x^2 + 3x - 6$, marking clearly the points where the graph intersects the x-axis.
b From the graph identify the values of x for which $0 \geqslant 3x^2 + 3x - 6$
Give your answer using set notation.

Q11b hint When is the graph below the x-axis? $\square \leqslant x \leqslant \square$

12 **Problem-solving** By sketching the graph of $y = 2x^2 + 4x - 6$, find the values of x which satisfy $0 \leqslant 2x^2 + 5x - 3$
Give your answer using set notation.

Q12 hint Rearrange the inequality to make one side equal to $2x^2 + 4x - 6$

13 **Reasoning** a Sketch the graph of $y = x^2 - x - 6$
b Hence find the values of x which satisfy the inequality $6 < x^2 - x$
Give your answer using set notation.

14 **Reasoning** Sketch graphs to find the values of x which satisfy these inequalities. Give your answers using set notation.
a $x^2 + x < 12$ b $2x^2 \geqslant 2x + 4$ c $x^2 \geqslant 9$

15.3 Graphs of quadratic functions

Objective

- Recognise and draw quadratic functions.

Did you know?

The ancient Egyptians left behind a scroll showing a solution to a quadratic equation. It may be 4000 years old.

Fluency

What shape is a quadratic graph?

Warm up

1 By factorising, solve the equations
a $0 = x^2 + 3x + 2$ b $0 = 2x^2 + 5x - 12$

2 Write in the form $a(x + p)^2 + q$
a $x^2 + 2x - 5$ b $2x^2 + 8x + 4$

3 Here is the graph of $y = -x^2 + 2x + 6$
a What are the coordinates of the maximum point?
b What is its line of symmetry?

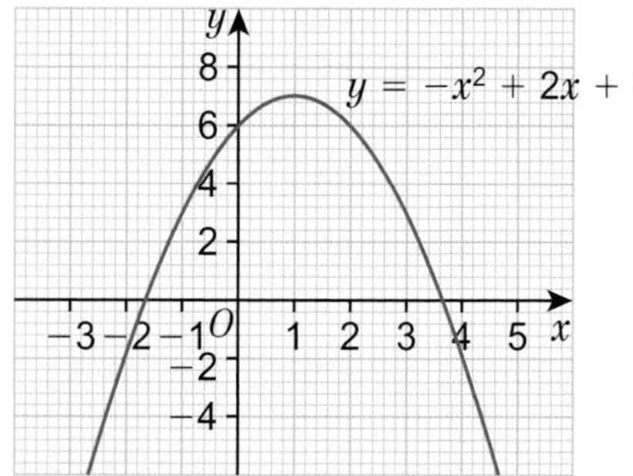

4 Sketch a $y = x^2$ b $y = -x^2$

Key point 4

The lowest or highest point of the parabola, where the graph turns, is called the **turning point**.
The turning point is either a minimum or maximum point.
The x-values where the graph intersects the x-axis are the solutions, or **roots**, of the equation $y = 0$.
⌣ is a minimum, ⌢ is a maximum

5 Here is the graph of $y = x^2 + 4x + 3$

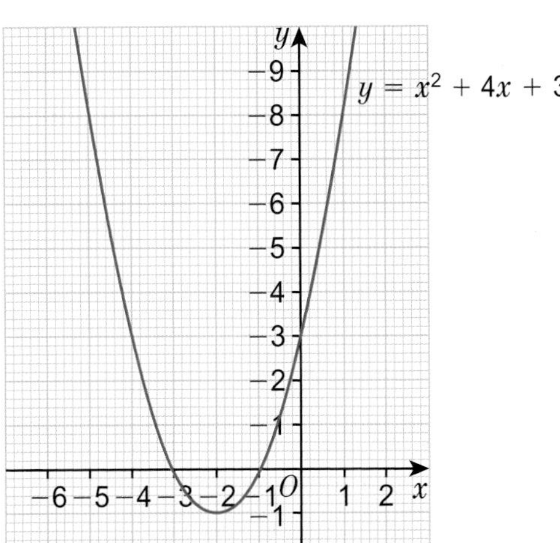

a Use the graph to find the roots of the equation $x^2 + 4x + 3 = 0$.

b Where does the graph intersect the y-axis?

c Is the turning point a maximum or a minimum?

d What are the coordinates of the turning point?

Discussion How can you tell from the equation whether a quadratic graph will have a maximum or minimum point?

6 a Plot the graph of $y = x^2 + 8x + 15$

b Use your graph to find the solutions to the equation $x^2 + 8x + 15 = 0$.

c Where does the graph intersect the y-axis?

d Does the graph have a maximum or minimum point?

e What are the coordinates of the turning point?

Q6a hint Copy and complete this table.

x	−6	−5	−4	−3	−2	−1	0
y							

7 a Solve the equations

i $0 = x^2 + 4x + 3$ ii $0 = x^2 + 8x + 15$

b Find the value of y when $x = 0$ for the equations

i $y = x^2 + 4x + 3$ ii $y = x^2 + 8x + 15$

Discussion How can you find the roots of an equation and where a graph crosses the y-axis without drawing an accurate graph? (Compare your graphs in **Q5** and **Q6** with your calculations in **Q7**.)

8 **Reasoning / Communication** Match the graphs to their equations, explaining your reasoning.

a $y = x^2 + 5x + 6$ b $y = x^2 + 5x - 6$ c $y = -x^2 + 2x + 8$ d $y = -2x^2 + 6x - 8$

i

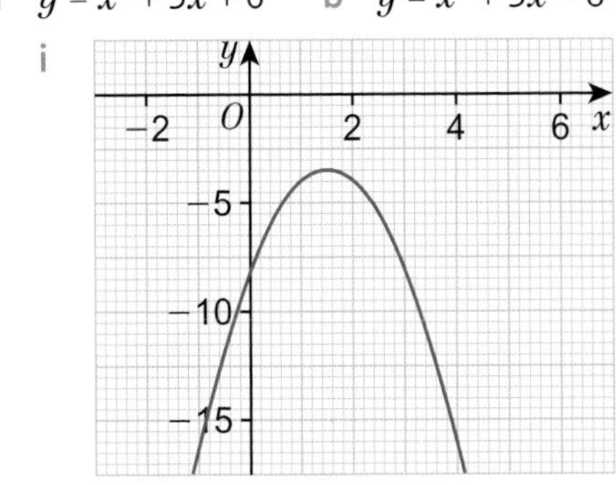

ii

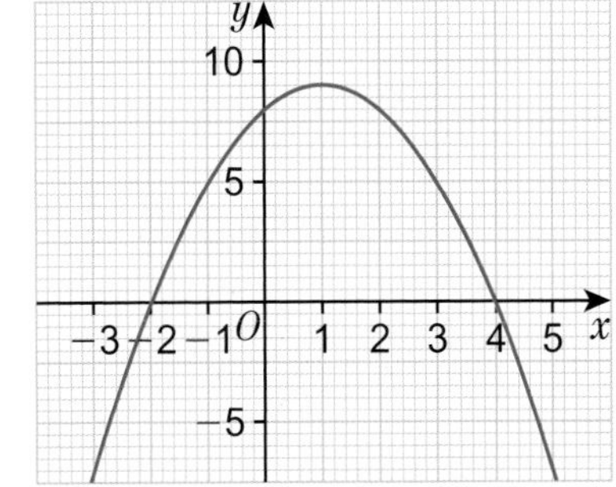

iii

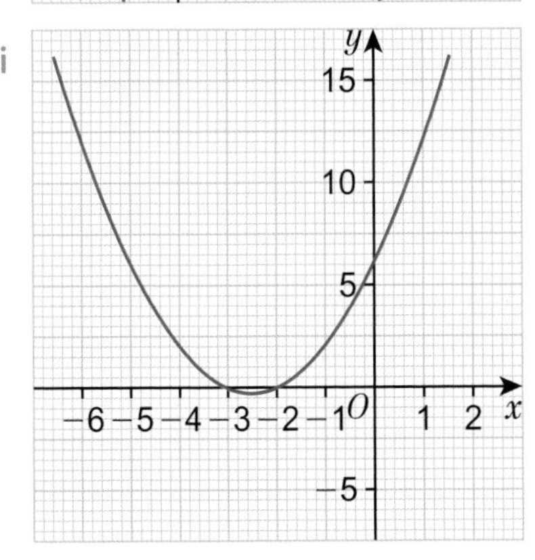

iv

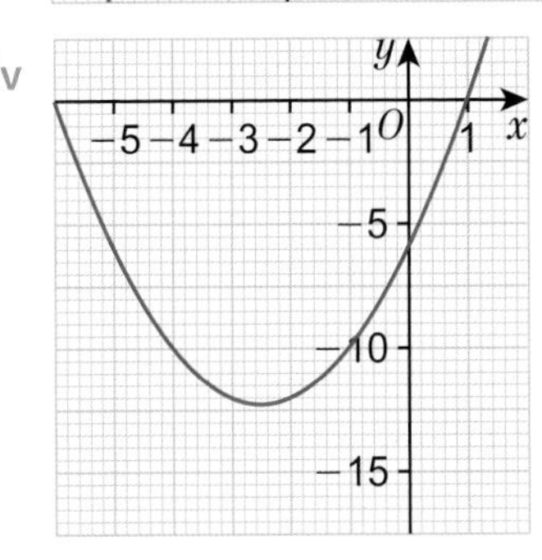

Key point 5

To find the coordinate of the turning point, write the equation in completed square form:
$y = a(x + b)^2 + c$
$(x + b)^2 \geqslant 0$, so the minimum for y is when $x + b = 0$ and $y = c$

Example 3

a Does the graph of $y = x^2 + 8x + 15$ have a maximum or a minimum point?
b Find the coordinates of the turning point.

a Minimum — The coefficient of x^2 is positive, so the turning point is a minimum.

b $y = x^2 + 8x + 15$
$y = (x + 4)^2 - 16 + 15$ — Write the quadratic function in completed square form.
$y = (x + 4)^2 - 1$
$x + 4 = 0$, so $x = -4$
Minimum at $(-4, -1)$

The smallest value that y can take is −1. This occurs when $(x + 4)^2 = 0$. $(x + 4)^2$ cannot be less than 0 because a square is always positive.
Solve the equation to find the x-coordinate.

9 **Reasoning** For each quadratic function, work out the coordinates of the turning point and state whether it is a maximum or a minimum.

a $y = x^2 - 2x + 4$
b $y = -x^2 - 6x - 11$
c $y = x^2 - 10x + 23$
d $y = 2x^2 + 12x + 13$
e $y = 3x^2 - 12x + 13$
f $y = -2x^2 - 4x + 2$

Discussion What do you notice about the completed square form and the coordinates of the turning point?

Q9b hint $y = -(x^2 + 6x + 11)$
$= -((x + 3)^2 + 2)$
$= -(x + 3)^2 - 2$ ← y-coordinate of the turning point

Key point 6

When a quadratic is written in completed square form $y = a(x + b)^2 + c$
the coordinate of the turning point is $(-b, c)$

10 a Factorise the expression $x^2 - 2x - 8$
b Hence write down the coordinates of the roots of $y = x^2 - 2x - 8$
c Where does the graph of $y = x^2 - 2x - 8$ cross the y-axis?
d Write $x^2 - 2x - 8$ in completed square form.
e Hence write down the coordinate of the turning point of $y = x^2 - 2x - 8$
f Is the turning point a maximum or a minimum?
Explain your answer.
g Using your answers to parts **a** to **f**, sketch the graph of $y = x^2 - 2x - 8$

Q10b hint Solve $x^2 - 2x - 8 = 0$

Key point 7

To sketch a quadratic function
- Calculate the solutions to the equation '$y = 0$' (points of intersection with the x-axis).
- Calculate the point at which the graph crosses the y-axis.
- Find the coordinates of the turning point and whether it is a maximum or a minimum.

11 Use the method in **Q10** to sketch these graphs.

a $y = x^2 - 2x - 3$
b $y = -x^2 - 2x + 8$
c $y = 2x^2 - 4x - 6$
d $y = 3x^2 - 3$

12 **Exam-style question**

a Solve the equation $x^2 - 4x + 3 = 0$ **(2 marks)**

b On a copy of the grid, sketch the graph of $y = x^2 - 4x + 3$, marking clearly the coordinates of the points of intersection with the axes and the coordinates of the turning point. **(4 marks)**

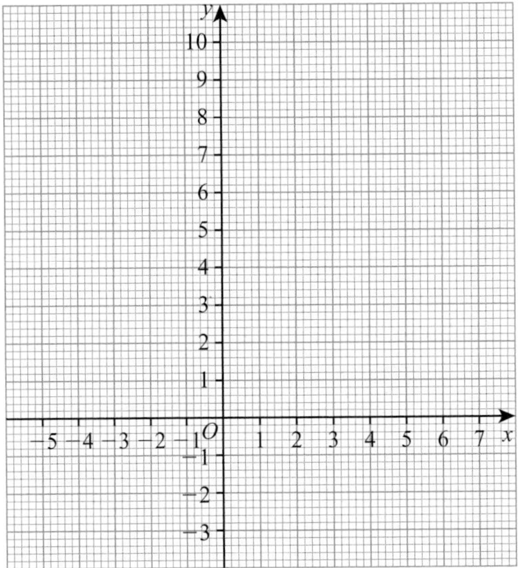

Q12 strategy hint You can use your answer to part **a** to help you draw the graph.

13 a Write down the coordinates of the turning point of the graph of $y = (x + 3)^2 - 5$

b Substitute $y = 0$ into the equation $y = (x + 3)^2 - 5$ and hence find the coordinates of the roots.

Q13b hint Give your answers in surd form.

14 Find the roots of these equations given in completed square form, giving your answers in surd form.

a $(x + 1)^2 - 3 = 0$ b $2(x - 1)^2 - 18 = 0$ c $3(x + 2)^2 - 4 = 0$

15 By writing the equations in completed square form, calculate the roots of the equations. Give your answers in surd form.

a $x^2 + 4x - 3 = 0$ b $2x^2 - 8x - 1 = 0$ c $3x^2 + 18x - 12 = 0$

16 **Reasoning / Communication** Give three reasons why the graph shown is **not** $y = -2x^2 + 4x + 6$

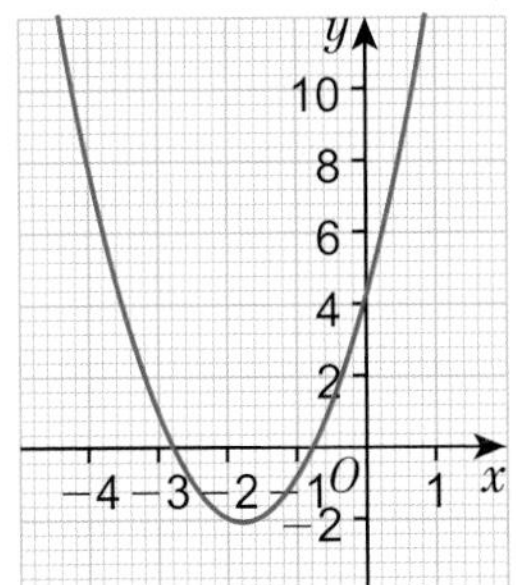

17 **Problem-solving** Find the equation of this quadratic graph.

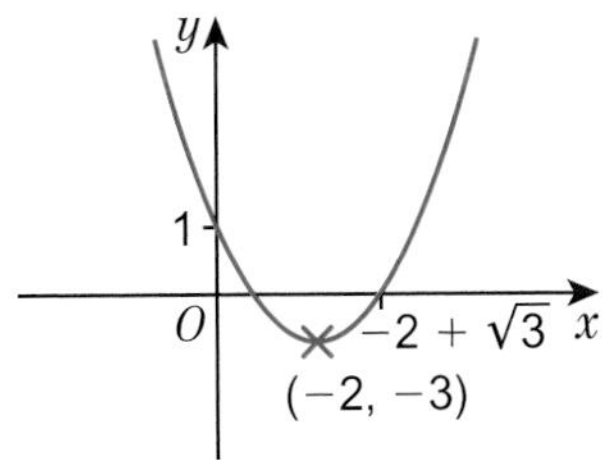

Q17 hint Substitute the value of the turning point and the y-intercept into $y = (x + b)^2 + c$

15.4 Solving quadratic equations graphically

Objectives

- Find approximate solutions to quadratic equations graphically.
- Solve quadratic equations using an iterative process.

Did you know?

In 3000 BC the ancient Babylonians used quadratic equations to work out how much tax to pay.

Fluency

Will the graph of each quadratic have a maximum or a minimum point?

- $y = 2x^2 + 7x - 9$
- $y = -3x^2 + 2x + 5$
- $y = (x - 3)(x - 2)$
- $y = (5 - x)(2 + x)$

Warm up

1 Find the roots of the equation by writing these equations in completed square form.

a $x^2 + 6x + 5 = 0$ b $2x^2 + 4x - 1 = 0$ c $3x^2 + 6x - 1 = 0$

Q1 hint Give your answer in surd form where necessary.

2 Use the quadratic formula to calculate the roots of each equation. Give your answers to 1 decimal place.

a $x^2 - 4x - 7 = 0$ b $2x^2 - 3x - 4 = 0$ c $-3x^2 + 2x + 5 = 0$

3 **Reasoning** Match each graph to its equation. Hence estimate the solutions to the equations.

a $x^2 - 4x - 7 = 0$ b $3x^2 - x - 5 = 0$

c $-x^2 + 2x + 4 = 0$ d $-2x^2 - 2x + 1 = 0$

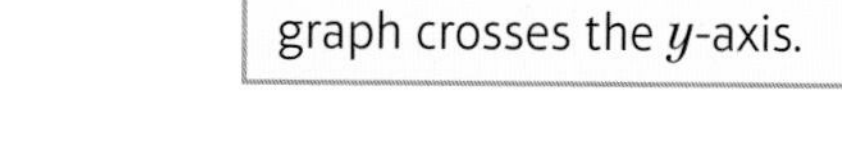

Q3 hint Work out where the graph crosses the y-axis.

i

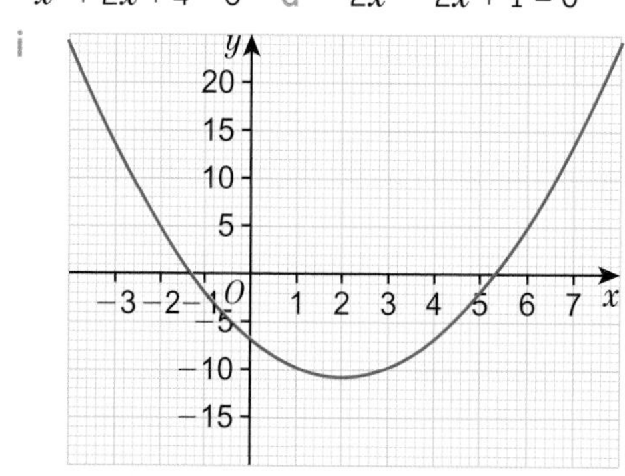

ii

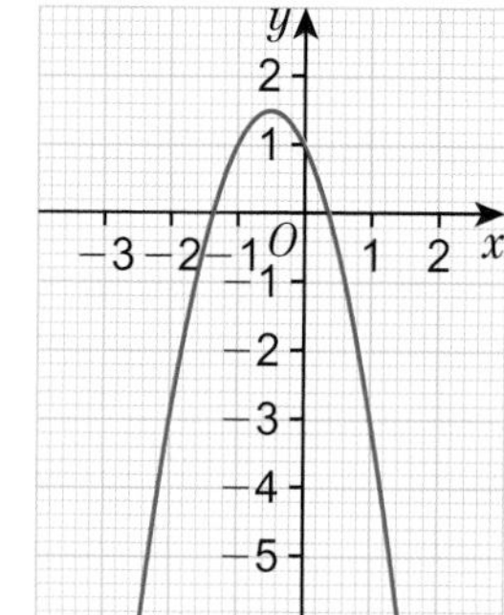

iii

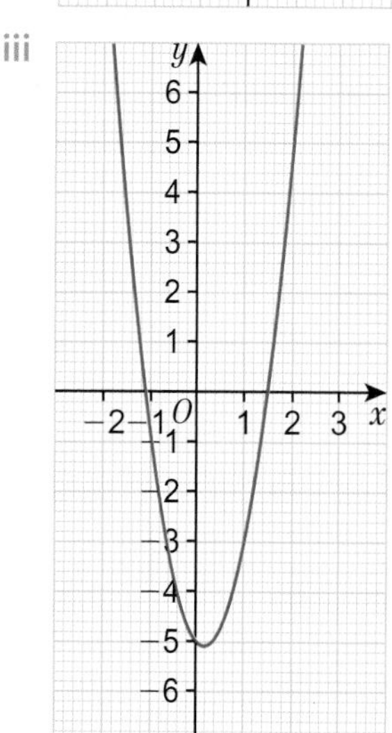

iv

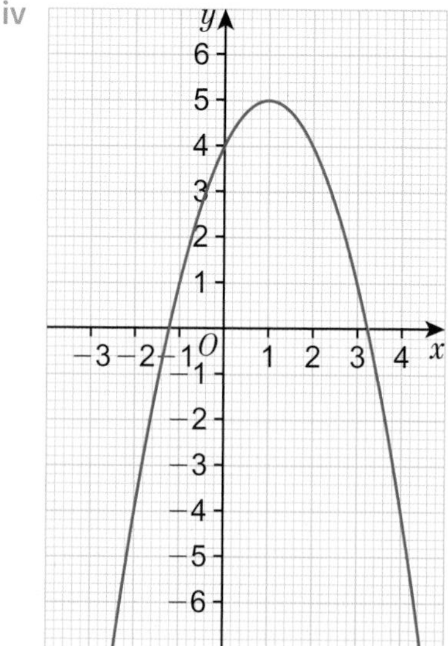

4 a Copy and complete the table of values for $y = 2x^2 - 8x + 7$

x	−2	−1	0	1	3	5
y		17				

b Plot the graph of $y = 2x^2 - 8x + 7$ on a suitable grid.

c From the graph estimate the roots to the equation $2x^2 - 8x + 7 = 0$.

5 a Plot the graphs of the following functions.

i $y = x^2 + 4x - 7$ ii $y = -x^2 - 3x + 2$ iii $y = 2x^2 - 4x - 2$ iv $y = -3x^2 + 6x + 4$

b Hence estimate the solutions to the equations

i $x^2 + 4x - 7 = 0$ ii $-x^2 - 3x + 2 = 0$ iii $2x^2 - 4x - 2 = 0$ iv $-3x^2 + 6x + 4 = 0$

c Use the quadratic formula to find the roots of the equations in part **a** to 3 significant figures. Check your answers to part **a**.

6 **Exam-style question**

a Complete the table of values for $y = 2x^2 + 4x - 2$ **(2 marks)**

x	−4	−3	−2	−1	0	1	2
y							

b Plot the graph of $y = 2x^2 + 4x - 2$

c Use your graph to estimate the roots of the equation $2x^2 + 4x - 2 = 0$ **(3 marks)**

d Write the expression $2x^2 + 4x - 2$ in the form $a(x + b)^2 + c$ **(2 marks)**

Exam hint
For greater accuracy use a ruler and draw lines on the graph to read off values.

7 **Reasoning** For each graph

i find the coordinates of the turning point

ii find the y-intercept

iii sketch the graph.

a $y = x^2 + 2x + 2$ b $y = -x^2 - 4x - 7$ c $y = 2x^2 - 4x + 3$

Discussion Do all these equations have solutions when $y = 0$?

8 **Reasoning** Ali is sketching the graph of $y = 2(x - 3)^2 + 5$
She is finding it difficult to identify the roots of the equation $2(x - 3)^2 + 5 = 0$.
Explain why.

Key point 8

The quadratic equation $ax^2 + bx + c = 0$ is said to have no real roots if its graph does not cross the x-axis. If its graph just touches the x-axis, the equation has one repeated root.

9 **Reasoning** By completing the square, decide whether these quadratic equations have

- no roots
- two roots
- one repeated root

a $x^2 + 6x + 11 = 0$ b $x^2 + 4x - 3 = 0$ c $x^2 - 6x - 12 = 0$ d $3x^2 + 12x + 12 = 0$

e $-x^2 + 2x - 5 = 0$ f $2x^2 - 8x + 5 = 0$ g $-2x^2 - 12x - 18 = 0$ h $-3x^2 - 2x + 1 = 0$

10 **Exam-style question**

a By completing the square, find the roots of the equation $x^2 - 4x - 3 = 0$, giving your answer in surd form. **(3 marks)**

b Show algebraically that $x^2 - 7x + 13 = 0$ has no real roots. **(3 marks)**

Exam hint
Write down all your working. Even if your final answer is wrong, you might still get some marks.

11 Write the equation for each quadratic graph.

Q11 hint Write $y = (x - a)(x - b)$ and expand.

a

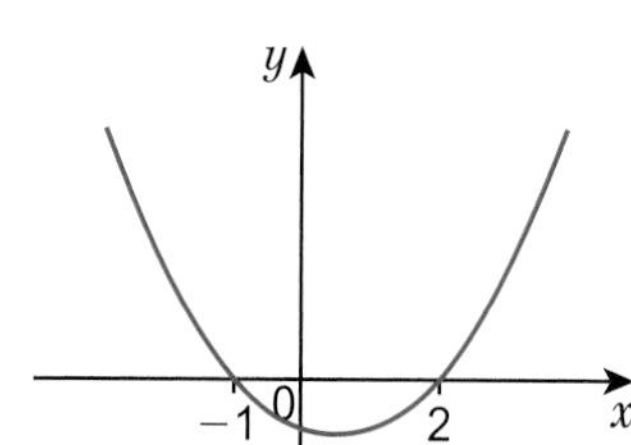

b

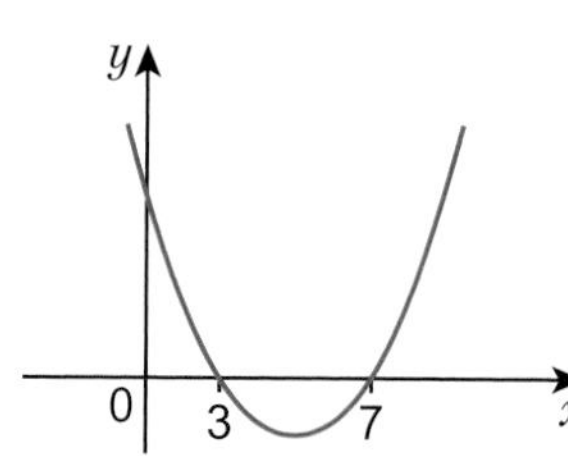

c

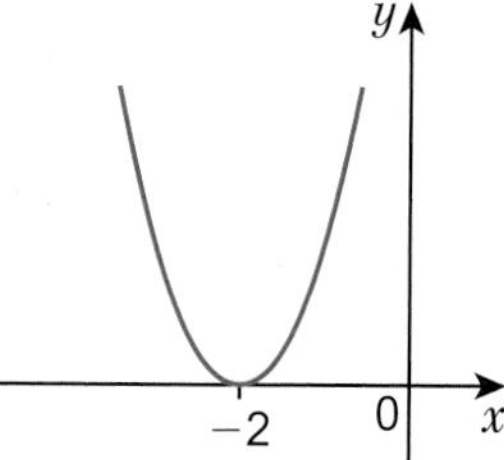

Key point 9

To find an accurate root of a quadratic equation you can use an **iterative** process.
Iterative means carrying out a repeated action.

Example 4

Use an iterative formula to find the positive root of the equation $y = x^2 + x - 5$ to 5 decimal places.

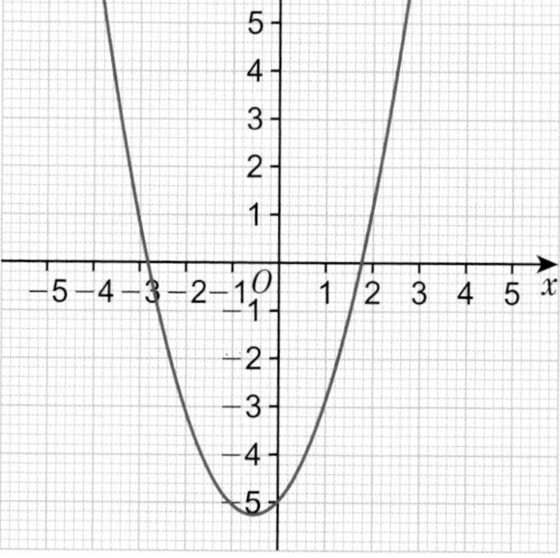

$0 = x^2 + x - 5$ — Rearrange the equation to make the highest power of x the subject.

$x^2 = 5 - x$

$x = \sqrt{5 - x}$ — Take the square root of each side to write $x = \ldots$

$x_1 = \sqrt{5 - x_0}$ so $x_{n+1} = \sqrt{5 - x_n}$ — Starting with initial value x_0 on the RHS gives a new value x_1. Repeating over and over again gives a sequence. So value x_n on the RHS gives the new value x_{n+1} on the LHS.

$x_0 = 2$ — From the graph you can see that the positive root is approximately 2. Use this as the value of x_0

$x_1 = \sqrt{5 - x_0} = \sqrt{5 - 2} = \sqrt{3} = 1.73200808$ — Find the value of x_2 by substituting x_1 into the iterative formula.

$x_2 = \sqrt{5 - x_1} = \sqrt{5 - ANS} = 1.807746993$ — Use the ANS key on your calculator so that the EXACT value is used.

$x_3 = 1.786687719$ — You can use the '=' key to produce the next iteration.

$x_4 = 1.792571416$

$x_5 = 1.790929531$

$x_6 = 1.791387861$

$x_7 = 1.791259931$

$x_8 = 1.79129564$

$x_9 = 1.791285672 = 1.79129$ (5 d.p.)

$x_{10} = 1.791288455 = 1.79129$ (5 d.p.) — Round all the answers to 5 d.p. until you get the same value twice. The answer is converging to $x = 1.79129$

$x \approx 1.79129$

12 Use the iterative equation and the starting point given to find one root for each quadratic equation. Give your answers correct to 5 decimal places.

a $y = x^2 - 2x - 4$ $\quad x = \sqrt{4 + 2x}$ $\quad x_0 = 3$

b $y = x^2 - 5x - 4$ $\quad x = \sqrt{5x + 4}$ $\quad x_0 = 5.5$

c $y = x^2 - x - 3$ $\quad x = \dfrac{3}{x - 1}$ $\quad x_0 = -1.5$

13 a Solve the quadratic equation $x^2 + x - 2 = 0$.

b Sketch the graph of $y = x^2 + x - 2$

c Write the set of values of x that satisfy $x^2 + x - 2 < 0$ (where the curve is below the x-axis)

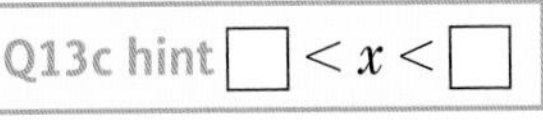

Q13c hint $\square < x < \square$

d Write the set of values of x that satisfy $x^2 + x - 2 > 0$ (where the curve is above the x-axis)

Q13d hint $x < \square$ and $x > \square$

Key point 10

To solve a quadratic inequality:

- Solve as a quadratic equation
- Sketch the graph
- Use the graph to find the values that satisfy the inequality.

14 Find the set of values that satisfy each inequality.

a $x^2 - 2x - 3 < 0$ b $x^2 + 3x - 10 < 0$ c $x^2 + 5x + 4 > 0$

d $x^2 + 7x + 10 < 0$ e $x^2 - 6x + 8 > 0$ f $x^2 - 6x + 5 < 0$

Q14a hint Solve $x^2 - 2x - 3 = 0$ first.

15.5 Graphs of cubic functions

Objectives

- Find the roots of cubic equations.
- Sketch graphs of cubic functions.
- Solve cubic equations using an iterative process.

Did you know?

The movements of tides can be represented by cubic equations.

Fluency

Where do the graphs cross the x-axis?

- $y = (x - 4)(x + 2)$ • $y = (x - 7)(x - 3)$ • $y = (x + 5)(x + 2)$

Warm up

1 Expand and simplify

a $(x + 2)(x + 3)$ b $(x - 3)(x + 4)$ c $(2x + 1)(x - 5)$ d $(3x - 4)(x - 2)$

2 Find the roots of each quadratic equation by factorising

a $x^2 - 3x - 10 = 0$ b $x^2 - 1 = 0$ c $2x^2 - 4x - 6 = 0$ d $6x^2 + 10x - 4 = 0$

3 Expand the expression $(x^2 + 4x + 1)(x + 2)$

Q3 hint Multiply each term in the first bracket by each term in the second bracket. Then simplify.

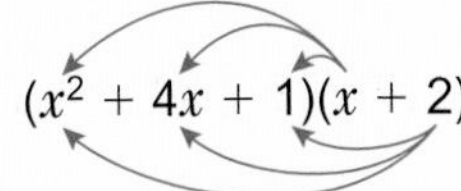

$(x^2 + 4x + 1)(x + 2)$

4 Copy and complete to expand the expression

$(x + 2)(x + 4)(x + 3) = (x^2 + \square x + \square)(x + 3)$

$= x^3 + \square + x^2 + \square x = \square$

5 Expand the expressions

a $(x + 2)(x + 5)(x + 1)$ b $(x - 3)(x + 4)(x - 2)$

c $(x + 2)(x - 1)(x + 3)$ d $x(x + 5)(x - 4)$

e $(x + 1)^2(x - 1)$ f $(x + 3)^3$

Key point 10

A **cubic** function is one whose highest power of x is x^3.
It is written in the form $\boldsymbol{y = ax^3 + bx^2 + cx + d}$

When $a > 0$ the function looks like

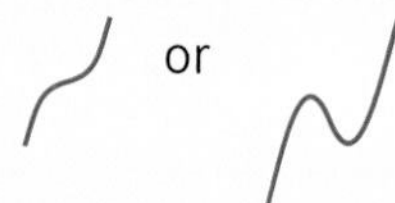

When $a < 0$ the function looks like

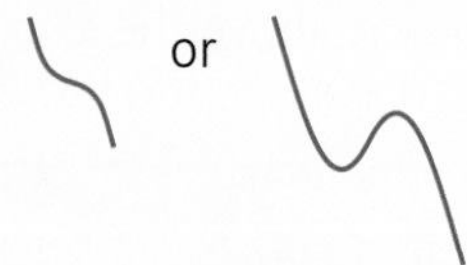

The graph intersects the y-axis at the point $y = d$
The graph's roots can be found by finding the values of x for which $y = 0$.

6 Here is the graph of $y = x^3 + 3x^2 - 6x - 8$
   a What are the roots of the equation $x^3 + 3x^2 - 6x - 8 = 0$?
   b Where does the graph cross the y-axis?

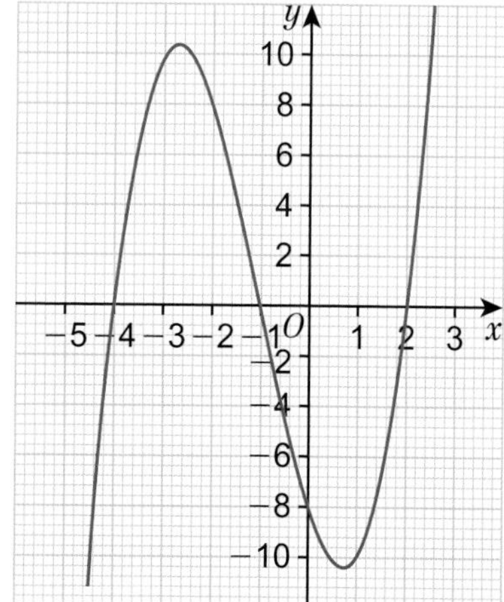

Example 5

Sketch the graph of $y = (x - 1)(x - 2)(x + 2)$

When $x = 0$, $(x - 1)(x - 2)(x + 2) = -1 \times -2 \times 2 = 4$ — The graph crosses the y-axis when $x = 0$

The graph crosses the y-axis at 4.

$(x - 1)(x - 2)(x + 2) = x^3 - x^2 - 4x + 4$ — Expand the expression.

The coefficient of x^3 is 1, so $a = 1$

Since $a > 1$ the graph has the shape

$0 = (x - 1)(x - 2)(x + 2)$ — Find the roots of the equation.

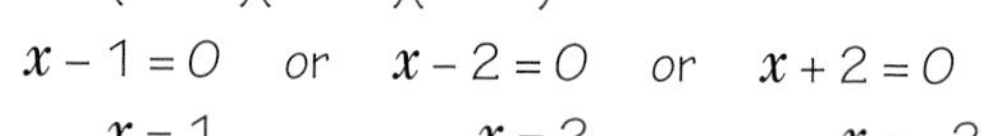

$x - 1 = 0$ or $x - 2 = 0$ or $x + 2 = 0$
$x = 1$ $\quad$ $x = 2$ $\quad$ $x = -2$ — If the product of any expressions is zero, one of them must itself be zero.

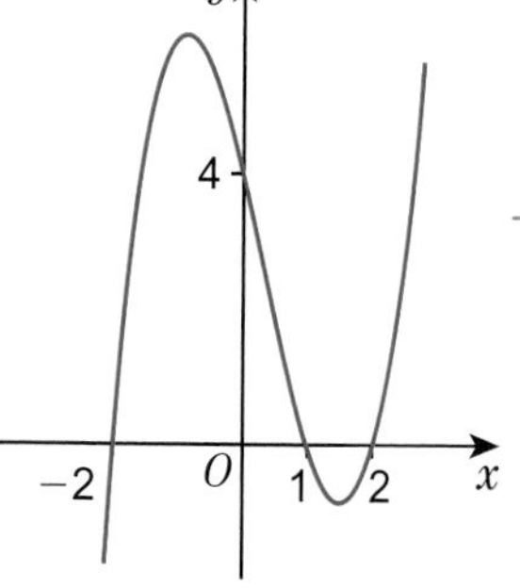

Sketch the graph, marking on the points of intersection with the x- and y-axes.

7 a What are the roots of the equation $y = (x + 1)(x + 2)(x + 5)$?

b Where does the graph of $y = (x + 1)(x + 2)(x + 5)$ cross the y-axis?

c Sketch the graph of $y = (x + 1)(x + 2)(x + 5)$

> **Q7b hint** You could multiply the constant terms to find where it crosses the y-axis.

8 **Reasoning** Match the equation of each graph to its image.

a $y = (x + 1)(x + 2)(x + 3)$ b $y = (x + 2)^2(x + 5)$

c $y = (x - 2)(x + 2)(x + 4)$ d $y = -x(x + 1)(x - 3)$

e $y = (1 - x)(x + 2)(x + 4)$ f $y = -x^2(x + 3)$

i

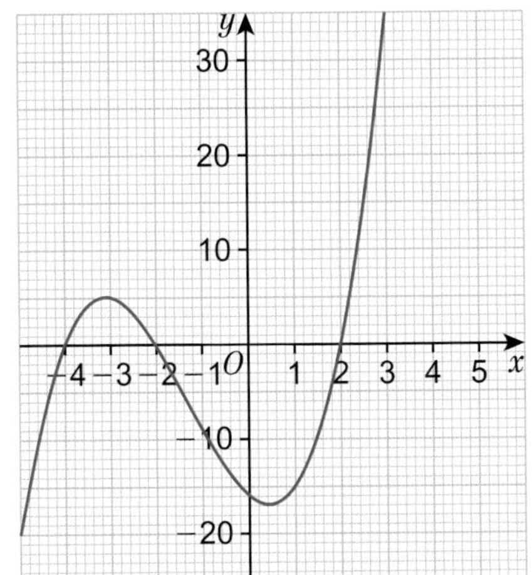

ii

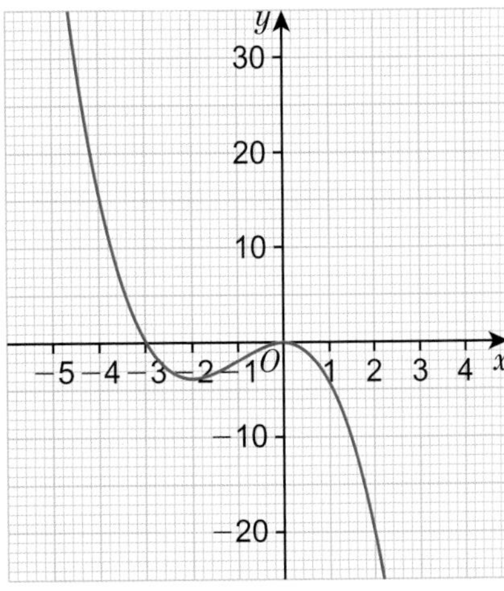

iii

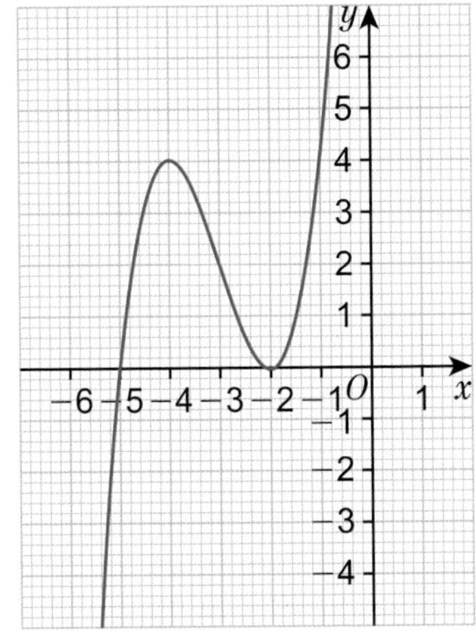

iv

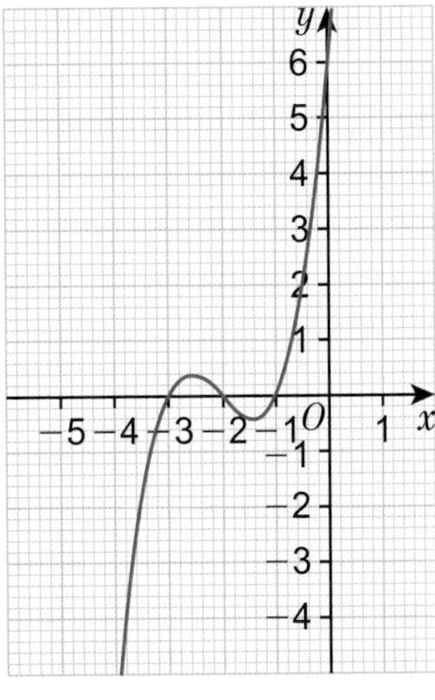

v

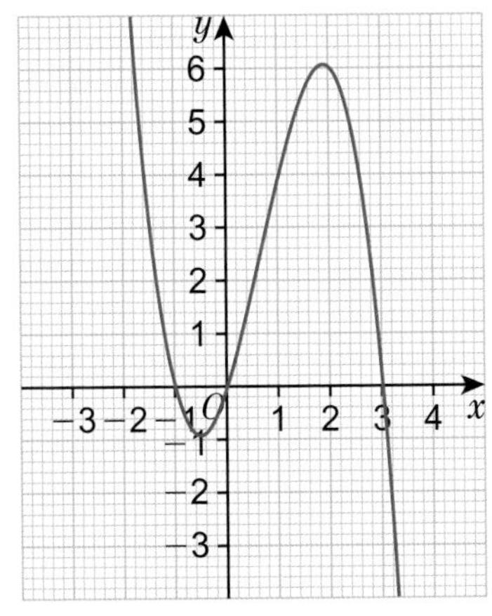

vi

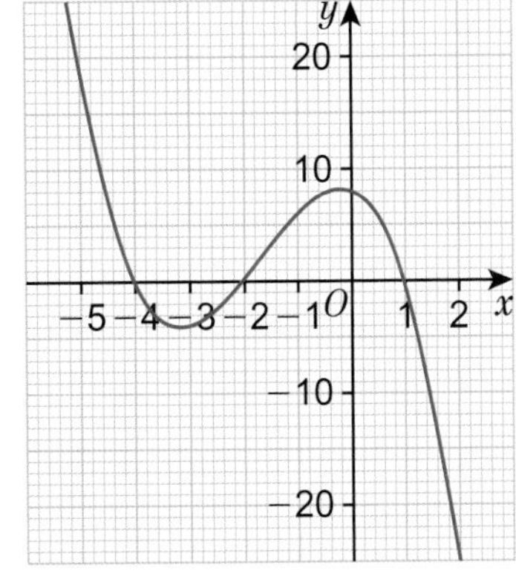

Key point 11

When the graph of a cubic function y crosses the x-axis three times, the equation $y = 0$ has three solutions.
For example $y = (x + 2)(x - 1)(x - 3)$

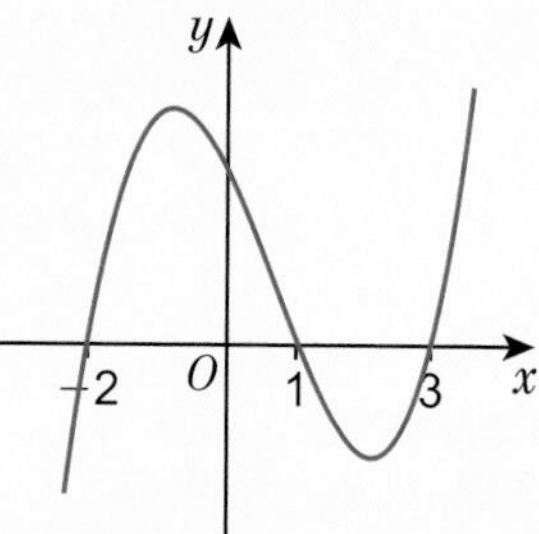

When the graph of a cubic function y crosses the x-axis once and touches the x-axis once, the equation $y = 0$ has three solutions but one of them is repeated.
For example $y = (x - 1)(x + 2)^2$.

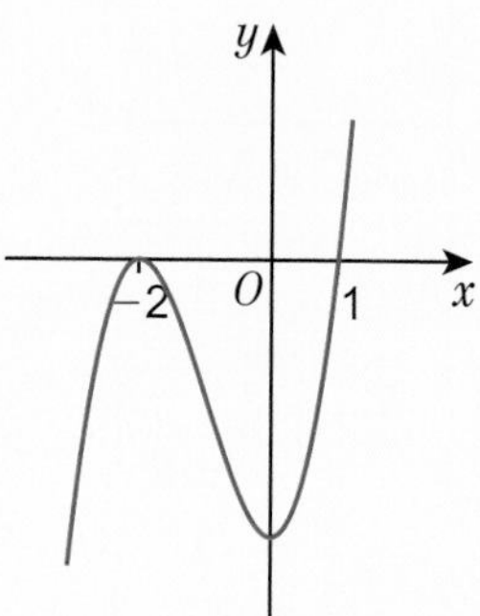

When the graph of a cubic function y crosses the x-axis once, the equation $y = 0$ can have

- one distinct, repeated solution, for example $y = (x - 1)^3$

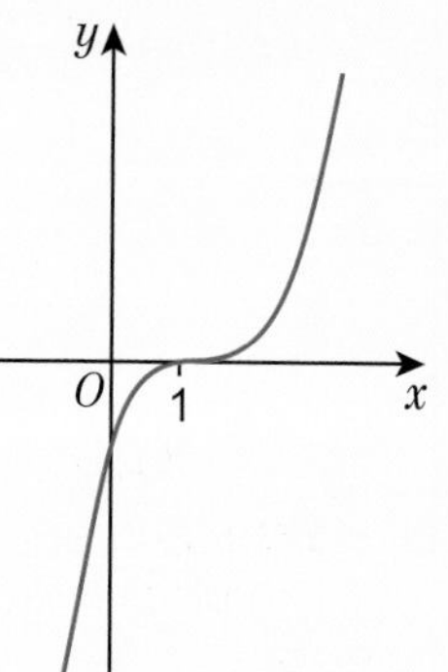

- or only one real solution, for example $(x + 2)(x^2 + x + 1)$
  The quadratic $(x^2 + x + 1)$ has no real solutions.

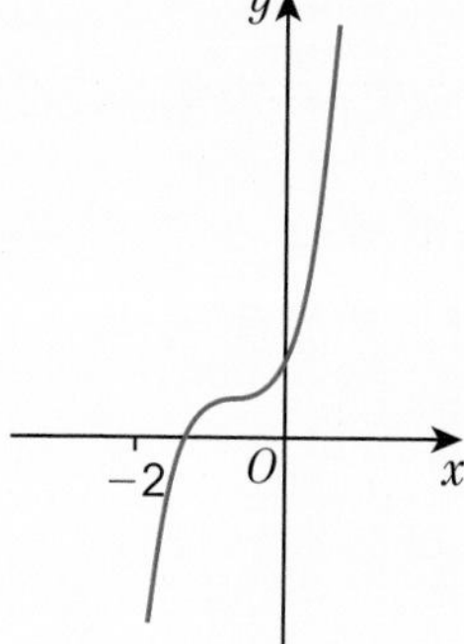

9 How many solutions does each cubic equation have?

a $(x + 1)(x - 3)(x + 4) = 0$
b $(x + 3)^3 = 0$
c $-x(x + 1)(x - 3) = 0$
d $x^2(x + 4) = 0$
e $(x^2 + 2x + 5)(x - 2) = 0$
f $(10 - x)(x + 4)(x - 1) = 0$

10 Sketch the graphs, marking clearly the points of intersection with the x- and y-axes.

a $y = (x - 3)(x + 2)(x - 1)$
b $y = x(x + 1)(x - 4)$
c $y = (-x + 1)(x + 3)(x - 1)$
d $y = (x + 1)^2(x + 3)$
e $y = (x + 2)^3$

11 **Exam-style question**

Sketch the graph of $y = -x(x + 2)^2$ marking clearly the points of intersection with the axes. **(3 marks)**

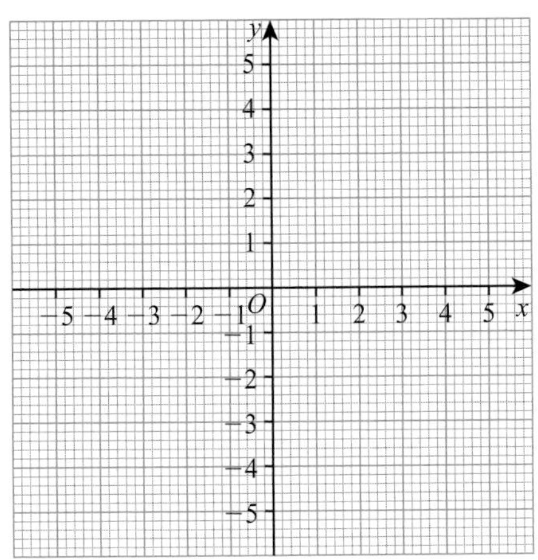

Exam hint
Work out the coordinates of the points of intersection of the axes before you plot them.

12 **Problem-solving** The graph has equation $y = x^3 + ax^2 + bx + c$

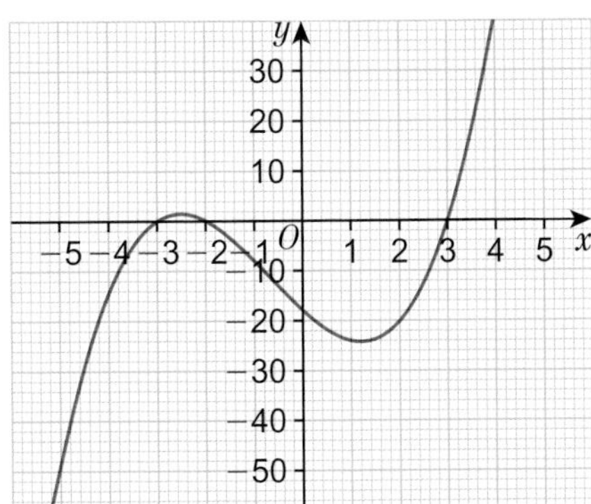

Work out the values of a, b and c.

13 **Problem-solving** A graph has equation $y = -x^3 + ax^2 + bx + c$
It crosses the x-axis at $x = 1$, $x = 4$ and $x = -2$.
Without drawing the graph, work out the values of a, b and c.

14 Use an iterative formula to find the one root of $x^3 + 2x^2 - 6 = 0$ to 4 d.p.
The first steps have been done for you:

$x^3 = -2x^2 + 6$

$x^1 = \sqrt[3]{-2x^2 + 6}$

$x_{n+1} = \sqrt[3]{-2x_n^{\ 2} + 6}$

$x_0 = \square$

$x_1 = \sqrt[3]{-2x_0 + 6} = \square$

$x_2 = \sqrt[3]{-2x_1 + 6} = \square$

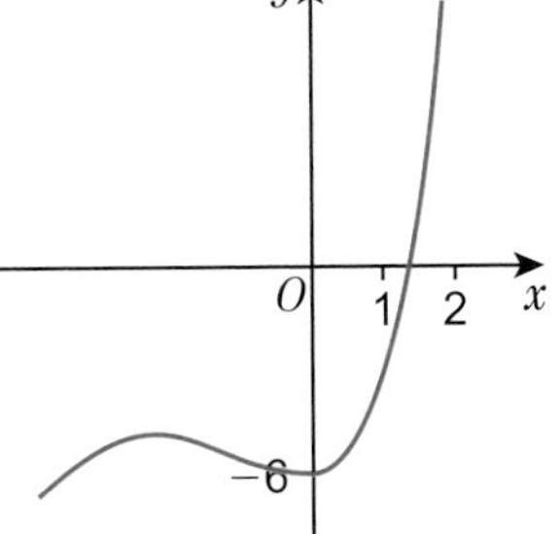

Q14 hint Estimate the root from the x-intercept on the graph.

15 Use an iterative formula to find the negative root of the equation $x^3 - x^2 - 4x + 3 = 0$ to 5 d.p.

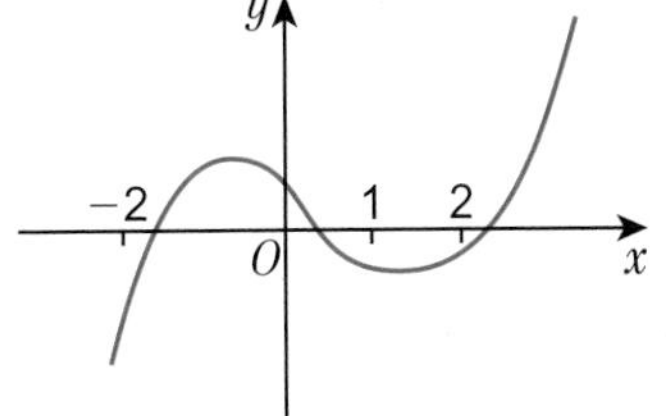

15 Problem-solving

Objective

- Use graphs to help you solve problems.

Example 6

The table shows data from a science experiment. The experiment measured the time in seconds it takes a steel sphere to fall from different heights.

Height (metres)	0	0.25	0.5	0.75	1
Time (seconds)	0	0.23	0.32	0.39	0.45

Estimate the time it would take the steel sphere to fall from 60 cm.

> Look at the table. Is there a pattern in the numbers in both rows? If not, then a graph is a good problem-solving strategy.

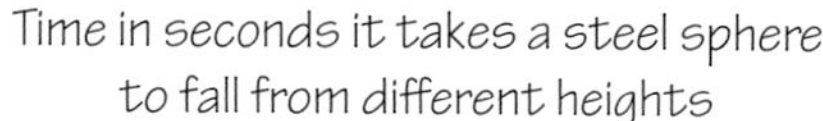

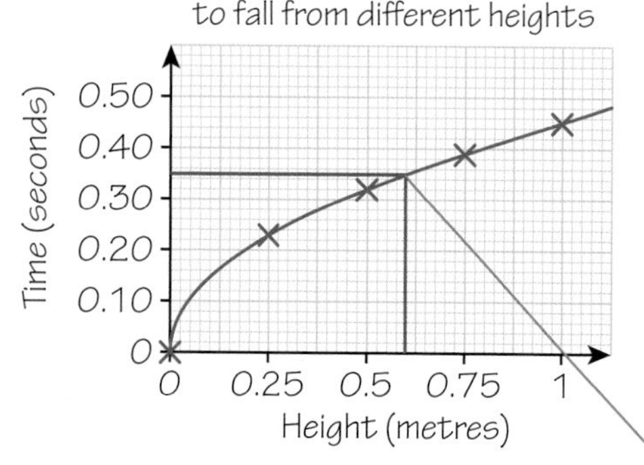

> Draw the axes for your graph. The first row in the table is usually on the horizontal axes. Decide on a scale for each axis by looking at the smallest and largest number in each row. Give your graph a title.

> Carefully plot the points from the table.

> The points lie on a curve, so join the points with a smooth curve.

> The question asks for the time to fall when the steel sphere is dropped from 60 cm, i.e. 0.6 m. Draw a line from 0.6m on the height axis to your graph. Draw a line across to the time axis and read the time in seconds.

Time it would take the steel sphere to fall from 60 cm = 0.35 seconds

> Check your answer against the table. Where would 0.6 m appear in the table? Would its likely time be 0.35 seconds?

1 The table shows the temperature in degrees Celsius at altitudes in 1000s of feet.

Altitude (1000s of feet)	15	20	25	30	35	40
Temperature (° Celsius)	4.5	−5.9	−16.1	−27.6	−39.8	−50.2

Estimate the temperature at an altitude of 28 000 feet.

2 **Modelling** The shape of a bridge is modelled by the equation $y = -\frac{1}{4}x^2 + x + \frac{5}{2}$, where x and y are measured in metres. How high is the highest point of the bridge?

> **Q2 hint** When drawing a graph of an equation, sketch the shape first.

3 **Finance** A British businessman in Abu Dhabi records distances (in kilometres) and fares in the local currency (Arab Emirate Dirhams) for three taxi rides in the city:

Distance (km)	2	3	7
Cost (AED)	6.90	8.60	15.40

a His fourth taxi ride is 13 km. What is the taxi fare for this trip?

b The exchange rate for converting British pounds to Arab Emirate Dirhams is £1 : 5.80 AED. How much British money can the businessman claim for the four taxi rides?

4 **Reasoning** The table shows the ratio of median house price to median earnings for a London borough.

Year	2004	2005	2006	2007	2008	2009	2010	2011
$\frac{\text{Median house price}}{\text{Median earnings}}$	9.91	10.31	10.83	12.15	12.06	10.82	11.89	12.17

a Explain what the increasing ratio shows about house prices in relation to people's earnings.

b Use the data to estimate the likely ratio in 2012.

c The ratio in 2012 was 13.87. What does this tell you about the median house price or the median earnings in this borough?

Q4b hint Draw a graph.

5 **Modelling** A boy throws a football upwards out of an upstairs window. The height, h, of the football, in metres, is modelled by the equation $h = -4t^2 + 7t + 6.5$

At the same time, his brother fires a toy paintball gun upwards at the football on a trajectory that can be modelled by the equation $h = \frac{1}{2}t + 2$

a How high is the upstairs window?

b Show that the paint from the paintball gun hits the football. At what height does this happen?

Q5 hint What is time, t, when the boy throws the football? What is the height at this time?

Q5b hint Where do the graphs intersect?

6 **Reflect** What clues could you look for in a question to tell you that a graph may be a good problem-solving strategy? What types of graphs could you draw?

15 Check up

Log how you did on your Student Progression Chart.

Simultaneous equations and inequalities

1 **Reasoning** A mobile phone company offers two different packages.

Package A	**Package B**
No monthly line rental.	Monthly line rental £20.
10p per minute calls.	5p per minute calls.

a For each package, form an equation to calculate the monthly cost with y = monthly cost and x = total minutes of calls.

b Use a graphical method to work out how many minutes of calls are used if the two packages cost the same.

2 Draw a coordinate grid with x-axis from −5 to 5 and y-axis from −5 to 11.

Draw graphs to solve the simultaneous equations

$y = x^2 - 2x - 4$

$y = -x - 2$

3 **Reasoning** Write down the three inequalities satisfied by the coordinates of all the points in the shaded region of the graph.

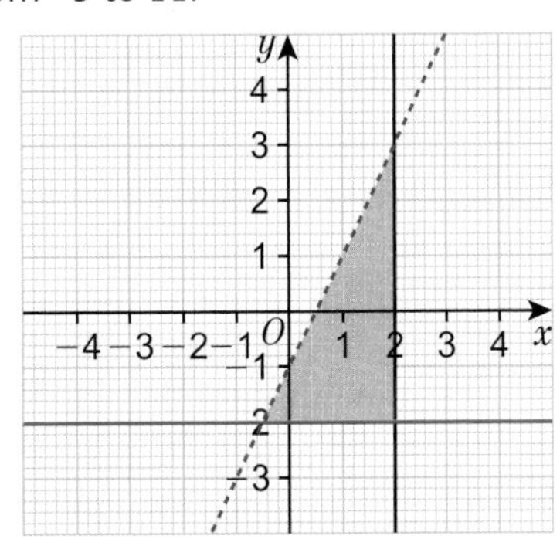

Graphs of quadratic functions

4 a Sketch the graph of $y = x^2 + x - 12$

b Showing all your workings, mark clearly the points of intersection with the y-axis, the solutions to the equation $x^2 + x - 12 = 0$ and the turning point.

c Find the set of values that satisfy $x^2 + x - 12 < 0$

5 a Here is the graph of $y = x^2 - 2x - 2$.
Showing all your workings, mark clearly the point of intersection with the y-axis and the turning point.

b Estimate the roots of the equation
$x^2 - 2x - 2 = 0$

c Use an iterative equation to find the positive solution correct to 5 d.p.

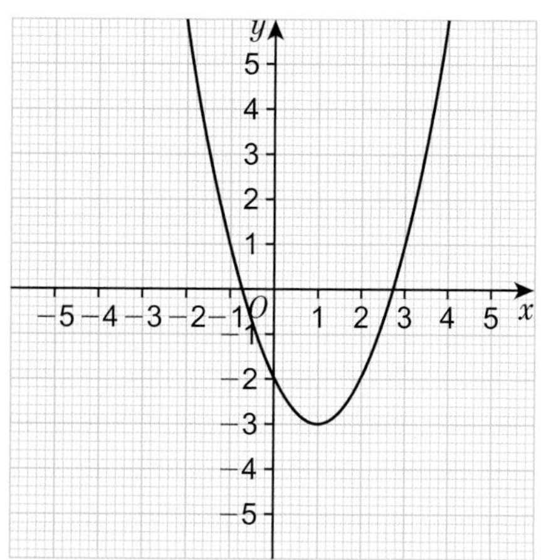

Graphs of cubic functions

6 Showing all your working, sketch the graph of $y = (x - 2)^2(x + 5)$, marking clearly the x- and y-intercepts.

Reflect

7 How sure are you of your answers? Were you mostly
Just guessing ☹ Feeling doubtful 😐 Confident ☺
What next? Use your results to decide whether to strengthen or extend your learning.

*** Challenge**

8 A function has roots −1 and 3. List as many possible equations of the function as possible.

Q8 hint The function could be quadratic OR cubic.

15 Strengthen

Simultaneous equations and inequalities

1 a Copy and complete the table of values for $y = x^2 - 3x + 1$

x	−2	−1	0	1	2	3	4	5
y		5		−1				

b On suitable axes plot the graph of $y = x^2 - 3x + 1$

c On the same axes draw the graph of $y = -x + 4$

d Where do the graphs intersect one another?

e Write down the solutions to the simultaneous equations
$y = x^2 - 3x + 1$
$y = -x + 4$

Q1e hint Look at your answer to part **d**.

2 Solve the simultaneous equations graphically.

a $y = 2x + 4$ and $y = -x + 7$

b $y = -x^2 + 3x + 4$ and $y = x + 1$

Q2a hint Plot both graphs on the same grid.

Q2b hint Use the method in **Q1**.

3 The sum of two numbers is 12.
The difference between them is 6.
Let the numbers be represented by x and y.

a Write an equation to show their sum is 12.

b Write an equation to show their difference is 6.

c On the same grid, plot the graphs of the equations you have given for parts **a** and **b**.

d Find x and y.

Q3 hint You can always check your answer to problems like this by checking the numbers work. Do they sum to 12? Is the difference 6?

Q3d hint Where do the graphs intersect one another?

4

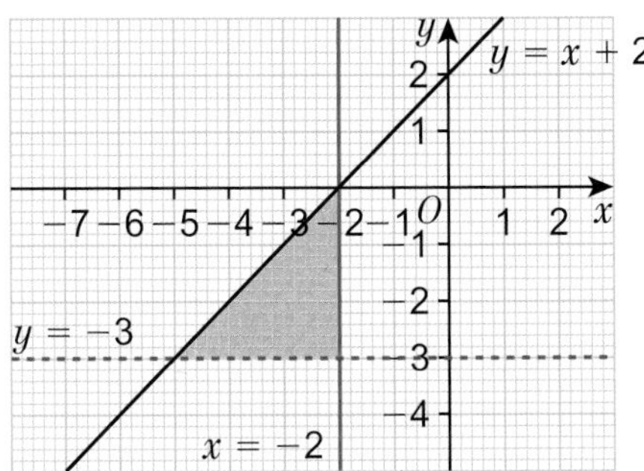

Choose one inequality from each row to describe the shaded area.

$y \leqslant x + 2$ $\quad$ $y < x + 2$ $\quad$ $y \geqslant x + 2$

$x \leqslant -2$ $\quad$ $x \geqslant -2$

$y < -3$ $\quad$ $y > -3$ $\quad$ $y \leqslant -3$

Q4 hint Choose a coordinate in the space, for example (–3, –2)
Which of the inequalities does it satisfy?

Q4 hint If the symbol is $\geqslant$ or $\leqslant$ a **solid** line is drawn, meaning that the points on the line itself are included.
If the symbol is $<$ or $>$ a **dotted** line is drawn, meaning that the points on the line itself are **not included**.

5 Write down the three inequalities that describe the shaded area;

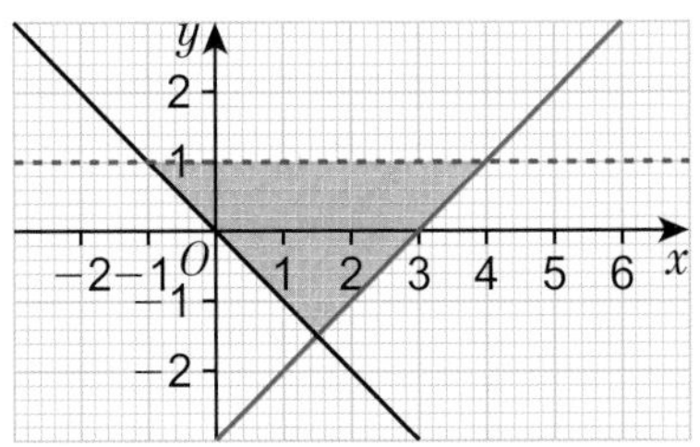

Q5 hint Write down the equation of the three lines. Use the method in **Q4** to work out which inequalities describe the shaded area.

6 a Draw a coordinate grid with –5 to 5 on both axes. Draw the line $y = x$.

b Mark on the coordinate (3, 5).

c For this coordinate is $y \geqslant x$?

d Mark on the coordinate (4, 3).

e For this coordinate is $y \geqslant x$?

f Use this information to shade the area that represents $y \geqslant x$.

Q6e hint $\geqslant$ means greater than OR equal to.

7 Use the method in **Q6** to draw a graph and shade the area that represents each inequality.

a $y \leqslant x$

b $y > 2x$

c $y \leqslant 2x + 2$

d $y > 10 - 2x$

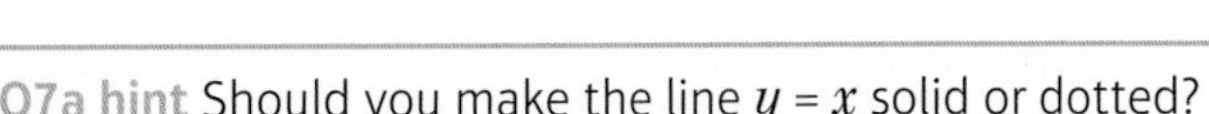
Q7a hint Should you make the line $y = x$ solid or dotted?

Q7b hint Choose a coordinate to test whether $y > 2x$. This will show you whether to shade that side of the line.

8 The diagram shows the lines $y = x$, $y = -2x$ and $y = 5$

On a copy of this diagram, shade the area represented by

$y \geqslant -2x$, $y > x$, $y < 5$

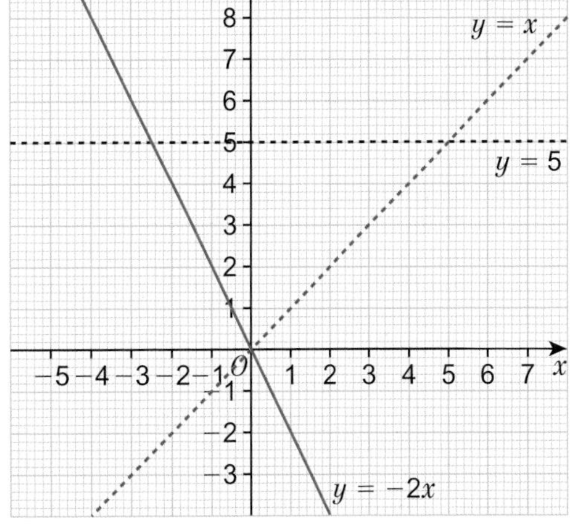

Q8 hint For each line, decide if the area to shade is above or below it. To do this, choose a coordinate pair and test it.

9 On a suitable coordinate grid, shade the region that satisfies the inequalities
$x \leqslant 5, x \geqslant 2, y < 3, y > -x + 3$

> **Q9 hint** Draw the four graphs on a suitable coordinate grid.
> Decide which area satisfies all four inequalities by testing coordinates.

Graphs of quadratic functions

1 For each of these graphs write down the
 i solutions to $y = 0$ ii intercept with the y-axis

a

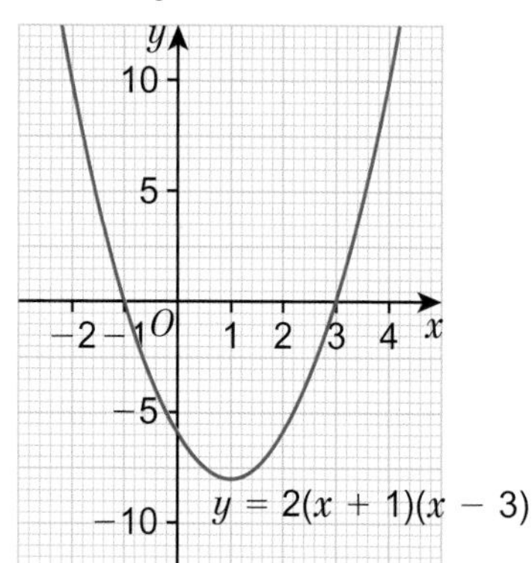

b

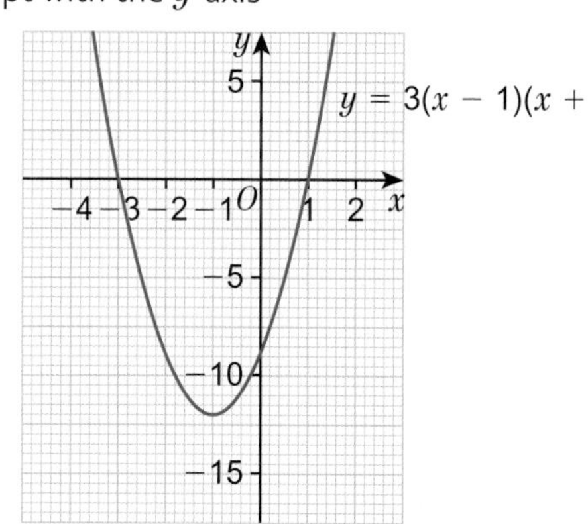

c

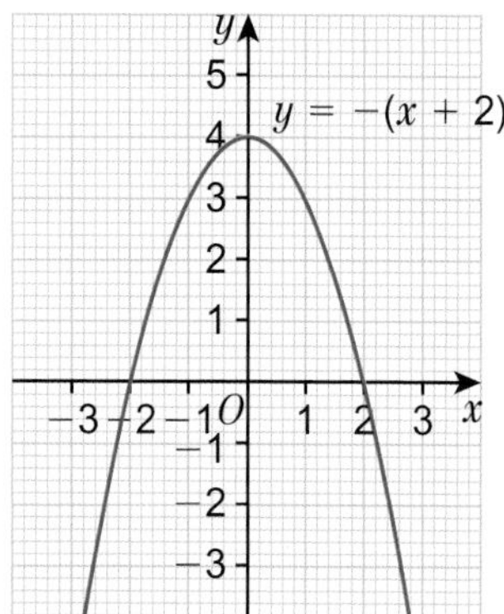

d

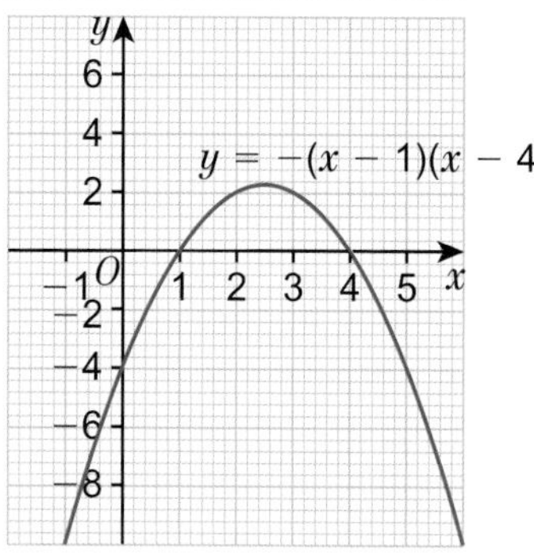

> **Q1a hint**
> Find the x values when $y = 0$.

> **Q1b hint**
>
> 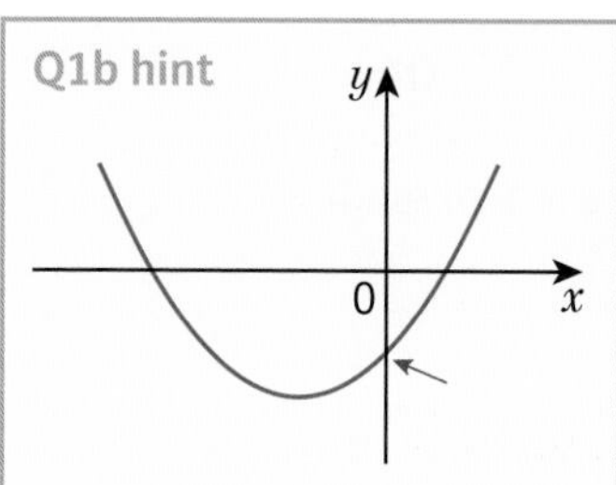
>
> Find the y-value when $x = 0$

2 A curve has the equation $y = x^2 + 2x - 8$. Copy and complete

a When $y = 0$
.... = 0
So $(x - \square)(x + \square) = 0$
There are two possible solutions:
$x - \square = 0$ hence $x = \square$
$x + \square = 0$ hence $x = \square$
So the roots are $x = \square$ and $x = \square$

b When $x = 0$
$y = x^2 + 2x - 8$
$y = 0^3 - \square - \square$
$y = \square$
So the intercept with the y-axis is at $y = \square$

3 Find the roots and y-intercept of

a $x^2 - 2x - 15 = 0$ b $-x^2 - 6x + 16 = 0$
c $2x^2 - 4x - 6 = 0$ d $3x^2 + 12x - 15 = 0$

> **Q3 hint** Use the method in **Q2**.

4 Copy and complete to find the turning point of $y = x^2 + 2x - 8$

a $y = (x + \square)^2 - \square - 8$

$y = (x + \square)^2 - \square$

b Write down the turning point of the graph $y = x^2 + 2x - 8$

c Decide if the turning point is a maximum or minimum.

Q4a hint To find the missing term in the bracket divide the coefficient of the x term by 2.

Q4b hint The turning point is the value of y when $(x + \square) = 0$

Q4c hint

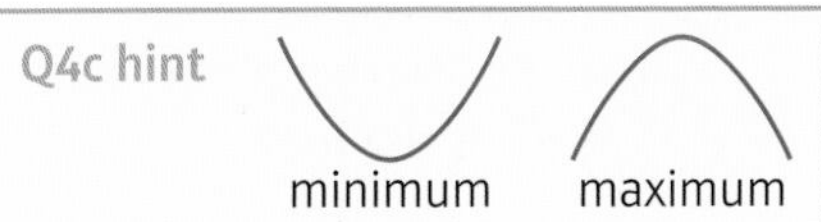

5 a Find the coordinates of the turning point of the graphs in **Q3**.

b Decide whether each turning point is a maximum or minimum.

Q5a hint Use the method in **Q4**.

Q5b hint Look at the coefficient of x^2.

6 Sketch quadratic graphs for

a $y = x^2 - 2x - 15$

b $y = -x^2 - 6x + 16$

c $y = 2x^2 - 4x - 6$

d $y = 3x^2 + 12x - 15$

Q6 hint Use your answers from **Q3** and **Q5** to work out these points:

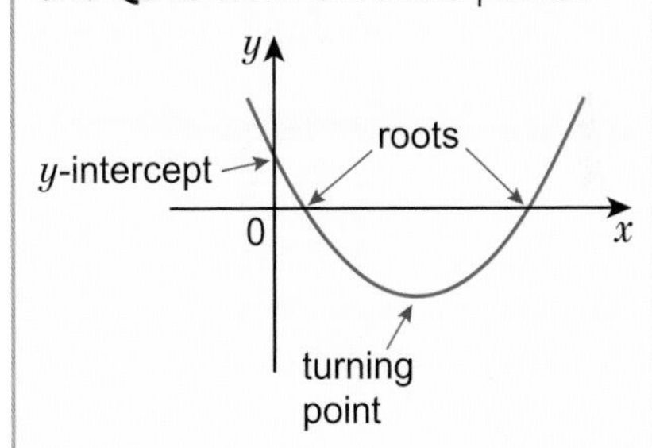

7 Use your sketch graph from **Q6a** to find the values of x that satisfy

a $x^2 - 2x - 15 < 0$

b $x^2 - 2x - 15 > 0$

Q7a hint

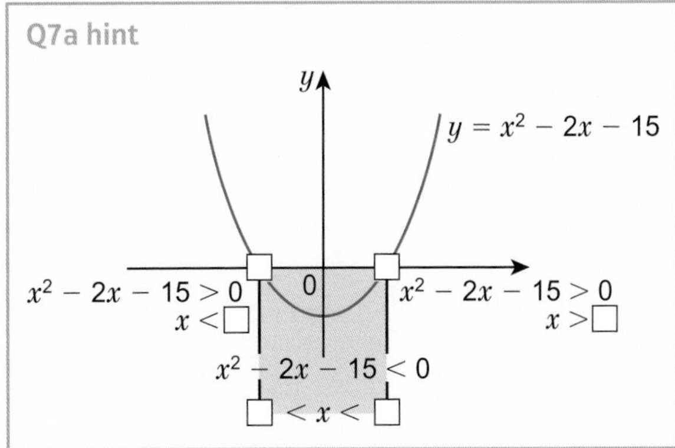

8 Sketch the graphs of

a $y = x^2 + 8x + 15$

b $y = -2x^2 + 12x - 10$

Q8 hint Find the solutions to $y = 0$ and the y-intercept, as in **Q3**.

9 For the graph of $y = x^2 + 4x - 3$ copy and complete the table of values.

x	−5	−4	−3	−2	−1	0	1	2
y	2							

a Plot the graph.

b Find the approximate roots of the equation $x^2 + 4x - 3 = 0$

Q9b hint Read the values of the roots from your graph.

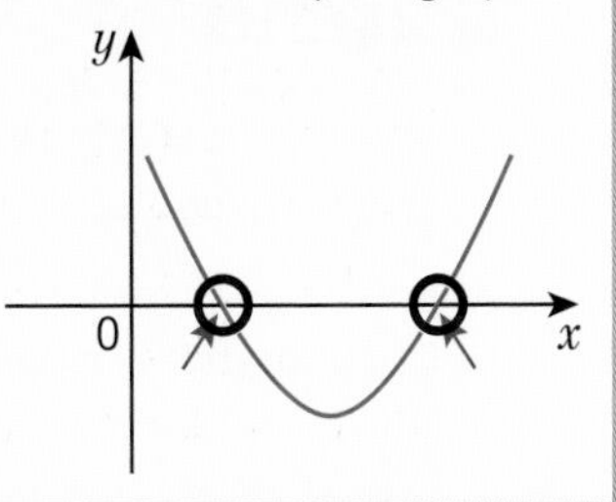

10 Plot the graphs and hence estimate the roots of

a $x^2 - 4x + 1 = 0$

b $2x^2 - 8x + 2 = 0$

c $-2x^2 + x + 4 = 0$

Q10 hint Use the method in **Q9**.

11 a Copy and complete to find an iterative formula to find the roots of the equation $y = x^2 - 4x + 1$

$0 = x^2 - 4x + 1$

$x^2 = 4x - \square$

$x = \sqrt{4x - \square}$

$x_{n+1} = \sqrt{4x_n - \square}$

b Use the starting point of $x_0 = 3.5$. Work out the values of x_1 to x_5.

c Work out the root to 3 decimal places.

Q11b hint $x_0 = 3.5$

$x_1 = \sqrt{4 \times 3.5 - \square} = \square$

Q11c hint Carry on repeating the iterative process until the first four decimal places do not change.

12 a Find an iterative formula that you could use to find the roots to the equation $y = x^2 - 8x + 1$

b Using the starting point of $x_0 = 7.5$, find the root to 3 decimal places.

Q12a hint Start with $0 = x^2 - 8x + 1$
Rearrange to give $x^2 = \square - \square$
Then $x =$ ....

Graphs of cubic functions

1 For each of these graphs write down the

i solutions of the equation $y = 0$ ii y-intercept.

a
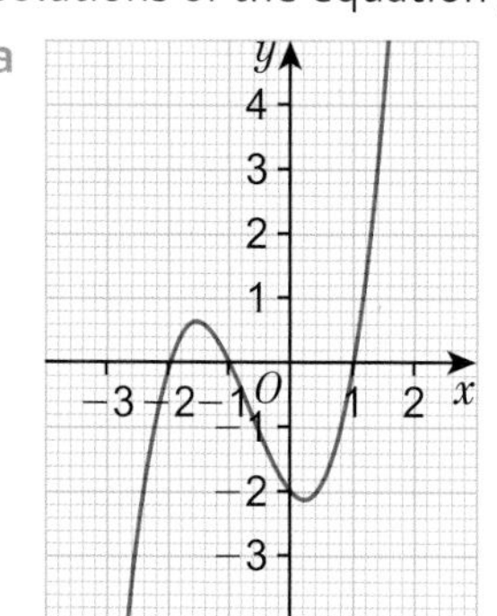

b
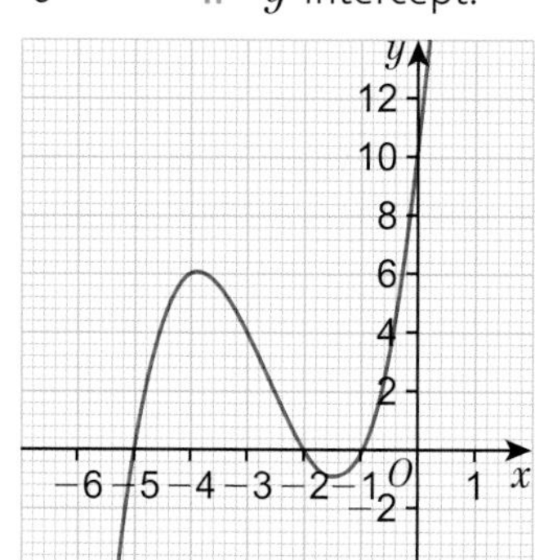

c
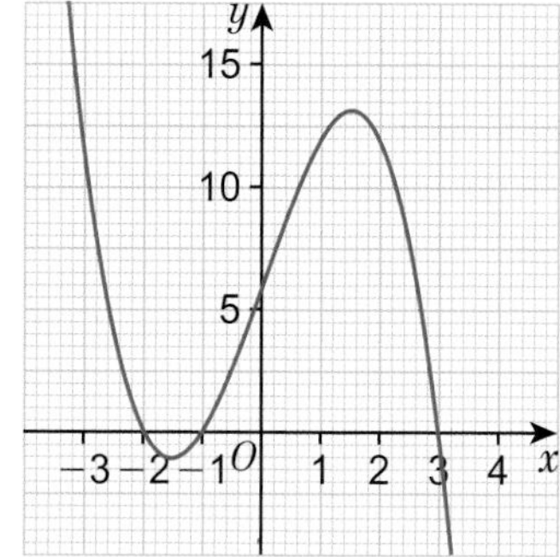

2 A curve has the equation $y = x^3 - 13x + 12$. Copy and complete

a When $y = 0$

.... $= 0$

$(x - \square)(x + \square)(x - \square) = 0$

So there are three possible solutions:

$x - \square = 0$ hence $x = \square$

$x + \square = 0$ hence $x = \square$

$x - \square = 0$ hence $x = \square$

b When $x = 0$

$y = (x - 3)(x + 4)(x - 1)$

$y = (0 - 3)(0 + 4)(\square - \square)$

$y = \square$

3 Find the roots and y-intercept of

a $y = (x + 3)(x - 7)(x + 2)$

b $y = (10 - x)(x + 2)(x + 1)$

c $y = (2x - 5)(x + 2)(x - 1)$

d $y = (x - 3)^2(3x - 6)$

Q3 hint Use the method in **Q2**.

15 Extend

1 a Shade the region that satisfies the inequalities

$y \leqslant 2x + 5 \qquad x \leqslant y + 1 \qquad y \leqslant -2$

b Calculate the area of the shaded triangle.

Q1a hint Rearrange the inequality $x \leqslant y + 1$

2 **Reasoning**

a What shape is made by the region that satisfies

$y \leqslant x \qquad y \geqslant -2x - 4 \qquad y > \frac{1}{2}x - 4 \qquad x < 2 \qquad y \leqslant 1$

b What is the sum of the interior angles of this shape?

3 **Reasoning** Calculate the exact area of the region satisfied by the inequalities
$x^2 + y^2 \leqslant 30 \qquad x \geqslant 0 \qquad y \geqslant 0$

4 **Reasoning** Work out how many real roots each equation has.
Show your workings.

a $x^2 + 2x + 1 = 0$ b $x^2 + 2x + 3 = 0$ c $x^2 - 1 = 0$
d $x^2 + 5x + 7 = 0$ e $10 + 3x - x^2 = 0$ f $-x^2 - 2x - 3 = 0$

5 **Reasoning** a Write down a quadratic function with
i a maximum at (−3, 4) ii a minimum at (4, −3)
b Write down a quadratic equation with roots at $x = -3$ and $x = 0$

6 **Exam-style question**

A is a curve with equation $y = x^2 + 4x - 3$
B is a line with equation $y = 2x + 5$
A intersects B at the points P and Q.
Work out the exact length of the straight line PQ. **(6 marks)**

Exam hint
Read the question carefully. What do you need to work out first?

7 a Expand the expression $x(x + 4)(x + 2)(x - 1)$
b How many roots does the equation $x(x + 4)(x + 2)(x - 1) = 0$ have?

8 **Exam-style question**

Show that
$x^3 - 4x = x(x - 2)(x + 2)$ **(2 marks)**

Exam hint
'Show that' means you must write down each stage of your working clearly.

9 **Problem-solving / Communication**
The general term of a sequence is $-n^2 + 4n + 20$
Explain why the largest term in the sequence occurs at $n = 2$

Q9 strategy hint Consider the graph of $y = -n^2 + 4n + 20$

10 **Problem-solving** Where do the graphs of
$y = (x + 1)(x - 3)(x - 5)$ and $y - x = 15$
intersect each other?

11 Expand the expression $(x^2 + 2x + 1)(x^2 + x + 2)$

Q11 hint Multiply each term in the second bracket by x^2, then $2x$, then 1.

12 **Reasoning** Use a graphical method to find approximate solutions to each pair of simultaneous equations.

a $x^2 + y^2 = 18 \qquad y = x^2 + 3x + 2$

b $y = \frac{1}{x} \qquad y = 4x - 1$

c $y = x^2 - x - 6 \qquad y = -x^2 + 16$

13 **Reasoning** How many whole number pairs of coordinates satisfy the inequalities
$y \geqslant 2x^2 - 4x - 16$ and $y < -15$?

14 Use a graphical method to find approximate solutions to each pair of simultaneous equations. Give all the solutions for values of x between 0 and 360°.

a $y = \sin x \qquad y = 0.6$
b $y = \tan x \qquad y = 0.2$

Discussion How could you check your answers to **Q14**?

15 Find the set of values of x that satisfy
a $2x^2 - x - 3 < 0$
b $-x^2 + 3x + 10 > 0$

Q15b hint

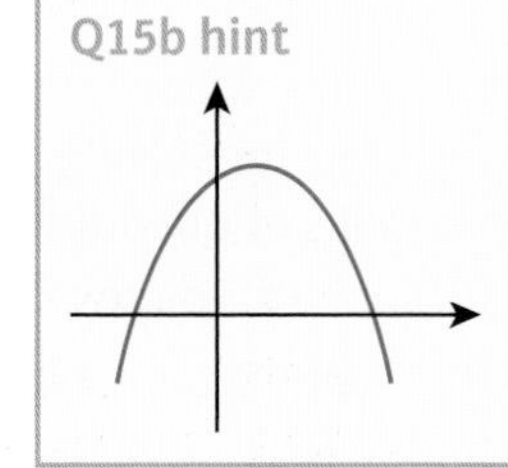

15 Knowledge check

- You can solve a pair of simultaneous equations by plotting the graphs and finding where they intersect one another. .... *Mastery lesson 15.1*
- The points that satisfy an inequality can be represented on a graph by shading the area to one side of the line.
  A dotted line is used to indicate $<$ or $>$
  A solid line is used to indicate $\geqslant$ or $\leqslant$ .... *Mastery lesson 15.2*
- A set of values that satisfy an inequality can be described using set notation, for example $\{x : x > 3\} \cup \{x < -2\}$ or $\{x : -2 \leqslant x \leqslant 3\}$ .... *Mastery lesson 15.2*
- The graph of a quadratic function is a smooth curve called a parabola. The lowest or highest point of the parabola, where the graph turns, is called the turning point. The turning point is either a minimum or maximum point. The x-values where the graph intersects the x-axis are the solutions, or roots, of the equation $y = 0$. .... *Mastery lesson 15.3*
- To find the coordinate of the turning point, write the equation in completed square form: $y = a(x + b)^2 + c$. .... *Mastery lesson 15.3*
- When a quadratic is written in completed square form $y = a(x + b)^2 + c$ the coordinate of the turning point is $(-b, c)$ .... *Mastery lesson 15.3*
- To sketch a quadratic function
  Calculate the solutions to the equation $y = 0$ (points of intersection with the x-axis). Calculate the point at which the graph crosses the y-axis. Find the coordinates of the turning point and whether it is a maximum or a minimum. .... *Mastery lesson 15.3*
- The quadratic equation $ax^2 + bx + c = 0$ is said to have no real roots if its graph does not cross the x-axis. If its graph just touches the x-axis, the equation has one repeated root. .... *Mastery lesson 15.4*
- To find an accurate root of a quadratic or cubic equation you can use an iterative process. Iterative means carrying out a repeated action. ... *Mastery lesson 15.4*
- To solve a quadratic inequality, solve as a quadratic equation then sketch the graph. Use the graph to find the values that satisfy the inequality. .... *Mastery lesson 15.5*
- To expand three pairs of brackets, first expand two of the brackets. .... *Mastery lesson 15.5*
- A cubic function is one whose highest power of x is x^3.
  It is written in the form $y = ax^3 + bx^2 + cx + d$
  The graph intersects the y-axis at the point $y = d$. The graph's roots can be found by finding the values of x for which $y = 0$. .... *Mastery lesson 15.5*
- When the graph of a cubic function y crosses the x-axis three times, the equation $y = 0$ has three solutions. When it crosses once and touches once it has three solutions but one is repeated. When it crosses once it can have one distinct, repeated solution or only one real solution. .... *Mastery lesson 15.5*

Reflect

In this unit, which was easiest, and which was hardest, to work with:

- Graphs for solving simultaneous equations?
- Graphs of inequalities?
- Graphs of quadratic functions?
- Graphs of cubic functions?

Copy and complete these sentences.

I find graphs________ easiest, because ____________________

I find graphs________ hardest, because ____________________

15 Unit test

Log how you did on your Student Progression Chart.

1 Use a graphical method to solve the simultaneous equations
$2y = x - 8$
$y = 8 - x$ *(4 marks)*

2 **Reasoning / Communication** Match the equations to their graphs.
Explain your reasoning.

a $y = x^2 - 4x + 4$ b $y = (x - 3)(x + 1)$
c $y = -x^2 + 5x + 6$ d $y = (2 - x)^2 - 3$ *(4 marks)*

i
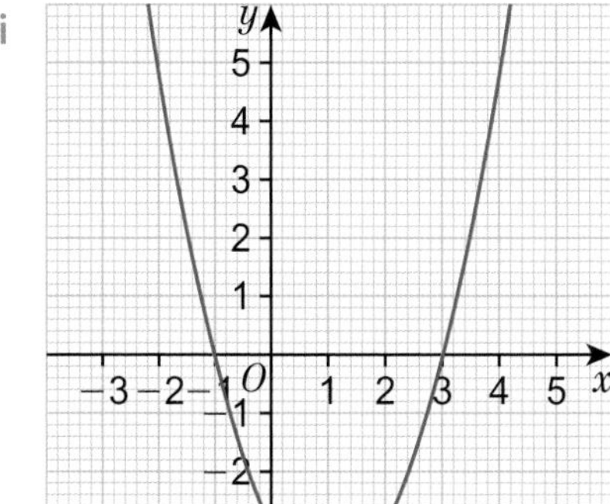

ii
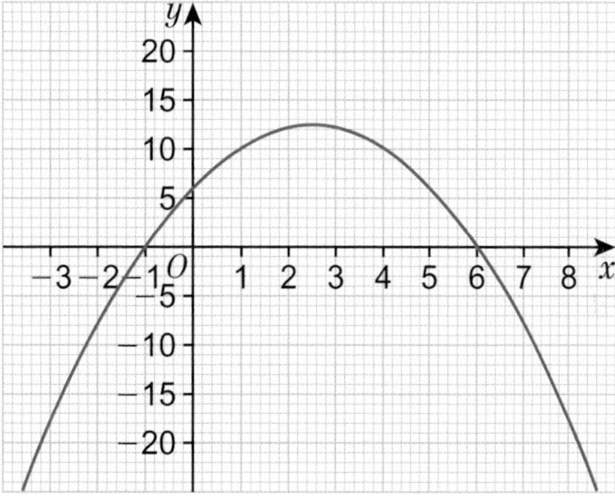

iii
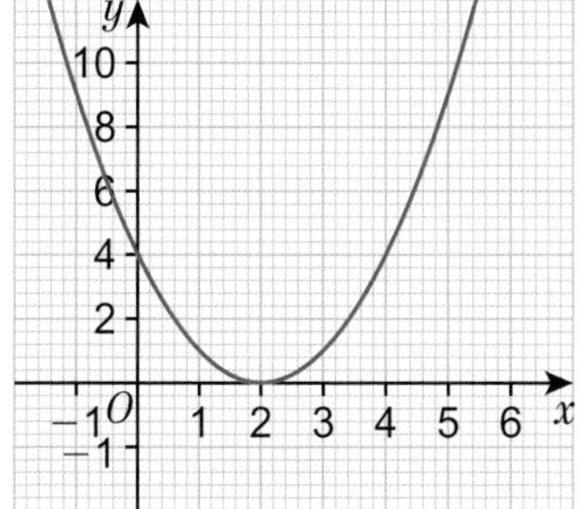

iv
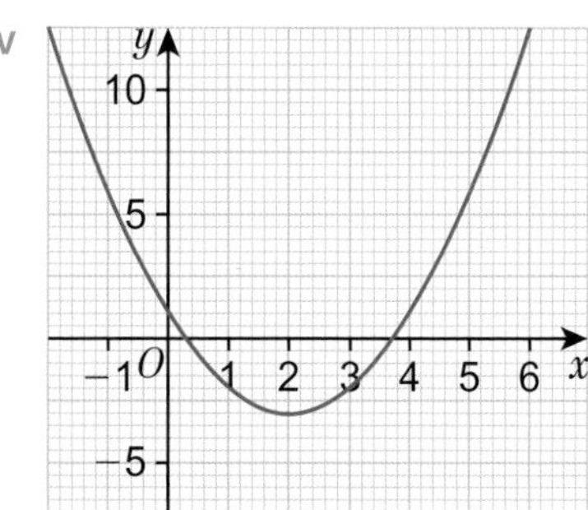

3 Draw a coordinate grid with −10 to +10 on both axes.
Shade the area that satisfies the inequalities
$y - 2x < 7$ and $y > 4$ *(3 marks)*

4 a Expand the expression $(x - 3)(x + 4)(x - 2)$
b What type of graph is $y = (x - 3)(x + 4)(x - 2)$?
c Write down the solutions to the equation $(x - 3)(x + 4)(x - 2) = 0$.
d Where does the graph intersect the y-axis? *(5 marks)*

5 Sketch the graph of $y = x^2 - 4x - 1$. *(3 marks)*

6 **Reasoning** Write down the equation of a graph with a maximum point at (3, −4). *(3 marks)*

7 **Reasoning** Calculate the exact area satisfied by the inequalities
$x^2 + y^2 \geqslant 25$, $x \leqslant 5$ and $y \leqslant 5$ *(4 marks)*

ActiveLearn Homework, practice and support: Higher 15 Unit test

8 **Problem-solving** A is the graph of $y = x^2 - 3x + 5$
B is the graph of $y - x = 5$. The graphs intersect at points P and Q.
Find the length of PQ. Give your answer as a surd. *(6 marks)*

9 Find the values of x that satisfy $x^2 + x - 6 > 0$ *(5 marks)*

10 **Reasoning** Which of these equations has real roots?

a $x^2 - 6x + 13 = 0$

b $2x^2 + 4x - 5 = 0$

c $-x^2 - 4x + 9 = 0$ *(5 marks)*

11 Use an iterative formula to find the only real solution to $x^3 - x + 8 = 0$ to 5 d.p. *(5 marks)*

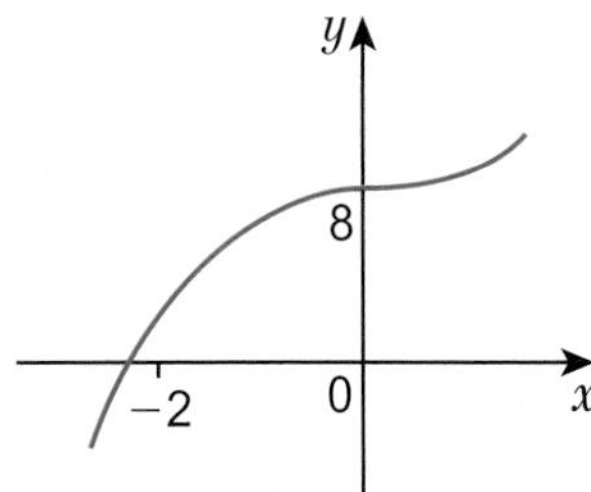

Sample student answers

Which student gives the best answer and why?

Exam-style question

The graph of $y = x^2 + bx + c$ is shown on the grid.

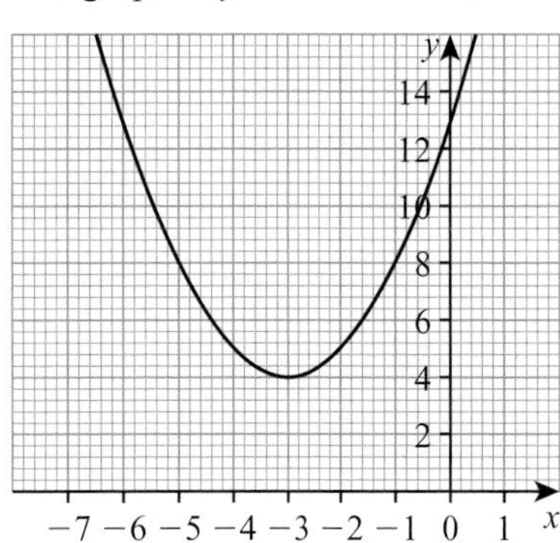

Work out the value of b and the value of c. **(4 marks)**

Student A answer

When $x = 0, y = 13$

$13 = 0^2 + 0 + c$

$c = 13$

$b = -3$ since the smallest value of y is when $x = -3$

Student B answer

The turning point is at $(-3, 4)$

$y = (x + 3)^2 + 4$

$y = x^2 + 6x + 9 + 4$

$y = x^2 + 6x + 13$

$b = 6, c = 13$

Student C answer

The turning point is at $(-3, 4)$.

$y = (x + 3)(x - 4)$

$y = x^2 + 3x - 4x - 12$

$y = x^2 - x - 12$

$b = -1, c = -12$

16 CIRCLE THEOREMS

The Ancient Greeks considered a circle to be the perfect shape – a symbol of symmetry and balance in nature. Many civilisations have built circular structures. Stonehenge is the largest and most famous stone circle in Britain. Archaeologists believe it was built between 3000 BC and 2000 BC, and its diameter is around 90 metres.

How do you think our ancestors marked out the circle for Stonehenge before positioning the stones?

16 Prior knowledge check

Numerical fluency

1 Work out the value of c.

a $5 = \frac{3}{4} \times 2 + c$

b $12 = -\frac{4}{3} \times 6 + c$

c $-9 = \frac{7}{8} \times -4 + c$

Algebraic fluency

2 Factorise: $4x + 4y$.

3 Work out the value of c.

a $y = \frac{3}{5}x + c$, when $x = 2$ and $y = 4$

b $y = \frac{2}{3}x + c$, when $x = 3$ and $y = -2$

Geometrical fluency

4 a Draw a circle with radius 6 cm.

b On your circle, draw and label a radius, an arc, a sector and a segment.

5 Find the size of angle x.

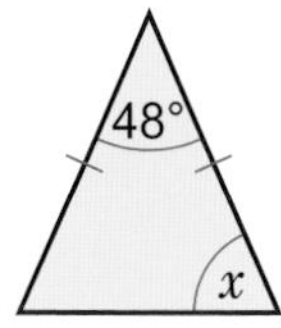

6 Work out the lengths of AC and PQ in these right-angled triangles. Give your answers correct to 1 d.p.

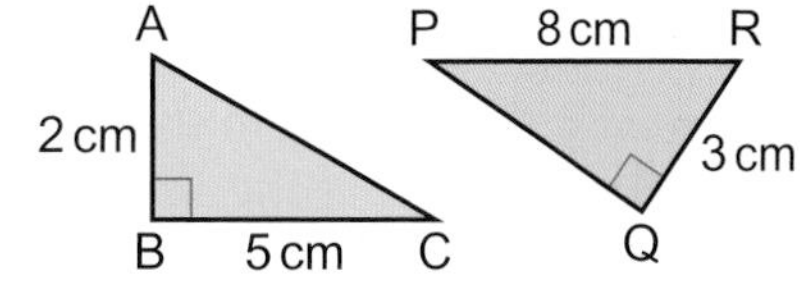

7 Prove that these two right-angled triangles are congruent.

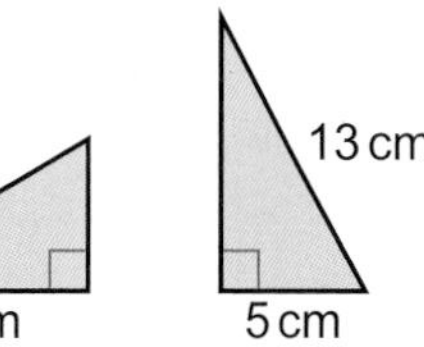

★ Challenge

8 Draw a circle with radius 5 cm. Then draw in a diameter and label it AC.

Choose any point on the circumference and label it B. Now draw a triangle between this point and the two ends of the diameter by drawing in lines AB and BC.

Measure angle B.

Repeat for other points on the circumference. What do you notice?

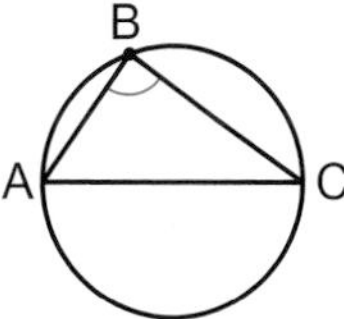

16.1 Radii and chords

Objectives

- Solve problems involving angles, triangles and circles.
- Understand and use facts about chords and their distance from the centre of a circle.
- Solve problems involving chords and radii.

Did you know?

A **theorem** is a rule that can be proved by a chain of reasoning.

Fluency

- What are the properties of an isosceles triangle?
- What is a chord in a circle?

Warm up

1 Work out the size of each angle marked with a letter. Give reasons for your answers.

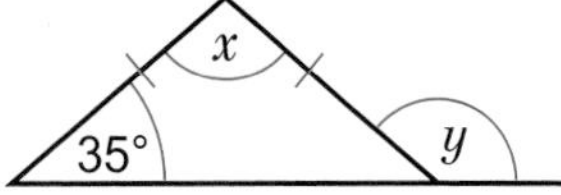

2 Prove that triangles ABD and ACD are congruent.

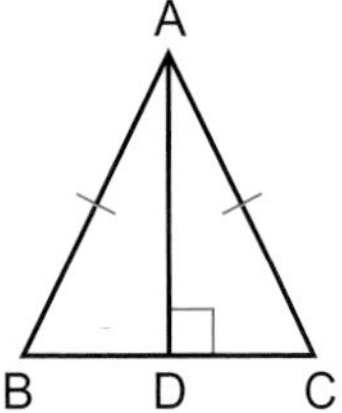

Questions in this unit are targeted at the steps indicated.

3 **Reasoning** Each diagram shows a circle with centre O. Work out the size of each angle marked with a letter.

Q3a hint Two sides of this triangle are radii of the circle. What sort of triangle is it?

a
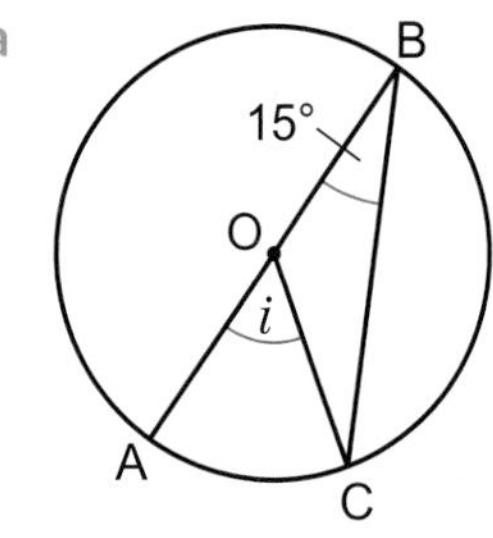

b
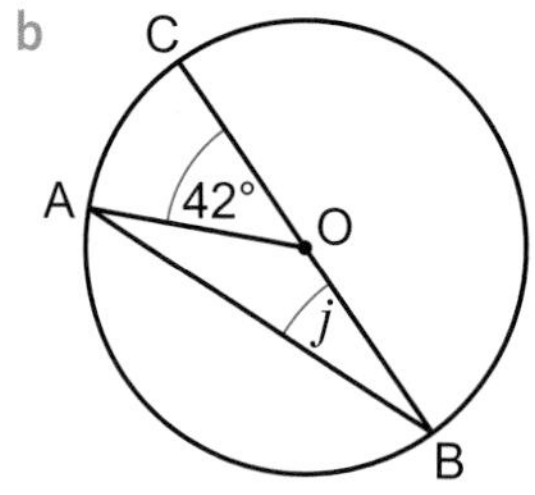

c
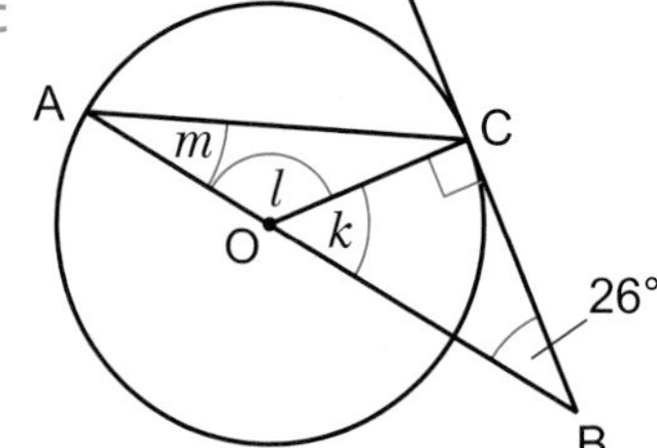

4 **Reasoning** Bill says that P is the centre of this circle. Explain how you know Bill must be wrong.

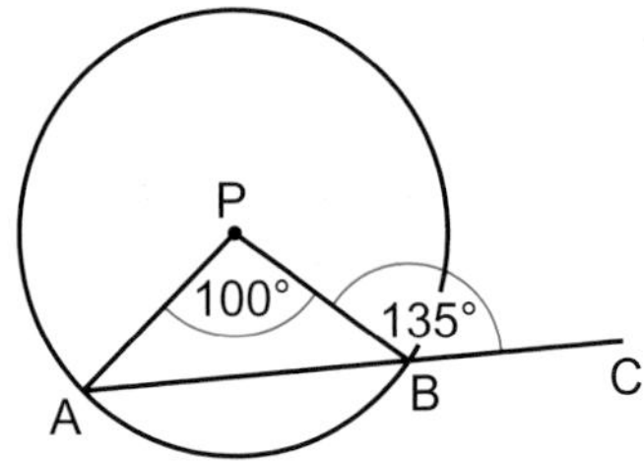

5 **Reasoning / Communication** O is the centre of a circle. OA and OB are radii. OM is perpendicular to AB.

a Prove that triangles OAM and OBM are congruent.

b Show that M is the midpoint of AB.

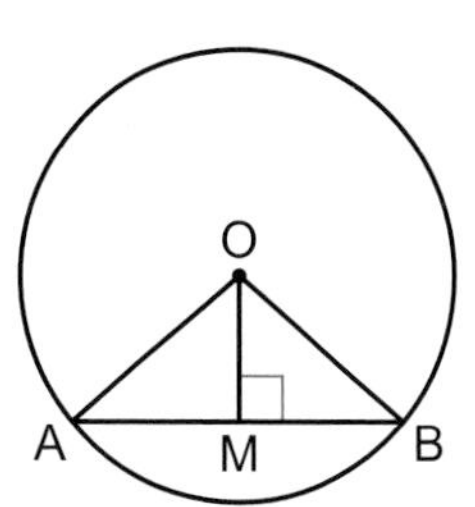

Q5b hint Show that AM = MB.

Discussion When you draw a line from the centre of a circle to the midpoint of a chord, at what angle does it meet the chord?

Key point 1

A **chord** is a straight line connecting two points on a circle.
The perpendicular from the centre of a circle to a chord bisects the chord and the line drawn from the centre of a circle to the midpoint of a chord is at right angles to the chord.

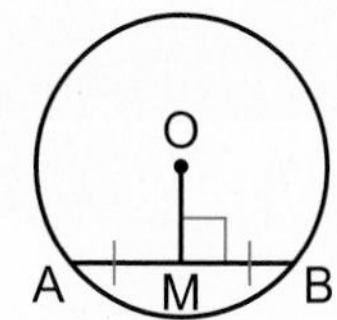

Example 1

O is the centre of a circle.

The length of chord AB is 18 cm.

OM is perpendicular to AB.

Work out the length of AM.

State any circle theorems that you use.

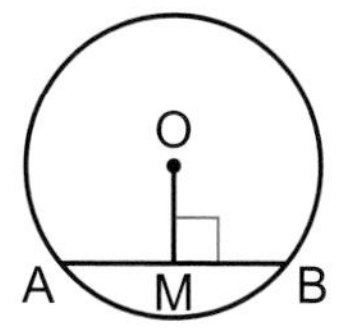

AB = 18 cm

So, AM = $\frac{18}{2}$ = 9 cm

You know that the perpendicular from the centre of a circle to a chord bisects the chord. So the length of AM will be exactly half the length of AB.

AM = 9 cm The perpendicular from the centre of a circle to a chord bisects the chord.

The question asks you to state any circle theorems that you use.

6 **Reasoning** O is the centre of a circle. M is a point on chord AB. The length of chord AB is 12 cm. OM is perpendicular to AB. OM is 8 cm.

a Work out the length of AM. State any circle theorems that you use.

b What is the length of the radius of the circle?

Q6a hint Start by drawing a diagram and mark on all the information you are given in the question. Your diagram should look like the one in **Example 1**.

Q6b hint Use Pythagoras' theorem.

7 **Reasoning** O is the centre of a circle. OA = 17 cm and AB = 16 cm. M is the midpoint of AB. Work out the length of OM.

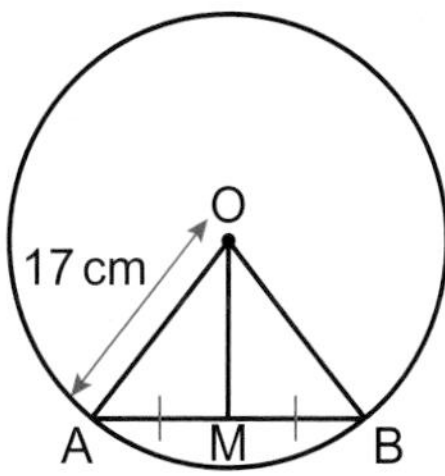

8 **Reasoning** O is the centre of a circle. The radius of the circle is 26 cm. The distance from O to the midpoint of chord AB is 24 cm. Work out the length of chord AB.

Q8 hint Draw a diagram and mark on all of the information that you know.

Reflect The hints for **Q6** and **Q8** suggested drawing a diagram to help with your answer. Did the diagram help you? Write a sentence explaining how.

9 **Reasoning** O is the centre of a circle.
M is the midpoint of chord AB.
Angle OAB = 25°.
a What is angle AMO?
b Work out angle AOM.
c Work out angle AOB.

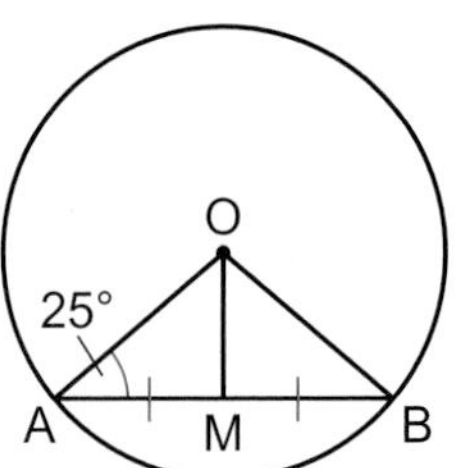

16.2 Tangents

Objectives

- Understand and use facts about tangents at a point and from a point.
- Give reasons for angle and length calculations involving tangents.

Did you know?

The word 'tangent' comes from the Latin verb 'tangere', which means to touch.

Fluency

When a line is drawn from the centre of a circle to the midpoint of a chord, at what angle does the line meet the chord?

Warm up

1 a Find the missing angle in this triangle.

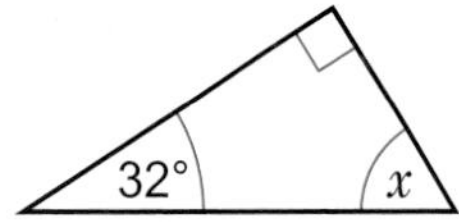

b Find the missing length in this triangle.

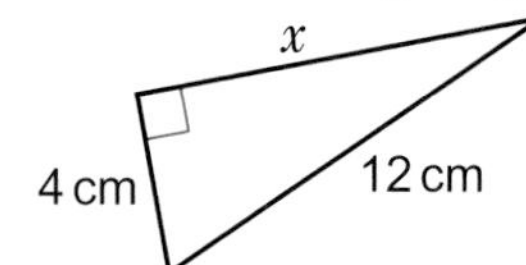

Key point 2

A **tangent** is a straight line that touches a circle at one point only.
The angle between a tangent and the radius is 90°.

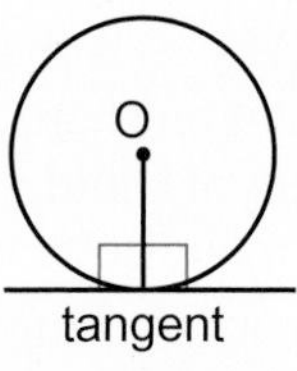

2 **Problem-solving / Communication**
PA and PB are two tangents to a circle with centre O.
Prove that triangles APO and BPO are congruent.
Discussion What can you say about the lengths PA and PB?

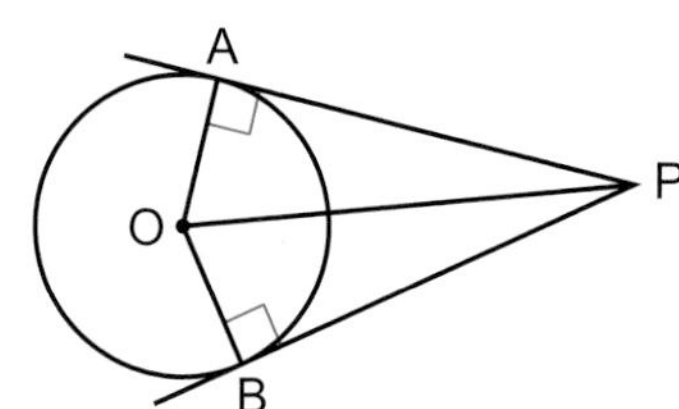

Key point 3

Tangents drawn to a circle from a point outside the circle are equal in length.
So AB = AC.

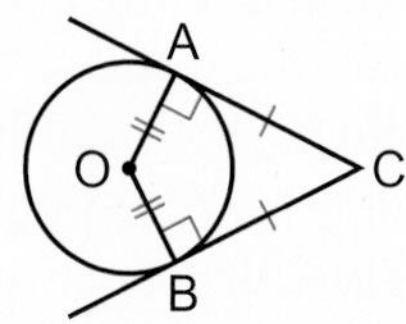

Example 2

In the diagram, O is the centre of the circle. Angle POQ is 130°.
PQ is a chord. PR and QR are tangents to the circle.

Work out the size of

a angle OPQ b angle QPR c angle PRQ.

Give reasons for your answers.

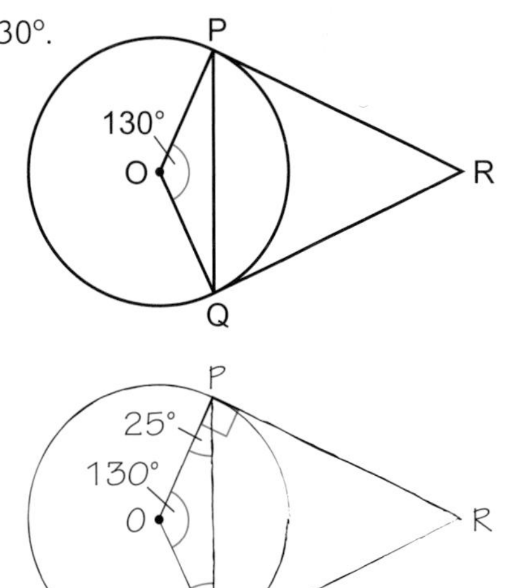

a OP = OQ (radii of same circle)

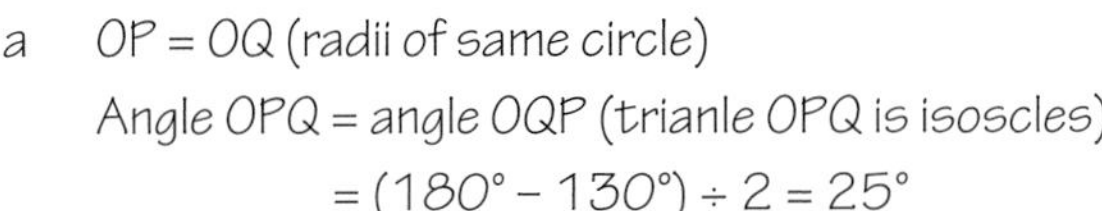
Angle OPQ = angle OQP (trianle OPQ is isoscles)
= (180° − 130°) ÷ 2 = 25°

b Angle OPR = angle OQR = 90°

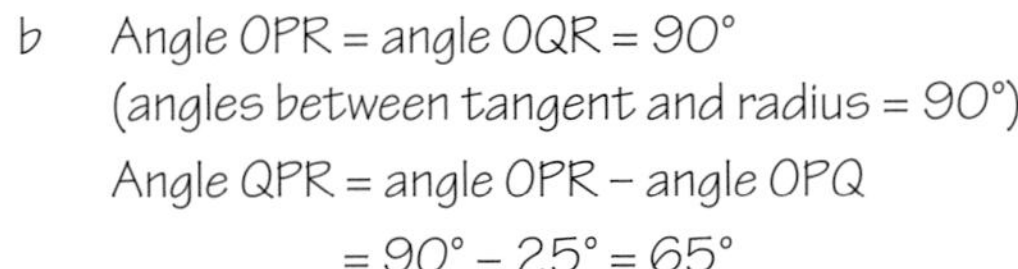
(angles between tangent and radius = 90°)
Angle QPR = angle OPR − angle OPQ
= 90° − 25° = 65°

c RPOQ is a quadrilateral.
Angle PRQ = 360° − 90° − 90° − 130° = 50°
(angles in a quadrilateral add up to 360°)

3 **Reasoning** The diagrams all show circles, centre O.
Work out the size of each angle marked with a letter. Give reasons for your answers.

a

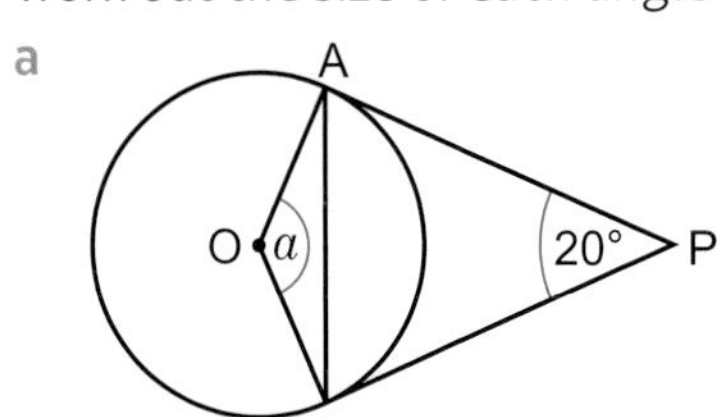

b (diagram: A, P, b, 136°, O, B)

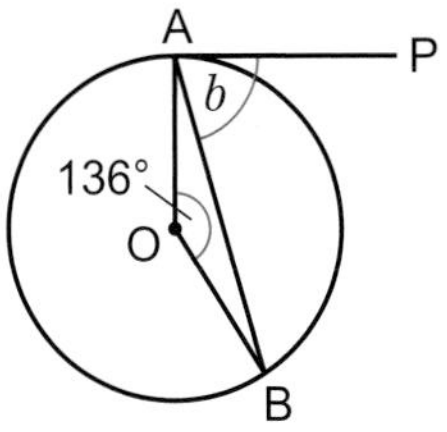

c

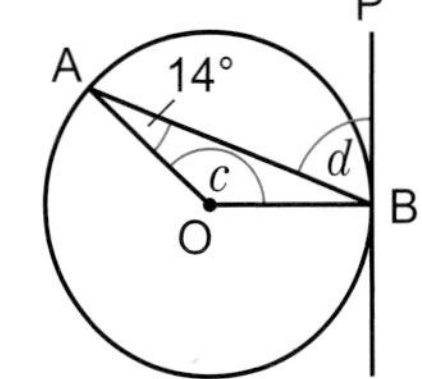

d (diagram: P, 34°, A, B, e, f, O)

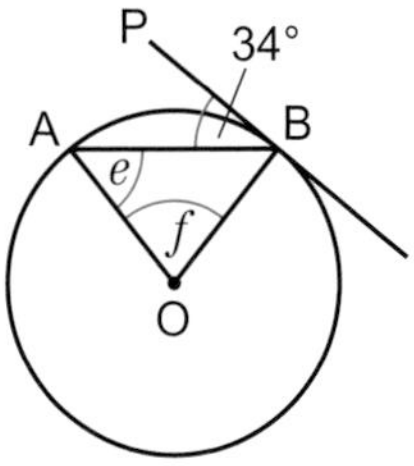

e

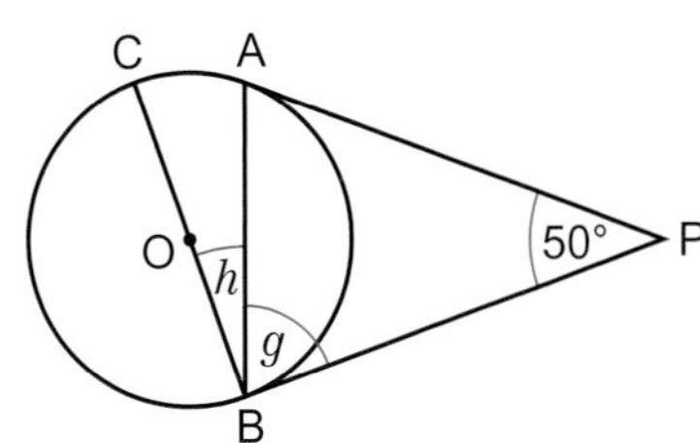

4 **Exam-style question**

S and *T* are points on the circumference of a circle, centre *O*.

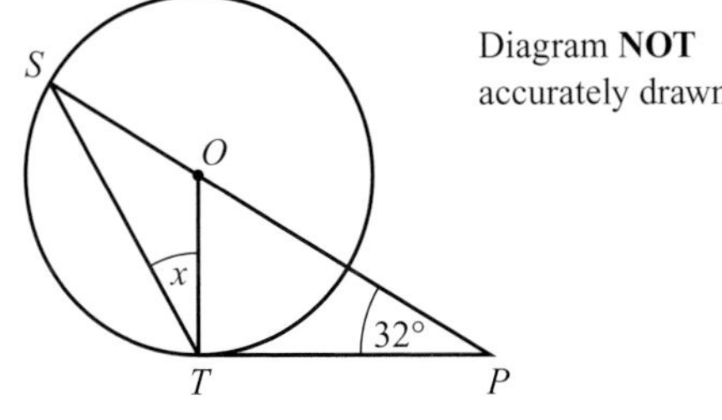

PT is a tangent to the circle.
SOP is a straight line.
Angle *OPT* = 32°.
Work out the size of the angle marked x.
Give reasons for your answer. **(5 marks)**

June 2013, Q16, 1MA0/2H

Exam hint
Mark the angles in the correct places on the diagram as you work them out.

5 **Problem-solving / Communication** A and B are points on the circumference of a circle, centre O. TA and TB are tangents to the circle.

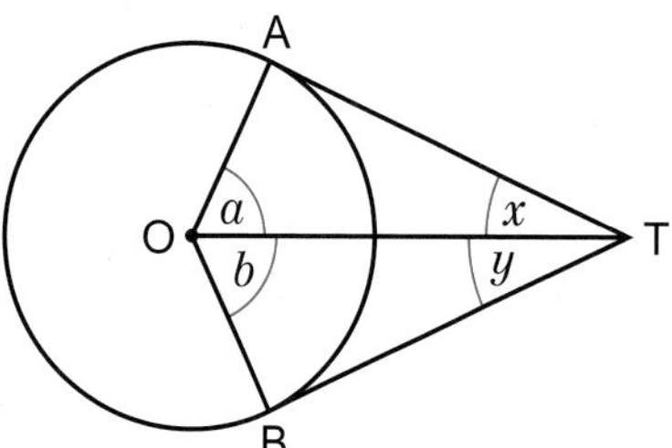

Q5 hint Use congruence.

Show that angles x and y are equal, and that angles a and b are equal.

6 **Reasoning** OA is the radius of a circle with diameter 18 cm. AT is a tangent to the circle from point T.
AT = 12 cm.
Calculate the distance from T to the centre of the circle. State any circle theorems that you use.

Q6 strategy hint Draw a diagram and mark the values on it.

16.3 Angles in circles 1

Objectives

- Understand, prove and use facts about angles subtended at the centre and the circumference of circles.
- Understand, prove and use facts about the angle in a semicircle being a right angle.
- Find missing angles using these theorems and give reasons for answers.

Did you know?

The Greek mathematician Euclid proved many results about circles in the 13 volumes of his *Elements*, which he wrote around 300 BC.

Fluency

Angles round a point add to .... degrees.

Warm up

1 Copy each diagram and colour the arc that the marked angle stands on.

a
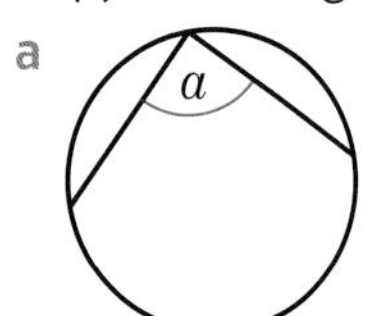

b
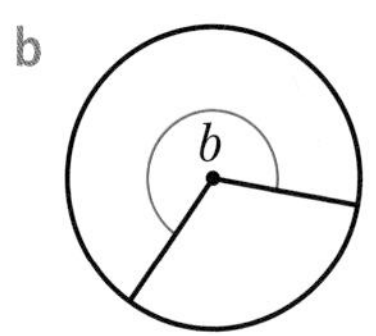

2 Work out the size of angle x in this diagram.
Give reasons for your answer.

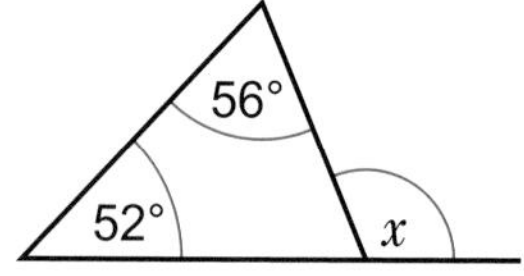

Key point 4

Circle theorem: The angle at the centre of a circle is twice the angle at the circumference when both are subtended by the same arc.

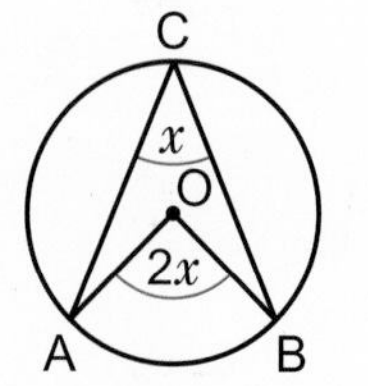

Communication hint
Subtended means that the arms of the angle start and finish at the ends of the arc. So angles ACB and AOB are both subtended by arc AB.

ActiveLearn Homework, practice and support: Higher 16.3

Example 3

Prove that the angle at the centre of a circle is twice the angle at the circumference when both are subtended by the same arc.

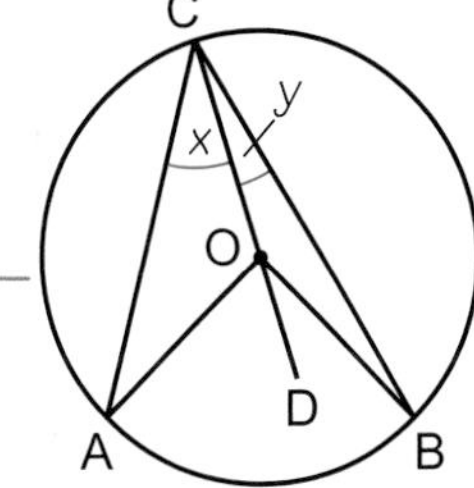

Draw the line CO and extend it to point D. Let angle ACO = x and angle BCO = y.

AO = OC (radii of same circle)

Angle ACO = angle OAC = x
(base angles of isosceles triangle)

Similarly, angle BCO = angle OBC = y

Angle AOD = $2x$ (exterior angle equals the sum of the two interior opposite angles)

Similarly, angle BOD = $2y$

Angle ACB = $x + y$

Angle AOB = $2x + 2y = 2(x + y)$ = 2(angle ACB)

3 **Reasoning** The diagrams show circles, centre O.
Work out the size of each angle marked with a letter. Give reasons for your answers.

a
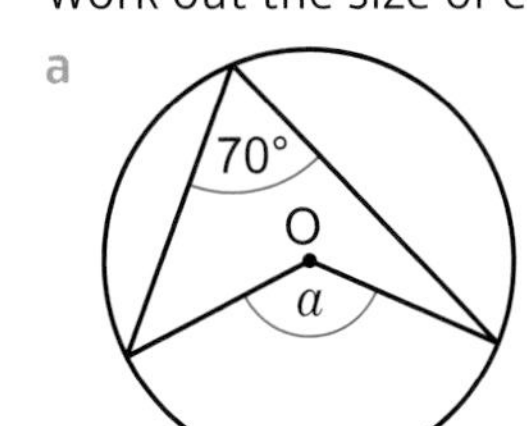

b 96°, O, b

c
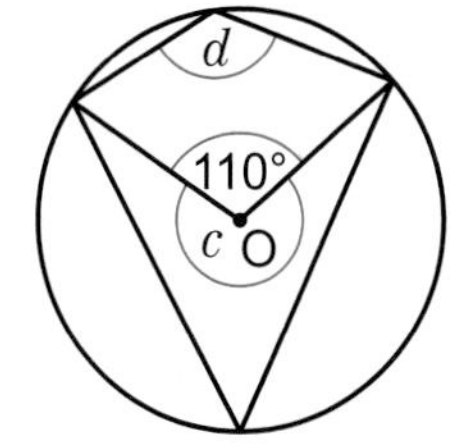

4 **Reasoning / Communication**
Prove that the angle in a semicircle is 90°.

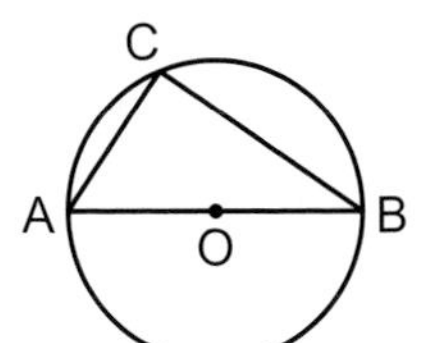

Q4 hint AOB is a straight line so what size is angle AOB? Angle ACB is half of angle AOB.

Key point 5

The angle in a semicircle is a right angle.
So angle ABC = 90°.

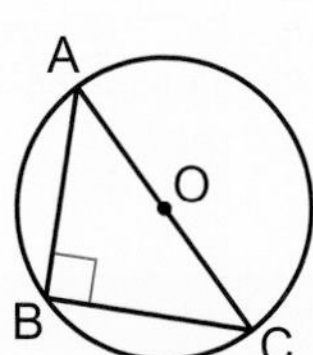

5 **Reasoning** The diagrams show circles, centre O.
Work out the size of each angle marked with a letter. Give reasons for your answers.

a
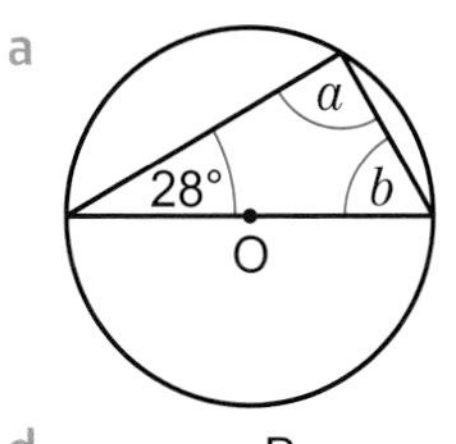

b
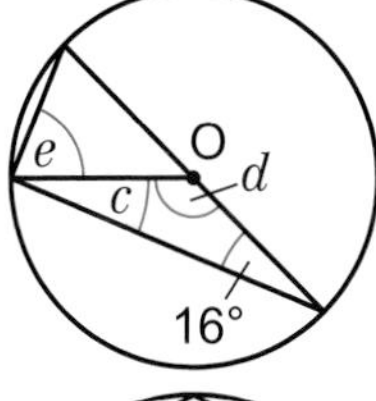

c
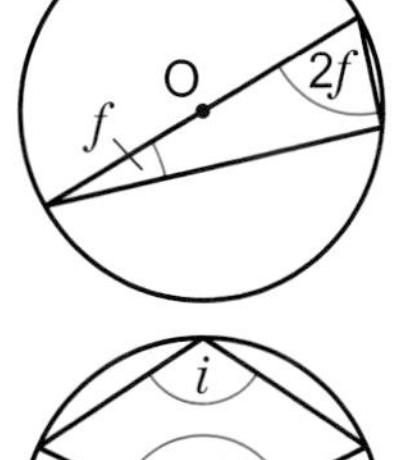

d
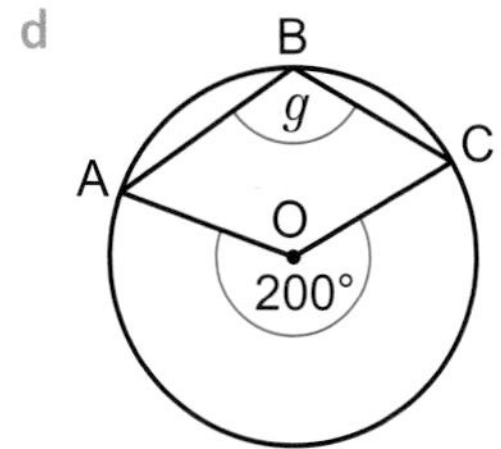

e
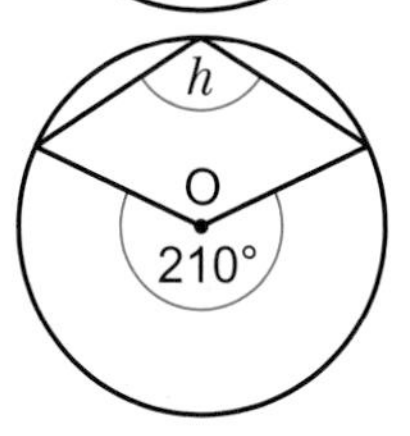

f
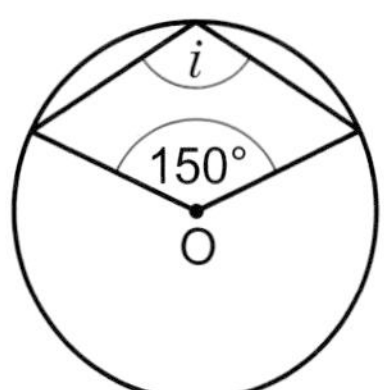

Q5d hint The reflex angle at point O and angle g are subtended by the same arc (AC). So the reflex angle AOC must be twice the size of angle g.

6 **Communication** Mario says the size of angle a is 65°.
Andy says the size of angle a is 115°.

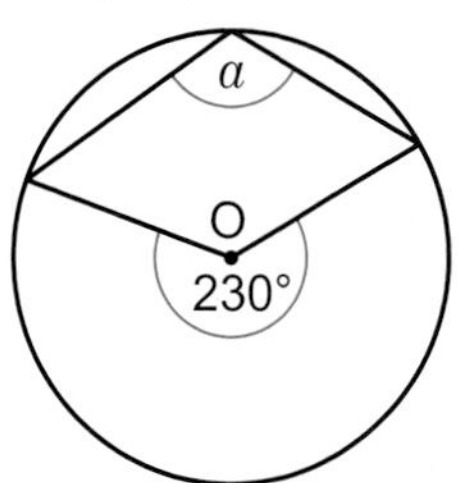

Q6 hint Copy the diagram and colour the arc subtending angle a and the arc subtending the 230° angle.

Show that Andy is correct.

Discussion What mistake has Mario made?

7 Find the size of each angle marked with a letter. The centres of the circles are marked O.

a

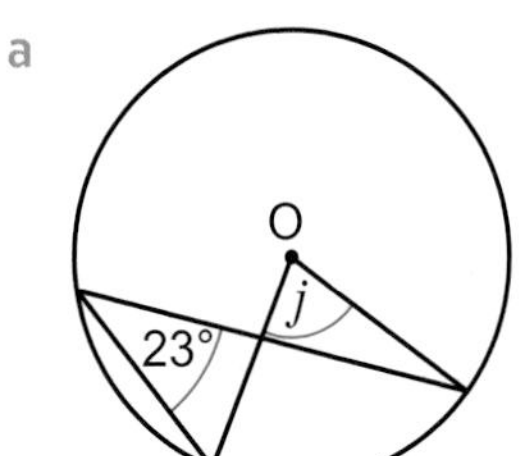

b

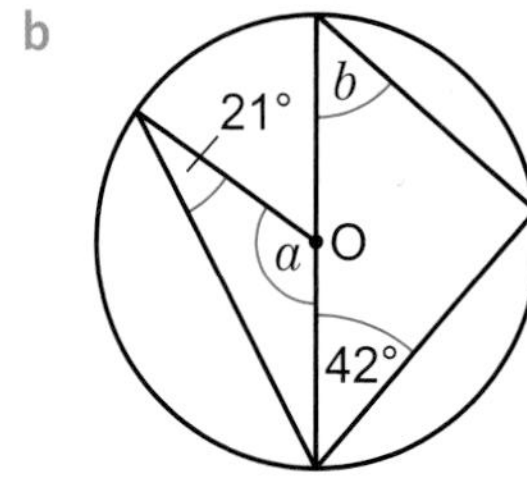

Q7a hint Look at the 23° angle and angle j. Are they subtended by the same arc?

8 **Communication** Lucy says the size of angle a is 15°.
Sue says the size of angle a is 60°.

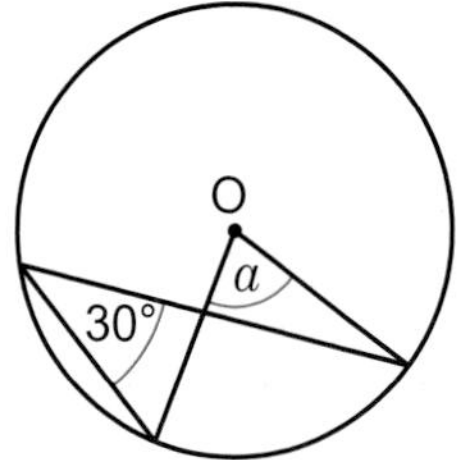

Show that Sue is correct.

Discussion What mistake has Lucy made?

9 **Exam-style question**

B, C and D are points on the circumference of a circle, centre O.

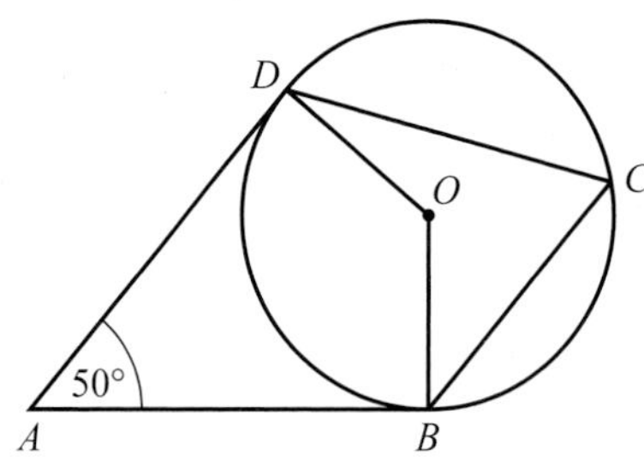

Diagram **NOT** accurately drawn

AB and AD are tangents to the circle.

Angle $DAB = 50°$.

Work out the size of angle BCD.

Give a reason for each stage in your working. **(4 marks)**

June 2012, Q21, 1MA0/1H

Exam hint
Each reason given must be a statement of a mathematical rule and not just the calculations you have done.

16.4 Angles in circles 2

Objectives

- Understand, prove and use facts about angles subtended at the circumference of a circle.
- Understand, prove and use facts about cyclic quadrilaterals.
- Prove the alternate segment theorem.

Did you know?

A **cyclic polygon** has all of its vertices on the circumference of a circle.

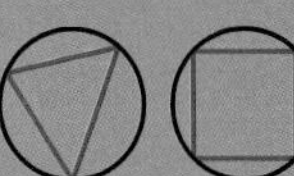

Fluency

- What do the angles in a quadrilateral add up to?
- What is the shaded part of this circle called?

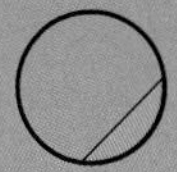

1 Write an expression in terms of x for angle BAC.

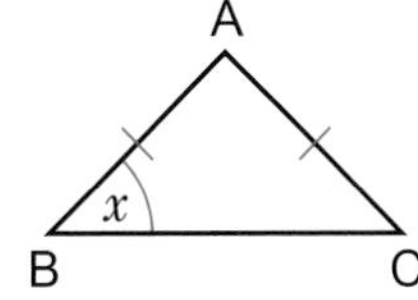

2 In each diagram, O is the centre of the circle.

a Work out the size of each angle marked with a letter.

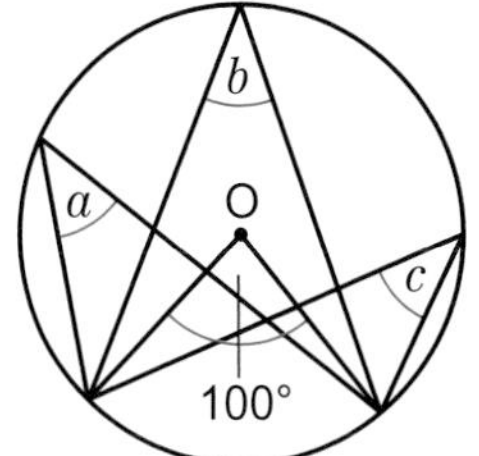

b Work out the size of angle d in terms of x.

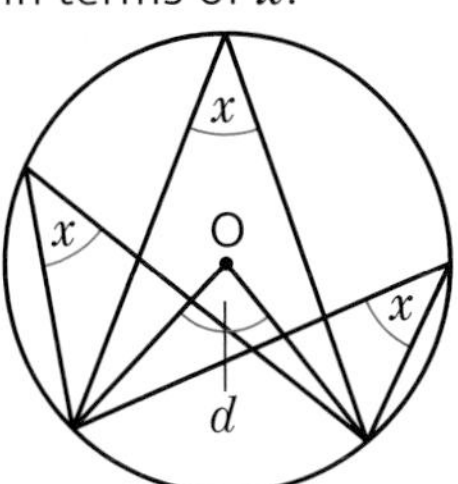

Discussion What do you notice about all the angles at the circumference in the same segment?

Key point 6

Angles subtended at the circumference by the same arc are equal. Another form of the same theorem is that angles in the same segment are equal.

3 Problem-solving / Communication

a Prove that angle ACB = angle ADB.

b **Reflect** Why does your answer to part **a** prove that *all* angles in the same segment are equal?

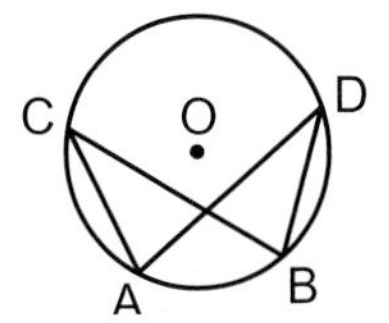

Q3 strategy hint Copy the diagram and draw in angle AOB. Angle AOB = 2 × angle ACB.

4 Reasoning In each diagram, O is the centre of the circle.

a

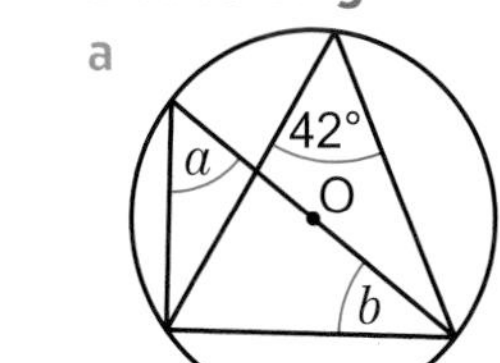

b

c

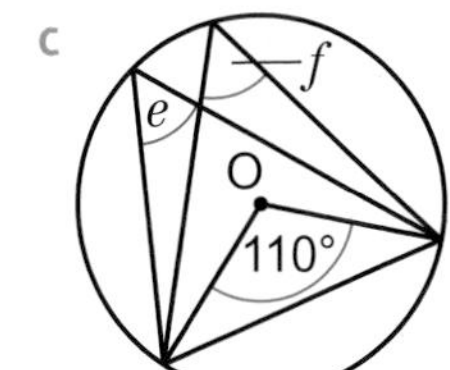

d

Work out the size of each angle marked with a letter. Give reasons for each step in your working.

Q4a hint Which angle is subtended by the same arc as the 42° angle? Look for an angle in a semicircle.

Key point 7

A **cyclic quadrilateral** is a quadrilateral with all four vertices on the circumference of a circle.

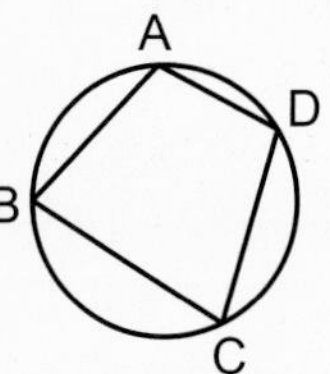

5 Reasoning In each diagram, O is the centre of the circle.
Work out the sizes of angles a, b and c in each diagram.

a

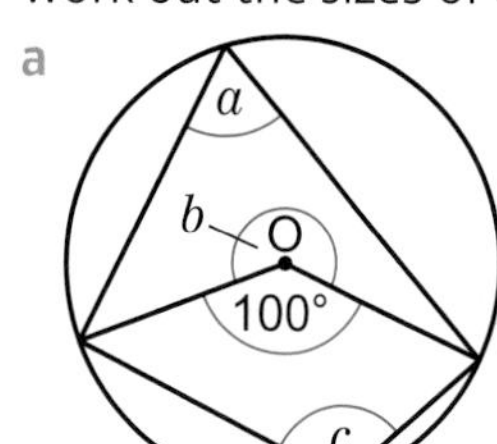

b

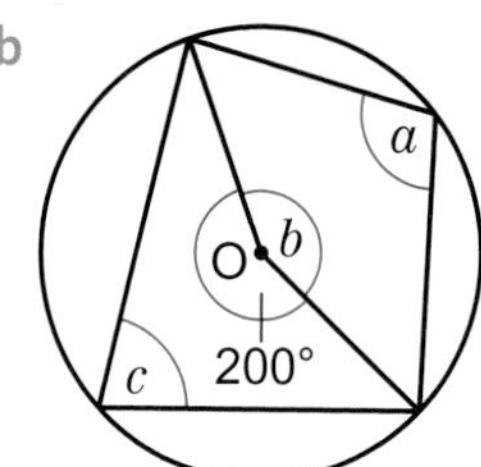

Discussion Work out angle a + angle c for each diagram. What do you notice?

Key point 8

Opposite angles of a cyclic quadrilateral add up to 180°:
So, $x + y = 180°$ and $p + q = 180°$.

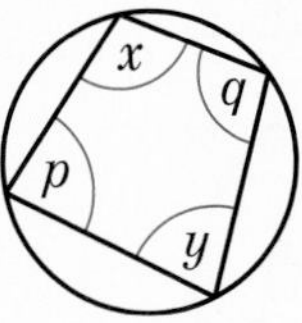

6 Reasoning ABCD is a cyclic quadrilateral.
O is the centre of the circle.

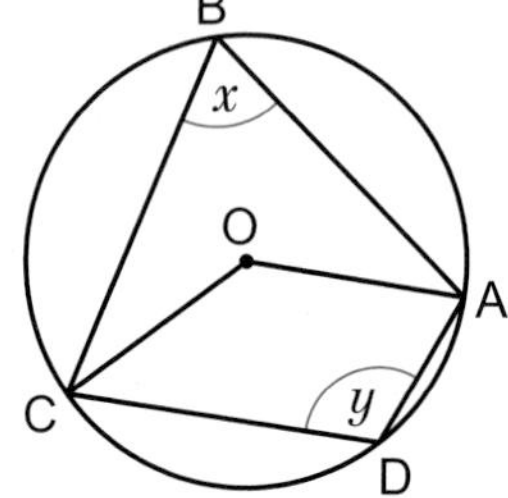

a What is the obtuse angle AOC in terms of x?

b What is the reflex angle AOC in terms of y?

c Copy and complete:
Obtuse angle AOC + reflex angle AOC = 360°

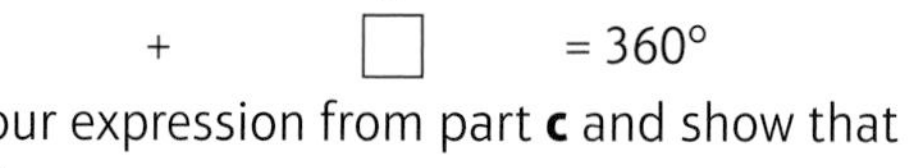

d Factorise your expression from part **c** and show that $x + y = 180°$

Discussion How does this prove the theorem about opposite angles in a cyclic quadrilateral?

7 Reasoning In each diagram, O is the centre of the circle.

a

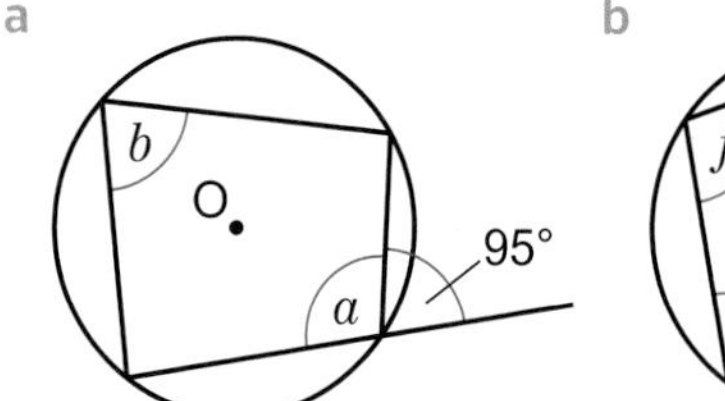

b

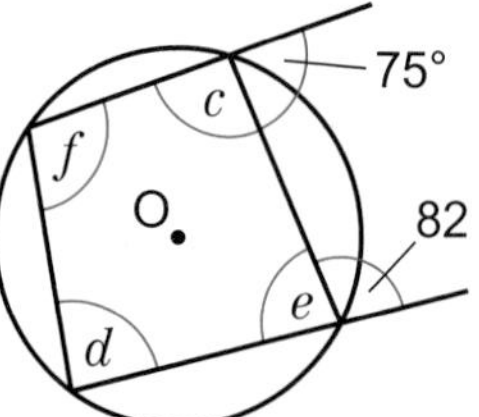

Work out the size of each angle marked with a letter.
Give reasons for each step in your working.

Discussion What do you notice about the exterior angle of a cyclic quadrilateral and the opposite interior angle?

8 **Communication** Prove that an exterior angle of a cyclic quadrilateral is equal to the opposite interior angle.

a Copy the diagram.

b Copy the working and complete the reasons.
Angle x + angle y = 180° because ....
Angle x + angle z = 180° because ....
So angle y = angle z.

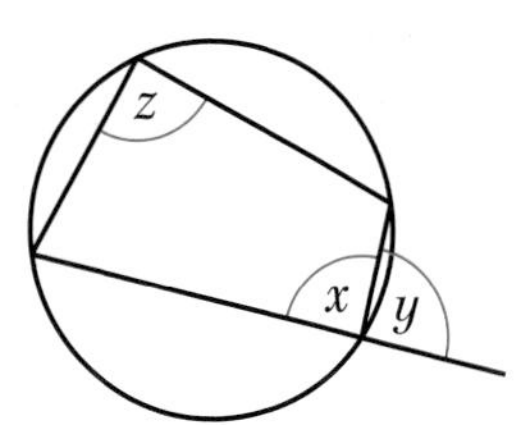

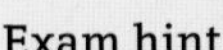
Exam hint
You need to be able to prove this theorem.

Key point 9

An exterior angle of a cyclic quadrilateral is equal to the opposite interior angle.

9 **Reasoning** Work out the size of each angle marked with a letter. O is the centre of the circle. Give reasons for each step in your working.

a
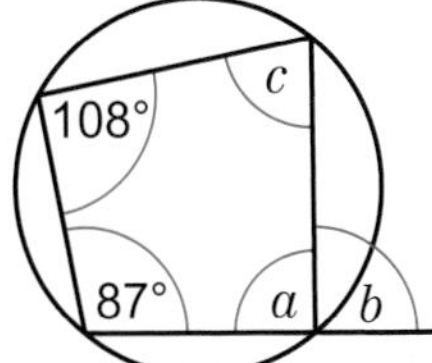

b
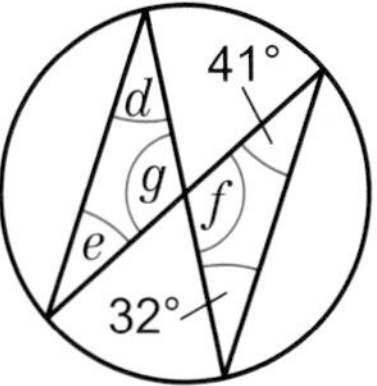

c
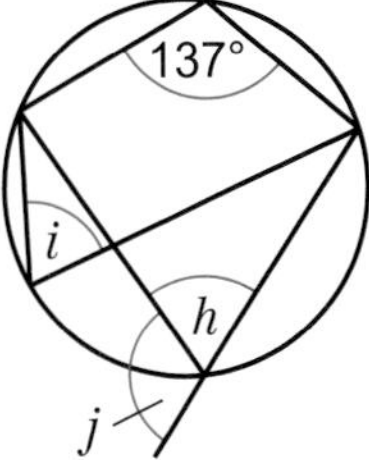

d
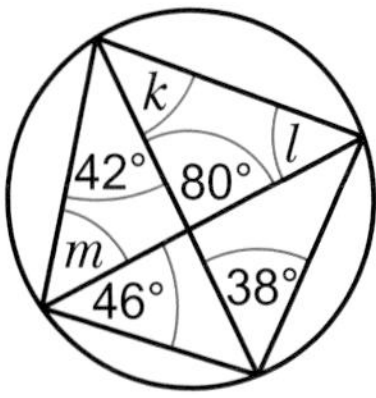

e
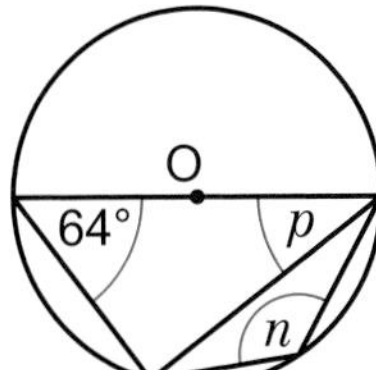

10 **Reasoning** O is the centre of a circle. AT is a tangent to the circle.

a Copy the diagram.

b Copy the working and complete the reasons.
Angle OAT = 90° because the angle between the tangent and the ...... = 90°.
Angle OAB = 90° − 58° = 32°.
OA = OB because radii ....
Angle OAB = angle OBA because the base angles of .... triangle are equal.
Angle AOB = 180° − 32° − 32° = 116° because angles in a .... add up to ......
Angle ACB = 116° ÷ 2 = 58° because the angle at the ...... is twice the ....

11 **Reasoning** Repeat **Q10** but this time with angle BAT = 72°.
What is the size of angle ACB?

Discussion What do you notice about angle BAT and angle ACB?

12 **Reasoning / Communication** Prove that angle BAT = angle ACB.

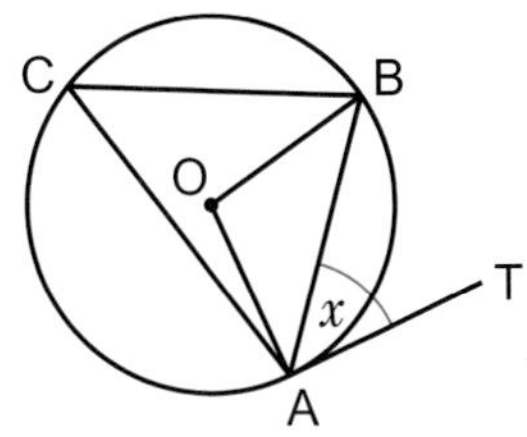

Q12 hint Repeat the steps in **Q10** but with angle BAT = x.

Key point 10

AT is a tangent to the circle. AB is a chord.
Angle BAT is the angle between the tangent and the chord in one segment.
The other segment made by the chord AB contains angle ACB.
This is called the **alternate segment**.
The angle between the tangent and the chord is equal to the angle in the alternate segment.
So angle BAT = angle ACB.

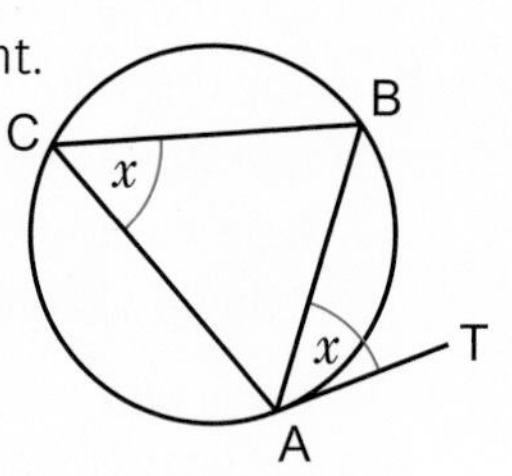

13 **Exam-style question**

M and *N* are two points on the circumference of a circle, centre *O*.

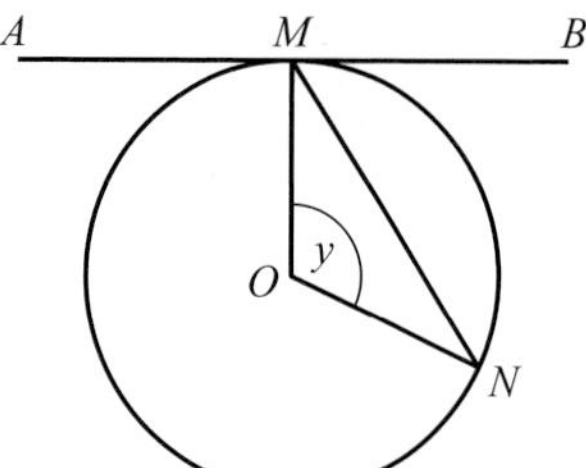

Diagram **NOT** accurately drawn

The straight line *AMB* is the tangent to the circle at *M*.
Angle *MON* = *y*.
Prove that angle $BMN = \frac{1}{2}y$. **(5 marks)**

Nov 2012, Q15, 5MB2H/01

Exam hint
Draw on the diagram the acute angle MPN, where P is a point on the circumference of the circle.

16.5 Applying circle theorems

Objectives

- Solve angle problems using circle theorems.
- Give reasons for angle sizes using mathematical language.
- Find the equation of the tangent to a circle at a given point.

Did you know?

The word 'circle' derives from the Greek word *κρικος* (**krikos**), meaning a hoop or a ring. Circles have been known since the earliest recorded history.

Fluency

- What does the graph of $x^2 + y^2 = 36$ look like?
- What is the gradient of a line perpendicular to $y = 2x + 3$?

Warm up

1 **a** Work out the size of each angle marked with a letter.
b Which angle is alternate to the angle marked 136°?
c Which angle is vertically opposite to the angle marked 136°?

2 AB is a tangent.
What is the size of angle BAO?

3 Find the equation of the line that is perpendicular to $y = 2x - 3$ and passes through the point (−1, 2).

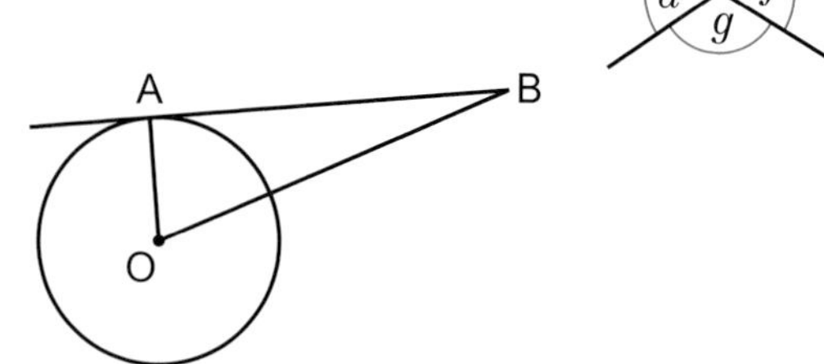

ActiveLearn Homework, practice and support: Higher 16.5

4 **Reasoning** In each diagram, O is the centre of the circle.

a

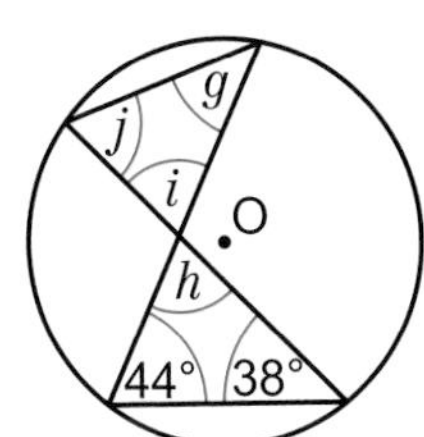

b

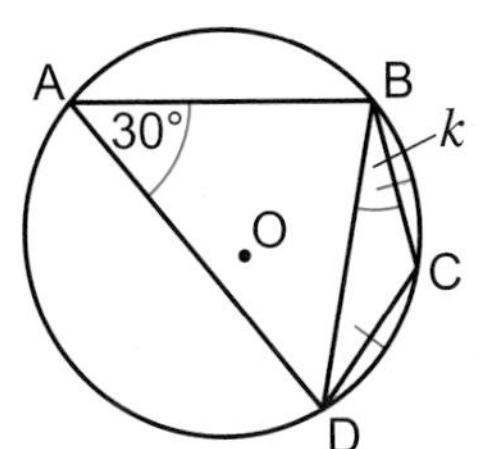

c

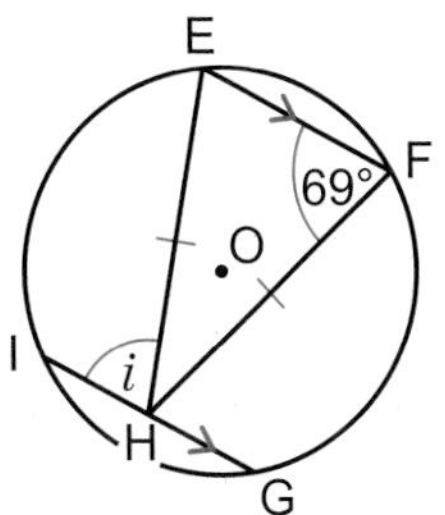

Work out the size of each angle marked with a letter. Give reasons for each step in your working.

5 **Reasoning** In each diagram, AT is a tangent to the circle.

a

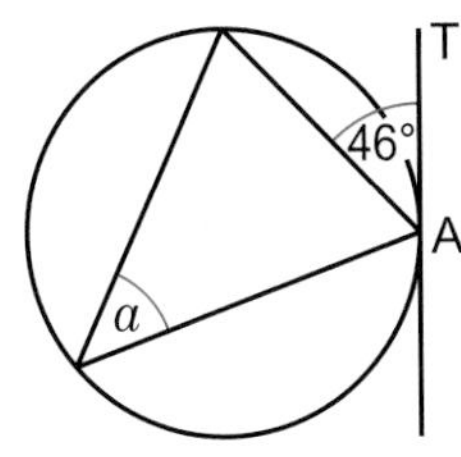

b

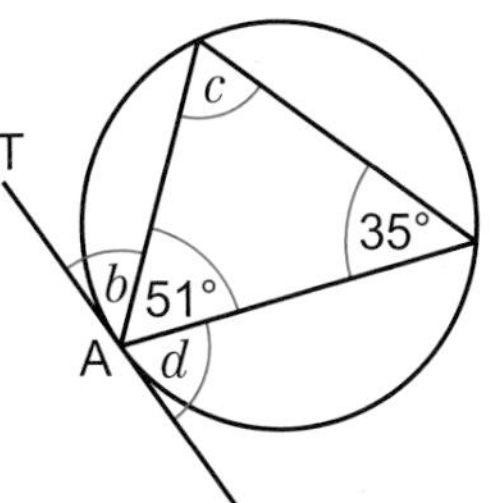

c

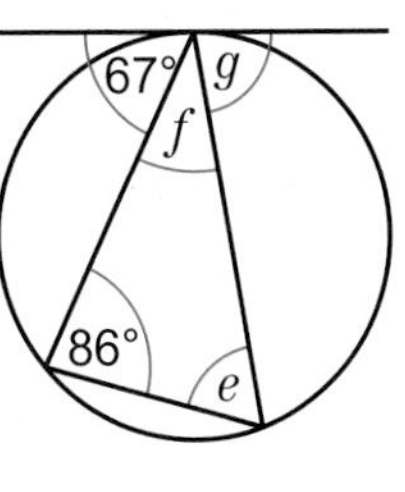

Work out the size of each angle marked with a letter. Give reasons for each step in your working.

Discussion Is there more than one way to get the answers?

6 **Reasoning** Work out the size of each angle marked with a letter.
Give reasons for each step in your working.

a

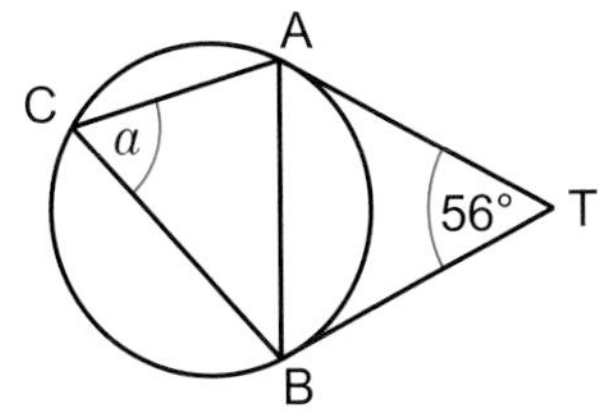

b

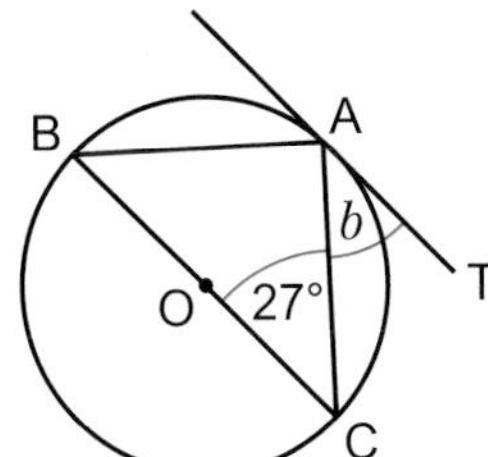

c

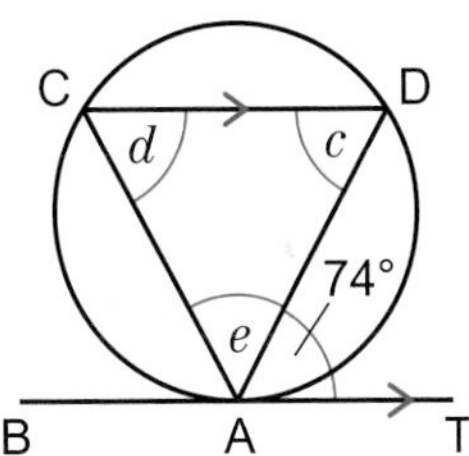

7 **Reasoning** O is the centre of the circle.
DAT and BT are tangents to the circle.
Angle CAD = 50° and angle ATB = 48°.
Work out the size of

a angle CAO
b angle AOB
c angle AOC
d angle COB
e angle CBO.

Give reasons for each step in your working.

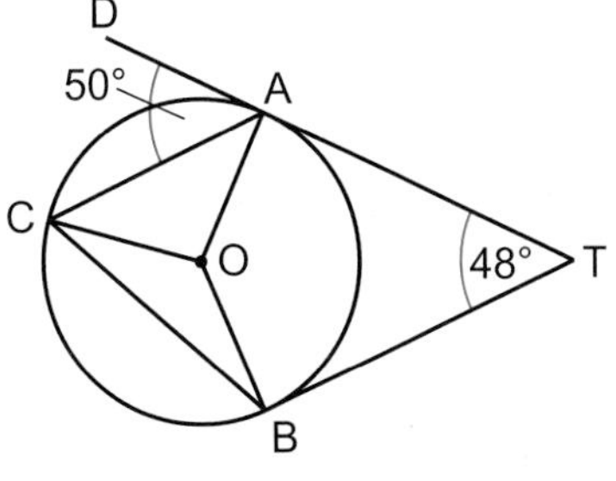

8 **Exam-style question**

B, C and D are points on the circumference of a circle, centre O.
ABE and ADF are tangents to the circle.
Angle $DAB = 40°$
Angle $CBE = 75°$
Work out the size of angle ODC.

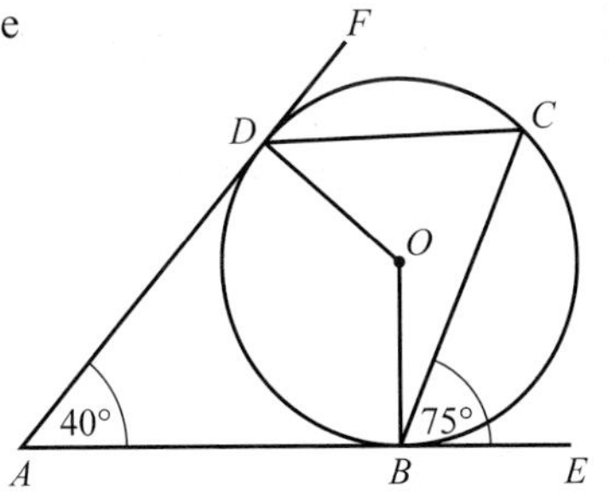

Diagram **NOT** accurately drawn

Exam hint
Remember that reasons are always words and not calculations.

(3 marks)

June 2014, Q21, 1MA0/1H

Example 4

Find the equation of the tangent to the circle $x^2 + y^2 = 25$ at the point A (3, −4).

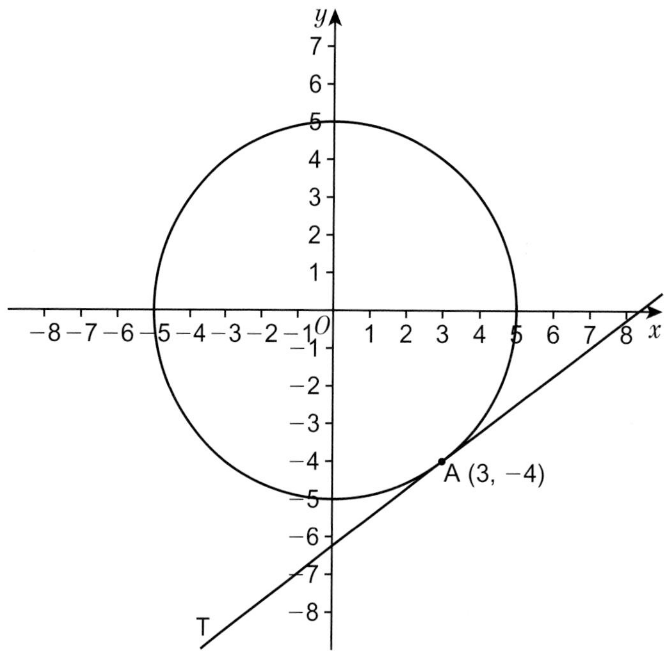

Gradient of line OA $= \frac{-4}{3}$

Gradient = $\frac{\text{change in } y}{\text{change in } x} = \frac{-4}{3}$

Gradient of line AT $= \frac{3}{4}$

Tangent is perpendicular to radius.

Equation of line AT is $y = \frac{3}{4}x + c$

$-4 = (\frac{3}{4} \times 3) + c$

Line passes through (3, −4) so substitute $x = 3$ and $y = -4$ in $y = mx + c$.

$c = -4 - \frac{9}{4}$

$= -\frac{25}{4}$

$y = \frac{3}{4}x - \frac{25}{4}$

Equation of line AT is $4y - 3x = -25$.

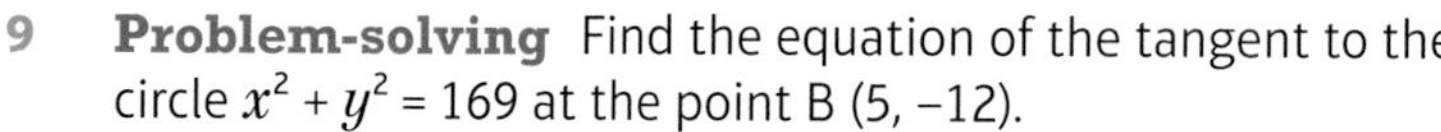

9 **Problem-solving** Find the equation of the tangent to the circle $x^2 + y^2 = 169$ at the point B (5, −12).

Reflect Did you follow the steps in **Example 4**? If so, how did it help you?

Q9 hint Draw the diagram and then follow the steps in **Example 4**.

10 **Problem-solving** Find the equation of the tangent to the circle $x^2 + y^2 = 225$ at the point C (9, 12).

11 **Problem-solving** Find the equation of the tangent to the circle $x^2 + y^2 = 100$ at the point D (−8, 6).

12 **Problem-solving** Find the equation of the tangent to the circle $x^2 + y^2 = 289$ at the point E (−8, −15).

16 Problem-solving

Objective

- Use logical reasoning to help you solve problems.

Example 5

AC and BD are chords of a circle. They intersect at point E.

a Prove that triangles AED and BEC are similar.

b AE = 3 cm, DE = 6 cm and BE = 4 cm. Show that CE = 8 cm.

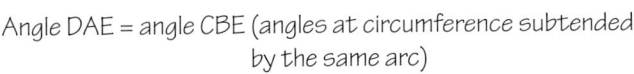

> State how you can show two triangles are similar.

a Shapes are similar if corresponding angles are equal.

Corresponding angles:

> Use the diagram to identify the corresponding angles in the two triangles.

Triangle AED		Triangle BEC
DAE	and	CBE
ADE	and	BCE
AED	and	BEC

Angle DAE = angle CBE (angles at circumference subtended by the same arc)

Angle ADE = angle BCE (angles at circumference subtended by the same arc)

Angle AED = angle BEC (vertically opposite angles)

> Look at the diagram. What reasons can you give for each pair of corresponding angles being equal?

All corresponding angles are equal. Therefore, triangles AED and BEC are similar.

b In similar shapes, all corresponding sides are in the same ratio.

> State what you know about the sides of similar shapes.

Corresponding sides:

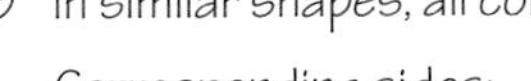

> Use the diagram to identify which sides in the question are corresponding sides.

Triangle AED		Triangle BEC
AE	and	BE
DE	and	CE

Corresponding sides:

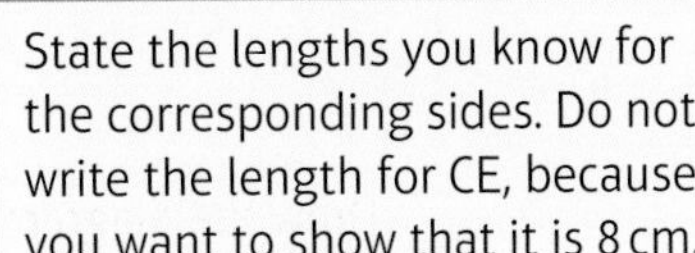

> State the lengths you know for the corresponding sides. Do not write the length for CE, because you want to show that it is 8 cm.

Triangle AED	Triangle BEC	Ratios
AE = 3 cm	BE = 4 cm	$\frac{3}{4}$
DE = 6 cm	CE	$\frac{6}{CE}$

> Find the ratios of corresponding sides.

As all pairs of corresponding sides are in the same ratio, $\frac{3}{4} = \frac{6}{CE}$

Therefore, CE = 8 cm

1 A, B and C are points on the circumference of a circle with centre O.
AC is the diameter of the circle.
AB = 4 cm and BC = 8 cm.
Show that the area of the circle is 20π cm^2.

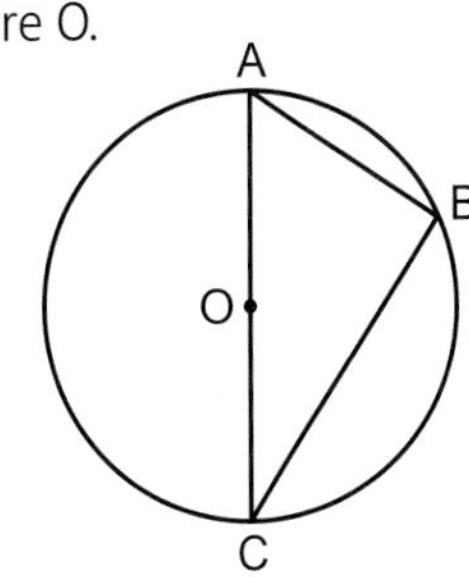

> **Q1 hint** What do you know about the angle in a semicircle? What theorem can you use to find AC?

2 You can write two consecutive numbers as n and $n + 1$.

a Prove that if you add the squares of two consecutive numbers and then add 1, the answer is an even number.

b Prove that if you add the squares of three consecutive numbers and then add 1, the answer is a multiple of 3.

> **Q2a hint** Square n and $n + 1$. Add them, then add 1. Factorise to show the result is divisible by 2.

3 a Show that $\frac{3x + 12}{3} = \frac{x^2 - 16}{x - 4}$

> **Q3a hint** Look at each side separately. Factorise the numerator. Then divide by the denominator.

b Prove that these triangles are similar.

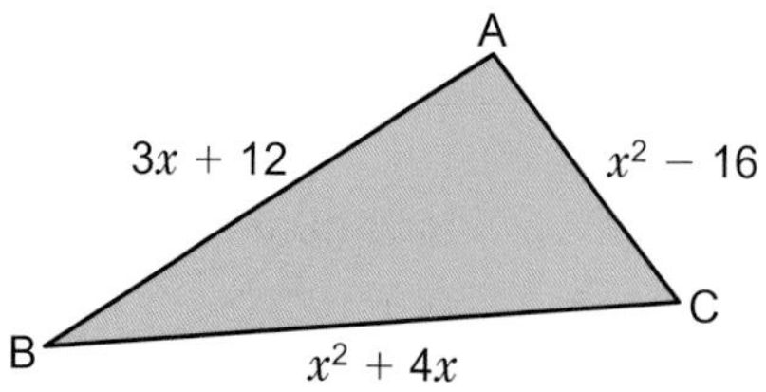

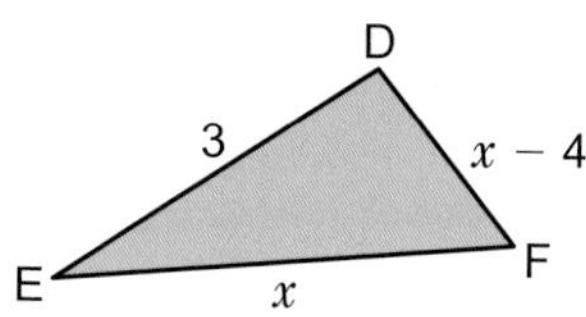

> **Q3b hint** Use your answer to part **a** to show that all pairs of corresponding sides are in the same ratio.

4 V, W, X and Y are points on the circumference of a circle.
WXZ and VYZ are straight lines.
Angle WYX = 30° and angle YXZ = 65°.

a Show that XWY = 35°.

> **Q4a hint** Don't forget to give reasons to accompany your working.

b Show that XVY = 35°.

c Given that VXY = 55°, show that VWX is a right angle.

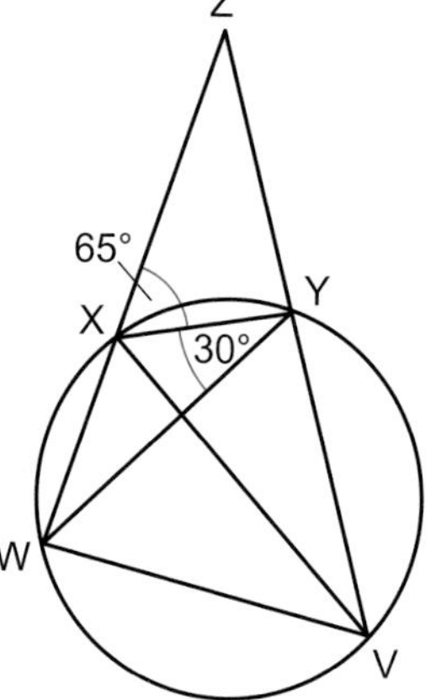

5 a Show that this triangle PQR, with an interior angle of 44° and an exterior angle of 112°, is isosceles.

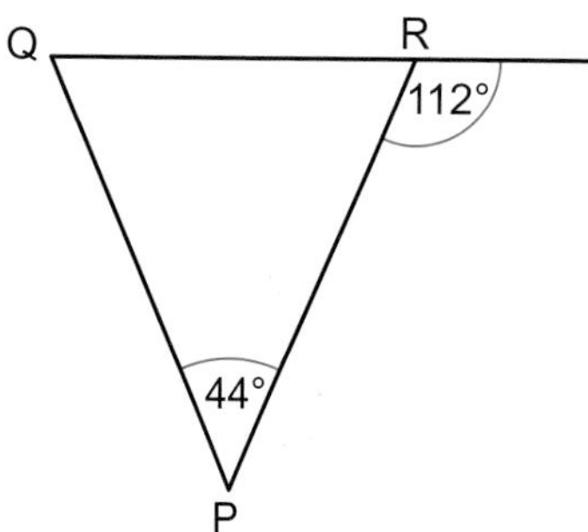

b Prove that this triangle XYZ, with an interior angle of $2x$ and an exterior angle of $x + 90°$, is isosceles.

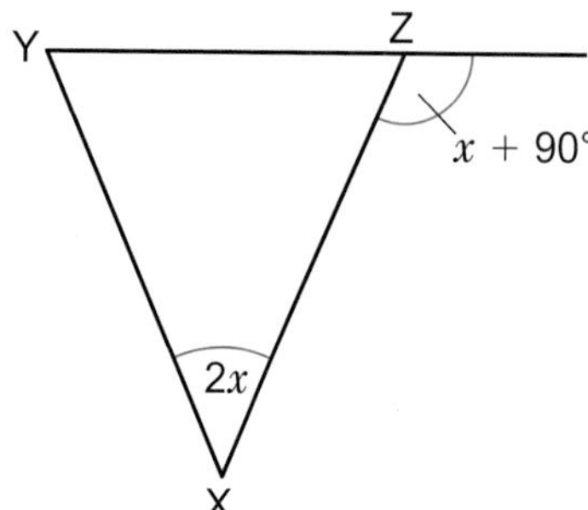

6 The formula for converting from degrees Fahrenheit (F) to degrees Celsius (C) is
$\frac{5}{9}(F - 32) = C$

Q6a communication hint **Verify** means substitute to show a statement is true.

a Verify that –40 °F = –40 °C
b Prove that the temperature –40° has the same value in both Fahrenheit and Celsius.

Q6b hint If F and C have the same value, then $F = C$. Therefore, substitute $F = C$ into the formula, and then solve to find C.

7 **Reflect**
Choose A, B or C.
Solving problems by logical reasoning is:
A always easy B sometimes easy, sometimes hard C always hard
Discuss with a classmate or your teacher what you find easy or hard.

16 Check up

Log how you did on your Student Progression Chart.

Chords, radii and tangents

1 Draw a circle with radius 6 cm. Draw and label clearly a chord, a tangent and a segment.

2 **Reasoning** In the diagrams, O is the centre of the circle.

a
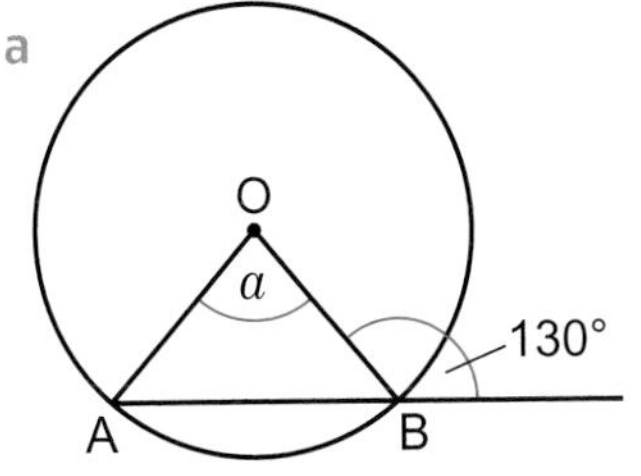

b
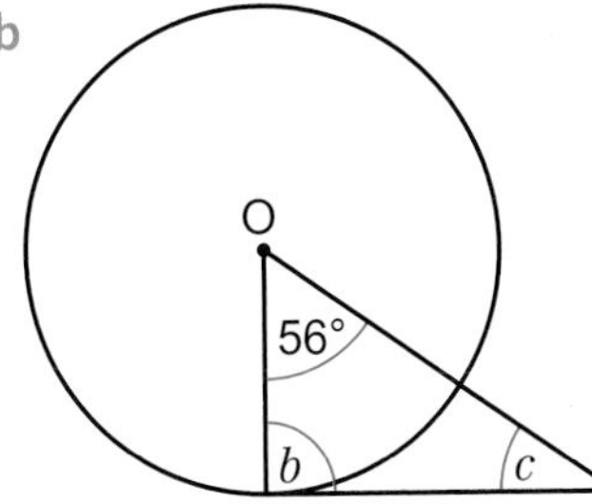

c
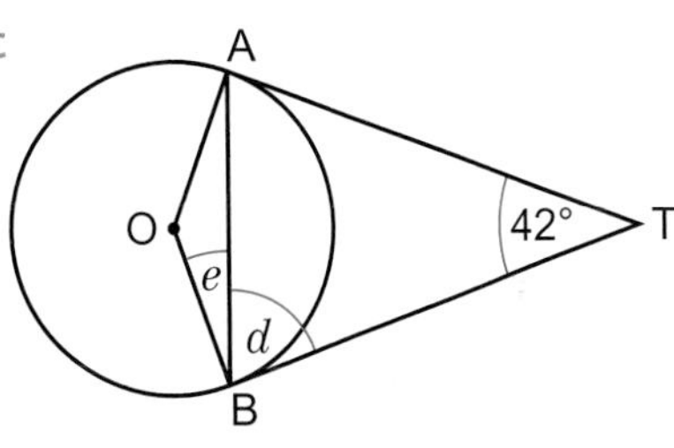

Work out the size of each angle marked with a letter.
Give reasons for each step in your working.

3 **Reasoning** O is the centre of a circle with radius 8.5 cm.
AB is a chord with length 15 cm. Angle OMB = 90°.
Work out the length of OM.

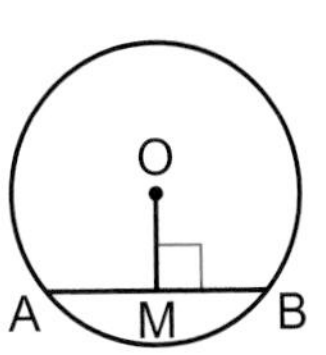

Circle theorems

4 **Reasoning** In the diagrams, O is the centre of each circle.
A, B, C and D are all points on the circumference of the circles.

a
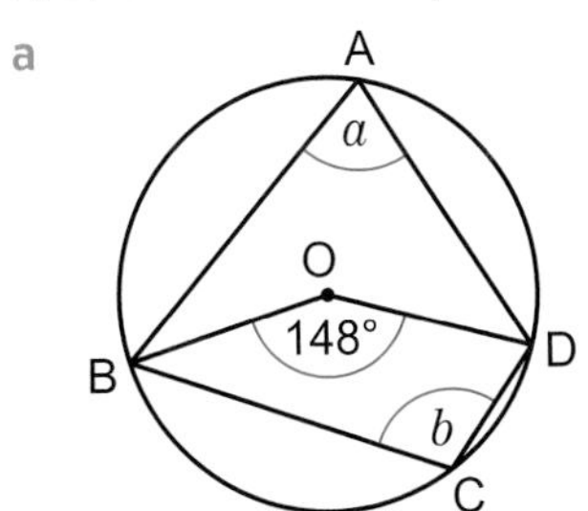

b
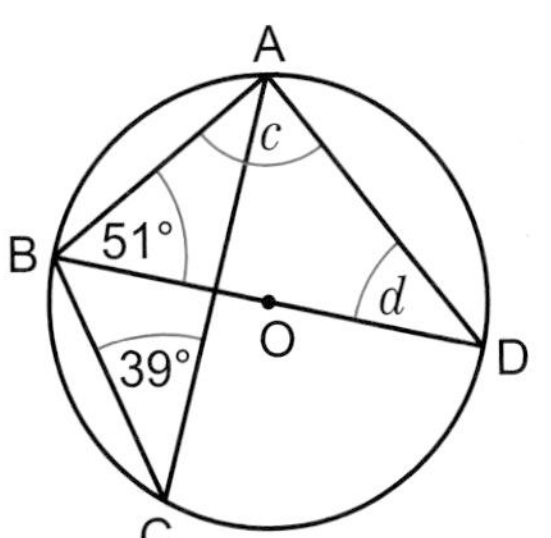

Work out the size of each angle marked with a letter.
Give reasons for each step in your working.

5 **Reasoning** Work out the size of

a angle a

b angle b.

Give reasons for each step in your working.

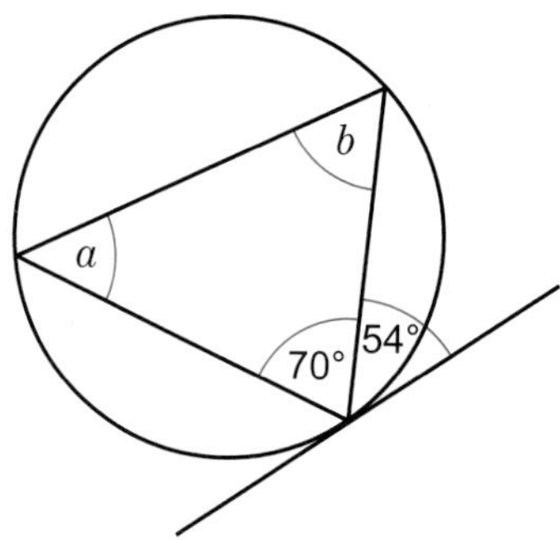

Proofs and equation of tangent to a circle at a given point

6 **Problem-solving** Find the equation of the tangent to the circle $x^2 + y^2 = 676$ at the point A (10, −24).

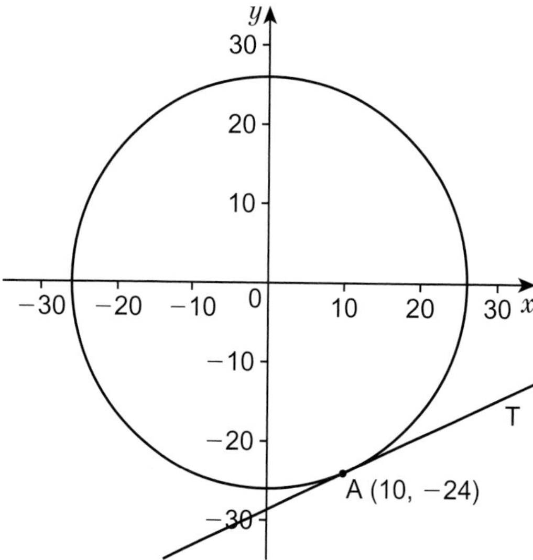

7 **Reasoning / Communication** Prove that the angle at the centre of a circle is equal to twice the angle at the circumference when both are subtended by the same arc.

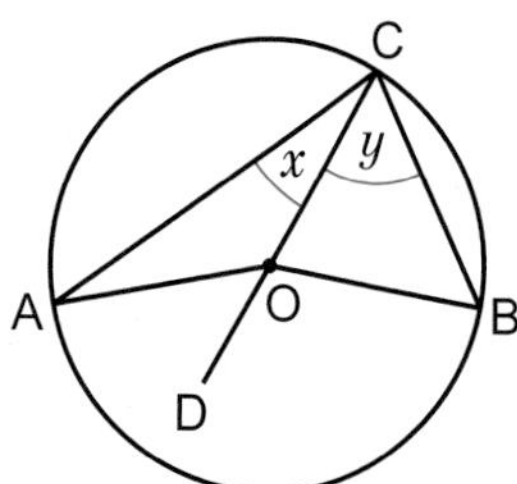

Reflect

8 How sure are you of your answers? Were you mostly

Just guessing Feeling doubtful Confident

What next? Use your results to decide whether to strengthen or extend your learning.

* Challenge

9 Copy this diagram of a circle, with centre O.

Choose a value for x and write in the size of as many angles as you can.

Repeat with a different value for x, if you have time.

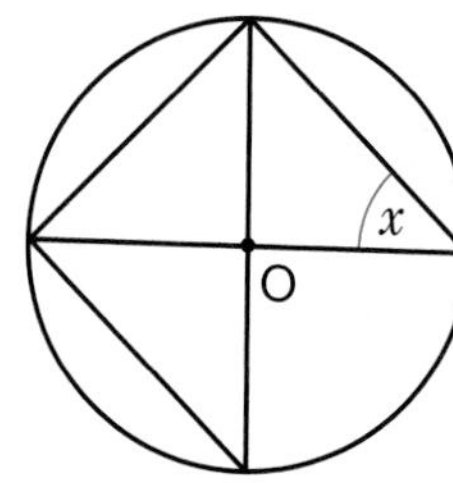

16 Strengthen

Chords, radii and tangents

1 Copy the diagram.
Use these words to label it.

chord	tangent	segment

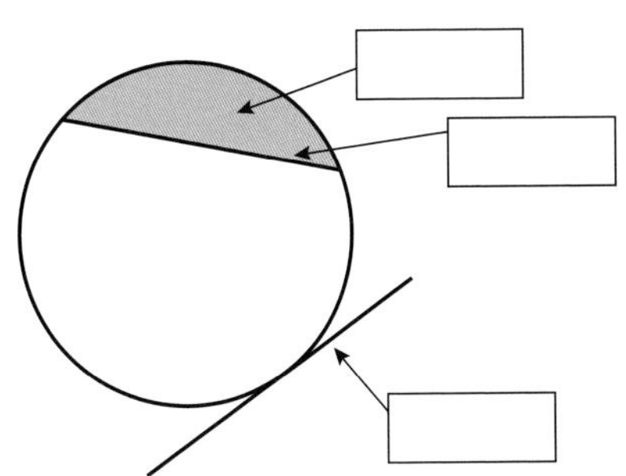

Q1 hint A chord joins two points on the circumference. A tangent touches the circle.

2 O is the centre of a circle. ABC is a straight line. Angle OBC = 160°.

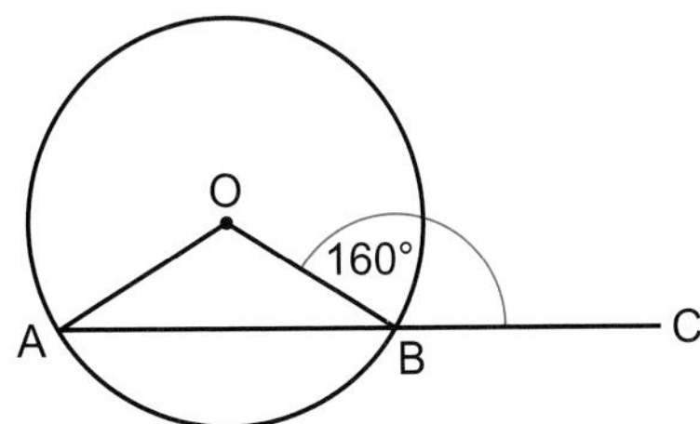

a Copy the diagram. Which two lines are equal length? Mark them.
b What type of triangle is OAB?
c Work out the sizes of angle OBA, angle OAB and angle AOB.

3 Draw a circle.
a Label the centre, O.
b Draw in a radius.
Label the point where it meets the circumference A.

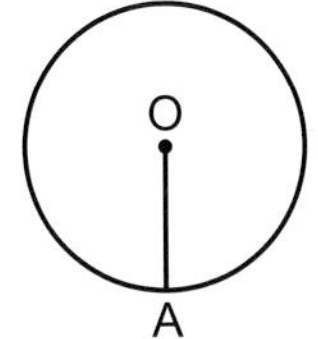

c Draw a line at 90° to your radius, through point A.
d Draw lines that meet your radius and the circumference at other angles.
Are they tangents?
e What is the angle between a tangent and a radius?

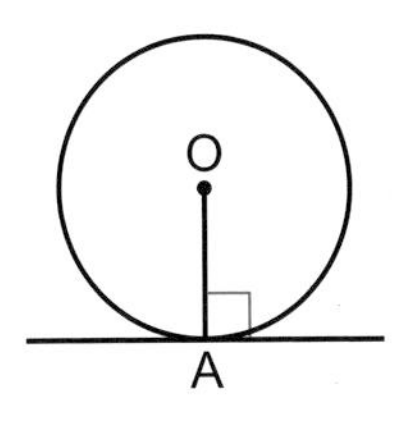

Q3d hint At how many points does it touch the circumference of the circle?

4 **Reasoning** O is the centre of a circle. AT and BT are tangents. AB is a chord. Angle ATB = 34°.

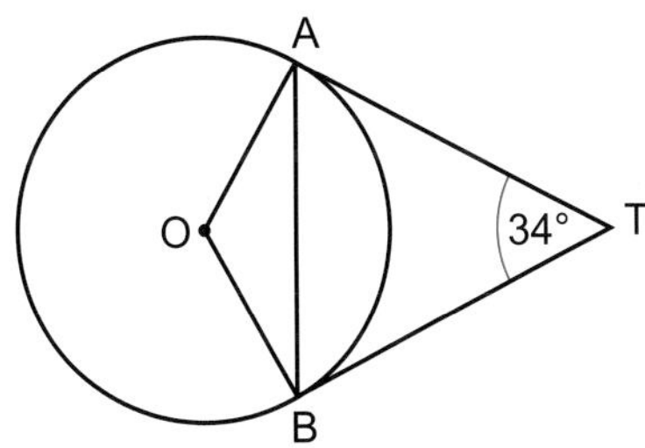

Q4a hint Tangents to a circle from the same external point are ______________.
The angle between a radius and a tangent is ______________°.

a Copy the diagram. Mark on any angles you know and any lines of equal length.
b What type of triangle is ABT?
c Work out the sizes of angle ABT, angle OBT and angle OBA.

5 **Reasoning** O is the centre of a circle with radius 6.5 cm.
AB is a chord with length 12 cm. Angle OMB = 90°.

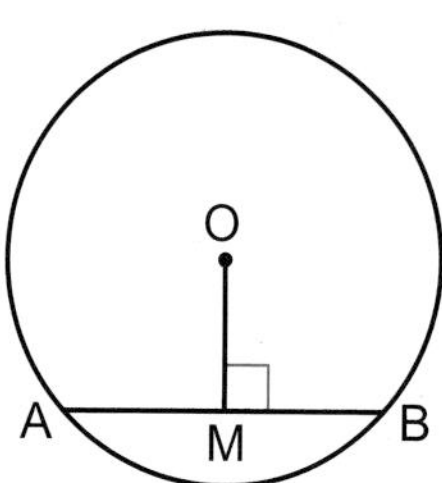

a Write down the length of OA.
b What is the length of AM?
c Work out the length of OM.

Q5a hint Copy the diagram and draw the radius OA.

Q5b hint M is the midpoint of AB, so AM = half of AB. Mark the lengths you know on the diagram.

Q5c hint Use Pythagoras' theorem.

Circle theorems

1 O is the centre of a circle.
P, Q and R are points on the circumference of the circle.
Angle QOR = 130°.
a Copy the diagram.
b Colour the arc that subtends angle QOR at the centre.
c What angle at the circumference is subtended by the same arc?
d Work out the size of angle QPR.

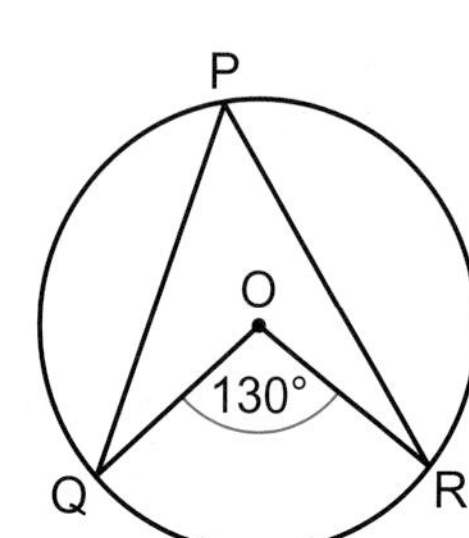

Q1d hint The angle at the centre of a circle is twice the angle at the circumference.

2 Work out the size of the angle marked with a letter in each of these diagrams.

a 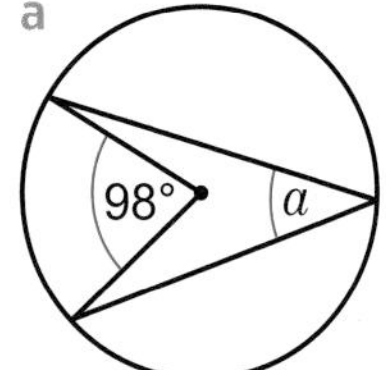

b 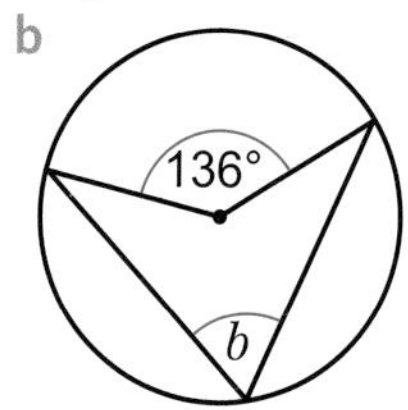

c 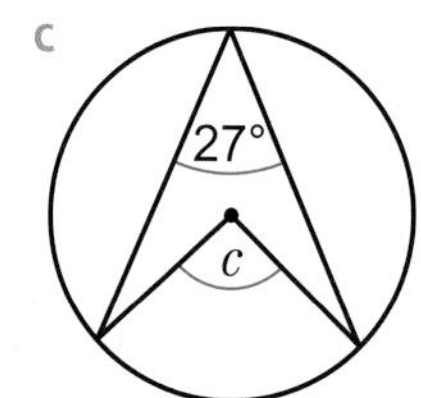

d 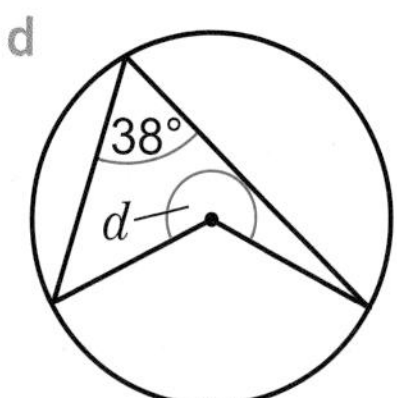

3 Draw a circle. Mark four points on the circumference and join them to make a cyclic quadrilateral.
Measure all of the angles in your cyclic quadrilateral.
Add the opposite angles. What do you notice?

4 P, Q, R and S are points on the circumference of a circle.
Angle QPS = 96° and angle PQR = 78°.
a What type of quadrilateral is PQRS?
b Which angle in the quadrilateral is opposite angle QPS?
c Work out the size of angle QRS.

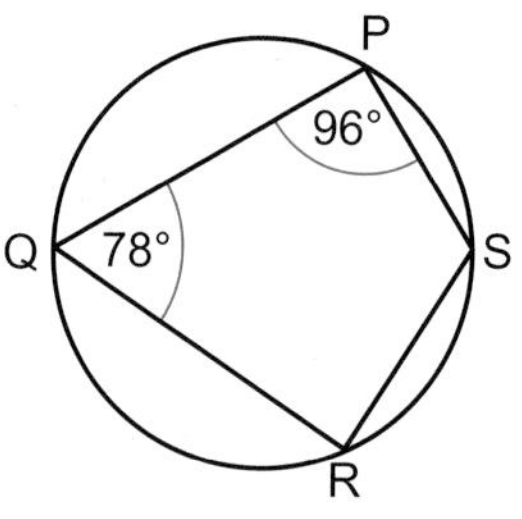

Q4c hint Opposite angles in a cyclic quadrilateral add up to ______°.

5 Which of these are cyclic quadrilaterals?

A
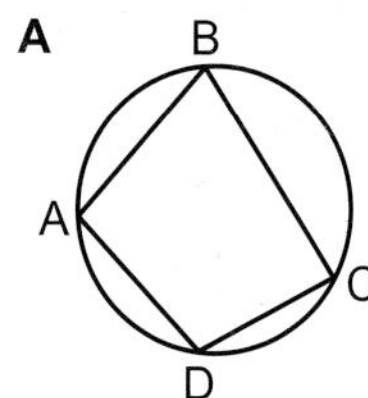

B
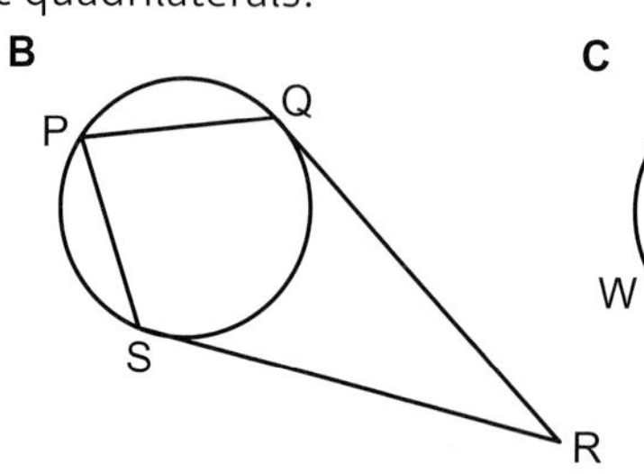

C

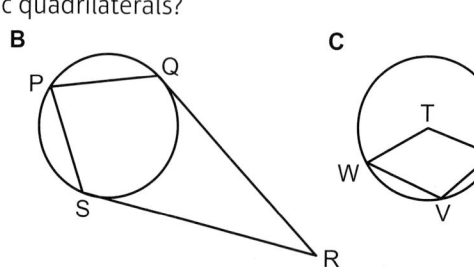

6 **Reasoning** O is the centre of a circle.
A, B, C and D are points on the circumference.
Angle BOD = 114°.

a Write down the vertices of the cyclic quadrilateral.

b Work out the size of angle BAD.

c What is the size of angle BCD?

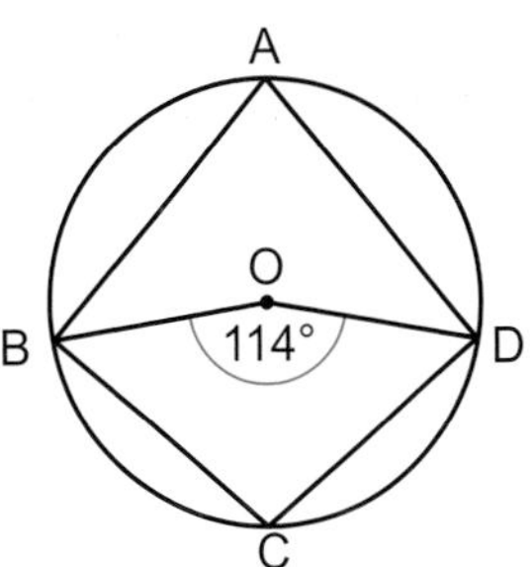

7 O is the centre of a circle.
A, B and C are points on the circumference.
Angle BAC = 36°.
Copy the diagram.
Then copy the working and complete the reasons.

AC is a ....

Angle ABC = ...° (angle in a ....)

Angle ACB = ...° (angles in a ....
add up to .... °)

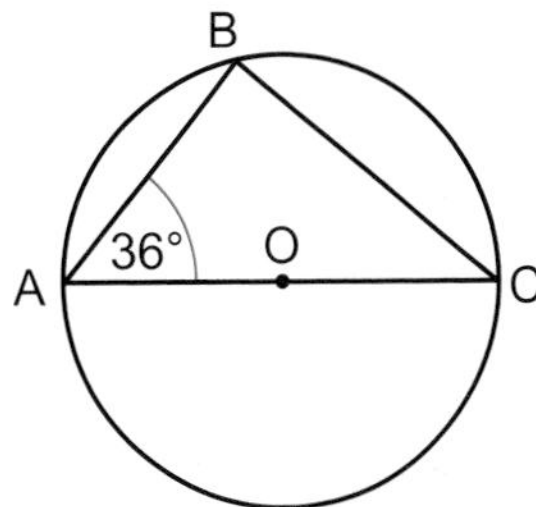

8 Draw a circle with diameter of at least 5 cm. Mark two points, A and B, on the circumference.
Draw four different triangles with base AB and the third vertex on the circumference.

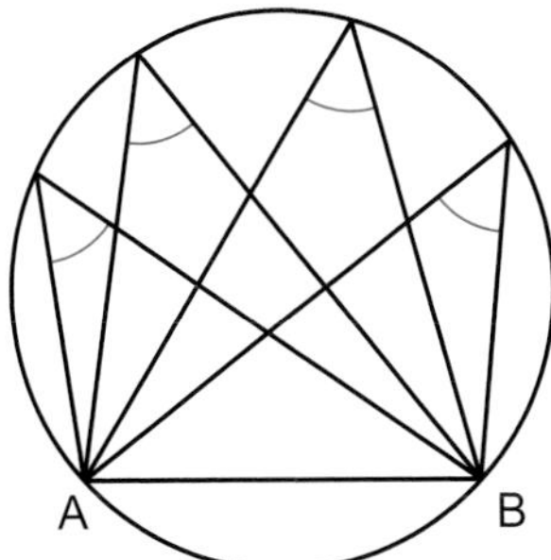

Measure each of the four angles at the circumference. What do you notice?

9 For each diagram, write down the pairs of equal angles.

a

b

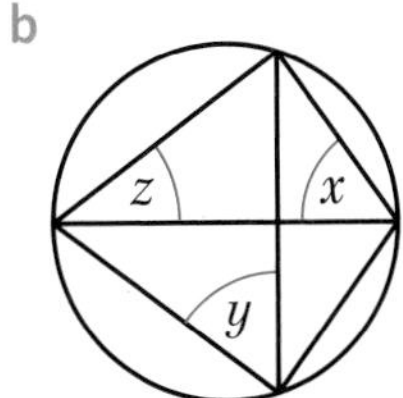

c

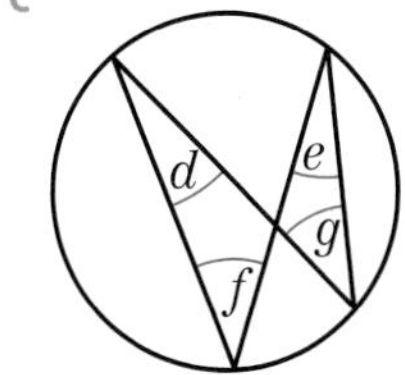

Q9 hint Turning the diagram round can often help.

10 O is the centre of a circle. A, B and C are points on the circumference of the circle.
TBD is a tangent to the circle at point B.

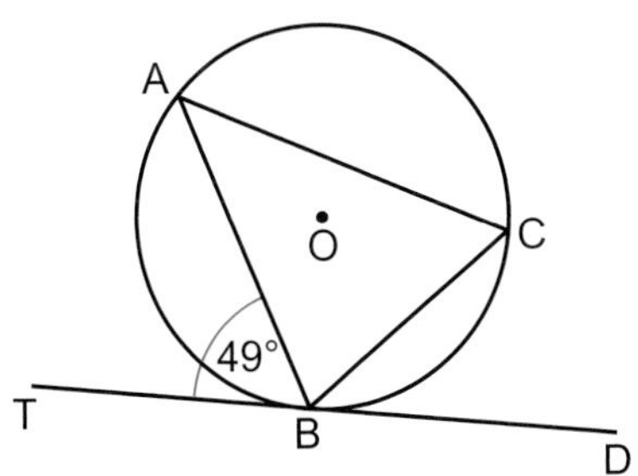

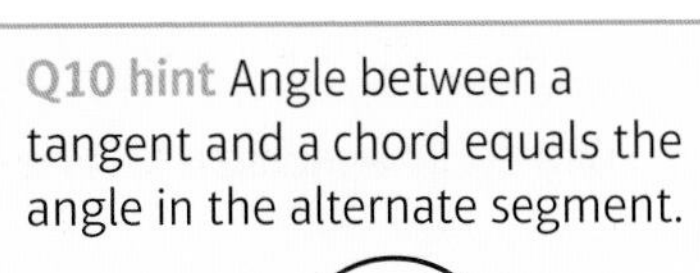

Q10 hint Angle between a tangent and a chord equals the angle in the alternate segment.

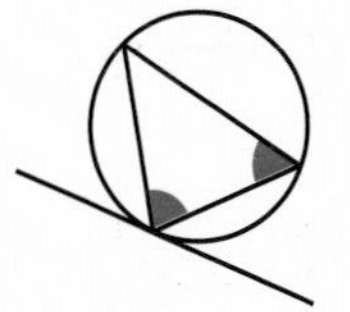

Angle ABT= 49°. What other angle is also 49°?

11 O is the centre of a circle.
A, B and C are points on the circumference.
TBD is a tangent to the circle at point B.
Angle ABC = 75° and angle ABT = 37°.
Work out the sizes of angle ACB and angle CAB.
Give reasons for each step in your working.

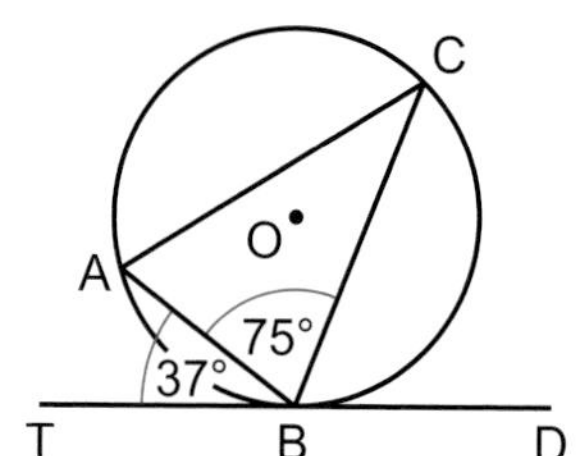

Proofs and equation of tangent to a circle at a given point

1 **Problem-solving** The diagram shows the tangent to the circle $x^2 + y^2 = 25$ at the point A (3, 4).

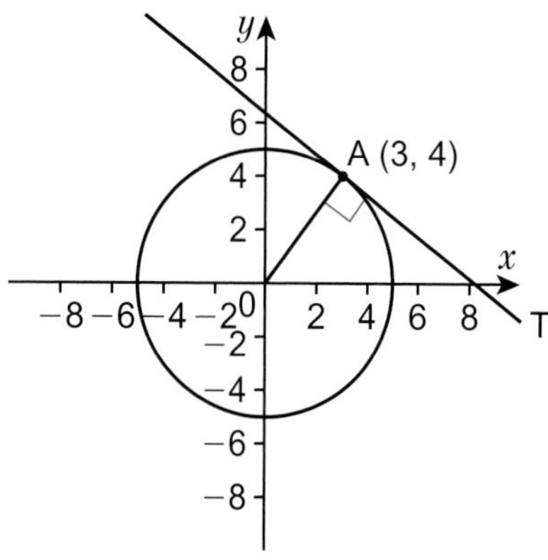

a What is the gradient of the radius?

b What is the gradient of the tangent AT?

c Complete for the equation of the tangent: $y = \square x + c$.

d Work out the value of c.

e Write the equation of line AT.

Q1b hint Gradient of perpendicular is $\frac{-1}{m}$.

Q1d hint Substitute $x = 3$ and $y = 4$ into your answer to part **c**.

2 **Reasoning** O is the centre of a circle. AOC is a straight line.

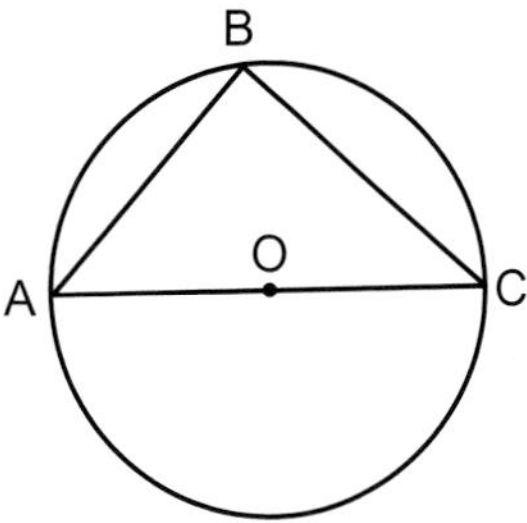

a What size is angle AOC?

b What arc subtends angle ABC?

c Work out the size of angle ABC.

d Which circle theorem have you proved?

3 **Problem-solving** Use other circle theorems to prove that an exterior angle of a cyclic quadrilateral is equal to the interior opposite angle.

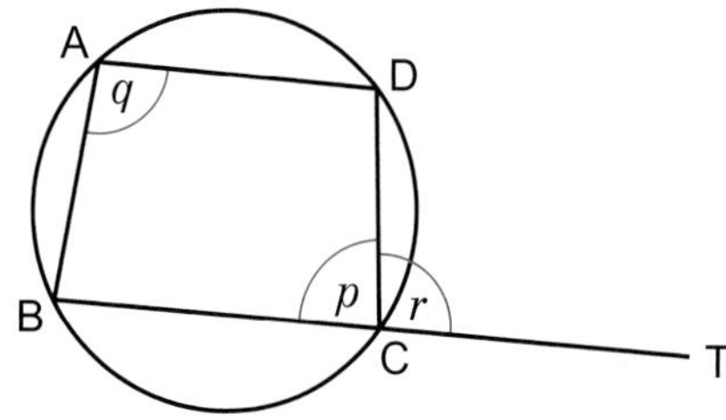

Q3 hint Angles on a straight line add to_______°.

16 Extend

1 **Reasoning** O is the centre of a circle.
OBC is an equilateral triangle.
Angle ABC = 130°.
Work out the sizes of angle a, angle b and angle c.
Give reasons for your answers.

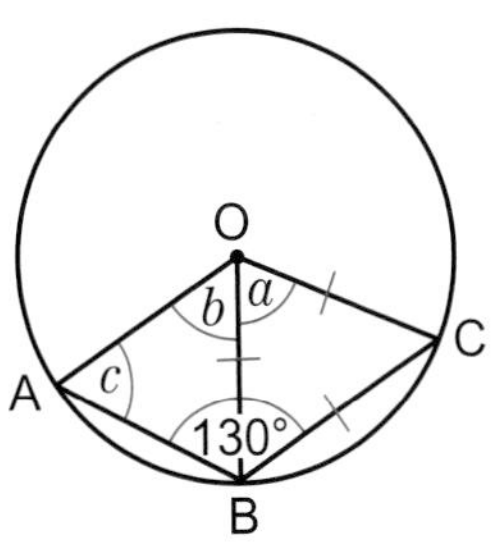

2 **Reasoning** O is the centre of a circle with radius 15.4 cm. PQ = 21.6 cm.

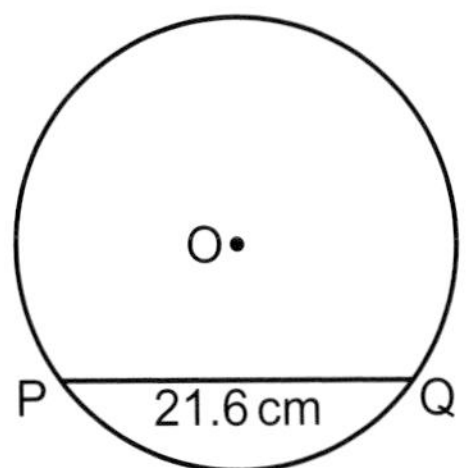

How far is the midpoint of PQ from the centre of the circle?
Give your answer correct to 1 d.p.

3 **Reasoning** O is the centre of a circle. Angle BAC = $3x$ and angle ACB = $2x$.

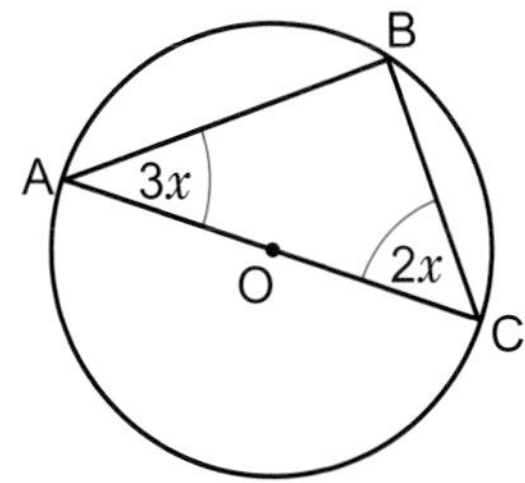

Q3 strategy hint Work out angle ABC first.

Work out the actual size of each angle in triangle ABC.

4 **Exam-style question**

A, *B*, *C* and *D* are points on the circumference of a circle, centre *O*.
Angle *AOC* = *y*.

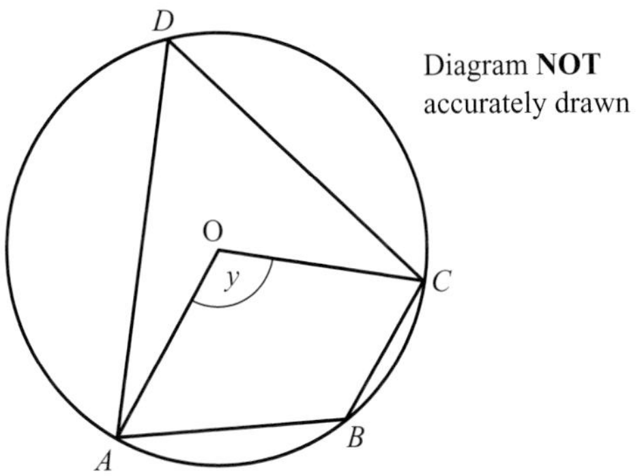

Find the size of angle *ABC* in terms of *y*.
Give a reason for each stage of your working. **(4 marks)**

Nov 2013, Q22, 1MA0/1H

Exam hint
Start by expressing angle *ABC* in terms of y.

5 **Reasoning** O is the centre of a circle.
A, B and C are points on the circumference.
Angle BOC = 40° and angle AOB = 70°.
Prove that AC bisects angle OCB.

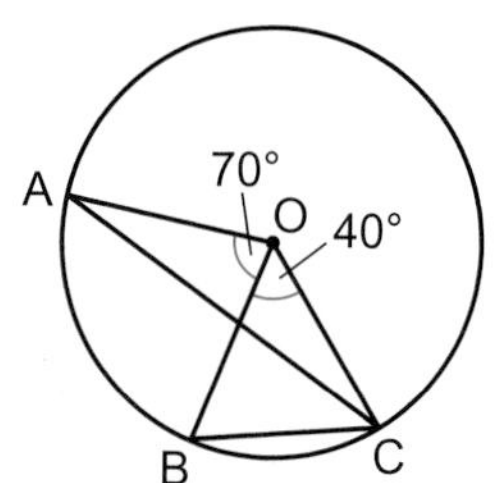

6 **Reasoning** O is the centre of a circle.
A, B, C and D are points on the circumference.
Angle ABC = 114°.
Work out the size of angle COD.
Give a reason for each step of your working.

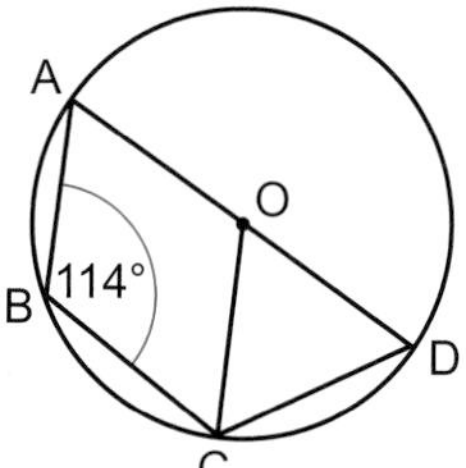

7 **Problem-solving** O is the centre of a circle.
A, B, C and D are points on the circumference.
Angle BAD = 150°.
Prove that triangle OBD is equilateral.

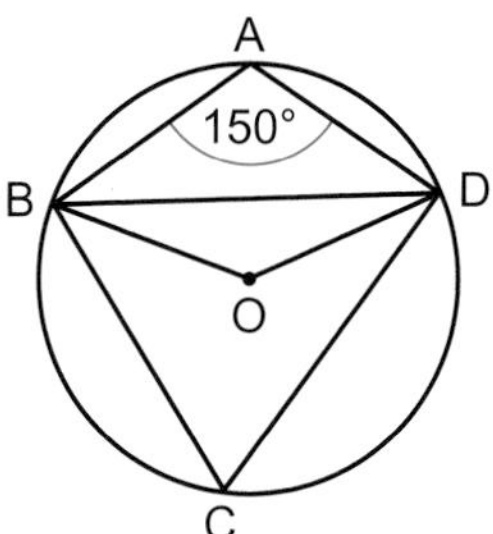

8 **Reasoning** O is the centre of a circle with radius 25 cm. AB and CD are parallel chords.
AB = 14 cm and CD = 40 cm. MON is a straight line.

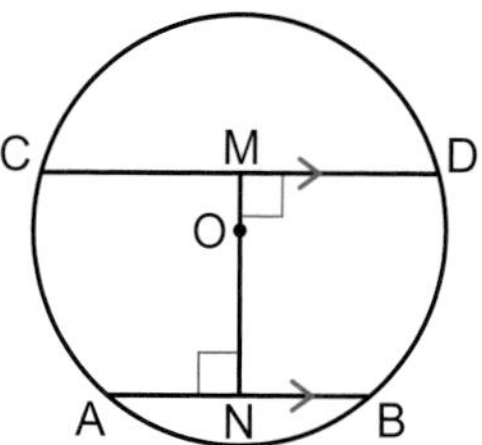

Q8 strategy hint First work out lengths OM and ON separately.

Work out the length of MN.

9 **Problem-solving** O is the centre of a circle.
Diameter AC is 40 mm.
Angle OMB = 90°, and OM = AM.
Work out the length of chord AB.
Give your answer correct to a suitable degree of accuracy.

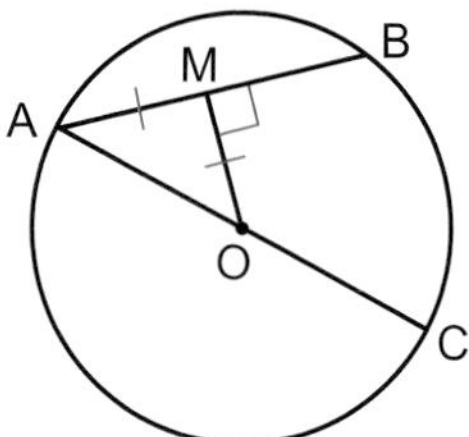

10 **Problem-solving / Communication** CD is parallel to BT. BT is a tangent to the circle.

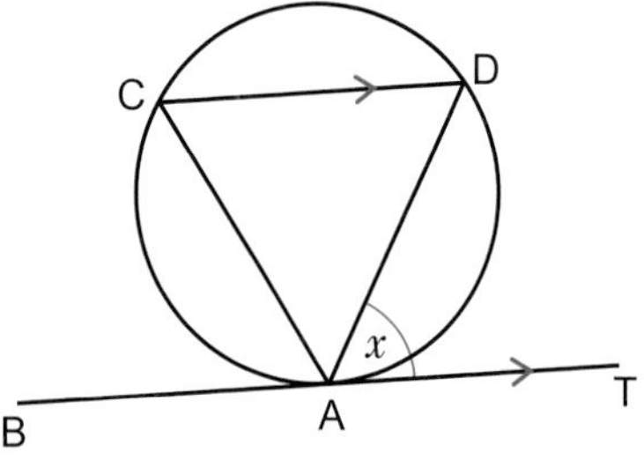

Prove that triangle ACD is isosceles.

11 **Problem-solving / Communication** AB and BC are tangents to a circle, centre O.

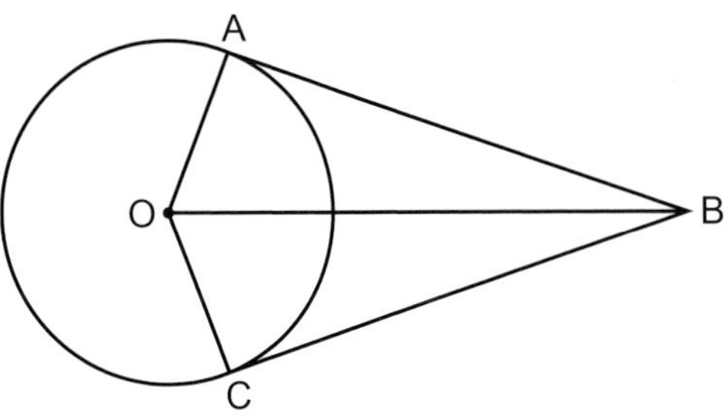

Q11 hint Prove that triangles ABO and OBC are congruent.

Prove that AB = BC.

12 **Problem-solving** O is the centre of a circle.
P, Q and S are points on the circumference.
RST is a tangent touching the circle at point S.
Angle RSP = 62°. Reflex angle QOS = 220°.

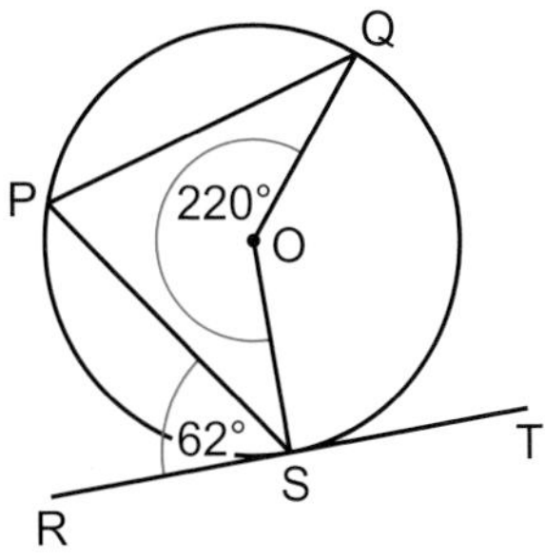

Work out the size of angle PQO.
Give a reason for each step of your working.

13 **Problem-solving / Communication**
Prove that the line drawn from the centre of a circle to the midpoint of a chord is perpendicular to the chord.

Q13 strategy hint Draw a diagram. You can use SSS to prove two triangles are congruent.

14 **Exam-style question**

AOC and *BOD* are diameters of a circle, centre *O*.

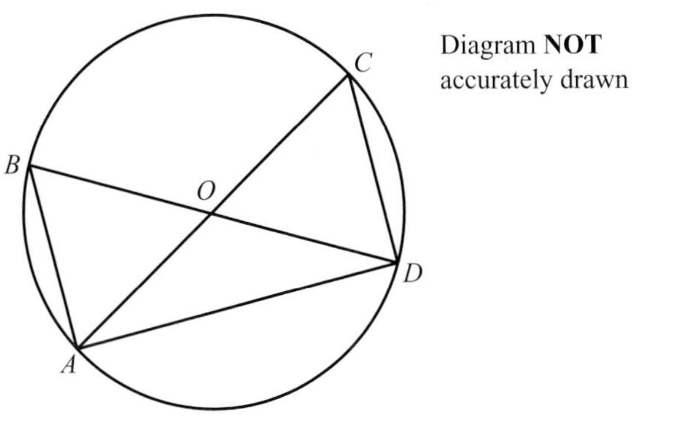

Prove that triangle *ABD* and triangle *DCA* are congruent. **(3 marks)**

Nov 2013, Q28, 1MA0/2H

Exam hint
Start by drawing triangles *ABD* and *DBA* separately, marking on the diagrams which sides and angles are equal. You must give a reason at each stage of your proof and a reason for congruence.

16 Knowledge check

- The angle between a **tangent** and the radius is 90°.

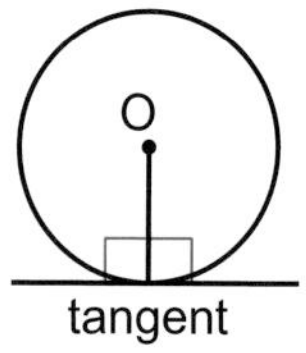

O

tangent .... *Mastery lesson 16.2*

- Tangents drawn to a circle from a point outside the circle are equal in length.
  So AB = AC.

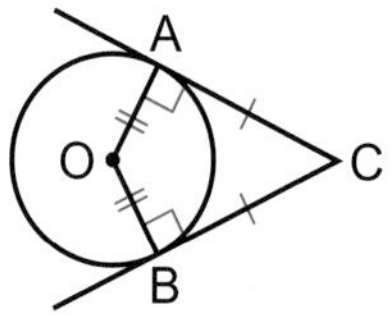

A O C B .... *Mastery lesson 16.2*

- You must learn all the circle theorems.
  You could be asked to prove any of the facts below.

- A **chord** is a straight line connecting two points on a circle.
  The perpendicular from the centre of a circle to a chord bisects the chord and the line drawn from the centre of a circle to the midpoint of a chord is at right angles to the chord.

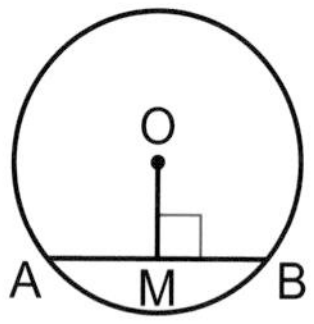

O A M B .... *Mastery lesson 16.1*

- The angle at the centre of a circle is twice the angle at the circumference when both are subtended by the same arc.

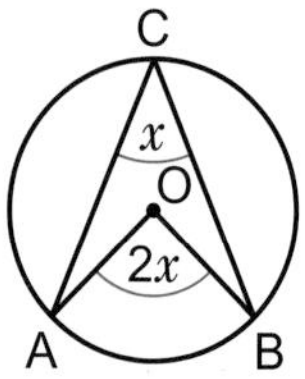

C x O $2x$ A B .... *Mastery lesson 16.3*

- The angle in a semicircle is a right angle.
  So angle ABC = 90°.

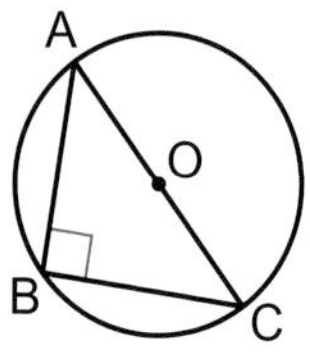

A O B C .... *Mastery lesson 16.3*

- Angles subtended at the circumference by the same arc are equal; or angles in the same segment are equal. .... *Mastery lesson 16.4*

- A **cyclic quadrilateral** is a quadrilateral with all four vertices on the circumference of a circle.

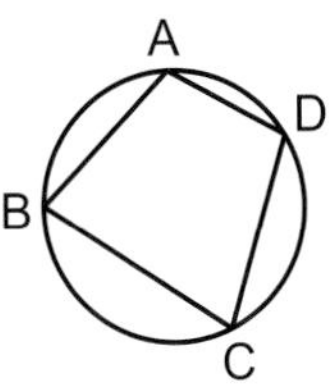

Mastery lesson 16.4

- Opposite angles of a cyclic quadrilateral add up to 180°:
  So, $x + y = 180°$ and $p + q = 180°$.

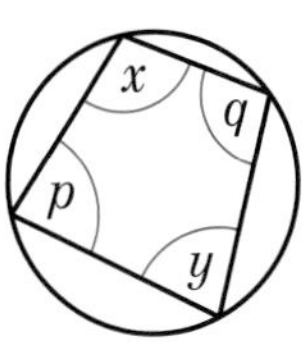

Mastery lesson 16.4

- An exterior angle of a cyclic quadrilateral is equal to the opposite interior angle. ... Mastery lesson 16.4

- AT is a tangent to the circle. AB is a chord.
  Angle BAT is the angle between the tangent and the chord in one segment.
  The other segment made by the chord AB contains angle ACB.
  This is called the **alternate segment**.

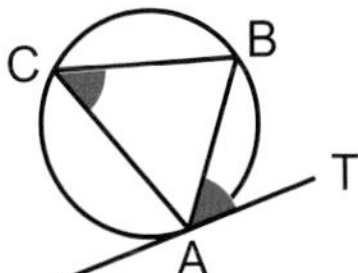

The angle between the tangent and the chord is equal to the angle in the alternate segment. ... Mastery lesson 16.4

Reflect

Look back at this unit.
Which lesson did you like most? Write a sentence to explain why.
Which lesson did you like least? Write a sentence to explain why.
Begin your sentence with: I liked lesson ____ most/least because ____

16 Unit test

Log how you did on your Student Progression Chart.

1 O is the centre of a circle. ABC is a straight line. Angle OBC = 146°.

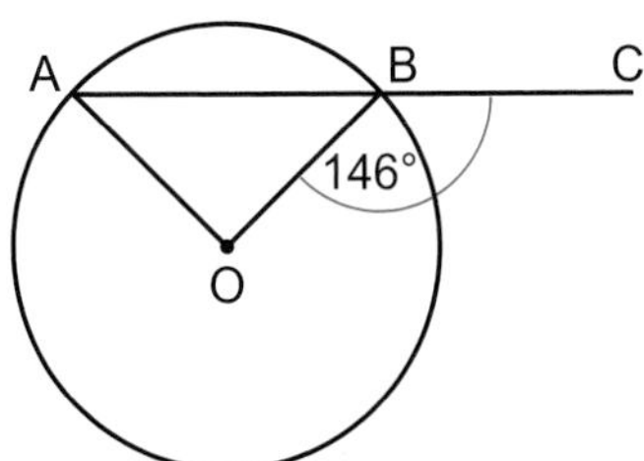

Work out the size of angle AOB. Give reasons for each step in your working. *(4 marks)*

ActiveLearn Homework, practice and support: Higher 16 Unit test

2 **Reasoning** O is the centre of a circle.
AT is a tangent and AB is a chord.
Angle AOB = 124°.

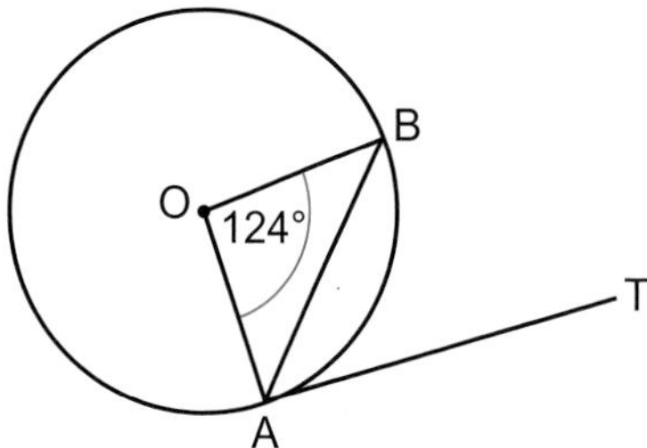

Work out the size of angle BAT.
Give reasons for each step in your working. *(4 marks)*

3 O is the centre of a circle with radius 8 cm.
AB is a chord with length 10 cm.
M is the midpoint of AB.

a Write down the size of angle OMB.

b Work out the length of OM (to 1 d.p.)

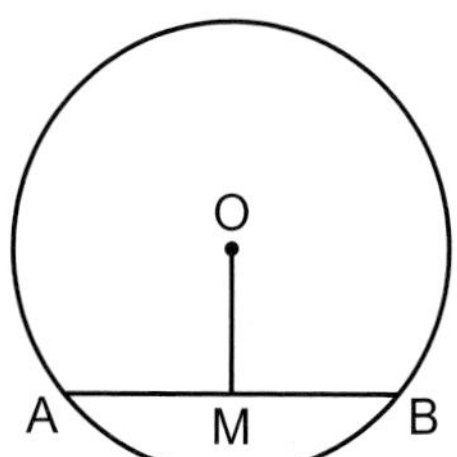

(2 marks)

4 **Reasoning** O is the centre of a circle.
A, B, C and D are points on the circumference of the circle.
Angle BCD = 110°.

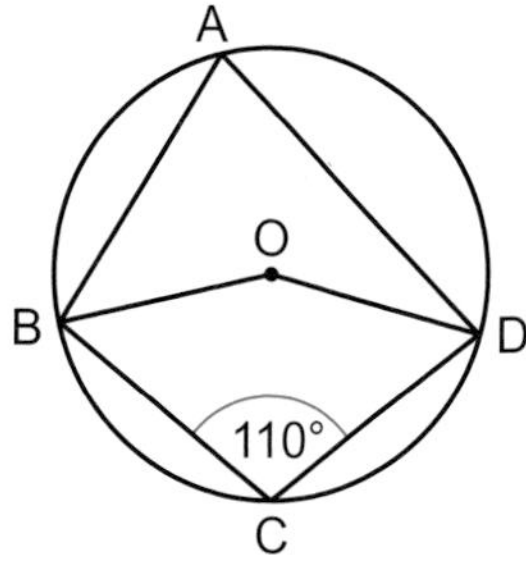

Work out the size of angle BOD.
Give reasons for each step in your working. *(3 marks)*

5 **Reasoning** O is the centre of a circle.
A, B, C and D are points on the circumference of the circle.
Angle ADB = 19°.
Work out the size of

a angle ABD

b angle ACB.

Give reasons for each step in your working. *(4 marks)*

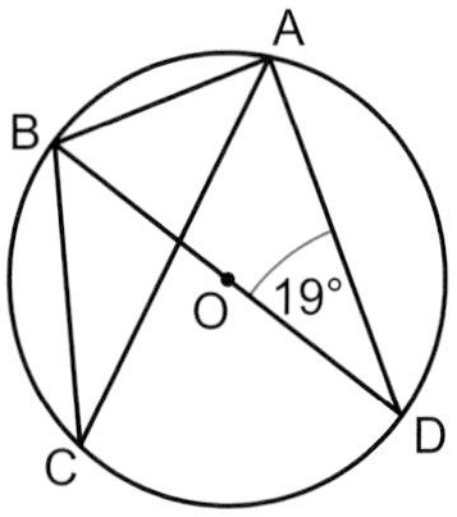

6 **Reasoning** O is the centre of a circle.
AC is a diameter.
Work out the actual size of angle BAC.

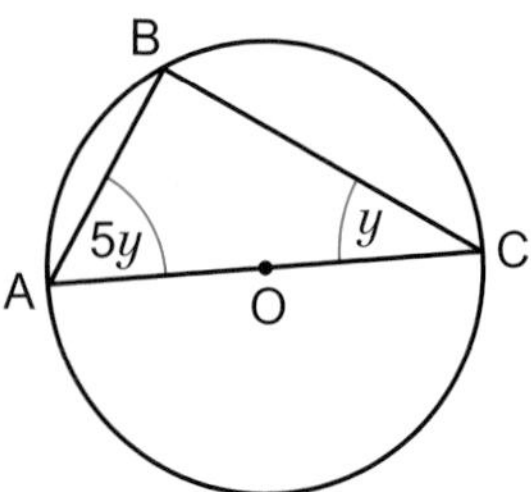

(3 marks)

7 **Exam-style question**

A and *B* are points on the circumference of a circle, centre *O*.
AC and *BC* are tangents to the circle.
Angle $ACB = 36°$.

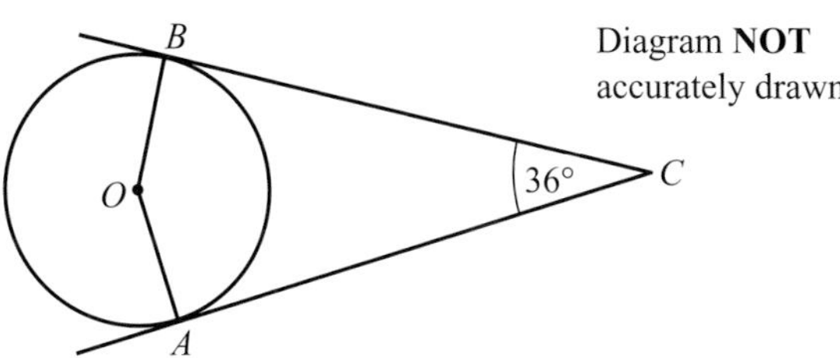

Diagram **NOT** accurately drawn

Find the size of angle *OBA*.
Give reasons for your answer. **(4 marks)**

June 2013, Q13, 5MB2H/01

8 **Reasoning** O is the centre of a circle.
A, B and C are points on the circumference of the circle.
DCT is a tangent to the circle at point C.
Angle BAC = 53° and angle ACB = 62°.
Work out the size of angle ACT.
Give reasons for each step in your working.

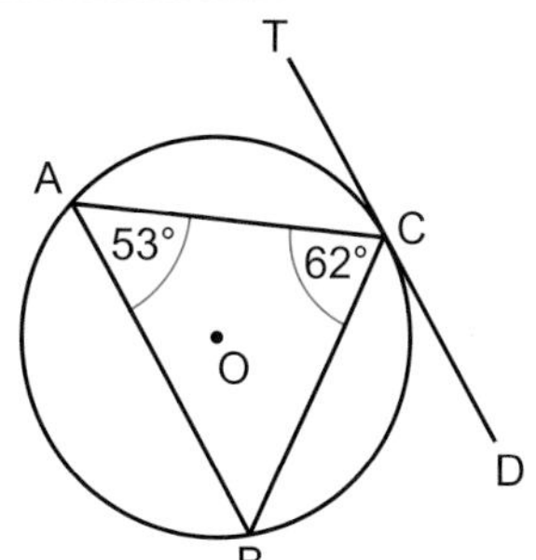

(3 marks)

9 **Communication / Problem-solving** Prove that triangle PQR is isosceles.

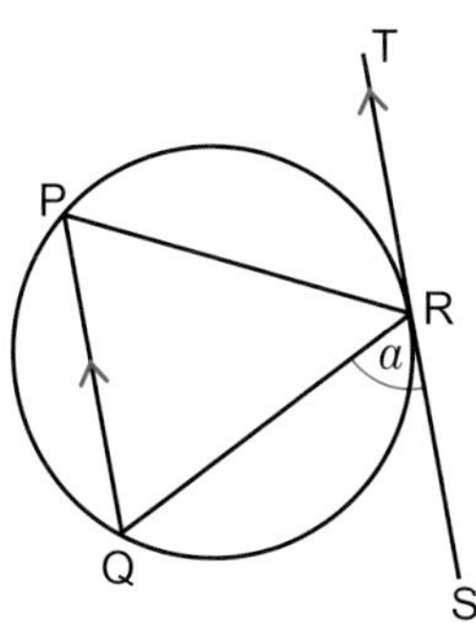

(3 marks)

10 **Problem-solving** A circle has equation $x^2 + y^2 = 25$.
P is the point (–4, 3) on its circumference.

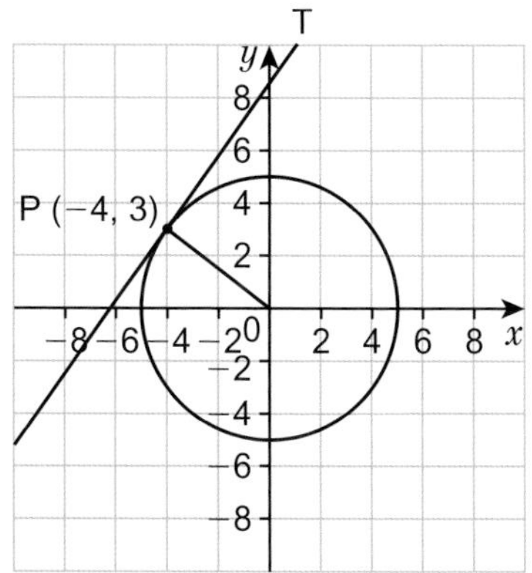

Find the equation of the tangent to the circle at P. *(5 marks)*

Sample student answers

Which student gave the best answer and why?

Exam-style question

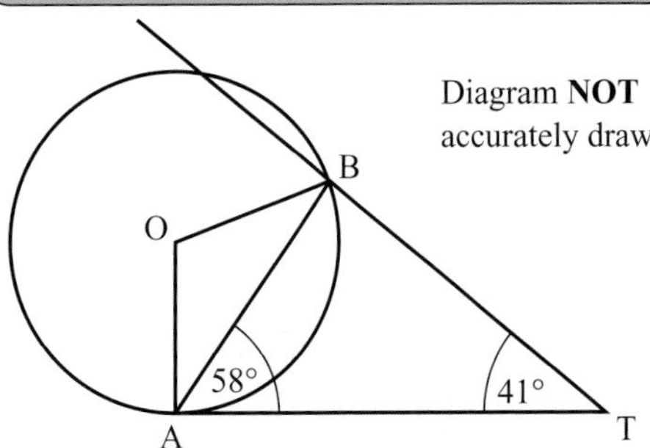

A and *B* are points on the circumference of a circle, centre *O*.
AT is a tangent to the circle.
Angle $TAB = 58°$.
Angle $BTA = 41°$.
Calculate the size of angle *OBT*.
You must give reasons at each stage of your working. **(5 marks)**

Nov 2013, Q14, 5MB2H/01

Student A

Angle TAO = 90° (angle between tangent and radius = 90°)
Angle OAB = angle OBA = 32° (base angles of isosceles triangle are equal)
Angle ABT = 81° (angles in a triangle add up to 180°)
Angle OBT = 32° + 81° = 113°

Student B

Angle OAB = 32° because 90° − 58° = 32°
Angle ABT = 81° because 180° − 58° − 41° = 81°
Angle OBT = 113° because 81° + 32° = 113°

17 MORE ALGEBRA

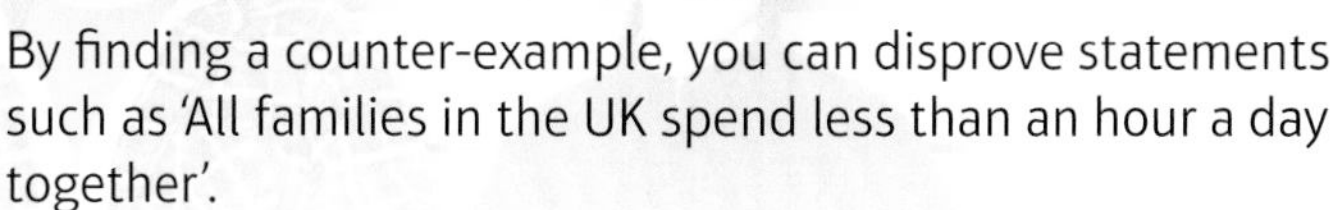
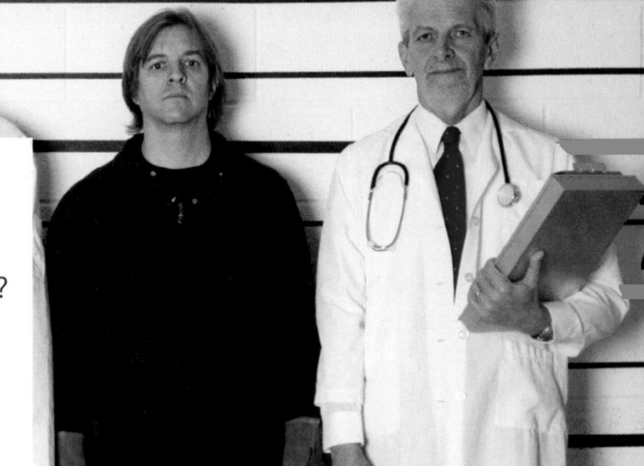

By finding a counter-example, you can disprove statements such as 'All families in the UK spend less than an hour a day together'.
Which person provides a counter-example to each statement?
'All people wear a hat.'
'None of the people wear glasses.'
Give a counter-example to show that this statement is false.
'All square numbers are even.'

17 Prior knowledge check

Numerical fluency

1 Find the LCM of 21 and 28.

2 Work out
a $\frac{7}{11} - \frac{3}{5}$ b $\frac{4}{9} + \frac{3}{4}$ c $\frac{5}{8} \times \frac{6}{10}$ d $\frac{5}{6} \div \frac{9}{10}$

3 Simplify these surds
a $\sqrt{50}$ b $\sqrt{80}$

Algebraic fluency

4 Expand and simplify.
a $7(2 - 5x)$ b $(x + 4)(2x - 3)$
c $(2x - 1)^2$

5 Solve these equations.
a $\frac{(n - 4)}{3} = 12$ b $7p - 3 = 3p + 17$
c $4(d + 5) = 7(d - 1)$

6 Make x the subject of each formula.
a $y = 4x + 7$ b $W = h + 3hx$
c $y = 4(x + 1)$ d $P = \frac{(6x + 1)}{3}$

7 Simplify
a $y^3 \times y^5$ b $4y^2 \times 7y^4$
c $y^8 \div y$ d $10y^7 \div 25y^2$

8 Factorise
a $x^2 + 6x + 5$ b $x^2 - 7x - 30$
c $x^2 - 5x + 6$ d $x^2 - 36$

9 Solve these equations by factorising.
a $x^2 + 11x + 30 = 0$
b $x^2 - 12x + 11 = 0$
c $2x^2 + 9x + 7 = 0$

10 Solve $x^2 + 5x + 2 = 0$ by using the quadratic formula.
Leave your answer in surd form.

11 Solve $x^2 + 8x + 10 = 0$ by completing the square.
Leave your answer in surd form.

* Challenge

12 a Write down any five consecutive integers.
b Work out their sum.
c Repeat parts **a** and **b** for four more sets of consecutive integers.
What do you notice?
d Predict the missing number in this sentence.
The sum of five consecutive numbers is a multiple of ____.
e Use algebra to show why this happens.

Q12e hint Let the numbers be $n, n + 1, n + 2, \ldots$

17.1 Rearranging formulae

Objectives

- Change the subject of a formula where the power of the subject appears.
- Change the subject of a formula where the subject appears twice.

Why learn this?

Physicists rearrange complex formulae in order to find important measures.

Fluency

$y = 5x - 2$. Find the value of x when

- $y = 3$
- $y = -7$
- $y = 0$

Warm up

1 In each formula change the subject to the letter given in brackets.

a $v = u + at$ (a) b $C = 2\pi r$ (r) c $A = \frac{1}{2}bh$ (h)

d $A = \pi r^2$ (r) e $x = \sqrt{t}$ (t) f $r = \sqrt{3s}$ (s)

2 Factorise

a $xy + 2y$ b $pq - q$ c $ak - 4k$

Questions in this unit are targeted at the steps indicated.

3 Make v the subject of the formula
$E = \frac{1}{2}mv^2$

Q3 hint First multiply both sides by 2.

4 Make x the subject of the formula
$H = \sqrt{x - y}$

Q4 hint First square both sides.

Example 1

Make x the subject of the formula $P = d\sqrt{\frac{x}{y}}$

$\frac{P}{d} = \sqrt{\frac{x}{y}}$ — Divide both sides by d.

$\frac{P^2}{d^2} = \frac{x}{y}$ — Square both sides.

$\frac{yP^2}{d^2} = x$ or $x = \frac{yP^2}{d^2}$

5 Make x the subject of each formula.

a $T = 2p\sqrt{\frac{x}{k}}$ b $y = 4\sqrt{\frac{1}{x}}$

c $P = \sqrt{\frac{xy}{z}}$ d $L = 3(1 + x)^2$

Q5d hint First divide both sides by 3. Then square root both sides.

6 In each formula change the subject to the letter given in brackets.

a $V = \frac{4}{3}pr^3$ (r) b $V = 4x^3$ (x)

c $y = \sqrt[3]{5x}$ (x) d $z = \sqrt[3]{\frac{x}{y}}$ (y)

Q6a hint First make r^3 the subject. Finally take the cube root to give r as the subject.

Q6c hint First cube both sides.

ActiveLearn Homework, practice and support: Higher 17.1

Key point 1

When the letter to be made the subject appears twice in the formula you will need to factorise.

Example 2

Make w the subject of the formula $A = wh + lh + lw$

$A - lh = wh + lw$ — w appears twice in this formula. Subtract lh from both sides to get the terms in w together on one side of the equals sign.

$A - lh = w(h + l)$ — Factorise the right-hand side, so w appears only once.

$w = \frac{A - lh}{h + l}$ — Divide both sides by $(h + l)$.

7 Make y the subject of the formula $h = 3y + xy$

8 Make d the subject of the formula $H = ad - ac - bd$

9 **Reasoning** $5xy + 2 = w + 3xy$

a Make y the subject.

b Make x the subject.

Discussion What do you notice about your answers?

10 **Reasoning** $H = xy + 2x + 7$

Zoe rearranges the formula to make x the subject.

Her answer is $x = \frac{(H - 7 - xy)}{2}$

a Explain why this cannot be the correct answer.

b What mistake has Zoe made?

c Work out the correct answer.

11 Make x the subject of the formula $V = \frac{1 + 7x}{x}$

Q11 hint First multiply both sides by x.

12 **Exam-style question**

Make k the subject of the formula $t = \frac{k}{k - 2}$ **(4 marks)**

June 2011, Q23, 1380/3H

Exam hint
First multiply both sides by $(k - 2)$.
Then expand the bracket on the left-hand side.

17.2 Algebraic fractions

Objectives

- Add and subtract algebraic fractions.
- Multiply and divide algebraic fractions.
- Change the subject of a formula involving fractions where all the variables are in the denominators.

Why learn this?

Bridge designers use algebraic fractions when making sure their designs are structurally safe.

Fluency

Simplify
- $\frac{7}{28}$
- $\frac{4x}{2}$
- $\frac{5x^2}{35}$
- $\frac{x^2}{x}$

1 Work out

a $\frac{5}{12}+\frac{7}{18}$ b $\frac{7}{11}+\frac{2}{9}$ c $\frac{9}{12}-\frac{4}{5}$

2 Work out

a $\frac{5}{7}\times\frac{2}{11}$ b $\frac{6}{7}\div\frac{5}{9}$ c $\frac{25}{32}\div\frac{35}{14}$

3 Write as a single fraction in its simplest form. The first one has been started for you.

a $\frac{x}{2}\times\frac{x}{3}=\frac{x\times x}{2\times 3}=$

b $\frac{2x}{5}\times\frac{3y}{4}$ c $\frac{4}{9y}\times\frac{3}{5y}$

Q3b hint First cancel any common factors.

4 Write as a single fraction in its simplest form. The first one has been started for you.

a $\frac{4x^2}{y^3}\times\frac{3y}{8x}=\frac{{}^1\cancel{4}x^{\cancel{2}}\times 3y}{y^3\times{}^2\cancel{8}\cancel{x}}=$

b $\frac{14x^3}{10y^2}\times\frac{25y^6}{21x^5}$ c $\frac{12y^2}{7x}\times\frac{14x^5}{16y^4}$

5 Write as a single fraction in its simplest form.

a $\frac{4}{x}\div\frac{3}{x}$ b $x^3y\div\frac{1}{xy}$

c $\frac{2y^3}{3x^5}\div\frac{8y^7}{15x^3}$ d $\frac{y}{2}\div\frac{y-7}{10}$

Q5a hint Dividing by $\frac{3}{x}$ is equivalent to multiplying by $\frac{x}{3}$.

Q5b hint Write x^3y as $\frac{x^3y}{1}$.

Example 3

Simplify $\frac{x}{5}+\frac{x}{3}$

LCM of 5 and 3 is 15 — Find the LCM of the denominators.

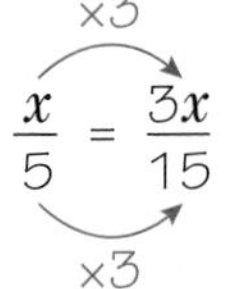

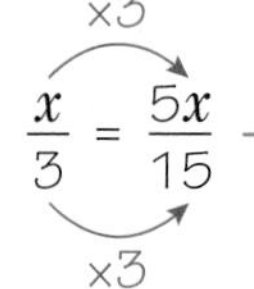

Write both fractions with the same denominator.

$\frac{3x}{15}+\frac{5x}{15}=\frac{8x}{15}$ — Add the fractions.

6 Write as a single fraction in its simplest form.

a $\frac{3x}{10}+\frac{x}{2}$ b $\frac{4x}{3}-\frac{x}{4}$ c $\frac{6x}{7}-\frac{x}{2}$

7 Write down the LCM of

a $2x$ and $5x$ b $3x$ and $6x$

c $4x$ and $7x$ d $4x$ and $3x$

Q7a hint Multiples of $2x$: $2x$, $4x$, …
Multiples of $5x$: $5x$, $10x$, …

8 a Write $\frac{1}{4x}$ and $\frac{1}{3x}$ as equivalent fractions with denominator the LCM of $4x$ and $3x$.

b Simplify $\frac{1}{4x}+\frac{1}{3x}$

9 Write as a single fraction in its simplest form.

a $\frac{1}{9x}+\frac{1}{2x}$ b $\frac{1}{4x}-\frac{1}{5x}$ c $\frac{1}{6x}+\frac{5}{9x}$

10 a Copy and complete.

$\frac{x-4}{2}=\frac{\square(x-4)}{5\times 2}=\frac{\square x-\square}{10}$

b Copy and complete.

$\frac{x+7}{5}=\frac{\square(x+7)}{2\times 5}=\frac{\square x+\square}{10}$

c Use your answers to parts **a** and **b** to work out $\frac{x-4}{2}+\frac{x+7}{5}$

11 Write as a single fraction in its simplest form.

a $\frac{x+2}{2} + \frac{x+1}{3}$ b $\frac{x+5}{2} - \frac{x-3}{7}$ c $\frac{x+7}{4} - \frac{2x-1}{9}$

12 **Exam-style question**

Write as a single fraction in its simplest form

$\frac{x+6}{2} + \frac{2x-3}{5}$ **(3 marks)**

Q12 strategy hint Start by rewriting each fraction so that the denominator of each is the same.

13 Make a the subject of the formula $\frac{1}{a} + \frac{1}{b} = 1$.

The working has been started for you.

$\frac{1}{a} + \frac{1}{b} = 1$

$\frac{1}{a} = 1 - \frac{1}{b}$

$\frac{1}{a} = \frac{\square}{\square} - \frac{1}{b} =$

Q13 hint Write the right-hand side as a single fraction using the common denominator of 1 and b: b. Then find the reciprocal to find a.

14 **STEM** Scientists use the lens formula to solve problems involving light.

The lens formula is $\frac{1}{f} = \frac{1}{u} + \frac{1}{v}$, where f = focal length, u = object distance and v = image distance.

Make u the subject of the formula.

17.3 Simplifying algebraic fractions

Objective

- Simplify algebraic fractions.

Why learn this?

Aerospace engineers use and simplify algebraic fractions when designing planes.

Fluency

Factorise

- $6x + 18$
- $x^2 + 3x$
- $x^3 + 4x^2$
- $3x^3 - 15x$

1 Simplify

a $\frac{x}{x^3}$ b $\frac{5x^3}{x}$ c $\frac{10x^4}{2x^2}$

2 Fully factorise

a $x^2 - 9x + 18$ b $x^2 - 81$ c $5x^2 + 21x + 4$

3 Simplify

a $\frac{x}{xy}$ b $\frac{x+6}{3(x+6)}$

c $\frac{x-7}{(x-7)^2}$ d $\frac{(x+2)(x-1)}{(x-1)(x-5)}$

e $\frac{(x+9)(x-3)}{x(x+9)}$ f $\frac{x^2(x-1)}{x(x-1)^2}$

Q3 hint You can simplify an algebraic fraction in the same way as simplifying a normal fraction. Cancel any common factors in the numerator and denominator.

Q3e hint You can only cancel whole brackets.

Warm up

Key point 2

You may need to factorise before simplifying an algebraic fraction:

(1) Factorise the numerator and denominator.

(2) Divide the numerator and denominator by any common factors.

4 a Factorise $x^2 - 6x$

b Use your answer to part **a** to simplify $\frac{x^2 - 6x}{x - 6}$

> **Q4b hint** Replace the numerator with your factorisation from part **a**. Cancel common factors.

5 Simplify fully

a $\frac{x^2 + 8x}{x}$ b $\frac{12x^2 + 15x}{4x + 5}$

c $\frac{10x - 25}{4x^2 - 10x}$

> **Q5c hint** Factorise the numerator and denominator.

6 **Reasoning** Simplify $\frac{x^2 + 2x}{x^2 + 2}$

Sally says, '$(x + 2)$ is a factor of the numerator and the denominator.'

a Is Sally correct? Explain.

b Can the fraction be simplified? Explain your answer.

7 Simplify fully

a $\frac{2(x + 3)}{x^2 + 8x + 15}$ b $\frac{x^2 - x - 6}{5(x + 2)}$

> **Q7a hint** Do not expand the numerator.

Example 4

Simplify fully $\frac{x^2 + 5x + 4}{x^2 - 3x - 28}$

$$\frac{x^2 + 5x + 4}{x^2 - 3x - 28} = \frac{(x + 1)(x + 4)}{(x - 7)(x + 4)}$$

Factorise the numerator and denominator.

$$= \frac{x + 1}{x - 7}$$

Divide the numerator and denominator by the common factor $(x + 4)$.

8 Simplify fully

a $\frac{x^2 + 8x + 15}{x^2 + 2x - 15}$ b $\frac{x^2 - 11x + 30}{x^2 + x - 42}$ c $\frac{x^2 - 25}{(x + 5)^2}$

> **Q8c hint** Factorise $(x^2 - 25)$ using the difference of two squares.

9 **Exam-style question**

Simplify fully

$\frac{x^2 + 14x + 49}{x^2 - 49}$ **(3 marks)**

> **Exam hint**
> First factorise the numerator and denominator.
> Use the fact that $a^2 - b^2 = (a + b)(a - b)$.

10 Simplify fully

a $\frac{2x^2 - x - 3}{3x^2 + x - 2}$ b $\frac{5x^2 + 14x - 3}{6x^2 + 23x + 15}$ c $\frac{25x^2 - 1}{25x^2 + 10x + 1}$

11 **Exam-style question**

Simplify fully

$\frac{x^2 + 3x - 4}{2x^2 - 5x + 3}$ **(3 marks)**

June 2012, Q23a, 1MA0/1H

> **Exam hint**
> 1 mark is awarded for correctly factorising the numerator; 1 mark for factorising the denominator; and 1 mark for the correct final answer.

12 a Copy and complete.

$(6 - x) = -(\square - \square)$

b Simplify

i $\frac{(6 - x)}{(x - 6)}$ ii $\frac{(36 - x^2)}{(x^2 - 3x - 18)}$

13 Simplify fully

a $\dfrac{16 - x^2}{x^2 - 4x}$ b $\dfrac{x^2 - 12x + 36}{2x^2 - 72}$ c $\dfrac{6x^2 - 10x}{6x^2 - 19x + 15}$

14 **Communication** Show that $\dfrac{(x^2 + x - 12)(x^2 + 2x - 3)(10x^2 + 12x)}{(9 - x^2)(5x^2 + 26x + 24)(7x - 7)} = -\dfrac{2x}{7}$

> **Q14 hint** Start with the numerator and then the denominator. Use factorising and simplifying to work towards $-\dfrac{2x}{7}$.

17.4 More algebraic fractions

Objectives

- Add and subtract more complex algebraic fractions.
- Multiply and divide more complex algebraic fractions.

Why learn this?

Opticians use algebraic fractions when working out a lens prescription.

Fluency

Simplify • $\dfrac{4x - 4}{x - 1}$ • $\dfrac{(x + 2)(x - 3)}{(x - 3)(x + 4)}$ • $\dfrac{(x + 7)^2}{(x - 3)(x + 7)}$

Warm up

1 Write as a single fraction.

a $\dfrac{3x^2}{y^2} \times \dfrac{5y}{4x}$ b $\dfrac{5y}{2} \div \dfrac{2y}{15}$ c $\dfrac{x}{4} \div \dfrac{x - 2}{12}$

2 Write as a single fraction in its simplest form.

a $\dfrac{2x}{3} + \dfrac{x}{5}$ b $\dfrac{1}{3x} - \dfrac{1}{8x}$ c $\dfrac{x - 1}{3} + \dfrac{x + 5}{4}$

3 Write as a single fraction in its simplest form.

a $(x + 3)^2 \times \dfrac{x - 4}{x + 3}$ b $\dfrac{x + 2}{x - 1} \times \dfrac{x - 1}{x + 5}$ c $\dfrac{x - 4}{6} \times \dfrac{2}{3x - 12}$

d $\dfrac{5}{x + 2} \div \dfrac{15}{8x + 16}$ e $\dfrac{2x + 6}{x + 7} \div \dfrac{x + 3}{x - 1}$ f $\dfrac{(x + 4)^2}{x - 2} \div \dfrac{(x + 4)}{x}$

> **Q3c hint** First factorise $3x - 12$.

Key point 3

You may need to factorise the numerator and/or denominator before you multiply or divide algebraic fractions.

4 a Factorise $x^2 - 9$

b Factorise $x^2 + 5x + 6$

c Write $\dfrac{x^2 - 9}{4} \times \dfrac{8}{x^2 + 5x + 6}$ as a single fraction in its simplest form.

5 Write as a single fraction in its simplest form.

a $\dfrac{x^2 - 7x + 10}{x^2 + 4x + 3} \times \dfrac{x^2 - 9}{x^2 - x - 20}$ b $\dfrac{14x + 21}{2x^2 + 7x + 6} \div \dfrac{x^2 - 10x + 21}{x^2 + 9x + 14}$

6 Write down the LCM of

a x and $x + 2$ b $x + 2$ and $x + 3$ c $x + 4$ and $x + 5$

d $x + 1$ and $x - 1$ e $2x - 3$ and $2x - 4$

ActiveLearn Homework, practice and support: Higher 17.4

Example 5

Write $\frac{7}{x+2} - \frac{3}{x+3}$ as a single fraction in its simplest form.

Common denominator $= (x+2)(x+3)$ — Find a common denominator.

$\frac{7(x+3)}{(x+2)(x+3)} - \frac{3(x+2)}{(x+2)(x+3)}$ — Convert each fraction to an equivalent fraction with the common denominator $(x+2)(x+3)$.

$= \frac{7(x+3) - 3(x+2)}{(x+2)(x+3)}$ — Subtract the fractions.

$= \frac{7x + 21 - 3x - 6}{(x+2)(x+3)} = \frac{4x+15}{(x+2)(x+3)}$ — Expand the brackets in the numerator, then simplify.

7 Simplify fully

a $\frac{1}{x+4} + \frac{1}{x+5}$

b $\frac{3}{x+1} + \frac{4}{x-1}$

c $\frac{7}{x-5} - \frac{1}{x+3}$

d $\frac{1}{2x-3} - \frac{1}{2x+4}$

8 **Exam-style question**

Write as a single fraction in its simplest form

$\frac{2}{x-4} - \frac{1}{x+3}$ **(3 marks)**

Nov 2011, Q23c, 1380/4H

Exam hint
Take care when multiplying out a bracket which has a negative sign in front of it.

9 a Factorise

i $3x + 9$

ii $4x + 12$

b Write down the LCM of $3x + 9$ and $4x + 12$.

c Write $\frac{1}{3x+9} + \frac{1}{4x+12}$ as a single fraction in its simplest form.

Q9b hint Look at the factorised form of each expression:
$a(x + y)$
$b(x + y)$
LCM $= ab(x + y)$

10 a Factorise $x^2 - 16$

b Write $\frac{1}{x+4} + \frac{1}{x^2-16}$ as a single fraction in its simplest form.

11 Write as a single fraction in its simplest form.

a $\frac{1}{3x^2+8x+5} - \frac{1}{3x+5}$

b $\frac{1}{x^2+7x+6} - \frac{1}{2x+12}$

c $\frac{1}{x^2+6x+8} + \frac{3}{x^2-3x-28}$

d $\frac{4}{25-x^2} - \frac{3}{5-x}$

Q11b hint Factorise $(x^2 + 7x + 6)$ and $(2x + 12)$.

12 Write $\frac{1}{5x} + \frac{1}{5(x-1)} + \frac{1}{10}$ as a single fraction in its simplest form.

Q12 hint Work out the lowest common denominator of $5x$, $5(x - 1)$ and 10.

13 **Communication** Show that

$\frac{1}{x^2+5x+6} + \frac{1}{5x+10} = \frac{x+8}{A(x+3)(x+2)}$ and find the value of A.

17.5 Surds

Objectives

- Simplify expressions involving surds.
- Expand expressions involving surds.
- Rationalise the denominator of a fraction.

Did you know?

Surds occur in nature. An example is the Golden Ratio $\frac{(1 + \sqrt{5})}{2}$, which is also used in architecture.

Fluency

Are these numbers rational or irrational?

-7 $\quad \frac{4}{9}$ $\quad \sqrt{6}$ $\quad 0.\dot{2}$ $\quad \sqrt{\frac{25}{49}}$

Warm up

1 Work out

a $\sqrt{5} \times \sqrt{5}$ b $7\sqrt{3} - 4\sqrt{3}$ c $3\sqrt{2} + 5\sqrt{2}$

2 Copy and complete.

a $\sqrt{6} = \sqrt{2} \times \sqrt{\square}$

b $\sqrt{\square} = \sqrt{5} \times \sqrt{6}$

c $\sqrt{\frac{\square}{\square}} = \frac{\sqrt{5}}{\sqrt{7}}$

Q2a hint Use $\sqrt{m} \times \sqrt{n} = \sqrt{mn}$

Q2c hint Use $\frac{\sqrt{m}}{\sqrt{n}} = \sqrt{\frac{m}{n}}$

3 Find the value of the integer k.

a $\sqrt{50} = \sqrt{\square} \times \sqrt{2} = k\sqrt{2}$ b $\sqrt{18} = k\sqrt{2}$ c $\sqrt{48} = k\sqrt{3}$

4 Rationalise the denominators. Simplify your answers if possible.

a $\frac{1}{\sqrt{10}}$ b $\frac{3}{\sqrt{15}}$ c $\frac{8}{\sqrt{32}}$

5 Simplify

a i $\sqrt{45}$ ii $\sqrt{20}$

b Use your answers to part **a** to simplify $3\sqrt{45} + 7\sqrt{20}$

6 Simplify

a $2\sqrt{75} - 3\sqrt{27}$

b $\sqrt{200} + 3\sqrt{32}$

c $5\sqrt{18} - \sqrt{128} + 4\sqrt{8}$

Q6a hint First simplify each surd. $\sqrt{75} = k\sqrt{3}$, $\sqrt{27} = l\sqrt{3}$

7 Factorise these expressions. The first one has been started for you.

a $\sqrt{12} + 2 = 2\sqrt{\square} + 2 = 2(\square + \square)$

b $9 + \sqrt{54}$ c $18 - \sqrt{45}$ d $\sqrt{75} - \sqrt{50}$

8 Expand and simplify

a $\sqrt{5}(4 + \sqrt{5})$ b $(\sqrt{7} + 1)(4 + \sqrt{7})$

c $(6 - \sqrt{2})(4 + \sqrt{2})$ d $(2 - \sqrt{2})^2$

e $(4 - \sqrt{10})^2$ f $(7 + \sqrt{3})^2$

Q8d hint $(2 - \sqrt{2})^2 = (2 - \sqrt{2})(2 - \sqrt{2})$ Your answer should be in the form $a - b\sqrt{2}$.

9 **Exam-style question**

Expand $(5 - \sqrt{5})^2$. Write your answer in the form $a + b\sqrt{c}$, where a, b and c are integers. **(2 marks)**

Exam hint
Make sure your answer is in the form $a + b\sqrt{c}$.

10 a Work out the area of each shape.
Write your answers in the form $a + b\sqrt{2}$.

i

$11 + \sqrt{2}$

$5 - \sqrt{2}$

ii

$2 + \sqrt{8}$

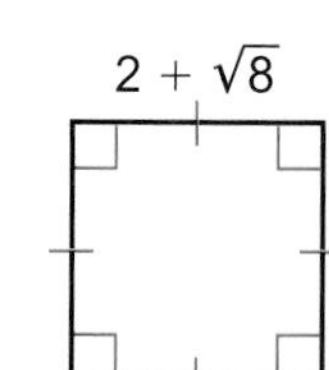

b **Reasoning** Would the perimeter of each shape be rational or irrational? Explain.

11 Rationalise the denominators. The first one has been started for you.

a $\dfrac{3 \times \sqrt{2}}{\sqrt{2}} \times \dfrac{\sqrt{2}}{\sqrt{2}} = \dfrac{3 \times \sqrt{2} + \sqrt{2} \times \sqrt{2}}{\sqrt{2} \times \sqrt{2}} =$

b $\dfrac{6 - \sqrt{3}}{\sqrt{3}}$ c $\dfrac{19 - \sqrt{7}}{\sqrt{7}}$ d $\dfrac{5 + \sqrt{5}}{\sqrt{5}}$

12 **Exam-style question**

Given that
$\dfrac{8 - \sqrt{18}}{\sqrt{2}} = a + b\sqrt{2}$, where a and b are integers,
find the value of a and the value of b. **(3 marks)**

June 2011, Q22b, 1380/3H

Exam hint
Make sure you multiply both parts of the expression in the numerator by $\sqrt{2}$.

13 **Reasoning**

a Expand and simplify $(3 + \sqrt{5})(3 - \sqrt{5})$

b Is your answer rational or irrational?

c How can you tell if your answer will be rational or irrational?

d Which of these will have rational answers when expanded?

i $(7 + \sqrt{2})(2 - \sqrt{2})$ ii $(7 + \sqrt{2})(7 + \sqrt{2})$ iii $(7 + \sqrt{2})(7 - \sqrt{2})$

Check by expanding the brackets.

e Rationalise the denominator of $\dfrac{1}{(7 + \sqrt{2})}$

Q13e hint Multiply the numerator and denominator by $(7 - \sqrt{2})$.

Key point 4

To rationalise the fraction $\dfrac{1}{a\sqrt{b}}$, multiply by $\dfrac{\sqrt{b}}{\sqrt{b}}$

To rationalise the fraction $\dfrac{1}{a \mp \sqrt{b}}$, multiply by $\dfrac{a \pm \sqrt{b}}{a \pm \sqrt{b}}$

14 Rationalise the denominators. Give your answers in the form $a \pm \sqrt{b}$ or $a \pm b\sqrt{c}$ where a, b and c are rational.

a 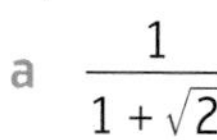 $\dfrac{1}{1 + \sqrt{2}}$ b $\dfrac{1}{5 - \sqrt{3}}$ c $\dfrac{7}{4 - \sqrt{5}}$

d $\dfrac{4}{1 + \sqrt{6}}$ e $\dfrac{\sqrt{5}}{1 - \sqrt{5}}$ f $\dfrac{6 + \sqrt{2}}{8 - \sqrt{2}}$

Q14a hint Multiply the numerator and denominator by $(1 - \sqrt{2})$.

15 a Solve $x^2 - 6x + 1 = 0$ by using the quadratic formula.

b Solve the equation $x^2 + 10x + 13 = 0$ by completing the square.

c Solve the equation $x^2 - 16x + 8 = 0$.

Write all your answers in surd form.

Q15a hint Simplify your surd answer.

17.6 Solving algebraic fraction equations

Objective

- Solve equations that involve algebraic fractions.

Why learn this?

Pharmacists use algebraic fraction equations to calculate the correct dosage when issuing medication.

Fluency

Find the LCM of
- x and 4
- $4x$ and x
- $x + 3$ and $x + 2$

Warm up

1 Simplify

a $(x + 3)(x - 2) \times \frac{2}{(x - 2)}$ b $(x + 6)(x + 4) \times \frac{4}{(x + 6)}$

2 Write as a single fraction in its simplest form.

a $\frac{6}{x} - \frac{1}{x}$ b $\frac{7}{2x} - \frac{3}{2x}$ c $\frac{8}{x - 6} + \frac{2}{x - 6}$

3 Solve by factorising.

a $x^2 + 6x + 8 = 0$ b $2x^2 - 13x + 11 = 0$ c $5x^2 - 25x + 20 = 0$

4 Solve $3x^2 + 8x - 17 = 0$ by using the quadratic formula.
Give your answers correct to 2 decimal places.

5 Solve these equations. Give your answer as a simplified fraction.

a $\frac{3}{x} + \frac{2}{x} = 4$ b $\frac{6}{x - 1} - \frac{2}{x - 1} = 7$ c $8 = \frac{3}{x + 5} - \frac{7}{x + 5}$

Q5a hint First simplify the LHS of the equation.

6 Solve these quadratic equations.

a $\frac{4}{x} = \frac{3x - 7}{5}$ b $\frac{2x + 1}{3} = \frac{2}{x}$

c $\frac{5x - 3}{2} = \frac{7}{x}$ d $\frac{10}{x} = \frac{2x + 3}{2}$

Q6a hint First multiply both sides by the LCM ($5x$) and simplify. Then multiply out the bracket and solve by factorising.

Example 6

Solve $\frac{3}{2x - 1} + \frac{4}{x + 2} = 2$

$\frac{3(x + 2)}{(2x - 1)(x + 2)} + \frac{4(2x - 1)}{(x + 2)(2x - 1)} = 2$ — Rewrite the LHS using the common denominator $(2x - 1)(x + 2)$.

$\frac{3(x + 2) + 4(2x - 1)}{(2x - 1)(x + 2)} = 2$ — Add the fractions.

$\frac{3x + 6 + 8x - 4}{(2x - 1)(x + 2)} = \frac{11x + 2}{(2x - 1)(x + 2)} = 2$ — Expand the brackets in the numerator and simplify.

$11x + 2 = 2(2x - 1)(x + 2)$ — Multiply both sides by $(2x - 1)(x + 2)$.

$11x + 2 = 4x^2 + 6x - 4$ — Multiply out the brackets and simplify the right-hand side.

$4x^2 - 5x - 6 = 0$ — Rearrange into the form $ax^2 + bx + c = 0$.

$(4x + 3)(x - 2) = 0$, so either $4x + 3 = 0$ or $x - 2 = 0$ — Solve by factorising.

The solutions are $x = -\frac{3}{4}$ and $x = 2$.

7 Copy and complete Sioned's working to solve $\frac{3}{x+1} + \frac{2}{2x-3} = 1$

$$\frac{3(x+1)(2x-3)}{x+1} + \frac{2(x+1)(2x-3)}{2x-3} = 1(x+1)(2x-3)$$

Multiply all the terms by the common denominator $(x + 1)(2x - 3)$ and simplify.

$$3(2x-3) + 2(x+1) = (x+1)(2x-3)$$

Simplify both sides and expand the brackets.

$$6x - 9 + 2x + 2 =$$

Rearrange into the form $ax^2 + bx + c = 0$. Solve by factorisation.

Reflect Sioned has used a different method to the example.
Which method do you prefer? Why?

8 **Communication**

a Show that the equation $\frac{x}{2x-3} + \frac{4}{x+1} = 1$ can be rearranged to give $x^2 - 10x + 9 = 0$

b Solve $x^2 - 10x + 9 = 0$

Discussion How can you check your solution is correct?

9 Solve these quadratic equations.

a $\frac{1}{x-1} + \frac{1}{5-x} = 1$

b $\frac{5}{x+2} + \frac{3}{x-2} = 1$

c $\frac{4}{x} - \frac{3}{2x-1} = 1$

d $\frac{3}{x+1} - \frac{2}{x+3} = 1$

e $\frac{4}{x} - \frac{3}{2x+3} = 1$

10 Solve these quadratic equations.
Give your answers correct to 2 decimal places.

a $\frac{4x-1}{2-x} = \frac{x}{3}$

b $\frac{1}{x-1} + \frac{1}{x+2} = 5$

c $\frac{4}{x} - \frac{2}{1-x} = 1$

d $\frac{2}{x-5} + \frac{1}{x+1} = 3$

Q10 communication hint 'Give your answers correct to 2 decimal places' shows that you will need to use the quadratic formula.

11 **Exam-style question**

Find the exact solutions of $x + \frac{5}{x} = 12$ **(3 marks)**

Exam hint
'Find the exact solutions' means that you should not use a calculator.
You should give your answers using simplified surds.

17.7 Functions

Objectives

- Use function notation.
- Find composite functions.
- Find inverse functions.

Why learn this?

Function notation is an easy way to distinguish different equations; each can be labelled using different letters.

Fluency

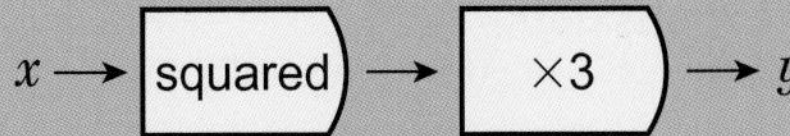

What is the output when

- $x = 3$
- $x = -2$
- $x = 5$?

Warm up

1 Write each expression using function machines.

a $2x + 5$ b $\frac{x}{2} - 6$ c $3(x + 1)$

2 Find the value of x when

a $5x - 3 = 4$ b $7x - 8 = 8$

3 a $H = 4x$ and $x = 3t$. Write H in terms of t.

b $P = \frac{x}{3}$ and $x = \frac{1}{2}y$. Write P in terms of y.

c $y = x^2$ and $x = h + 3$. Write y in terms of h.

Q3a hint Substitute $x = 3t$ into $H = 4x$.

Key point 5

A function is a rule for working out values of y for given values of x.
For example, $y = 3x$ and $y = x^2$ are functions. The notation $f(x)$ is read as 'f of x'. f is the function.
$f(x) = 3x$ means the function of x is $3x$.

4 $f(x) = \frac{10}{x}$. Work out

a $f(5)$ b $f(-2)$
c $f(\frac{1}{2})$ d $f(-20)$

Q4a hint Substitute $x = 5$ into $\frac{10}{x}$.

5 **Reasoning** $h(x) = 5x^2$. Alice says that $h(2) = 100$.

a Explain what Alice did wrong. b Work out $h(2)$.

6 $g(x) = 2x^3$. Work out

a $g(3)$ b $g(-1)$
c $g(\frac{1}{2})$ d $g(-5)$

Q6 hint Use the priority of operations.

7 $f(x) = x + x^3$, $g(x) = 3x^2$. Work out

a $f(1) + g(1)$ b $f(4) - g(2)$ c $f(2) \times g(4)$
d $\frac{g(5)}{f(3)}$ e $2g(10)$ f $3f(-1) - g(3)$

Q7e hint First work out g(10) and then multiply the answer by 2.

8 $g(x) = 5x - 3$. Work out the value of a when

a $g(a) = 12$ b $g(a) = 0$ c $g(a) = -7$

Q8a hint $g(a) = 5a - 3 = 12$
Solve for a.

9 $f(x) = x^2 - 8$. Work out the values of a when

a $f(a) = 17$ b $f(a) = -4$
c $f(a) = 0$ d $f(a) = 12$

Q9c hint Write your answer as a surd in its simplest form.

10 $f(x) = x(x + 3)$, $g(x) = (x - 1)(x + 5)$. Work out the values of a when

a $f(a) = 0$ b $g(a) = 0$

c $f(a) = -2$ d $g(a) = -8$

Q10a hint $f(a) = a(a + 3) = 0$. Solve for a.

11 $f(x) = 5x - 4$. Write out in full

a $f(x) + 5$ b $f(x) - 9$

c $2f(x)$ d $7f(x)$

e $f(2x)$ f $f(4x)$

Q11a hint $f(x) + 5 = 5x - 4 + 5 =$ ______

Q11c hint $2f(x) = 2(5x - 4) =$ ______

Q11e hint Replace x by $2x$.

12 $h(x) = 3x^2 - 4$. Write out in full

a $h(x) + 7$ b $2h(x)$ c $h(2x)$ d $h(-x)$

Discussion What do you notice about your answer to part **d**? Explain why this happens.

Key point 6

fg is a composite function. To work out $fg(x)$, first work out $g(x)$ and then substitute your answer into $f(x)$.

13 **Reasoning** $f(x) = 6 - 2x$, $g(x) = x^2 + 7$. Work out

a $gf(2)$ b $gf(7)$

c $fg(4)$ d $fg(5)$

Q13a hint First work out f(2) and then substitute your answer into $g(x)$.

14 **Reasoning** $f(x) = 4x - 3$, $g(x) = 10 - x$, $h(x) = x^2 + 7$. Work out

a $gf(x)$ b $fg(x)$

c $fh(x)$ d $hf(x)$

e $gh(x)$ f $hg(x)$

Q14a hint $gf(x)$ means substitute $f(x)$ for x in $g(x)$.
$gf(x) = g(4x - 3) = 10 - (4x - 3) =$ ______

Key point 7

The inverse function reverses the effect of the original function.

Example 7

Find the inverse function of $x \to 5x - 1$

Communication hint $x \to 5x - 1$ is another way of showing $f(x) = 5x - 1$

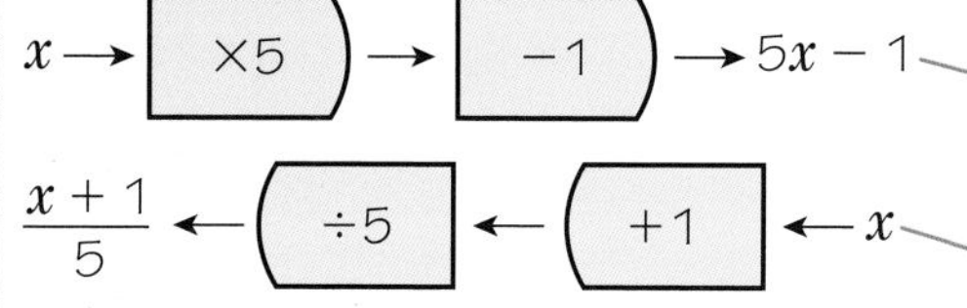

Write the function as a function machine.

Reverse the function machine to find the inverse function. Start with x as the input.

The inverse function of $x \to 5x - 1$ is $x \to \frac{x + 1}{5}$

15 Find the inverse of each function.

a $x \to 4x + 9$

b $x \to \frac{x}{3} - 4$

c $x \to 2(x + 6)$

d $x \to 7(x - 4) - 1$

Q15a hint You can check your answer by substituting e.g. $x = 2$ into the original function and then the answer into the inverse.

Q15d hint Simplify the function first. $x \to 7(x - 4) - 1$ is the same as $x \to 7x - 29$

Key point 8

$f^{-1}(x)$ is the inverse of $f(x)$.

16 **Reasoning** $f(x) = 4(x - 1)$, $g(x) = 4(x + 1)$

a Find $f^{-1}(x)$. b Find $g^{-1}(x)$. c Work out $f^{-1}(x) + g^{-1}(x)$.

d If $f^{-1}(a) + g^{-1}(a) = 1$ work out the value of a.

17.8 Proof

Objective

- Prove a result using algebra.

Did you know?

In the 1990s, Andrew Wiles spent over seven years trying to prove Fermat's Last Theorem. He received a knighthood for his successful proof.

Fluency

What type of number is **a** $2n$ **b** $2n + 1$ for any n?

Warm up

1 Which sequences contain
 i only even numbers ii only odd numbers iii even and odd numbers?
 a $n + 2$ b $2n$ c $5n$ d $2n - 1$ e n^2

2 Expand and simplify.
 a $x(x - 1)$ b $(x + 3)^2$ c $2x(2x + 1)$

3 Are these equations or identities?
 a $2(n + 3) = 2n + 6$ b $5n - 7 = 8$ c $\frac{1}{2}(4n + 10) = 2n + 5$ d $2(3n - 5) = 4$

Key point 9

To show a statement is an identity, expand and simplify the expressions on one or both sides of the equals sign, until the two expressions are the same.

Example 8

Show that $(x + 4)^2 - 7 \equiv x^2 + 8x + 9$

LHS $= (x + 4)^2 - 7 \equiv (x + 4)(x + 4) - 7 = x^2 + 8x + 16 - 7$
$= x^2 + 8x + 9$

Expand the brackets on the left-hand side (LHS).

RHS $= x^2 + 8x + 9$

So LHS = RHS and $(x + 4)^2 - 7 \equiv x^2 + 8x + 9$

Aim to show that LHS = RHS.

4 **Communication** Show that
 a $(x - 3)^2 + 6x \equiv x^2 + 9$
 b $x^2 + 8x + 49 \equiv (x + 7)^2 - 6x$
 c $(x - 5)^2 - 4 \equiv (x - 3)(x - 7)$
 d $16 - (x + 2)^2 \equiv (6 + x)(2 - x)$
 Reflect For part **c** can you think of a different method than the one given in the hint?

Q4b hint Start with the RHS.

Q4c hint First expand and simplify the LHS. Then factorise.

5 **Communication / Reasoning** a Show that $(x - 1)(x + 1) \equiv x^2 - 1$
 b Use your rule to work out
 i 99×101 ii 199×201

6 **Reasoning** The blue card is a rectangle of length $x + 5$ and width $x + 2$.
 a Write an expression for the area of the blue card.
 A rectangle of length $x + 1$ and width x is cut out and removed.
 b Write an expression for the area of the rectangle cut out.
 c Show that the area of the remaining card is $6x + 10$.

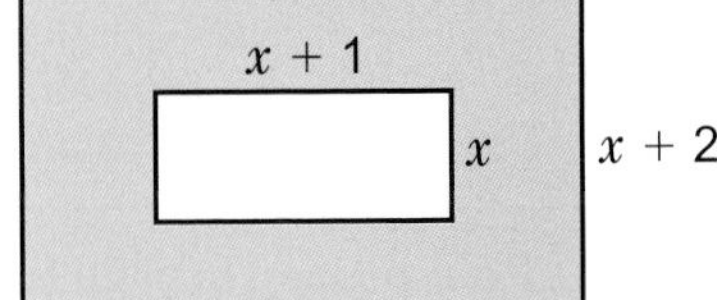

Q6c hint Subtract your expression from part **b** from your expression from part **a**.

7 **Exam-style question**

The diagram shows a large rectangle of length $(3x + 4)$ cm and width x cm.
A smaller rectangle of length x cm and width 5 cm is cut out and removed.
The area of the shape that is left is 70 cm^2.
Show that $3x^2 - x - 70 = 0$. **(3 marks)**

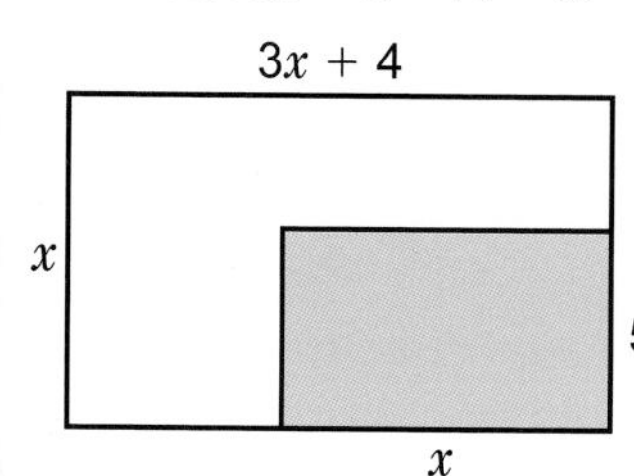

Exam hint
'Show that…' means you need to write down every stage of your working.

8 Give a counter-example to prove that these statements are *not* true.

a All prime numbers are odd.
b The cube of a number is always greater than its square.
c The difference between two numbers is always less than their sum.
d The difference between two square numbers is always odd.

Q8a hint List some prime numbers.

Q8c hint Try some negative numbers.

Key point 10

A **proof** is a logical argument for a mathematical statement. To prove a statement is true, you must show that it will be true in *all* cases.
To prove a statement is not true you can find a **counter-example** – an example that does not fit the statement.

9 a **Communication / Reasoning** Prove that the sum of any odd number and any even number is always odd.
b **Reasoning** Explain why any odd number can be written as $2n + 1$ or $2n - 1$.

Q9a hint Let $2n$ be any even number. Let $2n + 1$ be any odd number.

10 **Communication / Reasoning**

a The nth even number is $2n$. Explain why the next even number is $2n + 2$.
b Prove that the product of two consecutive even numbers is a multiple of 4.

11 **Communication / Reasoning** Prove that the product of any two odd numbers is odd.

12 **Communication / Reasoning** Given that $2(x - a) = x + 5$, where a is an integer, show that x must be an odd number.

13 **Communication / Reasoning**

a Work out
i $\frac{1}{5} - \frac{1}{6}$ ii $\frac{1}{3} - \frac{1}{4}$ iii $\frac{1}{7} - \frac{1}{8}$

b Use your answers to part **a** to write down the answer to $\frac{1}{9} - \frac{1}{10}$.

c **Reasoning** Explain how you can quickly calculate $\frac{1}{99} - \frac{1}{100}$.

d i Simplify $\frac{1}{x} - \frac{1}{x+1}$
ii **Reasoning** Explain how this proves your answer from part **c**.

14 Communication / Reasoning

Show that $\frac{1}{x^2 - x} - \frac{1}{x^2 + 3x} = \frac{A}{x(x - 1)(x + 3)}$ and find the value of A.

15 Communication / Reasoning Prove that $n^2 + n$ is a multiple of 2 for all values of n.

Q15 hint Factorise first.

16 Communication

a Write an expression for the product of three consecutive integers, $n - 1$, n and $n + 1$.

b Hence show that $n^3 - n$ is a multiple of 2.

Q16b hint Consider when n is even and when n is odd.

17 Exam-style question

Prove algebraically that the difference between the squares of any two consecutive integers is equal to the sum of these two integers. **(4 marks)**

February 2013, Q21, 1MA0/1H

Q17 strategy hint Start by using algebra to write down expressions for the squares of two numbers that are consecutive.

17 Problem-solving: Surface gravity

Objectives

- Be able to rearrange equations that involve powers.
- Be able to use standard form in calculations.

Big objects, like planets, create gravity that pull objects towards them. This is strongest on the surface of the planet.
To find the surface gravity of a planet we can use the formula:

$g = \frac{GM}{r^2}$ where M = mass of the planet (kg),
r = radius of planet (m),
G = gravitational constant

The formula uses a gravitational constant. This is a number that was calculated many years after the formula was constructed. $G = 6.67 \times 10^{-11}$

1 Earth's surface gravity is approximately 9.81 m/s^2. The mass of Earth is 5.97×10^{24} kg. Find the radius of the Earth in kilometres (to 3 significant figures).

Q1 hint Rearrange the equation to make r the subject.

2 To find the volume of a planet we can assume that it is spherical and use the formula $V = \frac{4}{3}\pi r^3$. The volume of Mars is 1.63×10^{20} m^3.

a Find the radius of Mars (to 3 significant figures).

Q2a hint Rearrange the equation for the volume of the sphere to make r the subject.

b Given that the mass of Mars is 6.42×10^{23} kg, calculate the gravity on the surface of Mars.

3 The density of Saturn is 687 kg/m^3. Its mass is 5.68×10^{26} kg. Calculate the gravity at the surface.

Q3 hint Remember that Density = mass ÷ volume

4 Imagine that the Earth starts growing. Assuming that the Earth's density remained constant, what radius would it need to grow to in order to have the same surface gravity as on Saturn?

Q4 hint Start by combining the equation for the volume of a sphere, the equation for gravity and the equation for density. Remember that as the radius increases so will the mass. Your equation will only involve values that will remain constant (G, D, g and r).

17 Check up

Log how you did on your Student Progression Chart.

Surds

1 Simplify

a $\sqrt{200} + 2\sqrt{50}$ b $(4 - \sqrt{7})^2$

2 Rationalise the denominators.

a $\frac{3 - \sqrt{2}}{\sqrt{5}}$ b $\frac{3}{2 - \sqrt{3}}$

Formulae and functions

3 Find $f^{-1}(x)$ for each function.

a $f(x) = \frac{x - 5}{2}$ b $f(x) = 3x + 4$

4 $f(x) = 9 - 2x$, $g(x) = x^2 + 4x$. Work out

a $f(2) + g(3)$ b $f(2) - g(3)$ c $f(3) \times g(4)$ d $\frac{g(6)}{f(6)}$

5 Make y the subject of the formula $z = \sqrt[3]{\frac{x + 1}{y}}$

6 Make y the subject of $5xy + 3x = 9 - 2y$

7 Make k the subject of the formula $T = 2p\sqrt{\frac{x}{k}}$

8 $f(x) = 4x^2 - 7$

a Work out f(3). b Find the value of a where $f(a) = 0$.

Algebraic fractions

9 Simplify fully

a $\frac{x^2 - 4}{3x + 6}$ b $\frac{x^2 + 4x - 32}{x^2 + 9x + 8}$

10 Write as a single fraction in its simplest form.

a $\frac{5}{2x} - \frac{7}{3x}$ b $\frac{3}{x + 4} + \frac{1}{x - 5}$ c $\frac{4}{x^2 - 7x + 6} - \frac{2}{x - 1}$

11 Write as a single fraction in its simplest form.

a $\frac{16x^3}{21y^9} \times \frac{14y^4}{12x}$ b $\frac{x^2 + 9x - 10}{x^2 + 5x + 4} \div \frac{4x - 4}{3x + 12}$

12 Solve the equation $\frac{2}{x + 1} - \frac{1}{x + 2} = 1$

Proof

13 **Communication** Show that $23 - (x + 1)^2 \equiv (6 + x)(4 - x)$

14 **Communication / Reasoning** Prove that this statement is not true:
The sum of two cubed numbers is always odd.

Reflect

15 How sure are you of your answers? Were you mostly

Just guessing ☹ Feeling doubtful 😐 Confident ☺

What next? Use your results to decide whether to strengthen or extend your learning.

* Challenge

16 Prove that

a the sum of two consecutive odd numbers is a multiple of 4

b the sum of three consecutive even numbers is a multiple of 6

c the sum of four consecutive odd numbers is a multiple of 8.

Q16 hint Let $2n$ be any even number and $2n + 1$ be any odd number.

ActiveLearn Homework, practice and support: Higher 17 Check up

17 Strengthen

Surds

1 Copy and complete.

a $\sqrt{3} \times \sqrt{3} = \square$

b $\sqrt{7} \times \sqrt{\square} = 7$

c $2\sqrt{2} \times \sqrt{2} = \square$

d $\sqrt{5}(6 - \sqrt{5}) = \sqrt{5} \times 6 - \sqrt{5} \times \sqrt{5} = 6\sqrt{5} - \square$

e $\sqrt{180} + \sqrt{45}$

Q1d hint $\sqrt{5} \times 6 = 6 \times \sqrt{5} = 6\sqrt{5}$
Always write the whole number before the surd.

Q1e hint $\sqrt{180} = \sqrt{9 \times 4 \times 5}$ and $\sqrt{45} = \sqrt{\square \times \square}$

2 Rationalise the denominators.

a $\frac{12}{\sqrt{3}} = \frac{12}{\sqrt{3}} \times \frac{\sqrt{3}}{\sqrt{3}} = \frac{\square\sqrt{3}}{\square} = \square\sqrt{3}$

b $\frac{4 + \sqrt{11}}{\sqrt{11}} = \frac{4 + \sqrt{11}}{\sqrt{11}} \times \frac{\sqrt{11}}{\sqrt{11}} = \frac{4 \times \sqrt{11} + \square}{\sqrt{11} \times \square} = \frac{\square}{\square}$

c $\frac{8 - \sqrt{5}}{\sqrt{5}}$

Q2a hint $\sqrt{3} \times \sqrt{3} = 3$
To get rid of $\sqrt{3}$ in the denominator, multiply the fraction by $\frac{\sqrt{3}}{\sqrt{3}}$.

Q2b hint Multiply both parts of the expression in the numerator by $\sqrt{11}$.

Q2 communication hint 'Rationalise the denominator' means get rid of any surds in the denominator, so it is a rational number.

3 Expand and simplify.
The first one has been started for you.

a $(4 - \sqrt{7})(2 + \sqrt{7}) = 8 + 4\sqrt{7} - \square\sqrt{7} - \square$

b $(5 - \sqrt{2})^2 = \square + \square\sqrt{7}$

c $(3 - \sqrt{5})(3 + \sqrt{5}) = 9 + 3\sqrt{5} - \square\sqrt{5} - \square$

d $(2 + \sqrt{11})(2 - \sqrt{11})$

e $(4 - \sqrt{7})(4 + \sqrt{7})$

f **Reasoning** Look at your answers to parts **c** to **e**. What do you notice? Why does this happen?

g What would you multiply these expressions by to get an integer answer?
i $(6 + \sqrt{8})$ ii $(3 - \sqrt{11})$

Q3 hint Multiply each term in the second bracket by each term in the first bracket.
FOIL: Firsts, Outers, Inners, Lasts.

Q3b hint $(5 - \sqrt{2})^2 = (5 - \sqrt{2})(5 - \sqrt{2})$

4 Rationalise the denominators.

a $\frac{8}{5 - \sqrt{2}} = \frac{8 \times (\square)}{(5 - \sqrt{2})(5 + \sqrt{2})} =$

b $\frac{7}{2 + \sqrt{3}}$

c $\frac{6}{7 - \sqrt{10}}$

Q4a hint To get rid of $(5 - \sqrt{2})$ in the denominator, multiply the fraction by $\frac{5 + \sqrt{2}}{5 + \sqrt{2}}$.

Formulae and functions

1 a Copy and complete.

i $y = \sqrt{3}$, so $y^2 = \square$

ii $y = \sqrt{x}$, so $y^2 = \square$

iii $y = \sqrt{(3x - 1)}$, so $y^2 = \square$

b Use your answer to part **a iii** to make x the subject of the formula $y = \sqrt{(3x - 1)}$

Q1b hint Your answer will be $x =$ __

2 Here are all the steps to make y the subject of $x = \frac{7 + y}{y}$
Match each step to one of these rearrangements.

Rewrite the formula so there is no fraction.	$y(x - 1) = 7$
Get all the terms containing y on the left-hand side and all other terms on the right-hand side.	$xy = 7 + y$
Factorise so that y appears only once.	$y = \frac{7}{x - 1}$
Get y on its own on the left-hand side.	$xy - y = 7$

3 Make y the subject of the formula $F = \frac{1 - 5y}{y}$

Q3 hint Follow the steps in **Q2**.

4 a $y = 5x - 9$. Work out the value of y when $x = 2$.
b $f(x) = 5x - 9$. Work out f(2).
c Work out
i f(5) ii f(−3) iii f(0)

Q4b hint f(2) means substitute $x = 2$ in $5x - 9$.

5 a Solve these equations.
i $8x - 1 = 0$ ii $2 - 7x = 0$
b $f(x) = 8x - 1$ and $g(x) = 2 - 7x$.
Use your answers to part **a** to find the value of a where
i $f(a) = 0$ ii $g(a) = 0$
c $p(x) = 9x - 4$. Find the value of a where $p(a) = 0$.

Q5b i hint $f(x) = 8x - 1$
$f(a) = 0$ means that $8a - 1 = 0$

6 **Reasoning** $f(x) = \frac{10}{x}$, $g(x) = x^2 - 1$, $h(x) = x(x - 5)$.
Work out
a f(5) b g(6) c f(5) × g(6)
d 4f(5) e 2g(6) f $g(2x)$
g i g(3)
ii hg(3)
iii f(2)
iv hf(2)

Q6c hint Multiply your value for f(5) by your value for g(6).

Q6d hint 4f(5) means 4 × f(5).

Q6f hint $g(x) = x^2 - 1$
$g(2x) = (2x)^2 - 1 = \_\_ - 1$

Q6g ii hint Substitute your value for g(3) into $h(x)$.

7 Jake draws a function machine to illustrate $y = 5x - 4$.
To find the inverse function he reverses the machine and replaces the functions with their inverse.

x → ×5 → −4 → y

y ← ÷5 ← +4 ← x

Copy and complete the inverse function.

$y = \frac{x + \square}{\square}$

8 Find $f^{-1}(x)$ for each function.
a $f(x) = 2x - 9$ b $f(x) = 3(x - 5)$
c $f(x) = \frac{(x + 4)}{2}$ d $f(x) = \frac{2(x + 1)}{5}$

Q8 hint Use the same method as **Q7**.

Algebraic fractions

1 Simplify

a $\frac{2 \times 15}{5 \times 12}$

b $\frac{x^2}{x}$

c $\frac{(x + 10)(x - 8)}{(x + 7)(x + 10)}$

d $\frac{x}{x - 2} \times \frac{x + 5}{x}$

e $\frac{(x + 4)}{(x - 3)} \times \frac{(x - 3)}{(x + 8)} \times \frac{(x - 2)}{(x - 4)}$

f $\frac{(x + 5)}{18} \times \frac{10}{(x - 1)} \times \frac{(x + 1)}{(x + 5)}$

Q1 hint Cancel common factors.

Q1a hint Look to group common factors.
$\frac{2 \times 15}{5 \times 12} = \frac{15}{5} \times \frac{2}{12}$

2 a Copy and complete.

i $\frac{15}{20} = \frac{\square}{\square}$ ii $\frac{9}{6} = \frac{\square}{\square}$

iii $\frac{x}{x^3} = \frac{\square}{\square}$ iv $\frac{y^6}{y^2} = \frac{\square}{\square}$

b Use your answers from part **a** to fully simplify $\frac{15y^6}{6x^3} \times \frac{9x}{20y^2}$

Q2b hint Regroup the terms and cancel common factors.
$\frac{15y^6}{6x^3} \times \frac{9x}{20y^2} = \frac{15}{6} \times \frac{9}{20} \times \frac{x}{x^3} \times \frac{y^6}{y^2}$

3 Write each of these as a single fraction in its simplest form.

a $\frac{4x^5}{15y^2} \times \frac{20y}{12x^3}$

b $\frac{12y^2}{21x^2} \div \frac{9y^5}{14x^3}$

Q3 strategy hint Use the same strategy as in **Q2**.

Q3b hint Multiply the first fraction by the reciprocal of the second fraction: $\frac{14x^3}{9y^5}$

4 a Factorise

i $3x + 18$

ii $x^2 + 6x$

b Use your answers from part **a** to simplify $\frac{3x + 18}{x^2 + 6x}$ fully.

c Simplify fully $\frac{x^2 - 25}{2x + 10}$

Q4b hint Rewrite the numerator and denominator in factorised form. Cancel common factors.

Q4c hint Use the difference of two squares.
$x^2 - 25 = (x + 5)(x - 5)$

5 Simplify fully.

a $\frac{8x + 32}{x^2 + 12x + 32} = \frac{8(\square + \square)}{(x + \square)(x + \square)} = \frac{\square}{\square}$

b $\frac{x^2 + 6x - 16}{x^2 - 11x + 18}$

c $\frac{x^2 - 3x - 40}{x^2 + 8x + 15}$

6 a Factorise

i $3x + 9$

ii $x^2 + 9x + 18$

iii $x^2 + 8x + 15$

iv $2x + 10$

b Use your answers to part **a** to write as a single fraction.

i $\frac{3x + 9}{x^2 + 9x + 18} \times \frac{x^2 + 8x + 15}{2x + 10}$

ii $\frac{2x + 10}{x^2 + 8x + 15} \div \frac{3x + 9}{x^2 + 9x + 18}$

7 Solve these quadratic equations.

a $(x - 8)(x + 7) = 0$

b $x^2 - 2x - 63 = 0$

c $x^2 + 3x + 3 = 4x + 9$

d $\frac{2x + 3}{x^2 + 5x - 7} = 1$

Q7b hint Factorise the equation.

Q7c hint Rearrange the equation into the form $x^2 + bx + c = 0$.

8 a Write down the LCM of x and $x - 1$.

b Copy and complete.

i $\frac{3}{x} = \frac{3(\square)}{x(x-1)}$

ii $\frac{2}{x-1} = \frac{2x}{(x-1)(\square)}$

c Copy and complete, using your answers to part **b**.

$\frac{3}{x} + \frac{2}{x-1} = \frac{\square + \square}{x(x-1)} = \frac{\square}{\square}$

d Use your answer to part **c** to solve $\frac{3}{x} + \frac{2}{x-1} = 1$

Q8d hint First set your fraction answer from part **c** equal to 1.

9 Write as a single fraction in its simplest form.

$\frac{3}{x^2 - 5x + 4} - \frac{2}{x - 4}$

Q9 hint First factorise $x^2 - 5x + 4$.

Proof

1 a Expand $(x - 4)^2$

b Expand and simplify $(x - 4)^2 - 9$

c Expand $(x - 7)(x - 1)$

d Use your answers to parts **b** and **c** to show that $(x - 4)^2 - 9 \equiv (x - 7)(x - 1)$

Q1d hint $\equiv$ means 'identical to'.

2 **Communication** Show that $(x - 1)^2 - 16 \equiv (x - 5)(x + 3)$

3 a List the first five cube numbers.

b Give a counter-example to prove this statement is *not* true: The difference between two cube numbers is always odd.

Q3 hint Look for a pair of numbers in your list from part **a** whose difference is even.

17 Extend

1 **Communication / Reasoning** Both Jack and Ruth make y the subject of the formula $1 - 2y = x$

Jack's answer is $y = \frac{x - 1}{-2}$

Ruth's answer is $y = \frac{1 - x}{2}$

a Show that both answers are correct.

b Explain why Ruth's answer might be considered a better answer.

c Make x the subject of the formula $\frac{P - 2x^2}{3} = d$

2 **STEM** The total resistance of a set of resistors in a parallel circuit is given by the formula

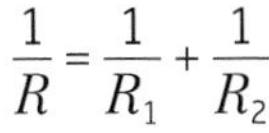

$\frac{1}{R} = \frac{1}{R_1} + \frac{1}{R_2}$

Make R_2 the subject of the formula.

3 **Communication** $\frac{1}{a} = \frac{1}{b} + \frac{1}{c} - \frac{1}{d}$

a Write down an expression for $\frac{1}{d}$.

b Show that $d = \frac{abc}{ac + ab - bc}$

4 Solve these equations.

a $\frac{4}{2x-3} = \frac{x}{5}$

b $\frac{4}{2-x} - \frac{1}{x-3} = 5$

5 **Communication**

a Show that $\frac{1}{1+\frac{1}{x}} = \frac{x}{x+1}$

b Work out the exact value of $\frac{1}{1+\frac{1}{9}}$

6 **Exam-style question**

The functions f and g are such that $f(x) = 3 - 4x$, $g(x) = 3 + 4x$

a Find f(6)

b Find $gf(x)$

c Find

i $f^{-1}(x)$

ii $g^{-1}(x)$

d Show that $f^{-1}(x) + g^{-1}(x) = 0$, for all values of x. **(7 marks)**

Q6 strategy hint $f^{-1}(x)$ is the inverse of $f(x)$.

7 **Exam-style question**

$f(x) = x + 7$, $g(x) = x^2 + 6$

a Work out

i $fg(x)$

ii $gf(x)$

b Solve $fg(x) = gf(x)$ **(6 marks)**

Q7 strategy hint Remember that fg means do g first and then f.

8 **Communication / Reasoning** $f(x) = \frac{x-7}{2}$, $g(x) = 2x + 7$

a Work out

i $fg(x)$

ii $gf(x)$

b Are $f(x)$ and $g(x)$ inverse functions? Explain your answer.

c Check whether $f(x) = \frac{1}{4}x - 1$ and $g(x) = 4(x + 1)$ are inverse functions.

Q8b hint Functions f and g are inverses of each other if $fg(x) = gf(x) = x$

9 **Communication / Reasoning** Show that $\frac{49 - x^2}{x^2 - 49} = -1$

10 **Communication / Reasoning** Show that $(3n + 1)^2 - (3n - 1)^2$ is a multiple of 12, for all positive values of n.

11 **Communication / Reasoning** Show that

$\frac{1}{5x^2 - 13x - 6} - \frac{1}{x^2 - 9} = \frac{Ax + B}{(x-3)(x+3)(5x+2)}$ and find the value of A and B.

12 **STEM** Newton's Law of Universal Gravitation can be used to calculate the force (F) between two different objects.

$F = \frac{Gm_1m_2}{r^2}$, where G is the gravitational constant (6.67×10^{-11} Nm^{2}kg^{-2}), m_1 and m_2 are the masses of the two objects (kg) and r is the distance between them (km).

a Rearrange the formula to make r the subject.

The gravitational force between the Earth and the Sun is 3.52×10^{22} N.

The mass of the Sun is 1.99×10^{30} kg and the mass of the Earth is 5.97×10^{24} kg.

b Work out the distance between the Sun and the Earth.

17 Knowledge check

- You can change the subject of a formula by isolating the terms involving the new subject. .... *Mastery lesson 17.1*
- When the letter to be made the subject appears twice in the formula you will need to factorise. .... *Mastery lesson 17.1*
- To add or subtract algebraic fractions, write each fraction as an equivalent fraction with a common denominator. .... *Mastery lesson 17.2*
- You may need to factorise before simplifying an algebraic fraction:
  - Factorise the numerator and denominator.
  - Divide the numerator and denominator by any common factors. .... *Mastery lesson 17.3*
- To find the lowest common denominator of algebraic fractions, you may need to factorise the denominators first. .... *Mastery lesson 17.4*
- You may need to factorise the numerator and/or denominator before you multiply or divide algebraic fractions. .... *Mastery lesson 17.4*
- To rationalise the fraction $\frac{1}{a \pm \sqrt{b}}$, multiply by $\frac{a \mp \sqrt{b}}{a \mp \sqrt{b}}$ .... *Mastery lesson 17.5*
- A function is a rule for working out values of y when given values of x e.g. $y = 3x$ and $y = x^2$. .... *Mastery lesson 17.7*
- The notation f(x) is read as 'f of x'. .... *Mastery lesson 17.7*
- fg is the composition of the function f with the function g. To work out fg(x), first work out g(x) and then substitute your answer into f(x). .... *Mastery lesson 17.7*
- The inverse function reverses the effect of the original function. $f^{-1}(x)$ is the inverse of f(x). .... *Mastery lesson 17.7*
- To show a statement is an identity, expand and simplify the expressions on one or both sides of the equals sign, until the two expressions are the same. .... *Mastery lesson 17.8*
- A **proof** is a logical argument for a mathematical statement. To prove a statement is true, you must show that it will be true in *all* cases. .... *Mastery lesson 17.8*
- To prove a statement is not true you can find a **counter-example** – an example that does not fit the statement.
- For an algebraic proof, use n to represent any integer. .... *Mastery lesson 17.8*

Even number	$2n$
Odd number	$2n + 1$ or $2n - 1$
Consecutive numbers	$n, n + 1, n + 2, \ldots$
Consecutive even numbers	$2n, 2n + 2, 2n + 4, \ldots$
Consecutive odd numbers	$2n + 1, 2n + 3, 2n + 5, \ldots$

Reflect

Look back at this unit.
Which lesson made you think the hardest? Write a sentence to explain why.
Begin your sentence with: Lesson ____ made me think the hardest because ____

17 Unit test

Log how you did on your Student Progression Chart.

1 Find the inverse of the function $f(x) = 5(x + 4)$ *(3 marks)*

2 Show that $(x + 4)^2 - (2x + 7) \equiv (x + 3)^2$ for all values of x. *(2 marks)*

3 Make x the subject of the formula $y = \frac{1}{2}(x + 3)^2$ *(2 marks)*

4 Simplify fully $\frac{9 - x^2}{x(x - 3)}$ *(2 marks)*

5 Write as a single fraction in its simplest form.

a $\frac{9x^3}{8y} \times \frac{4y^2}{15x^5}$ b $\frac{9}{4x} - \frac{2}{5x}$ *(4 marks)*

6 Make x the subject of the formula $V = \frac{1 + 5x}{x}$ *(2 marks)*

7 Expand and simplify.

a $(3 + \sqrt{2})(4 - \sqrt{2})$ b $(3 + \sqrt{5})^2$ *(4 marks)*

8 Write as a single fraction in its simplest form.

a $\frac{x^2 + x - 30}{x^2 + 10x + 24} \div \frac{x^2 - 12x + 35}{x^2 + 3x - 4}$ b $\frac{5x^2 - 4x - 12}{4x - 8} \times \frac{5x + 5}{5x^2 + 11x + 6}$ *(4 marks)*

9 Rationalise the denominator.

$\frac{9}{1 - \sqrt{3}}$ *(2 marks)*

10 Solve these quadratic equations.

a $\frac{5x - 1}{2} = \frac{3}{x}$ b $\frac{5}{x - 1} + \frac{7}{x - 1} = x$ *(6 marks)*

11 Solve the equation $\frac{1}{x + 2} - \frac{1}{x + 4} = 1$

Give your answers correct to 2 decimal places. *(3 marks)*

12 $f(x) = x^2 - 9$, $g(x) = 2x + 1$

a Work out

i $f(4) + g(2)$ ii $f(2) \times g(-1)$

b Find the value of a where $f(a) = 0$.

c Show that $fg(x) = 4x^2 + 4x - 8$ *(5 marks)*

Sample student answer

Explain what common mistake the student has made right at the very start of the answer.
Suggest a way to avoid making this mistake.

Exam-style question

Make b the subject of the formula $a = \frac{2 - 7b}{b - 5}$ **(4 marks)**

May 2008, Q22, 5540/3H

Student answer

$ab - 5 = 2 - 7b$

$ab + 7b = 2 + 5$

$b(a + 7) = 7$

$b = \frac{7}{a + 7}$

ActiveLearn Homework, practice and support: Higher 17 Unit test

Master p.558 | Problem-solve p.572 | Check p.574 | Strengthen p.576 | Extend p.579 | Test p.584

18 VECTORS AND GEOMETRIC PROOF

Vectors can be used to work out the adjustment in direction for a pilot landing a plane when there is a cross wind.

An airline pilot compensates for a cross wind by pointing her plane away from the runway. What angle did she make with the runway in order to make a perfect landing?

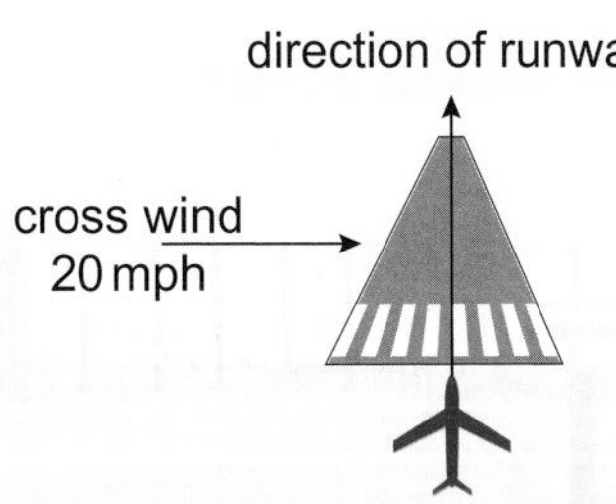

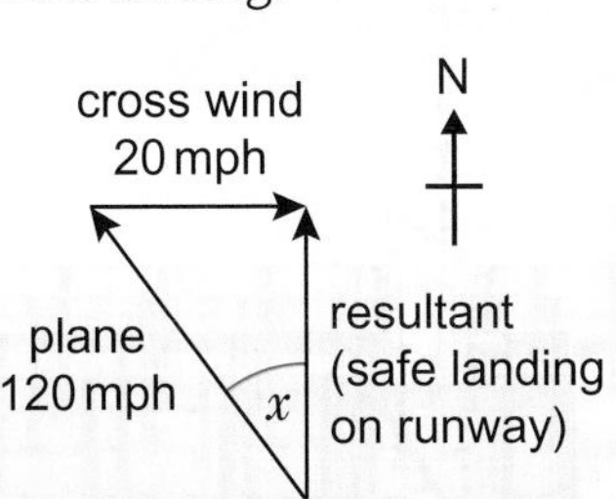

18 Prior knowledge check

Numerical fluency

1 $PQ = \frac{2}{3}PR$ P —— Q —— R

Work out

a PQ : PR b RQ : PR c PQ : QR

2 The point M divides the line LX in the ratio 5 : 2.

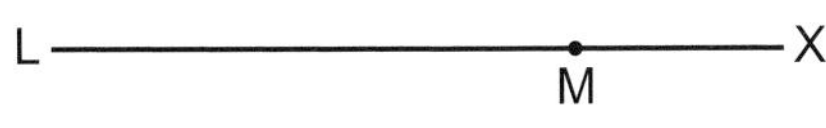

Copy and complete

a $LM = \frac{\square}{\square}LX$ b $MX = \frac{\square}{\square}LX$

c $LX = \frac{\square}{\square}MX$

3 Write each surd in its simplest form.

a $\sqrt{27}$ b $\sqrt{80}$

c $\sqrt{75}$ d $\sqrt{112}$

Algebraic fluency

4 Simplify

a $4x + 3y - x + 2y$ b $3(2x - 3)$

c $2(3a - b) + \frac{2}{3}(6a + 9b)$

Geometrical fluency

5 Sketch these shapes.

a isosceles triangle

b isosceles trapezium

c square

d rhombus

e parallelogram

Mark any

i equal sides ii parallel sides.

6 Find x. Give your answer to 3 significant figures (3 s.f.).

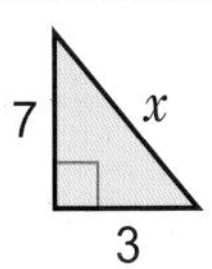

7 Copy the diagram.

Translate shape

a P by $\begin{pmatrix} 5 \\ -3 \end{pmatrix}$

b Q by $\begin{pmatrix} -4 \\ 3 \end{pmatrix}$

c R by $\begin{pmatrix} -5 \\ -2 \end{pmatrix}$

⁎ Challenge

8 Work out the bearing for the approach of each plane.

a

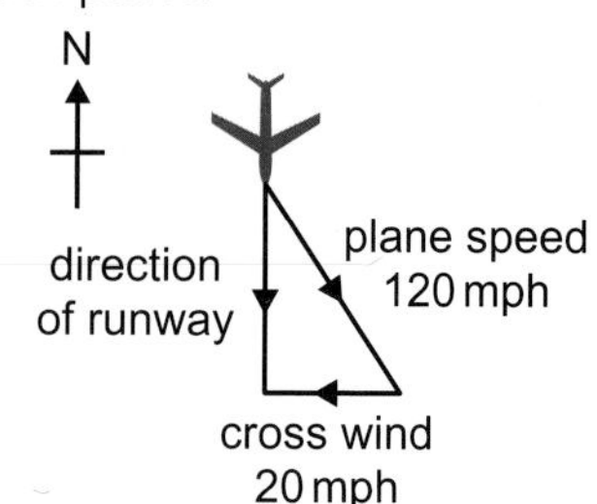

b

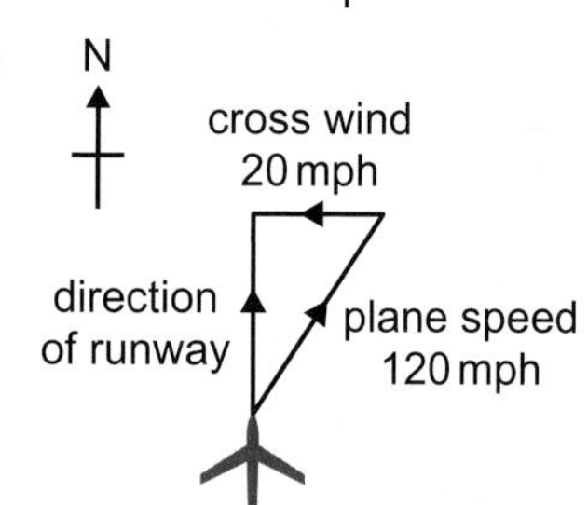

18.1 Vectors and vector notation

Objectives

- Understand and use vector notation.
- Work out the magnitude of a vector.

Why learn this?

You can describe journeys using vectors. For example, a flight from Bristol to Birmingham is a vector with magnitude 125 km and direction 021°.

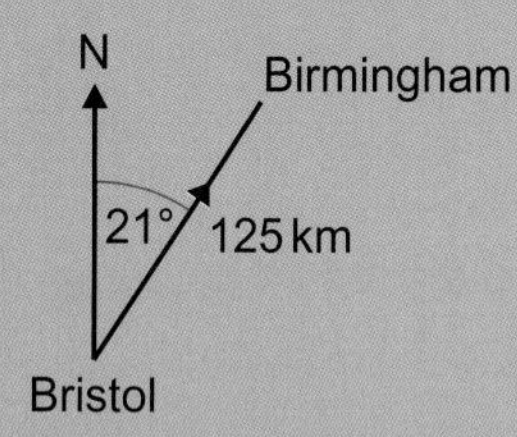

Fluency

How far does the translation $\begin{pmatrix} 3 \\ 5 \end{pmatrix}$ move an object in **a** the x direction **b** the y direction?

Warm up

1 Write the column vector for the translation of shape

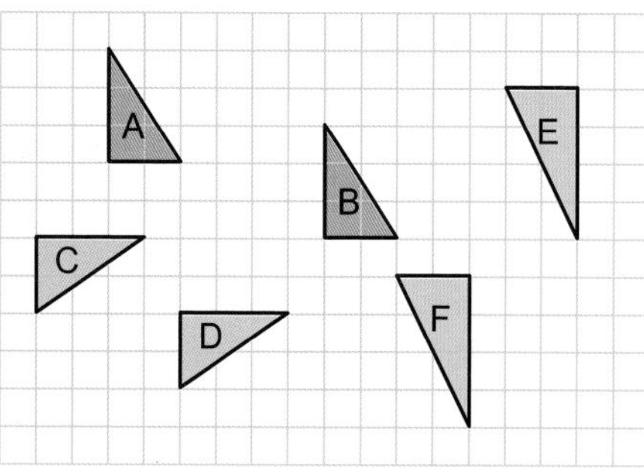

a A to B b C to D c E to F

2 Find x in this right angled triangle. Give your answers in surd form.

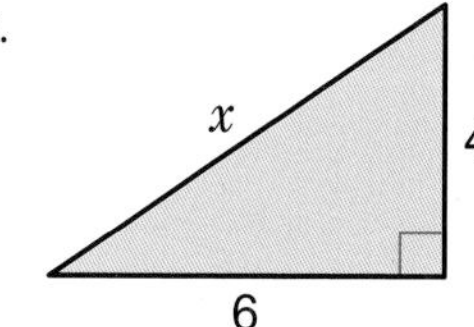

ActiveLearn Homework, practice and support: Higher 18.1

Key point 1

A **vector** is a quantity that has magnitude and direction.
The **magnitude** of a vector is its size.
Displacement is change in position. A displacement can be written as $\begin{pmatrix}3\\4\end{pmatrix}$ where 3 is the x component and 4 is the y component.
Examples of vectors are force (5 N acting vertically upwards) and velocity (15 km/h due north).

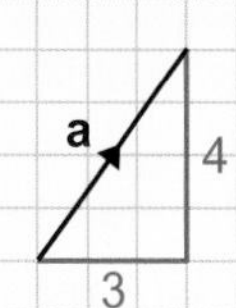

Questions in this unit are targeted at the steps indicated.

3 On squared paper draw and label these vectors.

a $\mathbf{a} = \begin{pmatrix}1\\3\end{pmatrix}$ b $\mathbf{b} = \begin{pmatrix}3\\-2\end{pmatrix}$ c $\mathbf{c} = \begin{pmatrix}-4\\-3\end{pmatrix}$ d $\overrightarrow{AB} = \begin{pmatrix}-5\\4\end{pmatrix}$ e $\overrightarrow{CD} = \begin{pmatrix}0\\5\end{pmatrix}$

Key point 2

The displacement vector from A to B is written $\overrightarrow{AB}$.
Vectors are written as **bold** lower case letters: **a**, **b**, **c**
When handwriting, underline the letter: a, b, c

Example 1

a Point A has coordinates (3, 5) and point B has coordinates (5, 1).
Write $\overrightarrow{AB}$ as a column vector.
b The point C is such that $\overrightarrow{BC} = \begin{pmatrix}-1\\2\end{pmatrix}$. Find the coordinates of C.

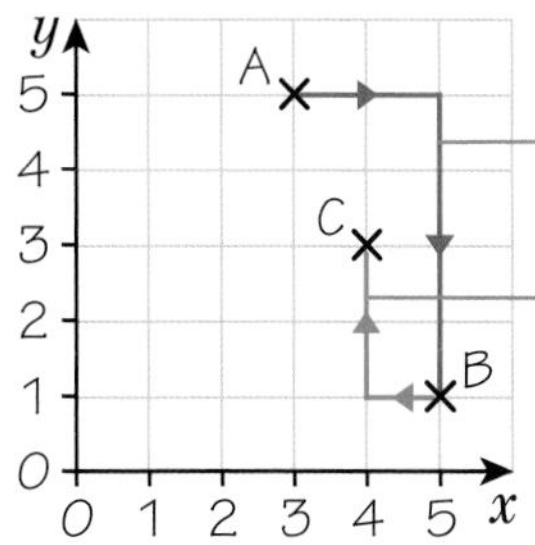

First mark the points A and B on a grid.
To move from A to B go 2 to the right and 4 down.

From B go 1 to the left and 2 up to find point C.

a $\overrightarrow{AB} = \begin{pmatrix}2\\-4\end{pmatrix}$

b *The coordinates of C are (4, 3)*

4 The point A is (1, 2), the point B is (3, 4) and the point C is (5, −1).
Write as column vectors
a $\overrightarrow{AB}$ b $\overrightarrow{BC}$ c $\overrightarrow{AC}$
Discussion What do you notice about your answers?

Q4 hint Mark the points on a grid.

Key point 3

Equal vectors have the same magnitude and the same direction.

5 Which of these vectors are equal?

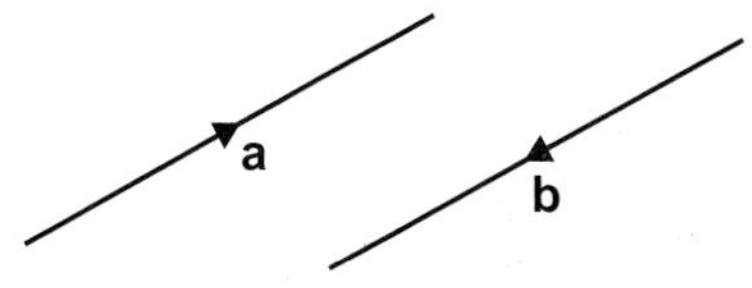

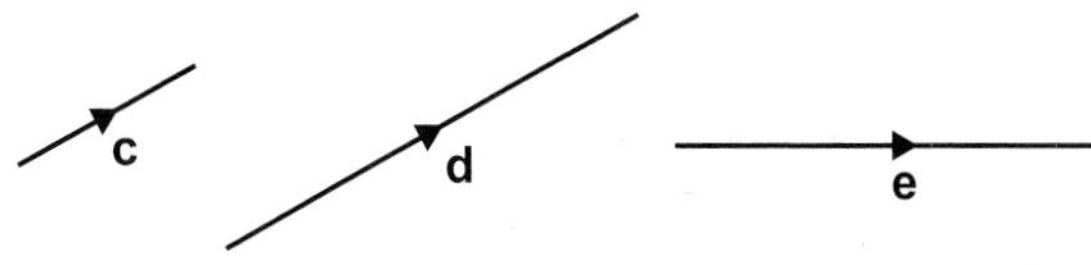

Discussion Are parallel vectors always equal?

Key point 4

The magnitude of the vector $\begin{pmatrix} x \\ y \end{pmatrix}$ is its length, i.e. $\sqrt{x^2 + y^2}$

|a| means the magnitude of vector **a**. |OA| means the magnitude of vector $\overline{\text{OA}}$.

6 Find the magnitude of the vector $\overrightarrow{AB} = \begin{pmatrix} 4 \\ -5 \end{pmatrix}$. Give your answer to 3 significant figures.

7 Work out the magnitude of each vector. Where necessary, leave your answer as a surd.

a $\mathbf{a} = \begin{pmatrix} 6 \\ 8 \end{pmatrix}$ b $\mathbf{b} = \begin{pmatrix} -5 \\ 12 \end{pmatrix}$ c $\mathbf{c} = \begin{pmatrix} -1 \\ -3 \end{pmatrix}$ d $\overrightarrow{AB} = \begin{pmatrix} 8 \\ 15 \end{pmatrix}$ e $\overrightarrow{CD} = \begin{pmatrix} 4 \\ -6 \end{pmatrix}$

8 **Reasoning / Communication** In triangle ABC, $\overrightarrow{AB} = \begin{pmatrix} 20 \\ -15 \end{pmatrix}$ and $\overrightarrow{AC} = \begin{pmatrix} -7 \\ 24 \end{pmatrix}$.

a Work out the length of the side AB of the triangle.

b Show that triangle ABC is isosceles.

Q8 hint Sketch A, B and C.

9 **Exam-style question**

A is the point (3, 4) and *B* is the point (–3, 0).

a Write $\overrightarrow{AB}$ as a column vector. **(1 mark)**

b Find the length of vector $\overrightarrow{AB}$. **(2 marks)**

Exam hint
Write vectors in column vector form: $\begin{pmatrix} p \\ q \end{pmatrix}$ not $\left(\frac{p}{q}\right)$ or (p, q)

10 **Reasoning** $\overrightarrow{AB} = \begin{pmatrix} 3 \\ 4 \end{pmatrix}$. B is the point (2, 3). Work out the coordinates of A.

18.2 Vector arithmetic

Objectives

- Calculate using vectors and represent the solutions graphically.
- Calculate the resultant of two vectors.

Why learn this?

The driver of a car going over a road-hump instinctively works out resultant forces.
A road-designer does the same, but in advance and with greater accuracy.

Fluency

The components of the column vector $\begin{pmatrix} 3 \\ -5 \end{pmatrix}$ are 3 units … and … units …

Warm up

1 Copy shape A on a coordinate grid.

Translate shape A by the vector $\begin{pmatrix} 0 \\ 2 \end{pmatrix}$. Label this new shape B.

Translate shape B by the vector $\begin{pmatrix} 2 \\ -1 \end{pmatrix}$. Label this new shape C.

What single translation maps

a shape A onto shape C b shape C onto shape A?

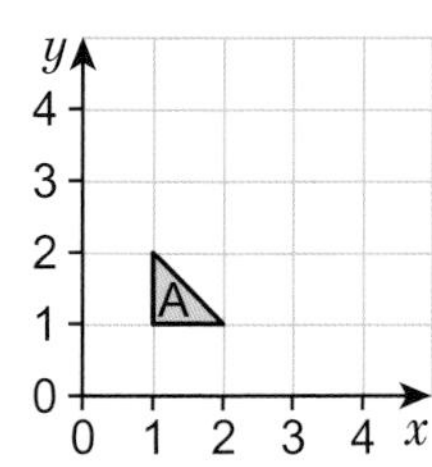

2 The vector $\begin{pmatrix} 3 \\ -2 \end{pmatrix}$ transforms shape A to shape B.

What vector transforms shape B to shape A?

Q2 hint Translate shape A from **Q1** by $\begin{pmatrix} 3 \\ -2 \end{pmatrix}$.

3 **Reasoning** The points A, B, C and D are the vertices of a quadrilateral where A has coordinates (1, 1).

$\overrightarrow{AB} = \begin{pmatrix} 1 \\ 2 \end{pmatrix}$, $\overrightarrow{BC} = \begin{pmatrix} 4 \\ 1 \end{pmatrix}$ and $\overrightarrow{CD} = \begin{pmatrix} 3 \\ -1 \end{pmatrix}$.

a Draw quadrilateral ABCD on squared paper.

b Write $\overrightarrow{AD}$ as a column vector.

c What type of quadrilateral is ABCD?

d What do you notice about $\overrightarrow{BC}$ and $\overrightarrow{AD}$?

4 **Reasoning** The points A, B, C and D are the vertices of a parallelogram.
A has coordinates (1, 1), $\overrightarrow{AB} = \begin{pmatrix} 2 \\ 3 \end{pmatrix}$ and $\overrightarrow{AD} = \begin{pmatrix} 4 \\ -1 \end{pmatrix}$.

a Draw parallelogram ABCD on squared paper.

b Write as a column vector i $\overrightarrow{CB}$ ii $\overrightarrow{BC}$

What do you notice?

c What do you notice about i $\overrightarrow{AB}$ and $\overrightarrow{DC}$ ii $\overrightarrow{AD}$ and $\overrightarrow{CB}$?

Key point 5

If $\overrightarrow{AB} = \overrightarrow{CD}$ then the line segments AB and CD are equal in length and are parallel.

$\overrightarrow{AB} = -\overrightarrow{BA}$

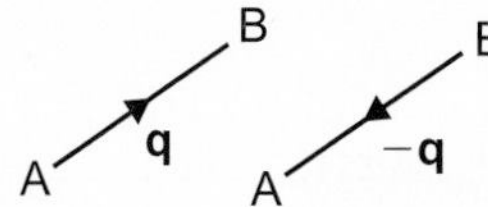

5 In quadrilateral ABCD, $\overrightarrow{AB} = \begin{pmatrix} 3 \\ 4 \end{pmatrix}$, $\overrightarrow{BC} = \begin{pmatrix} 5 \\ 0 \end{pmatrix}$, $\overrightarrow{CD} = \begin{pmatrix} -3 \\ -4 \end{pmatrix}$ and $\overrightarrow{DA} = \begin{pmatrix} -5 \\ 0 \end{pmatrix}$.
What type of quadrilateral is ABCD?

Q5 hint Look at the vectors for opposite sides.

6 **Exam-style question**

P is the point (0, 3). $\overrightarrow{PQ} = \begin{pmatrix} 2 \\ 3 \end{pmatrix}$

a Find the coordinates of Q. **(1 mark)**

R is the point (2, 4). QS is a diagonal of the parallelogram $PQRS$.

b Express $\overrightarrow{PR}$ as a column vector. **(3 marks)**

$\overrightarrow{RT} = \begin{pmatrix} 2 \\ -3 \end{pmatrix}$

c Calculate the length of PT **(3 marks)**

Exam hint
Sketch a diagram.

Key point 6

2**a** is twice as long as **a** and in the same direction.
−**a** is the same length as **a** but in the opposite direction.

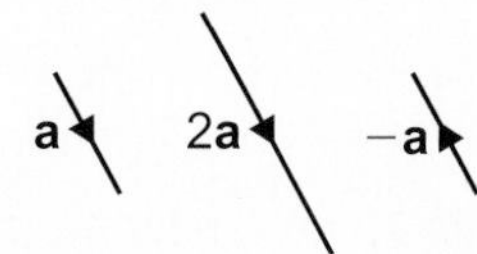

7 On squared paper draw vectors to represent

a 2**a** b −**a**

c −**b** d 3**b**

e −2**b**

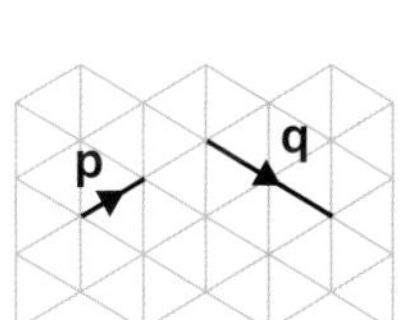

8 The vectors **p** and **q** are shown on an isometric grid.
Draw these vectors on an isometric grid.

a 2**p** b $\frac{1}{2}$**q**

c −**p** d −**q**

Key point 7

When a vector **a** is multiplied by a scalar k then the vector k**a** is parallel to **a** and is equal to k times **a**.
A **scalar** is a number, e.g. 3, 2, $\frac{1}{2}$, −1…

9 **Reasoning** $\overrightarrow{AB} = \begin{pmatrix} 2 \\ 1 \end{pmatrix}$

a Copy and complete to find the column vector for $2\overrightarrow{AB}$.

$$2\overrightarrow{AB} = 2 \times \begin{pmatrix} 2 \\ 1 \end{pmatrix} = \begin{pmatrix} 2 \times 2 \\ 2 \times 1 \end{pmatrix} = \begin{pmatrix} \square \\ \square \end{pmatrix}$$

b Write down the column vector for

i $3\overrightarrow{AB}$ ii $-4\overrightarrow{AB}$ iii $\frac{1}{2}\overrightarrow{AB}$

10 **Reasoning** $\overrightarrow{AB} = \begin{pmatrix} 1 \\ 2 \end{pmatrix}$ and $\overrightarrow{BC} = \begin{pmatrix} 3 \\ 1 \end{pmatrix}$.

Write down the vector $\overrightarrow{AC}$.

Discussion How can you find $\overrightarrow{AC}$ from $\overrightarrow{AB}$ and $\overrightarrow{BC}$?

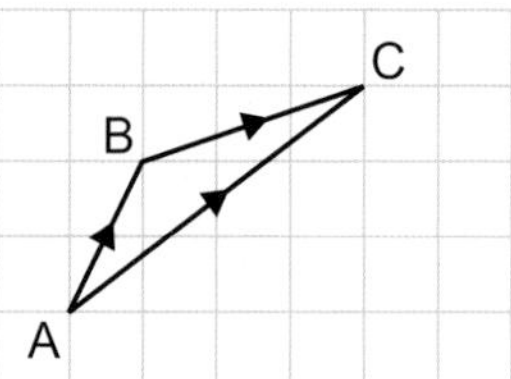

Key point 8

The two-stage journey from A to B and then from B to C has the same starting point and the same finishing point as the single journey from A to C. So A to B followed by B to C is equivalent to A to C.

$\overrightarrow{AB} + \overrightarrow{BC} = \overrightarrow{AC}$

Triangle law for vector addition

Let $\overrightarrow{AB} = \mathbf{a}$, $\overrightarrow{BC} = \mathbf{b}$ and $\overrightarrow{AC} = \mathbf{c}$.

Then $\mathbf{a} + \mathbf{b} = \mathbf{c}$ forms a triangle.

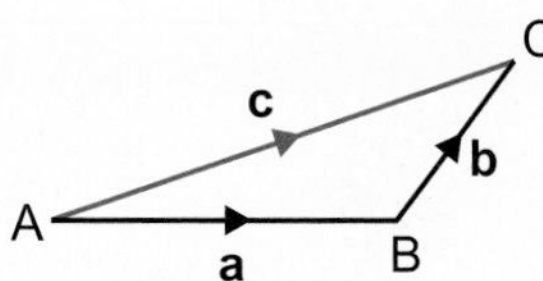

11 a Find, by drawing, the sum of the vectors **a** and **b**.

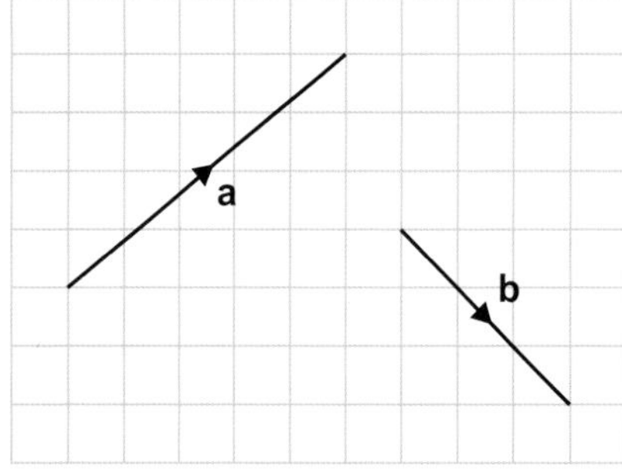

Q11a hint Use the triangle law of addition. Move vector **b** to the end of vector **a** so that the lines follow on. Draw and label the vector **a** + **b** to complete the triangle.

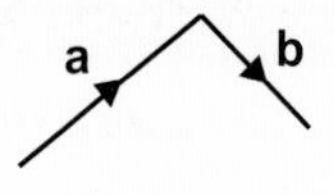

b Copy and complete this vector addition.

$$\mathbf{a} \quad + \quad \mathbf{b} \quad = \quad \mathbf{a} + \mathbf{b}$$

$$\begin{pmatrix} 5 \\ 4 \end{pmatrix} + \begin{pmatrix} 3 \\ -3 \end{pmatrix} = \begin{pmatrix} \square \\ \square \end{pmatrix}$$

12 a $\overrightarrow{AB} = \begin{pmatrix} 2 \\ 5 \end{pmatrix}$ and $\overrightarrow{BC} = \begin{pmatrix} 7 \\ -3 \end{pmatrix}$. Find $\overrightarrow{AC}$.

b $\mathbf{a} = \begin{pmatrix} -7 \\ 2 \end{pmatrix}$ and $\mathbf{b} = \begin{pmatrix} 8 \\ -3 \end{pmatrix}$. Find $\mathbf{a} + \mathbf{b}$.

13 $\mathbf{p} = \begin{pmatrix} 3 \\ 4 \end{pmatrix}$, $\mathbf{q} = \begin{pmatrix} -5 \\ 2 \end{pmatrix}$ and $\mathbf{r} = \begin{pmatrix} 2 \\ -6 \end{pmatrix}$.

a Work out

i **p** + **q** ii **q** + **p**

Discussion What do you notice about your answers to parts **a i** and **ii**?

b Work out

i (**p** + **q**) + **r** ii **p** + (**q** + **r**)

Discussion What do you notice about your answers to parts **b i** and **ii**?

Key point 9

$\mathbf{a} + \mathbf{b} = \mathbf{b} + \mathbf{a}$

14 $\mathbf{p} = \begin{pmatrix} 2 \\ 7 \end{pmatrix}$ and $\mathbf{q} = \begin{pmatrix} -2 \\ 3 \end{pmatrix}$.
Work out $\mathbf{p} - \mathbf{q}$

Q14 hint $\mathbf{p} - \mathbf{q} = \mathbf{p} + (-\mathbf{q})$
$-\mathbf{q} = \begin{pmatrix} \square \\ \square \end{pmatrix}$

15 $\mathbf{a} = \begin{pmatrix} 1 \\ 4 \end{pmatrix}$, $\mathbf{b} = \begin{pmatrix} -2 \\ 7 \end{pmatrix}$ and $\mathbf{c} = \begin{pmatrix} 0 \\ -3 \end{pmatrix}$.
Write down the column vector for

a $-\mathbf{a}$ b $\mathbf{a} + \mathbf{b}$ c $\mathbf{a} + \mathbf{b} + \mathbf{c}$ d $\mathbf{a} - \mathbf{b}$ e $\mathbf{b} - \mathbf{c}$

16 **Reflect** Write examples to help you remember how to
- a add two column vectors
- b subtract one column vector from another
- c multiply a column vector by a scalar, k.

Q16 hint Use the column vectors $\begin{pmatrix} a \\ b \end{pmatrix}$ and $\begin{pmatrix} c \\ d \end{pmatrix}$.

18.3 More vector arithmetic

Objectives
- Solve problems using vectors.
- Use the resultant of two vectors to solve vector problems.

Why learn this?

Civil engineers use vectors in road design to model the movement of a vehicle travelling along a curved section of road.

Fluency

In this parallelogram which line segments are
a parallel
b equal?

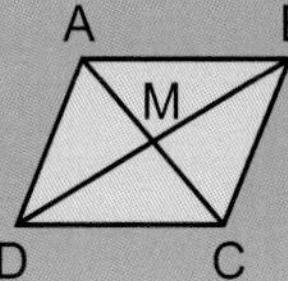

1 $\overrightarrow{AB}$ is the column vector $\begin{pmatrix} 4 \\ -3 \end{pmatrix}$. $\overrightarrow{BC}$ is the column vector $\begin{pmatrix} 2 \\ 4 \end{pmatrix}$.
- a Find the column vector $\overrightarrow{AC}$. Draw a diagram to show your answer.
- b Work out the magnitude of $\overrightarrow{AC}$.

2 $\mathbf{p} = \begin{pmatrix} -2 \\ 6 \end{pmatrix}$. What is $-\mathbf{p}$?

3 **Reasoning / Communication** $\overrightarrow{AB} = \begin{pmatrix} 2 \\ 3 \end{pmatrix}$, $\overrightarrow{BC} = \begin{pmatrix} 1 \\ -4 \end{pmatrix}$ and $\overrightarrow{CD} = \begin{pmatrix} -4 \\ 5 \end{pmatrix}$.
- a Find the column vector for $\overrightarrow{AD}$.
  Draw a diagram to show this.
- b Show that $\overrightarrow{AC} = \overrightarrow{DB}$.

4 $\mathbf{a} = \begin{pmatrix} 5 \\ 9 \end{pmatrix}$ and $\mathbf{b} = \begin{pmatrix} 3 \\ 3 \end{pmatrix}$. Work out the magnitude of

a $\mathbf{a}$ b $2\mathbf{b}$ c $\mathbf{a} + \mathbf{b}$ d $\mathbf{a} - \mathbf{b}$

5 In the quadrilateral ABCD, $\overrightarrow{AB} = \mathbf{a}$, $\overrightarrow{BC} = \mathbf{b}$ and $\overrightarrow{CD} = \mathbf{c}$.
Find in terms of **a**, **b** and **c**

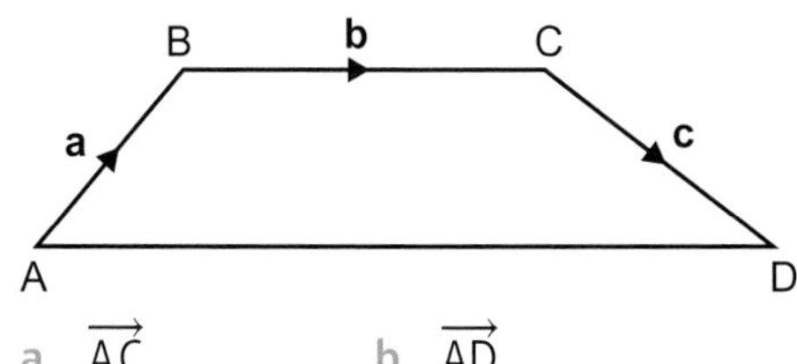

a $\overrightarrow{AC}$ b $\overrightarrow{AD}$

Q5a hint $\overrightarrow{AC} = \overrightarrow{A\square} + \overrightarrow{\square C} = \square + \square$

Q5b hint $\overrightarrow{AD} = \overrightarrow{A\square} + \overrightarrow{\square\square} + \overrightarrow{\square D} = \square + \square + \square$

6 $\overrightarrow{OA} = \mathbf{a}$

M is the midpoint of OA.

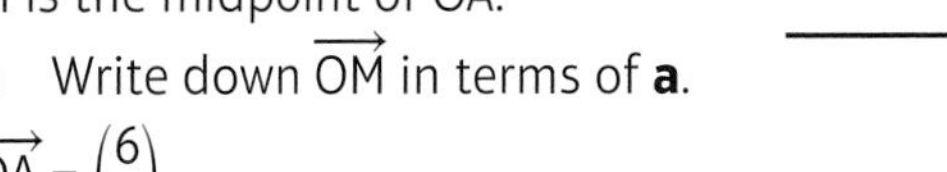

a Write down $\overrightarrow{OM}$ in terms of **a**.

$\overrightarrow{OA} = \begin{pmatrix} 6 \\ 4 \end{pmatrix}$

b Express as a column vector

i $\overrightarrow{AO}$ ii $\overrightarrow{OM}$

Q6a hint OM = $\frac{\square}{\square}$OA

7 In the diagram $\overrightarrow{BA} = \mathbf{a}$ and $\overrightarrow{AP} = \mathbf{b}$. P is the midpoint of AC.

Write down in terms of **a** and/or **b**

a $\overrightarrow{AC}$ b $\overrightarrow{BP}$ c $\overrightarrow{BC}$

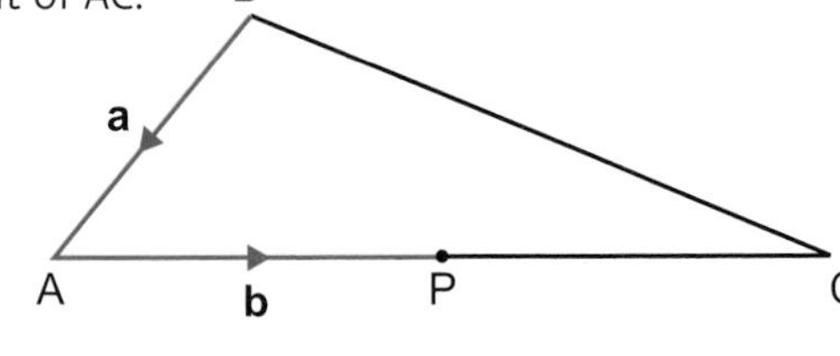

8 **Communication** PQRS is a parallelogram.

$\overrightarrow{PQ} = \mathbf{a}$ and $\overrightarrow{PS} = \mathbf{b}$.

a Explain why $\overrightarrow{SR} = \mathbf{a}$.

b Find

i $\overrightarrow{QR}$ ii $\overrightarrow{PR}$

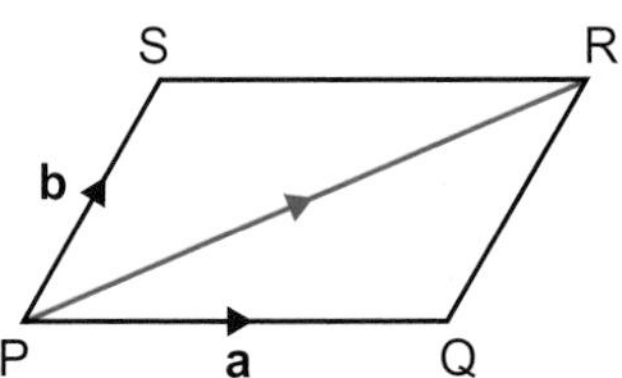

Key point 10

In parallelogram PQRS where $\overrightarrow{PQ}$ is **a** and $\overrightarrow{PS}$ is **b**, the diagonal $\overrightarrow{PR}$ of the parallelogram is **a** + **b**.
This is called the **parallelogram law for vector addition**.
When **c** = **a** + **b** the vector **c** is called the **resultant vector** of the two vectors **a** and **b**.

9 **Exam-style question**

Q9 strategy hint Use the parallelogram law for vector addition.

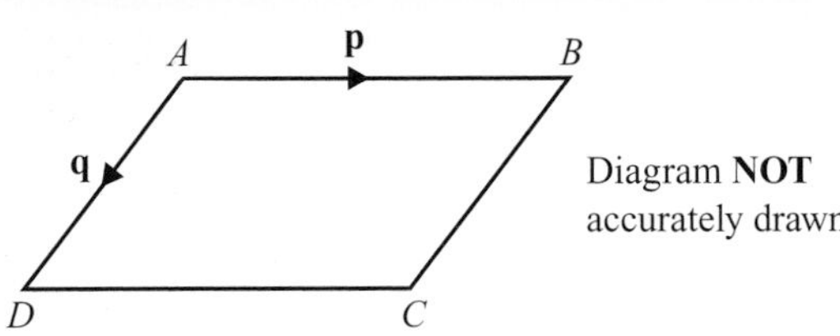

ABCD is a parallelogram.
AB is parallel to *DC*.
AD is parallel to *BC*.
$\overrightarrow{AB} = \mathbf{p}$ $\overrightarrow{AD} = \mathbf{q}$

a Express, in terms of **p** and **q**

i $\overrightarrow{AC}$ **ii** $\overrightarrow{BD}$ **(2 marks)**

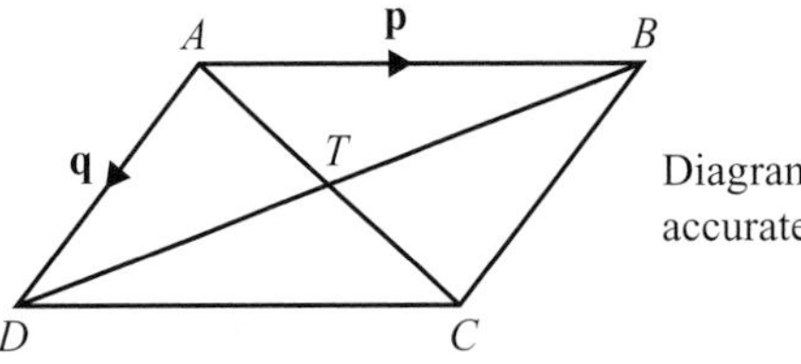

AC and *BD* are diagonals of parallelogram *ABCD*.
AC and *BD* intersect at *T*.

b Express $\overrightarrow{AT}$ in terms of **p** and **q**. **(1 mark)**

June 2006, Q13, 5525/06

10 Reasoning $\overrightarrow{PQ} = \mathbf{a}$ and $\overrightarrow{PR} = \mathbf{b}$.

a Write $\overrightarrow{QR}$ in terms of **a** and **b**.

b Where is the point S such that $\overrightarrow{PS} = \frac{1}{2}\mathbf{b}$?

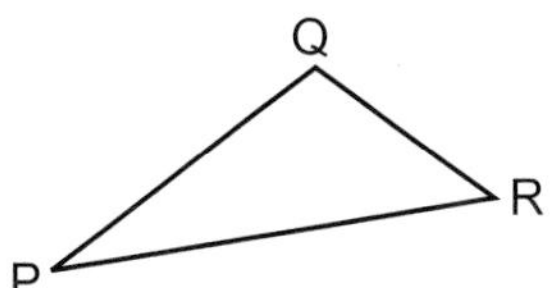

11 Reasoning ABCDEF is a regular hexagon.

$\overrightarrow{AB} = \mathbf{n}$

a Explain why $\overrightarrow{ED} = \mathbf{n}$.

$\overrightarrow{BC} = \mathbf{m}$ and $\overrightarrow{CD} = \mathbf{p}$.

b Find i $\overrightarrow{FE}$ ii $\overrightarrow{AF}$

c Find i $\overrightarrow{AC}$ ii $\overrightarrow{AD}$

d What is $\overrightarrow{FD}$?

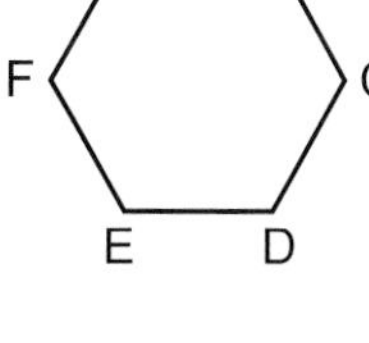

12 Reasoning ABCD is a square.

M is the midpoint of AB.

$\overrightarrow{DC} = \mathbf{r}$ and $\overrightarrow{DA} = \mathbf{s}$.

Write in terms of **r** and **s**

a $\overrightarrow{AB}$ b $\overrightarrow{BC}$

c $\overrightarrow{AM}$ d $\overrightarrow{DM}$

13 Reasoning Here are five vectors.

$\overrightarrow{AB} = 4\mathbf{a} - 2\mathbf{b}$ $\overrightarrow{CD} = 8\mathbf{a} + 12\mathbf{b}$ $\overrightarrow{EF} = 8\mathbf{a} - 4\mathbf{b}$ $\overrightarrow{GH} = -2\mathbf{a} + \mathbf{b}$ $\overrightarrow{IJ} = 12\mathbf{a} - 8\mathbf{b}$

a Three of these vectors are parallel. Which three?

b Simplify

i $4\mathbf{p} + 3\mathbf{q} - \mathbf{p} - 6\mathbf{q}$ ii $2(2\mathbf{a} - 3\mathbf{b}) + \frac{1}{2}(4\mathbf{a} - \mathbf{b})$

14 Reasoning In triangle PQR, $\overrightarrow{PQ} = \mathbf{a}$ and $\overrightarrow{PR} = \mathbf{b}$.

M is the midpoint of QR.

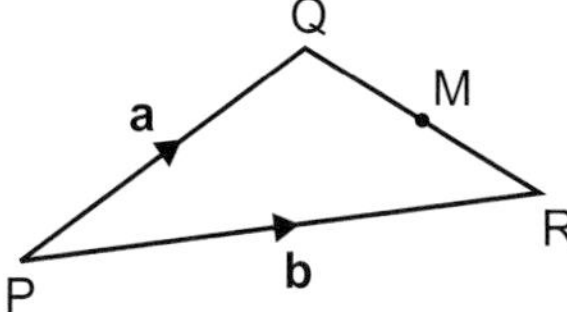

Write in terms of **a** and **b**

a $\overrightarrow{QR}$ b $\overrightarrow{QM}$ c $\overrightarrow{PM}$

Discussion Complete the parallelogram PQSR.

How can you use the parallelogram law to find $\overrightarrow{PM}$?

> **Q14c hint** Make use of the vectors you already know.
>
> $\overrightarrow{PM} = \overrightarrow{P\square} + \overrightarrow{\square M}$
>
> Simplify the expression by adding like vectors.

15 In triangle ABC, $\overrightarrow{AB} = \mathbf{a}$ and $\overrightarrow{AC} = \mathbf{b}$.

P is the midpoint of AB.

Q is the midpoint of AC.

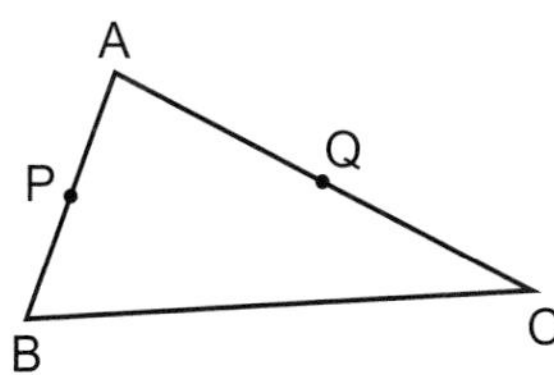

Write in terms of **a** and **b**

a $\overrightarrow{BC}$ b $\overrightarrow{AP}$ c $\overrightarrow{AQ}$ d $\overrightarrow{PQ}$

Discussion What do your answers show about the lines PQ and BC?

18.4 Parallel vectors and collinear points

Objectives

- Express points as position vectors.
- Prove lines are parallel.
- Prove points are collinear.

Why learn this?

Planes flying in formation follow parallel vector flight paths.

Fluency

Here are five vectors. Three of them are parallel. Which three?

$\begin{pmatrix}-2\\4\end{pmatrix}$ $\begin{pmatrix}6\\-12\end{pmatrix}$ $\begin{pmatrix}-1\\-2\end{pmatrix}$ $\begin{pmatrix}4\\-6\end{pmatrix}$ $\begin{pmatrix}1\\-2\end{pmatrix}$

Warm up

1 Here are vectors **a** and **b**.
On squared paper draw the vectors
a $-\mathbf{a}$
b $\mathbf{a} + \mathbf{b}$
c $2\mathbf{a} - 3\mathbf{b}$

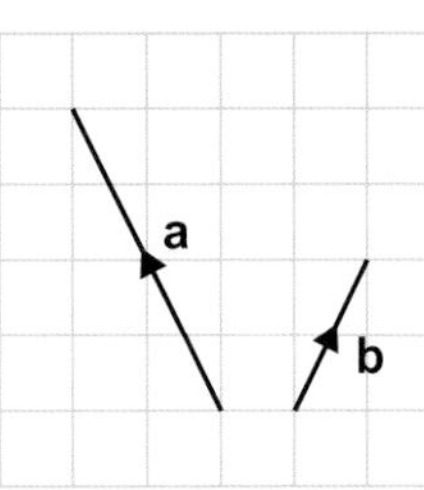

2 Work out
a $\begin{pmatrix}2\\-1\end{pmatrix} + \begin{pmatrix}-3\\4\end{pmatrix}$ b $\begin{pmatrix}5\\2\end{pmatrix} - \begin{pmatrix}3\\1\end{pmatrix}$ c $\begin{pmatrix}-2\\3\end{pmatrix} + \begin{pmatrix}2\\-3\end{pmatrix}$

3 $\mathbf{a} = \begin{pmatrix}-1\\3\end{pmatrix}$, $\mathbf{b} = \begin{pmatrix}-2\\4\end{pmatrix}$ and $\mathbf{a} + \mathbf{c} = \mathbf{b}$.
Calculate **c**.

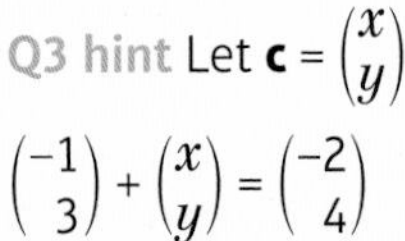

Q3 hint Let $\mathbf{c} = \begin{pmatrix}x\\y\end{pmatrix}$
$\begin{pmatrix}-1\\3\end{pmatrix} + \begin{pmatrix}x\\y\end{pmatrix} = \begin{pmatrix}-2\\4\end{pmatrix}$

4 $2\begin{pmatrix}x\\y\end{pmatrix} + \begin{pmatrix}3\\-1\end{pmatrix} = \begin{pmatrix}4\\-2\end{pmatrix}$
Find $\begin{pmatrix}x\\y\end{pmatrix}$.

Q4 hint $2x + 3 = \square$
$2y - 1 = \square$

5 $\mathbf{e} = \begin{pmatrix}5\\1\end{pmatrix}$ and $\mathbf{f} = \begin{pmatrix}-1\\4\end{pmatrix}$.
Calculate **g** given that $2\mathbf{e} - \mathbf{g} = \mathbf{f}$.

6 O is the origin (0, 0).
A has coordinates (1, 5) and B has coordinates (2, 4).
Find as column vectors
a $\overrightarrow{OA}$ b $\overrightarrow{AO}$ c $\overrightarrow{OB}$ d $\overrightarrow{AB}$

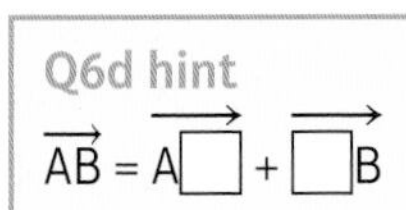

Q6d hint
$\overrightarrow{AB} = \overrightarrow{A\square} + \overrightarrow{\square B}$

Key point 11

With the origin O, the vectors $\overrightarrow{OA}$ and $\overrightarrow{OB}$ are called the **position vectors** of the points A and B.

In general, a point with coordinates (p, q) has position vector $\begin{pmatrix}p\\q\end{pmatrix}$.

7 $\overrightarrow{OA} = \mathbf{a}$ and $\overrightarrow{OB} = \mathbf{b}$.
Express $\overrightarrow{AB}$ in terms of **a** and **b**.

Q7 hint

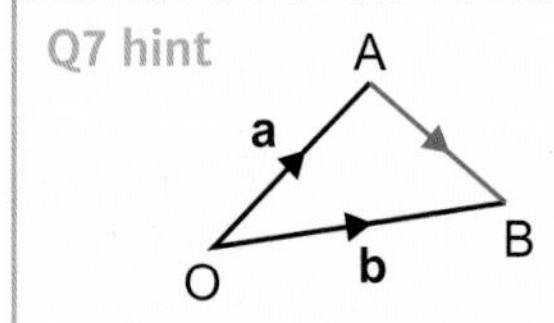

Key point 12

When $\overrightarrow{OA} = \mathbf{a}$ and $\overrightarrow{OB} = \mathbf{b}$, $\overrightarrow{AB} = \overrightarrow{AO} + \overrightarrow{OB} = \mathbf{b} - \mathbf{a}$.

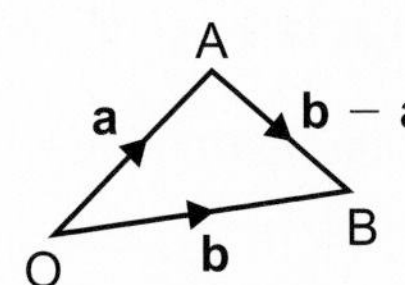

Example 2

The points A, B, C and D have coordinates (1, 3), (2, 7), (−6, −10) and (−1, 10) respectively. O is the origin.

a Write down the position vectors $\overrightarrow{OA}$ and $\overrightarrow{OB}$.

b Work out as column vectors

i $\overrightarrow{AB}$ ii $\overrightarrow{CD}$

c What do these results show about AB and CD?

a $\overrightarrow{OA} = \begin{pmatrix}1\\3\end{pmatrix}$ $\overrightarrow{OB} = \begin{pmatrix}2\\7\end{pmatrix}$

Position vector of (1, 3) is $\begin{pmatrix}1\\3\end{pmatrix}$

b $\overrightarrow{AB} = \overrightarrow{AO} + \overrightarrow{OB} = -\overrightarrow{OA} + \overrightarrow{OB} = \overrightarrow{OB} - \overrightarrow{OA}$

$= \begin{pmatrix}2\\7\end{pmatrix} - \begin{pmatrix}1\\3\end{pmatrix} = \begin{pmatrix}1\\4\end{pmatrix}$

$\overrightarrow{CD} = \overrightarrow{CO} + \overrightarrow{OD} = -\overrightarrow{OC} + \overrightarrow{OD} = \overrightarrow{OD} - \overrightarrow{OC}$

$= \begin{pmatrix}-1\\10\end{pmatrix} - \begin{pmatrix}-6\\-10\end{pmatrix} = \begin{pmatrix}5\\20\end{pmatrix}$

Use the triangle law:

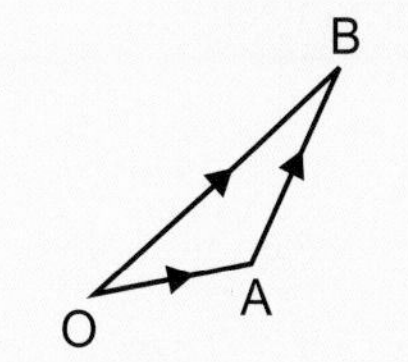

c $\overrightarrow{AB} = \begin{pmatrix}1\\4\end{pmatrix}$ $\overrightarrow{CD} = \begin{pmatrix}5\\20\end{pmatrix} = 5\begin{pmatrix}1\\4\end{pmatrix}$

$\overrightarrow{CD} = 5\overrightarrow{AB}$

The lines CD and AB are parallel and the length of the line CD is 5 times the length of the line AB.

This means CD is a multiple of AB. Explain clearly what $\overrightarrow{CD} = 5\overrightarrow{AB}$ means.

8 **Reasoning** The points P, Q, R and S have coordinates (−2, 5), (3, 1), (−6, −9) and (14, −25) respectively. O is the origin.

a Write down the position vectors $\overrightarrow{OP}$ and $\overrightarrow{OQ}$.

b Work out as a column vector

i $\overrightarrow{PQ}$ ii $\overrightarrow{RS}$

c What do these results show about the lines PQ and RS?

9 **Exam-style question**

P is the point (7, 5) and *Q* is the point (−3, 1).

a Find $\overrightarrow{PQ}$ as a column vector. **(1 mark)**

R is the point such that $\overrightarrow{QR} = \begin{pmatrix}4\\7\end{pmatrix}$

b Write down the co-ordinates of the point *R*. **(2 marks)**

X is the midpoint of *PQ*. *O* is the origin.

c Find $\overrightarrow{OX}$ as a column vector. **(2 marks)**

Exam hint
In this type of vector question it can be helpful to draw a sketch.

10 **Reasoning** The point A has coordinates (1, 3), the point B has coordinates (4, 5) and the point C has coordinates (−2, −4).

a Write $\overrightarrow{AB}$ as a column vector.

b $\overrightarrow{CD} = 6\overrightarrow{AB}$. Find $\overrightarrow{CD}$.

c Find the coordinates of D.

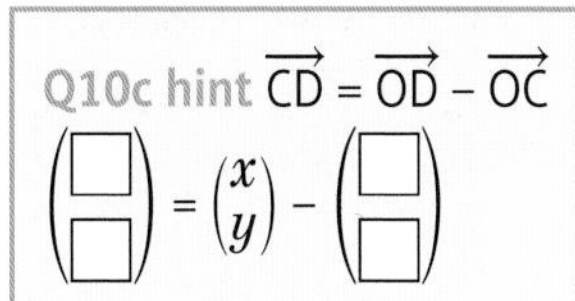

11 **Problem-solving** $\mathbf{a} = \begin{pmatrix} 4 \\ 1 \end{pmatrix}$ and $\mathbf{b} = \begin{pmatrix} -2 \\ 3 \end{pmatrix}$.
Find a vector **c** such that **a** + **c** is parallel to **a** – **b**.

Q11 hint Find **a** – **b** first.

12 **Reasoning** OABC is a quadrilateral in which $\overrightarrow{OA} = \mathbf{a}$, $\overrightarrow{OB} = \mathbf{a} + 2\mathbf{b}$ and $\overrightarrow{OC} = 2\mathbf{b}$.

a Find $\overrightarrow{AB}$ in terms of **a** and **b**.
What does this tell you about $\overrightarrow{AB}$ and $\overrightarrow{OC}$?

b Find $\overrightarrow{BC}$ in terms of **a** and **b**.
What does this tell you about $\overrightarrow{OA}$ and $\overrightarrow{BC}$?

c What type of quadrilateral is OABC?

Q12 hint Sketch a quadrilateral OABC.

Key point 13

$\overrightarrow{PQ} = k\overrightarrow{QR}$ shows that the lines PQ and QR are parallel. Also they both pass through point Q so PQ and QR are part of the same straight line.
P, Q and R are said to be **collinear** (they all lie on the same straight line).

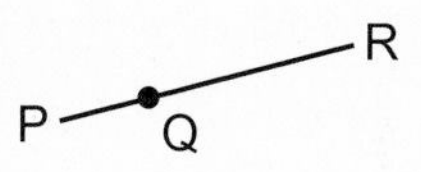

13 **Problem-solving** The points A, B and C have coordinates (2, 13), (5, 22) and (11, 40) respectively.

a Find as column vectors
i $\overrightarrow{AB}$ ii $\overrightarrow{AC}$

b What do these results show you about the points A, B and C?

Discussion How can you show that three points A, B, C are collinear?

14 **Problem-solving / Communication** The point P has coordinates (1, 3).
The point Q has coordinates (4, 6). The point R has coordinates (10, 12).
Show that points P, Q and R are collinear.

15 **Reflect** Explain in your own words what 'Points X, Y and Z are collinear' means.

18.5 Solving geometric problems

Objectives

- Solve geometric problems in two dimensions using vector methods.
- Apply vector methods for simple geometric proofs.

Why learn this?

Programmers use vectors to calculate collisions between objects and/or people in computer games.

Fluency

The point M is on AB such that AM : MB = 1 : 3.
What fraction of AB is AM?

Warm up

1 Find three pairs of parallel vectors.

2**p**	**p** – **q**	3**q** – **p**	5**p**
4**p** – 8**q**	4**q** – 4**p**	2**p** – 6**q**	3**p** – **q**

2 $\overrightarrow{PQ} = 3\mathbf{a} - 2\mathbf{b}$ and $\overrightarrow{PR} = 9\mathbf{a} - 6\mathbf{b}$.
What does this tell you about

a PQ and PR b the points P, Q and R?

ActiveLearn Homework, practice and support: Higher 18.5

3 **Reasoning / Communication** In triangle OAB, the point M is the midpoint of OA and the point N is the midpoint of OB.

$\overrightarrow{OA} = 2\mathbf{a}$ and $\overrightarrow{OB} = 2\mathbf{b}$.

a Express in terms of **a** and/or **b**

i $\overrightarrow{OM}$ ii $\overrightarrow{ON}$ iii $\overrightarrow{MO}$ iv $\overrightarrow{NO}$

b Express $\overrightarrow{AB}$ in terms of **a** and **b**.

c Express $\overrightarrow{MN}$ in terms of **a** and **b**.

d Explain what the answers to parts **b** and **c** show about AB and MN.

Q3a i hint M is the midpoint of OA so $\overrightarrow{OM} = \square$

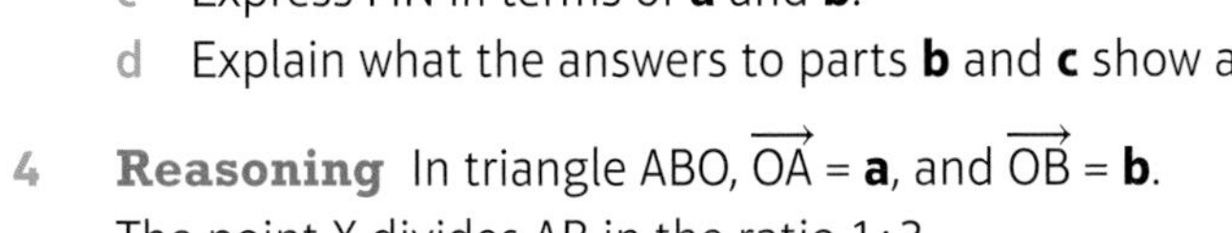

4 **Reasoning** In triangle ABO, $\overrightarrow{OA} = \mathbf{a}$, and $\overrightarrow{OB} = \mathbf{b}$.

The point X divides AB in the ratio 1:2.

Express in terms of **a** and **b**

a $\overrightarrow{AX}$ b $\overrightarrow{OX}$

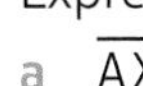

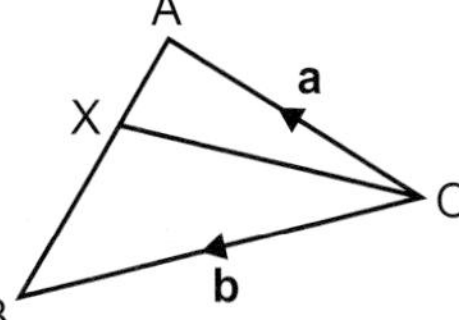

Q4b hint
$\overrightarrow{OX} = \overrightarrow{OA} + \overrightarrow{\square}$

5 **Problem-solving** In triangle OAB, $\overrightarrow{OA} = \mathbf{a}$ and $\overrightarrow{OB} = \mathbf{b}$.

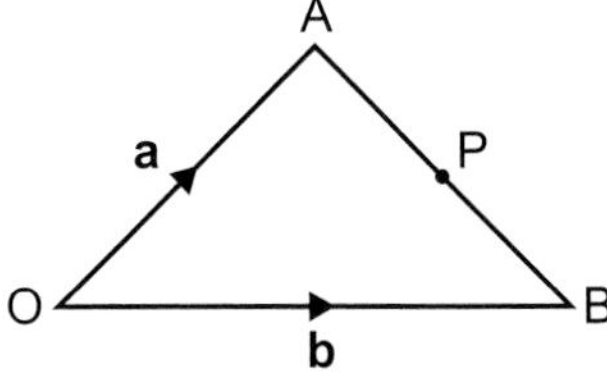

a Find in terms of **a** and **b** the vector $\overrightarrow{AB}$.

P is the midpoint of AB.

b Find in terms of **a** and **b** the vector $\overrightarrow{AP}$.

c Find in terms of **a** and **b** the vector $\overrightarrow{OP}$.

6 **Exam-style question**

$\overrightarrow{OX} = 2\mathbf{a} + \mathbf{b}$ and $\overrightarrow{OY} = 4\mathbf{a} + 3\mathbf{b}$

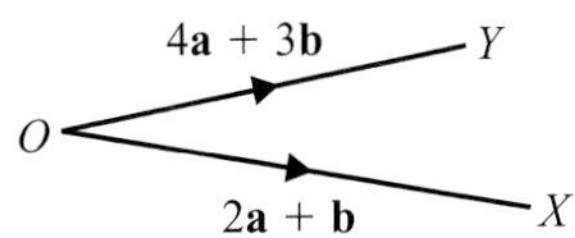

a Express the vector $\overrightarrow{XY}$ in terms of **a** and **b**.
Give your answer in its simplest form. **(2 marks)**

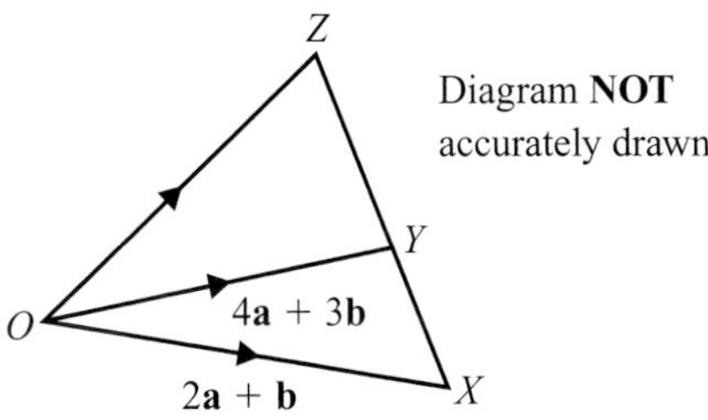

XYZ is a straight line.

$XY : YZ = 2 : 3$.

b Express the vector $\overrightarrow{OZ}$ in terms of **a** and **b**.
Give your answer in its simplest form. **(3 marks)**

Nov 2008, Q26, 5540H/4H

Q6b strategy hint
$XY : YZ = 2 : 3$

$YZ = \frac{\square}{\square} XZ$

Example 3

OABC is a quadrilateral in which $\overrightarrow{OA} = \mathbf{a}$, $\overrightarrow{OB} = \mathbf{a} + 2\mathbf{b}$ and $\overrightarrow{OC} = 4\mathbf{b}$.
D is the point such that $\overrightarrow{BD} = \overrightarrow{OC}$ and X is the midpoint of BC.

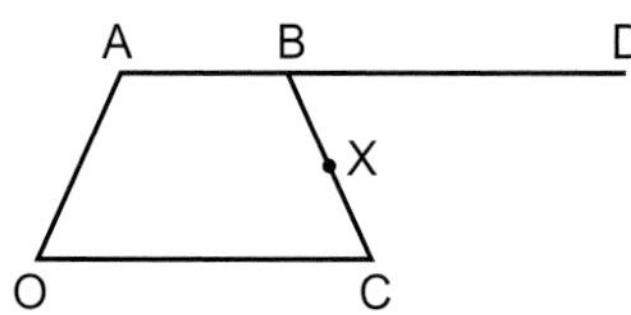

a Find in terms of **a** and **b**

i $\overrightarrow{OD}$ ii $\overrightarrow{OX}$

b Explain what your answers to parts **a i** and **a ii** mean.

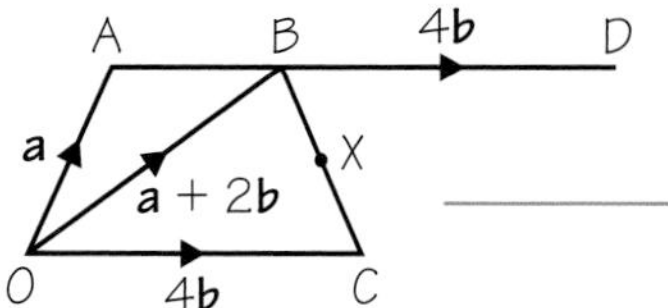

Copy the diagram and mark on all the vectors.
$\overrightarrow{BD} = \overrightarrow{OC}$ so BD and OC are parallel and have the same length.

i $\overrightarrow{OD} = \overrightarrow{OB} + \overrightarrow{BD}$

$= \mathbf{a} + 2\mathbf{b} + 4\mathbf{b} = \mathbf{a} + 6\mathbf{b}$

ii $\overrightarrow{OX} = \overrightarrow{OC} + \overrightarrow{CX}$ $\quad \overrightarrow{CX} = \frac{1}{2}\overrightarrow{CB}$

To find $\overrightarrow{CX}$ you first need to find $\overrightarrow{CB}$.

$\overrightarrow{CB} = \overrightarrow{CO} + \overrightarrow{OB} = \overrightarrow{OB} - \overrightarrow{OC} = \mathbf{a} + 2\mathbf{b} - 4\mathbf{b} = \mathbf{a} - 2\mathbf{b}$

$\overrightarrow{CX} = \frac{1}{2}(\mathbf{a} - 2\mathbf{b}) = \frac{1}{2}\mathbf{a} - \mathbf{b}$

$\overrightarrow{OX} = 4\mathbf{b} + \frac{1}{2}\mathbf{a} - \mathbf{b} = \frac{1}{2}\mathbf{a} + 3\mathbf{b}$

b $\overrightarrow{OD} = \mathbf{a} + 6\mathbf{b} = 2(\frac{1}{2}\mathbf{a} + 3\mathbf{b})$

$\overrightarrow{OD} = 2\overrightarrow{OX}$

Compare $\overrightarrow{OD}$ and $\overrightarrow{OX}$ to see if one is a multiple of the other.

So OD and OX are parallel with a point in common. This means that O, X and D lie on the same straight line. The length of OD is 2 times the length of OX. So X is the midpoint of OD.

7 **Exam-style question**

The diagram shows a regular hexagon *ABCDEF* with centre *O*.

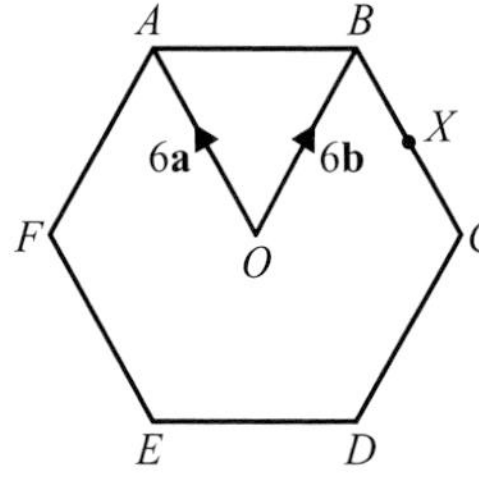

$\overrightarrow{OA} = 6\mathbf{a}$ $\quad \overrightarrow{OB} = 6\mathbf{b}$

a Express in terms of **a** and/or **b**

i $\overrightarrow{AB}$ **ii** $\overrightarrow{EF}$ **(2 marks)**

X is the midpoint of *BC*.

b Express $\overrightarrow{EX}$ in terms of **a** and/or **b**. **(2 marks)**

Y is the point on *AB* extended, such that $AB : BY = 3 : 2$

c Prove that *E*, *X* and *Y* lie on the same straight line. **(3 marks)**

June 2003, Q23, 5505/05

8 **Problem-solving / Communication** OACB is a parallelogram with $\overrightarrow{OA} = \mathbf{a}$ and $\overrightarrow{OB} = \mathbf{b}$.
E is the point on AC such that AE = $\frac{1}{4}$AC.
F is the point on BC such that BF = $\frac{1}{4}$BC.

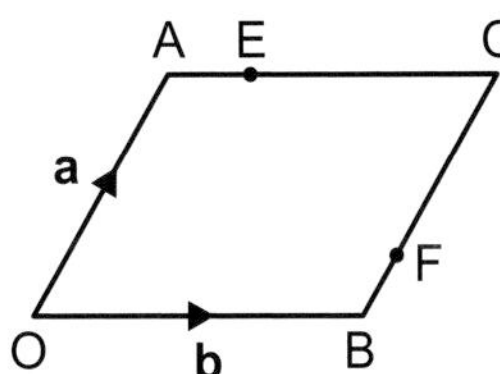

a Find in terms of **a** and/or **b**
i $\overrightarrow{AB}$ ii $\overrightarrow{AE}$ iii $\overrightarrow{OE}$ iv $\overrightarrow{OF}$

b Show that EF is parallel to AB.

9 **Problem-solving / Communication** In triangle OMN, $\overrightarrow{OM} = \mathbf{m}$ and $\overrightarrow{ON} = \mathbf{n}$.
The point P is the midpoint of MN and Q is the point such that $\overrightarrow{OQ} = \frac{3}{2}\overrightarrow{OP}$.

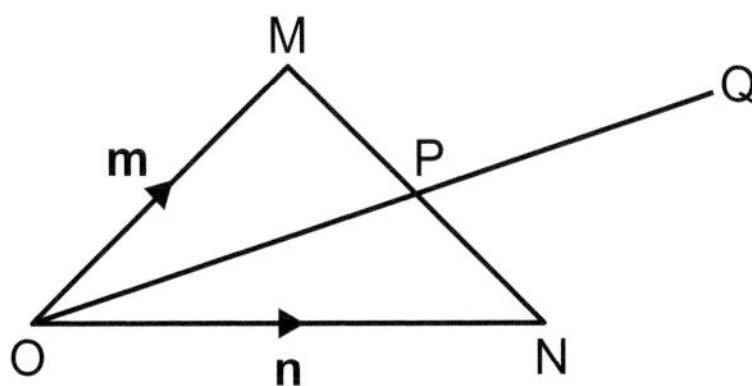

a Find in terms of **m** and **n**
i $\overrightarrow{OP}$ ii $\overrightarrow{OQ}$ iii $\overrightarrow{MQ}$

The point R is such that $\overrightarrow{OR} = 3\overrightarrow{ON}$.

b Find in terms of **m** and **n** the vector $\overrightarrow{MR}$.

c Explain why MQR is a straight line and give the value of $\frac{MR}{MQ}$.

10 **Problem-solving / Communication** In the diagram, $\overrightarrow{OR} = 6\mathbf{a}$, $\overrightarrow{OP} = 2\mathbf{b}$ and $\overrightarrow{PQ} = 3\mathbf{a}$.

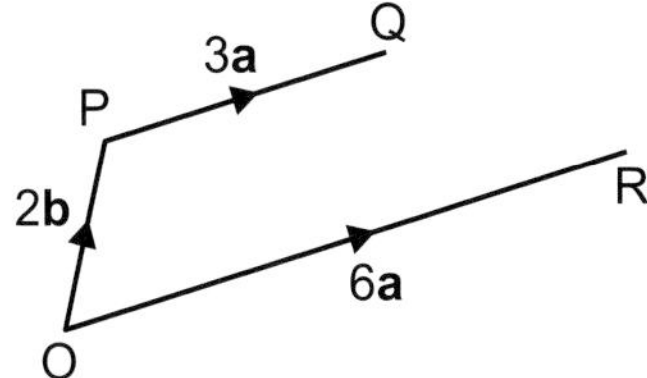

The point M is on PQ such that $\overrightarrow{PM} = 2\mathbf{a}$.
The point N is on OR such that $\overrightarrow{ON} = \frac{1}{3}\overrightarrow{OR}$.
The midpoint of MN is the point S.

a Find in terms of **a** and/or **b** the vector $\overrightarrow{NM}$.

b Find in terms of **a** and/or **b** the vector $\overrightarrow{OS}$.

c T is the point such that $\overrightarrow{QT} = \mathbf{a}$. Find in terms of **a** and **b** the vector $\overrightarrow{OT}$.

d Show that S lies on the line OT.

e When $\mathbf{a} = \begin{pmatrix} 8 \\ 2 \end{pmatrix}$ and $\mathbf{b} = \begin{pmatrix} 3 \\ 15 \end{pmatrix}$ find the length of QR.

18 Problem-solving

Objective • Use different problem-solving strategies and then 'explain'.

Key point 14

There are many different problem-solving strategies. Here are some you can use:

- pictures
- lists
- smaller numbers
- bar models
- x for the unknown
- flow diagrams
- arrow diagrams
- geometric sketches
- graphs
- logical reasoning

Example 4

The diagram shows triangle OPQ.

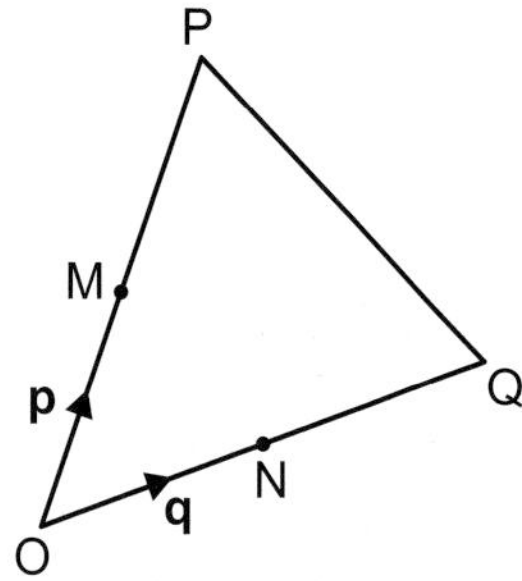

M is the midpoint of OP and N is the midpoint of OQ.
$\overrightarrow{OM} = \mathbf{p}$ and $\overrightarrow{ON} = \mathbf{q}$.

a i Find $\overrightarrow{OP}$ and hence $\overrightarrow{PO}$.
 ii Find $\overrightarrow{OQ}$ and hence $\overrightarrow{QO}$.

b Find $\overrightarrow{PN}$ and $\overrightarrow{QM}$.

c X lies on PN such that PX = $\frac{2}{3}$PN.
 Y lies on QM such that QY = $\frac{2}{3}$QM.
 Find $\overrightarrow{OX}$ and $\overrightarrow{OY}$.
 Explain what your answer means.

a i $\overrightarrow{OP} = 2\mathbf{p}$ hence $\overrightarrow{PO} = -2\mathbf{p}$

M is the midpoint of OP. This means OP = 2 × OM.
$\overrightarrow{PO}$ is in the opposite direction and so is equal to $-\overrightarrow{OP}$.

 ii $\overrightarrow{OQ} = 2\mathbf{q}$ hence $\overrightarrow{QO} = -2\mathbf{q}$

N is the midpoint of OQ. This means OQ = 2 × ON.
$\overrightarrow{QO}$ is in the opposite direction and so is equal to $-\overrightarrow{OQ}$.

b

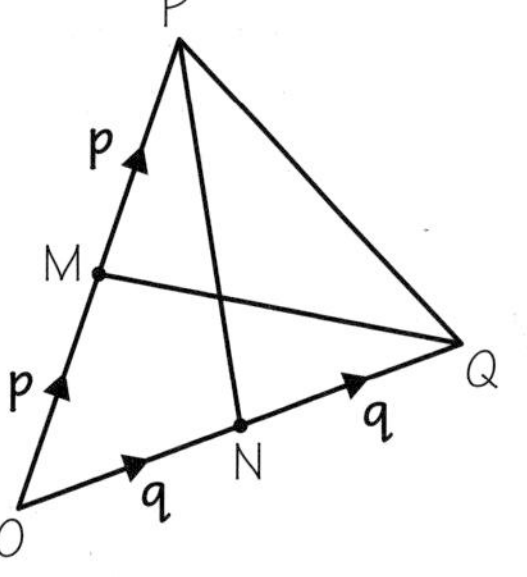

Sketch the diagram. Write on your diagram all the vectors you know. Join PN. Join QM.

$\overrightarrow{PN} = \overrightarrow{PO} + \overrightarrow{ON} = -2\mathbf{p} + \mathbf{q}$
$\overrightarrow{QM} = \overrightarrow{QO} + \overrightarrow{OM} = -2\mathbf{q} + \mathbf{p} = \mathbf{p} - 2\mathbf{q}$

Look at the diagram. Move from P to N and from Q to M along vectors you know.

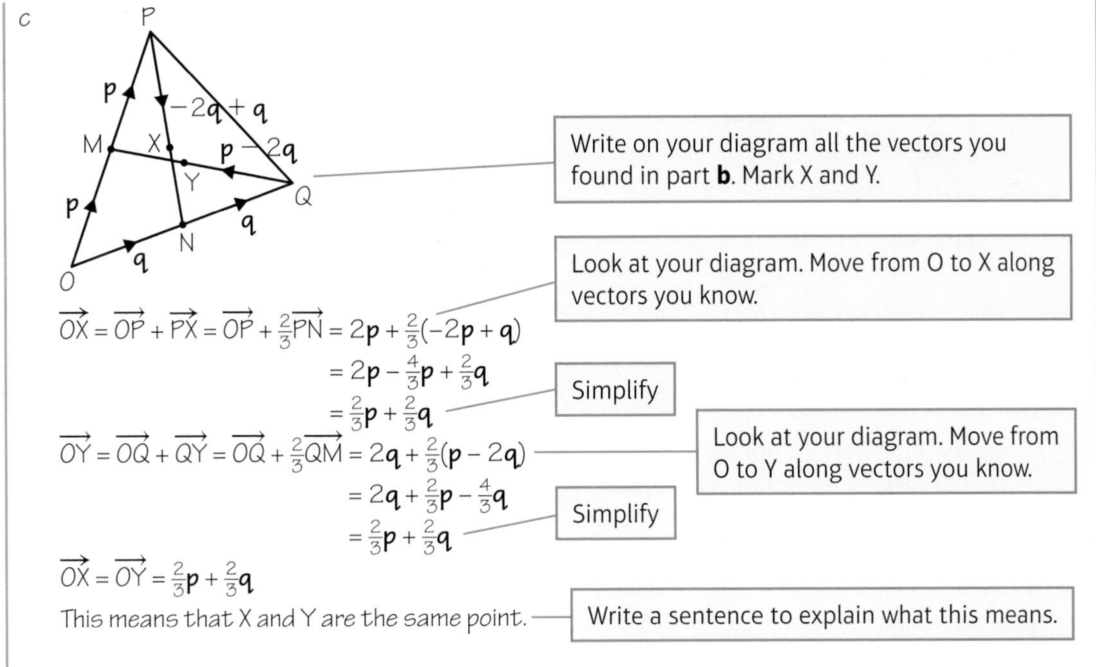

c
P
p
−2q + q
M
X
p − 2q
Y
Q
p
q
N
q
O
Write on your diagram all the vectors you found in part **b**. Mark X and Y.
Look at your diagram. Move from O to X along vectors you know.
$\overrightarrow{OX} = \overrightarrow{OP} + \overrightarrow{PX} = \overrightarrow{OP} + \frac{2}{3}\overrightarrow{PN} = 2\mathbf{p} + \frac{2}{3}(-2\mathbf{p} + \mathbf{q})$
$= 2\mathbf{p} - \frac{4}{3}\mathbf{p} + \frac{2}{3}\mathbf{q}$
$= \frac{2}{3}\mathbf{p} + \frac{2}{3}\mathbf{q}$
Simplify
$\overrightarrow{OY} = \overrightarrow{OQ} + \overrightarrow{QY} = \overrightarrow{OQ} + \frac{2}{3}\overrightarrow{QM} = 2\mathbf{q} + \frac{2}{3}(\mathbf{p} - 2\mathbf{q})$
Look at your diagram. Move from O to Y along vectors you know.
$= 2\mathbf{q} + \frac{2}{3}\mathbf{p} - \frac{4}{3}\mathbf{q}$
$= \frac{2}{3}\mathbf{p} + \frac{2}{3}\mathbf{q}$
Simplify
$\overrightarrow{OX} = \overrightarrow{OY} = \frac{2}{3}\mathbf{p} + \frac{2}{3}\mathbf{q}$
This means that X and Y are the same point.
Write a sentence to explain what this means.

1 In triangle OAB, $\overrightarrow{OA} = \mathbf{a}$ and $\overrightarrow{OB} = \mathbf{b}$.
X lies on OA such that OX = $\frac{1}{4}$OA.
Y lies on OB such that OY = $\frac{1}{4}$OB. Find

a $\overrightarrow{AB}$

b i $\overrightarrow{OX}$ ii $\overrightarrow{OY}$ iii $\overrightarrow{XY}$

c Explain what your answers to parts **a** and **b iii** show about AB and XY.

> **Q1 hint** Sketch a diagram of the triangle. Write on your diagram all the vectors you know. Mark X and Y.

> **Q1b hint** Write on your diagram all the vectors you have found so far.

2 Anna, Beth, Cara and Dana are in the final of a solo singing competition.
Anna says, 'There are twelve possible results for first and second place.'
Dana says, 'There are four of us in the competition and two places for first and second, so there are only eight possible results.'
Who is correct? Explain.

> **Q2 hint** You could list all possible results to help you solve the problem. Work logically so you don't miss any.

3 An aircraft pilot records the air temperature, in °C, at different heights (in metres) above sea level. The table shows three of his results.

Height above sea level (m)	500	1400	2000
Temperature (°C)	13	4	−2

The pilot predicts that the temperature at 2500 m above sea level will be −7 °C. Explain.

> **Q3 hint** You could use a graph to help you solve the problem.

4 Two dog walkers meet at the corner of a park, 350 m × 235 m. They are both heading to the opposite corner. They set off at the same time.
The spaniel walker takes a path along two adjacent sides of the park, and walks at an average pace of 5.2 km/h.
The labrador walker takes a different path across the diagonal of the park and walks at an average pace of 4.3 km/h.
Which dog walker gets to the opposite corner first? Explain.

> **Q4 hint** You could use a geometric sketch to help you solve the problem.

5 These urns of sunflower oil are mathematically similar.

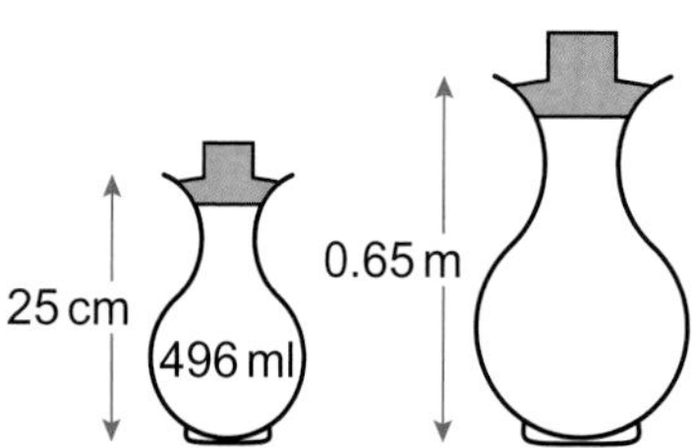

> **Q5 hint**
> Volume scale factor = (linear scale factor)^3
> You could use an arrow diagram to help you solve the problem.
>
> 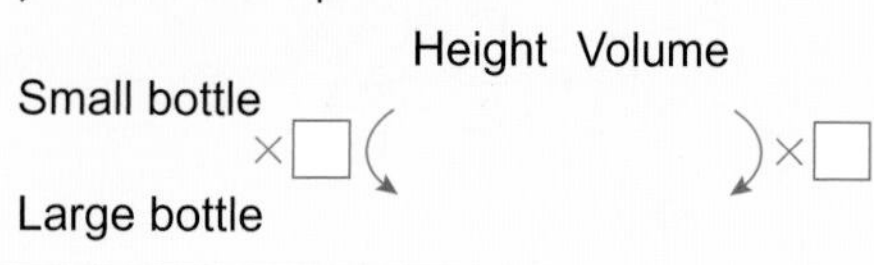
>

Maria says, 'The large urn holds 1 litre 290 millilitres (to the nearest millilitre)'.
Alexandra says, 'The large urn holds 8 litres 718 millilitres (to the nearest millilitre)'.
Who is correct? Explain.

6 **Finance** Ross earns a bonus. It is 15% of his annual salary. He spends 40% of his bonus on a holiday and $\frac{1}{4}$ of the remainder on a laptop. £1890 of his bonus is left.
Antony also earns a bonus. It is 12.5% of his annual salary. He spends 45% of his bonus on a holiday and $\frac{1}{5}$ of the remainder on a lawnmower. £1650 of his bonus is left.
Who earns the bigger annual salary? Explain.

> **Q6 hint** You could use bar models (one for Ross and one for Antony) to help you solve the problem.

7 Every autumn, orchard owners employ apple pickers. Each day, an apple picker picks, on average, 15 bags of apples. Each bag of apples weighs, on average, 18 kg. The apples are packed into trays that, on average, contain 620 g of apples.
One orchard owner's target is to have 10 800 trays of apples at the end of every day.
He says that to achieve this he must employ at least 25 apple pickers. Explain.

> **Q7 hint** You could use x for the unknown. Read each sentence, one at a time, to build an equation that you can solve for x. Beware of different units.

8 **Reflect** What clues are there in a question that help you decide which problem-solving strategy to use?

18 Check up

Log how you did on your Student Progression Chart.

Vector notation

1 The diagram shows two vectors **a** and **b**.

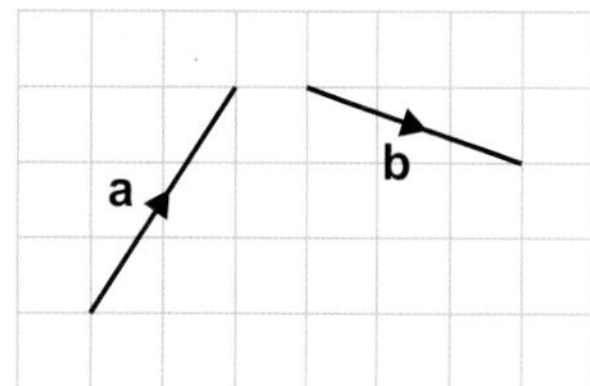

Write **a** and **b** as column vectors.

2 a A is the point (1, 4) and $\overrightarrow{AB} = \begin{pmatrix} 2 \\ 3 \end{pmatrix}$.

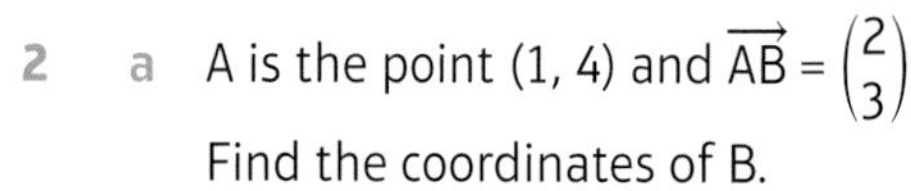

Find the coordinates of B.

b C is the point (4, 3) and D is the point (7, −2). Express $\overrightarrow{CD}$ as a column vector.

3 Find the magnitude of the vector $\begin{pmatrix} -3 \\ 5 \end{pmatrix}$. Give your answer in surd form.

Vector arithmetic

4 The diagram shows two vectors **p** and **q**.
On squared paper draw vectors to represent

a $2\mathbf{q}$ b $\mathbf{p} + \mathbf{q}$ c $\mathbf{p} - \mathbf{q}$

5 $\overrightarrow{AB} = \begin{pmatrix} 3 \\ -1 \end{pmatrix}$ and $\overrightarrow{BC} = \begin{pmatrix} -5 \\ 4 \end{pmatrix}$. Find $\overrightarrow{AC}$.

6 $\mathbf{a} = \begin{pmatrix} 4 \\ -2 \end{pmatrix}$ and $\mathbf{b} = \begin{pmatrix} -1 \\ 6 \end{pmatrix}$. Find

a $\mathbf{a} + \mathbf{b}$ b $\mathbf{b} - \mathbf{a}$ c $3\mathbf{a}$

7 $\mathbf{p} = \begin{pmatrix} 5 \\ -1 \end{pmatrix}$ and $\mathbf{q} = \begin{pmatrix} 9 \\ -3 \end{pmatrix}$.

$\mathbf{p} + 2\mathbf{r} = \mathbf{q}$

Find **r** as a column vector.

Geometrical problems

8 Which of these vectors are parallel to

a $\mathbf{a} - \mathbf{b}$ b $\mathbf{a} + \mathbf{b}$

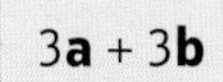

$3\mathbf{a} + 3\mathbf{b}$ $2\mathbf{a} - \mathbf{b}$ $3\mathbf{a} - 3\mathbf{b}$ $\frac{1}{2}\mathbf{a} - \frac{1}{2}\mathbf{b}$

9 P is the point (4, −3) and Q is the point (−2, 7).

a Write down the position vector, **p**, of the point P.

b Write down the position vector, **q**, of the point Q.

c Work out $\overrightarrow{PQ}$.

10 **Reasoning** The points A, B and C have coordinates (2, 13), (5, 22) and (11, 40) respectively.

a Find as column vectors

i $\overrightarrow{AB}$ ii $\overrightarrow{AC}$

b What do these results show about the points A, B and C?

11 **Exam-style question**

OABC is a parallelogram.
M is the midpoint of *CB*.
N is the midpoint of *AB*.
$\overrightarrow{OA} = \mathbf{a}$
$\overrightarrow{OC} = \mathbf{c}$

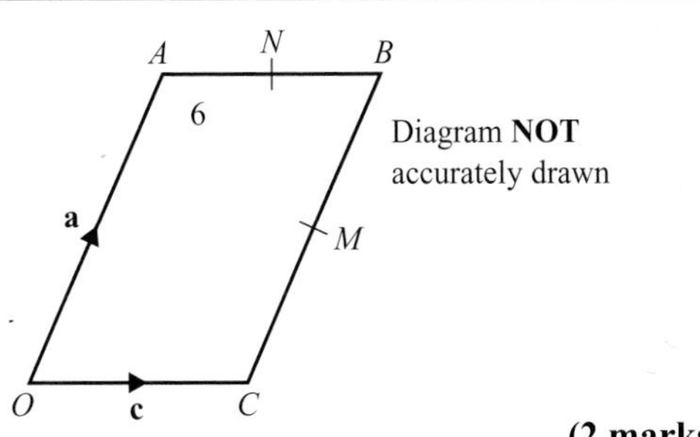

a Find, in terms of **a** and/or **c**, the vectors

i $\overrightarrow{MB}$

ii $\overrightarrow{MN}$ **(2 marks)**

b Show that *CA* is parallel to *MN*. **(2 marks)**

May 2008, Q25, 5540H/3H

12 How sure are you of your answers? Were you mostly

Just guessing  Feeling doubtful  Confident 

What next? Use your results to decide whether to strengthen or extend your learning.

* Challenge

13 **p** and **q** are two vectors such that

- $\mathbf{p} \neq \mathbf{q}$
- magnitude of $\mathbf{p} + \mathbf{q}$ = magnitude of $\mathbf{p} - \mathbf{q}$

Show that **p** and **q** are perpendicular vectors.

18 Strengthen

Vector notation

1 Write down the column vector that describes each transformation.

a A to B b B to A c A to C d A to D

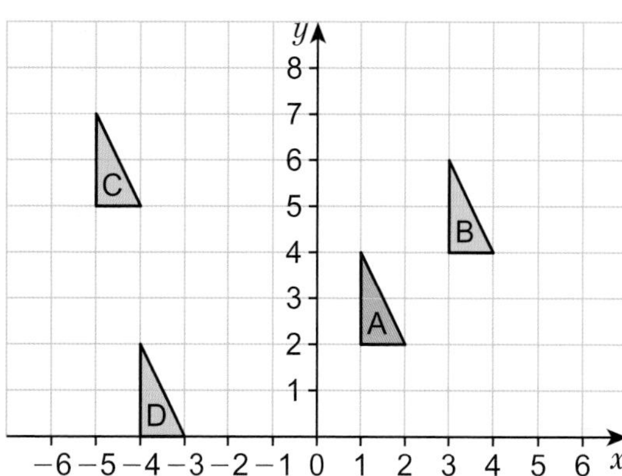

2 Write these as column vectors.

a $\overrightarrow{OA}$ b $\overrightarrow{OB}$

Q2 hint $\overrightarrow{OA}$ is the vector that translates O to A.

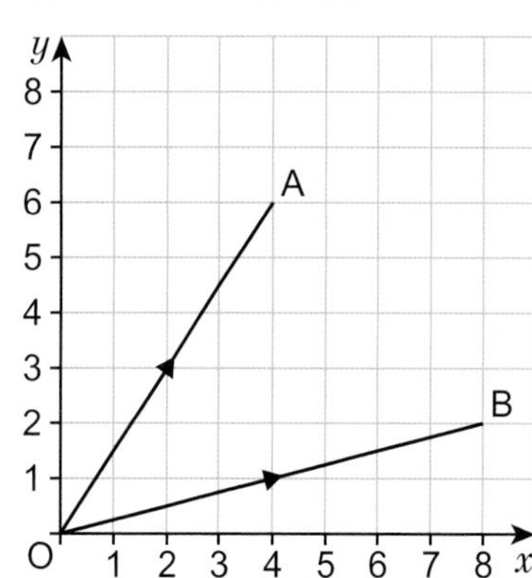

3 P is the point (2, 3) and $\overrightarrow{PQ} = \begin{pmatrix} 1 \\ 5 \end{pmatrix}$.

Find the coordinates of Q.

Q3 hint Plot the point P on a grid. From P move 1 unit across and 5 units up to find the position of Q.

4 A is the point (4, 2).
B is the point (7, 1).
Express $\overrightarrow{AB}$ as a column vector.

Q4 hint Plot the points A and B on a grid. Write the column vector that translates A to B.

5 a Copy and complete the diagram to show the vector $\begin{pmatrix} 2 \\ -5 \end{pmatrix}$.

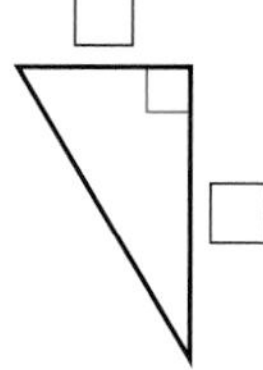

b Use Pythagoras' theorem to work out the magnitude of the vector $\begin{pmatrix} 2 \\ -5 \end{pmatrix}$ (the length of the hypotenuse of the triangle).
Give your answer in surd form.

6 Find the magnitude of each vector, giving your answers in surd form.

a $\begin{pmatrix} 3 \\ -4 \end{pmatrix}$ b $\begin{pmatrix} -9 \\ 5 \end{pmatrix}$ c $\begin{pmatrix} 9 \\ 7 \end{pmatrix}$ d $\begin{pmatrix} -3 \\ -5 \end{pmatrix}$

Q6 hint Use the method from **Q5**.

Vector arithmetic

1 The diagram shows the vector **a**.

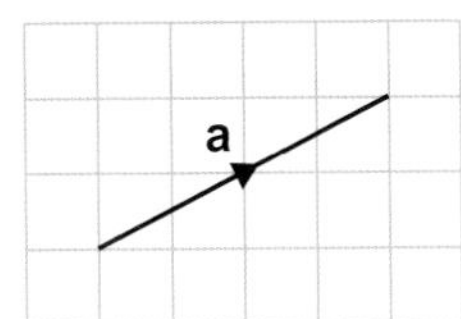

Q1a hint Draw vector **a** twice end to end.

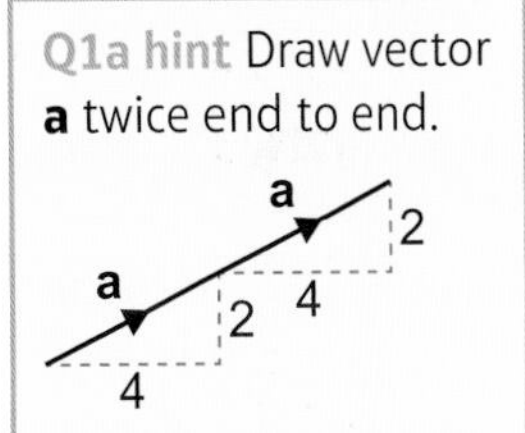

On squared paper draw vectors to represent

a 2**a** b 3**a**

c $\frac{1}{2}$**a** d −**a**

Q1d hint −**a** is in the opposite direction to **a**.

2 The diagram shows the vectors **b** and **c**.

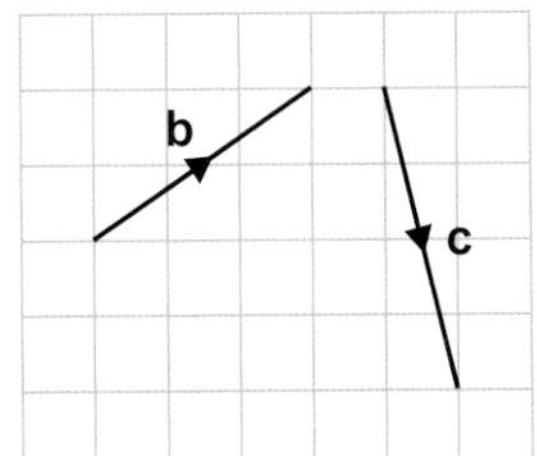

Q2a hint

start, b, c, finish

On squared paper draw vectors to represent

a **b** + **c** b **b** − **c**

3 $\mathbf{a} = \begin{pmatrix} 3 \\ 4 \end{pmatrix}$ and $\mathbf{b} = \begin{pmatrix} 1 \\ 2 \end{pmatrix}$.

a On squared paper draw the vectors

i **a** ii **b** iii **a** + **b** iv **a** − **b** v 2**a** + **b**

b Use your answers to part **a** to write as column vectors

i **a** + **b** ii **a** − **b** iii 2**a** + **b**

c Work out

i $\mathbf{a} + \mathbf{b} = \begin{pmatrix} 3 \\ 4 \end{pmatrix} + \begin{pmatrix} 1 \\ 2 \end{pmatrix} = \begin{pmatrix} 3 + 1 \\ 4 + 2 \end{pmatrix}$

ii $\mathbf{a} - \mathbf{b} = \begin{pmatrix} 3 \\ 4 \end{pmatrix} - \begin{pmatrix} 1 \\ 2 \end{pmatrix}$

iii $2\mathbf{a} + \mathbf{b} = 2\begin{pmatrix} 3 \\ 4 \end{pmatrix} + \begin{pmatrix} 1 \\ 2 \end{pmatrix}$

d What do you notice about your answers to parts **b** and **c**?

4 $\mathbf{a} = \begin{pmatrix} 3 \\ 2 \end{pmatrix}$ and $\mathbf{b} = \begin{pmatrix} -1 \\ 4 \end{pmatrix}$.

Calculate

a 2**a** + **b** b **a** − 3**b** c 4**a** + 2**b** d 3(**a** − **b**)

5 $\overrightarrow{AB} = \begin{pmatrix} 2 \\ 4 \end{pmatrix}$, $\overrightarrow{BC} = \begin{pmatrix} 4 \\ -1 \end{pmatrix}$ and $\overrightarrow{CD} = \begin{pmatrix} -5 \\ 3 \end{pmatrix}$.

Work out

a $\overrightarrow{AB} + \overrightarrow{BC}$ b $\overrightarrow{BC} - \overrightarrow{CD}$ c $2\overrightarrow{AB} + \overrightarrow{CD}$

6 $\mathbf{a} = \begin{pmatrix} 2 \\ 7 \end{pmatrix}$, $\mathbf{b} = \begin{pmatrix} -1 \\ -3 \end{pmatrix}$ and **a** − **x** = **b**.

Find **x** as a column vector.

Q6 hint **a** − **x** = **b**

$x = \begin{pmatrix} x \\ y \end{pmatrix}$ $\begin{pmatrix} 2 \\ 7 \end{pmatrix} - \begin{pmatrix} x \\ y \end{pmatrix} = \begin{pmatrix} -1 \\ -3 \end{pmatrix}$ $2 - x$

Geometrical problems

1 Draw two parallel lines of the same length on a squared grid. Write their column vectors.
What do you notice about the vectors of parallel lines?
Draw some more pairs of parallel lines to check your findings.

2 Which vectors are parallel?

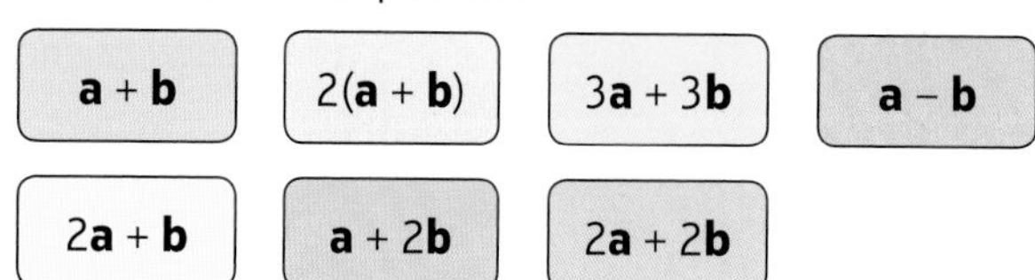

Q2 hint For any number k, $-\mathbf{a}$ and $k\mathbf{a}$ are parallel to **a**.

3 Which vectors are parallel?

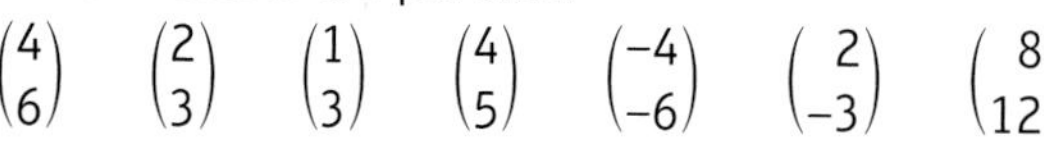

Q3 hint For any number k, $-\overrightarrow{AB}$ and $k\overrightarrow{AB}$ are parallel to $\overrightarrow{AB}$.

4 ABCD is a trapezium.
$\overrightarrow{AB} = \mathbf{a}$ and $\overrightarrow{AD} = \mathbf{b}$.

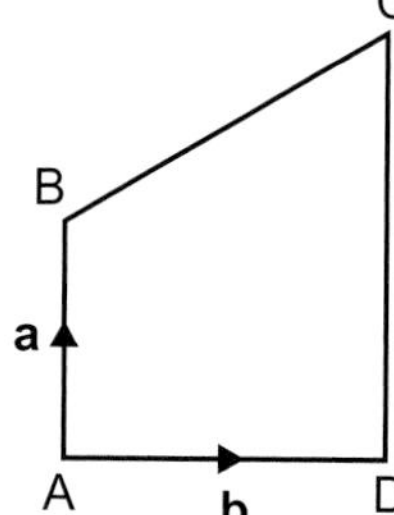

a CD is parallel to AB.
What does this tell you about the vectors $\overrightarrow{AB}$ and $\overrightarrow{CD}$?

b CD = 2AB
Copy and complete
$\overrightarrow{CD} = \square\overrightarrow{AB} = \square$

c Copy and complete
$\overrightarrow{BC} = \overrightarrow{BA} + \overrightarrow{\square} + \overrightarrow{\square}$
$= -\mathbf{a} + \square + \square$
Simplify your answer by collecting like vectors.

Q4c hint Trace around the diagram.

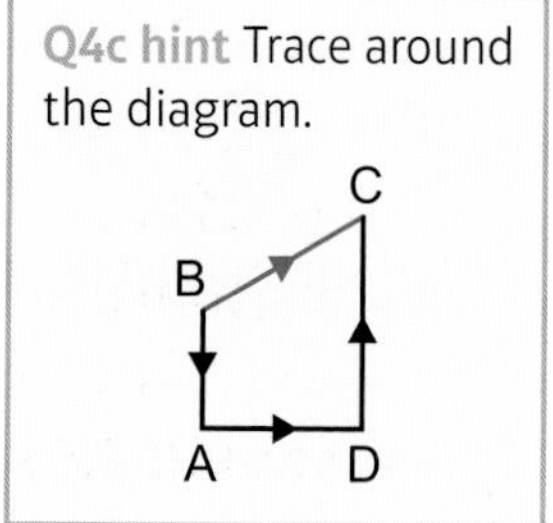

5 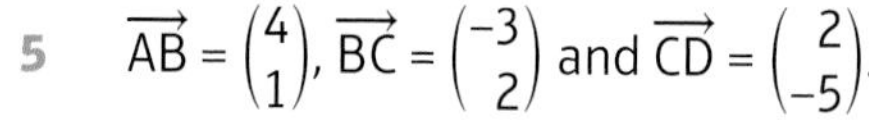

a On squared paper draw a diagram to show ABCD.

b Find the column vector for $\overrightarrow{AD}$.

c Show that $\overrightarrow{AC} = \overrightarrow{DB}$.
What does this tell you about the vectors $\overrightarrow{AC}$ and $\overrightarrow{DB}$?

Q5a hint

6 **Problem-solving** A, B and C are collinear.

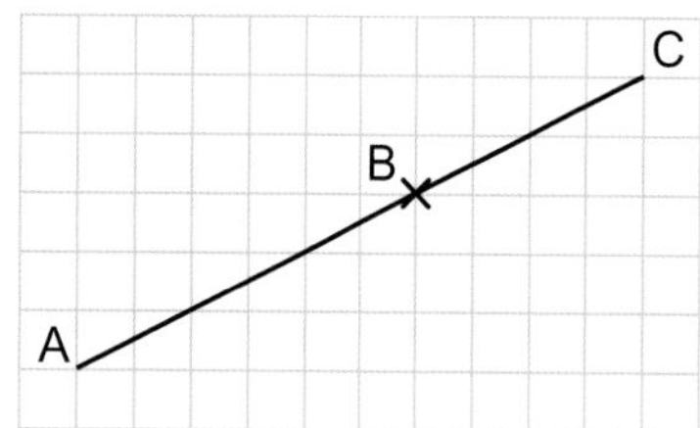

a Write down the column vectors
i $\overrightarrow{AB}$ ii $\overrightarrow{BC}$

b How do these vectors show that AB and BC are collinear?

7 P is the point with coordinates (3, –2).
Q is the point with coordinates (5, –1).

a Write down the position vector, **p**, of the point P.

b Write down the position vector, **q**, of the point Q.

c Work out $\overrightarrow{PQ}$.

Q7 hint Plot O(0, 0), P and Q on a coordinate grid. $\overrightarrow{OP}$ is the vector that translates O to P.

8 A is the point (4, 1), B is the point (8, 4) and C is the point (20, 13).

a Find $\overrightarrow{AB}$ and $\overrightarrow{BC}$.

b Show that AB and BC are parallel.

c Copy and complete:

AB and BC are _________ and both pass through the point _____

So, ABC is a __________ line and A, B, C are collinear.

9 OAB is a triangle. C is the midpoint of OA. D is the midpoint of AB.

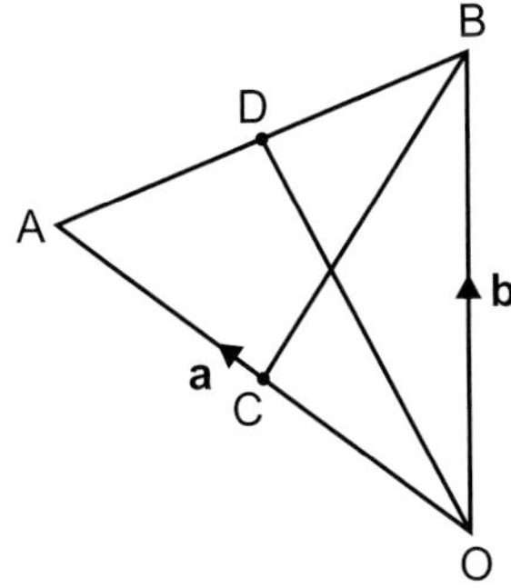

$\overrightarrow{OA} = \mathbf{a}$ and $\overrightarrow{OB} = \mathbf{b}$.

a Write $\overrightarrow{OC}$ in terms of **a**.

b Find $\overrightarrow{BA}$ in terms of **a** and **b**.

c Find $\overrightarrow{BD}$ in terms of **a** and **b**.

d Find $\overrightarrow{OD}$ in terms of **a** and **b**.

e Find $\overrightarrow{CD}$ in terms of **a** and **b**.

f Show that CD is parallel to OB.

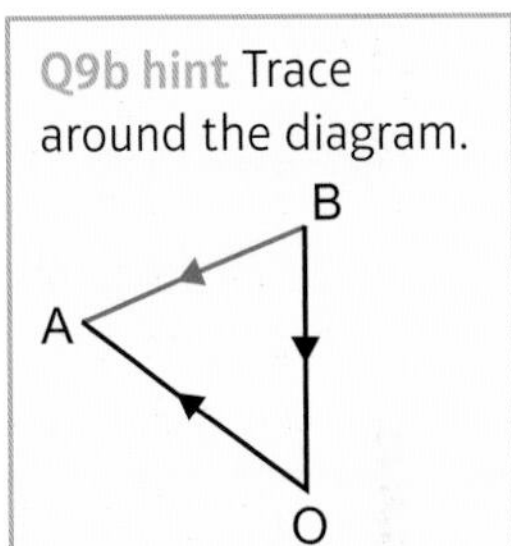

Q9b hint Trace around the diagram.

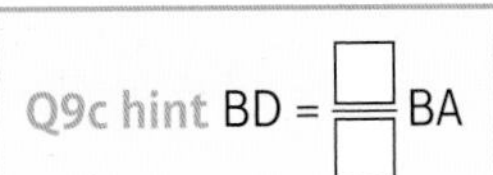

Q9c hint BD = $\frac{\square}{\square}$ BA

10 **Reasoning** $\overrightarrow{OA} = \mathbf{a}$, $\overrightarrow{OB} = \mathbf{b}$ and AP : PB = 3 : 2.

a Copy and complete

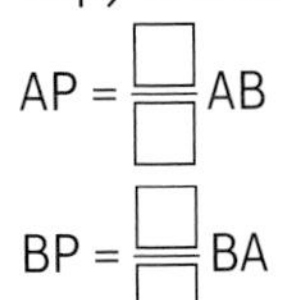

AP = $\frac{\square}{\square}$ AB

BP = $\frac{\square}{\square}$ BA

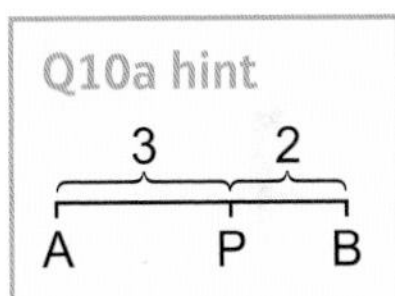

Q10a hint

b Express in terms of **a** and **b**

i $\overrightarrow{AB}$ ii $\overrightarrow{AP}$ iii $\overrightarrow{BA}$ iv $\overrightarrow{PB}$ v $\overrightarrow{OP}$

18 Extend

1 The diagram shows vectors **a** and **b**.
On isometric paper draw the vectors

a $\mathbf{a} + \mathbf{b}$ b $\mathbf{b} - \mathbf{a}$ c $2\mathbf{b} + \mathbf{a}$

d $\frac{1}{2}\mathbf{a}$

e $\frac{1}{2}\mathbf{a} + 2\mathbf{b}$

2 On this grid, $\overrightarrow{OA} = \mathbf{a}$ and $\overrightarrow{OB} = \mathbf{b}$.
Write in terms of **a** and **b**

a $\overrightarrow{OM}$ b $\overrightarrow{OH}$ c $\overrightarrow{MN}$ d $\overrightarrow{KE}$

e $\overrightarrow{AB}$ f $\overrightarrow{CM}$ g $\overrightarrow{DI}$ h $\overrightarrow{ME}$

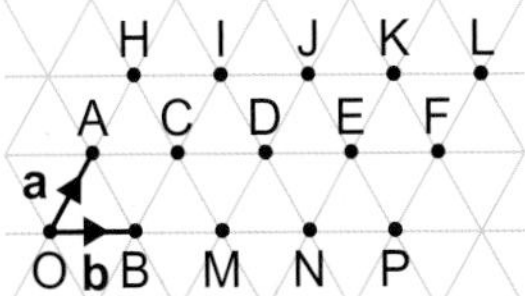

3 **Reasoning** a A is the point (1, 3) and $\overrightarrow{AB} = \begin{pmatrix} 3 \\ 2 \end{pmatrix}$.
Find the coordinates of B.

b C is the point (4, 3). BD is a diagonal of the parallelogram ABCD.
Express $\overrightarrow{BD}$ as a column vector.

c $\overrightarrow{CE} = \begin{pmatrix} 1 \\ -3 \end{pmatrix}$. Find the coordinates of E.

4 P is the point (2, 2) and Q is the point (6, 1).

a Write down the vector $\overrightarrow{PQ}$ as a column vector.

PQRS is a parallelogram. $\overrightarrow{PR} = \begin{pmatrix} 2 \\ 5 \end{pmatrix}$

b Find the vector $\overrightarrow{QS}$ as a column vector.

c Find the magnitude of vector $\overrightarrow{QS}$.

5 ORS is a triangle.
$ST = \frac{2}{3}SR$
$\overrightarrow{OS} = \mathbf{a}$ and $\overrightarrow{OR} = \mathbf{b}$.

a Write down an expression for $\overrightarrow{SR}$ in terms of **a** and **b**.

b Express $\overrightarrow{OT}$ in terms of **a** and **b**.
Give your answer in its simplest form.

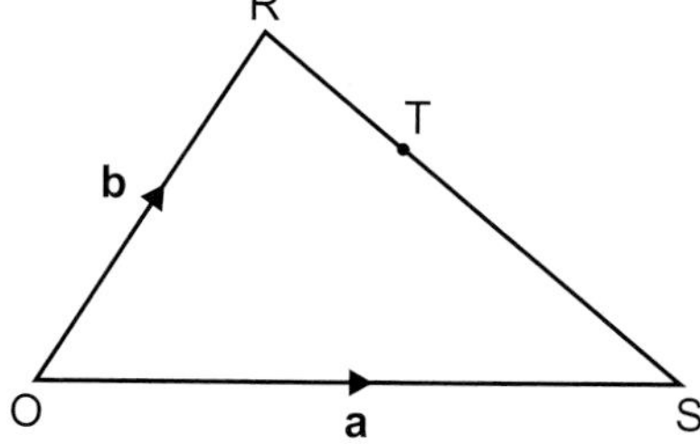

6 **Reasoning** OABC is a trapezium.
$\overrightarrow{OA} = \mathbf{a}$ and $\overrightarrow{OC} = \mathbf{c}$.
CB is parallel to OA and CB = 2OA.
M is the midpoint of AB and X divides CB in the ratio 2:3.
Write in terms of **a** and **c**

a $\overrightarrow{CB}$ b $\overrightarrow{OM}$ c $\overrightarrow{AB}$ d $\overrightarrow{OX}$

7 **Problem-solving** JKLM is a parallelogram.

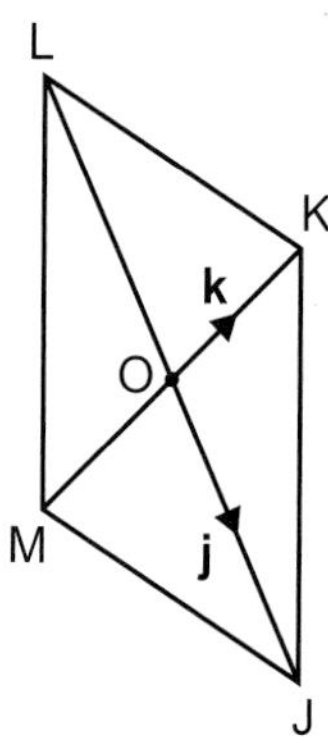

> **Q8 strategy hint** To do this question you need to remember the properties of a parallelogram, i.e. the diagonals bisect each other.

The diagonals of the parallelogram intersect at O.
$\overrightarrow{OJ} = \mathbf{j}$ and $\overrightarrow{OK} = \mathbf{k}$.

a Write an expression, in terms of **j** and **k**, for

i $\overrightarrow{LJ}$ ii $\overrightarrow{KJ}$ iii $\overrightarrow{KL}$

X is the point such that $\overrightarrow{OX} = 2\mathbf{j} - \mathbf{k}$.

b i Write down an expression, in terms of **j** and **k**, for $\overrightarrow{JX}$.

ii Explain why J, K and X lie on the same straight line.

8 **Exam-style question**

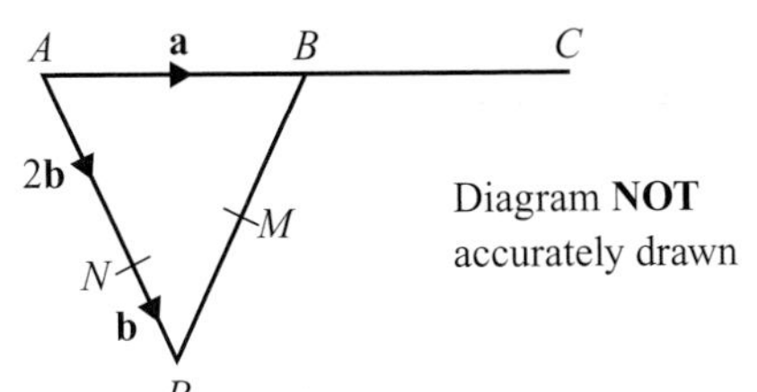

Diagram **NOT** accurately drawn

APB is a triangle. *N* is a point on *AP*.

$\overrightarrow{AB} = \mathbf{a}$ $\overrightarrow{AN} = 2\mathbf{b}$ $\overrightarrow{NP} = \mathbf{b}$

a Find the vector $\overrightarrow{PB}$, in terms of **a** and **b**. **(1 mark)**

B is the midpoint of *AC*.

M is the midpoint of *PB*.

b Show that *NMC* is a straight line. **(4 marks)**

Nov 2012, Q28, 1MA0/1H

Q9b strategy hint Show that *N*, *M* and *C* are collinear.

9 **Problem-solving** OPQ is a triangle.

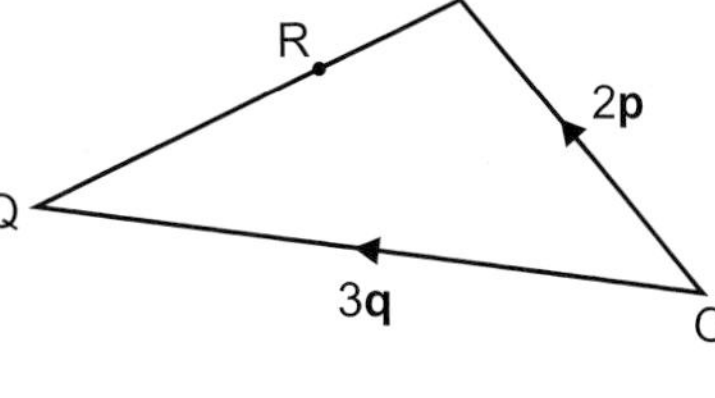

$\overrightarrow{OP} = 2\mathbf{p}$

$\overrightarrow{OQ} = 3\mathbf{q}$

a Find $\overrightarrow{PQ}$ in terms of **p** and **q**.

R is the point on PQ such that PR : RQ = 2 : 3.

b Show that $\overrightarrow{OR}$ is parallel to the vector **p** + **q**.

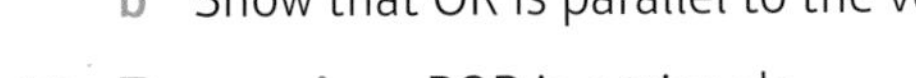

10 **Reasoning** PQR is a triangle.

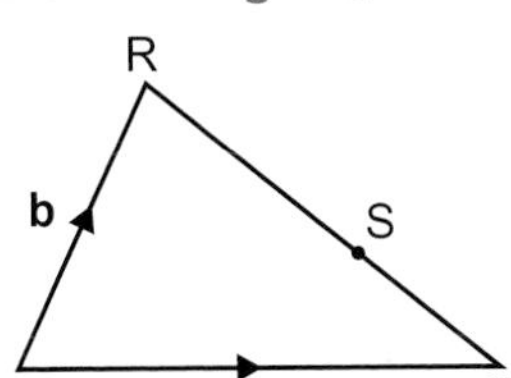

S is the point on RQ such that RS : SQ = 3 : 2.

$\overrightarrow{PQ} = \mathbf{a}$ and $\overrightarrow{PR} = \mathbf{b}$.

a Write down an expression for $\overrightarrow{RQ}$ in terms of **a** and **b**.

b Express $\overrightarrow{SP}$ in terms of **a** and **b**.

Q11b hint $\overrightarrow{SP} = \overrightarrow{SQ} + \overrightarrow{QP}$ Work out what SQ is as a fraction of RQ.

11 **Problem-solving** OPQ is a triangle. B is the midpoint of PQ.

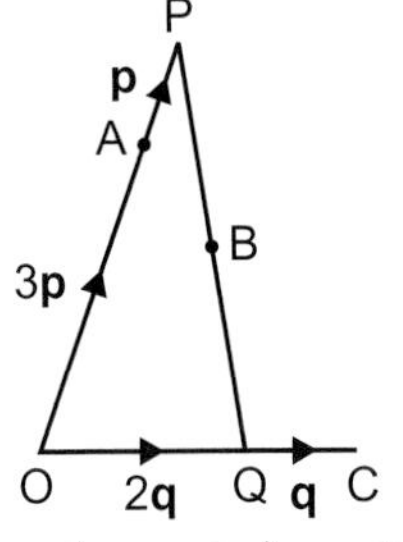

$\overrightarrow{OA} = 3\mathbf{p}$, $\overrightarrow{AP} = \mathbf{p}$, $\overrightarrow{OQ} = 2\mathbf{q}$ and $\overrightarrow{QC} = \mathbf{q}$.

a Find, in terms of **p** and **q**, the vectors

i $\overrightarrow{PQ}$ ii $\overrightarrow{AC}$ iii $\overrightarrow{BC}$

b Hence explain why ABC is a straight line.

The length of AB is 3 cm.

c Find the length of AC.

Q12b Communication hint 'Hence' means you should use your answers to part **a** to help you answer part **b**.

12 **Exam-style question**

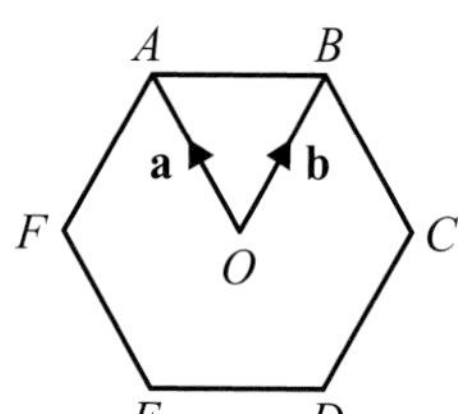

Diagram **NOT** accurately drawn

ABCDEF is a regular hexagon, with centre *O*.

$\overrightarrow{OA} = \mathbf{a}$, $\overrightarrow{OB} = \mathbf{b}$.

a Write the vector $\overrightarrow{AB}$ in terms of **a** and **b**. **(1 mark)**

The line *AB* is extended to the point *K* so that $AB : BK = 1 : 2$

b Write the vector $\overrightarrow{CK}$ in terms of **a** and **b**.

Give your answer in its simplest form. **(3 marks)**

March 2012, Q23, 1380/3H

Q12b strategy hint
Copy the diagram.
Extend *AB* to *K*.
Draw the line *DK*.

13 **Exam-style question**

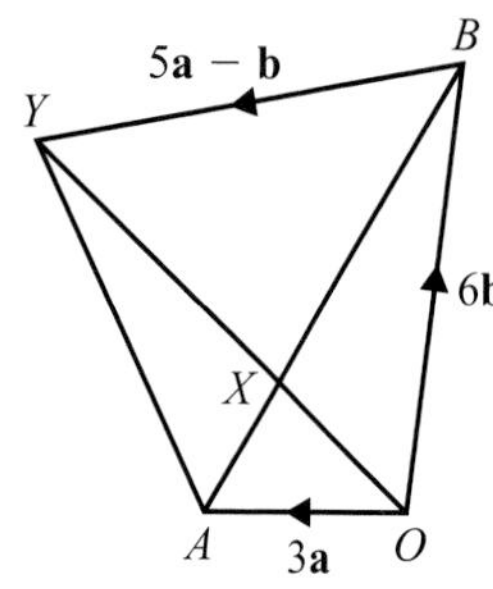

Diagram **NOT** accurately drawn

OAYB is a quadrilateral.

$\overrightarrow{OA} = 3\mathbf{a}$

$\overrightarrow{OB} = 6\mathbf{b}$

a Express $\overrightarrow{AB}$ in terms of **a** and **b**. **(1 mark)**

X is the point on *AB* such that $AX : XB = 1 : 2$ and $\overrightarrow{BY} = 5\mathbf{a} - \mathbf{b}$

b Prove that $\overrightarrow{OX} = \frac{2}{5}\overrightarrow{OY}$ **(4 marks)**

March 2013, Q26, 1MA0/1H

Q13b strategy hint
Find expressions for $\overrightarrow{OX}$ and $\overrightarrow{OY}$ in terms of **a** and **b**.

18 Knowledge check

- A **vector** is a quantity that has magnitude and direction. For example, velocity is a vector because it describes how fast something is moving and in which direction. The **magnitude** of a vector is its size. **Displacement** is change in position. .... *Mastery lesson 18.1*
- A displacement can be written as $\begin{pmatrix} 3 \\ 4 \end{pmatrix}$, where 3 is the x component and 4 is the y component.
  The displacement vector from A to B is written $\overrightarrow{AB}$. .... *Mastery lesson 18.1*
- Vectors are written as **bold** lower case letters: **a**, **b**, **c**. When handwriting, underline the letter: $\underline{a}$, $\underline{b}$, $\underline{c}$ .... *Mastery lesson 18.1*
- Equal vectors have the same magnitude and the same direction. .... *Mastery lesson 18.1*

- The magnitude of the vector $\begin{pmatrix} x \\ y \end{pmatrix}$ is its length, i.e. $\sqrt{x^2 + y^2}$. .... *Mastery lesson 18.1*
- If $\overrightarrow{AB} = \overrightarrow{CD}$ then the line segments AB and CD are equal in length and are parallel. $\overrightarrow{AB} = -\overrightarrow{BA}$

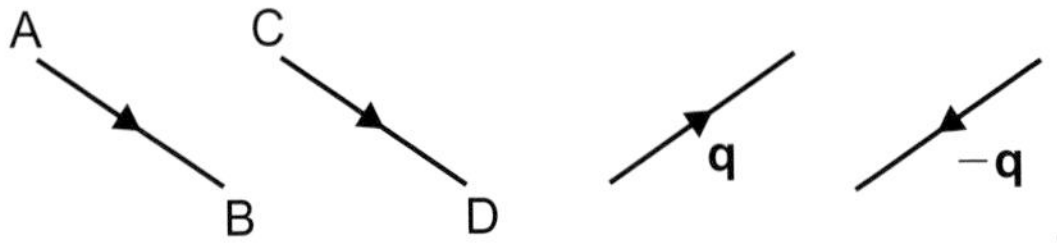

A C q −q B D

.... *Mastery lesson 18.2*

- 2**a** is twice as long as **a** and in the same direction.
  −**a** is the same length as **a** but in the opposite direction.

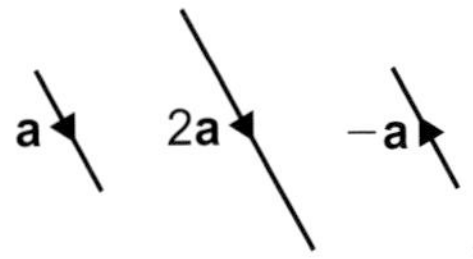

a 2a −a

.... *Mastery lesson 18.2*

- When a vector **a** is multiplied by a scalar k then the vector k**a** is parallel to **a** and is equal to k times **a**.
  A scalar is a number, e.g. 3, 2, $\frac{1}{2}$, −1, ... *Mastery lesson 18.2*
- The two-stage journey from A to B and then from B to C has the same starting point and the same finishing point as the single journey from A to C. So A to B followed by B to C is equivalent to A to C.
  $\overrightarrow{AB} + \overrightarrow{BC} = \overrightarrow{AC}$ .... *Mastery lesson 18.2*
- **Triangle law for vector addition**: Let $\overrightarrow{AB}$ = **a**, $\overrightarrow{BC}$ = **b** and $\overrightarrow{AC}$ = **c**.
  Then **a** + **b** = **c** forms a triangle.

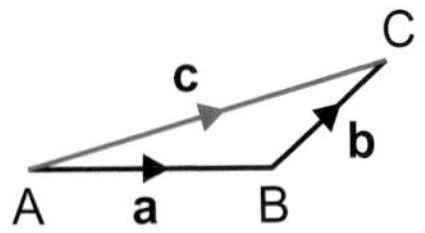

C c b A a B

.... *Mastery lesson 18.2*

- **a** + **b** = **b** + **a**. .... *Mastery lesson 18.2*
- **Parallelogram law for vector addition**: In parallelogram PQRS where $\overrightarrow{PQ}$ is **a** and $\overrightarrow{PS}$ is **b**, the diagonal $\overrightarrow{PR}$ of the parallelogram is **a** + **b**. .... *Mastery lesson 18.3*
- When **c** = **a** + **b** the vector **c** is called the **resultant vector** of the two vectors **a** and **b**. .... *Mastery lesson 18.3*
- With the origin O, the vectors $\overrightarrow{OA}$ and $\overrightarrow{OB}$ are called the **position vectors** of the points A and B. In general, a point with coordinates (p, q) has position vector $\begin{pmatrix} p \\ q \end{pmatrix}$. .... *Mastery lesson 18.4*
- When $\overrightarrow{OA}$ = **a** and $\overrightarrow{OB}$ = **b**, $\overrightarrow{AB} = \overrightarrow{AO} + \overrightarrow{OB}$ = **b** − **a**.

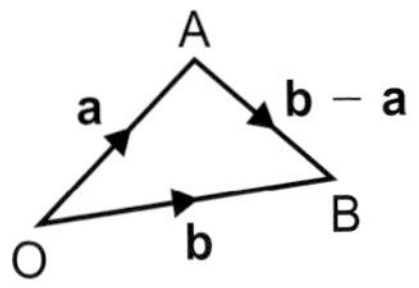

A a b − a B O b

.... *Mastery lesson 18.4*

- $\overrightarrow{PQ} = k\overrightarrow{QR}$ shows that the lines PQ and QR are parallel. Also they both pass through point Q so PQ and QR are part of the same straight line. P, Q and R are said to be **collinear** (they all lie on the same straight line).

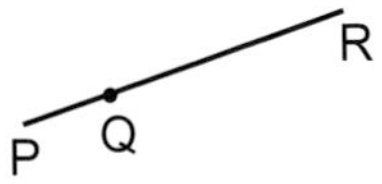

R P Q

.... *Mastery lesson 18.4*

Reflect

Think back to all units where you have been asked to prove things.

Choose A B or C to complete each statement:

I am	**A** good at proof	**B** OK at proof	**C** not very good at proof
I think proof is …	**A** easy	**B** OK	**C** hard
When I think about doing a proof, I feel	**A** confident	**B** OK	**C** unsure

Did you answer mostly As and Bs?
Are you surprised by how you feel about proof?
Why?

Did you answer mostly Cs? Find the three questions about proof that you found the hardest.
Ask someone to explain them to you.
Then complete the statements above again.

Hint You may choose proof questions from this unit or another unit. You could look back at Units 12 and 16 too.

18 Unit test

Log how you did on your Student Progression Chart.

1 $\mathbf{a} = \begin{pmatrix} 2 \\ -3 \end{pmatrix}$ $\mathbf{b} = \begin{pmatrix} 5 \\ 4 \end{pmatrix}$

Write as a column vector

a $5\mathbf{a}$ *(1 mark)*

b $\mathbf{a} + \mathbf{b}$ *(1 mark)*

c $2\mathbf{a} - 3\mathbf{b}$ *(2 marks)*

2 The diagram shows two vectors **a** and **b**.

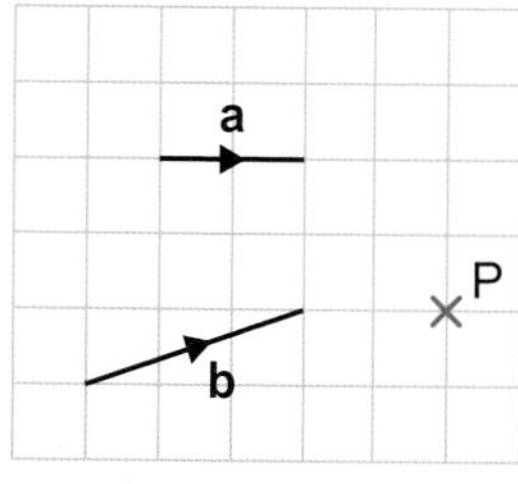

$\overrightarrow{PQ} = \mathbf{a} + 2\mathbf{b}$

Draw the vector $\overrightarrow{PQ}$ on squared paper. *(3 marks)*

3 ABCD is a parallelogram. D is the midpoint of AE.

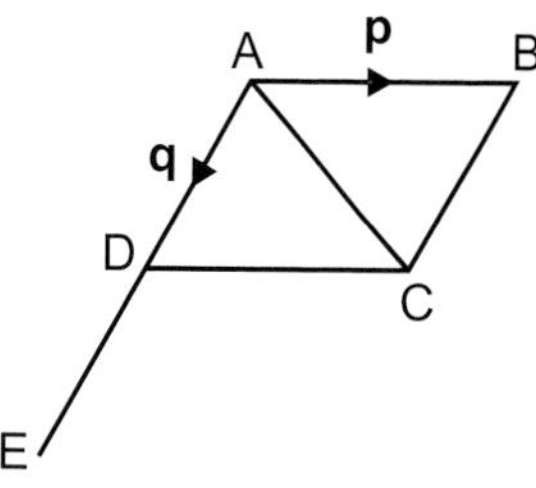

$\overrightarrow{AB} = \mathbf{p}$ and $\overrightarrow{AD} = \mathbf{q}$.

Write down in terms of **p** and/or **q**

a $\overrightarrow{AE}$ *(1 mark)*

b $\overrightarrow{AC}$ *(1 mark)*

ActiveLearn Homework, practice and support: Higher 18 Unit test

4 The diagram is a sketch.
P is the point (1, 4). Q is the point (3, 6).

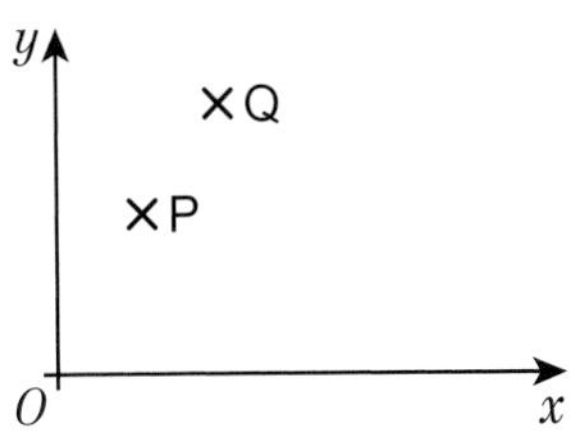

a Find the vector $\overrightarrow{PQ}$.
Give your answer as a column vector. *(2 marks)*

$\overrightarrow{QR} = \begin{pmatrix} 4 \\ -2 \end{pmatrix}$

M is the midpoint of PQ.
N is the midpoint of QR.

b Find the vector $\overrightarrow{MN}$.
Give your answer as a column vector. *(3 marks)*

5 **Reasoning** ABCDEF is a regular hexagon.

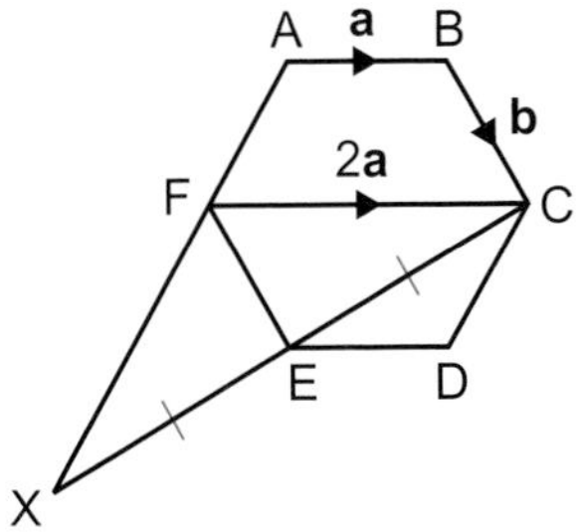

$\overrightarrow{AB} = \mathbf{a}$, $\overrightarrow{BC} = \mathbf{b}$ and $\overrightarrow{FC} = 2\mathbf{a}$.

a Find in terms of **a** and **b**
i $\overrightarrow{FE}$ ii $\overrightarrow{CE}$ *(2 marks)*

$\overrightarrow{CE} = \overrightarrow{EX}$

b Prove that FX is parallel to CD. *(3 marks)*

6 **Reasoning** a A is the point (2, 5) and $\overrightarrow{AB} = \begin{pmatrix} 6 \\ -2 \end{pmatrix}$.
Find the coordinates of B. *(1 mark)*

b C is the point (5, 1).
BD is a diagonal of the parallelogram ABCD.
Express $\overrightarrow{BD}$ as a column vector. *(2 marks)*

7 **Reasoning** OPQR is a trapezium.

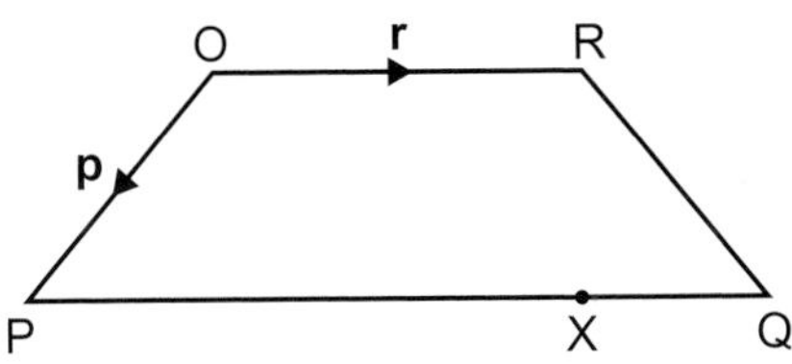

OR is parallel to PQ.
$\overrightarrow{OP} = \mathbf{p}$ and $\overrightarrow{OR} = \mathbf{r}$.
PQ = 2OR
X is the point on PQ such that PX : XQ = 3 : 1.
Express $\overrightarrow{OX}$ in terms of **p** and **q**. *(3 marks)*

8 **Exam-style question**

OACB is a parallelogram.

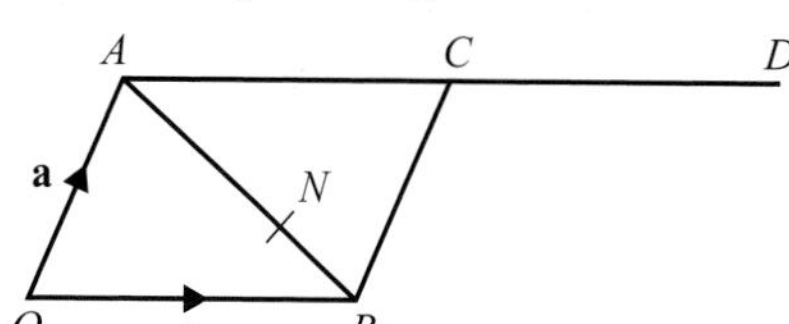

Diagram **NOT** accurately drawn

$\overrightarrow{OA} = \mathbf{a}$ and $\overrightarrow{OB} = \mathbf{b}$

D is a point such that $\overrightarrow{AC} = \overrightarrow{CD}$

The point *N* divides *AB* in the ratio 2 : 1

a Write an expression for $\overrightarrow{ON}$ in terms of **a** and **b**. **(3 marks)**

b Prove that *OND* is a straight line. **(3 marks)**

Nov 2013, Q24, 1MA0/1H

9 A is the point (3, 4) and B is the point (−1, 0).

a Express $\overrightarrow{AB}$ as a column vector. *(1 mark)*

$\overrightarrow{BC} = \begin{pmatrix} 2 \\ 5 \end{pmatrix}$

b Write down the coordinates of point C. *(1 mark)*

X is the midpoint of AB. O is the origin.

c Find $\overrightarrow{OX}$ as a column vector. *(2 marks)*

d What is the magnitude of $\overrightarrow{AB}$? *(2 marks)*

Sample student answer

a What mistake has the student made?

b How could you improve this answer?

Exam-style question

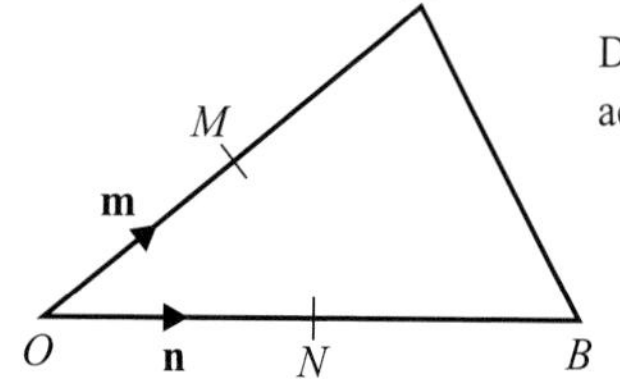

Diagram **NOT** accurately drawn

OAB is a triangle.

M is the midpoint of *OA*.

N is the midpoint of *OB*.

$\overrightarrow{OM} = \mathbf{m}$

$\overrightarrow{ON} = \mathbf{n}$

Show that *AB* is parallel to *MN*. **(3 marks)**

June 2014, Q24, 1MA0/1H

Student answer

$\overrightarrow{OA} = 2\mathbf{m}$

$\overrightarrow{OB} = 2\mathbf{n}$

$\overrightarrow{AB} = 2\mathbf{m} + 2\mathbf{n} = 2(\mathbf{m} + \mathbf{n})$

$\overrightarrow{MN} = \mathbf{m} + \mathbf{n}$

So $\overrightarrow{AB} = 2\overrightarrow{MN}$

Master p.588 | Problem-solve p.610 | Check p.611 | Strengthen p.613 | Extend p.615 | Test p.621

19 PROPORTION AND GRAPHS

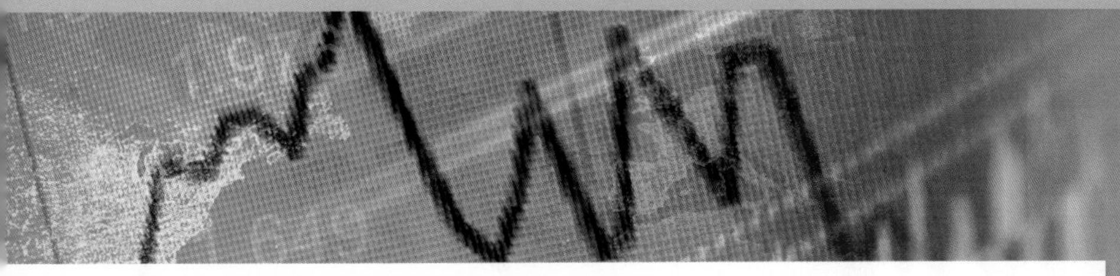

Economists and scientists use graphs to help visualise the relationship between different variables. The diagram shows a common supply and demand graph used in economics.

The demand graph shows that as the price of goods falls, the more we want to buy.

The supply graph shows that as the price of goods falls, the less we want to sell.

What is represented by the coordinates where the graphs intersect?

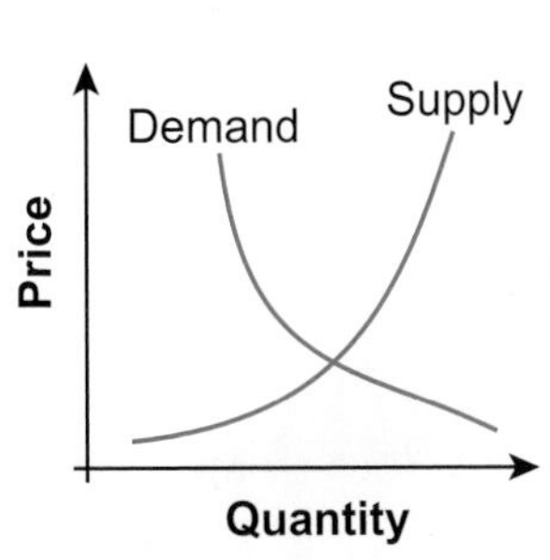

19 Prior knowledge check

Graphical fluency

1 The velocity–time graph shows a car's test drive.

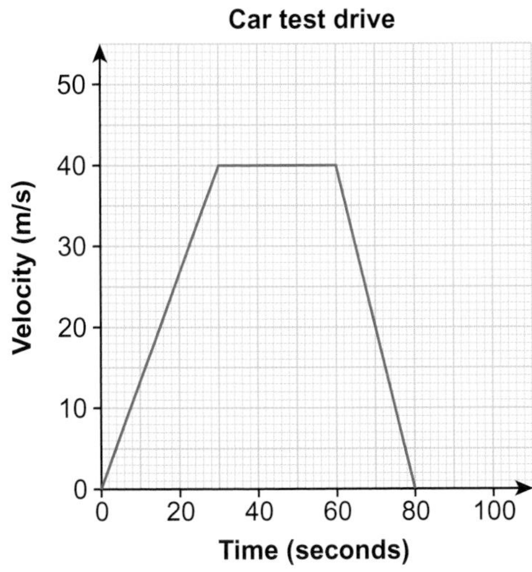

The car travelled in a straight line.
Work out

a its acceleration in m/s^2 for the first 30 seconds

b its deceleration in m/s^2 at the end of the drive

c the distance travelled.

2 Which graph shows direct proportion?

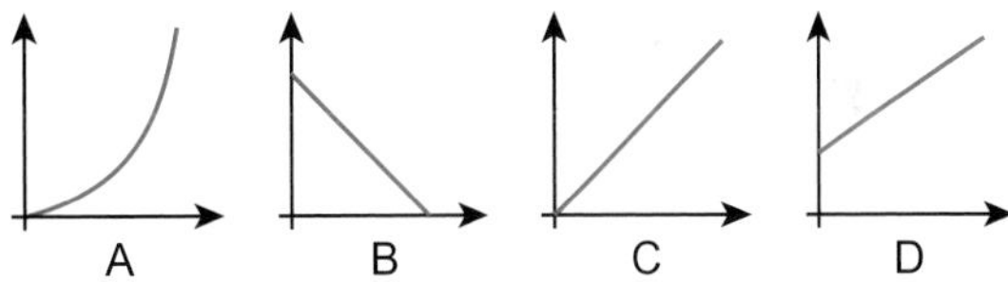

3 Sketch the graphs of

a $y = x^2$

b $y = x^3$

c $y = -x^2$

d $y = x^2 - 4x + 7$

4 Water is poured into each container at a steady rate. The graphs show how the depth of water changes. Match each graph to a container.

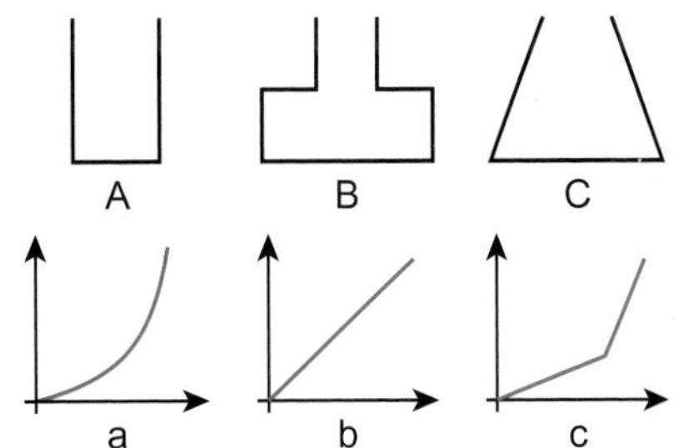

5 Here is the graph of $y = \cos x$

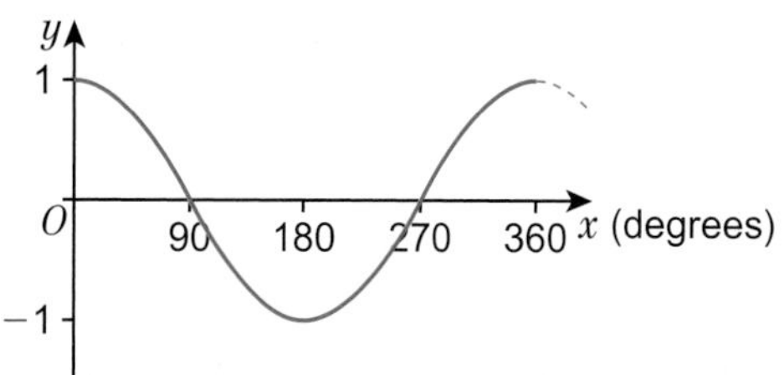

Write the equation of each graph.

a

b

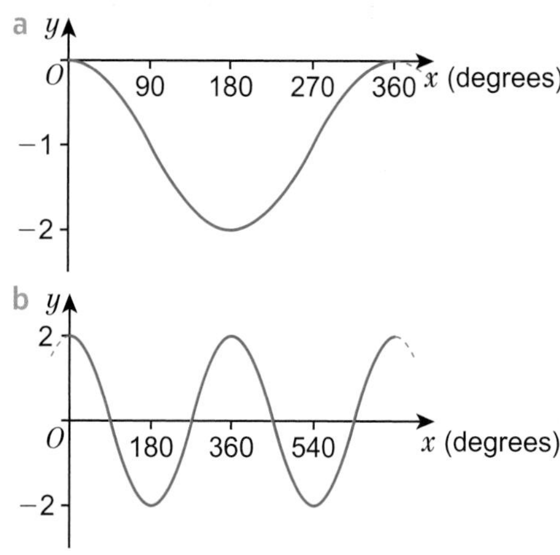

Numerical fluency

6 A boat sails a distance of 6 km at a constant speed of 15 km/hour.
How many minutes does the journey take?

7 Write the reciprocal of

a 3 b 0.5 c x

8 Work out

a 2^{-2} b 2^{-1} c 2^0
d 3^2 e 3^{-1}

9 The tables show information about two pairs of variables.

a	2	6	8	20	100
b	10	30	35	95	450

c	3	7	9	14	21
d	21	49	63	98	147

a Which pair of variables are directly proportional?

b Find the formula linking the pair of proportional variables.

Algebraic fluency

10 $f(x) = 2x^2 - 8$

a Work out f(3).

b Find the values of a where $f(a) = 0$.

c Write out in full $f(x + 3)$

* Challenge

11 Any number multiplied by its reciprocal is always equal to 1. For example, $\frac{1}{2} \times 2 = 1$

a Pair these numbers up so their product is 1.

4	8	5	0.25
0.2	0.125	10	0.1
2.5	0.4	1.6	0.625

b Write four more product pairs that equal 1.

Q11 hint Convert the decimals into fractions.

19.1 Direct proportion

Objective

- Write and use equations to solve problems involving direct proportion.

Why learn this?

Direct proportion is used to calculate exchange rates and compare prices of goods from different countries.

Fluency

Are x and y in direct proportion? How can you tell?

x	3	5	8
y	12	20	40

Warm up

1 Is cost in direct proportion to the quantity?
If so, write an equation linking cost and quantity.
  a 5 apples cost £4.20 and 15 apples cost £12.60.
  b 500 Mb of data costs £47 and 50 Mb of data costs £6.
  c 750 screws cost £12 and 1000 screws cost £14.
  d 120 units of gas costs £40 and 80 units of gas costs £30.

> **Q1 hint** Compare ratios cost : quantity

Questions in this unit are targeted at the steps indicated.

2 **Finance / Reasoning** The tables show the price paid for different quantities of euros from two currency exchange websites in July.

travelcash.com

Sterling (£)	50	210	120	400	380	300	250	280
Euros (€)	60	260	150	520	470	390	300	350

currencyexchange.co.uk

Sterling (£)	400	150	30	250	350	300	200	450
Euros (€)	440	160	32	280	390	330	210	500

  a Draw a scatter graph for both sets of information on the same axes.
  b Draw a line of best fit for each set of data.
  c Write a formula for euros, E, in terms of sterling, S, for
    i travelcash.com
    ii currencyexchange.co.uk
  d Which currency exchange website offers the best value for money?
  Explain your answer.

> **Q2c hint** $E = \square S$

Discussion How can you tell that two quantities are in direct proportion from their graph?
How can you tell that two quantities are in direct proportion from an equation?

Key point 1

The symbol $\propto$ means 'is directly proportional to'.
$y \propto x$ means y is directly proportional to x.
In general if y is directly proportional to x,
$y \propto x$ and $y = kx$
where k is a number, called the **constant of proportionality**.

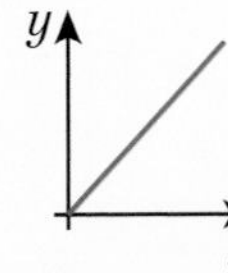

Example 1

y is directly proportional to x.
When $y = 20$, $x = 8$
  a Express y in terms of x.
  b Find x when $y = 35$.

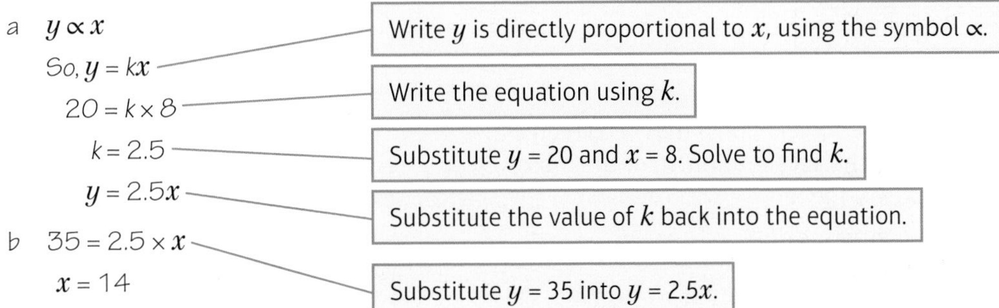

a $y \propto x$
So, $y = kx$
$20 = k \times 8$
$k = 2.5$
$y = 2.5x$
b $35 = 2.5 \times x$
$x = 14$
Write y is directly proportional to x, using the symbol $\propto$.
Write the equation using k.
Substitute $y = 20$ and $x = 8$. Solve to find k.
Substitute the value of k back into the equation.
Substitute $y = 35$ into $y = 2.5x$.

3 y is directly proportional to x.
$y = 15$ when $x = 3$
a Express y in terms of x. b Find y when $x = 10$
c Find x when $y = 65$

Q3a hint Start with the statement $y \propto x$, then the equation $y = kx$. Use the values of x and y to find the value of k.

4 y is directly proportional to x.
$y = 52$ when $x = 8$
a Write a formula for y in terms of x.
b Find y when $x = 14$
c Find x when $y = 143$

5 **Problem-solving** y is directly proportional to x.
a $y = 6$ when $x = 4$. Find x when $y = 7.5$
b $y = 31.5$ when $x = 7$. Find x when $y = 45.5$
c $y = 8$ when $x = 5$. Find x when $y = 13$

6 **Exam-style question**

y is directly proportional to x.
When $x = 600$, $y = 10$
a Find a formula for y in terms of x. **(3 marks)**
b Calculate the value of y when $x = 540$. **(1 mark)**
June 2012 Q13, 5MB3H/01

Exam hint
Most of the marks are for part **a**. Show all of the working you do to get your formula.

19.2 More direct proportion

Objectives
- Write and use equations to solve problems involving direct proportion.
- Solve problems involving square and cubic proportionality.

Why learn this?
Scientists use statements of proportionality to write equations for different variables.

Fluency

$y \propto x$. What are the values of A and B?

x	3	4	B
y	A	18	63

1 $y \propto x$
Use the table of values to find the constant of proportionality.

x	6	10	12
y	15	25	30

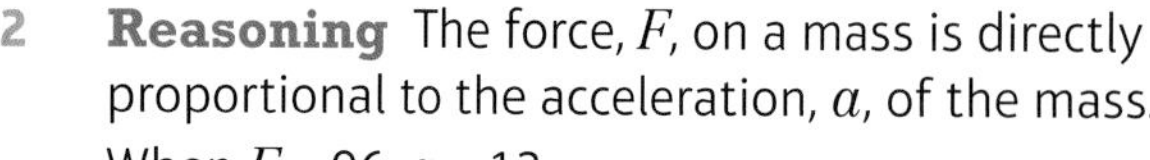

2 **Reasoning** The force, F, on a mass is directly proportional to the acceleration, a, of the mass.
When $F = 96$, $a = 12$
a Express F in terms of a.
b Find F when $a = 20$.
c Find a when $F = 112$.

Q2a hint Start with the statement $F \propto a$, then equation $F = ka$. Use the values of F and a to find the value of k.

3 **Reasoning / Communication** The table gives information about the perimeter, P, of a shape and the length, l, of one of its sides.

Perimeter, P (cm)	12	24	30	48
Length, l (cm)	5	10	12.5	20

a Show that P is directly proportional to l.
b Given that $P = kl$, work out the value of k.
c Write a formula for P in terms of l.
d Use your formula to work out
  i the value of P, when $l = 18$ ii the value of l, when $P = 42$

Q3a hint Use equivalent ratios.
$12:5 = 24:\square = \square:\square = \square:\square$

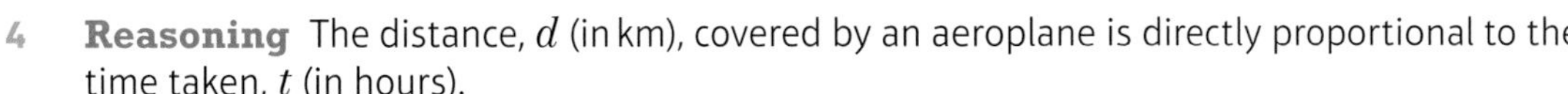

4 **Reasoning** The distance, d (in km), covered by an aeroplane is directly proportional to the time taken, t (in hours).
The aeroplane covers a distance of 1600 km in 3.2 hours.
a Find a formula for d in terms of t.
b Find the value of d, when $t = 5$
c Find the value of t, when $d = 2250$
d What happens to the distance travelled, d, when the time, t, is
  i doubled ii halved?

5 **Real / Reasoning** The cost, C (in £), of a newspaper advert is directly proportional to the area, A (in cm^2), of the advert.
An advert with an area of 40 cm^2 costs £2000.
a Sketch a graph of C against A.
b Write a formula for C in terms of A.
c Use your formula to work out the cost of an 85 cm^2 advert.

Key point 2

A quantity can be directly proportional to the *square*, the *cube*, or the *square root* of another quantity. For example:
- If y is proportional to the square of x then $y \propto x^2$ and $y = kx^2$
- If y is proportional to the cube of x then $y \propto x^3$ and $y = kx^3$
- If y is proportional to the square root of x then $y \propto \sqrt{x}$ and $y = k\sqrt{x}$

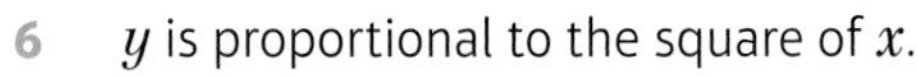

6 y is proportional to the square of x.
When $x = 3$, $y = 36$
a Write the statement of proportionality.
b Write an equation using k.
c Work out the value of k.
d Find y when $x = 5$
e Find x when $y = 25$

Q6a hint $\square \propto \square$

7 **Problem-solving** y is proportional to the cube of x.
When $x = 2$, $y = 28.8$
a Write a formula for y in terms of x.
b Find y when $x = 4$
c Find x when $y = 450$

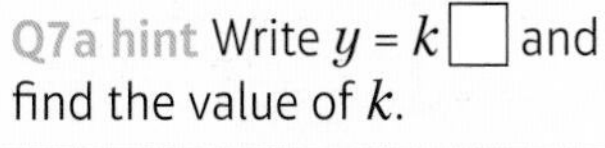

Q7a hint Write $y = k\square$ and find the value of k.

8 y is proportional to the square root of x.
When $x = 4$, $y = 50$
a Find a formula for y in terms of x.
b Find y when $x = 9$
c Find x when $y = 250$

9 **Exam-style question**

y is directly proportional to the square of x.
When $x = 5$, $y = 150$
Find the value of y when $x = 3$. **(4 marks)**

Exam hint
Work through the first three steps in **Q6**. Make sure you write everything down to get maximum marks.

10 **STEM / Reasoning** The kinetic energy, E (in joules, J), of an object varies in direct proportion to the square of its speed, s (in m/s).
An object moving at 5 m/s has 125 J of kinetic energy.

a Write a formula for E in terms of s.
b How much kinetic energy does the object have if it is moving at 2 m/s?
c What speed is the object moving at if it has 192.2 J of kinetic energy?
d What happens to the kinetic energy, E, if the speed of the object is doubled?

11 **Problem-solving** The cost of fuel per hour, C (in £), to propel a boat through the water is directly proportional to the cube of its speed, s (in mph).
A boat travelling at 10 mph uses £50 of fuel per hour.

a Write a formula for C in terms of s.
b Calculate C when the boat is travelling at 5 mph.

12 **STEM / Reasoning** In a factory, chemical reactions are carried out in spherical containers. The time, T (in minutes), the chemical reaction takes is directly proportional to the square of the radius, R (in cm), of the spherical container.
When $R = 120$, $T = 32$

a Write a formula for T in terms of R.
b Find the value of T when $R = 150$

13 **Problem-solving** In an experiment, measurements of g and h were taken.

h	2	5	7
g	24	375	1029

Which of these relationships fits the result?

$g \propto h$ $\quad g \propto h^2$ $\quad g \propto h^3$ $\quad g \propto \sqrt{h}$

19.3 Inverse proportion

Objectives

- Write and use equations to solve problems involving inverse proportion.
- Use and recognise graphs showing inverse proportion.

Why learn this?

You can use inverse proportion to work out how long it will take different numbers of people to complete a task.

Fluency

It takes 2 people 30 minutes to deliver 200 leaflets.
How long will it take 3 people?

Warm up

1 A is directly proportional to B.
$A = 10$ when $B = 40$

a Write a formula for A in terms of B.
b Use your formula to work out the value of A when $B = 460$

ActiveLearn Homework, practice and support: Higher 19.3

2 Match each statement of proportionality to the correct graph.

$y \propto x$ $\quad y \propto x^2$ $\quad y \propto \frac{1}{x}$

a

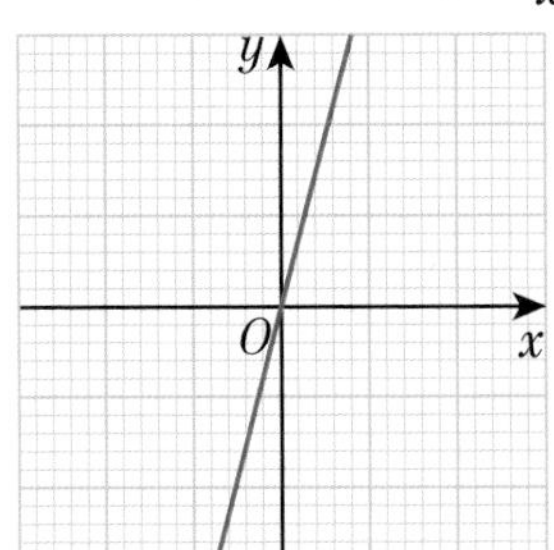

b

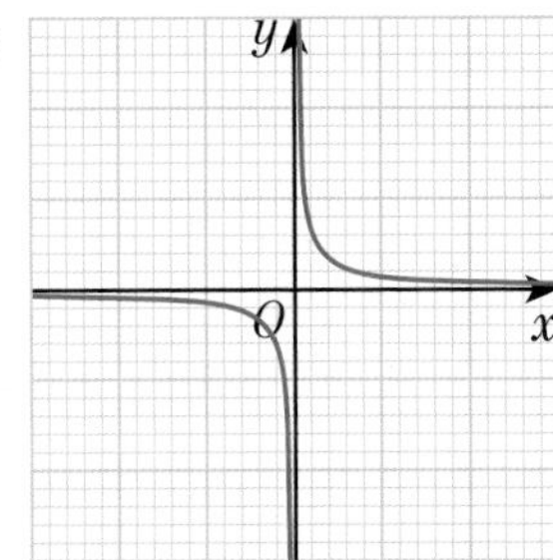

c

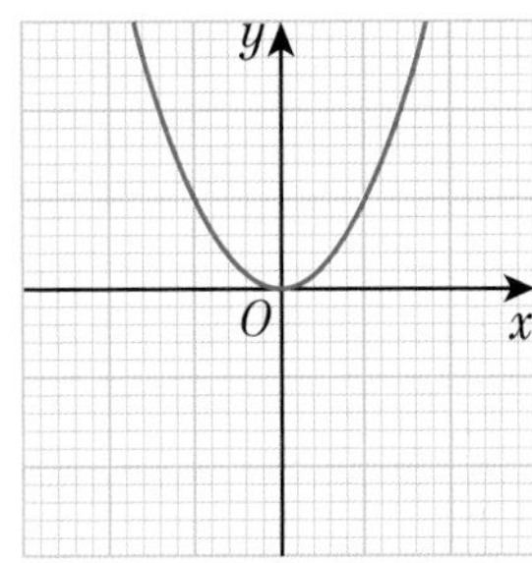

Key point 3

When y is **inversely proportional** to x

$y \propto \frac{1}{x}$

$y = \frac{k}{x}$

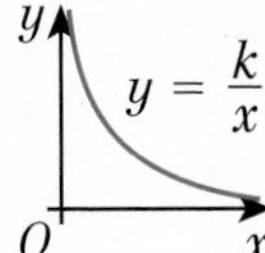

Example 2

y is inversely proportional to x.
When $y = 2$, $x = 3$

a Write a formula for y in terms of x.

b Calculate the value of y when $x = 8$

a $y \propto \frac{1}{x}$ so $y = \frac{k}{x}$ — Write y is inversely proportional to x using the $\propto$ symbol. Then write the equation using k.

$2 = \frac{k}{3}$ so $k = 6$ — Substitute $y = 2$ and $x = 3$. Solve to find k.

$y = \frac{6}{x}$ — Substitute k back into the equation.

b $y = \frac{6}{8} = \frac{3}{4}$ — Substitute $x = 8$ into your formula.

3 y is inversely proportional to x.
When $y = 5$, $x = 2$

a Write a formula for y in terms of x.

b Calculate the value of y when $x = 20$

c Calculate the value of x when $y = 4$

4 **STEM / Reasoning** The pressure, P (in N/m^2), of a gas is inversely proportional to the volume, V (in m^3).
$P = 1500\,\text{N/m}^2$ when $V = 2\,\text{m}^3$

a Write a formula for P in terms of V.

b Work out the pressure when the volume of the gas is 1.5 m^3.

c Work out the volume of gas when the pressure is 1200 N/m^2.

d What happens to the volume of the gas when the pressure doubles?

5 **Problem-solving / STEM** The time taken, t (in seconds), to boil water in a kettle is inversely proportional to the power, p (in watts), of the kettle.
A full kettle of power 1500 W boils the water in 400 seconds.

a Write a formula for t in terms of p.

b A similar kettle has a power of 2500 W. Can this kettle boil the same amount of water in less than 3 minutes?

6 **Reasoning** y is inversely proportional to x.

x	0.25	0.5	1	2	4	8	16	32
y	32	16	8	4	2	1	0.5	0.25

a Draw a graph of y and x. What type of graph is this?

b $y = \frac{k}{x}$ where k is the constant of proportionality. Find k.

c Work out $x \times y$ for each pair of values in the table. What do you notice?

Discussion How can you tell from a table of values if two variables are inversely proportional?

7 **Reasoning** Which graph shows variables in inverse proportion?

A
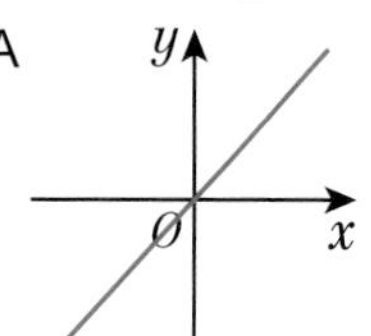

B
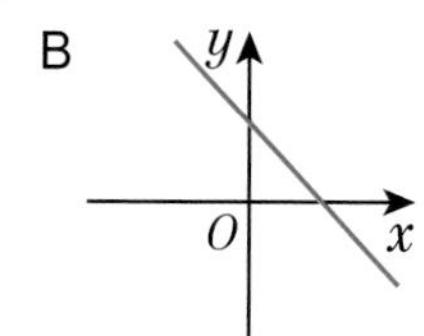

C
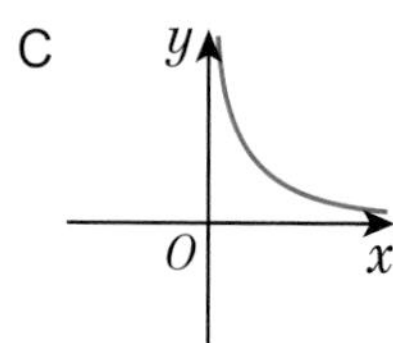

8 **STEM / Reasoning** The time, t (in seconds), it takes an object to travel a fixed distance is inversely proportional to the speed, s (in m/s), at which the object is travelling.
When travelling at 20 m/s it takes an object 40 seconds to travel from A to B.

a Write a formula for s in terms of t.

b Copy and complete the table of values for s and t.

Speed, s (m/s)	4		20	40		160
Time, t (seconds)		80	40		10	

c Sketch a graph to show how s varies with t.

Discussion What happens to the time taken as the speed approaches 0 m/s?

9 **Problem-solving** The graph shows two variables that are inversely proportional to each other.
Find the values for a and b.

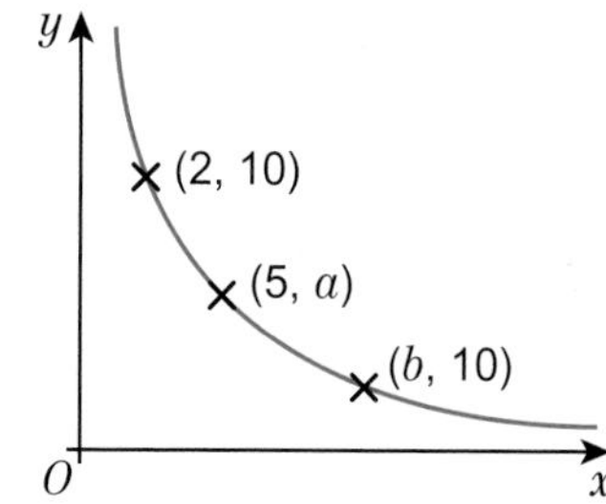

10 **Reasoning** A farmer employs fruit pickers to harvest his apple crop.
The fruit pickers work in different-sized teams.
The farmer records the times it takes different teams to harvest the apples from 10 trees.

Number of people in team, n	3	2	5	8	9	10	6	4	7
Time taken, t (minutes)	95	155	60	40	30	25	40	85	40

a Draw a scatter graph of t against n.

b Draw a curve of best fit.

c Write a formula for estimating t when given n.

d Use your formula to estimate the time it would take a team of 15 people to harvest the apples from 10 trees.

Q10d hint $t = \frac{k}{\square}$

Use a coordinate on the line of best fit to find the value of k.

11 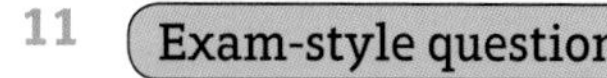

Exam-style question

h is inversely proportional to the square of r.
When $r = 5$, $h = 3.4$
Find the value of h when $r = 8$ **(3 marks)**

June 2013, Q22, 1MA0/2H

Exam hint
Read the question thoroughly. Note that you need to use *inverse* proportion.

12 y is inversely proportional to the cube of x.
When $y = 2$, $x = 3$

a Write a formula for y in terms of x.
b Calculate y when $x = 5$
c Calculate x when $y = 6.75$

13 y is inversely proportional to the square root of x.
When $y = 2$, $x = 9$

a Write a formula for y in terms of x.
b Calculate y when $x = 4$
c Calculate x when $y = 6$

14 **Reasoning** When 20 litres of water are poured into any cylinder, the depth, D (in cm), of the water is inversely proportional to the square of the radius, r (in cm), of the cylinder.
When $r = 15$ cm, $D = 28.4$ cm

a Write a formula for D in terms of r.
b Find the depth of the water when the radius of the cylinder is 25 cm.
c Find the radius of the cylinder when the depth is 64 cm.
d Cylinder A has radius x cm and is filled with water to a depth of d cm.
This water is poured into cylinder B with radius $2x$ cm.
What is the depth of water in cylinder B?

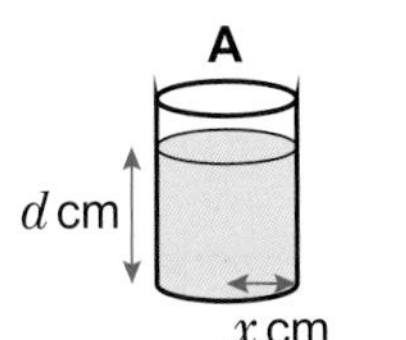

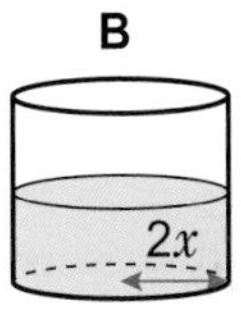

15 **Reasoning** The speed, s (in revolutions per minute), at which each cog in a machine turns is inversely proportional to the square of the radius, r (in cm).
When $r = 4$ cm, $s = 212.5$ revolutions per minute.

a Write a formula for s in terms of r.
b Work out the value of s when $r = 4.2$. Round your answer to 2 decimal places.

19.4 Exponential functions

Objectives

- Recognise graphs of exponential functions.
- Sketch graphs of exponential functions.

Why learn this?

Exponential graphs are used by scientists to describe population growth and radioactive decay.

Fluency

Work out the value of 2^3, 6^{-1}, 3^{-2}, 5^0

Warm up

1 Calculate

a 2^4 b 9^0 c 5^{-1} d 4^{-2}

2 A man places a grain of rice on the first square of a chess board. He then places two grains on the second square, four on the third square, eight on the fourth square, and so on, doubling the number of grains each time.
How many grains of rice does he place on the

a 10th square b 15th square c 20th square?

Discussion What happens to the amount of rice as he moves further around the chessboard?

3 Find the value of x for each of these equations.

a $2^x = 8$ b $3^x = 81$ c $10^x = 10\,000$

Q3a hint $2^{\square} = 8$

Key point 4

Expressions of the form a^x, where a is a positive number, are called **exponential functions**.

4 a Copy and complete the table of values for $y = 2^x$.
Give the values correct to 2 decimal places.

x	−4	−3	−2	−1	0	1	2	3	4
y									

b Draw the graph of $y = 2^x$ for $-4 \leqslant x \leqslant 4$.

c Use the graph to find an estimate for

i the value of y when $x = 3.5$ ii the value of x when $y = 10$.

Discussion What happens to the value of y as the value of x decreases?

Key point 5

The graph of an exponential function has one of these shapes.

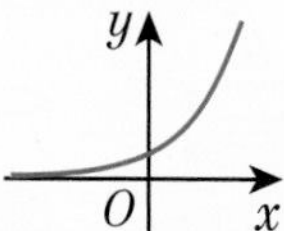

$y = a^x$ where $a > 1$ or
$y = b^{-x}$ where $0 < b < 1$
exponential growth

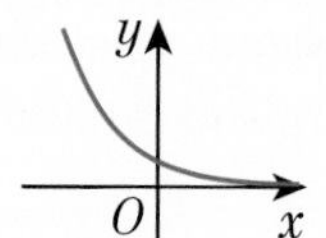

$y = a^{-x}$ where $a > 1$ or
$y = b^x$ where $0 < b < 1$
exponential decay

Hint The graph of $y = a^x$ when $a = 1$ is just the graph of $y = 1$.

5 **Reasoning**

a Draw the graph of each function.
Use a grid with x-axis −2 to 2 and y-axis 0 to 30.

i $y = 3^x$ ii $y = 5^x$

Q5a hint Create a table of values similar to **Q4a**.

b Predict where the graph of $y = 4^x$ would be. Sketch it on the same axes.

c At which point do all the graphs intersect the y-axis?

Discussion Why do exponential graphs always cross the y-axis at the same point?

6 a Copy and complete the table of values for $y = 2^{-x}$.
Give the values correct to 2 decimal places.

x	−4	−3	−2	−1	0	1	2	3	4
y									

b Draw the graph of $y = 2^{-x}$ for $-4 \leqslant x \leqslant 4$

c Use the graph to find an estimate for

i the value of y when $x = 3.5$ ii the value of x when $y = 10$

7 **STEM / Reasoning** The table gives information about the count rate of seaborgium-266.
The count rate is the number of radioactive emissions per second.

Time (seconds)	0	30	60	90	120	150	180
Count rate	800	400	200	100	50	25	12.5

a Draw a graph of the data. Plot time on the horizontal axis and count rate on the vertical axis.

b Is this an example of exponential growth or exponential decay?

c The half-life of a radioactive material is the time it takes for the count rate to halve.
What is the half-life of seaborgium-266?

Example 3

The sketch shows part of the graph $y = ab^x$
The points with coordinates (0, 3) and (2, 12) lie on the graph.
Work out the values of a and b.

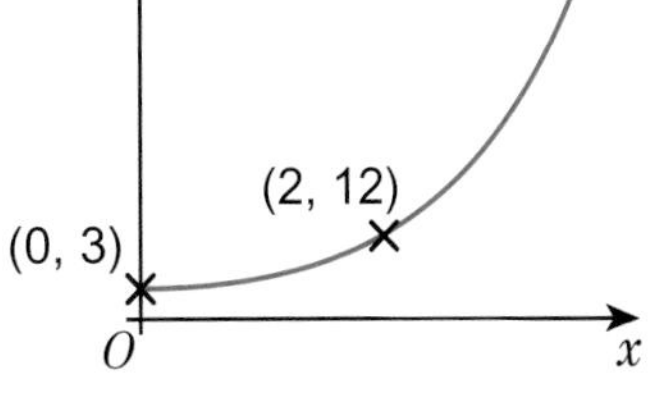

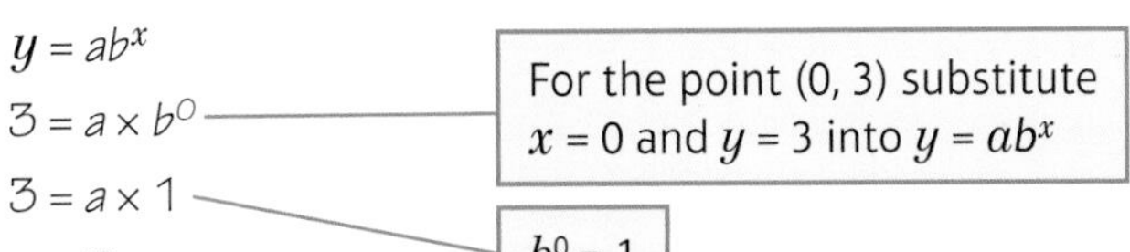

$y = ab^x$
$3 = a \times b^0$ — For the point (0, 3) substitute $x = 0$ and $y = 3$ into $y = ab^x$
$3 = a \times 1$ — $b^0 = 1$
$a = 3$

$y = 3b^x$ — $a = 3$, so the equation is $y = 3b^x$
$12 = 3b^2$ — For the point (2, 12), substitute $x = 2$ and $y = 12$ into $y = 3b^x$
$4 = b^2$
$b = 2$

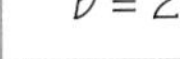

8 **Exam-style question**

The sketch shows a curve with equation $y = ka^x$, where k and a are constants and $a > 0$.

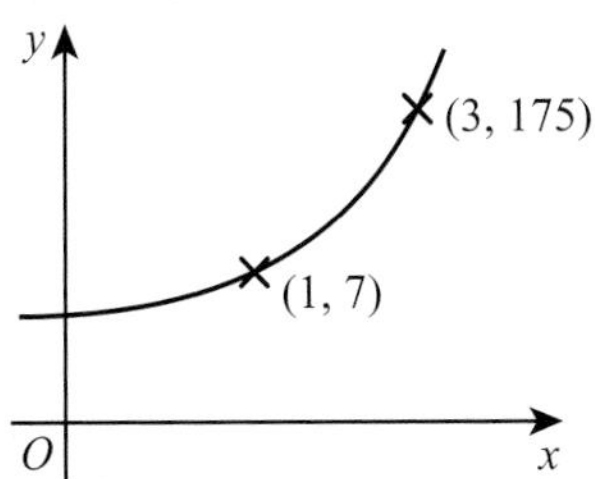

The curve passes through the points (1, 7) and (3, 175).
Calculate the value of k and the value of a. **(3 marks)**

June 2008, Q18, 5544H/15H

Exam hint
Start by using the point (1, 7) to express k in terms of a.

9 **Finance / Reasoning** The value, V (in £), of a car depreciates exponentially over time.
The value of the car on 1 January 2015 was £20 000.
The value of the car on 1 January 2017 was £16 200.
The sketch graph shows how the value of the car changes over time.
The equation of the graph is $V = ab^t$
where t is the number of years after 1 January 2015, and a and b are positive constants.

a Use the information to find the value of a and b.
b Use your values of a and b in the formula $V = ab^t$ to estimate the value of the car on 1 January 2018.
c By what percentage does the car depreciate each year?

10 **Reasoning / Modelling** The population of a country is currently 4 million, and is growing at a rate of 5% a year.
The expected population, p (in millions), in t years' time, is given by the formula $p = 4 \times 1.05^t$

a Use a table of values to draw the graph of p against t for the next 6 years.
b Use your graph to estimate
   i the size of the population after 2.5 years
   ii the time taken for the population to reach 5 million.

Reflect How do you know that the growth in population is exponential?

11 **Reasoning / Finance** £10 000 is invested in a savings account paying 4% compound interest a year.

a Write a formula for the value of the savings account (V) and the number of years (t).

b Draw a graph of V against t for the first 10 years.

c Use the graph to estimate when the investment will reach a value of £11 000.

19.5 Non-linear graphs

Objectives

- Calculate the gradient of a tangent at a point.
- Estimate the area under a non-linear graph.

Why learn this?

Formula 1 engineers use curved speed–time graphs to track the performance of their cars.

Fluency

Calculate the area of

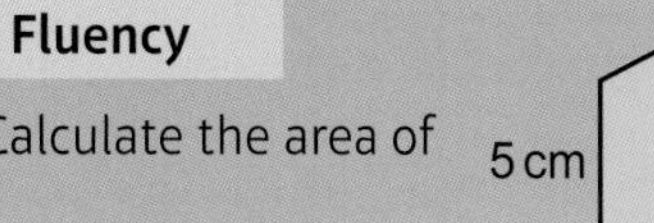

Warm up

1 The diagrams show four flasks filling up with water. h is the height of water after time t.

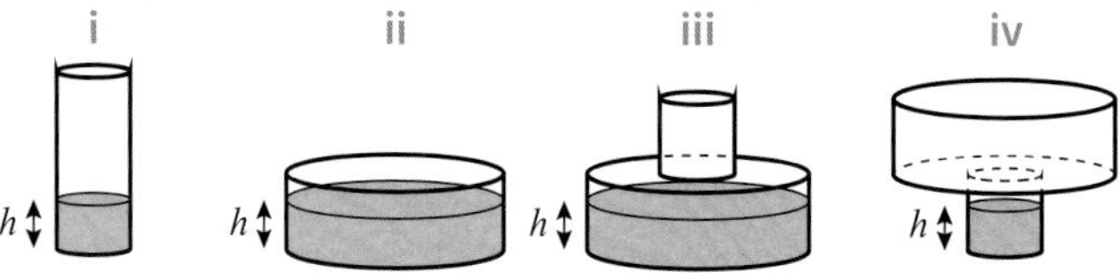

These graphs each show a relationship between h and t.
Match each flask to a graph.

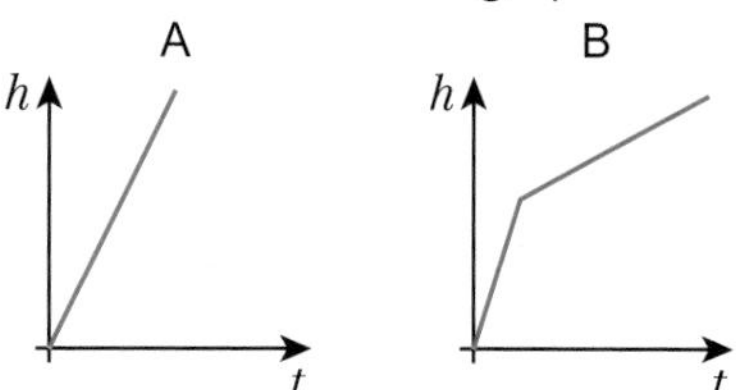

C D

2 The velocity–time graph shows the motion of a particle.
Given that the particle travelled in a straight line, calculate the distance travelled between $t = 0$ and $t = 12$.

3 Points A and B are connected by a straight line.
Write the gradient of the line AB for

a A(0, 0) and B(2, 6) b A(3, 5) and B(7, 11) c A(−2, 4) and B(2, 0)

Key point 6

The **tangent** to a curved graph is a straight line that touches the graph at a point.
The gradient at a point on a curve is the gradient of the tangent at that point.

Example 4

Water is poured into the container at a constant rate.
The graph shows the height, h (in cm), of the water after time, t (in seconds).
Estimate the rate at which h is increasing after 10 seconds.

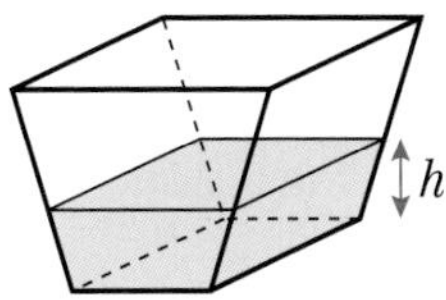

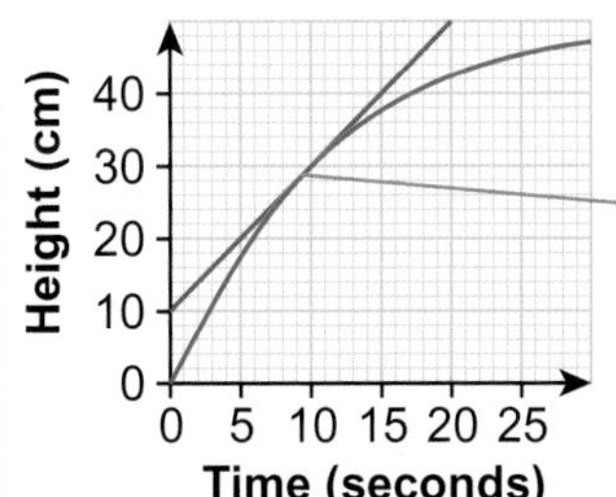

Draw a tangent to the curve at $t = 10$

$$\text{Gradient} = \frac{\text{change in } h}{\text{change in } t} = \frac{50 - 10}{20 - 0} = \frac{40}{20} = 2$$

Calculate the gradient of the tangent.

At $t = 10$ the height of the water is increasing at 2 cm per second.

4 **Communication** Water is poured into a flask at a constant rate.
h is the height of water after time, t.

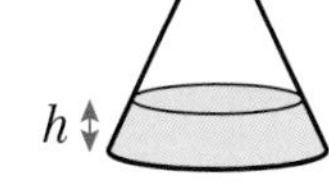

a Describe how the rate at which the height increases changes over time.

b Which graph best describes the relationship between h and t?

A
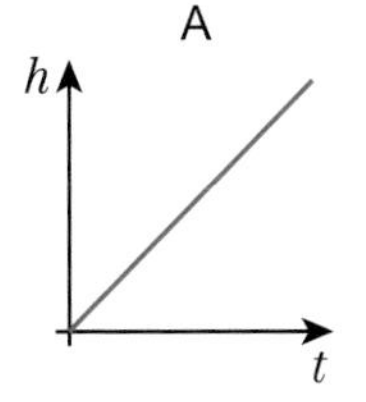

B
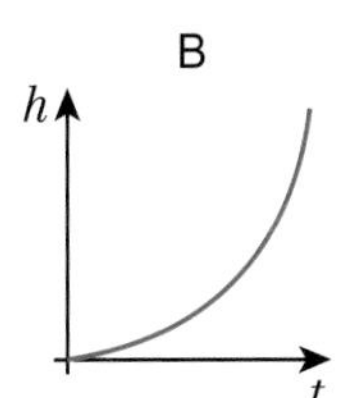

C
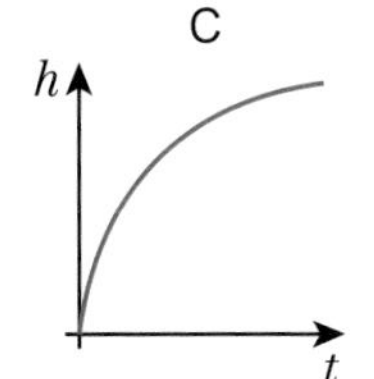

D
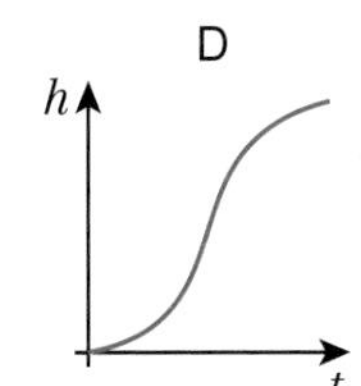

5 **Reasoning / Communication** The graph shows the relationship between the temperature, T, of a cup of coffee and time, t.

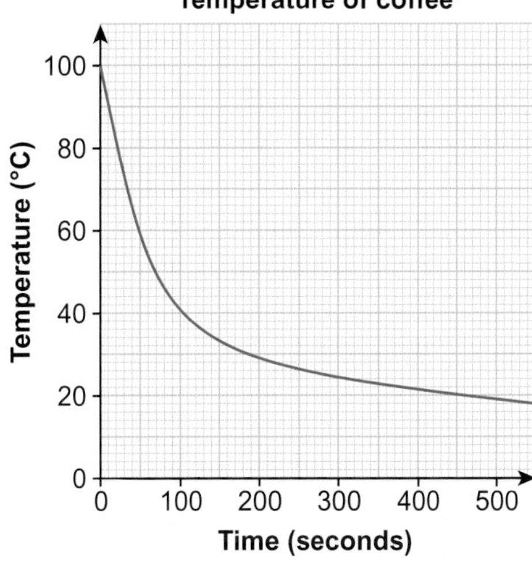

a What is the temperature, T, of the coffee after 50 seconds?

b Describe the rate at which the coffee cools down.

c Calculate the drop in temperature between 100 and 200 seconds.

d Calculate the average rate of temperature reduction between 100 and 200 seconds.

Q5d hint Find the temperature at $t = 100$ and $t = 200$.

e Becky says, 'The coffee cools twice as quickly between 100 and 200 seconds as between 0 and 400 seconds.' Is Becky correct?
Explain your answer.

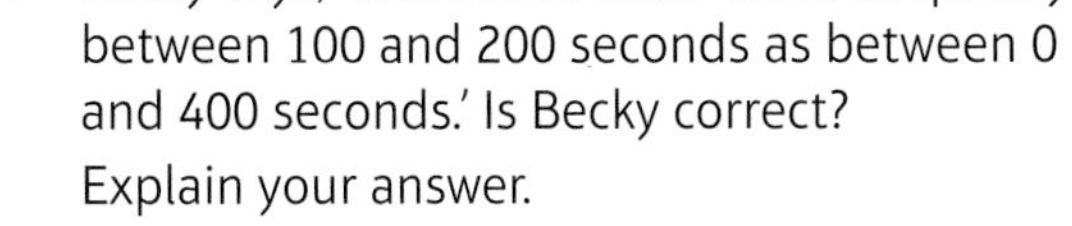

f Compare the average rate of temperature reduction over the first 300 seconds with the rate of temperature reduction at exactly 300 seconds.

Key point 7

On a distance–time graph, the gradient of the tangent at any point gives the speed.

6 **Reasoning** The distance–time graph shows information about a runner in a 100 metre race.

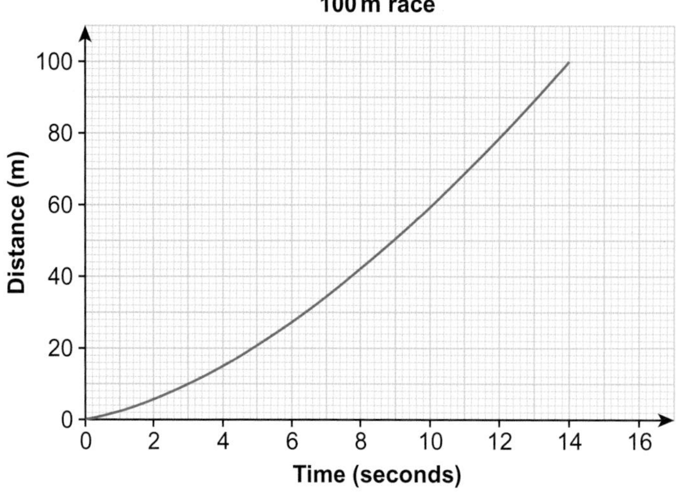

a Estimate the speed of the runner 12 seconds into the race.

b After how many seconds is the runner running at full speed?

c If the runner could maintain this speed, how long would it take to run a 200 m race?

Key point 8

The straight line that connects two points on a curve is called a **chord**. The gradient of the chord gives the average rate of change and can be used to find the average speed on a distance–time graph.

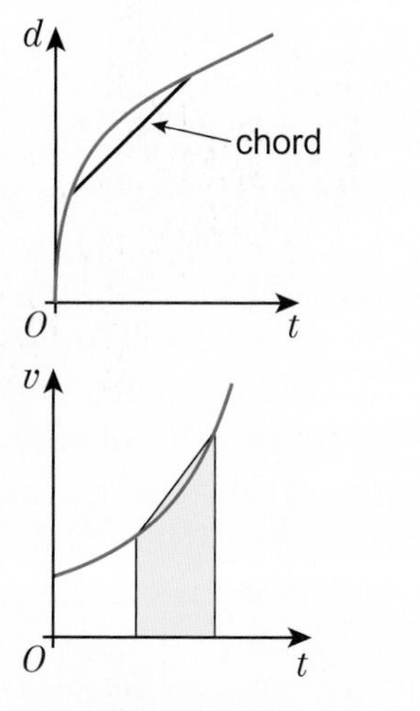

The area under a velocity–time graph shows the displacement, or distance from the starting point. To estimate the area under a part of a curved graph, draw a chord between the two points you are interested in, and straight lines down to the horizontal axis to create a trapezium. The area of the trapezium is an estimate for the area under this part of the graph.

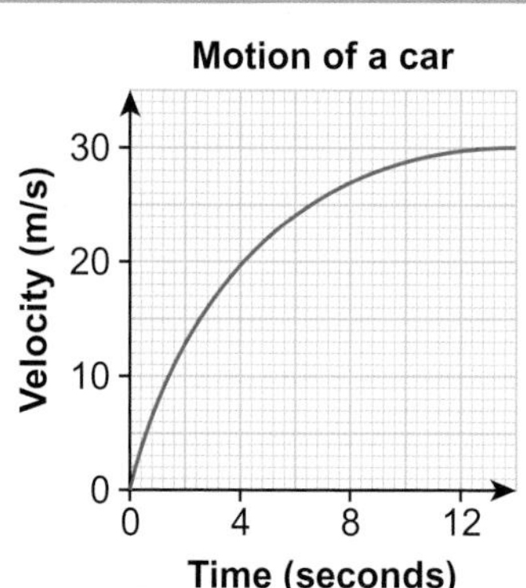

7 **Reasoning / Communication** A car drives away from a set of traffic lights. The velocity–time graph gives some information about the motion of the car.

Motion of a car

Velocity (m/s) — 0, 10, 20, 30; Time (seconds) — 0, 4, 8, 12

a Copy the graph. Draw a chord from $t = 0$ to $t = 10$.

b Calculate the average acceleration of the car over the first 10 seconds.

Q7b hint Find the gradient of the chord.

c Estimate the acceleration at time $t = 4$ seconds.

d Describe how the acceleration changes over the 12 seconds.

e Estimate the displacement of the car from $t = 4$ to $t = 8$.

Discussion Is the displacement always the same as the distance travelled?

8 a Draw the graph of $y = x^2 + 3$ for $0 \leqslant x \leqslant 4$
b Draw in a chord from $x = 0$ to $x = 1$ and use it to make a trapezium under the graph.
c Repeat with chords from $x = 1$ to $x = 2$, $x = 2$ to $x = 3$ and $x = 3$ to $x = 4$.
d Calculate the areas of your trapezia to estimate the area under the graph of $y = x^2 + 3$ from $x = 0$ to $x = 4$

Q8 hint

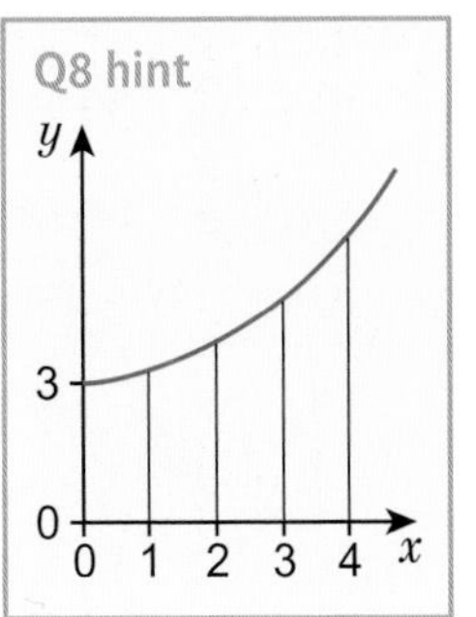

9 **Reasoning / Communication** The distance–time graph shows information about a 400 m race between Amy and Clare.

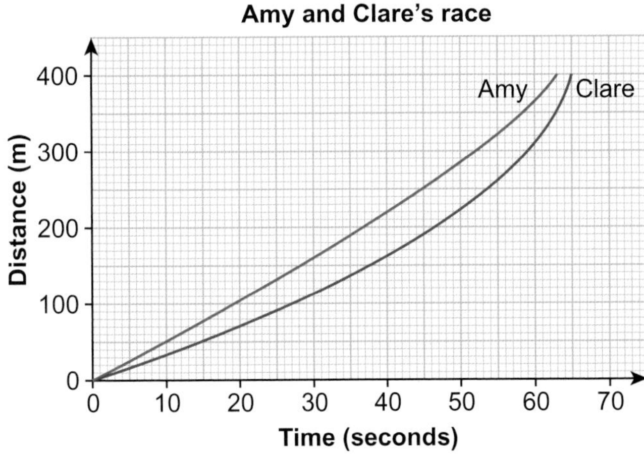

a Describe the race between Amy and Clare.
b Compare Clare's speeds for the first and second halves of the race.
c Estimate the difference in their speeds 50 seconds into the race.

10 **Exam-style question**

The velocity–time graph describes the motion of a car. Velocity, v, is measured in metres per second (m/s) and time, t, is measured in seconds. The car travels in a straight line.

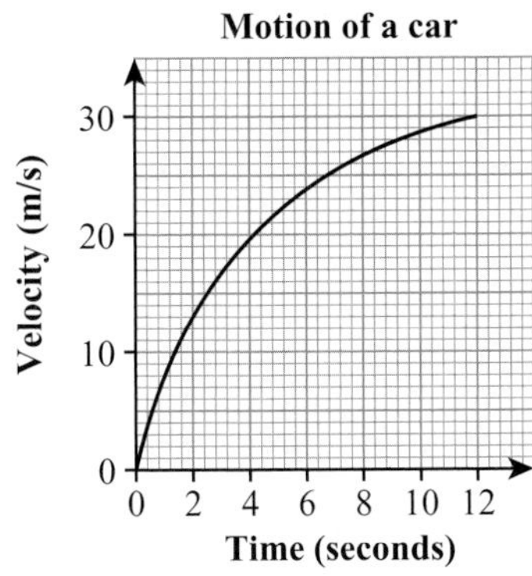

a Estimate the acceleration at $t = 4$. **(2 marks)**
b Estimate the distance travelled between $t = 6$ and $t = 8$. **(3 marks)**
c The instantaneous acceleration at time T is equal to the average acceleration for the first 8 seconds. Find an estimate for the value of T. **(3 marks)**

Q10a hint For motion, draw a tangent to the curve.

Q10c hint Calculate the average acceleration for the first 8 seconds.

Q10c communication hint Instantaneous acceleration is the acceleration at a given moment (or instant) in time.

19.6 Translating graphs of functions

Objective

- Understand the relationship between translating a graph and the change in its function notation.

Why learn this?

Architects, designers and artists all use different types of transformations to create inspirational designs.

Fluency

The point (1, 2) is translated by $\begin{pmatrix}3\\0\end{pmatrix}$ Find the new coordinate.

The point (−3, 4) is translated by $\begin{pmatrix}0\\-2\end{pmatrix}$ Find the new coordinate.

Warm up

1 $f(x) = 3x + 1$, $g(x) = 2x^2$
Find the values of
a f(2) b f(−3) c g(1) d f(0) e g(0)

2 $g(x) = 5x + 2$
a Find the value of i $g(3) + 2$ ii $g(3 + 2)$
b Write out in full i $y = g(x) + 2$ ii $y = g(x + 2)$

3 Here is the graph of $y = \sin x$

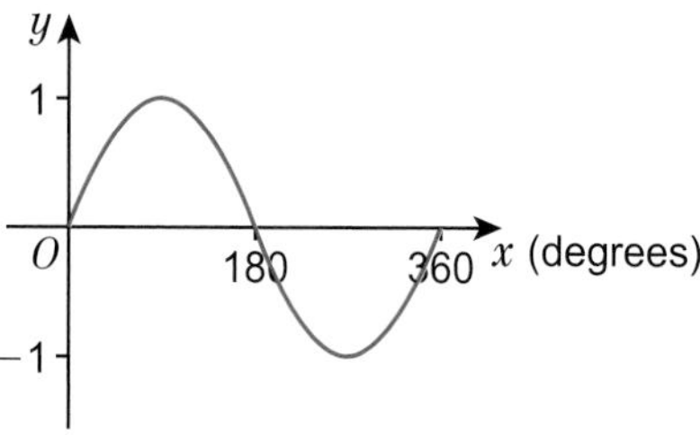

Match the equations to these graphs.
a $y = \sin x + 0.5$ b $y = \sin(x + 30)$

A

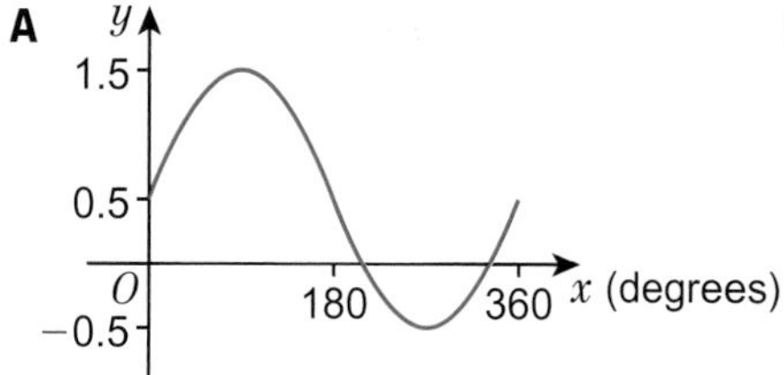

B

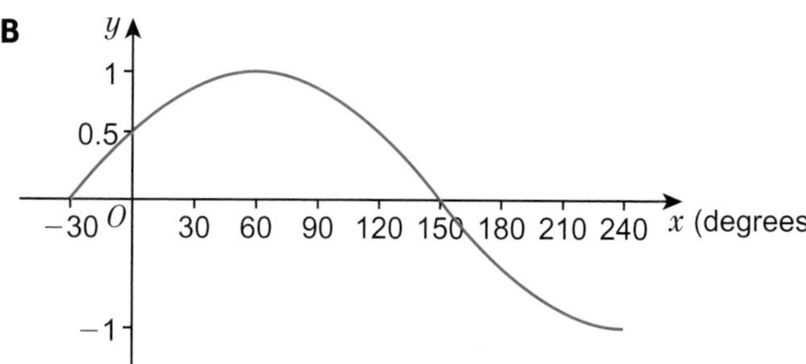

4 **Communication** Draw a coordinate grid with −5 to +5 on the x-axis and with −10 to +30 on the y-axis.
a On the same set of axes draw the graphs of
 i $y = f(x) = x^2$
 ii $y = f(x) - 5 = x^2 - 5$
 iii $y = f(x + 1) = (x + 1)^2$

Q4a hint Create a table of values for each graph.

b The minimum point of $y = f(x)$ is (0, 0).
Write the coordinates of the minimum point of
 i $y = f(x) - 5$ ii $y = f(x + 1)$
c Describe the transformation that maps the graph of $y = f(x)$ onto the graph of
 i $y = f(x) - 5$ ii $y = f(x + 1)$

Key point 9

The graph of $y = \mathrm{f}(x)$ is transformed into the graph of $y = \mathrm{f}(x) + a$ by $\begin{pmatrix} 0 \\ a \end{pmatrix}$.

The graph of $y = \mathrm{f}(x)$ is transformed into the graph of $y = \mathrm{f}(x + a)$ by $\begin{pmatrix} -a \\ 0 \end{pmatrix}$.

5 Here is the graph of $y = \mathrm{f}(x) = x^2$

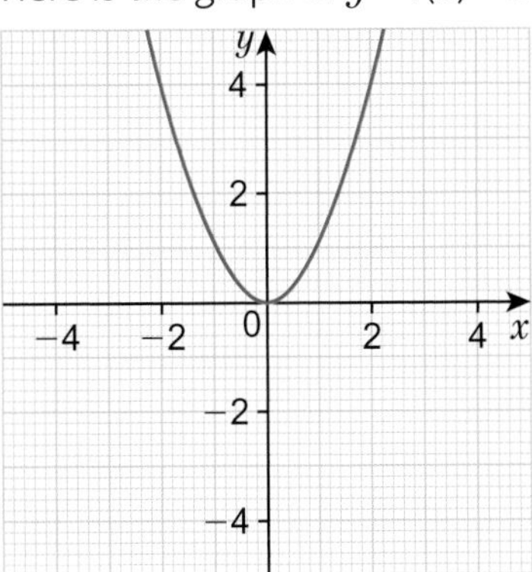

Q5 hint Label the coordinates of the turning point and the y-intercept on each of your sketch graphs.

Copy the axes and sketch the graphs of

a $y = \mathrm{f}(x) + 1$ b $y = \mathrm{f}(x) - 2$ c $y = \mathrm{f}(x + 2)$ d $y = \mathrm{f}(x - 4)$

6 Write the vector that translates $y = \mathrm{f}(x)$ onto

a $y = \mathrm{f}(x) + 2$ b $y = \mathrm{f}(x) - 3$ c $y = \mathrm{f}(x + 1)$ d $y = \mathrm{f}(x - 4)$ e $y = \mathrm{f}(x + 5) - 2$

Example 5

Graph A is a translation of the graph of $y = \mathrm{f}(x)$.
Write the equation of graph A.

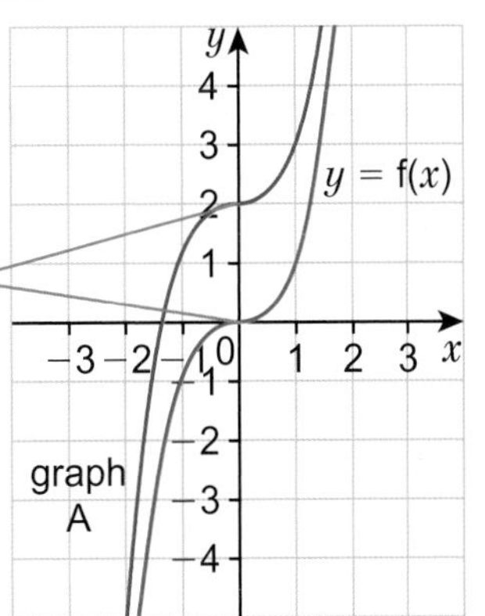

Find corresponding points on the graph, for example where the graph is horizontal.

Describe the translation to the corresponding point.

The graph of $y = \mathrm{f}(x)$ has been translated by $\begin{pmatrix} 0 \\ 2 \end{pmatrix}$

The equation of graph A is $y = \mathrm{f}(x) + 2$

Write your final answer in function notation.

7 **Exam-style question**

The graph of $y = \mathrm{f}(x)$ is shown on the grid.

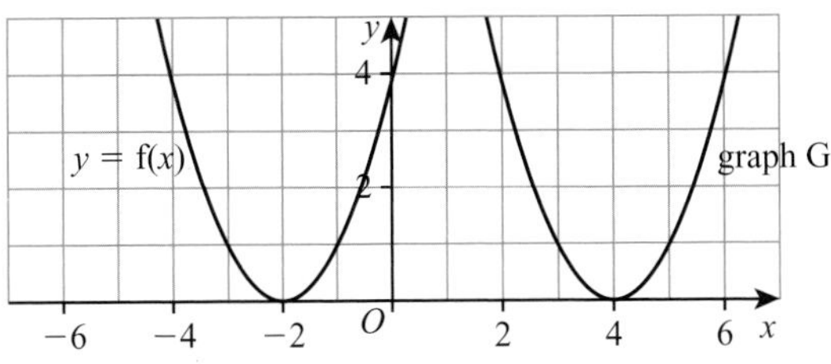

The graph G is a translation of the graph of $y = \mathrm{f}(x)$.
Write down the equation of graph G. **(1 mark)**

March 2013, Q25b, 1MA0/1H

Exam hint
First see if the translation is to the right or to the left or up or down and by how many squares.

8 **Exam-style question**

The graph of $y = f(x)$ is shown on the grid.
Copy the diagram and sketch the graph of $y = f(x - 3)$. **(2 marks)**

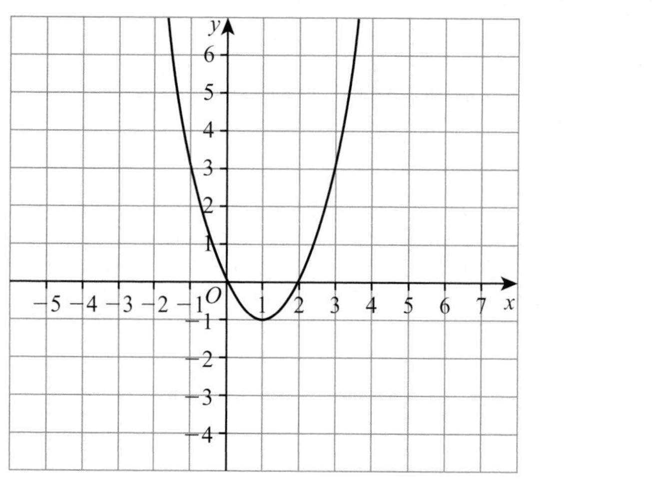

June 2012, Q26a, 1MA0/1H

Q8 strategy hint
Make sure you translate the points that have integer coordinates such as (–1, 3) and (2, 0) exactly three squares in the correct direction.

9 Here is a sketch of $y = f(x) = x^3$

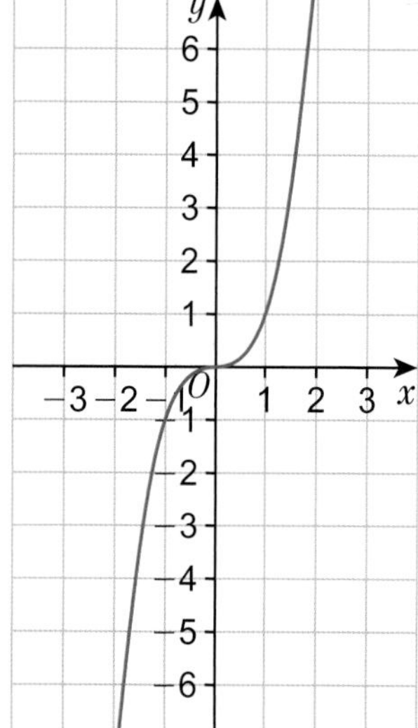

a Draw sketches of the graphs
   i $y = f(x) + 3$ ii $y = f(x - 3)$

b Write the coordinates of the point which (0, 0) is mapped to for both graphs.

10 **Reasoning** $f(x) = 3x + 2$

a Draw the graph of $y = f(x)$

b Draw the graph of $y = f(x + 1)$

c Write the algebraic equation of $y = f(x + 1)$

11 **Problem-solving** $f(x) = \frac{1}{x}$

a Sketch the graph of $y = f(x + 2) - 3$

b Write the equation of each asymptote.

Q11 communication hint
An **asymptote** is a line that a curve approaches but never reaches.

19.7 Reflecting and stretching graphs of functions

Objectives

- Understand the effect stretching a curve parallel to one of the axes has on its function form.
- Understand the effect reflecting a curve in one of the axes has on its function form.

Why learn this?

Function transformations are used in computer animation software.

Fluency

Paul draws a curve by connecting coordinates. What happens to the curve if he doubles all the y-values, but keeps the x-values the same?

Warm up

1 $f(x) = 6x - 4$
Write out in full
a $f(-x)$ b $-f(x)$

2 $g(x) = 2x + 5$
Find
a $-g(-3)$ b $2g(10)$

3 Here is the graph of $y = \cos x$

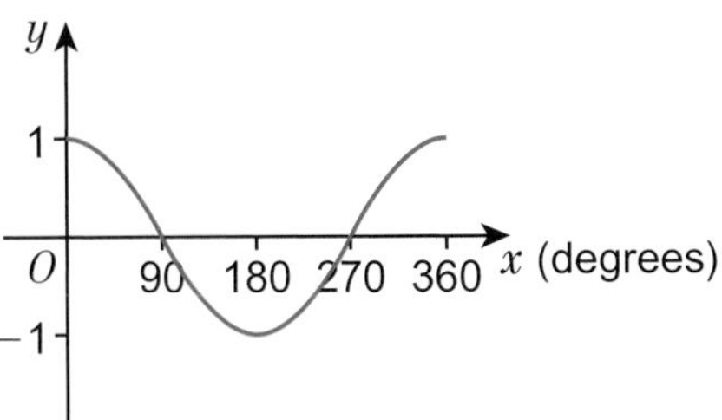

Sketch the graph of
a $y = -\cos x$ b $y = 2\cos x$

4 $f(x) = 4x - 2$
a Copy and complete the table.

x	−2	−1	0	1	2
f(x)					
−f(x)					
f(−x)					

b On the same set of axes, draw the graphs of
i $y = f(x)$ ii $y = -f(x)$ iii $y = f(-x)$
c Describe the transformation that maps $f(x)$ onto $-f(x)$.
d Describe the transformation that maps $f(x)$ onto $f(-x)$.

Key point 10

The transformation that maps the graph $y = f(x)$ onto the graph $y = f(-x)$ is a reflection in the y-axis.
The transformation that maps the graph $y = f(x)$ onto the graph $y = -f(x)$ is a reflection in the x-axis.

5 **Reasoning** The diagram shows the graph of $y = f(x)$

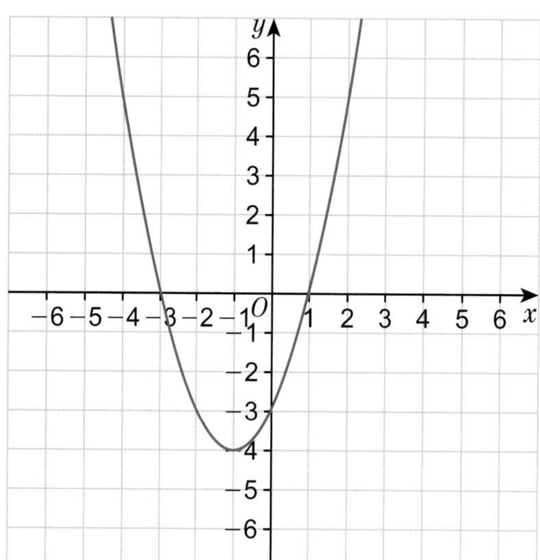

a Copy the sketch. On the same axes sketch the graphs of
   i $y = -f(x)$ ii $y = f(-x)$

b Finley says, 'The graphs of $y = f(x)$ and $y = -f(x)$ always intersect the y-axis in the same place. The graphs of $y = f(x)$ and $y = f(-x)$ always intersect the x-axis in the same place.' Is Finley right? Explain your answer.

6 **Reasoning** The diagram shows the graph of $y = f(x)$
The turning point of the curve is A(2, 4).
Write the coordinates of the turning points of the curves with these equations.

a $y = -f(x)$

b $y = f(-x)$

c $y = -f(-x)$

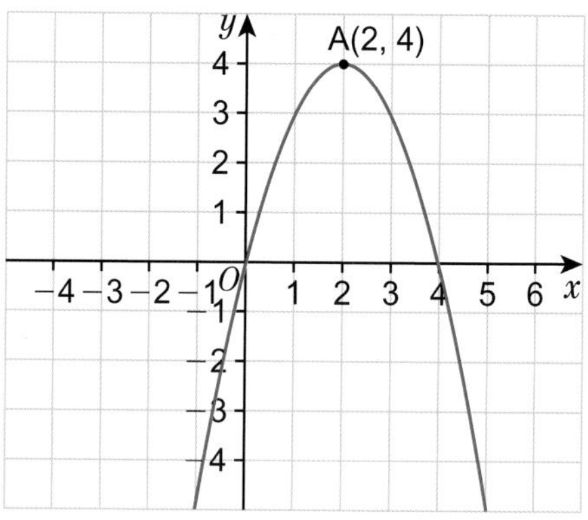

7 **Reasoning** Here is the graph of $y = f(x) = x^3 + 2$

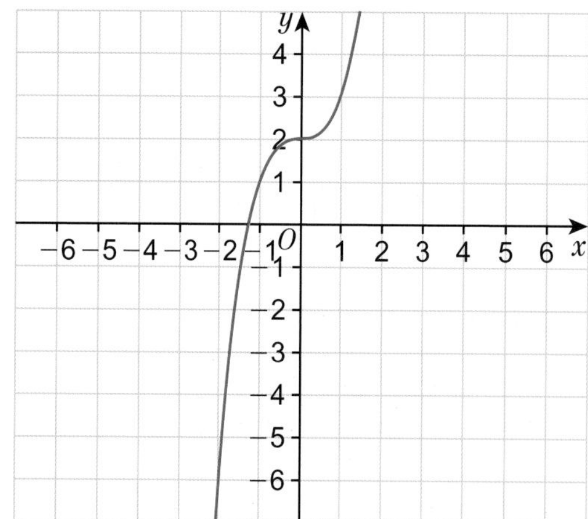

a Copy the sketch and on the same axes sketch the graphs of
   i $-f(x)$ ii $f(-x)$ iii $-f(-x)$

b Describe the transformation that maps $f(x)$ onto $-f(-x)$.

8 $f(x) = x^2 - 4$

a Copy and complete the table.

x	−2	−1	0	1	2
f(x)					
2f(x)					
f(2x)					

b On the same set of axes draw the graphs of

i $y = f(x)$ ii $y = 2f(x)$ iii $y = f(2x)$

Key point 11

For any function, f, the transformation which maps the graph of $y = f(x)$ onto the graph of $y = af(x)$ is a stretch of scale factor a parallel to the y-axis.

For any function, f, the transformation which maps the graph of $y = f(x)$ onto the graph of $y = f(ax)$ is a stretch of scale factor $\frac{1}{a}$ parallel to the x-axis.

Example 6

The diagram shows the graph of $y = f(x)$, where $f(x) = x^2 - 2x + 1$

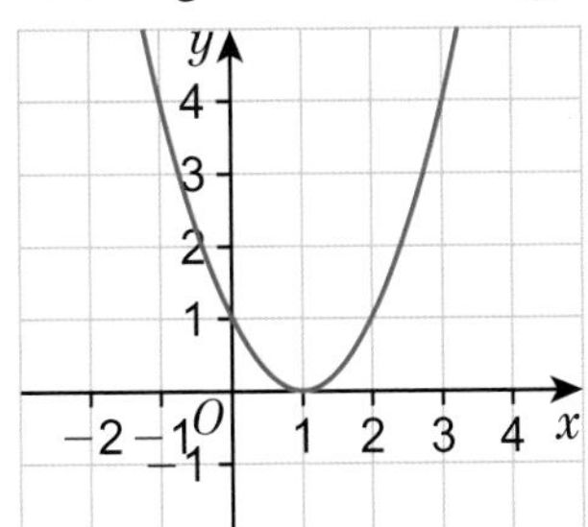

a Sketch the graph of $y = 3f(x)$

b Sketch the graph of $y = f(2x)$

a

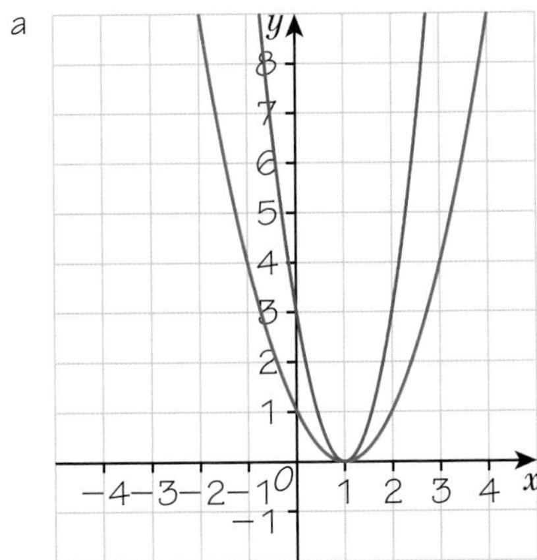

The transformation that maps $y = f(x)$ onto $y = 3f(x)$ is a stretch of scale factor 3 parallel to the y-axis. The x-coordinate of each point on the graph will remain the same, but each y-coordinate will be multiplied by 3.
(1, 0) → (1, 0)   (0, 1) → (0, 3)
(2, 1) → (2, 3)

b

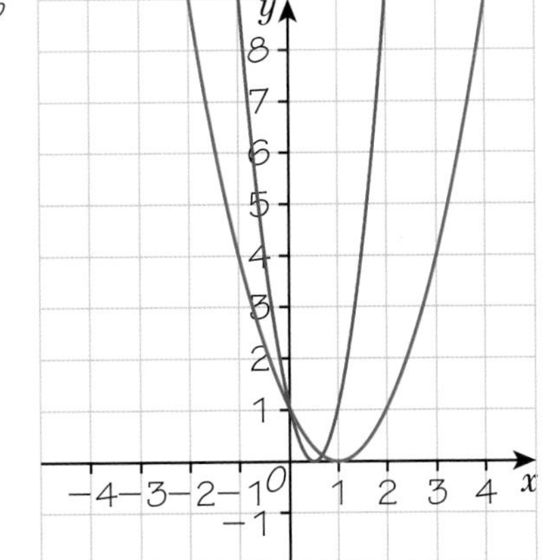

The transformation that maps $y = f(x)$ onto $y = f(2x)$ is a stretch of scale factor $\frac{1}{2}$ parallel to the x-axis. The y-coordinate on each graph will remain the same, but each x-coordinate will be multiplied by $\frac{1}{2}$.
(1, 0) → (0.5, 0)   (0, 1) → (0, 1)
(2, 1) → (1, 1)

9 Here is the graph of $y = f(x)$
Draw the graphs of
a $y = 2f(x)$
b $y = 3f(x)$
c $y = \frac{1}{2}f(x)$

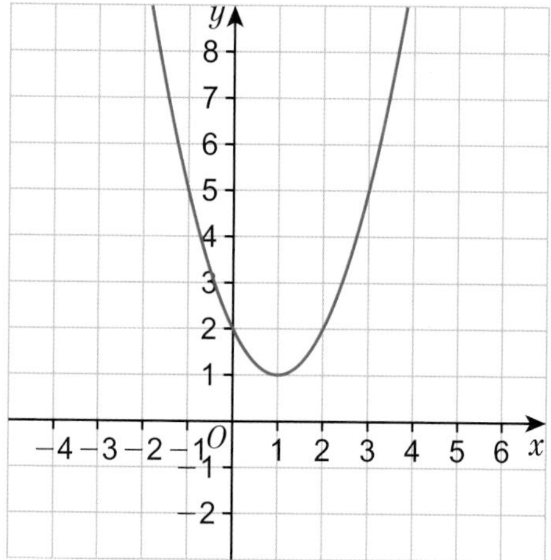

10 Here is the graph of $y = f(x)$

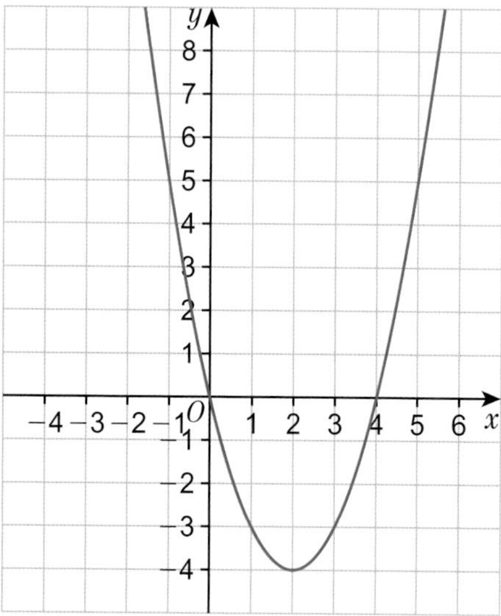

Draw the graphs of
a $y = f(2x)$ b $y = f(\frac{1}{3}x)$

11 **Exam-style question**

The graph of $y = f(x)$ is shown on the grid.
Copy the diagram and sketch the graph of $y = 2f(x)$. **(2 marks)**

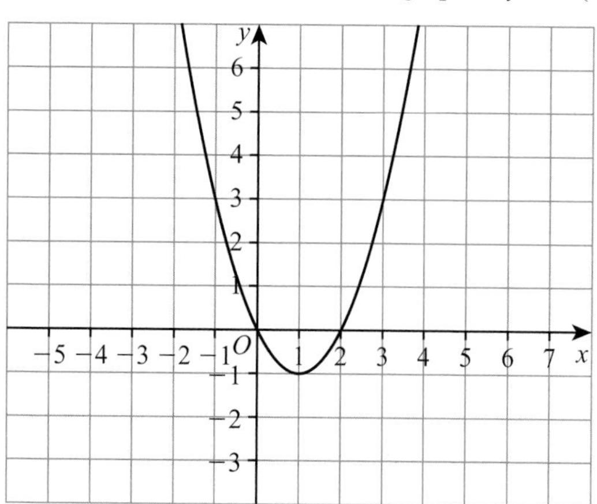

June 2012, Q26b, 1MA0/1H

Exam hint
Calculate some y-values for $y = 2f(x)$ to make sure your graph passes through the correct points.

12 Reasoning The diagram shows the graph of $y = f(x) = 3x + 2$

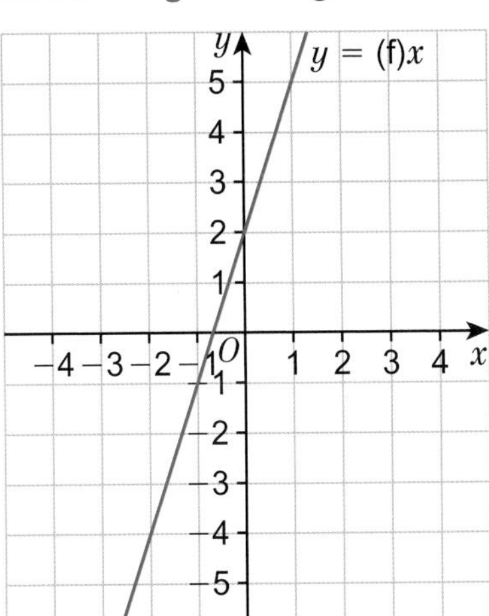

The graph has been transformed in different ways.
Match the function notation to the graphs.

a $f(2x)$ b $2f(x)$ c $f(-x)$ d $-f(x)$

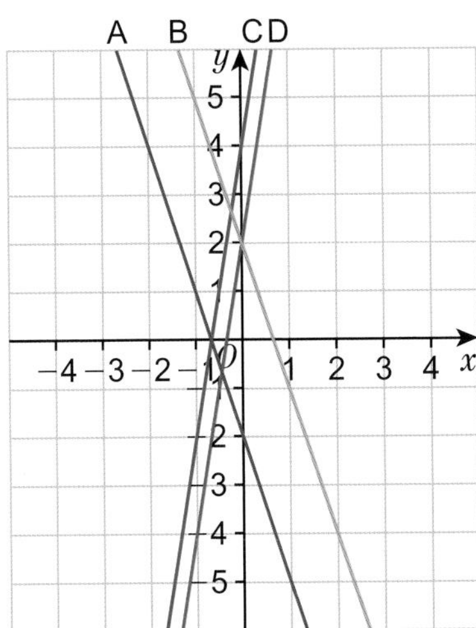

13 Reasoning Here is a sketch of $y = f(x) = -x^2 + 4$

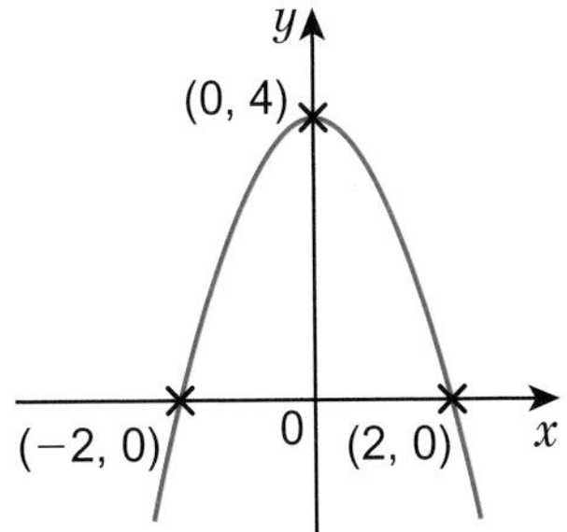

The graph has a maximum value at (0, 4).
It intersects the x-axis at (−2, 0) and (2, 0).

a Sketch the graph of $y = (2x)$
b Write the maximum value of $f(2x)$.
c Write the coordinates of the points where $y = f(2x)$ intersects the x-axis.
d Why does $f(2x)$ have a maximum value?

19 Problem-solving: Modelling outbreaks

Objectives
- Plot and interpret exponential graphs.
- Use and evaluate mathematical models.

Mathematics can be used to model disease outbreaks. This helps organisations to plan ahead and respond to emergencies.
One simple model for N, the number of infected people, is $N = A \times B^t$, where A is the initial number of infected people, t is the time in days since the start of the outbreak, and B is a measure of how contagious the disease is.

1 One outbreak is modelled using the equation $N = 7 \times 2^t$
Calculate the number of people who are infected when $t = 0$, $t = 1$, $t = 2$, $t = 3$ and $t = 4$.
When will the number of infected people exceed 100?

Q1 hint You could draw and complete a table of values, and then plot a graph of N against t.

2 A different outbreak of a long-lasting disease has begun in a nearby city. The local authorities have recorded the reported total number of infected people in the first five days in the table at the bottom of the page.

a Fit a model of the form $N = A \times B^t$ to this data, by finding approximate values of A (to the nearest integer) and B (to 1 decimal place).

Q2a hint To estimate B, find the ratios between the numbers of infected people on successive days

The local authorities are worried about two things:
- City doctors can only treat 500 newly infected people each day.
- The hospitals currently only have enough supplies to treat 3000 infected people in total.

b Use your model from part **a** to decide whether it is more urgent to send extra supplies or extra doctors. Justify your decision.

3 $N = A \times B^t$ is not appropriate for modelling diseases in the long term. Write down two reasons why the number of infected people would be unlikely to continue to increase exponentially.

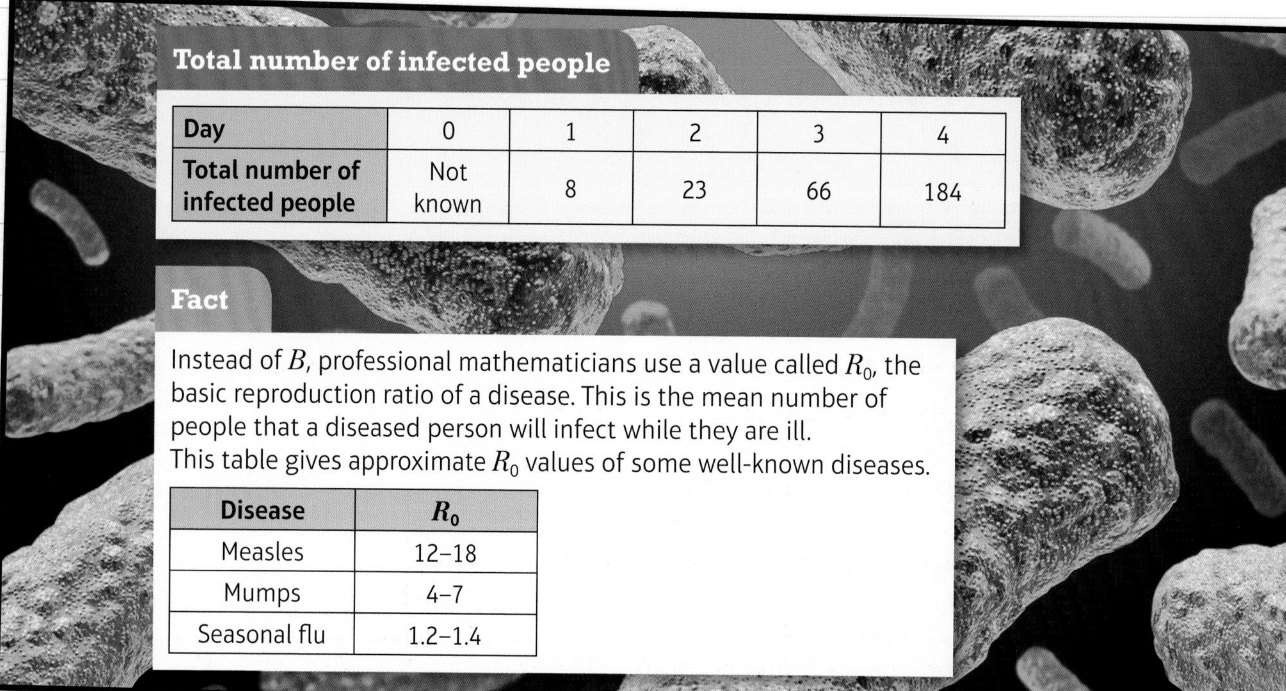

Total number of infected people

Day	0	1	2	3	4
Total number of infected people	Not known	8	23	66	184

Fact

Instead of B, professional mathematicians use a value called R_0, the basic reproduction ratio of a disease. This is the mean number of people that a diseased person will infect while they are ill.
This table gives approximate R_0 values of some well-known diseases.

Disease	R_0
Measles	12–18
Mumps	4–7
Seasonal flu	1.2–1.4

19 Check up

Log how you did on your Student Progression Chart.

Proportion

1 In a circuit with a fixed resistance, the current, I, is directly proportional to the voltage, V. When the current is 10 amps, the voltage is 4 volts.

a Write a formula for I in terms of V.

b Calculate the current when the voltage is 10 volts.

When the voltage is constant and the resistance is allowed to vary, the current is inversely proportional to the resistance, R.

When the resistance is 20 ohms, the current is 2 amps.

c Write a formula for I in terms of R.

d Calculate the current when the resistance is 4 ohms.

2 y is directly proportional to x^2.
When $y = 48$, $x = 2$

a Write a formula for y in terms of x.

b Find y when $x = 3$.

c Find x when $y = 300$.

3 c is inversely proportional to d^3.
When $c = 5.5$, $d = 4$

a Write a formula for c in terms of d.

b Find c when $d = 5$.

Exponential and other non-linear graphs

4 On the same axes, sketch the graphs of

a $y = 2^x$ b $y = 2^{-x}$ c $y = 3^x$

5 **Reasoning** The diagram shows a sketch of the curve $y = ab^x$

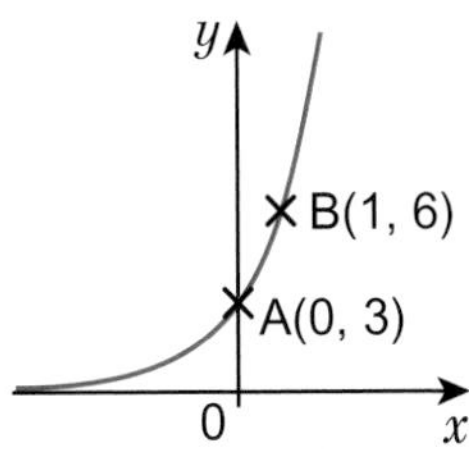

The curve passes through the points A(0, 3) and B(1, 6).

a Find the values of

i a ii b

b Find the value of y when $x = 4$.

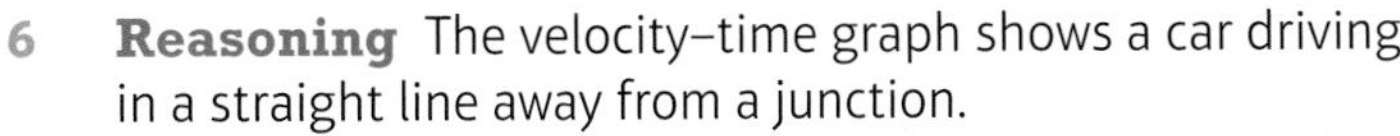

6 **Reasoning** The velocity–time graph shows a car driving in a straight line away from a junction.

The time after the junction, t, is measured in seconds, s.

The velocity, v, is measured in metres per second, m/s.

a Calculate the average acceleration of the car between $t = 20$ and $t = 30$.

b Estimate the acceleration at $t = 40$.

c Estimate the distance travelled between $t = 40$ and $t = 60$

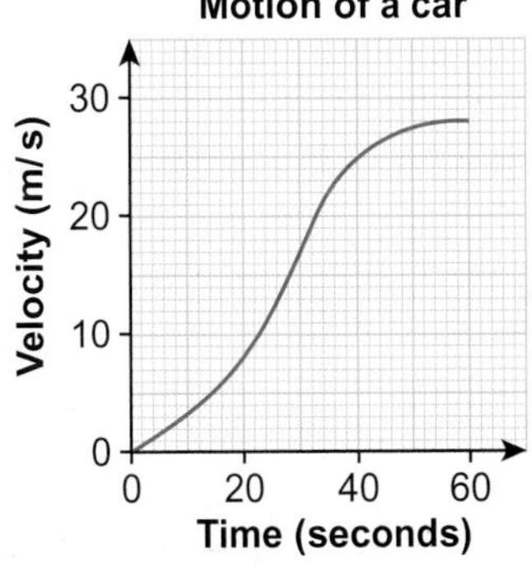

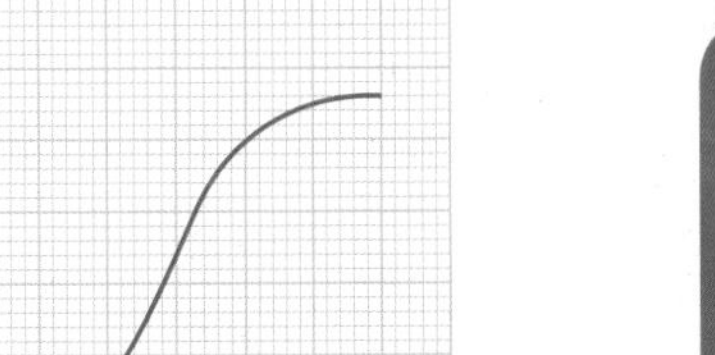

Transformations of graphs functions

7 The function $y = f(x)$ is shown in the diagram. Sketch the graph of

a $y = f(x) + 3$

b $y = f(x - 2)$

c $y = 2f(x)$

d $y = f(-x)$

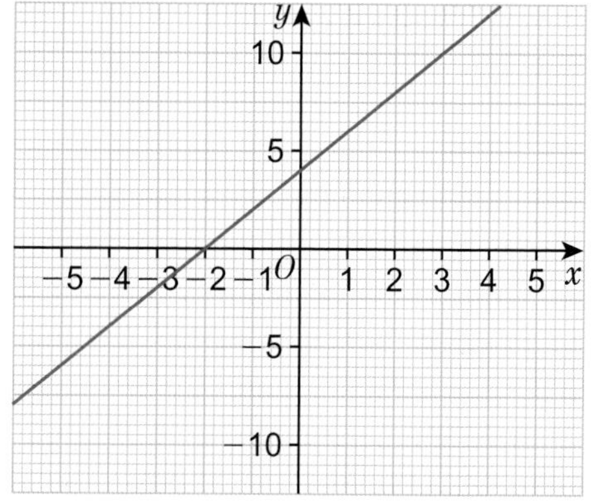

8 $y = f(x)$

The graph of $y = f(x)$ is shown on the grid.

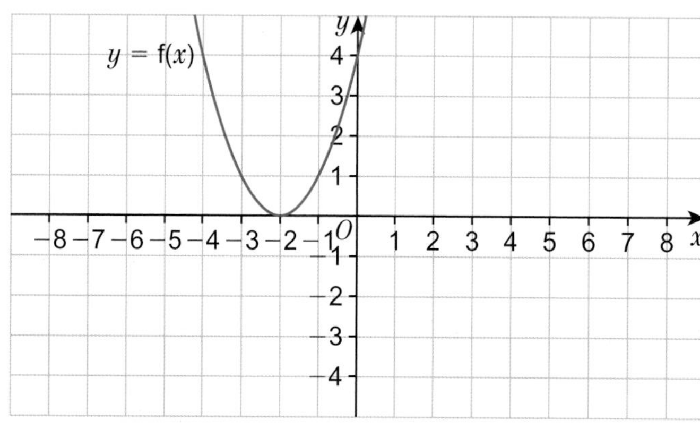

Copy the diagram and sketch the graph of $y = -f(x)$

Reflect

9 How sure are you of your answers? Were you mostly

Just guessing ☹ Feeling doubtful 😐 Confident ☺

What next? Use your results to decide whether to strengthen or extend your learning.

★ Challenge

10 Finley transforms the graph $y = 4x + 3$ using the transformations shown on the cards.

Reflect in the x-axis	Translate $\begin{pmatrix}0\\4\end{pmatrix}$	Reflect in the y-axis	Translate $\begin{pmatrix}1\\0\end{pmatrix}$

a Does the order of the transformations change the final result?

> **Q10a hint** Draw a sketch and try different orders.

There is more than one order that maps the graph onto itself.

b Find the different orders of transformations that map the graph onto itself.

c Explain why applying the transformations in different orders can lead to the same outcome.

> **Q10c hint** Use a combination of algebraic and function notation.

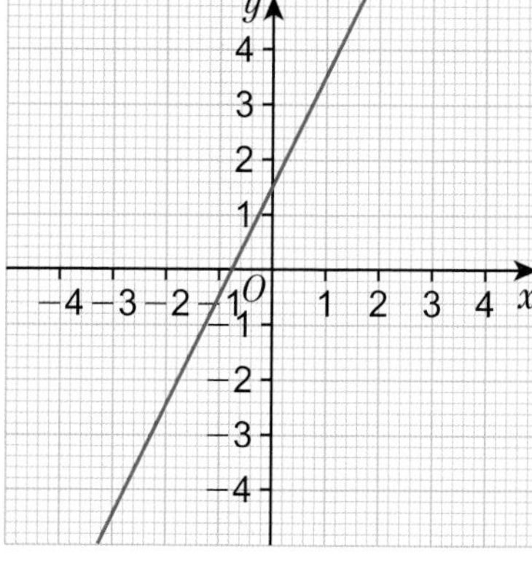

19 Strengthen

Proportion

1 **Real** The cost, c, of diesel is directly proportional to the number of litres, l, purchased. Diesel costs £1.32 a litre.
a Write a formula for the cost of diesel.
b Use your formula to calculate the amount of diesel you can purchase for £12.

2 **Reasoning**
a Write the lengths, l, and widths, w, of different rectangles with an area of $24\,cm^2$.
b Is l directly or inversely proportional to w?
c Draw a set of axes with l on the horizontal axis and w on the vertical axis. Draw a graph of possible lengths and widths of a $24\,cm^2$ rectangle.

Q2b hint $l = \square w$ or $l = \frac{\square}{w}$?

3 The graphs show how two pairs of variables relate to each other. Which graph shows
a direct proportion
b inverse proportion?

i
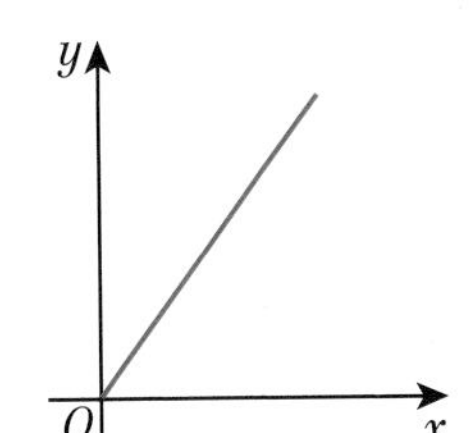

ii
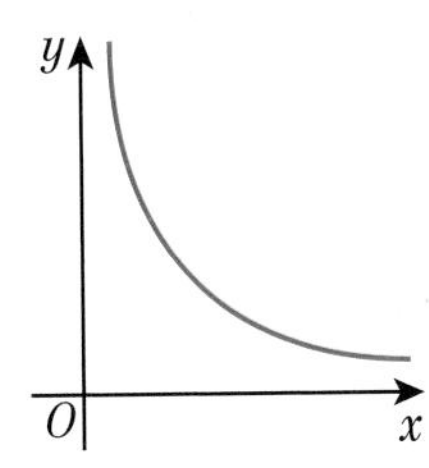

4 Write
i the statement of proportionality ii the formula for each of these.
a A is directly proportional to B.
b C is inversely proportional to D.
c M is directly proportional to the square of N.
d F is inversely proportional to the cube of G.
e H is inversely proportional to the square root of T.
f R is directly proportional to the cube of S.

Q4 hint Use $\propto$ to write a statement of proportionality, $y \propto x$
Use k (the constant of proportionality) to create a formula, $y = kx$

5 F is directly proportional to a.
$F = 20$ when $a = 2$
$F \propto a$ so $F = ka$, where k is the constant of proportionality.
a Find the value of k.
b Write a formula for F in terms of a.
c Work out the value of F when $a = 4$.
d Work out the value of a when $F = 60$.

Q5a hint Substitute $F = 20$ and $a = 2$ into $F = ka$

Q5b hint Use $F = ka$ and your value for k.

Q5c hint Use your formula from part **b**.

6 a is inversely proportional to b.
When $a = 10$, $b = 2$
$a \propto \frac{1}{b}$ so $a = \frac{k}{b}$ where k is the constant of proportionality.
a Find the value of k.
b Write a formula for a in terms of b.
c Calculate a when $b = 5$.
d Calculate b when $a = 5$.

Q6a hint Substitute $a = 10$ and $b = 2$ into the formula $a = \frac{k}{b}$

7 d is directly proportional to the square of t.
$d = 80$ when $t = 4$
$d \propto t^2$ so $d = kt^2$, where k is the constant of proportionality.
a Find the value of k.
b Write a formula for d in terms of t.
c Calculate the value of d when $t = 7$.
d Calculate the positive value of t when $d = 45$.

Q7 hint The 'square of t' means t^2.

Exponential and other non-linear graphs

1 **STEM** The number of bacteria, n, in a Petri dish doubles every minute. At time $t = 0$ minutes there is 1 bacterium in the Petri dish.

a Copy and complete the table of values.

t	0	1	2	3	4	5	6
n	1						

b Draw a set of axes, with n on the vertical axis from 0 to 70 and t on the horizontal axis from 0 to 6.

c Sketch a graph of t and n on your axes.

Q1c hint Roughly plot the values from part **a**.

2 **Reasoning** The diagram shows a sketch of the curve $y = ab^x$

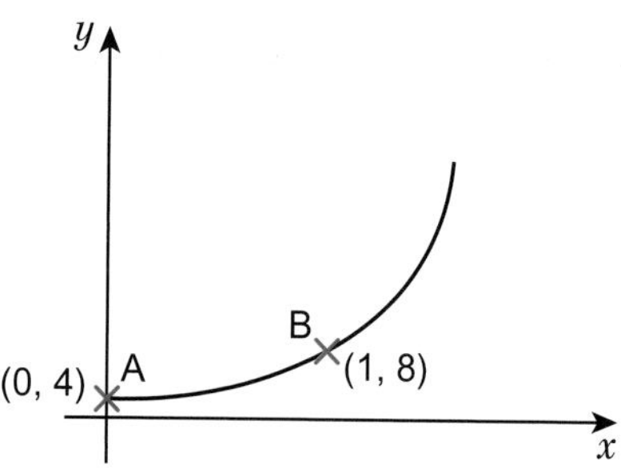

The curve passes through the points A(0, 4) and B(1, 8).

a Substitute the values of x and y from point A into $y = ab^x$

b Find the value of a.

c Substitute the values of x and y from point B and your value of a into $y = ab^x$

d Find the value of b.

e Substitute your values of a and b into $y = ab^x$

f Find the value of y when $x = 3$.

Q2b hint $b^0 = 1$

Q2d hint $b^1 = b$

Q2e hint $b^3 = b \times b \times b$

3 **Reasoning** The velocity–time graph shows a train pulling away from a station. The train travels in a straight line.

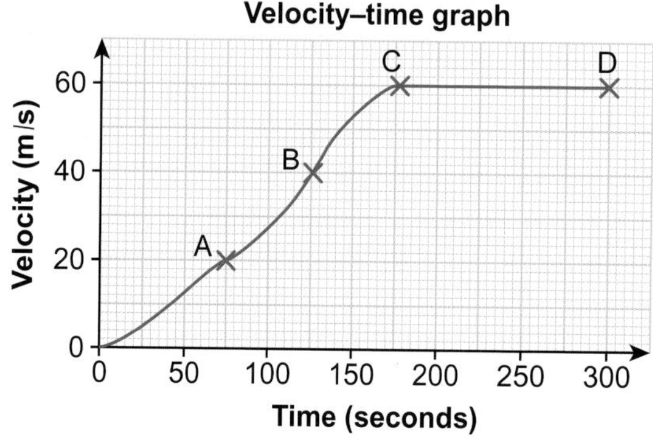

a What is the velocity of the train at point
i A ii B iii C?

b Estimate the acceleration between points A and B.

c Estimate the distance travelled between
i C and D ii B and C.

Q3b hint Acceleration = $\frac{\text{change in velocity}}{\text{time taken}}$

Q3c i hint Distance travelled = area under the graph

Q3c ii hint Read off the values and draw a trapezium. Find the area of the trapezium underneath the line BC.

Transformations of graphs functions

1 $f(x) = x^2$

a Copy and complete the table of values for $f(x)$.

x	−4	−3	−2	−1	0	1	2	3	4
$f(x)$									

b Sketch the graph of $y = f(x)$

c Copy and complete the table of values for $2f(x)$.

x	−4	−3	−2	−1	0	1	2	3	4
$2f(x)$									

d Copy and complete the sentence.
For the same values of x, the values of $2f(x)$ are always .... the values of $f(x)$.

e Sketch the graph of $y = 2f(x)$

f Describe how $y = f(x)$ is transformed into $y = 2f(x)$

g Copy and complete the table of values for $f(2x)$

x	−4	−3	−2	−1	0	1	2	3	4
$f(2x)$									

h Sketch the graph of $y = f(2x)$

i Describe how $y = f(x)$ is transformed into $y = f(2x)$

2 $f(x) = 2x - 4$

a Copy and complete the table of values for $f(x)$.

x	−3	−2	−1	0	1	2	3
$f(x)$							

b Sketch the graph of $y = f(x)$

c Copy and complete the table of values for $f(x) + 3$

x	−3	−2	−1	0	1	2	3
$f(x) + 3$							

d Copy and complete the sentence.
For the same values of x, the values of $f(x) + 3$ are always .... than the values of $f(x)$

e Sketch the graph of $y = f(x) + 3$

f Describe how $y = f(x)$ is transformed into $y = f(x) + 3$

g Copy and complete the table of values for $f(x + 2)$

x	−3	−2	−1	0	1	2	3
$f(x + 2)$							

h Sketch the graph of $y = f(x + 2)$

i Describe how $y = f(x)$ is transformed into $y = f(x + 2)$

19 Extend

1 A ball falls vertically after being dropped. It falls a distance, d metres, in a time of t seconds.
d is directly proportional to the square of t.
The ball falls 20 metres in a time of 2 seconds.

a Find a formula for d in terms of t.

b Calculate the distance the ball falls in 3 seconds.

c Calculate the time the ball takes to fall 605 m.

d Sketch a graph of how d varies with t.

e Describe the motion of the ball.

Q1e hint Describe any changes in its speed.

2 The diagram shows the graphs of $y = 2^x$, $y = 6^x$, $y = 0.5^x$ and $y = 3^{-x}$.
Match each graph to its equation.

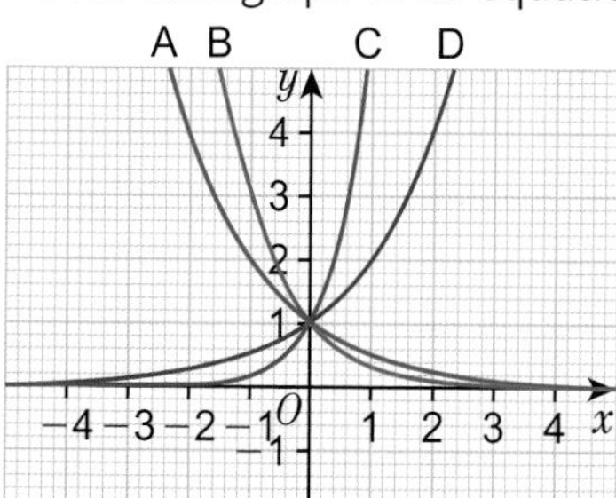

3 **Problem-solving** Choose the graphs that best match the following descriptions.

a The more petrol you buy, the higher the cost.

b The FTSE is rising more slowly than it has done for the last 6 months.

c Someone tells a joke. At first no one laughs, but then a couple of his friends do and before long everyone in the class is in hysterics.

d If you sell goods at a low price you don't make much profit, but if you charge too much people won't buy your product.

e The smaller the radius of the balls, the more balls you can fit in a ball pool.

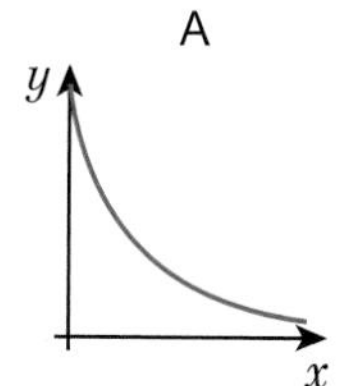

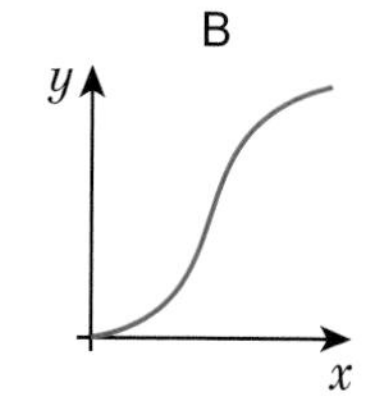

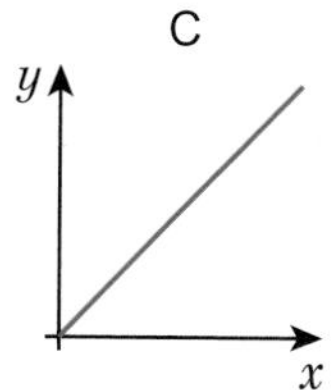

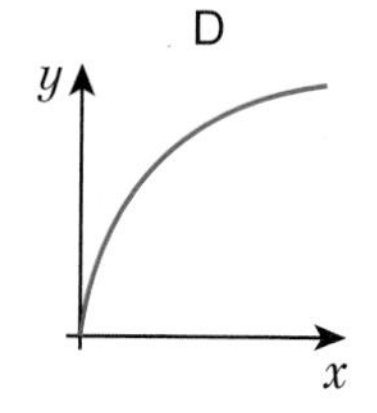

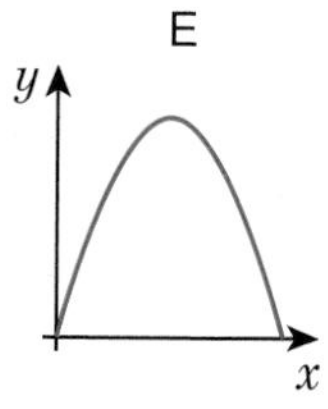

4 **Reasoning** The distance–time graph describes the distance, d, of a tennis ball from a fixed point as it is thrown vertically upwards.

a Describe the motion of the ball at points A, B and C.

b Compare the speed of the ball at A and C.

c Compare the velocity of the ball at A and C.

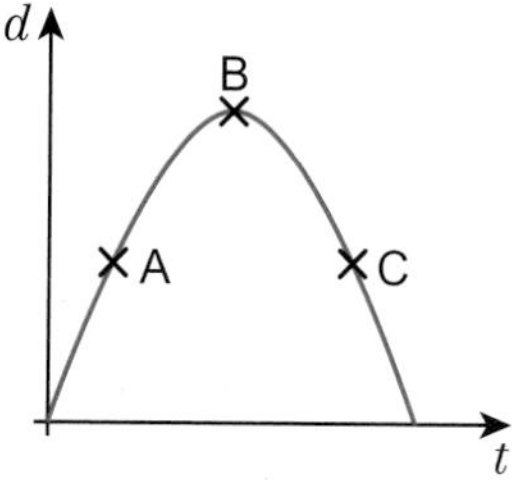

Q4 hint Velocity is a vector. It is speed in a given direction.

5 **Exam-style question**

Jane invests £6000 for 3 years at 3% per annum compound interest.

a Calculate the value of her investment at the end of 3 years.

Lee invests a sum of money for 30 years at 4% per annum compound interest.

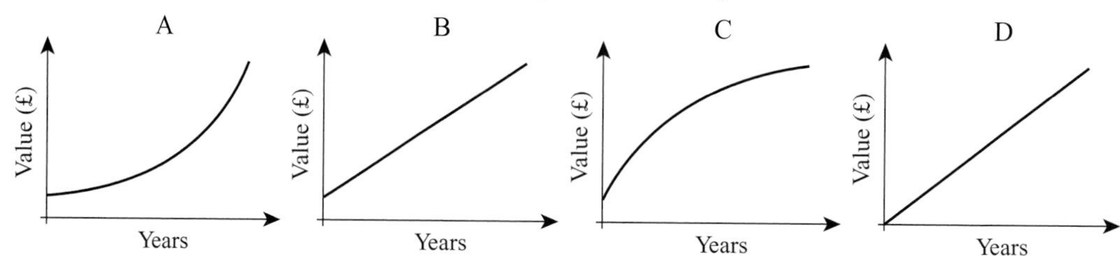

b Which graph best shows how the value of Lee's investment changes over the 30 years.

Hamish invested an amount of money in an account paying 5% per annum compound interest.
After 1 year the value of his investment was £2775.

c Work out the amount of money that Hamish invested. **(5 marks)**

Q5c strategy hint This is *not* the same as finding 5% of £2775 and then subtracting the answer from £2775.

6 The graph shows a company's profits over a 6-month period.

a Between which two months did the profit increase the fastest?
Explain how the graph shows this.

b Describe how the level of profit changed over this period.

> Q6b hint Use your answer to part **a** and describe the rate of change over the rest of the period.

c The area under the graph represents the total profit in a period of time.
Estimate the total profit between months 2 and 4.

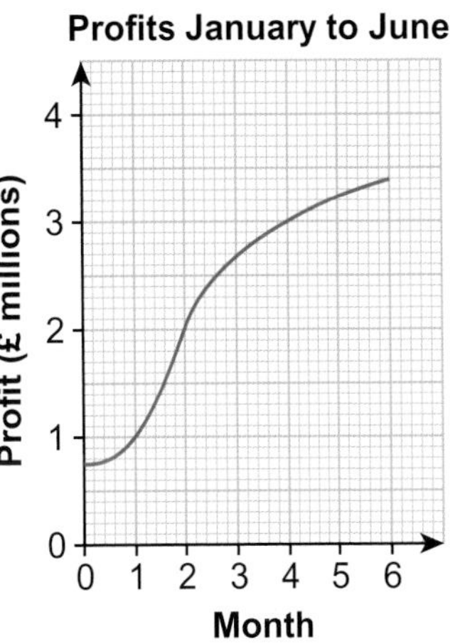

7 **STEM / Problem-solving** As part of a science experiment, Michael places different-sized spheres into a measuring jug of water.
He estimates the radius, r, of the spheres and measures the amount of water displaced, W.
The table shows the results from his experiment.

Radius, r (cm)	2	8	6	10	5	1	3	4.5
Water displaced, W (litres)	0.03	2.1	0.9	4.1	0.5	0.004	0.11	0.38

a Draw a scatter graph of Michael's results.

b Which rule best describes the relationship between r and W?

$W \propto r$ | $W \propto r^2$ | $W \propto r^3$

$W \propto \frac{1}{r}$ | $W \propto \frac{1}{r^2}$ | $W \propto \frac{1}{r^3}$

c Write a formula for estimating the relationship between r and W.

d Estimate the amount of water displaced by a sphere with radius 16 cm.

8 **Problem-solving** The diagram shows the graph of $y = \text{f}(x)$
The graph intersects the x-axis at points A (–2, 0), B (0, 0) and C (2, 0).
The graph intersects the y-axis at (0, 0).

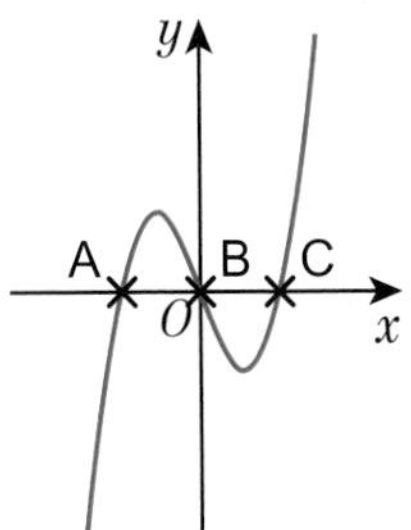

a Write the y-intercept of the graphs
i $y = 4\text{f}(x)$ ii $y = \text{f}(2x)$ iii $y = \text{f}(x) + 5$

b Write the x-intercepts of the graphs
i $y = \text{f}(x - 3)$ ii $y = \text{f}\left(\frac{x}{4}\right)$

9 **Reasoning** Here is the graph of $y = \text{f}(x) = \sin x$

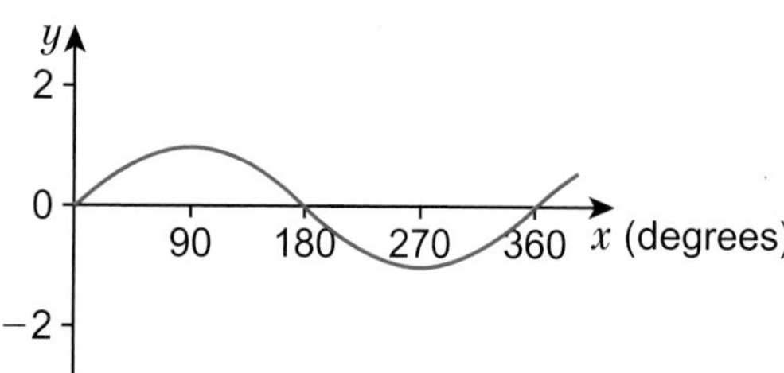

a Sketch the graph with equation $y = \text{f}\left(\frac{x}{2}\right)$

b How many solutions does the equation $\text{f}\left(\frac{x}{2}\right) = 0.5$ have in the range $0 < x < 360$?

10 **Reasoning** $\text{f}(x) = x^2 + 2x - 8$

a Write the coordinates of the points where $y = \text{f}(x)$ intersects the x-axis.

> **Q10a hint** Factorise $x^2 + 2x - 8$

b Write the minimum value of y.

> **Q10b hint** Complete the square.

c Sketch the graph of

i $y = \text{f}(x)$ ii $y = x^2 + 2x - 2$ iii $y = (x + 1)^2 + 2(x + 1) - 8$

11 The expression $x^2 + 6x + 5$ can be written in the form $(x + a)^2 + b$

a Find the values of a and b.

b Sketch the graph $y = x^2$

c On the same axes sketch the graph $y = x^2 + 6x + 5$

> **Q11 hint** When written in the form $y = (x + a)^2 + b$ you can use the values of a and b to transform the curve $y = x^2$. You can do this in two steps.

12 **Problem-solving** The diagram shows the graph of $y = a - b\sin(cx)$.

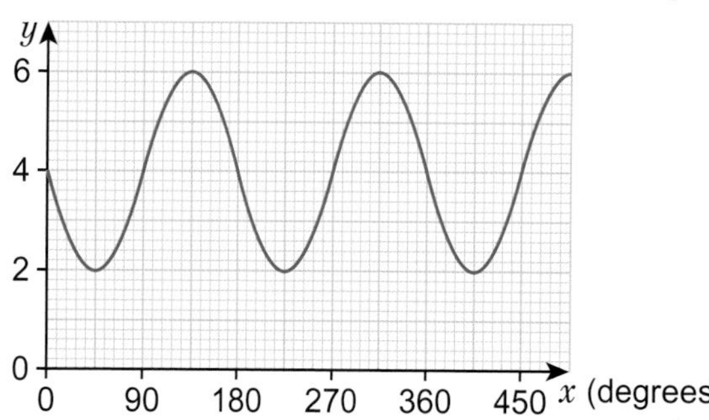

Use the graph to find the values of a, b and c.

13 **Problem-solving** The diagram shows the graphs of C_1 and C_2.
C_1 is the graph of $y = \text{f}(x)$
Write the equation of C_2 in function form.

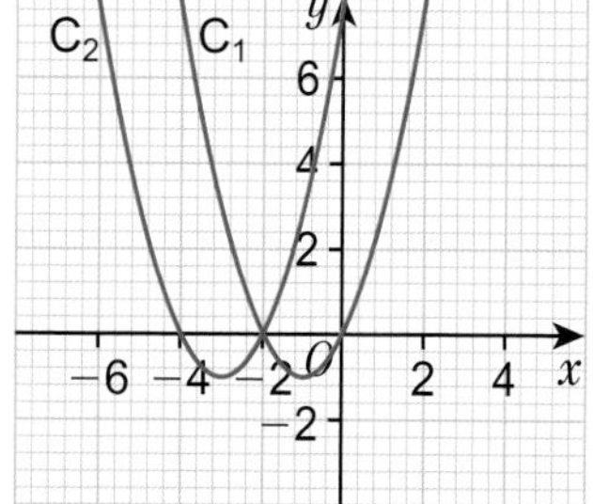

14 **Problem-solving** Here is the graph of $y = f(x) = \frac{1}{x}$

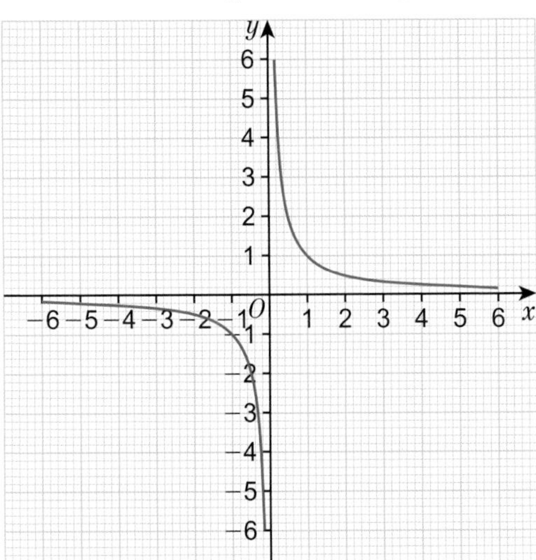

a Draw the graph of $y = f(x + 2) - 3$

b Write the equations of the two asymptotes.

15 At the start of an experiment, the count rate of a sample of the radioactive substance nobelium-259 is 840 counts per second. The radioactive decay for nobelium-259 can be modelled by the iterative equation $C_{t+1} = \frac{1}{2}C_t$ where C = count rate and t = time in hours. Calculate the count rate after 3 hours.

Q15 hint $C_0 = 840$. After 3 hours, $t = 3$.

16 Barry invests some money in a long-term savings account. The value of his investment after t years can be modelled by the iterative equation $M_{t+1} = 1.06\,M_t$

After two years the value of his investment is £16 854

a Find the value of his investment after 3 years.

b Find the value of his initial investment.

Q16 hint Work backwards to find M_1 and M_0

19 Knowledge check

- When a graph of two quantities is a straight line through the origin, one quantity is directly proportional to the other. .... *Mastery lesson 19.1*

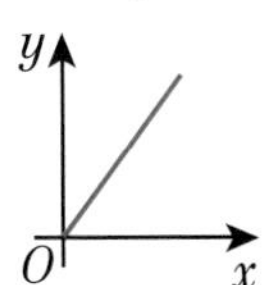

- The symbol ∝ means 'is directly proportional to'. .... *Mastery lesson 19.1*
- If y is directly proportional to x, $y \propto x$ and $y = kx$, where k is a number, called the **constant of proportionality**. .... *Mastery lesson 19.1*
- Where k is the constant of proportionality:
    - if y is proportional to the square of x then $y \propto x^2$ and $y = kx^2$
    - if y is proportional to the cube of x then $y \propto x^3$ and $y = kx^3$
    - if y is proportional to the square root of x then $y \propto \sqrt{x}$ and $y = k\sqrt{x}$ .... *Mastery lesson 19.2*

- When y is **inversely proportional** to x, $y \propto \frac{1}{x}$ and $y = \frac{k}{x}$ .... *Mastery lesson 19.3*

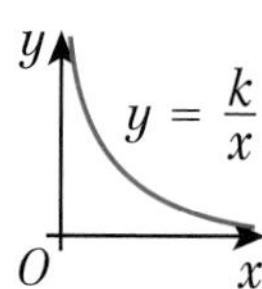

- To solve problems involving proportion:
  - write the statement of proportionality
  - write the formula
  - substitute given values into the formula to find the solution. .... *Mastery lesson 19.3*
- Expressions of the form a^x or a^{-x}, where $a > 1$, are called **exponential functions**. .... *Mastery lesson 19.4*
- The graph of an exponential function has one of these shapes. .... *Mastery lesson 19.4*

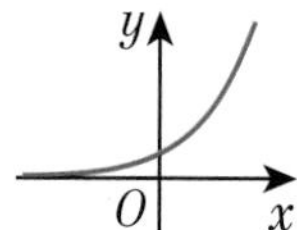

$y = a^x$ where $a > 1$ or
$y = b^{-x}$ where $0 < b < 1$
exponential growth

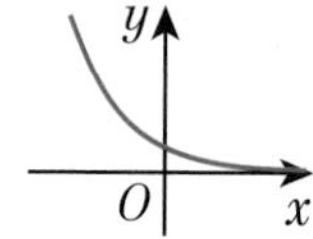

$y = a^{-x}$ where $a > 1$ or
$y = b^x$ where $0 < b < 1$
exponential decay

- Exponential graphs intersect the y-axis at (0, 1) because $a^0 = 1$ for all values of a. .... *Mastery lesson 19.4*
- The tangent to a curved graph is a straight line that touches the graph at a point. The gradient at a point on a curve is the gradient of the tangent at that point. .... *Mastery lesson 19.5*
- The straight line that connects two points on a curve is called a **chord**. The gradient of the chord gives the average rate of change and can be used to find the average rate of change between two points. ...... *Mastery lesson 19.5*

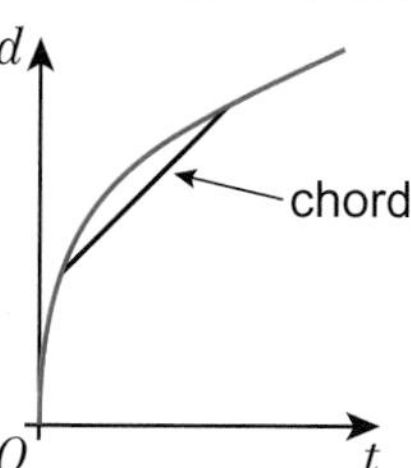

- The area under a velocity–time graph shows the displacement, or distance from the starting point. To estimate the area under a part of a curved graph, draw a chord between the two points you are interested in, and straight lines down to the horizontal axis to create a trapezium. The area of the trapezium is an estimate for the area under this part of the graph. .... *Mastery lesson 19.5*

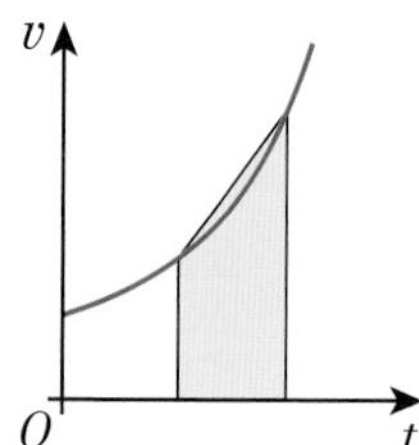

- The graph of $y = \mathrm{f}(x)$ is transformed into the graph of:
  - $y = \mathrm{f}(x) + a$ by a translation of a units parallel to the y-axis or a translation by $\begin{pmatrix} 0 \\ a \end{pmatrix}$ .... *Mastery lesson 19.6*
  - $y = \mathrm{f}(x + a)$ by a translation of $-a$ units parallel to the x-axis or a translation by $\begin{pmatrix} -a \\ 0 \end{pmatrix}$ .... *Mastery lesson 19.6*
  - $y = \mathrm{f}(-x)$ by a reflection in the y-axis .... *Mastery lesson 19.7*
  - $y = -\mathrm{f}(x)$ by a reflection in the x-axis .... *Mastery lesson 19.7*
  - $y = a\mathrm{f}(x)$ by a stretch of scale factor a parallel to the y-axis .... *Mastery lesson 19.7*
  - $y = \mathrm{f}(ax)$ by a stretch of scale factor $\frac{1}{a}$ parallel to the x-axis .... *Mastery lesson 19.7*

Reflect

Write down a word that describes how you feel:

a before a maths test

b during a maths test (when you know how to answer a question)

c during a maths test (when you don't immediately know how to answer a question)

d after a maths test

e discuss with a classmate what you could do to change ☹ feelings to ☺ feelings.

> **Hint** You might choose one of these (or a different word): OK, worried, excited, happy, focussed, panicked, calm.
> Beside each word, draw a face to show if it is a good or a bad feeling: ☺ ☹

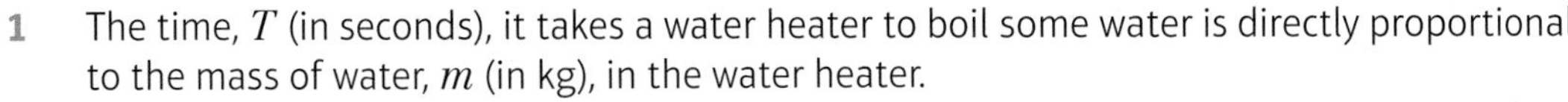

19 Unit test

Log how you did on your Student Progression Chart.

1 The time, T (in seconds), it takes a water heater to boil some water is directly proportional to the mass of water, m (in kg), in the water heater.

When $m = 250$, $T = 600$

a Find T when $m = 400$.

The time, T (in seconds), it takes a water heater to boil a constant mass of water is inversely proportional to the power, P (in watts), of the water heater.

When $P = 1400$, $T = 360$ *(3 marks)*

b Find the value of T when $P = 900$. *(3 marks)*

2 Write the letters of the graphs that could have equations

a $y = 3x - 2$ b $y = -2x^2$ c $y = \frac{4}{x}$ d $y = 3^x$ *(4 marks)*

A
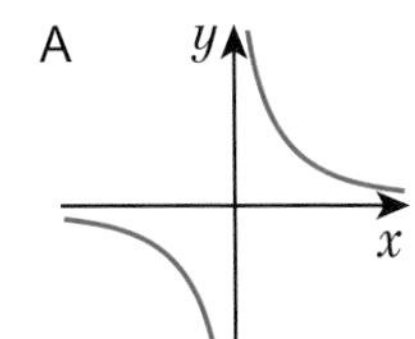

B
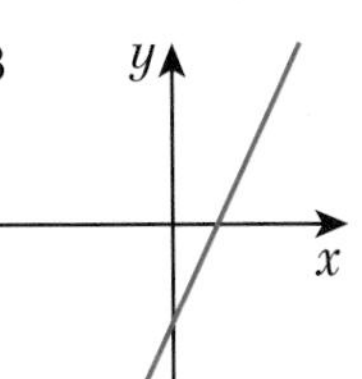

C
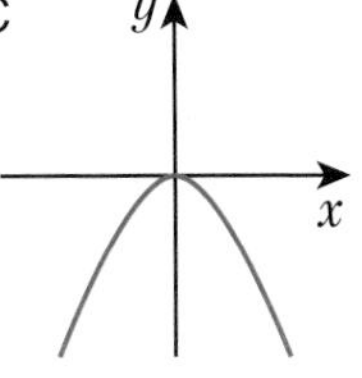

D
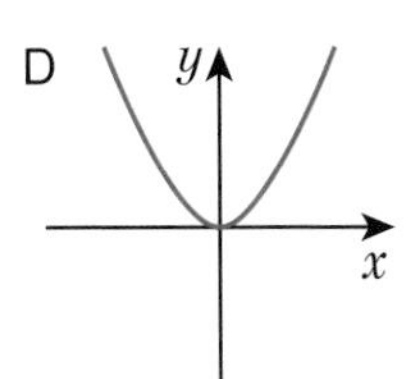

E
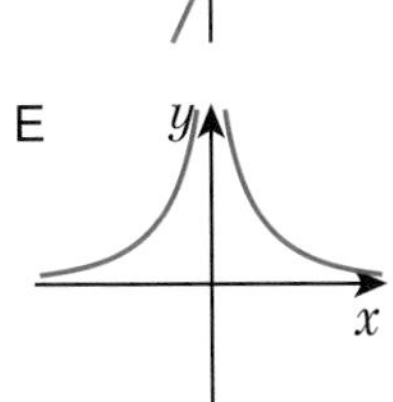

F
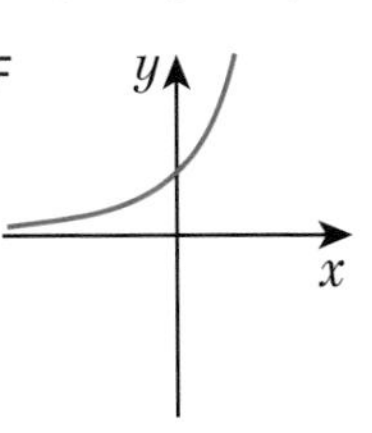

3 The distance, D, travelled by a particle is directly proportional to the square of the time taken, t.
When $t = 40$, $D = 30$

a Find a formula for D in terms of t *(3 marks)*

b Calculate the value of D when $t = 64$ *(1 mark)*

c Calculate the value of t when $D = 12$
Give your answer correct to 3 significant figures. *(1 mark)*

4 y is inversely proportional to x^2.
Given that $y = 2.5$ when $x = 24$

a find an expression for y in terms of x *(3 marks)*

b find the value of y when $x = 20$ *(1 mark)*

c find a value of x when $y = 1.6$ *(1 mark)*

5 **Reasoning** The sketch shows the graph of $y = f(x)$

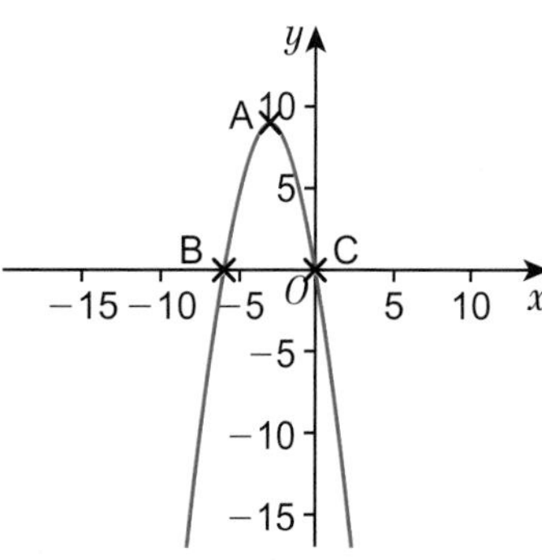

The maximum turning point at A has coordinates (−3, 9).
The graph intersects the x-axis at B(−6, 0) and C(0, 0).
Write the coordinates of A, B and C for the graph of

a $y = -f(x)$ b $y = f(-x)$ c $y = f(x - 2)$

d $y = 2f(x)$ e $y = f(3x)$ *(5 marks)*

6 **Reasoning** The velocity–time graph shows the first 60 seconds of a space shuttle flight.
Time, t, is measured in seconds. Velocity, v, is measured in metres per second.

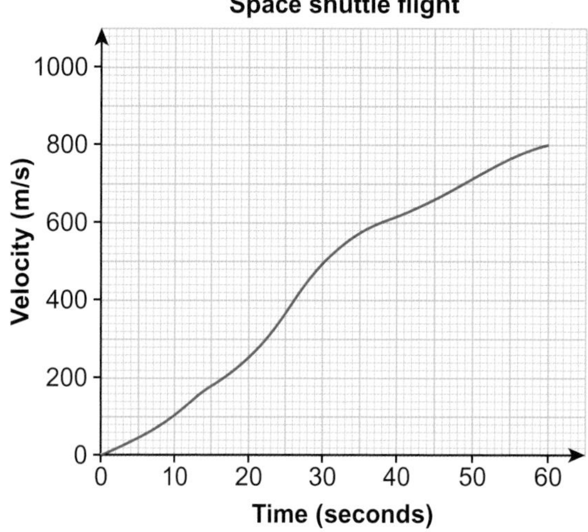

a Calculate the average rate of acceleration between $t = 40$ and $t = 50$. *(3 marks)*

b Estimate the rate of acceleration at $t = 30$. *(2 marks)*

c Given that the shuttle travelled in a straight line, estimate the distance travelled between $t = 20$ and $t = 30$. *(2 marks)*

7 The diagram shows a sketch of $y = f(x)$

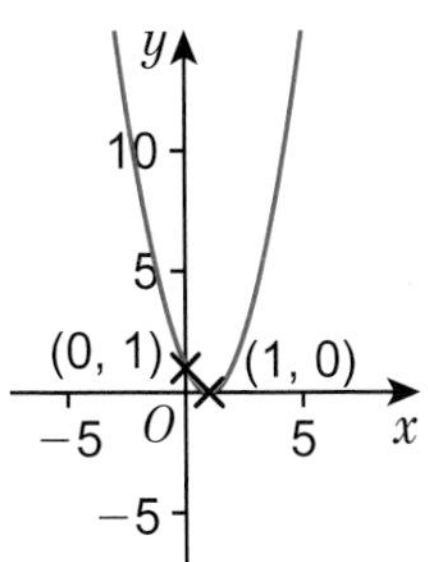

Sketch the graphs of

a $y = f(x) - 1$ b $y = 3f(x)$ *(2 marks)*

8 **Reasoning** The diagram shows a sketch of the curve $y = ab^x$

The curve passes through the points A(1, 12) and B(3, 108).

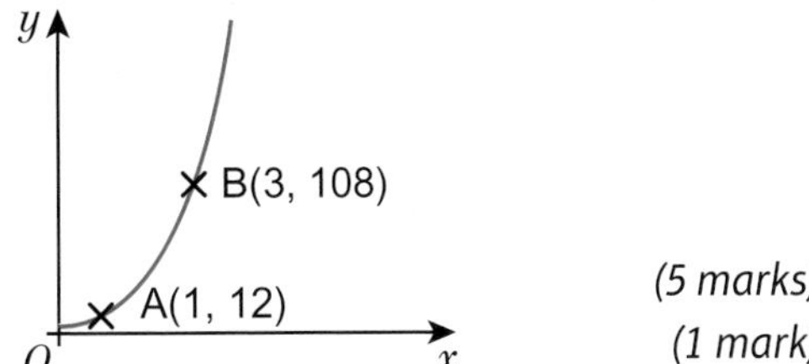

a Find the value of i a ii b *(5 marks)*

b Find the value of y when $x = 4$ *(1 mark)*

Sample student answer

a Describe the key points to look for on a graph to help transform and sketch a curve correctly.

b Which key point has the student got incorrect?

c What could they draw on the graph to help make sure they count correctly when transforming?

Exam-style question

This is a graph of the function $y = f(x)$

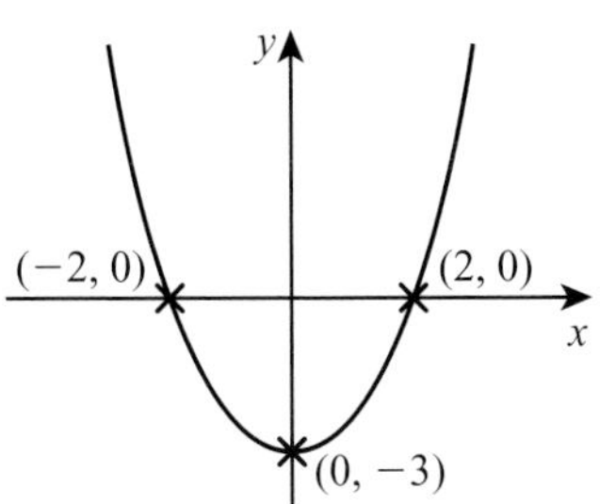

a Sketch the curve of the function $y = f(x + 2)$ **(2 marks)**

b Write the coordinates of the new minimum point **(1 mark)**

Student answer

a

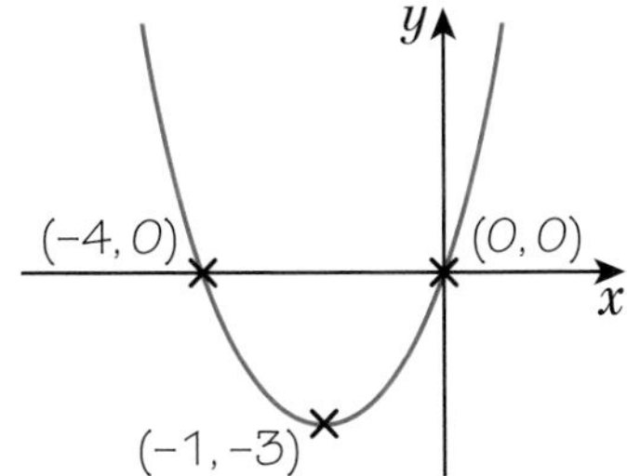

b (−1, −3)

ANSWERS

UNIT 1

1 Prior knowledge check

1 a 1.5 b 1.94 c 30 d 300
e 42 f 0.24 g 0.018 h 0.0081
i 2 j 30 k 1.5 l 22

2 a > b < c < d >

3 a Factors of 12: 1, 2, 3, 4, 6, 12
Factors of 18: 1, 2, 3, 6, 9, 18
b 1, 2, 3, 6
c 6

4

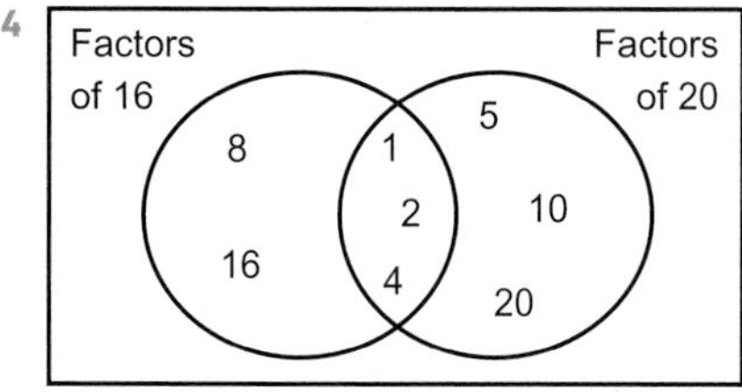

b 4

5 a Multiples of 6: 6, 12, 18, 24, 30, 36, 42, 48, 54, 60
Multiples of 9: 9, 18, 27, 36, 45, 54, 63, 72, 81, 90
b 18, 36, 54
c 18

6 a
Multiples of 4
Multiples of 10
4 8 12 16 24 28 32 36
20 60 80 100 40
10 30 50 70 90

b 20

7 a 2 b 18 c −11
d 4 e 17 f 36

8 (9 + 18) ÷ 3 = 9

9 a 63 b 25 c 10 d 6

10 a 6, −6 b 1, −1 c 8, −8

11 a 900 b 4896 c 18018 d 270

12 a 2 b 4

13 a 16 b 1 c 1331 d 128

14 6, 24

1.1 Number problems and reasoning

1 a H, 1 H, 2 H, 3 H, 4 H, 5 H, 6
T, 1 T, 2 T, 3 T, 4 T, 5 T, 6
b 12

2 a 2, 1 4, 1 6, 1
2, 3 4, 3 6, 3
2, 5 4, 5 6, 5
2, 7 4, 7 6, 7
2, 9 4, 9 6, 9
b 15

3 a 6 b 2 c 3 d 5

4 a VP, VB, VC, VL, SP, SB, SC, SL, MP, MB, MC, ML
b Students' own answer.
c 15
d 3 starters and 4 mains: 12 combinations
3 starters and 5 mains: 15 combinations
n starters and m mains: $n \times m$ combinations
e 24

5 a 10 b 10000 c 5000

6 a ABC, ACB, BAC, BCA, CAB, CBA
b i 24 ii 720 iii 3628800

7 a i 1000000 ii 6760000 iii 118813760
b i 151200 ii 3276000 iii 78936000

1.2 Place value and estimating

1 a i 900000 ii 870000
b i 2000 ii 2000
c i 0.007 ii 0.0071

2 a 99 b 29 c 27
d −10 e 63 f 5

3 10

4 a 192 b 192 c 192 d 192

5 a 364.82 b 0.36482 c 0.36482
d 3.7 e 37 f 0.986

6 a Students' own answer.
b Students' own answer.
c Students' own answer.
d It must use the digits 399492, so it must end in a 2.

7 a 2, 3 b Students' own answer
c 2.2, 2.4, 2.6, 2.8

8 The correct answers are given here. 0.1 out in either direction is acceptable.
a 6.9 b 4.7 c 9.2
d 11.3 e 3.2 f 6.3

9 a 4.5 cm, accept 4.4 or 4.6

10 a 64, 81
b 69, 77 (accept 1 out in either direction)

11 The correct answers are given here. 1 out in either direction is acceptable.
a 10 b 22 c 3
d 50 e 40 f 96

12 a i 16 ii 4 iii 11 iv 1
b i 16.8 ii 4.4 iii 11.2 iv 1.1
c Students' own answer.

13 23

14 $9.2^2 \approx 85$; $85 \times 6 = 510\,\text{cm}^2$

15 $\sqrt{80} \approx 8.9$; $8.9 \times 4 = 35.6\,\text{cm}$

16 a i £160 ii £96 iii £64
b i £174.64 ii £89.29 iii £70.04

17 a Students should use estimates to get an answer larger than 11.8.
b 18.4

1.3 HCF and LCM

1 a 1, 2, 4, 5, 10, 20 b 2 and 5

2 a 2, 3, 5, 7, 11, 13, 17, 19
b 1, 2, 3, 4, 6, 8, 12, 24
c

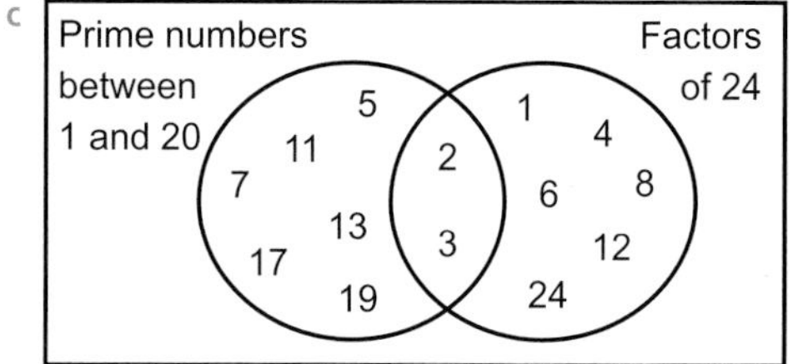

3 a

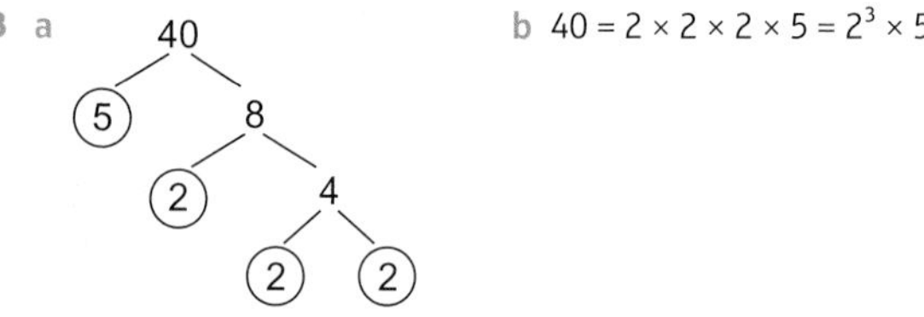

b $40 = 2 \times 2 \times 2 \times 5 = 2^3 \times 5$

4 $75 = 3 \times 5 \times 5 = 3 \times 5^2$

5 a $60 = 2 \times 2 \times 3 \times 5 = 2^2 \times 3 \times 5$
b $60 = 2 \times 2 \times 3 \times 5 = 2^2 \times 3 \times 5$
c $48 = 2 \times 2 \times 2 \times 2 \times 3 = 2^4 \times 3$

6 Students' own answers

7 a 2×3^2 b $2 \times 3 \times 7$ c 5^2
d 6^2 e $2^3 \times 3$ f $2^4 \times 5$

8 $m = 3, n = 3, p = 5$

9 a HCF = 6; LCM = 120 b HCF = 2; LCM = 420
c HCF = 2; LCM = 72 d HCF = 15; LCM = 45
e HCF = 9; LCM = 108 f HCF = 33; LCM = 66

10 3:30pm

11 12 cm tiles

12 2 numbers with an HCF of 2, e.g. 8 and 10.

13 a 1, 2, 3, 6, 9
b The factors of 18 other than 18.

14 a $2^2 \times 3$ b $2^4 \times 3^2$

15 a $2^2 \times 3^4 \times 5$ b $2^3 \times 3^6 \times 5^2$

16 $2^4 \times 5$

17 a 3×5^2 b $2^3 \times 3$ c $2^2 \times 3$ d $2 \times 3 \times 5$

18 a No, Yes, Yes
b $792 = 2^3 \times 3^2 \times 11$; so $12 = 2^2 \times 3$ divides into 792 exactly.
c Yes. $132 = 2^2 \times 3 \times 11$
d No. $27 = 3^3$ and 792 only contains 3^2

19 a $2^2 \times 5 \times 7$ b $2^3 \times 5^2 \times 7^2$ c **i and iii** d **ii**

1.4 Calculating with powers (indices)

1 a 27 b −1 c 64 d 45
e 800 f 0.008 g 12 h 72

2 a 16 b 27

3 a 4 b 4 c 2 d 3

4 a 3 b −1 c 10 d −5

5 a 5 b 5 c 46 d 19
e 4 f 15 g −72 h −0.1

6 a 64 b 8 c 2

7 a 2 b 3 c 10

8 a i 100 000 ii 100 000
iii 100 000 000 iv 100 000 000
b Add the indices together.
c i 10^7 ii 10^6 iii 10^{-5}

9 a 3^6 b 4^{10} c 9^7

10 a 3 b 2 c 7

11 a 3^8 b 4^6 c 5^4
d 2^7 e 8^3 f 3^6

12 a i 5^3 ii $5^5 \div 5^2 = 5^3$
b 4^4
c $6^1 = 6$

13 a 7^4 b 4^2 c $3^1 = 3$

14 a 2 b 4 c −3

15 a i 1, 2, 9; 1, 3, 8; 1, 4, 7; 1, 5, 6; 2, 3, 7; 2, 4, 6; 3, 4, 5
ii 4, 4, 4
b i Any two numbers that are both greater than 20 where one is 6 more than the other, e.g. 41, 35
ii 12, 6 or −6, −12

16 a 5^6 b 6^8 c 5^2 d 8^1

17 a 16 b $\frac{1}{2}$

18 a 2^{15} b 6^{12} c 8^{14}

19 a 2^{12} b 6^{10} c 4^{-6} d 5^{12}

20 a 2^{11} b 5^5 c $4^3 = 2^6$

1.5 Zero, negative and fractional indices

1 a 36 b 8 c 81 d 125

2 a 3^{10} b 2^2 c 2^7 d 7^8

3 a $\frac{1}{5}$ b $-\frac{2}{3}$ c −1 d 3

4 a 3 b 4 c 6 d 3

5 a i 0.5 ii 0.25 iii 0.2 iv 0.1
b i $\frac{1}{2}$ ii $\frac{1}{4}$ iii $\frac{1}{5}$ iv $\frac{1}{10}$
c i 0.25 ii 0.0625 iii 0.04 iv 0.01
d i $\frac{1}{4}$ ii $\frac{1}{16}$ iii $\frac{1}{25}$ iv $\frac{1}{100}$
e i 2 ii $\frac{16}{9} = 1.\dot{7}$

6 a $\left(\frac{1}{4}\right)^{-2} = 16, \frac{1}{3^5} = 3^{-5}, \frac{3}{2} = \left(\frac{2}{3}\right)^{-1}, \frac{1}{2^4} = 2^{-4}, \frac{1}{5^3} = 5^{-3}$
b $3^{-8} = \frac{1}{3^8}, \frac{1}{8^3} = 8^{-3}$ c $\left(\frac{2}{3}\right)^{-1} = \frac{3}{2}$ so $\left(\frac{a}{b}\right)^{-1} = \frac{b}{a}$

7 a 6^{12} b 5^{-4} c 8^{-1}

8 a $2^3 \div 2^3 = 2^0$ b 8 c $2^3 \div 2^3 = 8 \div 8 = 1$
d $2^3 \div 2^3 = 2^0 = 1$
e $7^5 \div 7^5 = 7^0 = 16\,807 \div 16\,807 = 1$
f $a^0 = 1$

9 a $\frac{1}{3}$ b $\frac{1}{16}$ c $\frac{1}{100000}$ d $\frac{4}{3}$
e $\frac{125}{64}$ f $\frac{4}{5}$ g $\frac{16}{121}$ h $\frac{10}{7}$
i 100 000 j $\frac{125}{8}$ k $5^0 = 1$ l $7^1 = 7$

10 a i 7 ii 4 iii 11 iv $\frac{2}{5}$
b Square root
c i 3 ii 10 iii −1 iv $\frac{1}{10}$
d Cube root
e i 5 ii 2

11 a 6 b 9 c $\frac{1}{3}$ d $\frac{4}{5}$
e $\frac{8}{7}$ f −2 g $\frac{1}{3}$ h $-\frac{4}{5}$

12 a $\frac{1}{5}$ b $\frac{1}{4}$ c $\frac{5}{3}$

13 a 16 b 1000 c 64
d $\frac{8}{27}$ e $\frac{1}{9}$ f $-\frac{1}{27}$

14 a 9 b $\frac{125}{12}$ c $\frac{125}{8}$

15 a 4 b $\frac{1}{3}$ c −2
d $-\frac{1}{2}$ e $\frac{7}{2}$ f $\frac{7}{4}$

16 a $\frac{1}{5} \times 16 = \frac{16}{5}$ b He said $25^{-\frac{1}{2}} = 5$ but it's $\frac{1}{5}$

17 $8^{\frac{4}{3}} = 16, 16^{\frac{3}{4}} = 8, 32^{-\frac{2}{5}} = \frac{1}{4}, \frac{1}{64} = 16^{-\frac{3}{2}}, \frac{9}{4} = \left(\frac{8}{27}\right)^{-\frac{2}{3}},$
$\left(\frac{1}{64}\right)^{\frac{2}{3}} = \frac{1}{16}, \left(\frac{81}{16}\right)^{-\frac{1}{2}} = \frac{4}{9}$

1.6 Powers of 10 and standard form

1 a 1 b $\frac{1}{10} = 0.1$ c $\frac{1}{100} = 0.01$
d $\frac{1}{1000} = 0.001$ e $\frac{1}{10000} = 0.000\,01$
f $\frac{1}{100000} = 0.000\,001$

2 a 3 b 100 000 c 8
d $\frac{1}{10}$ e −4 f 0.000 001

3 a 5.67 b 15.8 c 4.908 34

4 Answers in bold

Prefix	Letter	Power	Number
tera	T	10^{12}	1 000 000 000 000
giga	G	10^9	**1 000 000 000**
mega	M	$\mathbf{10^6}$	1 000 000
kilo	k	10^3	**1000**
deci	d	$\mathbf{10^{-1}}$	0.1
centi	c	10^{-2}	**0.01**
milli	m	$\mathbf{10^{-3}}$	0.001
micro	μ	10^{-6}	**0.000 001**
nano	n	$\mathbf{10^{-9}}$	0.000 000 001
pico	p	10^{-12}	**0.000 000 000 001**

5 a 0.015 g b 0.000 000 007 m
c 0.0017 kg d 0.000 000 000 0073 s

6 a $0.000\,0012$ m b $0.000\,000\,000\,0025$ m
c $0.000\,000\,000\,9$ m
7 a $4.5 \times 10\,000$ b 10^4 c 4.5×10^4
8 A, D, F
9 a 8.7×10^4 b 1.042×10^6 c 1.394×10^9
d 7×10^{-3} e 2.84×10^{-6} f 1.003×10^{-4}
10 a 400 000 b 350 c 6780
d 0.062 e 0.000 0893 f 0.004 04
11 a i 4.5×10^{12}
ii Calculator says 4.5 E +12, for example.
b i 7×10^{-5} ii Calculator says 7 E −05.
12 a 6×10^7 b 2×10^{11} c 4.8×10^6 d 2×10^3
e 3×10^{-8} f 2.5×10^{-5} g 2.5×10^7 h 6.4×10^{-5}
13 500 seconds
14 0.51 kg
15 a i 80 000 ii 300
b 8.03×10^4
16 a 4.07×10^5 b 9.778×10^4
c 7.2062×10^2 d 8.299993×10^5
17 $x = 5, y = 1, z = -2$

1.7 Surds

1 a $\frac{3}{10}$ b $\frac{7}{12}$
2 a $\frac{3}{5}$ b $\frac{17}{20}$ c $\frac{13}{8}$
d $\frac{17}{4}$ e $\frac{1}{3}$ f $\frac{14}{9}$
3 a 2.24 b 2.65 c 4.36 d 7.28
4 a i 2.44948… ii 2.44948…
b i 3.87298… ii 3.87298…
c Answers to parts **i** and **ii** are the same.
d i 12 ii 5 iii 5
5 a 5 b 2 c 8 d 6
6 a $2\sqrt{5}$ b $10\sqrt{3}$ c $2\sqrt{11}$
d $5\sqrt{10}$ e $20\sqrt{2}$ f $12\sqrt{14}$
7 a $5\sqrt{3}$ b 8.66 (2 d.p.)
8 a Students' own answers, e.g. $\sqrt{80}$
b Students' own answers
9 a $\frac{\sqrt{7}}{2}$ b $\frac{\sqrt{5}}{3}$ c $\frac{2\sqrt{3}}{7}$ d $\frac{3\sqrt{2}}{5}$
10

Rational	Irrational
$\frac{3}{8}$ $\sqrt{6.25}$ -4 $1.\dot{4}$ $\sqrt{\frac{4}{49}}$ 0.3	$\sqrt[3]{6}$ $\sqrt{17}$ $-\sqrt{8}$

11 $x = \pm 3\sqrt{10}$
12 a $x = \pm 5\sqrt{2}$ b $x = \pm 4\sqrt{10}$ c $x = \pm 2\sqrt{3}$ d $x = \pm 2\sqrt{7}$
13 $2\sqrt{15}$ cm
14 a i $60\sqrt{6}$ ii $48\sqrt{15}$
iii $180\sqrt{2}$ iv 144
15 a $\frac{\sqrt{7}}{7}$ b $\frac{\sqrt{3}}{3}$ c $\frac{\sqrt{5}}{5}$ d $\frac{\sqrt{5}}{10}$
e $\frac{\sqrt{2}}{2}$ f $\frac{\sqrt{15}}{5}$ g $\frac{8\sqrt{10}}{5}$ h $\sqrt{11}$
16 a $\frac{1}{\sqrt{7}} = \frac{1 \times \sqrt{7}}{\sqrt{7} \times \sqrt{7}} = \frac{\sqrt{7}}{\sqrt{49}} = \frac{\sqrt{7}}{7}$
17 $4\sqrt{5}$ cm
18 a $\frac{\sqrt{3}}{2}$ b $\frac{16}{7}$ c $\frac{15}{4}$

1 Problem-solving

1 a 2 tables with 4 chairs and 2 tables with 6 chairs; 5 tables with 4 chairs; 4 tables with 5 chairs; 1 table with 4 chairs, 2 tables with 5 chairs, and 1 table with 6 chairs.
b 5 tables
2 44
3 80 cm and 65 cm
4 2 starters, 2 mains and 3 desserts
5 a 2 tandems, 3 road bikes b 7 people
6 12 noon
7 Use the lowest common multiple

1 Check up

1 a 1536.4 b 0.92
2 7.3
3 a i 4 ii $\frac{1}{2}$
b i 3.8 ii 0.5
4 $2 \times 3^2 \times 5$
5 HCF = 2; LCM = 126
6 a $2 \times 5^2 \times 7^2$ b $2^3 \times 5^2 \times 7^3$
7 a $10^3 = 1000$ b $4^3 = 64$ c $2^4 = 16$ d $5^0 = 1$
8 a 2 b 144 c 1
9 a 9^4 b 3^8 c 5^5
d 2^7 e 2^{12} f 4^{-2}
10 a $\frac{1}{16}$ b 125 c $\frac{8}{27}$ d $\frac{1}{4}$
11 a $3\sqrt{6}$ b $50\sqrt{10}$
12 a $\frac{\sqrt{10}}{10}$ b $\sqrt{2}$
13 a 3.204×10^7 b 7×10^{-4}
14 a 56 000 b 0.001 09
15 a 4.5×10^{12} b 5×10^2 c 8.6×10^3
17 a WEABCDW or WDCBAEW; 135 minutes
b Students' own answers

1 Strengthen

Calculations, factors and multiples

1 a 11.172, 111.72, 1117.2, 11 172, 111 720
b 63.5, 635, 6350, 63 500, 635 000
2 a 641.69 b 64 169 c 0.641 69
d 0.064 169 e 0.89 f 890
3
$\sqrt{1}$ $\sqrt{4}$ $\sqrt{\boxed{9}}$ $\sqrt{16}$ $\sqrt{\boxed{25}}$ $\sqrt{\boxed{36}}$ $\sqrt{\boxed{49}}$ $\sqrt{64}$ $\sqrt{\boxed{81}}$
$\boxed{1}$ $\boxed{2}$ 3 $\boxed{4}$ 5 6 7 $\boxed{8}$ 9
4 a 7.2 b 7.7 c 8.7
5 a i 6 ii 20 iii 12 iv 4
b i 6.1 ii 19.7 iii 12.0 iv 4.3
6 a $2^3 \times 3^2$ b $2^2 \times 3 \times 5$ c $3^4 \times 7^2$
7 a
60
6 $\boxed{10}$
(2) (3) (2) (5)
b $60 = 2 \times 2 \times 3 \times 5$ c $60 = 2^2 \times 3 \times 5$
8 a $2^3 \times 3$ b $2^4 \times 5$ c $3^2 \times 5$
d $2 \times 3 \times 5$ e 2^4 f $2^3 \times 3^2$
9 a $18 = 2 \times 3 \times 3$ b $45 = 3 \times 3 \times 5$
c
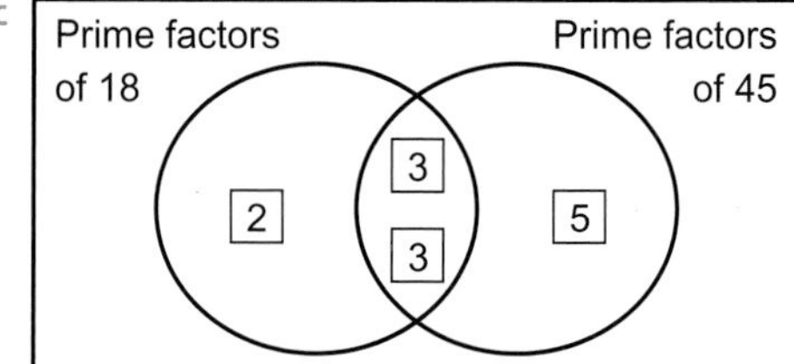

d 9
e 90

10 a HCF = 10; LCM = 60 b HCF = 7; LCM = 84
c HCF = 5; LCM = 75 d HCF = 4; LCM = 396

Indices and surds

1 a $2^3 = 8$ b $5^2 = 25$ c $(-3)^3 = -27$

2 a 32 b 125 c 3 d 3

3 a 27 b 32 c −24
d −20 e 63 f 150

4 a 27 b 169 c 4

5 a $3^5 \times 3^3 = 3^8$ b $4^4 \times 4^6 = 4^{10}$
c 6^2 d $\frac{7^6}{7^3} = 7^3$
e To multiply powers, **add** the indices.
To divide powers, **subtract** the indices.

6 a 5^9 b 7^{11} c 5^5
d 9^6 e 8^{-2} f 7^7

7 a 2^7 b 5^7 c 2^{13} d 3^8

8 a 4^6 b 6^{12} c 7^{10} d 8^{21}
e To work out a power raised to a power, **multiply** the indices.

9 a i 1 ii 1 iii 1 iv 1
b i 1 ii 1 iii 1 iv 1

10 a i 13 ii 13
b i 8 ii 5 iii 9 iv 12
c i 8 ii 8
d i 5 ii 3 iii 10 iv 2
e $\sqrt[4]{16} = 2$

11 a i 4 ii 16
b i 25 ii 9 iii 100 iv 4
c 8

12 a i $\frac{1}{4^3}$ ii 10^{-5} iii 2^{-1}
iv $\frac{1}{3^{\frac{1}{3}}}$ v $\left(\frac{7}{6}\right)^{-2}$
b i $\frac{1}{4}$ ii $\frac{1}{100}$ iii $\frac{1}{6}$ iv $\frac{1}{5}$

13 a $50 = 25 \times 2$ so $\sqrt{50} = \sqrt{25} \times \sqrt{2} = 5\sqrt{2}$
b $84 = 4 \times 21$ so $\sqrt{84} = \sqrt{4} \times \sqrt{21} = 2\sqrt{21}$
c $4\sqrt{6}$ d $5\sqrt{7}$ e $8\sqrt{2}$

14 a 4 b 25 c 17 d 21

15 a $\frac{\sqrt{17}}{17}$ b $\frac{\sqrt{21}}{7}$ c $\frac{\sqrt{2}}{4}$ d $\frac{3\sqrt{5}}{5}$

Standard form

1 a Yes
b No; 32 is not between 1 and 10.
c No; cannot have millions.
d No; 0.8 is not between 1 and 10.

2 a 6.8×10^4 b 9.4×10^7 c 8.01×10^5
d 4×10^{-6} e 3.9×10^{-3} f 5.3×10^{-8}

3 a 8×10^9 b 6×10^{14} c 6×10^2
d 4.8×10^{13} e 5.6×10^{10} f 4.8×10^{-5}

4 a 25 000, 0.013 b 25 000.013

1 Extend

1 a Square A b Square B

2 $27 = 3^3$; $(3^3)^2 = 3^6$
$9 = 3^2$; $(3^2)^3 = 3^6$

3 36 and 108

4 a $48 = 2 \times 2 \times 2 \times 2 \times 3$
$90 = 2 \times 3 \times 3 \times 5$
$150 = 2 \times 3 \times 5 \times 5$
b HCF = 6; LCM = 3600

5 300 minutes = 5 hours

6 a Numbers ending in zero are multiples of 10. $10 = 2 \times 5$
b 4 c $2^5 \times 9 \times 5^5 = 900\,000 = 9 \times 10^5$

7 a $2^8 \times 5^9$ b $2^7 \times 3^6$ c $2^6 \times 3^5 \times 5^7$ d $2^{10} \times 3^9 \times 5^7$

8 625

9 a i 120 536 km ii 1.20536×10^5 km
b i 227 900 000 000 m ii 2.279×10^{11} m
c i 0.000 004 m ii 4×10^{-6} m
d i 0.000 000 001 m ii 1×10^{-10} m

10 a 11 881 376 b 4 084 101

11 1.7962×10^8 kg

12 a $\frac{7\sqrt{2}}{2}$ b $-\frac{\sqrt{2}}{6}$

13 $\frac{\sqrt{3}}{3}$

14 One course : 5 + 7 + 3 = 15
Two courses: 35 + 15 + 21 = 71
Three courses: 105
Total: 15 + 71 + 105 = 191

15 a 9 000 000 b 800 c 0.000 008

16 a 20 b 3 c 19 d 3
e 2 f 2

1 Unit test

Sample student answer

The student has not used her answer to conclude whether or not the 500 sheets of paper will fit in the printer.

UNIT 2

2 Prior knowledge check

1 a 6 b 5 c 6 d 22

2 a 12 b −2 c −11
d 8 e 16 f 81

3 a $\frac{1}{2}$ b $\frac{2}{5}$ c $\frac{3}{4}$ d 10

4 a $2x$ b y^2 c $2w$
d t e 5 f $2z$

5 a p^4 b c^2d^3 c $14m^2$
d $-18f^2$ e $36x^3$ f y

6 a 32 b 20 c 37 d 1

7 16

8 a $7x + 21$ b $2x - 6$ c $3y^2 + 21$
d $18x - 9y + 9$

9 a $2(4x - 1)$ b $5(4y + 3)$ c $c(c - 2)$ d $n(1 + 2n)$

10 a −2 b 4 c 5 d 8

11 5

12 $3x + 90 = 180$; $x = 30$

13 a $x = y + 5$ b $x = \frac{y}{4}$

14 a 22 b 17

15 a 9, 16, 23, 30 b 14, 8, 2, −4

16 a add 9; 29, 38 b multiply by 3; −27, −81
c subtract 4; −6, −10 d divide by 10; 0.002, 0.0002

17 a a and c b b and d c a d b, c and d

18 a 4, 11, 18, 25 b 3, 6, 12, 24

19 a i 3 ii 6 iii 10 iv 15
b Students' own answers c 5050
d i 9 ii 36 iii 100 iv 225
e $\frac{1}{4}n^2(n + 1)^2$

2.1 Algebraic indices

1 a 2^7 b 2^3 c 2^{12} d 2^{-1}

2 a 10^5 b 5^3 c 3^3

3 a x^7 b x^7 c a^{11}
d y^9 e m^2

4 a $6a^8$ b $8c^6$ c $40n^7$ d $7v^5$
e $15s^5t^6$ f $30p^6q^6$

5 a x^2 b x^3 c p^3 d y^6
e r f t^2

6 a $2g^2$ b $3f^4$ c $3x^2$ d $3w^2$

7 a x^6 b x^{18} c t^9 d j^{18}

8 a $8r^6$ b $9f^8$ c $\frac{b^6}{8}$

9 Clockwise for multiply: $3x^4y^3$, $18x^7y^5$, $72x^8y^5$, $12x^5y^3$.
Divide: $4x$, $6x^3y^2$

10 a $8x^6y^9$ b $36x^{10}y^4$ c $81x^8y^4$ d $\frac{4x^6y^4}{9}$

11 a $x^3 \div x^3 = x^{3-3} = x^0$

$x^3 \div x^3 = \frac{x^3}{x^3} = 1$

Therefore $x^0 = 1$

b $x^3 \div x^4 = x^{3-4} = x^{-1}$

$x^3 \div x^4 = \frac{x \times x \times x}{x \times x \times x \times x} = \frac{1}{x}$

Therefore $x^{-1} = \frac{1}{x}$

c $x^3 \div x^5 = x^{3-5} = x^{-2}$

$x^3 \div x^5 = \frac{x \times x \times x}{x \times x \times x \times x \times x} = \frac{1}{x^2}$

Therefore $x^{-2} = \frac{1}{x^2}$

12 a $\frac{1}{b}$ b $\frac{1}{h^3}$ c 1 d $\frac{1}{r^6}$

13 a $12c^3d$ b 3

14 a $\frac{1}{t^6}$ b x^2 c 1 d w

15 a 1 b $\frac{1}{e^2f^3}$ c $\frac{1}{4p^{10}q^2}$ d $\frac{5v^3}{2u^4}$

16 a x^2 b $3x$ c $2x^4$ d $4x^2y^3$

17 a $x^{\frac{1}{2}} \times x^{\frac{1}{2}} = x^{\frac{1}{2}+\frac{1}{2}} = x^1 = x$

$\sqrt{x} \times \sqrt{x} = x$

Therefore $x^{\frac{1}{2}} = \sqrt{x}$

b $x^{\frac{1}{3}} \times x^{\frac{1}{3}} \times x^{\frac{1}{3}} = x^{\frac{1}{3}+\frac{1}{3}+\frac{1}{3}} = x^1 = x$

$\sqrt[3]{x} \times \sqrt[3]{x} \times \sqrt[3]{x} = x$

Therefore $x^{\frac{1}{3}} = \sqrt[3]{x}$

18 a $\frac{q^8}{9p^2}$ b $4c^3$ c $\frac{x}{2y^4}$ d $\frac{y}{2x^2}$

2.2 Expanding and factorising

1 a $4x + 8$ b $3q - 15$ c $14m + 7$ d $-2y - 12$

2 a $9a + 5$ b $x + 4$ c $5s$

3 a 2 b 9 c 3

4 a $2(x + 5)$ b $2x$, 10

c The answers to part **b** add up to give the answer to part **a**.

5 **a**, **b** and **d** are all identities; **c** is an equation.

a $x \times x \equiv x^2$ b $3x + 4x - x \equiv 6x$ d $\frac{6x}{3} \equiv 2x$

6 a

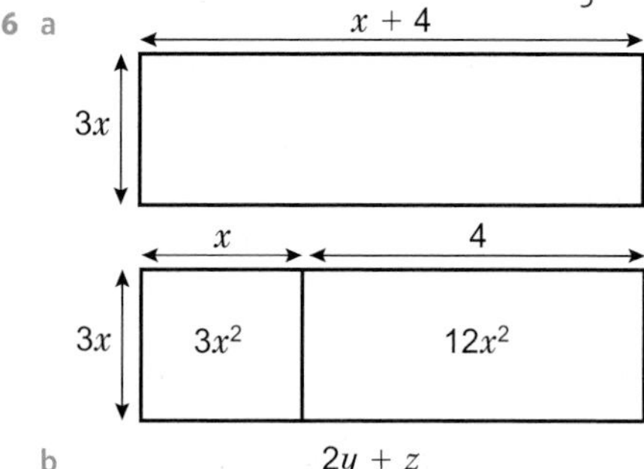

b

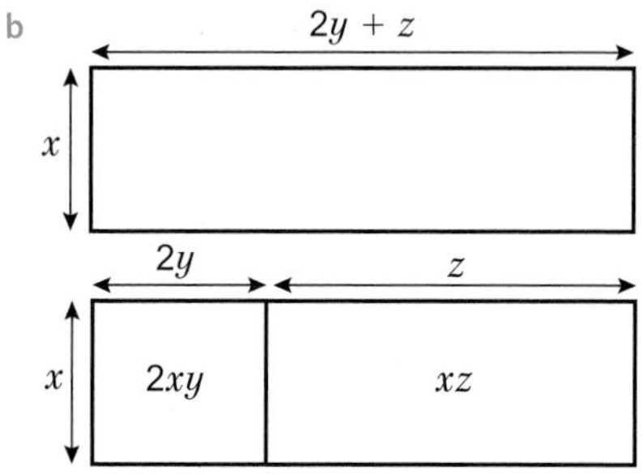

c

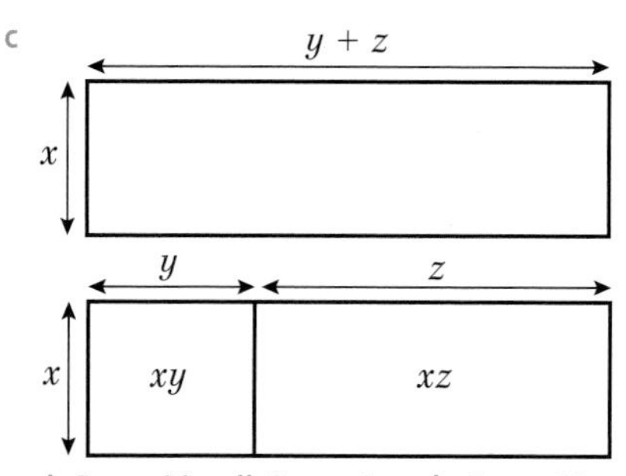

7 a i $5xy + 20x$ ii $3xy + 6y$ b $8xy + 20x + 6y$

8 a $8e + 18$ b $8y + 14$ c $4x + 27$ d $9m + 27$
e $5a + 8b$ f $13x + 8y$

9 a $x^2 - 2x$ b $11y - 12$ c $10t - 6$
d $2p^2 + pq + q^2$ e $w + 3w^2$ f $5e^2 + 3ef + 2f^2$

10 a $2x$ b x c $4y$ d $5xy$

11 a $2(x + 6)$ b $2x(2 + 3y)$ c $b(3a - 5)$ d $7x(y + z)$
e $ab(1 - c)$ f $t^2(t + 2)$ g $3pq(2p - 3)$ h $3xz(x + 4)$
i $5jk(4k - 3j)$ j $2pq(6r - 5s)$

12 a $4(s + 2t)$

b $= 4(s + 2t)[(s + 2t) - 2]$

$= 4(s + 2t)(s + 2t - 2)$

13 a $7(p + 1)(2p + 5)$ b $5(c + 1)(c - 1)$
c $4(y + 4)(3y + 10)$ d $(a + 3b)(a + 3b - 2)$
e $5(f + 5)(1 + 2f)$ f $5(a + b)(a + b - 2)$

14 One of the numbers is even so can be written as $2m$.

One of the other numbers is a multiple of three so can be written as $3n$.

If the other number is p their product is $2m \times 3n \times p = 6mnp$ so is divisible by 6.

15 a $8x^2 - 20xy$ b $2cp(2 - 3p)$ c $3m^2n^3$

2.3 Equations

1 a $x = 7$ b $x = 1$ c $x = 2$

2 a 6 b 24 c 12

3 $3^3 - 2 \times 3 = 27 - 6 = 21$

4 a $8x + 6$ b $21x - 4$ c $-8x + 17$

5 a $3x + 1 - 3x = 5x - 9 - 3x$

$1 = 2x - 9$

b $x = 5$

6 a 5 b 4 c 10 d -2

7 a i $12x - 16$ ii $7x - 21$ b -1

8 a $3x + 16$ b $x = 3$

9 a 1 b -1

10 a $x = -\frac{3}{11}$ b $x = \frac{22}{7}$ c $x = \frac{29}{2}$ d $x = -\frac{4}{3}$
e $x = \frac{21}{25}$ f $x = \frac{33}{37}$

11 a $2x$ b $\frac{y}{2}$ c $9z$ d $6w$

12 a $\frac{7x - 1}{4} \times 4 = 5 \times 4$

$7x - 1 = 20$

b 3

13 a $\frac{10}{x - 4} \times (x - 4) = 3 \times (x - 4)$

$10 = 3x - 12$

b $\frac{22}{3}$

14 a $-\frac{8}{5}$ b $\frac{7}{2}$

15 a $b = 9$ b $n = 1$ c $c = \frac{13}{3}$ d $x = -\frac{1}{5}$
e $x = \frac{41}{6}$

16 30°

17 a $\frac{x}{60}$ b $\frac{x}{45}$ c $\frac{x}{60} + \frac{x}{45} = 7$ d 180 miles

2.4 Formulae

1 a 4 b 13 c 32 d 17

2 a 7.5×10^7 b 300 000 000

3 1.22

4 a Formula b Equation c Expression d Identity
e Expression f Formula g Expression h Equation
i Identity j Equation

5 a 2000 b −2

6 a 350 b 8

7 a 130 minutes b $T = 30 + 40m$

8 a $A = \frac{BH}{2}$ b i 9 ii 10

9 £12 521.56

10 a 320 m b $22.4\,\text{ms}^{-2}$

11 a $a = \frac{v-u}{t}$ b $n = \frac{m-E}{2}$ c $G = \frac{WH}{3}$
d $Q = 7(R - C)$ e $V = 3T + W$ f $a = \frac{2}{t^2}(s - ut)$

12 a 82.4 °F b $C = \frac{5F - 160}{9}$ c 40 °C

13 a $T = \frac{D}{S}$ b 192 seconds

14 a $a = \frac{c-9}{6b}$ b 6

15 a 4654 m b 179107 m

2.5 Linear sequences

1 a 6 b 16 c 26

2 4, 25

3 a 2, 4, 6, 8, 10 b 4, 7, 10, 13, 16
c −4, −8, −12, −16, −20 d 1, −1, −3, −5, −7

4 a 10, 13, 16, 37, 307 b 98, 96, 94, 80, −100
c 6, 6, 6, 6, 6,

5 a 0.02 and 0.67 b $\frac{1}{2}$ and $\frac{5}{4}$
c −5 and −8 d 1 and 2.569

6 a 3, 2, 3, −4, −2 b In front of n
c i 5 ii −3
d i 3, 8, 13 ii 1, −2, −5

7 a $2n + 1$ b $4n + 10$ c $10n - 8$ d $-3n + 16$
e $5n$

8 a nth term = $3n + 2$.
The solution to the equation $3n + 2 = 596$ is $n = 198$; which is a whole number.
Therefore 596 is a term in the arithmetic sequence.
b The solution to the equation $7n - 3 = 139$ is $n = \frac{142}{3}$ which is not an whole number.
Therefore 139 is not a term in the arithmetic sequence.

9 a $6n - 3$
b No Ben is not right. The solution to the equation $6n - 3 = 150$ is $n = 25.5$ which is not a whole number.
Therefore 150 is not a term in the arithmetic sequence.

10 a 125.375
b From part **a** it is the 126th term; this is $8 \times 126 - 3 = 1005$

11 a 4007 b 49

12 a i 99.6 kg ii 99.2 kg iii 98.8 kg b 28 weeks

13 28 weeks

14 a 10, 17, 24, 31 b 7 c 3

15 a i 9, 21, 33, 45, 57 ii 41, 81, 121, 161, 201
b

Sequence	Input difference	Output difference
i	3	12
ii	10	40

The differences are scaled by 4 which is the multiplier in the function machine.

16 a 4 b 8 c 2 d $q = 4$

17 $p = 7, q = 6$

2.6 Non-linear sequences

1 a £1248 b £153

2 a 48, 96. Multiply by 2. b 27, $\frac{1}{3}$. Divide by 3.
c 18. Multiply by −3.

3 a 5, 8, 13 b 5, 9, 14 c −1, 0, −1

4 a $1, \frac{1}{2}, \frac{1}{3}, \frac{1}{4}$ b 2, 4, 8, 16 c 0.3, 0.09, 0.027, 0.0081

5 a $\sqrt{2}, 2, 2\sqrt{2}, 4, 4\sqrt{2}$ b $3, 6\sqrt{3}, 36, 72\sqrt{3}$

6 8 months

7 a £8400
b i £8820 ii £9261
c 5 years

8 Option 1 gives £10 400 and Option 2 gives £14 486.54.
Option 2 gives more money.

9 a 1, 4, 9, 16, 25, 36
b i $n^2 + 1$ ii $n^2 - 1$ iii $(n + 1)^2$

10

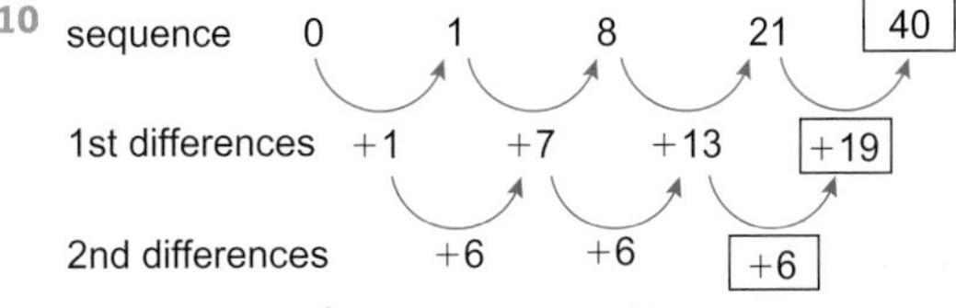

11 a 126 b 46 c 10

12 a

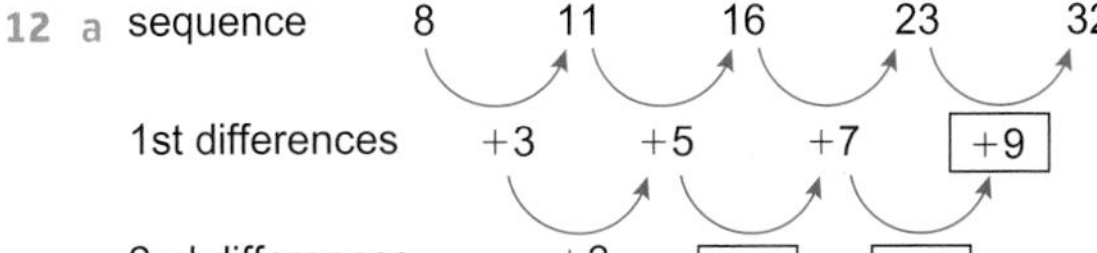

b

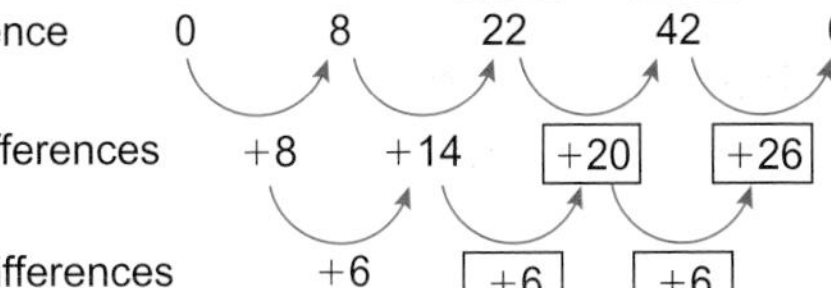

13 a $2n^2 + 1$ b $3n^2 - 5$ c $\frac{1}{2}n^2 + 4$

14 a 1, 5, 10, 10, 5, 1
b

Row, n	0	1	2	3	4	5
Sum	1	2	4	8	16	32

c 2^n

15 a 2. Halve the second difference to find the coefficient of n^2. Therefore $a = 1$.
b 3, 5, 7 c $2n - 1$ d $n^2 + 2n - 1$

16 a $n^2 + 3n$ or $n(n + 3)$ b $(n - 1)^2$
c $2n^2 + n + 2$ d $3n^2 - n + 1$

17 $u_5 \times u_8 = 10^5 \times 10^8 = 10^{5+8} = 10^{13} = u_{13}$

18 a 2, 4, 8, 16 b Multiply by 2
c $u_m \times u_n = 2^m \times 2^n = 2^{m+n} = u_{m+n}$

2.7 More expanding and factorising

1 a 2, 3 b −4, −1

2 a $4x^2$ b $25y^2$

3 a $(x + 2)(x + 1)$ b $x^2 + x + 2x + 2$; $x^2 + 3x + 2$

4 a $x^2 + 16x + 60$ b $x^2 + 3x - 18$
c $x^2 + 6x - 40$ d $x^2 - 7x + 12$

5 a $(x + 2)(x + 3) = x^2 + 5x + 6$
b $(x - 3)(x + 8) = x^2 + 5x - 24$

6 a $x^2 + 4x + 4$ b $x^2 - 6x + 9$
c $x^2 + 10x + 25$ d $x^2 - 8x + 16$

7 a $(51 + 49)(51 - 49) = 2 \times 100 = 200$
b i 400 ii 0.12

8 a $x^2 - 16$ b $x^2 - 4$

9 a $(x - 5)(x + 5)$
b $(y - 7)(y + 7)$
c $(t - 9)(t + 9)$

10 a $(x + 1)(x + 7)$ b $(x + 3)(x + 4)$
c $(x + 3)(x + 5)$ d $(x + 3)(x - 1)$
e $(x - 3)(x + 1)$ f $(x - 2)(x - 4)$
g $(x - 7)(x + 1)$ h $(x - 3)(x - 4)$
i $(x - 2)^2$ j $(x - 12)(x - 2)$
k $(x - 8)(x + 2)$ l $(x + 1)^2$

11 a $(x + 5)(x + 2) = x^2 + 7x + 10$
b $7x + 10$ c $x = 3$

12 $x = 1.5$

13 a $4x^2 - 9 = (2x)^2 - 3^2 = (2x - 3)(2x + 3)$
b $16y^2 - 1 = (4y)^2 - 1^2 = (4y - 1)(4y + 1)$

14 a $(3m - 5)(3m + 5)$
b $(5c - 9)(5c + 9)$
c $(x - 7y)(x + 7y)$

15 a $(x + 5)(x + 6)$ b $9u^2 - 24uv + 16v^2$

2 Problem-solving

1 a 3500 b $7xyz + m$

2 a 64 b $\frac{pq}{st}$

3 a 115 b $xy - mn$

4 a 462 b $x(x - 1)$

5 a 273 b $2n - 1$

6 500

2 Check up

1 a $20p^4$ b $5x^2$ c b^{-6}

2 $5q$

3 a $2y(x - 3)$ b $3a(b - 2a)$

4 a $x^2 - 2x - 24$ b $x^2 + 10x + 25$

5 a $\frac{2}{x^2}$ b 4 c $3c$ d $4p^{-5}$

6 $2s^2 + 5rs - 3r^2$

7 a $(x - 9)(x + 9)$ b $(x - 2)(x - 7)$

8 a Formula b Identity c Expression d Equation

9 $x = \frac{9}{2}$

10 $x = -17$

11 40

12 $1.1^3 + 4 \times 1.1 = 5.731 < 6$ and $1.2^3 + 4 \times 1.2 = 6.528 > 6$

13 $C = 25 + 36n$

14 a $y = \frac{4 - 2x}{3}$ b $b = \frac{S - 4a^2}{6a}$

15 $x = 10$

16 18, 29

17 a $9n - 7$
b The equation $9n - 7 = 167$ has solution $\frac{174}{9}$ which is not a whole number.
Therefore 167 is not a term in the sequence.
c 173

18 $3n^2 + 7$

20 a Clockwise from top left:
$x^2 + x - 6$, $3x^2 - 7x + 2$, $6x^2 + 10x - 4$, $2x^2 + 10x + 12$.
b $12x^2 + 14x + 4$. Any order would give the same result, multiplication is not commutative.
c $2(2x + 1)(3x + 2)$

2 Strengthen

Simplifying, expanding and factorising

1 a t^5 b t^7 c t^4 d t^2
e t^{-7} f t^2

2 a $18p^5$ b $72z^5$ c $14b^8$ d $8r^3$
e $6x^2$ f $10s^{-6}$

3 a t^4 b t^3 c $t^0 = 1$

4 a $5p^4$ b $3a^5$ c $3y^{-3}$ d $2p$

5 a $(x^2)^2 = x^2 \times x^2 = x^4$
b $(x^2)^3 = x^2 \times x^2 \times x^2 = x^6$
c $(x^2)^4 = x^2 \times x^2 \times x^2 \times x^2 = x^8$
d When you find the power of a power you multiply the powers together.

6 a a^2 b r^{-2} c $8g$

7 a $6x + 3y$ b $6x - 8y$ c $12x - 5y$

8 a $11c + d$ b $14m + 10n$

9 a $ab(3b - 2)$ b $2x(4y + 3)$
c $3st(t - 2)$ d $7b(2ab + 3)$

10 a

×	x	+5
x	x^2	$5x$
+4	$4x$	+20

b $(x + 4)(x + 5) = x^2 + 5x + 4x + 20$
$= x^2 + 9x + 20$

11 a $x^2 - 12x + 36$

×	x	−6
x	x^2	$-6x$
−6	$-6x$	+36

b $x^2 - 16$

×	x	+4
x	x^2	$+4x$
−4	$-4x$	−16

12 a $3x^2 + 26x + 16$

×	$3x$	+2
x	$3x_2$	$+2x$
+8	$+24x$	+16

b $10x^2 + 11x + 3$

×	$2x$	+1
$5x$	$10x^2$	$+5x$
+3	$+6x$	+3

c $3x^2 + 5x - 28$

×	$3x$	−7
x	$3x^2$	$-7x$
+4	$+12x$	−28

13 $(x - 3)^2 = x^2 - 6x + 9$
$(x + 1)(x + 5) = x^2 + 6x + 5$
$(x - 3)(x + 3) = x^2 - 9$
$(x + 2)(x - 3) = x^2 - x - 6$
$(x + 3)(x + 2) = x^2 + 5x + 6$

14 a 3 and 4, 2 and 6 b 2 and 6
c $(x + 2)(x + 6)$

15 a $(x + 12)(x + 1)$ b $(x + 3)(x + 4)$

16 a 2 and −5, 1 and −10, −1 and 10
b i $(x - 10)(x + 1)$ ii $(x + 10)(x - 1)$
iii $(x + 5)(x - 2)$ iv $(x - 5)(x + 2)$

17 a −24 and −1, −4 and −6, −2 and −12
b i $(x - 1)(x - 24)$ ii $(x - 2)(x - 12)$
iii $(x - 4)(x - 6)$ iv $(x - 3)(x - 8)$

Equations and formulae

1 a Expression b Identity c Formula d Equation

2 a 9 b 36 c 41

3 23

4 −5

5 $x = \frac{y + 4}{2}$ or $\frac{1}{2}(y + 4)$ or $\frac{1}{2}y + 2$

6 $Q = aP - ab$ or $a(P - b)$

7 a $b = \frac{4c}{3}$ b $s = \frac{v^2 - u^2}{2a}$

8 $x = 4$

9 a i $14x - 28$ ii $6x + 10$

b $x = \frac{19}{4}$

10 a $2x - 2$ b $x = \frac{7}{2}$

11 a x b $2x$ c $4x$

12 a $x = 20$ b $x = \frac{10}{3}$

13 a $x = 18$ b $x = \frac{7}{4}$

c Find the LCM of the denominators

14 a $x = 60$ b $x = 1$

Sequences

1 a 13, 21 b 50, 81 c 42, 68

2 a 5, 7, 9 b 48, 46, 44 c 2, 5, 10 d 10, 40, 90

3 a 3 b 36

c The general term is $3n$

4 a $10n$ b $7n$ c $12n$

5 a 6

b The general term is $n + 6$

6 a $n + 2$ b $n + 12$ c $n - 4$

7 a $4n$ b 3 c $4n + 3$

8 a i 30, 36 ii 9, 11 iii 16, 19 iv 5, 0

b i $6n$ ii $2n - 1$ iii $3n + 1$ iv $-5n + 30$

9 a 14, 18, 22, 26, 30

b The numbers in the sequence are even but 351 is odd.

c 23rd

10 a 20th b 112 c 152

11 a 53, 14, 18, 4, 4 b $2n^2 + 3$

12 a $4n^2 + 5$ b $n^2 - 10$

2 Extend

1 a i 6, 7, 8 ii 20, 10, 5 iii 3, −1, −5 iv −3, 9, −27

b **i** and **iii** are arithmetic; **ii** and **iv** are geometric

2 a 0.473 b 11.5 c 13 d 15.7

3 a In-store: 12800, 10240; Online: 3240, 4860

b 2017

4 £838.76

5 13

6 a £950 b $P = \frac{10(D - B)}{N}$ c £700

7 a $a = \frac{v^2 - u^2}{2s}$ b $h = \frac{3V}{\pi r^2}$

c $a = \frac{(r - 1)S}{r^n - 1}$ d $y = \frac{a^2x - c}{b^2}$

8 a $2c^4d^4$ b $12x^2y$ c $16m^{-1}n^3$ d $2p^{-1}q^4$

9 $x^2 + 20x + 19$

10 a $p^2 + 5p - 36$ b $w = -2$

c $(x + 3)(x - 3)$ d $3x^4y^{\frac{3}{2}}$

11 a It is not lower as the temperature in Mrs Smith's home is 25 °C.

b $F = \frac{9C + 160}{5}$

12 a The odd numbers are one greater than the even numbers which are multiples of 2.

b $(2m + 1)(2n + 1) = 4mn + 2m + 2n + 1$
$= 2(2mn + m + n) + 1$ which is odd

13 a $(x - 4)(x - 8)$ b $(x - 6)^2$

c $(x - 2)(x + 1)$ d $\left(\frac{x}{5} - \frac{y}{7}\right)\left(\frac{x}{5} + \frac{y}{7}\right)$

14 a $x = -\frac{5}{6}$ b 26 c $-\frac{24}{5}$ d −1

15 If the consecutive numbers are n and $n + 1$ then the difference between their squares is $(n + 1)^2 - n^2 = n^2 + 2n + 1 - n^2 = 2n + 1$, which is an odd number.

16 a $-2n^2 + 3$

b $-n^2 + 2n - 1$ or $-(n - 1)^2$

2 Unit test

Sample student answer

a You can label the dimensions you know.

b From the formulae sheet.

c It separates the sphere from the cone.

d So it doesn't get confused with the radius of the sphere.

UNIT 3

3 Prior knowledge check

1 a 22 b $\frac{27}{2}$ c $\frac{59}{20}$

2 a 11 b 14 c 3.5

3 a 2 hours 12 minutes b 4:11pm c A

4 a Mean = 2, median = 1, mode = 1, range = 5

b Mean = 5.375, median = 4, mode = 3, range = 12

c Mean = 3.1 (1 d.p.), median = 4, mode = 5 and 1, range = 6

5 a 10% b 20%

c i CT ii T iii T

6

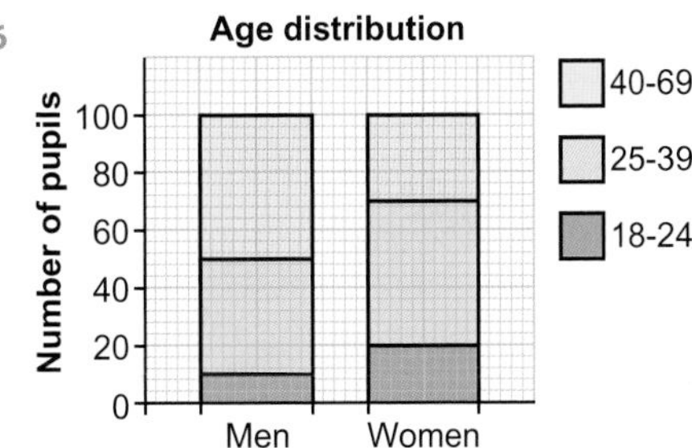

7 a 10 b 5 c March and June

d January e May

f Boys given 6 more detentions than girls (133 versus 127).

8

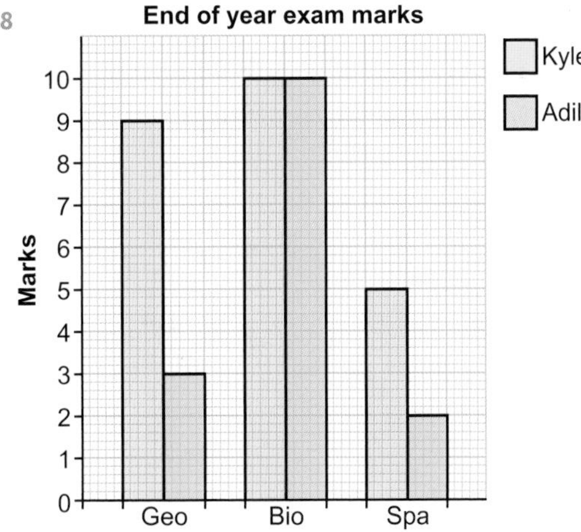

9 a 6 b 80

10 Correctly drawn pie chart with UK 165°, France 45°, Spain 90°, USA 60°.

11 a 5 b 25 c 5

12 a 5 b 35 c 2.1

d On average families in rural communities have 1 child more than those living in the city.

13 a

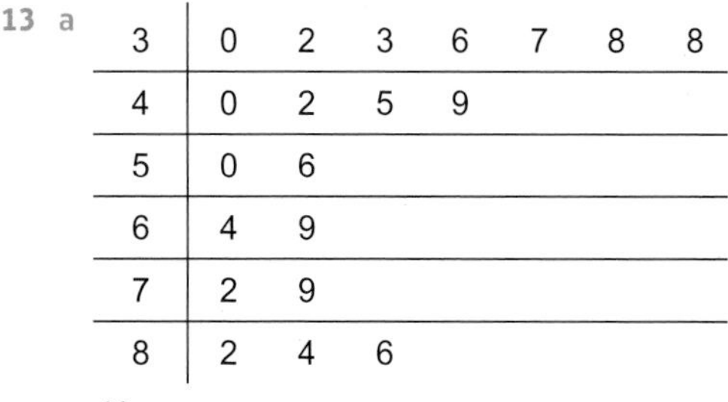

3	0 2 3 6 7 8 8
4	0 2 5 9
5	0 6
6	4 9
7	2 9
8	2 4 6

Key

3 | 0 means 30

b 20 **c** 4

14 Any suitable data collection sheet.

15 a Continuous **b** $x = 5$, $y = 4$

c

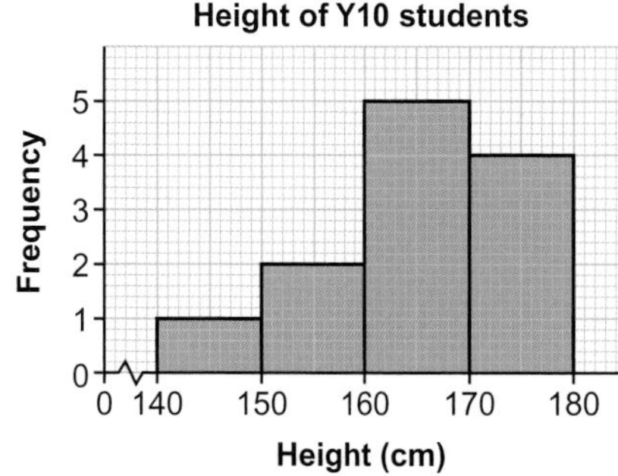

16 a 4, 4, 6, 7, 14; no

b 4, 4, 4, 8, 8, 14 or 4, 4, 5, 7, 8, 14 or 3, 4, 4, 8, 10, 13

3.1 Statistical diagrams 1

1 On average Sophie has a better score than Celia because her median is lower. She is also more consistent because her range is lower.

2 a 750 **b** Theatre

c 20000 × (45 ÷ 360) = 2500 at festival but only 1500 × (90 ÷ 360) = 375 at theatre

3 a 10 **b** 89 kg **c** 35 kg **d** 71 kg

e Yes; the number of people is 10 < 12 and total mass is 716 kg < 800 kg.

f 71.6 kg

4 Maths range = 54, English range = 37
Maths median = 60, English median = 58
The Maths marks are more spread out since the range for maths is far greater than the range for English. The average marks are very similar since the median mark for Maths is only 2 marks higher than the median English mark.

5 a

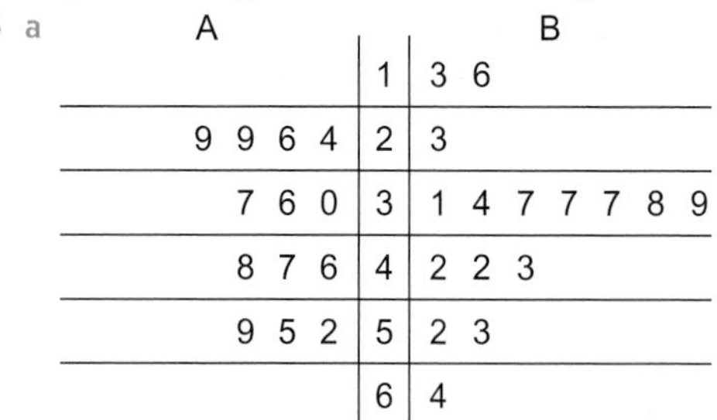

A		B
	1	3 6
9 9 6 4	2	3
7 6 0	3	1 4 7 7 7 8 9
8 7 6	4	2 2 3
9 5 2	5	2 3
	6	4

Key

A: 4 | 2 represents 24 cm   B: 1 | 3 represents 13 cm

b For type A there is an even spread of tulips in the range 20 to 59 cm because the numbers on the left-hand side appear as a block. For type B most tulips are between 30 and 39 cm, with a few shorter and a few taller, because the numbers show a distribution that is peaked at the centre and diminishes at either side of the centre.

6 a 2 **b** 37 **c** 27 **d** 27

7

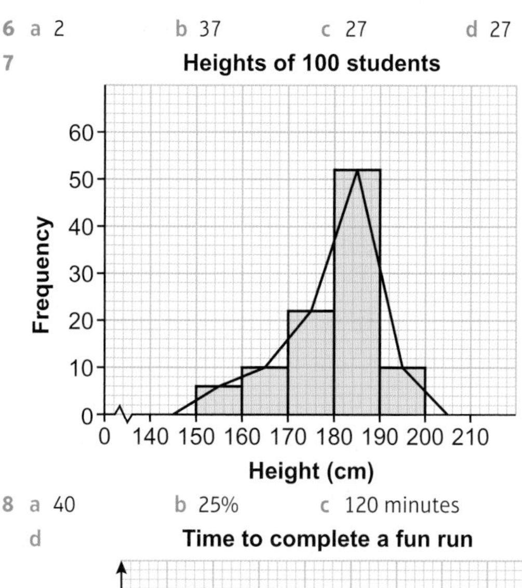

8 a 40 **b** 25% **c** 120 minutes

d

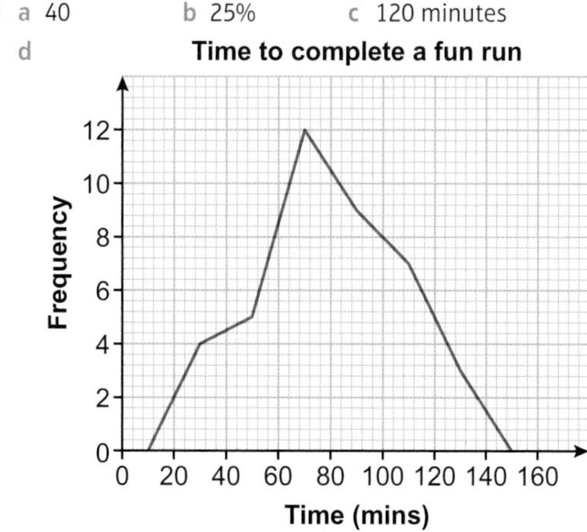

9 a B **b** About the same **c** A

3.2 Time series

1 a ii **b** iv **c** i

d iii **e** v

2 a 4, 6 **b** 39, 48

3 25%

4 a 37.3 °C **b** 38.3 °C, 10 am **c** 37.4 °C

d

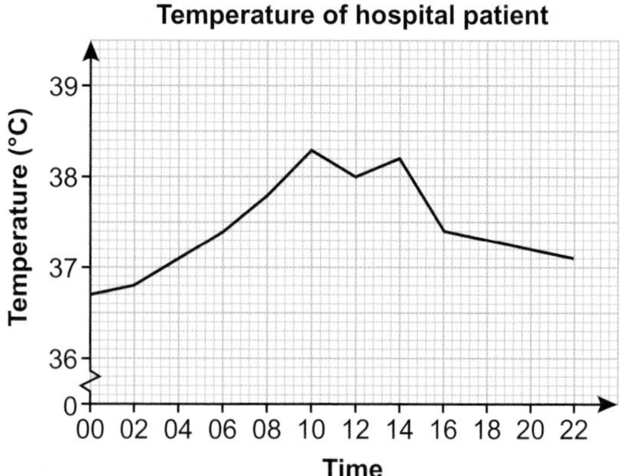

The temperature increased steadily between midnight and 10 am, then fluctuated around 38 °C for four hours. Between 14:00 and 16:00 it decreased sharply, then continued to decrease slowly until 22:00.

5 a
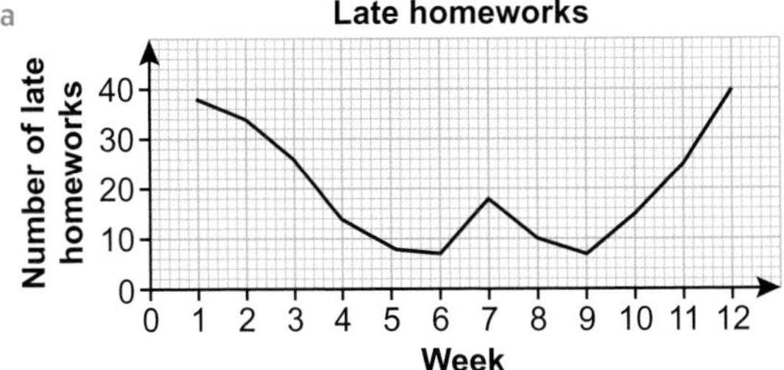

The number of late homeworks starts off very high, decreases mid-term but then gets steadily higher during the last three weeks of term. There is small blip near half-term.

6 a £4.90
b Yes; Magazine A has gone up by £3.40 whereas Magazine B has only risen by 90p.
c Yes; the graph is bending downwards or the slope of the graph is going down.
d e.g. Both magazines are likely to be the same price of about £7.50.

7 a 66 000
b Q1 of 2012
c
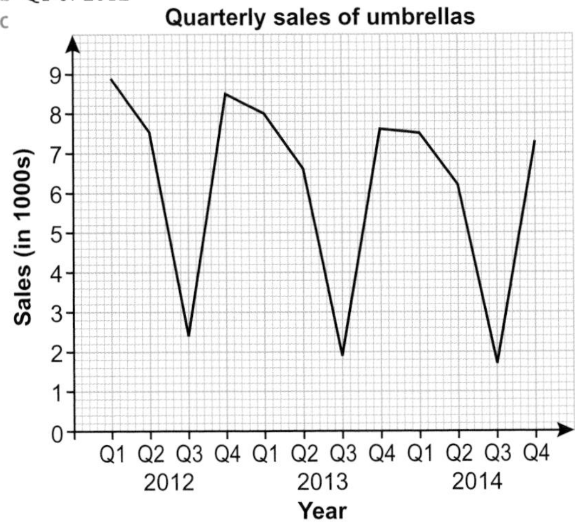

d Due to seasonal variations, sales fluctuate wildly. The overall trend is a decrease in sales.

8 a
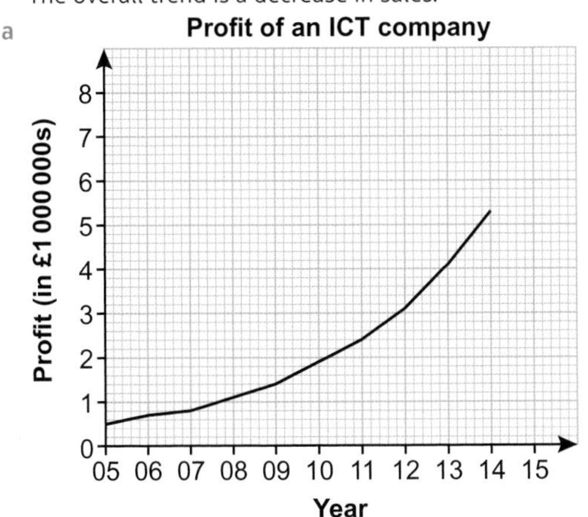

b Profit increases at an increasing rate.
c About £1.9 million
d About £6.9 million

9 a
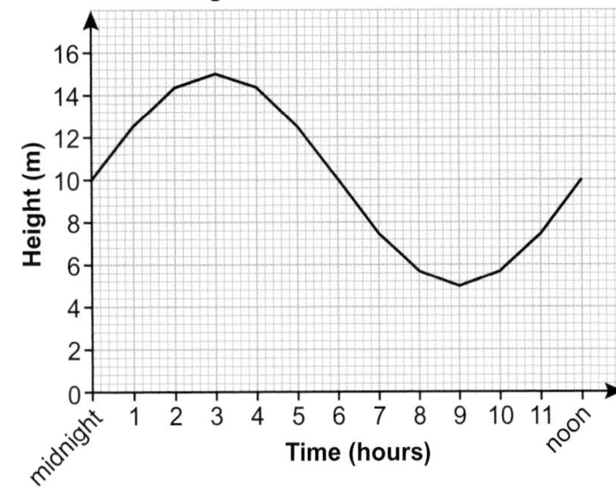

b 3 am
c 15 m
d Between 7 am and 11 am and again between 7 pm and 11 pm.

3.3 Scatter graphs

1 Point A

2 a
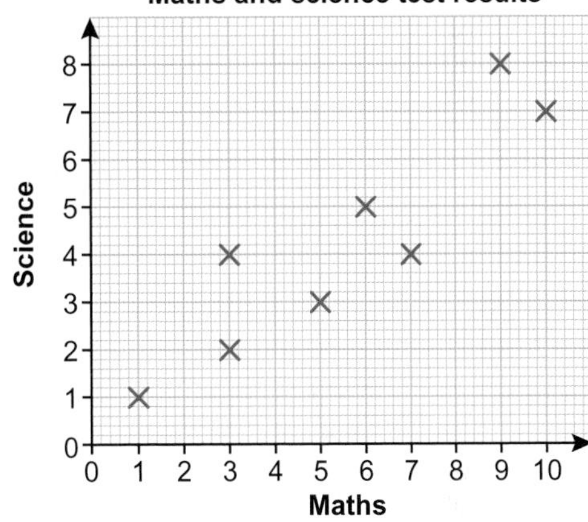

b In general, students with higher maths scores got **higher** science scores and students with lower maths scores got **lower** science scores.

3 a No correlation; there is no relationship between price and temperature.
b Negative correlation; as the price of ice creams increase, sales decrease.
c Positive correlation; as the temperature increases sales of ice creams increase.

4 a
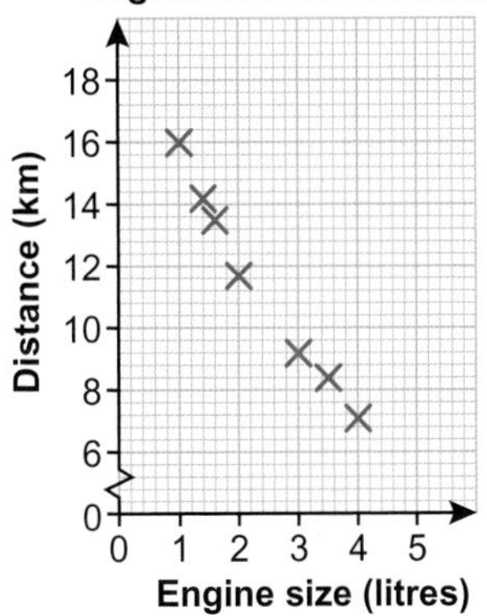

b Negative correlation. The larger the engine the shorter the distance travelled on a litre of petrol.

5 a

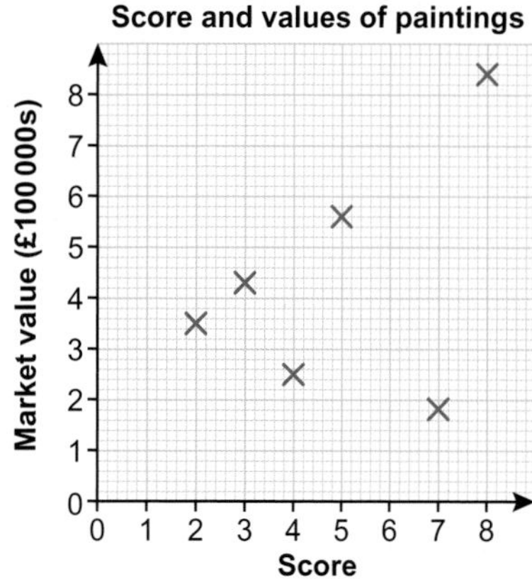

No correlation; there is no relationship between the value of painting and the mark awarded.

6 a

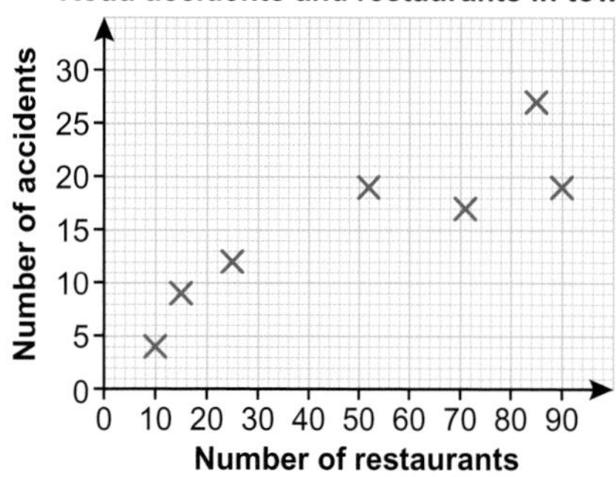

b Positive correlation; cities with a larger number of takeaway restaurants tend to have a larger number of road accidents.

c The data provides no support for the councillor's views; although there is positive correlation it may not be one of cause and effect. Large, busy towns are likely to have both more takeaway restaurants and more accidents than small, quiet towns.

7 a Negative correlation

b Positive correlation

c No correlation

8 a

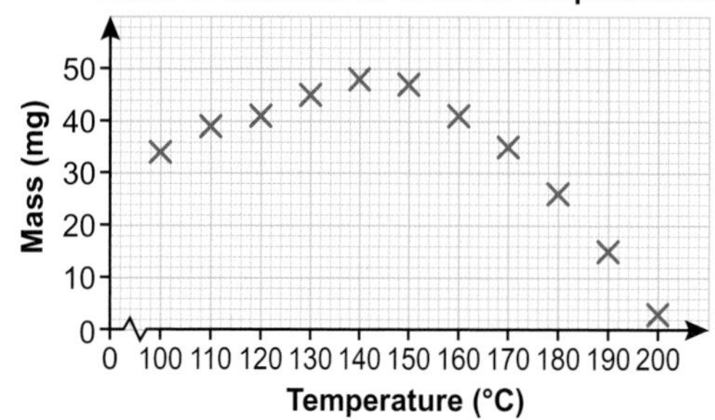

b Positive correlation at temperatures up to about 145 °C; negative correlation at temperatures over 145 °C.

c The mass increases to a maximum and then steadily decreases.

d Approximately 50 mg at about 145 °C

9 a Negative correlation. As money spent on quality control goes up the proportion of faulty mp3 players goes down.

b 4.2%

c £73 000

3.4 Line of best fit

1 a 4 b 3

2 C

3 a, b

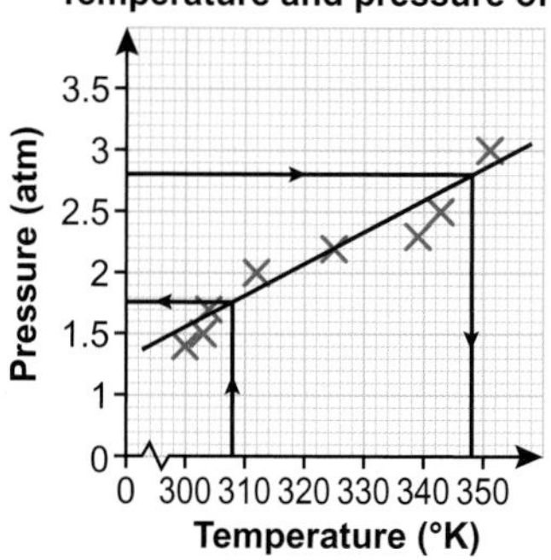

c About 85 kg d About 163 cm

4

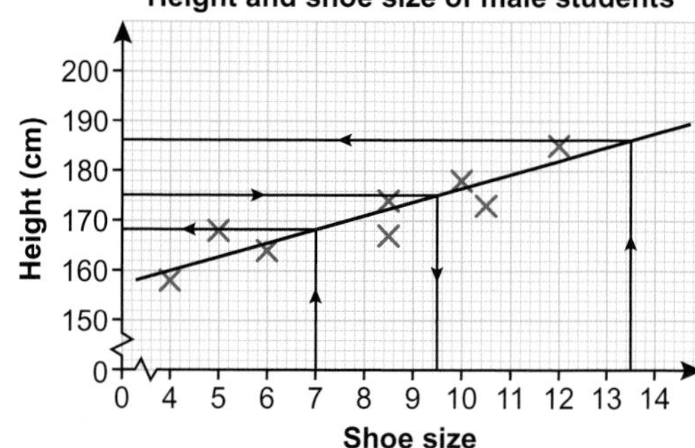

a About 349 °K b About 1.7 atm

5

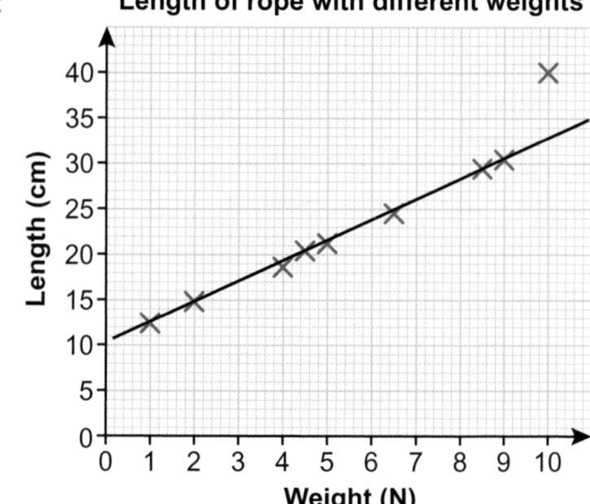

a i About 9.5 ii About 168 cm iii About 185 cm

b **iii** is the least reliable because it lies outside the range of the data points.

6 a Jack 2.9 and Joe 2.3

b Jack's estimate is more reliable. He uses more points and the points on his diagram are much closer to the line of best fit.

7 a, c

Length of rope with different weights

Length (cm)

0 5 10 15 20 25 30 35 40

0 1 2 3 4 5 6 7 8 9 10

Weight (N)

b The last point does not fit into the pattern of the other points. This might be due to an experimental misread or even a change in behaviour of the elastic when it is subject to a larger weight.

d About 26 cm e About 10 cm

8 a

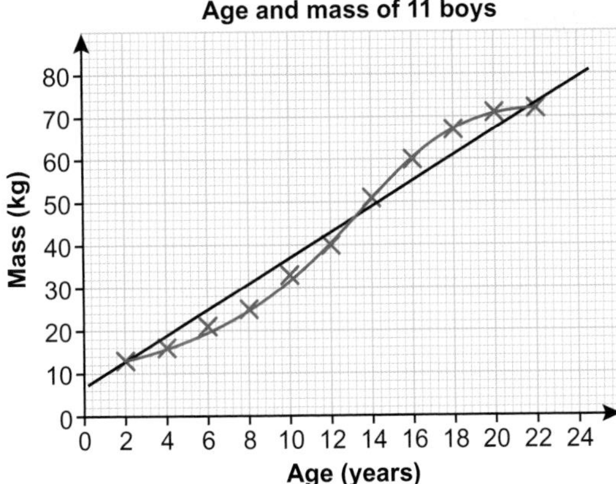

b i About 53 kg ii About 81 kg

c **i** is more reliable because the age of 15 is inside the data points whereas 24 is outside.

d i About 55 kg ii About 72 kg

e **d** is more reliable. The points nearly lie exactly on a smooth curve. This is a better fit of the data points. People have a growth spurt during their teenage years, this slows down till it stops at the age of 24 for boys so the graph should flatten off.

9 a

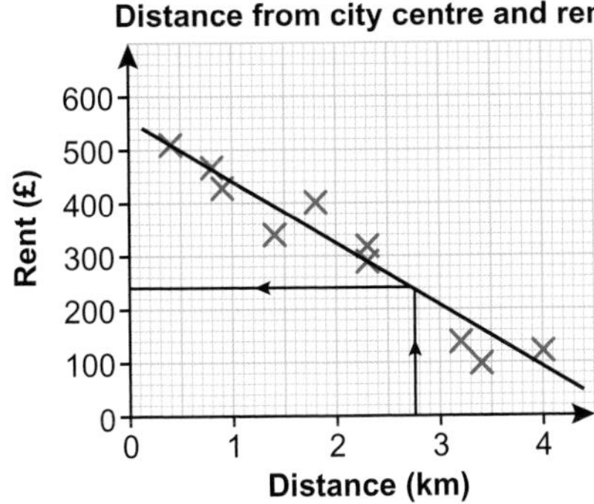

b There is negative correlation between rent and distance. The further you live from the city the lower the monthly rent.

c About £240

3.5 Averages and range

1 a 14 b 1

2 a 15.5 b 30

3 a Mean £19 400, median £15 000, mode £12 000

b Median; mean is distorted by one high salary and mode is lowest salary so neither of these give a typical salary.

4 a Mean 8.375, median 7.75, mode 7

b Mode; this is the popular shoe size so it makes sense to order what customers are likely to want to buy. The values of mean and median aren't proper shoe sizes.

5 a Mean £212, median £190, mode £180

b Mean which takes into account all five values and could be used to work out the total bill.

6 a Median; the low value of 6s distorts the mean, making it too low, and the mode gives the longest time so neither of these are typical.

b Mode; the data is qualitative so you cannot work out the mean or median.

7 a Outlier 7 kg, range 25 kg

b Outlier £38,000, range £24 000

8 a 2.9 and 500 are outliers, which are probably misreadings so should be ignored; 18 °C

b −£250 000 is an outlier and without any information to the contrary is probably correct and just a bad year so should be included; £400 000.

9 a

Time, T (minutes)	Frequency, f	Mid-point, x	xf
$0 \leqslant T < 4$	27	2	2 × 27 = 54
$4 \leqslant T < 10$	34	**7**	**7 × 34 = 238**
$10 \leqslant T < 20$	15	**15**	**15 × 15 = 225**
$20 \leqslant T < 60$	4	**40**	**40 × 4 = 160**
Total	**80**		**677**

Mean = 8.4625

b £33.85

10 a 3 b 8 c 16 d $20 \leqslant t < 30$

e There are 21 data values, so the median is in position $\frac{21+2}{2} = 11$

f $20 \leqslant t < 30$

11 a 36

b The total number of items is 36, so the median is at item $\frac{36+1}{2} = 18.5$

c $7.5 \leqslant d < 8.0$ d $7.0 \leqslant d < 7.5$

e Ben f $7.5 \leqslant d < 8.0$

g Jamie, because he has done more jumps in training over 8.0.

12 a 6

b Mean

$= \frac{(1 \times 1) + (3 \times 3) + (5 \times 4) + (7 \times 2) + (9 \times 3) + (11 \times 2) + (13 \times 1) + (15 \times 2) + (17 \times 4) + (19 \times 0)}{22}$

$= \frac{1 + 9 + 20 + 14 + 27 + 22 + 13 + 30 + 68}{22}$

= 9.27, so no compensation

3.6 Statistical diagrams 2

1 Pie chart with Tea 72°, Coffee 180°, Cola 45°, Water 63°

2

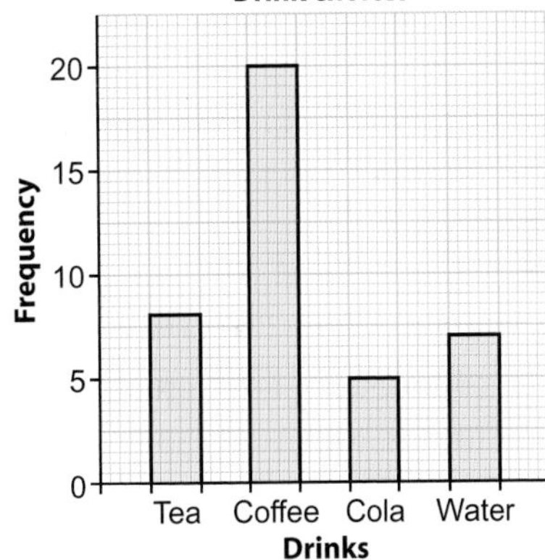

3

	Yes	No	Total
Boys	**50**	30	**80**
Girls	25	**75**	100
Total	**75**	**105**	180

4 a

	No change	Improved	Much improved	Total
Drug A	10	**45**	**5**	60
Drug B	**7**	**20**	13	**40**
Total	17	65	**18**	100

b $\frac{2}{5}$

c Students' own answer, e.g. Drug B had a greater proportion much improved but Drug A had a greater proportion improved.

5 a Any suitable two-way table.

b No since only 56% are in favour.

6 a The vertical scale does not start from zero. This makes a small difference look like a big difference.

b

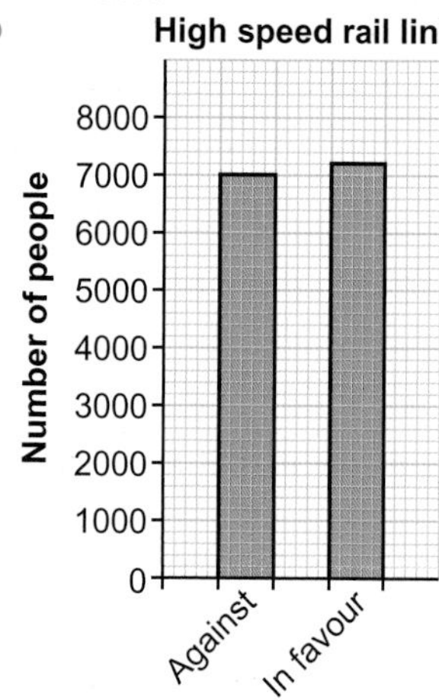

Very similar numbers in favour of and against the rail link.

7 The bar for 'Nutty Oats' is wider than the rest. The vertical axis has no scale or units.

8 a £4
b £300
c Buy in April and sell in September

9 a The data is not quantitative
b i Pie chart ii Dual bar chart
c 198

10 a It is possible to see the actual marks in a stem and leaf.

b

Boys		Girls
7 7 6 5 3	0	1 6
5 0 0	1	1 5 6 7
3 1	2	0 4 5 8
9 4 1	3	9 9

Key Boys Girls
0|1 represents 10 marks 1|5 represents 15 marks

c boy's median = 10; girl's median = 18.5 so girls have a higher average than the boys

11 a Frequency polygon
b Stem and leaf plots cannot be used with grouped data. Scatter diagrams are for data pairs so cannot be used. The researcher wants to compare the performance of several hospitals so could plot two or three frequency polygons on the same diagram. Not easy to compare waiting times across several hospitals by drawing several pie charts.

3 Problem-solving

1 The annual mean of PM10 is 34.1 mg/m^3 and the mean for PM2.5 is 23.4 mg/m^3. Both annual means are below the legal limit for the each type of particulate.

2 When we estimate the mean we use the midpoints of the groups. There is therefore a chance we are underestimating. To find out the maximum possible value of the mean we can use the upper limits for each group. This would produce a maximum value of the mean of 39.1 mg/m^3 for PM10 and 25.9 mg/m^3 for PM2.5. This would suggest that there is a chance the company is over the PM2.5 legal limit.

3 Check up

1

	Party A	Party B	Total
Men	120	**80**	200
Women	**130**	**50**	**180**
Total	**250**	130	380

a 180 b 130

2 a 24
b No; the proportions are the same but the numbers are different. 12 adults and 8 children chose maths.

3 a 15
b The range for cod is 39 which is greater than the range for plaice which is 28.
c 28.5 cm d 164 cm

4

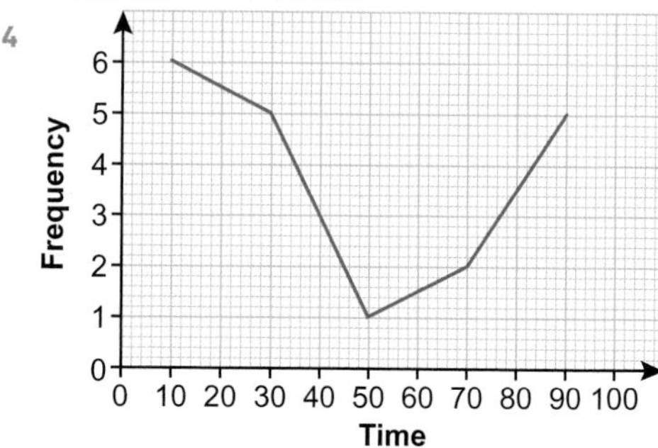

5 a Outlier £4800; range £4350
b Outliers 22, 24; range 29

6 a 9.625 b $5 \leqslant x < 10$ c 25 d $5 \leqslant x < 10$

7 a 5000
b Seasonal variations in weather
c Sales of sun cream are generally increasing.

8 a Positive
b i 42 ii 96
c Part **i**, because part **ii** is outside the data points.

10 a, b Students' own answers
c The new mean is 3 more than the old mean.
d Old mean = $\frac{w+x+y+z}{4}$

New mean = $\frac{(w+3)+(x+3)+(y+3)+(z+3)}{4}$

$= \frac{w+x+y+z+12}{4}$

$= \frac{w+x+y+z}{4} + 3$

= old mean + 3

e Mean is multiplied by c

New mean = $\frac{cw+cx+cy+cz}{4}$

$= \frac{c(w+x+y+z)}{4} = c \times$ old mean

3 Strengthen

Statistical diagrams

1 a 36 b 10
c No; the proportions are the same but the numbers are different. 5 boys and 9 girls chose badminton.

2 a 35 b 20 c $\frac{1}{2}$

3 a 2 b 6 c 14

d

	Yes	No	Total
Boys	**2**	4	**6**
Girls	3	**11**	**14**
Total	5	**15**	20

4 a 10
b 14 minutes
c 2
d 46 minutes by a girl

5 a

Set A		Set B
7 0	2	6 8 8
8 2 0 0	3	2 3
9	4	0

Key Set A: 7 | 2 represents 27   Set B: 2 | 6 represents 26

b i Both sets have a median of 30
ii Sets A and B have a range of 29 and 14 respectively.
c Set A has a much higher range than Set B, but the medians are the same.

6 a

Time	$0 \leqslant t < 2$	$2 \leqslant t < 4$	$4 \leqslant t < 6$	$6 \leqslant t < 8$	$8 \leqslant t < 10$
Frequency	1	3	9	5	2
Mid-points	1	3	**5**	**7**	**9**

b

Time spent on internet

7

Mid-point	0.5	1.5	**2.5**	**3.5**	**4.5**
Interval	$0 \leqslant t < 1$	$1 \leqslant t < 2$	$\mathbf{2 \leqslant t < 3}$	$\mathbf{3 \leqslant t < 4}$	$\mathbf{4 \leqslant t < 5}$
Frequency	1	2	**4**	**5**	**3**

Averages and range

1 a 46 cm
b He has probably misread the length during the experiment or has forgotten to put the decimal point in.
c 2.9 cm

2 a, b

Lengths (L cm)	Frequency	Mid-point	Frequency × midpoint
$0 \leqslant L < 10$	7	5	7 × 5 = 35
$10 \leqslant L < 20$	12	15	12 × 15 = 180
$20 \leqslant L < 30$	20	**25**	**20 × 25 = 500**
$30 \leqslant L < 40$	8	**35**	**8 × 35 = 280**
$40 \leqslant L < 50$	3	**45**	**3 × 45 = 135**
Total	50		**1130**

c 22.6 cm

3 a 300   b 6%
c

Speed (x mph)	Frequency	Mid-point	Frequency × mid-point
$0 \leqslant x < 20$	8	10	8 × 10 = 80
$20 \leqslant x < 25$	90	**22.5**	**90 × 22.5 = 2025**
$25 \leqslant x < 30$	184	**27.5**	**184 × 27.5 = 5060**
$30 \leqslant x < 40$	18	**35**	**18 × 35 = 630**
Total	**300**		**7795**

Mean speed = 26 mph (to the nearest whole number)

4 a $20 \leqslant L < 30$ b $25 \leqslant x < 30$

5 a $20 \leqslant L < 30$ b $25 \leqslant x < 30$

Scatter graphs and time series

1 a Zero correlation   b Negative correlation
c Positive correlation

2 a, d

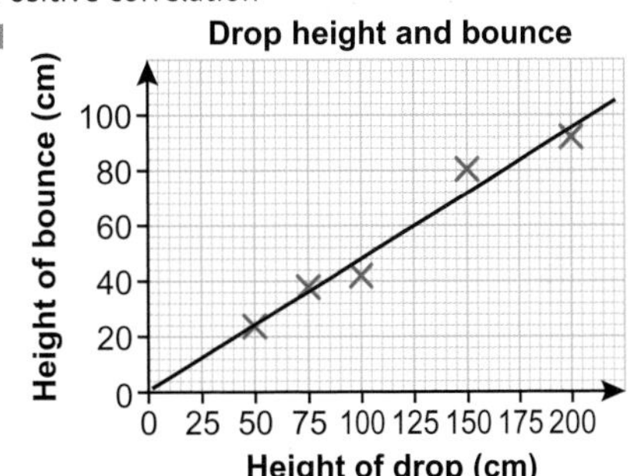

b As the height of the drop increases the height of the bounce increases.
c Positive correlation   e About 60 cm

3 a, d

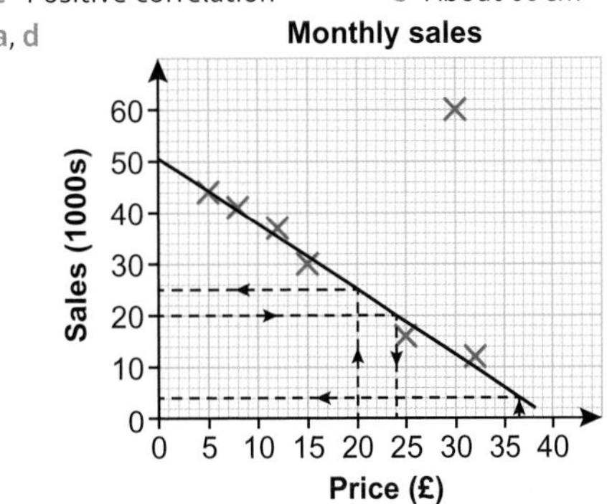

b (30, 60)
c Negative correlation; as price increases sales decrease.
e i About 25 000   ii About 4000
f **i** is more reliable because it is inside the range of data points whereas **ii** is outside
g About £24

4

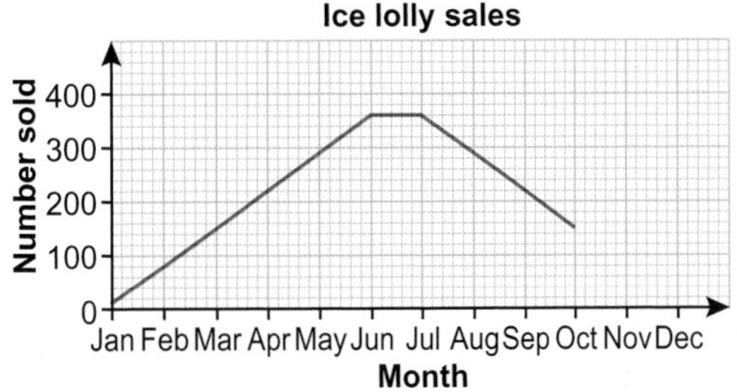

b Sales rise steadily until June when they remain constant for a month before decreasing at the same steady rate.
c November 80, December 10

3 Extend

1 a C and D; both points have the same y-coordinate.
  b B; D is a lot more expensive than B but only a little heavier.
  c A and B; they lie on a straight line from the origin so they are in direct proportion.
  d E; it is the most expensive but almost the smallest bag.
  e No correlation

2 a Negative b Positive c Positive

3 a 7 b 12

4 a 8 kg b 4
  c Yes, greenhouse mean yield is 6.5 kg and the mean yield outdoors is 3 kg.

5 a 28.46; the raw marks are unknown.
  b $\frac{50+1}{2}$ = 25.5 so the median is halfway between the 25th and 26th items. The first two groups contain 16 and the next one has 22 so the median is in the third group, 26–30.
  c 20
  d i Decrease ii Increase

6 a 36
  b There might be more students in Year 11.

7 a 40
  b

Women		Men
9	0	8
6	1	2 8
3 3	2	1 6 7
1	3	1 4 7 9
8 5	4	0 0 0 2 5
8 3	5	0 7 7 8
4 2	6	2 3 4
7	7	0
2 1	8	3

Key
Women: 6 | 1 represents 16 years
Men: 1 | 2 represents 12 years

  c Women's ages are uniformly spread out whereas men's ages are concentrated in the 30-59 age range.

8 a (15, 22) and (55, 15) correctly plotted
  b Negative correlation; as temperature increases, time decreases.
  c Answer in 18-20 range
  d Either 'Outside range of data points' or 'Line of best fit gives negative time'.

9 a Negative correlation

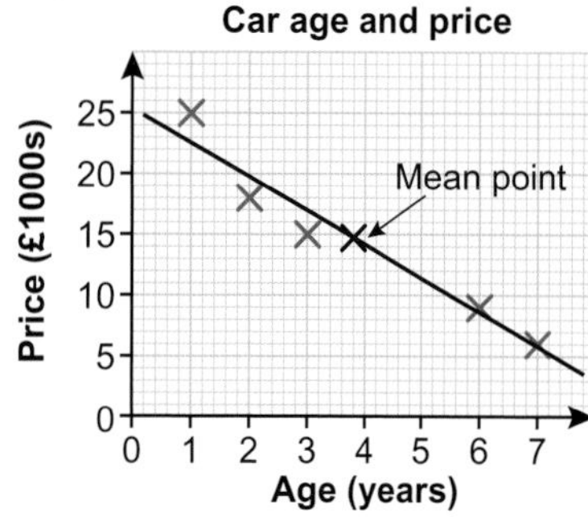

  b 3.8 c 14.6 e About £11 000
  f It is outside the range of existing data points; new cars depreciate significantly in the first year.

10 a

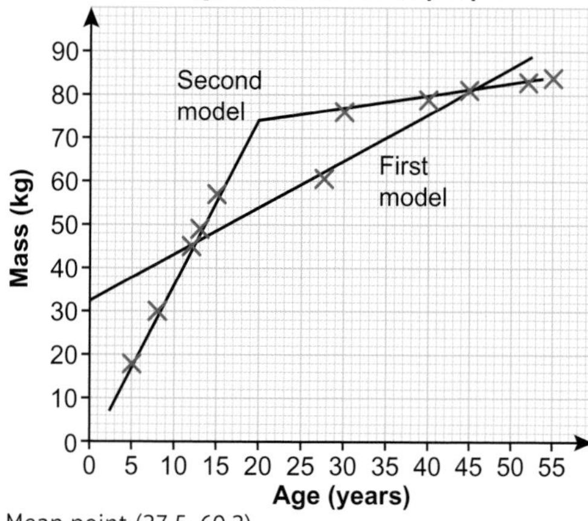

  b Mean point (27.5, 60.2)
  c About 56 kg d About 75 kg

11 a

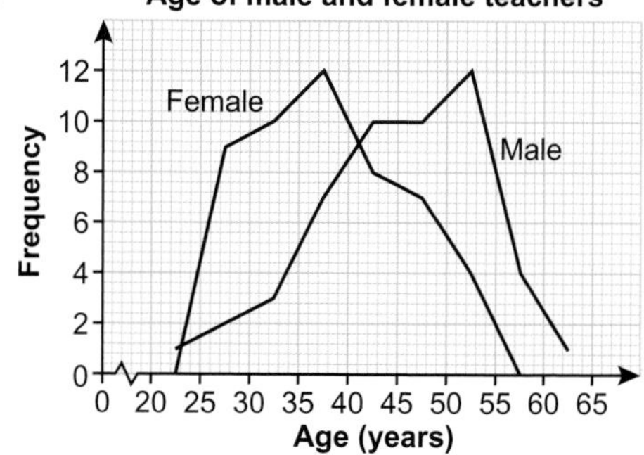

  b The mean age of male teachers is 45.2 and the mean age of female teachers is 38.1 showing that male teachers are, on average, 7 years older than the female teachers.
  c The male frequency polygon is to the right of the female frequency polygon.

12 a

	Under £30 000	At least £30 000	Total
Men	60	30	90
Women	60	50	110
Total	120	80	200

  b 55% c 66.7% (1 d.p.) d 62.5%

13

	Music	Drama	Sport	Total
Boys	35	21	39	95
Girls	30	28	32	90
Total	65	49	71	185

14 a i 50% ii 23% b 29.6% (1 d.p.)

15 Pie chart, because the other options cannot be used to display qualitative data.

16 a

  b $\frac{45+35+20}{3} = \frac{100}{3} = 33.3$(1 d.p.)

c $\frac{35+20+40}{3} = \frac{95}{3} = 31.7$(1 d.p.)

d

Months	1–3	2–4	3–5	4–6	5–7
Moving Average (1 d.p.)	33.3	31.7	**30**	**28.3**	**25**

e Decreasing

17 a 29, 35, **36**, **38**
b The trend is for an increase in prices.

18 $x = 15$

3 Unit test

Sample student answer

a Student A has got the answer correct.

b Student B gets more marks, because they have shown correct working and only got the final calculation wrong. Student A has shown no working so gets no method marks.

UNIT 4

4 Prior knowledge check

1 a $\frac{3}{10}$ b $\frac{7}{10}$

2 a 3 kg b £7.50 c 25 litres d 56 m

3 a $\frac{3}{20}$ b $\frac{3}{5}$

4 a £3 b 45 g

5 a $\frac{3}{4}$ b $\frac{9}{4}$

6 a $\frac{2}{15}$ b $\frac{21}{20}$ c $\frac{7}{10}$

7 a $\frac{5}{8}$ b $\frac{11}{30}$ c $\frac{7}{45}$

8 a $2\frac{2}{5}$ b $7\frac{7}{8}$

9 a 12 b $9\frac{1}{3}$

10 a 2 : 5 b 9 : 1

11 8 : 5

12 $\frac{1}{4}$

13 a 3 eggs b 75 g

14 a £15.60 b 36.4 kg c 153 ml d 40.8 m

15 a 60% b 81.7%
c 49 out of 60 is better as the score is a higher percentage.

16 £72

17 a 74% b 150%

18 a 0.65 = 65% b 0.429 = 42.9%
c 1.38 = 138% d 2.75 = 275%

19 a 1.17 b 1.38 c 1.06 d 2.1

20 £31.36

21 4.6 m

22 3.242 424

23 a $0.\dot{3}$ b $0.\dot{2}$

24 $\frac{3}{4}$ of £120 is £90. This is the largest amount.

4.1 Fractions

1 a $\frac{27}{8}$ b $2\frac{5}{6}$

2 a 10 b 9 c $3\frac{1}{3}$ d $4\frac{2}{5}$

3 a 54 b 40 c 27

4 $1\frac{17}{24}$

5 a $\frac{1}{8}$ b $\frac{1}{0.145}\left(=\frac{200}{29}\right)$ c $\frac{1}{4.8}\left(=\frac{5}{24}\right)$ d $\frac{3}{2}$

6 a 2 b $\frac{5}{2}$ c $\frac{4}{15}$ d $\frac{3}{17}$

7 a $\frac{5}{24}$ b $\frac{14}{15}$ c $4\frac{1}{8}$ d 6

8 a $\frac{1}{2}$ b $\frac{10}{3}$

9 a $2\frac{5}{14}$ b $1\frac{37}{80}$ c $7\frac{1}{2}$ d $1\frac{1}{7}$

10 Yes. Students' own answers, e.g. compare the answers of $3 \div \frac{1}{2}$ and 3×2.

11 $4\frac{7}{10} + 3\frac{1}{2} = 7\frac{7}{10} + \frac{1}{2} = 7\frac{7}{10} + \frac{5}{10} = 7\frac{12}{10} = 8\frac{2}{10} = 8\frac{1}{5}$

12 a $4\frac{1}{2}$ b $6\frac{1}{8}$ c $11\frac{19}{30}$ d $9\frac{7}{36}$

13 Yes, the part will fit because it is $7\frac{1}{9}$ cm, which is within the acceptable range.

14 a $4\frac{1}{2}$ b $3\frac{29}{40}$ c $-\frac{17}{18}$ d $-1\frac{1}{8}$

15 10 000 m^2

16 a 4 hours 5 minutes b 5 hours

4.2 Ratios

1 a 1 : 2 b 3 : 5 c 4 : 7 d 1 : 5
e 1 : 4 f 15 : 2

2 a 1 : 5 b 1 : 0.5 c 1 : 6 d $1 : \frac{7}{12}$

3 a 3 : 1 b $\frac{2}{3} : 1$ c 15 : 1 d $\frac{5}{6} : 1$

4 a 1 : 0.2 b 1 : 0.016 c $1 : \frac{3}{8}$ d 1 : 36.5

5 a 11.5 : 1 b The first school

6 Julie (Julie uses 5 parts of water to 1 part squash, Hammad uses 5.7 parts water)

7 £84

8 a 5 : 2 b 22.5 g of resin c 4.8 g of hardener

9 a 300 : 1 b 81 cm

10 350 students

11 a $\frac{3}{5}$ b $\frac{2}{5}$
c Sally gets £21 and David gets £14.

12 Benji gets 217 bricks and Freddie gets 248

13 8.16 m

14 a £68 : £136 : £170
b £5.84 : £17.51 : £23.35
c 26.1 m : 8.7 m : 52.2 m
d 129 kg : 451.5 kg : 193.5 kg

15 No, he needs 20 kg of cement.

16 a 40 : 73 b 142 : 241 c 3 : 7 d 15 : 1

17

Size	Blue	Green	Yellow
1 litre	0.6	0.375	0.025
2.5 litres	1.5	0.9375	0.0625
5.5 litres	3.3	2.0625	0.1375

4.3 Ratio and proportion

1 2 : 5 and 6 : 15

2 a $360 b £420

3 Cheaper in HK by £2 or HK$24.80

4 a 1 : 1.6 b Adrian by 3.4 km or 2.13 miles.

5 Yes; the ratio 4 : 5 is the same as 16 : 20

6 a $\frac{7}{12}$ b £36.75
c He is under 21. We do not know any more than this.

7 $s = b \times \frac{4}{3} = \frac{4}{3}b$
$b = s \times \frac{3}{4} = \frac{3}{4}s$

8 a $c = \frac{n}{250}$ b 11 chillies d $c = \frac{n}{125}$

9 a No b Yes c No

10 Yes

11 a Yes, because Q is 1.5 × the value of P
b $Q = 1.5 \times P$ c 2 : 3

12 $P = 48$, $Q = 56$, $R = 12.5$, $S = 45$

13 £5.44

14 2.4 m

15 The cheese is cheaper in Switzerland. 160 g would cost 3.58 SFr in England.

4.4 Percentages

1 a 1.15 b 1.3 c 1.05

2 a 0.75 b 0.90 c 0.94

3 £379.26
4 £550
5 £770
6 a £7680 b £6144
7 a i £30 ii £56
b £17 436.25
8 a £3300 b £3752 c £8427 d £19 167
9 a £128 b 4%
10 8.5%
11 23%
12 250%
13 Jo
14 £43.80
15 1.0506, £1.0506x
16 a 7.1% b £160 000
17 a Students' own answers, e.g. Let original value = n; 20% increase → $1.2n$; 20% of $1.2n = 0.24n$; $1.2n - 0.24n = 0.96n$, which is a decrease of 4%
b The final amount will be the same.

4.5 Fractions, decimals and percentages

1 a $m = \frac{1}{3}$ b $n = -4$ c $p = 32$
2

Fraction	Decimal	Percentage
$\frac{1}{8}$	0.125	12.5%
$\frac{9}{20}$	0.45	45%
$\frac{2}{3}$	$0.\dot{6}$	$66.\dot{6}$%
$\frac{4}{5}$	0.8	80%
$\frac{3}{2}$	1.5	150%

3 a 3.75 b 10 c 12.8 d 168
e 800 f £47.50
4 a £17.10 b 158.3%
5 24
6 30%
7 $\frac{2}{7}$
8 Students' own answers
9 £13.60
10 $0.\dot{6} = \frac{2}{3}$, students' own answer.
11 a Yes, because s = a constant × t.
b $t = \frac{s}{4}$ c 1 : 4
12 a $\frac{2}{3}$ b $\frac{1}{9}$ c $\frac{52}{99}$ d $\frac{2}{11}$
e $\frac{743}{999}$ f $\frac{29}{111}$
13 b and d are recurring

4 Problem-solving

1 a £430 b £375
2 Caroline pays £615 and Naomi pays £410.
3 6
4 No, because their weights are 75 kg and 90 kg.
5 £2000
6 Accountant 2 days, book-keeper 3 days, clerk 5 days.
7 27

4 Check up

1 a $4\frac{9}{40}$ b $\frac{7}{18}$
2 a $\frac{35}{12}$ b 4
3 a 7 : 40 b 7 : 32
4 a 1 : 7 b $1 : \frac{18}{5}$
5 a €57.15 b £385
6 a 200 : 75 = 8 : 3 b $f = \frac{8b}{3}$
c 80 biscuits
7 £66, £44, £22.
8 a £483.75 b 849.1 kg
9 20%
10 £720
11 a 9.3% b £15 300.43
13 2 and students' own answers.

4 Strengthen

Fractions

1 a $\frac{4}{5}$ b $\frac{4}{9}$ c $\frac{8}{3}$ d $\frac{15}{8}$
e $\frac{9}{4}$ f $\frac{7}{10}$
2 a $4\frac{1}{12}$ b $6\frac{1}{3}$ c $7\frac{1}{3}$ d $5\frac{1}{5}$
3 a $\frac{13}{12}$ b $6\frac{19}{24}$ c $4\frac{1}{45}$ d $10\frac{59}{70}$
4 a 5 b $2\frac{7}{8}$ c $1\frac{24}{35}$ d $3\frac{13}{24}$
e $\frac{9}{40}$
5 a $\frac{36}{5}$ b $\frac{56}{15}$ c 4 d $\frac{15}{2}$

Ratio and proportion

1 a D only b A and C
2 a i 1 : 5 ii 0.2 : 1
b i 1 : 0.25 ii 4 : 1
c i 1 : 8 ii 0.125 : 1
d i 1 : 0.57 ii 1.75 : 1
3 a 13 : 8 b 10 : 17 c 7 : 10 d 7 : 11
4 a i \$8 ii \$10 iii Double the number of £
b i £3 ii £5 iii Halve the number of \$
5 $X = 18$, $Y = 20$
6 $5.4 = \frac{27}{5}$
7 Kiran = £14.20, Stephen = £28.40, Jane = £49.70

Fractions, decimals and percentages

1 a 1.04 b 1.265 c 0.983
2 a 103.5% b 1.035 c £46.58
3 a 95.8% b 0.958 c £34.49
4 a 4.5% b Student's own answers.
5 a 2.25, 8, 28.125% b 25, 145, 17.2%
c 115, 615, 18.7%
6 a 112.5 g b £3.75 c £12
7 a 1.28 b x → ×1.28 → 13.44
c £10.50
8 a £780 b £185 000
9 a 5000 × 1.025 = £5125 b £5283.88

4 Extend

1 £48
2 $\frac{279}{560}$
3 a $\frac{1}{4}$ b $\frac{1}{8}$ c $\frac{1}{4}$ d $\frac{3}{8}$
4 50%
5 $\frac{17}{42}$
6 £1.92
7 43.5%
8 a 2 : 1 b 25%
9 43%
10 a 1 b 1 c $\frac{10}{81}$
11 Actual = 9227 bottles. Estimate = 9600 = 16 × 150 × 4
12 £164.00
13 £960.68
14 50%
15 51

4 Unit test

Sample student answer

The question has asked them to 'Compare' the costs of the handbags. They must state, using the costs found, which handbag is cheaper/more expensive. For example, 'The handbag is cheaper in Manchester as it costs £52.50, whereas in Paris it costs £54.'

UNIT 5

5 Prior knowledge check

1 a 25 b 5 c 8 d 10
2 a i 3.32 ii $\sqrt{11}$ b i 7.21 ii $2\sqrt{13}$
3 a Trapezium b Rectangle, square
c Rhombus, rectangle, square d Kite
e Kite, rhombus, parallelogram
4 a Equilateral triangle b Square
5 B and C
6 $a = 73°$ angles are alternate
$b = 73°$ angles are vertically opposite or corresponding
$c = 125°$ angles are co-interior, $d = 125°$ angles are vertically opposite
7 a

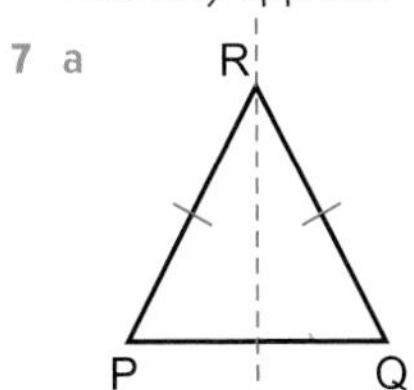

b Using symmetry ∠RPQ = ∠RQP
8 a 92° b 71° c 55°
9 a 130 b 112
10 $\frac{1}{3}$
11 a $y = \frac{x}{2}$ b $y = 6x$ c $y = \frac{x}{4}$
12 a $2x + 20 = 90$; $x = 35$
Angles are $x = 35°$ and $x + 20 = 55°$
b $4y = 180$; $y = 45°$
Angles are: 45°, 45° and 90°
c $6z - 30 = 180$; $z = 35$
Angles are: 70° and 110°
13 45°, 45° and 90° or 36°, 72° and 72°

5.1 Angle properties of triangles and quadrilaterals

1 60°
2 25°
3 75° (corresponding angles are equal and angles on a straight line sum to 180°)
4 a 75° b 68° c 68° d 7°
5 a Students' own drawings
b ∠BAD = 80°, ∠ADC = 100°, ∠DCB = 80°
c Opposite angles are equal
d Yes
e Opposite angles are equal
6 a 180° (angles on a straight line)
b i ∠CAB = x (alternate angles)
ii ∠ABC = z (alternate angles)
c $x + y + z = 180°$ (the angle sum of a triangle is 180°)
7 ∠AED = 38° (alternate angles are equal)
ADE = $\frac{180 - 38}{2} = 71°$ (the angle sum of a triangle is 180°)
∠EAD and ∠ADE are equal (base angles of a isosceles)
∠ADC = 180 − 71 = 109° (angles on a straight line sum to 180°)
8 $a + b + c = 180°$ and $e + d + f = 180°$
$a + b + c + d + e + f = 360°$
Therefore the angle sum of a quadrilateral = 360°
9 a $y = 103°$ b $y = 148°$ c $y = 111°$ d $y = z = 120°$
10 ∠BCE + ∠CBE = 132° (exterior angles of a triangle is equal to the sum of the two interior angles at the other two vertices)
∠BCE = ∠CBE (triangle BEC is isosceles)
∠BCE = ∠CBE = $\frac{132}{2}$ = 66°
∠CBA = 66° (alternate angles)
∠DAB = ∠CBA (trapezium is an isosceles trapezium)
∠DAB = 66°
11 ∠CBE = 110° (corresponding angles)
∠CBA = 70° (angles on a straight line sum to 180°)
∠ACB = 180 − (74 + 70) = 36° (angles in a triangle sum to 180°)
12 $a + 2a - 30 + a - 10 + 90 = 360$; $a = 77.5°$
So ∠CBD = 180 − (a − 10) = 112.5° (angles on a straight line)

5.2 Interior angles of a polygon

1 a 180 b 540 c 900 d 1080
2 a $x = 45°$ b $y = 81°$ c $z = 60°$
3 720°
4

Polygon	Number of sides (n)	Number of triangles formed	Sum of interior angles
Triangle	3	1	180°
Quadrilateral	4	**2**	**360°**
Pentagon	5	3	540°
Hexagon	6	**4**	**720°**
Heptagon	7	**5**	**900°**

5 a 3240° b 162°
6 a 108° b 135° c 128.6° d 156°
7 a $a = 130°$
b $x = 60°$, angles are: 60°, 180°, 60°, 120°, 120°
8 19
9 a 72° b 54° c 54°
10 Rhombus and hexagon
11 105°

5.3 Exterior angles of a polygon

1 a 900° b 540° c 1440°
2 a $a = 96°$; $b = 156°$; $c = 108°$
b $a = 140°$; $b = 140°$; $c = 80°$
c $a = 120°$; $b = 120°$; $c = 120°$
3 a $a = 95°$; $b = 85°$; $c = 85°$; $d = 95°$
b $a = 103°$; $b = 77°$; $c = 103°$; $d = 77°$
4 a Pentagon: $a = 60°$; $b = 60°$; $c = 90°$; $d = 100°$; $e = 50°$
Hexagon: $a = 50°$; $b = 70°$; $c = 70°$; $d = 70°$; $e = 60°$; $f = 40°$
b Sum of exterior angles is 360°
c Both sets of exterior angles sum to 360°
5 60°
6 $a = 80°$; $c = 66°$
7 138°, 70°, 167°, 113°, 125°, 169°, 127°, 171°
8 $x = 50°$ giving exterior angles of 70°, 50°, 90°, 50° and 100°
9 a 36 sides b 5 sides c 18 sides
10 a 6 sides b 12 sides c 9 sides
11 No; when you divide 360 by 70 you do not get a whole number.
12 8
13 6

5.4 Pythagoras' theorem 1

1 a 10 b 5 c 3 d 7
2 a 58.7 b 7.7
3 a 3.46 b 14.4
4 a Eleri's; she does not round the value before she finds the square root.
  b Yes
5 a 9.2 cm b 9.8 cm c 8.0 m
6 4.86 m
7 6.1 cm (1 d.p.)
8 37.7 feet (3 s.f.)
9 22.6 miles (3 s.f.)
10 30.4 m (3 s.f.)
11 33.2 cm
12 a No; $4^2 + 5^2 \neq 8^2$ b Yes; $9^2 + 12^2 = 15^2$
  c Yes; $5^2 + 12^2 = 13^2$

5.5 Pythagoras' theorem 2

1 8.1
2 a $a = 3$ b $b = 8$ c $c = 12$
3 11.5 cm
4 2.7 m (1 d.p.)
5 2.98 m
6 4.5 cm
7 a 8.49 cm (3 s.f.) b 12 cm
8 a $\sqrt{12} = 2\sqrt{3}$ cm b $\sqrt{27} = 3\sqrt{3}$ cm
  c $\sqrt{125} = 5\sqrt{5}$ cm
9 $\sqrt{3}$ cm
10 a 7.21 cm b 10.8 cm c 31.0 cm

5.6 Trigonometry 1

1 a 28° b 55° c 45°, 45°
2 a $x = 20$ b $x = 2$
  c $x = 7.35$ d $x = 1.05$ (3 s.f.)
3 a Accurate drawing of triangle ABC with sides correctly labelled.
  b Opposite (AC) = 2.9 cm; hypotenuse (BC) = 5.8 cm
  c i $\frac{2.9}{5.8} = 0.5$ ii $\frac{5}{5.8} = 0.9$ iii $\frac{2.9}{5} = 0.6$
  d i Opposite = 4.0 cm; hypotenuse = 8.1 cm
  $\frac{\text{opposite}}{\text{hypotenuse}} = 0.5$; $\frac{\text{adjacent}}{\text{hypotenuse}} = 0.9$; $\frac{\text{opposite}}{\text{adjacent}} = 0.6$
  ii Opposite = 4.6 cm; hypotenuse = 9.2 cm
  $\frac{\text{opposite}}{\text{hypotenuse}} = 0.5$; $\frac{\text{adjacent}}{\text{hypotenuse}} = 0.9$; $\frac{\text{opposite}}{\text{adjacent}} = 0.6$
4 a 0.6 b 1.0 c 7.1 d 0.3
  e 0.2 f 1.2
5 a $x = 4.17$ cm b $x = 9.66$ cm c $x = 1.88$ cm
6 Students' own answers
7 35.5 cm
8 a 8.5 cm b 9.1 cm c 5.6 cm
9 27.7 cm
10 2.6 m
11 6.7 m (1 d.p.)

5.7 Trigonometry 2

1 a 1.2 b 1.0 c 1.0
2 2.83 cm
3 a 34.2° b 36.4° c 13.8° d 53.1°
  e 38.2° f 36.5°
4 a 53.1° b 64.6° c 32.0°
5 48.2°
6 48.6°
7 26.6°
8 a 5.8 km b 31.0°
9 57.1°
10 11.6 cm²
11 31.8°
12 a 1 b $\sqrt{2}$ cm c i $\frac{1}{\sqrt{2}}$ ii $\frac{1}{\sqrt{2}}$
13 a i $\frac{1}{2}$ ii $\frac{1}{2}$ b $\sqrt{3}$
  c i $\frac{\sqrt{3}}{2}$ ii $\sqrt{3}$ iii $\frac{\sqrt{3}}{2}$ iv $\frac{1}{\sqrt{3}}$
14 a 45° b 30° c 2 d 45° e $\sqrt{3}$

5 Problem-solving

1 ABC = 120°; isosceles triangles
2 2.5, 10, 6
3 Length = 6 cm; width = 3 cm
4 90°, 51.32°, 38.68°
5 Angle E = 40°, Angle F = 160°, Angle G = 120°

5 Check up

1 a 144° b 72°
2 8
3 140°
4 ∠ABE = 90° (angles in a rectangle are all 90°)
  ∠BEA = 180 − (90 + 36) = 54° (angles in a triangle sum to 180°)
  ∠DEB = 180 − 54 = 126° (angles on a straight line sum to 180°)
5 $14x + 22x = 180$ (angles on a straight line sum to 180°); $x = 5°$
  ∠ABE = ∠BED = 70° (alternate angles are equal)
6 Any quadrilateral can be split into two triangles by joining one vertex to all the other vertices. The sum of angles in a triangle is 180°. Therefore the sum of the angles in a quadrilateral is 360°.

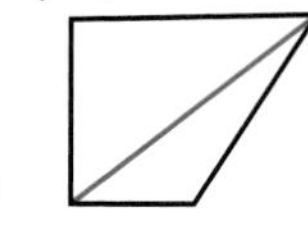

7 ∠DBC = 62° since it is corresponding to ∠EAB.
  $x = 180 - (2 \times 62) = 56°$ (since angles in a triangle sum to 180° and ∠CDB = ∠DBC as the triangle is isosceles)
8 a 6.3 cm b 2.2 m
9 No. $6^2 + 3^2 \neq 7^2$ and for a triangle to be right angled the square of the longest side must equal the sum of the square of the two shorter sides.
10 $3\sqrt{5}$ cm
11 a 8.63 cm b 8.43 m
12 21.8°
13 56.7°
14 a 1 b $\frac{1}{2}$ c $\frac{1}{2}$
16 4

5 Strengthen

Angles and polygons

1 a

Polygon	Quadrilateral	Pentagon	Hexagon	Heptagon
Number of sides (n)	4	**5**	**6**	**7**
Number of triangles	2	**3**	**4**	**5**
Sum of interior angles	2 × 180° = 360°	**3 × 180° = 540°**	**4 × 180° = 720°**	**5 × 180 = 900°**

  b Number of triangles = $n - 2$;
  Sum of interior angles = $(n - 2) \times 180°$
  d 1440°
2 a 140° b 150° c 162°

3 a 90° b 36° c 20°

4 a $n = \dfrac{360°}{\text{exterior angle}}$

b i 4 ii 6 iii 12 iv 30

5 a Interior angle DCB = 162°, exterior angles = x or DCA

b x = 18° c 20

6 a 720° b 165°

7 174°

8 ∠ACB = 56° (alternate angles are equal)

∠CBA = $\dfrac{180 - 56}{2}$ = 62° (angles in a triangle sum to 180°

and base angles on an isosceles are equal)

y = 180 − 62 = 118° (angles on a straight line sum to 180°)

9 a False. Angles on a straight line sum to 180°.

b False. Angles in a triangle sum to 180°.

c True. Angles in a triangle sum to 180° as do angles on a straight line.

d True. Angles on a straight line sum to 180°.

e True. Angles in a triangle sum to 180° as do angles on a straight line.

10 i

ii

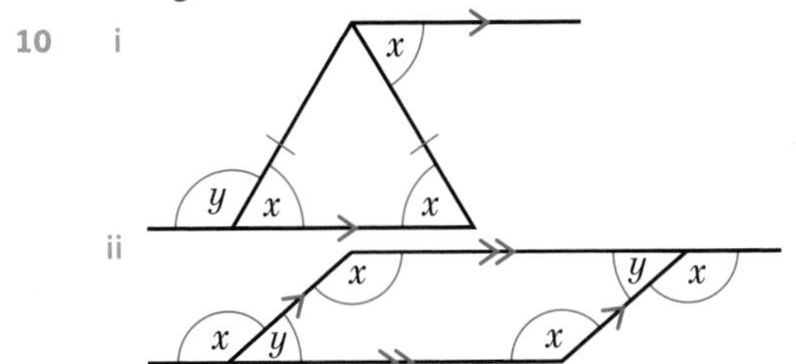

c Vertically opposite angles are equal

Pythagoras' theorem

1 a AB b PR c LN

2 $c^2 = 8^2 + 6^2$; $c^2 = 100$; $c = \sqrt{100}$; c = 10 cm

3 a 17 cm b 25.5 m

4 a $10^2 = 6^2 + b^2$; $100 = 36 + b^2$; $b^2 = 100 - 36$; $b = \sqrt{64}$; b = 8 cm

5 a 12 cm b 5.53 m

6 No. $18^2 \neq 7^2 + 16^2$

7 a $\sqrt{5}$ cm b $3\sqrt{3}$ c $4\sqrt{2}$

Trigonometry

1 a

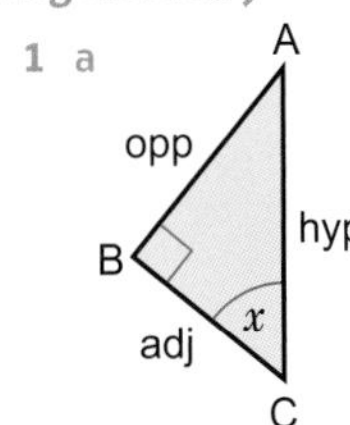

b c

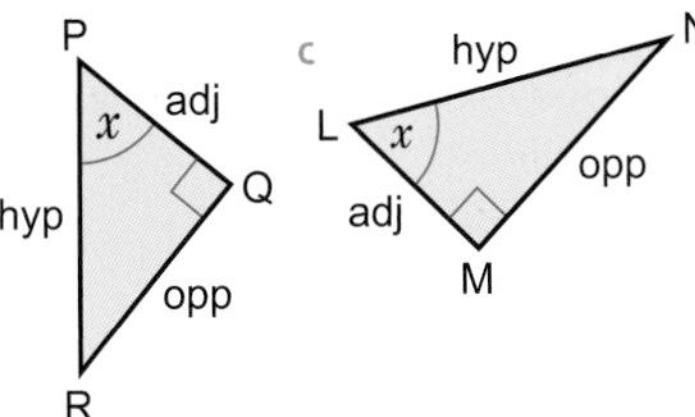

2 a $\sin x = \frac{8}{10}$ or $\frac{4}{5}$ b $\cos x = \frac{6}{10}$ or $\frac{3}{5}$

c $\tan x = \frac{8}{6}$ or $\frac{4}{3}$

3 a 0.4 b 0.7 c 0.3 d 1.6

4 a 19° b 53.7° c 40.3° d 41.8° e 19.7°

5 $x = 5 \times \tan 36°$; x = 3.6 cm

6 a 10.5 cm b 0.2 cm

7 $\cos = \frac{\text{adj}}{\text{hyp}}$; $\cos x = \frac{12}{15}$; $x = \cos^{-1}\left(\frac{12}{15}\right)$; x = 36.9°

8 68.0°

9 b is angle of elevation; a is angle of depression.

10 a

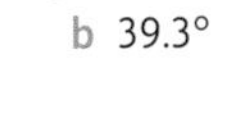

b 39.3°

11

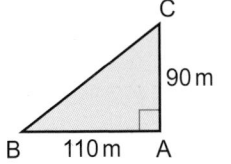

	30°	45°	60°	0°	90°
sin	$\frac{1}{2}$	$\sqrt{2}$	$\frac{\sqrt{3}}{2}$.	0	1
cos	$\frac{\sqrt{3}}{2}$.	$\frac{1}{\sqrt{2}}$	$\frac{1}{2}$	1	0
tan	$\sqrt{3}$	1	$\sqrt{3}$	0	

5 Extend

1 30°

2 Angles in a pentagon sum to 540°

Each angle in a regular pentagon = $\frac{540}{5}$ = 108°

∠BCD = 108°

∠FCD = 180 − 108 = 72° (angles on a straight line sum to 180°)

∠FCD = ∠CDF (triangle CDF is isosceles)

∠CFD = 180° − 2 × 72 = 36° (angles in a triangle sum to 180°)

3 $2x - 20 + x + 5 = 2x + 35$ so x = 50° (the exterior angle of a triangle is equal to the sum of the two interior angles at the other vertices)

∠QSR = $2x + 35$ = 135°

∠QSP = 180 − 135 = 45° (angles on a straight line)

4 12.0 cm

5 122.02 m

6 36°, 72°, 108°, 144°, 180°

7 6 sides

8 32.5°

9 48 700 feet

10 a i AB = 1.50 cm ii AC = 2.60 cm b 7.1 cm

11 $(4\sqrt{3} + 12)$ cm

12 7.1 cm

13 a 3.4 cm b 22.3°

5 Unit test

Sample student answers

Student C gives the best answer. Student A rounded too early and Student B's algebra was incorrect.

UNIT 6

6 Prior knowledge check

1 a 21 b −22

2 a $\frac{1}{7}$ b 4 c $-\frac{1}{3}$ d $-\frac{5}{2}$

3 5 km/h

4 a 9 b 7 c $\frac{1}{5}$ d 25 e 125

5 a $\frac{4}{3}$ b 3

6 A: $y = 3$, B: $x = -3$, C: $y = x$, D: $y = -x$

7 a

x	−3	−2	−1	0	1	2	3
y	**−5**	**−3**	**−1**	**1**	**3**	**5**	**7**

b Students' graph of $2x + 1$

8 a i 3 ii −1 b i $\frac{1}{2}$ ii 2 c i −1 ii 1

9 a i £55 ii £140

b i £50 ii 45 minutes iii £20 per hour

10 (−1, 3) and (3, 11)

6.1 Linear graphs

1 $y = 2x - 5$

2 a Any line with gradient = 5

b Any line with gradient = 0.5

c Any line with gradient = −3

3

Equation of line	Gradient	y-intercept
$y = 2x + 4$	**2**	**4**
$y = 2x$	**2**	**0**
$y = 2x - 3$	**2**	**−3**

4 a A: $y = 2x - 2$, B: $y = 3x + 1$, C: $y = 2x + 1$ D: $y = -x$
b D c B d B and C e A and C

5 A: $y = 2x$, B: $y = 3x - 4$, C: $y = \frac{1}{2}x + 2$, D: $y = -x - 1$, E: $y = 2x + 2$, F: $y = x - 1$

6 a ii and iv b i and iv

7 a i

x	0	−3
y	$1\frac{1}{2}$	0

ii

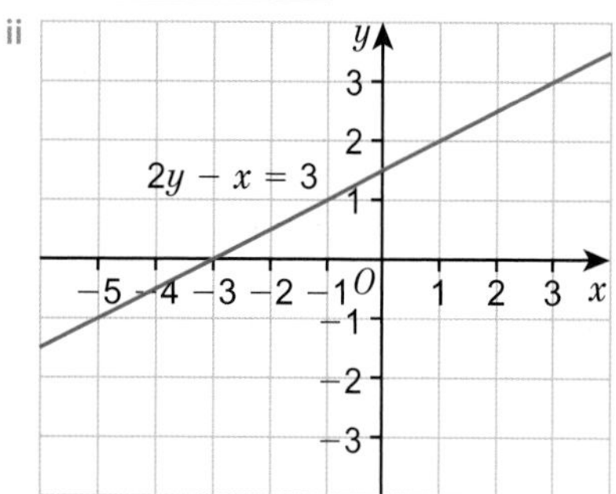

b i

x	0	4
y	4	0

ii

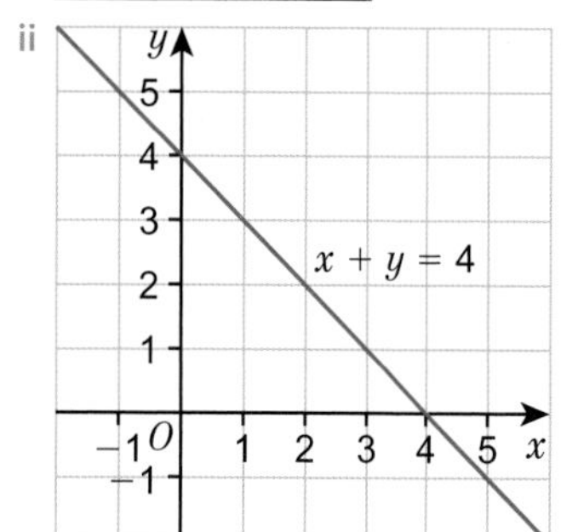

c i

x	0	7
y	7	0

ii

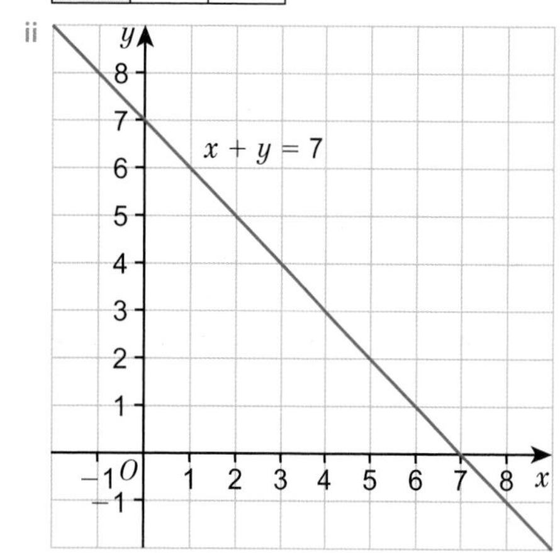

8 a $y = \frac{3 + x}{2}$; $y = 4 - x$; $y = 7 - x$
b Gradient $\frac{1}{2}$, y-intercept $\frac{3}{2}$; gradient −1, y-intercept 4; gradient −1, y-intercept 7
c Students' own check

9 Line c is steepest.

10 B and E

6.2 More linear graphs

1 A: $y = 2x + 2$, B: $-\frac{1}{3}x - 1$, C: $y = -2x + 1$

2 $c = 3$

3 a

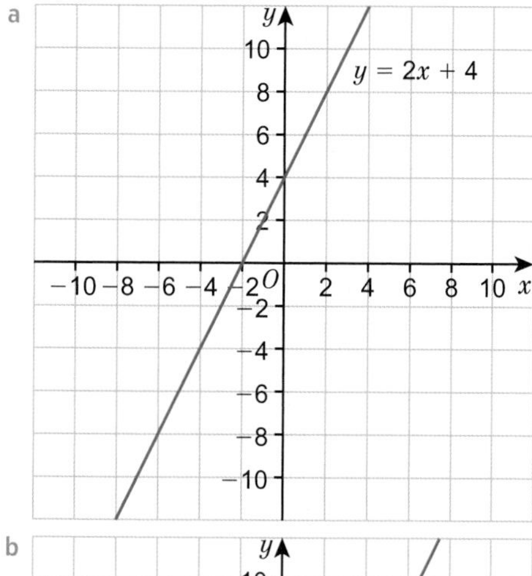

b

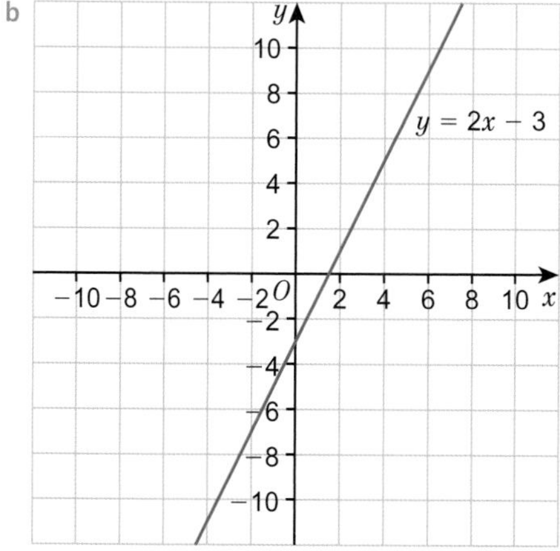

c

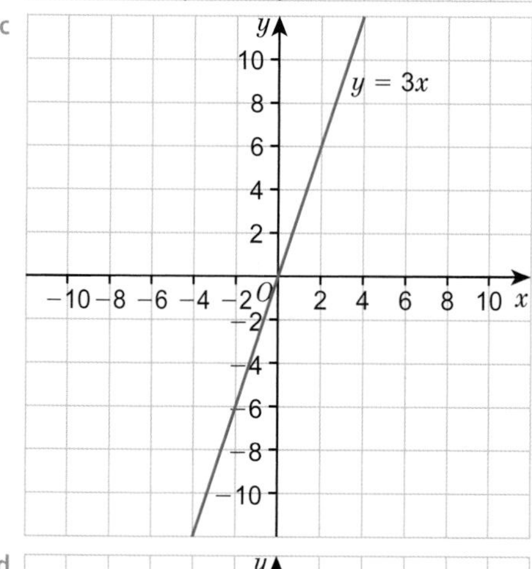

d

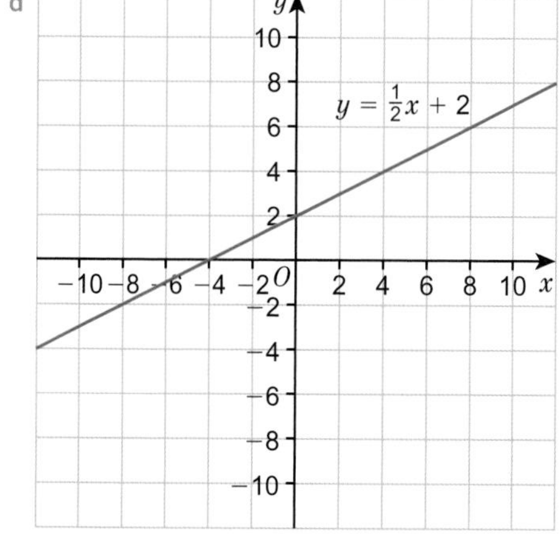

e
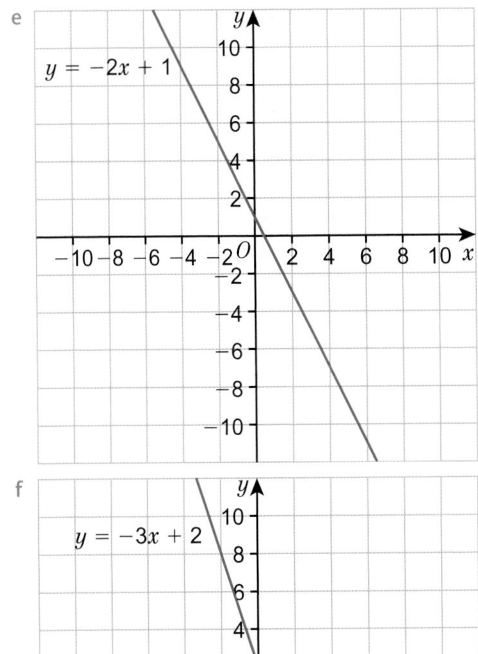

f
$y = -3x + 2$

4 A: $y = \frac{1}{2}x + 3$, B: $y = -x + 4$, C: $y = 2x + 3$, D: $y = 5x + 1$, E: $y = -3x$

5 a Sketch graph of $y = 2x$
b Sketch graph of $y = 3x + 1$
c Sketch graph of $x + y = 5$

6 a i x-intercept = 3, y-intercept = 3
ii x-intercept = −2, y-intercept = −6
iii x-intercept = −2 , y-intercept = 2
iv x-intercept = −2, y-intercept = 4
b Students' sketches of graphs

7 a, b, c, d

8 Students' own answers

9 a No b Yes c No

10 $y = 2x - 3$

11 a $y = 3x + 5$ b $y = -x + 3$
c $y = \frac{1}{2}x - 2$ d $y = -2x + 6$

12 a Students' own answers b $\frac{3}{4}$

13 a $-\frac{1}{2}$ b $y = -\frac{1}{2}x + c$; $c = 5$
c $y = -\frac{1}{2}x + 5$

14 a $4x - 3 = -x + 12$ b $x = 3$
c $y = 9$ d (3, 9)

15 (1, 1)

6.3 Graphing rates of change

1 a $10\,\text{mm}^2$ b $10.5\,\text{cm}^2$

2 a 6 km
b 5.45 pm
c 30 minutes
d 2 hrs 30 minutes
e 12 km/h
f 12

3 a
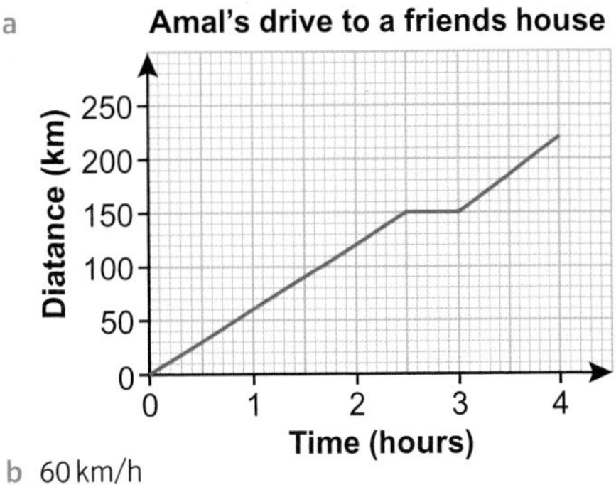

b 60 km/h

4 a
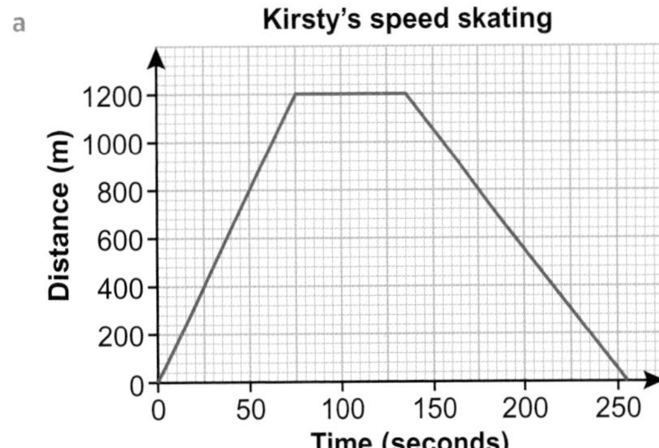

b 16 m/s

5 a 30 minutes b 22 km
c
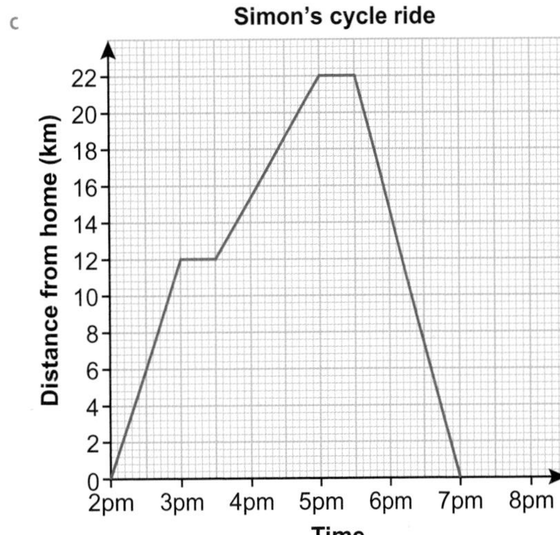

6 a
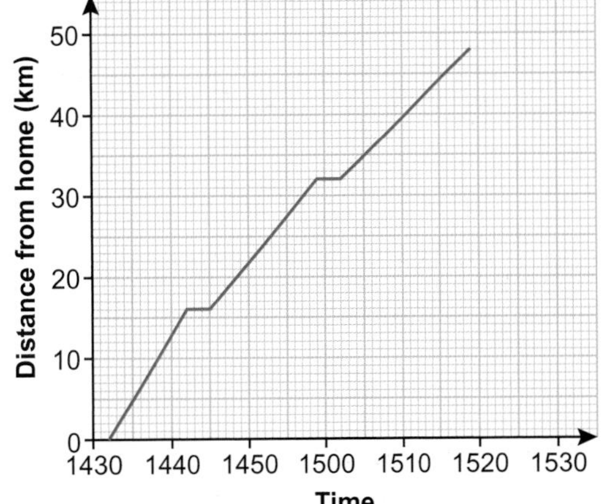

b 96 mph c 48 mph d 61.3 mph

7 55 km/h

8 a 85 miles approximately
  b 100 mph, 109 mph
  c Between 3:00 and 3:40; the line is steeper
  d Train A

9 a C b iii
  c Aii, Bi, Ciii

10 Students' own sketches, showing B is steepest and A is the least steep.

11 a 1.6 m/s b 5 minutes
  c 0.0092 m/s² d 96 m
  e He accelerated at 0.0092 m/s² for the first 2 minutes, then ran at a constant velocity of 1.1 m/s for 5 minutes. Next, he accelerated at 0.0083 m/s² for 1 minute, then ran at a constant velocity of 1.6 m/s for 8 minutes. Then he decelerated at 0.013 m/s² for the last 2 minutes.

6.4 Real-life graphs

1 a **Charge for Electricity**

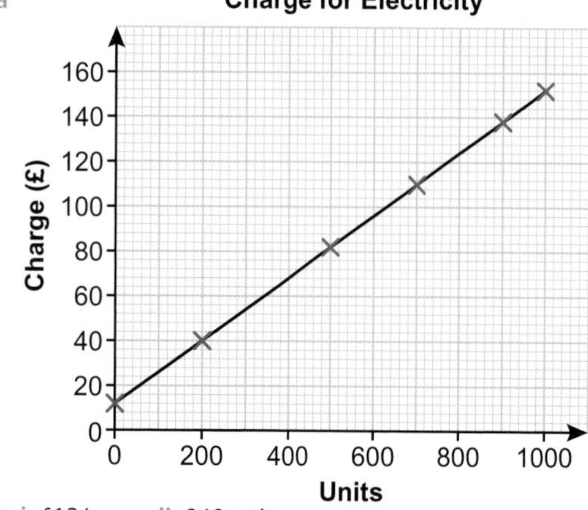

  b i £124 ii 340 units

2 a £1.00 b £54 c 149 pens

3 a C$14 b €0.7 c 1.4

4 a £24 per day b £40
  c $y = 24x + 40$ d 17 days

5 B and C

6 Question 3

7 a

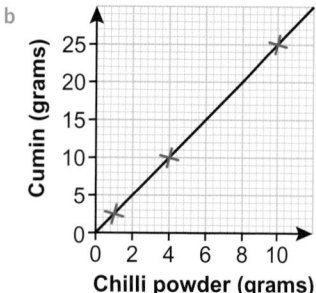

Chilli powder (grams)	1	4	10
Cumin (grams)	2.5	10	25

  b

  Cumin (grams): 0, 5, 10, 15, 20, 25
  Chilli powder (grams): 0, 2, 4, 6, 8, 10

  c $y = 2.5x$ d 34 g

8 a The y-intercept tells you the initial temperature of the freezer.
  The x-intercept tells you at what time the temperature reaches 0 °C.
  b 10 °C
  c 20 °C
  d Yes, because it is a straight line.

9 a **Sugar dissolving into coffee**

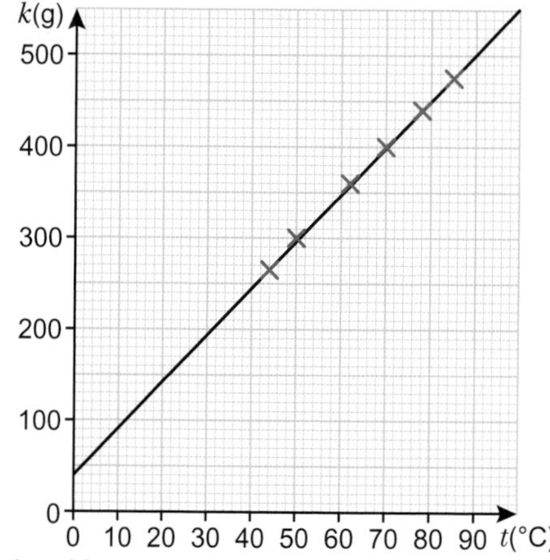

  b i 16 °C ii 455 g
  c $a = 5$, $b = 50$
  d Yes. 4 teaspoons is much less than 500 g.

10 a i £18 ii £14.50
  b This is the minimum monthly spend – you have to pay this even if you don't use the phone that month.
  c The point of intersection is where the two plans charge the same amount for the same number of minutes.
  d Molly – Plan C; Theo – Plan B

11 a,c **Age and price of car**

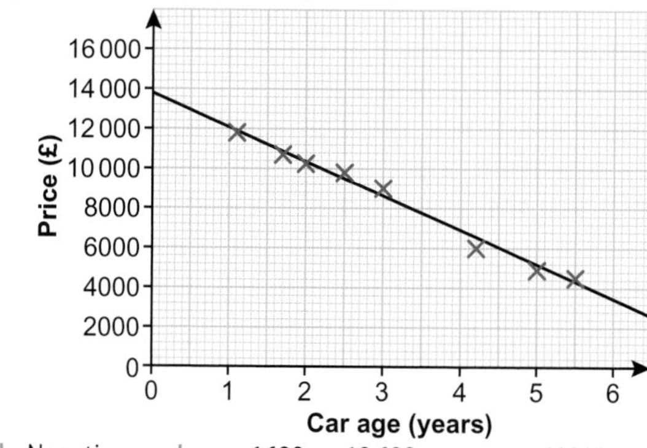

  b Negative d $y = -1600x + 13\,600$ e £8000

12 a i 84.6 ii 91.4
  b 84.6 in 2020, because 2020 is closer to the years covered by the data.

6.5 Line segments

1 a 7 b 1 c 5.5 d −3.5

2 13 cm

3 Gradient: 2, y-intercept: −3

4 a (2, 5) b (5.5, 5.5) c (1, 7) d (−2, −0.5)

5 a (1.5, 5.5) b (4, −1) c (2, 0.5) d (−1, −2)

6 a 3 b −2 c $-\frac{1}{2}$ d $\frac{1}{3}$

7 a 5 b 10 c 13.6

8 a $y = 2x + c$ b $y = x - 9$

9 $y = 15x + 180$

10 $y = \frac{1}{3}x - 5$

11 $y = 3x + 4$

12 a A_1: 2, A_2: $-\frac{1}{2}$; B_1: 3, B_2: $-\frac{1}{3}$; C_1: $\frac{1}{4}$, C_2: −4
  b They all equal −1

13 a $-\frac{1}{3}$ b 4 c $-\frac{5}{2}$

14 a $y = -2x + 5$ b $y = x - 4$

6.6 Quadratic graphs

1 Line A is $y = -2$ Line B is $x = 5$
  Line C is $x = -3$ Line D is $y = 0$

2

x	–4	–3	–2	–1	0	1	2	3	4
y	**16**	**9**	**4**	**1**	**0**	**1**	**4**	**9**	**16**

3

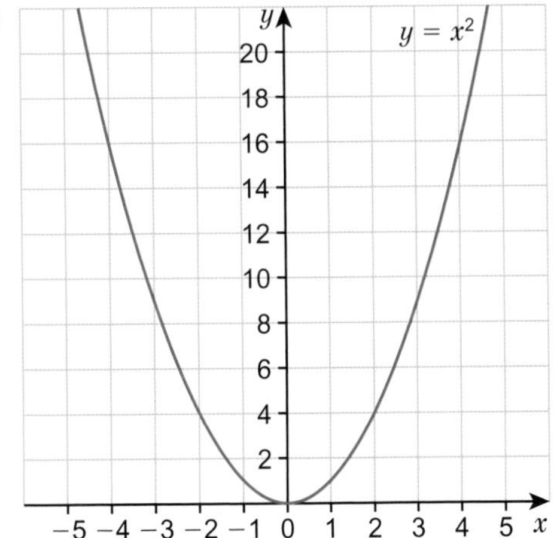

4 a

x	–3	–2	–1	0	1	2	3
x^2	**9**	**4**	**1**	**0**	**1**	**4**	**9**
–3	–3	–3	–3	–3	–3	–3	–3
y	**6**	**1**	**–2**	**–3**	**–2**	**1**	**6**

b

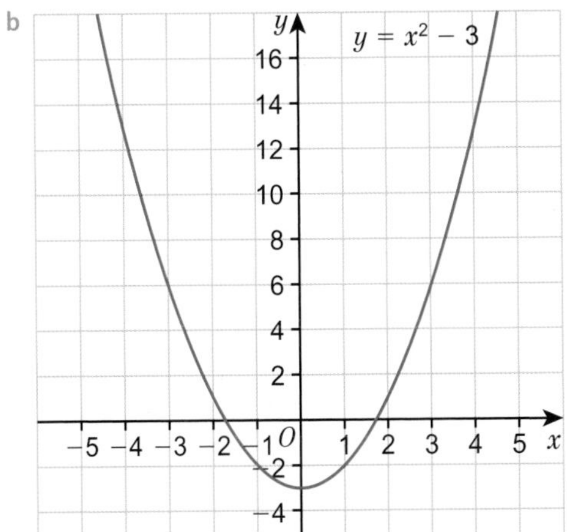

5

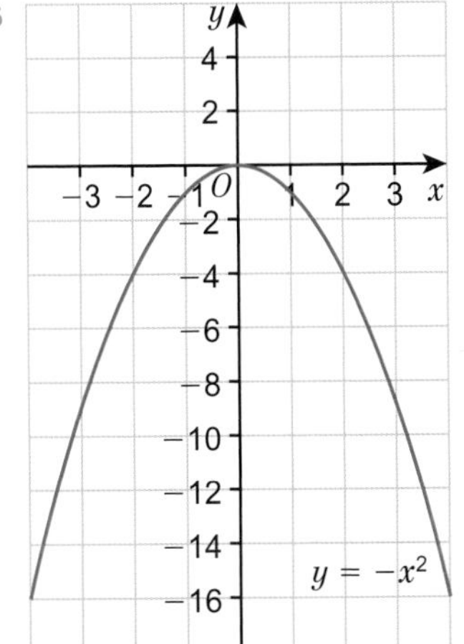

6 a

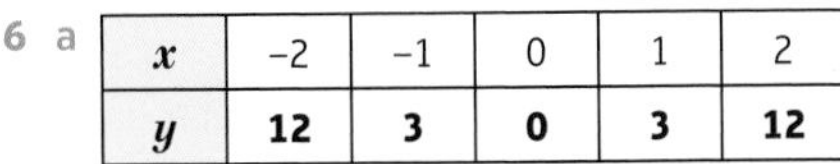

x	–2	–1	0	1	2
y	**12**	**3**	**0**	**3**	**12**

b

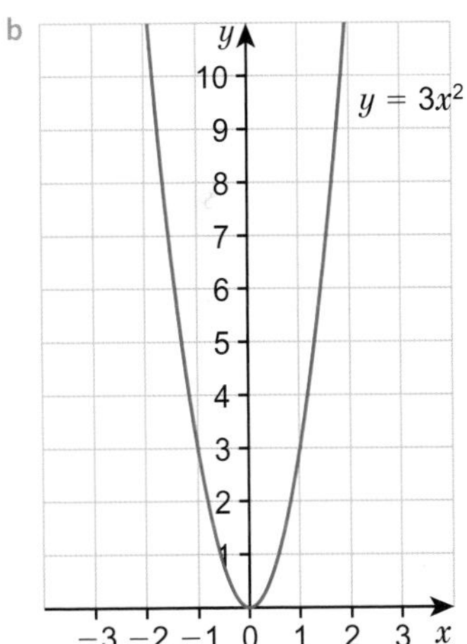

7 a Same parabola shape
b **Q3**, **Q4** and **Q6** have minimum, **Q5** has maximum
c **Q3**: (0,0), **Q4**: (0, –3), **Q5**: (0,0), **Q6**: (0, 0)
d $x = 0$ for all four graphs

8 a Quadratic b Between 3.6 and 5.2 seconds
c 2.6 seconds d 32 m e 5.2 seconds

9 $x = 1$ or -1

10 a $x = 0$ or 2 b $x = 1$
c $x = -3$ or 0.5 d $x = -3$ or 2

11 a $x = -0.7$ or 2.7 b $x = -1$ or 3
c $x = -2.6$ or 0.6 d $x = -3.3$ or 0.3
e The curve does not meet or cross the x-axis (or $y = 0$)

12 a

x	–2	–1	0	1	2	3	4
y	10	1	–4	–5	–2	5	16

b

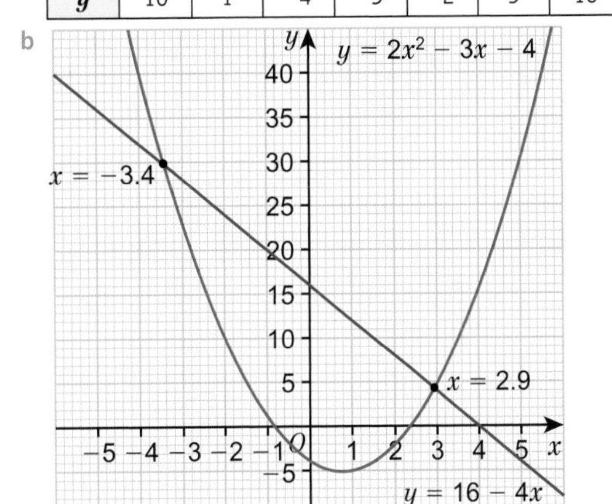

c $x = -3.4$ or 2.9

13 a

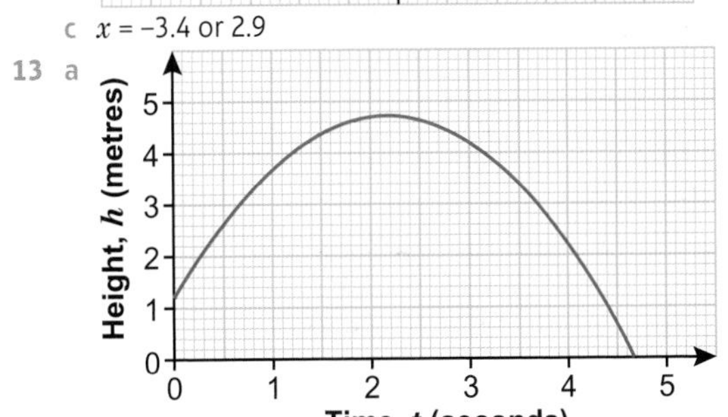

b Around 4.65 seconds

6.7 Cubic and reciprocal graphs

1

x	–3	–2	–1	0	1	2	3
y	**–27**	**–8**	**–1**	**0**	**1**	**8**	**27**

2 a 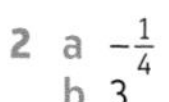 $-\frac{1}{4}$
b 3

3

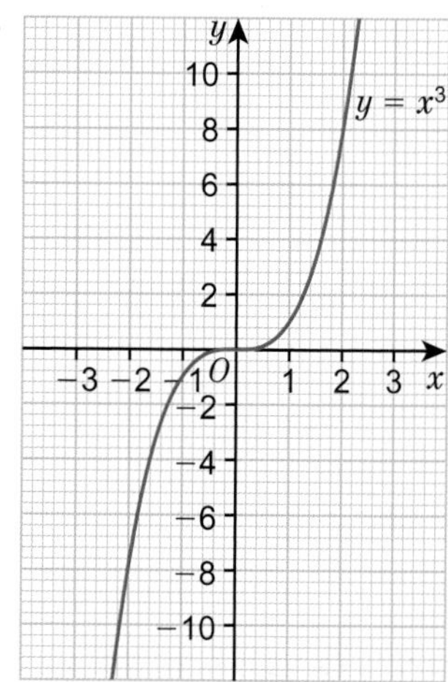

4 a Students' own estimate
b Students' own estimate
c 4.91
d −2.22

5

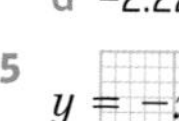
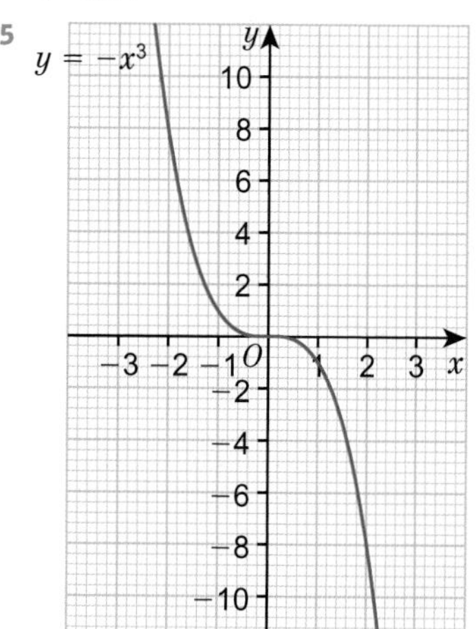

6 a

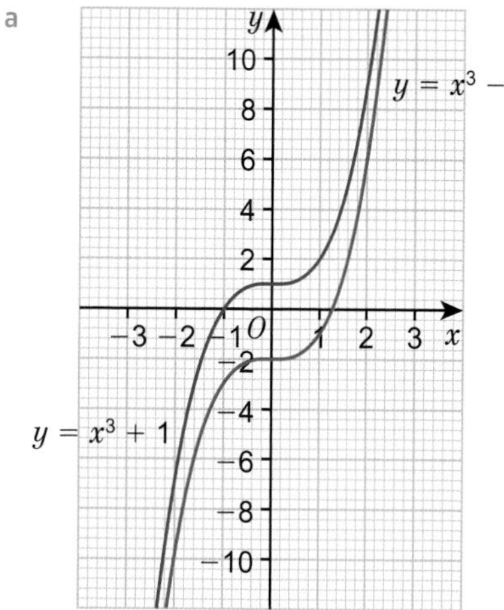

b Students' own answers

7 a

x	−3	−2	−1	$-\frac{1}{2}$	$-\frac{1}{4}$	$\frac{1}{4}$	$\frac{1}{2}$	1	2	3
y	$\frac{1}{3}$	$\frac{1}{2}$	1	2	4	−4	−2	−1	$-\frac{1}{2}$	$-\frac{1}{3}$

b

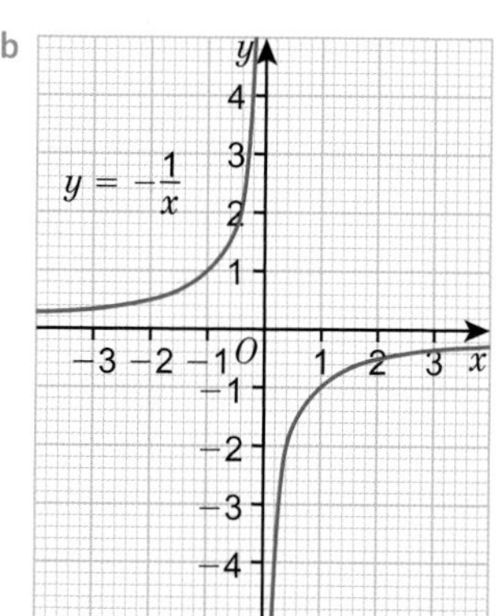

c Students' own answers, e.g. The two graphs have the same shape but in different quadrants.

8 a

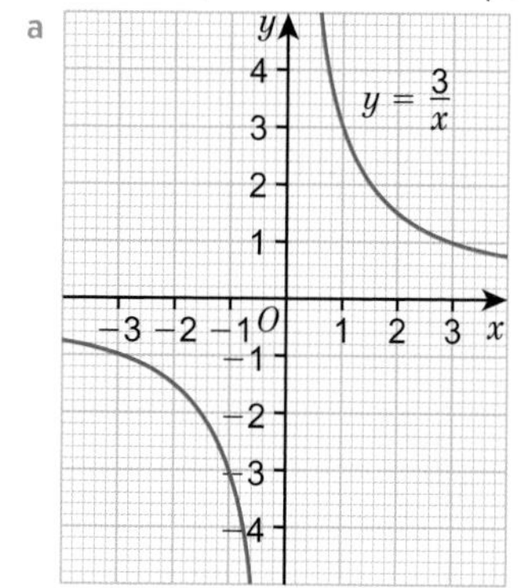

b i 1 ii −3 iii −1.2

9 a E b F c B d D
e C f A

11 a $x = 1.3$ b $x = -1$ c $x = -1$

12 a

x	−3	−2	−1	0	1	2	3
y	**−12**	**2**	4	0	**−4**	**−2**	12

b

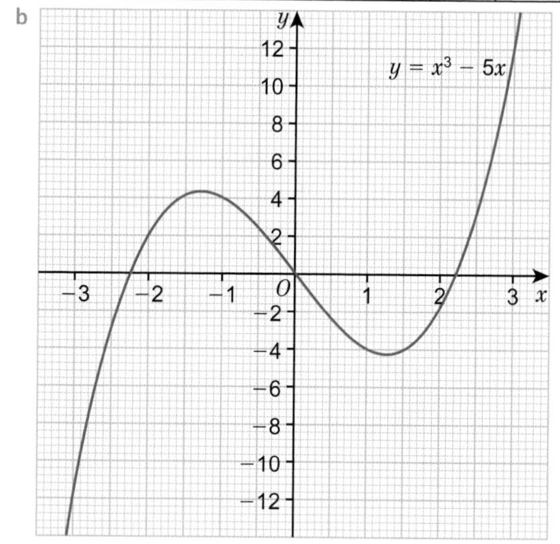

c $x = -2, -0.4$ or 2.4

13 a $x = -1.3$ or 0 or 1.7 b $x = -1.6$ or 1

6.8 More graphs

1 a 3.5 b 5.7

2 Students' circle with radius 5 cm

3 a 120 miles b $3\frac{1}{4}$ hours c 70 mph d 21 30
e 37 mph f The bus's speed increasing

4 2000 m as this is when she starts to descend at a constant speed.

5 a Data set A

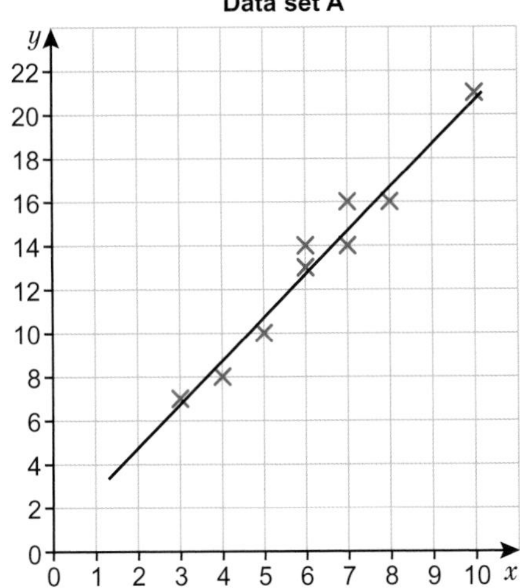

Data set B

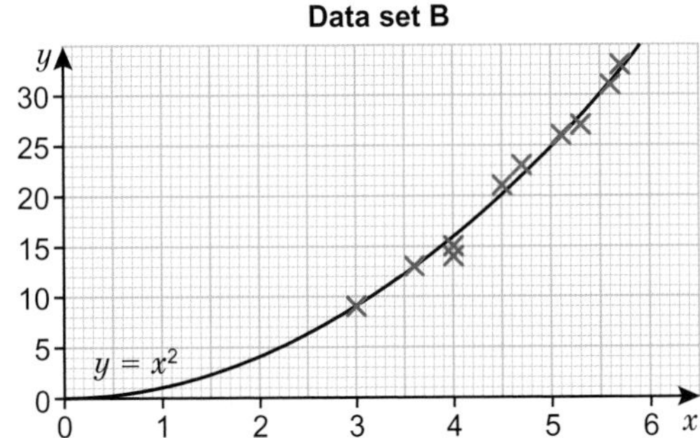

b Both graphs show strong positive correlation.

c x is proportional to y

d

x	3.0	3.2	3.5	3.7	4.0
y	**9.00**	**10.24**	**12.25**	**13.69**	**16.00**

x	4.5	4.8	5.0	5.5	6.0
y	**20.25**	**23.04**	**25.00**	**30.25**	**36.00**

e $y = x^2$ is very close to the line of best fit, so y is proportional to x^2.

6 a Petrol consumption of a car

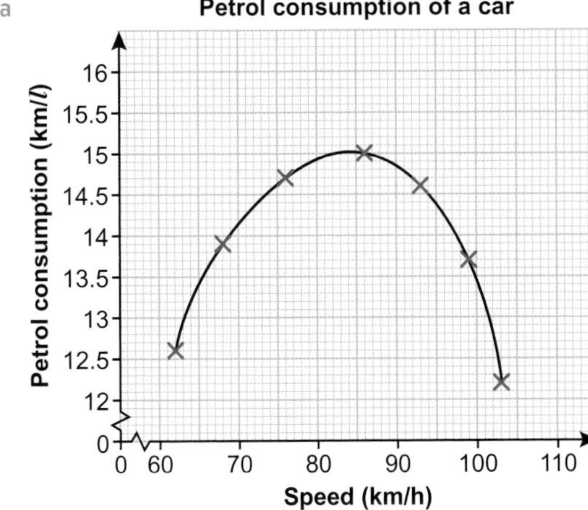

b 14.6 km/l c 65 km/h and 100 km/h

7 a 4 rats b Found the y-intercept
c i 14 rats ii 32 rats
d The number of rats increased (at a faster rate)

8 a 17 counts per second b 15 minutes
c 9 minutes d No

9 a £1000 b £1126 c £24
d 2.4% over the 5 years

10 a

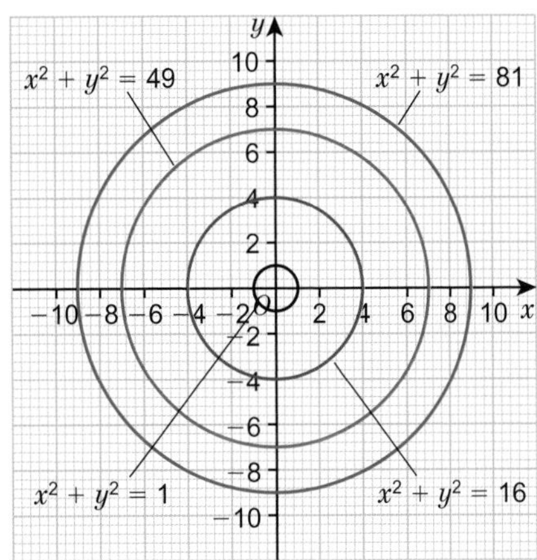

6 Problem-solving

1 Tom makes (T – 25) profit on each pair of trainers. His total profit will be this expression multiplied by the number of trainers he sells, Q.

2 When Tom sells his trainers at £40 then $Q = 10(40 - 40) = 0$. When he sells them at £39 then $Q = 10(40 - 39) = 10$ and this will increase by 10 for every £1 he lowers.

3 $P = -10T^2 + 650T - 10000$

4 One appropriate method would be to draw a graph.

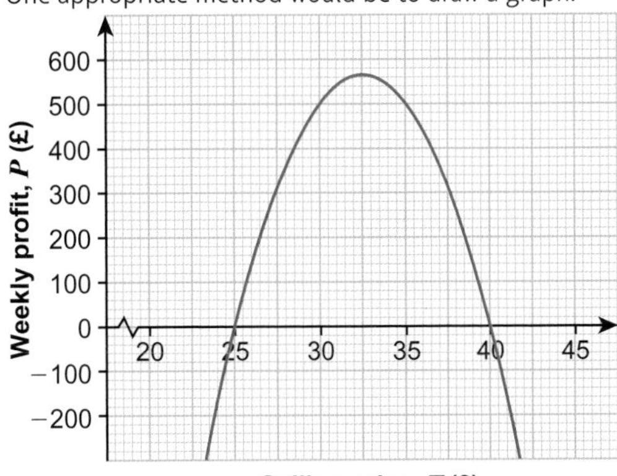

This shows the optimum result is at £32.50, giving a profit of £562.50.

5 The new formula would be $P = (400 - 10T)(T - 26) = -10T^2 + 660T - 10400$. The new optimum result is at £33, giving a profit of £490.

6 Check up

1 Gradient: $-\frac{2}{3}$, y-intercept: $\frac{7}{3}$

2 A: $2x + 3$, B: $3x + 2$, C: $-2x + 3$, D: $\frac{1}{2}x + 5$

3

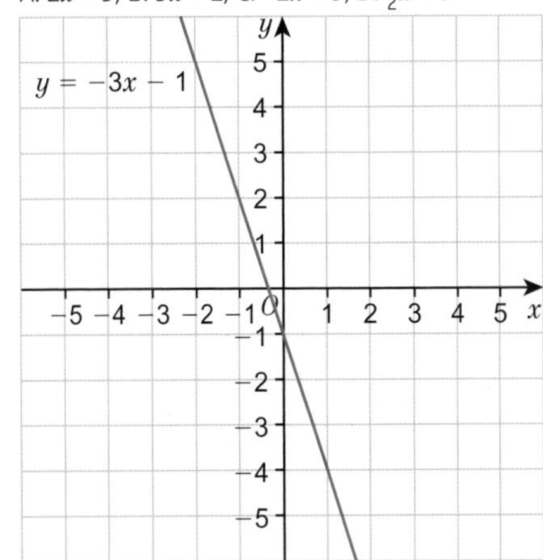

4 a $-\frac{1}{2}$ b 3
5 a 12, Hamzah is cycling at 12 mph b 14.7 mph
c 2.75 to 3 hours (or final 15 minutes) d 40 mph
6 a Price per cupcake b Minimum price c No
7 a (−0.5, −3) b 6.4
8 a $y = 3x - 13$
b Any equation with $y = -\frac{1}{3}x$ plus a constant, e.g. $y = -\frac{1}{3}x + 8$
9 a A b E c B d G e C f F g D
10 $x = -1.5, -0.4, 1.9$
11 a (0.75, 2) b 2 m c 2.1 seconds
d The height that the water comes out of the hosepipe.
12 a D b A c C d B
14 Students' own answers

6 Strengthen

Linear graphs

1 a i, iv
b ii $y = 2x + 3$ iii $y = -x + 4$ v $y = -5x + \frac{1}{2}$
2 a B and C b C and D c A and B
d i D ii B iii C iv A
3 a B b C c A
4 a e.g. b e.g.

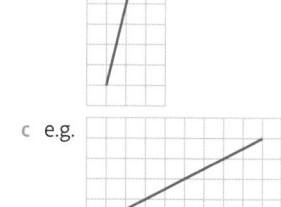

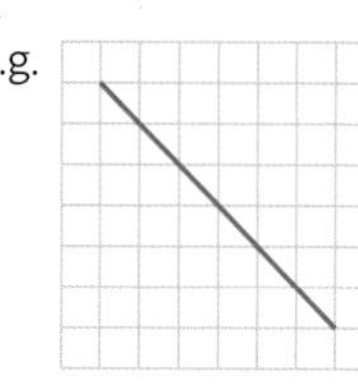

c e.g.

5 a,b,c

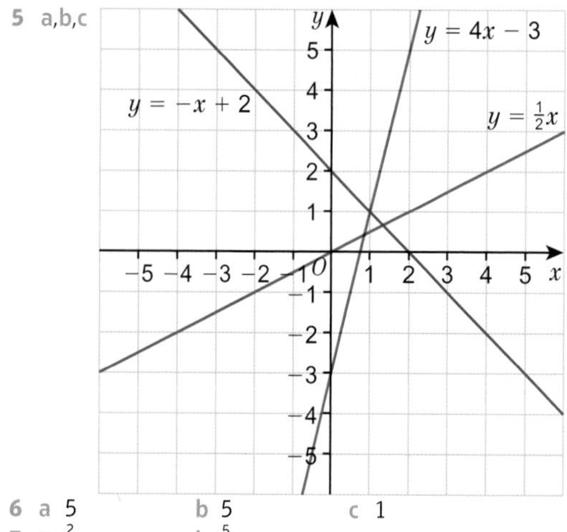

6 a 5 b 5 c 1
7 a $\frac{2}{3}$ b $\frac{5}{4}$
8 $\sqrt{13}, \sqrt{41}$
9 (−0.5, 4), (−2, 1.5)
10 a A and F, B and D
b $y = 2x$ plus a constant e.g. $y = 2x - 8$, $y = -x$ plus a constant e.g $y = -x + 5$
11 a $-\frac{1}{4}$ b −3 c $\frac{1}{10}$ d $\frac{5}{3}$
12 A and B, C and F, D and E
13 a L and P
b No, only points L and P satisfy the equation.
14 a 3 b $y = 3x + 10$

Non-linear graphs

1 (0, 0), (−5, 23), (−2, 7), (3, 10)
2 a (3, −26) b (−2, −7), (3, −26) c (−2, −9), (3, 26)
3 a A: $y = x^2 + 1$, F: $y = -x^2 + 4$ b B: $y = x^3 - 2$, E: $y = -x^3$
c C: $y = \frac{1}{x}$ d D: $y = x - 2$

Real-life graphs

1 a Quadratic b (0, 5)
c 4.6 m d 3.5 m and −3.5 m
e Distance in front of the goal posts
f 5 m g Yes
2 a The water level in Vase 1 rises faster first, then slower. The water in Vase 2 rises slower first, then faster.
b Vase 1 – Graph B; Vase 2 – Graph A
3 a i £1.25 ii 260 iii £8.75 iv 95 (or 96)
b They are in direct proportion

6 Extend

1 a C and D b A and C
2 a (3, 3.5) b $-\frac{4}{5}$
3 a (11, 13) b $\frac{4}{3}$
4 Number of seeds = −17 × seed size + 49
5 a $2t = 5a$
b

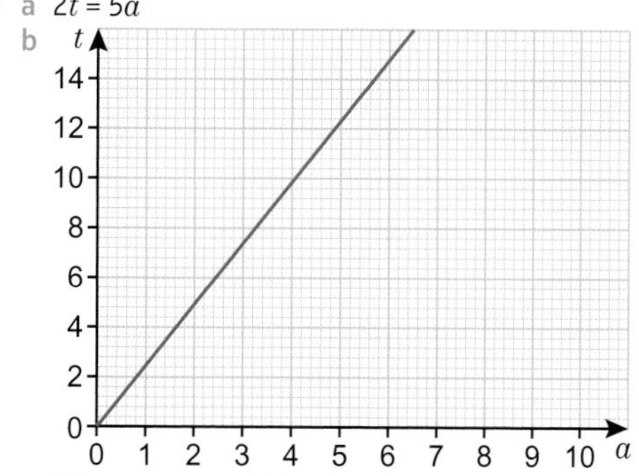

c 2.5 e 240 aubergines
6 a C b A c B
7 a $x^2 - 3x - 2x - 1 + 1 = x^2 - 5x = 0$
b $y = 2x - 1$ c $x = 0, 5$
8 a

x	−3	−2	−1	−0.5	−0.1	0.1	0.5	1	2	3
y	1.7	**2**	3	5	**21**	−19	**−3**	**−1**	**0**	0.3

b

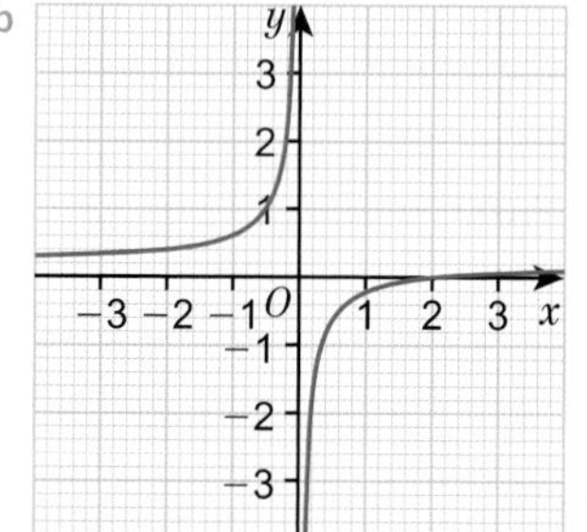

c $x = 0$ and $y = 1$
9 a graph plotted with points at (0, 0), (1, 0.6), (2, 2.4), (2, 2.4) (3, 5.4), (4, 9.6), (5, 15), (6, 21.6), (7, 29.4), (8, 38.4), (9, 48.6), (10, 60)
b 12.2 m c 8.2 seconds
10 a $3x + 5y = c$ where $c \neq -10$ (or $y = -\frac{3}{5}x + c$ where $c \neq -2$)
b $2y - 4x = 2$ (or $y = 2x + 1$)
c $y = \frac{3}{4}x + 2.5$ (or $4y = 3x + 10$)
11 a $y = -0.5x - 2$ b (−2,−1) c 5.8
12 $k = 7$; D = (4, 7); point (4, 7) satisfies the equation $y = 2x - 1$
13 PD = 7.5
14 a i C ii D iii A iv B
b A: $x = 0$, B: no line symmetry, C: $x = 0$, D: $y = x$
15 a A b B c C d D

6 Unit test

Sample student answers

Students' own answers

UNIT 7

7 Prior knowledge check

1 400

2 a 3.6 b 2.1 c 5.0

3 a 9.40 b 13.98

4 Teaspoon 5 ml, drink can 330 ml, bucket 5 litres, juice carton 1 litre

5 a 520 cm b 240 mm c 1000 mm d 3410 m
e 327 ml f 2.4 litres

6 a 18 b 52 c 48 d 2

7 a $x = \frac{y}{c}$ b $x = \frac{a}{bz}$ c $x = \frac{2m}{y}$

8 Suitable circle with correct centre, radius and diameter labelled.

9 a

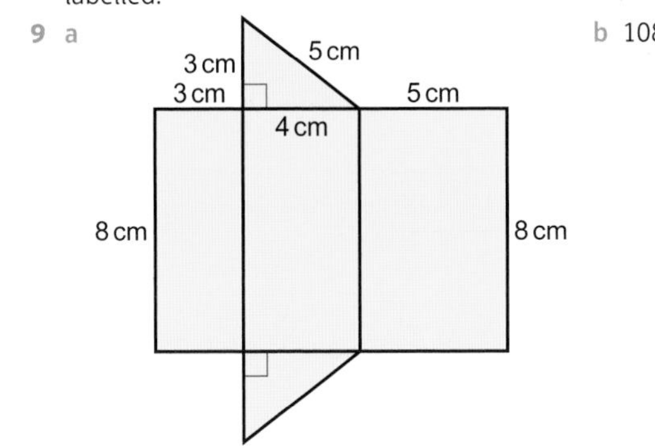

b 108 cm²

10 a

2 cm
6 cm
4 cm

b

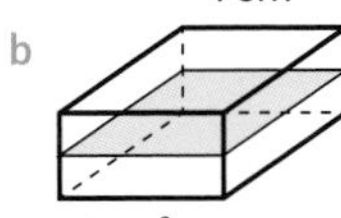
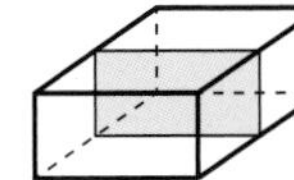
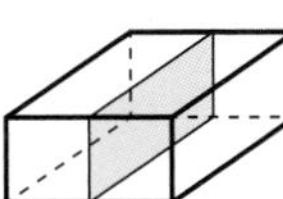

c 48 cm³

11 40 cm²

12 370 cm³

13 Students' own answers. The boxes will fit exactly into a cuboid box with dimensions 9 cm × 12 cm × 12 cm.

7.1 Perimeter and area

1 a 12 cm², 16 cm b 30 cm², 30 cm c 96 cm², 44 cm

2 a $x = 4$ b $x = 12$ c $x = 5$

3 a Area = 46.5 m², perimeter = 29 m b 1548 mm²
c 1406 mm²

4 a 60 cm² b 30 cm² c 26.6 m² d 373.5 cm²

5 Area = 276 cm², perimeter = 72 cm

6 £34.93

7 4350 m²

8 a $96 = \frac{1}{2}(9 + 15)h$ b $96 = 12h$ c $h = 8$ cm

9 4 cm

10 a $a = 6$ cm b $b = 2.8$ m (1 d.p.)

11 100 cm²

7.2 Units and accuracy

1 a 3.6 b 320 c 8.50 d 15.7

2 a i 2.5 kg ii 22.5 kg iii 27.5 kg
b i 2 m ii 38 m iii 42 m

3 a 1 cm = 10 mm, so they have same side length
b 1 cm² and 100 mm²
c 1 cm² = 100 mm²

4 a Suitable sketch of the squares
b 1 m² = 10 000 cm² c Divide by 10 000

5 a 2.5 cm² b 52 000 cm² c 0.7 m² d 340 mm²
e 88 500 cm² f 12.46 cm² g 370 000 mm² h 2.8 m²

6 a 0.6 m² b 544 mm² c 468 mm²
d 21 600 cm²

7 24.2 ha

8 1 600 000

9 a 33 mm, 27 mm b 27 mm ⩽ length ⩽ 33 mm

10 19 g ⩽ mass ⩽ 21 g

11 a i 35.5 cm ii 111.5 cm b i 2.45 cm ii 6.65 kg

12 a 17.5 m ⩽ x < 18.5 m b 24.45 kg ⩽ x < 24.55 kg
c 1.35 m ⩽ x < 1.45 m d 5.255 km ⩽ x < 5.265 km

13 a i 7.5 cm ii 8.5 cm b i 5.25 kg ii 5.35 kg
c i 11.35 m ii 11.45 m d i 2.245 litres ii 2.255 litres
e i 4500 m ii 5500 m f i 31.5 mm ii 32.5 mm
g i 1.525 kg ii 1.535 kg

14 a 14.5 cm, 15.5 cm, 27.5 cm, 28.5 cm
b Lower bound 84 cm, upper bound 88 cm

15 Upper bound 80.798 m², lower bound 79.008 m²

16 a Height 6.15 cm, 6.25 cm; area 23.5 cm², 24.5 cm²
b i 3.92 ii 3.98 (2 d.p.)
c 3.98 cm (2 d.p.)

7.3 Prisms

1 72 cm³

2 a $b = 8$ b $h = 4$

3 a Students' own sketch
b, c Areas are 20 cm², 28 cm² and 35 cm². The identical pairs are (top,bottom) (front,back) and (left side,right side).
d Surface area is 166 cm²

4 72 cm²

5 a Yes, because it has the same cross section all along its length.
b 12 cm²
c 72 cm³, same value as for volume calculated in **Q1**.

6 a 80 cm³ b 204 cm³ c 81 cm³

7 a 4 cm b 108 cm²

8 4 cm

9 a Suitable sketch of cube b 1 cm³ = 1000 mm³
c Divide by 1000

10 a 1 m³ and 1 000 000 cm³ b Multiply by 1 000 000

11 a 4 500 000 cm³ b 52 000 mm³
c 9.5 m³ d 3.421 cm³
e 5.2 litres f 700 cm³
g 175 cm³ h 3000 litres

12 a 9.44 m² b 3

13 9.2 cm

14 a 0.05 m³
b Estimated volume of leaf mould in wood is
20 000 × 0.2 = 4000 m³
$\frac{4000}{0.05}$ = 80 000, 12 × 80 000 = 960 000 worms

15 Volume = $\frac{1}{2} \times 4x \times 2x \times 5x = 20x^3$

16 Upper bound: 5.5 × 3.5 × 8.5 = 163.63 cm³,
Lower bound: 4.5 × 2.5 × 7.5 = 84.38 cm³

7.4 Circles

1 a $r = 5$ $r = \pm 5$

2 a $x = \frac{y}{m}$ b $x = \pm t$ c $x = \pm\sqrt{p}$

3 a All ratios are 3.14 to 2 d.p. b 3.14159265

4 a 28.3 cm b 14.8 m c 75.4 mm d 21.4 cm

5 a 50.3 cm² b 4.5 m² c 38.5 m² d 21.2 cm²

7 5 boxes

8 a 38.5 mm b 1164 mm²

9 Circumference = 201 cm, $\frac{1000}{2.01} = 497$

10 a 10π cm, 25π cm^2 b 14π cm, 49π cm^2
c 20π cm, 100π cm^2 d 24π cm, 144π cm^2

11 a i area 36π cm, circumference 12π cm
ii 110 cm^2 (2 s.f.), 38 cm
b The answers in terms of π because they have not been rounded

12 a $104 = \pi d$ b $d = 33.1$ cm

13 3.8 cm (1 d.p.)

14 a 12.87 m b 28.3 cm

15 a $\frac{A}{\pi} = r^2$ $\sqrt{\frac{A}{\pi}} = r$
b X: 3.6 cm Y: 2.8 cm Z: 4.7 cm

16 8 × 66 circles = 528
Total area of circles = 528 × 9π = 14 929 cm^2 (to nearest cm^2)
Area thrown away = 20 000 − 14 929 = 5071 cm^2
Percentage thrown away = $\frac{5071}{20000}$ = 0.25 or 25%

7.5 Sectors of circles

1 a 16π cm, 64π cm^2 b 50.3 cm, 201 cm^2

2 a 4π b $2\pi + 10.2$ c $3\pi + 7$

3 a i 18π cm^2 ii 56.5 cm^2 b i 25π cm^2 ii 78.5 cm^2

4 a i $(3\pi + 6)$ cm ii 15.4 cm b i $(5\pi + 10)$ cm ii 25.7 cm

5 a $(16\pi + 16)$ cm b 66.3 cm

6 a 4.4 m^2 b 8.8 m

7 21.5 cm^2

8 a 3.090 193 616 cm b 3.1 cm

9 58.9 cm^2

10 Arc length = 7.85 cm, perimeter = 37.9 cm

11 a 24.4 cm, 85.5 cm^2 b 73.7 cm^2 $\leqslant$ area $<$ 98.2 cm^2

12 a $10 = \frac{x}{360} \times \pi \times 3^2$ b 127° (to the nearest degree)

13 74°

14 8.56 m

15 5.0 cm (1 d.p.)

16 $(16\pi - 32)$ cm^2

7.6 Cylinders and spheres

1 Students' sketches

2 a ±6 b 4.3 c ±1.6 d 2.2

3 a πr^2 b $V = \pi r^2 h$

4 a 197.9 cm^3 b 167283.5 mm^3 c 2.6 m^3

5 0.267 m^3

6 3.2 cm

7 a 188.5 cm^2 b 16 889.2 mm^2 c 41.3 m^2

8 26 mm

9 300 cm^3

10 a SA = 324π mm^2, V = 972π mm^3
b SA = 100π cm^2, V = $\frac{500\pi}{3}$ cm^3

11 191 mm^3 $\leqslant$ volume $\leqslant$ 348 mm^3

12 a Total volume = $\frac{1088}{3}\pi$ = 1140 mm^3 (3 s.f.)
b Total SA = 208π = 653 mm^2

13 17 mm

14 6.31 m

15 3.1 cm

7.7 Spheres and composite solids

1 a 20 cm^2 b 21 cm^2

2 9.4 cm

3 a Net of square-based pyramid, square 4 cm side, height of each triangle 6 cm
b Triangular face 12 cm^2, square 16 cm^2 c 64 cm^2

4 213 cm^3

5 a x = 8.7 cm (1 d.p.) b Total volume = 950 cm^3

6 a 96π cm^3 b 302 cm^3

7 a 25π cm^2 b 65π cm^2 c 90π cm^2

8 $l = \sqrt{97}$ = 9.85 cm (2 d.p.), area = 123.8 cm^2

9 Volume = 63 363 mm^3 (to nearest mm^3)
Surface area = 9694 mm^2 (to nearest mm^2)

10 10.6 cm

11 Radius of sphere = 5.2322… cm, height of cone = 20.9 cm

12 Volume of whole cone = 144π, volume of smaller cone = $\frac{128}{3}\pi$
Volume of frustum = $\frac{304}{3}\pi$

13 a $10\pi x^3$ b $20\pi x^2$
c $10\pi x^3 + 20\pi x^2 = 10\pi x^2(x + 2)$

14 51π = 160 cm^3 (3 s.f.)

7 Problem-solving

1 6.6 cm

2 8.6 cm

3 72.2%

4 £935.03

5 864 cm^3

7 Check up

1 a 32 cm^2 b 25 cm

2 7 cm

3 a 50.3 cm b 64π = 201.1 cm^2

4 25.7 cm

5 a 15.7 cm^2 b 5.2 cm

6 a 40 000 cm^2 b 0.56 m^2
c 9.5 m^3 d 3000 ml = 3000 cm^3

7 9.5 cm^3 $\leqslant$ volume $\leqslant$ 10.5 cm^3

8 a 35.5 m $\leqslant$ 36 m $<$ 36.5 m
b 9.15 cm $\leqslant$ 9.2 cm $<$ 9.25 cm
c 23.55 km $\leqslant$ 23.6 km $<$ 23.65 km

9 36 cm^3

10 492.9 cm^2

11 36π cm^3

12 301.6 cm^3

14 Students' own answers

7 Strengthen

2D shapes

1 a 2 cm b 10.3 cm

2 a $55 = \frac{1}{2}(7 + b) \times 10$ b $55 = 35 + 5b$ c $b = 4$ cm

3 a i 4π cm ii 12.6 cm b i 12π cm ii 37.7 cm

4 a $2\pi r$ = 19.5 cm, πr^2 = 34 cm^2 b $A = \pi r^2$ c $C = 2\pi r$

5 a i 4π cm^2 ii 12.6 cm^2 b i 36π cm^2 ii 113.1 cm^2

6 a 153.9 cm^2 b 77.0 cm^2 c 44.0 cm d 22.0 cm
e 14.0 cm f 36.0 cm

7 a $\frac{1}{4}$ b $\frac{1}{8}$ c $\frac{100}{360} = \frac{5}{18}$

8 a $\frac{1}{8}$ b 201.1 cm^2 c 25.1 cm^2 d 50.3 cm
e 6.3 cm f 16 cm g 22.3 cm

Accuracy and measures

1 a 10 000 cm^2, 20 000 cm^2
b
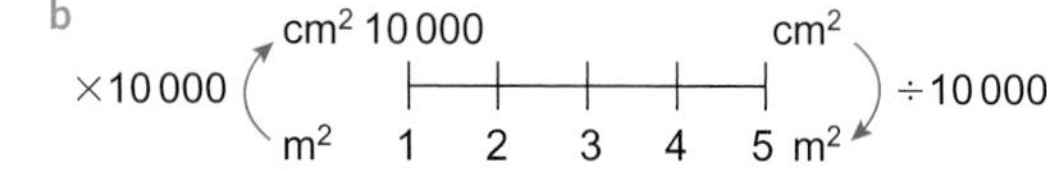

2 a 1 000 000 cm^3, 2 000 000 cm^3
b

3 a 2.5 b 27.5 and 22.5
c 22.5 $\leqslant$ 25 $\leqslant$ 27.5

4 a 22.5 $\leqslant l <$ 23.5 cm b 31.5 $\leqslant l <$ 32.5 mm

3D solids

1 a $60\,cm^3$ b $63\,cm^3$ c $240\pi = 754\,cm^3$

2 a Students' sketches
  b i $113.1\,cm^2$ ii $37.7\,cm$ iii $301.6\,cm^2$ iv $527.8\,cm^2$

3 a Volume = $340\,cm^3 = \frac{4}{3}\pi r^3$, surface area = $746\,cm^2 = 4\pi r^2$
  b Surface area = $4\pi r^2$ c Volume = $\frac{4}{3}\pi r^3$
  d i $452.39\,cm^2$ ii $904.78\,cm^3$

4 a 4 cm b 5 cm c $37.7\,cm^3$ d $75.4\,cm^2$

7 Extend

1 $82.5\,cm^2$

2 a $50xy$ b $500xyz$

3 a Split the garden with a line parallel to the wall to form a rectangle and a right-angled triangle. The hypotenuse of the triangle is 5 m. The right hand side of the triangle is 11 – 8 = 3 m. These are two sides of a Pythagorean triple, so the third side is 4 m. This side is the same length as the wall, so the wall is 4 m long.
  b Area of lawn = $38\,m^2$; 2 bottles

4 210 m

5 Capacity of tank = $942\,477.8\,cm^3 = 942\,477.8\,ml$
  Time to fill = 3142 seconds = 52 minutes

6 a Shade rectangle a by x
  b Shade rectangle a by x and rectangle b by x
  c Shade one of the end triangles

7 a 3.5 cm b $28\,cm^2$

8 Area of trapezium = $\frac{1}{2}(a + b)h$
  So $144 = \frac{1}{2}((x - 6) + (x + 2)) \times 3x$
  $144 = \frac{1}{2}(2x - 4) \times 3x$
  $144 = (x - 2) \times 3x$
  Therefore $3x^2 - 6x = 144$

9 6 litres

10 $3x$

11 a 19.9 b 24.8625 c 389.098125 d 3.7710…

12 a 528.75 b 265 225 c 37.184…

13 $3.85 \times 10^{13}\,m^2$

7 Unit test

Sample student answer

The question asks for the answer correct to 3 s.f. The student has rounded to 4 s.f. Although all the maths is correct they must make sure they write the answer in the requested way.

UNIT 8

8 Prior knowledge check

1 a 248 b 5 c 30 d 4

2 a 15 b 21 c 16 d 63

3 a 36 b 100 c 0.23 d 100

4 1 m = 100 cm
  1 km = 1000 m
  1 km = 100 000 cm

5 a 400 cm b 6200 m

6 a B and E b C and F, A and D

7 a

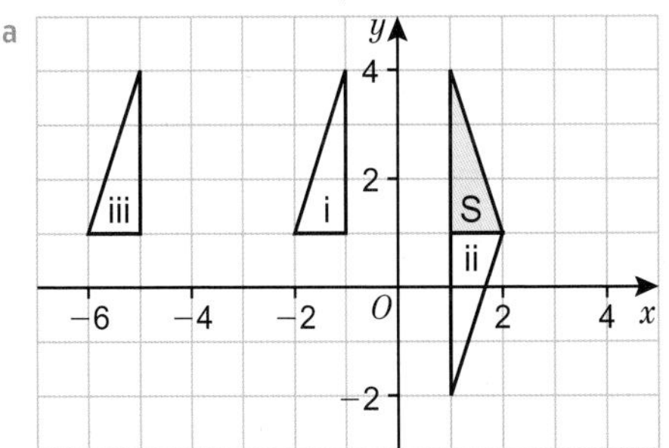

b, c

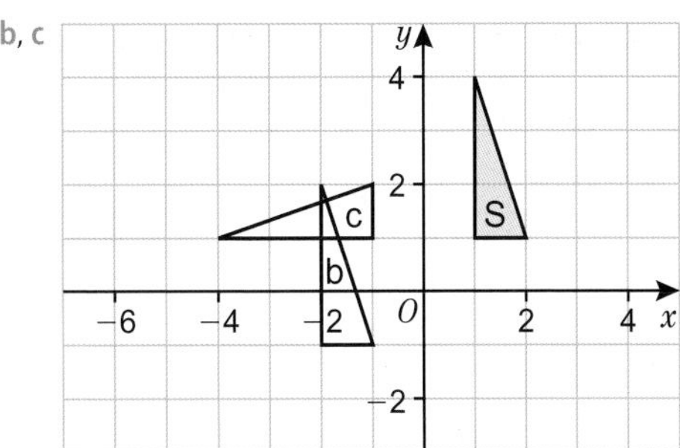

8

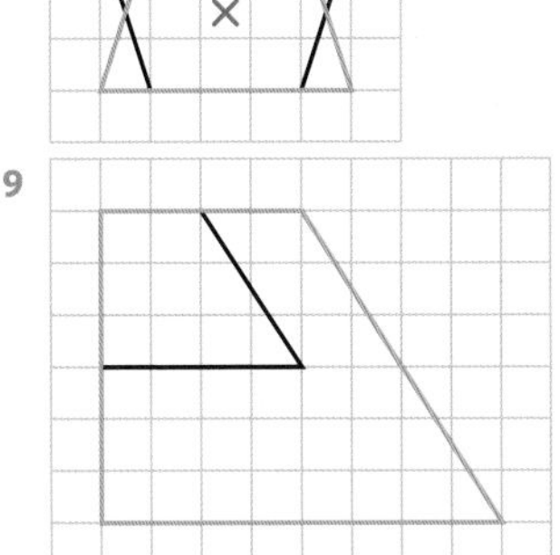

9

10 Similar

11 Students' own accurate triangle.

12 Students' own answers

8.1 3D Solids

1 a e.g. b e.g. c e.g.

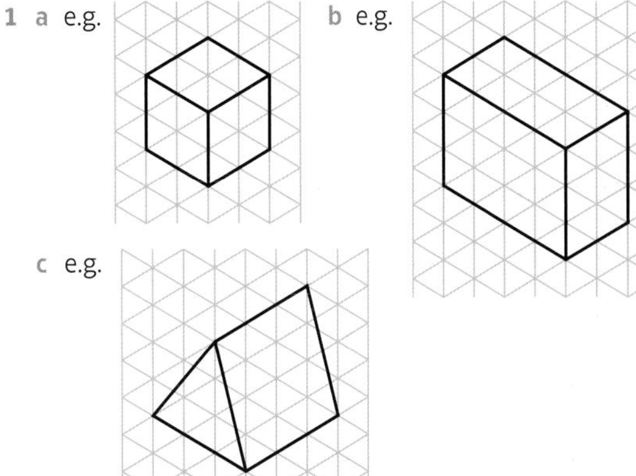

2 a

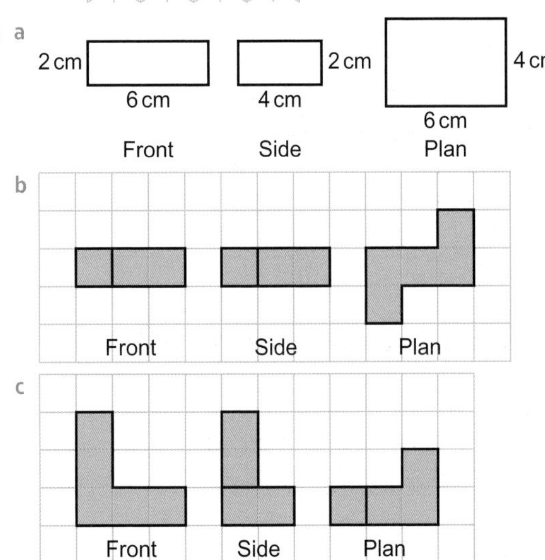

3 a b

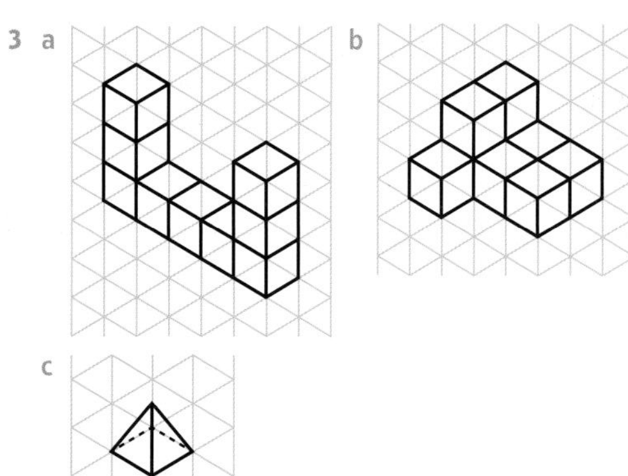

c

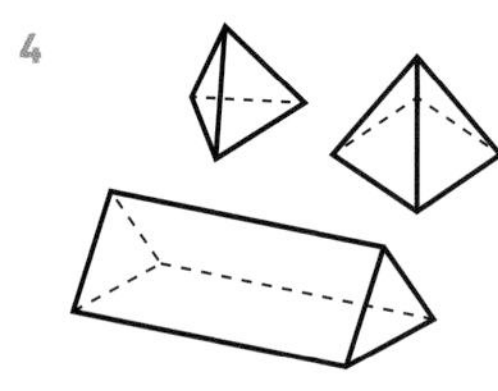

4

5 a $96\,cm^2$

b

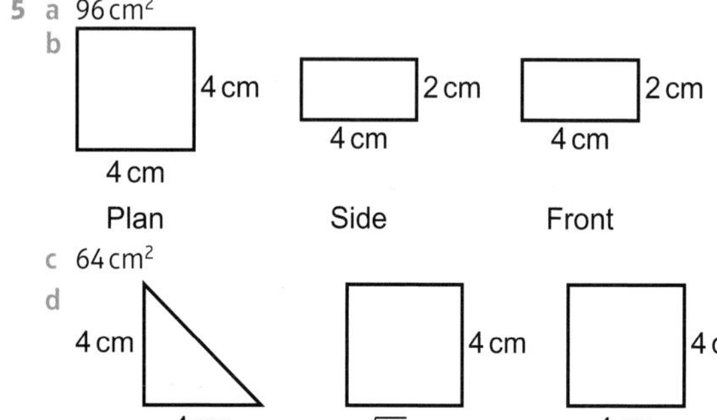

c $64\,cm^2$

d

4 cm, 4 cm, Plan; 4 cm, $\sqrt{32}$ cm, Side; 4 cm, 4 cm, Front

Surface area $70.6\,cm^2$

6

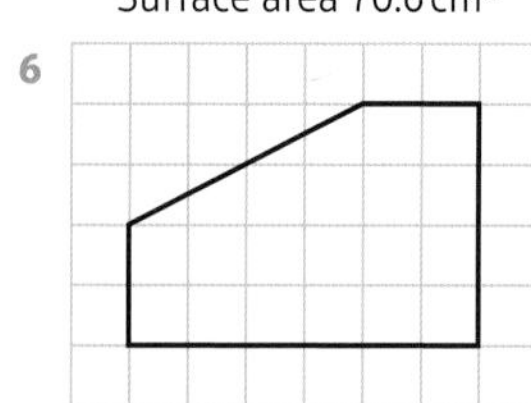

8.2 Reflection and rotation

1

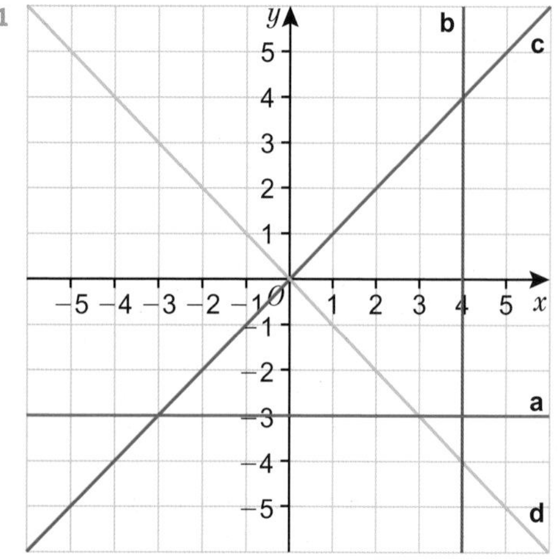

2 a, b, c, d

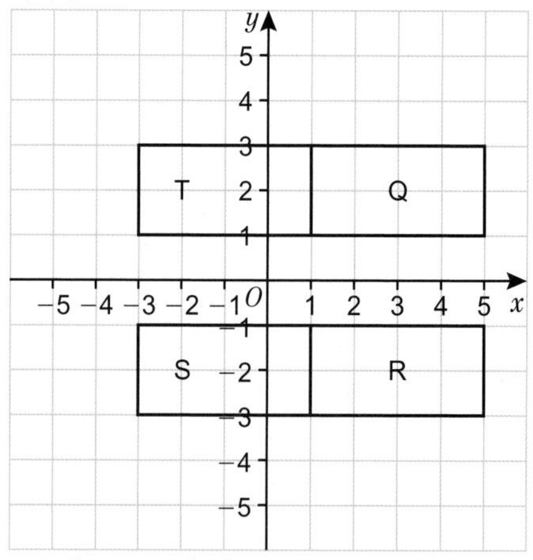

e Reflection in the line $x = 1$

3

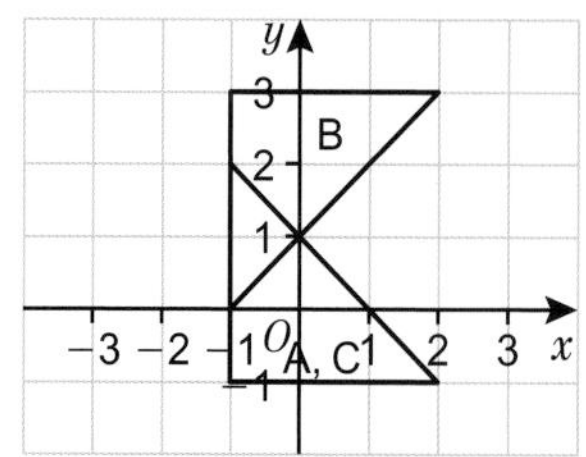

4 a Reflection in the y-axis or the line $x = 0$

b Reflection in the line $y = -1$

c Reflection in the line $y = -x$

d Reflection in the line $y = 1$

5 a Reflection in the line $x = 3$

b Reflection in the line $y = -x$

6 a, b, c, d

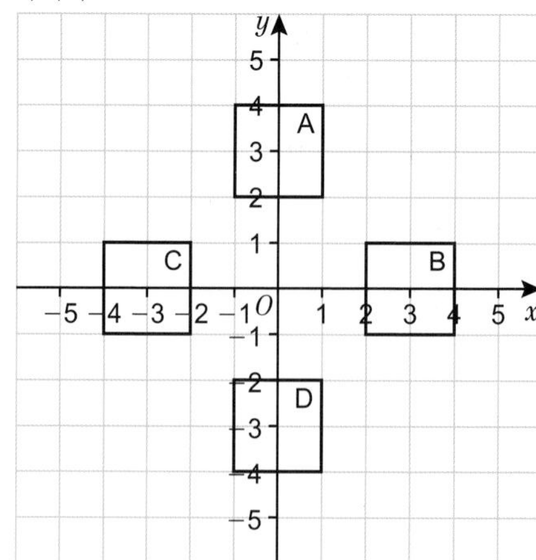

e Reflection in the line $y = -x$

7

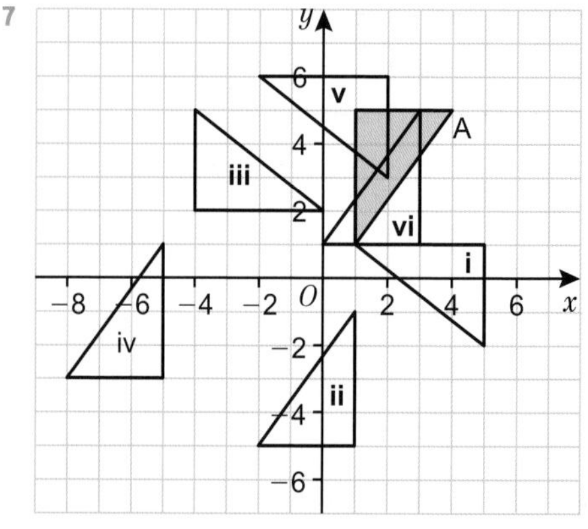

8 a Rotation 90° clockwise about (–3, 3)
b Rotation 90° clockwise about (1, 4)
c Rotation 90° anticlockwise about (3, –3)
d Rotation 180° about (–3, –3)

9 a, b, c

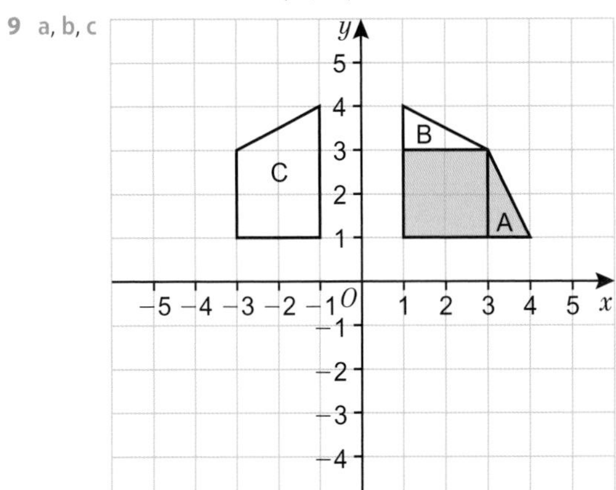

d Rotation 90° anticlockwise about (0, 0)

10 Always true. The image of point (a, b) reflected in the x- and y-axis is $(-a, -b)$, which is equivalent to a rotation of 180° about the origin.

11 Rotation 180° about (3, 3)

8.3 Enlargement

1 a, b

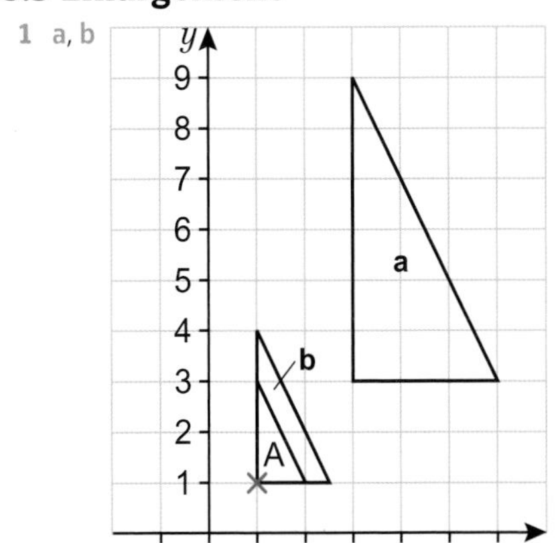

2 a, b, c

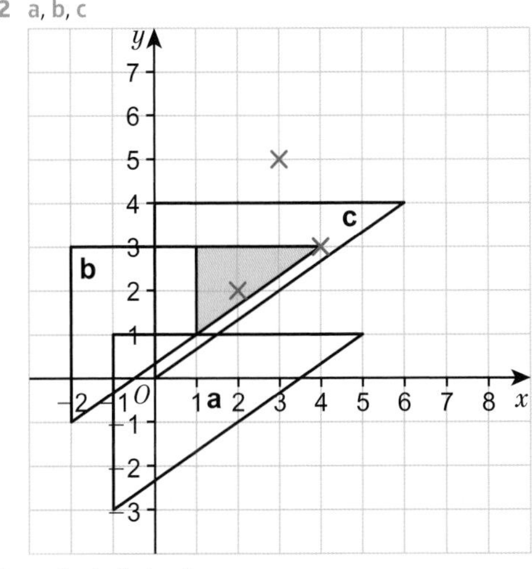

3 a Scale factor 3
b Correct construction lines
c (–5, –2)
d Enlargement by scale factor 3, centre (–5, –2)

4 a 6 cm^2 b 24 cm^2 c 54 cm^2 d 96 cm^2

e

Shape	Scale factor	$\frac{\text{Area of enlarged shape}}{\text{Area of A}}$
B	2	4
C	3	9
D	4	16

5 a e.g.

b e.g.

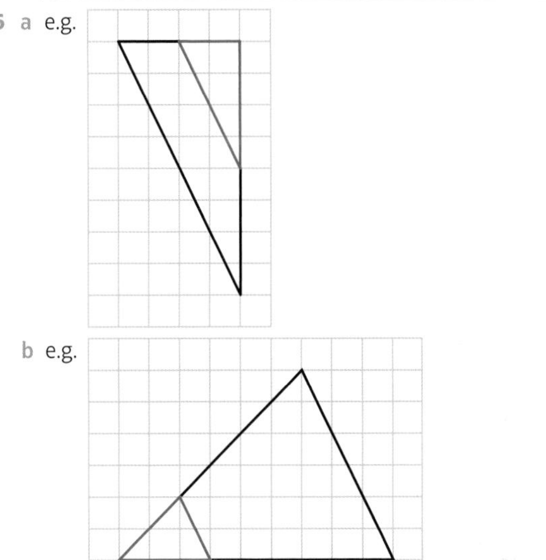

6 a i

ii

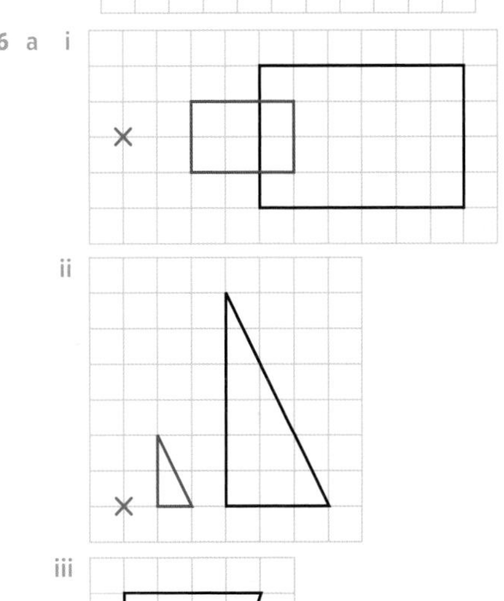

iii

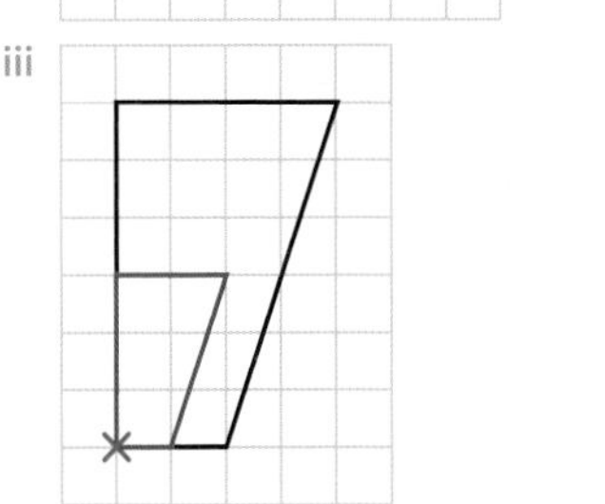

b Yes, e.g. An area is the product of two dimension (e.g. $a \times b$). Each dimension is multiplied by a common scale factor (s), so the enlarged area is $sa \times sb = s^2(a \times b)$. This means that the original area has been multiplied by s^2 (the scale factor squared). Therefore, if the scale factor is $\frac{1}{2}$, then the area is enlarged by $(\frac{1}{2})^2$.

7 a Enlargement, scale factor $\frac{1}{3}$, centre (–5, –2)
b Enlargement, scale factor $\frac{1}{3}$, centre (2, –5)

8

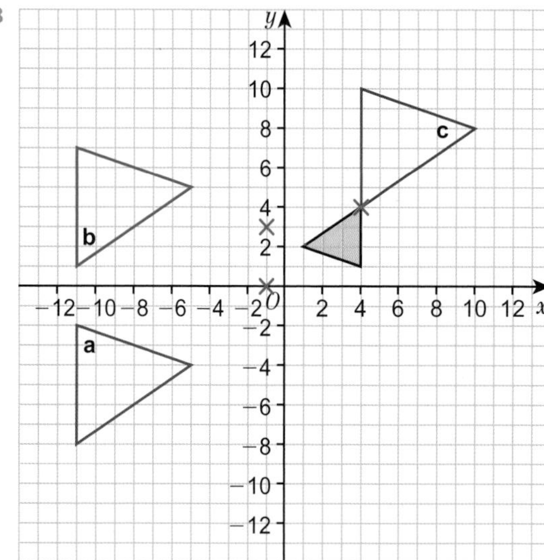

9 Enlargement, scale factor −3, centre (−2, −2)

10 Jamie is incorrect. Enlargements with scale factors between −1 and 1 make a shape smaller than the original shape.

11 Enlargement, scale factor −3, centre (−2, −2)

8.4 Translations and combinations of transformations

1 A 2 square right, 2 squares up
B 2 squares left, 2 squares up
C 2 square left, 2 squares up
D 1 square left, 2 squares up
E −2 squares left, 5 squares down

2

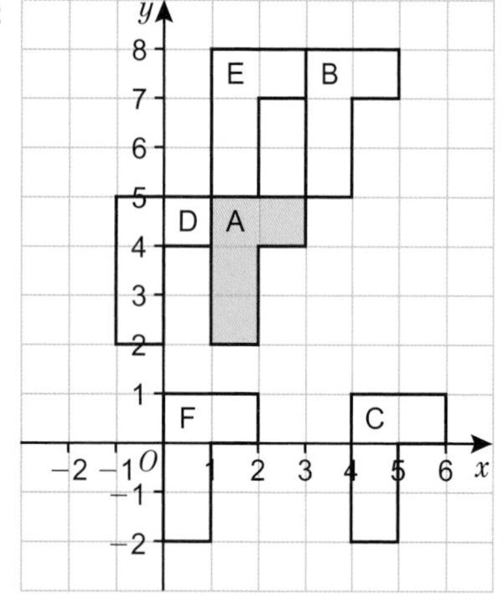

3 a $\begin{pmatrix}4\\2\end{pmatrix}$ b $\begin{pmatrix}0\\-4\end{pmatrix}$ c $\begin{pmatrix}4\\-6\end{pmatrix}$ d $\begin{pmatrix}5\\1\end{pmatrix}$ e $\begin{pmatrix}-5\\-1\end{pmatrix}$

4 $\begin{pmatrix}-a\\-b\end{pmatrix}$; with correct explanation

5 a, b

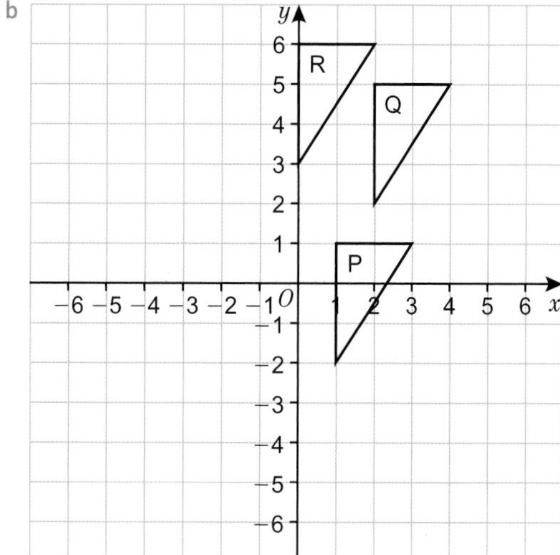

c $\begin{pmatrix}-1\\5\end{pmatrix}$

6 a $\begin{pmatrix}4\\1\end{pmatrix}$

b $\begin{pmatrix}a+c\\b+d\end{pmatrix}$, e.g. because this is the total horizontal movement and total vertical movement.

7 a, b

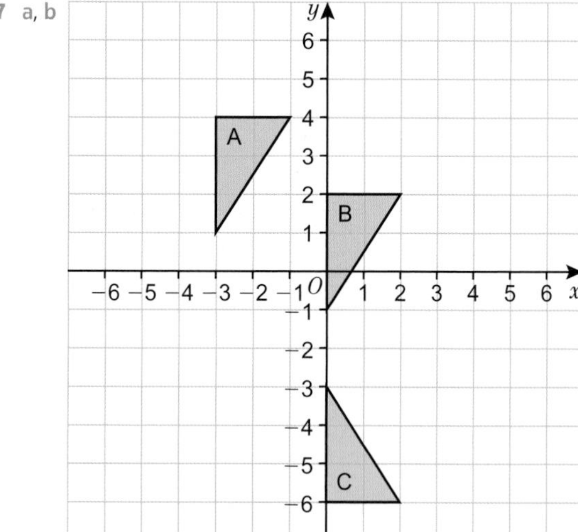

8 a, b, c

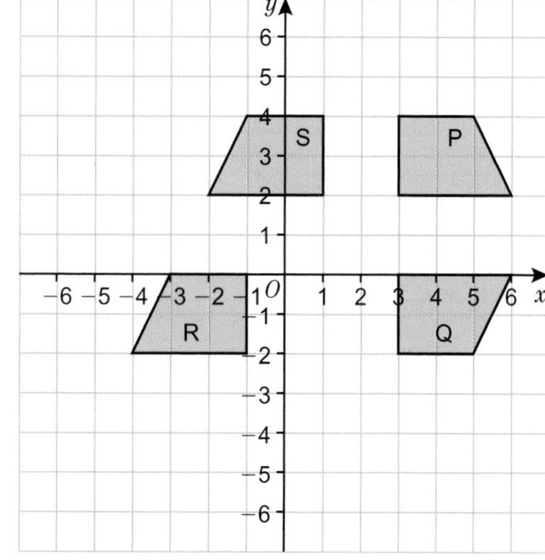

d Reflection in the line $x = 2$

9 a, b, c

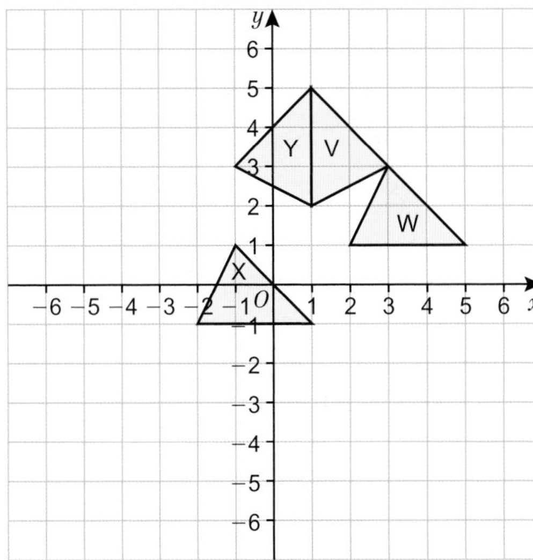

d Reflection in the line $x = 1$

10 a, b, c

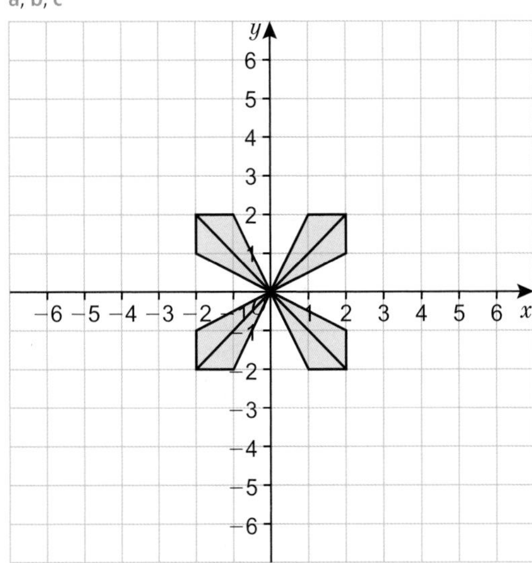

d Enlargement scale factor 3, centre (0, 0)

11 a $\begin{pmatrix} 6 \\ -1 \end{pmatrix}$

b

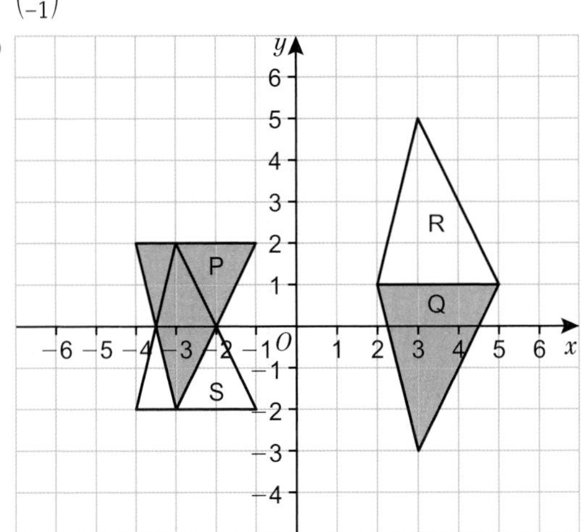

c Reflection in line $y = 1$

12 Adam's statement is not always true. When a shape is enlarged, the image and the original are similar but not congruent.

8.5 Bearings and scale drawings

1 a 20 m b 2.5 cm

2 a Accurate scale drawing of right-angled triangle with base 6 cm, hypotenuse 10 cm
b 36 m

3 a 072° b 255°

4 Bearing of 285° accurately drawn

5 a 1 cm : 120 m b i 360 m ii 600 m
c Answers between 4 and 4.24 minutes

6 a 5 km b 3 km c 1.25 km d 0.25 km

7 a 10 cm b 5 cm c 4 cm

8 a 2.5 km b 16 cm

9 a i 130 km ii 136 km b Sligo

10 a Accurate scale drawing, with AB = 5 cm
b AC = 67 m, CB = 78 m

11

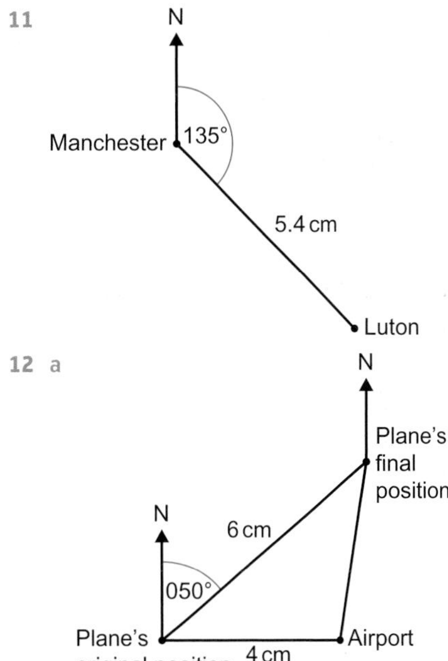

b 188°

13 a

N
160°
N
8 cm
060°
6 cm

b 27 km c 280°

14 323°

15 a 260° b 050°

16

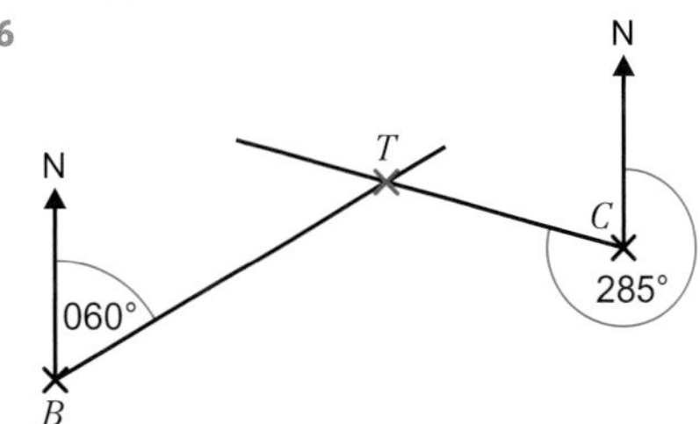

8.6 Constructions 1

1 Accurate drawing of the triangle
2 Accurate drawing of the triangle
3 Accurate drawing of the triangle
4 a, b, c Accurate drawings of the triangles
5 Accurate drawing of an equilateral triangle with side 6.5 cm
6 The sum of the two shorter sides is less than the longest side so the triangle will not be possible.
7 Accurate drawing of triangle with sides of length 5 cm, 15 cm and 17 cm (for real-life sides of 100 cm, 300 cm and 340 cm respectively)
8 Accurate drawing of triangle with sides of length 5.5 cm, 5.5 cm and 10 cm (for real-life sides of 11 m, 11 m and 20 m respectively)
9 Accurate net with sides of length 6 cm.

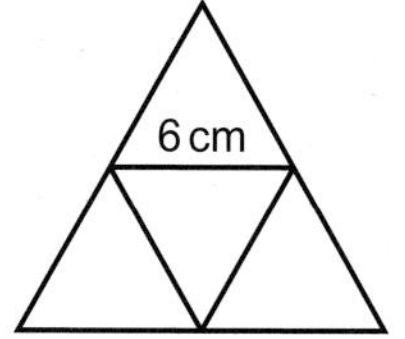

or

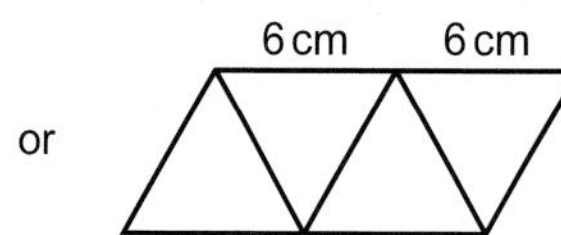

10 a, b Perpendicular bisector of line segment AB of length 7 cm drawn accurately
c AP is the same distance as BP.
11 a, b Perpendicular bisector accurately constructed of 2 points, S and T, 10 cm apart. The perpendicular bisector shows possible positions of the lifeboat.
12 a, b, c Perpendicular bisector from point P to the line AB accurately constructed.
13 a, b, c Perpendicular at point P on a line accurately constructed.
14 a Shortest distances to sides accurately drawn.
b 2.5 m
c 10 seconds

8.7 Constructions 2

1 a Accurate drawing of a triangle with sides 10 cm, 8 cm, 6 cm
b Right-angled triangle
2 a Accurate construction of an equilateral triangle with sides 5 cm
b 60°
3 Angles accurately drawn and bisected
4 a Accurate construction of 90° angle
b Accurate construction of 45° angle
5 a Accurate construction of 60° angle
b Accurate construction of 30° angle
6 Accurate construction of 120° angle
7 a Accurate scale drawing with sides 3 cm and 5 cm
b Accurate construction of angle of 30°
c $115\,m^2$
8 a Accurate scale drawing with pole height 4 cm, top guy length 4.5 cm and angle CBA 63°
b 2.4 m
c 27.6 m
9 a Accurate construction of triangle
b Accurate construction of line perpendicular to AB that passes through C
c $16\,cm^2$
10 Accurate scale drawing
11 a Accurate construction of triangle

b, c, d, e

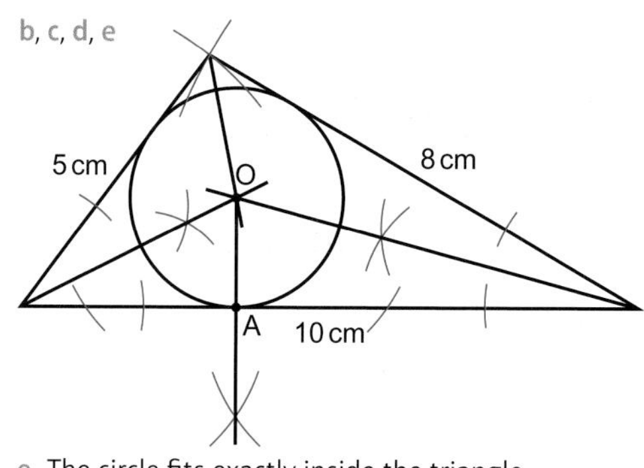

e The circle fits exactly inside the triangle.
12 a, b, c Construction of regular hexagon in a circle
d Hexagon
13 Construction of regular octagon in a circle with radius 5 cm

8.8 Loci

1 Circle
2 Sketch of a circle with radius marked 5 m
3

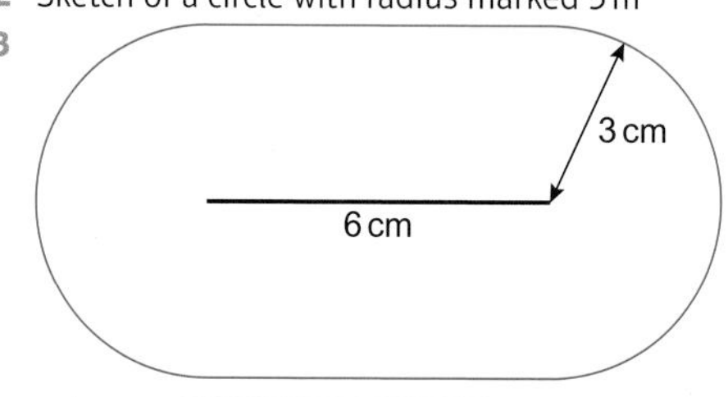

4 a, b

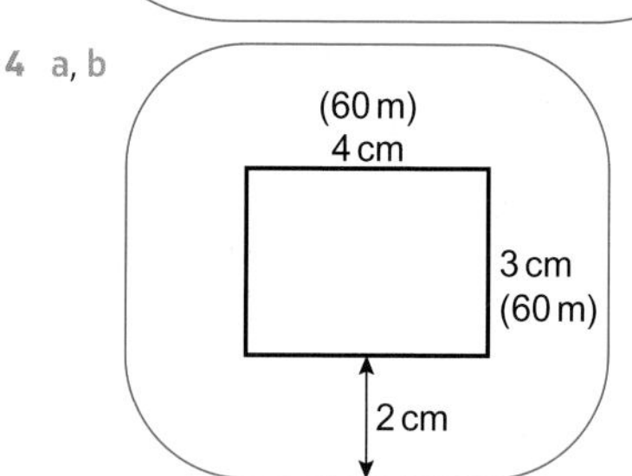

5 a, b, c

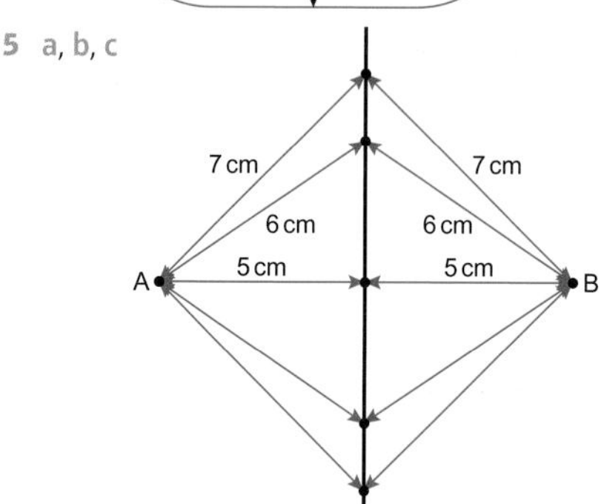

6

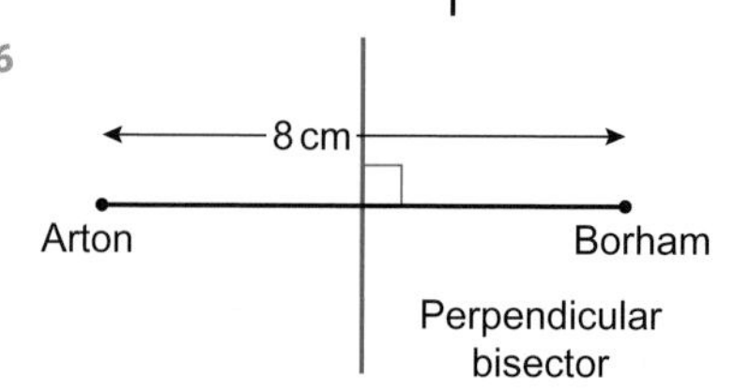

7

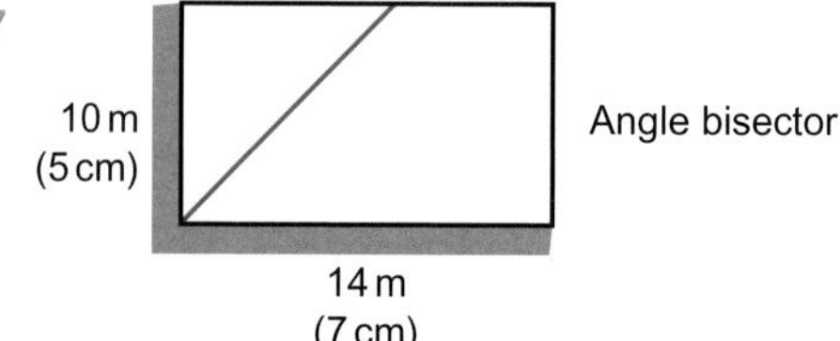

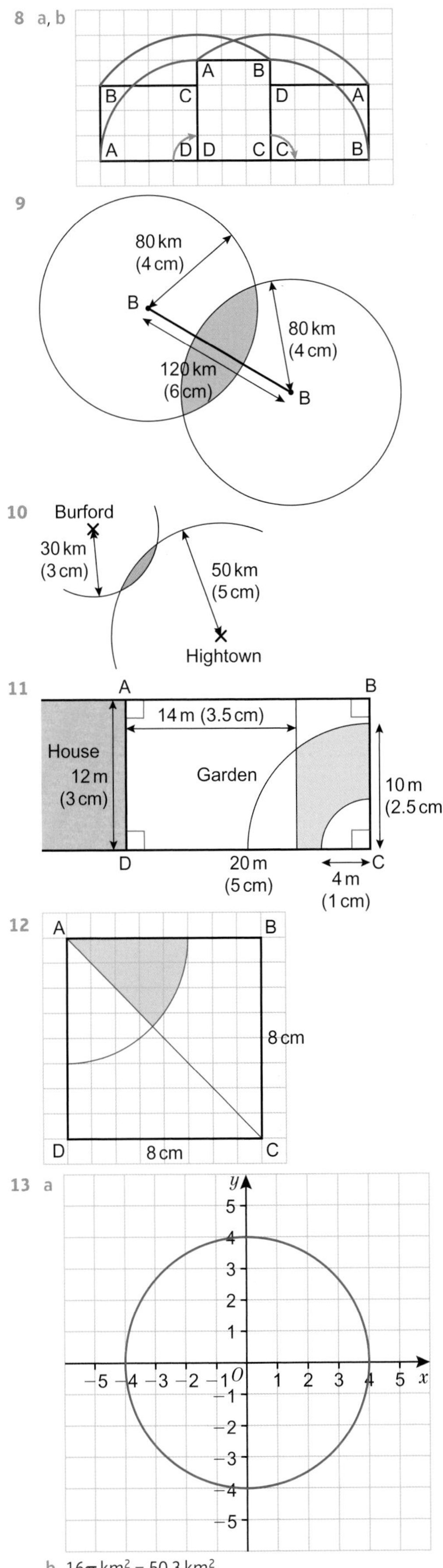

b $16\pi\ \text{km}^2 = 50.3\ \text{km}^2$

8 Problem-solving

1 One method would be as follows:
Construct a perpendicular bisector to establish 0° and 90°.
Construct an equilateral triangle with a vertex at the origin to find 60°.
Bisect the 60° angle to create a 30° angle.
Bisect the existing angles to create 15°, 45° and 75°.
Repeat on the other side of 90° to extend this up to 180°.

2 Construct the perpendicular bisectors of the sides of the regular hexagon and place a point where each of these bisectors meets the circle. Join these additional six points with the vertices of the hexagon to construct a regular dodecagon.

3 Square correctly constructed.

4 Start by constructing a square, then find the centre of the square and draw the circle which goes through all of its vertices. Repeat the method described above on the square to construct a regular octagon.

5 If you start with a square you can construct any polygon where the number of sides is a power of 2 (8 sides, 16 sides, 32 sides…). If you start with a hexagon you can construct any polygon where the number of sides is 3 multiplied by a power of 2 (12 sides, 24 sides, 48 sides…).

8 Check up

1

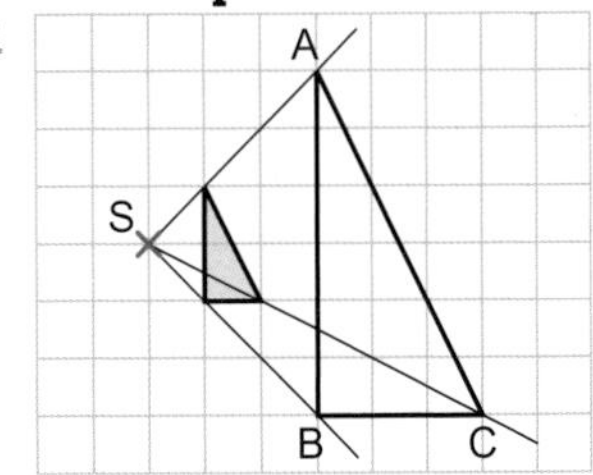

2 a Reflection in the line $y = x$
b Rotation 90° anticlockwise about (4, −1)
c Translation $\begin{pmatrix} -4 \\ 4 \end{pmatrix}$

3 a Enlargement scale factor 2, centre (−3, 5)
b Enlargement scale factor $\frac{1}{2}$, centre (−3, 3)
c Enlargement scale factor −1, centre (0, −2.5)

4 a, b, c

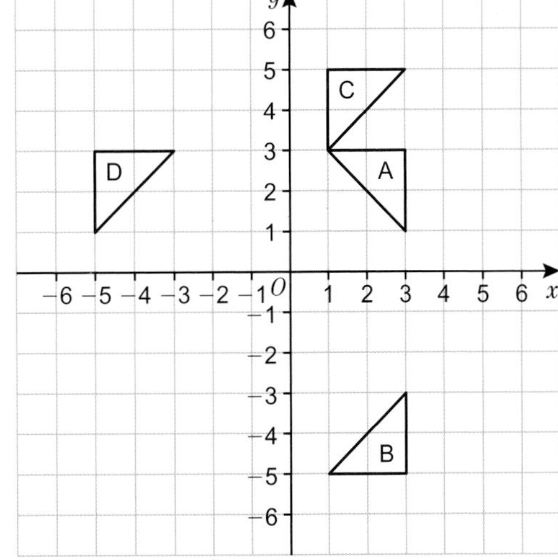

d Reflection in the line $x = -1$

5

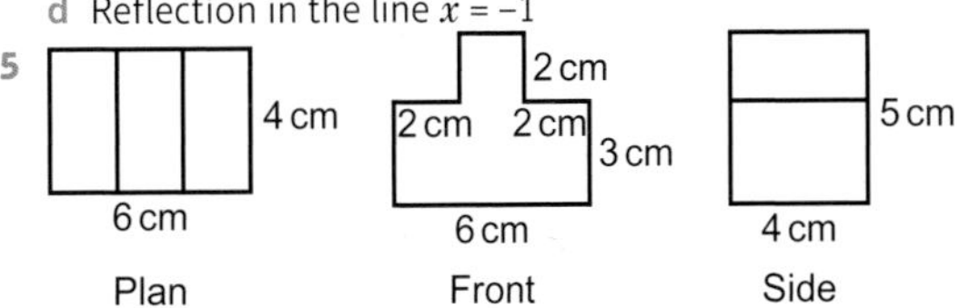

6 4 cm

7 290°

8 a

N

Airport 130°

50 km (5 cm)

N

050°

40 km (4 cm)

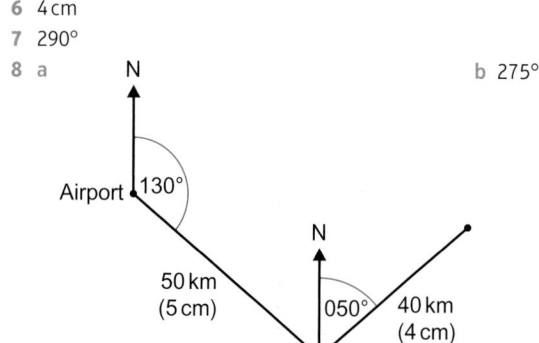

b 275°

9 Perpendicular bisector accurately constructed on a line of length 10 cm

10 Angle accurately bisected

11 a, b

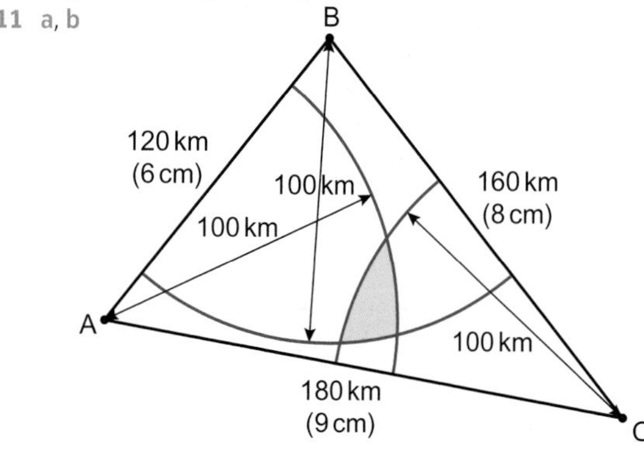

13 Regular polygons with these numbers of sides: 3, 4, 5, 6, 8, 10, 12, 15, 16, 17, 20, 24, 30, 32, 34, 40, 48, 51, 60, 64, 68 …

8 Strengthen

Transformations

1 a, b

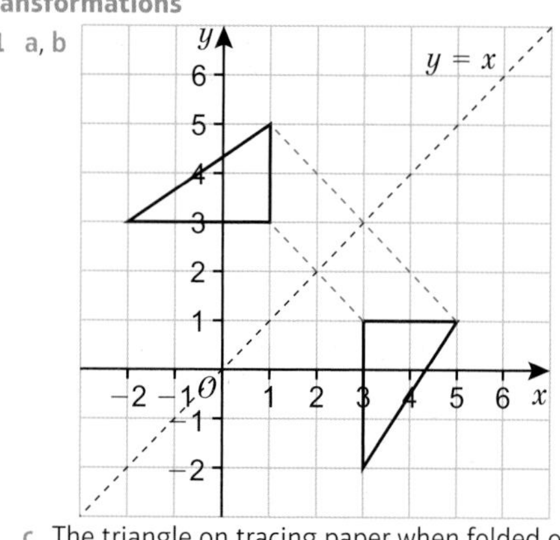

c The triangle on tracing paper when folded on the line of reflection becomes the image.

2 a, b, c

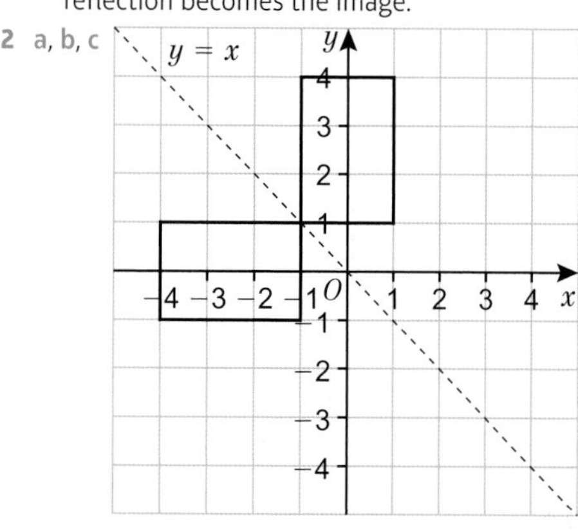

3 Shape A is reflected in the line $y = x$ to give image B.

4 a $\begin{pmatrix}3\\2\end{pmatrix}$ b $\begin{pmatrix}5\\1\end{pmatrix}$ c $\begin{pmatrix}2\\-1\end{pmatrix}$ d $\begin{pmatrix}-3\\2\end{pmatrix}$

e $\begin{pmatrix}-6\\-3\end{pmatrix}$ f $\begin{pmatrix}3\\-4\end{pmatrix}$

5 a 3 right, 1 up or $\begin{pmatrix}3\\1\end{pmatrix}$ b 5 left or $\begin{pmatrix}-5\\0\end{pmatrix}$

c 2 left, 3 down or $\begin{pmatrix}-2\\-3\end{pmatrix}$ d 6 left, 6 down or $\begin{pmatrix}-6\\-6\end{pmatrix}$

6 C

7 a Rotation of 90° clockwise about (0, 0)

b Rotation of 180° about (0, 2)

c Rotation of 90° anticlockwise about (4, −1)

8 a, b, c

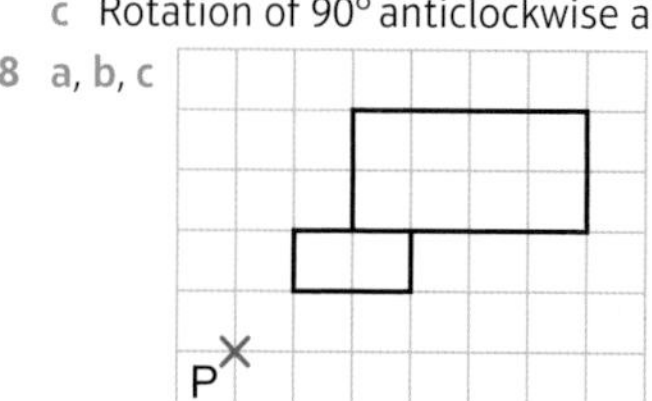

9

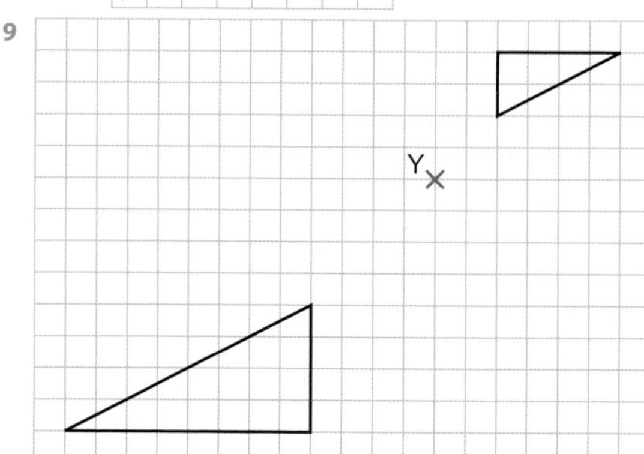

10

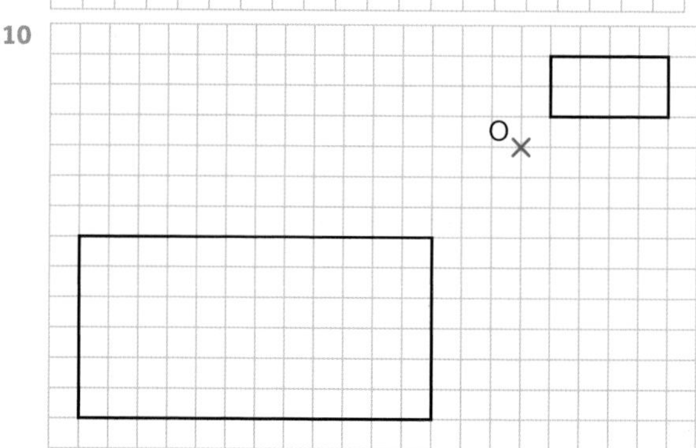

11

Rotation	angle, direction and centre of rotation
Translation	horizontal movement and vertical movement (or translation vector)
Enlargement	scale factor and centre of enlargement

Drawings and bearings

1 a B b A c C

2

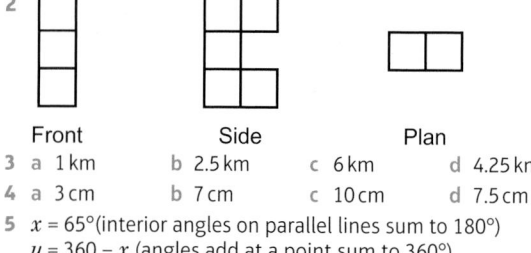

Front Side Plan

3 a 1 km b 2.5 km c 6 km d 4.25 km

4 a 3 cm b 7 cm c 10 cm d 7.5 cm

5 x = 65°(interior angles on parallel lines sum to 180°)
y = 360 − x (angles add at a point sum to 360°)
= 360 − 65 = 295°
Bearing of A from B is 295°

6 a, b

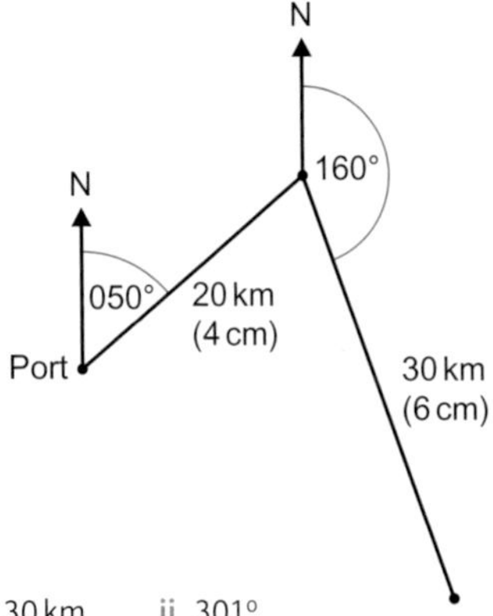

c i 30 km ii 301°

Constructions and loci

1 Accurate construction of a triangle with sides of length 10 cm, 7 cm, 6 cm
2 Accurate construction of a triangle with sides of length 7 cm, 8 cm, 9 cm
3 Accurate construction of the perpendicular bisector of a line of length 12 cm
4 Accurate construction of the perpendicular bisector of a line of length 7 cm
5 Accurate construction of angle bisector of 70° angle
6 Accurate construction of angle bisector of 100° angle

7 a

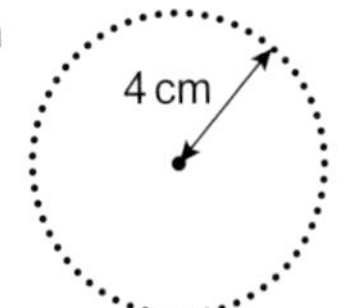

b Circle

c Points that are equidistant from a centre make a circle.

8 a, b, c

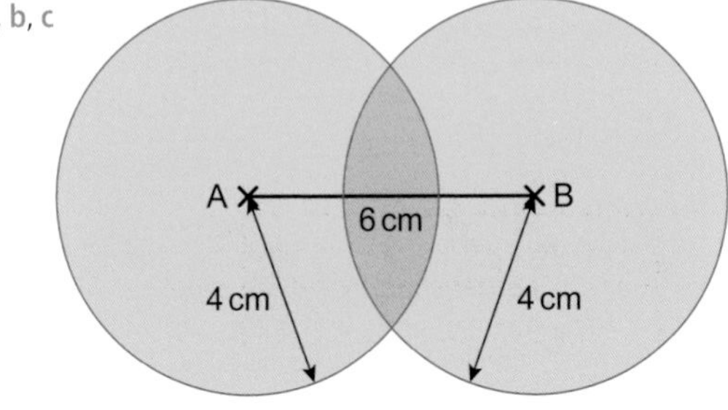

8 Extend

1 a

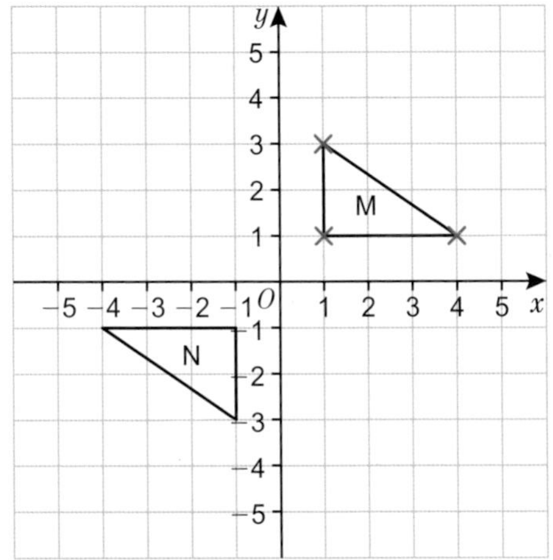

b Rotation 180° about the origin
c Yes, it always works.

2 Enlargement, scale factor −2, centre of enlargement (−1, 0)

3 a, b, d

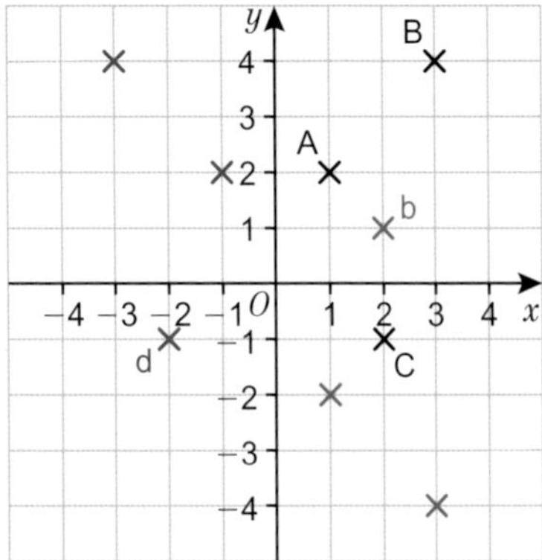

c When points are reflected in the x-axis, the y-coordinate is multiplied by −1
d When points are reflected in the y-axis, the x-coordinate is multiplied by −1
e i $(p, -q)$ ii $(-p, q)$ iii $(-p, -q)$

4 a, b, c

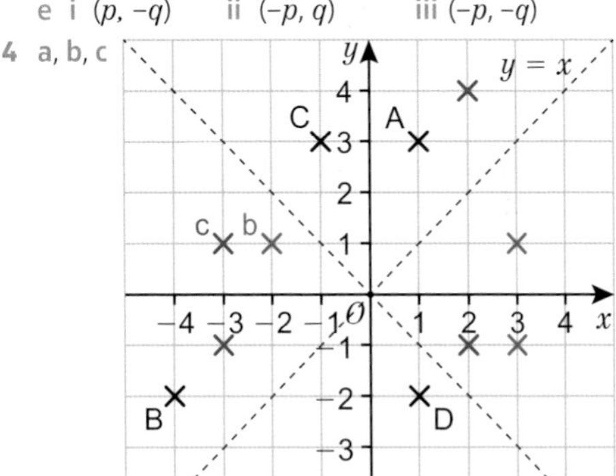

d i (q, p) ii $(-q, -p)$ iii $(-p, -q)$

5 (−2, −1), (−4, −1), (−3, −5), (−5, −5)

6 Reflection in $y = x$ followed by reflection in the y-axis
Reflection in $y = -x$ followed by reflection in the x-axis

7 a, d

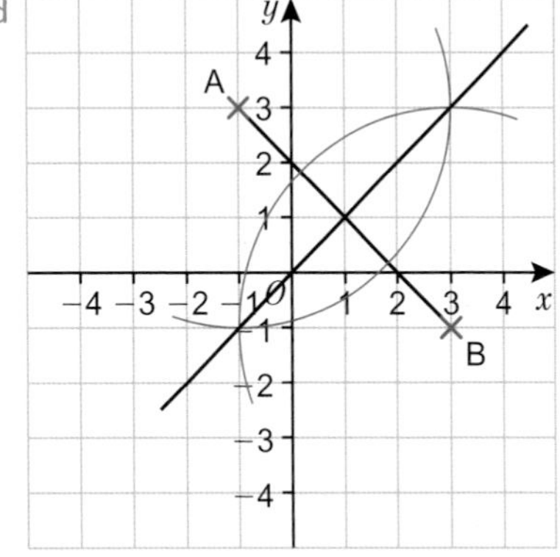

b $y = -x + 2$ c 1

8 a

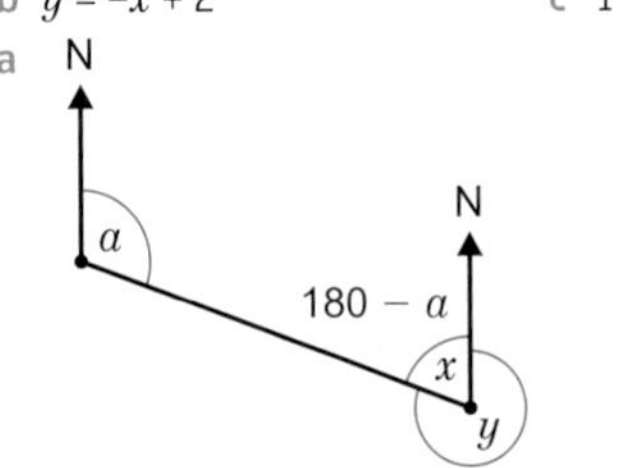

$x = 180 - a$ (interior angles on parallel lines sum to 180°)

$y = 360 - (180 - a)$ (angles at a point sum to 360°)
$= 180 + a$

b $(a - 180)°$

9 a $60°$
b, c Accurate construction of a hexagon

10 Accurate construction of a triangle with area $30\,cm^2$

11 a Sphere b Circle

12 a

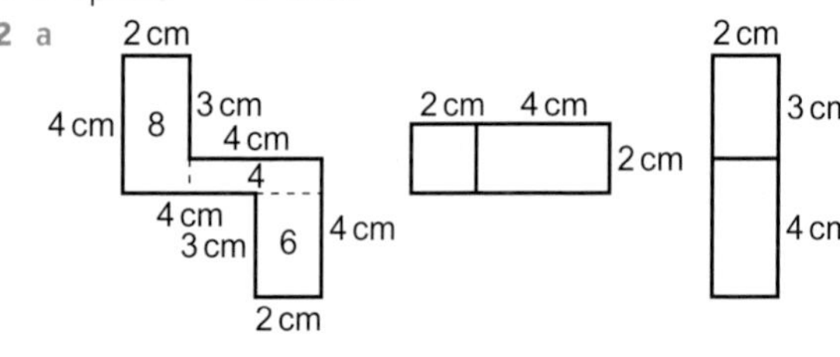

b $88\,cm^2$

13 Two correct transformations, e.g. 90° clockwise rotation about (0.5, 1.5), followed by a reflection in $x = -0.5$, or translation (–2, 0) followed by reflection in $y = -x$

14 a, b

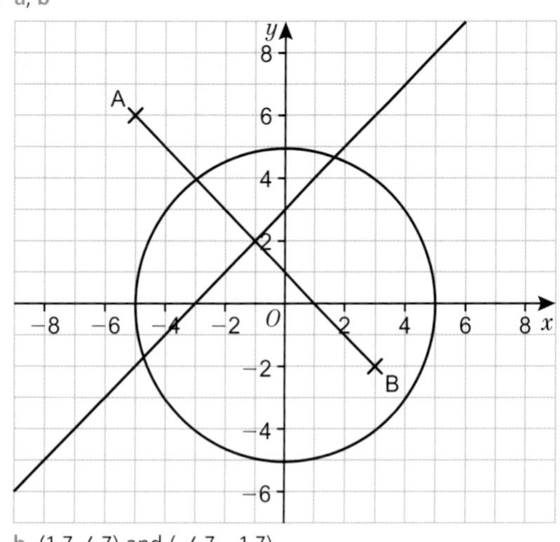

b (1.7, 4.7) and (–4.7, –1.7)

8 Unit test

Sample student answer

Student B gives the best answer. Student A has not given the direction or centre of rotation. Student C has given two transformations. The question asks for a single transformation.

UNIT 9

9 Prior knowledge check

1 a No b Yes c No d No

2 a No b No c No d Yes e Yes

3 36

4 +12 and –12

5 a $2\sqrt{3}$ b $2\sqrt{5}$

6 0

7 $x^2 - x - 12$

8 a $x = 2$ b $x = 5$ c $x = -3$

9 a $x(x + 8)$ b $(x + 3)(x + 1)$
c $(y - 5)(y + 2)$ d $(x + 5)(x - 5)$
e $(2 + y)(2 - y)$

10 $2x + 3y - 53 = 0$

11 $x = -1$ or $x = 4$

12 –2 and –8

9.1 Solving quadratic equations 1

1 6 and –2 or –6 and 2

2 a $x(x - 5)$ b $(y - 2)(y + 2)$ c $(x + 5)(x - 2)$

3 a $z = \pm 6$ b $z = \pm 4$ c $z = \pm 5$

4 a $x = \pm 4$ b $x = \pm 7$ c $x = \pm 5$

5 a $x = 4$ and $x = 6$ b $x = 5$ and $x = -6$
c $y = -2$ and $y = -1$ d $b = 2$ and $b = -5$

6 a $x = 0$ or $x = 2$ b $x = 4$ or $x = -4$
c $y = -2$ or $y = 2$

7 a $x = -3$ or $x = 2$ b $(x + 3)(x - 2)$

8 a $x = -1$ and $x = -6$ b $x = -4$ and $x = 3$
c $x = 2$ and $x = 4$ d $x = 0$ and $x = 7$

9 $(x - 4)(x + 6)$ or any multiple e.g. $(2x - 8)(x + 6)$; $(x - 4)(2x + 12)$

9.2 Solving quadratic equations 2

1 a $x = 1$ and $x = 3$ b $x = -1$ and $x = -4$
c $x = 3$ and $x = -2$

2 a $2x^2 + 7x + 3$ b $3x^2 + 5x - 2$ c $2x^2 - 6x - 8$

3 a 0.382 b 0.154

4 a $2\sqrt{6}$ b $2\sqrt{7}$ c $2\sqrt{10}$
d $-1+\sqrt{2} = \sqrt{2} - 1$ e $\frac{\sqrt{3}}{2} - 1$

5 $x(x + 1) = 30$; $x = 5$ m

6 $2a(a + 1) = 12$
$2a^2 + 2a = 12$
$2a^2 + 2a - 12 = 0$
$(2a + 6)(a - 2) = 0$
Therefore $a = -3$ or $a = 2$
Since a cannot be –3, small rug is 2 m × 2 m

7 $(3x - 1)(x + 2)$

8 a $5(x + 1)(x + 2)$ b $(2x + 5)(x - 1)$
c $2(2x + 1)(x - 2)$ d $(3x - 4)(x + 3)$
e $(2x + 3)(x - 5)$

9 a $a = -2.5$ and $a = 8$ b $x = 4.5$ and $x = -4$
c $y = \frac{4}{3}$ and $y = -2.5$ d $b = \frac{3}{4}$ or 0.75 and $b = 2\frac{2}{3}$ or 2.67

10 a $(2x - 5)(x + 1)$ so either $x = 2.5$ or $x = -1$
b $(3x - 4)(x + 3)$ so either $x = 1.33 = \frac{4}{3}$ or $x = -3$
c $(4x + 6)(x - 3) = 0$ so either $x = -1.5$ or $x = 3$
d $(2x + 5)(3x - 3) = 0$ so either $x = -2.5$ or $x = 1$

11 a $4x^2 + 18x = 10$
b $2x^2 + 9x - 5 = 0$
So $(2x - 1)(x + 5) = 0$
Therefore $x = 0.5$ or $x = -5$; since –5 is not a realistic solution, border should be 0.5 m wide.

12 $(4x - 7)(2x + 3)$, giving $x = 1.75$ or $x = -1.5$

13 a $x = -2.5 + \frac{\sqrt{5}}{2}$ or $x = -2.5 - \frac{\sqrt{5}}{2}$
b $x = -3.5 + \frac{\sqrt{41}}{2}$ or $x = -3.5 - \frac{\sqrt{41}}{2}$
c $x = -1 + \sqrt{3}$ or $x = -1 - \sqrt{3}$
d $x = -1 + \sqrt{7}$ or $x = -1 - \sqrt{7}$
e $x = -1.5 + \frac{\sqrt{21}}{6}$ or $x = -1.5 - \frac{\sqrt{21}}{6}$

14 a $x = 1.36$ or $x = -7.36$ b $x = 3.39$ or $x = -0.89$
c $x = 0.55$ or $x = -1.22$ d $x = 1.39$ or $x = -2.89$

15 a, b $x = -1.5$ or $x = 5$

16 $x = 0.27$ or $x = -1.47$

9.3 Completing the square

1 a $x^2 + 8x + 16$ b $x^2 - 6x + 9$ c $4x^2 + 12x + 9$
d $x^2 + 4x + 8$ e $x^2 + 2x + 8$

2 a $3\sqrt{5}$ b $4\sqrt{2}$ c $4\sqrt{3}$ d $3\sqrt{10}$

3 a $x = 1 + \sqrt{3}, x = 1 - \sqrt{3}$ b $x = -2 + \sqrt{2}, x = -2 - \sqrt{2}$
c $x = 7 + \sqrt{5}, x = 7 - \sqrt{5}$ d $x = 5 + \sqrt{3}, x = 5 - \sqrt{3}$

4 $(x + 2)^2 = x^2 + 4x + 4$

5 a $(x + 2)^2 + 1$ b $(x + 2)^2 + 2$ c $(x + 2)^2 - 5$

6 a $(x + 3)^2$ b $(x + 4)^2$ c $(x + 5)^2$ d $(x + 6)^2$

7 a $(x+1)^2-2$ b $(x+4)^2-16$ c $(x+6)^2-36$
d $(x+3)^2+2$ e $(x-2)^2+2$

8 $(x+2)^2-4+1=0$
$(x+2)^2=3$
$(x+2)=\pm\sqrt{3}$
$x=-2-\sqrt{3}$ or $x=-2+\sqrt{3}$

9 a $x=-3-\sqrt{2}$ or $x=-3+\sqrt{2}$
b $x=-1-\sqrt{6}$ or $x=-1+\sqrt{6}$
c $x=-4-\sqrt{7}$ or $x=-4+\sqrt{7}$

10 $3x^2-12x-1=3(x^2-4x)-1$
$=3[(x-2)^2-4]-1$
$=3(x-2)^2-12-1$
$=3(x-2)^2-13$

11 a $2(x+3)^2-16$ b $3(x-1)^2+2$
c $5(x+1)^2+20$ d $4(x+\frac{3}{2})^2-16$

12 a $x=3-2\sqrt{2}$ or $x=3+2\sqrt{2}$
b $x=-2-\sqrt{5}$ or $x=-2+\sqrt{5}$

13 $4x^2-8x-12=0$
$x^2-2x-3=0$
$(x-1)^2-(-1)^2-3=0$
$(x-1)^2=(-1)^2+3$
$x-1=\pm\sqrt{4}$
$x=1+\sqrt{4}$ or $x=1-\sqrt{4}$
$x=3$ or $x=-1$

14 a $x=1.24$ or $x=-3.24$ b $x=0.88$ or $x=-0.38$
c $x=1.08$ or $x=-3.08$ d $x=3.25$ or $x=-0.25$
e $x=0.60$ or $x=-2.10$

15 $x=1.24$ or $x=-1.74$

9.4 Solving simple simultaneous equations

1 a $b=2a+12$ b $b=5-3c$ c $b=\dfrac{5a-5}{3}$

2 a $x+y=12$ b $x-y=4$ or $y-x=4$

3 c and d

4 a $x=4, y=3$ b $x=3, y=5$
c $x=14, y=-6$ d $x=1, y=4$
e $x=9, y=11$ f $x=2, y=5$
g $x=3, y=9$ h $x=2, y=4$

5 Meal = £11 and wine = £14

6 50p

7 a $x=3, y=2$ b $x=4, y=-5$ c $x=4, y=2$

8 Students' own answers. $x=3, y=2$

9 a $6x+2y=34$ b $10x+0=50, x=5$
c $y=2$

10 a $x=2, y=4$ b $x=-2, y=4$
c $x=3, y=2$ d $x=8, y=3$

11 $x=2, y=-1$

12 $x=0, y=0$

13 $x+y=23$; $x-y=5$; $x=14, y=9$

14 a £12 b £24

15 $x=26$ and $y=12$. Perimeter = 76 cm

9.5 More simultaneous equations

1 a $8x+12y=24$ b $4x-24y=28$

2 a $x=2, y=4$ b $x=4, y=2$

3 a e.g. $y=2x+1$ b e.g. $y=2x+2$
c $m=3, c=-1$ d $y=3x-1$

4 $y=3-x$

5 a $x=2, y=1$ b $x=3, y=2$
c $x=-5, y=3$ d $x=2, y=4$
e $x=4, y=-6$

6 Coffee = 200p (£2) and scone = 90p

7 Adult = £1.80 and child = £1.00

8 Pear = 25p and banana = 60p

9 $x=0.67, y=-1.5$

10 Sand = 20 kg and cement = 40 kg

11 $x=£15$ and $y=£0.50=50$p. So a 50-mile hire would be £15 + 50 × £0.50 = £40

12 a $4a-4b=-2$ and $2a+b=8$ b $a=2.5, b=3$
c Rectangle is 6 m by 10 m

9.6 Solving simultaneous linear and quadratic equations

1 a $x=1$ or $x=-4$ b $x=1.5$ or $x=-1$
c $x=\frac{1}{3}$ or $x=-2$

2 a $x=1.19$ or $x=-4.19$ b $x=1.18$ or $x=-0.85$
c $x=0.50$ or $x=-3.00$

3 10

4 a $x=-4, y=-4$ or $x=3, y=3$
b $x=-2, y=-11$ or $x=4, y=1$
c $x=-0.5, y=4$ or $x=1, y=10$
d $x=-1.33, y=-9.67$ or $x=1, y=2$
e $x=-1.89, y=-0.66$ or $x=-1.11, y=1.66$

5 a $x=0.70, y=5.90$ or $x=-5.70, y=25.10$
b $x=3.37, y=9.74$ or $x=-2.37, y=-1.74$

6 Points are (1, 0) and (−1, 8)

7 a $x=1, y=1$ or $x=-2, y=4$
b $x=1.41, y=0.59$ or $x=-1.41, y=3.41$
c $x=-5, y=0$ or $x=0, y=5$
d $x=1.82, y=0.63$ or $x=0.18, y=-2.63$
e $x=-1, y=-3$ or $x=3, y=5$
f $x=-3, y=1$ or $x=9, y=5$

8 a $y=2x+3$ b $2x^2+x-3=0$
c $x=1, y=5$. Width = 5 m

9 (0, −1) and (6, 11)

10 AB = $2\sqrt{5}$ or 4.47

11 a $x^2+y^2=4$ b (1.27,1.54) and (−0.47,−1.94)

12 a $x=-4.93, y=-1.93$ or $x=1.93, y=4.93$
b $x=-1.54, y=-5.71$ or $x=0.77, y=5.87$
c $x=-4.30, y=-5.61$ or $x=1.90, y=6.81$

9.7 Solving linear inequalities

1 a $x=\frac{1}{2}$ b $x=2$ c $x=\frac{2}{3}$ d $x=-2$

2 a 5, 10, 12 b 3, 0, −2 c 5, 3, 0, −2 d −2, 0, 3

3 a 2, 3, 4, 5… b 0, −1, −2, −3…
c 2, 3, 4, 5… d 2 ,3, 4, 5…
e −4, −3, −2, −1, 0, 1, 2 f 1, 2, 3, 4
g −3, −2, −1, 0, 1, 2, 3

4 a $x \geqslant 2$ b $x \leqslant 0$ c $x < 2$ d $x > -1$
e $1 < x \leqslant 5$ f $-4 \leqslant x < 2$

5 a −3, −2, −1, 0, 1 b $-2 \leqslant x < 4$

6 a
−5 −4 −3 −2 −1 0 1 2 3 4 5 x

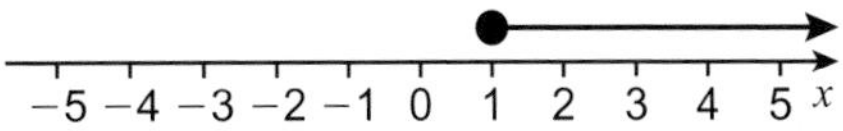

b
−5 −4 −3 −2 −1 0 1 2 3 4 5 x

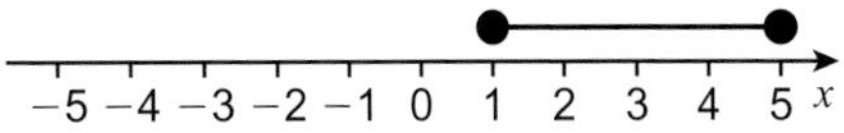

c
−5 −4 −3 −2 −1 0 1 2 3 4 5 x

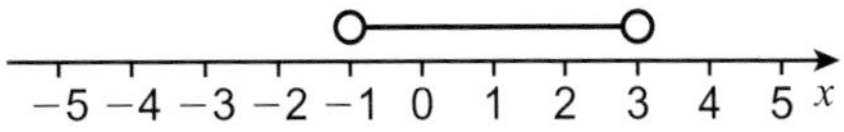

d
−5 −4 −3 −2 −1 0 1 2 3 4 5 x

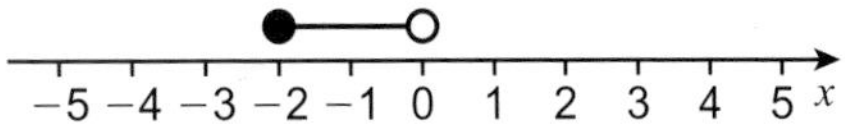

7 a The set of all x values such that x is less than 2
b The set of all x values such that x is less than or equal to −2
c The set of all x values such that x is greater than or equal to 0
d The set of all x values such that x is less than or equal to 0
e The set of all x values such that x is greater than −1

8 a b c d e

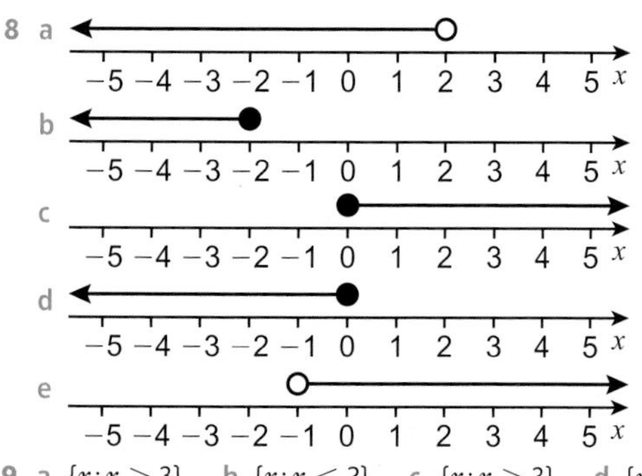

9 a $\{x : x > 3\}$ b $\{x : x < 2\}$ c $\{x : x > 3\}$ d $\{x : x \leqslant 2\}$

10 The smallest value of x is 4

11 a $\{x : x > 4\}$ b $\{x : x > -3\}$
c $\{x : x > -5.5\}$ d $\{x : x > 1.8\}$

12 a $-4 < x \leqslant 2$ b $-3 < x \leqslant 4$
c $-3 \leqslant x \leqslant 9$ d $-1 < x \leqslant 3$

13 a No b No

14 a 3, 4, 5, 6, 7 b 1, 0, −1, −2, −3, −4
c −2, −1, 0, 1, 2, 3 d −2, −1, 0, 1, 2, 3, 4, 5

15 a $\{x : x \geqslant -3\}$ b $\{x : x > -1\}$
c $\{x : -2 < x \leqslant 4\}$ d $\{x : 1\frac{1}{2} < x \leqslant 3\}$

9 Problem solving

1 $s = 5 \times 10 + \frac{1}{2} \times 8 \times 10^2 = 50 + 400 = 450$ m

2 $20 = 3t + \frac{1}{2} \times 2 \times t^2$ $2t^2 + 3t - 20 = 0$
$(2t - 5)(t + 4) = 0$ $t = 2.5$ or $t = -4$
The answer is $t = 2.5$ seconds as we cannot have a negative time in this situation.

3 6 seconds

4 a $\frac{25a}{2} = 5v$
b Yes. e.g. the motorbike travelling at 10 m s^{-1}, while the car accelerates at a constant rate of 4 m s^{-2} or the motorbike travelling at 5 m s^{-1} while the car accelerates at 2 m s^{-2}.

9 Check up

1 a $x = 0$ and $x = -3$ b $x = 2$ and $x = -3$

2 $(3x + 2)(x - 2) = 0$

3 $(x - 2)(x - 4) = 3$
$x^2 - 6x + 5 = 0$
$x = 1$ and $x = 5$, but $x = 5$ is the only sensible answer

4 $x = 1 + \sqrt{7}$ and $x = 1 - \sqrt{7}$

5 $(x + 3)^2 - 6$

6 $x = -3 + 2\sqrt{3}$ or $x = -3 - 2\sqrt{3}$

7 a $x = -5, y = -2$ b $x = 2, y = 0$

8 $y = 2x - 3$

9 $x = -5, y = 15$ or $x = 1, y = 3$

10 1, 0, −1, −2, −3, −4 and −5

11 a $0 < x \leqslant 9$
b 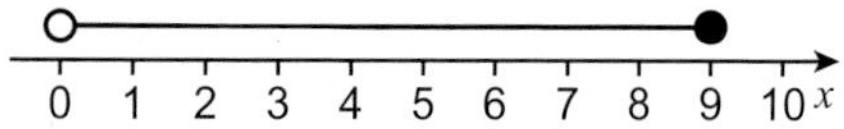
c $\{x : 0 < x \leqslant 9\}$

13 a 6 years old
b Students' own answers

9 Strengthen

Quadratic equations

1 $x = 0, x = -7$

2 a $x(x + 5)$ b $x = 0, x = -5$

3 a −4 and 3 b $(x + 3)(x - 4)$ c $x = -3, x = 4$

4 a $x = -6, x = 3$ b $x = 4, x = 3$
c $x = -5, x = 3$

5 a −1, 6; 1, −6; −2, 3; 2, −3 b $(2x - 3)(x + 2)$
c $x = 1.5$ or $x = -2$

6 a $x = \frac{4}{3}, x = -3$ b $x = -\frac{4}{3}, x = 2$
c $x = 1.5, x = 2$

7 a $x(x - 2)$
b $(x + 4)^2 = x^2 + 8x + 16$

8 a $(x + 1)(x - 2) = x^2 - x - 2$ b $x^2 - x - 2 = 4$
c $x^2 - x - 6 = 0$
d $x = 3, x = -2$; $x = 3$ m is the only sensible answer

9 a $a = 2, b = 3, c = 1$ b $a = 2, b = -4, c = -6$
c $a = 3, b = 4, c = -1$

10 a $9 - 8 = 1$ b $16 + 48 = 64$ c $16 + 12 = 28$

11 a $x = -1, x = -\frac{1}{2}$ b $x = 3, x = -1$
c $x = -\frac{2}{3} - \frac{\sqrt{7}}{3}, x = -\frac{2}{3} + \frac{\sqrt{7}}{3}$

12 a i $x^2 + 6x + 9$ ii $x^2 - 10x + 25$
b $12x$ c $x^2 + 12x + 36$

13 a $(x + 2)^2 = x^2 + 4x + 4$
$(x + 2)^2 - 4 = x^2 + 4x$
b $x = -2 - 2\sqrt{3}$ or $x = -2 + 2\sqrt{3}$

14 a $(x - 2)^2 = x^2 - 6x + 9$
$(x - 2)^2 - 6 = x^2 - 9x$
b $x = 3 + 3\sqrt{2}$ or $x = 3 - 3\sqrt{2}$

Simultaneous equations

1 a $x = \frac{7}{3}$ b $x = 2$

2 $x = 2, y = 4$

3 b $3 \times 2 + 4 \times -3 = -6$

4 $x = -4, y = 2$

5 a $y = 4 - 5x$ b $x^2 - 2x = 6 + 4 - 5x$
c $x^2 + 3x - 10 = 0$ d $x = 2$ or $x = -5$
e $y = -6$ or $y = 29$

6 $x = 1, y = 1$ or $x = -6, y = 8$

Inequalities

1 a $-2 \leqslant x \leqslant 1$ b $4 < y < 10$ c $3 \leqslant z < 8$

2 a i 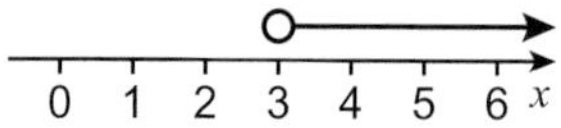
ii 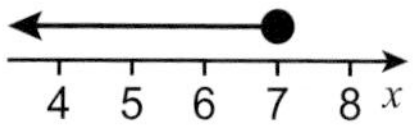
iii 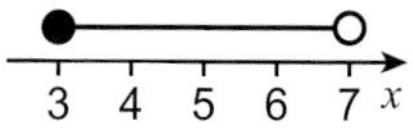
b i e.g. 4, 5 ii e.g. 6, 5 iii e.g. 4, 5, 6

3 a 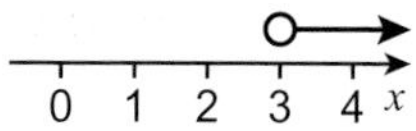
b 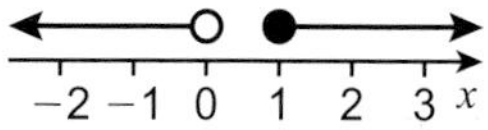
c 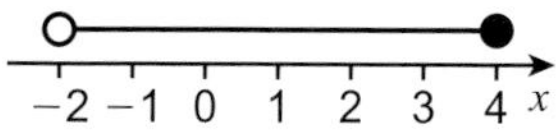
d 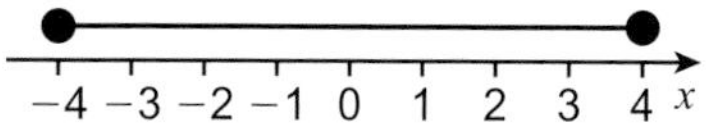
e Values for c are −1, 0, 1, 2, 3, 4
Values for d are −4, −3, −2, −1, 0, 1, 2, 3, 4

4 a $x > 8$ b $x > 3$
c $-3 < x < 4$ d $4 \leqslant x \leqslant 12$

5 a $x > 1$ b $x < 4$ c 2, 3

6 0, 1, 2

7 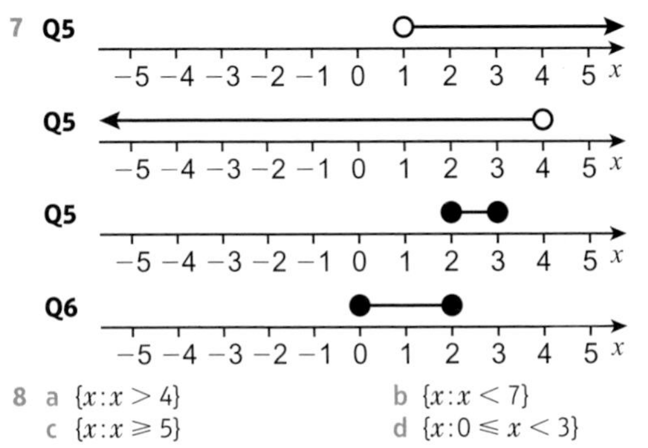

8 a $\{x : x > 4\}$ b $\{x : x < 7\}$
c $\{x : x \geqslant 5\}$ d $\{x : 0 \leqslant x < 3\}$

9 Extend

1 a $x(x + 1) = 30$ b 5 and 6

2 23.32 m

3 $3(x + 0.4)^2 - 0.48$
$a = 3$, $p = 0.4$, $q = -0.48$

4 68 m × 125 m

5 a Area $= x(x + 4) + x(x + 1) = x^2 + 4x + x^2 + x$
$2x^2 + 5x = 75$
So $2x^2 + 5x - 75 = 0$
b $x = 5$ or $x = -7.5$

6 Charging lead = £3.50; phone case = £7.50

7 $x = -5$ or $x = \frac{8}{3}$. Since −5 is not a valid solution then in this case $x = 2\frac{2}{3}$ m.

8 Points are (1.1, 4.9) and (−3.6, 9.6)

9 5 goals

10 $1.25 \leqslant t \leqslant 3.75$ seconds

11 1.52 m

12 $x = 0, y = 5$ or $x = -4, y = -3$

13 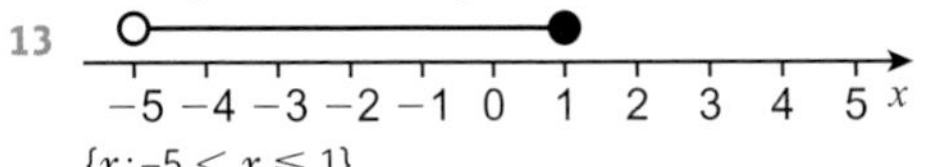

$\{x : -5 < x \leqslant 1\}$

14 a $(x + p)^2 + q = (x + p)(x + p) + q = x^2 + 2px + p^2 + q$
b i $(x + 3)^2 + 6$ ii $(x + 4)^2 - 19$ iii $(x - 2)^2 - 2$
iv $(x + \frac{3}{2})^2 + \frac{19}{4}$

9 Unit test

Sample student answer

1 There could end up being 3 or 4 different equations, so it is a good idea to label them to avoid confusion.

2 The letters s and l could look like a 5 or 1, which might lead to a mistake.

3 The student needs to make it clear how many paper clips are in each box, with a statement like, 'There are 60 paper clips in the small box and 175 paper clips in the large box'.

UNIT 10

10 Prior knowledge check

1 a 0.97 b 0.85 c 0.35 d 0.38

2 a 0.78 b 0.24 c $\frac{2}{5}$ d $\frac{7}{12}$
e 73% f 32%

3 a $\frac{2}{8}, \frac{3}{12}, \frac{4}{16}$, etc. b $\frac{4}{10}, \frac{6}{15}, \frac{8}{20}$, etc
c $\frac{10}{12}, \frac{15}{18}, \frac{20}{24}$, etc d $\frac{14}{20}, \frac{21}{30}, \frac{28}{40}$, etc

4 $\frac{2}{7}$ $\frac{1}{3}$ $\frac{3}{8}$ $\frac{2}{5}$ $\frac{5}{12}$

5 a 40 b 24 c 26.4 d 50
e 300 f 126

6 a 0.25, 25% b 0.3, 30%
c 0.6, 60% d 0.375, 37.5%
e 0.85, 85% f 0.4625, 46.25%

7 a 68 b 81.9 c 205 d 4.2

8 a $\frac{11}{12}$ b $\frac{23}{30}$ c $\frac{7}{12}$ d $\frac{13}{84}$

9
impossible unlikely even chance likely certain
0 $\frac{1}{2}$ = 0.5 1
0% 50% 100%
E D B A C

10 b 100 c $\frac{39}{50}, \frac{17}{100}, \frac{1}{20}$ d 234

11 a 1, 2, 3, 4, 5, 6 b $\frac{1}{6}$ c 1

12 a $\frac{1}{4}$ b 25 times c $\frac{7}{25}$

13 a

	Glasses	No glasses	Total
Boys	4	10	14
Girls	6	12	18
Total	10	22	32

b $\frac{5}{16}$ c $\frac{9}{16}$

14 0.55

10.1 Combined events

1 a 8 b 10 c 12 d 18 e mn

2 a BA, BP, BS, LA, LP, LS, CA, CP, CS, HA, HP, HS, MA, MP, MS
b 15 c $\frac{1}{15}$ d $\frac{2}{15}$

3 a 10 b $\frac{2}{5}$ c $\frac{2}{5}$ d $\frac{1}{10}$

4 a HH, HT, TH, TT b 4
c i $\frac{1}{4}$ ii $\frac{1}{2}$

5 a

		Dice					
		1	**2**	**3**	**4**	**5**	**6**
Spinner	**1**	2	3	4	5	6	7
	2	3	4	5	6	7	8
	3	4	5	6	7	8	9
	4	5	6	7	8	9	10

b i $\frac{1}{8}$ ii $\frac{5}{24}$ iii 0

6 a

13	15
15	17

b (4, 7), (6, 5), (8, 3) c $\frac{3}{20}$

7 a

		Dice 1					
		1	**2**	**3**	**4**	**5**	**6**
Dice 2	**1**	2	3	4	5	6	7
	2	3	4	5	6	7	8
	3	4	5	6	7	8	9
	4	5	6	7	8	9	10
	5	6	7	8	9	10	11
	6	7	8	9	10	11	12

b 36 c i $\frac{1}{18}$ ii $\frac{1}{2}$ iii $\frac{5}{18}$ d 7

8 a

		Bag A			
		S	**O**	**L**	**B**
Bag B	**S**	SS	OS	LS	BS
	B	SB	OB	LB	BB
	L	SL	OL	LL	BL

b i $\frac{1}{12}$ ii $\frac{1}{4}$ iii $\frac{3}{4}$

9 $\frac{1}{8}$

10 $\frac{3}{5}$

10.2 Mutually exclusive events

1 $\frac{1}{3}$

2 a $\frac{1}{3}$ b $\frac{1}{2}$ c $\frac{1}{3}$

3 $\frac{5}{9}$

4 **a** and **c** (a square number and a multiple of 3)

5 a $\frac{1}{2}$ b $\frac{5}{6}$ c $\frac{2}{3}$

6 a $\frac{1}{2}$ b $\frac{15}{52}$

7 23%

8 $\frac{4}{5}$

9 a 0.4 b 0.9

10 a $\frac{1}{6}$ b $\frac{5}{6}$

11 a $\frac{1}{4}$ b $\frac{3}{4}$

12 0.12

13 0.15

14 $\frac{7}{9}$

15 0.35

16 0.62

10.3 Experimental probability

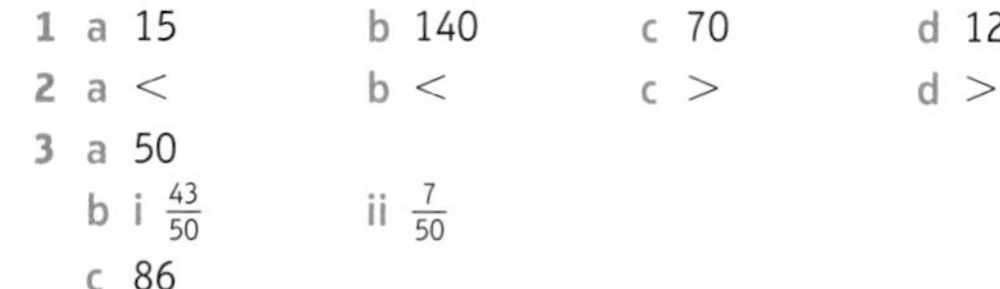

1 a 15 b 140 c 70 d 120

2 a < b < c > d >

3 a 50

b i $\frac{43}{50}$ ii $\frac{7}{50}$

c 86

4 Betty; the greater the number of trials, the better the estimate.

5 a $\frac{3}{8}$ b 150

6 a $\frac{5}{12}$ b 75

7 18

8 a 0.23, 0.22, 0.21, 0.18, 0.09, 0.07

b 0.07 c 35

d No, a fair dice has a theoretical probability of 0.17 for each outcome. For this dice, the estimated probability of rolling a 1 is more than three times more likely than rolling a 6.

9 Yes, because the estimated probabilities of 0.23, 0.195, 0.185, 0.2, 0.19 are all close to the theoretical probability of 0.2

10 40

11 No. Assuming there are more than 200 tickets in the draw, there will be more than 200 tickets that do not win, so buying 200 tickets will not guarantee a prize.

12 a 5 b 30 c 75

13 The dentist's estimate is a little high. The results from the 160 patients suggest a probability of 0.156

10.4 Independent events and tree diagrams

1 a $\frac{7}{45}$ b $\frac{5}{12}$ c $\frac{13}{28}$ d 0.08

e 0.42 f 0.44

2

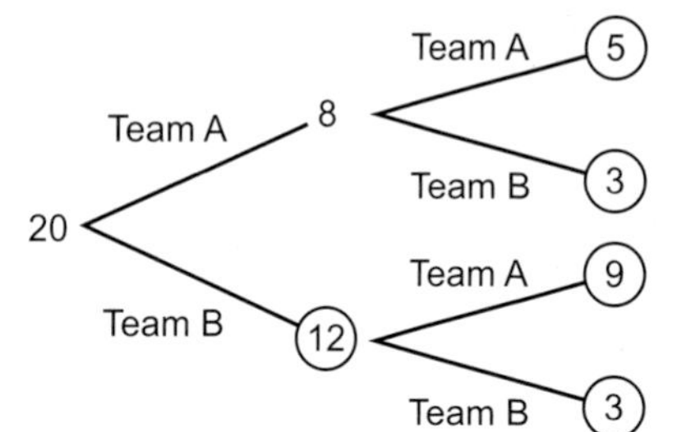

3 a

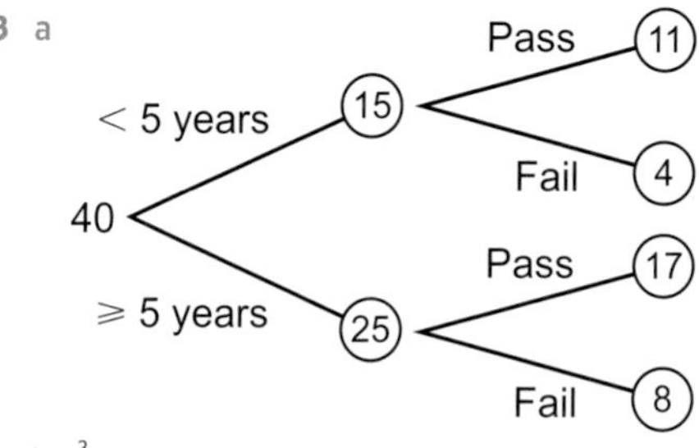

b $\frac{3}{10}$

4 a

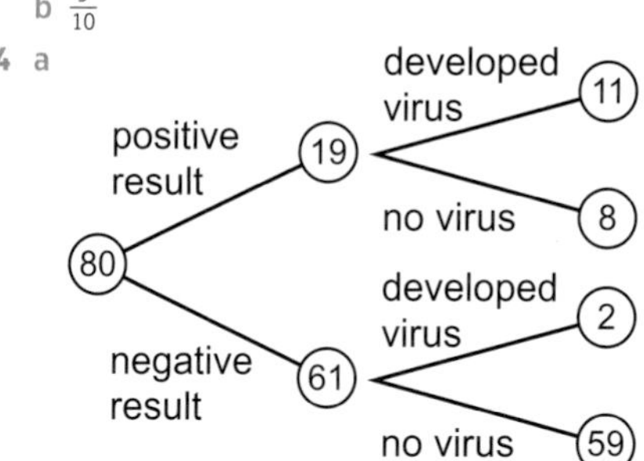

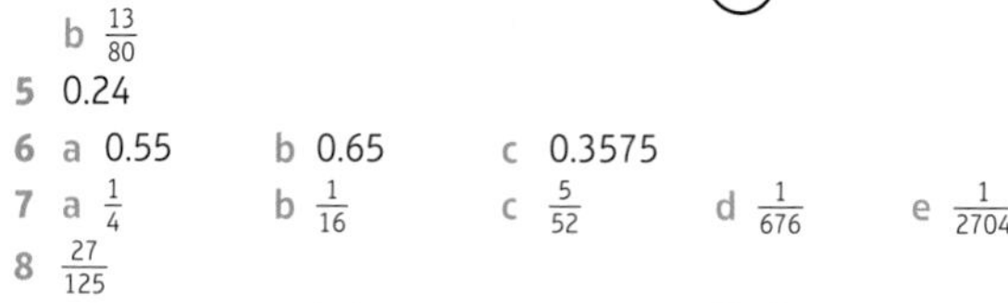

b $\frac{13}{80}$

5 0.24

6 a 0.55 b 0.65 c 0.3575

7 a $\frac{1}{4}$ b $\frac{1}{16}$ c $\frac{5}{52}$ d $\frac{1}{676}$ e $\frac{1}{2704}$

8 $\frac{27}{125}$

9 a

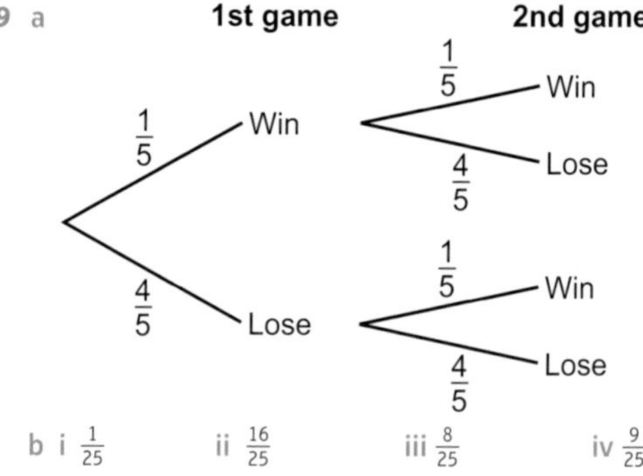

b i $\frac{1}{25}$ ii $\frac{16}{25}$ iii $\frac{8}{25}$ iv $\frac{9}{25}$

10 a

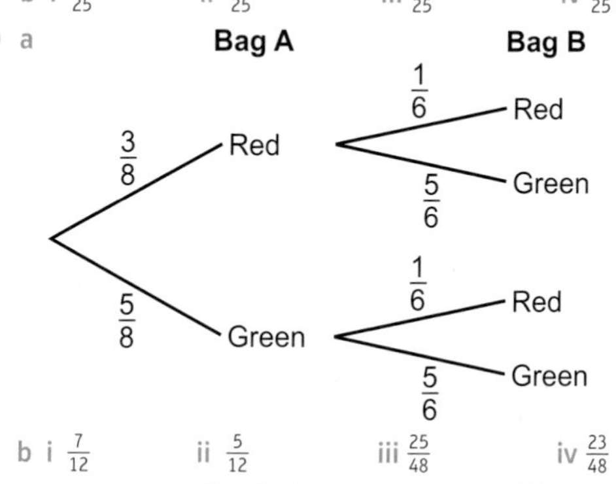

b i $\frac{7}{12}$ ii $\frac{5}{12}$ iii $\frac{25}{48}$ iv $\frac{23}{48}$

11 a

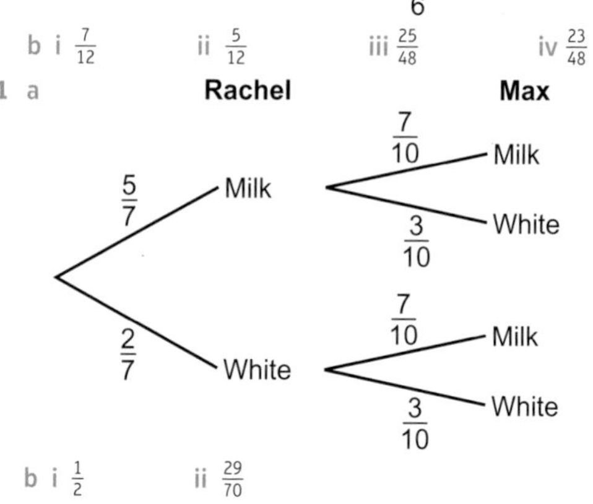

b i $\frac{1}{2}$ ii $\frac{29}{70}$

12 a

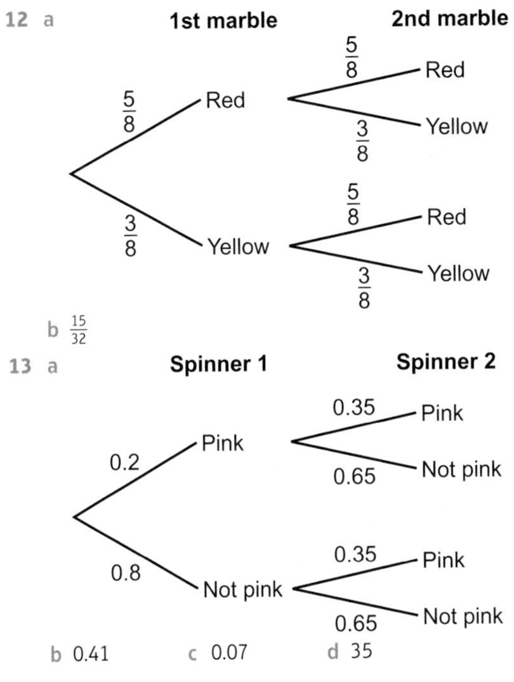

b $\frac{15}{32}$

13 a

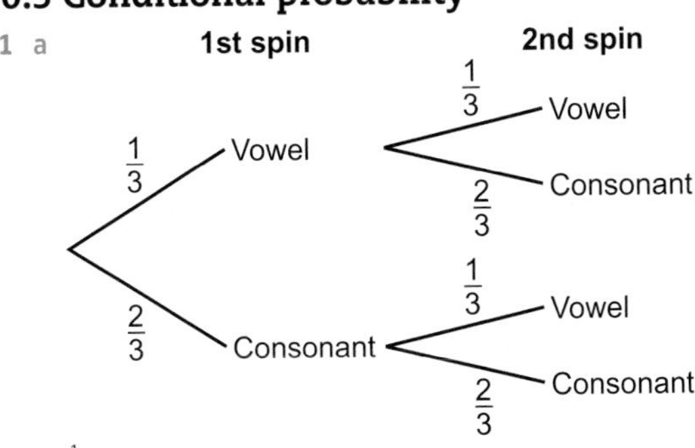

b 0.41 c 0.07 d 35

10.5 Conditional probability

1 a

1st spin 2nd spin

$\frac{1}{3}$ Vowel — $\frac{1}{3}$ Vowel, $\frac{2}{3}$ Consonant

$\frac{2}{3}$ Consonant — $\frac{1}{3}$ Vowel, $\frac{2}{3}$ Consonant

b $\frac{1}{9}$

2 a $\frac{47}{130}$ b $\frac{87}{130}$ c $\frac{23}{60}$

3 a dependent
b independent
c independent
d dependent
e independent

4 a 0.04 b 0.03 c 0.91

5 a

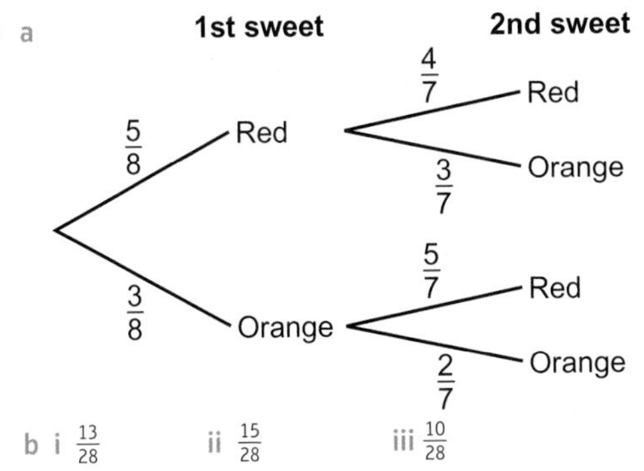

b i $\frac{13}{28}$ ii $\frac{15}{28}$ iii $\frac{10}{28}$

6 11.25%

7 0.7875

8 a

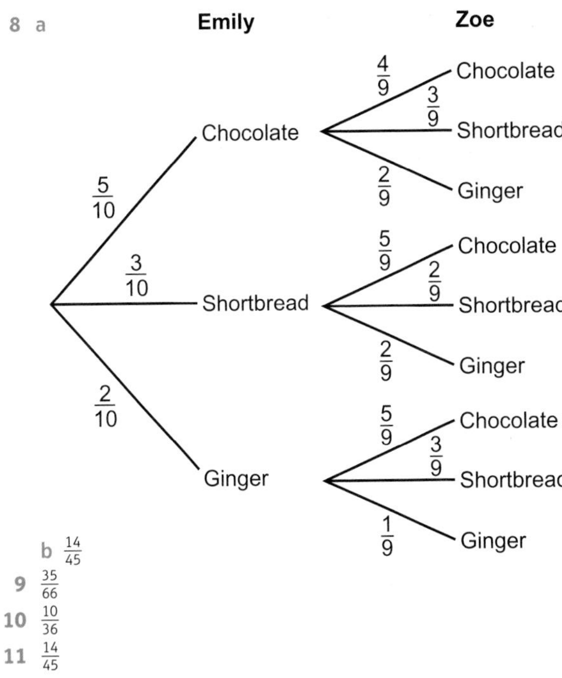

b $\frac{14}{45}$

9 $\frac{35}{66}$

10 $\frac{10}{36}$

11 $\frac{14}{45}$

10.6 Venn diagrams and probability

1 a 2 b 32 c 17 d 100

2 a A = {2, 4, 6, 8}; B = {2, 3, 5, 7}
b i true ii false iii true

3 a {1, 3, 5, 7, 9, 11, 13, 15}
b {1, 4, 9}
c {1, 2, 3, 4, 5, 6, 7, 8, 9, 10, 11, 12, 13, 14, 15}
d Q
e P: odd numbers < 16;
ξ: positive numbers < 16

4 a {1, 3, 4, 5, 7, 9, 11, 13, 15}
b {1, 9}
c {2, 4, 6, 8, 10, 12, 14}
d {2, 3, 5, 6, 7, 8, 10, 11, 12, 13, 14, 15}
e {4}
f {3, 5, 7, 11, 13, 15}

5 a $\frac{5}{12}$ b $\frac{1}{2}$ c $\frac{1}{12}$ d $\frac{5}{6}$
e $\frac{7}{12}$ f $\frac{1}{2}$ g $\frac{1}{3}$ h $\frac{2}{3}$

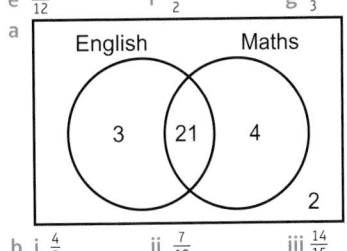

6 a

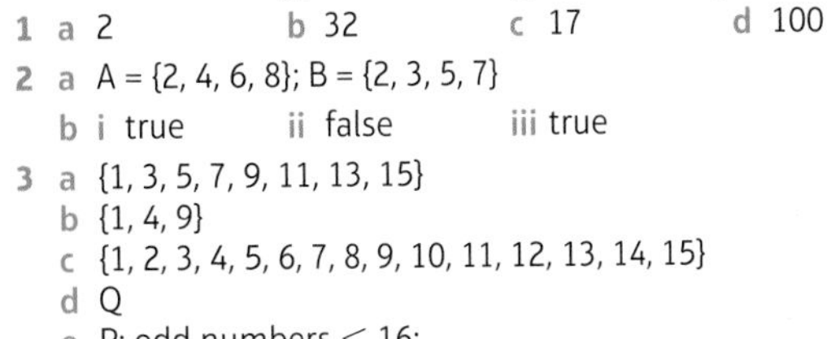

b i $\frac{4}{5}$ ii $\frac{7}{10}$ iii $\frac{14}{15}$ iv $\frac{2}{15}$

7 a

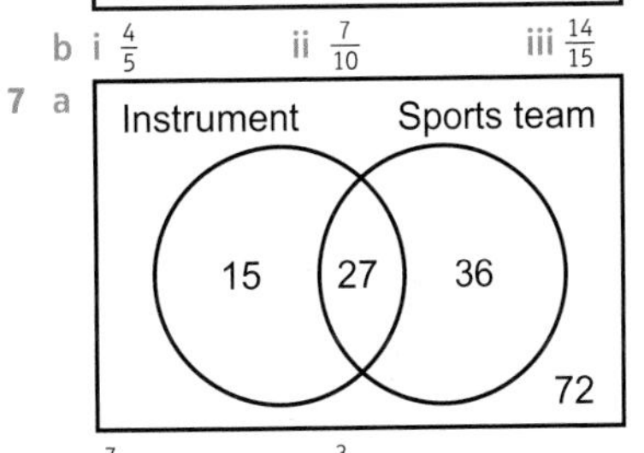

b $\frac{7}{25}$ c $\frac{3}{7}$

8 a 120
b i $\frac{31}{60}$ ii $\frac{7}{24}$ iii $\frac{22}{47}$

9 a 7 b 40
c i $\frac{7}{40}$ ii $\frac{11}{40}$ iii $\frac{11}{13}$

10 a

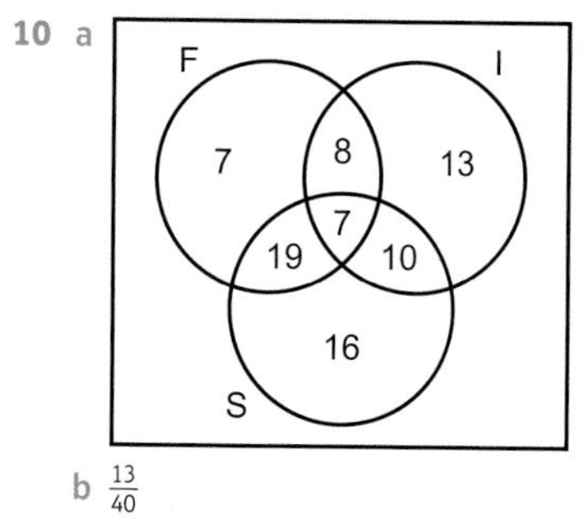

b $\frac{13}{40}$ **c** $\frac{1}{2}$

10 Problem solving

1 and **2**

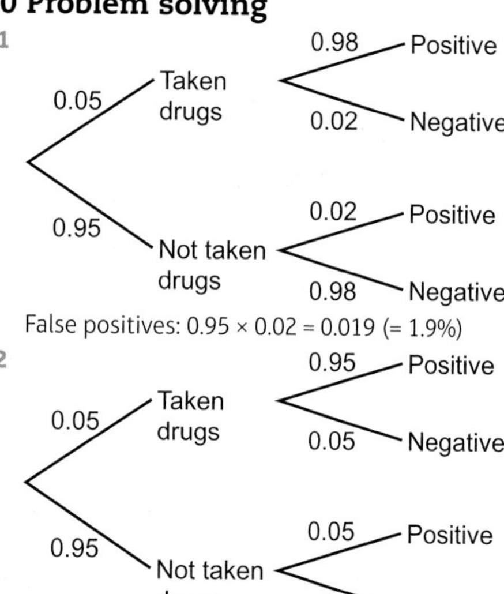

True positive: 0.05 × 0.95 = 0.0475
False positive: 0.95 × 0.05 = 0.0475
There would be the same amount of false positives as genuine positive results. This means that half the people that received positive results would be innocent.

3 Students may have different arguments to make.
Probability of positive for test A = 0.05 × 0.98 + 0.95 × 0.02 = 0.068
Number of retests = 600 × 0.068 = 41
Cost = 641 × £52 = £33 332
Probability of positive for test B = 0.05 × 0.95 + 0.95 × 0.05 = 0.095
Number of retests = 600 × 0.095 = 57
Cost = 657 × £40 = £26 280
It would be significantly cheaper to use test B. However we have shown that for test B, 4.75% of results are a false positive. So this could result in at least one of the retests giving a false positive.

10 Check up

1 a 0.75 **b** 0.075

2 a $\frac{4}{5}$ **b** $\frac{7}{10}$

3 a

		Dice					
		1	**2**	**3**	**4**	**5**	**6**
Spinner	**2**	3	4	5	6	7	8
	4	5	6	7	8	9	10
	6	7	8	9	10	11	12
	8	9	10	11	12	13	14

b i $\frac{1}{8}$ **ii** $\frac{1}{3}$ **iii** $\frac{3}{8}$

4 $\frac{2}{17}$

5 0.15

6 a

Number	1	2	3	4	5	6
Probability	0.2	0.3	0.1	0.15	0.1	0.15

b 45
c No, the probabilities are different. If the spinner was fair the probabilities would all be the same.

7

Spinner A — Spinner B

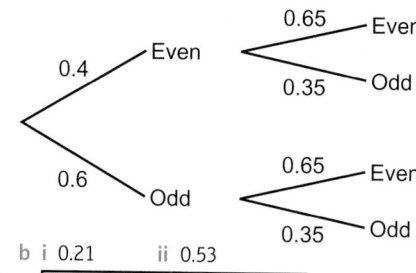

- 0.4 Even
  - 0.65 Even
  - 0.35 Odd
- 0.6 Odd
  - 0.65 Even
  - 0.35 Odd

b i 0.21 **ii** 0.53

8 a

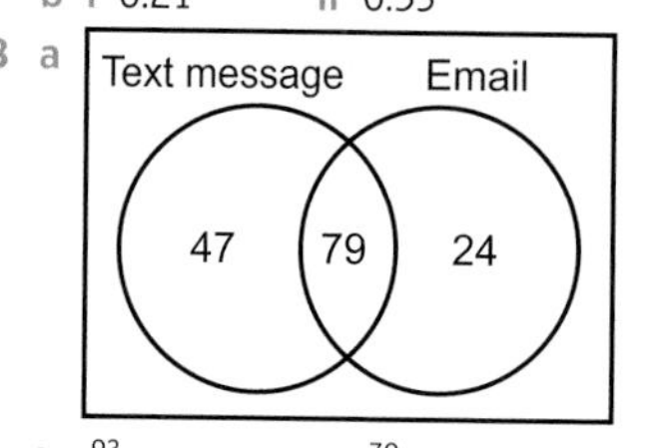

b $\frac{93}{140}$ **c** $\frac{79}{126}$

9 a R = {2, 3, 5, 8}
b R′ = {1, 4, 6, 7, 9, 10}
c $R \cap S$ = {3, 5}
d ξ = {1, 2, 3, 4, 5, 6, 7, 8, 9, 10}
e Yes

11 Students' own answers

10 Strengthen

Calculating probability

1 a $\frac{1}{5}$ **b** $\frac{4}{5}$ **c** $\frac{4}{5}$
d The answers are the same.
e i $\frac{3}{10}$ **ii** $\frac{7}{10}$

2 a $\frac{1}{10}$ **b** $\frac{3}{10}$ **c** $\frac{2}{5}$
d The total of answers to parts **a** and **b** is the answer to part **c**.
e $\frac{1}{2}$

3 0.15

4 a

		1st Match		
		Win	**Draw**	**Lose**
2nd Match	**Win**	W, W	D, W	L, W
	Draw	W, D	D, D	L, D
	Lose	W, L	D, L	L, L

b 9 **c** $\frac{1}{9}$
d (W, W), (D, W), (L, W), (W, D) or (W, L) **e** $\frac{5}{9}$

5 a

		Spinner A				
		2	**2**	**4**	**4**	**6**
Spinner B	**1**	2, 1	2, 1	4, 1	4, 1	6, 1
	2	2, 2	2, 2	4, 2	4, 2	6, 2
	2	2, 2	2, 2	4, 2	4, 2	6, 2
	3	2, 3	2, 3	4, 3	4, 3	6, 3
	3	2, 3	2, 3	4, 3	4, 3	6, 3

25 outcomes

b i $\frac{2}{5}$ **ii** $\frac{8}{25}$ **c** two even numbers

d

		Spinner A				
		2	**2**	**4**	**4**	**6**
Spinner B	**1**	3	3	5	5	7
	2	4	4	6	6	8
	2	4	4	6	6	8
	3	5	5	7	7	9
	3	5	5	7	7	9

e 5 f $\frac{13}{25}$

6 a 77 b $\frac{41}{77}$

7 a $\frac{10}{19}$; multiply and add b $\frac{3}{10}$; add

c $\frac{14}{95}$; multiply d $\frac{3}{10}$; multiply

Experimental probability

1 a $\frac{1}{20}$ b 25 c 36

2 a 30

b No, as the expected number is double the amount that Dylan did get.

3 a $\frac{1}{5}, \frac{3}{20}, \frac{7}{40}, \frac{17}{80}, \frac{21}{80}$ b 30

Tree diagrams and Venn diagrams

1 a without replacement b with replacement

c without replacement

2 a and b

3 a **1st counter** **2nd counter**

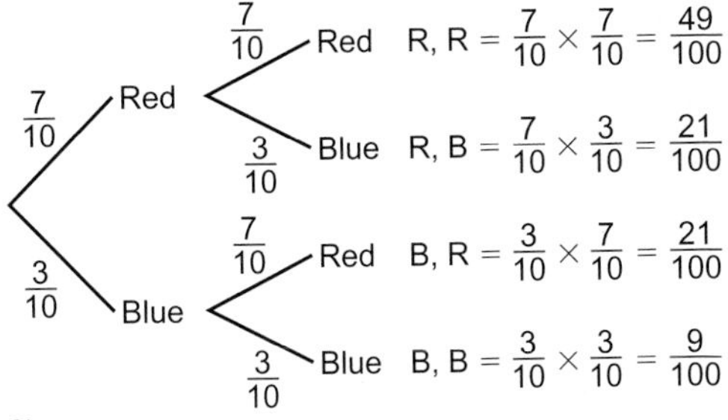

b $\frac{21}{50}$

4 a **1st crayon** **2nd crayon**

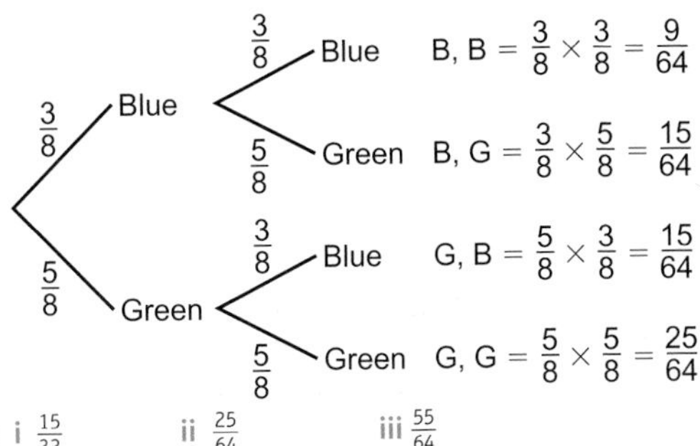

b i $\frac{15}{32}$ ii $\frac{25}{64}$ iii $\frac{55}{64}$

5 a **1st card** **2nd card**

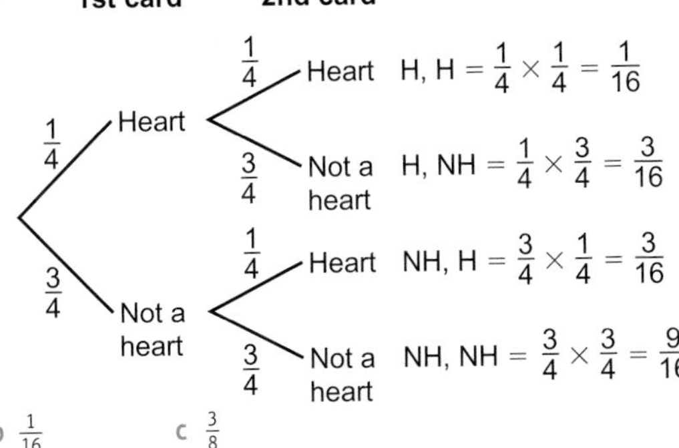

b $\frac{1}{16}$ c $\frac{3}{8}$

6 a $\frac{3}{5}$ b $\frac{5}{9}$

c **1st chocolate** **2nd chocolate**

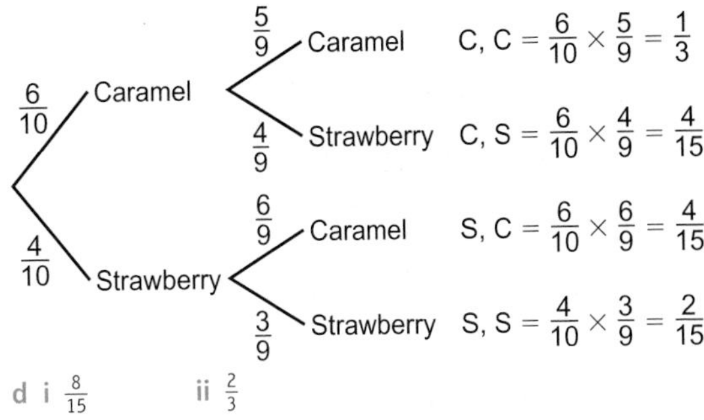

d i $\frac{8}{15}$ ii $\frac{2}{3}$

7 a **1st cartoon** **2nd cartoon**

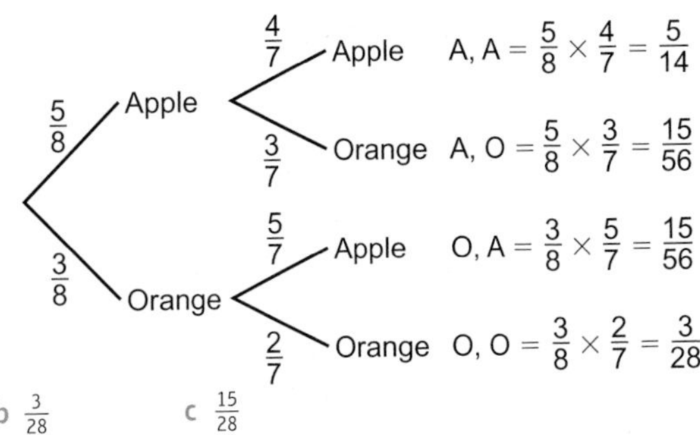

b $\frac{3}{28}$ c $\frac{15}{28}$

8 a **1st balloon** **2nd balloon**

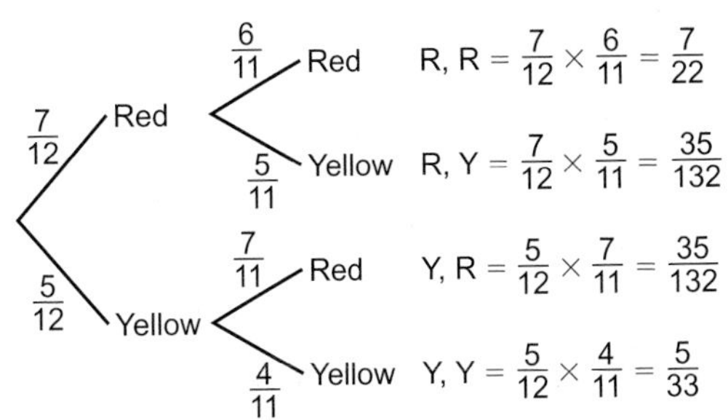

b $\frac{31}{66}$ c $\frac{35}{132}$

9 a i, ii and iii

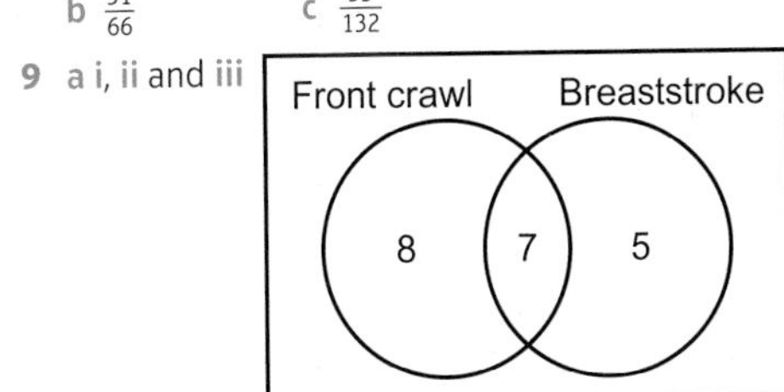

b 20

c i $\frac{7}{20}$ ii $\frac{2}{5}$ d $\frac{7}{15}$

10 a

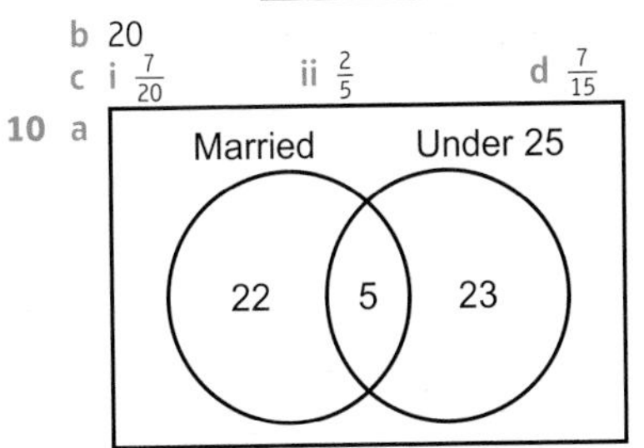

b $\frac{27}{50}$ c $\frac{5}{28}$

11 a i {50, 75, 100, 150}

ii {100, 150, 200, 250, 300}

iii {50, 75, 100, 150, 200, 250, 300}

iv {100, 150}

v {50, 75, 100, 125, 150, 175, 200, 225, 250, 275, 300}

b i $200 \in B$

ii $175 \in \xi$

iii $100 \in A \cap B$

10 Extend

1 Any multiple of: 2 red, 2 green, 5 blue and 1 yellow.

2 Example: If both cards are the same colour, both players turn over the next card. The winner is the first person to turn over a red card when the other player has turned over a black card.

Check that any rule given by the student gives the same probability for player A and player B.

3 a 0.35 b 20

4 a i 24% ii 32% b 111 days c 19.36%

5 $\frac{1904}{4495}$

6 $\frac{8}{25}$

7 a $\frac{5}{14}$ b $\frac{1}{28}$ c $\frac{2}{7}$

8 a $\frac{1}{12}$ b $\frac{1}{12}$ c $\frac{5}{96}$ d $\frac{5}{96}$

9 51

10 $\frac{27}{45}$

11 a $A \cup B$ b $B' \cup A$ c $B \cap C \cap A'$

10 Unit test

Sample student answer

a Labels to show the flavour each branch represents are missing from the tree diagram.

b Labels to show the combination that each calculation represents are missing.

c There should be a sentence to clearly state the answer to the question.

UNIT 11

11 Prior knowledge check

1 £2.60

2 £67.50

3 5 minutes

4 2000 g

5 6 pint bottle is cheaper, e.g. cost of 2 pints: 4 pint bottle 49p, 6 pint bottle 48p

6 a 6 days b 3 days

7 a 1:1000 b 1:10 c 1:1000 d 1:60
e 1:60

8 a 1.8 m b 280 m c 54.6 km

9 a 48 inches b 15 feet c 4 feet 10 inches

10 a 80 fluid ounces b 40 pints
c 2 gallons 4 pints

11 a 64 km b 30 miles

12 a 90 minutes b 3000 seconds
c 3 hours 45 minutes

13 a $806.50 b £49.60

14 1 cm = 10 mm; 1 m = 100 cm
1 cm^2 = 100 mm^2; 1 m^2 = 10 000 cm^2
1 cm^3 = 1000 mm^3; 1 m^3 = 1 000 000 cm^3

15 a $t = \frac{v - u}{a}$ b $M = DV$ c $A = \frac{F}{P}$

16 a 50 b 104

11.1 Growth and decay

1 a 1.3 b 0.86 c 1.072 d 0.975

2 a 0.65 b £4225 c 0.85 d £3591.25
e 0.5525

3 a $1.12^3 = 1.404$ (3 d.p.) b $0.85^4 = 0.522$ (3 d.p.)

4 £36 949.50

5 No; 1.15 × 1.22 = 1.403, which is equivalent to a 40.3% increase.

6 a 1.0815 b 0.68 c 1.0246

7 £38 024

8 £5412

9 £2719.62 (to the nearest penny)

10 £3792.88 (to the nearest penny)

11 Students' own answers

12 £232.33 (to the nearest penny)

13 a £209.70
b The cost of her train ticket before the increase was £225 ÷ 1.125 = £200, so her train ticket has gone up by £25. Her pay before the increase was £535.50 ÷ 1.05 = £510, so her pay has gone up by £25.50. Her pay increase is greater than the increase in the cost of the train ticket.

14 £3753.67

15 5 years

16 a 1263.5 b 21 hours

17 a 301
b The nearest whole number

18 449

11.2 Compound measures

1 a 6 b −30 c −0.125

2 a 16 km/h b 30 km c 3 hours

3 a 7 hours 30 minutes b 6 hours 12 minutes

4 a £399.50 b 3 hours

5 a i 1.5 litres ii 3.75 litres b 40 hours

6 a 16 km/litre b 4.1 litres (1 d.p.)

7 a 0.65 km/h b 7.8 km/h c 256 km/h d 188 km/h

8 a 3600 m/h b 43 200 m/h c 28 800 m/h
d 16 200 m/h e More

9

metres per second	kilometres per hour
15	54
20	72
30	**108**
45	**162**

10 900 km/h

11 Falcon is fastest. Car: 350 km/h = 97.2 m/s (1 d.p.); Falcon: 388.8 km/h = 108 m/s

12 a $\frac{1000x}{3600}$ b $\frac{3600y}{1000}$

13 53.8 km/h (1 d.p.)

14 26.4 km

15 1.8 km/h

16 44.7 m/s (1 d.p.)

17 2 m/s^2

18 2.5 m/s

11.3 More compound measures

1 a 7500 g b 6.25 m^2 c 0.095 m^3

2 a $m = 30$ b $v = 16$

3 8.3 g/cm^3

4 2.4 g/cm^3

5 2047.5 g

6 675 cm^3

7 8 940 000 g/m^3

8 2700 kg/m^3

9 $1000x$ kg/m^3

10 Platinum is denser: gold density = 19.32 g/cm^3; platinum density = 21.45 g/cm

11 1.01 g/cm³ (2 d.p.)
12 17.3 N/m² (1 d.p.)
13 90 N
14

Force	Area	Pressure
60 N	2.6 m²	**23.1** N/m²
73.0 N	4.8 m²	15.2 N/m²
100 N	**8.33** m²	12 N/m²

15 0.153 N/cm²
16 12 000 N (3 s.f.)
17 a 784 N b 49 N/cm² c 18 375 N/m² d Sitting
18 a 500 000 N/m² b $\frac{x}{10000}$ N/m²

11.4 Ratio and proportion

1 a B b A c C
2 a 2 b 3 c 5 d 9
3 a A and D
b Graph of data in table A with points plotted at (2, 8) (4, 16) (6, 24) and (8, 32)
Graph of data in table D with points plotted at (2, 10) (4, 20) (6, 30) and (8, 40)
c Straight lines d A, $y = 4x$; B, $y = 5x$
4 a $A = \frac{3}{5}B$ b $P = \frac{7}{4}Q$ c $X:Y = 9:5$
5 a 1:1.6
b Graph with miles on the horizontal axis and kilometres on the vertical axis. Points plotted at (8, 10) (10, 16) (15, 24) and (20, 32) and joined with a straight line.
c Yes, they are in direct proportion. When plotted the graph is a straight line from origin.
d Gradient = 1.6 e Kilometres = 1.6 × miles
6 a Yes, s is in direct proportion to t as $\frac{8}{10} = \frac{16}{20} = \frac{24}{30} = \frac{32}{40} = \frac{40}{50}$
b $t = 1.25s$ c 20 miles
7 Students' own answers
8 a Students' own answers, e.g.
Table of values:

x	0	5	10	20
C	0	6.5	13	26

Graph plotted from the table of values is a straight line, from origin so in direct proportion.
b $C = 1.3x$ c £71.50
9 2 hours 9 minutes
10 a 6 hours 40 minutes b 13 hours 20 minutes
c For both parts **a** and **b**, $H \times N = 40$
11

A	10	20	14	**2**	**5**
B	14	**7**	**10**	70	28

12 a Direct b Indirect c Neither d Indirect
e Direct
13 15 amps
14 a $r = \frac{4.5}{t}$ b 1.125
15 a

x	1	2	5	10
y	**10**	**5**	**2**	**1**

b
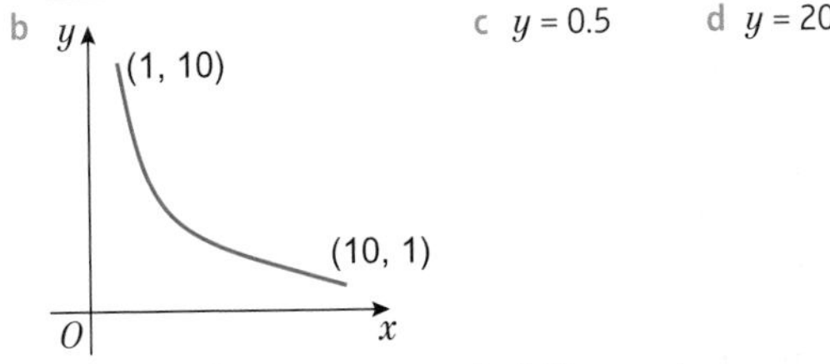

c $y = 0.5$ d $y = 20$
16 a 960 seconds b 560 seconds

11 Problem-solving

1 Profit of £23.07
2 60 cm
3 528 inches per second
4 3 hours 30 minutes
5 5.472 tonnes
6 3p

11 Check up

1 £7200
2 £3869.28
3 6059
4 5 years
5 a £342.23 b 6 hours
6 320 seconds
7 0.8 g/cm³
8 a 8050 kg/m³
b 1006.25 g or 1.00625 kg
9 9.375 N/m²
10 Usain Bolt is faster: Usain Bolt: 12.3 m/s = 44.2 km/h; White shark: 11.1 m/s = 40 km/h
11 a Yes. Values are in same ratio.
b $E = 1.3P$ c €32.50
12 When d = 8, P = 0.8, so $P = 0.1d$
When d = 75, the pressure on the watch will be 75 × 0.1 = 7.5 bars.
This is less than 8.5 bars, so the watch will still work.
13 10.5 amps
15 5 km race leader board: Allia, Chaya, Hafsa, Billie
10 km race leader board: Fion, Daisy, Gracie, Ellie

11 Strengthen

Percentages

1 a 1.2 b 1.09 c 1.037
2 a 0.77 b 0.94 c 0.925
3 a 1.308 b 1.265 c 0.7238 d 0.8099
e 1.0304
4 £605.63
5

Year	Amount at start of year	Amount plus interest	Total amount at end of year
4	£437.09	437.09 × 1.03 = **400 × 1.03⁴**	**£450.20**
5	**£450.20**	**£450.20 × 1.03 = 400 × 1.03⁵**	**£463.71**
6	**£463.71**	**£463.71× 1.03 = 400 × 1.03⁶**	**£477.62**

6 £6144
7 557
8 5 years

Compound measures

1 £302.60
2 a 3 l/min b 4 minutes
3

Metal	Mass (g)	Volume (cm³)	Density (g/cm³)
Copper	**1090**	122	8.96
Lead	450	**39.8**	11.3
Mercury	110	8.15	**13.5**

4

Force (N)	Area (cm²)	Pressure (N/cm²)
104	13	8
48	12	**4**
65	**5**	13

5 a **1000** g = 1 kg b **10 000** cm² = 1 m²
c **1 000 000** cm³ = 1 m³

6 a i 12 kg ii 15 000 g
b i 27 g/cm² ii 450 kg/m²
iii 50 000 kg/m³ iv 0.02 g/cm³

7

km/h	m/h	m/min	m/s
18	**18 000**	**300**	**5**
36	**36 000**	**600**	10
24	**24 000**	**400**	**6.67**
57.6	**57 600**	**960**	16

Ratio and proportion

1 $W = 24, X = 22.5, Y = 30, Z = 18$
2 200 seconds
3 a Table of values:

P	10	25	50	100
E	8	20	40	80

Graph plotted from the table of values; points joined with a straight line through the origin.
b $P = 1.25E$
4 64 N
5 $W = 6, X = 3, Y = 4, Z = 4$

11 Extend

1 a $T = 25x$ b 375 N c 24 cm
2 1100 N
3 a 162 km/h b 16 m/s² c 10 m/s²
d $a = \frac{v}{20}$ m/s²
4 $n = 4$
5 a $0.91x$ b 8 years
6 £10787.82
7 1.025 × 1.015 = 1.040375 is equivalent to just over 4% after 2 years, so 2.5% then 1.5% is preferable to 3.5%.
8 363 g
9 a

Number of years, n	0	1	2	3	4
Value, y	3000	3100	3200	3300	3400

b Graph plotted from the table of values; points joined with a line.
c 3.5 years
10 C
11 a Sam is correct. Exterior angle × number of sides = constant, so if number of sides is doubled exterior angle is halved
b 18°
12 148
13 1 hour 45 minutes

11 Unit test

Sample student answers

Student B gives the better answer as they have written a sentence at the end answering the question. It is also easier to follow Student B's working as they have labelled their working as 'International Bank' and 'Friendly Bank'.

UNIT 12

12 Prior knowledge check

1 a 1.5 b 6.4
2 a 36 b 3 c $\frac{4}{7}$ d 8 e 4 f $\frac{3}{2}$
3 a $\frac{2}{3}$ b $\frac{1}{5}$
4 a $x = \frac{4}{3} = 1\frac{1}{3}$ b $x = \frac{32}{3} = 10\frac{2}{3}$
5 C and G
6 a 2 b $\frac{1}{2}$
7 a 24 cm b 20 cm²
8 a $p = r, q = s$ (vertically opposite)
b $u = w, t = v$ (corresponding angles)
9 Accurate construction of the triangle. (If correct, angle between 6 cm line and 8 cm line will be 90°.)
10 Accurate constructions of the triangles
11

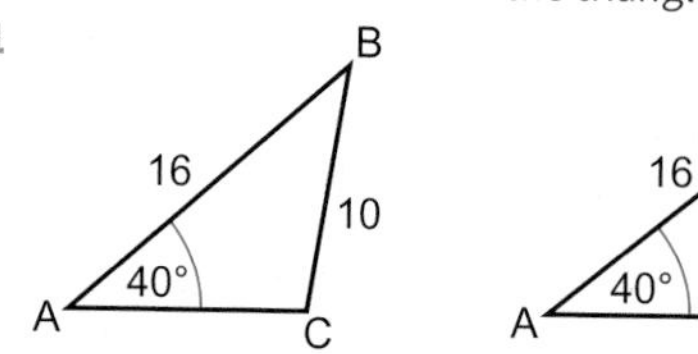

12.1 Congruence

1 A and D; C and F
2 $a = 134°$ (vertically opposite); $b = 134°$ (alternate with 134°); $c = 46°$ (angles on a straight line)
3 12 cm
4 a 112° b 5 cm
5 a SAS b RHS c SSS d AAS e AAS
6 a Congruent, SAS. A corresponds to R, B corresponds to Q, C corresponds to P.
b Not congruent
7 DEF (SSS) and GHI (SAS)
8 Yes, congruent, SSS (using Pythagoras to find the missing sides)
9 No. All the triangles with a 12 cm hypotenuse will be congruent, and all the triangles where 12 cm is not the hypotenuse will be congruent.
10 a ∠EBA = 50°, ∠EAB = 20°, ∠EDC = 20°, ∠CED = 110°
b Two angles and a corresponding side are equal (AAS).
11 a 119° b 119° c 35°
d Suitable proof, e.g. Two angles and a corresponding side are equal (AAS), so JKL and JML are congruent.

12.2 Geometric proof and congruence

1

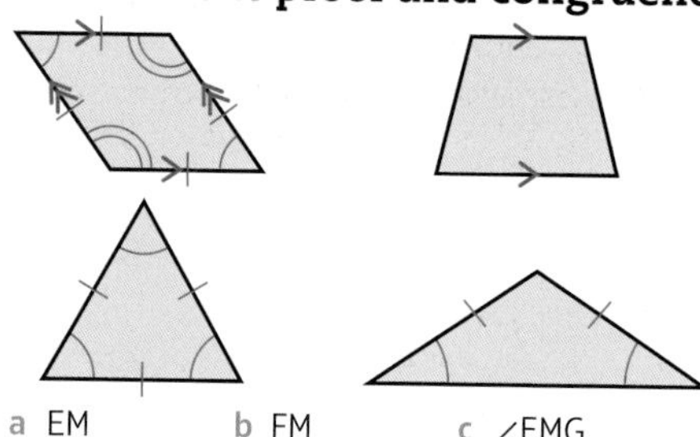

2 a EM b FM c ∠EMG
3 a Suitable proof, e.g. ∠WXT = ∠WZY = 90°; WY is common (hypotenuse); WZ = XY. Therefore the triangles are congruent (RHS).
b ∠WYZ
4 a Suitable proof, e.g. SQ is common; PQ = SR; PS = QR. Therefore the triangles are congruent (SSS).
b i 123° ii 28.5°
5 Suitable proof, e.g. LX = XM; XK = XL; ∠JXK = ∠LXM (vertically opposite). Therefore the triangles are congruent (SAS).

6 Suitable proof, e.g. ABCD is a rhombus, so ∠ABC = ∠ADC, BA = AD and BC = CD. Therefore the triangles are congruent (SAS).

7 a FG = GH; EG is common; ∠FEG = ∠GEH = 90°. Therefore the triangles are congruent (RHS).
  b If the triangles are congruent, then FE = EH, therefore FE = $\frac{1}{2}$FH.

8 SM is common, RS = ST; ∠SMR = ∠SMT = 90°. Therefore triangles RSM and MST are congruent (RHS). If the triangles are congruent, then RM and MT are the same length and the line SM bisects the base.

9 PQ = ST; ∠QPR = ∠RTS (alternate); ∠PQR = ∠RST (alternate). Therefore triangles PQR and RST are congruent (AAS). If the triangles are congruent, then PR = RT and R is the midpoint for PT.

10 a Suitable proof, e.g. ∠GDE = ∠GFC (alternate); DE = CF; ∠DEG = ∠FCG (alternate). Therefore triangles DEG and CFG are congruent (AAS).
  b Suitable proof, e.g. ∠DCG = ∠GEF (alternate); ∠GDC = ∠GFE (alternate); DC = EF. Therefore triangles CDG and EFG are congruent (AAS).
  c Triangles DEG and CFG are congruent, so DG = FG. Triangles CDG and EFG are congruent, so CG = GE. Therefore G is the midpoint of CE and DF.

11 a, b

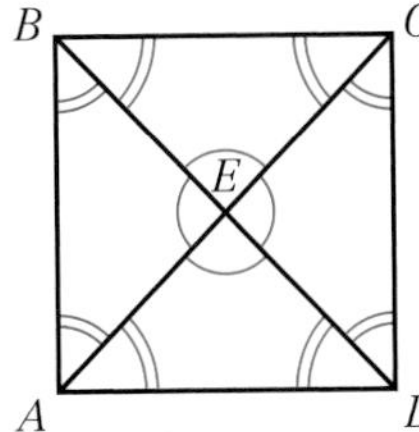

  c BEC, CED, DEA and AEB are congruent. BCD, ACD, ABD and ABC are congruent.
  d They are all equal, therefore they are all 90°.
  e Since all 4 triangles are congruent, AE = EC and BE = ED, therefore E is the midpoint of both AC and BD.

12

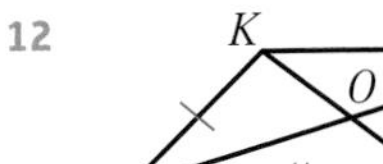

Suitable proof, e.g. ∠NKO = ∠MLO (alternate angles); ∠KNO = ∠MPK (alternate angles); LM = KN. Therefore triangles KON and LMO are congruent (AAS).
∠OKL = ∠OMN (alternate angles); ∠KLO = ∠ONM (alternate angles); KL = NM. Therefore triangles OMN and KOL are congruent (AAS).
Since the triangles are congruent, KO = MO, NO = LO, so O is the midpoint of both diagonals.

13 XY = XZ (XYZ is isosceles); XA = XB; ∠BXA is common. Therefore triangle XBY and XAZ are congruent (SAS).

12.3 Similarity

1 a 2 b $\frac{1}{2}$

2 PQ and XY, PR and XZ, QR and YZ

3 a PQ = 1.5 cm; XY = 3 cm; PR = 1.9 cm; XZ = 3.8 cm; QR = 1.6 cm; YZ = 3.2 cm
  ∠QPR = ∠YXZ = 54°; ∠PQR = ∠XYZ = 76°; ∠QRP = ∠YZX = 50°
  b All are $\frac{1}{2}$.

4 a i TU ii UV b $\frac{4}{9}$ c $\frac{5}{12}$
  d The ratios of corresponding sides are not the same, therefore the parallelograms are not similar.

5 a $\frac{10.5}{7} = \frac{3}{2}$ b $\frac{FG}{BA} = \frac{6}{4} = \frac{3}{2}$
  c Yes; the ratios of corresponding sides are the same.

6 a Similar b Similar c Not similar

7 ∠DCB = ∠YXW = 155°
All the angles are the same, therefore the shapes are similar.

8 55 cm

9 20 cm

10 10 m

11 a The corresponding angles are all equal. b 15 cm
  c All right-angled triangles with one other angle the same are similar.
  d 0.5 e sine

12 14 cm

13 a 6 cm b 2.8 m

14 a Yes; corresponding sides are all in the same ratio.
  b No; corresponding sides are not in the same ratio.

15 a Each shape has 6 equal sides and 6 equal angles, so they are similar.
  b Yes

16 a 12.7 cm b 17.7 cm

12.4 More similarity

1 a $\frac{3}{2}$ b $\frac{2}{3}$

2 a ∠EDC = EBA (alternate); ∠DCE = ∠EAB (alternate); ∠CED = ∠AEB (vertically opposite). Therefore all angles are equal and the triangles are similar.

3 a ∠RPQ =∠RTS (alternate); ∠PQR = ∠RST (alternate); ∠PRQ = ∠SRT (vertically opposite). Therefore all angles are equal and the triangles are similar.
  b 10 cm

4 a ∠F is common; ∠FGH =∠FJK (corresponding); ∠FHG = ∠FKJ (corresponding). Therefore all angles are equal and the triangles are similar.
  b 60 mm c 64 mm

5 a ∠PQN = 52°; ∠LMN = 102°
  b ∠L is common; ∠MNL = ∠PQN (corresponding); ∠LMN = ∠MPQ (corresponding). Therefore all angles are equal and the triangles are similar.
  c 44 cm d 22 cm e 18 cm

6 308 m tall

7 Perimeter = 54 m; area = 135 m^2

8 28 800 cm^2

9 Perimeter = 5 cm; area = 2.1 cm^2

10 a 18 cm b 7.5 cm

11 5 cm

12 a 24 cm^2 b 54 cm^2

13 12 cm, 15 cm, 19.2 cm

14 a 4 b 2 c 21 cm

15 30 cm

12.5 Similarity in 3D solids

1 a 5 b $\frac{4}{3}$

2 1500 cm^3

3

Linear scale factor	Volume A	Volume B	Volume scale factor
2	2	16	8
k	24	$24k^3$	k^3

4 96 cm^3

5 405 cm^3

6 60 cm^3

7 7.5 cm

8 a 21 cm b 6 cm

9 a 125 b 5 c 25 d 1500 cm^2

10 563 cm^2

11 4220 cm^2

12 Area scale factor = $\frac{92}{207} = \frac{4}{9}$.
So linear scale factor = $\frac{2}{3}$ and volume scale factor = $\frac{8}{27}$.
Volume of cone B = 837 × $\frac{8}{27}$ = 248 cm^3

13 a 640 cm^3 b 40 cm^2

14 8.44 litres

15 All lengths in a cube are the same, so all cubes have sides in the same ratio. Cuboids may vary.

12 Problem-solving

1

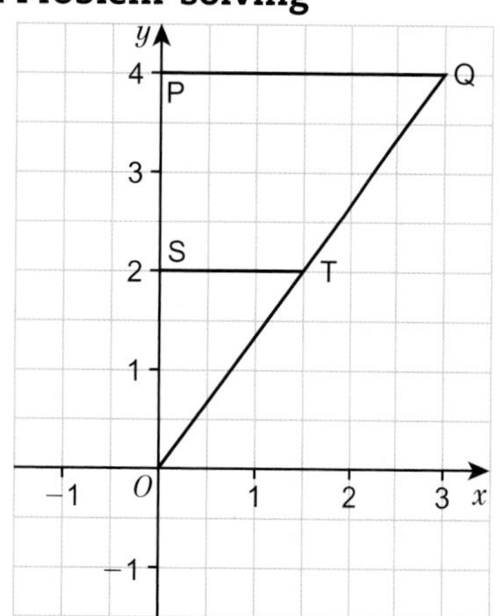

$\frac{PQ}{ST} = \frac{OP}{OS} = \frac{OQ}{OT} = 2$. All corresponding sides are in the same ratio, therefore, triangles OPQ and OST are similar.

2 7.2 m

3 1 m

4 a 60°

b Proof using SSS or SAS that the triangles are congruent.

5 Interior angles of triangle are 84°, 48° and 48°. Two angles are equal, so the triangle is isosceles.

6 An 8 m ladder reaching a 7.8 m high gutter would be 1.77 m (2 d.p.) away from the base of the wall. The 4 in 1 rule requires it to be 1.95 m away.

12 Check up

1 A and C are congruent (SAS)

2 a, b Suitable proof using any of SSS, SAS, RHS, e.g. ∠JHK = ∠HKL; ∠JKH = ∠LHK; HK is common. Therefore the triangles are congruent (AAS).

3 a All corresponding sides are in the same ratio (scale factor 2).

b Missing angles are 58° and 35°, therefore all angles are equal and the triangles are similar.

4 a 20 cm b 18 cm

5 a ∠A is common; ∠EBA = ∠DCA (corresponding); ∠EDC = ∠AEB (corresponding). All angles are equal so the triangles are similar.

b CD = 32 cm

6 a ∠PQR = ∠RST (alternate); ∠QPR = ∠RTS (alternate); ∠PRQ = ∠SRT (vertically opposite). All angles are equal so the triangles are similar.

b $x = 40$ cm; $y = 19.5$ cm

7 Area = 200 cm^2; perimeter = 55.2 cm

8 2580.5 cm^3

9 519 cm^3

11 Right-angled triangles with an angle of 45° are similar, and tan 45 = 1

Right-angled triangles with an angle of 60° are similar, and cos 60 = 0.5

12 Strengthen

Congruence

1 D

2 A and C

3 Rectangle, parallelogram (×2 ways), kite, isoscles triangle (×2 ways)

4 a SSS b RHS c AAS

5 a Suitable proof, e.g. KM is common, LM = NM, KN = KL. Therefore the triangles are congruent (SSS).

b LN is common but the other sides do not correspond.

Similarity in 2D shapes

1 a i AB and ED, DF and BC, AC and EF

ii ∠ABC = ∠EDF, ∠ACB = ∠EFD, ∠CAB = ∠FED

b i IK and IJ, GI and HI, GK and HJ

ii ∠IJH = ∠IKG, ∠KIG = ∠JIH, ∠IHJ = ∠IGK

c i LM and OP, MN and PQ, LN and OQ

ii ∠OPQ = ∠LMN, ∠POQ = ∠MLN, ∠PQO = ∠MNL

d i SR and WU, ST and WT, RT and UT

ii ∠RST = ∠UWT, ∠SRT = ∠WUT, ∠STR = ∠WTU

2 a

C	D	$\frac{C}{D}$
3	6	$\frac{1}{2}$
5	x	$\frac{5}{x}$
y	8	$\frac{y}{8}$

c $x = 10$, $y = 4$

3 1.5 cm

4 $a = 4$ cm, $b = 20$ cm, $c = 12$ cm, $d = 6$ cm

5 C and E

6 a i Alternate angles ii Alternate angles

iii Vertically opposite angles

b The triangles are similar.

c The paired angles are the same size.

d $x = 10$ cm, $y = 11$ cm

7 a

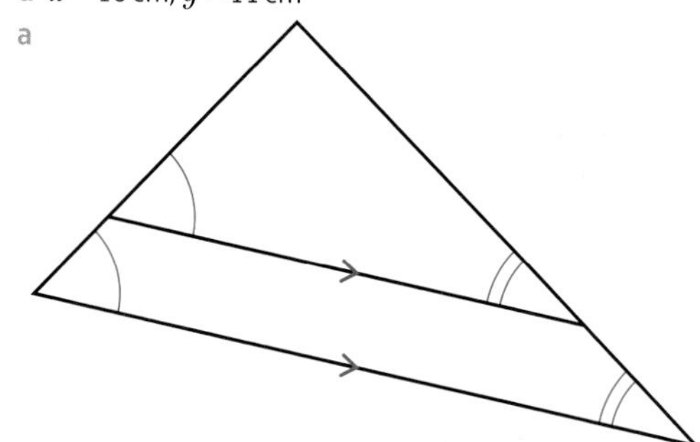

b ∠ACB = ∠AED (corresponding); ∠ABC = ∠ADE (corresponding); ∠CAB = ∠EAD (angle is common). All three angles are equal, so the triangles are similar.

c Students' sketches.

d Scale factor = $\frac{5}{4}$. BD = 13 cm, ED = 80 cm

8 a Drawing of rectangle A (2 by 5)

b Drawing of rectangle B (4 by 10)

c Perimeter of A = 14, perimeter of B = 28

d Scale factor is 2.

e Drawings of rectangle C (6 by 15)

f 42

9 a 5 b 60 cm

10 Scale factor is $4 = 2^2$

11 150 cm^2.

Similarity in 3D solids

1 a 8 b Scale factor is $8 = 2^3$ c 3

d Scale factor is $27 = 3^3$ e Students' own answers

2 a 9 b 3 c 27 d 337.5 cm^3

12 Extend

1 a Students' drawings b 20 cm

2 a ∠JAK = ∠BAC (angle is common); ∠AKJ = ∠ACB; ∠AJK = ∠ABC (from sum of angles in triangle = 180°). All three angles are the same, so the triangles are similar.

b 18 cm c 5.4 cm

3 $1.5\,cm^2$

4 41 cm

5 $675\,cm^2$

6 53.76 kg

7 1.5 cm

8 a $13291.25\,cm^3$ b 75

9 Suitable proof, e.g. AEB is isosceles (AE = BE) therefore ∠EAB = ∠EBA.
By alternate angles, ∠EAB = ∠ECD and ∠EBA = ∠EDC so triangle CDE is isosceles.
So AC = BD (corresponding lengths are the same)
By alternate angles, ∠AGF = ∠FDC and ∠BHJ = ∠JCD, so ∠FDE = ∠JCE.
Therefore, by AAS, we know that triangles ACH and BDG are congruent.

10 30.8 cm

11 a 18 cm b $1350\pi\,cm^3$ c $400\pi\,cm^3$
d $950\pi = 2980\,cm^3$ to 3 s.f.

12 $3500\,cm^3$

13 AD = AB and AE = AG; ∠EAB = ∠DAG = 90° + ∠DAE.
Therefore the triangles are congruent (SAS).

14 Linear scale factor $= \frac{(x^2-1)}{2(x-1)} = \frac{(x+1)(x-1)}{2(x-1)} = \frac{x+1}{2}$

Area scale factor $= \left(\frac{x+1}{2}\right)^2$

Area of B $= 8 \times \left(\frac{x+1}{2}\right)^2 = \frac{8(x^2+2x+1)}{4} = 2x^2+4x+2$

12 Unit test

Sample student answer

a Drawing the relevant triangles next to each other, the same way up, makes it easier to match the corresponding angles to see if they are the same. It also avoids confusion with the other parts of the diagram.

b The student has explained each step of the answer clearly and separately to show how each angle was calculated, and has summarised a final proof statement.

UNIT 13

13 Prior knowledge check

1 a 1.5 b 5.8

2 13.3

3 5.62

4 22.0

5 104.5°

6 a 2 b $\sqrt{2}$

7 a 30.5° b 38.7°

8 a 35.0 cm b 12.4 cm

9 $21\,cm^2$

10 a

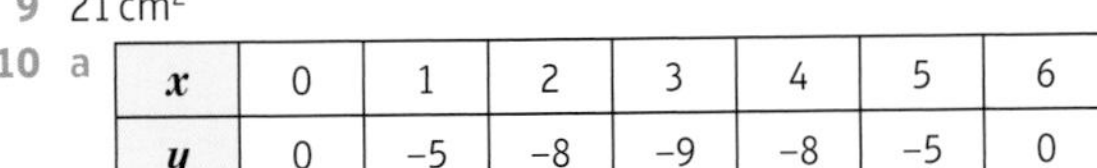

x	0	1	2	3	4	5	6
y	0	–5	–8	–9	–8	–5	0

b

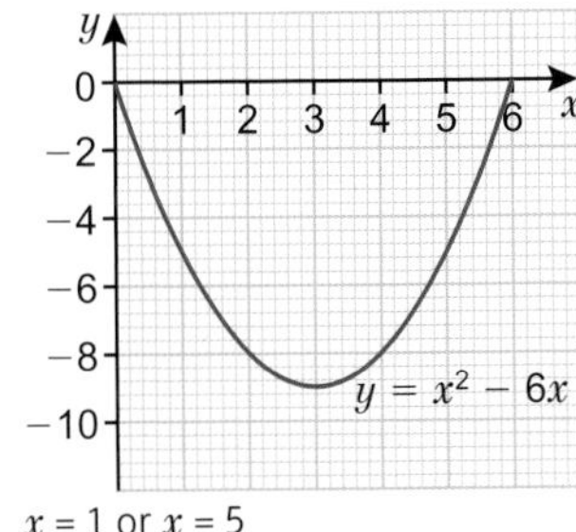

c $x = 1$ or $x = 5$

11 16

13.1 Accuracy

Key: UB = upper bound; LB = lower bound

1 a UB: $y = 3.65$; LB: $y = 3.55$ b UB: $z = 9.25$; LB: $z = 9.15$
c UB: $x = 33.7625$ d LB: $x = 32.4825$

2 a UB: $y = 1.25$; LB: $y = 1.15$ b UB: $z = 0.45$; LB: $z = 0.35$
c UB: $x = \frac{25}{7}$ d LB: $x = \frac{23}{9}$

3 a i 7.45 cm ii 8.65 cm b 11.416 (3 d.p.)
c i 7.35 cm ii 8.55 cm d 11.274 (3 d.p.)

4 a LB = 42.922° (3 d.p.); UB = 45.136° (3 d.p.)
b LB for x gives UB for cos x and UB for x gives LB for cos x

5 a UB: $x = 29.949$; LB: $x = 29.846$
b $x = 30°$ (to the nearest degree)

6 a UB: $x = 30.172$; LB: $x = 29.155$
b $x = 30°$ (to the nearest 10 degrees)

7 UB: $x = 7.938$; LB: $x = 7.757$

8 UB: $x = 275.213$; LB: $x = 223.811$

9 UB: $x = 3.187$; LB: $x = 3.110$

13.2 Graph of the sine function

1 a 1 b 0.6

2 0.28

3 a 0.6 b 0.4 c –0.2 d –0.8

4 a 1 b –1 c 0°, 180°, etc.

5 150°

6 a Decreases from 1 to 0 b Decreases from 0 to –1
c Increases from –1 to 0

7 a i 1 ii 0.96
b Reflection symmetry with mirror line $x = 90°$
d i 120° ii 135° iii 180° iv 150°
e Answers from 12° to 18°

8 Rotational symmetry of order 2 about (180, 0)

9 a Students' graph of $y = \sin x$ for the interval $0° \leqslant x \leqslant 540°$
b i 0 ii 1 c i $\frac{\sqrt{3}}{2}$ ii $\frac{\sqrt{3}}{2}$
d The graph repeats, so sin 420° is the same as sin 60°. The graph is symmetrical between $x = 360°$ and $x = 540°$, so sin 480° = sin 420°.

10 a 210°, 330°, 570°, 690° b 240°, 300°, 600°, 660°

11 A(90, 1), B(180, 0), C(270, –1), D(360, 0)

12 18.2°, 161.8°, 378.2°, 521.8°

13 56.4°, 123.6°, 416.4°, 483.6°

13.3 Graph of the cosine function

1 a 0.8 b 0.96

2 a 0.4 b –0.6

3 300°

4 330°

5 a Decreases from 0 to –1 b Increases from –1 to 0
c Increases from 0 to 1

6 a i –0.5 ii –1
b Reflection symmetry. Mirror line is $x = 180$.
c i 300° ii 270° iii 240° iv 360°

7 a Students' graph of $y = \cos x$ for the interval $0° \leqslant x \leqslant 720°$
b i 0.5 ii –0.5 c i $\frac{\sqrt{3}}{2}$ ii $-\frac{\sqrt{3}}{2}$

8 a 60°, 300°, 420°, 660° b 150°, 210°, 510°, 570°

9 A(90, 0), B(180, –1), C(360, 1)

10 a 35.9°
b Students' graph of $y = \cos x$ for the interval $0° \leqslant x \leqslant 720°$
c 35.9°, 324.1°, 395.9°, 684.1°

11 117.0°, 243.0°, 477.0°, 603.0°

13.4 The tangent function

1 a 1 b 0.225

2 a 0.6 b 0.6 c −1.2 d 1.5
3 240°
4 a 315° b 165°
5 a decreases from 0 to minus infinity
b increases from 0 to infinity
c decreases from 0 to minus infinity
6 a Every 180° b i 1.7 ii −1.7
c Rotational symmetry of order 2 about (180,0)
d i 240° ii 280° iii 300°
7 a Students' graph of $y = \tan x$ for the interval $0° \leqslant x \leqslant 540°$
b i 0 ii 1 c i $\sqrt{3}$ ii $-\sqrt{3}$
8 a 45°, 225°, 405°, 585° b 135°, 315°, 495°, 675°
9 a 74.7°
b Students' graph of $y = \tan x$ for the interval $0° \leqslant x \leqslant 720°$
c 74.7°, 254.7°, 434.7°
10 75.7°, 255.7°, 435.7°, 615.7°
11 a Students' graph of $y = \tan x$ for the interval $0° \leqslant x \leqslant 720°$
b 30°, 210°, 390°, 570°

13.5 Calculating areas and the sine rule

1 a $A = \pi r^2$ b $A = \frac{\pi r^2}{2}$ c $A = \frac{\pi r^2}{4}$ d $A = \frac{\pi r^2}{3}$
2 3.38 cm (3 s.f.)
3 a $h = p\sin\theta$ b $A = \frac{1}{2}pq\sin\theta$
4 a 36.8 cm² b 1.54 m²
5 7.99 cm (3 s.f.)
6 a 12.45 cm² (2 d.p.) b 20.73 cm² (2 d.p.)
c 8.27 cm² (2 d.p.)
7 119 m² (3 s.f.)
8 a 118.0° b 28.8 mm²
9 164 cm²
10 a 22.9 cm b 25.5 cm c 47.6 m d 14.7 m
11 a 48.2° b 19.8° c 68.9° d 55.2°
12 a 11.3 cm b 38.7°
13 59.0° or 121.0°
14 75.4° or 104.6°

13.6 The cosine rule and 2D trigonometric problems

1 a 067° b 247°
2 5.05
3 23.4 cm²
4 a 8.43 cm b 8.15 cm c 21.1 cm d 12.5 m
5 a 59.6° b 151.3° c 99.1° d 82.4°
6 106.3°
7 a 15.4 cm b 26.6° c 93.7 cm²
8 113°
9 a 16.6 km b 291°
10 a 56.3° b 110.8°
11 12.7 cm

13.7 Solving problems in 3D

1 a 65.0° b 36.7° c 60.9°
2 a 6.37 cm b 65.6 cm
3 a i 15 cm ii 20.5 cm iii 16.6 cm iv 20.5 cm
b 43.0° c 43.0° d 35.8°
4 19.1°
5 a 10.3 cm b 6.65 cm c 109.6°
6 a i 22.6 cm (3 s.f.) ii 11.3 cm (3 s.f.)
iii 21.2 cm (3 s.f.)
b 62° c 21°
7 a 19.8 cm b 238 cm²
8 32°

13.8 Transforming trigonometric graphs 1

1 a $\frac{\sqrt{3}}{2}$ b $\frac{1}{\sqrt{3}}$ c $\frac{1}{\sqrt{2}}$ d 0
e 0 f $\sqrt{3}$
2 a Students' graph of $y = \sin x$ for the interval $0° \leqslant x \leqslant 360°$
b Students' graph of $y = \cos x$ for the interval $0° \leqslant x \leqslant 360°$
c Students' graph of $y = \tan x$ for the interval $0° \leqslant x \leqslant 360°$
3 a

	x	$\sin x$	$-\sin x$
A	−180°	0	0
B	−150°	−0.5	0.5
C	−90°	−1	1
D	−30°	−0.5	0.5
E	30°	0.5	−0.5
F	90°	1	−1
G	150°	0.5	−0.5
H	180°	0	0

b
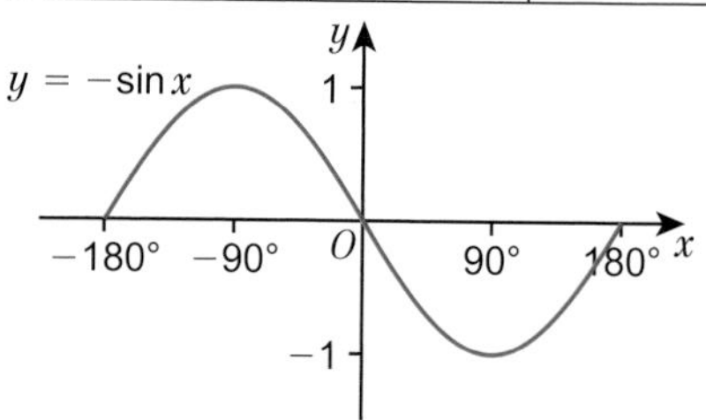

c Reflection in the x-axis
4 a

	x	$\sin x$	$-\sin x$	$\sin(-x)$
A	−180°	0	0	0
B	−150°	−0.5	0.5	0.5
C	−90°	−1	1	1
D	−30°	−0.5	0.5	0.5
E	30°	0.5	−0.5	−0.5
F	90°	1	−1	−1
G	150°	0.5	−0.5	−0.5

b
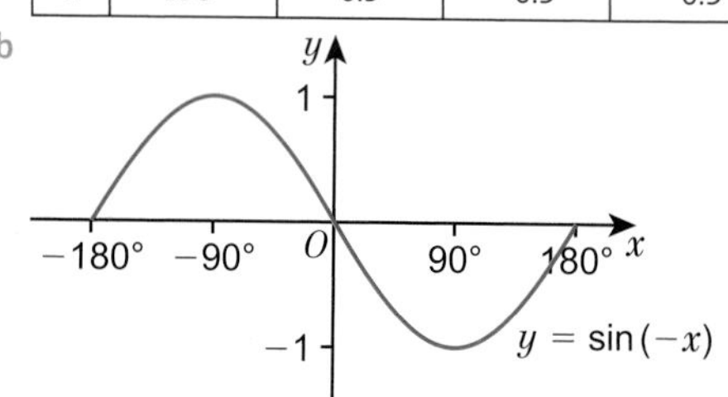

c Reflection in the y-axis
5
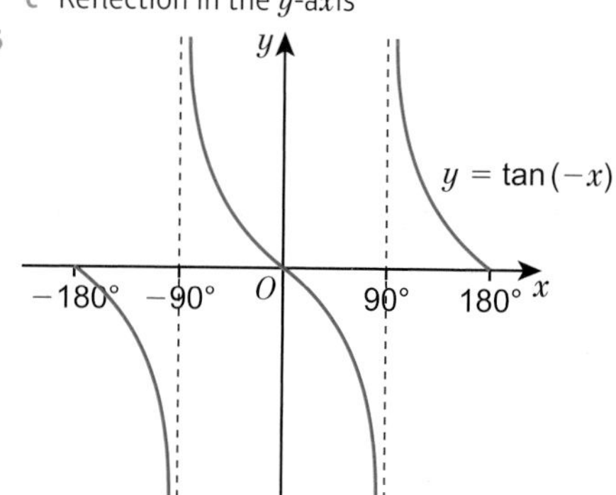

6 a Reflection in y-axis b Same as graph of $y = \sin x$
7 The graph has rotational symmetry of order 2 about the origin.

8 a

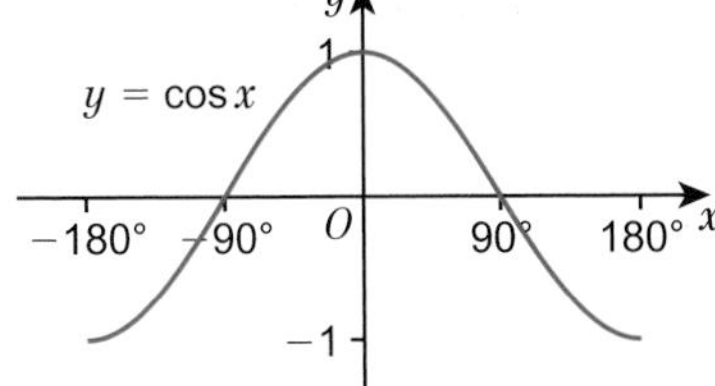

b

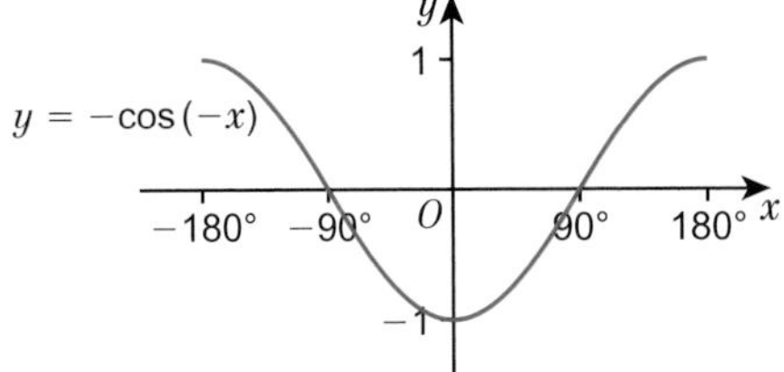

9 a Reflection in y-axis

b

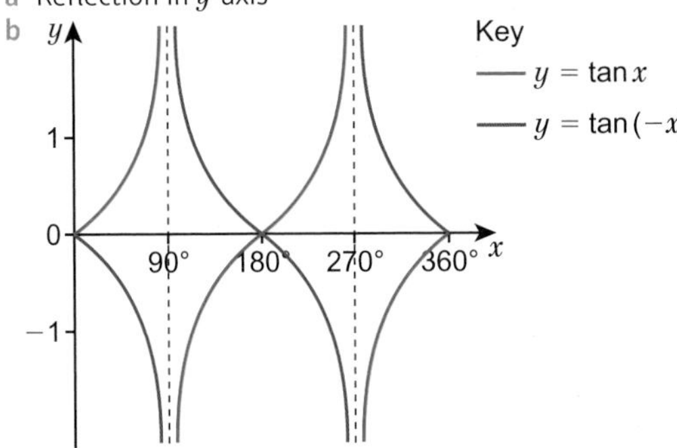

10 a Reflection in x-axis

b

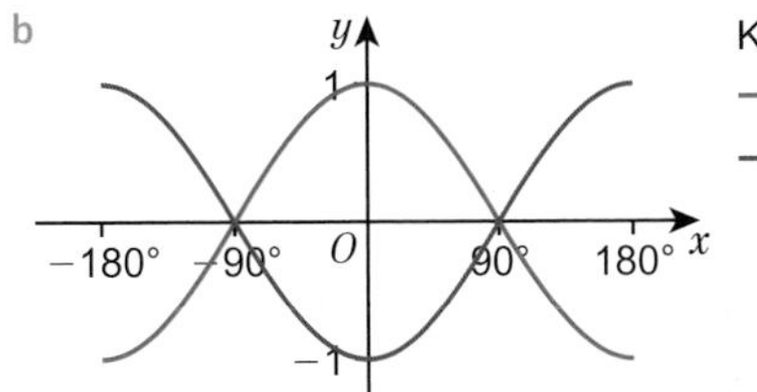

11 a Rotation through 180° about the origin

b

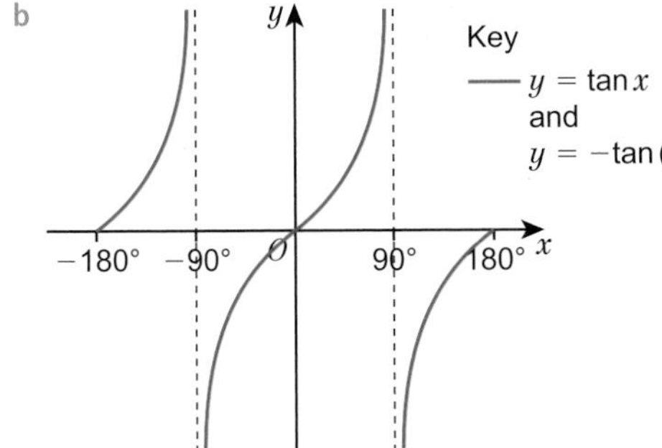

12 P(90, −1), Q(360, 0), R(450, −1)

13.9 Transforming trigonometric graphs 2

1 a (30, 2.5) b (30, 2)

2 a, b, c Students' graphs d Translation by $\begin{pmatrix} 0 \\ 0.5 \end{pmatrix}$

e, f Students' graphs

g Translation by $\begin{pmatrix} 0 \\ -0.5 \end{pmatrix}$ h $y = \sin x - 0.5$

3 a $y = \cos x - 1$ b $y = \sin x + 1$ c $y = \tan x + 2$

4 a

x	0°	30°	60°	90°
cos (x + 30°)	$\frac{\sqrt{3}}{2}$	$\frac{1}{2}$	0	$-\frac{1}{2}$

b

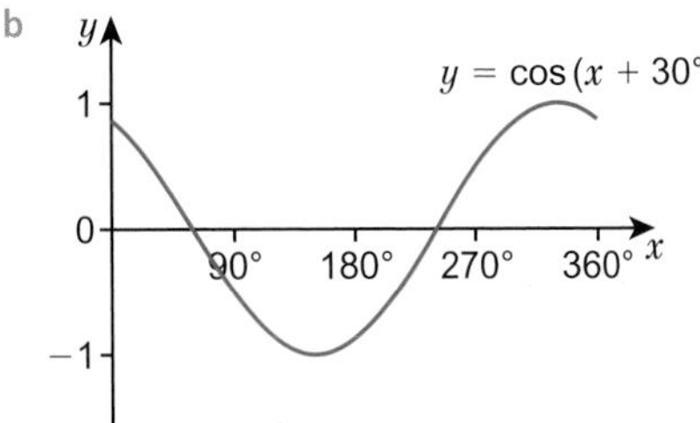

c Translation by $\begin{pmatrix} -30 \\ 0 \end{pmatrix}$

5 a Translation by $\begin{pmatrix} -60 \\ 0 \end{pmatrix}$ b Translation by $\begin{pmatrix} -20 \\ 0 \end{pmatrix}$

c Translation by $\begin{pmatrix} 30 \\ 0 \end{pmatrix}$

6 a Translation by $\begin{pmatrix} -40 \\ 0 \end{pmatrix}$ b Translation by $\begin{pmatrix} -30 \\ 0 \end{pmatrix}$

d Translation by $\begin{pmatrix} 60 \\ 0 \end{pmatrix}$

7 a C b B c A

8 a, c

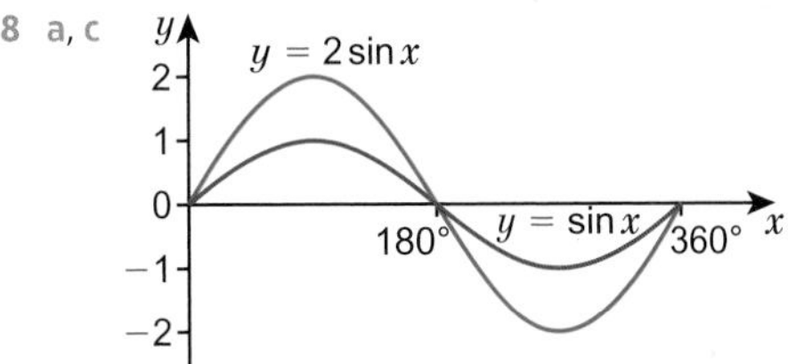

b

x	0°	30°	60°	90°	120°
sin x	0	0.5	$\frac{\sqrt{3}}{2}$	1	$\frac{\sqrt{3}}{2}$
2sin x	0	1	$\sqrt{3}$	2	$\sqrt{3}$

9 a

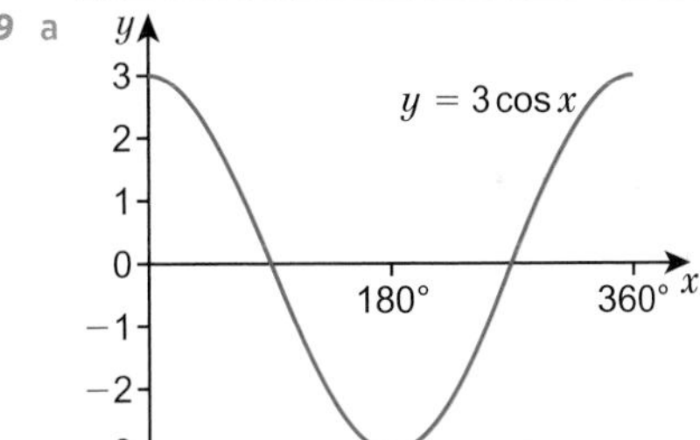

b

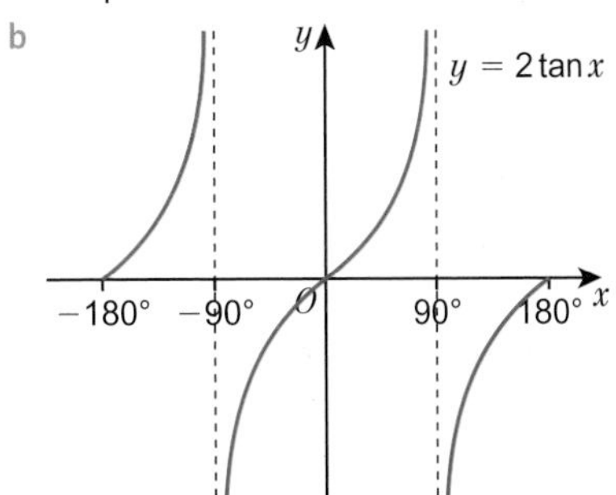

c

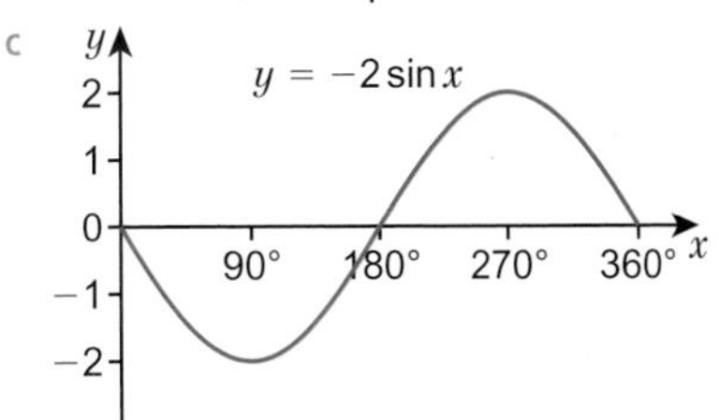

10 a, c

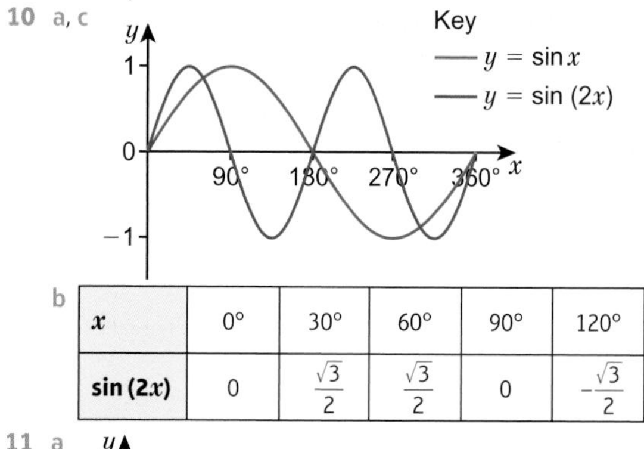

b

x	0°	30°	60°	90°	120°
$\sin(2x)$	0	$\frac{\sqrt{3}}{2}$	$\frac{\sqrt{3}}{2}$	0	$-\frac{\sqrt{3}}{2}$

11 a

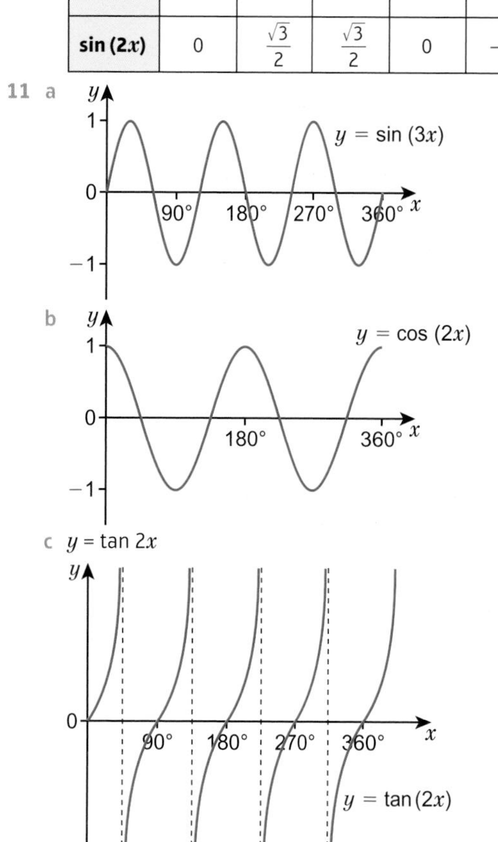

12 a (180, 0) b (270, −1) c $a = 2, b = 3, c = 1$

13 Problem-solving

1 The distance from A (or B) to the centre of the square is $10\sqrt{2}$ km. The time taken in hours is $\frac{10\sqrt{2}}{5}$ over the field and $\frac{10\sqrt{2}}{2}$ through the wood. This is a total of $7\sqrt{2} = 9.90$ hours, or 9 hours and 54 minutes, to the nearest minute.

2 Students' own answers, e.g. the route shown would take 9 hours and 10 minutes to the nearest minute. ($12.88\text{ km} < x < 19.57\text{ km}$)

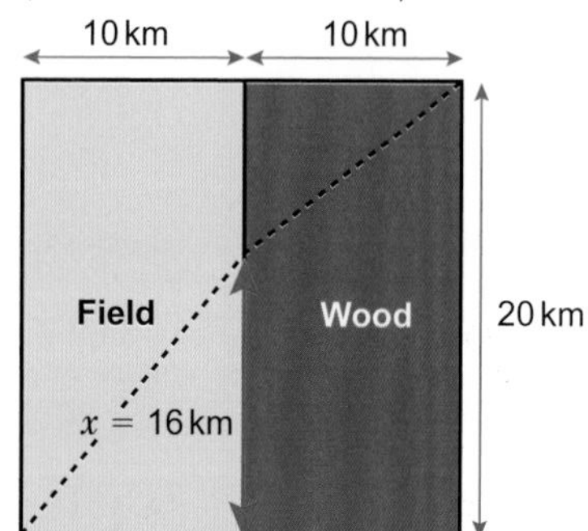

3 The quickest possible time is 9 hours and 8 minutes to the nearest minute.

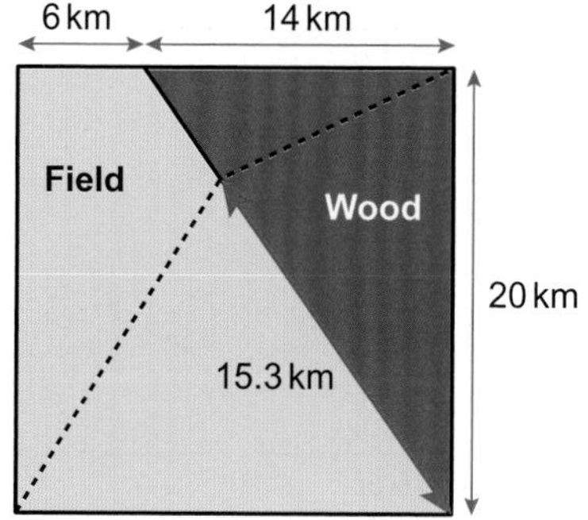

13 Check up

Key: UB = upper bound; LB = lower bound

1 7.96 m^2

2 a 16.0 cm b 4.61 cm

3 a 71.7° b 20.0°

4 UB: x = 4.5939 m (4 d.p.); LB: x = 4.4151 m (4 d.p.); x = 4.5 m (to the nearest 0.5 metres)

5 Students' graph of $y = \tan\theta$ for the interval $-360° \leqslant \theta \leqslant 360°$

6 66°, 294°

7 a B b C c A

8 19.5°, 160.5°, 379.5°, 520.5°

9 23.3 cm (1 d.p.)

10 a 15.2 cm b 11.3 cm

12 e.g. $\tan x = \frac{1}{\sqrt{3}}$

13 Strengthen

Accuracy and 2D problem-solving

Key: UB = upper bound; LB = lower bound

1 a

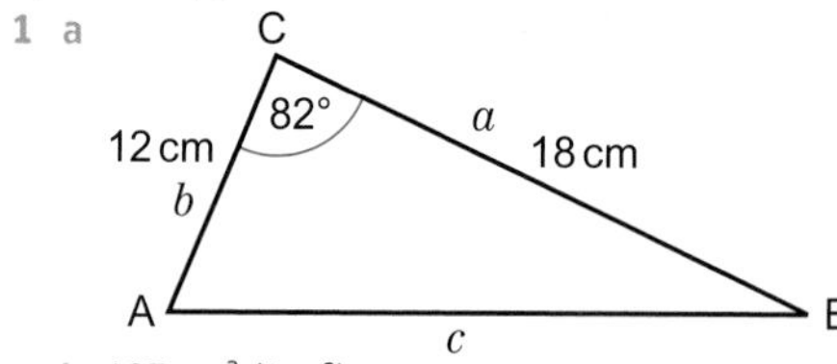

b 107 cm^2 (3 s.f.)

2 a 39.6 cm^2 b 166 cm^2 (3 s.f.)

3 a Triangle correctly labelled
b $\frac{x}{\sin 35°} = \frac{16}{\sin 56°}$ c 11.1 (3 s.f.)

4 a 25.4 m b 13.3 m

5 a Triangle correctly labelled
b $\frac{\sin\theta}{31} = \frac{\sin 146°}{74}$ c 13.5° (1 d.p.)

6 a 50.9° (1 d.p.) b 25.0° (1 d.p.)

7 a Triangle correctly labelled
b $x^2 = 23^2 + 37^2 - 2 \times 23 \times 37 \times \cos 48°$
c 27.6 cm (3 s.f.)

8 a 68.0 m (3 s.f.) b 61.6 m (3 s.f.)

9 a Triangle correctly labelled
b $\cos\theta = \frac{25^2 + 32^2 - 41^2}{2 \times 25 \times 32}$ c 91.1°

10 a 120.9° b 27.7°

11 a

	Upper bound	Lower bound
5.7	5.75	5.65
23	23.5	22.5

b UB = 0.3987; LB = 0.3827
c 5.65, 23.5 d 5.75, 22.5

12 UB = 10.7294; LB = 9.6295

Trigonometric graphs

1 a

x	0°	10°	20°	30°	40°	50°	60°	70°	80°	90°
sin x	0	0.17	0.34	0.5	0.64	0.77	0.87	0.94	0.98	1

b Students' graph of $y = \sin x$ for the interval $0° \leqslant x \leqslant 90°$

c, d Students' graph of $y = \sin x$ for the interval $0° \leqslant x \leqslant 360°$

2 a

x	0°	10°	20°	30°	40°	50°	60°	70°	80°
tan x	0	0.18	0.36	0.58	0.84	1.19	1.73	2.75	5.67

b Students' graph of $y = \tan x$ for the interval $0° \leqslant x \leqslant 80°$

c $\tan x$ increases rapidly towards infinity

d, e, f Students' graph of $y = \tan x$ for the interval $0° \leqslant x \leqslant 360°$

3 126° or 234° (from graph)

4

x	0°	30°	90°
sin x	0	0.5	1
sin2x − 1	−1	$\frac{\sqrt{3}}{2} - 1$	−1
2sin x − 1	−1	0	−1

b i $y = \sin x$, stretch factor 2 in the y direction, translation $\begin{pmatrix} 0 \\ -1 \end{pmatrix}$

ii $y = \cos x$, stetch factor 2 in the y direction

iii $y = \sin x$, stretch factor $\frac{1}{2}$ in the x direction, translation $\begin{pmatrix} 0 \\ -1 \end{pmatrix}$

c i B ii A iii C

5 a $\cos x = \frac{3}{4}$

b Students' graph of $y = \cos x$ for the interval $0° \leqslant x \leqslant 720°$

c 41.4°

d 41.4°, 318.6°, 401.4°, 678.6°

3D Problem solving

1 a Students' sketch of triangle CDG: CG = 25 cm; CD = 21 cm; ∠GCD = 90°

b 32.6 cm (3 s.f.)

c Students' sketch of triangle DFG: FG = 7 cm, DG = 32.6 cm, ∠FGD = 90°

d 12.1°

2 a Students' sketch of triangle ABC: AB = 11 cm, AC = 11 cm, ∠BAC = 56°

b 10.3 cm (3 s.f.)

c Students' of sketch triangle BCD: BC = 10.3 cm, CD = 10 cm, BD = 12 cm

d 72.3°

13 Extend

1 a 11.7 cm (3 s.f.)

b It is the same as Pythagoras' theorem.

2 a $\frac{1}{2}ab\sin C$

b $\frac{1}{2}ac\sin B$, $\frac{1}{2}bc\sin A$

c Each expression must have the same value so,

$\frac{1}{2}ab\sin C = \frac{1}{2}ac\sin B$

$ab\sin C = ac\sin B$

$b\sin C = c\sin B$

$\frac{b}{\sin B} = \frac{c}{\sin C}$

3 $d^2 = a^2 + b^2 + c^2$

4 a $a^2 = h^2 + b^2 - 2bx + x^2$ b $c^2 = h^2 + x^2$

c Substituting for $h^2 + x^2$ in part **a** gives $a^2 = b^2 + c^2 - 2bx$

d $x = \cos A$, so $a^2 = b^2 + c^2 - 2bc\cos A$

5 116°

6 a Students' graph of $y = \sin x$

b i 0.7 (from graph) ii 0.7 (from graph)

c The answers are the same.

d The graph is symmetrical about $x = 90°$.

7 $\frac{\pi r^2}{3} - \frac{\sqrt{3}r^2}{4}$

8 $1 + \sqrt{5}$

9 a, c

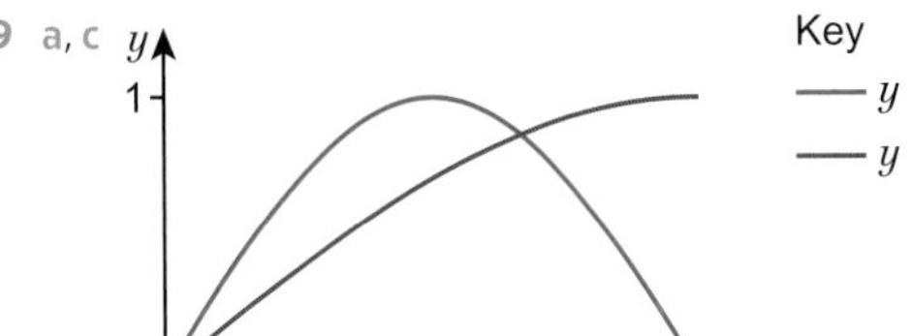

b

x	0°	60°	90°	120°
$\sin\left(\frac{x}{2}\right)$	0	0.5	0.7	0.9

10

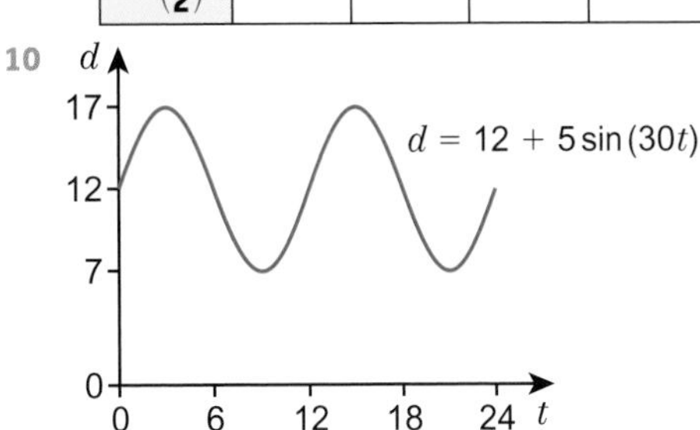

11 1130 cm³

12 33.6°

13 a $x = 5$ b $k = 10$

13 Unit test

Sample student answer

The student has assumed wrongly that the triangle is right-angled.

UNIT 14

14 Prior knowledge check

1 a 5% b 112 c 50

2 5:25

3 81

4 a Continuous b Discrete c Categorical

d Continuous e Discrete

5 a Mean = 5.4; median = 6; range = 7; mode = 4

b Mean = 6.5; median = 6; range = 5; mode = 6

c Mean = 3.0125; median = 2.65; range = 1.8; mode = 3.8

d Mean = 5.375; median = 5; range = 5; mode = 3

6 a $10 < m \leqslant 11$ b 10.75 c $10 < m \leqslant 11$

7 a 10 b 7 c 7

14.1 Sampling

1 a 620 b $\frac{13}{62}$ c 19.4% d 13

e $\frac{1}{6}$

2 a No; it is biased to people who shop in the supermarket.

b Yes; each person is equally likely to be selected.

3 a Yes, limited time frame.

b Yes, biased to people who do recycle.

c Yes, not everyone is equally likely.

d Not biased, but a very small sample.

4 a 13, 48, 09, 32, 02, 31, 50 b 86, 13, 60, 78, 48, 80

5 46, 12, 48, 06, 24, 14, 37, 39

6 a 100
  b List the members alphabetically.
  Generate 100 random numbers between 1 and 1000.
  c $\frac{1}{8}$
  d $\frac{1}{12}$

7 a Total number of students = 1000. $\frac{100}{1000} = \frac{1}{10} = 10\%$
  b 21, 19, 18, 20, 21
  c 21 + 19 + 18 + 20 + 21 = 100

8 a There are different proportions of male and female in the club.
  b 35 women, 45 men c 21 women, 27 men

9 7, 19, 15, 29

10 a $\frac{1}{8}$ b 320

11 2

14.2 Cumulative frequency

1 a 21 b i 15 ii 26

2

Mass, m (kg)	Cumulative frequency
$3 < m \leqslant 4$	4
$3 < m \leqslant 5$	**16**
$3 < m \leqslant 6$	**33**
$3 < m \leqslant 7$	**43**
$3 < m \leqslant 8$	**50**

3

Height, h (m)	Cumulative frequency
$4.0 < h \leqslant 4.2$	2
$4.2 < h \leqslant 4.4$	5
$4.4 < h \leqslant 4.6$	10
$4.6 < h \leqslant 4.8$	18
$4.8 < h \leqslant 5.0$	30
$5.0 < h \leqslant 5.2$	48
$5.2 < h \leqslant 5.4$	63
$5.4 < h \leqslant 5.6$	70

4 a

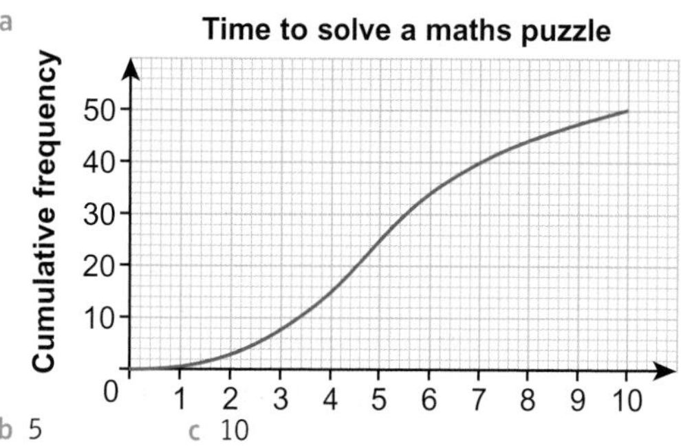

b 5 c 10

5 a

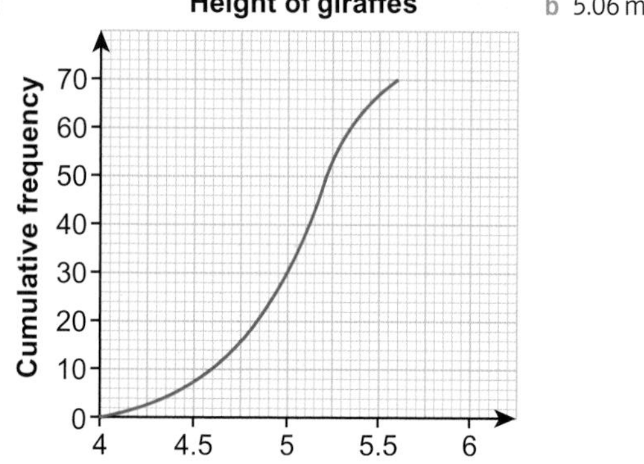

b 5.06 m

6 a

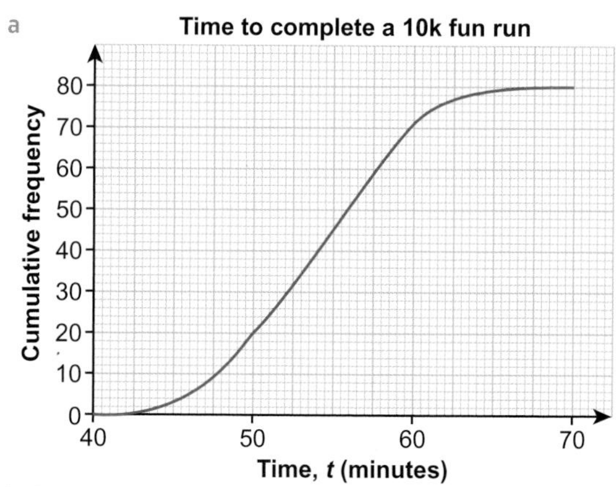

b 54
c 50
d 58
e 8

7 a

Masses of hippos

Cumulative frequency; Mass, m (tonnes)

b Median = 1.59, LQ = 1.52, UQ = 1.67, IQR = 0.15
c 22
d 1.64

8 a

Time, t (seconds)	Cumulative frequency
$0 < t \leqslant 10$	16
$10 < t \leqslant 20$	50
$20 < t \leqslant 30$	82
$30 < t \leqslant 40$	96
$40 < t \leqslant 50$	100

b

Time taken to drive to work

Cumulative frequency; Time, t (minutes)

c 20 minutes
d 55 days

14.3 Box plots

1 a 25.5 b 38

2

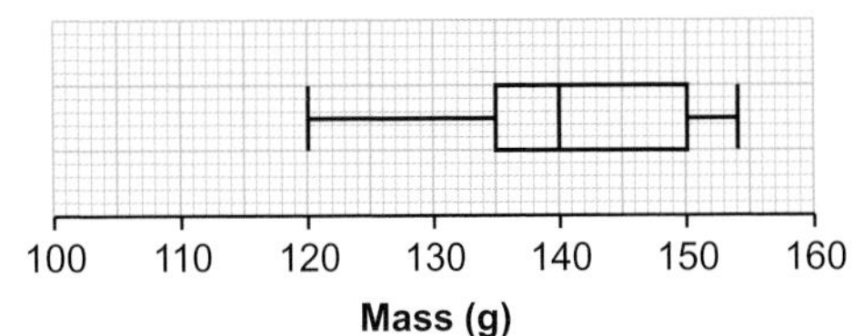

3 a 25 b LQ = 19, UQ = 28

c

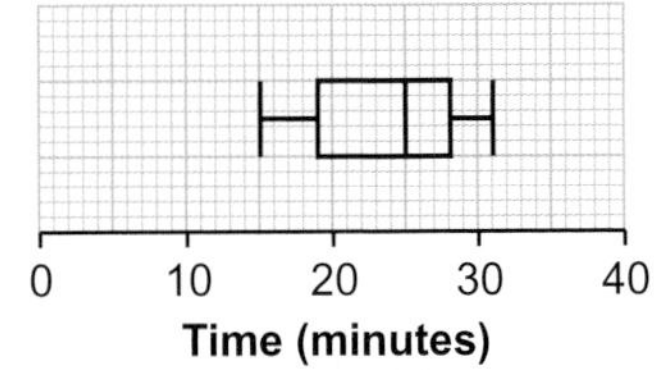

4

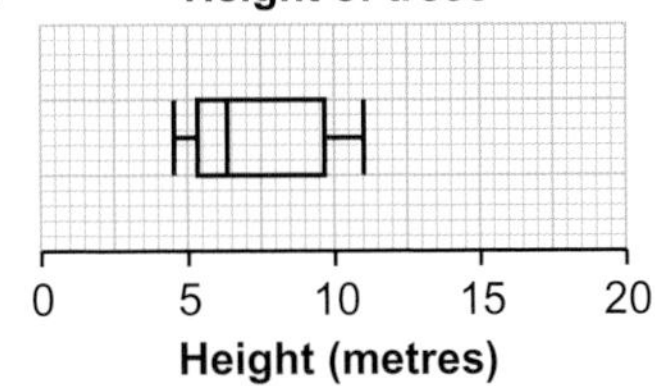

5 a 16 b 40 c LQ = 26.5, UQ =49.5 d 23

e

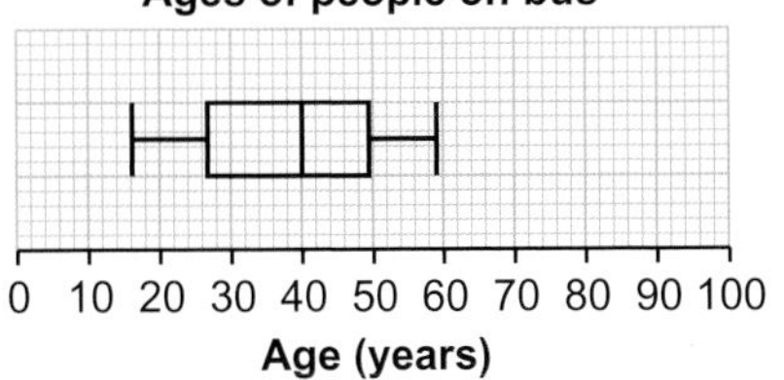

6 a Club A b Club A: 6; Club B: 3

c Club A: 12; Club B: 8

7 a

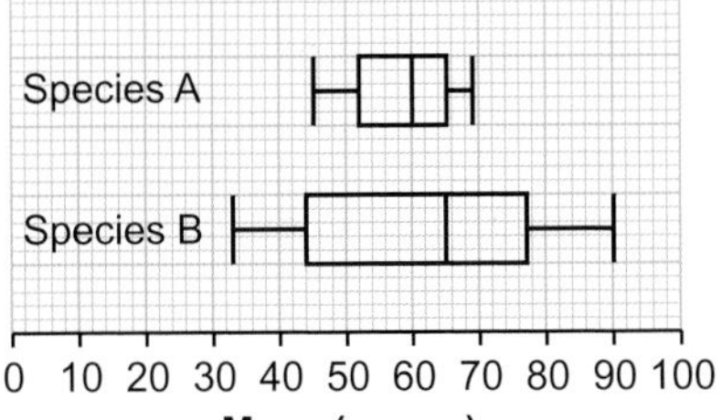

b Species B has a higher median mass and a greater spread of masses.

8 a Medians: 6.6 female, 7.2 male
LQs: 6 female, 6.2 male
UQs: 7.2 female, 7.9 male

b

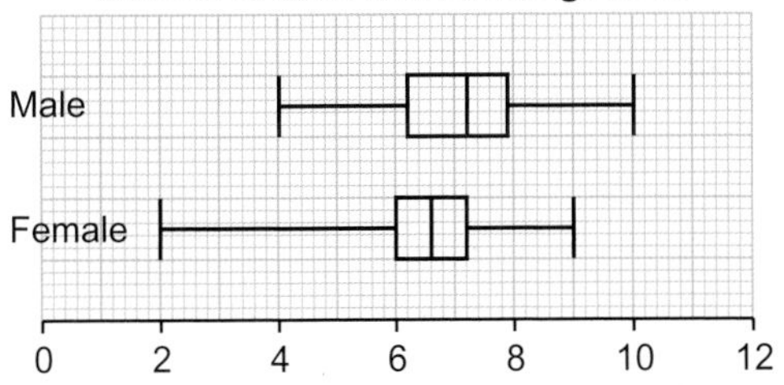

c Females had a lower median mass and a smaller spread of masses.

9 a 33

b

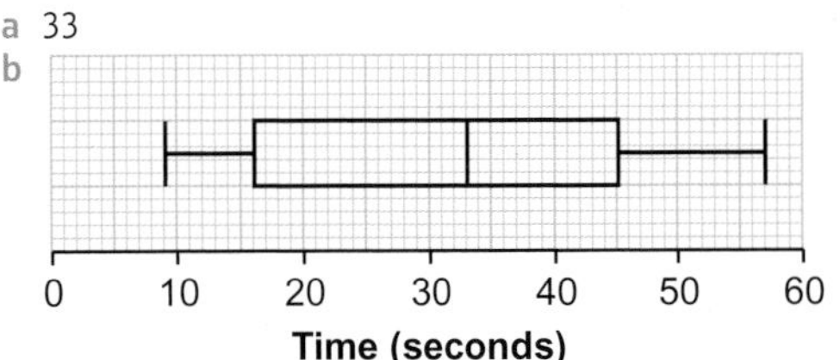

c Boys took longer on average and had a greater spread of times.

14.4 Drawing histograms

1 a

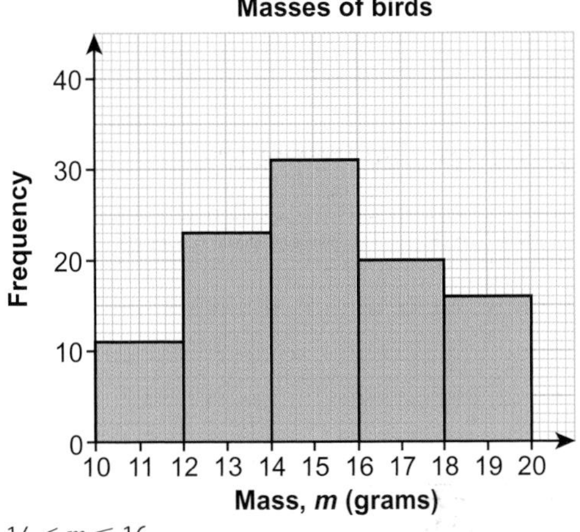

b $14 < m \leqslant 16$

c 15.14

2 a 5, 10, 10

b 2.4, 3.5, 1.5

3 5.5, 11.5, 15.5, 10, 8

4

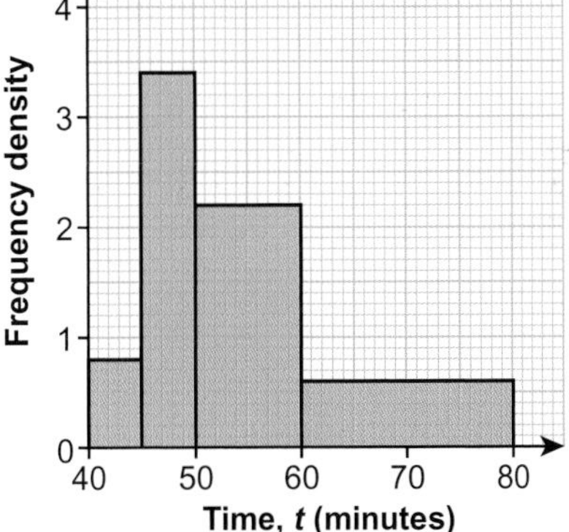

5

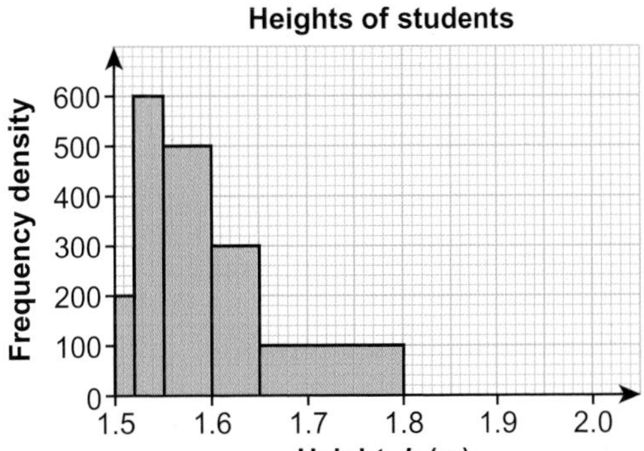

6 a 27.3

b

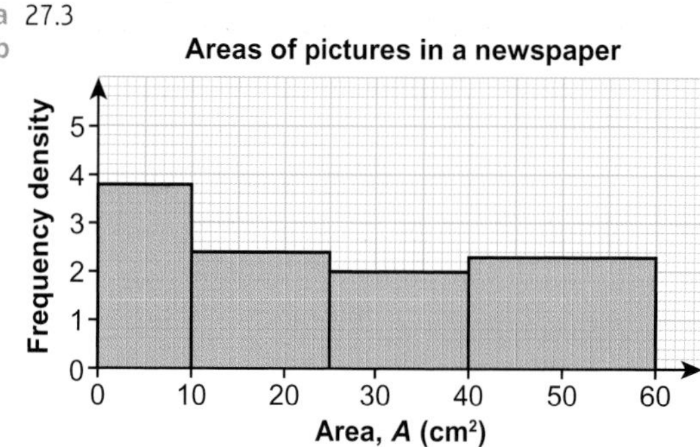

14.5 Interpreting histograms

1 a 119.875 b $120 < m \leqslant 130$

2 a 15 b 100 c 135

3 a 40 b 230 c 94

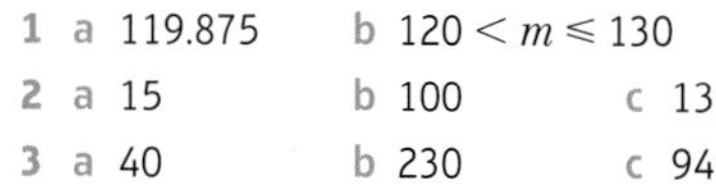

4 a 60, 40

b

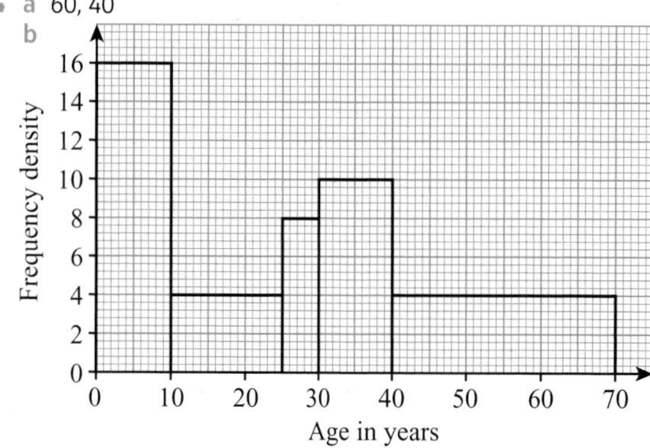

5 a

Time	Frequency
$15 < t \leqslant 16$	6
$16 < t \leqslant 18$	14
$18 < t \leqslant 20$	20
$20 < t \leqslant 25$	15
$25 < t \leqslant 30$	5

b 19.77 mins c 30

6 240.5th value = 20.5th value in the $25 < x \leqslant 30$ class.
$\frac{20.5}{80} \times 5 = 1.28$ so median = 26.28

7 a 100 b 85 c 1.51

d

Height (m)	Frequency
$1.4 < m \leqslant 1.45$	5
$1.45 < m \leqslant 1.48$	15
$1.48 < m \leqslant 1.5$	20
$1.5 < m \leqslant 1.55$	20
$1.55 < m \leqslant 1.6$	15
$1.6 < m \leqslant 1.7$	10

e 1.524 f 36

8 a 46 b 23.5th = 4.9375 c 15

14.6 Comparing distributions

1 a mean = 1.38, median = 1.4, mode = 1.5, range = 0.3

b mean = 4.9, median = 4.5, mode = 3, range = 5

2 a 294.1, 267.33 b 30, 14

3 Males weigh on average more and have a larger spread of masses

4 a, b African elephants are on average taller and have a greater spread of heights

5 a 23.55 b 18 c 85, 8

6 a Median and IQR (unaffected by extreme values: 120 in males, for example)

b Males completed the race quicker on average (medians are 78 and 84.5) and had a smaller spread (IQRs are 8 and 14)

7 a

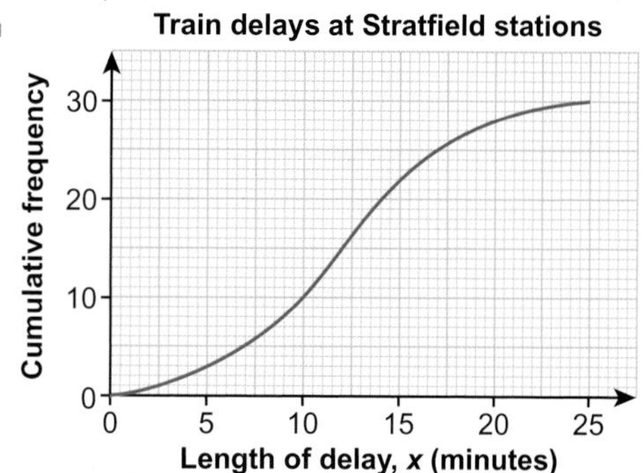

b median = 12, IQR = 6.8

c On average the delays at Westford were longer and had a larger spread

8 Checkpoint B higher average (median 38 compared to 32) and same spread (IQR 11 compared to 11)

9 Females had a higher average age and a larger spread

10 Women had a higher average but a lower spread

14 Check up

1 a 16, 18, 14, 17, 15

b 112, 283, 185, 191, 255

2 a

Mass, m (kg)	Cumulative frequency
$20 \leqslant m \leqslant 23$	1
$23 < m \leqslant 26$	5
$26 < m \leqslant 29$	13
$29 < m \leqslant 32$	34
$32 < m \leqslant 35$	66
$35 < m \leqslant 38$	84
$38 < m \leqslant 41$	90

b

Masses of emperor penguins

c 33 kg

d LQ = 30.5 kg, UQ = 35.2 kg, IQR = 4.7 kg

e i 74 ii 71

3 a

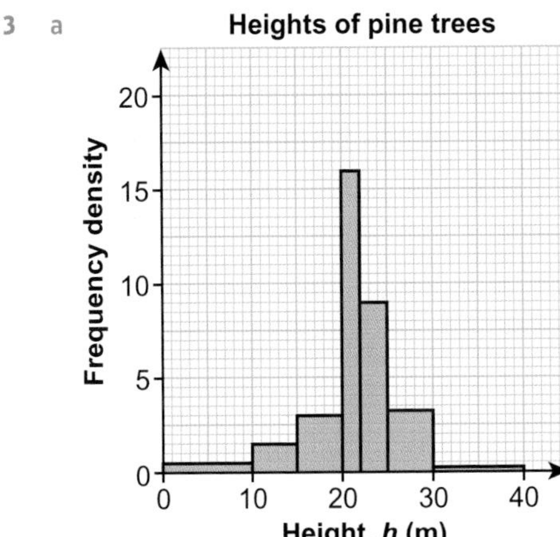

b 36

4 Girls have a higher average time and a bigger spread of times

5 a 28 b 22
c First party had a higher average age and a greater spread of ages

7 56

14 Strengthen

Sampling

1 a B b i 15 ii 8 c 23 d Equal

2 a 120
b 20%
c 3.2%, 5.6%, 6.4%, 4.8%, 4%

3 a 02, 79, 21, 51, 21, 08, 01, 57, 01, 87, 33, 73, 17, 70, 18, 40, 21, 24, 20, 66, 62
b 02, 21, 21, 08, 01, 01, 17, 18, 21, 24, 20
c 02, 21, 08, 01, 17, 18, 24, 20
d 02, 21, 08, 01, 17

4 02, 21, 51, 08, 01, 57, 33, 17

5 027, 108, 015, 018, 124

Graphs and charts

1 a

Mass, m (kg)	Cumulative frequency
$70 \leqslant m \leqslant 75$	1
$75 < m \leqslant 80$	9
$80 < m \leqslant 85$	28
$85 < m \leqslant 90$	43
$90 < m \leqslant 95$	53
$95 < m \leqslant 100$	60

b

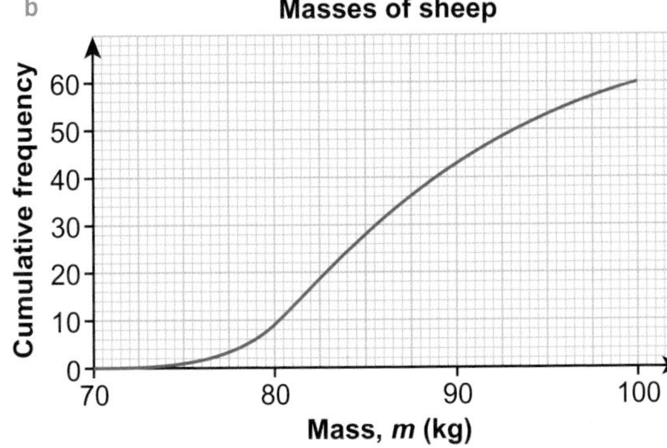

2 a median = £4.40, LQ = £3.50, UQ = £5.10
b £1.60

3 a 60 b 30 c 45 cm
d LQ = 35 cm, UQ = 54 cm e 19 cm

4 a

Time, t (mins)	Cumulative frequency
$65 \leqslant t \leqslant 68$	4
$68 < t \leqslant 71$	11
$71 < t \leqslant 74$	28
$74 < t \leqslant 77$	41
$77 < t \leqslant 80$	50

b

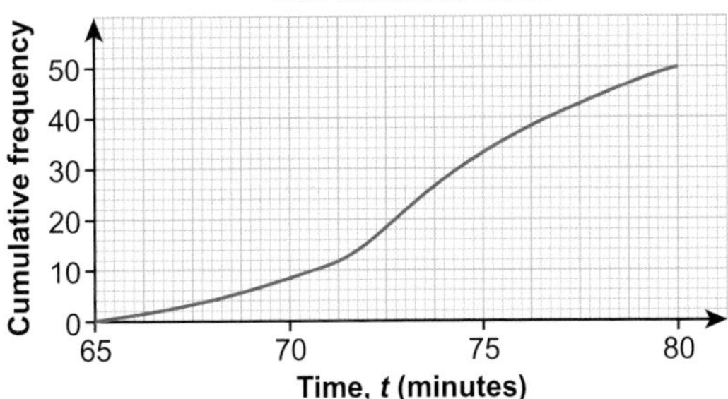

c Median = 73.4 mins, LQ = 71.3 mins, UQ = 76.2 mins
d 16 e 34

5

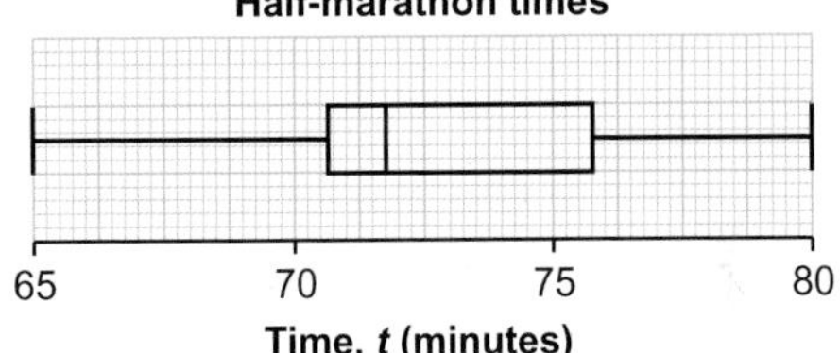

6 a i 5 ii 1.6

b

Length, l (mm)	Frequency	Class width	Frequency density
$10 \leqslant l \leqslant 15$	2	15 − 10 = 5	2 ÷ 5 = 0.4
$15 < l \leqslant 20$	8	20 − 15 = 5	8 ÷ 5 = 1.6
$20 < l \leqslant 30$	15	30 − 20 = 10	15 ÷ 10 = 1.5
$30 < l \leqslant 40$	12	40 − 30 = 10	12 ÷ 10 = 1.2
$40 < l \leqslant 60$	5	60 − 40 = 20	5 ÷ 20 = 0.25

c

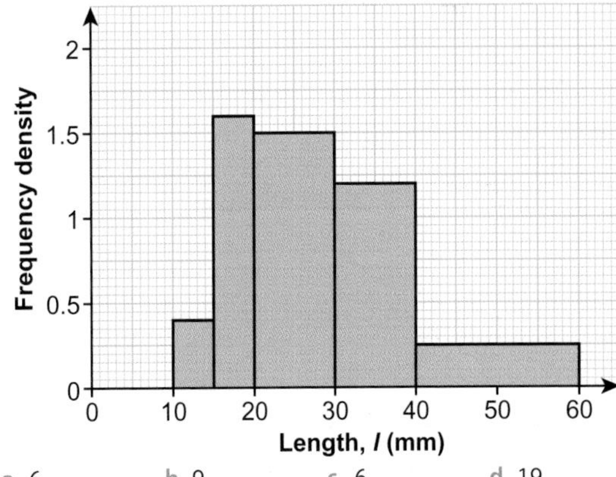

7 a 6 b 9 c 6 d 19

Comparing data

1 a

	Lower quartile	Median	Upper quartile	Interquartile range
Boys	4	6	7	3
Girls	3	5	8	5

b i higher ii Boys, girls

2 a 23 b 6, 12, 18 c 59 kg, 67 kg, 75 kg
d 67 kg e 16 kg

3 a 69 kg b LQ = 63 kg, UQ = 78.5 kg c 15.5 kg
d Female wild boars have a smaller mass on average and a higher spread of masses.

14 Extend

1 a Speeds of cheetahs

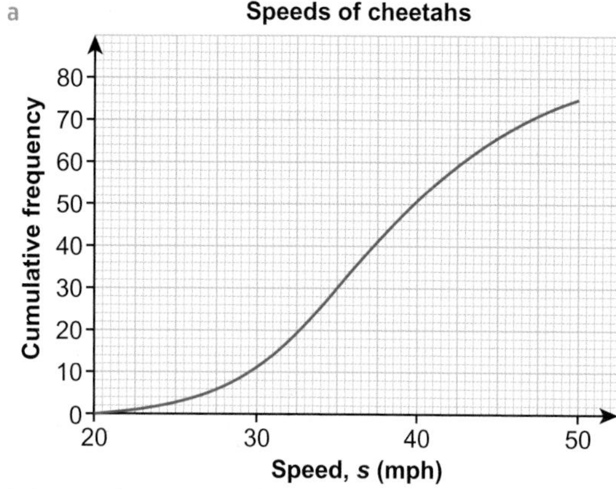

b i 37 mph ii 10 mph

2 75 and 45

3 Stratified: $\frac{1}{20}$ of each group – cranes: 22.5 (so 23 or 22), forklift: 31, dump: 6.5 (so 6 or 7)

4 Stratified: 25% - male: 30 builders, 10 electricians, 5 plumbers; Female: 17.5 (so 18) builders, 9 electricians, 8.5 (so 8) plumbers

5 a Number of clients with: rabbits = 16, guinea pigs = 18, hamsters = 9, gerbils = 7
b Numbered alphabetical list and random number generation

6 a 4
b 4 minutes and 9.6 minutes

7 a Less variation in temperatures
b Higher average temperature

8 a Discrete
b

Peas	Cumulative frequency
≤ 3	7
≤ 4	18
≤ 5	36
≤ 6	70
≤ 7	90
≤ 8	100

c, d, e Number of peas in a pod

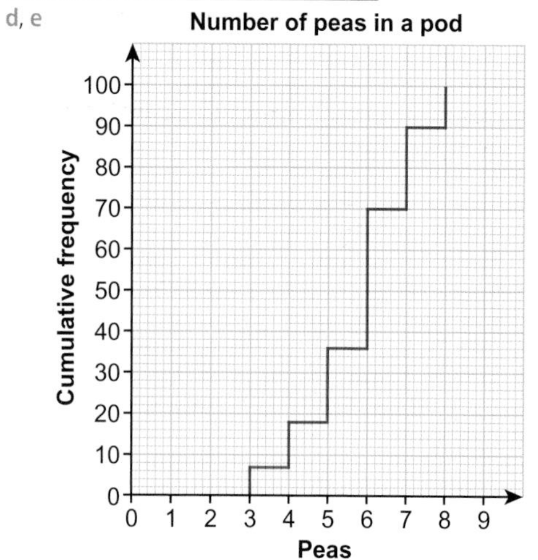

f Discrete data: no such thing as 4.5 peas g 6

9 a

Rabbits	Cumulative frequency
2	8
3	23
4	46
5	87
6	112
7	120

b Number of rabbits born

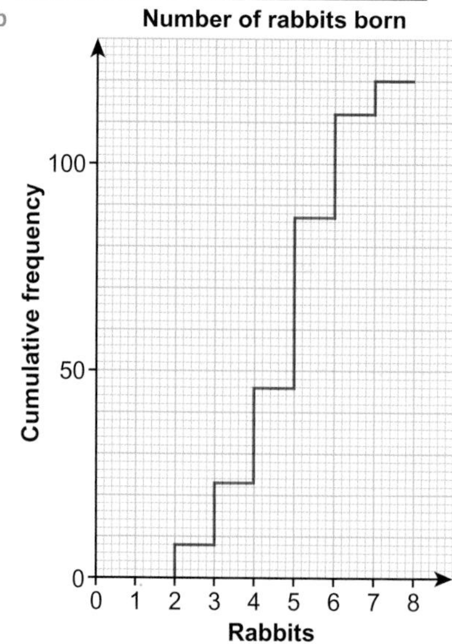

c 5 d 2

10 a 14.9 b LQ = 14.3, UQ = 15.5, IQR = 1.2
c 1.8 d 28.6

11 a 4 b 8 – 24 c 2
d Ages of people at party

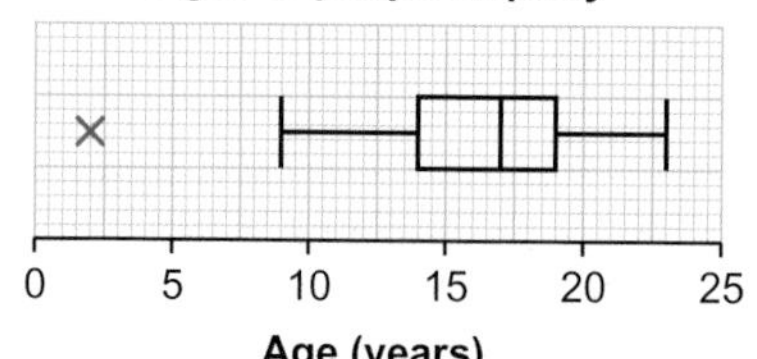

12 a Time taken to play a round of golf

Frequency density
Time, t (hours)

b 23

13 a

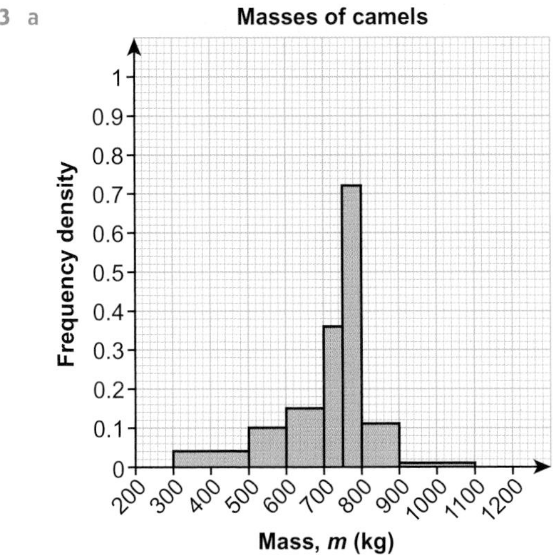

b 707.5 kg c 747 kg d $700 < m \leqslant 750$
e 53 f 760 kg – 770 kg

14 a

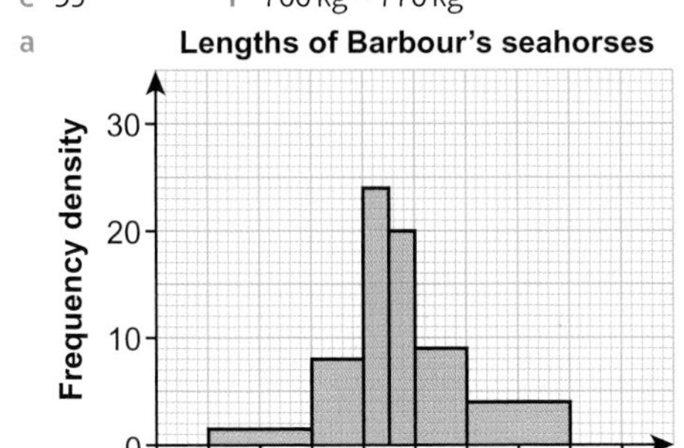

b 13.76 cm c 13.6 cm d 13.4 cm

14 Unit test

Sample student answer

a A ruler has been used making it neater and more accurate to read off the values. The group B lines have been drawn differently to group A to distinguish them on the graph and make it less likely to read off the wrong values.

b The lower quartile value is wrong because the scale has been read incorrectly. The student has tried to read '25' but has just counted up 5 squares, which really is 30.

UNIT 15

15 Prior knowledge check

1 a $3\sqrt{3}$ b $10\sqrt{2}$ c $2\sqrt{5}$

2 a Quadratic b Cubic c Quadratic d Linear
e Linear f Cubic g Linear

3 a −7 b 6 c 7 d −3 e 12

4 a $x \geqslant 3$
b $2 < x$
c $x > 7$
d $x \leqslant 5$

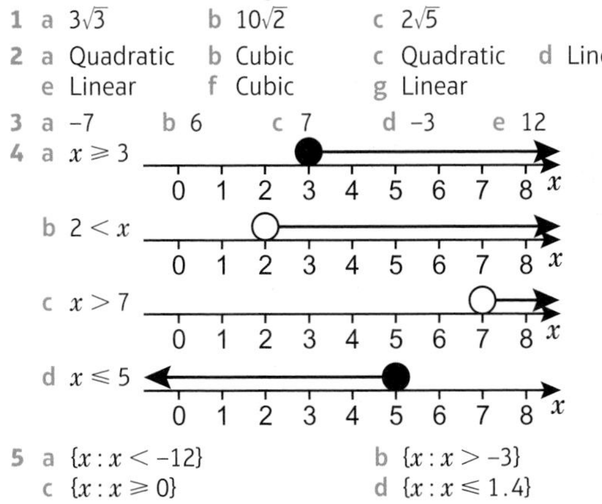

5 a $\{x : x < -12\}$ b $\{x : x > -3\}$
c $\{x : x \geqslant 0\}$ d $\{x : x \leqslant 1.4\}$

6 a −2, −1, 0, 1, 2, 3 b −2, −1, 0
c 0, 1 d 15, 16, 17, 18, 19, 20, 21, 22

7 a $x^2 + 10x + 21$ b $x^2 + 3x - 40$
c $x^2 - 5x + 6$ d $x^2 - 8x + 16$
e $2x^2 + 13x + 15$ f $3x^2 - 11x + 10$
g $9x^2 - 1$

8 a $(x + 5)(x + 2)$ b $(x - 3)(x + 1)$
c $(x + 5)(x - 3)$ d $(x - 7)(x + 1)$
e $(x - 1)(x + 1)$ f $3(x + 1)(x + 4)$
g $(2x + 5)(x - 2)$

9 $x = -2$ or $x = \frac{5}{3}$

10 a $(x - 3)^2 + 3$ b $(x - 4)^2 - 5$ c $3(x - 1)^2 + 6$

11 a $x = 5$ or $x = -1$ b $x = 0$ or $x = -2$
c $x = -\frac{3}{2} \pm \frac{\sqrt{29}}{2}$ d $x = 1$ or $x = -3$

12 a $y = 7, x = -1$ b $y = 5, x = 3$
c $x = 3$ and $y = -5$ or $x = -1$ and $y = 3$

13 a $x = -0.4$ or $x = -4.6$ b $x = 1.9$ or $x = -1.1$

14

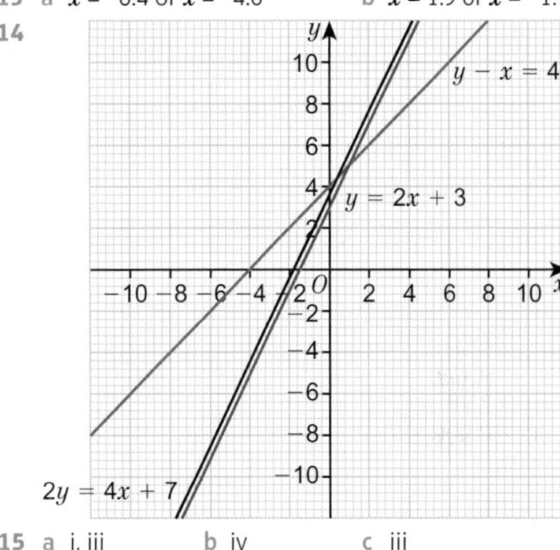

15 a i, iii b iv c iii

16 a

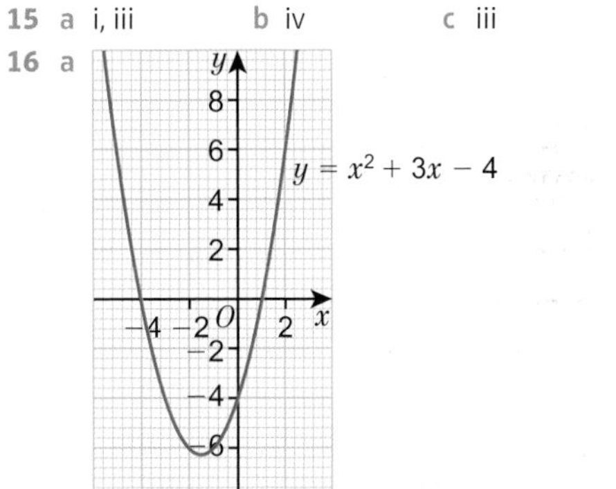

b i $x = 1$ or $x = -4$ ii $x = -4.4$ or $x = 1.4$

17 Any of the form $ax^2 - 5ax - 14a$

15.1 Solving simultaneous equations graphically

1

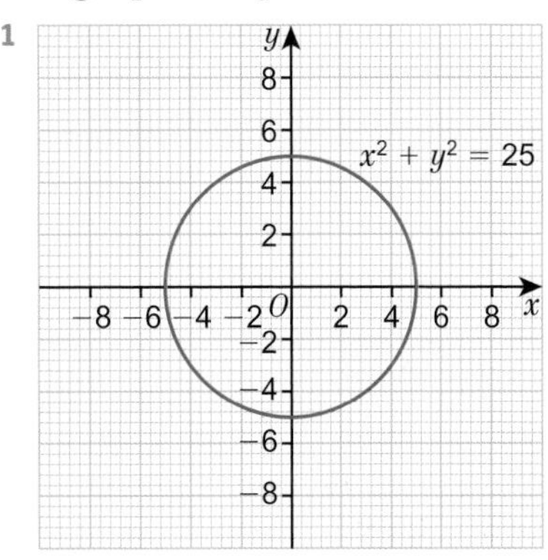

2 a

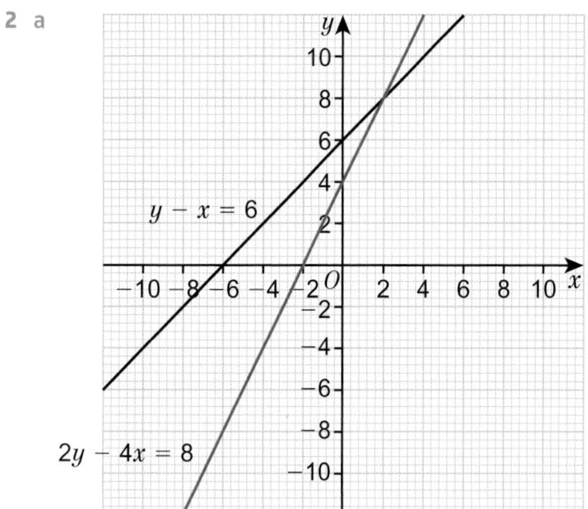

b (2, 8)

3 a i B ii A iii C
b i (1, 4) ii (4, 1) iii (0, 2)

4 a $x = 1, y = 6$
b $x = 3, y = 9$
c $x = -1, y = 7$
d $x = -2, y = -3$

5 $4x + 2y = 258$
$3x + 3y = 249$
a 37p b 46p

6 a Stream Speed: $y = 20 + 1.5x$, ONline: $y = 2x$
b 40 GB

7 $x = 1.3, y = 0.3$ or $x = -2.3, y = -3.3$

8 a $x = 3, y = -2$
b $x = -1.5, y = 4.75$

9 a, b

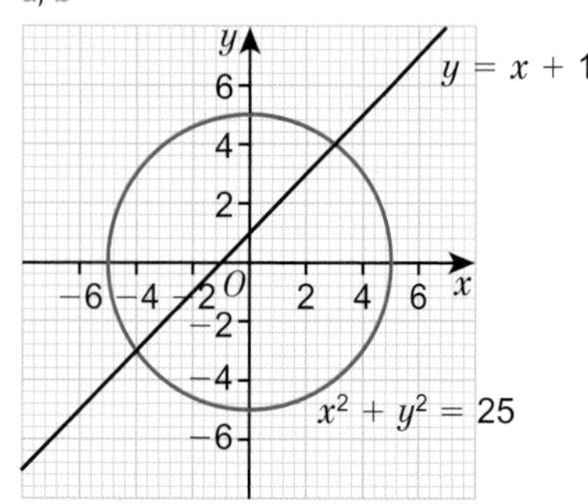

c (3, 4) and (−4, −3)

10 $x = -1.6, y = 2.6$ or $x = 2.6, y = -1.6$

15.2 Representing inequalities graphically

1 a $\{x : x \leqslant 6\}$ b $\{x : x < -2\}$ c $\{x : x \leqslant 4\}$

2 a i $x < 2$ ii $y \leqslant 4$ iii $-1 \leqslant x \leqslant 3$
b i

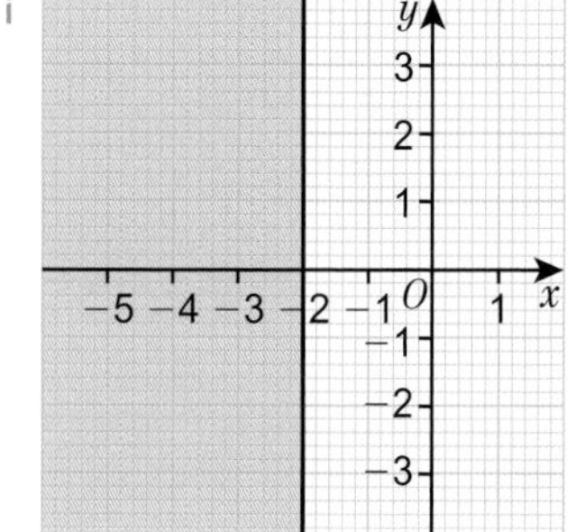

ii

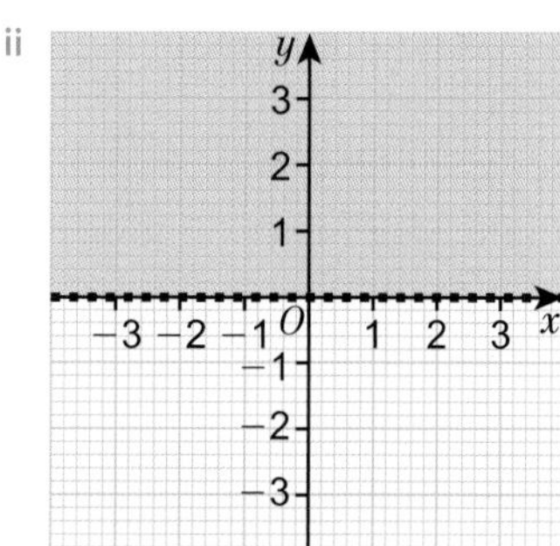

iii

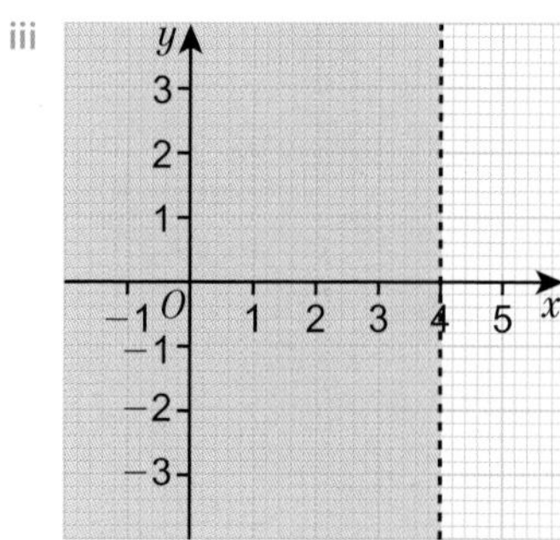

iv

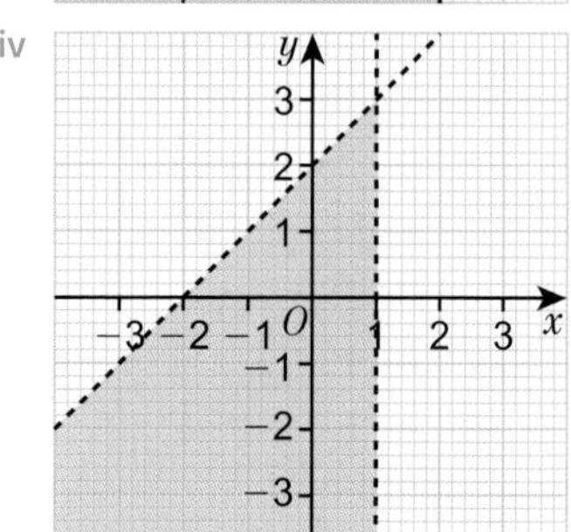

v

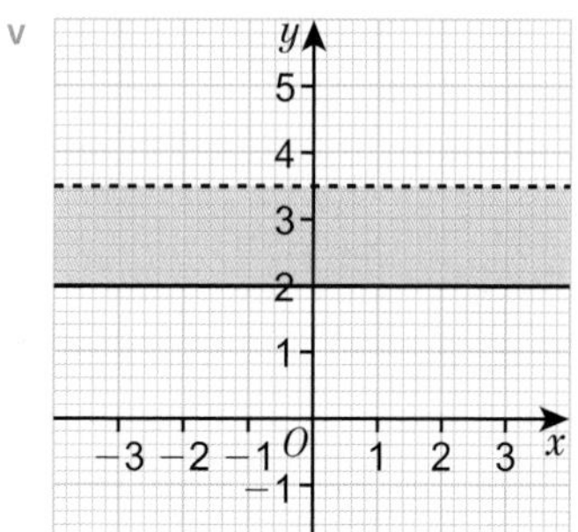

vi

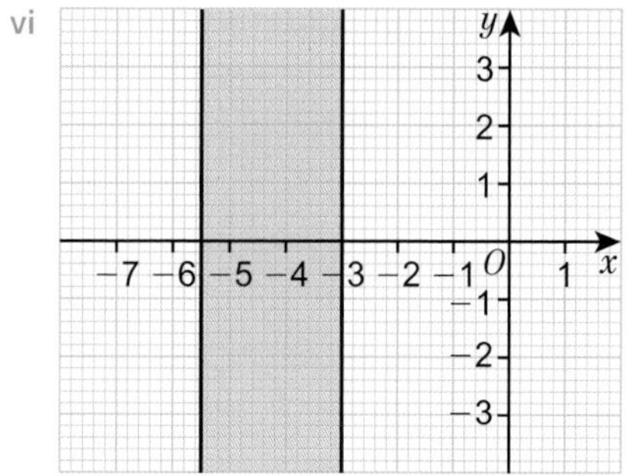

vii

3 a, b, d

c Yes

4 a

b

c

d

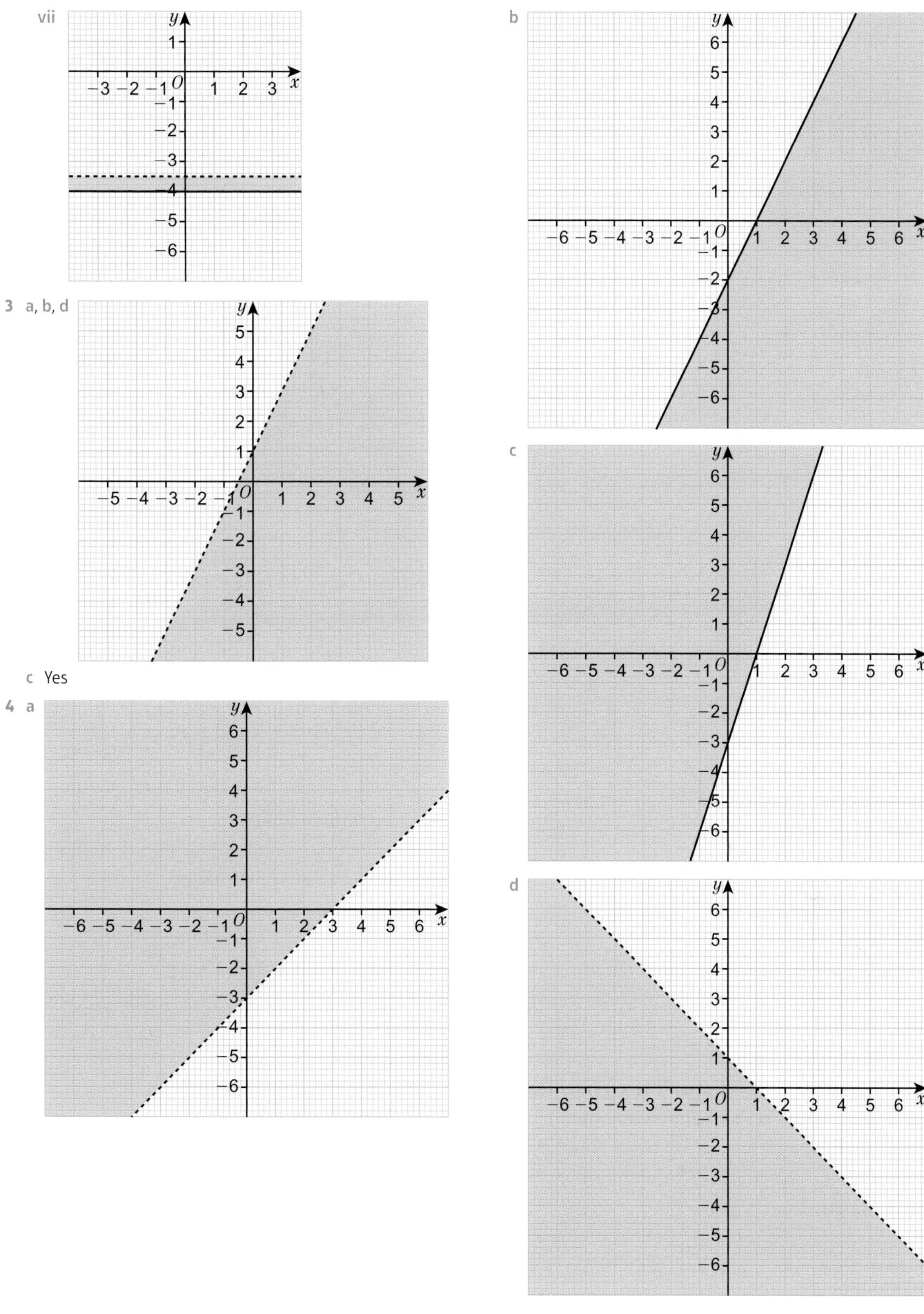

5
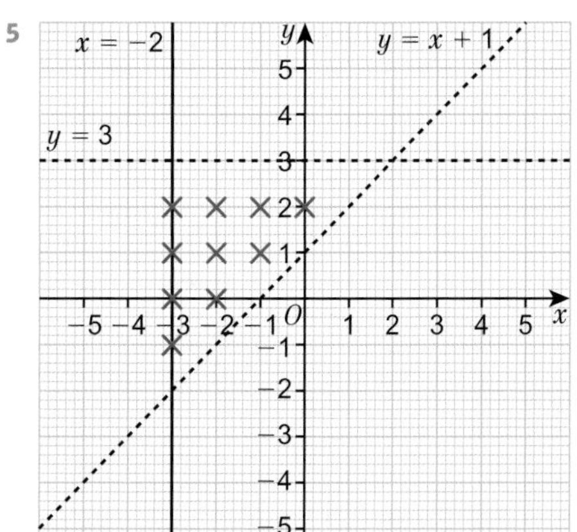

6
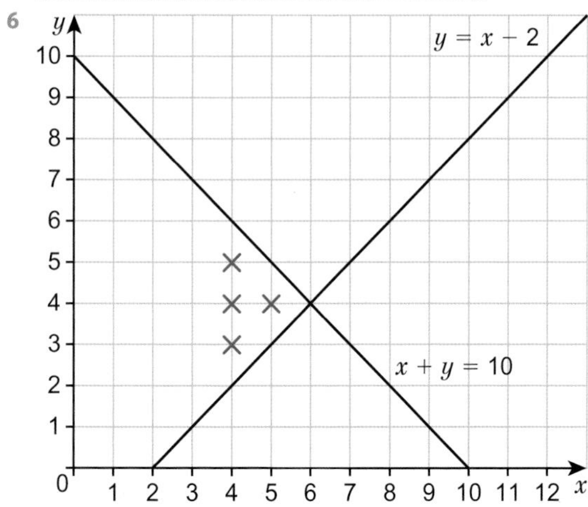

7 a i $y = 3, y = 3x - 2, x = -2$
ii $y < 3, y > 3x - 2, x \geqslant 2$
b i $x = 2, y = x + 2, y = -x - 2$
ii $y \leqslant x + 2, y \geqslant -x - 2, x < 2$
c i $y = 2x - 3, y = -3x, y = 5$
ii $y > 2x - 3, y > -3x, y \leqslant 5$

8 3 points

9
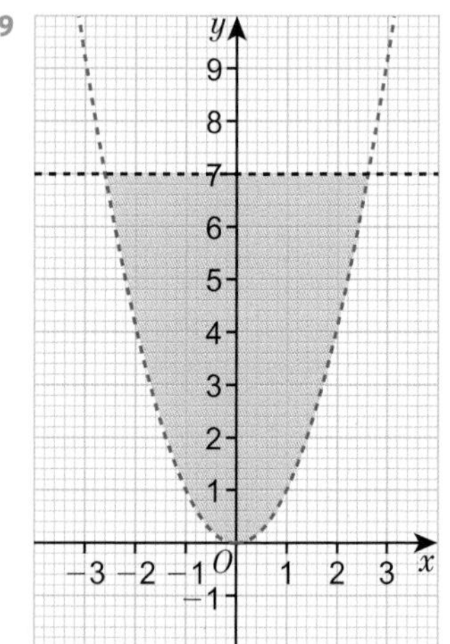

10 a $x < -1, x > 2$
b $x < -1, x > 2$

11 a
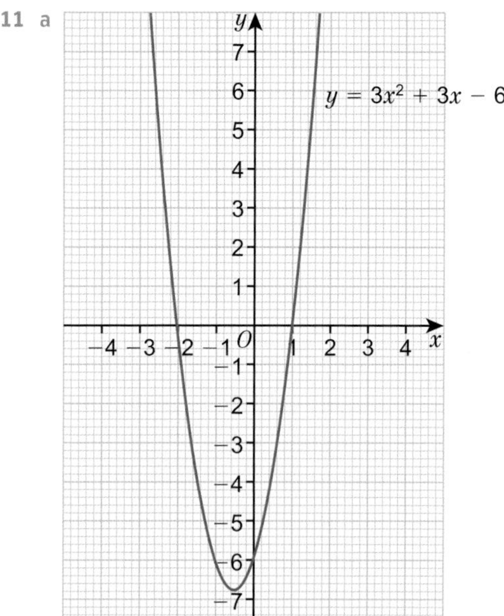

b $\{x : -2 \leqslant x \leqslant 1\}$

12 $\{x : x \leqslant -3\} \cup \{x : x > \frac{1}{2}\}$

13 a
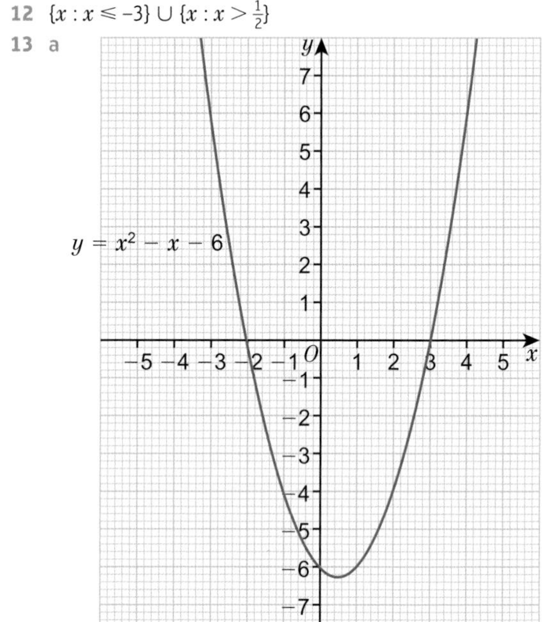

b $\{x : x > 3\} \cup \{x : x < -2\}$

14 a $\{x : -4 < x < 3\}$
b $\{x : x \geqslant 2\} \cup \{x : x \leqslant -1\}$
c $\{x : x \geqslant 3\} \cup \{x : x \leqslant -3\}$

15.3 Graphs of quadratic functions

1 a $x = -2, x = -1$
b $x = 1.5, x = -4$

2 a $(x + 1)^2 - 6$
b $2(x + 2)^2 - 4$

3 a $(1, 7)$
b $x = 1$

4

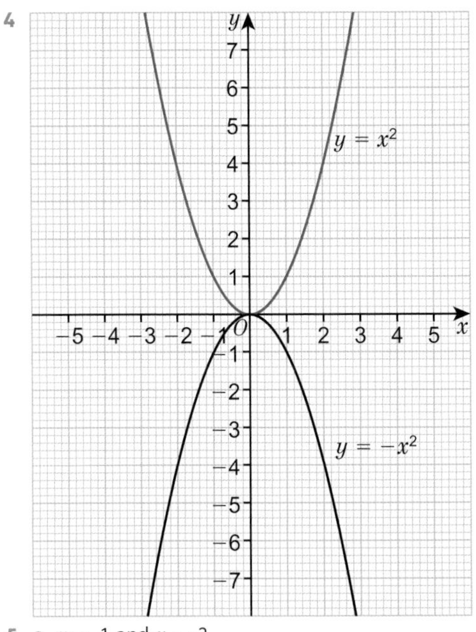

5 a $x = -1$ and $x = -3$
b (0, 3)
c Minimum
d (−2, −1)

6 a

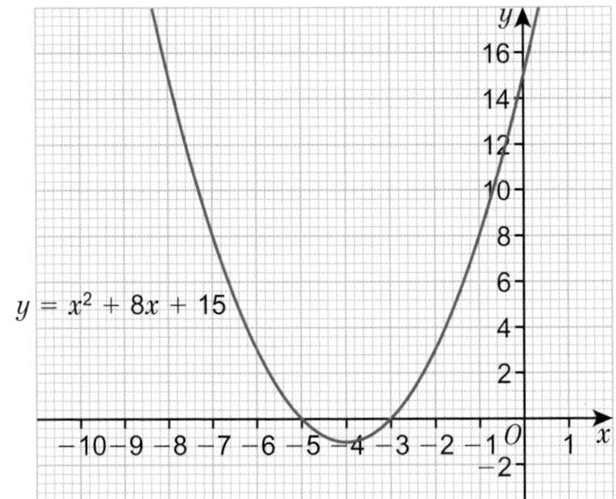

b $x = -3$ and $x = -5$
c (0, 15)
d Minimum
e (−4, −1)

7 a i $x = -3$ and $x = -1$ ii $x = -5$ and $x = -3$
b i 3 ii 15

8 a iii b iv c ii d i

9 a Minimum (1, 3)
b Maximum (−3, −2)
c Minimum (5, −2)
d Minimum (−3, −5)
e Minimum (2, 1)
f Maximum (−1, 4)

10 a $(x - 4)(x + 2)$
b (4, 0) and (−2, 0)
c (0, −8)
d $(x - 1)^2 - 9$
e (1, −9)
f Minimum, coefficient of x is positive.

g

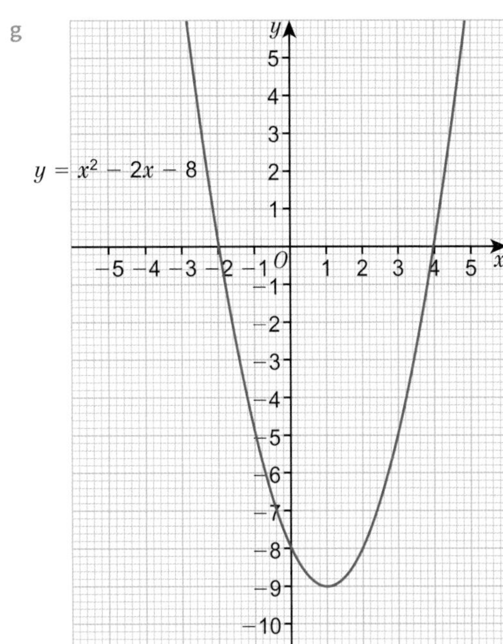

11 a

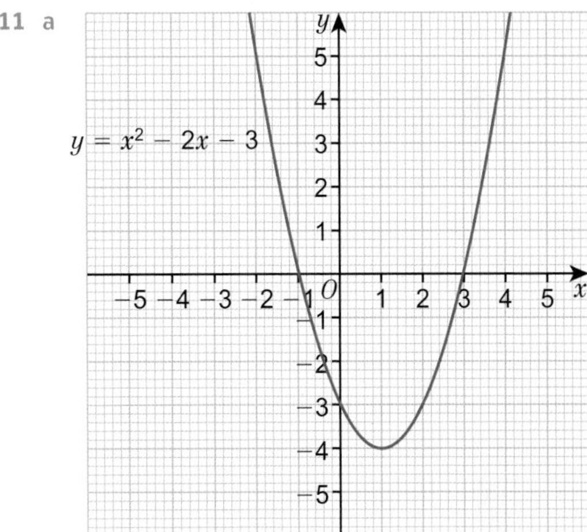

b

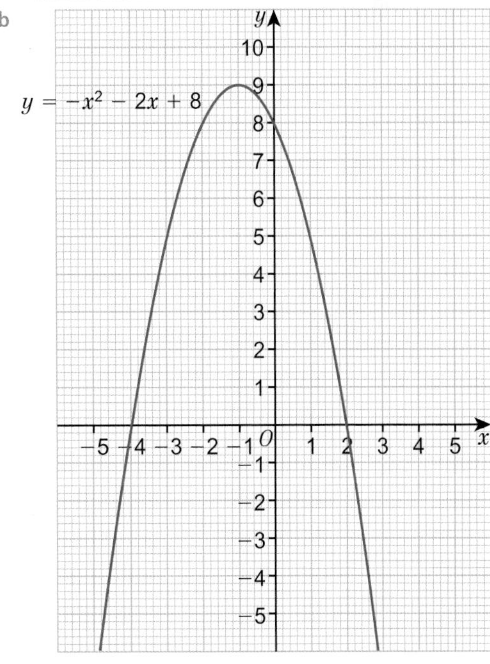

c

$y = 2x^2 - 4x - 6$

d

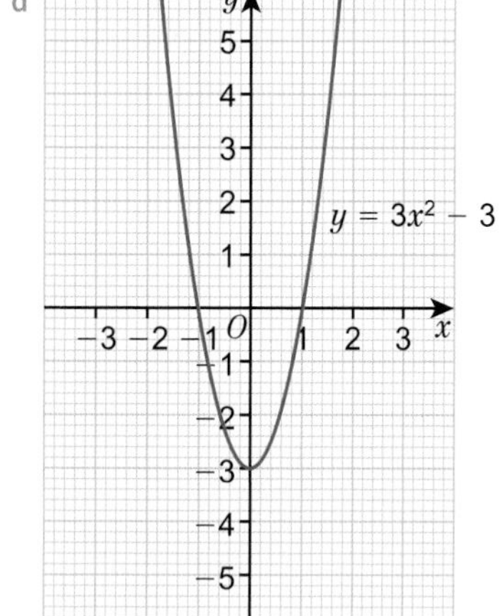

12 a $x = 3$ or $x = 1$

b

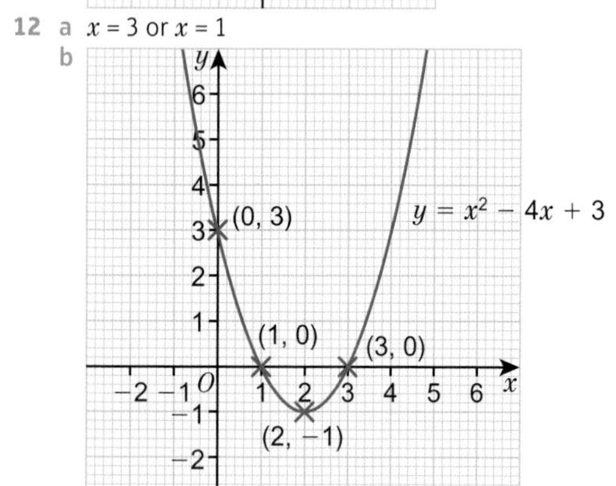

13 a (−3, −5) b $(-3 \pm \sqrt{5}, 0)$

14 a $x = -1 \pm \sqrt{3}$ b $x = -2$ or $x = 4$ c $x = -2 \pm \frac{2}{\sqrt{3}}$

15 a $x = -2 \pm \sqrt{7}$ b $x = 2 \pm \frac{3}{\sqrt{2}}$ c $x = -3 \pm \sqrt{13}$

16 The graphs should have a maximum since the coefficient of x is negative.
The graph should cross the y-axis at (0, 6).
The graph should have roots at $x = 3$ and $x = -1$.
The graph should have a turning point at (1, 8).

17 $y = x^2 + 4x + 1$

15.4 Solving quadratic equations graphically

1 a $x = -5$ and $x = -1$ b $-1 \pm \frac{\sqrt{6}}{2}$ c $-1 \pm \frac{2}{\sqrt{3}}$

2 a $x = 5.3$ and $x = -1.3$ b $x = 2.4$ and $x = -0.9$
c $x = -1.0$ or $x = 1.7$

3 a Graph i, $x = -1.3$ and $x = 5.3$
b Graph iii, $x = 1.5$ and $x = 1.1$
c Graph iv, $x = -1.2$ and $x = 3.2$
d Graph ii, $x = 0.4$ and $x = -1.4$

4 a

x	−2	−1	0	1	3	5
y	31	17	7	1	1	17

b

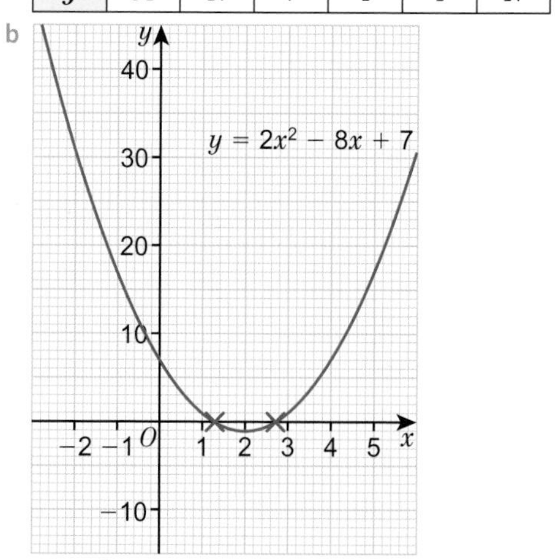

c 2.7 and 1.3

5 a Students' own graphs
b i $x = -5.3$ and $x = 1.3$ ii $x = -3.6$ and $x = 0.6$
iii $x = 2.4$ and $x = -0.4$ iv $x = 2.5$ and $x = -0.5$

6 a

x	−4	−3	−2	−1	0	1	2
y	14	4	−2	−4	−2	4	14

b

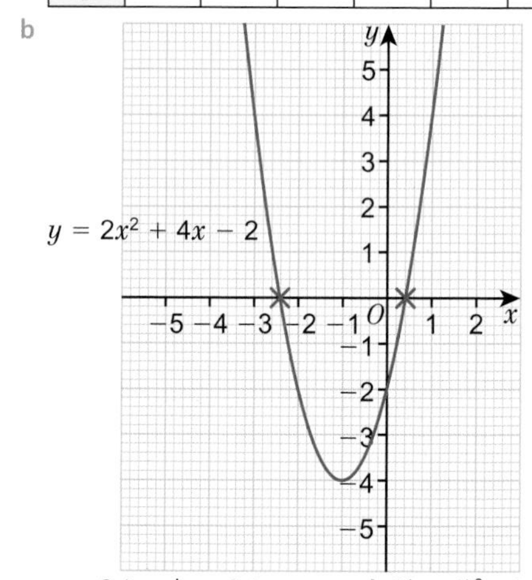

c $x = -2.4$ and $x = 0.4$ d $2(x + 1)^2 - 4$

7 a i (−1, 1) ii (0, 2)
iii

$y = x^2 + 2x + 2$

(0, 2)

(−1, 1)

b i (−2, −3) ii (0, −7)
iii

(−2, −3)
(0, −7)
$y = -x^2 - 4x - 7$

c i (1, 1) ii 3
iii

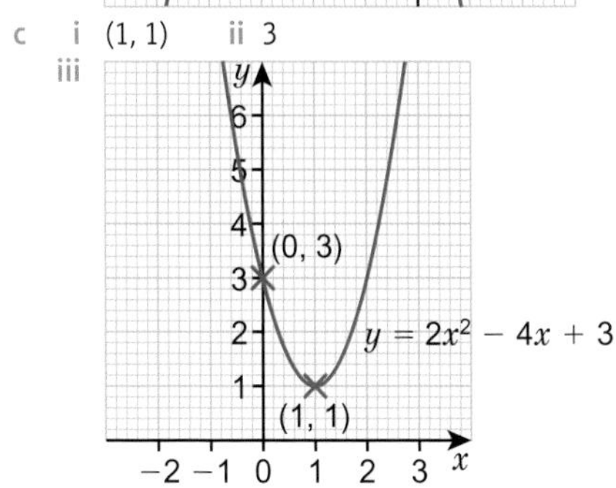

8 The minimum is at (3, 5) therefore the graph has no roots.

9 a 0 roots b 2 roots c 2 roots d 1 root
e 0 roots f 2 roots g 1 root h 2 roots

10 a $x = 2 \pm \sqrt{7}$
b $x^2 - 7x + 13 = 0$
$\left(x - \frac{7}{2}\right)^2 - \frac{49}{4} + 13 = 0$
$\left(x - \frac{7}{2}\right)^2 = -\frac{3}{4}$
There are no real roots of a negative number.
OR
The graph has a turning point at $(\frac{7}{2}, \frac{3}{4})$. Since this is a minimum the whole graph is above the x-axis.

11 a $y = x^2 - x - 2$
b $y = x^2 - 10x + 21$
c $y = x^2 + 4x + 4$

12 a 3.23607 b 5.70156 c −1.30278

13 a $x = -2$ and $x = 1$
b

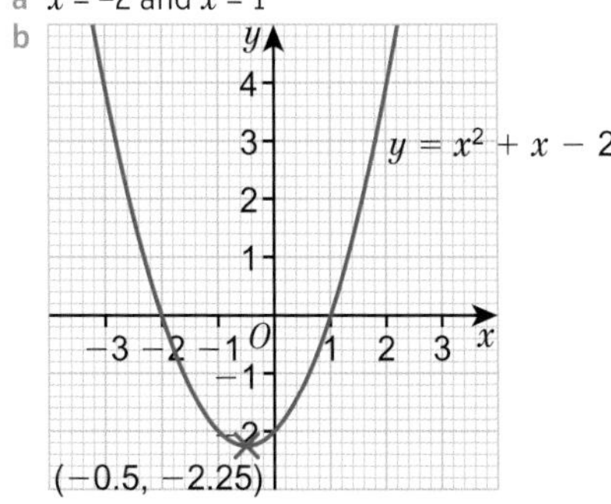

c $\{x : -2 < x < 1\}$
d $\{x : x < -2\} \cup \{x : x > 1\}$

14 a $\{x : -1 < x < 3\}$
b $\{x : -5 < x < 2\}$
c $\{x : x < -4\} \cup \{x : x > -1\}$

15.5 Graphs of cubic functions

1 a $x^2 + 5x + 6$ b $x^2 + x - 12$
c $2x^2 - 9x - 5$ d $3x^2 - 10x + 8$

2 a $x = 5$ or $x = -2$ b $x = 1$ or $x = -1$
c $x = 3$ or $x = -1$ d $x = -2$ or $x = \frac{1}{3}$

3 $x^3 + 6x^2 + 9x + 2$

4 $x^3 + 9x^2 + 26x + 24$

5 a $x^3 + 8x^2 + 17x + 10$ b $x^3 - x^2 - 14x + 24$
c $x^3 + 4x^2 + x - 6$ d $x^3 + x^2 - 20x$
e $x^3 + x^2 - x - 1$ f $x^3 + 9x^2 + 27x + 27$

6 a $x = -4, x = -1, x = 2$ b (0, −8)

7 a $x = -1, x = -2, x = -5$ b (0, 10)
c

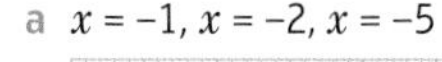

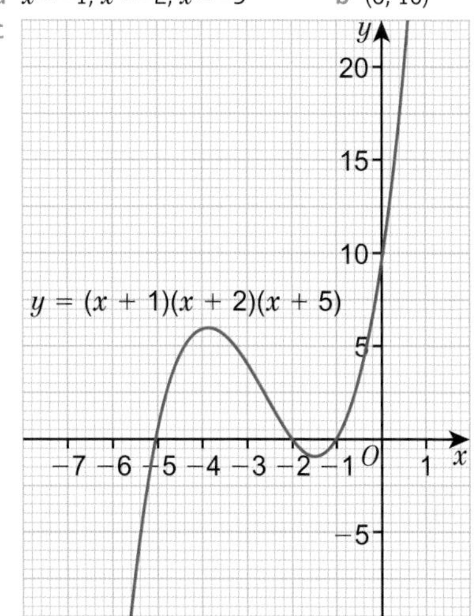

8 a iv b iii c i d v e vi f ii

9 a 3 b 1 repeated root
c 3 d 2 (one repeated)
e 1 f 3

10 a

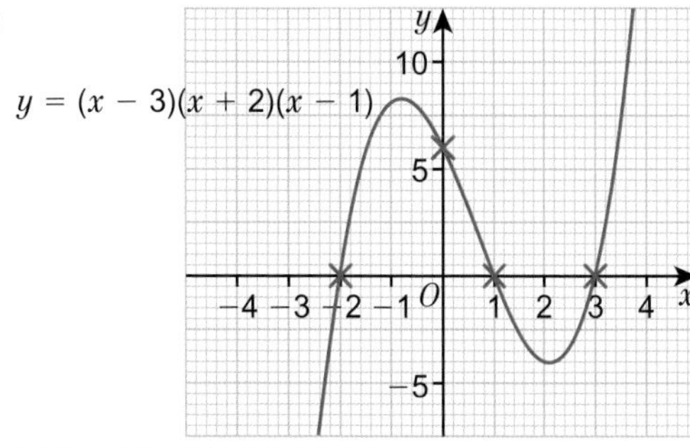

b

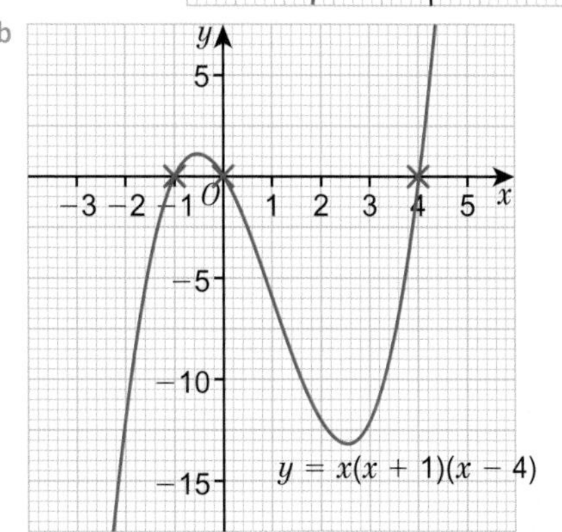

c

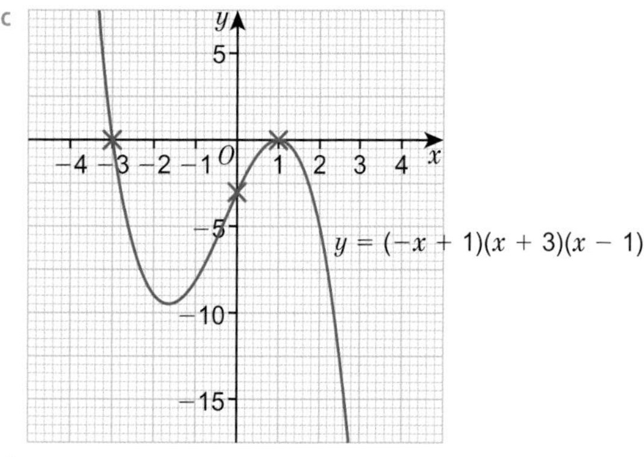

d

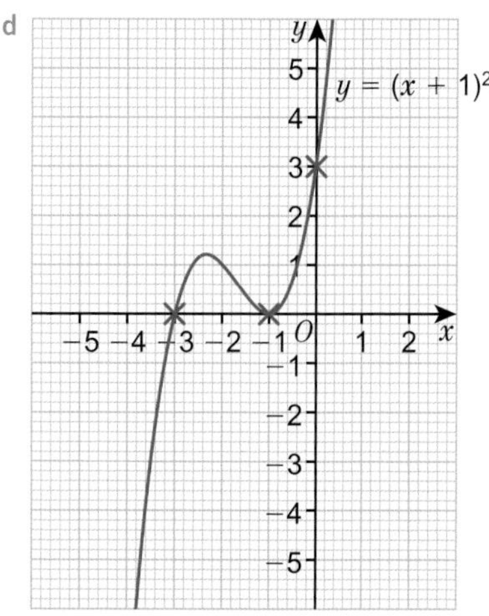

e

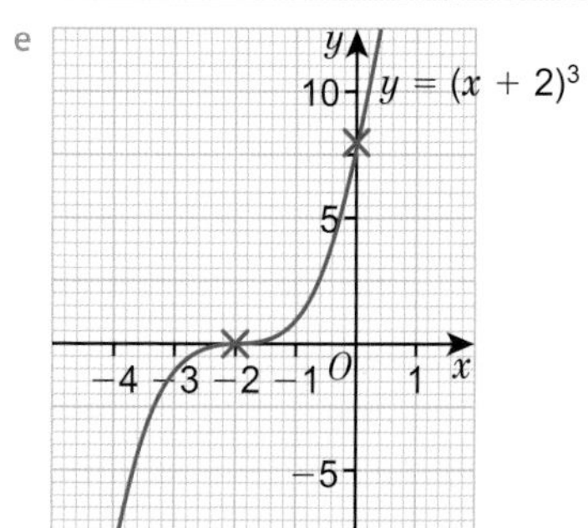

11

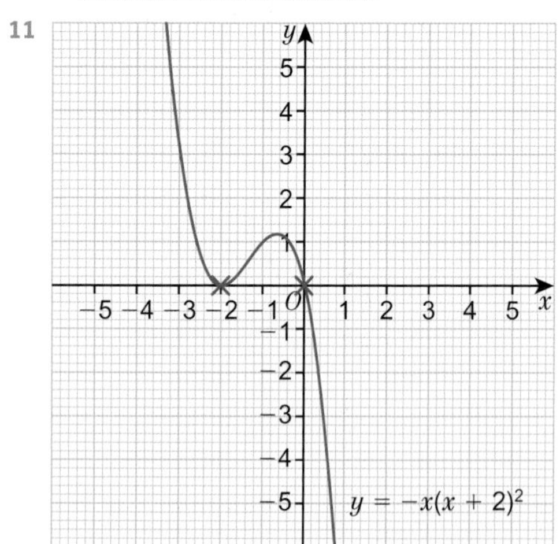

12 $a = 2, b = -9, c = -18$

13 $a = 3, b = 6, c = -8$

14 $x = 1.4562$

15 $x = -1.9122$

15 Problem-solving

1 Approximately −24 °C

2 3.5 metres

3 a 25.60 AED b £9.74

4 a House prices are increasing faster than earnings.
 b Approximately 12.5
 c House prices have increased, or earnings have decreased, more than expected.

5 a 6.5 metres b 3.1 metres

15 Check up

1 Package A: $y = 0.1x$
 Package B: $y = 20 + 0.05x$
 b 400 minutes

2

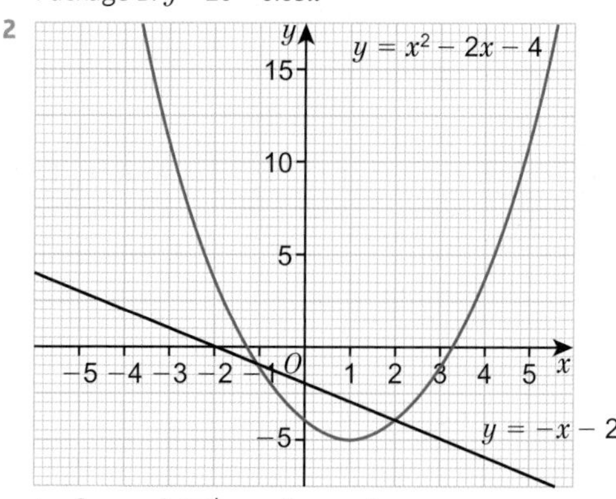

$x = 2, y = -4$ and $x = -1, y = -1$

3 $y < 2x - 1, x \leqslant 2, y \geqslant -2$

4 a, b

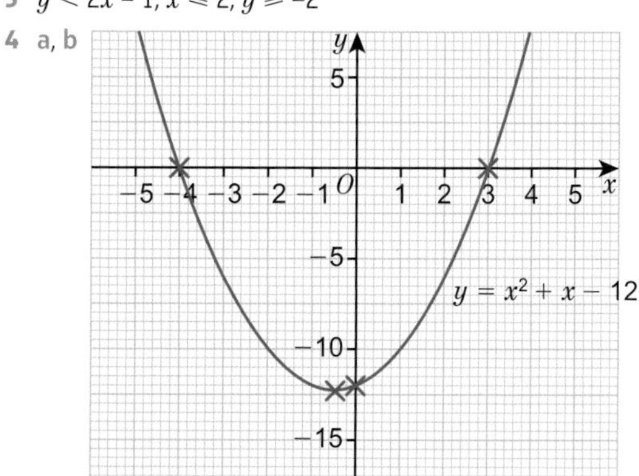

c $\{x : -4 < x < 3\}$

5 a

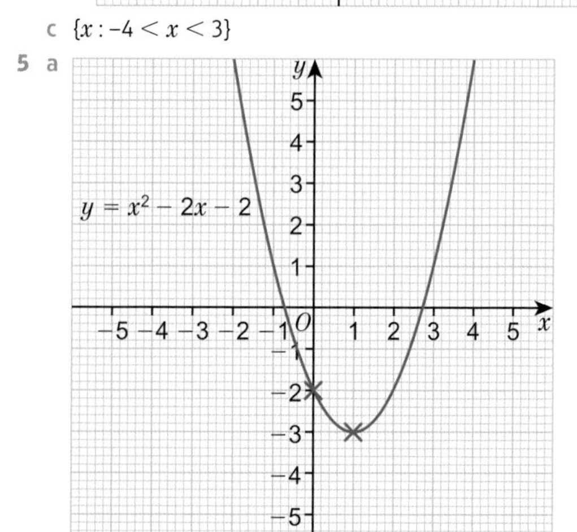

b 2.7 and −0.7

c Iterative equation:
 $x_{n+1} = \sqrt{2x_n + 2}$
 $x = 2.73205$

6
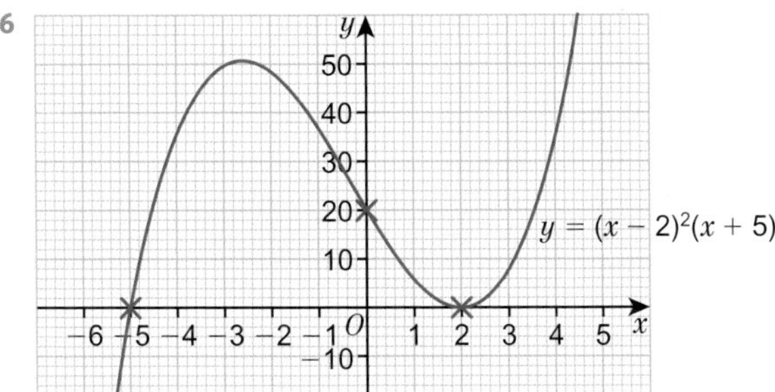

8 Functions of the form:
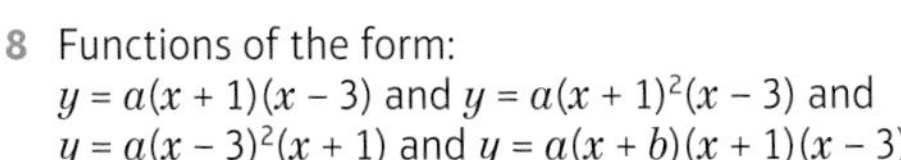
$y = a(x + 1)(x - 3)$ and $y = a(x + 1)^2(x - 3)$ and
$y = a(x - 3)^2(x + 1)$ and $y = a(x + b)(x + 1)(x - 3)$

15 Strengthen

Simultaneous equations and inequalities

1 a

x	−2	−1	0	1	2	3	4	5
y	11	5	1	−1	−1	1	5	11

b, c

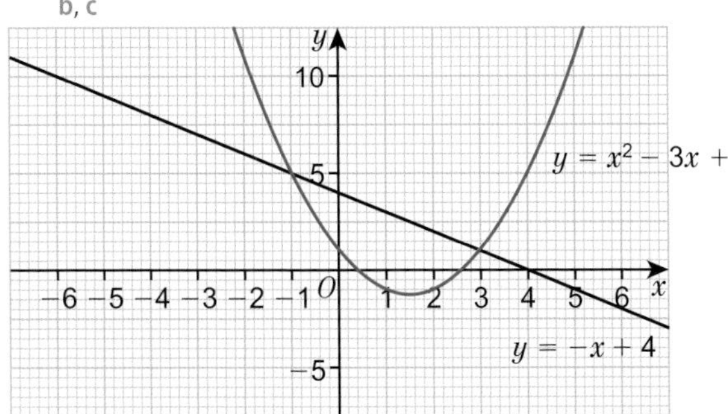

d (3, 1) and (−1, 5)
e $x = 3$, $y = 1$, and $x = -1$, $y = 5$.

2 a $x = 1$, $y = 6$
b $x = 3$, $y = 4$ or $x = -1$, $y = 0$

3 a $x + y = 12$
b $x - y = 6$
c
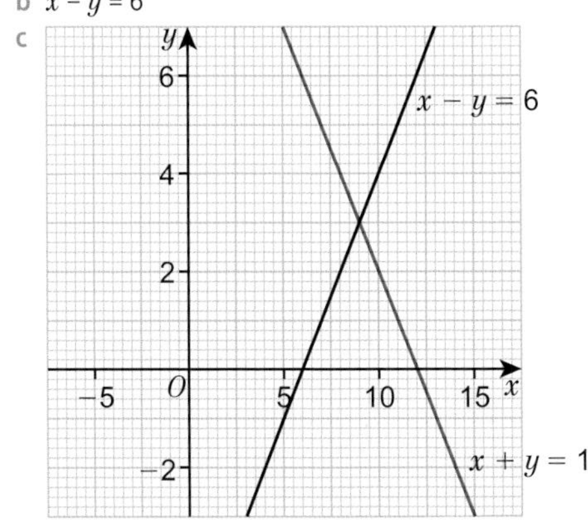

d $x = 9$, $y = 3$ or $y = 3$, $x = 9$, depending on how they have written the original functions.

4 $y \leqslant x + 2$, $x \leqslant 2$, $y > -3$

5 $y < 1$, $y \geqslant x - 3$, $y \geqslant -x$

6 a, b, d, f
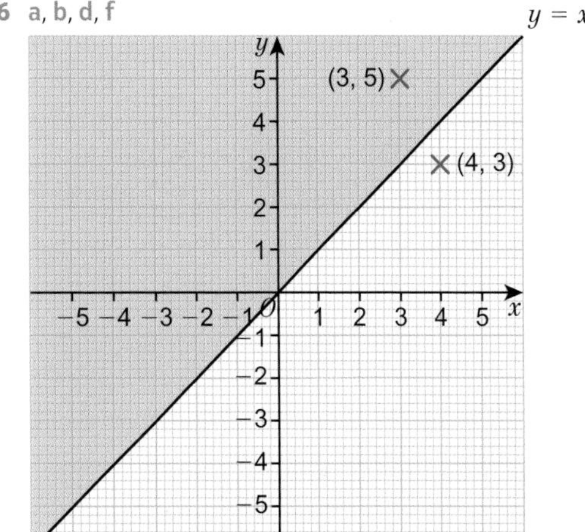

c Yes e No

7 a
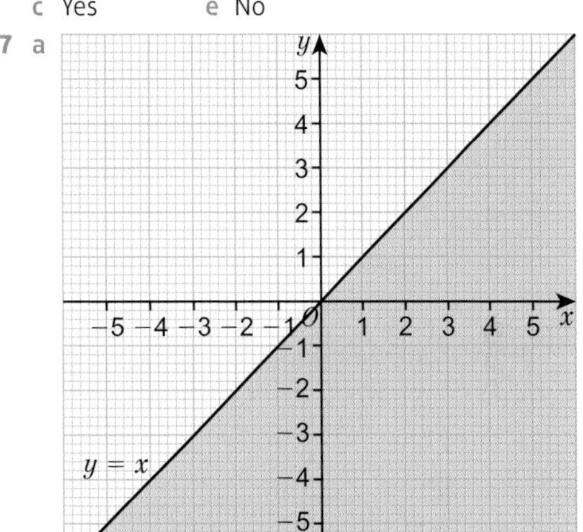

b
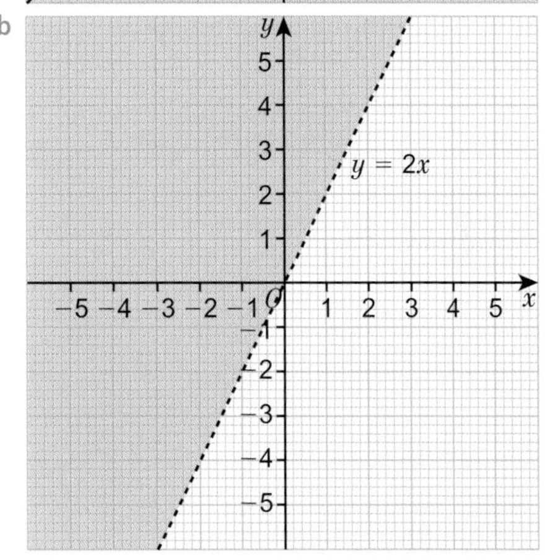

c
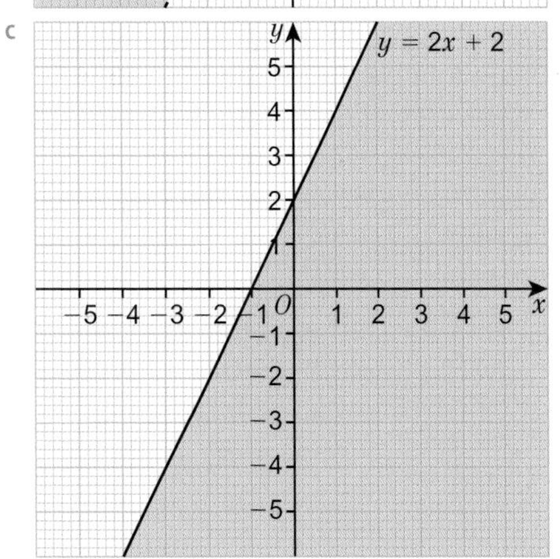

d

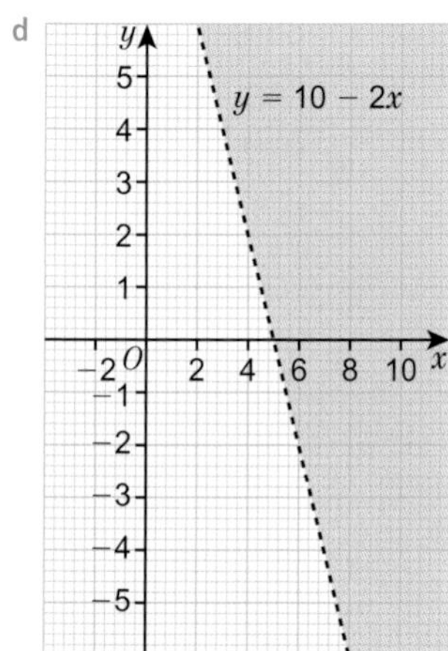

8

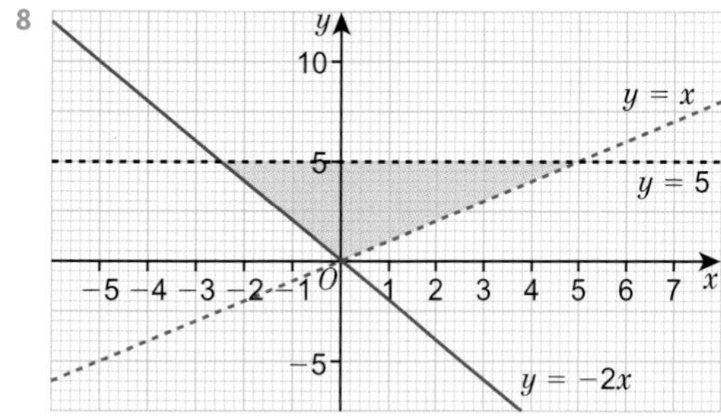

9

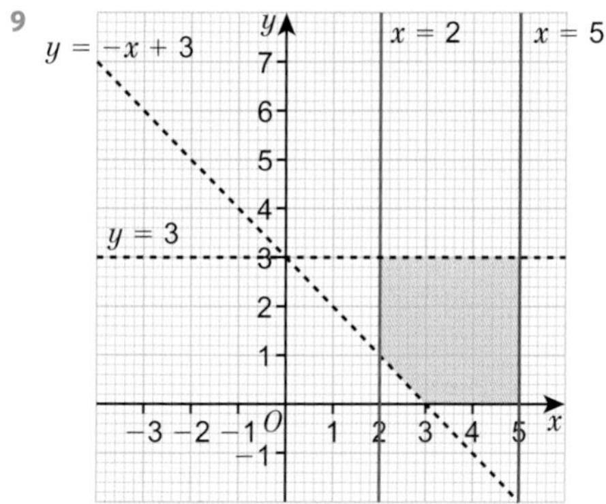

Graphs of quadratic functions

1 a Roots: −1, 3 y-intercept: −6
  b Roots: −3, 1 y-intercept: −9
  c Roots: −2, 2 y-intercept: 4
  d Roots: 1, 4 y-intercept: −4

2 a When $y = 0$
  $x^2 + 2x - 8 = 0$
  $(x - 2)(x + 4) = 0$
  There are two possible solutions:
  $x - 2 = 0$ hence $x = 2$
  $x + 4 = 0$ hence $x = -4$
  So the roots are $x = 2$ and $x = -4$.
  b When $x = 0$
  $y = x^2 + 2x - 8$
  $y = 0^3 - 0 - 8$
  $y = -8$
  So the intercept with the y-axis is $y = -8$.

3 a roots: 5 or − 3, y-intercept: −15
  b roots: −8 and 2, y-intercept: 16
  c roots: −0.5 and 3, y-intercept: −6
  d roots: 1 or −5, y-intercept: −15

4 a $y = (x + 1)^2 - 1 - 8$
  $y = (x + 1)^2 - 9$
  b (−1, −9)
  c Minimum

5 a (1, −16), minimum
  b (−3, 25), maximum
  c (1, −8), minimum
  d (−2, −27), minimum

6 a

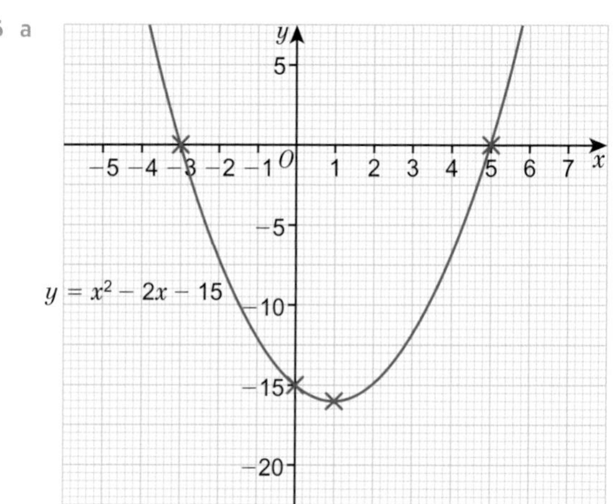

b

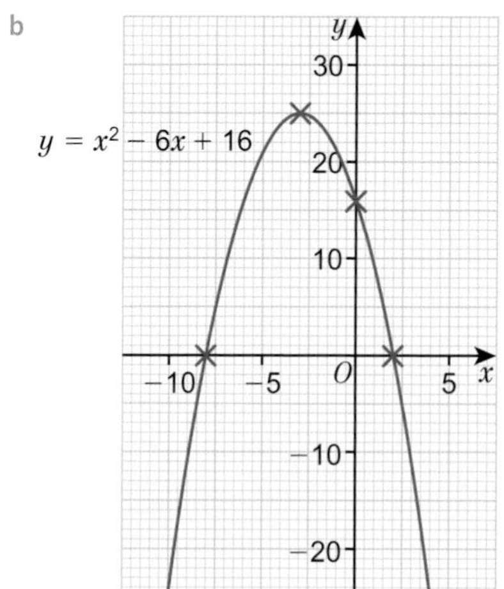

c

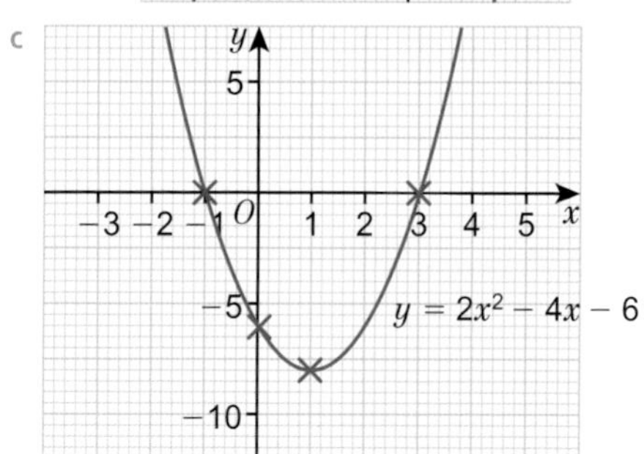

d

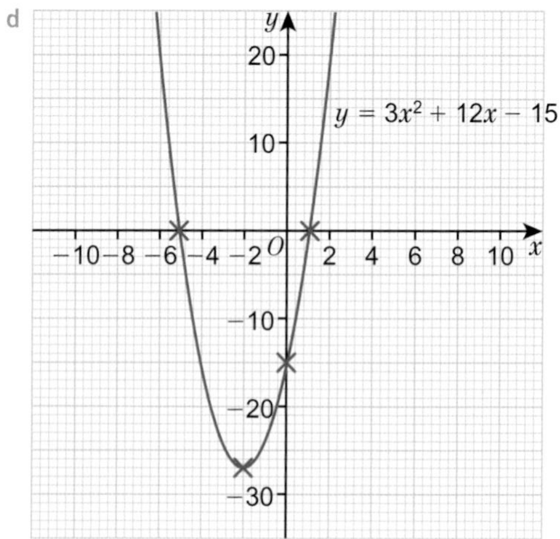

7 a $-3 < x < 5$
  b $x < -3$ and $x > 5$

8 a

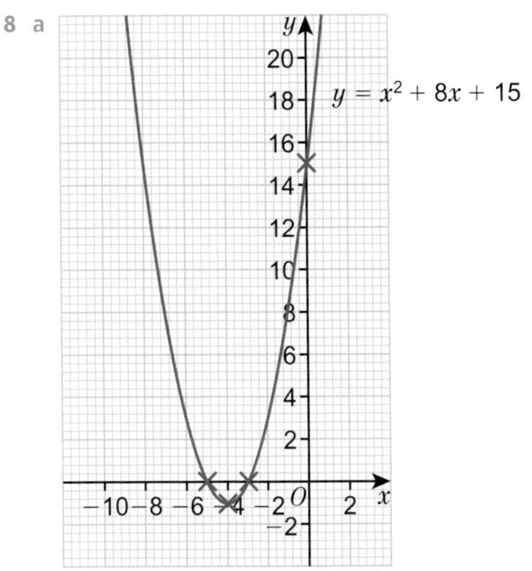

b

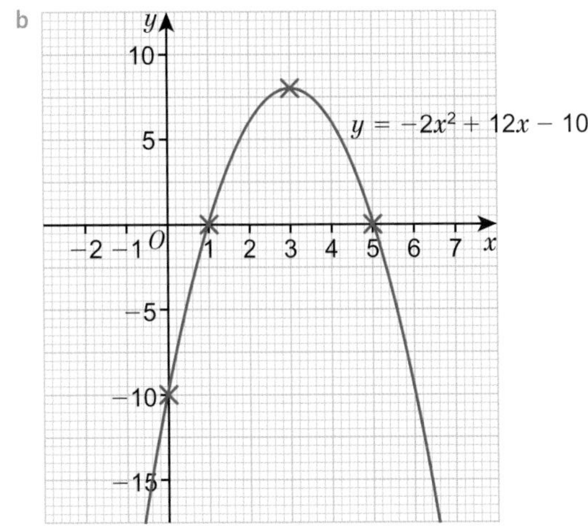

9

x	-5	-4	-3	-2	-1	0	1	2
y	2	-3	-6	-7	-6	-3	2	9

a

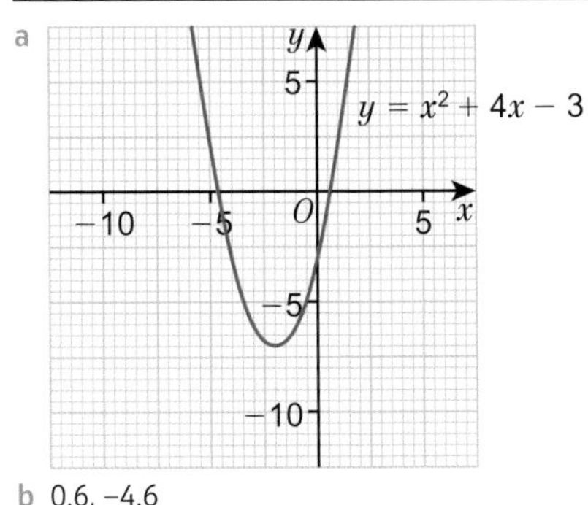

b 0.6, –4.6

10 a

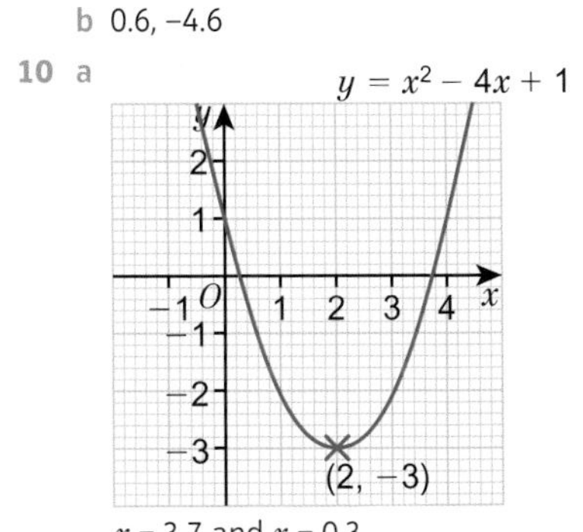

$x = 3.7$ and $x = 0.3$

b

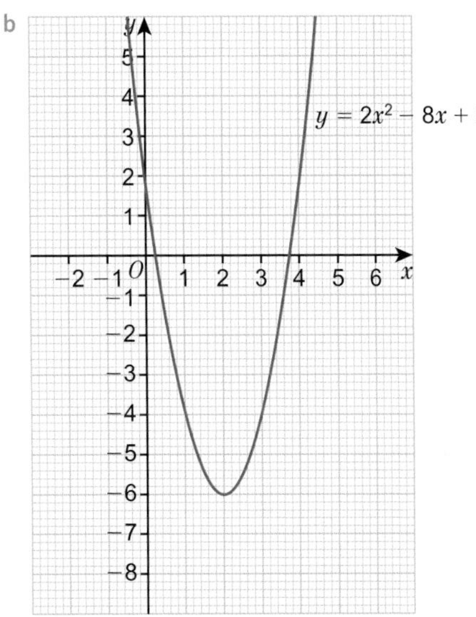

$x = 3.7$ and $x = 0.3$

c

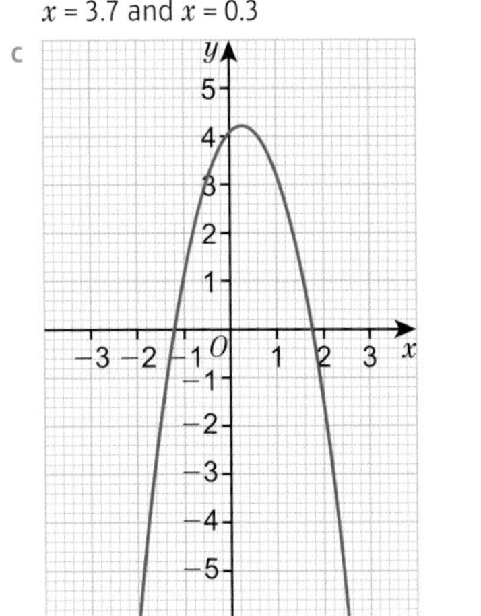

$x = 1.7$ and $x = -1.2$

11 a $y = x^2 - 4x + 1$.
$0 = x^2 - 4x + 1$
$x^2 = 4x - 1$
$x = \sqrt{4x - 1}$
$x_{n+1} = \sqrt{4x_n - 1}$

b $x_1 = 3.605551275$
$x_2 = 3.663632774$
$x_3 = 3.695203796$
$x_4 = 3.712252037$
$x_5 = 3.721425553$

c $x = 3.732$

12 a $x_{n+1} = \sqrt[3]{8x_n - 1}$ b 7.873

Graphs of cubic functions

1 a i $x = -1, x = 1, x = -2$ ii (0, –2)
b i $x = -1, x = -2, x = -5$ ii (0, 10)
c i $x = -1, x = -2, x = 3$ ii (0, 6)

2 a When $y = 0$
$(x - 3)(x + 4)(x - 1) = 0$
So there are three possible solutions:
$x - 3 = 0$ hence $x = 3$
$x + 4 = 0$ hence $x = -4$.
$x - 1 = 0$ hence $x = 1$

b When $x = 0$
$y = (x - 3)(x + 4)(x - 1)$
$y = (0 - 3)(0 + 4)(0 - 1)$
$y = 12$

3 a Roots: –3, 7, –2 y-intercept: –42
b Roots: 10, –2, –1 y-intercept: 20
c Roots: 2.5, –2, 1 y-intercept: 10
d Roots: 3, 2 y-intercept: –54

15 Extend

1 a

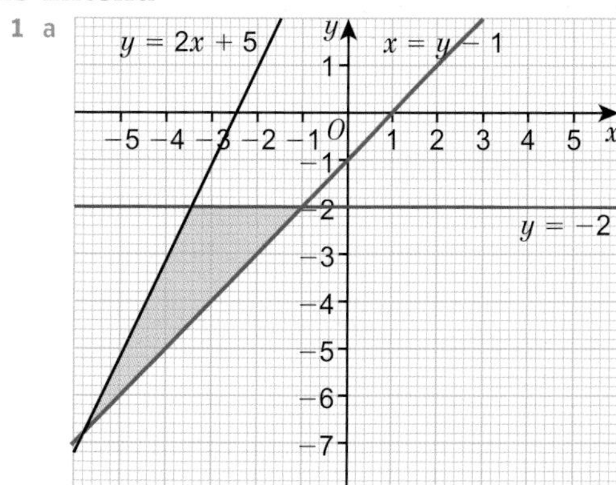

b Area = 6.25 units^2

2 a Pentagon b 540°

3 7.5π

4 a 1 b 0 c 2 d 0 e 2 f 0

5 a i $y = -a(x + 3)^2 - 4$ ii $y = a(x - 4)^2 - 3$
b $y = ax(x + 3)$

6 $6\sqrt{5}$

7 a $x^4 + 5x^3 + 2x^2 - 8x$ b 4

8 RHS $= x(x - 2)(x + 2)$
$= x(x^2 - 2x + 2x - 4)$
$= x(x^2 - 4)$
$= x^3 - 4x$

9 $-n^2 + 4n + 20 = -(n - 2)^2 + 24$
The largest value this can take is when $n - 2 = 0$, when $n = 2$.

10 (0, 15), (1, 16) and (6, 21)

11 $x^4 + 3x^3 + 5x^2 + 5x + 2$

12 a $x = 0.6, y = 4.2$
$x = -3.2, y = 2.7$
b $x = -2.6, y = -0.4$
$x = 1.6, y = 0.6$
c $x = -3.1, y = 6.5$
$x = 3.6, y = 3.2$

13 5

14 a 37° and 143° b 11°, 191°

15 a $\{x : -1 < x < 1.5\}$ b $\{x : -2 < x < 5\}$

15 Unit test

Sample student answers

Student B gives the best answer as they have used the information on the graph and the equation given.
Student A has calculated c correctly, but has incorrectly identified b.
Student C has incorrect values for b and c.

UNIT 16

16 Prior knowledge check

1 a $3\frac{1}{2}$ b 20 c $-5\frac{1}{2}$

2 $4(x + y)$

3 a 2.8 b −4

4

radius
arc
sector
segment

5 66°

6 AC = 5.4 cm; PQ = 7.4 cm

7 Using Pythagoras, both triangles have side lengths 12 cm, 13 cm and 5 cm, therefore the triangles are congruent (SSS).

8 The angle is always 90°.

16.1 Radii and chords

1 $x = 110°$ (angles in an isosceles triangle)
$y = 145°$ (angles on a straight line)

2 AD is common; ∠ABD = ∠ACD and AB = AC (isosceles triangle); ∠BAD = ∠CAD = 180° − 90° − ∠ABD (angles in a triangle). Therefore the triangles are congruent (SAS).

3 a $i = 30°$
b $j = 21°$
c $k = 64°$; $l = 116°$; $m = 32°$

4 ∠PBA = 45° (angles on a straight line add to 180°);
∠PAB = 35° (angles in a triangle).
The triangle is not isosceles so AP and PB are not radii.

5 a OM is common; OA = OB (radii of same circle);
∠OMA = 90° (angles on a straight line).
Therefore the triangles are congruent (RHS).
b OAM and OBM are congruent, so AM = AB.
Therefore M is the midpoint of AB.

6 a AM = 6 cm The perpendicular from the centre of a circle to a chord bisects the chord.
b AO = 10 cm

7 OM = 15 cm

8 AB = 20 cm

9 a 90° b 65° c 130°

16.2 Tangents

1 a 58° b 11.3 cm

2 OP is common; AO = OB (radii of same circle);
∠OAP = ∠OBP = 90°. Therefore the triangles are congruent (RHS)

3 a ∠OAP = ∠OBP = 90° (angle between tangent and radius is 90°)
a = 360° − 90° − 90° − 20° = 160° (angles in a quadrilateral add to 360°)
b OA = OB (radii of same circle)
∠OAB = (180° − 136°) ÷ 2 = 22° (isosceles triangle)
∠OAP = 90° (angle between tangent and radius is 90°)
So b = 90° − 22° = 68°
c OB = OA (radii of same circle)
∠OBA = ∠OAB = 14° (isosceles triangle)
c = 180° − 14° − 14° = 152° (angles in a triangle)
∠OAP = 90° (angle between tangent and radius is 90°)
So d = 90° − 14° = 76°
d ∠OBP = 90° (angle between tangent and radius = 90°)
∠ABO = 90° − 34° = 56°
e = ∠ABO = 56° (isosceles triangle)
f = 180° − 56° − 56° = 68° (angles in a triangle)
e g = ∠BAP (isosceles triangle)
g = (180° − 50°) ÷2 = 65°
∠OBP = 90° (angle between tangent and radius is 90°)
So h = 90° − 65° = 25°

4 ∠OTP = 90° (angle between tangent and radius is 90°)
∠TOP = 180 − (90 + 32) = 58° (angles in a triangle)
∠SOT = 180 − 58 = 122° (angles on a straight line)
OS = OT (radii of same circle)
x = (180 − 122) ÷ 2 = 29° (angles in a triangle)

5 OA = OB (radii of same circle); OT is common;
∠TAO = ∠TBO = 90° (angle between tangent and radius is 90°). Therefore triangles OAT and OBT are congruent (RHS).
In congruent triangles, equivalent angles are equal, so $a = b$ and $x = y$.

6 OT = 15 cm (angle between tangent and radius is 90°)

16.3 Angles in circles 1

1 a 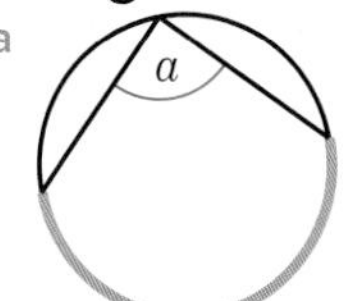

b 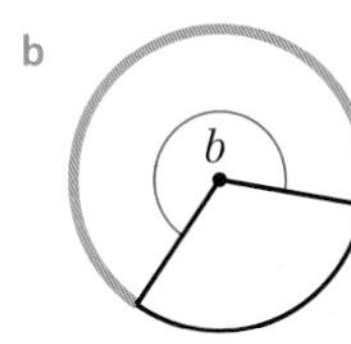

2 $x = 56° + 52° = 108°$ (exterior angle equals the sum of the two interior opposite angles)

3 a $a = 140°$ (angle at the centre is twice the angle at the circumference subtended by the same arc)
b $b = 48°$ (angle at the centre is twice the angle at the circumference subtended by the same arc)
c $c = 250°$ (angles round a point add to 360°)
$d = 125°$ (angle at the centre is twice the angle at the circumference subtended by the same arc)

4 ∠AOB =180°; ∠ACB = 180° ÷ 2 = 90° because the angle at the centre is twice the angle at the circumference subtended by the same arc.

5 a $a = 90°$ (angle in a semicircle is 90°)
$b = 62°$ (angles in a triangle)
b $c = 16°$ (isosceles triangle)
$d = 180 - 16 = 74°$ (angle in a semicircle is 90°)
c $f = 30°$ (angle in a semicircle is 90° and angles in a triangle add to 180°)
$2f = 60°$
d $g = 100°$ (angle at the centre is twice the angle at the circumference subtended by the same arc)
e $h = 105°$ (angle at the centre is twice the angle at the circumference subtended by the same arc)
f $i = 105°$ (angles round a point add to 360°, and the angle at the centre is twice the angle at the circumference subtended by the same arc)

6 The angle at the centre is twice the angle at the circumference subtended by the same arc, so $a = 230 \div 2 = 115°$

7 a $j = 46°$ (the angle at the centre is twice the angle at the circumference subtended by the same arc)
b $a = 138°$ (isosceles triangle)
$b = 48°$ (angle in a semicircle is 90° and angles in a triangle add to 180°)

8 The angles are on the same arc, so the angle at the centre is twice the angle at the circumference.
This means that $a = 60°$.
Lucy has worked out 30 ÷ 2 instead of 30 × 2.

9 ∠ABO = ∠ADO = 90° (angle between tangent and radius is 90°)
∠DOB = 130° (angles in a quadrilateral add to 360°)
∠BCD = 65° (the angle at the centre is twice the angle at the circumference subtended by the same arc)

16.4 Angles in circles 2

1 $180 - 2x$

2 a $a = b = c = 50°$ b $d = 2x$

3 ∠AOB = 2 × ∠ACB (the angle at the centre is twice the angle at the circumference subtended by the same arc). Similarly, ∠AOB = 2 × ∠ADB. So ∠ACB = ∠ADB

4 a $a = 42°$ (angles at the circumference subtended by the same arc are equal)
$b = 180 - 90 - 42 = 48°$ (angle in a semicircle is 90° and angles in a triangle add to 180°)
b $c = 56°$ (angles at the circumference subtended by the same arc are equal)
$d = 34°$ (angle in a semicircle is 90° and angles in a triangle add to 180°)
c $e = f = 55°$ (the angle at the centre is twice the angle at the circumference subtended by the same arc)
d $g = h = (360° - 170°) \div 2 = 95°$ (angles around a point add to 360°, the angle at the centre is twice the angle at the circumference subtended by the same arc)

5 a $a = 50°$; $b = 260°$; $c = 130°$
b $a = 100°$; $b = 160°$; $c = 80°$

6 a $2x$ b $2y$ c $2x + 2y = 360°$
d $2(x + y) = 2 \times 180$, so $x + y = 180$

7 a $a = 85°$ (angles on a straight line)
$b = 95°$ (opposite angles of a cyclic quadrilateral add to 180°)
b $c = 105°$ and $e = 98°$ (angles on a straight line)
$d = 75°$ and $f = 82°$(opposite angles of a cyclic quadrilateral add to 180°)

8 a Students' own drawing
b Angle x + angle y = 180° because angles on a straight line add to 180°.
Angle x + angle z = 180° because opposite angles of a cyclic quadrilateral add to 180°.

9 a $a = 72°$ (opposite angles of a cyclic quadrilateral add to 180°)
$b = 108°$ (angles on a straight line add to 180°)
$c = 93°$ (angles in a quadrilateral add to 360°)
b $d = 41°$ and $e = 32°$ (angles at the circumference subtended by the same arc)
$f = g = 107°$ (angles in a triangle add to 180°)
c $h = 43°$ (opposite angles of a cyclic quadrilateral add to 180°)
$i = 43°$ (angles at the circumference subtended by the same arc)
$j = 137°$ (angles on a straight line add to 180°)
d $k = 46°$ and $m = 38°$ (angles subtended by the same arc)
$l = 54°$(angles in a triangle add to 180°)
e $n = 116°$ (opposite angles of a cyclic quadrilateral add to 180°)
$p = 26°$ (angle in a semicircle is 90°)

10 a Students' own drawing
b Angle OAT = 90° because the angle between the tangent and the radius = 90°.
Angle OAB = 90° − 58° = 32°.
OA = OB because radii of the same circle.
Angle OAB = angle OBA because the base angles of an isosceles triangle are equal.
Angle AOB = 180° − 32° − 32° = 116° because angles in a triangle add to 180°.
Angle ACB = 116° ÷ 2 = 58° because the angle at the centre is twice the angle at the circumference when both are subtended by the same arc.

11 72°.

12 ∠OAT = 90° (angle between the tangent and the radius is 90°)
∠OAB = 90° − x
OA = OB (radii of the same circle)
∠OAB = ∠OBA (base angles of an isosceles triangle are equal)
∠AOB = 180° − (90° − x) − (90° − x) = $2x$ (angles in a triangle add to 180°)
∠ACB = $2x$ ÷ 2 = x (angle at the centre is twice the angle at the circumference when both are subtended by the same arc)
So ∠BAT = ∠ACB = x

15 OM = ON (radii of the same circle)
∠OMN = $\frac{1}{2}(180 - y)$ (angles in an isosceles triangle)
∠OMB = 90° (angle between the tangent and the radius is 90°)
∠BMN = 90 − ∠OMN = $90 - \frac{1}{2}(180 - y) = \frac{1}{2}y$

16.5 Applying circle theorems

1 a $a = c = d = f = 44°$; $b = e = g = 136°$
b e c b

2 90°

3 $y = -\frac{1}{2}x + \frac{3}{2}$

4 a g = 38° (angles at the circumference subtended by the same arc)
h = 98° (angles in a triangle add to 180°)
i = 98° (vertically opposite angles)
j = 44° (angles at circumference subtended by the same arc)
b ∠BCD = 150° (opposite angles of a cyclic quadrilateral add to 180°)
k = (180° – 150°) ÷ 2 = 15° (angles in a triangle add to 180° and base angles of isosceles triangle are equal)
c ∠FEH = 69° (base angles isosceles triangle are equal)
i = 69° (alternate angles)

5 a a = 46° (angle between the tangent and chord equals the angle in the alternate segment)
b b = 35° (angle between the tangent and chord equals the angle in the alternate segment)
c = 94° (angles in a triangle add to 180°)
d =94° (angle between the tangent and chord equals the angle in the alternate segment)
c e = 67° (angle between the tangent and chord equals the angle in the alternate segment)
f = 27° (angles in a triangle add to 180°)
g = 86° (angle between the tangent and chord equals the angle in the alternate segment)

6 a AT = BT (tangents to circle from same external point are equal)
∠TAB = (180° – 56°) ÷ 2 = 62°
a = 62° (angle between the tangent and chord equals the angle in the alternate segment)
b ∠BAC = 90° (angle in a semicircle is 90°)
∠ABC = 63° (angles in a triangle add to 180°)
b = 63° (angle between the tangent and chord equals the angle in the alternate segment)
c c = 74° (alternate angles are equal)
d = 74° (angle between the tangent and chord equals the angle in the alternate segment)
e = 32° (angles in a triangle add to 180°)

7 ∠OAT = 90° (angle between tangent and radius is 90°)
a ∠CAO = 180° – 50° – 90° = 40° (angles on a straight line add to 180°)
b ∠AOB = 360° – 90° – 90° – 48° = 132° (angles in a quadrilateral add to 360°)
c ∠AOC = 180° – 40° – 40° = 100° (angles in a triangle add to 180° and base angles isosceles triangle are equal)
d ∠COB = 360° – 132° – 100° = 128° (angles round a point add to 360°)
e ∠CBO = (180° – 128°) ÷ 2 = 26° (base angles of an isosceles triangle are equal)

8 ∠ABO = ∠ADO 90° (angle between tangent and radius is 90°)
∠BOD = 360 – 90 – 90 – 40 = 220° (angles in quadrilateral ADOB add to 360°)
∠EBO = 90 – 75 = 15°
∠BCD = (360 – 90 – 90 – 40) ÷ 2 = 70° (angle at the centre is twice the angle at the circumference subtended by the same arc)
∠ODC = 360 – 220 – 15 – 70 = 55° (angles in quadrilateral ODCB add to 360°)

9 $12y = 5x - 169$

10 $4y = -3x + 75$

11 $3y = 4x + 50$

12 $15y = -8x - 289$

16 Problem-solving

1 ∠ABC = 90° (angle in a semicircle is 90°)
Using Pythagoras, AC = $\sqrt{80}$ so $r = \frac{1}{2}\sqrt{80}$
Area = $\pi r^2 = \pi \times \frac{1}{4} \times 80 = 20\pi$

2 a $n^2 + (n + 1)^2 + 1 = n^2 + n^2 + 2n + 1 + 1 = 2n^2 + 2n + 2 = 2(n^2 + n + 1)$
b $n^2 + (n + 1)^2 + (n + 2)^2 + 1 = n^2 + n^2 + 2n + 1 + n^2 + 4n + 4 + 1 = 3n^2 + 6n + 6 = 3(n^2 + 2n + 2)$

3 a Both sides simplify to $x + 4$ so they are equal.
b Ratios of corresponding sides are all $x + 4$, so the triangles are similar.

4 a ∠YXW = 180 – 65 = 115° (angles on a straight line)
∠XWY = 180 – 115 – 30 = 35° (angles in a triangle add to 180°)
b ∠XVY = ∠XWY = 35° (angles at the circumference subtended by the same arc are equal)
c ∠VWY = ∠VXY = 55° (angles at the circumference subtended by the same arc are equal)
∠VWX = ∠VWY + ∠XWY = 55 + 35 = 90° (a right angle)

5 a ∠PRQ = 68° (angles on a straight line)
∠PQR = 180 – 68 – 44 = 68° (angles in a triangle add to 180°)
The triangle has two equal angles, and so it is isosceles.
b ∠XZY = 180 – (90 + x) = 90 – x (angles on a straight line)
∠XYZ = 180 – 2x – (90 – x) = 90 – x (angles in a triangle add to 180°)
The triangle has two equal angles, and so it is isosceles.

6 a $\frac{5}{9}(-40 - 32) = \frac{5}{9}(-72) = -40$
b If F = C, then $\frac{5}{9}(C - 32) = C$.
So $\frac{5}{9}C - \frac{160}{9} = C$; $-\frac{4}{9}C = \frac{160}{9}$; C = –40 °C

16 Check up

1

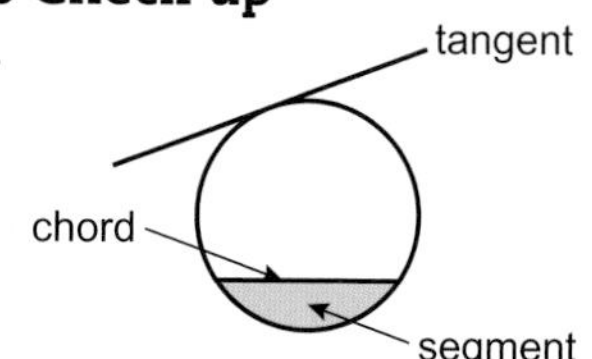

2 a ∠OBA = 50° (angles on a straight line add to 180°)
OA = OB (radii of same circle)
∠OAB = ∠OBA (base angles of isosceles triangle are equal)
a = 180 – 50 – 50 = 80° (angles in a triangle add to 180°)
b b = 90° (angle between radius and tangent is 90°)
c = 180 – 56 – 90 = 34° (angles in a triangle add to 180°)
c AT = BT (tangents to circle from same external point are equal)
∠TBA = ∠TAB (base angles isosceles triangle are equal)
d = (180° – 42°) ÷ 2 = 69°
∠OBT = 90° (angle between radius and tangent is 90°)
e = 90° – 69° = 21°

3 OM = 4 cm

4 a a = 74° (the angle subtended by an arc at the centre of a circle is twice the angle subtended at any point on the circumference)
b = 106° (opposite angles of a cyclic quadrilateral sum to 180°)
b c = 90° (the angle in a semicircle is a right angle)
d = ∠ACB = 39° (angles subtended at the circumference by the same arc are equal)

5 a = 54° (angle between tangent and chord is equal to the angle in the alternate segment)
b = 56° (angles in a triangle add to 180°)

6 $12y = 5x - 338$

7 AO = OC (radii of same circle)
∠ACO = ∠OAC = x and ∠BCO = ∠OBC = y (base angles of an isosceles triangle are equal)
∠AOD = 2x and ∠BOD = 2y (exterior angle of a triangle equals the sum of the 2 interior opposite angles)
∠ACB = $x + y$
∠AOB = 2x + 2y = 2(x + y) = 2(∠ACB)

9 Students' own answers

16 Strengthen

Chords, radii and tangents

1 Students' correctly labelled diagrams

2 a Students' diagrams with OA and OB marked as equal lengths.
 b Isosceles
 c ∠OBA = 20°; ∠OAB = 20°, ∠AOB = 140°

3 a–d Students' diagrams
 e 90°

4 a Students' diagrams with OA & OB and AT & BT marked as equal lengths, and angles OAT and OBT marked as right angles.
 b Isosceles
 c ∠ABT = 73°; ∠OBT = 90°; ∠OBA = 17°

5 a OA = 6.5 cm b AM = 6 cm c OM = 2.5 cm

Circle theorems

1 a, b Students' diagrams with arc QR coloured
 c ∠QPR d 284°

2 a 49° b 68° c 54° d 288°

3 They add to 180°.

4 a Cyclic b ∠QRS c 84°

5 A

6 a ABCD b ∠BAD = 57° c ∠BCD = 123°

7 AC is a diameter.
Angle ABC = 90° (angle in a semicircle)
Angle ACB = 54° (angles in a triangle add to 180°)

8 The angles at the circumference are equal.

9 a a and b b x and y c d and e; f and g

10 ∠ACB

11 ∠ACB = 37° (angle between tangent and chord is equal to the angle in the alternate segment)
∠CAB = 68° (angles in a triangle add to 180°)

Proofs and equation of tangent to a circle at a given point

1 a $\frac{4}{3}$ b $-\frac{3}{4}$ c $y = -\frac{3}{4}x + c$
 d $\frac{25}{4}$ e $y = -\frac{3}{4}x + \frac{25}{4}$ or $4y + 3x = 25$.

2 a ∠AOC = 180° b AC c ∠ABC = 180° ÷ 2 = 90°
 d The angle in a semicircle is a right angle.

3 $p + r = 180°$, so $r = 180 - p$ (angles on a straight line)
$p + q = 180°$, so $q = 180 - p$ (opposite angles of a cyclic quadrilateral sum to 180°)

16 Extend

1 $a = 60°$ (OBC is an equilateral triangle)
OA = OB so OAB is isosceles (radii same circle)
c = ∠OBA = 130° − 60° = 70° (base angles of an isosceles triangle are equal)
b = 180 − 70 − 70 = 40° (angles in a triangle add to 180°)

2 11.0 cm

3 ∠ABC =90°; ∠ACB = 36°; ∠BAC = 54°

4 ∠ADC = $\frac{y}{2}$ (angle at centre is twice the angle at the circumference)
∠ABC = 180 − $\frac{y}{2}$ (opposite angles in cyclic quadrilateral add to 180°)

5 OB = OC (radii same circle)
∠OCB = ∠OBC = (180° − 40°) ÷ 2 =70° (base angles isosceles triangle are equal and angles in a triangle add to 180°)
OA = OC (radii same circle)
∠OAC = ∠OCA = (180° −110°) ÷ 2 = 35° (base angles isosceles triangle are equal and angles in a triangle add to 180°)
∠BCA = ∠OCB − ∠OCA = 35°
So AC bisects angle OCB.

6 ∠ODC = 66° (opposite angles of a cyclic quadrilateral add to 180°)
OC = OD (radii same circle)
∠ODC = ∠OCD (base angles isosceles triangle are equal)
∠COD = 180° − 66° − 66° = 48° (angles in a triangle add to 180°)

7 ∠BCD = 30° (opposite angles of a cyclic quadrilateral add to 180°)
∠BOD = 60° (angle at the centre is twice angle at the circumference when both are subtended by the same arc)
OB = OD (radii same circle)
∠OBD = ∠ODB = (180° − 60°) ÷ 2 = 60° (base angles isosceles triangle are equal and angles in a triangle add to 180°)
In triangle OBD all the angles are 60° so it is equilateral.

8 MN = 39 cm

9 28.3 mm

10 ∠CDA = x (alternate angles are equal)
∠DCA = x (angle between the tangent and chord equals the angle in the alternate segment)
In triangle ACD, ∠CDA = ∠DCA, therefore AC = AD and the triangle is isosceles.

11 OB is common; OA = OC (radii same circle);
∠OAB = ∠OCB = 90° (angle between radius and tangent is 90°). Therefore triangles OAB and OCB are congruent (RHS) and AB = BC (corresponding sides).

12 ∠OSR = 90° (angle between radius and tangent is 90°)
∠PSO = 90 − 62 = 28°
∠QOS = 140° (angles round a point add to 360°)
∠SPQ = 70° (angle at centre is twice angle at circumference when both subtended by same arc)
∠PQS = 360° − 220° − 28° − 70° = 42° (angles in a quadrilateral add to 360°)

13 OA = OB (radii same circle); OM is common;
AM = MB given M is midpoint of AB;
Triangles OAM and OMB are congruent (SSS)
AMB is a straight line, ∠AMO = ∠OMB = 180° ÷ 2 = 90°

14 Proof that ABD and DCA are congruent using ASA, RHS or SAS

16 Unit test

Sample student answer

Both students worked out the correct value for angle OBT. Student A gave the best answer as they clearly stated the reasons for each part of their calculation. However, Student A's answer could have been improved by showing the calculations so that they didn't lose any method marks if they made an error in calculating.

UNIT 17

17 Prior knowledge check

1 84

2 a $\frac{2}{55}$ b $\frac{43}{36}$ or $1\frac{7}{36}$ c $\frac{3}{8}$ d $\frac{25}{27}$

3 a $5\sqrt{2}$ b $4\sqrt{5}$

4 a $14 - 35x$ b $2x^2 + 5x - 12$ c $4x^2 - 4x + 1$

5 a $n = 40$ b $p = 5$ c $d = 9$

6 a $x = \frac{y-7}{4}$ b $x = \frac{W-h}{3h}$ c $x = \frac{y}{4} - 1$ d $x = \frac{3P-1}{6}$

7 a y^8 b $28y^6$ c y^7 d $\frac{2y^5}{5}$

8 a $(x + 1)(x + 5)$ b $(x - 10)(x + 3)$
 c $(x - 2)(x - 3)$ d $(x - 6)(x + 6)$

9 a $x = -5, -6$ b $x = 1, 11$ c $x = -1, -\frac{7}{2}$

10 $x = \frac{-5 \pm \sqrt{17}}{2}$

11 $x = -4 \pm \sqrt{6}$

12 a Students' own answers, e.g. 1, 2, 3, 4 and 5.
b Students' own answers, e.g. 15.
c Students' own answers, e.g. 2 + 3 + 4 + 5 + 6 = 20, 3 + 4 + 5 + 6 + 7 = 25, 4 + 5 + 6 + 7 + 8 = 30, 5 + 6 + 7 + 8 + 9 = 35.
d The sum of 5 consecutive numbers is a multiple of **5**.
e $n + (n + 1) + (n + 2) + (n + 3) + (n + 4) = 5n + 10$

17.1 Rearranging formulae

1 a $a = \frac{v - u}{t}$ b $r = \frac{C}{2\pi}$ c $h = \frac{2A}{b}$ d $r = \sqrt{\frac{A}{\pi}}$
e $t = x^2$ f $s = \frac{r^2}{3}$
2 a $y(x + 2)$ b $q(p - 1)$ c $k(a - 4)$
3 $v = \sqrt{\frac{2E}{m}}$
4 $x = H^2 + y$
5 a $x = \left(\frac{T^2k}{4p^2}\right)$ b $x = \frac{16}{y^2}$ c $x = \frac{zP^2}{y}$
d $x = -1 + \sqrt{\frac{L}{3}}$
6 a $r = \sqrt[3]{\frac{3V}{4p}}$ b $x = \sqrt[3]{\frac{V}{4}}$ c $x = \frac{y^3}{5}$ d $y = \frac{x}{z^3}$
7 $y = \frac{h}{3 + x}$
8 $d = \frac{H + ac}{a - b}$
9 a $y = \frac{w - 2}{2x}$ b $x = \frac{w - 2}{2y}$
10 a There cannot be an x on both sides of the formula.
b Zoe should have factorised the x first.
c $x = \frac{H - 7}{y + 2}$
11 $x = \frac{1}{V - 7}$
12 $k = \frac{2t}{t - 1}$

17.2 Algebraic fractions

1 a $\frac{29}{36}$ b $\frac{85}{99}$ c $-\frac{1}{20}$
2 a $\frac{10}{77}$ b $\frac{54}{35}$ c $\frac{5}{16}$
3 a $\frac{x^2}{6}$ b $\frac{6xy}{20} = \frac{3xy}{10}$ c $\frac{12}{45y^2} = \frac{4}{15y^2}$
4 a $\frac{3x}{2y^2}$ b $\frac{5y^4}{3x^2}$ c $\frac{3x^4}{2y^2}$
5 a $\frac{4}{3}$ b x^4y^2 c $\frac{5}{4x^2y^4}$ d $\frac{5y}{y - 7}$
6 a $\frac{8x}{10} = \frac{4x}{5}$ b $\frac{13x}{12}$ c $\frac{5x}{14}$
7 a $10x$ b $6x$ c $28x$ d $12x$
8 a $\frac{3}{12x}$ and $\frac{4}{12x}$ b $\frac{7}{12x}$
9 a $\frac{11}{18x}$ b $\frac{1}{20x}$ c $\frac{13}{18x}$
10 a $\frac{x - 4}{2} = \frac{5(x - 4)}{5 \times 2} = \frac{5x - 20}{10}$ b $\frac{x + 7}{5} = \frac{2(x + 7)}{2 \times 5} = \frac{2x + 14}{10}$
c $\frac{7x - 6}{10}$
11 a $\frac{5x + 8}{6}$ b $\frac{5x + 41}{14}$ c $\frac{x + 67}{36}$
12 $\frac{9x + 24}{10}$
13 $a = \frac{b}{b - 1}$
14 $u = vf(v - f)$

17.3 Simplifying algebraic fractions

1 a $\frac{1}{x^2}$ b $5x^2$ c $5x^2$
2 a $(x - 6)(x - 3)$ b $(x - 9)(x + 9)$ c $(5x + 1)(x + 4)$
3 a $\frac{1}{y}$ b $\frac{1}{3}$ c $1(x - 7)$
d $\frac{x + 2}{x - 5}$ e $\frac{x - 3}{x}$ f $\frac{x}{x - 1}$
4 a $x(x - 6)$ b x
5 a $x + 8$ b $3x$ c $\frac{5}{2x}$
6 a Sally is incorrect because the two terms on the denominator have nothing in common. The denominator cannot be factorised.
b The expression cannot be simplified because the denominator cannot be factorised.
7 a $\frac{2}{x + 5}$ b $\frac{x - 3}{5}$
8 a $\frac{x + 3}{x - 3}$ b $\frac{x - 5}{x + 7}$ c $\frac{x - 5}{x + 5}$
9 $\frac{x + 7}{x - 7}$
10 a $\frac{2x - 3}{3x - 2}$ b $\frac{5x - 1}{6x + 5}$ c $\frac{5x - 1}{5x + 1}$
11 $\frac{x + 4}{2x - 3}$
12 a $(6 - x) = -(x - 6)$ b i -1 ii $-\frac{6 + x}{x + 3}$
13 a $-\frac{4 + x}{x}$ b $\frac{x - 6}{2(x + 6)}$ c $\frac{2x}{2x - 3}$
14 Numerator: $(x + 4)(x - 3)(x + 3)(x - 1)(2x)(5x + 6)$
Denominator: $(3 - x)(3 + x)(5x + 6)(x + 4)(7)(x - 1)$
$\frac{x - 3}{3 - x} = -1$; other factors cancel to leave $-\frac{2x}{7}$

17.4 More algebraic fractions

1 a $\frac{15x}{y}$ b $\frac{75}{4}$ c $\frac{3x}{x - 2}$
2 a $\frac{13x}{15}$ b $\frac{5}{24x}$ c $\frac{7x + 11}{12}$
3 a $(x + 3)(x - 4)$ b $\frac{x + 2}{x + 5}$ c $\frac{1}{9}$
d $\frac{8}{3}$ e $\frac{2x - 2}{x + 7}$ f $\frac{x(x + 4)}{x - 2}$
4 a $(x - 3)(x + 3)$ b $(x + 2)(x + 3)$ c $\frac{2x - 6}{x + 2}$
5 a $\frac{(x - 2)(x - 3)}{(x + 1)(x + 4)}$ b $\frac{7(x + 7)}{(x - 3)(x - 7)}$
6 a $x(x + 2)$ b $(x + 2)(x + 3)$
c $(x + 4)(x + 5)$ d $(x + 1)(x - 1)$
e $(7x - 3)(2x - 4)$
7 a $\frac{2x + 9}{(x + 4)(x + 5)}$ b $\frac{7x + 1}{(x + 1)(x - 1)}$
c $\frac{6x + 26}{(x - 5)(x + 3)}$ d $\frac{7}{(2x - 3)(2x + 4)}$
8 $\frac{x + 10}{(x - 4)(x + 3)}$
9 a i $3(x + 3)$ ii $4(x + 3)$ b $12(x + 3)$ c $\frac{7}{12(x + 3)}$
10 a $(x - 4)(x + 4)$ b $\frac{x - 3}{(x + 4)(x - 4)}$
11 a $\frac{-x}{(3x + 5)(x + 1)}$ b $\frac{1 - x}{2(x + 1)(x + 6)}$
c $\frac{4x - 1}{(x + 2)(x + 4)(x - 7)}$ d $\frac{-11 - 3x}{(5 - x)(5 + x)}$
12 $\frac{x^2 + 3x - 2}{10x(x - 1)}$
13 Students' own answer. $A = 5$

17.5 Surds

1 a 5 b $3\sqrt{3}$ c $8\sqrt{2}$

2 a $\sqrt{3}$ b $\sqrt{30}$ c $\sqrt{\frac{5}{7}}$

3 a $\sqrt{50} = \sqrt{25} \times \sqrt{2} = 5\sqrt{2}$, $k = 5$ b $k = 2\ 3$ c $k = 4$

4 a $\frac{\sqrt{10}}{10}$ b $\frac{\sqrt{15}}{5}$ c $\sqrt{2}$

5 a i $3\sqrt{5}$ ii $2\sqrt{5}$ b $23\sqrt{5}$

6 a $\sqrt{3}$ b $22\sqrt{2}$ c $15\sqrt{2}$

7 a $\sqrt{12} + 2 = 2\sqrt{3} + 2 = 2(\sqrt{3} + 1)$
b $3(3 + \sqrt{6})$ c $3(6 - \sqrt{5})$ d $5(\sqrt{3} - \sqrt{2})$

8 a $4\sqrt{5} + 5$ b $11 + 5\sqrt{7}$ c $22 + 2\sqrt{2}$ d $6 - 4\sqrt{2}$
e $26 - 8\sqrt{10}$ f $52 + 14\sqrt{3}$

9 $30 - 10\sqrt{5}$. $a = 30$, $b = -10$, $c = 5$

10 a i $53 - 6\sqrt{2}$ ii $12 + 4\sqrt{8} = 12 + 8\sqrt{2}$
b The perimeter of the first shape would be 32 units, which is rational. The perimeter of the second shape would be $8 + 4\sqrt{8}$ or $8 + 8\sqrt{2}$, which is irrational.

11 a $\frac{(3\sqrt{2} + 2)}{2}$ b $2\sqrt{3} - 1$ c $\frac{(19\sqrt{7} - 7)}{7}$ d $\sqrt{5} + 1$

12 $a = -3$, $b = 4$

13 a 4 b Rational
c The answer is rational as it is just a whole number/integer.
d i No. $12 - 5\sqrt{2}$ ii No. $51 + 14\sqrt{2}$ iii Yes. 47
e $\frac{(7 - \sqrt{2})}{47}$

14 a $\frac{(1 - \sqrt{2})}{-1}$ b $\frac{5 + \sqrt{3}}{22}$ c $\frac{7(4 + \sqrt{5})}{11}$ d $\frac{4(1 - \sqrt{6})}{-5}$
e $\frac{5 + \sqrt{5}}{-4}$ f $\frac{25 + 7\sqrt{2}}{31}$

15 a $x = 3 \pm 2\sqrt{2}$ b $x = 5 \pm 2\sqrt{3}$ c $x = 8 \pm 2\sqrt{14}$

17.6 Solving algebraic fraction equations

1 a $2(x + 3)$ b $4(x + 4)$

2 a $\frac{5}{x}$ b $\frac{4}{2x} = \frac{2}{x}$ c $\frac{10}{x - 6}$

3 a $x = -2$, $x = -4$ b $x = \frac{11}{2}$, $x = 1$
c $x = 4$, $x = 1$

4 $x = 1.40$, $x = -4.06$

5 a $x = \frac{5}{4}$ b $x = \frac{11}{7}$ c $x = -\frac{11}{2}$

6 a $x = -\frac{5}{3}$, $x = 4$ b $x = \frac{3}{2}$, $x = -2$
c $x = -\frac{7}{5}$, $x = 2$ d $x = \frac{5}{2}$, $x = -4$

7 $6x - 9 + 2x + 2 = 2x^2 + 2x - 3x - 3$
$8x - 7 = 2x^2 - x - 3$
$-2x^2 + 9x - 4 = 0$
$(2x - 1)(-x + 4)$
$x = \frac{1}{2}$, $x = 4$

8 a Students' own answer b $x = 1$, $x = 9$

9 a $x = 3$ b $x = 0$, $x = 8$
c $x = 1$, $x = 2$ d $x = -4$, $x = 1$
e $x = 3$, $x = -2$

10 a $x = 0.29$, $x = -10.29$ b $x = 1.21$, $x = -1.81$
c $x = 6.37$, $x = 0.63$ d $x = 5.70$, $x = -0.70$

11 $x = 6 \pm \sqrt{31}$

17.7 Functions

1 a $x \rightarrow \times 2 \rightarrow + 5 \rightarrow y$ b $x \rightarrow \div 2 \rightarrow - 6 \rightarrow y$
c $x \rightarrow + 1 \rightarrow \times 3 \rightarrow y$

2 a $x = \frac{7}{5}$ b $x = \frac{16}{7}$

3 a $H = 12t$ b $P = \frac{y}{6}$ c $y = (h + 3)^2 = h^2 + 6h + 9$

4 a 2 b −5 c 20 d $-\frac{1}{2}$

5 a Alice first multiplied 5 by 2 to get 10. Then she worked out 10 squared, which is 100.
b 20

6 a 54 b −2 c $\frac{1}{4}$ d −250

7 a 5 b 56 c 480 d 2.5
e 600 f −33

8 a $a = 3$ b $a = \frac{3}{5}$ c $a = -\frac{4}{5}$

9 a $a = \pm 5$ b $a = \pm 2$ c $a = \pm 2\sqrt{2}$ d $a = \pm 2\sqrt{5}$

10 a $a = 0$, $a = -3$ b $a = 1$, $a = -5$
c $a = -1$, $a = -2$ d $a = -1$, $a = -3$

11 a $5x + 1$ b $5x - 13$ c $10x - 8$ d $35x - 28$
e $10x - 4$ f $20x - 4$

12 a $3x^2 + 3$ b $6x^2 - 8$ c $12x^2 - 4$ d $3x^2 - 4$

13 a 11 b 71 c −40 d −58

14 a $-4x + 13$ b $37 - 4x$
c $4x^2 + 25$ d $16x^2 - 24x + 16$
e $-x^2 + 3$ f $107 - 20x + x^2$

15 a $x \rightarrow \frac{x - 9}{4}$ b $x \rightarrow 3(x + 4)$
c $x \rightarrow \frac{x}{2} - 6$ d $x \rightarrow \frac{x + 1}{7} + 4$

16 a $x \rightarrow \frac{x}{4} + 1$ b $x \rightarrow \frac{x}{4} - 1$ c $x \rightarrow \frac{x}{2}$ d $a = 2$

17.8 Proof

1 a even and odd b only even
c even and odd d only odd
e even and odd

2 a $x^2 - x$ b $x^2 + 6x + 9$ c $4x^2 + 2x$

3 a Identity b Equation ($n = 3$)
c Identity d Equation ($n = \frac{7}{3}$)

4 Students' own answers

5 a Students' own answer
b i 9999 (use $100^2 - 1$) ii 39 999 (use $200^2 - 1$)

6 a $(x + 5)(x + 2) = x^2 + 7x + 10$
b $x(x + 1) = x^2 + x$
c $x^2 + 7x + 10 - (x^2 + x) = 6x + 10$

7 $x(3x + 4) - 5x = 70$
$3x^2 + 4x - 5x - 70 = 0$
$3x^2 - x - 70 = 0$

8 a 2 is a prime.
b Any number less than 1 gives a cube that is less than its square.
c For example −5 − −2 = −3, −5 + −2 = −7
d For example 16 − 4 = 12.

9 $2n + 1 + 2n = 4n + 1 =$ odd.

10 a The next even number will be two more (because the next number, which is one more, will be odd).
b $(2n)(2n + 2) = 4n^2 + 4n = 4(n^2 + n)$. This is divisible by 4.

11 $(2n + 1)(2n - 1) = 4n^2 - 1$. $4n^2$ must be even, so $4n^2 - 1$ must be odd.

12 $2x - 2a = x + 5$
$x = 2a + 5$
$2a$ is even, even + odd = odd

13 a i $\frac{1}{30}$ ii $\frac{1}{12}$ iii $\frac{1}{56}$ b $\frac{1}{90}$
c It will be 1 divided by 99 times 100.
d i $\frac{1}{x(x + 1)}$
ii This shows that the difference between two fractions with 1 on the numerators and consecutive numbers on the denominator will be 1 divided by the denominators multiplied together.

14 $A = 4$

15 $n^2 + n = n(n + 1)$
When n is even, even × odd = even
When n is odd, odd × even = even

16 b $n^3 - n = (n-1)\ n\ (n + 1)$
= even × odd × even = even
Or = odd × even × odd = even

17 $(n+1)^2 - n^2 = n^2 + 2n + 1 - n^2 = 2n + 1$
$(n+1) + n = 2n + 1$

17 Problem-solving

1 6370 km
2 a 3390 km b 3.73 m/s^2
3 11.29 m/s^2
4 7320 km (3 s.f.)

17 Check up

1 a $20\sqrt{2}$ b $23 - 8\sqrt{7}$
2 a $\frac{3\sqrt{5} - \sqrt{10}}{5}$ b $6 + 3\sqrt{3}$
3 a $x \to 2x + 5$ b $x \to \frac{x - 4}{3}$
4 a 26 b −16 c 96 d −20
5 $y = \frac{x + 1}{z^3}$
6 $y = \frac{9 - 3x}{5x + 2}$
7 $k = \frac{4p^2x}{T^4}$
8 a 29 b $a = \pm\sqrt{\frac{7}{2}}$
9 a $\frac{x - 2}{3}$ b $\frac{x - 4}{x + 1}$
10 a $\frac{1}{6x}$ b $\frac{4x - 11}{(x + 4)(x - 5)}$ c $\frac{16 - 2x}{(x - 6)(x - 1)}$
11 a $\frac{8x^2}{9y^5}$ b $\frac{3(x + 10)}{4(x + 1)}$
12 $x = -1 \pm \sqrt{2}$
13 Students' own answer.
14 Students' own answer, e.g. $1^3 + 3^3 = 28$ or $2^3 + 4^4 = 72$
16 a $(2n + 1) + (2n + 3) = 4n + 4$
b $2n + (2n + 2) + (2n + 4) = 6n + 6$
c $(2n + 1) + (2n + 3) + (2n + 5) + (2n + 7) = 8n + 16$

17 Strengthen

Surds

1 a 3 b $\sqrt{7}$ c 4 d $6\sqrt{5} - 5$ e $9\sqrt{5}$
2 a $4\sqrt{3}$
b $\frac{4 + \sqrt{11}}{\sqrt{11}} = \frac{4 + \sqrt{11}}{\sqrt{11}} \times \frac{\sqrt{11}}{\sqrt{11}} = \frac{4 \times \sqrt{11} + \sqrt{11} \times \sqrt{11}}{\sqrt{11} \times \sqrt{11}} = \frac{4\sqrt{11} + 11}{11}$
c $\frac{8\sqrt{5} - 5}{5}$
3 a $1 + 2\sqrt{7}$ b $27 - 10\sqrt{2}$ c 4 d −7
e 9 f Students' own answer
g i $6 - \sqrt{8}$ ii $3 + \sqrt{11}$
4 a $\frac{40 + 8\sqrt{2}}{23}$ b $14 - 7\sqrt{3}$ c $\frac{42 + 6\sqrt{10}}{39} = \frac{14 + 2\sqrt{10}}{13}$

Formulae and functions

1 a i $y^2 = 3$ ii $y^2 = x$ iii $y^2 = 3x - 1$ b $x = \frac{y^2 + 1}{3}$
2 Rewrite the formula so there is no fraction. $xy = 7 + y$
Get all the terms containing y on the left-hand side and all other terms on the right-hand side. $xy - y = 7$
Factorise so that y appears only once. $y(x - 1) = 7$
Get y on its own on the left hand side. $y = \frac{7}{x - 1}$
3 $y = \frac{1}{F + 5}$
4 a $y = 1$ b $f(2) = 1$
c i $f(5) = 16$ ii $f(-3) = -24$ iii $f(0) = -9$
5 a i $x = \frac{1}{8}$ ii $x = \frac{2}{7}$
b i $a = \frac{1}{8}$ ii $a = \frac{2}{7}$
c $a = \frac{4}{9}$

6 a 2 b 35 c 70 d 8
e 70 f $4x^2 - 1$
g i 8 ii 24 iii 5 iv 0
7 $y = \frac{x + 4}{5}$
8 a $y = \frac{x + 9}{2}$ b $y = \frac{x}{3} + 5$ c $y = 2x - 4$ d $y = \frac{5}{2}x - 1$

Algebraic fractions

1 a $\frac{1}{2}$ b x c $\frac{x - 8}{x + 7}$ d $\frac{x + 5}{x - 2}$
e $\frac{(x + 4)(x - 2)}{(x + 8)(x - 4)}$ f $\frac{5(x + 1)}{9(x - 1)}$
2 a i $\frac{3}{4}$ ii $\frac{3}{2}$ iii $\frac{1}{x^2}$ iv $\frac{y^4}{1}$
b $\frac{9y^4}{8x^2}$
3 a $\frac{4x^2}{9y}$ b $\frac{8x}{9y^3}$
4 a i $3(x + 6)$ ii $x(x + 6)$ b $\frac{3}{x}$ c $\frac{x - 5}{2}$
5 a $\frac{8x + 32}{x^2 + 12x + 32} = \frac{8(x + 4)}{(x + 4)(x + 8)} = \frac{8}{(x + 8)}$
b $\frac{x + 8}{x - 9}$ c $\frac{x - 8}{x + 3}$
6 a i $3(x + 3)$ ii $(x + 6)(x + 3)$ iii $(x + 3)(x + 5)$
iv $2(x + 5)$
b i $\frac{3(x + 3)}{2(x + 6)}$ ii $\frac{2(x + 6)}{3(x + 3)}$
7 a $x = 8, x = -7$ b $x = 9, x = -7$
c $x = 3, x = -2$ d $x = -5, x = 2$
8 a $x(x - 1)$
b i $\frac{3(x - 1)}{x(x - 1)}$ ii $\frac{2x}{(x - 1)x}$ c $\frac{(5x - 3)}{x(x - 1)}$ d $x = 3 \pm \sqrt{6}$
9 $\frac{5 - 2x}{x^2 - 5x + 4}$

Proof

1 a $x^2 - 8x + 16$ b $x^2 - 8x + 7$ c $x^2 - 8x + 7$
d Students show that 1 side of the identity is the same as the other side of the identity.
2 Students show work to prove that the right-hand side is equal to the left hand side.
3 a 1, 8, 27, 64, 125
b Students' own answer, e.g. 64 − 8 = 56

17 Extend

1 a Students should show that both answers are correct.
b Ruth's answer is considered the better answer because the denominator is positive.
c $x = \sqrt{\frac{P - 3d}{2}}$
2 $R_2 = \frac{RR_1}{(R_1 - R)}$
3 a $\frac{1}{d} = \frac{1}{b} + \frac{1}{c} - \frac{1}{a}$
b Students should show their own work to make d the subject.
4 a $x = -\frac{5}{2}, x = 4$ b $x = \frac{10 \pm 2\sqrt{5}}{5}$
5 a Students' own answer
b Using the equivalent expression, it is $\frac{9}{10}$.
6 a −21 b $15 - 16x$
c i $f^{-1}(x) = \frac{x - 3}{-4} = \frac{x}{-4} + \frac{3}{4}$ ii $g^{-1}(x) = \frac{x - 3}{4} = \frac{x}{4} - \frac{3}{4}$
d When the two functions are added together, the x's and the number term cancel, leaving zero. Students should show this.
7 a i $x^2 + 13$ ii $x^2 + 14x + 55$
b $x = -3$

8 a i x ii x
  c Yes, because $fg(x) = gf(x) = x$, the functions are inverses.
  d These two functions are inverses. Students should show that $fg(x) = gf(x) = x$.

9 Students show their own work.
  The factors cancel to leave −1 as $(7 - x) = -(x - 7)$.

10 Students should show that the expression simplifies to $12n$, which is a multiple of 12.

11 $A = -4$ and $B = 1$

12 a $r = \sqrt{\frac{Gm_1m_2}{F}}$
  b 1.5×10^{11} m

17 Unit test

Sample student answer

The student has made an error in multiplying both sides by $(b - 5)$: $a \times (b - 5) = ab - 5a$, not $ab - 5$.

The student could avoid this mistake by putting the terms in a bracket before multiplying them up, which would remind them to multiply BOTH terms in the brackets by a.

UNIT 18

18 Prior knowledge check

1 a 2:3 b 1:3 c 2:1

2 a $LM = \frac{5}{7}LX$ b $MX = \frac{2}{7}LX$ c $LX = \frac{7}{2}MX$

3 a $3\sqrt{3}$ b $4\sqrt{5}$ c $5\sqrt{3}$ d $4\sqrt{7}$

4 a $3x + 5y$ b $6x - 9$ c $10a + 4b$

5 a

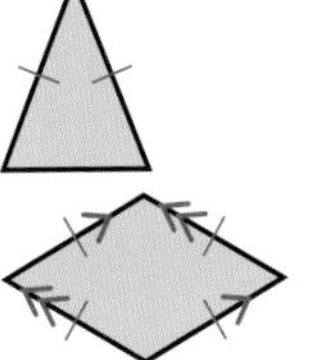

b

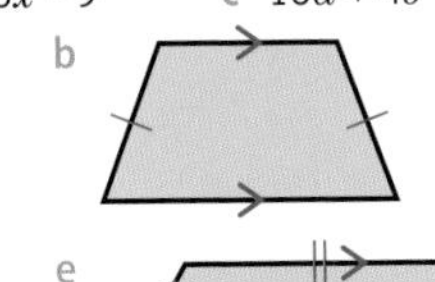

c

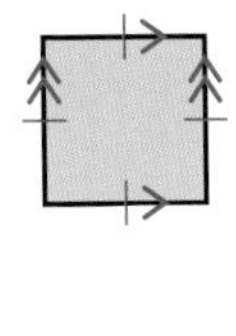

d e

6 7.62 (3 s.f.)

7

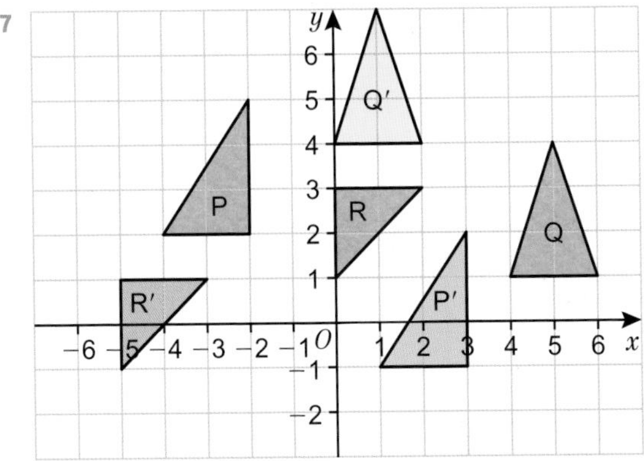

8 a 9.59° b 9.59°

18.1 Vectors and vector notation

1 a $\begin{pmatrix} 6 \\ -2 \end{pmatrix}$ b $\begin{pmatrix} 4 \\ -2 \end{pmatrix}$ c $\begin{pmatrix} -3 \\ -5 \end{pmatrix}$

2 $2\sqrt{13}$

3

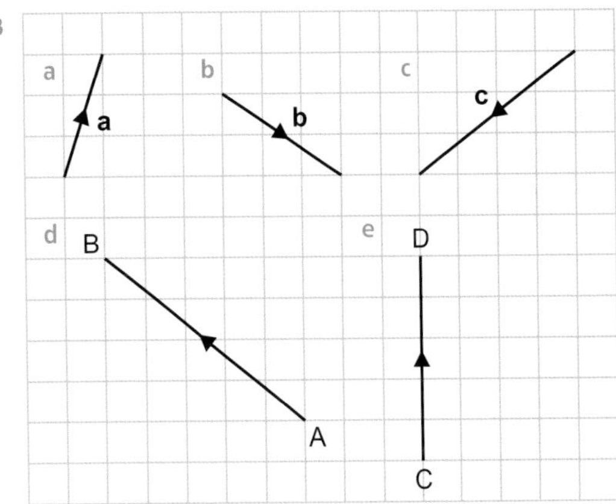

4 a $\begin{pmatrix} 2 \\ 2 \end{pmatrix}$ b $\begin{pmatrix} 2 \\ -5 \end{pmatrix}$ c $\begin{pmatrix} 4 \\ -3 \end{pmatrix}$

5 **a** and **d**

6 6.40

7 a 10 b 13 c $\sqrt{10}$ d 17 e $2\sqrt{13}$

8 a AB = 25
  b AC = 25, as AB = AC = 25 triangle ABC is isosceles

9 a $\begin{pmatrix} -6 \\ -4 \end{pmatrix}$ b $2\sqrt{13}$ or 7.21

10 (−1, −1)

18.2 Vector arithmetic

1 a $\begin{pmatrix} 2 \\ 1 \end{pmatrix}$ b $\begin{pmatrix} -2 \\ -1 \end{pmatrix}$

2 $\begin{pmatrix} -3 \\ 2 \end{pmatrix}$

3 a

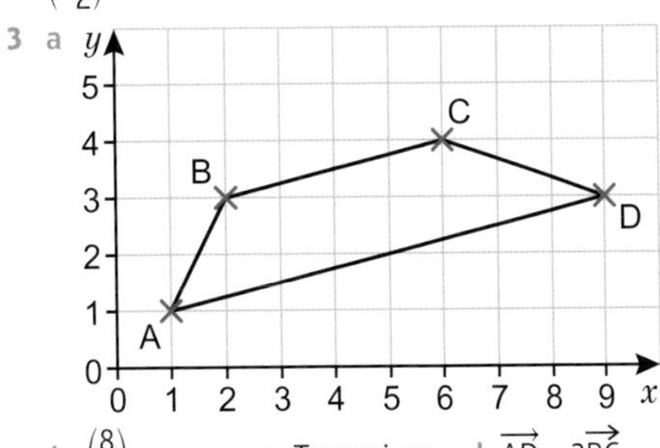

  b $\begin{pmatrix} 8 \\ 2 \end{pmatrix}$ c Trapezium d $\overrightarrow{AD} = 2\overrightarrow{BC}$

4 a

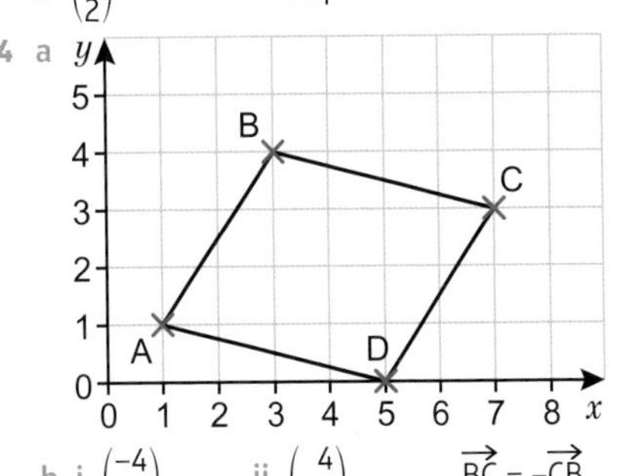

  b i $\begin{pmatrix} -4 \\ 1 \end{pmatrix}$ ii $\begin{pmatrix} 4 \\ -1 \end{pmatrix}$ $\overrightarrow{BC} = -\overrightarrow{CB}$
  c i $\overrightarrow{AB} = \overrightarrow{DC}$ ii $\overrightarrow{AD} = -\overrightarrow{CB}$

5 Parallelogram

6 a (2, 6) b $\begin{pmatrix} 2 \\ 1 \end{pmatrix}$ c $2\sqrt{5}$

7
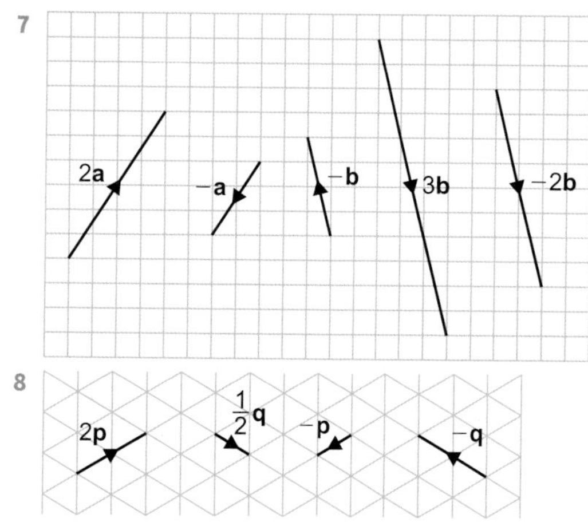

8

2p $\frac{1}{2}$q −p −q

9 a $\begin{pmatrix}4\\2\end{pmatrix}$ b $\begin{pmatrix}6\\3\end{pmatrix}$ c $\begin{pmatrix}-8\\-4\end{pmatrix}$ d $\begin{pmatrix}1\\0.5\end{pmatrix}$

10 $\begin{pmatrix}4\\3\end{pmatrix}$

11 a Students' drawings b $\begin{pmatrix}8\\1\end{pmatrix}$

12 a $\begin{pmatrix}9\\2\end{pmatrix}$ b $\begin{pmatrix}1\\-1\end{pmatrix}$

13 a i $\begin{pmatrix}-2\\6\end{pmatrix}$ ii $\begin{pmatrix}-2\\6\end{pmatrix}$ b i $\begin{pmatrix}0\\0\end{pmatrix}$ ii $\begin{pmatrix}0\\0\end{pmatrix}$

14 $\begin{pmatrix}4\\4\end{pmatrix}$

15 a $\begin{pmatrix}-1\\-4\end{pmatrix}$ b $\begin{pmatrix}-1\\11\end{pmatrix}$ c $\begin{pmatrix}-1\\8\end{pmatrix}$ d $\begin{pmatrix}3\\-3\end{pmatrix}$ e $\begin{pmatrix}-2\\10\end{pmatrix}$

16 Students' own answers

18.3 More vector arithmetic

1 a $\begin{pmatrix}6\\1\end{pmatrix}$ b 6.1

2 $\begin{pmatrix}2\\-6\end{pmatrix}$

3 a $\begin{pmatrix}-1\\4\end{pmatrix}$ b $\overrightarrow{DB} = \overrightarrow{AC} = \begin{pmatrix}3\\-1\end{pmatrix}$

4 a $\sqrt{106}$ b $6\sqrt{2}$ c $4\sqrt{13}$ d $2\sqrt{10}$

5 a **a** + **b** b **a** + **b** + **c**

6 a $\frac{1}{2}$**a** b i $\begin{pmatrix}-6\\-4\end{pmatrix}$ ii $\begin{pmatrix}3\\2\end{pmatrix}$

7 a 2**b** b **a** + **b** c **a** + 2**b**

8 a SR is parallel to PQ so $\overrightarrow{SR} = \overrightarrow{PQ}$ = **a**
b i **b** ii **a** + **b**

9 a i **p** + **q** ii **q** − **p** b $\frac{1}{2}$(**p** + **q**)

10 a **b** − **a** b S is the midpoint of PR.

11 a ED is parallel to and equal to AB, so $\overrightarrow{ED} = \overrightarrow{AB}$ = **n**
b i **m** ii **p**
c i **n** + **m** ii **n** + **m** + **p**
d **n** + **m**

12 a **r** b − **s** c $\frac{1}{2}$**r** d **s** + $\frac{1}{2}$**r**

13 a $\overrightarrow{AB}$, $\overrightarrow{EF}$ and $\overrightarrow{IJ}$ b i 3**p** − 3**q** ii 6**a** − $6\frac{1}{2}$**b**

14 a **b** − **a** b $\frac{1}{2}$(**b** − **a**) c $\frac{1}{2}$(**a** + **b**)

15 a **b** − **a** b $\frac{1}{2}$**a** c $\frac{1}{2}$**b** d $\frac{1}{2}$(**b** − **a**)

18.4 Parallel vectors and collinear points

1

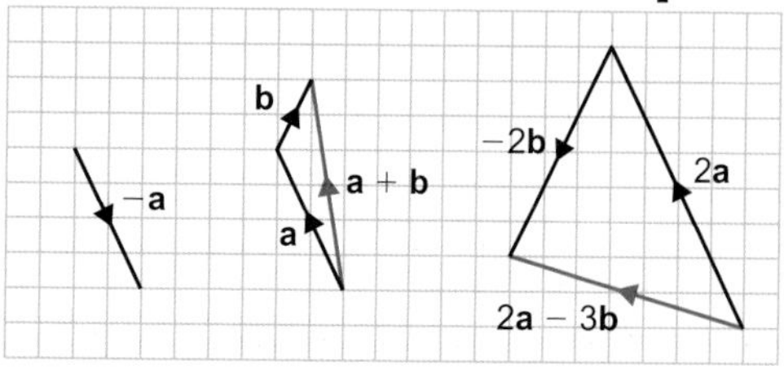

−a b a a + b −2b 2a 2a − 3b

2 a $\begin{pmatrix}-1\\3\end{pmatrix}$ b $\begin{pmatrix}2\\1\end{pmatrix}$ c $\begin{pmatrix}0\\0\end{pmatrix}$

3 $\begin{pmatrix}-1\\1\end{pmatrix}$

4 $\begin{pmatrix}0.5\\-0.5\end{pmatrix}$

5 $\begin{pmatrix}11\\-2\end{pmatrix}$

6 a $\begin{pmatrix}1\\5\end{pmatrix}$ b $\begin{pmatrix}-1\\-5\end{pmatrix}$ c $\begin{pmatrix}2\\4\end{pmatrix}$ d $\begin{pmatrix}1\\-1\end{pmatrix}$

7 **b** − **a**

8 a $\overrightarrow{OP} = \begin{pmatrix}-2\\5\end{pmatrix}$; $\overrightarrow{OQ} = \begin{pmatrix}3\\1\end{pmatrix}$
b i $\begin{pmatrix}5\\-4\end{pmatrix}$ ii $\begin{pmatrix}20\\-16\end{pmatrix}$
c $\begin{pmatrix}20\\-16\end{pmatrix} = 4\begin{pmatrix}5\\-4\end{pmatrix}$ so RS is parallel to PQ and is 4 times the length.

9 a $\begin{pmatrix}-10\\-4\end{pmatrix}$ b (1, 8) c $\begin{pmatrix}2\\3\end{pmatrix}$

10 a $\begin{pmatrix}3\\2\end{pmatrix}$ b $\begin{pmatrix}18\\12\end{pmatrix}$ c (16, 8)

11 $\begin{pmatrix}2\\-3\end{pmatrix}$

12 a $\overrightarrow{AB}$ = 2**b**, so $\overrightarrow{OC}$ is parallel to $\overrightarrow{AB}$ and the same length.
b $\overrightarrow{BC}$ = −**a**, so $\overrightarrow{BC}$ is parallel to, and the same length as, $\overrightarrow{OA}$ but the opposite direction.
c Parallelogram

13 a i $\begin{pmatrix}3\\9\end{pmatrix}$ ii $\begin{pmatrix}9\\27\end{pmatrix}$
b $\overrightarrow{AC} = 3 \times \overrightarrow{AB}$ as $\begin{pmatrix}9\\27\end{pmatrix} = 3\begin{pmatrix}3\\9\end{pmatrix}$ so the lines are parallel.
Both lines pass through point A so A, B and C are collinear.

14 $\overrightarrow{PQ} = \begin{pmatrix}3\\3\end{pmatrix}$, $\overrightarrow{QR}$. $= \begin{pmatrix}6\\6\end{pmatrix} = 2\begin{pmatrix}3\\3\end{pmatrix}$ so the lines are parallel.
Both lines pass through point Q so P, Q and R are collinear.

15 Students' own answers

18.5 Solving geometric problems

1 2**p** and 5**p**; **p** − **q** and 4**q** − 4**p**; 3**q** − **p** and 2**p** − 6**q**

2 a $\overrightarrow{PR}$ = 9**a** − 6**b** = 3(3**a** − 2**b**) so PR is parallel to PQ and is 3 times its length.
b P, Q and R are collinear.

3 a i **a** ii **b** iii −**a** iv −**b**
b 2**b** − 2**a** c **b** − **a**
d AB is parallel to MN and equal to twice its length.

4 a $\frac{1}{3}$(**b** − **a**) b $\frac{1}{3}$(**b** + 2**a**)

5 a **b** − **a** b $\frac{1}{2}$(**b** − **a**) c $\frac{1}{2}$(**b** + **a**)

6 a 2**a** + 2**b** b 7**a** + 6**b**

7 a i 6**b** − 6**a** ii 6**a**
b 12**b** − 3**a**
c $\overrightarrow{EX}$ = 12**b** − 3**a** = 3(4**b** − **a**); $\overrightarrow{EY}$ = 16**b** − 4**a** = 4(4**b** − **a**). The lines are parallel, and they both pass through point X so E, X and Y are collinear.

8 a i **b** − **a** ii $\frac{1}{4}$**b** iii **a** + $\frac{1}{4}$**b** iv **b** + $\frac{1}{4}$**a**
b $\overrightarrow{EF} = \frac{3}{4}$(**b** − **a**); $\overrightarrow{AB}$ = **b** − **a**. $\overrightarrow{EF}$ is a multiple of $\overrightarrow{AB}$, therefore the lines are parallel.

9 a i $\frac{1}{2}$(**m** + **n**) ii $\frac{3}{4}$(**m** + **n**) iii $\frac{3}{4}$**n** − $\frac{1}{4}$**m**
b 3**n** − **m**
c $\overrightarrow{MQ} = \frac{3}{4}$**n** − $\frac{1}{4}$**m** = $\frac{1}{4}$(3**n** − **m**). $\overrightarrow{MQ}$ is a multiple of $\overrightarrow{MR}$, therefore the lines are parallel. Both lines pass through point M, so MQR is a straight line.
$\frac{MR}{MQ} = 4$

10 a $2\mathbf{b}$ b $2\mathbf{a} + \mathbf{b}$ c $4\mathbf{a} + 2\mathbf{b}$

d $\overrightarrow{OS} = \frac{1}{2}(4\mathbf{a} + 2\mathbf{b})$ so S is the midpoint of OT.

e $\overrightarrow{QR} = 3\mathbf{a} - 2\mathbf{b} = \begin{pmatrix} 18 \\ -24 \end{pmatrix}$. QR is 30.

18 Problem-solving

1 a $\mathbf{b} - \mathbf{a}$ b i $\frac{1}{4}\mathbf{a}$ ii $\frac{1}{4}\mathbf{b}$ iii $\frac{1}{4}(\mathbf{b} - \mathbf{a})$

c AB and XY are parallel as one is a multiple of the other. AB is four times the length of XY.

2 Anna is correct (AB, AC, AD, BA, BC, BD, CA, CB, CD, DA, DB, DC)

3 The data gives the graph $y = -0.01 \quad + 18$, which shows the temperature at 2500 m is −7 °C.

4 The spaniel walker takes 6.75 minutes; the Labrador walker takes 5.88 minutes. The Labrador makes it to the opposite corner of the park first.

5 Alexandra is correct: $0.496 \times 2.6^3 = 8.7176$, i.e. 8 litres 718 m$l$ (to the nearest ml)

6 Antony earns the biggest annual salary. Ross earns £28 000 and Antony earns £30 000.

7 $\frac{15 \times 18000}{620}x = 10800$, so $x = 24.8$.

Hence the owner must employ at least 25 apple pickers.

18 Check up

1 a $\begin{pmatrix} 2 \\ 3 \end{pmatrix}$ b $\begin{pmatrix} 3 \\ -1 \end{pmatrix}$

2 a (3, 7) b $\begin{pmatrix} 3 \\ -5 \end{pmatrix}$

3 $\sqrt{34}$

4

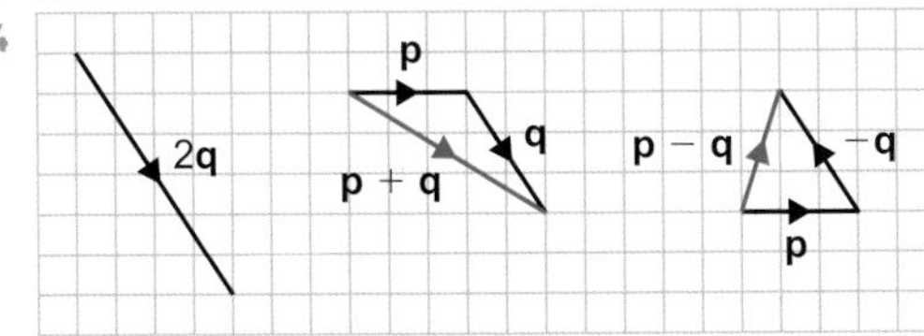

5 $\begin{pmatrix} -2 \\ 3 \end{pmatrix}$

6 a $\begin{pmatrix} 3 \\ 4 \end{pmatrix}$ b $\begin{pmatrix} -5 \\ 8 \end{pmatrix}$ c $\begin{pmatrix} 12 \\ -6 \end{pmatrix}$

7 $\begin{pmatrix} 2 \\ -1 \end{pmatrix}$

8 a $3\mathbf{a} - 3\mathbf{b}$; $\frac{1}{2}\mathbf{a} - \frac{1}{2}\mathbf{b}$ b $3\mathbf{a} + 3\mathbf{b}$

9 a $\begin{pmatrix} 4 \\ -3 \end{pmatrix}$ b $\begin{pmatrix} -2 \\ 7 \end{pmatrix}$ c $\begin{pmatrix} -6 \\ 10 \end{pmatrix}$

10 a i $\begin{pmatrix} 3 \\ 9 \end{pmatrix}$ ii $\begin{pmatrix} 9 \\ 27 \end{pmatrix}$

b $\overrightarrow{AC} = 3\begin{pmatrix} 3 \\ 9 \end{pmatrix}$ so A, B and C are collinear.

11 a i $\frac{1}{2}\mathbf{a}$ ii $\frac{1}{2}\mathbf{a} - \frac{1}{2}\mathbf{c}$

b $\overrightarrow{CA}$ is a multiple of $\overrightarrow{MN}$ so the lines are parallel.

13 From Pythagoras' theorem: $\sqrt{p^2 + q^2} = \sqrt{p^2 + (-q)^2}$

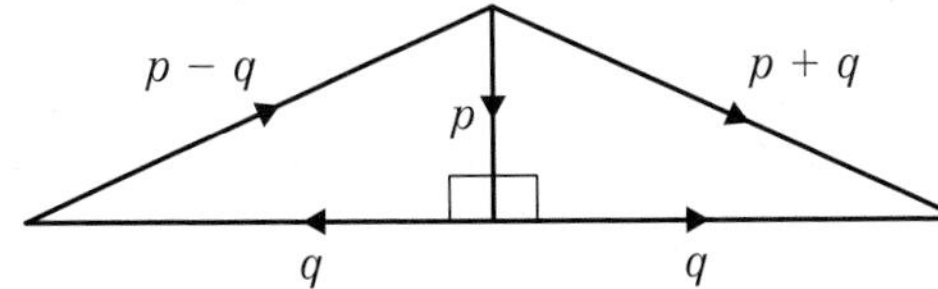

18 Strengthen

Vector notation

1 a $\begin{pmatrix} 2 \\ 2 \end{pmatrix}$ b $\begin{pmatrix} -2 \\ -2 \end{pmatrix}$ c $\begin{pmatrix} -6 \\ 3 \end{pmatrix}$ d $\begin{pmatrix} -5 \\ -2 \end{pmatrix}$

2 a $\begin{pmatrix} 4 \\ 6 \end{pmatrix}$ b $\overrightarrow{OB} = \begin{pmatrix} 8 \\ 2 \end{pmatrix}$

3 (3, 8)

4 $\begin{pmatrix} 3 \\ -1 \end{pmatrix}$

5 a

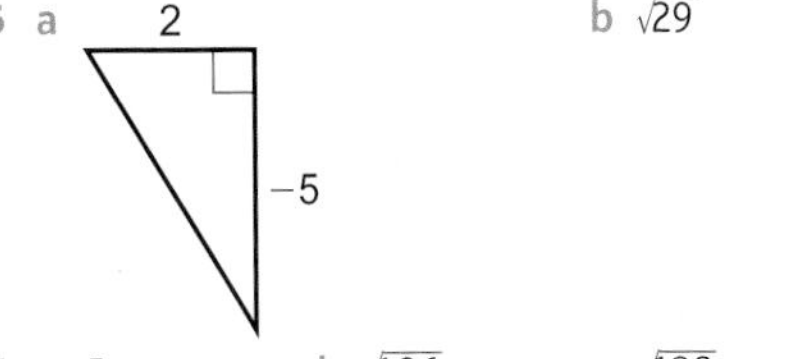

b $\sqrt{29}$

6 a 5 b $\sqrt{106}$ c $\sqrt{130}$ d $\sqrt{34}$

Vector arithmetic

1

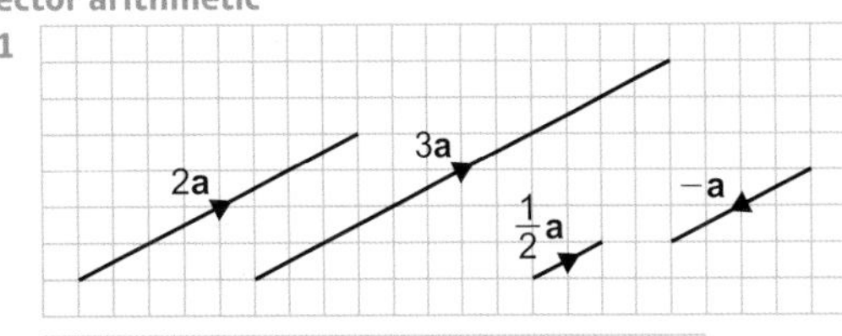

2

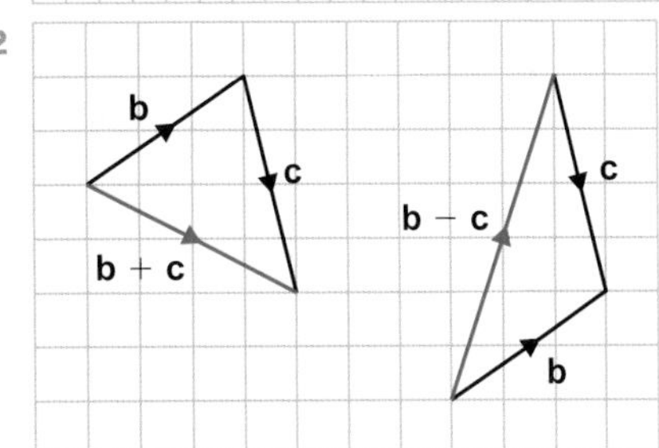

3 a

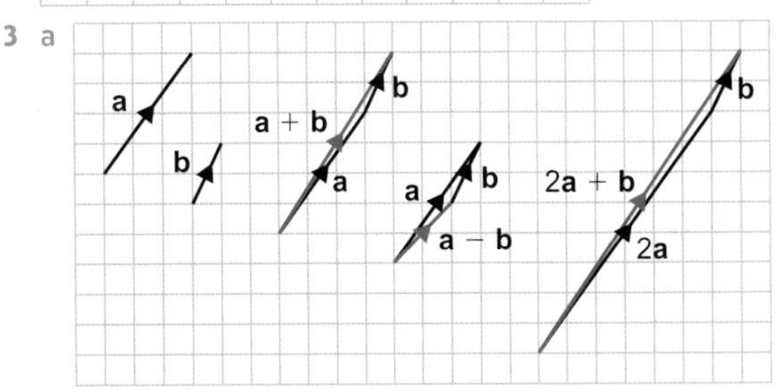

b i $\begin{pmatrix} 4 \\ 6 \end{pmatrix}$ ii $\begin{pmatrix} 2 \\ 2 \end{pmatrix}$ ii $\begin{pmatrix} 7 \\ 10 \end{pmatrix}$

c i $\begin{pmatrix} 4 \\ 6 \end{pmatrix}$ ii $\begin{pmatrix} 2 \\ 2 \end{pmatrix}$ ii $\begin{pmatrix} 7 \\ 10 \end{pmatrix}$

4 a $\begin{pmatrix} 5 \\ 8 \end{pmatrix}$ b $\begin{pmatrix} 6 \\ -10 \end{pmatrix}$ c $\begin{pmatrix} 10 \\ 16 \end{pmatrix}$ d $\begin{pmatrix} 12 \\ -6 \end{pmatrix}$

5 a $\begin{pmatrix} 6 \\ 3 \end{pmatrix}$ b $\begin{pmatrix} 9 \\ -4 \end{pmatrix}$ c $\begin{pmatrix} -1 \\ 11 \end{pmatrix}$

6 $\begin{pmatrix} 3 \\ 10 \end{pmatrix}$

Geometrical problems

1 Parallel lines of the same length have the same column vector.

2 $\mathbf{a} + \mathbf{b}$; $2(\mathbf{a} + \mathbf{b})$; $3\mathbf{a} + 3\mathbf{b}$; $2\mathbf{a} + 2\mathbf{b}$

3 $\begin{pmatrix} 4 \\ 6 \end{pmatrix}$, $\begin{pmatrix} 2 \\ 3 \end{pmatrix}$, $\begin{pmatrix} -4 \\ -6 \end{pmatrix}$, $\begin{pmatrix} 8 \\ 12 \end{pmatrix}$

4 a $\overrightarrow{CD}$is a multiple of $\overrightarrow{AB}$. b $\overrightarrow{CD} = 2\overrightarrow{AB} = 2\mathbf{a}$

c $\overrightarrow{BC} = \overrightarrow{BA} + \overrightarrow{AD} + \overrightarrow{DC} = -\mathbf{a} + \mathbf{b} + 2\mathbf{a} = \mathbf{b} + \mathbf{a}$

5 a

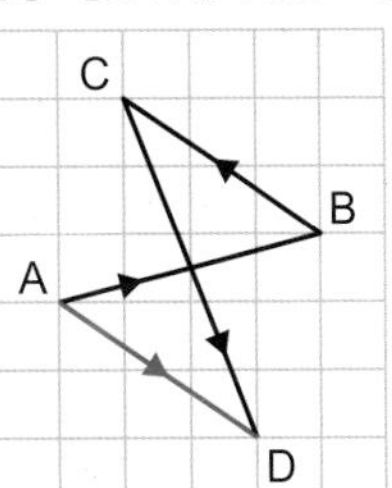

b $\begin{pmatrix} 3 \\ -2 \end{pmatrix}$

c $\overrightarrow{AC} = \begin{pmatrix} 4 \\ 1 \end{pmatrix} + \begin{pmatrix} -3 \\ 2 \end{pmatrix} = \begin{pmatrix} 1 \\ 3 \end{pmatrix}$; $\overrightarrow{DB} = \begin{pmatrix} -2 \\ 5 \end{pmatrix} + \begin{pmatrix} 3 \\ -2 \end{pmatrix} = \begin{pmatrix} 1 \\ 3 \end{pmatrix}$

AC is parallel to DB and they are equal in length.

6 a $\begin{pmatrix}6\\3\end{pmatrix}$ b $\begin{pmatrix}4\\2\end{pmatrix}$

c $\overrightarrow{AB} = 3\begin{pmatrix}2\\1\end{pmatrix}$ and $\overrightarrow{BC} = 2\begin{pmatrix}2\\1\end{pmatrix}$. Both are multiples of the same vector and so are parallel. They both pass through B, so they are collinear.

7 a $\begin{pmatrix}3\\-2\end{pmatrix}$ b $\begin{pmatrix}5\\-1\end{pmatrix}$ c $\begin{pmatrix}2\\1\end{pmatrix}$

8 a $\overrightarrow{AB} = \begin{pmatrix}4\\3\end{pmatrix}$; $\overrightarrow{BC} = \begin{pmatrix}12\\9\end{pmatrix}$

b $\overrightarrow{BC} = \begin{pmatrix}12\\9\end{pmatrix} = 3\begin{pmatrix}4\\3\end{pmatrix}$. Both are multiples of the same vector and so are parallel.

c $\overrightarrow{AB}$ and $\overrightarrow{BC}$are **parallel** and both pass through point B. So ABC is a **straight** line and A, B and C are collinear.

9 a $\frac{1}{2}\mathbf{a}$ b $\mathbf{a} - \mathbf{b}$ c $\frac{1}{2}(\mathbf{a} - \mathbf{b})$ d $\frac{1}{2}(\mathbf{a} + \mathbf{b})$

e $\frac{1}{2}\mathbf{b}$

f $\overrightarrow{OB}$ and $\overrightarrow{CD}$ are both multiples of **b**, so are parallel.

10 a AP = $\frac{3}{5}$AB; BP = $\frac{2}{5}$BA

b i $\mathbf{b} - \mathbf{a}$ ii $\frac{3}{5}(\mathbf{b} - \mathbf{a})$ iii $\mathbf{a} - \mathbf{b}$ iv $\frac{2}{5}(\mathbf{b} - \mathbf{a})$

v $\frac{2}{5}\mathbf{a} + \frac{3}{5}\mathbf{b}$

18 Extend

1

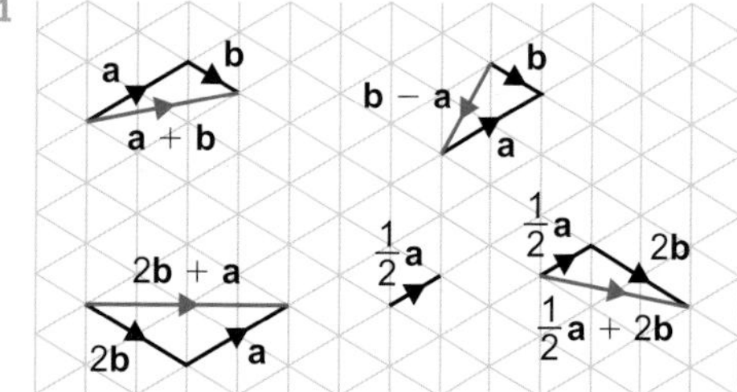

2 a 2b b 2a c b d −a

e b − a f b − a g a − b h b + a

3 a (4, 5) b $\begin{pmatrix}-3\\-4\end{pmatrix}$ c (5, 0)

4 a $\begin{pmatrix}4\\-1\end{pmatrix}$ b $\begin{pmatrix}2\\5\end{pmatrix}$ c $\sqrt{29}$

5 a b − a b $\frac{1}{3}$(a + 2b)

6 a 2a b $\frac{1}{2}$(3a + c) c a + c d $\frac{1}{5}$(4a + 5c)

7 a i 2j ii j − k iii −k − j

b i j − k ii $\overrightarrow{JX} = \overrightarrow{KJ}$, and point J is common.

8 a a − 3b

b $\overrightarrow{NM} = \frac{1}{2}$(a − b); $\overrightarrow{NC}$ = 2(a − b); $\overrightarrow{NC}$ is a multiple of $\overrightarrow{NM}$ and point N is common, so NMC is a straight line.

9 a 3q − 2p

b $\overrightarrow{OR} = \frac{6}{5}$(p + q); $\overrightarrow{OR}$ is a multiple of p + q and so it is parallel to **p** + q.

10 a a − b b $-\frac{1}{5}$(3a + 2b)

11 a i 2q − 4p ii 3(q − p) iii 2(q − p)

b $\overrightarrow{AB}$ and $\overrightarrow{AC}$ are multiples of q − p.
Point A is common, so ABC is a straight line.

c 9 cm

12 a b − a b 2b − a

13 a 6b − 3a

b $\overrightarrow{AX} = \frac{1}{3}\overrightarrow{AB}$ = 2b − a

$\overrightarrow{OX} = \overrightarrow{OA} + \overrightarrow{AX}$ = 2(b + a)

$\overrightarrow{OY} = \overrightarrow{OB} + \overrightarrow{BY}$ = 5(b + a)

So $\overrightarrow{OX} = \frac{2}{5}\overrightarrow{OY}$

18 Unit test

Sample student answer

a The answer should say $\overrightarrow{AB}$ = 2n − 2m = 2(n − m) and $\overrightarrow{MN}$ = n − m. The student has forgotten the direction of the vectors.

b The answer could be improved by adding a sentence at the end, e.g. 'This means AB is parallel to MN and is twice the length.'

UNIT 19

19 Prior knowledge check

1 a 1.33 m/s² b 2 m/s² c 2200 m or 2.2 km

2 C

3 a

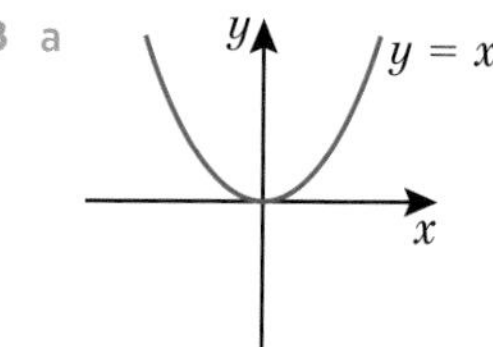

b $y = x^3$

c

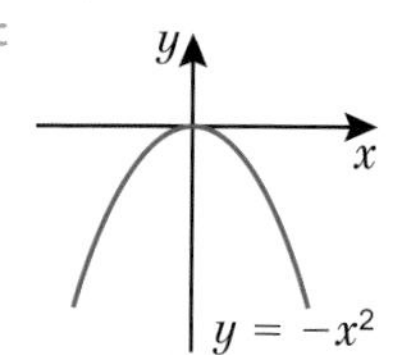

d

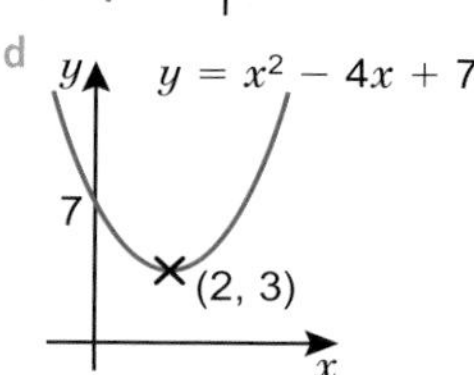

4 Ab, Bc, Ca

5 a $y = (\cos x) - 1$ b $y = 2\cos x$

6 24 minutes

7 a $\frac{1}{3}$ b 2 c $\frac{1}{x}$

8 a $\frac{1}{4}$ b $\frac{1}{2}$ c 1 d 9 e $\frac{1}{3}$

9 a c and d b $d = 7c$

10 a 10 b ±2 c $2(x^2 + 6x - 5)$

11 a 4 and 0.25, 8 and 0.125, 5 and 0.2, 10 and 0.1, 2.5 and 4, 1.6 and 0.625

b Students' own answers

19.1 Direct proportion

1 a Yes; $C = 0.84q$ b No

c No d No

2 a, b

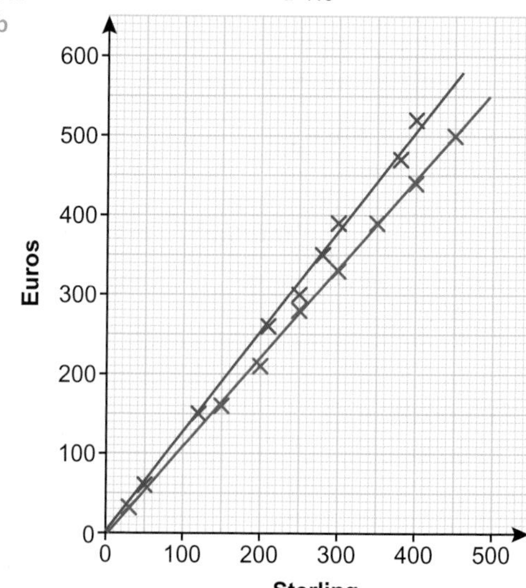

c i $E = 1.25S$ (where gradient is taken from students' graphs)
ii $E = 1.1S$ (where gradient is taken from students' graphs)
d travelcash.com

3 a $y = 5x$ b $y = 50$ c $x = 13$

4 a $y = 6.5x$ b $y = 91$ c $x = 22$

5 a $x = 5$ b $x = 10.1$ (1 d.p.)
c $x = 8.125$

6 a $y = \frac{x}{60}$ b $y = 9$

19.2 More direct proportion

1 $k = 2.5$

2 a $F = 8a$ b 160 c 14

3 a The ratio of $P:l$ simplifies to 12:5 for all pairs of values.
b $k = 2.4$
c $P = 2.4l$
d i $P = 43.2$ cm ii $l = 17.5$ cm

4 a $d = 500t$
b $d = 2500$ km
c $t = 4.5$ hours
d i The distance doubles. ii The distance halves.

5 a

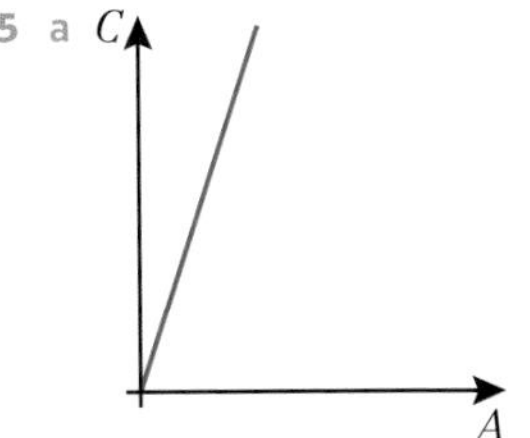

b $C = 50A$ c £4250

6 a $y \propto x^2$ b $y = kx^2$ c $k = 4$ d $y = 100$
e $x = 2.5$

7 a $y = 3.6x^3$ b $y = 230.4$ c $x = 5$

8 a $y = 25\sqrt{x}$ b $y = 75$ c $x = 100$

9 $y = 54$

10 a $E = 5s^2$ b $E = 20$ J c $s = 6.2$ m/s
d The kinetic energy, E is multiplied by 4.

11 a $C = 0.05s^3$ b $C = £6.25$

12 a $T = \frac{R^2}{450}$ b $T = 50$ minutes

13 $g \propto h^3$

19.3 Inverse proportion

1 a $A = \frac{B}{4}$ b $A = 115$

2 a $y \propto x$ b $y \propto \frac{1}{x}$ c $y \propto x^2$

3 a $y = \frac{10}{x}$ b $y = 0.5$ c $x = 2.5$

4 a $P = \frac{3000}{V}$ b $P = 2000$ N/m^2 c $V = 2.5$ m^3
d When the pressure doubles, the volume halves.

5 a $t = \frac{600\,000}{p}$
b No, it takes 4 minutes. When $p = 2500$ W, $t = 240$ seconds.

6 a A graph showing inverse proportion. b $k = 8$
c The product of x and y is always 8, the constant of proportionality.

7 C

8 a $t = \frac{800}{s}$

b

Speed, s (m/s)	4	**10**	20	40	**80**	160
Time, t (seconds)	**200**	80	40	**20**	10	**5**

c

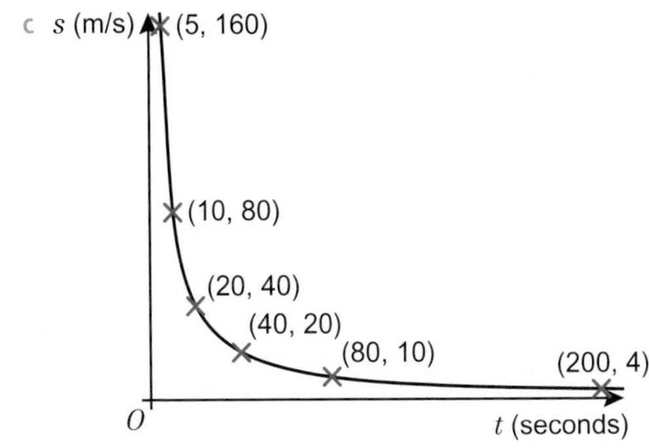

9 $a = 4, b = 2$

10 a, b

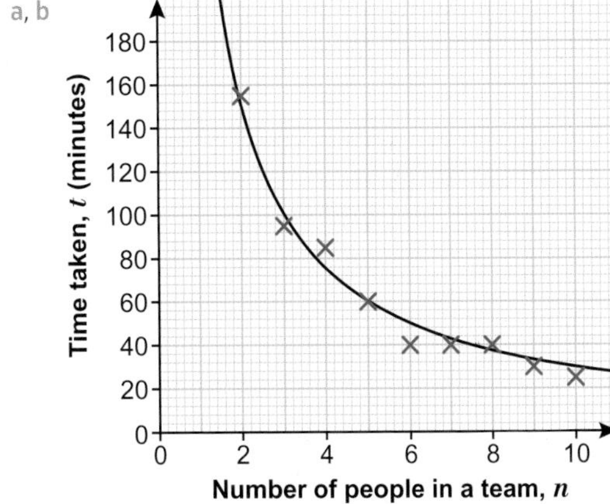

c Answers close to $t = \frac{300}{n}$
d Answer using students' formula from part c
$t = \frac{33}{n}$ gives $t = 20$ minutes.

11 1.33 (2 d.p.)

12 a $y = \frac{54}{x^3}$ b $y = 0.432$ c $x = 2$

13 a $y = \frac{6}{\sqrt{x}}$ b $y = 3$ c $x = 1$

14 a $D = \frac{6390}{r^2}$ b $D = 10.2$ cm (1 d.p.)
c $r = 10.0$ cm (1 d.p.) d $\frac{1}{4}d$ cm

15 a $s = \frac{3400}{r^2}$ b $s = 192.74$

19.4 Exponential functions

1 a 16 b 1 c $\frac{1}{5}$ d $\frac{1}{16}$

2 a 512 b 16384 c 524288

3 a $x = 3$ b $x = 4$ c $x = 4$

4 a

x	−4	−3	−2	−1	0	1	2	3	4
y	0.06	0.13	0.25	0.5	1	2	4	8	16

b

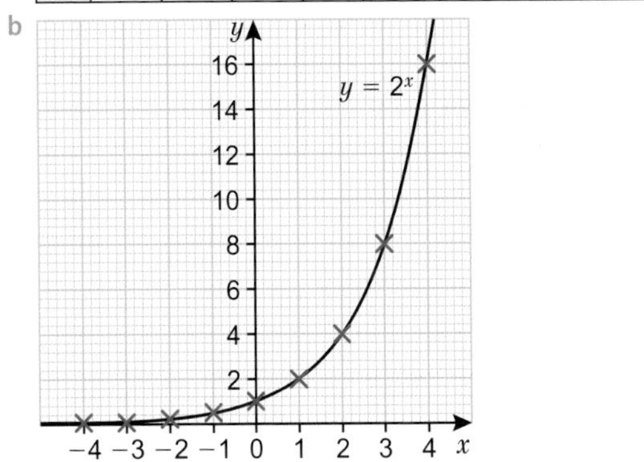

c i $y \approx 11$ ii $x \approx 3.3$

5 a, b

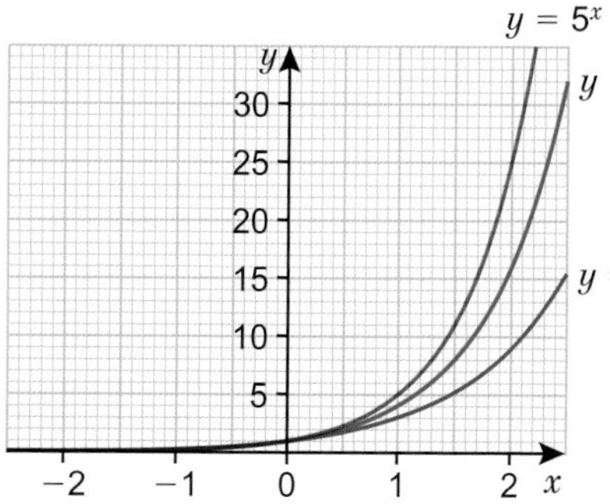

c (0, 1)

6 a

x	−4	−3	−2	−1	0	1	2	3	4
y	16	8	4	2	1	0.5	0.25	0.13	0.06

b

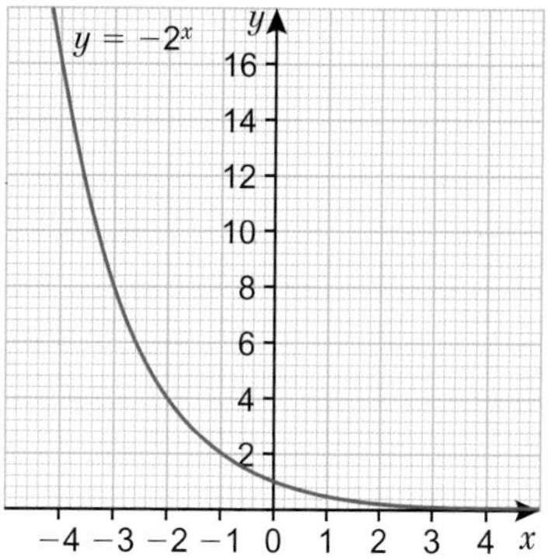

c i $y \approx 0.1$ ii $x \approx -3.3$

7 a

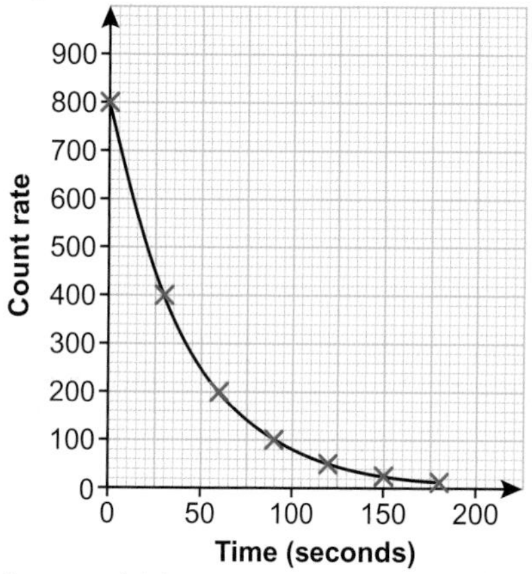

b Exponential decay c 30 seconds

8 $a = 5, k = \frac{7}{5}$

9 a $a = 20000, b = 0.9$ b £14 580 c 10%

10 a

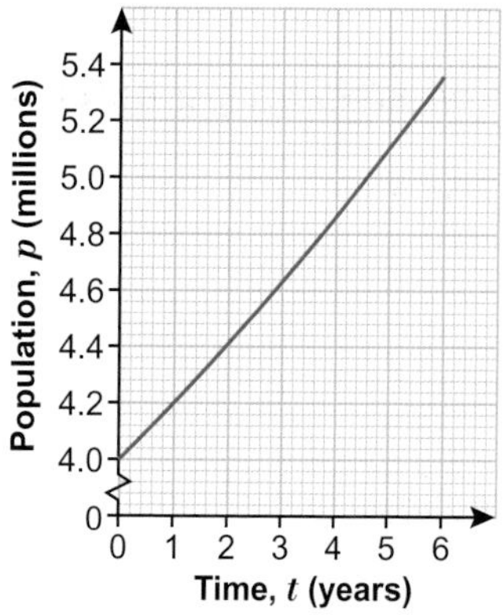

b i $p \approx 4.5$ million ii $t \approx 4.6$ years

11 a $V = 10000 \times 1.04^t$

b

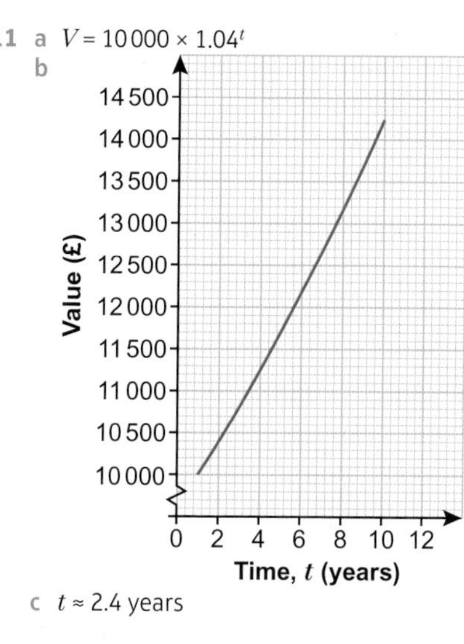

c $t \approx 2.4$ years

19.5 Non-linear graphs

1 i = A, ii = D, iii = C, iv = D

2 180 m

3 a 3 b 1.5 c −1

4 a As time increases, the height increases at a faster rate.
b B

5 a $T = 60\,°C$
b The rate of temperature change decreases over time.
c 10 °C
d ≈ 0.1 °C/sec
e No, the average rate of temperature reduction ≈ 0.2 °C/sec between 0 and 400 seconds.
f The average rate of temperature reduction over the first 300 seconds ≈ 0.25 °C. This is faster than the rate of temperature reduction at exactly 300 seconds ≈ 0.08 °C/sec.

6 a 10 m/s^{-1} b 10 seconds c 24 seconds

7 a

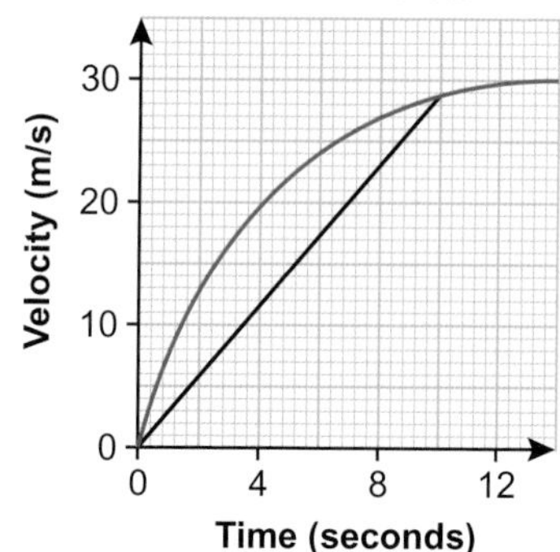

b 2.8–2.9 m/s^{-2}
c 2.5–3.0 m/s^{-2}
d The rate of acceleration decreases over time.
e ≈ 94 m

8 a, b, c

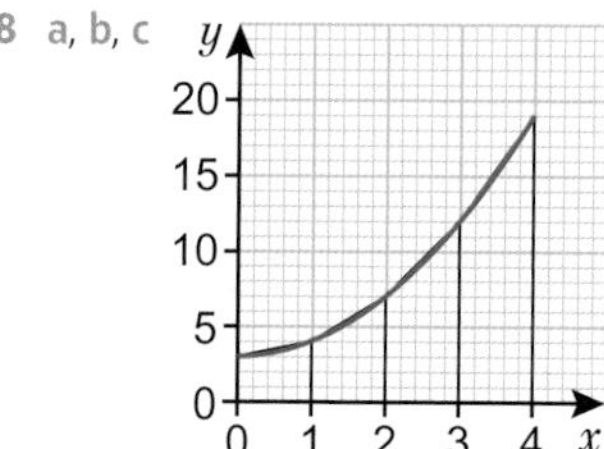

d 34

9 a Amy is leading the race throughout. The distance between Amy and Clare increases for the first 45 seconds, then Clare accelerates and begins to close the gap.

b Her average speed in the second half of the race is more than double her average speed in the first half.

c Estimating from the graph: Amy = 7.25 m/s; Clare 7.5 m/s. Difference 0.25 m/s

10 a 2.5 m/s b 50 m c $T \approx 3$ sec

19.6 Translating graphs of functions

1 a 7 b −8 c 2 d 1 e 0

2 a i 19 ii 27

b i $y = 5x + 4$ ii $y = 5(x + 2) + 2$

3 a A b B

4 a

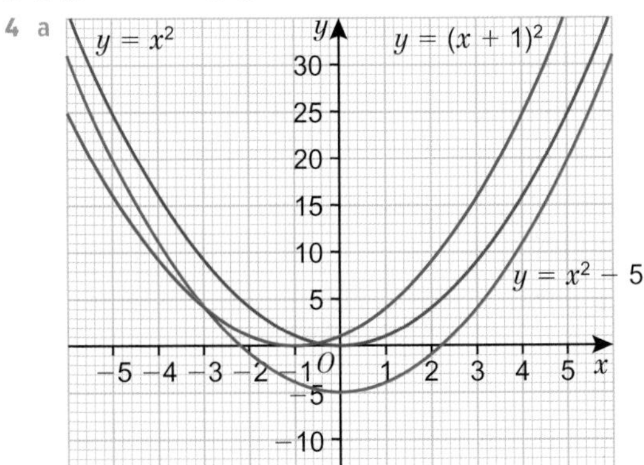

b i (0, −5) ii (−1, 0)

c i Translation by $\begin{pmatrix} 0 \\ -5 \end{pmatrix}$ ii Translation by $\begin{pmatrix} -1 \\ 0 \end{pmatrix}$

d i $y = x^2 - 5$ ii $y = (x + 1)^2$

5

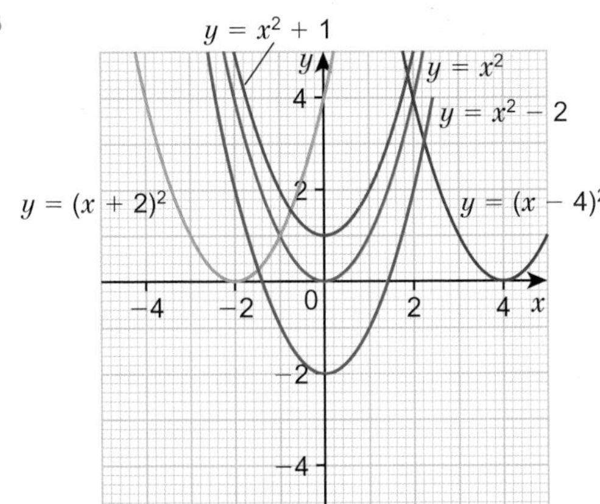

6 a $\begin{pmatrix} 0 \\ 2 \end{pmatrix}$ b $\begin{pmatrix} 0 \\ -3 \end{pmatrix}$ c $\begin{pmatrix} -1 \\ 0 \end{pmatrix}$ d $\begin{pmatrix} 4 \\ 0 \end{pmatrix}$ e $\begin{pmatrix} -5 \\ -2 \end{pmatrix}$

7 $y = f(x - 6)$

8

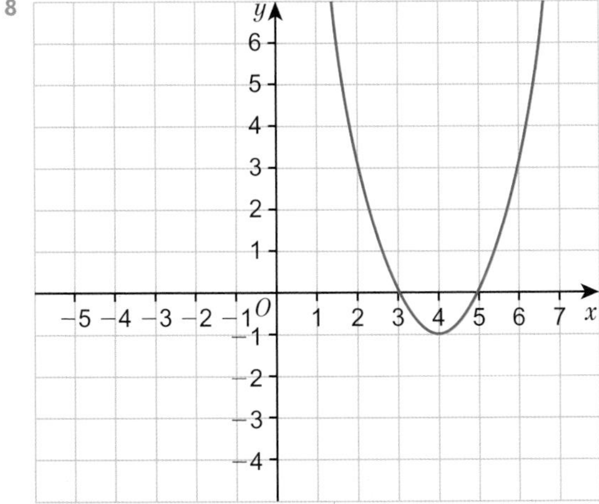

9 a

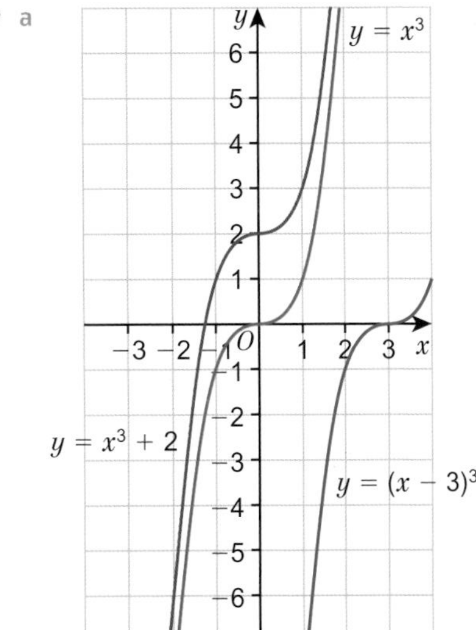

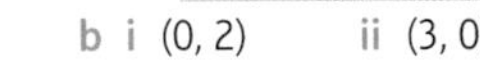

b i (0, 2) ii (3, 0)

10 a, b

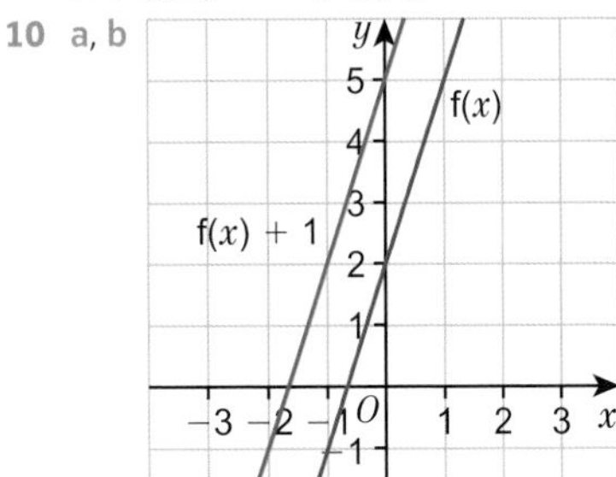

c $y = 3x + 5$

11 a

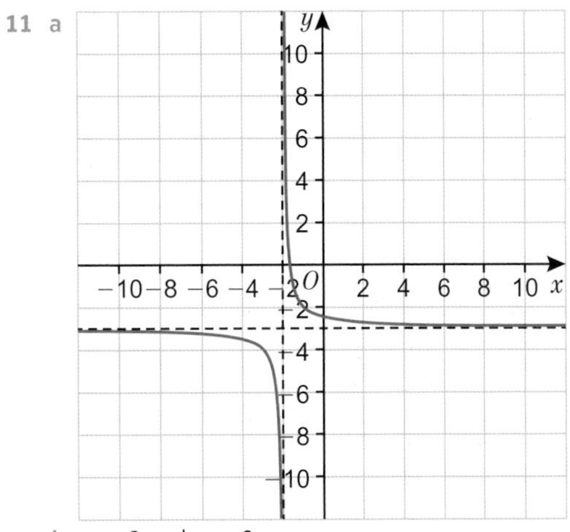

b $x = -2$ and $y = -3$

19.7 Reflecting and stretching graphs of functions

1 a $-6x - 4$ b $-6x + 4$

2 a 1 b 50

3 a

b

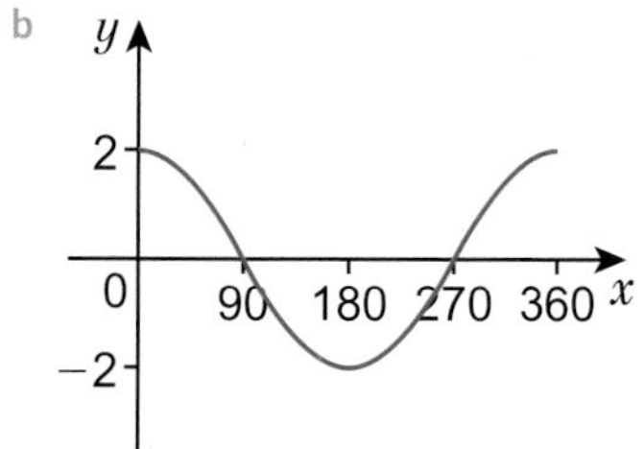

4 a

x	−2	−1	0	1	2
f(x)	−10	−6	−2	2	6
−f(x)	10	6	2	−2	−6
f(−x)	6	2	−2	−6	−10

b

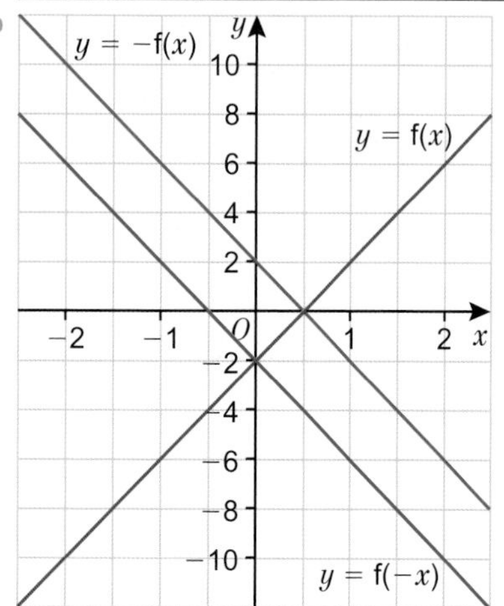

c Reflection in the x-axis

d Reflection in the y-axis

5 a

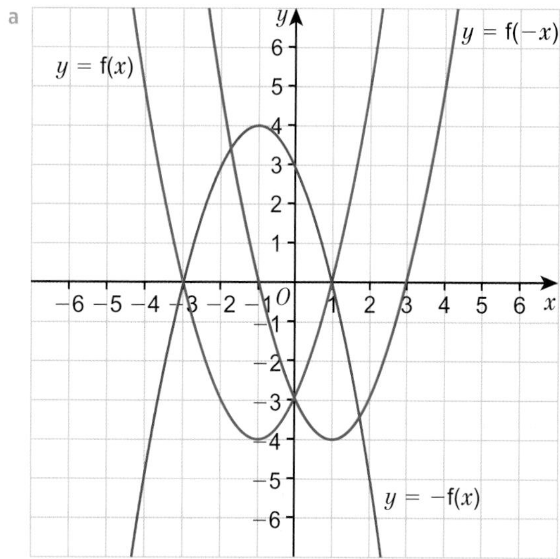

b No, the graphs $y = f(x)$ and $y = -f(x)$ always intersect the x-axis in the same place.
The graphs $y = f(x)$ and $y = f(-x)$ always intersect the y-axis in the same place.

6 a (2, −4) b (−2, 4) c (−2, −4)

7 a

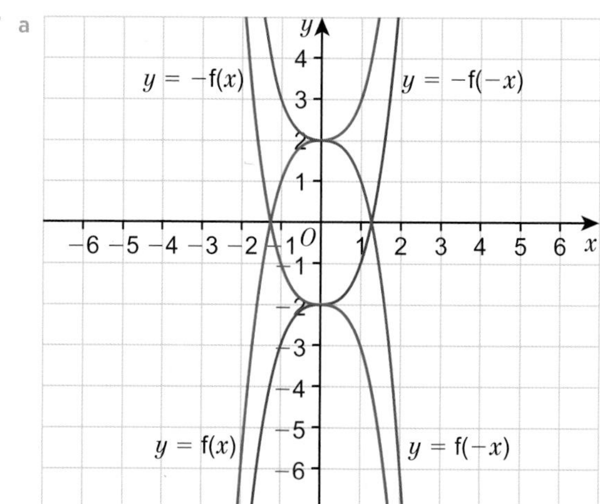

b Rotation of 180° around (0, 0)

8 a

x	−4	−3	−2	−1	0	1	2	3	4
f(x)	12	5	0	−3	−4	−3	0	5	12
2f(x)	24	10	0	−6	−8	−6	0	10	24
f(2x)	60	32	12	0	−4	0	12	32	60

b

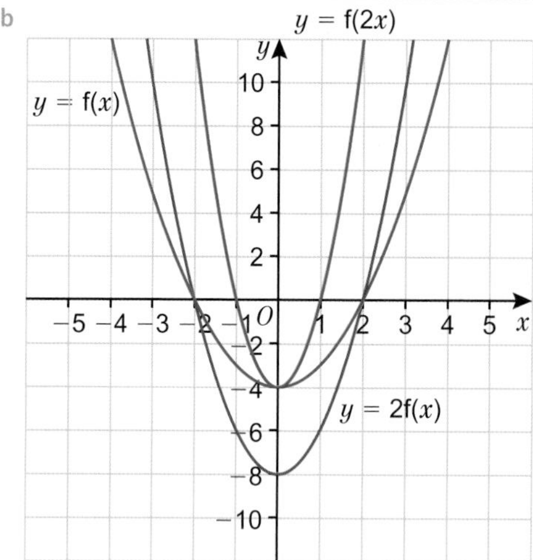

9

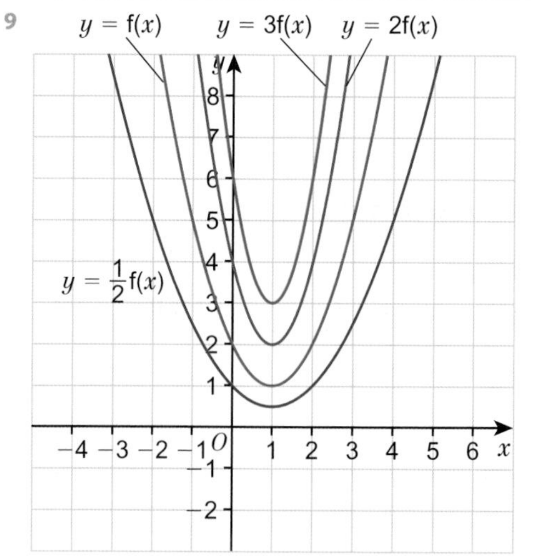

10

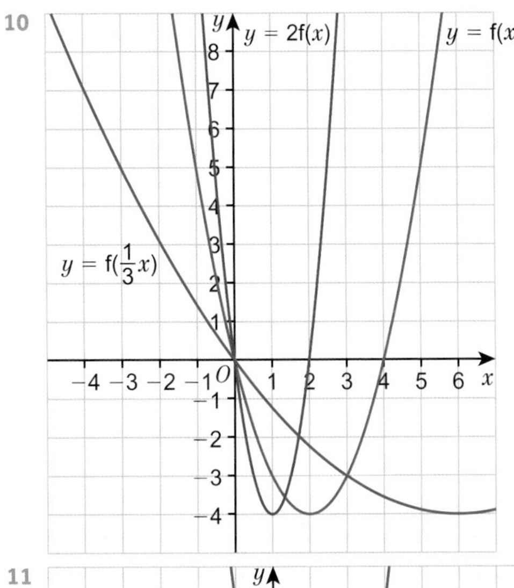

11

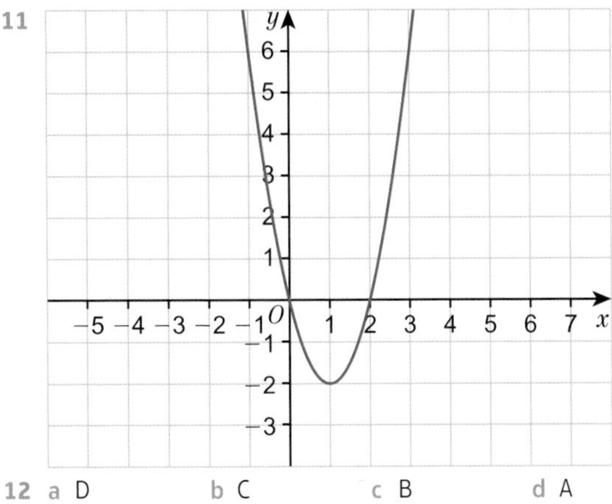

12 a D b C c B d A

13 a

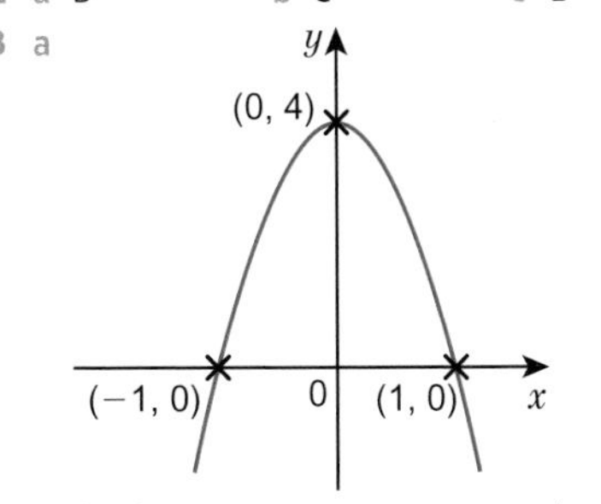

b (0, 4) c (−1, 0) and (1, 0)

d The x^2 term is negative.

19 Problem-solving

1 Total number of cases: 7, 14, 28, 56, 112. Day 4.

2 a $A = 3, B = 2.8$.

b It is more urgent to send more staff as they will be needed on day 6, whereas the extra supplies will not be needed until day 7.

3 Over time there will be fewer people who have not yet caught the disease. This model also does not include people who recover (or die) from the disease. Finally, the value of B might be influenced by human response to the illness (for example through the use of quarantine).

19 Check up

1 a $I = 2.5V$ b $I = 25$ amps c $I = \frac{40}{R}$ d $I = 10$ amps

2 a $y = 12x^2$ b $y = 108$ c $x = \pm 5$

3 a $c = \frac{352}{d^3}$ b $c = 2.816$

4

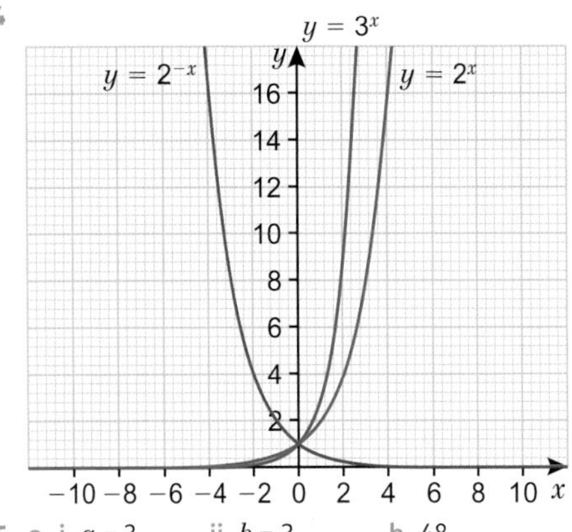

5 a i $a = 3$ ii $b = 2$ b 48

6 a 0.9 m/s^2 b 0.4 m/s^2 c 530 m

7

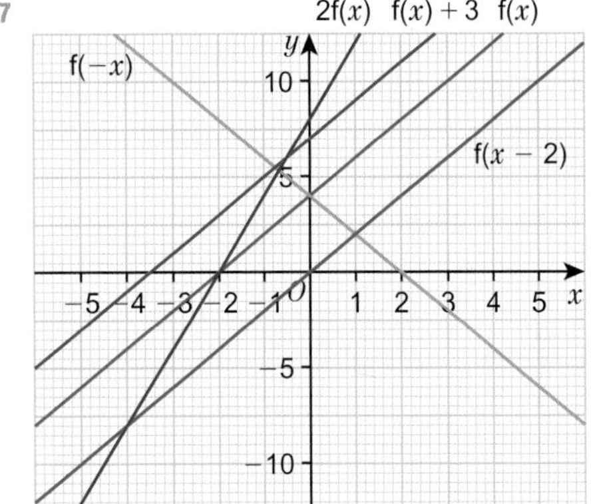

8

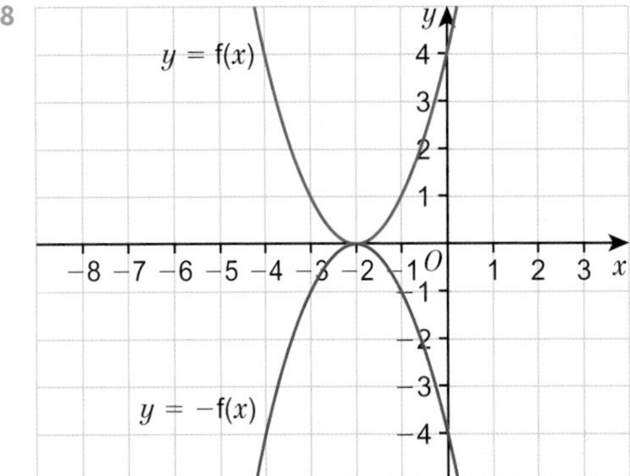

10 a Yes

b Students' own answers

c Students' own answers

19 Strengthen

Proportion

1 a $c = 1.32l$ b 9.09 litres (2 d.p.)

2 a e.g. $l = 24$ $w = 1$, $l = 12$ $w = 2$, $l = 8$, $w = 3$, $l = 6$, $w = 4$

b Indirectly proportional

c

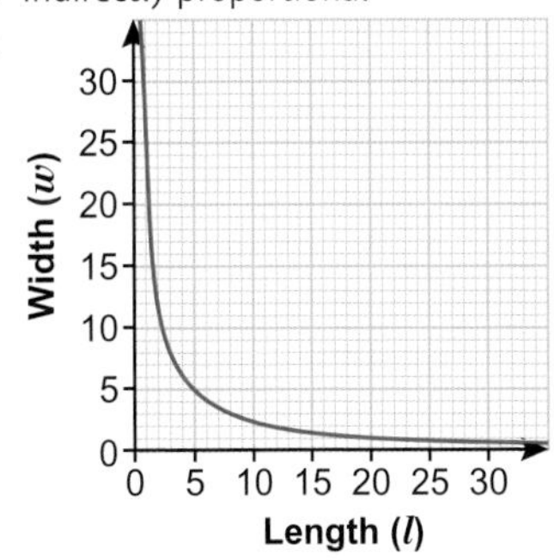

3 a i b ii

4 a i $A \propto B$ ii $A = kB$

b i $C \propto \frac{1}{D}$ ii $C = \frac{k}{D}$

c i $M \propto N^2$ ii $M = kN^2$

d i $F \propto \frac{1}{G^3}$ ii $F = \frac{k}{G^3}$

e i $H \propto \frac{1}{\sqrt{T}}$ ii $H = \frac{k}{\sqrt{T}}$

f i $R \propto S^3$ ii $R = kS^3$

5 a $k = 10$ b $F = 10a$ c $F = 40$ d $a = 6$

6 a $k = 20$ b $a = \frac{20}{b}$ c $a = 4$ d $b = 4$

7 a $k = 5$ b $d = 5t^2$ c $d = 245$ d $t = 3$

Exponential and other non-linear graphs

1 a

t	0	1	2	3	4	5	6
n	1	2	4	8	16	32	64

b, c

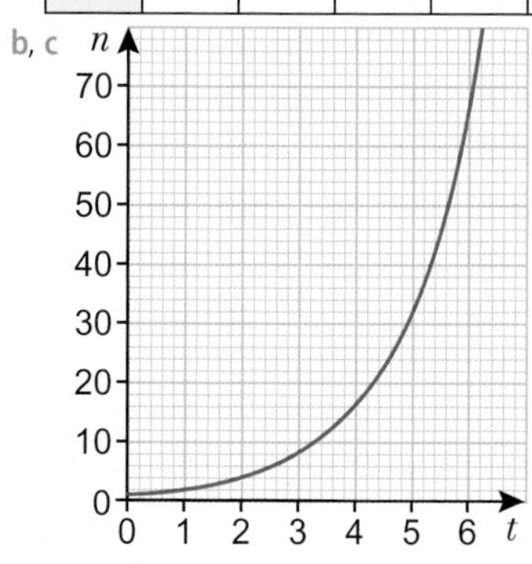

2 a $4 = ab^0$ b $a = 4$

c $8 = 4b$ d $b = 2$

e $y = 4 \times 2^x$ f $y = 32$

3 a i 20 m/s ii 40 m/s iii 60 m/s

b 0.4 m/s²

c i 7.5 km ii 2.5 km

Transformations of graphs of functions

1 a

x	−4	−3	−2	−1	0	1	2	3	4
f(x)	16	9	4	1	0	1	4	9	16

b, e, h

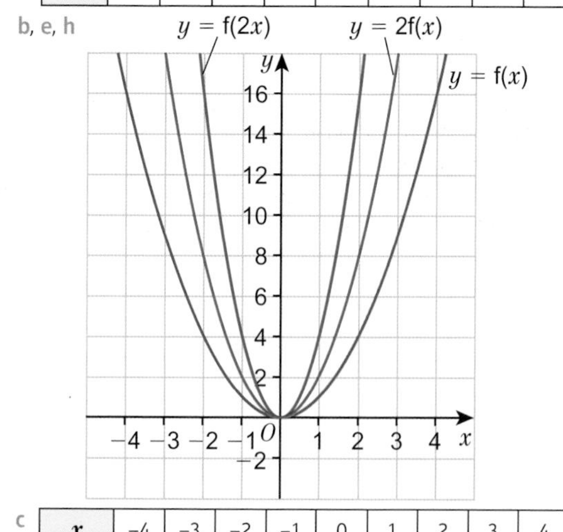

c

x	−4	−3	−2	−1	0	1	2	3	4
2f(x)	32	18	8	2	0	2	8	18	32

d Double

e See above

f $y = f(x)$ is stretched by a scale factor of 2 away from the x-axis.

g

x	−4	−3	−2	−1	0	1	2	3	4
f(2x)	64	36	16	4	0	4	16	36	64

h See above

i $y = f(x)$ is stretched by a scale factor of $\frac{1}{2}$ away from the y-axis.

2 a

x	−3	−2	−1	0	1	2	3
f(x)	−10	−8	−6	−4	−2	0	2

b, e, h

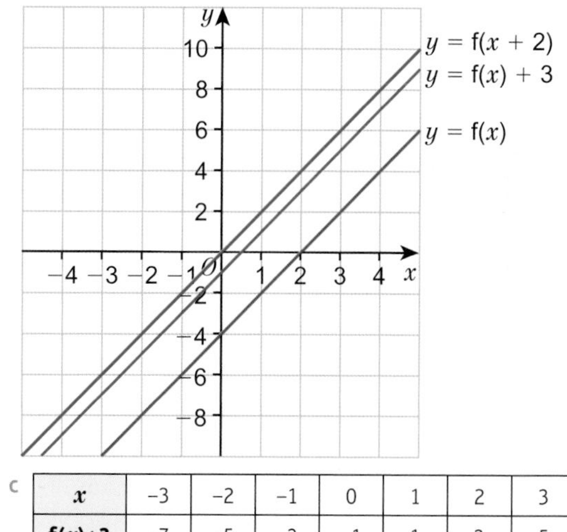

c

x	−3	−2	−1	0	1	2	3
f(x)+3	−7	−5	−3	−1	1	3	5

d 3 greater

e See above

f $y = f(x)$ is translated by $\begin{pmatrix}0\\3\end{pmatrix}$.

g

x	−3	−2	−1	0	1	2	3
f(x+2)	−6	−4	−2	0	2	4	6

h See above

i $y = f(x)$ is translated by $\begin{pmatrix}-2\\0\end{pmatrix}$.

19 Extend

1 a $d = 5t^2$

b 45 m

c 11 seconds

d

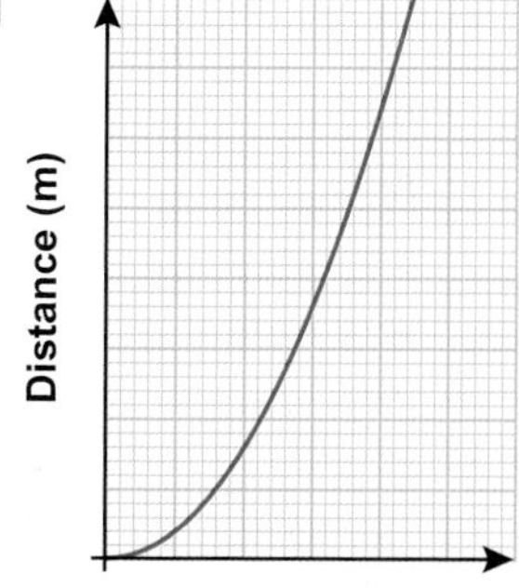

e It accelerates towards the ground.

2 $y = 2^x$ is D

$y = 6^x$ is C

$y = 0.5^x$ is A

$y = 3^{-x}$ is B

3 a C b D c B d E e A

4 a A = The ball is travelling upwards and decelerating
B = The ball has reached its maximum height
C = The ball is accelerating towards the ground
b The speed at A is the same as the speed at C.
c The velocity at A and C have the same magnitude, but one is positive and one is negative.

5 a £6556.36
b A
c £2642.86

6 a Month 1 and month 2. The graph is steepest gradient in this section.
b The profits increased over the period. The increase was greatest between months 1 and 2, and then at a slower but fairly steady rate for the next 4 months.
c £5 million

7 a

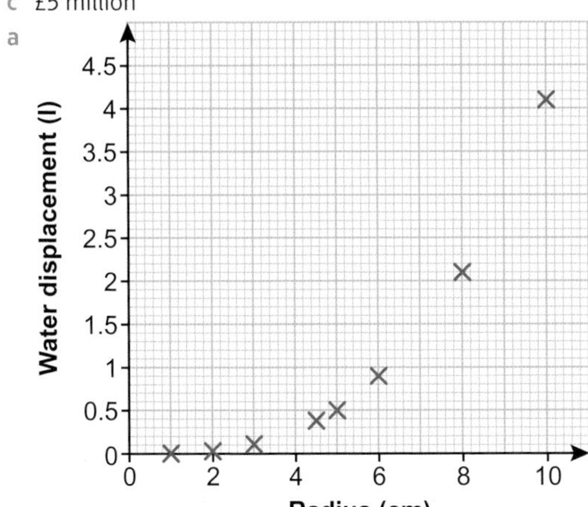

b $W \propto r^3$
c $W = 0.004r^3$
d 16.4 litres

8 a i (0, 0) ii (0, 0) iii (0, 5)
b i (1, 0), (3, 0), (5, 0) ii (−8, 0), (0, 0), (8, 0)

9 a

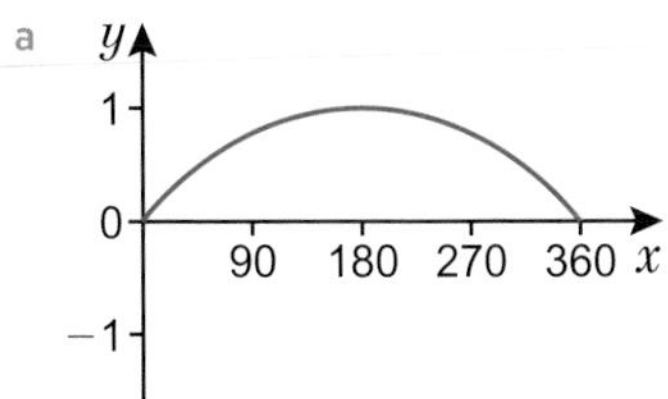

b 2

10 a (−4, 0), (2, 0) b (−1, −9)
c

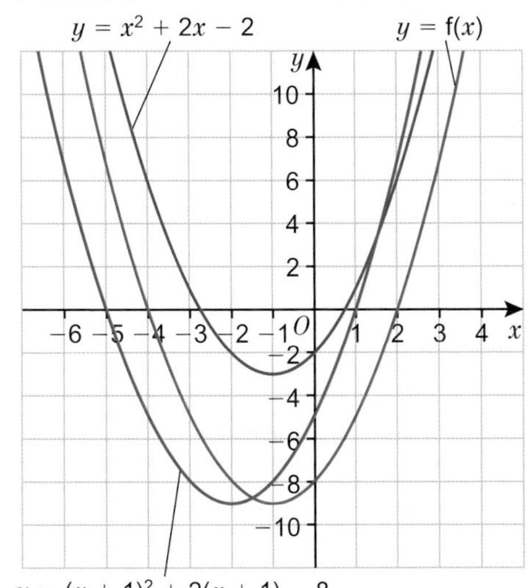

11 a $a = 3$ $b = -4$
b, c

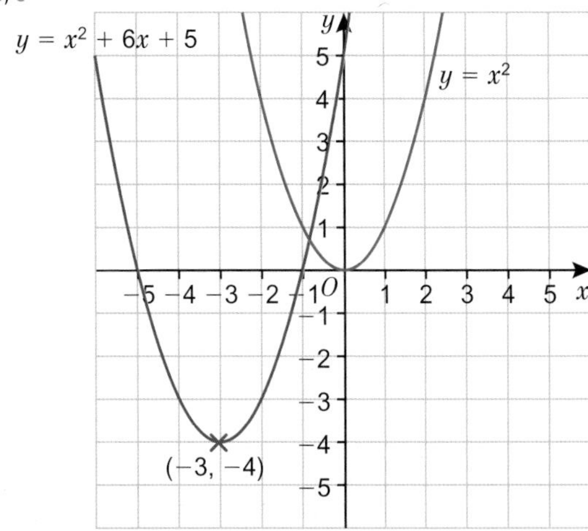

12 $a = 4$, $b = 2$, $c = 2$

13 C_2 $y = f(x + 2)$

14 a

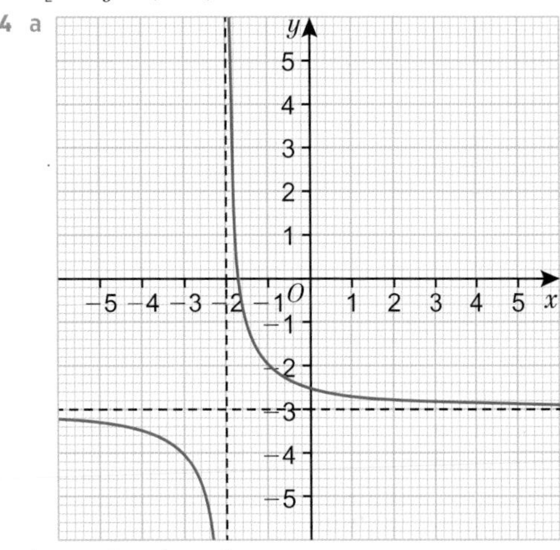

b $x = -2$ and $y = -3$

15 105 counts per second

16 a £178 658.24
b £15 000

19 Unit test

Sample student answer

a Where the curve crosses the y-axis and the x-axis, and where the minimum/maximum points are.

b The minimum point.

c The student could label the axes with an approximate scale to help count the number of units to be moved.

INDEX

1 : *n* *see* unit ratios
2D problems, trigonometry 415–18, 426, 427, 429–31, 433–5, 436
3D problems, trigonometry 418–21, 428, 432–3, 435, 436
2D shapes
  area 205–7, 213–19, 227, 229–30, 232–4, 235
  congruence 363–9, 382–3, 385–6, 390–3, 394
  perimeters 205–7, 213–19, 227, 229–30, 232–4, 235
  similarity 370–7, 382–4, 387–9, 390, 393, 394
3D solids
  cross-sections 211, 231, 235
  elevations 240–2, 267, 270–1, 275
  plans 240–2, 267, 270–1, 275
  similarity 378–81, 385, 389, 391–2, 394
  surface area 210–12, 220–6, 228, 231–2, 233–8
  volume 210–12, 220–6, 228, 231–2, 233–8

AAS triangles 365–6, 394
acceleration
  kinematics formulae 345–6, 352, 353, 357, 359
  surface gravity 548
  velocity-time graphs 169–70, 196
accuracy
  error intervals 208–10, 227, 230–1, 234, 235
  measurements 207–10, 227, 230–1, 234, 235
  upper and lower bounds 209–10, 227, 230–1, 234, 235, 400–2, 427, 429–31
addition
  algebraic fractions 537–8
  fractions 99–100, 109, 110–11, 112–13, 116
  indices 9–11, 21
  mixed numbers 99–100, 111, 112–13
  vectors 251–2, 562–5, 583
adjacent 136, 155–6
algebra 30–60, 531–56
  arithmetic sequences 40–2, 50, 54–6, 58
  composite functions 544, 549, 551, 554, 555
  equations 35–7, 38, 50, 53–4, 58
  equivalence 32
  expanding brackets 33–5, 46–8, 50, 51–2, 56, 57, 58
  expressions 38, 52, 57, 58, 59
  factorising 33–5, 46–8, 50, 52–3, 57, 58
  Fibonacci type sequences 43, 50, 54, 58
  formulae 37–9, 50, 53–4, 56, 58
  fractions 533–41, 549, 552–4, 555
  functions 543–4, 549–51, 553–4, 555
  geometric sequences 43, 50, 55–6, 58
  identity 34, 38, 53, 58
  indices 31–3
  inverse functions 544, 549, 551, 554–5
  linear sequences 40–2, 50, 54–6, 58
  non-linear sequences 42–5, 50, 54–6, 58
  problem-solving 48–9, 548
  proof 545–7, 549, 553–4, 555
  quadratic sequences 43–5, 50, 55–6, 58–9
  rearranging formulae 532–3, 548–51, 553–4, 555
  sequences 40–5, 50–1, 54–6, 58–9
  simplifying 31–3, 50, 51–3, 56–7
  subjects 39, 56, 58, 532–3, 548
  surds 539–40, 549, 553–4, 555
  variables 38, 56–8
algebraic fractions 533–41, 549, 552–4, 555
  addition, subtraction, multiplication and division 537–8
  simplifying 535–7
  solving 541–2
  surds 539–40, 549, 553–4, 555
alternative segment theorem 512, 521–2, 524–5, 527
angle bisectors 259–61, 267, 272–3, 276
angle of depression/elevation 139–41, 152, 156
angles 121–58
  circle theorems 501–30
  congruence 364–9, 382–3, 385–6, 393, 394
  cosine rule 415–18, 427, 430, 433, 435, 436
  cosines 137–8, 140, 142, 156, 405–7
  isosceles triangles 122, 124, 125, 127–8, 147
  parallelograms 123, 129, 145, 149
  polygons 126–31, 153–4, 155
  problem-solving 143–4
  Pythagoras' theorem 131–6, 146, 149–50, 154, 155
  quadrilaterals 122–7, 129–30, 145, 147–8
  sectors of circles 218–19, 227, 230
  semicircles 507, 526
  sine rule 413–14, 427, 429–30, 434, 436
  sines 137–8, 140, 142, 156, 402–5
  subtended in circles 506–7, 509–12, 513–14
  tangents 137–8, 140, 142, 156, 408–10
  tangents to circles 504–6, 512–14, 517, 519, 521–2, 524–7
  triangles 122–7, 128, 131–43, 145–7, 149–56
  trigonometry 121, 136–43, 146–7, 151–3, 154, 155–6
annual salaries, percentage increases 341, 357
approximation 4–5
  see also estimation
arcs, circles 218–19, 227, 230
area 203–38
  accuracy 207–10, 227, 230–1, 234, 235
  circles 213–19, 227, 229–30, 235
  circle segments, sine calculation 412–13
  2D shapes 204–9, 213–19, 226, 227, 229–30, 232–4, 235
  percentage increase 226
  perimeters 204–9, 213–19, 226, 227, 229–30, 232–4, 235
  pressure 339, 347–8, 353, 355, 356, 357, 359
  problem-solving 225
  scale factors 247, 376–81, 385, 389, 391–2, 394
  semicircles 217–18, 227, 230
  triangles 411–13, 429, 433–5, 436
  unit conversions 207–8, 227, 230–1, 234, 235
area under the curve 169–70, 233, 600–1, 620
arithmetic sequences, algebra 40–2, 50, 54–6, 58
asymptotes 196
averages 75–8, 81, 82–3, 85–7, 89, 90–3, 94
  grouped frequency tables 77–8, 82, 87, 90, 91–3, 94
  moving 93, 94
average speed 344–5, 352, 358, 359
  distance-time graphs 168, 187, 198
axes
  scaling 79
  zigzag lines 171

back-to-back stem and leaf diagrams 65–6, 82, 85, 94
bar models 110–11, 114
bearings 253–5, 267, 271, 274, 276, 415, 417, 426
bias 440–2, 458–9, 463, 466
bisecting 133
bisectors
  angle 259–61, 267, 272–3, 276
  perpendicular 258–9, 272, 274, 276
bivariate data, scatter graphs 70–5, 83, 87–8, 89, 91–2, 94
box plots (box-and-whisker diagrams) 446–8, 454, 456, 457, 461–2, 464–5, 467
  comparative 447–8, 456–7, 462, 464
  outliers 454, 464–5, 467
brackets
  expansion 33–5, 46–8, 50–2, 56–9
  squaring 46–7, 59

calculation with powers 8–13, 21, 23–5
capacity 212, 226, 233, 234, 235
capture-recapture sampling 442, 466
censuses 440, 466
centre of enlargement 246–9, 266–7, 269–70, 273–4, 276
centre of rotation 244–5, 252, 266–7, 268–9, 273–5, 276
chords
  circles 502–4, 512, 515–16, 517, 519–20, 523–4, 526
  graphs 600–1, 620
circles
  alternative segment theorem 512, 521–2, 524–5, 527
  arcs 218–19, 227, 230
  area 213–19, 227, 229–30, 235
  chords 502–4, 512, 515–16, 517, 519–20, 523–4, 526
  circumferences 213–16, 227, 229–30, 235
  cyclic polygons 509–12, 520–1, 522–4, 527
  graphs 185, 188, 199
  radii 502–4, 517, 519–20, 524, 526
  sectors 216–19, 227, 230
  segments 411–13, 417, 434
  subtended angles 506–7, 509–12, 520–1, 523–5, 526, 577–8
  tangents 504–6, 512–14, 517, 519, 521–2, 524–7
circle theorems 501–30
  alternative segment 512, 521–2, 524–5, 527
  application 512–17
  chords 502–4, 512, 515–16, 517, 519–20, 523–4, 526
  polygons 509–12, 520–1, 522–4, 527
  problem-solving 515–17
  quadrilaterals 510–11, 520–1, 522, 527
  radii 502–4, 517, 519–20, 524, 526
  semi-circles 507, 526
  subtended angles 506–7, 509–12, 520–1, 523–5, 526, 577–8
  tangents 504–6, 512–14, 517, 519, 521–2, 524–7
circumferences 213–16, 227, 229–30, 235
class widths 449–53, 461, 465, 467
collinear points 568, 578–9, 583
column vectors 250–2, 267–9, 276, 558–66
combinations of transformations 249–52, 266–7, 268–9, 273–5, 276
combined events, probability 307–9, 326, 328–9, 333–5
common difference 40–2, 50, 54–6, 58
comparative box plots 447–8, 456–7, 462, 464
comparing gradients 163
comparing populations 447–8, 453–6, 457–8, 462–5, 467
completing the square 284–6, 297–9, 301, 302, 478–85, 492, 495–6, 498
composite functions 544, 549, 551, 554, 555
compound interest 342–3, 352, 354, 357–8, 359, 598, 616, 619
compound measures
  density 346–7, 352, 353, 355, 357, 359
  kinematics formulae 345–6, 352, 353, 357, 359
  multiplicative reasoning 343–8, 355–6, 357–8, 359, 532–3
  pressure 339, 347–8, 353, 355, 356, 357, 359
  rates 344–6, 352, 355–8, 359
conditional probability 318–24, 325, 330–3, 334–5, 336
  Venn diagrams 321–4, 332–3, 335, 336
cones
  similarity/scale factors 379, 381, 392
  volume 222–4, 225
congruence 363–9, 382–3, 385–6, 390–3, 394
  geometric proof 367–9
consecutive integers 35
constant multipliers 43, 50, 55–6, 58
constant of proportionality 589–92, 611, 613, 616–18, 619
constant rates 598–9
  distance-time graphs 169, 189, 194, 196
  multiplicative reasoning 342–5, 350–1, 352–5, 352–6, 358
  velocity-time graphs 196
constructions 239–42
  angle bisectors 259–61, 267, 272–3, 276
  3D solids 240–2, 267, 270–1, 275
  locus 262–4, 267, 273, 276
  perpendicular bisectors 258–9, 272, 274, 276
  triangles 256–7, 261, 267, 272, 274
conversions
  currency 103, 105, 111, 113, 117, 120, 589
  decimals 108–11, 114–15, 116, 118
  fractions 108–11, 114–15, 116, 118
  metric speed measures 344–6, 352, 353, 355–6, 359
  percentages 108–11, 114–15, 116, 118
  units of area 207–8, 227, 230–1, 234, 235
  units of measurement 26
  units of temperature 57
correlation
  graphs 173–4, 183–4, 195, 199
  scatter graphs 70–5, 87–8, 183–4, 195
cosine 137–8, 140, 142, 156
cosine functions, graphs 405–7, 421–5, 427–8, 431–2, 435–6
cosine rule 415–18, 427, 430, 433, 435, 436
counter examples 546, 555
cross-sections, prisms 211, 231, 235
cube roots 9
cubes, similarity/scale factors 389
cubic equations 180–2, 188, 193, 198, 485–9, 492, 496, 498
cubic functions
  graphs 180–2, 188, 193, 198, 485–9, 492, 496, 498
  iteration 489
  roots 486–9, 492, 496, 498
  x/y-intercepts 486–9, 492, 496, 498

y-intersects 486–7, 492, 496, 498
cubic proportionality 591–2, 611, 613, 616–18, 619
cuboids
diagonals 419–20, 428, 432, 433
similarity/scale factors 378, 389
surface area 210–12, 231
trigonometry 419–20, 428, 432, 433
volume 210–12, 231
cumulative frequency 443–8, 454–7, 459–67
diagrams 444–5, 448, 455, 457, 459–61, 463–4, 466–7
interquartile range 445–8, 453–8, 460–5, 467
median 444–5, 448, 457–8, 460–5, 467
outliers 454, 464–5, 467
quartiles 445–8, 457, 460–2, 464–5, 467
step polygons 464
tables 443–5, 459–61, 464, 466–7
upper class boundaries 443–4, 466
currency, exchange rates 103, 105, 111, 113, 117, 120, 171, 589
curved surface area, cones 223–4, 225, 232, 236
cyclic polygons 509–12, 520–1, 522–4, 527
cyclic quadrilaterals 510–11, 520–1, 522, 527
cylinders
similarity/scale factors 379, 385, 391
surface area 220–2, 226, 228, 231, 234, 236
volume 220–2, 226, 228, 231, 234, 236

data
appropriate display methods 79–81, 93
averages 75–8, 81, 82–3, 85–7, 89, 90–3, 94
back-to-back stem and leaf diagrams 65–6, 82, 85, 94
comparing 447–8, 453–6, 457–8, 462–5, 467
cumulative frequency 443–8, 454–7, 459–67
frequency density 449–53, 458, 461, 465, 467
frequency polygons 66–7, 81, 86, 90, 92, 94
interpretation and representation 61–96
outliers 75, 77, 82, 86, 88, 94
pie charts 64, 82, 84, 90
range 75–8, 82–3, 85–7, 89, 90–3, 94
sampling 440–2, 457–9, 463–4, 466
scatter graphs 70–5, 83, 87–8, 89, 91–2, 94
statistical diagrams 63–7, 78–80, 81, 82, 84–5, 94
stem and leaf diagrams 64–6, 82, 85, 94
time series graphs 67–70, 80, 83, 87–9, 93, 94
see also statistics
decay 340–3, 351–2, 354, 357–8, 359
decimal multipliers, percentages 340–3, 351–2, 354, 357–8, 359
decimals
conversions 108–11, 114–15, 116, 118
indices 12–13
recurring 109, 115, 118
decomposition, prime factors 6–8, 21–3
decreases, percentages 106–7, 112, 114–16, 118
denominators, rationalising 18–19, 21, 25, 27, 28, 539–40, 549, 552–4, 555
density 346–7, 352, 353, 355, 357, 359
frequency 449–53, 458, 461, 465, 467
dependent events 318–24, 325, 330–3, 334–5, 336
depreciation 106–7, 114, 118, 341–3, 351–2, 354, 357–8, 597
diagonals, trigonometry 419–20, 428, 432, 433
diagrams
appropriate use 79–81, 93
cumulative frequency 444–5, 448, 455, 457, 459–61, 463–4, 466–7
misleading 79, 81
statistical 63–7, 78–80, 81, 82, 84–5, 94
see also representation of data
direct proportion 104–5, 109, 112, 115–16, 118
graphs 171–2, 199, 588–92, 611, 613, 616–18, 619
multiplicative reasoning 348–51, 352, 353, 356–8, 359
displacement, vectors 558–60, 582
distance–time graphs 167–9, 183, 187, 194, 196, 198, 600–1, 616
average speed 168, 187, 198
distance travelled, velocity-time graphs 169–70, 233
division
algebraic fractions 537–8
fractions 99–100, 111, 112–13, 115–16, 118
mixed numbers 99, 111, 113
powers 10
double brackets, expansion 46, 59

elements, sets 321, 322, 336
enlargement 245–9, 253–5, 266–7, 269–70, 273–4, 276
fractional scale factors 247–8, 266, 276
negative scale factors 248–9, 253–5, 270, 275
equal vectors 559–60, 582
equations 280–304
algebra 35–7, 38, 50, 53–4, 58
circle 185
graphs 471–500
inequalities 293–5, 297, 300, 301, 302
quadratic 281–6, 291–3, 296–301, 302
simultaneous 287–93, 296, 297, 299–301, 302
equidistant points, loci 262–4, 273, 276
equilateral triangles 136, 142
congruence 367–8
equivalence, algebraic expressions 32
error intervals 208–10, 227, 230–1, 234, 235
estimation 4–5, 20–1, 22, 27
graphs 172–4
place value 4–5
population size 442, 466
see also approximation
events
dependent 318–24, 325, 330–3, 334–5, 336
independent 314–18, 325, 330–1, 334–5
mutually exclusive 310–12, 326, 328–9, 333–5
exchange rates 103, 105, 111, 113, 589
expanding brackets
algebra 33–5, 46–8, 50, 51–2, 56, 57, 58
cubic equations 485–9, 498
double brackets 46, 59
surds 539, 549
experimental probability 312–14, 325, 326, 329, 333, 335
exponential functions 184, 343, 595–8, 610, 611, 614, 616, 620
expressions, algebra 38, 52, 57, 58, 59
exterior angles
cyclic quadrilaterals 511, 522, 527
hexagons 129, 131
pentagons 129–30
polygons 128–31, 144, 148–9, 154, 155
quadrilaterals 129–30, 149
triangles 123–5, 128, 155
extrapolation
scatter graphs 74–5, 91–2, 94
see also prediction

factorial 2–3, 20–1, 26
factorising
algebra 33–5, 46–8, 50, 52–3, 57, 58
algebraic fractions 535–7, 549, 552–4, 555
surds 539–40, 549, 553–4, 555
factors
decomposition 6–8, 20–3
number of outcomes 2–3, 20–1
fair games 313–14, 326, 329, 333
Fibonacci type sequences 43, 50, 54, 58
flow diagrams 225–6
forces
pressure 339, 347–8, 353, 355, 356, 357, 359
surface gravity 548
formulae
algebra 37–9, 50, 53–4, 56, 58
rearranging 532–3, 548–51, 553–4, 555
subjects 39, 56, 58, 532–3, 548–51, 553–4, 555
fractional indices 11–13, 21, 24–5
fractional scale factors, enlargement 247–8, 266, 276
fractions 97, 98–100, 108–11, 112–18
addition 99–100, 109, 110–11, 112–13, 116
algebraic 533–41, 549, 552–4, 555
conversions 108–11, 114–15, 116, 118
decimals and percentages 108–11
division 99–100, 111, 112–13, 115–16, 118
multiplication 98–100, 108–9, 110–11, 118
negative indices 11–12, 21
rationalising the denominator 18–19, 21, 25, 27, 28, 539–40, 549, 552–4, 555
reciprocals 99–100, 116, 118
recurring decimals 109, 118
subtraction 100, 113
surds 539–40, 549, 553–4, 555
frequency
cumulative 443–8, 454–7, 459–67
density 449–53, 458, 461, 465, 467
experimental probability 312–14, 325, 326, 329, 333, 335
frequency polygons 66–7, 81, 86, 90, 92, 94
frequency trees 314–15
front elevation, 3D solids 241–2, 267, 270–1, 275
frustrums, volume 224, 392
function machines 42
function notation 543–4, 549–51, 553–4, 555
functions
algebra 543–4, 549–51, 553–4, 555
composite 544, 549, 551, 554, 555
inverse 544, 549, 551, 554–5
reflections 605–6, 612, 615, 618, 621
scaling 606–9, 612, 615, 618, 621
translating 602–4, 612, 615, 618, 621

games, fairness 313–14, 326, 329, 333
geometric problems 568–74, 575, 578–82, 583
geometric proof of congruence 367–9
geometric sequences, algebra 43, 50, 55–6, 58
Golden Ratio 539
gradients
comparing 163
distance-time graphs 167–9, 187
linear graphs 161–3, 166, 190–2, 198
perpendicular lines 176
velocity-time graphs 169–70
graphs 159–202
area under the curve 600–1, 620
asymptotes 196
axes 162, 163, 171, 199
chords 600–1, 620
circles 185, 188, 199
correlation 173–4, 183–4, 199
cosine functions 405–7, 421–5, 427–8, 431–2, 435–6
cubic 180–2, 188, 193, 198, 485–9, 492, 496, 498
cumulative frequency diagrams 444–5, 448, 455, 457, 459–61, 463–4, 466–7
currency conversions 171
direct proportion 171–2, 199, 588–92, 611, 613, 616–18, 619
distance–time 167–9, 183, 187, 194, 196, 198, 600–1, 616
equations 471–500
estimation 172–4
exponential functions 184, 343, 595–8, 610, 611, 614, 616, 620
frequency polygons 66–7, 81, 86, 90, 92, 94
gradients 161–3, 166, 198–9
half-life 184, 343
histograms 449–53, 458, 461, 465, 467
inequalities 475–8, 491, 492–4, 496–7, 498
inverse proportion 592–5, 611, 613, 616–18, 620
linear 161–6, 170–6, 183, 187, 190–2, 193, 195–6, 198
line segments 174–6, 187
line of symmetry 197
misleading 79
non-linear 182–4, 192–3, 598–601, 611, 614, 616–17
parallel lines 162, 191–2, 197, 198
perpendicular lines 176, 191–2, 197
pie charts 64, 82, 84, 90
prediction 172–4
problem-solving 186, 490–1
profit 186, 194
proportion 587–623
quadratic 176–9, 188, 193, 196, 199, 482–5, 491–2, 494–6, 497, 498
rates of change 166–70
real-life 170–4, 188–9, 193–4
reciprocal 180–2, 188, 193, 199
reflections 422–3, 436, 605–6, 612, 615, 618, 621
rotations 422–3, 436
scale factors 606–9, 612, 615, 618, 621
scatter plots 70–5, 83, 87–8, 89, 91–2, 94, 183–4, 195
simultaneous equations 472–4, 491, 492–4, 497, 498
sine functions 402–5, 421–5, 428, 431–2, 434, 435–6
sketching 165
stretching functions 606–9, 612, 615, 618, 621
tangents 408–10, 421–5, 427–8, 431–2, 435–6, 598–600, 614, 616–17, 620
time series 67–70, 80, 83, 89, 93, 94
transformations 421–5, 428, 431–2, 434, 436, 602–9, 612, 615, 618, 621
translating functions 423–5, 436, 602–4, 612, 615, 618, 621
trigonometric 402–10, 421–5, 427–8, 431–2, 434, 435–6
velocity–time 169–70, 196, 198, 233, 598, 600–1, 611, 614, 620
gravity 178–9, 548
grouped frequency tables 77–8, 82, 87, 90, 91–3, 94
growth 340–3, 351–2, 354, 357–8, 359, 597

half-life 184, 343
HCF (highest common factor) 6–8, 21, 22–3, 26–7, 33–5
hectares 208, 226, 235
height
cone slants 223–4, 225, 232, 236
triangles 136
trigonometry 139–41, 152, 156
hexagonal prisms 380
hexagons 127, 128, 129, 131, 374
highest common factor (HCF) 6–8, 21, 22–3, 26–7, 33–5
histograms 449–53, 458, 461, 465, 467
drawing 449–50
interpreting 450–3
horizontal stretches, trigonometric graphs 425, 428, 432, 436
hypotenuse 131–3, 136, 146, 149–52, 154, 155–6

identities, algebra 34, 38, 53, 58, 545, 549, 553–4, 555
image, transformations 243–5
improper fractions 12–13, 21, 24–5, 98–100, 118
income 104, 106–7
independent events 314–18, 325, 330–1, 334–5
index form 7
indices 7, 8–16, 23–6, 27, 28–9
addition 9–11, 21, 23–4, 27
algebra 31–3
decimals 12–13
fractional 11–13, 21, 24–5
mixed numbers 12–13
multiplication 11, 23–4, 27
negative 11–13
powers of 10 14–16, 21, 27–8
prefixes 14
prime factor decomposition 7
subtraction 10, 21, 24, 27
zero 11–12
inequalities
set notation 478, 498
solving 293–5, 297, 300, 301, 302
solving graphically 475–8, 491, 492–4, 496–7, 498
initial velocity, kinematics formulae 345–6, 352, 353, 357, 359
integers 17, 35
interest
compound 342–3, 352, 354, 357–8, 359, 598, 616, 619
simple 106
interior angles
cyclic quadrilaterals 510–11, 520, 521–2, 527
hexagons 127
pentagons 126–7
polygons 126–8, 143–5, 147–8, 153–4, 155
quadrilaterals 126–7, 145, 147–8
triangles 123–5, 126–7, 131–43, 145–7, 149–56
interpolation, scatter graphs 74–5, 88
interpretation, data 61–96
averages 75–8, 81, 82–3, 86–7, 89, 90–3, 94
range 75–8, 82–3, 86–7, 89, 90–3, 94
see also statistics
interquartile range (IQR) 445–8, 453–8, 460–5, 467
intersections, set notation 323–4, 336
inverse functions 544, 549, 551, 554–5
inverse operations, percentage changes 107
inverse proportion
graphs 592–5, 611, 613, 616–18, 620
multiplicative reasoning 350–1, 353, 356, 358, 359
inverse trigonometric functions 140, 156
IQR *see* interquartile range
irrational numbers 539–40
surds 16–19, 21, 23–5, 28, 29
isosceles trapezia 125
area 206–7, 227, 229, 234
perimeters 206–7, 227, 229, 234
isosceles triangles 122, 124, 125, 127–8, 147
congruence 367–8, 369
similarity/scale factors 371, 373, 374
iteration
cubic function solutions 489
quadratic equation solutions 484–5, 492, 496, 497, 498

kinematics formulae 345–6, 352, 353, 357, 359

LCM (lowest common multiple) 6–8, 21, 22–3, 26–7, 33–5
length
of hypotenuse 131–3, 136, 146, 149–52, 154, 155–6
scale factors 370–85, 387–9, 390–2, 393, 394
linear equations, graphs 162–3, 187, 190–2, 198
linear functions, graphs 165
linear graphs 161–6, 170–6, 187, 190–2, 193
gradients 161–3, 166
real-life 170–4
linear inequalities, solving 293–5, 297, 300, 301, 302
linear relationships, graphs 183
linear scale factors 574
2D shapes 370–7, 382–4, 387–9, 390, 393, 394
3D solids 378–81, 385, 389, 391–2, 394
similarity 363–4, 370–85, 387–93, 394
linear sequences, algebra 40–2, 50, 54–6, 58
line of best fit 72–5, 91–2, 94
line segments, graphs 174–6, 187, 191, 199
line of symmetry, graphs 197
lists for problem-solving 19–20
loci (locus) 262–4, 267, 273, 276
losses, percentage 106–7, 112, 114–16, 118
lower bounds
measurements 209–10, 227, 230–1, 234, 235
trigonometry 400–2, 427, 429–31
lower quartiles (LQ) 445–8, 457, 460–2, 464–5, 467
lowest common multiple (LCM) 6–8, 21, 22–3, 26–7, 33–5
LQ *see* lower quartiles

magnitude, vectors 559–60, 582–3
maps *see* scale drawings
mass, density 346–7, 352, 353, 355, 357, 359
maximum turning points 177–9, 199, 478–85, 492, 495–6, 498
mean 75–8, 82, 89–91, 94
measurements
accuracy 207–10, 227, 230–1, 234, 235
error intervals 208–10, 227, 230–1, 234, 235
rounding up/down 209–10, 227, 230–1, 234, 235
units conversions 26
upper/lower bounds 209–10
median 76–8, 82, 85, 87, 90–1, 94
cumulative frequency 444–5, 448, 457–8, 460–5, 467
metric speed measurement conversions 344–6, 352, 353, 355–6, 359
midpoints, line segments 175–6, 187, 191, 199
minimum turning points 177–9, 199, 478–85, 492, 495–6, 498
misleading diagrams 79, 81
missing angles, within circles 507–8
mixed numbers 98–100, 111, 112–13, 118
addition 99–100, 111, 112–13
division 99, 111, 113
indices 12–13
multiplication 99, 111
subtraction 100, 113
modal class 77–8, 82, 87, 94
mode 76–8, 89
money *see* currency
motion, kinematics formulae 345–6, 352, 353, 357, 359
moving averages 93, 94
multiples, factors 2–3, 20–1
multiplication
algebraic fractions 537–8
linear scale factors 394
mixed numbers 99, 111
powers 9–11
simultaneous equations 289–91, 299, 301
vectors by scalars 561–2, 583
multiplicative reasoning 339–62
compound interest 342–3, 352, 354, 357–8, 359
compound measures 343–8, 355–6, 357–8, 359, 532–3
constant rates 342–5, 350–1, 352–5, 352–6, 358
density 346–7, 352, 353, 355, 357, 359
depreciation 341–3, 351–2, 354, 357–8
growth and decay 340–3, 351–2, 354, 357–8, 359
kinematics formulae 345–6, 352, 353, 357, 359
percentages 340–3, 351–2, 354, 357–8, 359
pressure 339, 347–8, 353, 355, 356, 357, 359
problem-solving 351–2
proportional change 341–3, 348–52, 353, 354, 356–8, 359
rates 344–6, 352, 355–8, 359
ratio and proportion 348–51, 352, 353, 356–8, 359
multipliers, percentages 340–3, 351–2, 354, 357–8
mutually exclusive events 310–12, 326, 328–9, 333–5
$m \times n$ 2–3

n : 1 *see* unit ratios
navigation 415, 417, 426
negative correlation, scatter graphs 71, 87–8
negative gradients 161, 190
negative indices 11–13, 21
negative numbers
cube roots 9
square roots 9
negative scale factors, enlargement 248–9, 253–5, 270, 275
Newtons, pressure 339, 347–8, 353, 355, 356, 357, 359
non-linear graphs 182–4, 192–3, 598–601, 611, 614, 616–17
non-linear relationships, graphs 182–4
non-linear sequences, algebra 42–5, 50, 54–6, 58
nth terms
arithmetic sequences 40–2
Fibonacci type sequences 43
geometric sequences 43
quadratic sequences 43–5
number 1–29
calculation 2–3, 8–13, 20–1, 22–3
estimation 4–5, 20–1, 22, 27
factors 2–3, 20–1
fractional indices 11–13, 21, 24–5
indices 7, 8–16, 21, 23–5
multiples 2–3, 20–1, 27
negative indices 11–13, 21
place value 4–5
powers of 10 14–16, 21, 27–8
powers 7, 8–16
powers to another power 11, 23–4
prefixes 14
prime factors 6–8, 21–3, 26–7
problem-solving 19–20
problems and reasoning 2–3
standard form 15, 21, 25, 27–8
surds 16–19, 21, 23–5, 28, 29
zero indices 11–12
number lines, square roots 4, 22
number of outcomes 2–3

objects, transformations 243–5
octagons 128
opposite 136, 155–6
opposite angles, cyclic quadrilaterals 510–11, 520, 521–2, 527
ordinary numbers, powers of 10 15, 21
origin, graphs 162
outliers 75, 77, 82, 86, 88, 94
cumulative frequency 454, 464–5, 467

parabolas 177–9, 186, 199, 478–85, 492, 495–6, 498
parallel lines 123, 145, 149, 162, 191–2, 197, 198
parallelogram law, vector addition 564–5, 583
parallelograms
angles 123, 129, 145, 149
congruence 367
similarity/scale factors 370–1
parallel vectors 566–8, 575, 578, 583
pentagonal prisms, similarity/scale factors 381, 391
pentagons 126–7, 129–30, 370, 371
percentage change 106–7, 112, 114–16, 118
areas 226
inverse operations 107
percentage loss (or profit) 106–8, 112, 114–16, 118
percentages 97, 105–11, 112, 114–16, 116–17, 118, 574
conversions 108–11, 114–15, 116, 118
multipliers 340–3, 351–2, 354, 357–8, 359
perfect squares 285
perimeters
area 204–9, 213–19, 226, 227, 229–30, 232–4, 235
circle sections 217–19, 227, 230
2D shapes 204–9, 213–19, 226, 227, 229–30, 232–4, 235
see also circumferences
perpendicular bisectors 258–9, 272, 274, 276
perpendicular lines, graphs 176, 191–2, 197
perpendicular planes 419–21, 436
phase, trigonometric graphs 423–5, 432
pictures, problem-solving 19–20
pie charts 64, 82, 84, 90
place value 4–5, 19–20
planes, 3D problems 419–21, 436
plans, 3D solids 240–2, 267, 270–1, 275
polygons 126–31, 133, 135, 142, 143–5, 147–9, 153–5
area 205–7, 227, 229–30, 232–4, 235
congruence 363–9, 382–3, 385–6, 390–3, 394

cyclic 509–12, 520–1, 522–4, 527
exterior angles 128–31, 144, 148–9, 154, 155
interior angles 126–8, 143–5, 147–8, 153–4, 155
perimeters 205–7, 227, 229–30, 232–4, 235
similarity 370–7, 382–4, 387–9, 390, 393, 394
sum of exterior angles 129, 155
sum of interior angles 127, 155
populations 597
comparison 447–8, 453–6, 457–8, 462–5, 467
definition 440
estimating size 442, 466
growth and decay 343, 351–2, 354, 358
sampling 440–2, 457–9, 463–4, 466
position vectors 566–73, 583
positive correlation, scatter graphs 71, 87–8
positive gradients 161, 190
possible outcomes, numbers of 2–3
powers of 10 14–16, 21, 27–8
powers *see* indices
powers
division 10
multiplication 9–11
to the power of another power 11, 23–4
prediction 43, 74–5, 91–2, 94, 172–4
prefixes, indices 14
pressure 339, 347–8, 353, 355, 356, 357, 359
prime factors 6–8, 21–3, 26–7
algebraic expansion 33–5
trees 6, 22
prisms
cross-sections 211, 231, 235
similarity/scale factors 378–81, 385, 389, 391–2, 394
trigonometry 420
volume 210–12, 228, 231, 233, 235
probability 305–38
combined events 307–9, 326, 328–9, 333–5
conditional 318–24, 325, 330–3, 334–5, 336
dependent events 318–24, 325, 330–3, 334–5, 336
experimental 312–14, 325, 326, 329, 333, 335
fair games 313–14, 326, 329, 333
independent events 314–18, 325, 330–1, 334–5
mutually exclusive events 310–12, 326, 328–9, 333–5
problem-solving 325
product rule 307–9, 313–14, 326, 327–9, 333–4, 335
sample space diagrams 308–9, 326, 327–8, 335
theoretical 313–14, 326, 329, 333
tree diagrams314–321 325, 327, 330–2, 335
two-way tables 318–19, 326, 328–9
Venn diagrams 321–4, 327, 332–3, 335, 336
product rule 307–9, 313–14, 326, 327–9, 333–4, 335
profit
calculation 107
graphs 186, 194
percentages 106–8, 112, 114–16, 118
proof
algebra 545–7, 549, 553–4, 555
congruence 367–9
proportion 103–5, 111–12, 113–14, 116–17, 118
constant of proportionality 589–92, 611, 613, 616–18, 619
direct 348–51, 352, 353, 356–8, 359, 588–92, 611, 613, 616–18, 619
exponential functions 184, 343, 595–8, 610, 611, 614, 616, 620
graphs 171–2, 587–623
inverse 350–1, 353, 356, 358, 359, 592–5, 611, 613, 616–18, 620
multiplicative reasoning 348–51, 352, 353, 356–8, 359
non-linear graphs 598–601, 611, 614, 616–17
to the square/cube/square root 591–2, 611, 613, 616–18, 619
proportional change
growth and decay 341–3, 351–2, 354, 357–8, 359
multiplicative reasoning 341–3, 348–52, 353, 354, 356–8, 359
pyramids
similarity/scale factors 379, 380, 392
trigonometry 420–1
volume 222–4, 236
Pythagoras' theorem 131–6, 146, 149–50, 154, 155
3D problems 418–21, 428, 432–3, 435, 436

quadratic equations 281–6, 291–3, 296–301, 302
algebraic fractions 541–2
completing the square 478–85, 492, 495–6, 498
formulae 283–4
graphs 176–9, 186, 199
iteration 484–5, 492, 496, 497, 498
pricing 186
quadratic formulae 283–4
roots 482–5, 492, 494–6, 497, 498
simultaneous 291–3, 296, 297–301, 302
solving 281
solving graphically 482–5, 491–2, 494–6, 497, 498
quadratic functions
completing the square 284–6, 298–9, 301, 302
graphs 478–85, 491–2, 494–6
perfect squares 285
roots 281–4, 291–2, 297–9, 301, 302, 478–85, 492, 494–6, 497, 498
turning points 478–85, 492, 495–6, 498
y-intercepts 479–85, 491–2, 494–6, 498
y-intersects 479–85, 491–2, 494–6, 498
quadratic graphs 176–9, 182, 184–6, 188, 193, 196, 199
quadratic sequences, algebra 43–5, 50, 55–6, 58–9
quadrilaterals
cyclic 510–11, 520–1, 522, 527
exterior angles 129–30, 149
interior angles 126–7, 145, 147–8
similarity/scale factors 384
quarter circles 217, 230
quartiles 445–8, 457, 460–2, 464–5, 467

radii, circles, theorems 502–4, 517, 519–20, 524, 526
random sampling 441, 459, 464, 466
range 75–8, 82–3, 85–7, 89, 90–3, 94
rates
compound measures 344–6, 352, 355–8, 359
constant 169, 189, 342–5, 350–1, 352–5, 352–6, 358
rates of change
graphs 166–70, 196
kinematics formulae 345–6, 352, 353, 357, 359
velocity-time graphs 169–70, 196
rationalising the denominator 18–19, 21, 25, 27, 28, 539–40, 549, 552–4, 555
rational numbers 17, 21
ratios 97, 101–5, 111–12, 113–14, 116–17, 118
multiplicative reasoning 348–51, 352, 353, 356–8, 359
proportion 103–5, 111–12, 113–14, 116–17, 118
scale drawings 253–61, 271
scale factors 370–7
real-life graphs 170–4, 188–9, 193–4
rearranging formulae 532–3, 548–51, 553–4, 555
reasoning
multiplicative 339–62
number 2–3
reciprocal functions 191, 199
reciprocal graphs 180–2, 188, 193, 199
reciprocals 99–100, 116, 118, 191
rectangles
congruence 367
similarity/scale factors 371–2
recurring decimals 109, 115, 118
reflections 242–5, 252, 266–7, 268–9, 270, 273–4, 275–6
graphs 605–6, 612, 615, 618, 621
trigonometric graphs 422–3, 436
regions, loci 262–4, 267, 273, 276
relative frequency, experimental probability 312–14, 325, 326, 329, 333, 335
representation of data 61–96
back-to-back stem and leaf diagrams 65–6, 82, 85, 94
frequency polygons 66–7, 81, 86, 90, 92, 94
pie charts 64, 82, 84, 90
scatter graphs 70–5, 83, 87–8, 89, 91–2, 94
statistical diagrams 63–7, 78–80, 81, 82, 84–5, 94
stem and leaf diagrams 64–6, 82, 85, 94
time series graphs 67–70, 80, 83, 87–9, 93, 94
see also statistics
resultant
parallelogram law 564–5, 583
triangle law 562–3, 583
vectors 562–5, 568–74, 575, 577, 579–82, 583
resultant vectors, transformations 251–2, 267, 273, 276
rhombus, congruence 367–9
RHS triangles 365–6, 394
right-angled triangles
hypotenuse 131–3, 136, 146, 149–52, 154, 155–6
Pythagoras' theorem 131–6, 146, 149–50, 154, 155
shorter sides 134–6, 155–6
similarity/scale factors 373, 375, 383, 387–90
surds 135–6, 142–3, 150
trigonometry 136–43, 146–7, 151–3, 154, 155–6
right-angles, semicircles 507, 526
roots
cubic functions 486–9, 492, 496, 498
quadratic equations 482–5, 492, 494–6, 497, 498
quadratic functions 281–4, 291–2, 297–9, 301, 302, 478–85, 492, 494–6, 497, 498
rotations 242–5, 252, 266–7, 268–9, 270, 273–4, 275–6
trigonometric graphs 422–3, 436
rounding up/down 209–10, 227, 230–1, 234, 235
route calculation 415, 417, 426

samples, definition 440, 466
sample space diagrams 308–9, 326, 327–8, 335
sampling 440–2, 457–9, 463–4, 466
bias 440–2, 458–9, 463, 466
capture-recapture method 442
random 441, 459, 464, 466
stratified 441–2, 457, 458–9, 463, 466
SAS triangles 365–6, 394
scalars, vectors 561–2, 583
scale drawings 253–61, 271
scale factors
area 247
enlargement 245–9, 253–5, 266–7, 269–70, 273–4, 276
graphs 606–9, 612, 615, 618, 621
maps 253–5, 271
trigonometric graphs 425, 428, 432, 434, 436
see also linear scale factors
scale models 102
scaling, axes 79
scatter graphs 70–5, 83, 87–8, 89, 91–2, 94, 195
correlation 71, 87–8, 183–4, 195
extrapolation 74–5, 91–2, 94
interpolation 74–5, 88
line of best fit 72–5, 91–2, 94
outliers 75, 88, 94
scientific notation 15, 21, 25, 27–8
sectors, circles 216–19, 227, 230
segments, circles 411–13, 434
semicircles 217–18, 227, 230, 507, 526
sequences
algebra 40–5, 50–1, 54–6, 58–9
arithmetic 40–2, 50, 54–6, 58
Fibonacci type 43, 50, 54, 58
function machines 42
geometric 43, 50, 55–6, 58
linear 40–1, 50, 54–6, 58
non-linear 42–5, 50, 55–6, 58–9
quadratic 43–5, 50, 55–6, 58–9
set notation 321–4, 336, 478, 498
shapes
congruence 363–9, 382–3, 385–6, 390–3, 394
similarity 363–4, 370–85, 387–93, 394
side elevation, 3D solids 241–2, 267, 270–1, 275
similarity 363–4, 370–85, 387–93, 394, 574
2D shapes 370–7, 382–4, 387–9, 390, 393, 394
3D solids 378–81, 385, 389, 391–2, 394
simple interest 106
simplifying
algebra 31–3, 50, 51–3, 56–7
algebraic fractions 535–7, 549, 552–4, 555
surd fractions 539–40, 549, 553–4, 555
simultaneous equations 287–93, 296, 297, 299–301, 302
linear 287–93, 296, 297, 299–301, 302
multiplication 289–91, 299, 301
quadratic 291–3, 296, 297–301, 302
solving graphically 472–4, 491, 492–4, 497, 498
straight lines 290–1
sine rule 413–14, 427, 429–30, 434, 436
sines 137–8, 140, 142, 156
area of circle segments 412–13
area of triangles 411–13, 429, 433–5, 436
graphs 402–5, 421–5, 428, 431–2, 434, 435–6
slant heights, cones 223–4, 225, 232, 236
SOHCAHTOA 138
speed
average 344–5, 352, 358, 359
distance time graphs 167–9
spheres
surface area 221–2, 228, 232, 236
volume 221–2, 228, 232, 236
spinners 2, 313
square-based pyramids, trigonometry 420–1

squared proportionality 591–2, 611, 613, 616–18, 619
square root proportionality 591–2, 611, 613, 616–18, 619
square roots 4, 9, 22
squares, perfect 285
squaring, brackets 46–7, 59
SSS triangles 365–6, 394
standard form 15, 21, 25, 27–8
statistics 61–96, 439–70
  back-to-back stem and leaf diagrams 65–6, 82, 85, 94
  bias, sampling 440–2, 458–9, 463, 466
  box plots 446–8, 454, 456, 457, 461–2, 464–5, 467
  class boundaries 443–4, 466
  class widths 449–53, 461, 465, 467
  comparative box plots 447–8, 456–7, 462, 464
  comparing data 447–8, 453–6, 457–8, 462–5, 467
  cumulative frequency 443–8, 454–7, 459–67
  diagrams 63–7, 78–80, 81, 82–5, 92–3, 94
  distance-time graphs 167–9, 183, 187, 194, 196, 198, 600–1, 616
  frequency density 449–53, 458, 461, 465, 467
  frequency polygons 66–7, 81, 86, 90, 92, 94
  histograms 449–53, 458, 461, 465, 467
  interquartile range 445–8, 453–8, 460–5, 467
  median 444–5, 448, 457–8, 460–5, 467
  outliers 75, 77, 82, 86, 88, 94, 454, 464–5, 467
  pie charts 64, 82, 84, 90
  problem-solving 81
  quartiles 445–8, 457, 460–2, 464–5, 467
  range 75–8, 82–3, 85–7, 89, 90–3, 94
  sampling 440–2, 457–9, 463–4, 466
  scatter graphs 70–5, 83, 87–8, 89, 91–2, 94, 195
  summary 446
  time series graphs 67–70, 80, 83, 87–9, 93, 94
  velocity-time graphs 169–70, 196, 198, 598, 600–1, 611, 614, 620
  see also data
steady speed, distance-time graphs 168
stem and leaf diagrams 64–6, 82, 85, 94
step polygons 464
straight lines, simultaneous equations 290–1
strata 441–2, 466
stratified sampling 441–2, 457, 458–9, 463, 466
stretches
  graphs 606–9, 612, 615, 618, 621
  trigonometric graphs 425, 428, 432, 434, 436
subjects of formulae 39, 56, 58, 532–3, 548–51, 553–4, 555
subtended angles, circles 506–7, 509–12, 520–1, 523–5, 526, 577–8
subtraction
  algebraic fractions 537–8
  fractions 100, 113
  mixed numbers 100, 113
sum of angles, triangles 123
sum of exterior angles, polygons 129
sum of interior angles
  polygons 127
  triangles 123
summary statistics 446
surds
  algebra 539–40, 549, 553–4, 555
  fractions 539–40, 549, 553–4, 555
  number 16–19, 21, 23–5, 28, 29
  right-angled triangles 135–6, 142–3, 150
surface area 210–12, 220–6, 228, 231–2, 233–8
  cones 222–4, 225, 228, 232, 236
  cylinders 220–2, 226, 228, 231, 234, 236
  prisms 210–12, 228, 231, 235–6
  pyramids 222–4, 236
  spheres 221–2, 228, 232, 236
surface gravity 548

tally counting 63
tangents 137–8, 140, 142, 156
  circle theorems 504–6, 512–14, 517, 519, 521–2, 524–7
  curved graphs 598–600, 614, 616–17, 620
  graphs 408–10, 421–5, 427–8, 431–2, 435–6
tax
  income 106
  Value Added 105, 118
temperature, unit conversion 57
tetrahedrons 379, 420, 428
theoretical probability 313–14, 326, 329, 333
three-dimensional... *see* 3D... (at start of index)
time series graphs 67–70, 80, 83, 87–9, 93, 94
  moving averages 93, 94
total interest 342–3, 352, 354, 357–8, 359
trajectories, graphs 177–9, 193–4, 198
transformations 239–79
  bearings 253–5, 267, 271, 274, 276
  column vectors 250–2, 267–9, 276
  combinations 249–52, 266–7, 268–9, 273–5, 276
  congruence 363–9, 382–3, 385–6, 390–3, 394
  cosine function graphs 421–5, 428, 432, 436
  enlargement 245–9, 253–5, 266–7, 269–70, 273–4, 276
  graphs 602–9, 612, 615, 618, 621
  reflections 242–5, 252, 266–7, 268–9, 273–4, 275–6
  resultant vectors 251–2, 267, 273, 276
  rotations 242–5, 252, 266–7, 268–9, 273–4, 275–6
  scale drawings 253–5, 271
  similarity 363–4, 370–85, 387–93, 394
  sine function graphs 421–5, 428, 432, 434, 436
  tangent function graphs 421–5, 431, 436
  translations 249–52, 266–7, 268–9, 270, 273–5, 276
  vector notation 558–60, 582
  vectors 250–2, 267–9, 273, 276, 558–60, 582
translations 249–52, 266–7, 268–9, 270, 273–5, 276
  graphs 602–4, 612, 615, 618, 621
  trigonometric graphs 423–5, 436
  vectors 558–60, 582
trapezia 125, 133, 135, 153
  area 205–7, 227, 229, 233–4, 235
  congruence 367
  perimeters 206–7, 227, 229, 233–4, 235
  similarity/scale factors 371
tree diagrams 314–21, 325, 327, 330–2, 335
trees, prime factors 6, 22
trends, time series graphs 67–70, 80, 83, 87–9, 93, 94
triangle law, vector additions 562–3, 583
triangles
  adjacent 136, 155–6
  angle properties 122–6
  area 411–13, 429, 433–5, 436
  congruence 364–9, 382–3, 385–6, 393, 394
  constructions 256–7, 261, 267, 272, 274
  cosine rule 415–18, 427, 430, 433, 435, 436
  exterior angles 123–5, 128, 155
  height 136
  hypotenuse 131–3, 136, 146, 149–52, 154, 155–6
  interior angles 123–5, 126–7, 131–43, 145–7, 149–56
  opposite 136, 155–6
  Pythagoras' theorem 131–6, 146, 149–50, 154, 155
  similarity 370–1, 373–5, 383–4, 387–90, 392–3, 394
  sine rule 413–14, 427, 429–30, 434, 436
  sum of interior angles 123
  surds 135–6, 142–3
  trigonometry 136–43, 146–7, 151–3, 154, 155–6, 399–438
triangular prisms 378
trigonometric graphs 402–10, 427–8, 431–2, 434, 435–6
  phase 423–5, 432
  reflections 422–3, 436
  rotations 422–3, 436
  scale factors 425, 428, 432, 434, 436
  transformations 421–5, 428, 432, 434, 436
  translations 423–5, 436
trigonometry 121, 136–43, 146–7, 151–3, 154, 155–6, 399–438
  accuracy 400–2, 427, 429–31
  angle of depression/elevation 139–41, 152, 156
  area of a triangle 411–13, 429, 433–5, 436
  bearings 415, 417, 426
  cosine function graphs 405–7, 421–5, 427–8, 431–2, 435–6
  cosine rule 415–18, 427, 430, 433, 435, 436
  diagonals 419–20, 428, 432, 433
  2D problems 418–21, 428, 432–3, 435, 436
  3D problems 418–21, 428, 432–3, 435, 436
  graphs 402–10, 421–5, 427–8, 431–2, 434, 435–6
  inverse functions 140, 156
  similar triangles 375, 383
  sine function graphs 402–5, 421–5, 428, 431–2, 434, 435–6
  sine rule 413–14, 427, 429–30, 434, 436
  SOHCAHTOA 138
  tangent function graphs 408–10, 421–5, 427–8, 431–2, 435–6
  upper and lower bounds 400–2, 427, 429–31
turning points 177–9, 199, 478–85, 492, 495–6, 498
two-dimensional... *see* 2D (at start of index)
two-way tables 79, 82–4, 92–3, 94
  conditional probability 318–19, 326, 328–9

unions, set notation 323–4, 336
unit ratios 101–2, 111, 113, 118
units
  accuracy 206–7, 227, 229, 233–4, 235
  area 207–8, 227, 230–1, 234, 235
  conversions 14, 26, 57, 589
  currency 103, 105, 111, 113, 117, 120
  prefixes 14
unknown angles
  cosine rule 415–18, 427, 430, 433, 435, 436
  sine rule 413–14, 427, 429–30, 434, 436
unknown sides
  cosine rule 415–18, 427, 430, 433, 435, 436
  sine rule 413–14, 427, 429–30, 434, 436
upper bounds
  measurements 209–10, 227, 230–1, 234, 235
  trigonometry 400–2, 427, 429–31
upper class boundaries, cumulative frequency tables 443–4, 466
upper quartiles (UQ) 445–8, 457, 460–2, 464–5, 467
UQ *see* upper quartiles

Value Added Tax (VAT) 105, 118
variables, algebra 38, 56–8
vector notation 558–60, 574, 576–7, 579, 581–2
vectors 557–86
  addition 562–5, 583
  arithmetic 560–5, 575, 577, 579–82, 583
  collinear points 568, 578–9, 583
  column-type 250–2, 267–9, 276, 558–66
  displacement 558–60, 582
  equal 559–60, 582
  geometric problems 568–74, 575, 578–82, 583
  magnitude 559–60, 582–3
  notation 558–60, 574, 576–7, 579, 581–2
  parallel 566–8, 575, 578, 583
  parallelogram law 564–5, 583
  position-type 566–73, 583
  resultant 251–2, 267, 276, 562–5, 568–74, 575, 577, 579–82, 583
  scalars 561–2, 583
  transformations 250–2, 267–9, 273, 276
  triangle law 562–3, 583
velocity, kinematics formulae 345–6, 352, 353, 357, 359
velocity-time graphs 169–70, 196, 198, 598, 600–1, 611, 614, 620
  acceleration 169–70, 196
  area under the curve 233
Venn diagrams 321–4, 327, 332–3, 335, 336
  prime factors 6–7, 23, 26, 28
vertical stretches, trigonometric graphs 425, 428, 432, 434, 436
volume 203, 210–12, 220–6, 228, 231–2, 233–8
  capacity 212, 226, 233, 234, 235
  cones 222–4, 225, 228, 232, 236
  cylinders 220–2, 226, 228, 231, 234, 236
  density 346–7, 352, 353, 355, 357, 359
  frustrums 224, 392
  prisms 210–12, 228, 231, 235–6
  problem-solving 225
  pyramids 222–4, 236
  scale factors 378–81, 385, 389, 391–2, 394
  spheres 221–2, 228, 232, 236

x-intercepts 163, 198
  cubic functions 486–9, 492, 496, 498
  quadratic functions 478–85, 492, 494–6, 497, 498

y-intercepts
  cubic functions 486–7, 492, 496, 498
  linear graphs 161–3, 190, 198
  quadratic functions 479–85, 491–2, 494–6, 498

zero correlation 71, 87–8
zero indices 11–12
zigzag lines, axes 171